Encyclopedia of
Applied Electrochemistry

Gerhard Kreysa • Ken-ichiro Ota
Robert F. Savinell
Editors

Encyclopedia of Applied Electrochemistry

Volume 3

P–Z

With 1250 Figures and 122 Tables

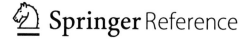

Editors
Gerhard Kreysa
Eppstein, Germany

Robert F. Savinell
Case Western Reserve University
Cleveland, OH, USA

Ken-ichiro Ota
Yokohama National University, Fac.
Engineering
Yokohama, Japan

ISBN 978-1-4419-6995-8   978-1-4419-6996-5 (eBook)
ISBN Bundle 978-1-4419-6997-2 (print and electronic bundle)
DOI 10.1007/978-1-4419-6996-5
Springer New York Heidelberg Dordrecht London

Library of Congress Control Number: 2014934571

© Springer Science+Business Media New York 2014
This work is subject to copyright. All rights are reserved by the Publisher, whether the whole or
part of the material is concerned, specifically the rights of translation, reprinting, reuse of
illustrations, recitation, broadcasting, reproduction on microfilms or in any other physical way,
and transmission or information storage and retrieval, electronic adaptation, computer software,
or by similar or dissimilar methodology now known or hereafter developed. Exempted from this
legal reservation are brief excerpts in connection with reviews or scholarly analysis or material
supplied specifically for the purpose of being entered and executed on a computer system, for
exclusive use by the purchaser of the work. Duplication of this publication or parts thereof is
permitted only under the provisions of the Copyright Law of the Publisher's location, in its
current version, and permission for use must always be obtained from Springer. Permissions for
use may be obtained through Rights Link at the Copyright Clearance Center. Violations are liable
to prosecution under the respective Copyright Law.
The use of general descriptive names, registered names, trademarks, service marks, etc. in this
publication does not imply, even in the absence of a specific statement, that such names are
exempt from the relevant protective laws and regulations and therefore free for general use. While
the advice and information in this book are believed to be true and accurate at the date of
publication, neither the authors nor the editors nor the publisher can accept any legal
responsibility for any errors or omissions that may be made. The publisher makes no warranty,
express or implied, with respect to the material contained herein.

Printed on acid-free paper

Springer is part of Springer Science+Business Media (www.springer.com)

Preface

Electrochemistry provides the opportunity to run chemical redox reactions directly with electrons as reaction partners and against the free energy gradient via external electrical energy input into the reaction system. It also serves as a way to efficiently generate energy from stored energy in chemical bonds. Applied electrochemistry has been the basis for many industrial processes ranging from metals recovery and purification to chemical synthesis and separations since the latter half of the 1800s when large-scale electricity generation became possible. Electrochemical processes have a large impact on energy as it has been estimated that these processes consume about 6–10 % of the world's electricity generation capacity. Applied electrochemistry is now impacting industry and society more and more with technologies for waste water treatment, efficient chemical separation, and environmental sensing and remediation. Electrochemistry is the foundation for electrochemical energy storage by batteries and electrochemical capacitors and energy conversion by fuel cells and solar cells. In fact, applied electrochemistry will play a major role in the world's ability to harness and use renewable energy sources. Electrochemistry is also fundamental to biological cell transport and many aspects of living systems and their activities. It is exploited for use in medical diagnostics to detect abnormalities and in biomedical engineering to relieve pain and deliver function.

The application of electrochemistry involves not just a fundamental understanding of the sciences, but also applying engineering principles to device and technology design by considering mass and energy balances, transport processes in the electrolyte and at electrode interfaces, and multi-scale modeling and simulation for predicting and optimizing performance. The interaction of the interfacial reactions, the transport driving forces, and the electric field defines the field of electrochemical engineering. The understanding of the scientific and engineering principles of electrochemical systems has driven advances in the application of electrochemistry especially during the last half century.

The purpose of this collection is to summarize the A–Z of the application of electrochemistry and electrochemical engineering for use by electrochemists and electrochemical engineers as well as nonspecialists such as engineers and scientists of all disciplines, economists, students, and even politicians. Electrochemical fundamentals, electrochemical processes and technologies, and electrochemical techniques are described by many experts in their fields from around the world, many from industry. Each entry is meant to be an introduction and also gives references for further study. With this collection, we hope that current technology and operating practices can be made available for future generations to learn from. We hope that this encyclopedia will stimulate understanding of the current state of the art and lead to advances in new and more efficient technologies with breakthroughs from new theory and materials. We hope you find it of value to your work.

Gerhard Kreysa
Ken-ichiro Ota
Robert F. Savinell
Editors-in-Chief

Acknowledgments

The Editors-in-Chief would like to acknowledge the backing of their institutions in supporting this work. Specifically we want to thank the Technical University of Dortmund (GK), Yokohama National University (KO), and Case Western Reserve University (RFS). We also want to thank the topical editors and authors; it is because of their dedication to their fields and hard work that this collection was possible. We thank our families for their understanding of the importance of this project to us and our profession. Finally, we want to thank the editorial staff of Springer, especially Barbara Wolf who worked long and hard on this project and Kenneth Howell specifically for encouraging us to embark on this project and nudging us along.

About the Editors

Prof. Dr.rer.nat. Dr.-Ing. E.h. Dr.tekn.h.c. Gerhard Kreysa Retired Chief Executive of DECHEMA e.V.

Gerhard Kreysa was born in 1945 in Dresden. He studied chemistry at the University of Dresden and received his Ph.D. in 1970. In 1973, he joined the Karl Winnacker Institute of DECHEMA in Frankfurt am Main. He developed new concepts for the utilization of three-dimensional electrodes, which became prominent for electrochemical waste water treatment in the process industry. He also played a leading role in the clarification of the "cold fusion" affaire in 1989. In 1985, he was appointed as professor in the Chemical Engineering Department at the University of Dortmund. In 1993, he was appointed as honorary professor at the University of Regensburg. From 1985 to 1995, he served as executive editorial board member of the *Journal of Applied Electrochemistry*. He was a recipient of the Chemviron Award in 1980, the Max-Buchner-Research-Award of DECHEMA and the Castner Medal of the Society of Chemical Industry in 1994, and the Wilhelm Ostwald Medal of the Saxon Academy of Sciences in Leipzig in 2006.

From 1992 to 2009, Dr. Kreysa served as chief executive of DECHEMA Society of Chemical Engineering and Biotechnology in Frankfurt am Main, Germany. During this time he also served as general secretary of the European Federation of Chemical Engineering and the European Federation of Biotechnology. He obtained many distinctions: Honorary doctor degrees of Technical University of Clausthal and of the Royal Institute of Technology in Stockholm, Foreign Member of the Royal Swedish Academy of Engineering Sciences, elected honorary fellow of the Institution of Chemical Engineers, and honorary member of the Czech Society of Chemical Engineering. In 2007 he was awarded with the Order of Merit of the Free State of Saxony, and in 2008 he became a member of the German Academy of Technical Sciences (acetech). He has 196 scientific publications and books and has given 312 scientific and public lectures. Despite the numerous duties and responsibilities of his former senior management position, he continues to have a lively interest in the further development of science and engineering and is highly regarded as an advisor on national and international issues.

Prof. Dr. G. Kreysa
Weingasse 22
D-65817 Eppstein

Ken-ichiro Ota

Ken-ichiro Ota is a professor at and the chairman of Green Hydrogen Research Center at Graduate School of Engineering, Yokohama National University, Japan. He received his B.S.E. in applied chemistry in 1968 and Ph.D. in engineering in 1973, both from the University of Tokyo. After graduation, he became a research associate at the university until 1979. In the same year, he became an associate professor at the Yokohama National University, and a professor in 1995. He has worked on hydrogen energy and fuel cell since 1974, focusing on materials science for fuel cells and hydrogen energy system including water electrolysis. In the fuel cell field he has worked on direct methanol fuel cell, molten carbonate fuel cell, and polymer electrolyte fuel cell. Recently, he is developing transition metal oxide–based cathode for polymer electrolyte fuel cell. He is also working on storage and transport of renewable energies by hydrogen technology. He has published more than 190 original papers, 80 review papers, and 50 scientific books. He received the Molten Salt Award in 1998 and the Industrial Electrolysis Award in 2002 from the Electrochemical Society of Japan. He received the Canadian Hydrogen Society Award in 2004 and the Society Award of the Electrochemical

Society of Japan in 2011. He is now the chairman of the National Committee for Standardization of the Stationary Fuel Cells. He was the president of the Hydrogen Energy Systems Society of Japan from 2000 to 2008 and also the president of the Electrochemical Society of Japan from 2008 to 2009. He is now the chairman of the Fuel Cell Development Information Center of Japan.

Dr. Robert F. Savinell Distinguish University Professor, George S. Dively Professor of Engineering, Department of Chemical Engineering, Case Western Reserve University, Cleveland, OH, USA

Dr. Robert F. Savinell received his B.Che. from Cleveland State University in 1973, and his M.S. (1974) and Ph.D. (1977), both in chemical engineering from University of Pittsburgh. He worked as a research engineer for Diamond Shamrock Corporation, then as a faculty member at the University of Akron before joining the faculty at Case Western Reserve University (CWRU) in 1986. Professor Savinell was the Director of the Ernest B. Yeager Center for Electrochemical Sciences at CWRU for ten years and served as Dean of Engineering at CWRU for seven years. Professor Savinell has been engaged in electrochemical engineering research and development for 40 years. Savinell's research is directed at fundamental science and mechanistic issues of electrochemical processes; and at electrochemical technology systems and device design, development, modeling and optimization. His research has addressed applications for energy conversion, energy storage, sensing, and electrochemical materials extraction and synthesis. Savinell has over 120 peer-reviewed and over 168 other publications, eight patents, and has been an invited and keynote speaker at hundreds of national and international conferences and workshops in the electrochemical field. He has supervised over 50 Ph.D./M.S. student projects.

Professor Savinell is the former North American editor of the *Journal of Applied Electrochemistry* and currently is the editor of the *Journal of the Electrochemical Society*. He is a Fellow of the Electrochemical Society, Fellow of the American Institute of Chemical Engineers, and Fellow of the International Society of Electrochemistry.

Section Editors

Electrocatalysis

Radoslav R. Adzic Chemistry Department, Brookhaven National Laboratory, Upton, NY, USA

Primary Batteries

George Blomgren Imara Corporation, Menlo Park, CA, USA

Environmental Electrochemistry

Christos Comninellis EPFL, Lausanne, Switzerland

Organic Electrochemistry

Toshio Fuchigami Department of Electrochemistry, Tokyo Institute of Technology, Midori-ku, Yokohama, Japan

Solid State Electrochemistry

Ulrich Guth Kurt-Schwabe-Institut für Mess- und Sensortechnik e.V. Meinsberg, Ziegra-Knobelsdorf, Germany

High-Temperature Molten Salts

Rika Hagiwara Graduate School of Energy Science, Kyoto University, Sakyo-ku, Kyoto, Japan

Inorganic Electrochemical Synthesis (Including Chlorine/Caustic, Chlorates, Hypochlorite)

Kenneth L. Hardee Research & Development Division, De Nora, Fairport Harbor, OH, USA

Bioelectrochemistry

Dirk Holtmann DECHEMA e.V., Frankfurt am Main, Germany

Electrochemical Instrumentation and Laboratory Techniques

Rudolf Holze Institut für Chemie, Technische Universität Chemnitz, Chemnitz, Germany

Fuel Cells

Minoru Inaba Department of Molecular Chemistry and Biochemistry, Doshisha University, Kyotanabe, Kyoto, Japan

Supercapacitors

Hiroshi Inoue Department of Applied Chemistry, Graduate School of Engineering, Osaka Prefecture University, Sakai, Osaka, Japan

High-Temperature Electrochemistry

Tatsumi Ishihara Department of Applied Chemistry, Kyushu University, Nishi-ku, Fukuoka, Japan

Secondary Batteries

Kiyoshi Kanamura Applied Chemistry Graduate School of Engineering, Tokyo Metropolitan University, Hachioji, Tokyo, Japan

Semiconductor Synthesis and Electrochemistry

Paul A. Kohl Georgia Institute of Technology, School of Chemical and Biomolecular Engineering, Atlanta, GA, USA

Electrolytes

Werner Kunz Institute of Physical and Theoretical Chemistry, Regensburg University, Regensburg, Germany

Electrodeposition (Electrochemical Metal Deposition and Plating)

Uziel Landau Chemical Engineering Department, Case Western Reserve University, Cleveland, OH, USA

Electrochemical Analysis and Sensors

Chung-Chiun Liu Case Western Reserve University, Cleveland, OH, USA

Electrocatalysis

Nebojsa Marinkovic Synchrotron Catalysis Consortium, University of Delaware, Newark, DE, USA

Environmental Electrochemistry

Yunny Meas Vong CIDETEQ, Parque Tecnológico Querétaro, Sanfandila, Pedro Escobedo, CP, México

Photoelectrochemistry

Tsutomu Miyasaka Graduate School of Engineering, Toin University of Yokohama, Kanagawa, Japan

Electrochemical Engineering

Trung Van Nguyen Department of Chemical and Petroleum Engineering, The University of Kansas, Lawrence, KS, USA

Fuel Cells

Thomas Schmidt Electrochemistry Laboratory, Paul Scherrer Institut, Villigen, Switzerland

Advisory Board

Richard C. Alkire Department of Chemical and Biomolecular Engineering, University of Illinois at Urbana-Champaign, Urbana, IL, USA

Jürgen Garche Zentrum für Sonnenenergie, Ulm, Germany

Angelika Heinzel Fakultät 5 / Abt. Maschinenbau Energietechnik, Universitat Duisburg-Essen, Duisburg, Germany

Zempachi Ogumi Office of Society-Academia Collaboration for Innovation, Center for Advanced Science and Innovation, Kyoto University, Uji, Japan

Tetsuya Osaka Department of Applied Chemistry, Waseda University, Tokyo, Japan

Mark Verbrugge Director, Chemical and Materials Systems Lab, General Motors Research & Development, Warren, MI, USA

Ralph E. White Department of Chemical Engineering, Swearingen Engineering Center, University of South Carolina, Columbia, SC, USA

Contributors

Abd El Aziz Abd-El-Latif Institute of Physical and Theoretical Chemistry, University of Bonn, Bonn, Germany

Luisa M. Abrantes Departamento de Química e Bioquímica, FCUL, Lisbon, Portugal

Radoslav R. Adzic Chemistry Department, Brookhaven National Laboratory, Upton, NY, USA

Sheikh A. Akbar The Ohio State University, Columbus, OH, USA

Francisco Alcaide Energy Department, Fundación CIDETEC, San Sebastián, Spain

Antonio Aldaz Instituto Universitario de Electroquímica, University of Alicante, Alicante, Spain

Leonardo S. Andrade Universidade Federal de Goiás, Catalão, Brazil

Juan Manuel Artés Institució Catalana de Recerca i Estudis Avançats (ICREA), Institute for Bioengineering of Catalonia (IBEC), Barcelona, Spain

Electrical and Computer Engineering, University of California Davis, Davis, CA, USA

Mahito Atobe Graduate School of Environment and Information Sciences, Yokohama National University, Yokohama, Japan

Nicola Aust BASF SE, Ludwigshafen, Germany

Arseto Bagastyo Advanced Water Management Centre (AWMC), The University of Queensland, Brisbane, QLD, Australia

Helmut Baltruschat Institute of Physical and Theoretical Chemistry, University of Bonn, Bonn, Germany

Cesar Alfredo Barbero Department of Chemistry, National University of Rio Cuarto, Rio Cuarto, Cordoba, Argentina

Scott Barnett Department of Materials Science and Engineering, Northwestern University, Evanston, IL, USA

Romas Baronas Faculty of Mathematics and Informatics, Vilnius University, Vilnius, Lithuania

Damien Batstone Advanced Water Management Centre (AWMC), The University of Queensland, Brisbane, QLD, Australia

Pierre Bauduin Institut de Chimie Separative de Marcoule, UMR 5257 – ICSM Site de Marcoule, CEA/CNRS/UM2/ENSCM, Bagnols sur Ceze, France

Dorin Bejan Department of Chemistry, Electrochemical Technology Centre, University of Guelph, Guelph, ON, Canada

Luc Belloni CEA Saclay, Gif-sur-Yvette, France

Henry Bergman Anhalt University, Anhalt, Germany

Sonia R. Biaggio Universidade Federal de Goiás, Catalão, Brazil

Salma Bilal National Centre of Excellence in Physical Chemistry, University of Peshawar, Peshawar, Pakistan

José M. Bisang Programa de Electroquímica Aplicada e Ingeniería Electroquímica (PRELINE), Facultad de Ingeniería Química, Universidad Nacional del Litoral, Santa Fe, Santa Fe, Argentina

George Blomgren Blomgren Consulting Services, Lakewood, OH, USA

Nerilso Bocchi Universidade Federal de Goiás, Catalão, Brazil

Pierre Boillat Paul Scherrer Institut, Villigen PSI, Switzerland

Nikolaos Bonanos Department of Energy Conversion and Storage, Technical University of Denmark, Roskilde, DK

Antoine Bonnefont Institut de Chimie de Strasbourg, CNRS-Université de Strasbourg, Strasbourg, France

Oleg Borodin Electrochemistry Branch, Sensor and Electron Devices Directorate, U.S. Army Research Laboratory, Adelphi, MD, USA

Stanko Brankovic Electrical and Computer Engineering Department, Chemical and Bimolecular Engineering Department, and Chemistry Department, University of Houston, Houston, TX, USA

Enric Brillas Laboratory of Electrochemistry of Materials and Environment, Department of Physical Chemistry, Faculty of Chemistry, University of Barcelona, Barcelona, Spain

Ralph Brodd Broddarp of Nevada, Inc., Henderson, NV, USA

Michael Bron Institut für Chemie, Technische Chemie, Martin-Luther-Universität Halle-Wittenberg, Halle, Germany

Nigel W. Brown Daresbury Innovation Centre, Arvia Technology Ltd., Daresbury, UK

Felix N. Büchi Paul Scherrer Institut, Villigen PSI, Switzerland

Richard Buchner Institute of Physical and Theoretical Chemistry, University of Regensburg, Regensburg, Germany

Ratnakumar V. Bugga Jet Propulsion Laboratory, Pasdena, CA, USA

Nigel J. Bunce Department of Chemistry, Electrochemical Technology Centre, University of Guelph, Guelph, ON, Canada

Andreas Bund FG Elektrochemie und Galvanotechnik, Institut für Werkstofftechnik, Technische Universität Ilmenau, Ilmenau, Germany

Erika Bustos Centro de Investigacióny Desarrollo Tecnológico en Electroquímica, S. C., Sanfandila, Pedro Escobedo, Querétaro, México

Julea Butt School of Chemistry, University of East Anglia, Norwich, UK

Yun Cai Material Science, Joint Center for Artificial Photosynthesis, Lawrence Berkeley National Laboratory, Berkeley, CA, USA

Claudio Cameselle Department of Chemical Engineering, University of Vigo, Vigo, Spain

Maja Cemazar Department of Experimental Oncology, Institute of Oncology Ljubljana, Ljubljana, Slovenia

Vidhya Chakrapani Department of Chemical and Biological Engineering, Rensselaer Polytechnic Institute, Troy, NY, USA

François Chellé Universite Catholique de Louvain, Louvain-la-Neuve, Belgium

Kaimin Chen MECC, Medtronic Inc., Minneapolis, MN, USA

Po-Yu Chen Department of Medicinal and Applied Chemistry, Kaohsiung Medical University, Kaohsiung, Taiwan

Kazuhiro Chiba Tokyo University of Agriculture and Technology, Fuchu, Tokyo, Japan

Masanobu Chiku Department of Applied Chemistry, Graduate School of Engineering, Osaka Prefecture University, Osaka, Japan

YongMan Choi Chemistry Department, Brookhaven National Laboratory, Upton, NY, USA

David E. Cliffel Department of Chemistry, Vanderbilt University, Nashville, TN, USA

Christos Comninellis Institute of Chemical Sciences and Engineering, Ecole Polytechnique Fédérale de Lausanne (EPFL), Lausanne, Switzerland

Ann Cornell School of Chemical Science and Engineering, Applied Electrochemistry, KTH Royal Institute of Technology, Stockholm, Sweden

Serge Cosnier Department of Molecular Chemistry, CNRS UMR 5250 CNRS-University of Grenoble, Grenoble, France

Vincent S. Craig Department of Applied Mathematics, Research School of Physical Sciences and Engineering, Australian National University, Canberra, ACT, Australia

Hideo Daimon Advanced Research and Education, Doshisha University, Kyotanabe, Kyoto, Japan

Manfred Decker Kurt-Schwabe-Institut fuer Mess- und Sensortechnik e.V. Meinsberg, Waldheim, Germany

Dario Dekel CellEra Inc., Caesarea, Israel

I. M. Dharmadasa Electronic Materials and Sensors Group, Materials and Engineering Research Institute, Sheffield Hallam University, Sheffield, UK

Petros Dimitriou-Christidis Environmental Chemistry Modeling Laboratory, Ecole Polytechnique Fédérale de Lausanne (EPFL), Lausanne, Switzerland

Pablo Docampo Clarendon Laboratory, Department of Physics, Oxford University, Oxford, UK

Deepak Dubal AG Elektrochemie, Institut für Chemie, Technische Universität Chemnitz, Chemnitz, Germany

Laurie Dudik Case Western Reserve University, Cleveland, OH, USA

Jean François Dufreche Institut de Chimie Séparative de Marcoule and Université Montpellier, Marcoule, France

Christian Durante Department of Chemical Sciences, University of Padova, Padova, Italy

Prabir K. Dutta The Ohio State University, Columbus, OH, USA

Ulrich Eberle Government Programs and Research Strategy, GM Alternative Propulsion Center, Adam Opel AG, Rüsselsheim, Germany

Obi Kingsley Echendu Electronic Materials and Sensors Group, Materials and Engineering Research Institute, Sheffield Hallam University, Sheffield, UK

Minato Egashira College of Bioresource Sciences, Nihon University, Fujisawa, Kanagawa, Japan

Takashi Eguro Frukawa Battery, Iwaki, Fukushima, Japan

Martin Eichler AG Elektrochemie, Institut für Chemie, Technische Universität Chemnitz, Chemnitz, Germany

Robert Eisenberg Department of Molecular Biophysics and Physiology, Rush University Medical Center, Chicago, IL, USA

Bernd Elsler Johannes Gutenberg-University Mainz, Mainz, Germany

Eduardo Expósito Instituto Universitario de Electroquímica, University of Alicante, Alicante, Spain

Emiliana Fabbri Electrochemistry Laboratory, Paul Scherrer Institute, Villigen, Switzerland

Yujie Feng Harbin Institute of Technology, Harbin, China

Rui Ferreira Instituto de Tecnologia Química e Biológica, Universidade Nova de Lisboa, Oeiras, Portugal

Sergio Ferro Department of Chemical and Pharmaceutical Sciences, University of Ferrara, Ferrara, Italy

Stéphane Fierro Institute of Chemical Sciences and Engineering, Ecole Polytechnique Fédérale de Lausanne (EPFL), Lausanne, Switzerland

Michael A. Filler School of Chemical and Biomolecular Engineering, Georgia Institute of Technology, Atlanta, GA, USA

Alanah Fitch Department of Chemistry, Loyola University, Chicago, IL, USA

Robert Forster School of Chemical Sciences National Center for Sensor Research, Dublin City University, Dublin, Ireland

György Fóti Institute of Chemical Sciences and Engineering, Ecole Polytechnique Fédérale de Lausanne (EPFL), Lausanne, Switzerland

Alejandro A. Franco Laboratoire de Réactivité et de Chimie des Solides (LRCS) - UMR 7314, Université de Picardie Jules Verne, CNRS and Réseau sur le Stockage Electrochimique de l'Energie (RS2E), Amiens, France

Matthias Franzreb Institute of Functional Interfaces, Karlsruhe Institute of Technology, Eggenstein-Leopoldshafen, Germany

Stefano Freguia Advanced Water Management Centre (AWMC), The University of Queensland, Brisbane, QLD, Australia

Bernardo A. Frontana-Uribe Centro Conjunto de Investigación en Química Sustentable, UAEMéx–UNAM, Toluca, Estado de México, Mexico

Instituto de Química UNAM, Mexico, Mexico

Albert J. Fry Weslayan University, Middletown, CT, USA

Toshio Fuchigami Department of Electrochemistry, Tokyo Institute of Technology, Midori-ku, Yokohama, Japan

Akira Fujishima Kanagawa Academy of Science and Technology, Takatsu–ku, Kawasaki, Kanagawa, Japan

Photocatalysis International Research Center, Tokyo University of Science, Noda, Chiba, Japan

Klaus Funke Institute of Physical Chemistry, University of Muenster, Muenster, Germany

Ping Gao AG Elektrochemie, Institut für Chemie, Technische Universität Chemnitz, Chemnitz, Germany

Vicente García-García Instituto Universitario de Electroquímica, University of Alicante, Alicante, Spain

Helga Garcia Instituto de Tecnologia Química e Biológica, Universidade Nova de Lisboa, Oeiras, Portugal

Darlene G. Garey CIDETEQ, Centro de Investigación y Desarrollo Tecnológico en Electroquímica Parque Tecnológico Querétaro, Pedro Escobedo, Edo. Querétaro, México

Armando Gennaro Department of Chemical Sciences, University of Padova, Padova, Italy

Abhijit Ghosh Advanced Ceramics Section Glass and Advanced Materials Division, Bhabha Atomic Research Centre, Mumbai, India

M. Mar Gil-Diaz IMIDRA, Alcalá de Henares, Madrid, Spain

Luc Girard Institut de Chimie Separative de Marcoule, UMR 5257 – ICSM Site de Marcoule, CEA/CNRS/UM2/ENSCM, Bagnols sur Ceze, France

Jean Gobet Adamant-Technologies, La Chaux-de-Fonds, Switzerland

Luis Godinez Centro de Investigación y Desarrollo Tecnológico en Electroquímica S.C., Querétaro, Mexico

Alan Le Goff Department of Molecular Chemistry, CNRS UMR 5250, CNRS-University of Grenoble, Grenoble, France

Muriel Golzio CNRS; IPBS (Institut de Pharmacologie et de Biologie Structurale), Toulouse, France

Ignacio Gonzalez Department of Chemistry, Universidad Autónoma Metropolitana-Iztapalapa, México, Mexico

Heiner Jakob Gores Institute of Physical Chemistry, Münster Electrochemical Energy Technology (MEET), Westfälische Wilhelms-Universität Münster (WWU), Münster, Germany

Pau Gorostiza Institució Catalana de Recerca i Estudis Avançats (ICREA), Barcelona, Spain

Lars Gundlach Department of Chemistry and Biochemistry and Department of Physics and Astronomy, University of Delaware, Newark, DE, USA

Ulrich Guth Kurt-Schwabe-Institut für Mess- und Sensortechnik e.V. Meinsberg, Waldheim, Germany

FB Chemie und Lebensmittelchemie, Technische Universität Dresden, Dresden, Germany

Geir Martin Haarberg Department of Materials Science and Engineering, Norwegian University of Science and Technology (NTNU), Trondheim, Norway

Jonathan E. Halls Department of Chemistry, The University of Bath, Bath, UK

Ahmad Hammad Research and Development Center, Saudi Aramco, Dhahran, Saudi Arabia

Achim Hannappel DECHEMA Research Institute of Biochemical Engineering, Frankfurt am Main, Germany

Falk Harnisch Institute of Environmental and Sustainable Chemistry, Technical University Braunschweig, Braunschweig, Germany

Akitoshi Hayashi Department of Applied Chemistry, Osaka Prefecture University, Sakai, Osaka, Japan

Christoph Held Department of Biochemical and Chemical Engineering, Technische Universität Dortmund, Dortmund, Germany

Wesley A. Henderson Department of Chemical and Biomolecular Engineering, North Carolina State University, Raleigh, NC, USA

Peter J. Hesketh School of Mechanical Engineering, Georgia Institute of Technology, Atlanta, GA, USA

Michael Heyrovsky J. Heyrovsky Institute of Physical Chemistry of the ASCR, Prague, Czech Republic

Takashi Hibino Graduate School of Environmental Studies, Nagoya University, Nagoya, Japan

Yoshio Hisaeda Department of Chemistry and Biochemistry, Kyushu University, Graduate School of Engineering, Fukuoka, Japan

Tuan Hoang University of Southern California, Los Angeles, CA, USA

Dirk Holtmann DECHEMA Research Institute of Biochemical Engineering, Frankfurt am Main, Germany

Rudolf Holze AG Elektrochemie, Institut für Chemie, Technische Universität Chemnitz, Chemnitz, Germany

Michael Holzinger Department of Molecular Chemistry, CNRS UMR 5250, CNRS-University of Grenoble, Grenoble, France

Dominik Horinek Institute of Physical and Theoretical Chemistry, University of Regensburg, Regensburg, Germany

Teruhisa Horita Fuel Cell Materials Group, Energy Technology Research Institute, National Institute of Advanced Industrial Science and Technology (AIST), Tsukuba, Ibaraki, Japan

Barbara Hribar-Lee Faculty of Chemistry and Chemical Technology, University of Ljubljana, Ljubljana, Slovenia

Chang-Jung Hsueh Electronics Design Center, and Chemical Engineering Department, Case Western Reserve University, Cleveland, OH, USA

Chi-Chang Hu Chemical Engineering Department, National Tsing Hua University, Hsinchu, Taiwan

Gary W. Hunter NASA Glenn Research Center, Cleveland, OH, USA

Jorge G. Ibanez Department of Chemical Engineering and Sciences, Universidad Iberoamericana, México, Mexico

Munehisa Ikoma Panasonic, Moriguchi, Japan

Nobuhito Imanaka Department of Applied Chemistry, Faculty of Engineering, Osaka University, Osaka, Japan

Shinsuke Inagi Tokyo Institute of Technology, Midori-ku, Yokohama, Japan

Hiroshi Inoue Osaka Prefecture University, Sakai, Osaka, Japan

György Inzelt Department of Physical Chemistry, Eötvös Loránd University, Budapest, Hungary

Tsutomu Ioroi AIST, Ikeda, Japan

Hiroshi Irie Yamanashi University, Yamanashi Prefecture, Japan

John Thomas Sirr Irvine School of Chemistry, University of St Andrews, St Andrews, UK

Manabu Ishifune Kinki University, Higashi-Osaka, Osaka, Japan

Akimitsu Ishihara Yokohama National University, Hodogaya-ku, Yokohama, Japan

Tatsumi Ishihara Department of Applied Chemistry, Faculty of Engineering, International Institute for Carbon Neutral Energy Research (WPI-I2CNER), Kyushu University, Nishi ku, Fukuoka, Japan

Masashi Ishikawa Kansai University, Suita, Osaka, Japan

Adriana Ispas FG Elektrochemie und Galvanotechnik, Institut für Werkstofftechnik, Technische Universitüt Ilmenau, Ilmenau, Germany

Gaurav Jain MECC, Medtronic Inc., Minneapolis, MN, USA

Metini Janyasupab Electronics Design Center, and Chemical Engineering Department, Case Western Reserve University, Cleveland, OH, USA

Fengjing Jiang Institute of Fuel Cells, School of Mechanical Engineering, Shanghai Jiao Tong University, Shanghai, People's Republic of China

Maria Jitaru Research Institute for Organic Auxiliary Products (ICPAO), Medias, Romania

Jakob Jörissen Chair of Technical Chemistry, Technical University of Dortmund, Germany

Pavel Jungwirth Institute of Organic Chemistry and Biochemistry, Academy of Sciences of the Czech Republic, Prague, Czech Republic

Yoshifumi Kado Asahi Kasei Chemicals Corporation, Tokyo, Japan

Heike Kahlert Institut für Biochemie, Universität Greifswald, Greifswald, Germany

Yijin Kang University of Pennsylvania, Philadelphia, PA, USA

Agnieszka Kapałka Institute of Chemical Sciences and Engineering, Ecole Polytechnique Fédérale de Lausanne (EPFL), Lausanne, Switzerland

Shigenori Kashimura Kinki University, Higashi-Osaka, Japan

Alexandros Katsaounis Department of Chemical Engineering, University of Patras, Patras, Greece

Jurg Keller Advanced Water Management Centre (AWMC), The University of Queensland, Brisbane, QLD, Australia

Geoffrey H. Kelsall Department of Chemical Engineering, Imperial College London, London, UK

Sangtae Kim Department of Chemical Engineering and Materials Science, University of California, Davis, CA, USA

Woong-Ki Kim Faculty of Electrical Engineering and Computer Science, Ingolstadt University of Applied Sciences, Ingolstadt, Germany

Axel Kirste BASF SE, Ludwigshafen, Germany

Naoki Kise Department of Chemistry and Biotechnology, Graduate School of Engineering, Tottori University, Tottori, Japan

Norihisa Kobayashi Chiba University, Chiba, Japan

Svenja Kochius DECHEMA Research Institute, Frankfurt am Main, Germany

Paul A. Kohl Georgia Institute of Technology, School of Chemical and Biomolecular Engineering, Atlanta, GA, USA

Ulrike I. Kramm Technical University Cottbus, Cottbus, Germany

Mario Krička AG Elektrochemie, Institut für Chemie, Technische Universität Chemnitz, Chemnitz, Germany

Nedeljko Krstajic Faculty of Technology and Metallurgy, University of Belgrade, Belgrade, Serbia

Akihiko Kudo Tokyo University of Science, Tokyo, Japan

Andrzej Kuklinski Fakultät Chemie, Biofilm Centre/Aquatische Biotechnologie, Universität Duisburg-Essen, Essen, Germany

Juozas Kulys Department of Chemistry and Bioengineering, Vilnius Gediminas Technical University, Vilnius, Lithuania

Werner Kunz Institut für Biophysik, Fachbereich Physik, Johann Wolfgang Goethe-Universität Frankfurt am Main, Frankfurt am Main, Germany

Manabu Kuroboshi Okayama University, Okayama, Japan

Jan Labuda Institute of Analytical Chemistry, Faculty of Chemical and Food Technology, Slovak University of Technology, Bratislava, Slovakia

Claude Lamy Institut Européen des Membranes, Université Montpellier 2, UMR CNRS n° 5635, Montpellier, France

Ying-Hui Lee Chemical Engineering Department, National Tsing Hua University, Hsinchu, Taiwan

Carlos A. Ponce de Leon Electrochemical Engineering Laboratory, University of Southampton, Faculty of Engineering and the Environment, Southampton, Hampshire, UK

Jean Lessard Universite de Sherbrooke, Quebec, Canada

Hans J. Lewerenz Joint Center for Artificial Photosynthesis, California Institute of Technology, Pasadena, CA, USA

Claudia Ley DECHEMA Research Institute, Frankfurt am Main, Germany

Meng Li Chemistry Department, Brookhaven National Laboratory, Upton, NY, USA

R. Daniel Little University of California, Santa Barbara, CA, USA

Chen-Wei Liu Institute for Material Science and Engineering, National Central University, Jhongli City, Taoyuan County, Taiwan

Chung-Chiun Liu Electronics Design Center, and Chemical Engineering Department, Case Western Reserve University, Cleveland, OH, USA

Ping Liu Brookhaven National Laboratory, Upton, NY, USA

Yoav D. Livney Faculty of Biotechnology and Food Engineering, The Technion, Israel Institute of Technology, Haifa, Israel

Leonardo Lizarraga Université Lyon 1, CNRS, UMR 5256, IRCELYON, Institut de recherches sur la catalyse et l'environnement de Lyon, Villeurbanne, France

M. Carmen Lobo IMIDRA, Alcalá de Henares, Madrid, Spain

Svenja Lohner Stanford University, Stanford, CA, USA

Manuel Lohrengel University of Düsseldorf, Düsseldorf, Germany

Reiner Lomoth Department of Chemistry - Ångström Laboratory, Uppsala University, Uppsala, Sweden

Daniel Lowy FlexEl, LLC, College Park, MD, USA

Roland Ludwig Department of Food Science and Technology, Vienna Institute of Biotechnology BOKU-University of Natural Resources and Life Sciences, Vienna, Austria

Dirk Lützenkirchen-Hecht Fachbereich C- Abteilung Physik, Wuppertal, Germany

Vadim F. Lvovich NASA Glenn Research Center, Electrochemistry Branch, Power and In-Space Propulsion Division, Cleveland, OH, USA

Johannes Lyklema Department of Physical Chemistry and Colloid Science, Wageningen University, Wageningen, The Netherlands

Hirofumi Maekawa Department of Materials Science and Technology, Nagaoka University of Technology, Nagaoka, Japan

Anders O. Magnusson DECHEMA Research Institute, Frankfurt am Main, Germany

J. Maier Max Planck Institute for Solid State Research, Stuttgart, Germany

Frédéric Maillard Laboratoire d'Electrochimie et de Physico-chimie des Matériaux et des Interfaces, Saint Martin d'Héres, France

Daniel Mandler Institute of Chemistry, The Hebrew University, Jerusalem, Israel

Klaus-Michael Mangold DECHEMA-Forschungsinstitut, Frankfurt am Main, Germany

Yizhak Marcus Department of Inorganic and Analytical Chemistry, The Hebrew University, Jerusalem, Israel

Nebojsa Marinkovic Synchrotron Catalysis Consortium, University of Delaware, Newark, DE, USA

Frank Marken University of Bath, Bath, UK

István Markó Universite Catholique de Louvain, Louvain-la-Neuve, Belgium

Marko S. Markov Research International, Williamsville, NY, USA

Jack Marple Research and Development, Energizer Battery Manufacturing Inc, Westlake, OH, USA

Virginie Marry Laboratoire Physicochimie des Electrolytes, Colloïdes et Sciences Analytiques, CNRS, ESPCI, Université Pierre et Marie Curie, Paris, France

Guillermo Marshall Laboratorio de Sistemas Complejos, Departamento de Ciencias de la Computación, Facultad de Ciencias Exactas y Naturales, Universidad de Buenos Aires, Buenos Aires, Argentina

Carlos Alberto Martinez-Huitle Institute of Chemistry, Federal University of Rio Grande do Norte, Lagoa Nova, Natal, RN - CEP, Brazil

Marco Mascini Dipartimento di Chimica "Ugo Schiff", Università degli Studi di Firenze, Sesto Fiorentino, Firenze, Italy

Rudy Matousek Severn Trent Services, Sugar Land, TX, USA

Hiroshige Matsumoto Kyushu University, Fukuoka, Japan

Kouichi Matsumoto Faculty of Science and Engineering, Kinki University, Higashi-osaka, Japan

Werner Mäntele Johann Wolfgang Goethe-Universität Frankfurt am Main, Institut für Biophysik, Fachbereich Physik, Frankfurt am Main, Germany

Steven McIntosh Department of Chemical Engineering, Lehigh University, Bethlehem, PA, USA

Jennifer R. McKenzie Department of Chemistry, Vanderbilt University, Nashville, TN, USA

Ellis Meng University of Southern California, Los Angeles, CA, USA

Pierre-Alain Michaud Institute of Chemical Sciences and Engineering, Ecole Polytechnique Fédérale de Lausanne (EPFL), Lausanne, Switzerland

Richard L. Middaugh Independent Consultant, formerly with Energizer Battery Co, Rocky River, OH, USA

Alessandro Minguzzi Dipartimento di Chimica, Università degli Studi di Milano, Milan, Italy

Shigenori Mitsushima Yokohama National University, Yokohama, Kanagawa Prefecture, Japan

Tsutomu Miyasaka Graduate School of Engineering, Toin University of Yokohama, Yokohama, Kanagawa, Japan

Kenji Miyatake University of Yamanashi, Kofu, Yamanashi, Japan

Junichiro Mizusaki Tohoku University, Funabashi, Chiba, Japan

Mogens Bjerg Mogensen Department of Energy Conversion and Storage, Technical University of Denmark, Roskilde, Denmark

Charles W. Monroe Department of Chemical Engineering, University of Michigan, Ann Arbor, MI, USA

Vicente Montiel Instituto Universitario de Electroquímica, University of Alicante, Alicante, Spain

Somayeh Moradi AG Elektrochemie, Institut für Chemie, Technische Universität Chemnitz, Chemnitz, Germany

J. Thomas Mortimer Case Western Reserve University, Cleveland, OH, USA

Hubert Motschmann Institute of Physical and Theoretical Chemistry, University of Regensburg, Regensburg, Germany

Christopher B. Murray University of Pennsylvania, Philadelphia, PA, USA

Katsuhiko Naoi Institute of Symbiotic Science and Technology, Tokyo University of Agriculture and Technology, Koganei, Tokyo, Japan

Hiroki Nara Faculty of Science and Engineering, Waseda University, Okubo, Shinjuku-ku, Tokyo, Japan

George Neophytides Electrochemistry Laboratory, Paul Scherrer Institut, Villigen, Switzerland

Roland Neueder Institute of Physical and Theoretical Chemistry, University of Regensburg, Regensburg, Germany

Jinren Ni Peking University, Beijing, China

Ernst Niebur Mind/Brain Institute, Johns Hopkins University, Baltimore, MD, USA

Branislav Ž. Nikolić Department of Physical Chemistry and Electrochemistry, University of Belgrade, Faculty of Technology and Metallurgy, Belgrade, Serbia

Yoshinori Nishiki Development Department, Permelec Electrode Ltd, Kanagawa, Japan

Shigeru Nishiyama Department of Chemistry, Keio University, Hiyoshi, Yokohama, Japan

Toshiyuki Nohira Graduate School of Energy Science, Kyoto University, Kyoto, Japan

Atusko Nosaka Department Materials Science and Technology, Nagaoka University of Technology, Nagaoka, Niigata, Japan

Naoyoshi Nunotani Department of Applied Chemistry, Faculty of Engineering, Osaka University, Osaka, Japan

Tsuyoshi Ochiai Kanagawa Academy of Science and Technology, Takatsu–ku, Kawasaki, Kanagawa, Japan

Photocatalysis International Research Center, Tokyo University of Science, Noda, Chiba, Japan

Wolfram Oelßner Kurt-Schwabe-Institut für Mess- und Sensortechnik e.V. Meinsberg, Kurt-Schwabe-Straße, Waldheim, Germany

Andreas Offenhäusser Institute of Complex Systems, Peter Grünberg Institute: Bioelectronics, Jülich, Germany

Ulker Bakir Ogutveren Anadolu University, Eskişehir, Turkey

Bunsho Ohtani Catalysis Research Center, Hokkaido University, Sapporo, Hokkaido, Japan

Yohei Okada Tokyo University of Agriculture and Technology, Fuchu, Tokyo, Japan

Osamu Onomura Nagasaki University, Nagasaki, Japan

Immaculada Ortiz Department of Chemical Engineering and Inorganic Chemistry, University of Cantabria, Santander, Cantabria, Spain

Juan Manuel Ortiz Instituto Universitario de Electroquímica, University of Alicante, Alicante, Spain

Tetsuya Osaka Faculty of Science and Engineering, Waseda University, Okubo, Shinjuku-ku, Tokyo, Japan

Ken-ichiro Ota Yokohama National University, Fac. Engineering, Yokohama, Japan

Lisbeth M. Ottosen Department of Civil Engineering, Technical University of Denmark, Lyngby, Denmark

Ilaria Palchetti Dipartimento di Chimica "Ugo Schiff", Università degli Studi di Firenze, Sesto Fiorentino, Firenze, Italy

Vladimir Panić University of Belgrade, Institute of Chemistry, Technology and Metallurgy, Belgrade, Serbia

Marco Panizza University of Genoa, Genoa, Italy

Juan Manuel Peralta-Hernández Centro de Innovación Aplicada en Tecnologías Competitivas, Guanajuato, Mexico

Cristina Pereira Instituto de Tecnologia Química e Biológica, Universidade Nova de Lisboa, Oeiras, Portugal

Araceli Pérez-Sanz IMIDRA, Alcalá de Henares, Madrid, Spain

Laurence (Laurie) Peter Department of Chemistry, University of Bath, Bath, UK

Marija Petkovic Instituto de Tecnologia Química e Biológica, Universidade Nova de Lisboa, Oeiras, Portugal

Ilje Pikaar Advanced Water Management Centre (AWMC), The University of Queensland, Brisbane, QLD, Australia

Antonio Plaza IMIDRA, Alcalá de Henares, Madrid, Spain

Dmitry E. Polyansky Chemistry Department, Brookhaven National Laboratory, Upton, NY, USA

Jinyi Qin Department of Environmental Microbiology, Helmholtz Centre for Environmental Research - UFZ, Leipzig, Germany

Jelena Radjenovic Advanced Water Management Centre (AWMC), The University of Queensland, Brisbane, QLD, Australia

Krishnan Rajeshwar The University of Texas at Arlington, Arlington, TX, USA

Nayif A. Rasheedi Research and Development Center, Saudi Aramco, Dhahran, Saudi Arabia

David Rauh EIC Laboratories, Inc, Norwood, MA, USA

Thomas B. Reddy Department of Materials Science and Engineering, Rutgers, The State University of New Jersey, Piscataway, NJ, USA

David Reyter INRS Energie, Matériaux et Télécommunications, Varennes, Quebec, Canada

Marcel Risch MIT, Cambridge, MA, USA

Vivian Robinson ETP Semra Pty Ltd, Canterbury, NSW, Australia

Romeu C. Rocha-Filho Universidade Federal de Goiás, Catalão, Brazil

Manuel A. Rodrigo Department of Chemical Engineering, Faculty of Chemical Sciences and Technology, Universided de Castille la Mandne, Ciudad Real, Spain

Paramaconi Rodriguez School of Chemistry, The University of Birmingham, Birmingham, UK

Alberto Rojas-Hernández Department of Chemistry, Universidad Autónoma Metropolitana-Iztapalapa, México, Mexico

Sandra Rondinini Dipartimento di Chimica, Università degli Studi di Milano, Milan, Italy

Benjamin Rotenberg Laboratoire Physicochimie des Electrolytes, Colloïdes et Sciences Analytiques, CNRS, ESPCI, Université Pierre et Marie Curie, Paris, France

Anna Joëlle Ruff Lehrstuhl für Biotechnologie, RWTH Aachen University, Aachen, Germany

Luís Augusto M. Ruotolo Department of Chemical Engineering, Federal University of São Carlos, São Carlos, SP, Brazil

Jennifer L. M. Rupp Electrochemical Materials, ETH Zurich, Zurich, Switzerland

Yoshihiko Sadaoka Ehime University, Matsuyama, Japan

Gabriele Sadowski Department of Biochemical and Chemical Engineering, Technische Universität Dortmund, Dortmund, Germany

Hikari Sakaebe Research Institute for Ubiquitous Energy Devices, National Institute of Advanced Industrial Science and Technology (AIST), Ikeda, Osaka, Japan

Hikari Sakaebe National Institute of Advanced Industrial Science and Technology (AIST), Ikeda, Osaka, Japan

Mathieu Salanne Laboratoire PECSA, UMR 7195, Université Pierre et Marie Curie, Paris, France

Wolfgang Sand Fakultät Chemie, Biofilm Centre/Aquatische Biotechnologie, Universität Duisburg-Essen, Essen, Germany

Shriram Santhanagopalan National Renewable Energy Laboratory, Golden, CO, USA

Hamidreza Sardary AG Elektrochemie, Institut für Chemie, Technische Universität Chemnitz, Chemnitz, Germany

Kotaro Sasaki Chemistry Department, Brookhaven National Laboratory, Upton, NY, USA

Richard Sass DECHEMA e.V., Informationssysteme und Datenbanken, Frankfurt, Germany

Shunsuke Sato Toyota Central Research and Development Laboratories, Inc., Nagakute, Aichi, Japan

André Savall Laboratoire de Génie Chimique, CNRS, Université Paul Sabatier, Toulouse, France

Elena Savinova Institut de Chimie et Procédés pour l'Energie, l'Environnement et la Santé, UMR 7515 CNRS, Université de Strasbourg-ECPM, Strasbourg, France

Natascha Schelero Stranski-Laboratorium, Institut für Chemie, Fakultät II, Technical University Berlin, Berlin, Germany

Günther G. Scherer Electrochemistry Laboratory, Paul Scherrer Institute, Villigen, Switzerland

Thomas J. Schmidt Electrochemistry Laboratory, Paul Scherrer Institut, Villigen, Switzerland

Jens Schrader DECHEMA Research Institute of Biochemical Engineering, Frankfurt am Main, Germany

Uwe Schroeder Institute of Environmental and Sustainable Chemistry, Technical University Braunschweig, Braunschweig, Germany

Brooke Schumm Eagle Cliffs, Inc., Bay Village, OH, USA

Ulrich Schwaneberg Lehrstuhl für Biotechnologie, RWTH Aachen University, Aachen, Germany

Hans-Georg Schweiger Faculty of Electrical Engineering and Computer Science, Ingolstadt University of Applied Sciences, Ingolstadt, Germany

Hisanori Senboku Hokkaido University, Sapporo, Hokkaido, Japan

Daniel Seo Department of Chemical and Biomolecular Engineering, North Carolina State University, Raleigh, NC, USA

Karine Groenen Serrano Laboratory of Chemical Engineering, University of Paul Sabatier, Toulouse, France

Anwar-ul-Haq Ali Shah Institute of Chemical Sciences, University of Peshawar, Peshawar, Pakistan

Yang Shao-Horn MIT, Cambridge, MA, USA

Hisashi Shimakoshi Department of Chemistry and Biochemistry, Kyushu University, Graduate School of Engineering, Fukuoka, Japan

Yasuhiro Shimizu Graduate School of Engineering, Nagasaki University, Nagasaki, Japan

Youichi Shimizu Department of Applied Chemistry, Graduate School of Engineering, Kyushu Institute of Technology, Kitakyushu, Fukuoka, Japan

Komaba Shinichi Department of Applied Chemistry, Tokyo University of Science, Shinjuku, Tokyo, Japan

Soshi Shiraishi Division of Molecular Science, Faculty of Science and Technology, Gunma University, Kiryu, Gunma, Japan

Pavel Shuk Rosemount Analytical Inc. Emerson Process Management, Solon, OH, USA

Jean-Pierre Simonin Laboratoire PECSA, UMR CNRS 7195, Université Paris, Paris, France

Subhash C. Singhal Pacific Northwest National Laboratory, Richland, WA, USA

Ignasi Sirés Laboratory of Electrochemistry of Materials and Environment, Department of Physical Chemistry, Faculty of Chemistry, University of Barcelona, Barcelona, Spain

Stephen Skinner Department of Materials, Imperial College London, London, UK

Marshall C. Smart Jet Propulsion Laboratory, Pasdena, CA, USA

Henry Snaith Clarendon Laboratory, Department of Physics, Oxford University, Oxford, UK

Stamatios Souentie Department of Chemical Engineering, University of Patras, Patras, Greece

Bernd Speiser Institut für Organische Chemie, Universität Tübingen, Tübingen, Germany

Jacob Spendelow Los Alamos National Laboratory, Los Alamos, NM, USA

Daniel Steingart Department of Mechanical and Aerospace Engineering, Andlinger Center for Energy, the Environment Princeton University, Princeton, NJ, USA

John Stickney University of Georgia, Athens, GA, USA

Margarita Stoytcheva Instituto de Ingenieria, Universidad Autonoma de Baja California, Mexicali, Baja California, Mexico

Svetlana B. Strbac ICTM-Institute of Electrochemistry, University of Belgrade, Belgrade, Serbia

Eric M. Stuve Department of Chemical Engineering, University of Washington, Seattle, WA, USA

Stenbjörn Styring Department of Chemistry - Ångström Laboratory, Uppsala University, Uppsala, Sweden

Seiji Suga Okayama University, Okayama, Japan

Wataru Sugimoto Faculty of Textile Science and Technology, Shinshu University, Nagano, Japan

I-Wen Sun Department of Chemistry, National Cheng kung University, Tainan, Taiwan

Jin Suntivich Cornell University, Ithaca, NY, USA

Hitoshi Takamura Department of Materials Science, Graduate School of Engineering, Tohoku University, Sendai, Japan

Prabhakar A. Tamirisa MECC, Medtronic Inc., Minneapolis, MN, USA

Shinji Tamura Department of Applied Chemistry, Faculty of Engineering, Osaka University, Osaka, Japan

Hideo Tanaka Okayama University, Okayama, Japan

Tadaaki Tani Society of Photography and Imaging of Japan, Tokyo, Japan

Akimasa Tasaka Doshisha University, Kyotanabe, Kyoto, Japan

Tetsu Tatsuma Institute of Industrial Science, University of Tokyo, Tokyo, Japan

Masahiro Tatsumisago Department of Applied Chemistry, Osaka Prefecture University, Sakai, Osaka, Japan

Pierre Taxil Laboratoire de Génie Chimique, Université de Toulouse, Toulouse, France

Justin Teissie CNRS; IPBS (Institut de Pharmacologie et de Biologie Structurale), Toulouse, France

Ingrid Tessmer Rudolf-Virchow-Zentrum, Experimentelle Biomedizin, University of Würzburg, Würzburg, Germany

Anders Thapper Department of Chemistry - Ångström Laboratory, Uppsala University, Uppsala, Sweden

Masataka Tomita Department of Applied Chemistry, Tokyo University of Science, Shinjuku, Tokyo, Japan

Marc Tornow Institut für Halbleitertechnik, Technische Universität München, München, Germany

Taro Toyoda Department of Engineering Science, The University of Electro-Communications, Chofu, Tokyo, Japan

Dimitrios Tsiplakides Department of Chemistry, Aristotle University of Thessaloniki, Thessaloniki, Greece

Pierre Turq Laboratoire Physicochimie des Electrolytes, Colloïdes et Sciences Analytiques, CNRS, ESPCI, Université Pierre et Marie Curie, Paris, France

Makoto Uchida Fuel Cell Nanomaterials Center, Yamanashi University, 6-43 Miyamaecho, Kofu, Yamanashi, Japan

Kai M. Udert Eawag, Dübendorf, Switzerland

Helmut Ullmann FB Chemie und Lebensmittelchemie, Technische Universität Dresden, Dresden, Germany

Soichiro Uno Nissan Research Center/EV System Laboratory, NISSAN MOTOR CO., LTD., Yokosuka-shi, Kanagawa-ken, Japan

Kohei Uosaki International Center for Materials Nanoarchitectonics (WPI-MANA), National Institute for Materials Science (NIMS), Tsukuba, Japan

Ane Urtiaga Department of Chemical Engineering and Inorganic Chemistry, University of Cantabria, Santander, Cantabria, Spain

Francisco J. Rodriguez Valadez Centro de Investigación y Desarrollo Tecnológico en Electroquímica S.C., Querétaro, Mexico

S. C. Sanfandila, Research Branch, Center for Research and Technological Development in Electrochemistry, Querétaro, Mexico

Vladimir Vashook Kurt-Schwabe-Institut für Mess- und Sensortechnik e.V. Meinsberg, Waldheim, Germany

FB Chemie und Lebensmittelchemie, Technische Universität Dresden, Dresden, Germany

Ruben Vasquez-Medrano Department of Chemical Engineering and Sciences, Universidad Iberoamericana, México, Mexico

Nicolaos Vatistas DICI, Università di Pisa, Pisa, Italy

Constantinos G. Vayenas Department of Chemical Engineering, University of Patras, Patras, Achaia, Greece

Danae Venieri Department of Environmental Engineering, Technical University of Crete, Chania, Greece

Philippe Vernoux Université Lyon 1, CNRS, UMR 5256, IRCELYON, Institut de recherches sur la catalyse et l'environnement de Lyon, Villeurbanne, France

Alberto Vertova Dipartimento di Chimica, Università degli Studi di Milano, Milan, Italy

Bernardino Virdis Advanced Water Management Centre (AWMC), The University of Queensland, Brisbane, QLD, Australia

Centre for Microbial Electrosynthesis (CEMES), The University of Queensland, Brisbane, QLD, Australia

Vojko Vlachy Faculty of Chemistry and Chemical Technology, University of Ljubljana, Ljubljana, Slovenia

Rittmar von Helmolt Government Programs and Research Strategy, GM Alternative Propulsion Center, Adam Opel AG, Rüsselsheim, Germany

Regine von Klitzing Stranski-Laboratorium, Institut für Chemie, Fakultät II, Technical University Berlin, Berlin, Germany

Winfried Vonau Kurt-Schwabe-Institut fuer Mess- und Sensortechnik e.V. Meinsberg, Waldheim, Germany

Yunny Meas Vong CIDETEQ, Parque Tecnológico Querétaro, México, Estado de Querétaro, México

Lj Vracar Faculty of Technology and Metallurgy University of Belgrade, Belgrade, Serbia

Miomir B. Vukmirovic Chemistry Department, Brookhaven National Laboratory, Upton, NY, USA

Vlastimil Vyskocil UNESCO Laboratory of Environmental Electrochemistry, Faculty of Science, Department of Analytical Chemistry, Charles University in Prague, Prague, Czech Republic

Jay D. Wadhawan Department of Chemistry, The University of Hull, Hull, UK

Siegfried R. Waldvogel Johannes Gutenberg-University Mainz, Mainz, Germany

Frank C. Walsh Electrochemical Engineering Laboratory, University of Southampton, Faculty of Engineering and the Environment, Southampton, Hampshire, UK

Jia X. Wang Brookhaven National Laboratory, Upton, NY, USA

Masahiro Watanabe University of Yamanashi, Kofu, Yamanashi, Japan

Takao Watanabe Central Research Institute of Electric Power Industry, Yokosuka, Kanagawa, Japan

Andrew Webber Technology, Energizer Battery Manufacturing Inc., Westlake, OH, USA

Adam Z. Weber Lawrence Berkeley National Laboratory, Berkeley, CA, USA

Hermann Weingärtner Lehrstuhl für Physikalische Chemie II, Ruhr-Universität Bochum, Bochum, Germany

Nina Welschoff Johannes Gutenberg-University Mainz, Mainz, Germany

Alan West Department of Chemical Engineering Columbia University, New York, NY, USA

Reiner Westermeier SERVA Electrophoresis GmbH, Heidelberg, Germany

Lukas Y. Wick Department of Environmental Microbiology, Helmholtz Centre for Environmental Research - UFZ, Leipzig, Germany

Andrzej Wieckowski University of Illinois, Urbana, IL, USA

Alexander Wiek AG Elektrochemie, Institut für Chemie, Technische Universität Chemnitz, Chemnitz, Germany

John P. Wikswo Department of Physics and Astronomy, Vanderbilt University, Nashville, TN, USA

Frank Willig Fritz-Haber-Institut der Max-Planck-Gesellschaft, Berlin, Germany

Yuping Wu Department of Chemistry, Fudan University, Shanghai, China

Naoaki Yabuuchi Department of Applied Chemistry, Tokyo University of Science, Shinjuku, Tokyo, Japan

Kohta Yamada Asahi Glass Co. Ltd., Research Center, Kanagawa-ku, Yokohama, Japan

Yoshiaki Yamaguchi Technical Development Division, Global Technical Headquarters, GS Yuasa International Ltd., Kyoto, Japan

Ichiro Yamanaka Department of Applied Chemistry, Tokyo Institute of Technology, Tokyo, Japan

Shigeaki Yamazaki Kansai University, Suita, Osaka, Japan

Harumi Yokokawa The University of Tokyo, Institute of Industrial Science, Tokyo, Japan

H.-I. Yoo Department of Materials Science and Engineering, Seoul National University, Seoul, Korea

Nobuko Yoshimoto Graduate School of Science and Engineering, Yamaguchi University, Yamaguchi, Japan

Akira Yoshino Yoshino Laboratory, Asahi Kasei Corporation, Fuji, Shizuoka, Japan

Jun-ichi Yosida Kyoto University, Kyoto, Japan

Zaki Yusuf Research and Development Center, Saudi Aramco, Dhahran, Saudi Arabia

Junliang Zhang Institute of Fuel Cells, School of Mechanical Engineering, Shanghai Jiao Tong University, Shanghai, People's Republic of China

Fengjuan Zhu Institute of Fuel Cells, School of Mechanical Engineering, Shanghai Jiao Tong University, Shanghai, People's Republic of China

Hongmin Zhu School of Metallurgical and Ecological Engineering, University of Science and Technology Beijing, Beijing, China

Roumen Zlatev Instituto de Ingenieria, Universidad Autonoma de Baja California, Mexicali, Baja California, Mexico

Jens Zosel Kurt-Schwabe-Institut für Mess- und Sensortechnik e.V. Meinsberg, Waldheim, Germany

Sandra Zugmann EVA Fahrzeugtechnik, Munich, Germany

Institució Catalana de Recerca i Estudis Avançats (ICREA), Institute for Bioengineering of Catalonia (IBEC), Barcelona, Spain

Electrical and Computer Engineering, University of California Davis, Davis, CA, USA

Andreas Züttel Empa Materials Science and Technology, Hydrogen and Energy, Dübendorf, Switzerland

Faculty of Applied Science, DelftChemTech, Delft, The Netherlands

P

Paired Electrosynthesis

Nicola Aust and Axel Kirste
BASF SE, Ludwigshafen, Germany

Introduction

Organic electrochemistry can be a very powerful synthetic tool [1–3]. Anodic oxidation as well as cathodic reduction processes are utilized [4]. If one wants to carry out a synthesis based on an anodic oxidation, the cathodic process is normally not of synthetic interest and vice versa. Nevertheless in some cases the counter-electrode process can be used also. An example is the cathodic evolution of hydrogen in protic solvents like water or methanol, while the anodic process renders the desired product. The evolving hydrogen can be utilized as fuel material. In these cases the counter-electrode reaction is of economical and not of synthetic use. But electrochemists dream of a paired electrosynthesis using cathodic and anodic process for synthesis to achieve the ultimate goal: a 200 % electrosynthesis.

There are several ways possible to perform such a paired electrosynthesis, e.g., in a parallel, convergent, divergent, or linear assembly [5, 6] (Fig. 1).

In a parallel electrosynthesis the products generated at anode and cathode do not interfere or react. In contrast stands the convergent paired electrosynthesis where the generated products react to one final product or where by anodic and cathodic process the same product is synthesized. Using one substrate and creating an oxidized and a reduced product is the main feature of a divergent paired electrosynthesis. The linear assembly has in common that also only one substrate is employed. But in the linear paired electrosynthesis, one product is synthesized from the substrate by one- or two-mediated electrode processes sometimes in a cascade-like path. All different forms of a paired electrosynthesis have in common that they are highly efficient and therefore resources preserving and waste/emissions minimizing. This is also summarized under the term "green synthesis" [6].

The benefit is high, but it is obvious that to conduct a successful paired electrosynthesis is not an easy task. Both electrode processes have to be compatible. Yield losses at the counter-electrode or a high effort to separate the electrode processes stand against a paired electrolysis. Product separation and isolation is an important factor especially in a parallel or divergent electrosynthesis [7]. The following electrolyses shall illustrate the important features of a paired electrosynthesis and show the broad range of application.

Example of a Parallel Paired Electrosynthesis

A successful industrial example of a parallel paired electrosynthesis is BASF's electrolysis of 4-*tert*-butylbenzaldehyde dimethylacetal and

G. Kreysa et al. (eds.), *Encyclopedia of Applied Electrochemistry*, DOI 10.1007/978-1-4419-6996-5,
© Springer Science+Business Media New York 2014

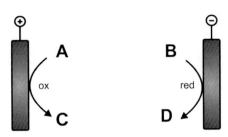
Parallel paired electrolysis

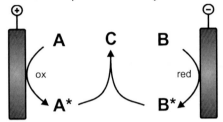
Convergent paired electrolysis

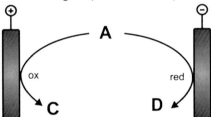
Divergent paired electrolysis

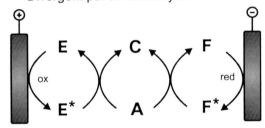
Linear paired electrolysis

Paired Electrosynthesis, Fig. 1 Different forms of a paired electrosynthesis (*A, B* = substrate; *C, D* = product; *E, F* = mediator)

phthalide simultaneously in an undivided cell [8] (Fig. 2).

Already in the 1980s the electrosynthesis of acetals of aromatic aldehydes has been established at BASF. Employing the anodic substitution with methanol as nucleophile on methyl-substituted aromatic compounds, one generates by double methoxylation first the intermediate ether and then the corresponding acetal [9, 10]. At the cathode hydrogen is generated (Fig. 3).

After the electrolysis the acetals can be hydrolyzed to their aldehydes and methanol is recovered. By this elegant way to avoid overoxidation to the acids, aromatic aldehydes are synthesized from toluene derivatives [11]. The electrosynthesis takes place in good yields for toluene derivatives with electron pushing para-substituents like the *tert*-butyl group. It is carried out in an undivided cell developed by BASF: the capillary gap cell which contains a stack of bipolar round graphite electrodes. The electrodes are separated by spacers and connected in series [12].

More than 10 years after the establishment of the acetalization process, electrochemists at BASF have searched for a compatible reductive process that can be run instead of the hydrogen evolution. What they have found is the reduction of phthalic acid dimethylester to phthalide [8] (Fig. 2). Phthalide is a compound that was up till then generated by classical catalytic hydrogenation from phthalic acid anhydride [7]. Part of the process development has been to fit the paired electrosynthesis in the same capillary gap cell like the acetalization of the toluene derivatives.

The methanol balance as well as the proton balance is remarkable: two moles of methanol that get released in the cathodic process are used for the acetalization at the anode and four protons used for the cathodic reduction are generated in the anodic acetalization. The electric energy consumption of the paired electrosynthesis is not increased compared to the non-paired process. Because hydrogen is avoided completely, the energy/fossil fuel to generate the hydrogen for the reduction step is economized [7].

But without the development of a skillful work-up, the paired electrosynthesis would not have become a successful industrial process up till now. In the non-paired electrosynthesis, 4-*tert*-butylbenzaldehyde dimethylacetal gets distilled for purification and this procedure has been retained. The main challenge has been to purify the rest of the electrolysis solution containing the phthalide. The BASF electrochemists have found a way to isolate

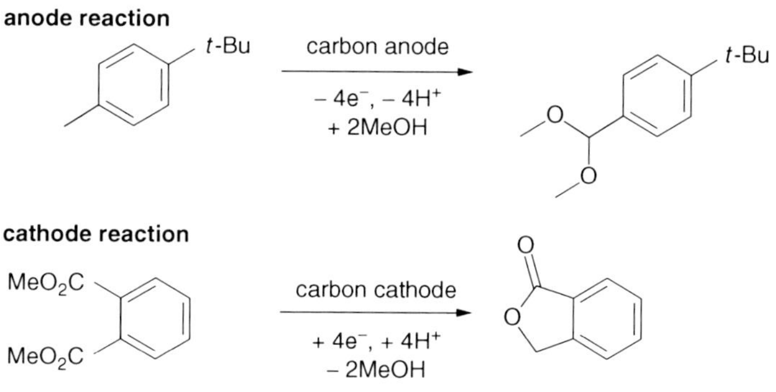

Paired Electrosynthesis, Fig. 2 Parallel electrosynthesis of 4-*tert*-butylbenzaldehyde dimethylacetal and phthalide

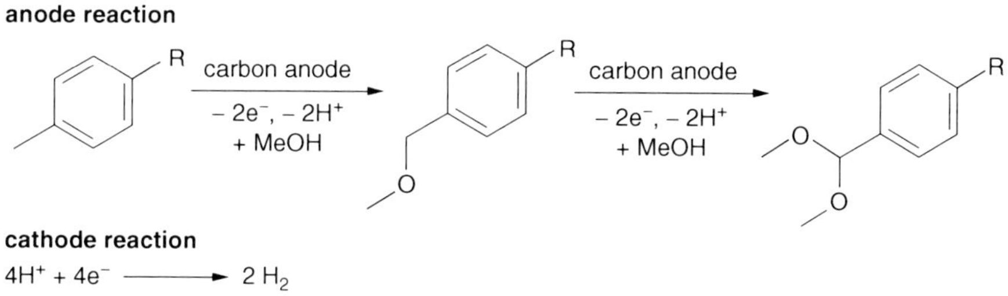

anode reaction

cathode reaction

Paired Electrosynthesis, Fig. 3 Electrosynthesis of acetals of aromatic aldehydes

and to purify phthalide with a skillful layer crystallization process [7].

Example of a Convergent Paired Electrosynthesis

The anodic and cathodic production of glyoxylic acid in aqueous solution is an interesting example of a convergent paired electrosynthesis [13, 14] (Fig. 4).

Industrially the main chemical route to glyoxylic acid is the oxidation of glyoxal with nitric acid [15]. A challenge in this synthesis is to prevent overoxidation to oxalic acid and CO_2. Electrochemically the anodic oxidation of glyoxal is possible as well as the cathodic reduction of oxalic acid. Both electrolyses have been investigated thoroughly.

An important aspect in the anodic oxidation of glyoxal that has been discovered is the clear improvement of the oxidation by hydrochloric acid as mediator [16] (Fig. 5).

First reports about the reduction of oxalic acid have already been published around 1900. In an acidic electrolyte the reduction of oxalic acid at

anode reaction

cathode reaction

Paired Electrosynthesis, Fig. 4 Convergent electrosynthesis of glyoxylic acid

a mercury or lead cathode is described [17]. Decades later, the reduction has been studied intensively [18, 19]. Preferred cathodes are still high overpotential cathodes like lead electrodes [18]. The reduction at graphite electrodes has also been of interest. To improve the reduction at graphite, the addition of metal salts which have a high hydrogen overpotential has been found [20]. This emphasizes the obvious: in aqueous solution it is crucial to oppress the

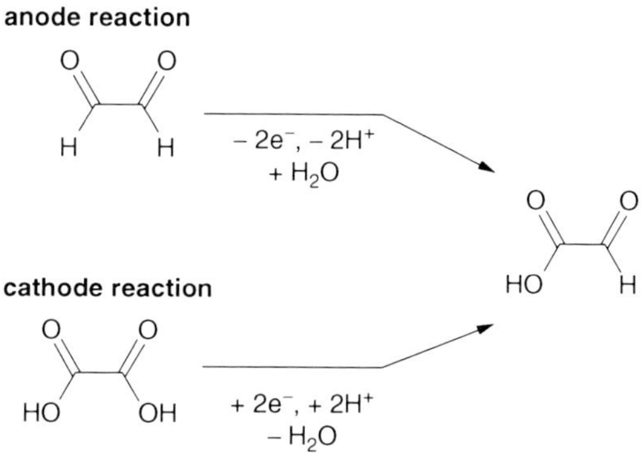

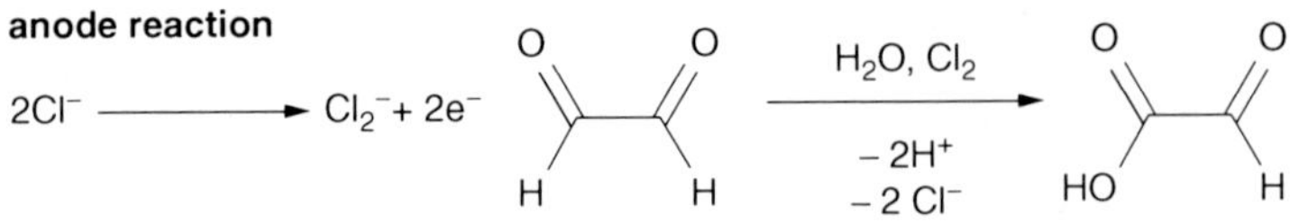

Paired Electrosynthesis, Fig. 5 Mediated anodic oxidation of glyoxal to glyoxylic acid

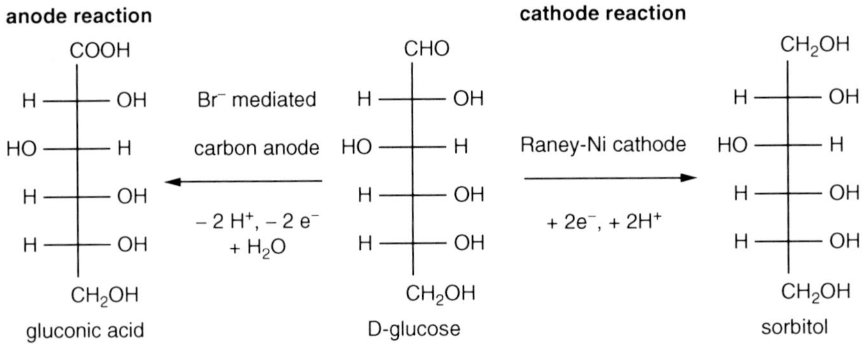

Paired Electrosynthesis, Fig. 6 Divergent electrosynthesis of gluconic acid and sorbitol from glucose

evolution of hydrogen as reductive process in order to be able to reduce the oxalic acid.

The combination of both processes to a paired synthesis bears the next improvement in the electrochemical synthesis of glyoxylic acid first in an undivided cell [13] and later in a divided cell [14]. In the example of the divided cell, the compartments are separated by a cationic exchange membrane. The use of hydrochloric acid renders also in the paired process a good selectivity for the anodic glyoxylic acid generation and guarantees a sufficient conductivity of the aqueous glyoxal solution. The cathodic reduction of oxalic acid takes place as well in an aqueous solution containing hydrochloric acid as supporting electrolyte. Graphite is used as anode and lead as cathode. The choice of temperature and pH is decisive. But also under optimized parameter, overoxidation of glyoxylic acid to oxalic acid and CO_2 is found. By precipitation of the glyoxylic acid, a pure product in good yields is obtained [14].

Example of a Divergent Paired Electrosynthesis

An established divergent paired electrosynthesis is the production of gluconic acid and sorbitol from glucose [21] (Fig. 6).

For the oxidation of glucose to gluconic acid, it is obviously important to avoid overoxidation to glucaric acid. The oxidation is halogenide mediated. In this case bromide is the mediator of choice. To control the selectivity to gluconic acid by the applied charge is an advantage of organic electrochemistry that is used in this case. pH and temperature are also crucial parameters for a sufficient selectivity. The sensitivity of a carbohydrate oxidation in general to these parameters is also shown later by Marsais [22] and Schäfer [23] who have investigated the (TEMPO)-mediated oxidation of carbohydrates (in non-paired processes). The cathodic reduction is improved by the choice of a Raney Nickel powder catalyst instead of a Zn(Hg) cathode. The paired process is carried out in aqueous solution in an undivided flow reactor that was adapted for this electrosynthesis [21].

Example of a Linear Paired Electrosynthesis

The synthesis of methyl ethyl ketone from 2,3-butanediol is a remarkable example of a linear paired electrosynthesis [24] (Fig. 7).

In a bromide-mediated reaction, acetoin is generated from 2,3-butanediol. Afterwards acetoin

Paired Electrosynthesis, Fig. 7 Linear electrosynthesis of methyl ethyl ketone from 2,3-butanediol

anode reaction — Br⁻ mediated carbon anode, $-2e^-$, $-2H^+$

cathode reaction — Zn(Hg) cathode, $+2e^-$, $+2H^+$, $-H_2O$

is reduced at a Hg/Zn cathode to methyl ethyl ketone. The use of an undivided cell is inevitable in this case. The reaction control is very important because the oxidation of 2,3-butanediol with the oxidized mediator is slow. To avoid a reduction of this species, a special flow cell has been developed. A reaction tube after the graphite anode renders sufficient time to oxidize 2,3-butanediol. This example emphasizes how crucial electrolysis conditions and setup is to avoid the undesired reaction at the counter-electrode. In this case the oxidized mediator has to be protected from reduction, and of course the formed methyl ethyl ketone should not be mediated oxidized at the anode to acetoin.

Summary and Future Directions

The examples illustrate the diversity as well as the common features of a paired electrosynthesis. One can start with one or two substrates to generate one or two products. Electrode processes can be mediated or direct. Undivided and divided cells are employed in paired electrosyntheses. But as in the BASF phthalide example, it is crucial for the synthesis of glyoxylic acid, sorbitol, and methyl ethyl ketone that the cathodic process is the reduction of the substrate and not the reduction of protons because in these cases protons are generated at the anode and the electrolysis takes place in a protic solvent. Therefore effects that minimize the overpotential of hydrogen have to be omitted. Reaction control is important in all described examples, and consequently the cell and the setup have to fit for each case. Work-up and product isolation are significant for a successful synthesis and can be even more challenging in a paired synthesis.

A skillful solution is demonstrated by the BASF process.

Future directions, respectively, particular processes, are hard to foresee though the development of more paired electrosynthesis examples is being expected because the essence of such an electrolysis is very attractive: it uses the mole of charge moved in the cell in the most efficient way. The work can be based on the knowledge about successful paired electrosyntheses that is generated for decades as it is partly reflected by the presented examples.

Cross-References

▶ Anodic Substitutions
▶ Electrochemical Functional Transformation
▶ Electrosynthesis Using Mediator
▶ Green Electrochemistry
▶ Organic Electrochemistry, Industrial Aspects

References

1. Lund H, Hammerich O (2001) Organic electrochemistry. Marcel Dekker, New York
2. Schäfer HJ, Bard AJ, Stratmann M (2004) Organic electrochemistry. In: Encyclopedia of electrochemistry, vol 8. Wiley-VCH, Weinheim
3. Eberson L, Nyberg K (1976) Synthetic uses of anodic substitution reactions. Tetrahedron 32:2185–2206
4. Schäfer HJ (1981) Anodic and cathodic CC-bond formation. Angew Chem Int Ed 20:911–934
5. Paddon CA, Atobe M, Fuchigami T, He P, Watts P, Haswell SJ, Pritchard GJ, Bull SD, Marken F (2006) Towards paired and coupled electrode reactions for clean organic microreactor electrosyntheses. J Appl Electrochem 36:617–634
6. Frontana-Uribe BA, Little RD, Ibanez JG, Palma A, Vasquez-Medrano R (2010) Organic

electrosynthesis: a promising green methodology in organic chemistry. Green Chem 12:2099–2119

7. Hannebaum H, Pütter H (1999) Elektrosynthesen Strom doppelt genutzt: Erste technische "Paired Electrosynthesis". Chemie in unserer Zeit 33:373–374
8. Hannebaum H, Pütter H (BASF) DE19618854
9. Wendt H, Bitterlich S (1992) Anodic synthesis of benzaldehydes – 1. Voltammetry of the anodic oxidation of toluene in non-aqueous solutions. Electrochim acta 37:1951–1958
10. Wendt H, Bitterlich S, Lodowicks E, Liu Z (1992) Anodic synthesis of benzaldehydes – 2. Optimization of the direct anodic oxidation of toluenes in methanol and ethanol. Electrochim Acta 37:1959–1969
11. Degner D (BASF) DE2848397. Degner D, Barl M, Siegel H (BASF) DE2848397
12. Beck F, Guthke H (1969) Entwicklung neuer Zellen für electro-organische Synthesen. Chem-Ing-Tech 41:943–950
13. Scott K (1991) A preliminary investigation of the simultaneous anodic and cathodic production of glyoxylic acid. Electrochim Acta 36:1447–1452
14. Jalbout AF, Zhang S (2002) New paired electrosynthesis route for glyoxalic acid. Acta Chim Slov 49:917–923
15. Mattioda G, Christidis Y (2000) Glyoxylic acid. In: Ullmann's encyclopedia of industrial chemistry, vol 17. Wiley-VCH, Weinheim, pp 89–92
16. Pierre G, El Kordi M, Cauquis G, Mattioda G, Christidis Y (1985) Electrochemical synthesis of glyoxylic acid from glyoxal. Part 1. Role of the electrolyte, temperature and electrode material. J Electroanal Chem 186:167–177
17. Tafel J, Friedrichs G (1904) Elektrolytische Reduction von Carbonsäuren und Carbonsäureestern in schwefelsaurer Lösung. Chem Ber 37:3187–3191
18. Goodridge F, Lister K, Plimley RE, Scott K (1980) Scale-up studies of the electrolytic recuction of oxalic to glyoxalic acid. J Appl Electrochem 10:55–60
19. Picket DJ, Yap KS (1974) A study of the production of glyoxylic acid by the electrochemical reduction of oxalic acid solution. J Appl Electrochem 4:17–23
20. Scharbert B, Dapperheld S, Babusiaux P (Hoechst) DE4205423
21. Park K, Pintauro PN, Baizer MM, Nobe K (1985) Flow rate studies of the paired electro-oxidation and electroreduction of glucose. J Electrochem Soc 132:1850–1855
22. Ibert M, Fuertès P, Merbouh N, Fiol-Petit C, Feasson C, Marsais F (2010) Improved preparative electrochemical oxidation of D-glucose to D-glucaric acid. Electrochim Acta 55:3589–3594
23. Schnatbaum K, Schäfer HJ (1999) Electroorganic Synthesis 66: Selective anodic oxidation of carbohydrates mediated by TEMPO. Synthesis 864–872
24. Li W, Nonaka T, Chou T-C (1999) Paired electrosynthesis of organic compounds. Electrochemistry 67:4–10

Permanent Electrochemical Promotion for Environmental Applications

Leonardo Lizarraga and Philippe Vernoux
Université Lyon 1, CNRS, UMR 5256, IRCELYON, Institut de recherches sur la catalyse et l'environnement de Lyon, Villeurbanne, France

Introduction

Electrochemical Promotion of Catalysis (EPOC), or Non-Faradaic Electrochemical Modification of Catalytic Activity (NEMCA) effect, is a promising concept for boosting catalytic processes and advancing the frontiers of catalysis. This innovative field aims to modify both the activity and the selectivity of catalysts in a controlled manner. EPOC utilizes solid electrolyte materials as catalytic carriers. Ions contained in these electrolytes are electrochemically supplied to the catalyst surface and act as promoting agents to modify the catalyst electronic properties in order to achieve optimal catalytic performance. EPOC can be considered as an electrically controlled catalyst-support interaction in which promoting ionic agents are accurately supplied onto the catalytic surface by electrical potential control. The main advantage of EPOC is that the electrochemical activation magnitude is much higher than that predicted by Faraday's law. Therefore, EPOC requires low currents or potentials. Moreover, promoting species such as O^{2-} and H^+ cannot be formed via gaseous adsorption and cannot be easily dosed by chemical means. EPOC has been investigated thoroughly for more than 70 catalytic reactions [1, 2] and is reversible, since the catalyst restores its initial activity, typically within a few minutes after current interruption. Recently, however, some studies have shown that, for specific operating conditions and catalyst-electrodes, the reversibility of this phenomenon strongly depends on the duration of the polarization.

For prolonged polarization times, the catalytic reaction rate after current interruption can remain higher than the value before current application [3]. This behavior has been reported as "permanent-electrochemical promotion of catalysis" (P-EPOC). This phenomenon was mainly observed over catalysts which have multiple valence states and the ability to form oxides, i.e., Pt/PtO$_x$ or Rh/Rh$_2$O$_3$.

State-of-the-Art

The phenomenon of P-EPOC has been mainly observed on yttria-stabilized zirconia (YSZ), an O^{2-} conducting electrolyte, interfaced with different catalysts: IrO$_2$ [4], RuO$_2$ [5], Rh [6–9], Pt [3, 10–12], and Pd [13]. Two recent studies have reported P-EPOC on Pt films interfaced with K–β″Al$_2$O$_3$, a K$^+$ conducting ceramic [14], or Ba$_3$Ca$_{1.18}$Nb$_{1.82}$O$_{9-a}$, a H$^+$ conductor [15]. Two parameters are commonly used to quantify the magnitude of the EPOC effect [2]: the rate enhancement ratio, r, defined as r = r/r$_0$, where r is the electropromoted catalytic rate and r$_0$ the open-circuit, i.e. non-promoted, catalytic rate; and the apparent Faradaic efficiency, Λ, defined as Λ = (r − r$_0$)/(I/nF). P-EPOC effect is quantified using a "permanent" rate enhancement ratio [3], y, which has been defined as y = r$_{P\text{-}EPOC}$/r$_0$, where r$_{P\text{-}EPOC}$ is the catalytic rate in an open-circuit state after current interruption.

A typical EPOC experiment setup consists of an electrochemical cell based on a dense pellet or tube of a solid electrolyte (YSZ, K–β″Al$_2$O$_3$, etc.). The pellet or tube is covered on both sides by three electrodes: working, counter and reference [2]. The catalytic activity of catalyst layers, used as working electrodes, can be modified by applying small potentials or currents between the working and the counter electrodes. Figure 1 shows two typical transients, i.e. variation of the catalytic rate versus time, for two different polarization times. At t = 0, a small positive current (300 μA) is applied across the IrO$_2$/YSZ interface exposed to a C$_2$H$_4$/O$_2$ atmosphere. Upon current application, the rate of ethylene catalytic

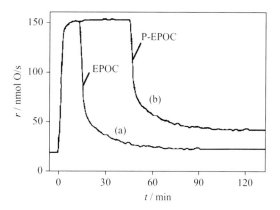

Permanent Electrochemical Promotion for Environmental Applications, Fig. 1 Polarization and relaxation transients of the rate of C$_2$H$_4$ combustion on IrO$_2$/YSZ catalyst due to an anodic current application (300 mA) for two different times: (a) short polarization to give reversible EPOC and (b) long polarization time to give P-EPOC. T = 380 °C, P$_{O2}$ = 17 kPa, P$_{C2H4}$ = 0.14 kPa (Reprint ref 3)

combustion on IrO$_2$, expressed in moles of O consumed per second, drastically increases by around a factor of 3.5 and reaches a plateau after a few minutes. This enhancement is non-Faradaic, with a Faradaic efficiency larger than 100. The reversibility of this NEMCA effect depends on the polarization time. For short current applications (a few minutes), the catalytic rate gradually decreases down to its initial value after the current interruption. For prolonged polarizations (1 h), the catalytic rate also slowly decreases when the current stops, but more than 1 h after this interruption, it reaches a stable value, called the P-EPOC state, higher that its initial one.

Table 1 lists all reactions and catalysts where the P-EPOC phenomenon was observed, along with operating conditions. The P-EPOC phenomenon was mainly observed (Fig. 1 and Table 1) when long anodic polarizations, i.e. high anodic charges, were applied at the catalyst/electrolyte interfaces. In addition, most of these studies, especially those using Pt, were performed in oxygen excess conditions in the temperature range where PtO$_x$ species are stable. Recent studies have shown that the y values have a $t^{1/2}$ dependence in respect to polarization time at constant

Permanent Electrochemical Promotion for Environmental Applications, Table 1 Exhaustive list of P-EPOC studies

Reactions	Catalyst (preparation method)	Solid electrolyte	T/°C	Preagent/kPa	P_{O_2}/kPa	Polarization time/min	γmax	Ref
NO reduction by C_3H_6	Rh (paste calcination)	YSZ	300–380	$P_{C3H6} = 0.1$ $P_{NO} = 0.1$	0.5	60–180	6	6–8
C_2H_4 deep oxidation	IrO_2 (thermal decomposition)	YSZ	380	$P_{C2H4} = 0.14$	17	120	3	4
	RuO_2	YSZ	360	$P_{C2H4} = 0.14$	12	180	1.3	5
	Pt (paste calcination)	YSZ	550	$P_{C2H4} = 0.25$	1	60–6,600	2	3
	Pt (sputtering)	YSZ	375	$P_{C2H4} = 0.2$	8.5	10–600	2.2	10
	Pt/FeOx (pulsed laser deposition)	YSZ	375-425	$P_{C2H4} = 0.19$	8.2	40–60	10	11
	Pt (paste calcination)	$Ba_3Ca_{1.18}Nb_{1.82}O_{9-\alpha}$	350	$P_{C2H4} = 1$	2-8	100	4	15
C_3H_8 deep oxidation	Pt (sputtering)	YSZ	350	$P_{C3H8} = 0.2$	4.5	15	1.3	12
C_3H_6 deep oxidation	Pt (wet impregnation)	$K-\beta''Al_2O_3$	310	$P_{C3H6} = 0.2$	1	300	1.35	14
CH_4 deep oxidation	Pd (wet impregnation)	YSZ CeO$_2$/YSZ	470-600	$P_{CH4} = 0.4$	1	150	2	13
CH_4 partial oxidation	Rh (paste impregnation)	TiO_2/YSZ	550	$P_{CH4} = 0.5–1$	0.5	15–55	11	9

current, indicating a diffusion controlled process [10, 12]. Besides, the Comninellis research group has performed different electrochemical investigations (cyclic voltammetry, double step chronopotentiometry, etc.) for the system O_2 (g), Pt/YSZ, which evidences the possible electrochemical formation of PtO_x at long polarization time, with a kinetic dependence of $t^{1/2}$ [3, 16–18]. All of these facts are indicative of the importance of the formation of oxide species, such as PtOx or Rh_2O_3, in the P-EPOC effect. Comninellis and collaborators (Fig. 2a) have proposed a mechanism for the P-EPOC phenomenon as a function of the three different locations of oxygen species when an anodic polarization is applied (O^{2-} ions pumped from YSZ to Pt) [3]. When starting anodic polarizations, oxygen species first generate a monolayer at the catalyst/YSZ interface (1) and then rapidly spillover to the catalyst surface in the form of promoting ionic species (2), i.e. partially discharged ions ($O^{\delta-}$), which are at the origin of the reversible EPOC phenomenon. The coverage of these

promoting species is maximal after a few minutes of polarization when reaching the plateau of catalytic activity. The third type of stored oxygen is attributed to the growth of an oxide layer starting at the Pt/YSZ interface (3). After current interruption, the promoters stored at the gas-exposed surface are consumed by the reaction, and the catalytic activity decreases. The oxide layer at the Pt/YSZ interface could then act as a tank of promoting ions which can spillover to the catalyst surface via solid state diffusion and enhance the catalytic rate (P-EPOC state).

The mechanism proposed by Comninellis and collaborators is well supported by the transients shown in Fig. 2b on thick Pt films, where the P-EPOC state disappears after some hours, depending on the anodic charge supplied (polarization time). After the release of all stored oxygen promoting species, the catalytic rate returns to its initial value. Therefore, the phenomenon is more persistent than permanent. However, long-term stable P-EPOC states were observed on thin Pt films (sputtering) interfaced

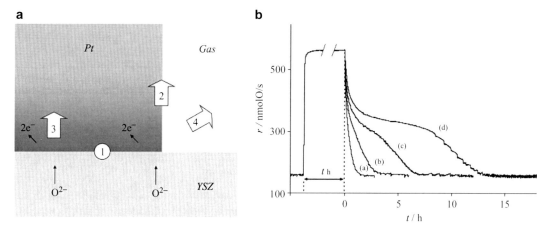

Permanent Electrochemical Promotion for Environmental Applications, Fig. 2 (a) Oxygen species produced electrochemically under anodic polarization. Storage at three different locations: (1) at the Pt/YSZ interface, (2) at the gas-exposed catalyst surface via backspillover and (3) in the bulk platinum at the vicinity of the Pt/YSZ interface. Leaving (via O_2 evolution and/or chemical reaction) toward the gas phase is indicated (4). (b) Dependence of the rate transients observed during open-circuit relaxation on the anodic polarization time: (a) 1 h, (b) 2 h, (c) 4 h, (d) 11 h. T = 525 °C. Feed composition: $P_{C2H4} = 0.25$ kPa, $P_{O2} = 1$ kPa. Reprinted from reference 3

on YSZ upon short anodic polarization times (a few minutes) [10, 12]. This interesting result can be explained through the electrochemical formation of a thermodynamically stable interfacial PtO_x species at the Pt/YSZ interface, in which it is not in contact with the reactive gas mixture. The P-EPOC effect, in this case, was only removed by applying cathodic currents that decompose PtO_x species. This interfacial PtO_x layer increases the oxygen chemical potential gradient across the Pt/YSZ interface and then promotes the thermal migration of $O^{\delta-}$ promoting species from the electrolyte to the metal/gas interface. This effect can be analyzed as a self-driven EPOC mechanism as permanent effects observed in wireless NEMCA transients [19].

Future Environmental Applications

Air pollution originating from automotive traffic and industries is an increasing problem, especially in urban areas. Major challenges are the treatment of gaseous streams from automobiles and plants, such as the abatement of NOx, soot, unburned CO and hydrocarbons, as well as low temperature catalytic combustion of traces such as volatile organic compounds (VOCs). VOCs present in buildings or cars are wide-ranging classes of chemicals, and currently over 300 compounds are considered as VOCs by the US Environmental Protection Agency (EPA). Their release has widespread environmental implications. The EPA has validated that indoor air pollution is one of the top human health risks. The studies on indoor air quality (IAQ) have been transited gradually to indoor volatile organic compounds. The removal of formaldehyde is vital for improving IAQ and human health due to a carcinogenic risk. The challenge addressed to the scientific catalysis community is to find smart, stable, efficient and selective catalysts with limited loadings of noble metals. EPOC is one of the ways to improve and control in-situ catalyst performance and stability. In addition, P-EPOC is particularly suitable for oxidation reactions occurring in oxygen excess (Table 1), for which noble metals are the most effective catalysts. Short pulse electrical polarizations could be sufficient to obtain high and long-term catalytic activity enhancements of noble metal-based catalysts (Pt, Pd, Rh), allowing use of extremely low loadings and electricity. In particular, P-EPOC is a promising and energy-saving solution for catalytic oxidation of VOCs and light hydrocarbon combustion. Nevertheless, additional fundamental and experimental studies are necessary to improve the understanding of the P-EPOC phenomena in order to optimize their efficiency, most particularly at low temperatures.

Cross-References

▶ Electrochemical Promotion for the Abatement of Gaseous Pollutants
▶ Non-Faradaic Electrochemical Modification of Catalytic Activity (NEMCA)
▶ Solid Electrolytes

References

1. Vayenas CG, Bebelis S, Ladas S (1990) Dependence of catalytic rates on catalyst work function. Nature 343:625–627
2. Vayenas CG, Bebelis S, Pliangos C, Brosda S, Tsiplakides D (2001) Electrochemical activation of catalysis: promotion, electrochemical promotion and metal–support interactions. Kluwer Academic/Plenum Publishers, New York
3. Falgairete C, Jaccoud A, Foti G, Comninellis C (2008) The phenomenon of "permanent" electrochemical promotion of catalysis (P-EPOC). J Appl Electrochem 38:1075–1082
4. Nicole J, Tsiplakides D, Wodiunig S, Comninellis C (1997) Activation of catalyst for gas-phase combustion by electrochemical pretreatment. J Electrochem Soc 144:L312–L314
5. Wodiunig S, Patsis V, Comninellis C (2000) Electrochemical promotion of RuO_2-catalysts for the gas phase combustion of C_2H_4. Solid State Ionics 136:813–817
6. Foti G, Lavanchy O, Comninellis C (2000) Electrochemical promotion of Rh catalyst in gas-phase reduction of NO by propylene. J Appl Electrochem 30:1223–1228
7. Pliangos C, Raptis C, Badas T, Vayenas CG (2000) Electrochemical promotion of NO reduction by C_3H_6 on Rh/YSZ catalyst-electrodes. Ionics 6:119–126
8. Pliangos C, Raptis C, Badas T, Vayenas CG (2000) Electrochemical promotion of NO reduction by C_3H_6 on Rh/YSZ catalyst-electrodes. Solid State Ionics 136:767–773
9. Baranova EA, Fóti G, Comnillis C (2004) Current-assisted activation of Rh/TiO_2/YSZ catalyst. Electrochem Commun 6:389–394

10. Souentie S, Xia C, Falgairete C, Li YD, Comninellis C (2010) Investigation of the "permanent" electrochemical promotion of catalysis (P-EPOC) by electrochemical mass spectrometry (EMS) measurements. Electrochem Commun 12:323–326
11. Mutoro E, Koutsodontis C, Luerssen B, Brosda S, Vayenas CG, Janek J (2010) Electrochemical promotion of Pt (111)/YSZ (111) and Pt-FeO$_x$/YSZ (111) thin catalyst films: Electrocatalytic, catalytic and morphological studies. Appl Catal B Environ 100:328–337
12. Souentie S, Lizarraga L, Papaioannou EI, Vayenas CG, Vernoux P (2010) Permanent electrochemical promotion of C_3H_8 oxidation over thin sputtered Pt films. Electrochem Commun 12:1133–1135
13. Jiménez-Borja C, Dorado F, Lucas-Consuegra A, García-Vargas JM, Valverde JL (2009) Complete oxidation of methane on Pd/YSZ and Pd/CeO_2/YSZ by electrochemical promotion. Catal Today 146:326–329
14. de Lucas-Consuegra A, Dorado F, Valverde JL, Karoum R, Vernoux P (2007) Low-Temperature propene combustion over Pt/K-bAl_2O_3 electrochemical catalys: Characterization, catalytic activity measurements, and investigation of the NEMCA effect. J Catal 251:474–484
15. Thursfield A, Brosda S, Pliangos C, Schober T, Vayenas CG (2003) Electrochemical Promotion of an oxidation reaction using a proton conductor. Electrochim Acta 48:3779–3788
16. Falgairete C, Fóti G (2009) Oxygen storage in O_2/Pt/YSZ cell. Catal Today 146:274–278
17. Fóti G, Jaccoud A, Falgairete C, Comninellis C (2009) Charge storage Pt/YSZ interface. J Electroceram 23:175–179
18. Falgairete C, Xia C, Li YD, Harbich W, Fóti G, Comninellis C (2010) Investigation of the Pt/YSz interface at low oxygen partial pressure by solid electrochemical mass spectroscopy under high vacuum conditions. J Appl Electrochem 40:1901–1907
19. Poulidi D, Rivas ME, Metcalfe IS (2011) Controlled spillover in a single catalyst pellet: rate modification, mechanism and relationship with electrochemical promotion. J Catal 281:188–197

Perovskite Proton Conductor

Nikolaos Bonanos
Department of Energy Conversion and Storage, Technical University of Denmark, Roskilde, DK

Introduction

Perovskite proton conductors belong to the class of high temperature proton conductors (HTPCs), solids that conduct electricity by transporting H⁺ ions (protons) at temperatures above ambient, typically 400–1,000 °C. Unlike more common proton conductors, such as ice or KHSO₄, in which hydrogen is part of the chemical structure, in HTPCs, it is taken up from traces of water vapor in the atmosphere via gas-solid equilibrium. Accordingly, the hydrogen does not normally appear in the chemical formulae of the compounds. Typical high temperature proton conductors are SrCeO₃, SrZrO₃, BaCeO₃, BaZrO₃, KTaO₃, LaNbO₄, suitably doped, and La₆WO₁₂ without a dopant; of these compounds, the first five are perovskites. Perovskite proton conductors are of interest as electrolytes in fuel cells, water vapor electrolysis cells, electrochemical hydrogen pumps, and hydrogen sensors. When they display electronic conduction in addition to the protonic, these materials can function as hydrogen semipermeable membranes of very high selectivity. In this article, perovskite proton conductors are describe with the non-specialist reader in mind. Detailed discussions of perovskite proton conductors and their applications can be found in review articles [1–5].

Since the 1960s, hydrogen has been known to exist in oxides, but only in trace amounts, and was not believed to alter the physical properties of the oxide significantly. In the early 1970s, it became clear that the role of hydrogen was important, and sometimes dominant. In the year 1981, in which the first international conference on Solid State Proton Conductors [6] was held in Paris, a landmark paper was published by H. Iwahara and coworkers [7] demonstrating proton conductivity in Sc-doped SrCeO₃ and operating a prototype water vapor electrolyzer at 900 °C. During the 30 years that have elapsed since then, the subject has moved from the marginal to the mainstream, and features strongly in the International Conferences on Solid State Ionics (SSI) and Solid State Proton Conductors (SSPC). Figure 1 shows the number of publications per year that include the words *proton* and *conduction* and *oxides* in their title or keywords. If the present trend were to continue, by 2020, the subject would attract about 1,000 publications per year!

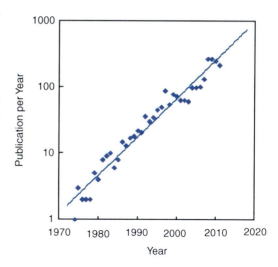

Perovskite Proton Conductor, Fig. 1 Number of scientific publications per year that include the words *proton*, *conduction*, and *oxides* in their title or keywords. The idea for this graph comes from ref. [3]

Generation of Protonic Carriers in Perovskite Proton Conductors

The chemical reaction that describes the incorporation of water molecules into an oxide is given in (Eq. 1), using Kröger-Vink notation:

$$H_2O(g) + O_O^\times + V_O^{\cdot\cdot} \leftrightarrow 2OH_O^{\cdot} \qquad (1)$$

This means that a water molecule reacts with an oxygen ion on a normal site ($O_O\times$) and an oxygen vacancy ($V_O^{\cdot\cdot}$) to produce two hydroxyl ions occupying oxygen sites; in other words, the protonic species. The oxygen vacancies are created by doping or, less commonly, are part of the structure. In the case where the host is tetravalent and the dopant trivalent, the dopant would be denoted as [M′]. Reaction (1) is characterized by an equilibrium constant K_W defined by the equation:

$$K_W = \frac{[OH_O^{\cdot}]^2}{[V_O^{\cdot\cdot}] P_{H_2O}} \qquad (2)$$

where

$$[V_O^{\cdot\cdot}] = \frac{1}{2}\left([M'] - [OH_O^{\cdot}]\right) \qquad (3)$$

In its "rest state," the protonic species resides on an oxygen ion, to be precise *within* the oxygen ion, since the ionic radius of O^{2-} is 1.40 Å, while that of OH^- is 1.37 Å (both values refer to sixfold coordination of these species). To enable ionic conduction, the hydrogen must be able to hop from one oxygen ion to another.

The requirements for a solid to be an HTPC can be summarized as follows:
- It must be an oxide (all systems known so far are oxides).
- It must contain oxygen vacancies, by nature or by doping.
- It must have the right chemical properties to favor the incorporation of water molecules.
- It must permit the momentary breaking of O–H bonds, if the protons are going to be mobile.
- If it is to be used as an electrolyte, it should not have predominant electronic conductivity.

The requirement for oxygen vacancies is strong, since without the reaction described by (Eq. 1), the $OH_O^{\cdot}$ species cannot form. Usually, the vacancies are created by doping with an element of valence lower than that of the host element. Perovskites (Fig. 2) are a structure type in which high temperature protonic conduction is commonly observed. Perovskites have two types of cation site: a larger 12-coordinated site (A) and a smaller 6-coordinated site (B). An example of a perovskite doped on the B-site would be $BaZr_{1-y}Y_yO_{-y/2}$. If the oxygen vacancies can be hydrated, as described above, the first three criteria for an HTPC will have been fulfilled.

Perovskites such as $BaZrO_3$ and $SrCeO_3$, doped as above, are able to absorb a quantity of water vapor, while others such as $CaTiO_3$ and $SrTiO_3$, are less able to do so. The difference in the ease of hydration has been linked to the basicity of the elements constituting the oxide [1]. Alternatively, the electronegativity has been implicated. More specifically, a correlation has been found between the standard enthalpy of hydration and the difference in electronegativity between the A-site and B-site elements in the perovskite (Fig. 3) [8]. The open points are from reference [8]; the solid points are for $BaCe_{0.9-x}Zr_xY_{0.1}O_{3-\delta}$ (BCZY) [9].

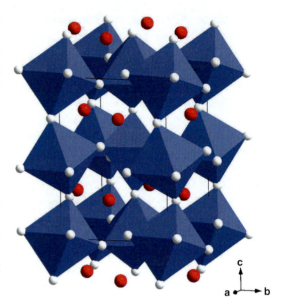

Perovskite Proton Conductor, Fig. 2 Structure of $SrCeO_3$, a typical perovskite proton conductor when doped with a trivalent element on the Ce site. The spheres represent the A-cation, Sr, while the octahedra have the B-cation, Ce at the center, and oxygen at the apices. A strong distortion from the cubic structure is visible the structure is shown in the *Pbnm* setting

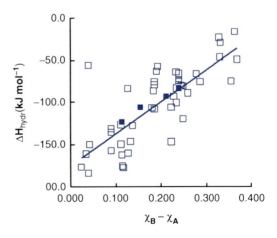

Perovskite Proton Conductor, Fig. 3 Correlation between standard enthalpy of hydration and the electronegativity difference of the A- and B-site elements in the perovskite (Fig. 3). The open points are tabulated by Norby et al. [8], while the solid points are for $BaCe_{0.9-x}Zr_xY_{0.1}O_{3-\delta}$ (BCZY) [9]

A corresponding plot of the standard entropy (not shown) displays no clear trend, suggesting that the entropy is determined by the conversion of the gaseous water molecules to solid species, rather than by which solid they enter into. The enthalpy and entropy of hydration, together with the temperature, determine the equilibrium constant of the reaction. For $\chi_B - \chi_A = 0.2$, midway in the plot, values of $\Delta H = -100$ kJ·mol^{-1} and $\Delta S = -110$ J·mol^{-1}·K^{-1} are interpolated, giving an equilibrium constant K_W of 1.31 for 873 K. Figure 4 shows the hydration isotherm for this value of K_W.

It is an unfortunate fact of chemistry that greater basicity implies greater affinity for CO_2, and this is one of the main disadvantages of the perovskite class of proton conductors. The problem is particularly acute at lower temperatures, which, ironically, is the range that can benefit most from perovskite proton conductors (the temperature range of 700 °C and above is already well served by oxide ion conductors). For this reason, researchers, especially at University of Oslo, believe that alternatives to perovskite systems are a *sine qua non* for cells that operate reliably in the presence of hydrocarbons. Forgoing the highly electropositive elements Sr and Ba and opting for La, compounds such as LaNbO$_4$ and LaTaO$_4$ are obtained, which are fully stable to CO_2 [10] These compounds require doping (usually Ca) and have low protonic conductivity (Fig. 5), while another, La$_6$WO$_{12}$ does not require doping and has higher conductivity [11]. On the downside, the tungstate's reactivity with NiO presents problems for the fabrication of fuel cell anodes in the conventional way.

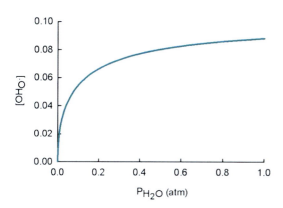

Perovskite Proton Conductor, Fig. 4 Concentration of protonic carriers as a function of partial pressure of water vapor for an oxide with an equilibrium constant K_W of 1.31 (Eq. 2). The plot is also called a hydration isotherm

Mechanism of Transport

The transport of protons occurs, as stated earlier, by a hopping mechanism. This can be modelled using large collections of atoms and applying

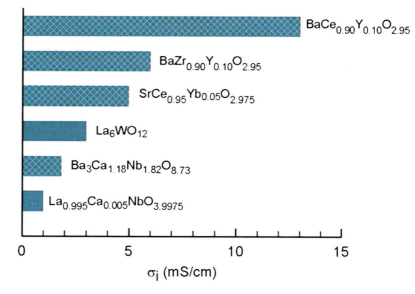

Perovskite Proton Conductor, Fig. 5 Ionic conductivity (predominantly protonic) of selected HTPCs at 800 °C (various sources)

experimentally determined interatomic potentials. Alternatively, the lattice can be modelled by quantum mechanical techniques, which do not require interatomic potentials as inputs, but are computationally more demanding. Simulations of this type demonstrate that the proton hopping is facilitated by vibrations of the oxygen ions; this leads to the counterintuitive conclusion that longer B-cation oxygen bonds lead to lower activation energies for transport – in line with experimental results. Molecular dynamics studies have followed the path of a single proton in the lattice of BaCeO$_3$ for a duration of *ca.* 100 ps, revealing two kinds of motion: fast rotation and slower transport from one oxygen to another [12].

Density Functional Theory (DFT) methods allow the ab initio calculation [13] of the energy of both molecules and lattices. By minimizing the energy, the positions of all atoms can be obtained and the paths of ions can be traced. These methods have opened a new window onto the process of proton transport in oxides. For example, a recent study has shown that, for SrTiO$_3$, when a proton is introduced into a lattice already containing another proton, it experiences an energy minimum at a short distance from the original species [14]. In other words, an attractive force arises that leads to defect pairing, despite the existing electrostatic repulsion between the two species. The experimental test of this prediction and its implications for hydration thermodynamics and conduction mechanisms would be promising areas of research.

Minority Conduction Processes

Considering (Eq. 1), it is apparent that the concentrations of oxide ion vacancies and hydroxyls are in competition and that the concentration of OH$_O^{\bullet}$ cannot exceed that of the dopant. When other point defects, namely, electrons (e$'$) and holes (h$_i^{\bullet}$), are allowed, the following electroneutrality condition is obtained:

$$2\left[V_O^{\bullet\bullet}\right] + \left[OH_O^{\bullet}\right] + \left[h_i^{\bullet}\right] - [e'] - [M'_B] = 0 \quad (4)$$

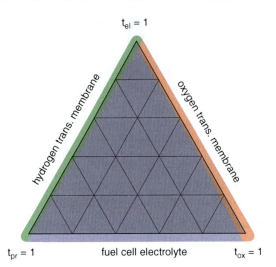

Perovskite Proton Conductor, Fig. 6 Diagram showing the possible ionic transport numbers for oxide ions, protons, and electron holes in a HTPC, following the relation: $t_{ox} + t_H + t_h = 1$. The preferred regimes for three types of function are shown in the figure

The conductivities due to each of these defects are determined by their concentrations multiplied by their mobilities. As a result of the electron–hole equilibrium, these defects cannot both be present in high concentration at the same time. Therefore, perovskite proton conductors normally display protonic, oxide ion and one type of electronic conductivity. For an HTPC in oxidizing conditions the main defects are V$_O^{\bullet\bullet}$, OH$_O^{\bullet}$, and h$_i^{\bullet}$. The situation can be visualized in the diagram of Fig. 6, which indicates the domains favorable for applications in fuel cell electrolytes and hydrogen transport membranes. The domain for oxygen transport membranes (i.e., no proton conduction) is included for the sake of completeness.

Even though perovskite proton conductors are selected for their protonic conductivity, possessing a minority electronic conductivity component is not necessarily a disadvantage. Figure 7a shows a plot of the total conductivity versus the oxygen partial pressure (P_{O2}) on a logarithmic scale. The curve follows an equation of the type:

$$\sigma_{tot} = \sigma_i + \sigma_p^{\circ} \cdot P_{O_2}^{n} + \sigma_n^{\circ} \cdot P_{O_2}^{n} \quad (5)$$

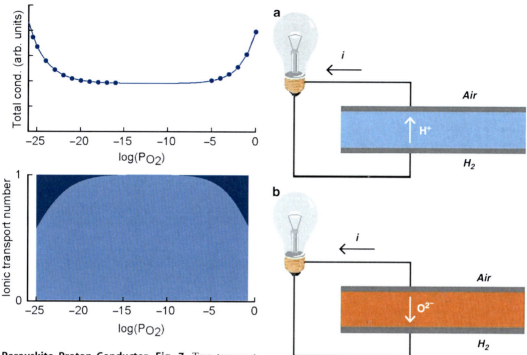

Perovskite Proton Conductor, Fig. 7 Two transport properties for a typical HTPC, plotted versus the logarithm of oxygen partial pressure (P_{O2}). (a) Total conductivity (b) Ionic transport number

Perovskite Proton Conductor, Fig. 8 Schematic diagram of two types of H_2/O_2 fuel cells: (a) PCFC based on high temperature proton conductor (b) SOFC, based on an oxide ion conductor

In this equation, σ_i is the ionic conductivity, assumed to be independent of P_{O2}, while σ_p° and σ_n° are parameters describing the other conductivity components at a P_{O2} of 1 atm [15]. In Fig. 7b, the ionic transport number σ_i/σ_{tot} is plotted on the same logarithmic scale. In a fuel cell, for example, an H_2/O_2 cell, the emf is dependent on the mean ionic transport number. In Fig. 7b, the mean ionic transport number is the area ratio of the light colored block to the whole block. As a result, partial electronic conductivity limited to the high and low P_{O2} regions is not a problem. This is also observed in practice, where H_2/O_2 or H_2/air cells based on perovskite proton conductors give open circuit voltages within a few percent of the theoretical values [16].

Future Directions

In the above, we have not specified whether the ionic conduction was via a protonic or an oxide ion mechanism – both allow the operation of a fuel cell. There is, however, a crucial difference between the two types of cell, illustrated in Fig. 8. In the proton conducting fuel cell (PCFC), the water vapor is evolved at the cathode, while in the solid-oxide fuel cell (SOFC) with oxide ion-conducting electrolyte, it is evolved at the anode. For a PCFC, the water vapor can be easily diluted by an excess of air, which also removes waste heat. In an SOFC, the fuel is recirculated to condense the water vapor, leading to a more complex auxiliary plant.

In the first years of work on perovskite proton conductors, many researchers, including the author, assumed that PCFCs would be useful only in connection with hydrogen fuel cells, while with hydrocarbon fuels, SOFCs would be the only choice. However, work by Coors and coworkers has demonstrated that perovskite proton conductors with minority oxide ion conduction can sustain stable currents for long periods with dry hydrocarbon fuels [17]. It appears that such cells work by internal reforming the hydrocarbon, mediated by oxygen transport. This is an interesting and promising area for future research, both applied and theoretical. Another area of interest is the PCFC cathode, for which there are unanswered questions concerning reaction mechanisms, for example, the role – if any – of oxide ion conduction in the cathode material. Cathodes for PCFCs also present material challenges in terms of transport properties and stability to high partial pressures of water vapor [18, 19].

The range of transport properties and the applications are likely to keep the subject of perovskite proton conductors in the spotlight for many years to come.

Acknowledgements S. Ricote is thanked for a critical reading of the text. The author is grateful to T. Norby and coworkers at the University of Oslo for providing data used in Fig. 3 and to K.S. Knight of the ISIS Facility, Rutherford Appleton Laboratory, for Fig. 2.

Cross-References

▶ Defect Chemistry in Solid State Ionic Materials
▶ DFT Screening and Designing of Electrocatalysts
▶ Kröger-Vinks Notation of Point Defects
▶ MIEC Materials
▶ Mixed Conductors, Determination of Electronic and Ionic Conductivity (Transport Numbers)
▶ Oxide Ion Conductor
▶ Solid Electrolytes
▶ Solid Oxide Fuel Cells, Introduction

References

1. Kreuer K-D (2003) Proton-conducting oxides. Annu Rev Mater Res 33:33–359
2. Malavasi L, Fischer CAJ, Islam MS (2010) Oxide-ion and proton conducting electrolyte materials for clean energy applications: structural and mechanistic features. Chem Soc Rev 39:4370–4387
3. Fabbri E, Pergolesi D, Traversa E (2010) Materials challenges toward proton-conducting oxide fuel cells: a critical review. Chem Soc Rev 39:4366–4369
4. Athanassiou C, Pekridis G, Kaklidis N, Kalimeri K, Vartzoka S, Marnellos G (2007) Hydrogen production in solid electrolyte membrane reactors (SEMRs). Int J Hydrog Energy 32(1):38–54
5. Iwahara H (1996) Proton conducting ceramics and their applications. Solid State Ion 86–88(1):9–15
6. Jensen J, Kleitz M (eds) (1982) Solid state protonic conductors I: Danish-French workshop on solid state materials for low to medium temperature fuel cells and monitors, with emphasis on proton conductors, Paris, France, 8–11 Dec 1981. Odense, Denmark
7. Iwahara H, Esaka T, Uchida H, Maeda N (1981) Proton conduction in sintered oxides and its application to steam electrolysis for hydrogen production. Solid State Ion 3/4:359
8. Norby T, Widerøe M, Glöckner R, Larring Y (2004) Hydrogen in oxides. Dalton Trans 2004:3012–3018
9. Ricote S, Bonanos N, Caboche G (2009) Water vapor solubility and conductivity study of the proton conductor $BaCe_{0.9-x}Zr_xY_{0.1}O_{3-\delta}$. Solid State Ion 180(14–16):990–997
10. Norby T, Haugsrud R (2006) Proton conduction in rare-earth ortho-niobates and ortho-tantalates. Nat Mater 5:193–196
11. Haugsrud R, Kjølseth C (2008) Effects of protons and acceptor substitution on the electrical conductivity of La_6WO_{12}. J Phys Chem Solids 69(7):1758–1765
12. Münch W, Seifert G, Kreuer K-D, Maier J (1996) A quantum molecular dynamics study of proton conduction phenomena in $BaCeO_3$. Solid State Ion 86–88:647–652
13. See for example Wikipedia, "Ab initio quantum chemistry methods"
14. Bork N, Bonanos N, Rossmeisl J, Vegge T (2011) Thermodynamic and kinetic properties of H defect pairs in $SrTiO_3$ from density functional theory. Phys Chem Chem Phys 13:15256–15263
15. Bonanos N (2001) Oxide-based protonic conductors, point defects and transport properties. Solid State Ion 145:265–274
16. Bonanos N, Knight KS, Ellis B (1995) Perovskite solid electrolytes: structure, transport properties and fuel cell applications. Solid State Ion 79:161–170
17. Coors WG (2004) Steam reforming and water-gas shift by steam permeation in a protonic ceramic fuel cell. J Electrochem Soc 51(7):A994
18. He F, Wu T, Peng R, Xia C (2009) Cathode reaction models and performance analysis of $Sm_{0.5}Sr_{0.5}CoO_{3-\delta}$–$BaCe_{0.8}Sm_{0.2}O_{3-\delta}$ composite cathode for solid oxide fuel cells with proton conducting electrolyte. J Power Sources 194:263–268
19. Ricote S, Bonanos N, Rørvik PM, Haavik C (2012) Microstr./perform. of $La_{0.58}Sr_{0.4}Co_{0.2}Fe_{0.8}O_{3-\delta}$ cathodes deposited on $BaCe_{0.2}Zr_{0.7}Y_{0.1}O_{3-\delta}$ by infiltration and spray pyrolysis. J Power Sources 209:172–179

pH Electrodes - Industrial, Medical, and Other Applications

Winfried Vonau
Kurt-Schwabe-Institut fuer Mess- und
Sensortechnik e.V. Meinsberg, Waldheim,
Germany

Introduction

The pH value of a solution is a measure of the activity of hydronium ions ($H_3O^+ \rightleftharpoons H^+ + H_2O$) in the solution and as such it is a measure of the acidity or basicity of that medium. On the one hand pH stands for power of hydrogen (H); on the other hand, the term pH is derived from p, the mathematical symbol of the negative logarithm. The pH value is often simply referred to as concentration of hydrogen ions. Acids and bases are connected with free hydronium and hydroxyl ions, respectively. The relationship between hydronium and hydroxyl ions in a given solution is constant for a given set of conditions and one can be determined by knowing the other. The usual range of pH values encountered is between 0 and 14, with 0 being the value for concentrated acids, 7 being the value for pure water, and 14 being the value for concentrated lyes. It is possible to get pH values <0 (e.g., 10 M HCl) and >14 (e.g., 10 M NaOH). In Fig. 1 examples of pH values of a selection of substances are given.

There is a wide variety of applications for pH measurement. For example, pH measurement and control is the key to the successful purification of drinking water, manufacturing of sugar, sewage treatment, food processing, electroplating, the effectiveness and safety of medicines, cosmetics, etc. Plants require the soil to be within a certain pH range in order to grow properly; animals or humans can sicken or die if their blood pH level is not within the correct limits.

Important Mathematical Correlations and Calculations

Definitions:

$$pH = -lg\ a_{H_3O^+} \tag{1}$$

$a_{H_3O^+} \cdots$ activity of hydronium ions

$$pOH = pK_w - pH \tag{2}$$

$K_w \cdots$ water ion product ($10^{-14}\mathrm{mol}^2\ \mathrm{L}^{-2}$) at 25 °C

For the calculation of pH values, it must be distinguished between strong and weak acid [HA(nion)] and base [C(ation)OH] solutions. Strong acids and bases are almost completely dissociated in water. Hydrochloric acid (HCl) is a good example of a strong acid and sodium hydroxide (NaOH) of a strong base.

For instance, for a 0.01 M solution of HCl, the activity of hydronium ions can be taken as 0.01 M and according to Eq. 1 the pH = 2.

For weak acids and bases, the situation is more complicated. Here, the acid and base dissociation constants ($K_{a(b)}$ or $pK_{a(b)}$ values) have to be considered. For such media there is no complete dissociation, but its degree is given by the equilibrium, Eq. 3. In the following, for reasons of simplification H_3O^+ is replaced by H^+ and [i] stands for the activity of a component i:

$$K_a = \frac{[H^+][A^-]}{[HA]} \tag{3}$$

Using the nominal concentration of the acid, C_a, which is equivalent to the amount of acid that is initially added to a solution

$$C_a = [HA] + [A^-] \tag{4}$$

for a strong acid which is completely dissociated the term [HA] can be neglected:

$$C_a = [A^-] \tag{5}$$

In the case of a weak acid, HA is not completely dissociated, and thus the assumption

pH Electrodes - Industrial, Medical, and Other Applications, Fig. 1 pH values of different substances at 25 °C [1]

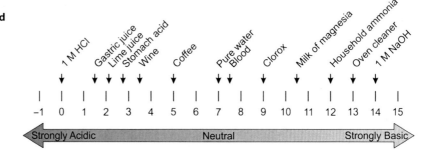

[A⁻] >> [HA] is not valid. Therefore the full mass balance Eq. 4 must be taken.

If the acid is not very diluted and not extremely weak, ions from water dissociation can be neglected; Eq. 6 can be formulated:

$$[H^+] \approx [A^-] \quad (6)$$

There are three equations with three unknowns, ($[H^+]$, $[A^-]$, and $[HA]$), which need to be solved for $[H^+]$. The mass balance equation allows to solve it for $[HA]$ in terms of $[H^+]$.

Rearrangements of above equations lead to Eqs. 7 and 8 resulting finally in Eq. 9 (with the positive square root) giving the expression for the H⁺ activity in weak acidic solutions:

$$K_a = \frac{[H^+]^2}{C_a - [H^+]} \quad (7)$$

$$[H^+]^2 + K_a[H^+] - K_a C_a = 0 \quad (8)$$

$$[H^+] = \frac{-K_a + \sqrt{K_a^2 + 4K_a C_a}}{2} \quad (9)$$

pH calculations in real media require a lot of knowledge of the composition of these systems and the ambient conditions (temperature, pressure, additional components, etc.). Unfortunately, this task is solvable only with additional effort in mathematics, where to a lot of literature is available, e.g., [2]. In most cases, a simple arithmetic determination of pH values of complex systems is not possible. For this reason it is necessary to have practicable measuring methods at one's disposal.

pH Measurement

In Fig. 2 the current procedures for pH determination are shown. As to be seen, electrochemical methods, which following solely will be reported in greater detail, dominate.

The measurement is executed potentiometrically using several pH and reference electrodes. The relation is based on the idealized Eq. 10:

$$E \cdots E^\ominus - (RT/F)\ln[H^+] \quad (10)$$

$E^\ominus$ ... standard electrochemical cell potential (voltage)
R ... gas constant
T ... temperature in Kelvins
F ... Faraday constant

Depending on the measurement task, indicator and reference electrode are chosen. For a long time, the utilization of hydrogen electrodes is known as a possibility to measure pH values. A hydrogen electrode can be made by coating a platinum electrode with a fresh layer of platinum black and passing a flow of hydrogen gas over it. The platinum-black coating allows the adsorption of hydrogen species onto the electrode. When dipped into a solution containing hydronium ions, according to Eq. 11 at $pH_2 = 101.3$ kPa in the hydrogen electrode turns up a potential which is proportional to the H⁺ activity. Hydrogen electrodes are mainly used under standard conditions as reference electrodes (see also subchapters Potentiometry and Reference electrodes) and in combination with these or with any other reference electrodes as indicator electrodes for the pH measurement:

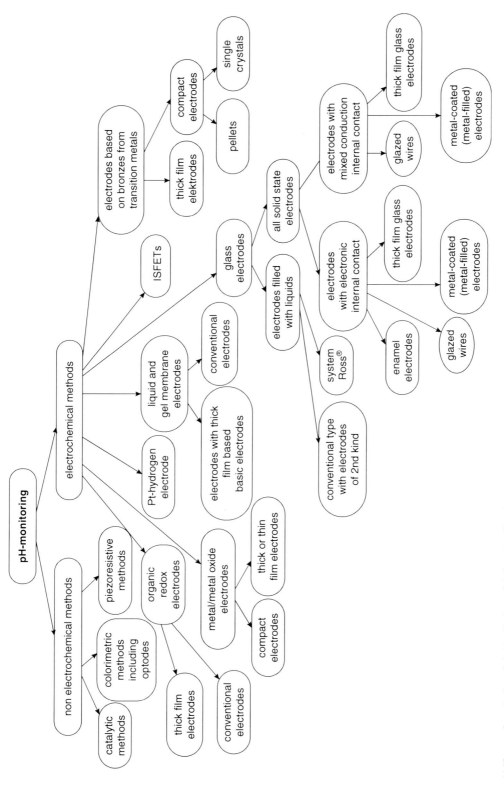

pH Electrodes - Industrial, Medical, and Other Applications, Fig. 2 Experimental methods to measure pH values

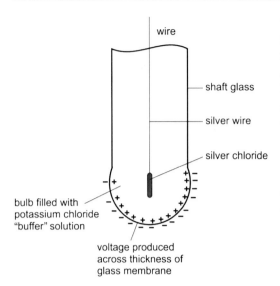

pH Electrodes - Industrial, Medical, and Other Applications, Fig. 3 Scheme of a glass electrode

$$E = E^\ominus + (RT/F)\ln [H^+] - (RT/2F)\ln pH_2 \qquad (11)$$

Due to their ungainliness, today hydrogen electrodes are only seldom in use, e.g., for the calibration of precision buffers. New concepts are focused on the use of miniaturized hydrogen electrodes with internal hydrogen source and gas-diffusion electrodes consisting of Pd and Pt to be applied for food quality control and for measurements in fluoride-containing solutions [3].

In most cases, pH glass electrodes serve as pH selective half-cell. This kind of electrode was introduced by Cremer [4] and is shown in principle in Fig. 3.

Main components are electrode glass, internal electrolytic solution, and internal reference electrode. Electrode glasses normally have a thickness of about 0.1–0.2 mm and their composition often is among the best-kept secrets of the manufacturers of the electrodes. However, the main components, e.g., SiO_2, Li_2O, Na_2O, or CaO, are known [5]. In most cases and in correspondence with the pH meter, the internal solution is a buffer of pH = 7 into which a silver/silver chloride electrode is immersed. A special type of pH glass electrodes uses homogeneous redox systems instead of electrodes of second kind as internal reference. Such electrodes deliver a better response behavior during temperature changes. As redox system is to be used for this purpose, $I_3^-/3I^-$ has been recommended in [6]. At $c_{I_3^-}/c_{I^-}^3 \approx 5 * 10^{-5}$ the potential between 0 and 70 °C changes by a maximum of 1 mV.

While in the past glass electrodes have been used only in a limited temperature range, an extensive development work had been done to enhance the working temperature range of the electrodes from −25 to 130 °C. However, there are no such glass electrodes, which cover the entire range of temperature. The traditional glass electrode with an internal liquid reference system, as all other electrodes of this kind, exhibits some drawbacks: position dependence when the electrodes are stored or used, pressure and temperature dependence, mechanical fragility, and several limitations concerning the miniaturization.

For these reasons, it is a goal to develop all-solid-state electrodes for pH measurement. A main problem to be solved was to connect the ionically conducting glass with an electronic conductor. Although the current flow is very low due to the high resistance of the glass, a reversible electrochemical reaction should take place at the phase boundary by transferring cations between the phases. The possibility to fabricate all-solid-state glass electrodes is in principle already mentioned by Kratz [7]. According to the electrochemical requirements, it is not sufficient to use metals instead of buffer solutions as internal reference electrodes. So, e.g., it was tried to coat bright platinum wires with pH-electrode glass. In the resulting system the electrochemical reaction was blocked. Improved approaches were carried out by arranging a silver plate on the inside of glass membranes, by filling of glass electrode bodies with mercury or several amalgams or by solidifying melts of alloys with low melting points within glass electrode bodies. Due to the solution of sodium or lithium and cadmium in the mercury or in the metal layer, the electrode reactions can become reversible. It was also practiced to use materials having both electronic and ionic conductivity as an internal reference system (interlayer of glass with mixed conductivity, tungsten bronze as a mixed conducting material, modified carbon fleeces, sandwich of pH-sensitive glass, and mixed conducting glass). It has been shown that the measuring behavior of such electrodes (a summarizing report is given in [8]), due to an enhanced reversibility, is much better than that of electrodes with directly metallized pH glass membranes.

Beside the functional optimization of the electrodes, also a further development of manufacturing technology for their production is a continuous challenge. Corresponding to the new knowledge about all-solid-state electrodes planar glass electrodes in thick film technology were developed, which e.g., are applicable as insertion electrode in meat or other solid foods (see Fig. 4).

Because of the good correlation of the coefficients of thermal expansion between pH-electrode glasses and steel, the last mentioned substance is a predestinated substrate material for thick film glass electrodes. This fact also causes the availability of pH enamel electrodes [9], which among others were developed for use in the severe, highly corrosive operating environments of agitated glassed steel reactors. It can be placed inside the chemical reactor and withstands the rigors of heat, pressure, and dynamic agitation to provide continuous pH measurement. No protective cage is needed or used. The pH probe is suited to situations where slurries are encountered, providing years of service with minimum maintenance. Typical applications include agitated reactors, pipelines with high flow rates, erosive and highly corrosive slurries, limestone recirculation lines, pulp and paper applications, pharmaceutical production, paint pigment manufacturing, fermentation, and neutralization processes. Sometimes, a constant concentration of one ionic component in the analyzed medium is given (e.g., Na^+). Then a second Na^+-sensitive enamel electrode can be used as external reference electrode as shown in Fig. 5.

There are a lot of other, partly niche applications where glass electrodes according to the national and international standards are not perfectly suited. Planar technologies open up opportunities for new multisensor arrays including pH electrodes. Depending on the special application, besides glass membrane electrodes, also other types of pH electrodes play a role as separate probe and as well as a part of multisensor systems.

Especially for medical and biotechnological applications miniaturized planar pH electrodes are needed. Here metal/metal oxide electrodes and electrodes based on ion-selective field effect

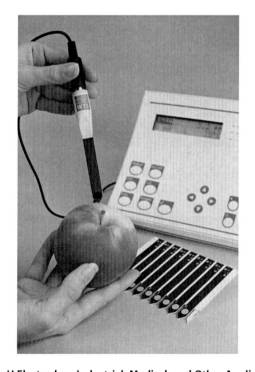

pH Electrodes - Industrial, Medical, and Other Applications, Fig. 4 Thick film pH glass electrode

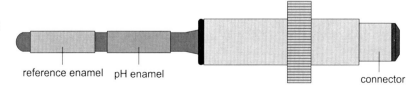

pH Electrodes - Industrial, Medical, and Other Applications, Fig. 5 All-solid-state enamel pH combination electrode

transistors (ISFET) are used. An example for the first mentioned type is Ru/RuO$_2$-electrodes that can be fabricated in thin as well as in thick film technology. They can for instance be used for the simultaneous in vitro measurement of metabolic, morphologic, and electrophysiologic parameters of living cells and tissue. pH-ISFET electrodes for the first time were mentioned by Bergveld [10]. In Fig. 6 an example for the usage of such type of electrodes as single probe is given. Figure 7 shows the layout of a multielectrode array including ISFETs for the pH determination.

For several applications in pH measurement older electrode concepts are still in use. For reasons of cost probes for the determination of pH values in gastroenterology (see Fig. 8) often consist of antimony electrodes with diameters <3 mm [12]. Their pH function as an electrode of the second kind is given by Eq. 12 which is formulated for a temperature of 25 °C:

pH Electrodes - Industrial, Medical, and Other Applications, Fig. 6 ISFET pH probes for measurements in stomatology (**a**) and gynecology (**b**)

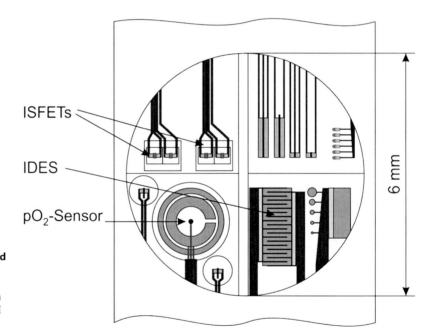

pH Electrodes - Industrial, Medical, and Other Applications, Fig. 7 Layout of a multiparametric silicon sensor chip including pH ISFETs [11]

pH Electrodes - Industrial, Medical, and Other Applications, Fig. 8 Antimony combination electrode with silver chloride reference electrode for pH measurements in the gastroenterology

$$E = E^{\ominus}_{(Sb)} + \frac{0.059}{3} \ln K_{L_{Sb(OH)_3}} + 0.059*14 - 0.059\, pH \quad (12)$$

The equilibrium of redox reactions where hydronium ions are taking part is dependent on the pH value. A favored redox system for pH measurements is the equilibrium between quinone and hydroquinone. The quinhydrone electrode based thereon [13] was the most commonly used device for the electrochemical pH determination in the laboratory until the introduction of the glass electrode. Here, the redox voltage is measured with a conventional reference and a blank platinum electrode. Due to the rapid and reproducible establishment of its chain voltage, this electrode possesses importance for several applications, e.g., for measurements in fluoride-containing media. They were advanced in recent years as pH thick film electrodes, too and among others, qualified for a usage in flow-through mode (see Fig. 9).

For measurements in emulsions, alkaline solutions, and solutions with only moderate concentrations of oxidizing or reducing agents, the well-known quinhydrone electrode was further developed as solid composite electrode.

An electrode based on this conception is also available for flow injection potentiometry [13].

Independent of the kind of the used pH electrode, especially under extreme operational conditions, high demands are placed on the stability of the selective membranes, reinforcing materials and tightness of the electrode housing. In Fig. 10 in this connection, a pressure and temperature stable antimony electrode is shown for the use in geothermal applications or in water steam circle of power stations. In Fig. 11 a glass electrode for precision measurement in the deep sea is presented establishing a corresponding internal pressure in the inner core of the pH electrode.

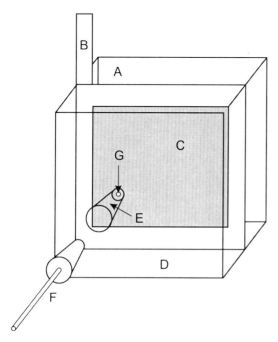

pH Electrodes - Industrial, Medical, and Other Applications, Fig. 9 Scheme of the detector cell: *A* and *D* poly (methyl methacrylate) plates, *B* copper spacer frame, *C* pH-sensitive layer, *E* hole for FIA fitting, *F* FIA tube fittings, *G* hole through the sensitive material [14]

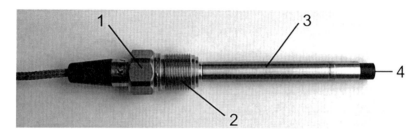

pH Electrodes - Industrial, Medical, and Other Applications, Fig. 10 pH electrode to be used under extreme application conditions. *1* ... electrode head, *2* ... thread M 22 × 1.5 mm, *3* ... stainless steel (or titanium) bushing (inner diameter: 12 mm), *4* ... antimony electrode

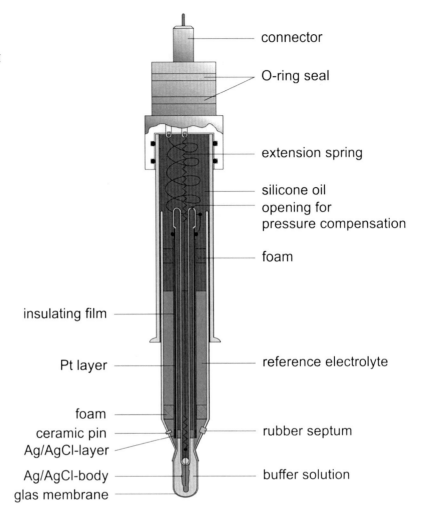

pH Electrodes - Industrial, Medical, and Other Applications, **Fig. 11** Pressure stable pH glass electrode in the deep sea

Future Directions

Because of the higher-than-average importance of pH electrodes in measurement technology, R&D in this field will be necessary also in the future. Major tasks will be improving selectivity and reducing cross sensitivities, respectively; further miniaturization while maintaining electrode performance of common electrodes; expansion of the area of application concerning pressure and temperature; reducing calibration efforts in connection with increasing the lifetime; inexpensive production using new fabrication methods; and further integration in multisensor systems.

Cross-References

► Potentiometry
► Reference Electrodes
► Sensors

References

1. Shipman J, Wilson J (1990) An introduction to physical science. D. C. Heath, Boston
2. Schwabe K (1969) pH-Fibel. Deutscher Verlag für Grundstoffindustrie, Leipzig
3. Vonau W, Oelßner W, Schwarz J, Hörig A, Kohnke HJ, Heller G, Kurzenknabe (2011) German Patent DE 10 2011 113 941

4. Cremer M (1906) Über die Ursachen der elektromagnetischen Eigenschaften der Gewebe, zugleich ein Beitrag zur Lehre von den polyphasischen Elektrolytketten. Z Biol 46:562–608
5. Solokow AI, Passinski AH (1932) Über Glaselektroden. Z Phys Chem A 160:366–377
6. Ross JW (1982) Potentiometrische Elektrode. German Patent, DE 146 066
7. Kratz L (1950) Die Glaselektrode und ihre Anwendungen. Dietrich Steinkopff, Frankfurt
8. Vonau W, Guth U (2006) pH monitoring: a review. J Solid State Electrochem 10(9):746–752
9. Emmerich B (1978) Die Emailelektrode, eine neue Lösung zur pH-Messung im Betrieb. Regelungstechn. Praxis 20:313
10. Bergveld P (1972) Development, operation and application of the ion-selective field-effect transistor as a tool for electrophysiology. IEEE Trans Biomed Eng 19:340–351
11. Brischwein M, Motrescu ER, Otto AM, Cabala E, Grothe H, Wolf B (2003) Functional cellular assays with multiparametric silicon sensor chips. Lab Chip 3(4):234–240
12. Vandenplas Y, Badriul H, Verghote M, Hauser B, Kaufman L (2004) Glass and antimony electrodes for oesophageal pH monitoring in distressed infants: how different are they? Eur J Gastroenterol Hepatol 16:1325–1330
13. Biilmann ME (1921) Sur I'hydrogenation des quinhydrones. Ann Chim 15:109–157
14. Kahlert H, Pörksen JR, Behnert J, Scholz F (2005) FIA acid-base titration with a new flow-through PH detector. Anal Bioanal Chem 382:1981–1986

Photocatalyst

Bunsho Ohtani
Catalysis Research Center, Hokkaido University, Sapporo, Hokkaido, Japan

Definition

Photocatalysts are materials that induce photocatalytic reaction under photoirradiation. A general definition of photocatalysis, a conceptual name of photocatalytic reactions, is a chemical reaction induced by photoabsorption of a solid material, or "photocatalyst," which remains unchanged during the reaction. Therefore, "photocatalyst" should act catalytically, i.e., without change, under light. Although molecules or metal complexes dissolved in solution or in the gas phase, not solid materials, can drive such photoinduced reactions without change during the reactions, they are called "photosensitizer" but not "photocatalyst." photoinduced reactions are, in principle, initiated by photoabsorption, i.e., excitation of electrons in a material, followed by electron transfer from/to a reaction substrate. The photoexcited electrons can reduce a substrate adsorbed on the surface of a photocatalyst, and positive holes, electron deficiencies, can oxidize an adsorbed substrate, if the electronic levels of photoexcited electrons and positive holes are higher and lower than the redox levels of substrates, respectively. This is one of the necessary conditions for photocatalysis and will be discussed later.

For photocatalysts, there are several modes of photoexcitation as shown below.

Band-Gap Excitation

For conventional photocatalysts such as titanium (IV) oxide (TiO_2; titania), band-gap excitation occurs when irradiated. In a schematic representation of the electronic structures of semiconducting (or insulating) materials, a band model, an electron in an electron-filled valence band (VB) is excited by photoabsorption to a vacant conduction band (CB), which is separated by a band gap (a forbidden band), from the VB, leaving a positive hole in the VB. An important point is that photoabsorption and electron-positive hole generation are inextricably linked; a VB electron is not excited after photoabsorption. These electrons and positive holes can induce reduction and oxidation, respectively, of chemical species adsorbed on the surface of a photocatalyst, unless they recombine with each other so as not to induce a redox reaction but to produce heat and/or photoemission. Such a mechanism accounts for the photocatalytic reactions of semiconducting (or insulating) materials absorbing photons by the bulk of materials.

Since light of energy greater than band gap can excite electrons in VB to CB, light of wavelength shorter than that corresponds to the band gap, i.e., longer-wavelength limit for photoabsorption (photoexcitation) is fixed by the band structure of a photocatalyst. For metal oxides, it has been known that VB is mainly composed of O_{2p} orbitals and thereby the position (top) of VB is independent of the kind of metal, i.e., only the CB (bottom) position is changed depending on the kind of metal. Therefore, in order to narrow the band gap, i.e., to use light of longer wavelength, metal oxides having lower CB level should be used. Considering the VB top level seems enough low to oxidize most of organic/inorganic compounds, the band-gap narrowing for metal oxides leads to disadvantageous lowering of the CB bottom level.

Transition Between an Electronic Level and a Band

Photocatalysts that can use visible light included in sunlight and indoor light have been looked for since the conventional photocatalysts such as titania can absorb, i.e., be excited only by ultraviolet light. The strategy to utilize visible light in photocatalysis can be roughly divided into three categories. The first one is to use the semiconductor with a narrow band gap. Metal nitride or sulfides often have colors and are expected to work as photocatalysts under visible-light irradiation when used. However, they tend to be oxidized and possibly lose their photocatalytic activity. Then, it was proposed to raise the top of VB of stable metal oxides to reduce the band-gap energy by doping various elements as the second strategy. Many papers have reported the band-gap narrowing by doping nitrogen, carbon or sulfur, etc., to titania and the introduction of visible-light responsibility. Actually, shift of the photoabsorption spectrum to the longer-wavelength side is observed by doing a variety of elements. It has been suggested in the recent studies, however, that the electronic structure of such doped material was not an expected one; levels of doped elements are separated from the VB to form an independent sub-bands.

In recent studies, clusters of ions such as copper or iron and their oxide were deposited (grafted) to induce photoexcitation of electrons in VB of base metal oxide such as titania to the electronic level of these loaded clusters or electrons in the electronic level to CB, i.e., interfacial charge transfer. When the electronic states of the grafted clusters are located between the CB bottom and the VB top of titania, the interfacial charge transfer can be driven under visible-light irradiation, and a new photoabsorption band appears in a wavelength region longer than that for band-gap excitation.

Excitation Through Surface-Plasmon Resonance Absorption

Another example of visible light-driven reaction through non-band-gap excitation is a photocatalytic reaction that uses surface-plasmon resonance (SPR) absorption of small metal particles loaded on base metal oxides. For example, gold particles of the size of ten to several ten nanometers, presenting purplish red color by the SPR absorption, loaded on titania particles have been used for photocatalytic reactions under visible-light irradiation at the wavelength of ca. 600 nm. Based on the results that titania or a related material is necessary for this visible light-driven reaction and that SPR absorption cannot induce electronic excitation of electrons, the mechanism of this kind of reaction seems complicated and is now under discussion.

Thermodynamics

As thermodynamics says, if ΔG is negative ($\Delta G < 0$ as in the case, e.g., in oxidative decomposition of organic compounds under aerated conditions) and if ΔG is positive ($DG > 0$, as in the case, e.g., in splitting of water into hydrogen and oxygen), energy is released and stored, respectively. Therefore, if the standard electrode potential of the compound to be reduced by electrons is higher, i.e., more negative (cathodic), than that of the compound to be oxidized by positive holes, ΔG is positive, i.e., the reaction stores energy and vice versa. A notable point is that both situations, energy release and storage, are possible in photocatalysis, while thermal catalyses are limited to only reactions of negative ΔG, i.e., spontaneous reactions. The reason why photocatalysts can drive even a reaction of positive ΔG, which does not proceed spontaneously, is that an overall redox reaction can proceed, even if its ΔG is positive, in a system in which reduction and oxidation steps are spatially or chemically separated; otherwise, the reaction between reduction and oxidation products proceeds to give no net products. Under these conditions, both partial Gibbs energy change for reactions of photoexcited electrons with oxidant (ΔG_e) and positive holes with reductant (ΔG_h) are required to be negative, i.e., reactions by photoexcited electrons and positive holes proceed spontaneously by photoexcitation. In other words, for the reaction through band-gap excitation, the CB bottom and VB top positions must be higher (more cathodic) and lower (more anodic) than standard electrode potentials of an electron acceptor (oxidant) and an electron donor (reductant), respectively, to make Gibbs energy change of both half reactions negative, as has often been pointed out as a necessary condition for photocatalysis.

Photocatalytic Activity

The widely used scientific term "activity" often appears in papers on photocatalysis as "photocatalytic activity." Although the author does not know who first started using this term in the field of photocatalysis, researchers in the field of conventional catalysis were using this term even before the 1980s, when photocatalysis studies had begun to be promoted by the famous work of the so-called Honda–Fujishima effect on photoelectrochemical decomposition of water into oxygen and hydrogen using a single-crystal titania electrode, as mentioned above. Most authors, including the present author, prefer to use the term "photocatalytic activity," but in almost all cases, the meaning seems to be the same as that of absolute or relative reaction rate. A possible reason why the term "photocatalytic activity" is preferably used is that the term may make readers think of "photocatalytic reaction rate" as a property or ability of a photocatalyst, i.e., photocatalysts have their own activity. On the other hand, "reaction rate" seems to be controlled by given reaction conditions including

a photocatalyst. In the field of conventional catalysis, "catalytic activity" has been used to show a property or performance of a catalyst, since an "active site," substantial or virtual, on a catalyst accounts for the catalytic reaction. The estimated reaction rate per active site can be called "catalytic activity." In a similar sense, the term "turnover frequency," i.e., number of turnovers per unit time of reaction, is sometimes used to show how many times one active site produces a reaction product(s) within unit time. On the contrary, it is clear that there are no active sites, as in the meaning used for conventional catalysis, in which rate of catalytic reaction is predominantly governed by the number or density of active sites, on a photocatalyst. The term "active site" is sometimes used for a photocatalytic reaction system with dispersed chemical species, e.g., metal complexes and atomically adsorbed species, on support materials. However, even in these cases, a photocatalytic reaction occurs only when the species absorb light, and therefore, species not irradiated cannot be active sites. A possible mechanism of photo induced reaction is that photoirradiation induces production of stable "active sites" that work as reaction centers of conventional catalytic reactions, though this is different from the common mechanism of photocatalysis by electron–positive hole pairs. Anyway, photocatalytic reaction rate strongly depends on various factors such as the irradiance of irradiated light that initiates a photocatalytic reaction. Considering that the dark (nonirradiated) side of a photocatalyst or suspension does not work for the photocatalytic reaction, the use of the term "active site" seems inappropriate.

Design of Active Photocatalysts

Since an ordinary photocatalysis is induced by photoexcited electrons and positive holes, rate of photocatalytic reaction must depend on photoirradiation irradiance (light flux) and efficiencies of both photoabsorption and electron–positive hole utilization. The efficiency of electron–positive hole utilization is called quantum efficiency, i.e., the number (or rate) ratio of product(s) and absorbed photons, and even if quantum efficiency is high, the overall rate should be negligible when the photocatalyst does not absorb incident light. This is schematically represented as

$$[\text{Rate}/\text{mol s}^{-1}] = [\text{Irradiance}/\text{mol s}^{-1}]$$
$$\times [\text{Photoabsorption efficiency}]$$
$$\times [\text{Quantum efficiency}].$$

Since all of the parameters in this equation must be functions of light wavelength, the overall rate can be estimated by integration of a product of spectra of photoirradiation, photoabsorption, and quantum efficiencies. When we discuss activity of a photocatalyst, it seems reasonable to evaluate a product of photoabsorption and quantum efficiencies, i.e., apparent quantum efficiency. Assuming that quantum efficiency does not depend on the irradiation (absorption) irradiance, the actual reaction rate can be estimated by multiplying with the irradiance. On the basis of these considerations, enhancement of photocatalytic activity can be achieved by increase in both efficiencies. For example, preparing visible-light absorbing photocatalysts, as a recent trend in the field of photocatalysis, and depositing noble metal particles onto the surface of photocatalysts lead to the improvement of these efficiencies, respectively. In this sense, the design of active photocatalysts seems simple and feasible, but we encounter the problem that both efficiencies are related to each other, and we do not know how we can improve the quantum efficiency since correlations between physical/ structural properties and photocatalytic activity have only partly been clarified.

Future Perspectives

Since most of researchers in the field of photocatalysis came from different fields of chemistry, catalysis chemistry, electrochemistry, materials chemistry, photochemistry, etc., there seemed no common concepts shared by them. It is necessary to understand photocatalysis

appropriately considering thermodynamics and kinetics of photocatalysis introduced in this section.

Cross-References

▶ TiO$_2$ Photocatalyst

References

1. Ohtani B (2008) Preparing articles on photocatalysis—beyond the illusions, misconceptions and speculation. Chem Lett 37(3):216–229
2. Ohtani B (2010) Photocatalysis A to Z—what we know and what we don't know. J Photochem Photobiol C Photochem Rev 11(4):157–178
3. Ohtani B (2011) Photocatalysis by inorganic solid materials: revisiting its definition, concepts, and experimental procedures. Adv Inorg Chem 63:395–430

Photochromizm and Imaging

Tetsu Tatsuma
Institute of Industrial Science, University of Tokyo, Tokyo, Japan

Introduction

Photochromism is the reversible color changes of a material, one way or both ways of which are induced by light irradiation. In relation to electrochemistry, photoinduced charge separation gives positive and negative charges, and the former and/or the latter cause redox reactions accompanied by color changes. There are three types of electrochemical photochromism (i.e., photoelectrochromism). (1) A material absorbs light and gives rise to charge separation. It reduces (or oxidizes) itself by separated charges, resulting in a color change. It is electrochemically or chemically reoxidized (or re-reduced) (single material systems). (2) Material A absorbs light and gives charge separation. It reduces (or oxidizes) material B by separated charges and changes its color. Reduced (or oxidized) material B is electrochemically or chemically reoxidized (or re-reduced) (composite material systems). (3) Material A absorbs light of wavelength λ_1 and gives charge separation. It reduces (or oxidizes) material B by separated charges and changes its color. Reduced (or oxidized) material B now absorbs light of wavelength λ_2 and gives charge separation. It reoxidizes (or re-reduces) itself to original material B (composite material systems with reversible photoinduced processes). Some examples are described in the following sections.

Single Material Systems

MoO_3, WO_3, and V_2O_5 films (amorphous in many cases) irradiated with UV light changes their color from colorless to blue [1, 2]. Electrons in the valence band of the metal oxide are excited to the conduction band and used for self-reduction of the film (Fig. 1a). The film is bleached gradually in dark by reoxidation in the presence of ambient oxygen. Slightly reduced MoO_3 are colored even by visible light [3].

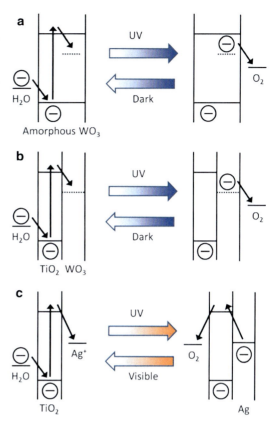

Photochromizm and Imaging, Fig. 1 Mechanisms of some typical photochromic materials. (**a**) Single material system. (**b**) Composite material system. (**c**) Composite material systems with reversible photoinduced processes

Composite Material Systems

A semiconductor photocatalyst such as TiO_2 or ZnO (Chapter 309535, 310654) is coupled with an electrochromic system. As electrochromic systems, metal ion/metal oxide systems (Tl^+/Tl_2O_3, Pb^{2+}/PbO_2, Mn^{2+}/MnO_2, and Co^{2+}/CoO_2) and metal ion/metal systems (Ag^+/Ag) can be used [4]. In the former system, metal ions are oxidized to colored metal oxides by holes generated in the valence band of the photo-irradiated semiconductor and bleached by electrochemical reduction of the oxide. In the latter, metal ions are reduced to colored metal particles by photoexcited electrons in the conduction band. A system with Prussian white/Prussian blue is also used. Prussian white on TiO_2 is oxidized to Prussian blue by UV irradiation [5]. Redox reactions of conducting polymers such as polypyrrole derivatives are also employed [6, 7]. Methylene blue can also be combined with polyaniline as a photocatalyst, and methylene blue is photoelectrochemically reduced to a colorless form [8]. WO_3 can be coupled with TiO_2, as a mixed suspension [9, 10], or a photoelectrochromic cell with a TiO_2 photoanode and a WO_3 electrochromic cathode [11]. A WO_3-TiO_2 composite film also works not only in an electrolyte [12] but also in air [13]. The blue film colored by photoelectrochemical reduction under UV irradiation can be bleached due to reoxidation by dissolved or ambient oxygen (Fig. 1b). In the coloration process in air, adsorbed water works as electrolyte so that a photoanodic reaction at TiO_2 and cathodic reaction at WO_3 can take place separately.

Composite Material Systems with Reversible Photoinduced Processes

In this type of photochromic materials, reversed processes are driven by light of different wavelengths. When TiO_2 with adsorbed Ag^+ ions is irradiated with UV light in air, electrons in the TiO_2 valence band are excited to the conduction band, and Ag^+ ions are reduced by the electrons to Ag nanoparticles that exhibit localized surface plasmon resonance (LSPR) (Fig. 1c) (Chapter 309533). The deposited Ag nanoparticles are different in size, so that they exhibit different colors since the LSPR wavelength redshifts as the particle size increases. If the Ag nanoparticle ensemble, which has absorption over the visible range, is irradiated with a monochromatic green light, for instance, particles resonant with green light are selectively excited and oxidized to Ag^+ ions due to plasmon-induced charge separation (Chapter 309533) [14]. After sufficient irradiation, there are no Ag particles that absorb green light, and the sample reflects green light [15]. As a result, the sample shows green color. Likewise, the material exhibits blue, red, or white color after irradiation with light of the corresponding color [16–18].

The multicolor photochromic material can be initialized by UV irradiation. A drawn image is gradually bleached under room light on the basis of its principle. To retain a drawn image, the Ag nanoparticles may be protected by an alkylthiol [19].

The multicolor photochromism can be extended to near-infrared region by using Ag nanorods so that invisible images, which are viewable with infrared cameras, can be drawn and overlaid with a visible image [20]. Polarization-selective imaging is also possible with nanorods.

Future Directions

Photochromic materials can be applied to smart windows and sunglasses. In view of thermal control, photochromism in the infrared region would be more important. Additional potential applications of photochromism are simple and inexpensive display materials for rewritable papers and book readers. Further development of multicolor photochromism would therefore be of significance. In particular, reproducibility of colors and drawing time for images would be important issues. It would also be applied to nanophotonic devices [21] exhibiting reversible responses.

Cross-References

- ▶ Photocatalyst
- ▶ Plasmonic Electrochemistry (Surface Plasmon Effect)
- ▶ TiO_2 Photocatalyst

References

1. Deb SK, Chopoorian JA (1966) J Appl Phys 37:4818
2. Colton RJ, Guzman AM, Rabalais JW (1978) Acc Chem Res 11:170
3. Yao JN, Hashimoto K, Fujishima A (1992) Nature 355:624
4. Inoue T, Fujishima A, Honda K (1980) J Electrochem Soc 127:1582
5. DeBerry DW, Viehbeck A (1983) J Electrochem Soc 130:249
6. Inganäs O, Lundström I (1984) J Electrochem Soc 131:1129
7. Yoneyama H, Wakamoto K, Tamura H (1985) J Electrochem Soc 132:2414
8. Kuwabata S, Mitsui K, Yoneyama H (1992) J Electrochem Soc 139:1824
9. Ohtani B, Atsumi T, Nishimoto S, Kagiya T (1988) Chem Lett 1988:295
10. Tennakone K, Ileperuma OA, Bandara JMS, Kiridena WCB (1992) Semicond Sci Technol 7:423
11. Bechinger C, Ferrere S, Zaban A, Sprague J, Gregg BA (1996) Nature 383:608
12. Tatsuma T, Saitoh S, Ohko Y, Fujishima A (2001) Chem Mater 13:2838
13. Tatsuma T, Saitoh S, Ngaotrakanwiwat P, Ohko Y, Fujishima A (2002) Langmuir 18:7777
14. Tian Y, Tatsuma T (2005) J Am Chem Soc 127:7632
15. Matsubara K, Tatsuma T (2007) Adv Mater 19:2802
16. Ohko Y, Tatsuma T, Fujii T, Naoi K, Niwa C, Kubota Y, Fujishima A (2003) Nat Mater 2:29
17. Naoi K, Ohko Y, Tatsuma T (2004) J Am Chem Soc 126:3664
18. Tatsuma T (2013) Bull Chem Soc Jpn 86:1

19. Naoi K, Ohko Y, Tatsuma T (2005) Chem Commun 1988:1288
20. Kazuma E, Tatsuma T (2012) Chem Commun 48:1733
21. Tanabe I, Tatsuma, T (2012) Nano Lett 12:5418

Photoelectrochemical CO$_2$ Reduction

Shunsuke Sato
Toyota Central Research and Development
Laboratories, Inc., Nagakute, Aichi, Japan

Introduction

The development of a photocatalytic system to produce useful organic chemicals by reducing CO_2 under sunlight is an increasingly important research area addressing global warming and fossil fuel shortages. If CO_2 can be reduced by utilizing water as an electron donor, such a reaction would mimic photosynthesis in plants. Development of the water oxidation catalysts and CO_2 reduction catalysts are important to achieve it. There are several methods to achieve CO_2 reduction using catalysts such as electrocatalysts, photoelectro-catalysts, and photocatalysts.

Electrocatalytic CO_2 reduction can be conducted by metal catalysts, semiconductor catalysts, and molecular catalysts (metal complexes) under electrical biases. The electrocatalyst requires a great electrical potential in order to achieve a catalytic CO_2 reduction, which may be regarded as activation energy for the reaction. Potentials for CO_2 reduction to various products are given in Table 1. Single-electron reduction of CO_2 to CO_2 radical anion occurs at -1.9 V versus NHE. The unfavorably high negative potential for the single-electron reduction of CO_2 is caused by the large reorganization energy between the linear CO_2 molecule and bent-bonded CO_2 radical anion. Therefore, it is difficult to reduce CO_2 molecule using an electrical energy only. The electrocatalysts are able to facilitate proton-coupled multi-electron reactions which require

Photoelectrochemical CO$_2$ Reduction, Table 1 CO$_2$ reduction potentials

	$E^0/$V vs. NHE
$CO_2 + e^- \longrightarrow CO_2^{-\cdot}$	< -1.9
$CO_2 + 2e^- + 2H^+ \longrightarrow HCOOH$	-0.61
$CO_2 + 2e^- + 2H^+ \longrightarrow CO + H_2O$	-0.53
$CO_2 + 4e^- + 4H^+ \longrightarrow HCHO + H_2O$	-0.48
$CO_2 + 6e^- + 6H^+ \longrightarrow CH_3OH + H_2O$	-0.38
$CO_2 + 8e^- + 8H^+ \longrightarrow CH_4 + 2H_2O$	-0.24

lower potentials than those for the single-electron reactions. The potentials for reduction of CO_2 to various organics show that the potentials required for reactions are lowered with increasing numbers of electrons and protons involved in the reactions. However, many electrocatalysts require much higher electrical energy (overpotential) than the theoretical values. Thus, developing electrocatalysts which can reduce the overpotential is necessary.

Advantages and Disadvantages of Photoelectrochemical Reaction

Photoelectrochemical (PEC) reaction is a similar to electrochemical reaction in regard to an experimental setup. However, PEC reaction uses semiconductor electrodes instead of conductor electrodes used in electrochemistry (Fig. 1). Semiconductors (electrodes) possess a band gap where no electron states can exist and generate pairs of electrons and holes upon absorption of light whose photon energy is larger than the band gap energy. Semiconductors have been used to convert solar photon energy into electrical energy by photovoltaic devices. Thus, there have been attempts to utilize semiconductor electrodes for PEC CO_2 reduction.

Advantage of the PEC reaction is the lower overpotential required for it than that for the electrochemical reaction, because photoexcitation of the semiconductor improves energy of electrons up to a level of conduction band from valence band in the semiconductor (Fig. 2). Here, reaction rate and catalytic ability depends on light

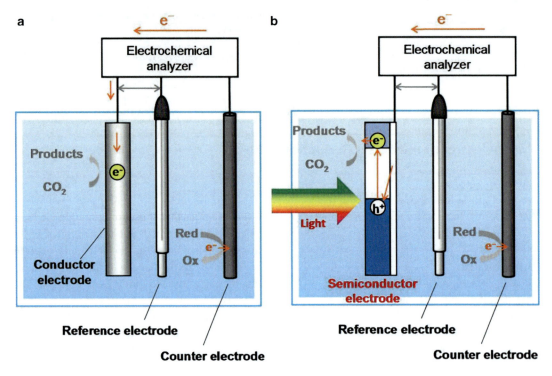

Photoelectrochemical CO₂ Reduction, Fig. 1 Reaction mechanism of electrochemical reaction (a) and PEC reaction (b)

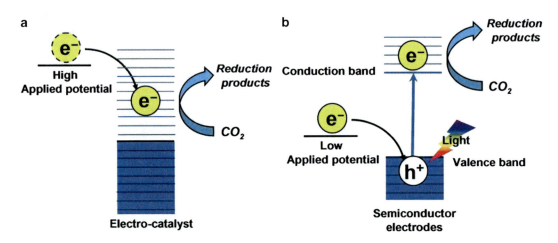

Photoelectrochemical CO₂ Reduction, Fig. 2 Schematic energy diagram of electrochemical reaction (a) and PEC reaction (b)

intensity and quality of semiconductor surface as a catalyst, respectively. When activity at a semiconductor surface for a specific reaction is poor, the corresponding PEC reaction rate is very slow, irrespective of a level of applied potential. Therefore, controlling surface chemistry of the semiconductor appropriate for each specific reaction is crucial for high-efficiency PEC reaction. The same is true of electrochemical reaction.

Photoelectrochemical CO₂ Reduction, Fig. 3 Schematic illustration of the tandem-type CO₂ photo-recycling cellreactor

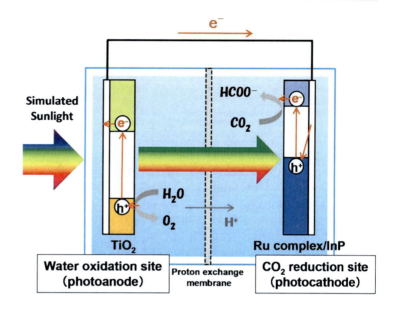

PEC CO₂ Reduction

In 1972, Honda and Fujishima reported first PEC hydrogen production reaction by water-splitting using Pt electrode (cathode) and TiO₂ semiconductor photoanode under ultraviolet light irradiation [1]. On the contrary, the first PEC reduction of CO₂ by InP photocathode was reported by Halmann in 1978 [2]. Then PEC CO₂ reduction was reported in many papers. However, the efficiency of the PEC CO₂ reduction was low even under irradiation, more intense than sunlight, and with a high electrical bias. Furthermore, most of the results suffer from low selectivity of products from CO₂ reduction, because semiconductor electrodes prefer to produce hydrogen, which is due to the catalytic nature of surface of semiconductors or co-catalysts deposited on the semiconductors. Therefore, majority of PEC reaction reported to date have been the solar hydrogen production [3–6].

Recently, there appeared reports on successful PEC CO₂ reduction reaction, which consists of combinations of semiconductor electrodes and molecular catalysts [7–9].

In 2008, Bocarsly et al. developed a CO₂ reduction system consisting of a GaP photocathode and a pyridinium ion dissolved in aqueous solution [7]. They insist that pyridinium ions which receive electron from visible-light-excited GaP acts as a catalyst for successive six-electron reduction of carbon dioxide to methanol in 0.1 M acetate buffer under an electrical bias. Reduction product selectivity ranges from 90 % to 51 % under an electric bias from −0.3 V to −0.7 V (vs. SCE).

In 2010, Arai et al. have developed a photocathode for CO₂ reduction in water which consists of a combination of Ru-complex catalyst and InP [8]. This system produces formic acid from CO₂ and H₂O under visible-light irradiation with a selectivity of ca. 62 % with an electrical bias of −0.6 V (vs. SCE). In this entry, by isotope tracer analyses using $^{13}CO_2$ and D₂O, it was verified that carbon and proton sources for formate formation were CO₂ and H₂O, respectively.

Future Directions

PEC CO₂ reduction technique can reduce electrical bias and overpotential compared to electrical CO₂ reduction. Development of semiconductor electrodes is very important in this research area, because PEC catalytic ability depends on the semiconductor electrodes in regard to photoabsorption and energy of photoexcited electrons. Development of co-catalysts such as molecule catalysts which can control selectivity for CO_2 reduction products is also necessary. The goal of the PEC reaction is to develop PEC-based system for selective CO_2 reduction that makes use of water molecules as both proton and electron sources under solar irradiation like the photosynthetic process in plant. Recently, it was reported that a tandem-type CO_2 photo-recycling cell was successfully constructed by combining the Ru-complex/InP photocathode for CO_2 reduction with a TiO_2 photoanode catalyst capable of for water oxidation and that the system utilized H_2O as both an electron donor and a proton source under simulated sunlight with no electrical bias (Fig. 3) [10]. Selectivity for formate formation was about 70 % and solar conversion efficiency for CO_2 reduction to formate was about 0.04 %. This is the first report of photocatalytic CO_2 reduction with high selectivity utilizing H_2O as both an electron donor and a proton source using sunlight as an only energy input, where the reaction pathway was completely proven by isotope tracer analyses. The future trend toward a significant and useful PEC CO_2 reduction would be the conjugation of the CO_2 reduction and the water oxidation.

Cross-References

▶ Photocatalyst
▶ Photoelectrochemistry, Fundamentals and Applications

References

1. Fujishima KH (1972) Electrochemical photolysis of water at a semiconductor electrode. Nature 238:37
2. Halmann M (1978) Photoelectrochemical reduction of aqueous carbon-dioxide on p-type gallium-phosphide in liquid junction solar-cells. Nature 275:115
3. Khaselev O, Turner JA (1998) A monolithic photo-voltaic-photoelectrochemical device for hydrogen production via water splitting. Science 280:425
4. Licht S (2001) Multiple band gap semiconductor/electrolyte solar energy conversion. J Phys Chem B 105:6281

5. Ingler WB Jr, Khan SUM (2006) A self-driven p/n-Fe_2O_3 tandem photoelectrochemical cell for water splitting. Electrochem Solid-State Lett 9:G144
6. Alexander BD, Kulesza PJ, Rutkowska I, Solarska R, Augustynski J (2008) Metal oxide photoanodes for solar hydrogen production. J Mater Chem 18:2298
7. Barton EE, Rampulla DM, Bocarsly AB (2008) ive solar-driven reduction of CO_2 to methanol using a catalyzed p-GaP based photoelectrochemical cell. J Am Chem Soc 130:6342
8. Arai T, Sato S, Uemura K, Morikawa T, Kajino T, Motohiro T (2010) Photoelectrochemical reduction of CO_2 in water under visible-light irradiation by a p-type InP photocathode modified with an electropolymerized ruthenium complex. Chem Commun 46:6944
9. Kumar B, Smieja JM, Kubiak CP (2010) Photoreduction of CO_2 on p-type Silicon using Re(bipy-Bu$_t$) (CO)$_3$Cl: photovoltages exceeding 600 mV for the selective reduction of CO_2 to CO. J Phys Chem C 114:14220
10. Sato S, Arai T, Morikawa T, Uemura K, Suzuki TM, Tanaka H, Kajino T (2011) Selective CO_2 conversion to formate conjugated with H_2O oxidation utilizing semiconductor/complex hybrid photocatalysts. J Am Chem Soc 133:15240

Photoelectrochemical Disinfection

Danae Venieri
Department of Environmental Engineering,
Technical University of Crete, Chania, Greece

Introduction

Maintenance of the microbiological quality and safety of water systems is imperative, as their fecal contamination may exact high risks to human health and result in significant economic losses. Disinfection of water and wastewater is important in the control of waterborne diseases, as it is the final barrier against human exposure to pathogenic microorganisms [1]. Waterborne diseases, which are transmitted through the ingestion of contaminated water that serves as the passive carrier of the infectious agent, illustrate the importance of effective inactivation of pathogens contained in water and wastewater [2]. Common disinfectants used in the drinking and wastewater industries include chlorine,

chlorine dioxide, ozone and ultraviolet light, which may act by many different means to kill organisms or prevent their growth (i.e. destruction of cellular components and inhibition of nucleic acid replication) [1, 3].

Over the last decades, there have been intensive efforts toward the development of efficient technologies for microbial inactivation in various aqueous matrices. In this perspective, advanced oxidation processes (AOPs) have been recognized as an emerging group of techniques, demonstrating high disinfection efficiency in aqueous samples [4]. These methods are based primarily on the in situ generation of highly reactive intermediate chemical species, like hydroxyl radicals, offering a simple and effective process for inactivating pathogenic bacteria and destroying organic compounds [3, 5, 6]. TiO_2 photocatalysis is an important member of AOPs, and its benefits regarding water and wastewater disinfection have been demonstrated with respect to *Escherichia coli*, *Staphylococcus aureus* and *Enterococcus faecalis* [2, 7–9]. In most photocatalytic applications, the catalyst is employed as a slurry of fine particles in a photochemical reactor, resulting in certain difficulties such as post-reaction catalyst recovery and low quantum efficiencies [10, 11]. These problems may be addressed by immobilization of TiO_2 on a conducting support and application of a potential bias, so as to reduce the recombination of charge carriers, which is the main limitation of the process photonic efficiency [12, 13]. These modifications have led to the development of photoelectrocatalysis (PEC), in which a small positive potential is applied to a TiO_2-based thin film in the form of a photoanode. The constant current density or bias potential applied to the semiconductor electrode promotes the efficient separation of electron–hole pairs from electron transfer by an external circuit and accelerates the production of photogenerated oxidizing chemical species on the catalyst surface [14].

According to the literature [15], illumination of a semiconductor-electrolyte interface with photons having energy greater than its band gap energy generates electron–hole pairs at the anode electrode surface. The simultaneous application

of a bias positive to the flat-band potential produces a bending of the conduction and valence bands which, in turn, causes a more effective separation of the photogenerated carriers within the space charge layer. In other words, the potential gradient forces the electrons towards the cathode, thus leaving the photogenerated holes to react at the anode with H_2O and/or OH^- to yield hydroxyl radicals, i.e.:

Anode (working electrode):

$$TiO_2 + hv \rightarrow TiO_2 - e^-_{cb} + TiO_2 - h^+_{vb} \quad (1)$$

$$TiO_2 - h^+_{vb} + H_2O_s \rightarrow TiO_2 - {}^{\bullet}OH_s + H^+ \quad (2)$$

$$TiO_2 - h^+_{vb} + OH^-_s \rightarrow TiO_2 - {}^{\bullet}OH_s \quad (3)$$

$$TiO_2 - e^-_{cb} + TiO_2 - h^+_{vb} \rightarrow \text{recombination} \quad (4)$$

Cathode (counter electrode):

$$2H_2O + 2e^- \rightarrow H_2 + 2OH^- \quad (5)$$

where the subscripts cb and vb denote the conduction and valence bands, respectively, h^+ and e^- denote the photogenerated holes and electrons, respectively, and the subscript s refers to species adsorbed onto the photoanode surface.

Microbial Inactivation During PEC

Given the importance of effective inactivation of pathogens contained in water and wastewater so as to control waterborne diseases, PEC seems to be a promising and efficient tool. Inactivation of microorganisms is a gradual process that involves a number of physicochemical and biochemical processes. The bactericidal function of this technique can be attributed to the oxidation properties of photocatalytically generated active oxygen species (AOS), which cause damage to cellular membranes and further destruction of bacterial structures [16]. The most commonly accepted photocatalytic inactivation mechanism is based on the attack of ROS to the bacterial cell wall, where the bacteria-catalyst contact takes place [17]. Furthermore, radiation during PEC is equally important, as it induces DNA lesions, damaging nucleic acids and making them functionless. The hydroxyl radical (and other AOS), directly generated by this process, is the main cause of DNA destruction, which in turn leads to cell death. Most studies dealing with UV irradiation of microbial cells have concluded that the extremely reactive hydroxyl radical, for which no defense exists, is able to damage DNA [18].

Factors Affecting Disinfection During PEC

The studies refer to PEC as means of disinfection, highlighting the importance of certain parameters like type of microorganism, applied voltage, bacterial concentration, treatment time, and the aqueous matrix, which are considered determining factors of microbial inactivation [14].

Type of Microorganism

Generally, most studies dealing with the evaluation of PEC as a disinfection method have mainly focused on *Escherichia coli* inactivation with the application of conventional culture techniques, while other pathogens have not been considered [2, 12, 16, 19]. The importance of the tested bacterial strain should be underlined when disinfection occurs, especially considering that the bactericidal effect of PEC involves loss of membrane integrity and peroxidation of its phospholipids. In this sense, bacteria other than coliforms with different resistance rates should be taken into account. Gram-negative bacteria possess an additional outer membrane containing two lipid bilayers, which provide them higher complexity. However, since it is strongly believed that the attack occurs on the bacteria outer cell wall, certain attention is paid to the differences of wall structure. In this sense, Gram-positive bacteria, which possess a thick peptidoglycan cell wall, have been reported to be photocatalytically more resistant than Gram-negative, and according to the literature, a higher number of ROS are needed for their complete inactivation [20]. On the other hand, several authors have suggested different inactivation mechanisms, showing that Gram-negative bacteria are more resistant to the photocatalytic process, due to their more complex structure given by the additional outer membrane [17]. Apart from bacterial species, certain considerations should be given to other microorganisms like protozoan parasites and viruses, whose presence in water and wastewater poses great risks for public health. Although these microorganisms have been recognized as highly resistant during disinfection treatments, they have been merely reported in studies dealing with PEC.

Applied Voltage

It is presumed that raising the anodic potential increases the depth of the space charge region and suppresses the extent of electron–hole recombination, thus enhancing photocatalytic rates [21]. The beneficial effect of increasing the applied potential on PEC bacteria killing has already been reported in studies dealing mainly with the inactivation of *E. coli* [12, 19]. However, it should been pointed out that increasing the anodic potential above a certain value will not result in an increased inactivation rate, depending on the microorganism tested and the experimental conditions [21].

Bacterial Concentration

PEC disinfection rate is inversely proportional to bacterial concentration in water samples. Total inactivation may be achieved in relatively short treatment time (i.e. within approximately 15 min) when bacterial inoculum contains 10^3 – 10^5 CFU/mL, depending on the bacterial strain. At higher concentrations, residual bacterial cells may reach a plateau, implying a threshold, under which no further inactivation occurs. It is important to note here that the overall effect of initial concentration on the disinfection process is highly dependent on the type of microorganism used. Gram-positive bacteria require longer treatment time than Gram-negative with equal cellular density, while even longer times have been recorded when dealing with protozoan parasites [22].

Aqueous Matrix

Disinfection efficiency of treatment methods should be performed in real conditions, i.e. using samples of complex composition that contain multiple bacterial populations. Very few studies have been performed applying PEC for the inactivation of microorganisms in wastewater samples. Generally, treatment of wastewater samples shows a lower degree of disinfection compared to water. When processing wastewater samples, the viability and culturability of bacteria is affected by the presence of other competitive microorganisms and the interaction amongst them. According to Hong et al. [23], samples containing multiple bacterial populations or biofilms show great resistance against conventional disinfection methods, requiring long treatment time or more stressed conditions [23]. Although PEC is capable of inactivating bacteria to a certain extent in real wastewater, the residual cells do not indicate a sample free of microbial indicators and, therefore, safe for public health. In addition, another parameter under consideration is the particulate matter present in wastewater, which aids in the resistance of microorganisms to disinfection, as it may interfere by physically shielding bacterial cells and protecting the integrity of the contained DNA. Furthermore, part of the photogenerated AOS may be wasted to attack the organic carbon of the wastewater (about 8 mg/L, which typically consists of highly resistant humic-type compounds and biomass-associated products) and/or scavenged by bicarbonates, sulfates and chlorides, rather than inactivate pathogens. This could be overcome by increasing AOS concentration through raising the applied potential [21].

Future Directions

PEC has been acknowledged as an emerging disinfection technique, as it shows satisfactory inactivation rates of microorganisms. However, challenges still remain for further improvement, considering the extended microbial variety and the different response of each microorganism during PEC treatment. Up to now, most studies have dealt with the common fecal indicator *Escherichia coli*, while other microbes/pathogens have been merely mentioned. Future directions regarding photoelectrocatalytic disinfection are referred to thorough study of multiple microorganisms, including protozoan parasites and viruses and determination of the disinfection mechanism which occurs during water and wastewater treatment. According to the results obtained, PEC has the potential to be applied for effective microbial inactivation in aqueous matrices, with the view of protecting public health.

Cross-References

▶ Disinfection of Water, Electrochemical
▶ Electrodisinfection of Urban Wastewater for Reuse
▶ Photoelectrochemical Disinfection of Air (TiO_2)

References

1. Maier RM, Pepper IL, Gerba CP (2009) Environmental microbiology. Academic Press/Elsevier, USA
2. Chen CY, Wu LC, Chen HY, Chung YC (2010) Inactivation of *Staphylococcus aureus* and *Escherichia coli* in water using photocatalysis with fixed TiO_2. Water Air Soil Pollut 212:231–238
3. Malato S, Fernández-Ibáñez P, Maldonado MI, Blanco J, Gernjak W (2009) Decontamination and disinfection of water by solar photocatalysis: recent overview and trends. Catal Today 147:1–59
4. Frontistis Z, Daskalaki VM, Katsaounis A, Poulios I, Mantzavinos D (2011) Electrochemical enhancement of solar photocatalysis: degradation of endocrine disruptor bisphenol-A on Ti/TiO2 films. Water Res 45:2996–3004

5. Chong MN, Jin B, Chow CWK, Saint C (2010) Recent developments in photocatalytic water treatment technology: a review. Water Res 44:2997–3027
6. Nissen S, Alexander BD, Dawood I, Tillotson M, Wells RP, Macphee DE, Killham K (2009) Remediation of a chlorinated aromatic hydrocarbon in water by photoelectrocatalysis. Environ Pollut 157:72–76
7. Chatzisymeon E, Droumpali A, Mantzavinos D, Venieri D (2011) Disinfection of water and wastewater by UV-A and UV-C irradiation: application of real-time PCR method. Photochem Photobiol Sci 10:389–395
8. Rémy SP, Simonet F, Cerda EE, Lazzaroni JC, Atlan D, Guillard C (2011) Photocatalysis and disinfection of water: identification of potential bacterial targets. Appl Catal B Environ 104:390–398
9. Venieri D, Chatzisymeon E, Gonzalo MS, Rosal R, Mantzavinos D (2011) Inactivation of *Enterococcus faecalis* by TiO$_2$-mediated UV and solar irradiation in water and wastewater: culture techniques never say the whole truth. Photochem Photobiol Sci 10:1744–1750
10. Egerton TA, Kosa SAM, Christensen PA (2006) Photoelectrocatalytic disinfection of *E. coli* suspensions by iron doped TiO$_2$. Phys Chem Chem Phys 8:398–406
11. Marugán J, Hufschmidt D, Sagawe G, Selzer V, Bahnemann D (2006) Optical density and photonic efficiency of silica-supported TiO$_2$ photocatalysts. Water Res 40:833–839
12. Egerton TA (2011) Does photoelectrocatalysis by TiO2 work? J Chem Technol Biotechnol 86:1024–1031
13. Marugán J, Christensen P, Egerton T, Purnama H (2009) Synthesis, characterization and activity of photocatalytic sol–gel TiO$_2$ powders and electrodes. Appl Catal B Environ 89:273–283
14. Martínez-Huitle CA, Brillas E (2009) Decontamination of wastewaters containing synthetic organic dyes by electrochemical methods: a general review. Appl Catal B Environ 87:105–145
15. Morrison SR (1980) Electrochemistry at semiconductors and oxidized metal electrodes. Plenum Press, New York
16. Li G, Liu X, Zhang H, An T, Zhang S, Carroll AR, Zhao H (2011) In situ photoelectrocatalytic generation of bactericide for instant inactivation and rapid decomposition of Gram-negative bacteria. J Catal 277:88–94
17. Pal A, Pehkonen SO, Yu LE, Ray MB (2007) Photocatalytic inactivation of Gram-positive and Gram-negative bacteria using fluorescent light. J Photochem Photobiol A Chem 186:335–341
18. Gogniat G, Dukan S (2007) TiO$_2$ Photocatalysis causes DNA damage via fenton reaction-generated hydroxyl radicals during the recovery period. Appl Environ Microbiol 73:7740–7743

19. Philippidis N, Nikolakaki E, Sotiropoulos S, Poulios I (2010) Photoelectrocatalytic inactivation of *E. coli* XL-1 blue colonies in water. J Chem Technol Biotechnol 85:1054–1060
20. Van Grieken R, Marugán J, Pablos C, Furones L, López A (2010) Comparison between the photocatalytic inactivation of Gram-positive E. faecalis and Gram-negative E. coli faecal contamination indicator microorganisms. Appl Catal B Environ 100:212–220
21. Baram N, Starosvetsky D, Starosvetsky J, Epshtein M, Armon R, Ein-Eli Y (2009) Enhanced inactivation of E. coli bacteria using immobilized porous TiO2 photoelectrocatalysis. Electrochim Acta 54:3381–3386
22. Cho M, Cates EL, Kim JH (2011) Inactivation and surface interactions of MS-2 bacteriophage in a TiO2 photoelectrocatalytic reactor. Water Res 45:2104–2110
23. Hong SH, Jeong J, Shim S, Kang H, Kwon S, Ahn KH, Yoon J (2008) Effect of electric currents on bacterial detachment and inactivation. Biotechnol Bioeng 100:379–386

Photoelectrochemical Disinfection of Air (TiO$_2$)

Tsuyoshi Ochiai and Akira Fujishima
Kanagawa Academy of Science and Technology, Takatsu–ku, Kawasaki, Kanagawa, Japan
Photocatalysis International Research Center, Tokyo University of Science, Noda, Chiba, Japan

Historical Overview

In the late 1960s, one of the present authors (AF) began to investigate the photoelectrolysis of water, using a single crystal TiO$_2$ electrode. Then, the first report on the efficient production of hydrogen from water by TiO$_2$ photocatalysis was published in Nature in 1972, at the time of the "oil crisis" [1]. Thus, TiO$_2$ photocatalysis drew the attention of many people as one of the promising methods to obtain this new energy source. However, even though the reaction efficiency is very high, TiO$_2$ can absorb only the UV light contained in solar light, which is only about 3 %. Therefore, from the viewpoint of hydrogen production technology, TiO$_2$ photocatalysis is not very attractive. Instead, research shifted in the 1980s to the utilization of the strong photoproduced oxidation power of TiO$_2$ for the decomposition of various contaminants in both water and air. In this case, the holes (h$^+$) generated in TiO$_2$ were highly oxidizing, and most contaminants were essentially oxidized completely. In addition, various forms of active oxygen, such as O$_2$$^{\bullet-}$, $^{\bullet}$OH, HO$_2$$^{\bullet}$, and O$^{\bullet}$, produced by the following processes, may be responsible for the decomposition reactions:

$$e^- + O_2 \rightarrow O_2^{\bullet-}(ad)$$

$$O_2^-(ad) + H^+ \rightarrow HO_2^{\bullet}(ad)$$

$$h^+ + H_2O \rightarrow {}^{\bullet}OH(ad) + H^+$$

$$h^+ + O_2^{\bullet-}(ad) \rightarrow 2O^{\bullet}(ad)$$

For the purpose of easy handling of photocatalysts, the immobilization of TiO$_2$ powders on supports was carried out in the late 1980s. Then, the novel concept of light cleaning materials coated with a TiO$_2$ film photocatalyst under UV light was investigated. In 1997, the marked change in water wettability of the TiO$_2$ surface before and after UV light irradiation was also reported in Nature as a novel phenomenon of TiO$_2$ photocatalysis [2]. With the discovery of this phenomenon, the application range of TiO$_2$ coatings has been largely widened, as mentioned in Fig. 1 and in the literature [3–5].

Figure 2 shows the market transition of industries related to photocatalysis, based on a survey by the Photocatalysis Industry Association of Japan. This data represents the sales volume for companies that are members of the Photocatalysis Industry Association of Japan. The sales volume greatly increased during the past decade, especially for the "cleanup" application involved in photocatalytic water and/or air purifiers. This trend indicates the increasing number of people who are interested in environmental issues such as food poisoning and sick house syndrome. Moreover, the swine influenza

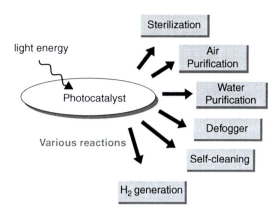

Photoelectrochemical Disinfection of Air (TiO$_2$), Fig.1 Schematic illustration of applications of photocatalysis

outbreak of 2009 raised serious fears of a global pandemic, suddenly increasing the sales volume of photocatalytic air purifiers. Therefore, photoelectrochemical disinfection of air is currently one of the most important applications of TiO$_2$ photocatalysis.

Mechanism of Photocatalytic Disinfection

Sunada et al. studied the photocatalytic disinfection process of *Escherichia coli* on TiO$_2$ film [6]. A typical experiment involves placing an *E. coli* suspension containing $\sim$2 $\times$ 10^5 colony forming units (CFU)/ml on an illuminated TiO$_2$-coated glass plate (1.0 mW cm^{-2} UV light). Under these conditions, there were no surviving cells after only 90 min of illumination. In contrast, no obvious changes in survival were observed when the TiO$_2$-coated glass plate was stored in the dark or when a glass plate was used as the substrate under UV illumination. Figure 3 shows AFM photographs of *E. coli* cells on a TiO$_2$-coated glass plate after different UV illumination times (Fig. 3a–c), along with a schematic illustration of the process (Fig. 3d–f). After illumination for 1 day, the outermost layer clearly seen in Fig. 3a disappeared (Fig. 3b). After illumination for 6 days, the cylindrical shape of the cells disappeared completely (Fig. 3c), suggesting the

Photoelectrochemical Disinfection of Air (TiO$_2$), Fig. 2 Market transition of industries related to photocatalysis (Source: Photocatalysis Industry Association of Japan)

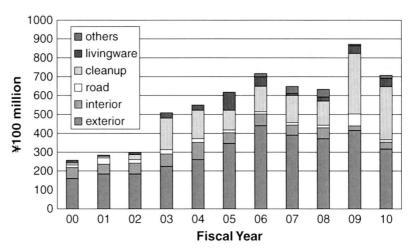

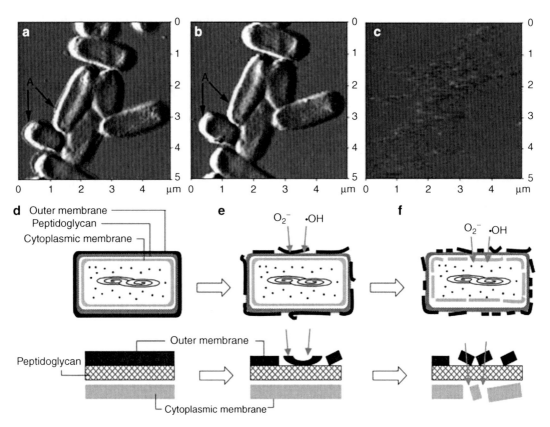

Photoelectrochemical Disinfection of Air (TiO$_2$), Fig. 3 AFM photographs of *E. coli* cells on TiO$_2$-coated glass plate after different UV illumination times (**a–c**) and a schematic illustration of the process (**d–f**) (Reprinted from Journal of Photochemistry and Photobiology A: Chemistry, 156, K. Sunada, T. Watanabe, K. Hashimoto, Studies on photokilling of bacteria on TiO$_2$ thin film, 227–233, 2003, with permission from Elsevier)

Photoelectrochemical Disinfection of Air (TiO$_2$), Fig. 4 The overview and the SEM image of TMiP

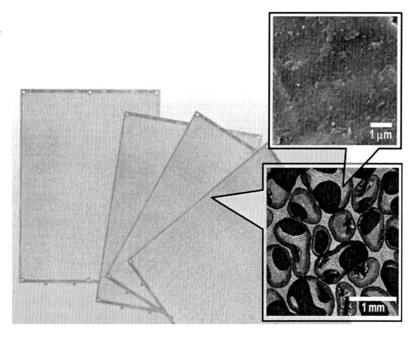

complete decomposition of dead cells. Taken together, they concluded that the photokilling of bacteria on the illuminated TiO$_2$ surface could occur by a two-step reaction mechanism (Fig. 3d–f): (1) the disordering of the outer membrane of cells on the illuminated TiO$_2$ surface and (2) disordering of the inner membrane (the cytoplasmic membrane) and killing of the cell. In the first stage, the outer cell membranes were decomposed partially by the reactive species (·OH, H$_2$O$_2$, O$_2$·$^-$) produced by the TiO$_2$ photocatalyst (Fig. 3e). During this stage, cell viability was not lost very efficiently. However, the permeability of reactive species will be increased. Consequently, reactive species easily reach and attack the inner membrane (Fig. 3f). The structural and functional disordering of the cytoplasmic membrane due to lipid peroxidation led to the loss of cell viability and to cell death. If the illumination continued for a sufficiently long time, the dead cells were found to be decomposed completely (Fig. 3c). Moreover, Kikuchi et al. reported that the *E. coli* could be killed even when the cells were separated from the TiO$_2$ surface by a 50 μm PTFE spacer under similar experimental conditions [7]. This result indicates that photocatalytic disinfection of bacteria can occur remotely. This results in important applications in the disinfection of air with TiO$_2$ photocatalysis and suggests that a combination of TiO$_2$ photocatalysts with antibacterial reagents that can permeate the outer membrane could show a far superior photokilling activity.

Future Directions

There are many approaches for solving the above mentioned problems. For the first problem, we have fabricated the easy-to-handle photocatalytic filter material, TiO$_2$ nanoparticles impregnated on titanium mesh (TMiP^{TM}), by collaboration with involved companies [8]. Figure 4 shows the overview and the SEM image of TMiP. The Ti mesh, obtained by controlled chemical etching of titanium foil of 0.2 mm thickness, was anodized in an acid solution. Then the Ti mesh was heated to make a TiO$_2$ layer on the surface. Heated Ti mesh was dip-coated with TiO$_2$ anatase sol and was then heated again to sinter TiO$_2$ nanoparticles onto the anodized Ti mesh surface. Because of its highly ordered three-dimensional structure with modified TiO$_2$ nanoparticles,

TMiP provides excellent air pass through while maintaining a high level of surface contact. In addition, TMiP is flexible and lightweight enough to design various air purification units with UV sources [9–11]. For solving the second problem, The Project to Create Photocatalyst Industry for Recycling-oriented Society (NEDO project) was conducted from 2007 to 2012 with an investment of ¥5.1 billion. This has given particular impetus to the research and development of highly sensitive visible light-responsive photocatalysts. The project team found that the visible light activity of a WO_3 photocatalyst with co-catalysts such as Pt, Pd, WC, CuO, or Cu(II) clusters was drastically enhanced via the efficient oxygen reduction process [12, 13] – in particular, Cu-modified WO_3-based photocatalyst, with a visible light reactivity 10 times higher than that of existing products made from N-doped TiO_2. This project is expected to lead to the development of photocatalyst effects of a magnitude sufficient for deodorization, VOC elimination, and sterilization and antibacterial scenarios for interior applications. Lastly, to eliminate "phony products," the standardization process in the photocatalyst industry is proceeding by the Japanese Industrial Standards Committee. At the same time, ISO standardization is also being introduced. Furthermore, the New Energy and Industrial Technology Development Organization (NEDO) has taken the lead in promoting standardization of performance evaluation methods for visible light-responsive photocatalysts since 2007. In addition, the Photocatalysis Industry Association of Japan is formulating standards for photocatalyst products tested by the standardized methods.

In conclusion, we can expect a large number of applications in the photoelectrochemical disinfection of air by using TiO_2 photocatalysts because of the expansion to new fields of research and development. In particular, we feel that achieving a healthy and comfortable living environment is becoming an important issue, and TiO_2 photocatalysis can fulfill an important role in the disinfection of air.

Cross-References

▶ Disinfection of Water, Electrochemical
▶ Organic Pollutants, Direct Electrochemical Oxidation
▶ Photocatalyst
▶ Photoelectrochemical Disinfection
▶ TiO_2 Photocatalyst

References

1. Fujishima A, Honda K (1972) Electrochemical photolysis of water at a semiconductor electrode. Nature 238:37–38
2. Wang R, Hashimoto K, Fujishima A, Chikuni M, Kojima E, Kitamura A, Shimohigoshi M, Watanabe T (1997) Light-induced amphiphilic surfaces. Nature 388:431–432
3. Fujishima A, Rao TN, Tryk DA (2000) Titanium dioxide photocatalysis. J Photochem Photobiol C Photochem Rev 1:1–21
4. Hashimoto K, Irie H, Fujishima A (2005) TiO_2 photocatalysis: a historical overview and future prospects. Jpn J Appl Phys 44:8269–8285
5. Fujishima A, Zhang X, Tryk DA (2008) TiO_2 photocatalysis and related surface phenomena. Surf Sci Rep 63:515–582
6. Sunada K, Watanabe T, Hashimoto K (2003) Studies on photokilling of bacteria on TiO_2 thin film. J Photochem Photobiol A Chem 156:227–233
7. Kikuchi Y, Sunada K, Iyoda T, Hashimoto K, Fujishima A (1997) Photocatalytic bactericidal effect of TiO_2 thin films: dynamic view of the active oxygen species responsible for the effect. J Photochem Photobiol A Chem 106:51–56
8. Ochiai T, Hoshi T, Slimen H, Nakata K, Murakami T, Tatejima H, Koide Y, Houas A, Horie T, Morito Y, Fujishima A (2011) Fabrication of TiO_2 nanoparticles impregnated titanium mesh filter and its application for environmental purification unit. Catal Sci Technol 1:1324–1327
9. Ochiai T, Niitsu Y, Kobayashi G, Kurano M, Serizawa I, Horio K, Nakata K, Murakami T, Morito Y, Fujishima A (2011) Compact and effective photocatalytic air-purification unit by using of mercury-free excimer lamps with TiO_2 coated titanium mesh filter. Catal Sci Technol 1:1328–1330
10. Slimen H, Ochiai T, Nakata K, Murakami T, Houas A, Morito Y, Fujishima A (2012) Photocatalytic decomposition of cigarette smoke by using TiO_2 impregnated titanium mesh filter. Ind Eng Chem Res 51:587–590
11. Ochiai T, Nakata K, Murakami T, Morito Y, Hosokawa S, Fujishima A (2011) Development of an air-purification unit using a photocatalysis-plasma hybrid reactor. Electrochemistry 79:838–841

12. Arai T, Horiguchi M, Yanagida M, Gunji T, Sugihara H, Sayama K (2008) Complete oxidation of acetaldehyde and toluene over a Pd/WO$_3$ photocatalyst under fluorescent- or visible-light irradiation. Chem Commun 2008:5565–5567
13. Abe R, Takami H, Murakami N, Ohtani B (2008) Pristine simple oxides as visible light driven photocatalysts: highly efficient decomposition of organic compounds over platinum-loaded tungsten oxide. J Am Chem Soc 130:7780–7781

Photoelectrochemical Processes, Electro-Fenton Approach for the Treatment of Contaminated Water

Luis Godinez[1] and Francisco J. Rodriguez Valadez[1,2]
[1]Centro de Investigación y Desarrollo Tecnológico en Electroquímica S.C., Querétaro, Mexico
[2]S. C. Sanfandila, Research Branch, Center for Research and Technological Development in Electrochemistry, Querétaro, Mexico

Introduction

Advanced Oxidation Processes (AOPs) are characterized by the generation and use of the •OH radical species for treating wastewaters, and among the different AOPs, the Fenton reaction is probably the most popular approach. The Fenton reaction was discovered in 1894 by H. J. Fenton, who reported that low concentrations of iron ions in H$_2$O$_2$ aqueous solutions could effectively promote the oxidation of tartaric acid. Later, in 1934, Haber & Weiss suggested that the ferrous ion was actually promoting the decomposition of H$_2$O$_2$ and thus the formation of the •OH radical species, whose presence was, in fact, the reason for the observed oxidation power of the Fenton mixture [1, 2].

The •OH radical species has many advantages over other common chemicals for the elimination of a wide variety of pollutants in water effluents. As can be seen in Table 1, its oxidation potential is very high (2.8 V), larger than ozone (2.42 V) and than H$_2$O$_2$ (1.78 V) itself. The oxidation kinetics of the ·OH radical species is also large, and, due to its inherent reactivity and molecular size, it is conveniently nondiscriminating for treating complex mixtures of organic pollutants in water. In addition, the oxidation by-products are hydroxylated compounds which, as opposed to the toxic by-products that may result from halogen-based oxidation processes, make these water treatment technologies environmentally friendly [3, 4].

The Fenton Process

The Fenton process involves the reaction of ferrous ions with hydrogen peroxide molecules in aqueous solution to generate a series of oxidizing species of which the hydroxyl radical is the most powerful and reactive. In this way, it is widely accepted that the •OH species is formed as described by Eq. 1 [5–8]:

$$Fe^{2+} + H_2O_2 \rightarrow Fe^{3+} + OH^- + \text{•}OH$$
$$k_1 = 70 \ M^{-1}s^{-1} \tag{1}$$

However, different studies have also suggested the presence of other oxidizing species that result from the coupled reactions 2–6 [8, 9]:

$$Fe^{3+} + H_2O_2 \rightarrow Fe - OOH^{2+} + H^+$$
$$k_2 = 0.001 - 0.01 \ M^{-1}s^{-1} \tag{2}$$

$$Fe - OOH^{2+} + H^+ \rightarrow HO_2^{\cdot} + Fe^{2+} + H^+ \tag{3}$$

$$\text{•}OH + H_2O_2 \rightarrow HO_2^{\cdot} + H_2O \tag{4}$$

$$Fe^{2+} + HO_2^{\cdot} + H^+ \rightarrow Fe^{3+} + H_2O_2 \tag{5}$$

$$Fe^{3+} + HO_2^{\cdot} \rightarrow Fe^{2+} + O_2 + H^+ \tag{6}$$

The Fenton process is therefore an attractive approach to treat a wide variety of contaminated effluents, in particular those that require fast kinetics, contain persistent pollutants, or impose limitations in terms of space requirements.

Photoelectrochemical Processes, Electro-Fenton Approach for the Treatment of Contaminated Water, Table 1 Oxidation potential of some chemicals commonly used in water treatment processes

Oxidant	Oxidation potential (V vs NHE)
Fluoride	3.0
•OH radical	2.8
Atomic oxygen	2.42
Ozone	2.42
Hydrogen peroxide	1.78
Permanganate	1.68
Chlorine dioxide	1.57
Hypochlorous acid	1.45
Chlorine	1.36
Bromine	1.09
Iodine	0.54

The Fenton mixture, however, has rarely been taken to industrial scale due to some inherent problems that result in its nonpractical application. Since the Fenton mixture requires a fixed concentration of H_2O_2, it is necessary to constantly dilute and control the amount of peroxide that needs to be fed into the reactor. The commercial H_2O_2 solutions needed for this process are expensive and require careful handling due to their reactive nature. The ferrous ions, on the other hand, need to be removed from the treated effluent, adding an additional operation that carries an associated cost. In order to solve these problems, some variations on the classical Fenton approach have been explored.

The Electro-Fenton Approach

Among these variations, the Electro-Fenton approach has recently gained much attention due to its potential advantages. As its name suggests, the Electro-Fenton approach is an electrochemical methodology in which one or more of the reagents needed for the Fenton mixture is formed through electrochemical reactions. The main advantage consists in the possibility to control their generation using electrical variables, i.e., current and potential, which are easily implemented and controlled [10, 11].

There are two main types of Electro-Fenton processes [1, 12]:

- Cathodic Electro-Fenton process
- Anodic Electro-Fenton process.

The main difference between these two methodologies is related to the way in which the iron ions are incorporated into the system. While in the cathodic Electro-Fenton process an Fe^{2+} or Fe^{3+} electrolyte is added to the reaction mixture, in the anodic Electro-Fenton system, an iron anode is employed as the source of the ferrous ions. In both processes, however, there is a continuous production of H_2O_2 which is formed from the electrochemical reduction of dissolved oxygen at the cathode surface as described by Eq. 7 [2, 12]:

$$O_{2(g)} + 2H^+ + 2e^- \rightarrow H_2O_2 \qquad (7)$$

As it was previously mentioned, the electro-generated H_2O_2 reacts with the ferrous ions to produce the radical species •OH. As described in Eq. 8, at the cathode surface, the reduction of ferric ions to their active ferrous form also takes place:

$$Fe^{3+} + e^- \rightarrow Fe^{2+} \qquad (8)$$

In this way, the Fe^{2+} species becomes available for further reaction with more electro-generated H_2O_2, thus sustaining the •OH electrochemical generation process.

In an anodic Electro-Fenton process, an iron anode is employed to electrochemically generate the Fe^{2+} ions, as described by Eq. 9:

$$Fe^0 \rightarrow Fe^{2+} + 2e^- \qquad (9)$$

In Fig. 1, a simplified scheme of the reactions that take place in an Electro-Fenton cell is presented.

As can be inferred by inspection of Fig. 1, the electrochemical generation of the Fenton mixture depends on a series of variables that must be properly defined and controlled. Among these, the saturation of oxygen must be maintained; the value of applied current, the conductivity of the solution, the pH, the material of the

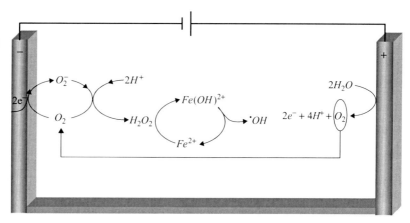

Photoelectrochemical Processes, Electro-Fenton Approach for the Treatment of Contaminated Water, **Fig. 1** Reactions in an Electro-Fenton cell [10]

electrodes, and the proper mixing of the electrogenerated reagents that must react with the pollutant species after being prepared are all of paramount importance.

Future Directions

As has been discussed in this brief presentation, the Electro-Fenton approach is characterized by a high potential for its practical application in water treatment processes. However, there still are a series of problems that must be dealt with before these systems can become widely used technologies in the field of water treatment. Among these, there is the need to properly understand and develop heterogeneous processes in which the iron is supported in the reactor so that the subsequent separation process can be avoided [13–16]. In this regard, it is also important to study and optimize the use of the Fenton mixture employing other transition metals which, as some reports suggest, may be more efficient than iron itself [17–19]. On the other hand, it is important to explore and incorporate the use of electromagnetic radiation in photo-assisted processes (in particular, the use of solar radiation, which would be a relatively cheap energy input to assist in the treatment process). It is well known that light-activated processes foster the regeneration of the Fe^{2+} species from its oxidized form and that the use of semiconductor anodes can trigger the electrochemical formation of •OH radicals by hole injection at the anode surface. In this regard, it is also important to point out that there are intensive research efforts preparing and studying semiconductor-doped materials for photo-assisted Electro-Fenton processes (thus defining another potentially important area of technology development called the Photo-Electro-Fenton approach) [20–22]. Depending on the specific nature of the technology to be developed for specific applications, it is also important to study the combination of the Fenton-based processes with traditional technologies such as biological and physicochemical reactors in order to fully optimize and take advantage of the full potential of the Fenton mixture [23].

Cross-References

▶ Electrochemical Bioremediation
▶ Electrochemical Treatment of Landfill Leachates
▶ Electro-Fenton Process for the Degradation of Organic Pollutants in Water
▶ Organic Pollutants, Direct Electrochemical Oxidation
▶ Photoelectrochemical Disinfection
▶ TiO_2 Photocatalyst

References

1. Parsons S (2004) Advanced oxidation process for water and wastewater treatment. IWA publishing, London
2. Goldstein G, Meyerstein D, Czapski G (1993) The Fenton reagents. Free Radic Biol Med 15:435–445

3. Bandala E, Corona B, Guisar R, Uscanga M (2007) Aplicación de procesos Avanzados de Oxidación en la desactivación secuencial de microorganismos Resistentes en Agua. Ciencia... Ahora 20:52–63
4. Kim JK, Metcalfe IS (2007) Investigation of the generation of hydroxyl radicals and their oxidative role in the presence of heterogeneous copper catalysts. Chemosphere 69:689–696
5. Peralta-Hernández JM, Meas-Vong Y, Rodríguez F, Chapman T, Maldonado I, Godínez LA (2006) In situ electrochemical and photo-electrochemical generation of the fenton reagent: a potentially important new water treatment technology. Water Res 40:1754–1762
6. Lucas MS, Peres JA (2006) Decolorization of the azo dye Reactive Black 5 by Fenton and photo-Fenton oxidation. Dye Pigment 71:236–244
7. Deng Y, Englehard JD (2009) Kinetics and oxidative mechanism for H_2O_2-enchanced iron-mediated aeration (IMA) treatment of recalcitrant organic compounds in mature landfill leachate. J Hazard Mater 169:370–375
8. Qiang Z, Chang J-H, Huang C-P (2003) Electrochemical regeneration of Fe_{2p} in Fenton oxidation processes. Water Res 37:1308–1319
9. Lu MC, Lin C-J, Liao C-H, Ting WP, Huang R-Y (2001) Influence of pH on the dewatering of activated sludge by Fenton's reagent. Water Sci Technol 44:327–332
10. Liou RM, Chen SH, Hung MY, Hsu CS (2004) Catalytic oxidation of pentachlorophenol in contaminated soil suspensions by Fe^{+3}- Resin/H_2O_2. Chemosphere 55:1271–1280
11. Duarte F, Maldonado-Hódar FJ, Pérez-Cardenas AF, Madeira LM (2009) Fenton-like degradation of azodye Orange II catalyzed by transition metal son carbón aerogels. Appl Catal B Environ 85:139–147
12. Pignatello JJ, Oliveros E, MacKay A (2006) Advanced oxidation processes for organic contaminant destruction based on the Fenton reaction and related chemistry. Crit Rev Environ Sci Technol 36:1–84
13. Qiuqiang Chen, Pingxiao Wu, Yuanyuan Li, Nengwu Zhu, Zhi Dang (2009) Heterogeneous photo-Fenton photodegradation of reactive brilliant orange X-GN over iron-pillared montmorillonite under visible irradiation. J Hazard Mater 168:901–908
14. Kasiri MB, Aleboyeh H, Aleboyeh A (2008) Degradation of Acid Blue 74 using Fe-ZSM5 zeolite as a heterogeneous photo-Fenton catalyst. Appl Catal Environ 84:9–15
15. Rey-May Liou, Shih-Hsiung Chen, Mu-Ya Hung, Chin-Shan Hsu a, Juin-Yih Lai (2005) Fe (III) supported on resin as effective catalyst for the heterogeneous oxidation of phenol in aqueous solution. Chemosphere 59:117–125
16. Balanosky E, Fernández J, Kiwi J, López A (1999) Degradation of membrane concentrates of the textile industry by Fenton like reactions in iron-free

solutions at biocompatible pH values (pH $\approx$ 7–8). Water Sci Technol 40(4–5):417–424
17. Verma P, Baldrian P, Gabriel J, Trnka T, Nerud F (2004) Copper-ligand complex for the decolorization of synthetic dyes. Chemosphere 57:107–1211
18. Bali U, Karagözoglu B (2007) Performance comparison of Fenton process ferric coagulation and H2O2/pyridine/Cu(II) system for Decolorization of Remazol Turquoise Blue G-133. Dye Pigment 74:73–80
19. Fathima N, Aravindhan R, Rao J, Nair B (2008) Dye house wastewater treatment through advanced oxidation process using Cu-exchanged Y Zeolite: a heterogeneous catalytic approach. Chemosphere 70:146–1151
20. Esquivel K, García MG, Rodríguez F, Ortiz-Frade L, Godínez L (2013) Study of the photo-electrochemical activity of cobalt- and nickel doped TiO2 photoanodes for the treatment of a dye-contaminated aqueous solution. Journal of Applied Electrochemistry 43:433–440
21. Zhao B, Meleb G, Piob I, Li J, Palmisano L, Vasapollob G (2010) Degradation of 4-nitrophenol (4-NP) using Fe-TiO2 as a heterogeneous photo-Fenton catalyst. Journal of Hazardous Materials 176:569–574
22. Zarei M, Khataee A, Fathinia M, Seyyednajafi F, Ranjbar H (2012) Combination of nanophotocatalysis with electro-Fenton-like process in the removal of phenol from aqueous solution: GC analysis and response surface approach. International Journal of Industrial Chemistry 3:1–11
23. Sheng H Lin, Chih C Chang (2000) Treatment of landfill leachate by combined electro-Fenton oxidation and sequencing batch reactor method. Water Res 34(17):4243–4249

Photoelectrochemistry, Fundamentals and Applications

Krishnan Rajeshwar
The University of Texas at Arlington, Arlington, TX, USA

Introduction

Light can influence an electrochemical system in variant ways. The basis for the photoeffect is *photoexcitation* either of a molecule located in the electrolyte phase or of the electrode material itself. The former constitutes the basis of either a *photogalvanic cell* or a *dye-sensitized solar cell* as discussed later. In the latter case of electrode photoexcitation a metal electrode can absorb the incident light and if the energy exceeds its work function threshold, can cause *photoemission* of electrons from the metal into the electrolyte phase where they become solvated and thus stabilized. On the other hand, if the electrode material is a semiconductor, photons can be absorbed in a quantized fashion if their incident energy is equal to or greater than the semiconductor band-gap energy. In these cases we have *photoelectrochemical cells*. Yet another situation may be distinguished where *colloidal* nanoparticles of an inorganic semiconductor are suspended in solutions and irradiated with photons to drive *photocatalytic* processes of interest. Finally a hybrid approach involves using a nanocrystalline semiconductor film as an electrode in a *photoelectrocatalytic* cell. The field of *photoelectrochemistry* encompasses all of the above variant scenarios, each one of which is examined in turn next. *Photosplitting* (*photoelectrolysis*) and *photocatalytic* systems are also discussed in what follows. The impetus for studies of these systems has been both fundamental (for example, to test electron transfer theories) and applied, with the applications largely directed toward solar energy conversion (specifically *storage*) and environmental remediation [1].

Approaches

Photogalvanic Cells

A homogeneous redox component, i.e., one that is dissolved in an electrolyte phase can be photochemically excited to yield excited state species that can undergo redox reactions. The prototypical molecule here is the metal complex, [Ru(bpy)$_3$]$^{2+}$ where bpy $=2,2'$-bipyridine ligand. Thus the excited state of this molecule is a fairly powerful reductant:

$$[Ru(bpy)_3{}^{2+}]^* + Fe^{3+} \rightarrow Fe^{2+} + Ru(bpy)_3{}^{3+}$$

The back reaction constitutes the back-reaction regenerating the original reactants:

$$Fe^{2+} + Ru(bpy)_3{}^{3+} \rightarrow Ru(bpy)_3{}^{2+} + Fe^{3+}$$

Thus the light energy simply is thermalized. However the kinetics of this reaction can be made slower (e.g., by manipulating local microenvironments) such that appreciable concentrations of the photoreaction products can be accumulated. By immersing electrodes into the solution one can force a Faradaic (photo)current to flow in response to the second step above with each electrode being made selective to one of the half-reactions. Since the original excitation energy can be harnessed (at least partially), these devices called *photogalvanic cells*, are relevant to applications related to solar energy conversion [2–4].

Much of the interest that was centered on these devices in the early 1970s have since largely subsided because of practical difficulties in designing electrode interfaces that are selective and have fast kinetics for the desired half-reaction.

Dye-sensitized Solar Cells (DSSCs)

These are conceptually very similar to the photogalvanic devices in that the initial photoexcitation occurs on a dye molecule that is either anchored or strongly adsorbed on an oxide semiconductor support (nominally TiO_2). Again using [Ru(bpy)$_3$]$^{2+}$ as an example, the photoexcited species, [Ru(bpy)$_3{}^{2+}$], inject a photoelectron into (acceptor) states in the oxide support thus getting oxidized. Regeneration of the oxidized state of the dye (using a redox shuttle such as iodide/polyiodide) accompanied by redox conversion of the resultant species at the countereelctrode completes the electrical circuit and results in a photocurrent flow in the external circuit of the cell. The result is light energy $\rightarrow$ electricity conversion in the device.

The breakthrough in the DSSC technology occurred in 1990 when strategies were developed (using corresponding developments in nanotechnology) to prepare *mesoporous* TiO_2 films so that appreciable amounts of dye could be sequestered on the oxide support phase thus enhancing the incident photon-to-electron conversion efficiency (IPCE) and the photocurrent yields [5–7]. These developments in turn had a marked influence on the health and viability of the field of photoelectrochemistry itself [8], and DSSCs continue to be an intensely studied research topic.

Recent developments in DSSC have turned toward adapting these devices to incorporate a solar energy storage component. Replacement of the liquid electrolyte with a *solid electrolyte* or a gel has also been an important area of research activity as are efforts directed to improving the long- term (several years!) stability of the DSSC components.

Photoelectrochemicalcells

Consider a metal working electrode first. The ejected electrons from a metal electrode surface will travel a few A into the electrolyte phase and then become solvated. These solvated species display interesting chemistry and electrochemistry if suitable electron scavengers are available to interact with them. The resultant *free radical species* can be probed spectroscopically in these photoelectrochemical or *spectrophotoelectrochemical* experiments. The irradiation can be either continuous or pulsed and with the advent of powerful laser sources a whole slew of experimental strategies open up [9, 10]. Of course with continuous irradiation the spectroscopic probe will have to be orthogonally placed to avoid interference with the incoming radiation [11].

On the other hand, solid-state physics principles teach that semiconductors are characterized by filled and empty states that are more appropriately termed as *bands*, specifically *valence* and *conduction bands* (VB and CB) respectively [12]. These are the solid states analogues of the corresponding energy levels in molecules called highest occupied molecular orbital (HOMO) and lowest unoccupied molecular orbital (LUMO) respectively; i.e., the E_g becomes the solid- state analog of the molecular HOMO-LUMO energy gap. While in a *metal*, the bands are half- filled (giving rise to its electrical conductivity when an electric field is applied), the VB and CB in a *semiconductor* are separated by an energy gap (E_g) that may range from $\sim$0.1 eV to $\sim$3.5 eV. In an *insulator* this gap is usually much larger. Now when photons are incident such that their energy, $hu > E_g$, electron–hole pairs are excited in the semiconductor, and in the presence of a built-in field at the interface (discussed in detail in other chapters in this encyclopedia), the *excitons* are dissociated into electrons and holes. Depending on whether the semiconductor is n- type or p-type, the minority carriers (holes for n-type, electrons for p-type) are driven to the semiconductor/electrolyte interface. Thus *photooxidation* and*photoreduction* reactions occur respectively between these carriers and redox species in the electrolyte phase.

Thus *semiconductor working electrode* materials and their interfaces with electrolytes form the basis for a very important class of *photoelectrochemical* cells [12]. In principle both *organic* and *inorganic* semiconductor materials may be utilized although we shall mainly focus on inorganic semiconductors for our present discussion.

If either an n-type photoanode or a p-type photocathode is used in conjunction with a metal counterelectrode and a reversible redox electrolyte (e.g., $Fe(CN)6^{3-/4-}$), we have the basis for a *regenerative* photoelectrochemical cell. Alternately both an n-type and a p-type semiconductor may be used in tandem in a twin-photoelectrode geometry for the cell, much like what plants do in photosynthesis (For example [13]). Note that in these case there is no *net chemistry occurring in the electrolyte phase in response tophotoexcitation*, i.e., what is photooxized (or photoreduced) at one terminal is re-reduced (or re-oxidized) back at the other. The result is conversion or transduction of photon energy to electrical energy.

One can well argue that there is no real advantage to be accrued from the use of such an energy conversion device relative to a *solid-state photovoltaic* device that is much more robust and simpler to implement. Who would want to put a (potentially toxic) liquid on a roof top, even in a sealed device with attendant risk of leakage?! Indeed aside from furnishing important fundamental insights into charge transfer at

Photoelectrolytic Cells

Consider an n-type (oxide) semiconductor in contact with a liquid electrolyte for specificity. If a reversible redox system is not present, then the photogenerated holes will attack the electrolyte species (or the solid lattice) instead. In an aqueous electrolyte adsorbed water molecules or hydroxide ions are always present. Photooxidation of these species will generate dioxygen if the energetics at the interface are satisfied [14]. At the counterelectrode the electrons will reduce protons in the electrolyte, assuming again that the interfacial energetics are optimal. The net result is the photosplitting of water into dioxygen and hydrogen in a *photoelectrolytic* cell. Note that unlike in the liquid-junction regenerative cell discussed above, there is next chemical change in the electrolyte phase.

As can be imagined, the implications in terms of solar energy storage are tremendously important and it is no wonder that the original paper on this topic [15] continues to be one of the most widely cited in the chemical literature. It is worth pointing out that in the original work a pH gradient was imposed in the electrochemical cell to serve as an additional "bias" to the insufficient energy of the conduction band electrons in rutile to photoreduce protons.

It is somewhat sobering that three decades have elapsed since the original study on the feasibility of solar water splitting and we are no closer to a satisfactory solution to this challenge [8] that has been aptly termed as the "Holy Grail" [16]. Issues such as process efficiency, photoelectrode stability, and potential materials cost continue to dog efforts aimed at overcoming this challenge. A detailed discussion of these issues is prohibited by space constraints here; instead, the reader is referred to review articles on this topic [14, 17]. Intense effort also has gone into the design and development of schemes that require no *external bias* voltage (over and above what is generated by the light) to sustain the water splitting process. That is, the entire process is then *spontaneous* under photoexcitation of the semiconductor and a photocurrent continues to flow as long as the light is on. Obviously this entails an (open-circuit) photovoltage in excess of $\sim$1.5 V (assuming operation at ambient temperature) to be generated in the system with no deleterious consequences in terms of electrode corrosion etc.!

Figure 1 provides a schematic illustration of the three types of devices, namely, regenerative, photoelectrolytic and DSSC.

Photocatalysis

An entirely related field of research, underpinned by very similar photophysics principles as those detailed above for photoelectrochemical cells, evolved around the same time as the first demonstration of solar water splitting using TiO_2 [18]. This involved the possibility of utilizing the (very energetic) holes available in a low-lying valence band in TiO_2 to photoxidize (organic) pollutants confined in the electrolyte phase [1, 19]. In this instance, unlike with the water photosplitting reaction above, the photooxidation of organics in thermodynamically *downhill* (negative free energy change) such that the terminology: *photocatalysis* is entirely appropriate. That is, light merely serves to speed up a sluggish electrochemical process in the dark [20]. By contrast the photogeneration of H_2 and O_2 from water represents a *photoelectrosynthetic* process.

This field of activity has enjoyed exponential growth since the 1970s spurred by environmental remediation applications and many semiconductors beyond TiO_2 have been examined as well as an astounding range of solution substrates [1, 19, 21]. It is worth noting that unlike advanced oxidation processes (AOPs) for combating environmental pollution, the semiconductor-based *heterogeneous photocatalysis* approach is compatible with both oxidizable and reducible solution species or even a combination of both [22]. For example AOPs clearly cannot be used to treat a waste stream consisting of phenols and toxic metal ions (e.g., Cd^{2+}, Hg^{2+}) unlike

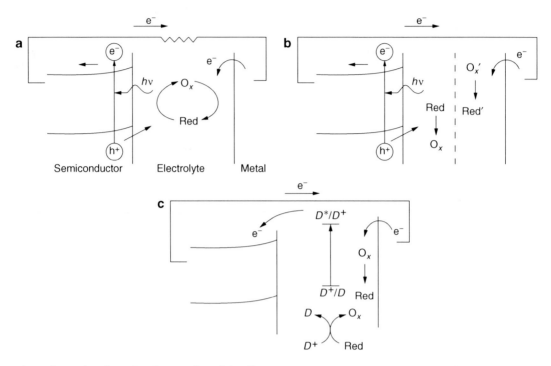

Photoelectrochemistry, Fundamentals and Applications, Fig. 1 Types of photoelectochemical devices for solar energy conversion (**a**), (**b**) and (**c**) depict regenerative, photoelectrolytic and dye-sentisitized configurations, respectively. As in the remainder of the chapter, and *n*-type semi-conductor is assumed in these cases for specificity

a remediation scheme based on the use of TiO$_2$ or another semiconductor. This is simply because both photogenerated holes and electrons can be utilized to oxidize and reduce the target pollutant respectively in the latter case. Incidentally many mechanism-oriented studies have been directed at the questions surrounding what the interfacial reactions involve: for example, direct oxidation by photogenerated holes or one mediated by hydroxyl radicals (or other reactive oxygen species, ROS), role of semiconductor surface states or traps etc. [19].

Finally the heterogeneous photocatalysis technique has been extended in scope to include the treatment of contaminated air and gas-phase pollutants. Photocatalyst suspensions, while they offer several advantages associated with huge surface area and consequently large reaction cross- sections, simplicity of reactor etc., are also limited in that a bias potential cannot be applied to further accelerate the photoprocess. Further, photocatalyst recovery after use is complicated. In contrast the use of *nanocrystalline photocatalyst* films affords effective solutions to both these issues in a*photoelectrocatalysis* cell geometry [21].

Note that the slurry-based approach can be used for water photosplitting purposes as well and has been a popular configuration by many researchers in Japan and elsewhere. In this scenario there is a crucial factor in a suspension-based vis-a-vis an electrode-based photoelectrolytic cell geometry that is worthy of attention. In the former a potential explosive mixture of H$_2$ and O$_2$ are photoegenrated in close proximity. Back-reactions involving the photoproducts (leading to a *photostationary state*) are also an issue in the former case. More simply one of the half-reactions (say the oxygen evolution reaction or OEC) is substituted with a (sacrificial) redox half-reaction such that the net reaction is thermodynamically down-hill and the reaction now becomes *photocatalytic* rather than*photosynthetic* [20].

Active Material Aspects

Early in the evolution of the field of photoelectrochemistry, the semiconductors were mostly used in single crystal form. While this configuration furnished a rich library of fundamental data in the 1970s and 1980s on the nature of semiconductor/electrolyte interfaces in the dark and under illumination, the use of single crystals is not compatible with practical deployment of photoelectrochemical devices. This then paved the way to the use of semiconductor *thin films* (usually polycrystalline but sometimes even amorphous) supported on metals or transparent conducting glass. More recently, the advent of nanotechnology has opened the door to the use of *nanocrystalline* semiconductor films. Currently considerable attention is focused on the photoelectrochemical behavior of inorganic semiconductors in the form of quantum (or Q) dots, either directly in suspended form in solution or in thin film form [22]. Their unique optical and electronic properties considerably expand the scope of photoelectrochemical devices that use Q- dots as the active material. The other aspect of the search for new families of semiconductor materials for photoelectrochemical applications has been to focus on compounds that contain earth-abundant and non-toxic elements such as Cu, Fe, W, Ti etc. instead of precious and toxic elements such as Ga, In, Se, Cd etc. This in turn has the added bonus of potentially lowering the cost associated with the ultimate device. In solar conversion devices, the active materials component(s) (i.e., semiconductors) constitute a healthy fraction of the overall device cost [23].

Further Application Possibilities

We close this discussion with the note that water splitting is not the only reaction of interest in solar energy conversion and environmental remediation. Splitting of CO_2 (to a fuel product such as methanol or methane) constitutes a value-added approach to combating the accumulation of this greenhouse gas molecule [24]. However, the kinetic bottlenecks to CO_2 splitting pose steep technical challenges in coming up with solutions that are being actively pursued in laboratories worldwide. Combining this value-added approach with the use of semiconductors such as CuO and Cu_2O are attractive from a translational perspective [25, 26]. Photoassisted water splitting to generate hydrogen from p-type semiconductors such as Cu_2O can also be combined with a (dark) oxidation reaction at the counterelectrode that is designed to break down an environmental pollutant (e.g., organic dye) thus adding value to the hydrogen photogeneration step [27].

Concluding Remarks

The field of photoelectrochemistry continues to make strides during its ca. 40 year lifespan. While our current knowledge of the fundamental aspects of charge transfer at semiconductor-electrolyte interfaces can be regarded as fairly mature, there are still unresolved aspects related to the specific *chemical* and *electrochemical* nature of traps and surface states at these interfaces. Much remains to be done in further development of experimental tools for studying the dynamic role of these charge transfer-mediating sites *under conditions typical of an operating solar cell*. On the applications side, the many challenges remaining for the development of stable, efficient, and cost-effective systems for solar water and CO_2 splitting will continue to keep future generations of researchers busily engaged for years to come. In particular solar conversion efficiencies routinely exceeding the $\sim$10 % threshold will be required to trigger the interest of the chemical engineering community and engage it in further concerted efforts aimed at designs of photoelectrolytic reactors and the like.

Cross-References

▶ Photocatalyst
▶ Photoelectrochemical CO_2 Reduction

- ▶ Photoelectrochemical Disinfection
- ▶ Photoelectrochemical Disinfection of Air (TiO$_2$)
- ▶ Photoelectrochemical Processes, Electro-Fenton Approach for the Treatment of Contaminated Water

References

1. Rajeshwar K (1997) Environmental electrochemistry. Academic, San Diego
2. Archer MD (1975) J Appl Electrochem 5:17
3. Lin CT, Sutin N (1976) J Phys Chem 80:97
4. Albery WJ, Archer MD (1978) J Electroanal Chem 86:1
5. Gratzel M (ed) (1983) Energy resources through photochemistry and catalysis. Academic, New York
6. O'Regan B, Gratzel M (1991) Nature 353:737
7. Gratzel M (2005) Inorg Chem 44:6841
8. Rajeshwar K (2011) J Phys Chem Lett 2:1301
9. Barker GC (1971) Ber Bunsenges Phys Chem 75:728
10. Kissinger PT, Heineman WR (eds) (1996) Laboratory techniques in electrochemistry, 2nd edn. Marcel Dekker, New York
11. Rajeshwar K, Lezna RO, de Tacconi NR (1992) Anal Chem 64:429A
12. Rajeshwar K (2001) In: Licht S (ed) Encyclopedia of electrochemistry, Chap. 1. Wiley-VCH, Weinheim, p 3–53
13. Nocera DG (2012) Acc Chem Res, 45:767 and references therein
14. Rajeshwar K (2007) J Appl Electrochem 37:765
15. Fujishima A, Honda K (1972) Nature 238:37
16. Bard AJ, Fox MA (1995) Acc Chem Res 28:141
17. Rajeshwar K (1985) J Appl Electrochem 15:1
18. Carey JH, Lawrence J, Tosine HM (1976) Bull Environ Contam Toxicol 16:897
19. Rajeshwar K (1995) J Appl Electrochem 25:1067
20. Rajeshwar K, Ibanez JG (1995) J Chem Educ 72:1044
21. Rajeshwar K, Osugi ME, Chanmanee W, Chenthamarakshan CR, Zanoni MVB, Kajitvichyanukul P, Krishnan-Ayer R (2008) J Photochem Photobiol C Photochem Rev 9:15
22. For example, Kamat PV (2008) J Phys Chem B 113:18737
23. Weaver N, Singh R, Rajeshwar K, Singh P, DuBow J (1981) Sol Cell 3:221
24. Boston DJ, Huang KL, de Tacconi NR, MacDonnell FM, Rajeshwar K (2013) In: Lewerenz HJ, Peter LM (eds) Photoelectrochemical water splitting: challenges and new perspectives. RSC Press (in press), UK
25. Ghadimkhani G, de Tacconi NR, Chanmanee W, Janaky C, Rajeshwar K (2013) Chem Commun 49:1297–1299
26. Rajeshwar K, de Tacconi NR, Ghadimkhani G, Chanmanee W, Janaky C (2013) Chem Phys Chem 14:2251–2259
27. Somasundaram S, Chenthamarakshan CR, de Tacconi NR, Rajeshwar K (2007) Int J Hydrogen Energ 32:4661

Photogalvanic Cells, Principles and Perspectives

Jonathan E. Halls[1] and Jay D. Wadhawan[2]
[1]Department of Chemistry, The University of Bath, Bath, UK
[2]Department of Chemistry, The University of Hull, Hull, UK

Introduction

In modern society, it is common knowledge that the levels of fossil fuels and the stored energy they provide are rapidly decreasing – as a finite resource, it is imperative that new methods of delivering energy to the global population are sought. Due to the abundance of radiant energy provided by the Sun – the solar energy falling on the surface of the Earth in two weeks is equivalent to the energy contained in the initial global supply of fossil fuels – and the environmental uncertainty surrounding nuclear power, much attention has been paid to the development of different systems for harnessing solar energy [1]. In particular, this area of interest has led to numerous investigations into photoelectrochemical (PEC) cells, due to their possible rôle as transducers of solar to electrical energy [2].

Photoelectrochemical Energy Conversion

A PEC cell can be defined as one in which one or both half cells exhibit a PEC effect – which can in turn be described as one in which the irradiation of an electrode or electrolyte system produces a change in the electrode potential (on open circuit) or in the current flowing (on closed circuit) [1].

Becquerel [3, 4] first proposed this theory when, in 1839, he observed the flow of current between two unsymmetrical illuminated metal electrodes in sunlight [5].

PEC effects, and therefore cells, can exist in two forms – photovoltaic (PV) and photogalvanic (PG) – the latter of which shall be discussed extensively within this review. A third type of PEC cell has been proposed by Tien et al. [6] which combines the effects of the photogalvanic and photovoltaic processes and is subsequently named the photogalvanovoltaic effect. However, the feasibility of using this process as a commercial application is questionable, and research into photogalvanovoltaic cells is extremely limited [7, 8].

A PV effect can be characterized as an effect due to photoreduction of electrons and holes, along with no accompanying chemical change. The PV cell absorbs photons from solar energy (sunlight), the energy of which is then transferred to electrons in the system, which consequently leave their respective positions to become part of the electrical current – forming holes [1]. PV cells are purely solid-state devices and have been found to have suitable properties for use as industrial and commercial solar cells – however, they are incredibly complex in terms of production due to the need for extensive purification of the silicon used within the semiconductor cells, which in turn leads to increased manufacturing costs. In contrast to PV systems, cells that exhibit a PEC effect such as the PG cells described in Table 6 further on exhibit a storage capacity and can consequently achieve in one "stage" what a combination of a PV cell with a storage battery achieves in two. This is a fundamental reason behind the extensive research into modifications of PG cells to increase their efficiency. While the abundance of solar radiation received by Earth is relatively constant over time, the exposure of a given surface at a given time to solar energy varies daily and seasonally – consequently, having a solar energy conversion device with an incorporated storage capability is extremely beneficial [1]. The values for mean total solar radiation received by Earth over a 24-h day are given in Fig. 1.

This review will focus solely on the characteristics, analysis, and developments of photogalvanic cells, which in contrast to PV cells are relatively cheap to construct – however, current research proves inconclusive as to whether they can exhibit efficiency and reliability properties as favorable as those found in PV devices.

The Photogalvanic Effect

The PG effect is described by Rabinowitch as "the change in the electrode potential of a galvanic system, produced by illumination and traceable to a photochemical process in the body of the electrolyte" [9, 10]. Cells exhibiting a PG effect have a higher storage capacity than PV cells, but a lower conversion efficiency (theoretically ~ 18 % but observed values are much lower) [5, 11]. Consequently, extensive research has been carried out concerning the manipulation of substances used in PG cells (i.e., reductants and photosensitizers) in order to maximize electrical output [5].

A typical PG cell is made up of two parallel electrodes (an illuminated and a dark) between which a thin layer of electrolyte solution is contained; see Fig. 2. A photochemical reaction is induced within the electrolyte as a result of solar energy entering through the semitransparent or transparent (illuminated) electrode – the resulting high energy products react, driving electrons around an external circuit [12, 13]. Two redox couples are dissolved within the electrolyte, and the reaction scheme is given in Eq. 1.

Absorption of solar radiation is undertaken by component A, a dye. A subsequent electron-transfer reaction between the excited A^* and Z yields energetic products B and Y [13]:

$$\frac{A + Zk}{\rightleftharpoons B + Y} \tag{1}$$

Photochemistry:

$$\frac{A h\nu}{\longrightarrow A^*} \tag{2}$$

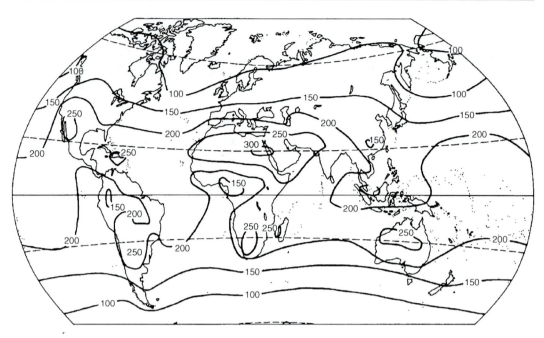

Photogalvanic Cells, Principles and Perspectives, Fig. 1 Mean annual intensity of solar radiation on a horizontal plane at the surface of the Earth. Numbers represent solar irradiance in W m^{-2} averaged over a 24 h day (Reproduced from reference Springer Science+Business Media, Journal of Applied Electrochemistry, 5, 1975, 17, Electrochemical aspects of solar energy conversion, M. D. Archer, Figure. With kind permission from Springer Science and Business Media)

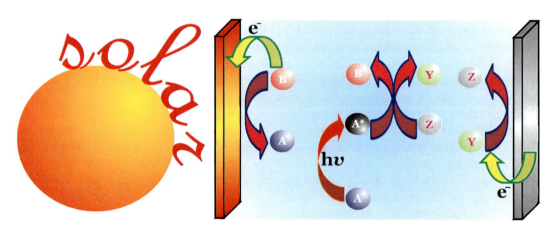

Photogalvanic Cells, Principles and Perspectives, Fig. 2 Schematic representation of a thin-layer photogalvanic cell. The illuminated electrode (*left-hand side*) is constructed from a semitransparent/transparent material (such as SnO_2, In_2O_3, or ZnO), with a photogalvanic electrode filling the space between the illuminated and dark (*right-hand side*) electrodes

$$\frac{A^* + ZK_q}{\longrightarrow B + Y} \quad (3)$$

A back reaction between **B** and **Y** yields original components **A** and **Z**:

$$\frac{B + Yk}{\longrightarrow A + Z} \quad (4)$$

Discussed in further detail later on, the rapidity of this reaction is of great hindrance to the performance of the cell. Component B must locate the illuminated electrode (where solar radiation permeates the cell) and react there before the back reaction occurs [13].

Electrode reactions:

$$B \pm e \longrightarrow A \qquad (5)$$

$$Y \pm e \longrightarrow Z \qquad (6)$$

Following on from Becquerel's aforementioned discovery [3], the PG effect was initially recognized by Rideal and Williams [14] before Rabinowitch [9, 10] subsequently examined the photochemical and photogalvanic properties of a thionine (Th)-iron system. The scheme is described below, while the mechanism and kinetics are described later.

$$\frac{1}{2}\text{Th} + \text{Fe}^{2+} \underset{\text{dark}}{\overset{hv}{\rightleftharpoons}} \frac{1}{2}\text{Leu} - \text{Th} + \text{Fe}^{3+} \qquad (7)$$

Rabinowitch discovered this redox equilibrium was the most light-sensitive known – thionine dyes absorb at 500–700 nm, comparable with values of maximal solar radiation – and as a result would possess a high level of PG sensitivity. As explained in the later sections, the thionine-iron (a dye and reductant, respectively) redox cell yielded a low energy conversion efficiency (<1 %), but the results were somewhat promising due to the ability to manipulate the dye and/or reductant – and subsequently established much research into optimization of the necessary components within the system [9, 15].

Photogalvanic Cell Overview

The Ideal Photogalvanic Cell

Albery and Archer analyzed the various factors which affect the ability of a PG cell to deliver the optimum power conversion efficiency (theoretically $\sim$18 %) [16, 17]. These authors discovered that the processes occurring within a PG cell were dependent on the system's photochemistry, homogeneous kinetics, mass transfer, and electrode kinetics – and therefore, the energy

converted depended on species concentration, hv intensity, the electrode kinetics, and their spacing and diffusion lengths [17].

Over a series of articles, Albery and Archer calculated the optimum length characteristics for an efficient PG cell [1, 12, 13, 16–20]. It should be noted that while Albery and Archer's work is perspicacious and seminal, it is imperative to realize that their considerations were only undertaken under the assumption of minimal electrical migration within the PG cell. Of course this is not true in many cases, and the generality of their results in conditions of low supporting electrolyte concentration must therefore be considered with caution.

To enhance solar energy conversion, it was discovered that the following conditions be obeyed:

$$10X_\varepsilon < X_g, X_k, X_l \qquad (8)$$

Although to produce the maximum theoretical power yield, the following PG cell conditions have been established – however, in practice this is unfeasible, as explained later:

$$10X_\varepsilon \cong X_G \cong \frac{1}{2}X_G < X_l \qquad (9)$$

Table 1 provides the definitions of these four X parameters (Fig. 3).

Equation 8 denotes that X_ε must be bigger than the three other length parameters controlling the efficiency of a PG cell. Initially, the absorbance length must be smaller than the generating length in order to ensure there is sufficient amounts of component A available to absorb the incoming photons of solar radiation and to inhibit the bleaching of the solution at the semitransparent electrode. In order for B to be able to diffuse to this electrode before it is broken down via the back reaction, the absorbance length must also be smaller than the reaction length. Finally, to ensure B reacts on the electrode interacting with hv, the absorbance length must be smaller than the cell length. The value of X_k is known to be approximately 10^{-3} cm; therefore by use of Eq. 8 and the equations described in Table 1,

Photogalvanic Cells, Principles and Perspectives, Table 1 Definition of characteristic lengths for a PG cell (Adapted from reference [12])

Name	Symbol and equation	Description
Cell length	X_l	The distance between electrodes
Absorbance length	$X_\varepsilon = (\varepsilon[A])^{-1}$	The distance over which light is absorbed
Generating length	$X_g = \left(\frac{D}{\phi I_0 \varepsilon}\right)^{\frac{1}{2}}$	The distance A diffuses in light of irradiance I_0 before being converted to B
Reaction length	$X_k = \left(\frac{D}{k[Y]}\right)^{\frac{1}{2}}$	The distance over which B diffuses before being broken down in reaction with Y

where D = diffusion coefficient of B, ϕ = quantum efficiency for generation of B and is tantamount to the ratio of photons absorbed to the amount produced, I = molar flux of photons, and ε = molar extinction coefficient of A (units cm^2 mol^{-1})

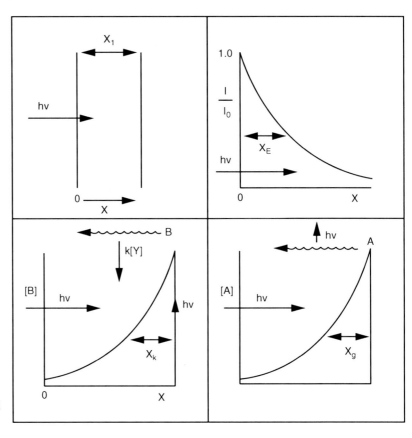

Photogalvanic Cells, Principles and Perspectives, Fig. 3 Schematic definition of the characteristic lengths for a PG cell (Reprinted with permission from W. John Albery (1982) Development of photogalvanic cells for solar energy conservation. Accounts of Chemical Research 15: 142. Copyright (1982) American Chemical Society)

Albery and Archer were able to calculate the values of the four parameters for an ideal cell [12, 13].

Electrode Selectivity and Kinetics

The selectivity of the electrodes and their kinetics are an essential consideration for the design of a practical PG cell. The B and Y components of the PG cell reaction must react at the illuminated and dark electrodes, respectively, in order to generate a current. If the illuminated electrode is not selective, the back reaction ($B + Y \rightarrow A + Z$) is catalyzed due to Y reacting at the site instead of B – however, a selective electrode facilitates diffusion of Y across to the dark electrode, allowing B to react at the necessary position.

Photogalvanic Cells, Principles and Perspectives, Table 2 Ideal electrochemical rate constants (Adapted from reference [12])

Couple	Illuminated electrode	Dark electrode
A, B	$k'_B > \dfrac{D}{X_k} \approx$ 5×10^{-3} cm s^{-1}	$k'_B[B] << k'_Y[Y]$
Y, Z	$k'_Y[Y] << k'_B[B]$	$k'_Y > \dfrac{D}{X_l} \approx$ 5×10^{-3} cm s^{-1}

Table 2 gives the electrochemical rate constants for an ideal PG cell. For a species to react at a specified electrode within an electrochemical reaction, the ratio D/X (where X is one of the length parameters outlined in Table 1) must be smaller than the rate constant k'. The opposite is true for a species to be lost via reaction in the electrolyte or diffusion to the opposite electrode. As B is required at the illuminated electrode, and as shown by Eq. 8, for an ideal PG cell it is necessary for reaction length to be smaller than cell length, B is less likely to diffuse to the dark electrode than it is to be destroyed by reaction with Y. Consequently, for component B the characteristic length is X_k. As explained further down, the concentration of Y and Z is assumed to be larger due to the diffusion distance, so for component Y the characteristic length is X_l [12].

Overall, it can be seen that if the electrode kinetics of the A,B and Y,Z couples are fast on the two electrodes (illuminated and dark, respectively) than the potentials of the respective electrodes will be similar to the standard electrode potentials of the two couples. Thus, the voltage generated by the PG system can be approximated by Eq. 10, and relies on differential electrode kinetics [12, 13].

$$E_\Delta \cong \left| E^{\ominus}{}_{A, B} - E^{\ominus}{}_{Y, Z} \right| \qquad (10)$$

If identical electrodes are selected for use in a PG cell, or if the A, B couple reacts via the dark electrode, then the system becomes a concentration cell – a severely inefficient solar energy conversion device, as explained later.

Effect of Species Concentration

Along with the optimization of electrode properties, the other major factor influencing the efficiency of a PG cell is the varying concentrations of the four components within the basic PEC reaction – A, B, Y, and Z. For the fundamental diffusion of components Y and Z across a PG cell in contrasting directions, the diffusive concentration gradient of B at the illuminated electrode is required to be in equality with Y and Z [12]. In addition, Eq. 8 denotes that the ideal distance between electrodes must be bigger than the absorbance length; thus, the variance in concentration across the system for components Y and Z must be greater than that of B [12]. Given that for an efficient PG cell, the flux of electrons of Y and Z should approximately equal I_0, the aforementioned variance in concentration (Δc) can be calculated as follows:

$$I_0 = \frac{D\Delta c}{X_l} \qquad (11)$$

where $\Delta c = [Y]_0 - [Y]_{X_l} = [Z]_{X_l} - [Z]_0$, with $[Y]$ and $[Z]$ at the midpoint of the cell being equal to their respective concentrations $[Y]_D$ and $[Z]_D$ in the absence of light [12].

For the diffusive transport of components Y and Z to the desired electrodes (dark and illuminated, respectively), the concentrations of the two need to be greater than $\frac{1}{2}\Delta c$ in the absence of light, which ensures their presence – and therefore their subsequent reaction – at the two electrodes. The presence of Y on the dark electrode ensures minimization of the back reaction, increases the electrochemical kinetics via inhibition of the Y, Z couple on the illuminated electrode, and produces a flow of current with B reacting on the opposite electrode [13, 17, 19, 20]. With Z located at the illuminated electrode, the excited A^* component (the dye) can be trapped, enabling formation of B. The following set of equations, established by Albery and Foulds, allow the calculation of optimum concentrations of Z, Y, and A [12].

For the formation of B, the quantum efficiency for production can only be maximized if

Photogalvanic Cells, Principles and Perspectives, Fig. 4 Graphical illustration of the polarization (*red*) and power (*blue*) characteristics of an ideal photogalvanic cell. Note that the fill factor as defined by Eq. 17 is 25 % for an ideal galvanic cell under discharge

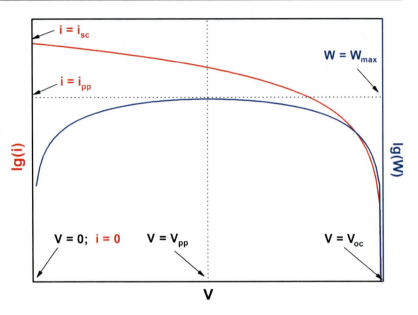

$$k_q[Z]_0 \tau > 10 \quad (12)$$

where τ = lifetime of excited state A^* if Z is not present.

As mentioned above, the presence of the Y, Z couple on the illuminated electrode must be as low as possible – therefore, by increasing $[Y]$ and minimizing $[Z]$, the potential difference produced by the cell improves via the resulting effects on the potential of the Y, Z half cell [17]. Taking into account these requirements, and the above equation, the subsequent condition for Z results

$$[Z]_D \approx \Delta c + \frac{10}{k_q \tau} \quad (13)$$

As stated, the concentration of Y must be increased to a value large enough to produce a beneficial voltage from the cell and to prevent concentration polarization, but not so large as to aid destruction of B via the back reaction [12, 17]:

$$[Y]_D > \frac{1}{2}\Delta c \quad (14)$$

From Table 1, we can therefore obtain

$$[Y]_D = \frac{40}{k} - \frac{1}{2}\Delta c \quad (15)$$

Finally, to shorten the absorbance length, $[A]$ should be maximized, ensuring photons of solar energy are retained in close proximity to the electrode undergoing irradiation of solar light, facilitating the PEC reaction [12].

Solar Energy Conversion and Practical Power Output

Following the manipulation of the above parameters within a PG cell, the practicality of the system as a solar energy converter can be assessed by calculating the values of conversion efficiency and fill factor (a measure of the device quality). Conversion efficiency indicates the amount of incoming solar irradiation converted to electrical energy, while the fill factor describes the ratio of the (actual) maximum obtainable power to the theoretical. Both can be expressed as percentages. Figure 4 gives a graphical illustration of fill factor.

Conversion efficiency:

$$\eta = \frac{V_{pp} i_{pp}}{AE} \times 100 \quad (16)$$

Fill factor:

$$F_f = \frac{V_{pp} i_{pp}}{V_{oc} i_{sc}} \quad (17)$$

where A = area of each electrode, E = input light irradiance, V_{pp} = value of the cell voltage at power point, i_{pp} = value of the current at power point, V_{oc} = open circuit voltage, and i_{sc} = current at short circuit.

To determine η, E is calculated as

$$E = \frac{W_{max}}{A} \qquad (18)$$

where W_{max} is the maximum power produced by the cell and, for a cell with differential electrode kinetics (i.e., the ideal cell explained earlier), can be calculated as follows:

$$W_{max} = 0.8AF\phi I_0 \Delta E \qquad (19)$$

where ΔE = difference in standard electrode potentials for the A, B and Y, Z couples.

For standard solar radiation with Sun elevation angle of $30°$ (AM 2.0), I_0 can be estimated to have a value of $\sim 1.6 \times 10^{-3}$ mol photons m^{-2} s^{-1}.[17] If $\phi = 1$ and $\Delta E = 1.1$ V [1] (the optimum value), Eq. 19 establishes W_{max} as having a value of 140 W m^{-2}, which corresponds to a power conversion efficiency of 18 % [12, 17, 19]. Albery and Archer demonstrated however that this theoretical conversion efficiency is, in practice, unrealistic due to the impossibility of optimizing all the available parameters [19].

If typical values of D and ε are taken, 10^{-5} cm^2 s^{-1} and 10^8 cm^2 mol^{-1}, respectively, the generating length is found to be

$$X_g \sim 10^{-3} \text{ cm} \qquad (20)$$

The ideal cell requires the generating length to be ten times larger than the optical absorbance length, so

$$X_\varepsilon \sim 10^{-4} \text{ cm} \qquad (21)$$

Therefore it follows that

$$[A] \sim 10^{-1} \text{ M} \qquad (22)$$

Consequently, the dye must prove to be relatively soluble in order to absorb the solar radiation as close to the illuminated electrode as possible. Taking the value of X_g above, along with Eq. 9, the following condition must be obeyed for the kinetics of the reaction B with Y:

$$k[Y] \sim 40\text{s}^{-1} \qquad (23)$$

To ensure the inhibition of concentration polarization (keeping the dark electrode unpolarized), $[Y]$ must be twice as large as $[B]$:

$$[Y] > 2[B] \sim 10^{-2}[A] \sim 10^{-3} \text{M} \qquad (24)$$

From Eqs. 8 and 9 it can be seen that the absorbance length must be smaller than the distance between electrodes, and as a result the rate constant, k', of the reaction of A and B at the illuminated electrode must be very fast:

$$k' > \frac{D}{X_\varepsilon} \sim 10^{-1} \text{cm s}^{-1} \qquad (25)$$

The required value of this rate constant proves a major disadvantage in obtaining the maximum theoretical conversion efficiency from a PG cell, as it is highly unusual for a redox couple to exhibit electrode kinetics as fast as the above value [19, 20].

It can be seen that obtaining maximum efficiency from a cell relies on the solubility of the dye (A), an extremely rapid rate constant for the A, B couple, and relatively slow kinetics for the electron-transfer reaction of B and Y. It is inconceivable that electron-transfer reactions obeying the above conditions while at the same time satisfying the Marcus theory could occur [16, 21–23].

As explained earlier, a PG system acting as a concentration cell proves to be severely inefficient as a solar energy converter, due to ΔE not having any bearing on the maximum obtainable power – both electrodes are identical. Therefore, W_{max} for a concentration cell is calculated as follows:

$$W_{max} = 0.28ART\phi I_0 \qquad (26)$$

Photogalvanic Cells, Principles and Perspectives, Fig. 5

Structures labeled: Thionine, Semithionine, Leucothionine

Photogalvanic Cells, Principles and Perspectives, Fig. 5 The three structures of the thionine dye within the iron-thionine PG system

Taking the same values of I_0 and ϕ as above, W_{max} is ~ 1.1 W m^{-2}, which corresponds to a maximum cell conversion efficiency of 0.15 %, i.e., practically useless [19, 20].

Early Developments of Molecular Systems for Photogalvanic Cells

Early Photogalvanic Systems

Hatchard and Parker [24] further investigated the kinetics of the initial iron-thionine system discovered by Rabinowitch and derived the following mechanism for the collection and conversion of solar energy within this system:

$$\frac{1}{2}\text{Th} + \text{Fe}^{2+} \underset{\text{dark}}{\overset{hv}{\rightleftharpoons}} \frac{1}{2}\text{Leu} - \text{Th} + \text{Fe}^{3+} \quad (27)$$

Forward reactions:

$$\text{Th} + hv + \text{Th}^* \quad (28)$$

$$\text{Th}^* + \text{Fe}^{2+} \longrightarrow \text{Th}^{\cdot -} \quad (29)$$

$$2\text{Th}^{\cdot -} \longrightarrow \text{Th} + \text{Leu} - \text{Th} \quad (30)$$

Back reactions:

$$\text{Leu} - \text{Th} + \text{Fe}^{3} \longrightarrow \text{Th}^{\cdot -} + Fe^{2+} \quad (31)$$

$$2\text{Th}^{\cdot -} \longrightarrow \text{Th} + \text{Leu} - \text{Th} \quad (32)$$

$$\text{Th}^{\cdot -} + \text{Fe}^{3+} \longrightarrow \text{Th} + \text{Fe}^{2+} \quad (33)$$

where Th^* = triplet state of thionine dye, $\text{Th}^{\cdot -}$ = semithionine half-reduced form, and Leu = leucothionine (Fig. 5).

The overall reaction for this type of PG cell differs from the basic reaction considered earlier, in that the thionine (photosensitive dye) couple is a two-electron redox system, and can be given by the following reaction:

$$2Z + A \underset{k_{-1}}{\overset{g}{\rightleftharpoons}} Z + B + Y \underset{k_{-2}}{\overset{k_2}{\rightleftharpoons}} C + 2Y \quad (34)$$

This two-electron thionine couple (A, C) and the one-electron outer sphere iron (reductant) couple (Y, Z) manifests itself as an ability to obtain the required electrode selectivity of B and Y on the illuminated and dark electrode, respectively. However, this mechanism also complicates the method of establishing the system's homogeneous kinetics, as explained further on.

Variations in Cell Parameters for an A, B, C System

Hatchard and Parker [24] and Ferreira and Harriman [25] established that the leucothionine product was formed via the disproportionation reaction of $B + B$, as opposed to the reaction with Z as seen earlier for an A, B cell [1]. B is therefore an intermediary product, the stability of which is not sufficient to reach the illuminated electrode. The quantum efficiency of leucothionine is consequently a deciding factor in determining ideal conditions for a cell of this type and is incorporated into efficiency calculations as follows:

$$\text{Flux producing } C = g\phi_c \quad (35)$$

where

$$\phi_c = \frac{1}{\left(\frac{2+k_{-1}[Y]}{k_3[B]}\right)} \tag{36}$$

$$g = \phi I_\varepsilon [A] \tag{37}$$

In order to realize the optimum value for the quantum efficiency of C (in this case, ½), B must be broken down by the aforementioned disproportionation reaction and half of the photogenerated B must react to form C. Thus, it follows that $k_3[B] \gg k_{-1}[Y]$ is an essential condition. If $k_3[B] \ll k_{-1}[Y]$, an inefficient system results, as the majority of the photogenerated B reacts with Y, forming the original reactant A (the back reaction).

An efficient cell therefore requires

$$\frac{2k_3[B]}{k_{-1}[Y]} > \frac{2k_3 g}{k_{-1}^2[Y]^2} > 1 \tag{38}$$

Subsequent substitution with Eq. 37 results in the conditions for the concentration of Y:

$$[Y] < \left(\frac{2\phi I_0}{X_\varepsilon} \times \frac{k_3}{k_{-1}^2}\right)^{\frac{1}{2}} \tag{39}$$

Referring back to Eq. 8, the above equation agrees with the requirement of a small absorbance length for an efficient cell – in this case, it increases the likelihood of B to disproportionate. In addition to this, in a similar fashion to the $[Y]_D$ parameters for the ideal A, B cell denoted earlier, $[Y]_D$ for this A, B, C cell can now be defined as

$$[Y]_D = \left(\frac{\phi I_0}{5X_\varepsilon} \times \frac{k_3}{k_{-1}^2}\right)^{\frac{1}{2}} - \left(\frac{1}{2}\Delta c\right) \tag{40}$$

Analysis of practical A, B, C photogalvanic cells has shown that the concentration of Y and Z (the inorganic iron couple) helps induce the required homogeneous kinetics [12, 23, 24]. The concentrations of the iron redox couple aid the prevention of concentration polarization at the dark electrode, while trapping A^* at the illuminated electrode ([Fe^{2+}]) and inhibiting destruction of B or C in the electrolyte ([Fe^{3+}]).

Photogalvanic Cells, Principles and Perspectives, Table 3 Conditions for an ideal PG cell (Adapted from references [12, 13])

Characteristic length	Approximate value for an ideal photogalvanic cell
X_l	10^{-2} cm, giving $[Y] \sim 10^{-2}$ M
X_ε	10^{-4} cm, giving $[A] \sim 10^{-1}$ M
X_k	10^{-3} cm, giving $k[Y] \sim 10$ s^{-1} and $k < 10^3$ M^{-1} s^{-1}

As explained in Table 3, for an ideal cell the absorbance length is approximately 10^{-4} cm, which corresponds to dominance of the k_{-2} parameter. However, thionine dye has a low solubility – meaning the absorbance length can be as long as 10^{-1} cm – and thus the k_{-1}^2/k_3 parameter becomes equally important. This is a considerable factor in the relative inefficiencies of iron-thionine PG cells – a high absorbance length means B struggles to disproportionate and C fails to reach the illuminated electrode [12].

Various modifications have been attempted in order to increase the solar energy conversion efficiency of the iron-thionine PG cell, including adding sulfonate groups onto the thionine dye in order to increase solubility [13] and irreversibly coating a thionine electrode with up to twenty thionine monolayers to increase absorbance of solar radiation and thus decrease ε [26]. However, research has broadened since early innovations by Rabinowitch, Albery, Archer, and Foulds, and the common approach is now motivated towards improvisations of the dye and reductant to induce wholesale changes in the efficiencies of PG cells.

Modern Approaches for Photogalvanic Cell Advancement

Modification of Photosensitizing Dye

Photosensitizing dyes, which in PG cells are almost always organometallic or heterocompounds, exhibit both the redox characteristics similar to the irreversible inorganic (Y, Z) couple and the ability to absorb light intensity of a certain wavelength. To aid the conversion of solar energy to electrical energy, a chosen dye in a PG cell must therefore absorb light at a wavelength of approximately 500–700 nm – while also having a relatively soluble structure and a suitable reduction potential [26–28]. A variety of dye structures have been used to investigate their suitability for use in PG cells, several of which are listed in Table 4 below.

Modification of Reductant

A survey of current literature indicates that the photosensitizing dye has been the subject of much modification in investigations to optimize PG cell efficiency. In a similar fashion, manipulation of the reductant has also received close attention, following on from the early realization that the use of sacrificial electron donors such as triethanolamine (TEA) and ethylenediaminetetraacetic acid (EDTA) with suitable dyes produces higher photopotentials and photocurrents than their early counterparts, e.g., iron and hydroquinone. Table 5 gives an illustration of this [2, 28]. However, EDTA and TEA are unable to produce higher photopotentials and photocurrents in cells with proflavin, rhodamine B, and rose Bengal due to these dyes having high negative potentials. One method of restricting this problem is by adding methyl viologen (MV^{2+}) to the system, which acts as a mediator – catalyzing the photoredox reduction (MV^{2+} is reduced – the process of which oxidizes the dye – thus forming a cation of the dye which reacts with the EDTA or TEA) [28].

Literature Results for a Variety of Molecular Photogalvanic Systems

Table 6 describes a selection of PG systems that have been investigated within current literature. The following table, Table 7, depicts the experimental results obtained from the systems listed. In terms of storage capacity, this is $t_{1/2}$ and is calculated by applying light radiation of a suitable (i.e., solar) wavelength until the potential reaches a constant value, at which point a load is applied to the cell to induce a current at power point. The time taken for the output of

the cell to fall, in the dark, to its half at power point is $t_{1/2}$ [15].

Tables 6 and 7 give an illustration of the variance within research to optimize PG cells. Given are systems with varying groups of dyes – thiazines (azure B, toluidine blue, methylene blue, azure A), phenazines (safranine, safranin-O), a xanthene (rhodamine 6G), and a triphenyl methyl derivative (malachite green), as well as differing reductants, including the aforementioned sacrificial electron donors EDTA and TEA. Much investigation has been undertaken concerning thiazine and phenazine dyes due to their suitable absorption wavelength maxima, solubility, and the ease in which different substituents can be modified on the basic structure of the compound, i.e., adding sulfonate groups to thiazine compounds to increase solubility [13].

However, the power produced, the fill factor, and the conversion efficiency of many engineered PG cells are all somewhat disappointing – a level has not yet been reached where PG cells could be considered a feasible replacement for the more costly PV cell. A number of factors contribute to this, many of which have been discussed within this review and can be related back to Eqs. 8 and 9 for an optimum PG cell, originally formulated by Albery and Archer [17, 18]. Without careful manipulation of electrode spacing, kinetics, and diffusion lengths at the electrodes – in conjunction with mass transfer, the photochemistry, and the homogeneous kinetics of the system – the power developed by a PG cell cannot be optimized and may consequently be a crucial determinant for current literature displaying disappointing results relative to the optimum theorized by the pioneering work of Albery and Archer [1, 12, 13, 16–20].

Effects of System Conditions

Lal [36] and Dube and Sharma [37] have shown it is possible to use a mixture of two dyes within the PG system in an attempt to maximize the use of the broad solar spectrum. Using EDTA as a reductant, Lal used a mixture of thionine and Azure B as a photosensitizing dye, which contributed to a solar energy conversion efficiency of 0.18 %. Dube and Sharma witnessed conversion

Photogalvanic Cells, Principles and Perspectives **1567**

Photogalvanic Cells, Principles and Perspectives, Table 4 A list of potential dyes for PG cells (Adapted from reference [2])

Dye	Class	Structure	λ_{max}/nm
Thionine	Thiazine		596
Toluidine blue	Thiazine		630
Methylene blue	Thiazine		665
Azure A	Thiazine		635
Azure B	Thiazine		647
Azure C	Thiazine		620
Phenosafranine	Phenazine		520
Safranin-O	Phenazine		520

(continued)

Photogalvanic Cells, Principles and Perspectives, Table 4 (continued)

Dye	Class	Structure	λ_{max}/nm
Fluorescein	Xanthene		490
Rhodamine B	Xanthene		551
Rhodamine 6G	Xanthene		524
Acridine orange	Acridine		492
Fuchsine	Triphenylmethane		545

efficiencies of 0.17 %, 0.18 %, and 0.19 % when using dye mixtures of Azure A and Azure B, Azure B and Azure C, and Azure A and Azure C, respectively, with mannitol as a reductant. In general, as can be seen by referral to Tables 6 and 7, the conversion efficiencies seen by Lal and Dube and Sharma do not correspond to a particularly efficient system, an analysis which is supported by the relatively low photocurrents and photopotentials also seen for these mixed dye systems. This and the additional complexity issue raised by using more than one dye are

Photogalvanic Cells, Principles and Perspectives, Table 5 The photocurrent and photovoltage of a selection of PG cells upon variation of the reductant (Adapted from references [2] and [28])

Dye	Reductant	Photocurrent/μA cm^{-2}	Photopotential/mV
Thionine	Fe^{2+}	3.30	140
	Hydroquinone	2.80	20.0
	EDTA	12.0	190
	TEA	18.0	250
	TEA + MV^{2+}	18.0	210
Riboflavine	Hydroquinone	0.830	20.0
	EDTA	58.0	720
	TEA	35.0	500
	TEA + MV^{2+}	37.0	430
Proflavin	Hydroquinone	0.430	1.90
	Hydroquinone + MV^{2+}	0.410	2.40
	EDTA	0.790	140
	EDTA + MV^{2+}	25.0	350
	TEA	0.250	60.0
	TEA + MV^{2+}	6.25	320
	MV^{2+}	0.038	5.00
Rhodamine B	EDTA	0.092	28.0
	EDTA + MV^{2+}	0.092	10.0
Rose bengal	EDTA	0.025	19.0
	EDTA + MV^{2+}	0.017	20.0

Photogalvanic Cells, Principles and Perspectives, Table 6 A selection of PG systems that have been investigated. Research is listed in chronological order

Reference	Dye	Reductant	Surfactant
[29]	Azure B	Nitrilotriacetic acid (NTA)	N/A
[30]	Safranine	EDTA	N/A
		Glucose	N/A
		NTA	N/A
[11]	Toluidine blue	Glucose	CTAB
[31]	Methylene blue	Oxalic acid	N/A
[32]	Azure A	Ascorbic acid	NaLS
[33]	Bromophenol red	EDTA	N/A
[5]	Rhodamine 6G	Oxalic acid	DSS
			CTAB
			TX-100
[15]	Safranine O	EDTA	Tween-80
[34]	Toluidine blue	Arabinose	NaLS
	Malachite green	Arabinose	NaLS
[35]	Safranine	DTPA	Brij 35
	Bismarck brown	DTPA	Brij 35
	Methyl orange	DTPA	Brij 35

Photogalvanic Cells, Principles and Perspectives, Table 7 Experimental results for selection of PG cells listed in Table 6

Reference	Photocurrent/μA	Photopotential/mV	Fill factor	Conversion efficiency/%	Storage capacity/min
[29]	60.0	340	0.290	0.180	50.0
[30]	50.0	760	0.580	0.262	19.0
	35.0	373	0.180	0.036	85.0
	35.0	415	0.370	0.084	8.00
[11]	51.0	268	0.410	0.058	6.00
[31]	110	312	0.280	0.121	35.0
[32]	160	770	0.340	0.546	110
[33]	45.0	581	0.270	0.036	35.0
[5]	425	1936	0.410	0.860	131
	215	1145	0.450	0.240	68.0
	310	1478	0.380	0.550	96.0
[15]	300	785	0.340	0.977	60.0
[34]	60	813	0.259	0.145	123
	36	348	0.024	0.059	32.0
[35]	155	842	0.410	0.644	122
	115	786	0.480	0.519	117
	95	625	0.400	0.207	94.0

possible reasons for the relatively small amount of research into mixed dye PG cells, with most engineered towards suitable selection and/or modification of a single photosensitizing dye [38].

It can be seen from Tables 6 and 7 that, in general, PG systems containing a surfactant produced higher relative values for photocurrent, photopotential, and conversion efficiency than systems studied in which a surfactant was absent. Fendler and Fendler [39] and Atwood and Florence [40] have attributed this to the ability of a surfactant to solubilize certain molecules (i.e., the photosensitizing dye) and the catalytic effect that carefully chosen surfactants induce on particular chemical reactions. Furthermore, Rohatgi-Mukherjee et al. theorized that addition of a surfactant into a PG system increases conversion efficiency via facilitating the separation of photogenerated products by hydrophobic-hydrophilic interaction of the products with the surfactant interface [27].

Genwa and Genwa provided a thorough example of the benefit of micellar PG systems, by analyzing the electrical output and efficiency of the same cell with three different surfactants – anionic, cationic, and nonionic (DSS, CTAB, and TX-100, respectively) [5]. The results are tabulated in Table 7, showing that the use of an anionic surfactant is more beneficial than nonionic, followed by cationic. Results for the same system containing no surfactant resulted in a photocurrent, photopotential, fill factor, and conversion efficiency of 152 μA, 985 mV, 0.39, and 0.22 %, respectively – thus illustrating the possibility of increasing the electrical output of a PG cell by more than double via the addition of a suitable surfactant.

Finally, in terms of pH of the PG system, it has been found that photopotential and photocurrent increase with increase in pH up to a certain maximum, after which it decreases at a similar rate. Several research articles, including Gangotri and Gangotri [15], Genwa and Chouhan [32], and Dube et al. [29], have observed that the pH for the optimum condition is related to the pK_a of the reductant, with the pH being equal to or slightly higher than the pK_a of the reductant. The aforementioned authors cite a possible reason for this as the availability of the reductant in its neutral or anionic form – a superior electron donor.

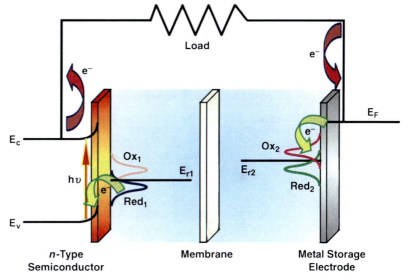

Photogalvanic Cells, Principles and Perspectives, Fig. 6 Light to electrical energy conversion and storage at a two-electrode semiconductor PG cell

Photogalvanic Cells, Principles and Perspectives, Fig. 7 A solar chargeable battery with n-type semiconductor using a three-electrode system (Reprinted from Electrochimica Acta, 36(7), Maheshwar Sharon, P. Veluchamy, C. Natarajan, Dhananjay Kumar, Solar rechargeable battery–principle and materials, 1107–1126, Copyright (1991), with permission from Elsevier)

Semiconductor-Based Photogalvanic Cells

In semiconductor photogalvanic cells, the semiconductor acts as the light energy harvester, allowing the formation of either weakly bound (Wannier) or strongly bound (Frenkel) excitons provided the incident photon energy is larger than the bandgap separation. For n-type semiconductors in contact with a redox electrolyte at open circuit, the band bending at the interface (tantamount to the formation of a Schottky barrier) allows for oxidation processes to take place via the valence band, causing the photoanode to acquire a negative potential. This in turn causes a shift in the Fermi level with consequent reduction in the band bending until the flatband potential is reached (at high illumination intensities).

Photogalvanic Cells, Principles and Perspectives, Table 8 Semiconductor-based PG systems (Reprinted from Electrochimica Acta, 36(7), Maheshwar Sharon, P. Veluchamy, C. Natarajan, Dhananjay Kumar, Solar rechargeable battery–principle and materials, 1107–1126, Copyright (1991), with permission from Elsevier)

System	V_{oc}/mV	Discharge current/μA
Three-electrode systems		
$BaTiO_3$ I Ce^{3+},Ce^{4+} ‖ Fe^{3+},Fe^{2+} I Pt	825	38
n-CdSe I S^{2-},S ‖ C ‖ storage electrode HO^-	430	N/A
$MoSe_2$ I Br^-,Br_3^- ‖ I_3^-,I^- I Pt	400	4,000
GaAs I S^{2-},Se_2^{2-} I Cd	250	8,000
CdSe I S^{2-},S_x^{2-} ‖ Se_2^{2-},Se,Se^{2-} I Pt	50–200	8,000
CdSe I S^{2-},S_x^{2-} ICd	300–400	10,000
n-$MoSe_2$ I HBr,Br_2 I Nafion®-315 I I_3^-,HI I Pt	490	15,000
n-MX_2 I I^-,I_3^- ‖ AQ,AQH_2 I C	260	1,000
$MX_2 = WSe_2,MoS_2$; AQ = anthraquinone-2,6-disodium disulfonate		
n-Pb_3O_4 I Fe^{2+},Fe^{3+}‖ IO_3^-,I^- I Pt	840	410
n-Cd(Se,Te) I S_x^{2-} I CoS I membrane I CsOH,CsSH I SnS I Sn	510	1,500
n-CdS I $Na_2S,S,NaOH$ ‖ NaOH I SbO_2^- I Sb	60	N/A
Four-electrode systems		
p-WSe_2 I MV^{2+} ‖ I^- I n-WSe_2	800	12,000
p-InP I $PEO,NaSCN,Na_2S,S$ I n-CdS	540	1
p-InP I porphyrin I Nafion-117 I $Ru(bpy)_3^{2+}$ I n-CdS	380	1
Without a membrane		
n-CdSe I $CdSO_4$ I p-CdTe	100	1

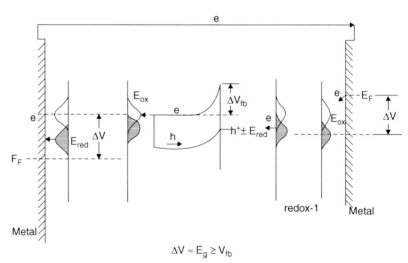

Photogalvanic Cells, Principles and Perspectives, Fig. 8 Schematic illustration of the requirements of the various energy levels for the Sharon-Schottky PG cell (Reprinted from Electrochimica Acta, 36(7), Maheshwar Sharon, P. Veluchamy, C. Natarajan, Dhananjay Kumar, Solar rechargeable battery–principle and materials, 1107–1126, Copyright (1991), with permission from Elsevier)

This system allows for light to electrical energy conversion, as illustrated in Fig. 6.

It follows that if a two-electrode configuration is used with a "storage electrode" acting as a dark electrode process and the illuminated electrode allowing for a redox transformation, Fig. 6, then the system may operate as a PG cell. However, this configuration holds the disadvantage that the dark reaction at the semiconductor (the reverse of the light-driven process) is necessarily sluggish

| Ag | Ag$_2$SO$_4$ | Li$_2$SO$_4$ 0.01 M
Na$_4$ZnTPPS
10^{-4} M
Tosylate$_4$ZnTMPP
10^{-4} M
(aq) ‖ | BTPPATPBCl
0.01 M
quencher
10^{-3} M
(DCE) | BTPPACl
0.001 M
LiCl
0.01 M
(aq) | AgCl | Ag′ |

Photogalvanic Cells, Principles and Perspectives, Fig. 9 Cell diagram for a liquid | liquid solar cell. The organic supporting electrolyte is bis(triphenyl-phosphoranylidene)ammonium tetrakis(4-chlorophenyl)borate, BTPPATPBCl (Reprinted from Electrochemistry Communications, David J Fermín, Hong D Duong, Zhifeng Ding, Pierre F Brevet, Hubert H Girault, Solar energy conversion using dye-sensitised liquid|liquid interfaces, 29–32, Copyright (1999), with permission from Elsevier)

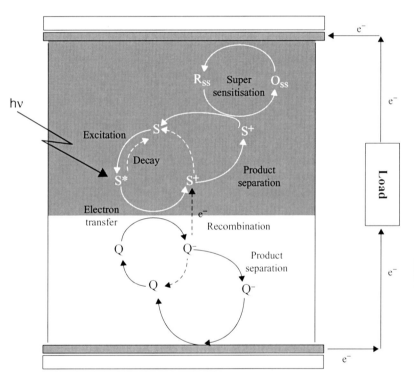

Photogalvanic Cells, Principles and Perspectives, Fig. 10 Schematic representation of the light energy conversion process based on the heterogeneous quenching of a sensitizer S by an electron acceptor Q at a liquid|liquid interface in the presence of a supersensitizer O$_{ss}$/R$_{ss}$ (Reprinted from Electrochemistry Communications, David J Fermín, Hong D Duong, Zhifeng Ding, Pierre F Brevet, Hubert H Girault, Solar energy conversion using dye-sensitised liquid|liquid interfaces, 29–32, Copyright (1999), with permission from Elsevier)

for efficient photoelectrochemistry; the formation of a barrier at the semiconductor | electrolyte interface opposes the dark reaction because of the depletion layer. This causes the need to use a semiconductor electrode for light-powered charging, but a third electrode made from a metal for the discharge process, Fig. 7 [41–43]. An example of this type of cell is the CdSe | polysulfide (S$_x^{2-}$) system, based on the cell diagram [42], CdSe | HS$^-$, HO$^-$, S$_x^{2-}$ | Membrane | HS$^-$, and HO$^-$ | SnS | Sn. Under illumination, holes (h$^+$) in the n-type CdSe electrode are able to oxidize HS$^-$ to sulfur:

$$HS^- + HO^- - 2h^+ \longrightarrow S + H_2O \quad (41)$$

The sulfur engages in further reaction within the electrolyte to form polysulfide complexes,

$$S + S_x^{2-} \longrightarrow S_{x+1}^{2-} \quad (42)$$

so that the bisulfide/sulfur equilibrium occurs (at open circuit) at the counter electrode,

$$S + H_2O + 2e^- \rightleftarrows HS^- + HO^- \quad (43)$$

and tin sulfide/sulfur equilibrium occurs (at open circuit) at the storage electrode,

$$SnS + H_2O + 2e^- \rightleftarrows Sn + HS^- + HO^- \quad (44)$$

Such cells can be readily extended to afford a four-electrode system containing an *n*-type anode and a *p*-type cathode for light-driven charging, with two metal electrodes for discharge [41], with the two compartments separated by a membrane. Consider the reaction

$$Fe^{3+} + Ce^{3+} \underset{dark}{\overset{light}{\rightleftarrows}} Ce^{4+} + Fe^{2+} \quad (45)$$

During charge, light-generated holes in the *n*-type semiconductor oxidize Ce^{3+} to Ce^{4+} with electrons passing through Ohmic contact with the semiconductor from the conduction band to the conduction band of a *p*-type semiconductor, where Fe^{3+} is reduced to Fe^{2+}, either in the dark or via a light-driven reaction. Under discharge, the continuum of electronic energy levels in the metal allows for the reverse processes

$$Ce^{4+} + e^- \longrightarrow Ce^{3+} \quad (46)$$

$$Fe^{2+} - e^- \longrightarrow Fe^{3+} \quad (47)$$

Several examples of successful cells of this type have been developed (Table 8), including all-solid-state devices [41–45].

A particularly exciting cell that has been developed is the Sharon-Schottky cell, Fig. 8 [41]. Here, the semiconductor light harvester is employed as a *membrane* between two redox electrolytes. One of these forms a Schottky barrier in front of the electrode and a second electrolyte is employed as an Ohmic contact. This allows complementary processes to occur at

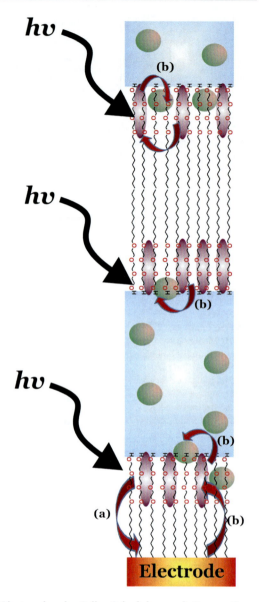

Photogalvanic Cells, Principles and Perspectives, Fig. 11 Cartoon illustrations of the development of a photo-Dember voltage (**a**) and a photogalvanic voltage (**b**) at the illuminated electrode. The aqueous subphase (23.1 Å) is *smaller* than the lipid bilayer (37.8 Å); the photoactive dye is indicated by the burgundy bananas, and redox species from the aqueous subphase are indicated by *green spheres*. The *arrows* representing *light* have been purposely drawn out of place to emphasize the photochemical electron-transfer nature of the process; the *red arrows* indicate the electron flow pathways (Reproduced with permission from reference [52]. Copyright (2012) Royal Society of Chemistry)

Photogalvanic Cells, Principles and Perspectives, Table 9 Performance of various galvanic cells constructed using the L_a phase of Brij 30/H_2O under 350 nm light (center band, bandwidth 30 nm, intensity of 1.8 mW cm^{-2}). Data presented were averaged over at least three different constructed cells (Copyright (2012) Royal Society of Chemistry)

Conditions	[a]E_{oc}/V	[b]j_{sc}/μA cm^{-2}	[c]E_m/V	[d]P_m/μW cm^{-2}	Fill factor/%	Maximum power conversion/%
Zn I L_α (0.7 mmol PMe, 98 μmol $ZnCl_2$, I ITO	1.11	17.2	0.31	2.73	14.2	0.15
Not degassed[e]						
Zn I L_α (0.1 mmol PMe, 0.1 mmol $ZnCl_2$, KCl[f]) I ITO	0.99	48.6	0.26	6.90	14.3	0.37
Not degassed[e]						
Zn I L_α (0.7 mmol PMe, 95 μmol $ZnCl_2$, KCl[f]) I ITO	0.99	118.7	0.67	20.81	17.6	1.11
Degassed[g]						
Zn I L_α (0.4 mmol PMe, 0.4 mmol PMeCl, 40 μmol $ZnCl_2$, KCl[f]) I ITO	1.19	172.3	0.61	34.4	16.3	1.83
Degassed[g]						
Al I L_α (0.5 mmol PMe, 3.0 mmol $AlCl_3$, KCl[f]) I ITO	0.85	264.8	0.26	18.5	8.21	0.99
Degassed[g]						

[a]Cell emf
[b]Current density of cell at short circuit
[c]Cell potential difference at maximum power
[d]Maximum power density
[e]*Not degassed* refers to the non-exclusion of molecular oxygen from the experiment
[f]*KCl* refers to the doping of the aqueous subphase with 0.1 M KCl
[g]*Degassed* refers to the rigorous exclusion of oxygen from the experiment

metal electrodes located a short distance from the semiconductor and essentially reduces the Ohmic loss within the device, since the semiconductor acts as an electrode for the majority carriers and a photoelectrode for the minority carriers. The advantage here is that the full bandgap of the semiconductor is used for energy conversion. An example of this system is [46] the cell diagram, Ag I Ag(CN)$_2^-$ I α-Fe$_2$O$_3$ (E_{CB} = −4.14 eV; E_{VB} = −6.14 eV) I Fe(CN)$_6^{4-}$, Fe(CN)$_6^{3-}$ I Pt in which Ohmic contact exits between the α-Fe$_2$O$_3$ and Ag(CN)$_2^-$, with a Schottky junction between α-Fe$_2$O$_3$ and ferri-/ferrocyanide. This cell generated 5.8 mA cm^{-2} at 0.9 V, with light-induced oxidation of ferrocyanide. It is important that, for efficient operation, the Fermi level of silver matches the conduction band energy of the semiconductor. The advantage of this system is that it is easy to fabricate and a large number of cells can be connected in series without using any conducting wire [41].

Future Directions

The use of the liquid I liquid interface as a means for charge separation after photochemically induced interfacial redox reaction is an exciting and relatively recent development for exploitation in PG cell systems [47–50], not least as a result of the similarity between electron transfer at the liquid I liquid interface and semiconductor I electrolyte junctions [22].

Fermín and Girault [51] developed the PG cell depicted in Fig. 9, employing Ru(bpy)$_3^{2+}$ as a supersensitizer (S) between a nonpolarized water | 1,2-dichloroethane interface, with a proposed reaction scheme illustrated in Fig. 10. The cell was found to provide an electron-to-photon ratio on the order of 0.1 % under green light (543 nm) and steady-state conditions. Analysis of the polarization and power characteristics for the experimental cell gave 0.7 nW cm^{-2} at 25 mV, which under linear extrapolation permits solar to electrical energy conversion on the order of 0.05 % under AM 1.0 illumination.

Our laboratory [52] extended this approach for the use of lyotropic liquid crystals ("liquid nanotechnology") doped with photoredox active materials in considering the cell Zn | Zn^{2+}(aq)/ N-methylphenothiazine (Brij 30) | ITO, where the dark reaction involves the use of a sacrificial electrode to keep the open circuit voltage high. The concept is illustrated in Fig. 11 and exploits the nanometric space and thus diffusion lengths available. Typical cell performance data are given in Table 9, illustrating that under violet (350 nm) light, approximately 2 % power conversion efficiency is possible with a 15 % fill factor. The cell was observed to behave as an electrochemical capacitor (voltage efficiency $\sim$85 %; power efficiency $\sim$80 %), with estimated maximum energy density of $\sim$1 W h kg^{-1} at a power density of $\sim$1 kW kg^{-1}, clearly demonstrating this approach to be of potential pragmatic use.

Conclusion

While it can be seen from Table 7 that the conversion efficiencies for many formulated PG cells are some way off the $\sim$18 % optimum theorized by Albery and Archer [17], the ability to vary a number of parameters within the system, and the relative ease of doing so, affords a customizable property that the PV cell cannot compete with. Providing a compromise can be made regarding the effects of the system photochemistry, mass transport and the homogeneous kinetics, illumination intensity, and species concentration, PG cells hold promise for a commercially and environmentally viable alternative to the common PV devices (e.g., solar panels) used today.

Lord Porter said that "I have no doubt that we will be successful in harnessing the Sun's energy... If sunbeams were weapons of war, we would have had solar energy centuries ago" [53]. Although unfortunate, it is perhaps human nature that matters concerning society are disregarded until they become distinctly personal – with the realization in the last five decades that Earth's finite energy resources are rapidly diminishing, the subject of discovering renewable, viable energy sources has become an increasingly important research area, the success of which can only increase as levels of understanding and modern technological advancement continue to grow.

Acknowledgements This work has been financed through EPSRC (grant number EP/G020833/1).

Cross-References

▶ Conductivity of Electrolytes
▶ Dye-Sensitization
▶ Dye-Sensitized Electrode, Photoanode
▶ Electrical Double-Layer Capacitors (EDLC)
▶ Electrode
▶ Membrane Processes, Electrodialysis
▶ Membrane Technology
▶ Photocatalyst
▶ Photoelectrochemistry, Fundamentals and Applications
▶ Polymer Electrolyte Fuel Cells, Mass Transport
▶ Quantum Dot Sensitization
▶ Semiconductor Electrode
▶ Solid State Dye-Sensitized Solar Cell

References

1. Archer MD (1975) Electrochemical aspects of solar-energy conversion. J Appl Electrochem 5:17
2. Jana AK (2000) Solar cells based on dyes. J Photochem Photobiol A 132:1
3. Becquerel E (1839) Recherches sur les effets de la radiation chimique de la lumiére solaire au moyen des courants électriques. C R Acad Sci Paris 9:145

4. Becquerel E (1839) Mémoire sur les effets électriques produits sous l'influence des rayons solaires. C R Acad Sci Paris 9:561
5. Genwa KR, Genwa M (2008) Photogalvanic cell: A new approach for green and sustainable chemistry. Sol Energy Mater Sol Cells 92:522
6. Mountz JM, Tien HT (1978) Photogalvanovoltaic cell. Sol Energy 21:291
7. Tien HT, Mountz JM (1978) Photo-galvano-voltaic cell - new approach to use of solar-energy. Int J Energy Res 2:197
8. Tien HT, Higgen J, Mountz JM (1979) Solar energy: chemical conversion and storage. Humana Press, Clifton, p 203
9. Rabinowitch E (1940) The photogalvanic effect I. The photochemical properties of the thionine-iron system. J Chem Phys 8:551
10. Rabinowitch E (1940) The photogalvanic effect II. The photogalvanic properties of the thionine-iron system. J Chem Phys 8:560
11. Gangotri KM, Meena RC, Meena R (1999) Use of micelles in photogalvanic cells for solar energy conversion and storage: cetyl trimethyl ammonium bromide-glucose-toluidine blue system. Photochem Photobiol A 123:93
12. Albery WJ, Foulds AW (1979) Photogalvanic cells. J Photochem 10:41
13. Albery WJ (1982) Development of photogalvanic cells for solar-energy conversion. Acc Chem Res 15:142
14. Rideal EK, Williams DC (1925) The action of light on the ferrous ferric iodine iodide equilibrium. J Chem Soc 127:258
15. Gangotri P, Gangotri KM (2009) Studies of the micellar effect on photogalvanics: solar energy conversion and storage in EDTA-safranine O-Tween-80 system. Energy Fuel 23:2767
16. Albery WJ, Archer MD (1976) Photogalvanic cells .1. potential of zero current. Electrochim Acta 21:1155
17. Albery WJ, Archer MD (1977) Optimum efficiency of photogalvanic cells for solar-energy conversion. Nature 270:399
18. Albery WJ, Archer MD (1977) Photogalvanic cells .2. current-voltage and power characteristics. J Electrochem Soc 124:688
19. Albery WJ, Archer MD (1978) Photogalvanic cells .3. maximum power obtainable from a thin-layer photogalvanic concentration cell with identical electrodes. J Electroanal Chem 86:1
20. Albery WJ, Archer MD (1978) Photogalvanic cells .4. maximum power from a thin-layer cell with differential electrode-kinetics. J Electroanal Chem 86:19
21. Marcus RA (1965) On the theory of electron-transfer reactions. VI. Unified treatment of homogeneous and electrode reactions. J Chem Phys 43:679
22. Smith BB, Halley JW, Nozik AJ (1996) On the Marcus model of electron transfer at immiscible liquid interfaces and its application to the semiconductor liquid interface. Chem Phys 205:245
23. Royea WJ, Fajardo AM, Lewis NS (1997) Fermi golden rule approach to evaluating outer-sphere electron-transfer rate constants at semiconductor/liquid interfaces. J Phys Chem B 101:11152
24. Hatchard CG, Parker CA (1961) The photoreduction of thionine by ferrous sulphate. Trans Faraday Soc 57:1093
25. Ferreira MIC, Harriman A (1977) Photoredox reactions of thionine. J Chem Soc Faraday Trans 1 73:1085
26. Hall DE, Eckert JA, Lichtin NN, Wildes PD (1976) Multilayer iron-thionine photogalvanic cell. J Electrochem Soc 123:1705
27. Rohatgi-Mukherjee KK, Choudhary R, Bhowmik BB (1985) Molecular interaction of phenosafranin with surfactants and its photogalvanic effect. J Colloid Interface Sci 106:45
28. Tsubomura H, Shimoura Y, Fujiwara S (1979) Chemical processes and electric-power in photogalvanic cells containing reversible or irreversible reducing agents. J Phys Chem 83:210
29. Dube S, Sharma SL, Ameta SC (1997) Photogalvanic effect in azur B-NTA system. Energy Convers Manage 38:101
30. Gangotri KM, Regar OP (1997) Use of azine dye as a photosensitizer in solar cells: Different reductants - Safranine systems. Int J Energy Res 21:1345
31. Gangotri KM, Meena RC (2001) Use of reductant and photosensitizer in photogalvanic cells for solar energy conversion and storage: oxalic acid-methylene blue system. J Photochem Photobiol A 141:175
32. Genwa KR, Chouhan A (2006) Role of heterocyclic dye (Azur A) as a photosensitizer in photogalvanic cell for solar energy conversion and storage: NaLS-ascorbic acid system. Sol Energy 80:1213
33. Ameta SC, Punjabi PB, Vardia J, Madhwani S, Chaudhary S (2006) Use of bromophenol red-EDTA system for generation of electricity in a photogalvanic cell. J Power Sources 159:747
34. Genwa KR, Kumar A, Sonel A (2009) Photogalvanic solar energy conversion: Study with photosensitizers toluidine blue and malachite green in presence of NaLS. Appl Energy 86:1431
35. Genwa KR, Khatri NC (2009) Comparative study of photosensitizing dyes in photogalvanic cells for solar energy conversion and storage: Brij-35-Diethylenetriamine Pentaacetic Acid (DTPA) system. Energy Fuel 23:1024
36. Lal C (2007) Use of mixed dyes in a photogalvanic cell for solar energy conversion and storage: EDTA-thionine-azur-B system. J Power Sources 164:926
37. Dube S, Sharma SL (1994) Studies in photochemical conversion of solar-energy - simultaneous use of 2 dyes with mannitol in photogalvanic cell. Energy Convers Manage 35:709

38. Jana AK, Bhowmik BB (1999) Enhancement in power output of solar cells consisting of mixed dyes. J Photochem Photobiol A 122:53
39. Fendler JH, Fendler EJ (1975) Catalysis in micellar and macromolecular systems. Academic Press, New York
40. Atwood D, Florence AT (1983) Surfactant systems. Chapman and Hall, New York
41. Sharon M, Veluchamy P, Natarajan C, Kumar D (1991) Solar rechargeable battery - principle and materials. Electrochim Acta 36:1107
42. Licht S (2002) Photoelectrochemical solar energy storage cells. In: Licht S, Bard AJ, Stratmann M (eds) Encyclopædia of electrochemistry, vol 6. Wiley-VCH, Weinheim, p 317
43. Licht S, Hodes G (2008) Photoelectrochemical storage cells. In: Archer MD, Nozik AJ (eds) Nanostructured and photoelectrochemical systems for solar photon conversion. Imperial College Press, London, p 591
44. Hada H, Takaoka K, Saikawa M, Yonezawa Y (1981) Energy-conversion and storage in solid-state photogalvanic cells. Bull Chem Soc Jpn 54:1640
45. Yonezawa Y, Okai M, Ishino M, Hada H (1983) A photochemical storage battery with an n-gap photo-electrode. Bull Chem Soc Jpn 56:2873
46. Sharon M, Rao GR (1986) Photoelectrochemical cell with liquid (ohmic) semiconductor liquid (schottky-barrier) system. Indian J Chem 25A:170
47. Volkov AG (1995) Energy-conversion at liquid/liquid interfaces - artificial photosynthetic systems. Electrochim Acta 40:2849
48. Mareček V, Armond AHD, Armond MKD (1989) Photochemical electron transfer in liquid/liquid solvent systems. J Am Chem Soc 111:2561
49. Brown AR, Yellowlees LJ, Girault HH (1993) Photoinitiated electron-transfer reactions across the interface between two immiscible electrolyte-solutions. J Chem Soc Faraday Trans 89:207
50. Fedoseeva M, Grilj J, Kel O, Koch M, Letrun R, Markovic V, Petkova I, Richert S, Rosspeintner A, Sherin PS, Villamaina D, Lang B, Vauthey E (2011) Photoinduced electron transfer reactions: From the elucidation of old problems in bulk solutions towards the exploration of interfaces. Chimia 65:350
51. Fermín DJ, Duong HD, Ding Z, Brevet PF, Girault HH (1999) Solar energy conversion using dye-sensitised liquid vertical bar liquid interfaces. Electrochem Commun 1:29
52. Halls JE, Wadhawan JD (2012) Photogalvanic cells based on lyotropic nanosystems: towards the use of liquid nanotechnology for personalised energy sources. Energy Environ Sci 5:6541
53. Rhodes BK, Odell R (1992) A dictionary of environmental quotations. John Hopkins University Press, Baltimore, p 265

Photography, Silver Halides

Tadaaki Tani
Society of Photography and Imaging of Japan, Tokyo, Japan

Introduction and Characteristics

Photography with silver halide (AgX) is based on the photodecomposition reaction of AgX under illumination and was invented by L. J. M. Daguerre in 1839. The sensitivity of a photographic material with AgX depends on the absorption spectrum of AgX, which extends from ultraviolet to blue region. Dye sensitization, which was invented by H. Vogel in 1873, makes photographic materials with AgX sensitive to light that sensitizing dyes on its surface absorb and thus to light in green, red, and near-infrared region [1]. Photographic materials are diversified to meet various uses and include color negative films, color reversal films, color papers, photographic films for medical examination, and printing.

Photographic materials are manufactured by coating aqueous gelatin suspensions of AgX grains and additives (i.e., photographic emulsions) on substrates and record images of light and high-energy irradiation through photographic processes of exposure, development, and fix. On exposure, an electron appears in the conduction band of an AgX grain as a result of the light absorption of the grain or a sensitizing dye molecule on it. An electron in the conduction band migrates and is captured by one of shallow traps on the grain surface (i.e., the electronic processes for photographic sensitivity). An electron at a trap attracts an interstitial silver ion and reacts with it to form an Ag atom (i.e., the ionic process). The repetition of the electronic and ionic processes at the same site leads to the formation of an Ag cluster and becomes to be a latent image center, which is composed of four or more Ag atoms and can initiate photographic development [1].

A latent image center that acts as a deep electron trap on an AgX grain receives an electron from a developing agent molecule in a developer and then attracts and reacts with an interstitial silver ion to add an Ag atom to a latent image center during the development process. The repetition of these processes results in the conversion of an AgX grain with a latent image center on it to an Ag grain. Silver halide grains that remain after the development process are dissolved during the fix process. Since only the AgX grains with latent image centers on their surfaces are converted to black Ag grains, a latent image is developed to give a negative black-and-white image after these processes. For color photography, a developing agent molecule gives an electron to a latent image center and becomes to be an oxidation product, which then reacts with a coupler molecule in a photographic material to form a dye molecule for a color image. Then, Ag and AgX grains are removed after color development [1].

Reactions associated with photographic development are based on electrochemistry [1]. The redox potential of a developer with respect to its silver potential is the driving force for photographic development and used for the examination of the development process, which it causes. The oxidation potential of a compound is used to examine its ability as a developing agent on the basis of the relationship between oxidation potentials and rates of development among a series of compounds [2]. Although the oxidation potential of a developing agent molecule is based on the oxidation reaction, in which two electrons are involved in the oxidation of one molecule, a latent image center is too small to receive two electrons at the same time. The potentials for one-electron oxidation of developing agents have been evaluated by use of their semiquinone formation constants and are in closer relationship with their development rates than those for two-electron oxidation as were measured by polarography [3].

Since a visible light image can be reproduced by the combination of three primary lights (i.e., blue, green, red), color films and papers are composed of a layered photo-sensor, which is

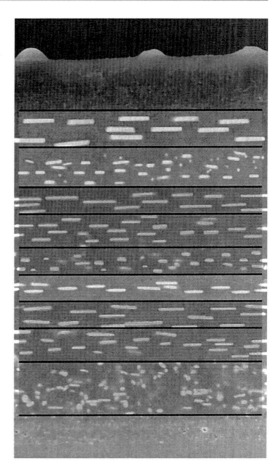

Photography, Silver Halides, Fig. 1 A scanning electron micrograph of the section of a layered photo-sensor in a color negative film, where white lines and spots are the sections of AgX grains. The photo-sensor is composed of 14 layers with different functions and $\sim$20 μm thick in total

sensitive to light in blue, green, and red regions separately owing to the light absorption of AgX grains and/or sensitizing dyes on the grains. A color negative film is composed of a layered photo-sensor coated on a TAC film base with thickness of $\sim$100 μm. Fig. 1 shows a scanning electron micrograph of the section of a layered photo-sensor in a color negative film, where white lines and spots are the sections of AgX grains. The photo-sensor in this figure is composed of 14 layers with different functions and $\sim$20 μm in total. Major layers are blue-, green-, and red-sensitive, while each major layer is

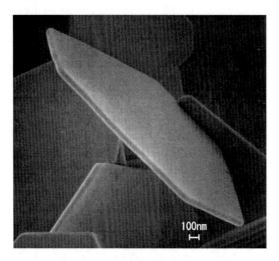

Photography, Silver Halides, Fig. 2 A scanning electron micrograph of tabular AgX grains developed for highly sensitive layers in a color negative film

composed of sub-layers with high, medium, and low sensitivities. Namely, a color negative film is designed to capture a light image in whole visible region with wide dynamic range of exposure [1].

As described above, photographic materials as represented by a color negative film are finely constructed hybrid systems composed of inorganic materials such as AgX grains that are well designed in nanoscale and more than 100 kinds of functional organic compounds. For example, monodispersed ultrathin tabular AgX grains as shown in Fig. 2 are designed from viewpoints of their structure, physical property, and photographic sensitivity, produced on a large scale and arranged with their main surfaces in parallel with the surface of a film base as seen in Fig. 1. The photo-sensor with 14 different layers in Fig. 1 is precisely coated all together on a large scale and at high speed. The coefficient of variation in thickness among arbitrary points in each layer is less than 2 %.

For the development of color films, extensive studies have been made on dye sensitization. After the electron transfer mechanism and the energy transfer one were proposed for it and examined extensively for many years from various viewpoints, the former has been accepted. Namely, a sensitizing dye molecule on an AgX grain is excited by an incident photon and injects an excited electron into the conduction band of the grain. An injected electron takes part in the formation of a latent image center on the grain. This model demands that the LUMO level of a sensitizing dye molecule should be higher than the bottom of the conduction band of AgX [1].

Thus, the height of LUMO of a dye with respect to that of the bottom of the conduction band of AgX together with the height of HOMO of a dye with respect to that of the top of the valence band of AgX is one of the most important properties for understanding and designing the dye sensitization of photographic materials and evaluated by organic electrochemistry in addition to ultraviolet photoelectron spectroscopy and molecular orbital method. The reduction and oxidation potentials (E_R and E_{OX}, respectively) are related to the heights of LUMO and HOMO (ε_{LU} and ε_{HO}, respectively), respectively, as follows [1].

$$\varepsilon_{LU} = E_R + C \qquad (1)$$

$$\varepsilon_{HO} = E_{OX} + C' \qquad (2)$$

where C and C' are constants. The values of E_R and E_{OX} of dyes could be measured by polarography, cyclic voltammetry, and phase-selective second-harmonic (PSSH) voltammetry. The values of E_R and E_{OX}, which were measured by means of PSSH voltammetry, could be free from the errors owing to the secondary reactions of reduced and oxidized dyes formed as results of electrode reactions [4]. Tremendous number of sensitizing dyes have been synthesized and subjected to the measurement of polarography and then PSSH voltammetry for precise evaluation of their ε_{LU} and ε_{HO}.

The electron transfer mechanism has enhanced the achievement of dye sensitization with high efficiency and the accumulation of the knowledge on the electron transfer process. The efficiency and rate of the electron transfer in dye sensitization have been analyzed in the framework of Marcus theory with respect to the value of E_R of a dye that is in linear relationship with the electronic energy gap between ε_{LU} of the dye and the bottom of the conduction band of AgX according to Eq. 1 [5]. The degree of the recombination between an injected electron and a positive hole remaining at HOMO of a dye has been analyzed with respect to the values of E_{OX} of the dye that is in linear relationship with ε_{HO} according to Eq. 2 [1]. The electron transfer mechanism for dye sensitization has been applied to dye-sensitized solar cells on the basis of electrochemistry [6]. J-aggregates with sharp and intense absorption bands have been indispensable for color films and have attracted wide interests owing to their unique characteristics [1]. The vacuum level shift as observed for the first time at interfaces between AgX and sensitizing dyes has become to be important knowledge for the electronic structure at interfaces between substrates and organic semiconductors in various devices [1].

Recently, digital still cameras with solid state devices (i.e., CCD and CMOS) as photo-sensors have achieved such sensitivity and image quality as to meet the demand of most consumers and overwhelmed color films owing to their convenience. Other photographic films have been similarly overwhelmed by electronic imaging technologies.

However, AgX photographic materials are characterized by high image quality with large frame area, natural image reproduction, durability for a very long period, and three-dimentional imaging with high resolution, being used in the future as color films with natural image reproduction, black-and-white films with high image information and durability for achives, and new nuclear films that are composed of nanoparticles of AgX and being developed for the detection of dark matters [1].

Cross-References

▶ Dye-Sensitization
▶ Polarography
▶ Solid State Dye-Sensitized Solar Cell

References

1. Tani T (2011) Photographic science advances in nanoparticles, J-aggregates, dye sensitization, and organic materials. Oxford University Press, Oxford
2. Bent RL, Dessloch JC, Duennebier FC, Fassett DW, Glass DB, James TH, Julian DB, Ruby WR, Smell JM, Sterner JH, Thirtle JR, Vittum PW, Weissberger A (1951) Chemical constitution, electrochemical, photographic and allergenic properties of p-amino-N-dialkylamines. J Am Chem Soc 73:3100–3125
3. Tong LKJ, Bishop CA, Glesmann MC (1964) Correlation between oxidation potentials and development rates of some hydroquinones and p-phenylenediamines. Photogr Sci Eng 8:326–328
4. Lenhard J (1986) Measurement of reversible electrode potentials for cyanine dyes by use of phase-selective second-harmonic AC voltammetry. J Imag Sci 30:27–35
5. Tani T, Suzumoto T, Ohzeki K (1990) Energy gap dependence of efficiency of photoinduced electron transfer from cyanine dyes to silver bromide microcrystals in spectral sensitization. J Phys Chem 94:1298–1301
6. Graetzel M (2007) Nanocrystalline injection solar cells. In: Poortmans J, Arkhipov V (eds) Thin film solar cells: fabrication, characterization and application. Wiley, West Sussex

Photolysis of Water

Akihiko Kudo
Tokyo University of Science, Tokyo, Japan

Introduction

Photocatalytic and photoelectrochemical water splitting are important from the viewpoint of energy and environmental issues in a global level because it enables an ideal hydrogen production from water using a renewable energy such as a solar energy. Once solar hydrogen is obtained, carbon dioxide can be converted to various organic compounds by the reaction with the hydrogen. Artificial photosynthesis is achieved through the hydrogenation of carbon dioxide using the solar hydrogen. The artificial photosynthesis is an attractive reaction as an ultimate green sustainable chemistry and will contribute to solving energy and environmental issues resulting in bringing an energy revolution.

The Honda-Fujishima effect of water splitting using a TiO$_2$ electrode was reported in the early stage of 1970s. When TiO$_2$ is irradiated with UV light, electrons and holes are generated as shown in Fig. 1 [1]. The photogenerated electrons reduce water to form H$_2$ on a Pt counter electrode while holes oxidize water to form O$_2$ on the TiO$_2$ electrode with some external bias by a power supply or pH difference between a catholyte and an anolyte. New photoelectrode and powdered photocatalyst materials for water splitting have been discovered one after another since this pioneer work.

Bases of Photocatalytic Water Splitting

Photoelectrochemical Cells and Powdered Photocatalysts

There are two methods for water splitting using photon energy as shown in Fig. 2. There are advantageous and disadvantageous points for each method. In photoelectrochemical cells represented by Honda-Fujishima effect shown in Fig. 1, n- and p-type photoelectrode materials can be use as an anode and cathode, respectively. Major advantageous point of this system is that H$_2$ can be obtained separately from O$_2$. Even if a material does not possess the suitable potential that is determined with the electronic band structure for water splitting, the water splitting may be possible with assistance of an external bias. Moreover, charge separation to suppress recombination between photogenerated electrons and holes is assisted by the external bias. However, materials for the photoelectrochemical cells are limited, because electrodes should be fabricated and possess an electric conductivity. In powdered photocatalysts, various kinds of materials can be used. The system is simple, because photocatalyst powder is just dispersed or dipped in water. So, it is advantageous for applying to a large-scale process.

Processes in Powdered Photocatalysts

Many heterogeneous photocatalysts have semiconductor properties. Figure 3 shows main processes in a photocatalytic reaction using a powdered system.

The first step is absorption of photons to form electron-hole pairs. Semiconductors have the band structure in which the conduction band is separated from the valence band by a band gap with a suitable width. When the energy of incident light is larger than that of a band gap, electrons and holes are generated in the conduction and valence bands, respectively.

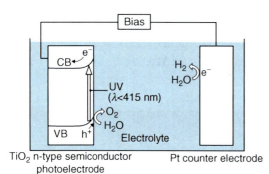

Photolysis of Water, Fig. 1 Honda-Fujishima effect for water splitting using an n-type TiO$_2$ photoelectrode

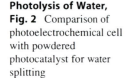

Photolysis of Water, Fig. 2 Comparison of photoelectrochemical cell with powdered photocatalyst for water splitting

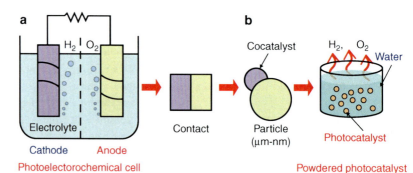

Photolysis of Water, Fig. 3 Processes of photocatalytic water splitting

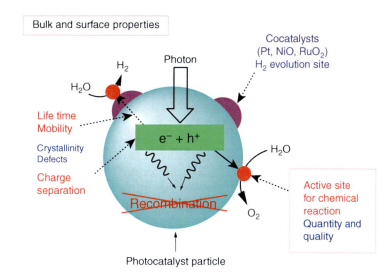

The photogenerated electrons and holes cause redox reactions similarly to electrolysis. Water molecules are reduced by the electrons to form H$_2$ and are oxidized by the holes to form O$_2$ for overall water splitting. Important points in the semiconductor photocatalyst materials are a width of the band gap and levels of the conduction and valence bands. The bottom level of the conduction band has to be more negative than the redox potential of H$^+$/H$_2$ (0 V vs. NHE), while the top level of the valence band be more positive than the redox potential of O$_2$/H$_2$O (1.23 V). Therefore, the theoretical minimum band gap for water splitting is 1.23 eV that corresponds to light of about 1,100 nm.

The second step consists of charge separation and migration of photogenerated carriers. Crystal structure, crystallinity, and particle size strongly affect the step. The higher the crystalline quality is, the smaller the amount of defects is. The defects operate as trapping and recombination centers between photogenerated electrons and holes, resulting in a decrease in the photocatalytic activity.

The final step involves the surface chemical reactions. The important points for this step are surface character (active sites) and quantity (surface area). Even if the photogenerated electrons and holes possess thermodynamically sufficient potentials for water splitting, they will have to recombine with each other if the active sites for redox reactions do not exist on the surface. Cocatalysts such as Pt, NiO, RuO$_2$, and Cr-Rh oxide are usually loaded to introduce active sites for H$_2$ evolution because the conduction band levels of many oxide photocatalysts are not high enough to reduce water to produce H$_2$ without catalytic assistance.

Water Splitting and Sacrificial H$_2$ or O$_2$ Evolution Using Powdered Photocatalysts

"Water splitting" means to split H$_2$O simultaneously giving H$_2$ and O$_2$ in a 2:1 ratio. On the other hand, there are sacrificial H$_2$ and O$_2$ evolution reactions as shown in Fig. 4. When the photocatalytic reaction is carried out in an aqueous solution including a reducing reagent, in other words, electron donors or hole scavengers, such as alcohol and a sulfide ion, photogenerated holes irreversibly oxidize the reducing reagent instead of water. It enriches electrons in a photocatalyst, and an H$_2$ evolution reaction is enhanced. This reaction will be meaningful for realistic hydrogen production if biomass and abundant compounds in nature and industries are used as the reducing reagents. On the other hand, photogenerated electrons in the conduction band are consumed by oxidizing reagents (electron acceptors or electron scavengers) such as Ag$^+$ and Fe^{3+} resulting in that an O$_2$

Photolysis of Water, Fig. 4 Sacrificial H$_2$ and O$_2$ evolution over photocatalysts

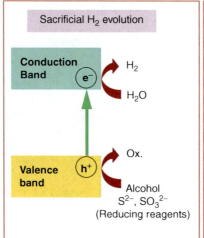
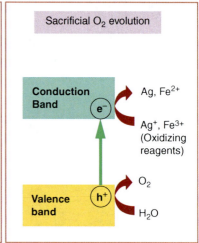

evolution reaction is enhanced. These reactions using sacrificial reagents are studied to evaluate if a certain photocatalyst satisfies the thermodynamic and kinetic potentials for H$_2$ and O$_2$ evolution. These reactions are regarded as half reactions of water splitting and are often employed as test reactions of photocatalytic H$_2$ or O$_2$ evolution. Even if a photocatalyst is active for these half reactions, the results do not guarantee a photocatalyst to be active for overall water splitting into H$_2$ and O$_2$ in the absence of sacrificial reagents.

Water splitting is an uphill reaction in which a light energy is converted to a chemical energy as shown in Fig. 5. In contrast, the sacrificial H$_2$ evolution is usually a downhill reaction. In this sense, the term of "water splitting" in the absence of sacrificial reagents should be distinguishably used toward H$_2$ or O$_2$ evolution from aqueous solutions in the presence of sacrificial reagents.

Water Splitting by Powdered Photocatalysts with Wide Band Gaps

Many metal oxide photocatalysts with wide band gaps can split water into H$_2$ and O$_2$ efficiently in a stoichiometric ratio under UV light irradiation

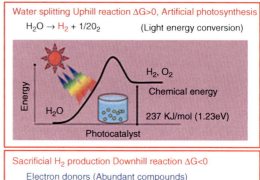

Photolysis of Water, Fig. 5 Photocatalytic water splitting and sacrificial H$_2$ evolution

as shown in Table 1 [2, 3]. These photocatalysts consist of metal cations with d^0 or d^{10} configuration. Representative photocatalysts with wide band gaps that respond to only UV are TiO$_2$ and SrTiO$_3$. ZrO$_2$, Ta$_2$O$_5$, and tantalates such as NaTaO$_3$ are also active for photocatalytic water splitting. K$_4$Nb$_6$O$_{17}$ and K$_2$La$_2$Ti$_3$O$_{10}$ with layered structures are unique photocatalysts. Ge$_3$N$_4$ and GaN: Zn are non-oxide photocatalysts [4]. NiO/NaTaO$_3$: La and RhyCr$_{2-y}$O$_3$/Ga$_2$O$_3$:

Photolysis of Water 1585

Photolysis of Water, Table 1 Wide band gap oxide photocatalysts consisting of metal cations with d^0 or d^{10} configuration for water splitting

Photocatalyst	Band gap/eV	Cocatalyst	Year
TiO_2	3.2	Rh	1985
$CaTiO_3$	3.5	NiO_x	2002
$SrTiO_3$	3.2	NiO_x	1980
$Rb_2La_2Ti_3O_{10}$	3.4–3.5	NiO_x	1997
$La_2Ti_2O_7$	3.8	NiO_x	1999
$La_2Ti_2O_7$:Ba		NiO_x	2005
$KLaZr_{0.3}Ti_{0.7}O_4$	3.91	NiO_x	2003
$La_4CaTi_5O_{17}$	3.8	NiO_x	1999
$KTiNbO_5$	3.6	NiO_x	1999
$Na_2Ti_6O_{13}$		RuO_2	1990
$Gd_2Ti_2O_7$	3.5	NiO_x	2006
$Y_2Ti_2O_7$	3.5	NiO_x	2004
ZrO_2	5	None	1993
$K_4Nb_6O_{17}$	3.4	NiO_x	1986
$Ca_2Nb_2O_7$	4.3	NiO_x	1999
$Sr_2Nb_2O_7$	4	NiO_x	1999
$Ba_5Nb_4O_{15}$	3.85	NiO_x	2006
$ZnNb_2O_6$	4	NiO_x	1999
$Cs_2Nb_4O_{11}$	3.7	NiO_x	2005
La_3NbO_7	3.9	NiO_x	2004
Ta_2O_5	4	NiO_x	1994
$K_3Ta_3Si_2O_{13}$	4.1	NiO	1997
$K_3Ta_3B_2O_{12}$	4	None	2006
$LiTaO_3$	4.7	None	1998
$NaTaO_3$	4	NiO	1998
$KTaO_3$	3.6	Ni	1996
$AgTaO_3$	3.4	NiO_x	2002
$NaTaO_3$:La	4.1	NiO	2000
$SrTa_2O_6$	4.4	NiO	1999
$Sr_2Ta_2O_7$	4.6	NiO	2000
$KBa_2Ta_3O_{10}$	3.5	NiO_x	1999
$PbWO_4$	3.9	RuO_2	2004
$RbWTaO_6$	3.8	NiO_x	2004
CeO_2:Sr		RuO_2	2007
$NaInO_2$	3.9	RuO_2	2003
$CaIn_2O_4$		RuO_2	2001
$SrIn_2O_4$	3.6	RuO_2	2001
$CaSb_2O_6$	3.6	RuO_2	2002
Ga_2O_3:Zn	4.6	Ni	2008

Zn photocatalysts have recently proven that highly efficient water splitting is possible using a powdered photocatalyst under UV light irradiation.

Water Splitting by Powdered Photocatalysts with Visible-Light Response

The goal of this research field is to produce hydrogen by usage of solar energy, the so-called solar hydrogen production. In order to utilize solar energy efficiently, it is indispensable to develop visible-light-driven photocatalysts [3].

There are two types of photocatalyst systems accompanied with one photon process and Z-schematic two-photon process for water splitting under visible-light irradiation as shown in Fig. 6. Some oxynitrides are active for water splitting as photocatalysts accompanied with one photon process [4]. Two-photon systems, as seen in photosynthesis by green plants (Z-scheme), are another way to achieve overall water splitting [5]. The Z-scheme is composed of an H_2-evolving photocatalyst, an O_2-evolving photocatalyst, and an electron mediator as shown in Table 2. Some Z-scheme photocatalysts work without electron mediators. Photocatalysts that are active only for half reactions of water splitting can be employed for the construction of the Z-scheme, that is, the merit of the Z-scheme.

Water Splitting by Photoelectrochemical Cells

It is important to see the stability of photocurrent and to determine the amounts of evolved H_2 and O_2 for photoelectrochemical water splitting, because we cannot always guarantee that obtained photocurrent is due to water splitting. Even if photocurrent is observed, it is sometimes due to some redox reactions rather than water reduction and oxidation. As mentioned above, an external bias can be applied to enhance photoelectrochemical water splitting. However, the applied external bias versus not a reference electrode but a counter electrode must be smaller than 1.23 V that was a theoretical voltage required for electrolysis of water, if one thinks of light energy conversion. Of course, non-bias is an ideal condition. WO_3, Fe_2O_3, and $BiVO_4$ have been extensively studied as photoanodes for O_2

Photolysis of Water, Fig. 6 Water splitting using single and Z-schematic photocatalysts

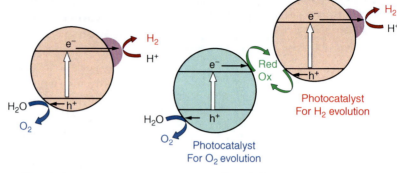

Photolysis of Water, Table 2 Z-scheme type photocatalyst systems working under VIS irradiation

H_2-evolving photocatalysts	H_2-evolving photocatalysts	Electron mediator
Pt/SrTiO$_3$:Cr,Ta	Pt/WO$_3$	IO$_3^-$/I$^-$
Pt/TaON	Pt/WO$_3$	IO$_3^-$/I$^-$
Pt/TaON	RuO$_2$/TaON	IO$_3^-$/I$^-$
Pt/CaTaO$_2$N	Pt/WO$_3$	IO$_3^-$/I$^-$
Pt/BaTaO$_2$N	Pt/WO$_3$	IO$_3^-$/I$^-$
Pt/ZrO$_2$/TaON	Pt/WO$_3$	IO$_3^-$/I$^-$
Coumarin/K$_4$Nb$_6$O$_{17}$	IrO$_2$-Pt/WO$_3$	IO$_3^-$/I$^-$
Pt/ZrO$_2$/TaON	Ir/R-TiO$_2$/Ta$_3$N$_5$	IO$_3^-$/I$^-$
Pt/ZrO$_2$/TaON	RuO$_2$/TaON	IO$_3^-$/I$^-$
Pt/SrTiO$_3$:Rh	BiVO$_4$	Fe$^{3+/2+}$
Pt/SrTiO$_3$:Rh	Bi$_2$MoO$_6$	Fe$^{3+/2+}$
Pt/SrTiO$_3$:Rh	WO$_3$	Fe$^{3+/2+}$
Ru/SrTiO$_3$:Rh	BiVO$_4$	Fe$^{3+/2+}$
Ru/SrTiO$_3$:Rh	PbWO$_4$:Cr	Fe$^{3+/2+}$
Ru/SrTiO$_3$:Rh	BiVO$_4$	[Co(bpy)$_3$]$^{3+/2+}$
Ru/BaTa$_2$O$_6$:Ir,La	BiVO$_4$	[Co(bpy)$_3$]$^{3+/2+}$
Pt/SrTiO$_3$:Rh	BiVO$_4$	None
Ru/SrTiO$_3$:Rh	BiVO$_4$	None
Ru/SrTiO$_3$:Rh	AgNO$_3$	None
Ru/SrTiO$_3$:Rh	Bi$_2$MoO$_6$	None
Ru/SrTiO$_3$:Rh	TiO$_2$:Cr,Sb	None
Ru/SrTiO$_3$:Rh	TiO$_2$:Rh,Sb	None
Ru/SrTiO$_3$:Rh	WO$_3$	None

evolution to combine an ordinary metal electrodes, dye sensitized solar cells, and p-type semiconductor photoelectrodes for H_2 evolution.

Summary

At the present stage, the materials with which solar water splitting can be carried out are limited and the efficiencies are not satisfying us. Various kinds of efficiencies have been reported to evaluate the performances of photoelectrochemical cells and powdered photocatalysts. A quantum efficiency is scientifically important. Moreover, energy conversion efficiency of solar to hydrogen (STH) will be a standard value to evaluate the efficiency of an artificial photosynthesis system. It is indicated on roadmaps in the world that the construction of solar fuel production systems by artificial photosynthesis is a conclusive solution for energy and environmental issues. Solar water splitting is the core reaction of the artificial photosynthesis. If we can produce abundant hydrogen by the artificial photosynthesis, we can use it for not only a clean energy but also a raw material in chemical industries. CO_2 and N_2 can be converted to useful compounds using the solar hydrogen. Of course, there are several processes to utilize solar energy. There are merits and demerits for each process. Characteristic point of photocatalytic water splitting is the simplicity. Therefore, it is advantageous to a large-scale application harvesting a wide area of sunlight. The final goal of this research area is to construct an artificial photosynthesis resulting in revolution for energy.

Cross-References

▶ Photocatalyst
▶ Photoelectrochemistry, Fundamentals and Applications

References

1. Fujishima A, Honda K (1972) Electrochemical photolysis of water at a semiconductor electrode. Nature 238:37
2. Osterloh FE (2008) Inorganic materials as catalysts for photochemical splitting of water. Chem Mater 20:35
3. Kudo A, Miseki Y (2009) Heterogeneous photocatalyst materials for water splitting. Chem Soc Rev 38:253
4. Maeda K, Domen K (2007) New non-oxide photocatalysts designed for overall water splitting under visible light. J Phys Chem C 111:7851
5. Kudo A (2011) Z-scheme photocatalyst systems for water splitting under visible light irradiation. MRS Bull 36:32

Photorechargeable Cell

Tsutomu Miyasaka
Graduate School of Engineering,
Toin University of Yokohama,
Yokohama, Kanagawa, Japan

Introduction

All kinds of existing devices for storage and charge–discharge of electric power (condensers, capacitor, secondary batteries, etc.) utilize electrochemical interfaces. On the other hand, electric power can be produced from light by photoelectrochemical cells which also use electrochemical interfaces bearing photosensitive materials (semiconductors and dyes). Combination of these phenomena into a single hybrid cell makes it possible to convert light energy into power and store it in the cell structure at the same time. Such kind of cell capable of reversible photo-charge and dark discharge is called photorechargeable cell, which is a type of solar cell endowed with power storage ability. The merit of this cell is its function of stabilizing and sustaining the output electric power against variation of light intensity and cutoff of light.

There have been many trials to combine a photovoltaic electrode and rechargeable materials for invention of photorechargeable cells [1–7]. The methods of power storage so far proposed are classified into two types. One is the cell incorporating charge–discharge reactions of redox materials that have been employed in the study of secondary batteries. The other is the cell employing electrostatic capacitance change for power storage and release.

Materials and Structures for Rechargeable Half Cell

Photorechargeable cells can be devised by electrochemical combination of photoelectrochemical half cell and rechargeable half cell. Because photovoltaic power generation, either by dye-sensitized photocells or conventional solid-state solar cells (Si, CIS, etc.), is characterized to produce low output voltage less than 1 V, it is desired that the function of power storage can realize high capacity and energy density within the low voltage range. Conductive polymer materials that perform charge–discharge reactions at low voltages with relatively high coulombic capacity meet the above requirements. Among these polymers are polypyrrole [4, 5, 7], polyaniline [6], and polythiophene [7]. In a typical cell structure, counter electrode of a dye-sensitized solar cell is loaded with a layer of polypyrrole [4]. Polypyrrole undergoes a charge–discharge reaction due to doping and undoping of anion such as ClO_4^- and BF_4^- at a potential around -0.3–0.4 V versus SCE. This doping reaction is also crucial for endowing the conductivity to the polymer. To construct photoanode of the photovoltaic half cell, metal oxide or compound semiconductor electrodes or dye-sensitized semiconductor electrodes have been employed. It is necessary for practical use that the photoanode has good sensitivity to visible light and performs rapid photo-charging with high photocurrent density. In addition, photoanode is desired to generate voltage as high as possible to drive charging of the storage materials. Dye-sensitized mesoporous semiconductors, sensitive to visible light (wavelengths up to 800–900 nm), meet these requirements. Photoexcitation of a dye-sensitized TiO_2 photoanode causes electron flow to the polypyrrole-loaded counter electrode where electrons are stored by reducing the polymer. In the optimum conditions for polymer structure and electrolyte composition, charging capacity can reach 90 mAh/g. A desired structure of the cell to operate efficient charge–discharge at high energy density is a three-electrode construction in which a middle electrode (inserted) separates electrolytes compositions for photovoltaic generation and charge storage. While counter electrode bears polypyrrole for the storage of photoexcited electrons, the middle electrode that is set opposed to the dye-sensitized anode bears a material for the storage of holes so that it is positively charged, i.e., oxidized, during light excitation of the anode. Polythiophene has been employed for a hole storage material [7]. The polymer-based photorechargeable cells achieve cyclic charge–discharge in the voltage range of 0.5–0.7 V [5, 7] with coulombic densities more than 50mAh g^{-1} [5]. Besides polymer materials, inorganic compounds such as MnO_2 and WO_3 [3, 8] also work as high-capacity storage materials to compose the rechargeable half cell. Carbonaceous materials are also considered to be useful for power storage materials. Especially, high energy density is achieved by lithium ion intercalation at charging voltages <4.2 V, which is however out of the range that photovoltaic generation concern. Soft carbonaceous materials have been successfully applied to various types of photoelectrochemical cells [9]. They can also be applied to the photorechargeable cell in the other way as they

also serve as excellent materials for capacitor electrodes that work in a wide range of voltage.

Photocapacitor

Besides the use of redox materials, the other type of photorechargeable cell is devised by using carbon materials for power storage half cell. Here, the charge–discharge mechanism differs from the above method and is based on the principle of capacitor that electrostatically stores electric power in terms of electric double layer capacitance. Activated carbon (AC) of large surface areas (>100 m g^{-1}) has been adopted for commercial capacitors of high charge–discharge capacity. In comparison with redox-type storage materials, these carbon-based capacitors are more stable against charge–discharge cycles and perform rapid charging reactions to photocurrent change (light variation). The photorechargeable capacitor is known as photocapacitor [10]. The device has been fabricated as a simple sandwich-type cell consisting of a dye-sensitized semiconductor photoanode, a redox-free liquid electrolyte, and a counter electrode in which two electrodes bear porous AC layers. In principle, charging proceeds by dye-sensitized electron injection to semiconductor (TiO_2) and simultaneous hole injection to the AC layer that is in junction with dye/TiO_2. The injected electron flows to the counter electrode and charges the AC layer. Figure 1 shows the structure and charge–discharge mechanism. The cell, which is free of redox reactions, is capable of charging up to 0.45 V [10]. The charging voltage, however, is lower than the expected range corresponding to the photovoltage of dye-sensitized solar cell, >0.7 V, which is due to the absence of redox system in electrolyte to favor the photovoltaic generation. Three-electrode type photocapacitor with a structure of separated electrolytes has been designed to overcome this subject [11]. Its structure is basically similar to those used in the above redox-type cells except for AC layers that replace polymers. Figure 2 illustrates the cell structure, in which the middle electrode can be a thin metal foil. Preparation of dye-sensitized

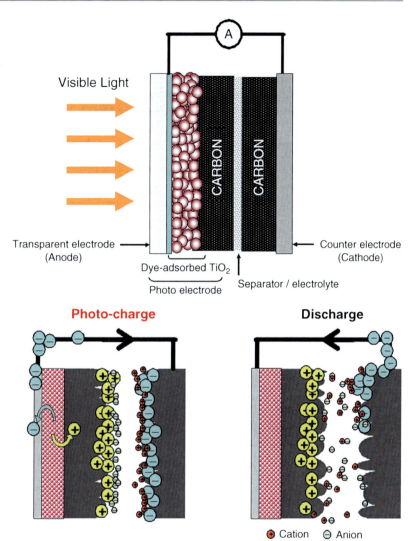

Photorechargeable Cell, Fig. 1 Two-electrode sandwich-type structure of photocapacitor and charge-transfer mechanism in the processes of photo-charge and dark discharge

photoanode follows the conventional method of fabricating a mesoporous TiO$_2$ electrode on a transparent conductive substrate (see keyword, dye-sensitized photoanode). The electrolyte of the photovoltaic unit consisting photoanode and middle electrode is an iodine-/triiodide-containing organic solution and that of the AC-coated capacitor unit is 1 M tetraethylamine BF$_4$-containing propylene carbonate. Various kinds of sensitizers (dyes, quantum dots) and semiconductors can be applied to the construction of photoanode.

To operate the photocapacitor, charging is done by visible light irradiation to the photoanode in the condition of short-circuiting the external circuit, and discharge is done by connecting the middle and counter electrodes in the circuit. Rate of charging is fully dependent on the light intensity and tends to decrease with time until charging is saturated. In discharge, current density can be regulated as constant by galvanostatic circuitry. Figure 3 shows a photo-charge and dark-discharge characteristics of the photocapacitor. Illuminated with white light of 100 mW cm^{-2}, charging voltage reaches as high as 0.8 V. On discharge in the dark, cell voltage exhibited a constant decrease with time under constant current. This linear decrease is a characteristic of the electric double layer capacitor, and the cell capacitance is determined by the reciprocal of the slope, Idt/dV, where I and t represent current and discharging time, respectively. Using a Ru complex-sensitized TiO_2 photoanode, three-electrode photocapacitor fabricated in optimized conditions shows a charging voltage of 0.85 V yielding a maximum capacitance per the amount of AC of 200 F g^{-1}, specific capacitance per electrode area of 6 F cm^{-2}, and energy density per area of 230 mWh cm^{-2}. This performance corresponds to 4 % as a global solar energy storage efficiency starting from the conversion of incident energy and ending with energy storage in the cell. Following the method of carbon-based photocapacitors, polythiophene derivatives have been applied to replace the AC layers as redox active materials. It gives an energy density of 21.3 μ Wh cm^{-2} with charging voltage of 0.75 V [12]. The photocapacitors are fairly stable against more than thousands of repeated charge–discharge cycles.

Photorechargeable cells work as the solar cell endowed with power-stabilizing function against light variation, the function being incorporated at the material level in a single sandwich-type device. Figure 3 displays data demonstrating that the photocapacitor stabilizes the output power under exposure to random fluctuation of outdoor sunlight [9] (Fig. 4).

Future Perspectives

Because photorechargeable devices have a flat shape, similar to solar cells, to receive light, the amount of charge storage material loaded on the electrode area is limited. In a single flat device, storage capacity and energy density can only be increased by

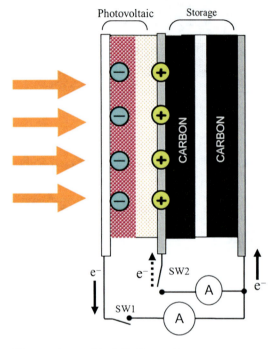

Photorechargeable Cell, Fig. 2 Three-electrode type photocapacitor. Photo-charge (*solid arrow*) and discharge (*dashed arrow*) processes are controlled on external circuit by switching (SW1, SW2)

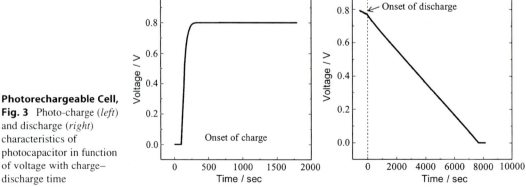

Photorechargeable Cell, Fig. 3 Photo-charge (*left*) and discharge (*right*) characteristics of photocapacitor in function of voltage with charge–discharge time

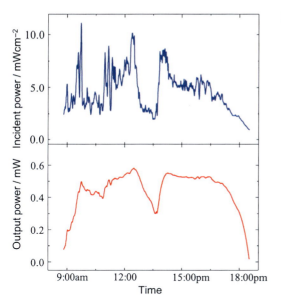

Photorechargeable Cell, Fig. 4 Output power of the photocapacitor in comparison with the change of incident solar irradiation power during a day

increasing the thickness of the material. For large power storage, conventional capacitors such as cylinder types are more useful by way of electrically connecting a solar cell module and a storage cell. In this case, however, circuitry of stabilizing the solar cell output is prerequisite to connection. Replacement of solar cell with photorechargeable cell may save such load of circuitry. For small power industry, which is related to consumer electronic devices, photorechargeable device is a more useful tool for continuous supply of photo-generated power to secondary batteries such as lithium ion batteries. High sensitivity of dye-sensitized power generation to weak indoor light is especially advantageous for versatile applications in power storage and supply to small power electronic devices such as computers and other IT equipment.

Cross-References

▶ Photocatalyst
▶ Photoelectrochemistry, Fundamentals and Applications
▶ TiO$_2$ Photocatalyst

References

1. Licht S, Hodes G, Tenne R, Manassen J (1987) A light-variation insensitive high efficiency solar cell. Nature 326:863–864
2. Kaneko M, Okada T, Minoura H, Sugiura T, Ueno Y (1990) Photochargeable multilayer membrane device composed of CdS film and prussian blue battery. Electrochim Acta 35:291–293
3. Hauch A, Georg A, Krasovec UO, Orel B (2002) Photovoltaically self-charging battery. J Electrochem Soc 149:A1208–A1211
4. Nagai H, Segawa H (2004) Energy-storable dye-sensitized solar cell with a polypyrrole electrode. Chem Commun 8:974–975
5. Saito Y, Ogawa A, Uchida S, Kubo T, Segawa H (2010) Energy-storable dye-sensitized solar cells with interdigitated nafion/polypyrrole-Pt comb-like electrodes. Chem Lett 39:488–489
6. Karami H, Mousavi MF, Shamsipur M (2003) New design for dry polyaniline rechargeable batteries. J Power Sources 117:255–259
7. Liu P, Yang HX, Ai XP, Li GR, Gao XP (2012) A solar rechargeable battery based on polymer charge storage electrodes. Electrochem Commun 16:69–72
8. Saito Y, Uchida S, Kubo T, Segawa H (2010) Surface-oxidized tungsten for energy-storable dye-sensitized solar cells. Thin Solid Films 518:3033–3036
9. Miyasaka T, Ikeda N, Murakami TN, Teshima K (2007) Light energy conversion and charge storage with soft carbonaceous materials that solidify mesoscopic electrochemical interfaces. Chem Lett 36:480–487
10. Miyasaka T, Murakami TN (2004) The photocapacitor: an efficient self-charging capacitor for direct storage of solar energy. Appl Phys Lett 85:3932–3934
11. Murakami TN, Kawashima N, Miyasaka T (2005) A high-voltage dye-sensitized photocapacitor of three-electrode system. Chem Commun 26:3346–3348
12. Hsu CY, Chen HW, Lee KM, Hu CW, Ho KC (2010) A dye-sensitized photo-supercapacitor based on PProDOT-Et2 thick films. J Power Sources 195:6232–6238

Plasmonic Electrochemistry (Surface Plasmon Effect)

Tetsu Tatsuma
Institute of Industrial Science, University of Tokyo, Tokyo, Japan

Introduction

Photoelectrochemistry has been applied to photocatalysis (see the entries ▶ Photocatalyst,

Plasmonic Electrochemistry (Surface Plasmon Effect), Fig. 1 Possible mechanisms for (**a**) plasmonic excitation of dye or semiconductor and (**b**) plasmon-induced charge separation

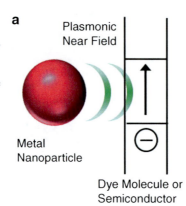
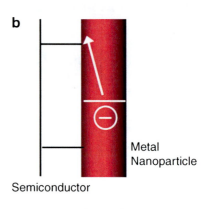

▶ TiO$_2$ Photocatalyst) and redox reaction-based solar cells such as dye-sensitized solar cells (see the entries ▶ Dye-Sensitized Electrode, Photoanode). In photoelectrochemistry, semiconductors such as TiO$_2$ or dye molecules such as Ru-complexes are used for light harvesting elements. Those elements are often used as thick deposits so that incident photons are absorbed extensively. An alternative approach to sufficient light absorption is the use of plasmonic metal nanoparticles. Nanoparticles of noble metal such as Au, Ag, and Cu absorb and scatter light due to localized surface plasmon resonance (LSPR), which is collective oscillation of conduction electrons at the metal surface by coupling with an incident electromagnetic field. LSPR is of interest for the following three aspects:
(1) strong light absorption, which is roughly proportional to particle volume;
(2) tunable optical properties, e.g., resonance intensity and wavelength dependent on particle size and shape, interparticle spacing, and dielectric environment; and
(3) non-propagating, oscillating electric field (optical near field) strongly localized in close proximity to the particle surface, beyond the diffraction limit.

There are two approaches to light harvesting based on LSPR:
(i) excitation of dye or semiconductor by plasmonic near field (Fig. 1a) and
(ii) charge separation at the interface between a plasmonic metal nanoparticle and a semiconductor (plasmon-induced charge separation) (Fig. 1b).

Plasmonic Excitation of Dye and Semiconductor

Plasmonic excitation of dye molecules and semiconductors has been used for surface-enhanced Raman scattering (SERS) and enhancement of fluorescence. It is also used for photoelectrochemistry such as dye-sensitized solar cells [1, 2] and photocatalysis [3, 4]. The LSPR band of metal nanoparticle antennae must overlap the absorption band of the dye or semiconductor used, so that the plasmonic near field around the metal nanoparticle can excite the dye or semiconductor. Since lifetime of LSPR is much longer than the time for which a photon passes by the metal nanoparticle, the excitation by near field can be more efficient than direct excitation by incident photons. The enhancement factor is defined by "photocurrent (or reaction rate) in the presence of the nanoparticles"/"photocurrent (or reaction rate) in the absence of the nanoparticles." Its value is typically from 2 to 20. The optimum spacing between the metal nanoparticle and the dye molecule or semiconductor is normally a few nm to a few tens nm [5, 6]. In the case of too small spacing or direct contact, energy transfer from the dye molecule or semiconductor to the metal nanoparticle may be so high that photoelectrochemical processes are suppressed. For the same reason, the use of plasmonic nanoparticles could have negative effect on systems with strong light absorption, whereas it is effective for enhancement of systems with weak light absorption.

Plasmon-Induced Charge Separation

Visible or near infrared light-induced charge separation [7] is observed for semiconductors (TiO_2, $SrTiO_3$, ZnO, CeO_2) in contact with plasmonic nanoparticles of Au, Ag, or Cu. The action spectra of the charge separation are close to LSPR-based absorption spectra of the corresponding metal nanoparticles. A possible mechanism is electron transfer from resonant metal nanoparticles to semiconductor due to an external photoelectric effect or hot electron injection. Actually, TiO_2 electrodes modified with plasmonic metal nanoparticles exhibit negative photopotential shifts and anodic photocurrents [8], whereas electrodes with metal nanoparticles coated with TiO_2 exhibit opposite responses, i.e., positive photopotential shifts and cathodic photocurrents [9]. On the other hand, the charge separation takes place preferentially at sites where the plasmonic electric field is localized [10].

In the case of Au, nanoparticles are stable both under light irradiation and in the dark. Therefore, the charge separation system is applied to photovoltaic cells and photocatalysis. Photovoltaic cells may be a wet type or solid state. In the wet-type cell, a photoanode such as electrode/nSC/MNP (nSC = n-type semiconductor; MNP = metal nanoparticle) [8] or a photocathode such as electrode/MNP/nSC [9] is used with an electrolyte containing a redox couple. Structures of solid-state cells are electrode/nSC/MNP/HTM/electrode (HTM = hole transport material or p-type semiconductor) [11] and electrode/nSC/MNP/electrode [12].

Photocatalytic systems based on the plasmon-induced charge separation can be used for oxidation of alcohols, aldehydes, and phenol [8, 13]; mineralization of carboxylic acids [14]; oxidation of benzene to phenol [15]; release of hydrogen from alcohols and ammonia [16]; and oxidation and reduction of water (but not water splitting) [17]. The photocatalytic system can also be applied to hydrophilic/hydrophobic patterning based on photocatalytic removal of a hydrophobic thiol adsorbed on metal nanoparticles [18].

In the case of Ag, nanoparticles are oxidized to Ag^+ ions by the charge separation [7, 19, 20]. The Ag^+ ions are reduced back to Ag nanoparticles by photocatalytic effects of photoexcited semiconductor (e.g., UV-irradiated TiO_2). This pseudo-reversible process is exploited for photochromism (see entry ▶ Photochromizm and Imaging) including multicolor photochromism [19], infrared and polarized photochromism, and fast photochromism. Holographic data storage [21] and photomorphing gels [20] are additional applications. Ag nanoparticles can be stabilized by chemical modification with a hydrophobic thiol. The electrode/MNP/TiO_2 system mentioned above can also protect Ag nanoparticles from oxidation so that it can be used for wet-type photovoltaic cells and photocatalysis.

In the case of Cu, nanoparticles are ready to be oxidized even in the absence of light. However, Cu nanoparticles coated with a thin protective layer (e.g., poly(vinyl alcohol)) are stable even under resonant light and exhibit plasmon-induced charge separation [22].

Future Directions

An increased number of papers have been published in recent years regarding electrochemistry based on the plasmonic excitation and the plasmon-induced charge separation. For further developments and practical applications, mechanisms and characteristics of those phenomena must be studied in further detail. These systems would make photovoltaic devices less expensive and simpler on the basis of efficient light absorption.

Cross-References

▶ Photocatalyst
▶ TiO_2 Photocatalyst
▶ Dye-Sensitized Electrode, Photoanode
▶ Photochromizm and Imaging

References

1. Wen C, Ishikawa K, Kishima M, Yamada K (2000) Effects of silver particles on the photovoltaic properties of dye-sensitized TiO_2 thin films. Sol Energ Mater Sol C 61:339
2. Standridge SD, Schatz GC, Hupp JT (2009) Distance dependence of plasmon-enhanced photocurrent in dye-sensitized solar cells. J Am Chem Soc 131:8407
3. Ingram DB, Linic S, (2011) Water splitting on composite plasmonic-metal/semiconductor photoelectrodes: evidence for selective plasmon-induced formation of charge carriers near the semiconductor Surface. J Am Chem Soc 133:5202
4. Gao H, Liu C, Jeong HE, Yang P (2012) Plasmon-enhanced photocatalytic activity of iron oxide on gold nanopillar. ACS Nano 6:234
5. Kawawaki T, Takahashi Y, Tatsuma T (2011) Enhancement of dye-sensitized photocurrents by gold nanoparticles: Effects of dye-particle spacing. nanoscale 3:2865
6. Torimoto T, Horibe H, Kameyama T, Okazaki K, Ikeda S, Matsumura M, Ishikawa A, Ishihara H, (2011) Plasmon-enhanced photocatalytic activity of cadmium sulfide nanoparticle immobilized on silica-coated gold particles. J Phys Chem Lett 2:2057
7. Tatsuma T, (2013) Plasmonic photoelectrochemistry: Functional materials based on photoinduced reversible Redox Reactions of Metal Nanoparticles. Bull Chem Soc Jpn 86:1
8. Tian Y, Tatsuma T (2005) Mechanisms and applications of plasmon-induced charge separation at TiO_2 films loaded with gold nanoparticles. J Am Chem Soc 127:7632
9. Sakai N, Fujiwara Y, Takahashi Y, Tatsuma T (2009) Plasmon resonance-based generation of cathodic photocurrent at electrodeposited gold nanoparticles coated with TiO_2 films. Chemphyschem 10:766
10. Kazuma E, Sakai N, Tatsuma T (2011) Nanoimaging of localized plasmon-induced charge separation. Chem Commun 47:5777
11. Yu K, Sakai N, Tatsuma T (2008) Plasmon resonance-based solid-state photovoltaic devices. Electrochemistry 76:161
12. Takahashi Y, Tatsuma T (2011) Solid state photovoltaic cells based on localized surface plasmon-induced charge separation. Appl Phys Lett 99:182110
13. Kowalska E, Abe R, Ohtani B (2009) Visible light-induced photocatalytic reaction of gold-modified titanium(IV) oxide particles: action spectrum analysis. Chem Commun 2009:241
14. Kominami H, Tanaka A, Hashimoto K (2010) Mineralization of organic acids in aqueous suspensions of gold nanoparticles supported on cerium(IV) oxide powder under visible light irradiation. Chem Commun 46:1287
15. Ide Y, Matsuoka M, Ogawa M (2010) Efficient visible-light-induced photocatalytic activity on gold-nanoparticle-supported layered titanate. J Am Chem Soc 132:16762
16. Tanaka A, Sakaguchi S, Hashimoto K, Kominami H (2012) Preparation of Au/TiO_2 exhibiting strong surface plasmon resonance effective for photoinduced hydrogen formation from organic and inorganic compounds under irradiation of visible light. Catal Sci Technol 2:907
17. Silva CG, Juarez R, Marino T, Molinari R, Garcia H (2011) Influence of excitation wavelength (UV or Visible Light) on the photocatalytic activity of titania containing gold nanoparticles for the generation of hydrogen or oxygen from water. J Am Chem Soc 133:595
18. Tian Y, Notsu H, Tatsuma T (2005) Visible-light-induced patterning of Au- and $Ag-TiO_2$ nanocomposite film surfaces on the basis of plasmon photoelectrochemistry. Photochem Photobiol Sci 4:598
19. Ohko Y, Tatsuma T, Fujii T, Naoi K, Niwa C, Kubota Y, Fujishima A (2003) Multicolor photochromism of TiO_2 films loaded with Ag nanoparticles. Nat Mater 2:29
20. Tatsuma T, Takada K, Miyazaki T (2007) UV light-induced swelling and visible light-induced shrinking of a TiO2-containing redox gel. Adv Mater 19:1249
21. Qiao Q, Zhang X, Lu Z, Wang L, Liu Y, Zhu X, Li J (2009) Formation of holographic fringes on photochromic Ag/TiO_2 nanocomposite films. Appl Phys Lett 94:074104
22. Yamaguchi T, Kazuma E, Sakai N, Tatsuma T (2012) Photoelectrochemical responses from polymer-coated plasmonic copper nanoparticles on TiO_2. Chem Lett 41:1340–1342

Platinum Monolayer Electrocatalysts

Radoslav R. Adzic[1] and Yun Cai[2]
[1]Chemistry Department, Brookhaven National Laboratory, Upton, NY, USA
[2]Material Science, Joint Center for Artificial Photosynthesis, Lawrence Berkeley National Laboratory, Berkeley, CA, USA

Introduction

Fuel cells are expected to be one of major sources of clean energy given their uniquely high energy conversion efficiency and low or zero emissions. Particularly important will be the application of proton exchange membrane fuel cell (PEMFC) in transportation. It will cause a substantial decrease of adverse effects of using fossil fuels on the environment. However, the slow kinetics of oxygen-reduction reaction (ORR) at fuel-cell cathodes cause a significant loss of the cell voltage even with the best Pt-based catalysts, which lowers the energy conversion efficiency. This drawback and the problem of Pt dissolution during long-term fuel-cell operations demand high Pt content at the cathode [1, 2]. The costs associated with high Pt contents hinder the commercialization of fuel-cell technology. Platinum monolayer electrocatalysts [3, 4] were developed to answer the challenges of slow kinetics of oxygen-reduction reaction (ORR) at fuel-cell cathodes, insufficient stability of Pt, and low CO tolerance Pt-based anode electrocatalysts [5].

Platinum monolayer electrocatalysts have a core–shell structure with a monolayer of Pt as the shell supported by various conductive nanostructures with single or multiple metal components as the core. This class of electrocatalysts possesses the following unique properties: (1) ultimately low Pt content, containing only one monolayer amount of Pt; (2) complete utilization of Pt since all Pt atoms are on the surface and directly participate in the reaction; and (3) tunable activity and stability from the modification of the structural and electronic properties of Pt_{ML} induced by substrate [3, 4, 6]. Based on the composition and structure of the substrates, the Pt_{ML} electrocatalysts can be divided into the following categories:
1. Pt_{ML} on metal or alloys
2. Pt_{ML} on core–shell cores

From the shape of substrates, in addition to the traditional 0D nanoparticles, 1D nanowires and 3D nanocrystals have also been used as substrates for Pt_{ML}. The unique features of Pt_{ML} electrocatalysts open various possibilities for designing electrocatalysts with specific catalytic properties by choosing appropriate substrates.

Synthesis of Pt_{ML} Electrocatalysts

A typical Pt_{ML} electrocatalyst is prepared by placing a monolayer of Pt atoms on metal nanoparticles using galvanic displacement of an underpotentially deposited (UPD) Cu monolayer by a Pt monolayer [3, 7]. Underpotential deposition describes the formation of a submonolayer or monolayer of a metal on a foreign metallic substrate at potentials positive to the reversible Nernst potential, that is, before bulk deposition can occur [8]. The experimental setup for Pt_{ML} deposition on an electrode consists of a cell with several compartments; one used for Cu UPD, one for rinsing the electrode after emersion from $CuSO_4$ solution, and one for displacement of a Cu monolayer upon electrode immersion in Pt^{2+} solution. The cell is under an inert gas (Ar or N_2) and facilitates all operations in an O_2-free environment.

The deposition of Pt_{ML} is a two-step process. First, a Cu UPD layer is deposited in a $CuSO_4$ solution via holding the potential positive to the bulk deposition for Cu. Then, after removing potential control, the electrode is transferred in a Pt^{2+} solution. Pt replaces the underpotentially deposited Cu by a simple redox exchange. A schematic diagram of the entire process is shown in Fig. 1.

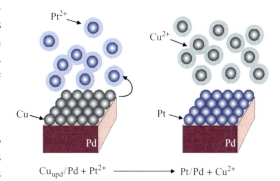

Platinum Monolayer Electrocatalysts, Fig. 1 Schematics of the galvanic displacement of a Cu UPD monolayer by Pt. *Blue* and *gray balls* represent Pt and Cu atoms, respectively. The *clouds* around balls indicate ions [9]

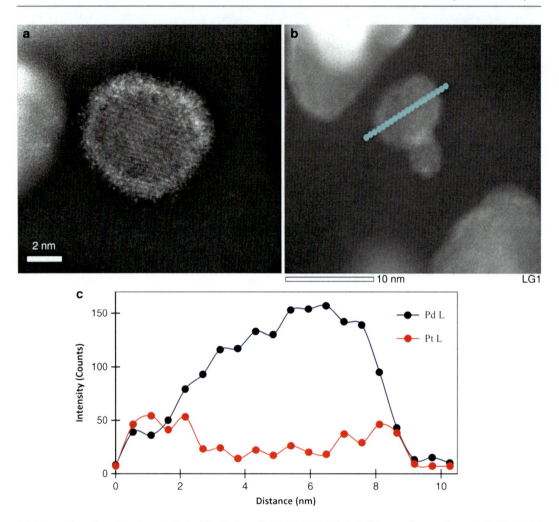

Platinum Monolayer Electrocatalysts, Fig. 2 Pt$_{ML}$/Pd/C: (**a**) HAADF-STEM image; line profile analysis (**b**) STEM image (**c**) the corresponding scanning EDS [10]

Scale-up synthesis of these electrocatalysts to produce kilogram quantities can be carried out in cells with larger cathode areas and cell volumes. A convenient cell consists of a Ti cylinder container (more than 10 cm in diameter) that also serves as the working electrode; the inside wall is coated with RuO$_2$, which is corrosion resistant, and the UPD of Cu is not taking place on this surface. A Pt black sheet used as counter electrode is in a separate compartment.

Structural and Electronic Characterizations of Pt$_{ML}$ Electrocatalysts

The morphology and structure of Pt$_{ML}$ electrocatalysts were characterized by high-angle annular dark-field (HAADF) imaging using scanning transmission electron microscopy (STEM). Figure 2a shows a typical HAADF-STEM Z-contrast image of Pt$_{ML}$/Pd/C nanoparticle, bright shell and relatively darker core, suggesting

Platinum Monolayer Electrocatalysts

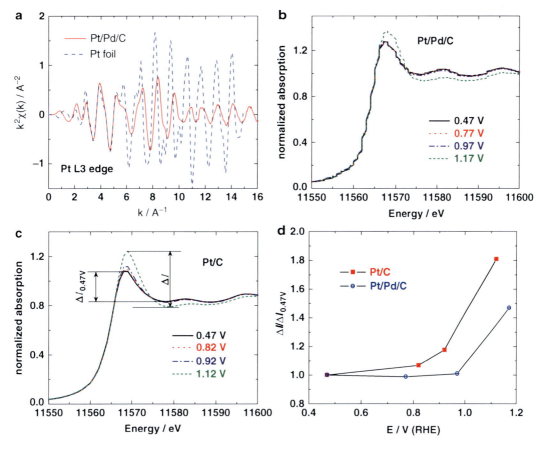

Platinum Monolayer Electrocatalysts, Fig. 3 (a) In situ EXAFS (k²-weighted) k-space spectra of Pt L3 edge obtained from Pt$_{ML}$/Pd/C and Pt foil in 1 M HClO$_4$ at a potential of 0.41 V/RHE. XANES spectra obtained with Pt$_{ML}$/Pd/C (**b**) and Pt/C (**c**) electrocatalysts at four different potentials in 1 M HClO$_4$. (**d**) A comparison of the change of the absorption peak as a function of potential for Pt$_{ML}$/Pd/C and Pt/C

the formation of a core–shell structure, i.e., a Pt (Z = 78) shell on a Pd (Z = 46) nanoparticle [10]. Figure 2b, c illustrates the line profile analysis by STEM/energy-dispersive X-ray spectrometry (EDS), showing the distribution of Pt and Pd component in a single nanoparticle. The analysis of Pt and Pd distribution exhibits the following three marked features:

1. Pt atoms fully cover the Pd nanoparticle surface,
2. The Pt intensity is fairly constant along the center of Pd nanoparticle.
3. At the both edges of Pd nanoparticle, the Pt intensity is approximately doubled compared with that at center.

These features demonstrate the formation of a core (Pd)–shell (Pt) structure. The elemental distribution of the catalysts can also be obtained from the chemical mapping from the EELS signal as discussed in ref [11].

The element-specific electronic properties, e.g., Pt 5d-band state, of Pt$_{ML}$ electrocatalysts and the local environment, such as bond distance and coordination number of individual component, were obtained using X-ray absorption near-edge structure (XANES), and extended X-ray absorption fine structure (EXAFS) spectroscopies, respectively.

Figure 3 shows the in situ EXAFS of Pt L3 edge from the Pt$_{ML}$/Pd/C electrocatalyst in 1 M HClO$_4$

at various potentials together with those from the reference foil [10]. Drastic difference between Pt_{ML}/Pd and the Pt foil was observed, especially the oscillatory behavior in k-space (Fig. 3a). The Pt–Pt and Pt–Pd bond lengths are determined to be 2.729 ± 0.005 Å and 2.724 ± 0.007 Å, respectively, which are smaller than that of bulk Pt (2.775 Å). The coordination numbers (CN) of Pt–Pt and Pt–Pd are 5.8 ± 0.8 and 2.7 ± 0.7, respectively. In the case of a complete Pt monolayer on a Pd(111), the CN of Pt–Pt is 6, while for Pt–Pd, it is 3. The slightly different CN for monolayers on nanoparticles, the CN of Pt–Pt and Pt–Pd from the fitting results, verify the Pt monolayer formation on Pd nanoparticle surfaces.

Strong evidence of delayed oxidation of a Pt monolayer on Pd nanoparticles in comparison with the oxidation of Pt nanoparticles was obtained from in situ XANES measurements as a function of potential. Figure 3a shows the Pt L3 edge spectra obtained on the $Pt_{ML}/Pd/C$ electrocatalysts at four different potentials. Only at the highest potentials is there an increase in the intensity of white line as a consequence of the PtOH formation causing a depletion of Pt's d-band. The increase in the intensity of the white line for the Pt/C electrocatalyst commences at considerably less positive potentials (Fig. 3b, c). This indicates that the oxidation of a Pt monolayer on palladium substrate requires higher potentials than that of platinum nanoparticles on a carbon substrate, which is in accord with the voltammetry data [3].

Factors Affecting the Activity of Pt Monolayer

Core–Shell Interaction
Monolayer of Pt atoms on a foreign metal will undergo compressive or tensile strain, depending on the difference in atomic radii of Pt and the other metal [4]. Thus, Pt atoms deposited on a Ru substrate would have a large compressive strain, but they would have only a small compressive strain on Pd and a tensile strain when deposited on Au. In addition to the strain effect, the electronic couplings between Pt_{ML} and its supporting substrates also affect the electronic property of surface Pt atoms. Both the surface strain and electronic modification generate a d-band center shift of the Pt monolayer [12, 13].

O/OH Binding
The studies of Pt monolayer electrocatalysts confirmed and elucidated the role of OH or O adsorption on the ORR kinetics [3]. This is in agreement with the density functional theory (DFT) calculations showing a strong correlation between the position of d-band center and binding energies of small adsorbates on strained surfaces and metal overlayers; the latter has a direct impact on the catalytic activity of the reactions [13]. For Pt_{ML} electrocatalysts, there is a volcano-type dependence of the ORR activity on the position of d-band center [6]. Figure 4 shows the experimentally determined activity (kinetic current) and the DFT calculation of binding energy of oxygen as functions of calculated d-band center ($\varepsilon_d-\varepsilon_F$) on various single-crystal substrates. The highest activity was observed on $Pt_{ML}/Pd(111)$ which has a slight upshift of d-band center and a weaker binding of O than that on Pt(111).

OH Coverage/OH–OH Repulsion Energy
The work on Pt_{ML} electrocatalysts revealed a new factor playing a role in determining activity for the ORR, which involves the OH–OH repulsion between OHs on Pt and other metals in the surface layer. In addition to the strength of O/OH binding with Pt, the coverage of OH is another important factor that affects the activity of Pt_{ML} electrocatalysts; the high OH coverage on Pt is known to inhibit ORR. The data for Pt_{ML} electrocatalysts, with an additional metal mixing with Pt to form a mixed monolayer, shows that the interaction between the Pt and the other metal in the shell induces a change of lateral repulsion energy of adsorbed OH or O and thus change the OH coverage on Pt [14].

Figure 5 depicts an experiment on a Pt–M mixed layer [14]. The lateral repulsion energy between adsorbed OH or between adsorbed OH and O changes when different M is used. A very

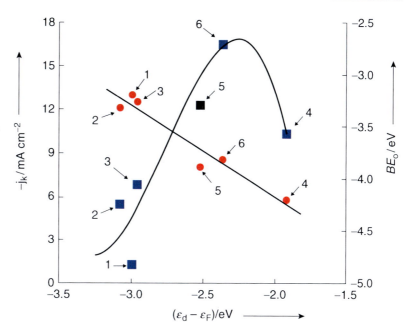

Platinum Monolayer Electrocatalysts, Fig. 4 Kinetic currents (j_K; *square symbols*) at 0.8 V for O_2 reduction on the platinum monolayers supported on different single-crystal surfaces in a 0.1 M $HClO_4$ solution and calculated binding energies of atomic oxygen (BE_O; *filled circles*) as functions of calculated d-band center ($\varepsilon_d - \varepsilon_F$); relative to the Fermi level of the respective clean platinum monolayersLabels. *1* Pt_{ML}/Ru(0001), *2* Pt_{ML}/Ir(111), *3* Pt_{ML}/Rh(111), *4* Pt_{ML}/Au(111), *5* Pt(111), *6* Pt_{ML}/Pd(111) [6]

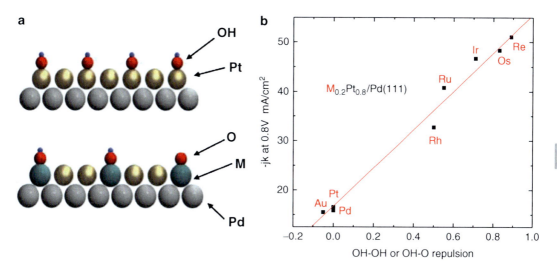

Platinum Monolayer Electrocatalysts, Fig. 5 (**a**) Model for the reduced OH adsorption on Pt atoms caused by a lateral repulsion with OH on another metal making a mixed monolayer. (**b**) Kinetic current at 0.80 V as a function of the calculated interaction energy between two OHs or OH and O on a $M_{0.2}Pt_{0.8}$ monolayer. *Positive energies* indicate more repulsive interaction compared to Pt_{ML}/Pd(111) [14]

good linear correlation is found between the measured kinetic current densities and the effective repulsion energy between two OH(a)s or an O(a) and an OH(a) calculated from first principles. The strong repulsion between these two adsorbates destabilizes the OH adsorption on Pt and accordingly reduces the OH coverage. Such good linear correlation suggests that the destabilization of OH on Pt and the resulted reduced OH coverage, due to the influence from the second metal, are responsible for the enhanced ORR kinetics.

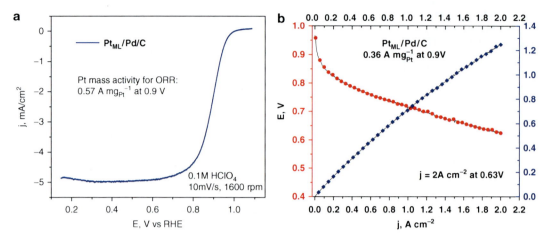

Platinum Monolayer Electrocatalysts, Fig. 6 Polarization curves for Pt$_{ML}$/Pd/C measured on RDE (**a**) and MEA (**b**). Other data are indicated in the graph

Rotating Disk Electrode and Fuel-Cell Measurements

Performance of Pt$_{ML}$ electrocatalysts has been determined using single-crystal surfaces [15], thin-film rotating disk electrode (RDE), and in fuel-cell membrane electrode assembly (MEA). Higher activities observed with rotating electrodes are ascribed to a better Pt utilization of Pt. Figure 6 displays the data on RDE and MEA prepared from Pt$_{ML}$/Pd/C nanoparticles: 0.57 and 0.36 A/mg$_{Pt}$ in Pt mass activities at 0.9 V were obtained from the measurements on RDE and MEA, respectively.

The Rational Design of Pt$_{ML}$ Electrocatalysts

The unique features of Pt$_{ML}$ electrocatalysts enable a wide selection of substrates to attain an electrocatalyst having low noble-metal content with enhanced catalytic activity and stability. A desired substrate should be able to affect the Pt monolayer by causing a weakening of the Pt–OH bond, a reduction of the OH coverage, and a delay of PtOH formation to more positive potentials than on pure Pt. The followings are some examples of the design of substrates to achieve the above goals.

Varying the Composition and Structure

Composition of the substrate has a significant impact on the catalytic activity of Pt$_{ML}$, through strain effect and electronic coupling. Pd has been found the most active metal as the substrate. To further reduce the content of noble metal and/or to enhance the catalytic activity and stability, non-noble metal, e.g., Co, Ni, and Fe, can be incorporated into the core as well as the components with higher dissolution potential, Au, Ir, for example. In addition to the random mix, a core–shell structure, with stable shell, is another effective way layer to preclude the dissolution of non-noble-metal core and/or to mediate the activity over a less reactive core.

For example, the addition of a small content of Au to Pd, e.g., a Pd$_9$Au$_1$ alloy core achieves remarkable enhancement of the stability while retaining the ORR activity as high as that on Pt$_{ML}$/Pd/C [16]. MEA fuel-cell tests showed that Pt$_{ML}$/Pd$_9$Au/C remains active even after 200,000 potential cycles. In addition to alloys, highly stable, inexpensive intermetallic compounds can also be attractive candidates as the supports for Pt$_{ML}$. Pt$_{ML}$ electrocatalyst with intermetallic Pd–Pb core exhibits an ORR activity superior to Pt/C [17].

Multimetallic alloy nanoparticle cores, obtained by in situ decomposition of a Prussian blue analogue, e.g., Pt$_{ML}$/AuNi0.5Fe, show a remarkable Pt mass activity as 1.38 A/mg$_{Pt}$

and all-noble-metal activity as 0.18 A/mg$_{Pt+Au}$ [18]. The interaction of Pt$_{ML}$ with the stratified structure of the core containing 3–5 atomic layers of Au – plays an essential role in determining the activity and stability of the catalyst (insignificant loss after 15,000 triangular-potential cycles) and the high electrochemical stability of the gold shell that precludes the exposure of the relatively active inner-core materials to the electrolyte.

The high-temperature-annealed Pd$_3$Fe(111) alloy core, i.e., Pd/Pd$_3$Fe, has a top layer with the structure same as Pd(111) but different electronic properties (i.e., a −0.25 eV downshift of the d-band center compared to Pd(111)) [19]. It is considerably more active than Pd(111), and the observed enhancement of ORR activity of Pt$_{ML}$/Pd/Pd$_3$Fe originates mainly from the destabilization of OH binding, leading to a decreased Pt–OH coverage on the Pt surface.

The IrNi core–shell nanoparticle core has a two-layer Ir shell around inner Ni core fabricated via thermally induced segregation [20]. The Ir shell completely protects the Ni atoms in the core from oxidation or dissolution under elevated potentials after 5,000 potential cycles. It is evident that Ir shell completely protects Ni core from oxidation or dissolution in acid electrolyte under elevated potentials. DFT calculations using a sphere-like model for the nanoparticle demonstrated that mixing Ni with Ir (Pt$_{ML}$/IrNi/C) induces geometric, electronic, and segregation effects, thus weakening the binding energy of oxygen and so resulting in higher activity than in Pt/C and Pt$_{ML}$/Ir/C electrocatalysts.

Iridium, given its stability, could be a good candidate as the substrate for Pt$_{ML}$. However, Ir as a core causes the Pt lattice to contract too much, significantly decreasing the d-band center, and it thus entails a very weak adsorption of O_2 on Pt, resulting in slow ORR kinetics. An addition of a Pd monolayer (interlayer), placed between the Pt monolayer and the Ir core, i.e., Pt$_{ML}$/Pd$_{ML}$/Ir/C, lowers the excessive effect of Ir as corroborated by DFT calculation and consequently ameliorates the ORR kinetics [21]. Xing et al. used the PdAu alloy as the interlayer to mediate the Pt$_{ML}$ catalytic properties [22]. Besides an enhanced activity, this catalyst showed an excellent stability that may be due to the stabilizing effect of Au [23].

Induced Lattice Contraction by Hollow Core

The study on Pt hollow nanoparticles shows that the hollow structure induces a lattice contraction in the Pt shell and exhibits higher activity and better stability than solid Pt nanoparticles for ORR [24]. The lattice contraction not only enhances the ORR kinetics but also prevents the instability caused by dissolution of core materials. Placing a Pt$_{ML}$ on Pd and Pd–Au hollow cores, obtained using Ni nanoparticles as sacrificial templates, improved its properties for the ORR. The hollow architecture results from the combination of galvanic replacement and the Kirkendall effect [25]. The larger ORR-active surface area and the significant savings of noble metal brought about by the hollow structure accordingly enhanced up to 0.57 A/mg of the total-metal mass activity of the electrocatalyst for the ORR.

The Effects of Facets and Shapes

Atoms at low-coordination sites, i.e., edges, kinks, and defects, have been shown to have a stronger binding with OH than those at 2D terrace sites and thus inhibit the oxygen reduction [11]. Those nonregistered atoms are also prone to stronger binding with adsorbates and thus prone to dissolution [26]. On the other hand, it has been shown that nanoparticles with a higher ratio of (111) facets have higher activity for the oxygen-reduction reaction [11]. Therefore, a desired nanoparticle core for Pt$_{ML}$ should possess a high ratio of atoms at (111) facets and less atoms at low-coordination sites. The following paragraphs show three examples to achieve this goal.

Reduction of Low-Coordination Sites

The removal of low-coordination sites via Br treatment on Pd/C core engenders nanoparticles with a smooth surface and increased content of (111) facets. As a result, 1.5 fold enhancement of the ORR activity is achieved on Pt$_{ML}$ catalyst with such core [27].

Platinum Monolayer Electrocatalysts, Table 1 Comparison of specific Pt mass and PGM mass activities of PtML electrocatalysts on different core nanoparticles for the ORR measured by the kinetic currents at 0.9 V. The values for conventional Pt/C electrocatalyst are given for comparison

Cores	Specific activity (mA cm^{-2})	Pt mass activity (A mg^{-1})	Noble-metal mass activity (A mg^{-1})
Pt/C	0.3	0.2	0.2
Pd	0.7	1.64	0.25
Pd$_{Hollow}$	0.9	1.5	0.45
Pd$_{20}$Au$_{Hollow}$	0.85	1.62	0.61
Pd/Ir$_{sublayer}$	0.94	2.17	0.13
Pd$_{NWe}$	0.85	1.85	0.22
PdAu	1.5	3.5	
Pd$_{NW}$	0.77	1.83	0.55
Pd$_{tetrahedral}$	0.64	0.92	0.14

Pd$_{NWe}$ Electrodeposited NW, *Pd$_{NW}$* wet chemistry synthesis. Other data are given in the graph

Pd Tetrahedron Core

The production of active facets can also be achieved by using nanocrystals with well-defined facets as substrates. The concave Pd tetrahedron (TH Pd) is a good example, which has a small number of low-coordination sites and defects and high content of the shape-determined high-coordinated facets [28, 29]. The ORR kinetics of Pt$_{ML}$/TH Pd/C is improved significantly.

Pd Nanowire Core

As it is generally observed, Pd nanowires (NW) have smooth surfaces with fewer low-coordination sites and edges and less surface imperfections than nanoparticles. Such surfaces are suitable for the ORR and make a good support for a Pt-ML since OH adsorption on them is shifted positively with reduced coverage. The power density of this catalyst is close to 1 W/cm^2 with only 40 μgPt/cm^2 (0.1 mg/cm^2 PGM) content, which compares favorably with reported data for other catalysts [30].

Table 1 displays a comparison of activities of several Pt$_{ML}$ electrocatalysts having different cores from the above rational designs. A considerable difference between some core–shell couples indicates a remarkable possibility for tuning the activity of these catalysts and the flexibility of this approach.

Pt Submonolayer Anode Electrocatalysts with High CO Tolerance

The presence of small concentration of CO is enough to strongly impair the performance of Pt-based catalysts. To increase CO tolerance, Ru is introduced to Pt to form a PtRu alloy for the anode. One of the roles played by Ru is to provide RuOH species for the oxidation of CO to CO$_2$, leaving Pt sites free for adsorption and oxidation of H$_2$ or MeOH (bifunctional mechanism [31]). Electroless (spontaneous) deposition of Pt submonolayer on Ru(0001) surface [32] and Ru nanoparticles [33] was used to synthesize such electrocatalyst with excellent CO tolerance for the oxidation of reformate H$_2$ [34]. The DFT calculation from Koper et al. found that CO binds most weakly on Pt$_{ML}$/Ru among various PtRu combinations in their study [35]. The example in Fig. 7a compares the CO tolerance in an accelerated test of the Pt submonolayer catalyst PtRu$_{20}$ with that of one commercial PtRu alloy catalyst in the oxidation of H$_2$ containing 1,000 ppm CO: the commercial PtRu lost almost all the activity after 100 min, while PtRu$_{20}$ remains active even after 350 min. The long-term fuel-cell performance stability test of the PtRu$_{20}$ electrocatalyst (Fig. 7b) shows no detectable loss in performance over 870 h with the Pt loading approximately 1/10th of the stand loading.

With the Pt monolayer or submonolayer concept, the DFT calculation from Nilekar et al. found a promising new category of electrocatalysts for anode with high CO–CO repulsion energy which leaves to low CO binding energy: mixed-metal Pt monolayer electrocatalysts [36]. Their results suggest that in addition to Ru, Ir, Os, and Re could also be good candidates for mixed-metal Pt monolayer.

Activity and Stability of Pt$_{ML}$ Electrocatalysts in Fuel-Cell Tests

Figure 8a displays the TEM image of the cross-section of MEA after 100,000 potential cycles from 0.6 to 1.0 V, and the corresponding distribution of Pt, Au, and Pd in the catalytic

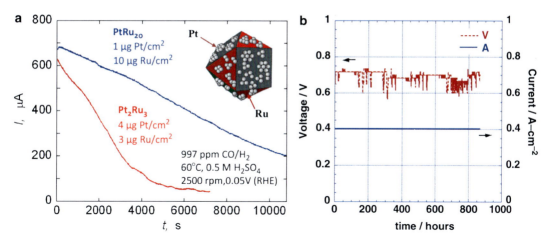

Platinum Monolayer Electrocatalysts, Fig. 7 (a) Accelerated CO tolerance tests of the PtRu$_{20}$ and of a commercial electrocatalyst for the oxidation of H$_2$ with 997 ppm of CO at 60 °C at 0.05 V on a rotating disk electrode (2,500 rpm) with the loadings indicated in the graph. The PtRu$_{20}$ (1 % Pt, 10 % Ru on C) electrocatalyst exhibits considerably higher CO tolerance. (b) Long-term performance stability test of the PtRu$_{20}$ electrocatalyst in an operating fuel cell. The fuel cell's voltage at constant current of 0.4 A/cm^2 is given as a function of time for the electrode of 50 cm^2 with an anode containing 180 μg/cm^2 of Ru, and 18 μg/cm^2 of Pt (or 0.063 g/kW of Pt that is approximately 1/10 of the standard Pt loading) and a standard air cathode with a Pt/C electrocatalyst. The fuel was clean H$_2$ or H$_2$ with 50 ppm of CO and 3 % air; temperature 80 °C. A change of fuel from the CO + H^2 mixture to H^2 caused an intermittent spike in the current trace

nanoparticles after the stability test is displayed in Fig. 8b [16]. A Pd band forms in the middle due to Pd dissolution and Pd^{2+} reduction by H$_2$ diffusing from the anode; Pt and Au remain in the cathode. Pd is a slightly more reactive metal than Pt and so dissolves at slightly lower potentials (0.92 (Pd) vs. 1.19 (Pt) V). The small dissolution limits the excursions of potential in an operating fuel cell, or at least minimizes it. Such partial dissolution of Pd entails a small contraction of the Pt$_{ML}$ shell, giving rise to a more stable structure with increased dissolution resistance and specific activity. This is the self-healing effect observed with this core–shell system as depicted in the model insert in Fig. 8b: the slow dissolution of Pd causes the decrease in the particle size, leading to some contraction of the Pt layer. The excess of Pt atoms from a monolayer shell can form a partial bilayered structure, or hollow particles may be formed due to the Kirkendall effect.

The distribution of Au in PdAu alloy has a great impact on the stabilization of Pd under potential cycling in the acid media. Figure 8c displays the cross-section of MEA and the posttest distribution of Pt, Au, and Pd in the Pt$_{ML}$/Pd$_9$Au catalyst based on highly uniform PdAu alloy. Such a uniform alloy causes a positive shift of Pd oxidation, in accord with its stabilization potential and reduced PdOH formation, as evidenced from voltammetry and in situ EXAFS studies, in particular confirming the changes in coordination number of Pd–O. Potential cycling did not entail any decrease in Pd, Pt, or Au. The Pt mass activity of the Pt/Pd$_9$Au/C electrocatalyst in a test involving 200,000 potential cycles decreased negligibly (Fig. 8d, red circles). The DOE's target for 30,000 potential cycles under the same protocol is a loss of 40 %. For comparison, the mass activity of a commercial Pt/C catalyst shows a terminal loss below 50,000 cycles (Fig. 8d, open triangles). The preparation of a highly compact Pt-ML, using the combined processes of H absorption and H adsorption on Pd to reduce Pt^{2+}, leads to the production of a new generation of Pt$_{ML}$/Pd/C catalysts with outstanding, unprecedented stability: there was no loss of activity over 200,000 potential

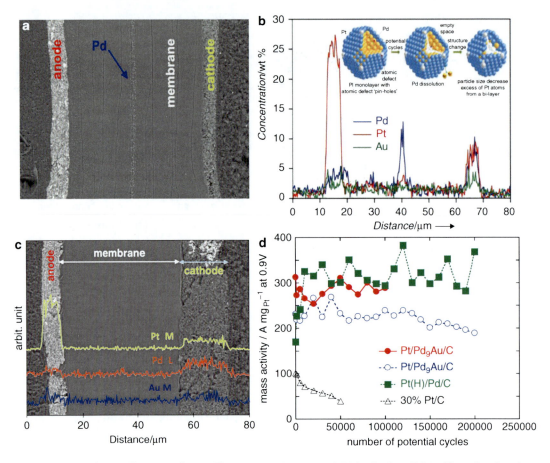

Platinum Monolayer Electrocatalysts, Fig. 8 (a) Cross-section of the MEA of Pt$_{ML}$/Pd$_9$Au/C after 100,000 potential cycles from 0.6 to 1.0 V. (b) Corresponding distribution of Pt, Au, and Pd, vs. distance after the test. The insert shows the model for the slow dissolution of Pd and the decrease in the particle's size, leading to some contraction of the Pt layer. (c) The cross-section of the MEA and overlaid posttest distribution of Pt, Au, and Pd in the Pt$_{ML}$/Pd$_9$Au/C catalyst based on a highly uniform PdAu alloy after 200,000 potential cycles. (d) Comparison of the Pt mass activity for Pt/Pd$_9$Au/C (*open circles*), Pt$_{ML}$/Pd$_9$Au catalyst based on highly uniform PdAu alloy (*red circles*) Pt$_{ML}$/Pd/C electrocatalyst with highly compact Pt-ML (*green squares*) and of a commercial Pt/C catalyst (*open triangles*) [16]

cycles (Fig. 8d green squares). Furthermore, under more severe conditions with a potential range of 0.6 –1.4 V, there were no significant losses of Pt and Au, although the dissolution of Pd was apparent.

Summary and Future Directions

Although Pt monolayer electrocatalysts are at early stage of development, significant improvements in their activity and stability have been achieved, and the accelerated stability tests in fuel cells established them as a viable practical concept. Moreover the well-established scaled-up synthesis has made Pt$_{ML}$ electrocatalysts ready for applications including the automotive one. In contrast to the content of platinum in the cathode catalysts currently being tested in fuel-cell vehicles, viz., 400 μgPt/cm^2, Pt$_{ML}$ electrocatalysts require only 40–80 μgPt/cm^2 and 60–100 μgPd/cm^2. Thus, the requirement for a 100 KW fuel cell in a medium-sized electric car, with the catalyst's performance of 1 W/cm^2, is 4–8 g of Pt and about 10 g of Pd. Currently, catalytic converters use 5 g of Pt per vehicle.

Therefore, there would be no need for considerable increase of the present rate of supply of Pt.

Owing to the unique features of Pt monolayer electrocatalysts, there is a broad range of possibilities for further improving the activity and stability and reducing the costs. The design of future Pt_{ML} electrocatalysts includes the change of composition, shape, and size of cores to optimize the core–shell interaction. In detail, the research can be performed from the following aspects: (1) Further reduction of the noble-metal (e.g., Pd) content by alloying it with non-noble metals, using its hollow nanoparticle counterparts, or by designing new inexpensive supports, such as IrNi described above. (2) Modification of the properties of the Pt-ML by selecting the appropriate top atomic layer of the cores to enable the employment of a variety of core materials. (3) Design and synthesis of core materials with defined facets, preferably (111) facets, like nanowires and nanorods which have their high resistance to surface oxidation and possess large fraction of (111) facets. Electrodeposition of nanowires is particularly promising since by its nature, this technique is capable of positioning the catalyst optimally in MEA, thus facilitating its highest utilization. (4) Incorporation of the synthesis of Pt_{ML} electrocatalysts with the fuel-cell setup to further reduce the production cost and better accommodation of the catalysts on the electrode. These studies would substantially reduce the technical barriers to produce durable, economical fuel cells. Given the limited resources of Pt, the concepts Pt_{ML} catalysts will have a broad impact on future catalysis research and technology.

Cross-References

▶ Electrocatalysts for the Oxygen Reaction, Core-Shell Electrocatalysts
▶ Hydrogen Oxidation and Evolution on Platinum in Acids
▶ Oxygen Reduction Reaction in Acid Solution
▶ Oxygen Reduction Reaction in Alkaline Solution
▶ Platinum-Based Anode Catalysts for Polymer Electrolyte Fuel Cells
▶ Platinum-Based Cathode Catalysts for Polymer Electrolyte Fuel Cells
▶ Polymer Electrolyte Fuel Cells (PEFCs), Introduction

References

1. Adžić RR (1998) Recent advances in the kinetics of oxygen reduction. In: Lipkowski J, Ross PN (eds) Electrocatal. Wiley-VCH, Weinheim, p 197
2. Gottesfeld S, Zawodzinski TA (2008) Polymer electrolyte fuel cells. In: Alkire RC, Gerischer H, Kolb DM, Tobias CW (eds) Advances in electrochemical science and engineering. Wiley-VCH, Weinham, pp 195–301
3. Zhang J, Mo Y, Vukmirovic MB, Klie R, Sasaki K, Adzic RR (2004) Platinum monolayer electrocatalysts for O_2 reduction: Pt Monolayer on Pd(111) and on carbon supported Pd nanoparticles. J Phys Chem B 108:10955–10964
4. Adzic R, Zhang J, Sasaki K, Vukmirovic M, Shao M, Wang J, Nilekar A, Mavrikakis M, Valerio J, Uribe F (2007) Platinum monolayer fuel cell electrocatalysts. Top Catal 46:249–262
5. Markovic NM (2003) Fuel cell electrocatalysis. In: Vielstich W, Lamm A, Gasteiger HA (eds) Handbook of fuel cells: fundamentals, technology applications. Wiley, Hoboken, pp 368–393
6. Zhang J, Vukmirovic MB, Xu Y, Mavrikakis M, Adzic RR (2005) Controlling the catalytic activity of platinum-monolayer electrocatalysts for oxygen reduction with different substrates. Angew Chem Int Ed 44:2132–2135
7. Brankovic SR, Wang JX, Adzic RR (2001) Metal monolayer deposition by replacement of metal adlayers on electrode surfaces. Surf Sci 474:L173–L179
8. Kolb DM (1978) In: Gerischer H, Tobias CW (eds) Advances in electrochemistry and electrochemical engineering. Wiley, New York, p 125
9. Vukmirović MB, Bliznakov ST, Sasaki K, Wang JX, Adzic RR (2011) Electrodeposition of metals in catalyst synthesis: the case of platinum monolayer electrocatalysts. Electrochem Soc Int 20:33–40
10. Sasaki K, Wang JX, Naohara H, Marinkovic N, More K, Inada H, Adzic RR (2010) Recent advances in platinum monolayer electrocatalysts for oxygen reduction reaction: scale-up synthesis, structure and activity of Pt shells on Pd cores. Electrochim Acta 55:2645–2652
11. Wang JX, Inada H, Wu L, Zhu Y, Choi Y, Liu P, Zhou W-P, Adzic RR (2009) Oxygen reduction on well-defined core – shell nanocatalysts: particle size, facet, and Pt shell thickness effects. J Am Chem Soc 131:17298–17302
12. Hammer B, Nørskov JK (2000) Theoretical surface science and catalysis – calculations and concepts. In: Bruce CG, Helmut K (eds) Advances in catalysis. Academic, New York, pp 71–129

13. Greeley J, Nørskov JK, Mavrikakis M (2002) Electronic structure and catalysis on metal surfaces. Annu Rev Phys Chem 53:319–348
14. Zhang J, Vukmirovic MB, Sasaki K, Nilekar AU, Mavrikakis M, Adzic RR (2005) Mixed-metal Pt monolayer electrocatalysts for enhanced oxygen reduction kinetics. J Am Chem Soc 127:12480–12481
15. Zhang J, Vukmirović MB, Sasaki K, Uribe F, Adžić RR (2005) Platinum monolayer electrocatalysts for oxygen reduction: effect of substrates, and long-term stability. J Serb Chem Soc 70:513–525
16. Sasaki K, Naohara H, Cai Y, Choi YM, Liu P, Vukmirovic MB, Wang JX, Adzic RR (2010) Core-protected platinum monolayer shell high-stability electrocatalysts for fuel-cell cathodes. Angew Chem Int Ed 49:8602–8607
17. Ghosh T, Vukmirovic MB, DiSalvo FJ, Adzic RR (2009) Intermetallics as Novel supports for Pt monolayer O2 reduction electrocatalysts: potential for significantly improving properties. J Am Chem Soc 132:906–907
18. Gong K, Su D, Adzic RR (2010) Platinum-monolayer shell on AuNi0.5Fe nanoparticle core electrocatalyst with high activity and stability for the oxygen reduction reaction. J Am Chem Soc 132:14364–14366
19. Zhou W-P, Yang X, Vukmirovic MB, Koel BE, Jiao J, Peng G, Mavrikakis M, Adzic RR (2009) Improving electrocatalysts for O_2 reduction by fine-tuning the Pt − support interaction: Pt monolayer on the surfaces of a Pd3Fe(111) single-crystal alloy. J Am Chem Soc 131:12755–12762
20. Sasaki K, Kuttiyiel KA, Barrio L, Su D, Frenkel AI, Marinkovic N, Mahajan D, Adzic RR (2011) Carbon-supported IrNi core–shell nanoparticles: synthesis, characterization, and catalytic activity. J Phys Chem C 115:9894–9902
21. Knupp S, Vukmirovic M, Haldar P, Herron J, Mavrikakis M, Adzic R (2010) Platinum monolayer electrocatalysts for oxygen reduction: Pt monolayer on carbon-supported PdIr nanoparticles. Electrocatal 1:213–223
22. Xing Y, Cai Y, Vukmirovic MB, Zhou W-P, Karan H, Wang JX, Adzic RR (2010) Enhancing oxygen reduction reaction activity via Pd − Au alloy sublayer mediation of Pt monolayer electrocatalysts. J Phys Chem Lett 1:3238–3242
23. Zhang J, Sasaki K, Sutter E, Adzic RR (2007) Stabilization of platinum oxygen-reduction electrocatalysts using gold clusters. Science 315:220–222
24. Wang JX, Ma C, Choi Y, Su D, Zhu Y, Liu P, Si R, Vukmirovic MB, Zhang Y, Adzic RR (2011) Kirkendall effect and lattice contraction in nanocatalysts: a new strategy to enhance sustainable activity. J Am Chem Soc 133:13551–13557
25. Yin Y, Rioux RM, Erdonmez CK, Hughes S, Somorjai GA, Alivisatos AP (2004) Formation of hollow nanocrystals through the nanoscale Kirkendall effect. Science 304:711–714
26. Budevski E, Staikov G, Lorenz WJ (1996) Electrochemical phase formation and growth – an introduction to the initial stages of metal deposition. VCH, Weinheim
27. Cai Y, Ma C, Zhu Y, Wang JX, Adzic RR (2011) Low-coordination sites in oxygen-reduction electrocatalysis: their roles and methods for removal. Langmuir 27:8540–8547
28. Gong K, Vukmirovic MB, Ma C, Zhu Y, Adzic RR (2011) Synthesis and catalytic activity of Pt monolayer on Pd tetrahedral nanocrystals with CO-adsorption-induced removal of surfactants. J Electroanal Chem 662:213–218
29. Huang X, Tang S, Zhang H, Zhou Z, Zheng N (2009) Controlled formation of concave tetrahedral/trigonal bipyramidal palladium nanocrystals. J Am Chem Soc 131:13916–13917
30. Wagner FT, Lakshmanan B, Mathias MF (2010) Electrochemistry and the future of the automobile. J Phys Chem Lett 1:2204–2219
31. Wang JX, Marinković NS, Zajonz H, Ocko BM, Adžić RR (2001) In situ x-ray reflectivity and voltammetry study of Ru(0001) surface oxidation in electrolyte solutions. J Phys Chem B 105:2809–2814
32. Brankovic SR, McBreen J, Adžić RR (2001) Spontaneous deposition of Pt on the Ru(0001) surface. J Electroanal Chem 503:99–104
33. Brankovic SR, Wang JX, Adzic RR (2001) Pt submonolayers on Ru nanoparticles: a novel low Pt loading, high CO tolerance fuel cell electrocatalyst. Electrochem Solid-State Lett 4:A217–A220
34. Sasaki K, Wang JX, Balasubramanian M, McBreen J, Uribe F, Adzic RR (2004) Ultra-low platinum content fuel cell anode electrocatalyst with a long-term performance stability. Electrochim Acta 49:3873–3877
35. Koper MTM, Shubina TE, van Santen RA (2002) Periodic density functional study of CO and OH adsorption on Pt − Ru alloy surfaces: implications for CO tolerant fuel cell catalysts. J Phys Chem B 106:686–692
36. Nilekar AU, Sasaki K, Farberow CA, Adzic RR, Mavrikakis M (2011) Mixed-metal Pt monolayer electrocatalysts with improved CO tolerance. J Am Chem Soc 133:18574–18576

Platinum-Based Anode Catalysts for Polymer Electrolyte Fuel Cells

Paramaconi Rodriguez[1] and Thomas J. Schmidt[2]
[1]School of Chemistry, The University of Birmingham, Birmingham, UK
[2]Electrochemistry Laboratory, Paul Scherrer Institut, Villigen, Switzerland

Introduction

Polymer electrolyte fuel cells (PEFCs) are considered to be a promising electric power source

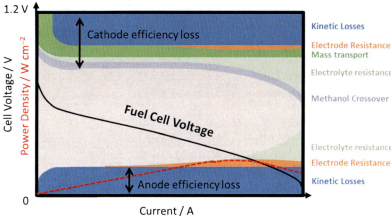

Platinum-Based Anode Catalysts for Polymer Electrolyte Fuel Cells, **Fig. 1** Contribution of anode-, cathode-, and electrolyte-related performance losses in a DMFC (Modified from [13, 27])

for stationary and automotive applications due to their potentially high energy density combined with exhaust-free operation. However, before the effective commercialization, several technical problems related specifically to durability issues must be resolved [1, 2]. The oxygen reduction reaction (ORR) is one of the main causes of the polarization loss in the cathodes and accounts for 80 % of the total loss [1–3]. The performance of a PEFC, hence, is limited by the ORR on the cathode side. Other significant challenges can be found on the anode side especially when operated with impure hydrogen derived from reformation reactions, which typically contain low levels of carbon monoxide, i.e., between 10 and 100 ppm of CO. Carbon monoxide is known to be a catalyst poison which strongly adsorbs on the commonly used Pt catalysts and can result in significant anodic polarization losses [4–8].

When operated with hydrogen feed gas, anode durability typically is not an issue provided that no starvation of hydrogen occurs, e.g., due to insufficient distribution of hydrogen inside the stack (so-called H_2 gross starvation). Under these undesired conditions, the anode potential is increased from below typically 0.1 V to values exceeding more than 1.5 V, resulting in instantaneous corrosion of the anode catalyst and support [9, 10].

Besides hydrogen as a fuel, also methanol or other organic liquid energy carriers can be oxidized in so-called direct oxidation fuel cells, e.g., direct methanol fuel cells (DMFC) [11–15] or direct alcohol fuel cells (DAFC) [16–18]. Methanol, however, is volatile and a toxic compound. As a replacement, other organic fuels, such as ethanol, ethylene glycol, propanol, and dimethyl oxalate, are being considered as fuels for DAFC [16, 17]. The most extensively studied alternative fuel is ethanol, considering that it is safer and has a higher energy density compared to methanol (8.01 kWh kg^{-1} versus 6.09 kWh kg^{-1}) [19–26].

Considering only the thermodynamics of the DMFC (used here as a representative of direct alcohol fuel cells), methanol should be oxidized spontaneously when the potential of the anode is above 0.05 V/SHE. Similarly, oxygen should be reduced spontaneously when the cathode potential is below 1.23 V/SHE, identical to a H_2–O_2 fuel cell. However, kinetic losses due to side reactions cause a deviation of ideal thermodynamic values and decrease the efficiency of the DMFC. This is presented in Fig. 1b, which includes various limiting effects as kinetics, ohmic resistance, alcohol crossover, and mass transport. The anode and cathode overpotentials for alcohol oxidation and oxygen reduction reduce the cell potential and together are responsible for the decay in efficiency of approximately 50 % in DMFCs [13, 27].

This essay briefly will summarize the different catalyst approaches for the underlying anode reactions for the oxidation of pure H_2, H_2-rich reformates including CO, and alcohols such as methanol and ethanol.

Anode Electrocatalysts for Hydrogen Oxidation Reaction (HOR)

The most effective catalyst for HOR at the anode in PEFCs are Pt nanoparticles supported on carbon-type supports. Since the HOR on Pt in acidic environment is very fast with values of the exchange current density of several hundreds of mA/cm$^2_{Pt}$ (80 °C, 100 kPa H$_2$ [28]) as compared to below 10^{-5} mA/cm$^2_{Pt}$ for the ORR at 80 °C and 100 kPa O$_2$ [29]. it is possible to reduce the anodic Pt loading to values of below 0.1 mg$_{Pt}$/cm^2 without increasing the polarization overpotential [30].

However, as was mentioned above that even small traces of CO can cause significant decrease in fuel cell performance. It was reported that only 5 ppm CO in the hydrogen stream leads to a drop of the maximum power density to less than half the value obtained for pure hydrogen [31]. The preparation of an efficient electrocatalyst with high activity for the HOR with good CO tolerance is related to the understanding of reaction mechanisms on its surface.

Several studies on platinum electrodes showed that the CO poisoning mechanism occurs because of strong adsorption of CO on the catalyst surface, which blocks the HOR [6, 32, 33]. Generally, in order to improve the CO tolerance of fuel cell anodes, several strategies can be employed [34]: (1) alloying of platinum with a second metal which either helps to reduce the adsorption energy of CO and therefore may help to reduce the effective CO coverage through a so-called ligand or electronic effect or which offers adsorption sites for oxygenated species in a bifunctional mechanism [7, 35–40]; (2) the use of catalysts which do not adsorb CO at all; or (3) the increase of the temperature in order to reduce the effective CO coverage (cf. Fig. 2, entry "▶ High-Temperature Polymer Electrolyte Fuel Cells", this volume).

Starting with the first approach, bimetallic PtM (M = Fe, Ru, Mo, Sn, or W) catalysts have been proposed [4, 7, 37–39, 41–51]. This first class has a (pseudo-)bifunctional mechanism of action, as in the case of Pt-Ru, where CO is adsorbed on both the Pt and Ru sites. The second metal does not

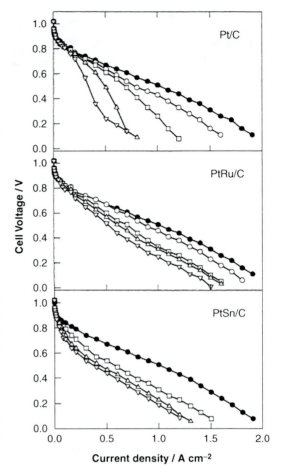

Platinum-Based Anode Catalysts for Polymer Electrolyte Fuel Cells, Fig. 2 Single cell performance plots for Pt/C, PtRu/C, and PtSn/C at 85 °C at several CO concentrations: (•) 0 ppm; () 5 ppm; (□) 20 ppm; (△) 50 ppm; and (▽) 100 ppm (Reproduced by permission from [8])

serve exclusively to adsorb OH$_{ad}$. The surface reaction for Pt-Ru can be formulated as

$$Pt(Ru) - CO_{ad} + Ru - OH_{ad} \rightarrow CO_2 + H^+ + e^-. \quad (1)$$

In the bifunctional mechanism, the presence of a second metal (normally oxophilic) favors the formation of the OH species promoting the early electrooxidation of the CO. As every oxidized CO opens up two sites for the hydrogen oxidation, Eq. 2,

$$H_2 \rightarrow 2H + 2e^-, \quad (2)$$

the strategy is clear that shifting the CO oxidation reaction to potentials as low as possible will significantly increase the rates for HOR.

The electronic effect claims that the second metal in the alloy modifies the H_2 and CO adsorption energies, reducing the CO coverage and by then creating Pt sites available for the H_2 oxidation. Quite significant efforts have been carried out to developing a CO-tolerant anode electrocatalyst at levels of 50 ppm of CO (with a noble metal loading lower than 0.1 mgcm^{-2} or less). Among the several binary alloy systems, PtRu and PtSn have shown promising performance for the HOR in the presence of CO, but the origin of this effect (bifunctional or electronic or both) remains under discussion [4, 37–39, 42–44, 46, 47]. Previous results on carbon-supported nanocrystalline PtRu and PtSn alloys, under in situ electrochemical conditions and using X-ray absorption spectroscopy, have shown changes in the electronic structure (Pt $5d$ band vacancy) but also a short-range atomic order (Pt–Pt bond distance and coordination number, the so-called lattice strain). Alloying Pt with Sn causes lowering of the Pt $5d$ band center and increase the Pt–Pt interatomic distance. On the other hand, the effect when alloying Ru with Pt is exactly opposite. Both effects induce a change in the CO poisoning characteristics in their systems, resulting in different activation energies, surface coverage, reaction orders, etc.

Figure 2 shows the single-cell polarization characteristics of CO-tolerant PtRu/C and PtSn/C relative to Pt/C for the HOR in PEFCs at 85 °C and different CO concentration.

Based on a simple kinetic model, Springer et al. [52] suggested that the polarization of the hydrogen electrode at low current density is limited essentially by the maximum rate of hydrogen dissociative chemisorption on a small fraction of the catalyst surface free of CO. It was also found that a rate of CO oxidation on Pt as low as 10 nAcm^{-2} could have a significant effect in lowering the CO steady-state coverage and thus in increasing the magnitude of the hydrogen electrooxidation current.

Other catalysts have shown similar properties, as it is the case of PtFe and PtMo. Watanabe and co-workers demonstrate that PtFe shows an excellent CO tolerance for the H_2 oxidation, similar to the PtRu catalyst [35, 50]. According to the authors, this effect is associated to a positive shift in the binding energy of the Pt $4f$ or $4d$ orbitals. The positive shift of the binding energies indicates an increase of the valence band ($5d$ orbital), resulting in a lower electron back-donation to the CO molecules and therefore a decrease in the CO coverage. Several investigations on PtMo report a good CO tolerance that it was ascribed to a bifunctional mechanism where oxygenated species on the molybdenum atoms react efficiently with the CO adsorbed on the Pt [53, 54].

The Pt/WO$_x$ system has also been subject of several studies although catalytic activity is not as strong as that for ruthenium. In this case, the catalytic activity was associated to the rapid change of the oxidation state of W, involving the redox couples $W^{(VI)}/W^{(IV)}$ or $W^{(VI)}/W^{(V)}$, which again promote the formation of hydrous oxide and therefore the oxidation of the CO [55–57].

The catalytic activity of a PtRu alloy catalyst, which is the most active system for oxidation of CO-contaminated H_2 fuel, can be further improved by using ternary alloy catalyst. In this regard, Chen et al. [58] reported a superior performance of a PtRuWO$_3$/C against CO poisoning in a stream of H_2/100 ppm CO at 80 °C at 220 mA cm^{-2} after a test period of 6 h. Other studies performed by Götz et al. demonstrated that under operation conditions of H_2/150 ppm CO (75 °C), the system PtRuW shows the highest activity at low current densities, while PtRuSn is the best catalyst at current densities higher than 300 mAcm^{-2} [51]. In addition, the activity of both catalysts at higher current densities exceeded the activity of the PtRu (E-TEK) commercial catalyst. Papageorgopoulos et al. [51] have reported that the addition of 10 at %Mo showed a significant improve in performance toward the H_2 oxidation in the presence of CO compared to PtRu/C. On the other hand, addition of Nb in a PtRu/C catalysts had a negative effect in PtRu by promoting CO poisoning [51].

As a result of a DFT calculations of a variety of model ternary PtRuM alloy catalysts yielded detailed adsorption energies and activation barriers and a combinatorial experimental electrocatalysis performed over 64-element electrode array with composition PtRuM (M = Co, Ni, and W), it is reported that PtRuCo ternary catalysts clearly outperformed the standard PtRu catalyst, followed by PtRuNi ternary catalysts [59].

In addition to the above mentioned catalyst, Schmidt et al. [60] reported the superior activity and CO tolerance of PdAu/C (Vulcan XC-72) (platinum-free electrocatalysts) at 60 °C compared to PtRu/C. This enhanced CO tolerance was explained by the significantly reduced CO adsorption energy on the PdAu surface, resulting in very low effective CO coverage and therefore in high HOR rates [60, 61].

Hydrogen Starvation

One of the main challenges for the commercialization of PEFCs is the long-term stability. As was explained above, under fuel cell operation, hydrogen is oxidized at the anode while oxygen is reduced at cathode catalyst layer.

$$2H^+ + 2e^- \rightarrow H_2 \qquad E^\circ = 0 \text{ V} \qquad (3)$$

$$O_2 + 4H^+ + 4e^- \rightarrow 2H_2O \quad E^\circ = 1.23 \text{ V} \quad (4)$$

In total, electric energy, water, and heat are produced. However, additional and undesired reactions can produce catalyst layer degradation and consequently can limit the lifetime of the fuel cell. Carbon corrosion and dissolution of the active metal catalyst are the main undesired side reactions in PEFCs [62–66].

Localized hydrogen starvation at a PEFCs anode can lead to the formation of local cells as presented on the Fig. 3a and takes place mainly during the start–stop cycling when a H_2/O_2 front is moving through the anode compartment of the cell [65]. In short, due to the presence of both H_2 and O_2 in the anode of a fuel cell, the cell becomes internally shorted with the result of a cathodic potential excursion to approx. 1.5 V [67].

The consequences of carbon corrosion at the *cathode* due to local hydrogen starvation have been confirmed with start–stop experiments and with experiments involving the deliberate blocking of hydrogen flow field channels. The main cause for local hydrogen starvation in real PEFCs is associated to water droplet formation inside flow field channels.

Gross hydrogen starvation occurs when the complete anode compartment of the PEFC is free of hydrogen. This situation occurs, e.g., when individual flow fields inside a fuel cell stack are not supplied by H_2 due to inhomogeneous hydrogen distribution when gas inlets of cells are blocked due to liquid water. Under operating conditions, i.e., when a current is drawn off the PEFC in the absence of fuel, the anode potential is increasing to a value where some other oxidation reactions can occur, typically to potentials when the carbon support of the anode catalyst is oxidized. In Fig. 3b, the effect of the hydrogen starvation mechanism on the cell potential of a PEFC anode is shown under a constant current of 20 mA/cm^2. In this model experiment, the anode potential increases as a function of time when the fuel gas is switched from hydrogen to Argon flow. In the presence of H_2, the potential exhibits a constant value of less than 0.1 V/ RHE (Eq. 3). However, when the H_2 flow is replaced by Ar, the potential gradually increases, at a rate that depends on the pseudocapacitance of the electrode material, to higher potentials than 1.6 V/RHE. This potential excursion leads to severe, rapid, and irreversible carbon corrosion [65, 66, 68–71].

Besides engineering solutions such as continuous H_2 flow in the anode to avoid accumulation of liquid water and/or nitrogen, different material mitigation strategies have been proposed in order to minimize effects of local hydrogen starvation which include the following: (1) the use of an oxygen evolution reaction (OER) catalyst in the cathode catalyst layer opening up an energetically more favorable reaction pathway during internal shorting as compared to carbon

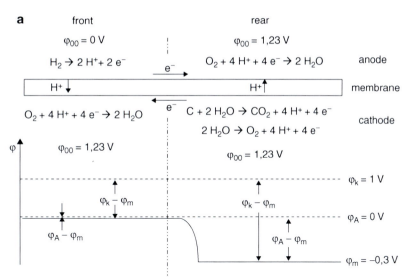
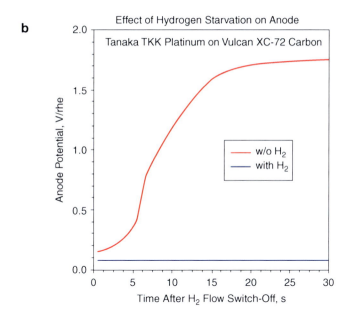

Platinum-Based Anode Catalysts for Polymer Electrolyte Fuel Cells, Fig. 3 (a) Reverse current decay mechanism at *local* hydrogen starvation in the rear part of a PEM cell (Reproduced by permission from [9]). (b) Effect of *gross* hydrogen starvation on the half-cell potential at the anode side (Reproduced by permission from [10])

oxidation; (2) the use of non-carbon supports which cannot be oxidized under local starvation conditions; and (3) the use of a highly selective anode catalyst which cannot reduce oxygen in the anode compartment [72], therefore avoiding the internal cell shorting.

More information regarding carbon corrosion and the use of novel support materials can be found in [3] and references therein.

Anode Electrocatalyst for Alcohol Oxidation

As in many other systems, the nature and the structure of the electrode material play an important role in the adsorption and electrooxidation of most organic fuels commonly used in DAFC. Many groups attempted to develop Pt-free electrocatalysts as anode materials for alcohol

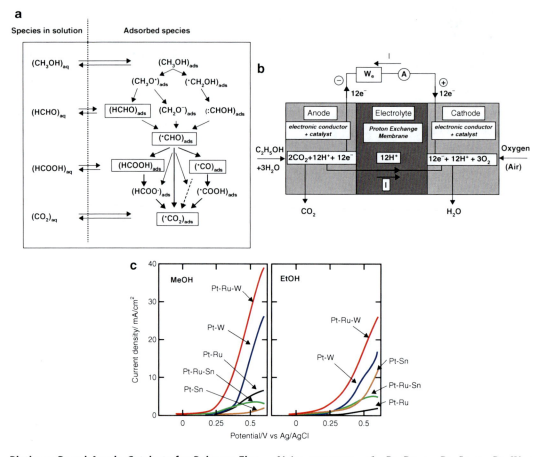

Platinum-Based Anode Catalysts for Polymer Electrolyte Fuel Cells, Fig. 4 (a) Reaction mechanism for the methanol electrochemical oxidation proposed by Lamy et al. (b) Schematic representation of a DEFC (Reproduced by permission from [16]). (c) Voltammograms of $Pt_{80}Ru_{20}$; $Pt_{80}Sn_{20}$; $Pt_{85}W_{15}$; $Pt_{60}Ru_{15}Sn_{25}$; $Pt_{65}Ru_{20}W_{15}$ supported nanoparticles electrodes in 0.5 M H_2SO_4 + 1 M methanol or ethanol solution at 30 °C; $v = 10$ mVs^{-1} (Reproduced by permission from [83])

oxidation, but only platinum seems to be able to adsorb alcohols and to break the C–H bonds efficiently. However, the dissociative chemisorption of an alcohol leads to the formation of strongly poisoning intermediate species (mainly adsorbed CO or COH), which block the electroactive sites of the catalysts, see discussion of CO tolerance above. During alcohol oxidation, many other intermediates species are formed, and different reaction mechanisms and poisoning species have been proposed. A temptative reaction mechanism proposed by Lamy et al. [16] is presented in Fig. 4a and is discussed in this volume, chapter "▶ Direct Alcohol Fuel Cells (DAFCs)".

As in the development of CO-tolerant catalysts for PEFC anodes, the main challenge for the development of catalysts for the oxidation of alcohols is to reduce or to avoid the formation of strongly adsorbed poisoning species (i.e., CO) or to favor their oxidation at low overpotentials.

As described before, a route to improve the electrocatalytic properties of platinum, and to decrease the poisoning of its surface by adsorbed CO or CO containing species, is to prepare alloys with a second metal (or third metal). Many different binary and ternary platinum-based anode catalysts, such as PtRu, PtSn, PtMo, PtRuMo, PtRh/SnO$_2$, or PtRuW, have been examined

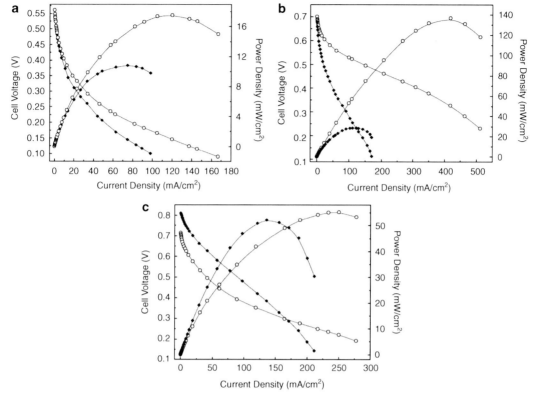

Platinum-Based Anode Catalysts for Polymer Electrolyte Fuel Cells, Fig. 5 The performance of single DAFC with (**a**) Pt/C; (**b**) PtRu/C (20–10 %), 1.33 mg$_{Pt}$/cm^2; and (**c**) PtSn/C, 1.33 mg$_{Pt}$/cm^2 as anode catalysts. Cathode: 1.0 mg$_{Pt}$/cm^2 (20 wt.% Pt/C, Johnson Matthey Co.). p$_{O2}$: 0.2 MPa (abs.); 90 °C; MeOH (or EtOH): 1 M and 1.0 ml/min. Nafion®-115 was used as electrolyte. (○) DMFC (♦) DEFC (Reproduced by permission from [91])

toward the methanol and ethanol electrooxidation [19, 20, 73–82].

Among all the alcohols, ethanol has being subject of extensive studies because it is a green fuel and can be easily obtained as biofuel. The schematic of direct ethanol fuel cell is presented in Fig. 4b, while Fig. 4c shows the onset on the oxidation potential of methanol and ethanol for different alloy catalyst.

Regarding the ethanol oxidation mechanism, adsorbed CO and CH$_x$ fragments have been identified as the main poisoning reaction intermediates [77, 84–90].

The cell performances reported by Zhou et al. for DMFCs and DEFCs with Pt/C, PtRu/C, and PtSn/C employed as anode catalyst are summarized in Fig. 5. The performances of single cells are significantly different when anode catalyst composition is the variable parameter to be considered, keeping the other parameter constant. In this evaluation, even though the Pt loading on anode catalyst layer is higher than that of other bimetallic anode catalysts (2.0 mg$_{Pt}$/cm^2 for Pt/C vs. 1.3 mg$_{Pt}$/cm^2 for PtRu/C and PtSn/C), the performance of the respective single cells is still poor. It can also be concluded that the performance of single DMFC with Pt/C and PtRu/C as anode catalyst is higher than that of single DEFC, due to the fact that methanol is more reactive than ethanol. On the other hand, PtSn/C catalyst has shown a better performance on a single DEFC than on a single cell of DMFC in the current density range from 0 to 168 mA/cm^2 [91].

Conclusions

Whereas the oxidation of pure hydrogen in H_2-PEFC is not challenging from the electrocatalytic point of view (the only challenge actually remaining is how much the Pt loading can be reduced), the oxidation of H_2-rich reformates obtained from hydrocarbon or alcohol reforming with even ppm levels of CO remains one of the biggest challenges for reformate-driven PEFC development. Although some Pt alloys, e.g., PtRu, show very high CO tolerance, from an operational system point of view, there is not a large range of CO levels which can be tolerated, and fuel processing to the desired level is a PEFC systems challenge. Although the many progresses have been made over the last years in the direct oxidation of hydrocarbons and alcohols in DAFCs, only hydrogen (both pure or reformate) fuelled systems offer the high power densities, e.g., necessary for stationary fuel cell systems. Direct alcohol systems may find a place in some applications in the portable electronics market, where power density requirements are not as high.

Acknowledgments P.R. acknowledges the financial support from NWO trough a Veni grant.

Cross-References

▶ Direct Alcohol Fuel Cells (DAFCs)
▶ Electrocatalysis - Basic Concepts, Theoretical Treatments in Electrocatalysis via DFT-Based Simulations
▶ Electrocatalysis of Anodic Reactions
▶ High-Temperature Polymer Electrolyte Fuel Cells
▶ Hydrogen Oxidation and Evolution on Platinum in Acids
▶ Platinum-Based Cathode Catalysts for Polymer Electrolyte Fuel Cells
▶ Polymer Electrolyte Fuel Cells (PEFCs), Introduction

References

1. Gasteiger HA et al (eds) (2011) Polymer electrolyte fuel cells 11. ECS transactions 41(1). The Electrochemical Society, Pennington
2. Vielstich W, Gasteiger HA, Yokokawa H (eds) Handbook of fuel cells – advances in electrocatalysis, materials, diagnostics and durability, vols 5 and 6. Wiley-VCH, Chichester
3. Rabis A, Rodriguez P, Schmidt TJ (2012) Electrocatalysis for polymer electrolyte fuel cells: recent advances and future challenges. ACS Catal 2:864–890
4. Pereira LGS, Paganin VA, Ticianelli EA (2009) Investigation of the CO tolerance mechanism at several Pt-based bimetallic anode electrocatalysts in a PEM fuel cell. Electrochim Acta 54(7):1992–1998
5. Wee J-H, Lee K-Y (2006) Overview of the development of CO-tolerant anode electrocatalysts for proton-exchange membrane fuel cells. J Power Sources 157(1):128–135
6. Chu HS et al (2006) Transient behavior of CO poisoning of the anode catalyst layer of a PEM fuel cell. J Power Sources 159(2):1071–1077
7. Arenz M et al (2005) Carbon-supported Pt-Sn electrocatalysts for the anodic oxidation of H_2, CO, and H_2/CO mixtures. Part II: the structure-activity relationship. J Catal 232(2):402–410
8. Lee SJ et al (1999) Electrocatalysis of CO tolerance in hydrogen oxidation reaction in PEM fuel cells. Electrochim Acta 44(19):3283–3293
9. Ohs JH et al (2011) Modeling hydrogen starvation conditions in proton-exchange membrane fuel cells. J Power Sources 196(1):255–263
10. Halalay IC et al (2011) Anode materials for mitigating hydrogen starvation effects in PEM fuel cells. J Electrochem Soc 158(3):B313–B321
11. Yan L et al (2007) Transition metal oxides as DMFC cathodes without platinum. J Electrochem Soc 154(7):B664–B669
12. Coutanceau C et al (2004) Preparation of Pt-Ru bimetallic anodes by galvanostatic pulse electrodeposition: characterization and application to the direct methanol fuel cell. J Appl Electrochem 34(1):61–66
13. Zainoodin AM, Kamarudin SK, Daud WRW (2010) Electrode in direct methanol fuel cells. Int J Hydrogen Energy 35(10):4606–4621
14. Ren X et al (2000) Recent advances in direct methanol fuel cells at Los Alamos National Laboratory. J Power Sources 86(1–2):111–116
15. Oedegaard A, Hentschel C (2006) Characterisation of a portable DMFC stack and a methanol-feeding concept. J Power Sources 158(1):177–187
16. Lamy C et al (2002) Recent advances in the development of direct alcohol fuel cells (DAFC). J Power Sources 105(2):283–296
17. Lamy C, Belgsir EM, Leger JM (2001) Electrocatalytic oxidation of aliphatic alcohols: application

to the direct alcohol fuel cell (DAFC). J Appl Electrochem 31(7):799–809

18. Braunchweig B et al (2013) Electrocatalysis: a direct alcohol fuel cell and surface science perspective. Catal Today 202:197–209

19. Kowal A et al (2009) Ternary $Pt/Rh/SnO_{(2)}$ electrocatalysts for oxidizing ethanol to $CO_{(2)}$. Nat Mater 8(4):325–330

20. Wang Z-B, Yin G-P, Lin Y-G (2007) Synthesis and characterization of PtRuMo/C nanoparticle electrocatalyst for direct ethanol fuel cell. J Power Sources 170(2):242–250

21. Rousseau S et al (2006) Direct ethanol fuel cell (DEFC): electrical performances and reaction products distribution under operating conditions with different platinum-based anodes. J Power Sources 158(1):18–24

22. Zhou WJ et al (2005) Direct ethanol fuel cells based on PtSn anodes: the effect of Sn content on the fuel cell performance. J Power Sources 140(1):50–58

23. Song SQ et al (2005) Direct ethanol PEM fuel cells: the case of platinum based anodes. Int J Hydrogen Energy 30(9):995–1001

24. Leger JM et al (2005) How bimetallic electrocatalysts does work for reactions involved in fuel cells? Example of ethanol oxidation and comparison to methanol. Electrochim Acta 50(25–26):5118–5125

25. Vigier F et al (2004) Development of anode catalysts for a direct ethanol fuel cell. J Appl Electrochem 34(4):439–446

26. Lamy C et al (2004) Recent progress in the direct ethanol fuel cell: development of new platinum-tin electrocatalysts. Electrochim Acta 49(22–23):3901–3908

27. Hogarth MP, Ralph TR (2002) Catalysis for low temperature fuel cells – Part III: challenges for the direct methanol fuel cell. Platinum Metals Rev 46(4):146–164

28. Neyerlin KC et al (2007) Study of the exchange current density for the hydrogen oxidation and evolution reactions. J Electrochem Soc 154(7):B631–B635

29. Neyerlin KC et al (2005) Effect of relative humidity on oxygen reduction kinetics in a PEMFC. J Electrochem Soc 152(6):A1073–A1080

30. Schwanitz B et al (2011) Stability of ultra-low Pt anodes for polymer electrolyte fuel cells prepared by magnetron sputtering. Electrocatalysis 2:35–41

31. Schmidt TJ, Gasteiger HA, Behm RJ (1999) Rotating disk electrode measurements on the CO tolerance of a high-surface area Pt/Vulcan carbon fuel cell catalyst. J Electrochem Soc 146(4):1296–1304

32. Camara GA, Ticianelli EA, Mukerjee S, Lee SJ, McBreen J (June 2002) CO poisoning of hydrogen oxidation reaction in PEMFCs. J Electrochem Soc 149(6) A748–753

33. Lemons RA (1990) Fuel cells for transportation. J Power Sources 29(1–2):251–264

34. Schmidt TJ, Markovic NM (2006) Electrocatalysis of inorganic reactions: bimetallic surfaces.

In: Somasundaran P (ed) Encyclopedia of surface and colloid science. Taylor and Francis, New York, pp 2017–2031

35. Igarashi H et al (2001) CO tolerance of Pt alloy electrocatalysts for polymer electrolyte fuel cells and the detoxification mechanism. Phys Chem Chem Phys 3(3):306–314

36. Hajbolouri F et al (2004) CO tolerance of commercial Pt and PtRu gas diffusion electrodes in polymer electrolyte fuel cells. Fuel Cells 4(3):160–168

37. Yajima T, Uchida H, Watanabe M (2004) In-situ ATR-FTIR spectroscopic study of electro-oxidation of methanol and adsorbed CO at Pt-Ru alloy. J Phys Chem B 108(8):2654–2659

38. Camara GA et al (2002) Correlation of electrochemical and physical properties of PtRu alloy electrocatalysts for PEM fuel cells. J Electroanal Chem 537(1–2):21–29

39. Lucas CA, Markovic NM, Ross PN (2000) Structural effects during CO adsorption on Pt-bimetallic surfaces. II. The Pt(111) electrode. Surf Sci 448(2–3):77–86

40. Nilekar AU et al (2011) Mixed-metal Pt mono layer electrocatalysts with improved CO tolerance. J Am Chem Soc 133(46):18574–18576

41. Zhang L et al (2011) A novel CO-tolerant PtRu core–shell structured electrocatalyst with Ru rich in core and Pt rich in shell for hydrogen oxidation reaction and its implication in proton exchange membrane fuel cell. J Power Sources 196(22):9117–9123

42. Yamanaka T et al (2010) Particle size dependence of CO tolerance of anode PtRu catalysts for polymer electrolyte fuel cells. J Power Sources 195(19):6398–6404

43. Lopes PP, Ticianelli EA (2010) The CO tolerance pathways on the Pt–Ru electrocatalytic system. J Electroanal Chem 644(2):110–116

44. Stolbov S et al (2009) High CO tolerance of Pt/Ru nanocatalyst: insight from first principles calculations. J Chem Phys 130(12):124714

45. Takeguchi T et al (2007) Preparation and characterization of CO-tolerant Pt and Pd anodes modified with SnO_2 nanoparticles for PEFC. J Electrochem Soc 154(11):B1132–B1137

46. Russell AE et al (2007) Unravelling the complexities of CO_2 tolerance at PtRu/C and PtMo/C. J Power Sources 171(1):72–78

47. Okanishi T et al (2006) Chemical interaction between Pt and SnO_2 and influence on adsorptive properties of carbon monoxide. Appl Catal A Gen 298:181–187

48. Stamenkovic V et al (2005) In situ CO oxidation on well characterized $Pt_3Sn(hkl)$ surfaces: a selective review. Surf Sci 576(1–3):145–157

49. Ioroi T et al (2003) Enhanced CO-tolerance of carbon-supported platinum and molybdenum oxide anode catalyst. J Electrochem Soc 150(9): A1225–A1230

50. Watanabe M, Zhu Y, Uchida H (2000) Oxidation of CO on a Pt–Fe alloy electrode studied by surface

50. enhanced infrared reflection-absorption spectroscopy. J Phys Chem B 104(8):1762–1768

51. Papageorgopoulos DC, Keijzer M, de Bruijn FA (2002) The inclusion of Mo, Nb and Ta in Pt and PtRu carbon supported 3 electrocatalysts in the quest for improved CO tolerant PEMFC anodes. Electrochim Acta 48(2):197–204

52. Springer T, Zawodzinski T, Gottesfeld S (1997) Modelling of polymer electrolyte fuel cell performance with reformate feed streams: effects of low levels of CO in hydrogen. The Electrochemical Society, Pennington

53. Grgur BN, Markovic NM, Ross PN Jr (1998) Electrooxidation of H_2, CO, and H_2/CO mixtures on a well-characterized Pt 70 Mo 30 bulk alloy electrode. J Phys Chem B 102(14):2494–2501

54. Mukerjee S et al (1999) Investigation of enhanced CO tolerance in proton exchange membrane fuel cells by carbon supported PtMo alloy catalyst. Electrochem Solid State Lett 2(1):12–15

55. Micoud F et al (2009) Unique CO-tolerance of Pt–WO_x materials. Electrochem Commun 11(3):651–654

56. Micoud F et al (2010) The role of the support in COads monolayer electrooxidation on Pt nanoparticles: Pt/WO_xvs. Pt/C. Phys Chem Chem Phys 12(5):1182–1193

57. Nagel T et al (2003) On the effect of tungsten on CO oxidation at Pt electrodes. J Solid State Electrochem 7(9):614–618

58. Chen KY, Shen PK, Tseung ACC (1995) Anodic oxidation of impure H_2 on teflon-bonded Pt-Ru/WO_3/C electrodes. J Electrochem Soc 142(10): L185–L187

59. Strasser P et al (2003) High throughput experimental and theoretical predictive screening of materials – a comparative study of search strategies for new fuel cell anode catalysts. J Phys Chem B 107(40):11013–11021

60. Schmidt TJ et al (2001) On the CO tolerance of novel colloidal PdAu/carbon electrocatalysts. J Electroanal Chem 501:132–140

61. Schmidt TJ et al (2003) Electrooxidation of H_2, CO, and CO/H_2 on well characterized Au(111)Pd surface alloys. Electrochim Acta 48:3823–3828

62. Ishigami Y et al (2012) Real-time visualization of CO_2 generated by corrosion of the carbon support in a PEFC cathode. Electrochem Solid State Lett 15(4):B51–B53

63. Kulikovsky AA (2011) A simple model for carbon corrosion in PEM fuel cell. J Electrochem Soc 158(8): B957–B962

64. Kim J, Lee J, Tak Y (2009) Relationship between carbon corrosion and positive electrode potential in a proton-exchange membrane fuel cell during start/stop operation. J Power Sources 192(2):674–678

65. Linse N et al (2012) Quantitative analysis of carbon corrosion during fuel cell start-up and shut-down by anode purging. J Power Sources 219:240–248

66. Linse N et al (2010) Start/stop induced carbon corrosion in polymer electrolyte fuel cells. In: Proceedings of the ASME 8th international conference on fuel cell

science, engineering, and technology, Brooklyn vol 22010. pp 357–362

67. Fuller T, Gray G (2006) Carbon corrosion induced by partial hydrogen coverage. ECS Trans 1(8):345–353

68. Tang H et al (2006) PEM fuel cell cathode carbon corrosion due to the formation of air/fuel boundary at the anode. J Power Sources 158(2):1306–1312

69. Schulenburg H et al (2011) 3D imaging of catalyst support corrosion in polymer electrolyte fuel cells. J Phys Chem C 115(29):14236–14243

70. Perry ML, Patterson TW, Reiser CA (2006) Systems strategies to mitigate carbon corrosion in fuel cells. ECS Trans 3(1):783–795

71. Liu ZY et al (2008) Characterization of carbon corrosion-induced structural damage of PEM fuel cell cathode electrodes caused by local fuel starvation. J Electrochem Soc 155(10):B979–B984

72. Genorio B et al (2011) Tailoring the selectivity and stability of chemically modified platinum nanocatalysts to design highly durable anodes for PEM fuel cells. Angew Chem Int Ed 50(24):5468–5472

73. García G, Tsiouvaras N, Pastor E, Peña MA, Fierro JLG, Martínez-Huerta MV (2012) Ethanol oxidation on PtRuMo/C catalysts: in situ FTIR spectroscopy and DEMS studies. Int J Hydrogen Energy 37(8):7131–7140

74. Li M et al (2010) Ethanol oxidation on the ternary Pt-Rh-$SnO_{(2)}$/C electrocatalysts with varied Pt:Rh:Sn ratios. Electrochim Acta 55(14):4331–4338

75. Lee E, Murthy A, Manthiram A (2010) Effect of Mo addition on the electrocatalytic activity of Pt-Sn-Mo/C for direct ethanol fuel cells. Electrochim Acta 56(3):1611–1618

76. Jian X-H et al (2009) Pt-Ru and Pt-Mo electrodeposited onto Ir-IrO_2 nanorods and their catalytic activities in methanol and ethanol oxidation. J Mater Chem 19(11):1601–1607

77. Giz MJ, Camara GA, Maia G (2009) The ethanol electrooxidation reaction at rough PtRu electrodeposits: a FTIRS study. Electrochem Commun 11(8):1586–1589

78. Li H et al (2007) Comparison of different promotion effect of PtRu/C and PtSn/C electrocatalysts for ethanol electro-oxidation. Electrochim Acta 52(24):6622–6629

79. Wang H, Jusys Z, Behm RJ (2006) Ethanol electrooxidation on carbon-supported Pt, PtRu and Pt3Sn catalysts: a quantitative DEMS study. J Power Sources 154(2):351–359

80. Colmati F, Antolini E, Gonzalez ER (2006) Effect of temperature on the mechanism of ethanol oxidation on carbon supported Pt, PtRu and Pt3Sn electrocatalysts. J Power Sources 157(1):98–103

81. Li M, Cullen DA, Sasaki K, Marinkovic NS, More K, Adzic RR (2013) Ternary electrocatalysts for oxidizing ethanol to carbon dioxide: making Ir capable of splitting C–C bond. J Am Chem Soc 135(1):132–141

82. Ammam M, Easton EB (2013) PtCu/C and Pt(Cu)/C catalysts: synthesis, characterization and catalytic activity towards ethanol electrooxidation. J Power Sources 222:79–87
83. Tanaka S et al (2005) Preparation and evaluation of a multi-component catalyst by using a co-sputtering system for anodic oxidation of ethanol. J Power Sources 152:34–39
84. Lai SCS, Koper MTM (2010) The influence of surface structure on selectivity in the ethanol electro-oxidation reaction on platinum. J Phys Chem Lett 1(7):1122–1125
85. Lai SCS et al (2008) Mechanism of the dissociation and electrooxidation of ethanol and acetaldehyde on platinum as studied by SERS. J Phys Chem C 112(48):19080–19087
86. Shao MH, Adzic RR (2005) Electrooxidation of ethanol on a Pt electrode in acid solutions: in situ ATR-SEIRAS study. Electrochim Acta 50(12):2415–2422
87. Vigier F et al (2004) On the mechanism of ethanol electro-oxidation on Pt and PtSn catalysts: electrochemical and in situ IR reflectance spectroscopy studies. J Electroanal Chem 563(1):81–89
88. de Souza JPI et al (2002) Electro-oxidation of ethanol on Pt, Rh, and PtRh electrodes. A study using DEMS and in-situ FTIR techniques. J Phys Chem B 106(38):9825–9830
89. Cantane DA et al (2012) Electro-oxidation of ethanol on Pt/C, Rh/C, and Pt/Rh/C-based electrocatalysts investigated by on-line DEMS. J Electroanal Chem 681:56–65
90. Detacconi NR et al (1994) In-situ FTIR study of the electrocatalytic oxidation of ethanol at iridium and rhodium electrodes. J Electroanal Chem 379(1–2):329–337
91. Zhou WJ et al (2004) Performance comparison of low-temperature direct alcohol fuel cells with different anode catalysts. J Power Sources 126(1–2):16–22

Platinum-Based Cathode Catalysts for Polymer Electrolyte Fuel Cells

Emiliana Fabbri and Thomas J. Schmidt
Electrochemistry Laboratory, Paul Scherrer Institute, Villigen, Switzerland

Introduction

Polymer electrolyte fuel cells (PEFCs) are electrochemical devices converting chemical energy of fuels into electrical energy with relatively high efficiencies and low emissions. For these reasons they have attracted great attention as promising power sources for small stationary, mobile, and portable applications. The heart of a PEFC is the membrane electrode assembly (MEA), which is composed of anodic and cathodic catalytic layers (CLs), where electrochemical reactions occur, gas diffusion layers (GDLs), and a proton-conducting membrane. For the CLs in hydrogen driven PEFCs, Pt has been early recognized and still remains the catalyst of choice. The development of Pt-based CLs has gone through a number of different stages in the last three decades. In the early days of PEFC development, the CLs relied on vey high loadings of Pt (about 4 mg cm^{-2}) to achieve reasonable fuel cell performance [1]. The CLs were made of unsupported Pt-black powders, with very low surface areas (10–30 m^2 g^{-1}) and thereby requiring high Pt loadings per unit area in order to attain reasonable performance. In the late 1980's and early 1990's, by developing CLs based on Pt supported on porous carbon and incorporating in the CL a proton-conducting phase, it was possible to achieve the same performance for the cathodic and anodic reactions reducing the noble metal loading down to 0.35 mg cm^{-2} [2, 3]. The introduction of a proton-conducting phase such as Nafion led to an extended reaction zone and, thereby, to a larger catalyst utilization. Supporting Pt particles on porous carbon allowed higher Pt dispersion and hence larger catalyst surface area compared to unsupported Pt-black. The carbon support not only provides high dispersion of Pt nanoparticles, but it also possess high electronic conductivity to minimize ohmic losses and adequate porosity to ensure efficient mass transport of reactants and products from and to the CLs. Table 1 provides a list of the most widely used carbon supports and their surface area. After the breakthrough of CLs based on carbon-supported Pt catalysts, the new generation of PEFCs emerged and the Pt loading could be progressively reduced. However, while at present the Pt anode loading can be as low as 0.05 mg cm^{-2} without measurable kinetic losses [5] (in the case of operation with pure H$_2$), the Pt

Platinum-Based Cathode Catalysts for Polymer Electrolyte Fuel Cells, Table 1 Brunauer–Emmett–Teller (BET) surface area of the most used carbon supports in Pt/C catalysts [4]

Carbon	BET surface area/m^{2}g$^{-1}_C$
Black pearl	1,600
Ketjenblack	800
Vulcan XC-72C	240
Black pearls graphitized	240
Ketjenblack graphitized	160
Vulcan XC-72C graphitized	80

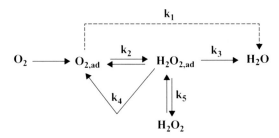

Platinum-Based Cathode Catalysts for Polymer Electrolyte Fuel Cells, Fig. 1 Possible reaction scheme for the ORR on Pt-based catalysts (Ref. [8])

loading required at the cathode is still about 0.4 mg cm^{-2} [6]. Besides the large Pt cathode loading, the slow ORR kinetics at the positive electrode are responsible for about two thirds of the overall voltage losses of a PEFC at high current density, and therefore large efforts are currently devoted to search for more active catalysts and reducing the Pt loading.

Performance: ORR Kinetics

As already mentioned above, the oxygen reduction reaction (ORR) at the PEFC cathode suffers from poor kinetics, resulting in overpotentials of about 0.3–0.4 V. In acidic environment, Pt-based catalysts are still recognized the materials of choice for the ORR which leads to water formation according to the following equation:

$$O_2 + 4H^+ + 4e^- \rightarrow 2H_2O$$
$$E_0 = 1.229\text{V vs. SHE at } 25\,°C \text{ in acid} \quad (1)$$

Besides direct 4e$^-$ reduction of adsorbed oxygen to water, ORR can also proceed by 2e$^-$ reduction of O$_2$ to adsorbed H$_2$O$_2$, which then either desorbs or undergoes a second 2e$^-$ reduction to water [7]:

$$O_2 + 2H^+ + 2e^- \rightarrow H_2O_{2(ads)}$$
$$E_0 = 0.67\text{V vs. SHE at } 25\,°C \text{ in acid} \quad (2)$$

$$H_2O_2 + 2H^+ + 2e^- \rightarrow 2H_2O$$
$$E_0 = 1.77\text{V vs. SHE at } 25\,°C \text{ in acid} \quad (3)$$

Figure 1 presents a simplified reaction scheme for the oxygen reduction reaction [8]: direct 4e$^-$ reduction of adsorbed oxygen to water (k1), 2e$^-$ reduction to adsorbed H$_2$O$_2$ as an intermediate (k2) which can be then further reduced to water (k3), chemically decompose to water and O$_2$ (k4), or desorbs from the surface (k5).

It was firstly reported that in absence of adsorbed impurities, the ORR on Pt proceeds principally through the direct pathway, both in acid and alkaline electrolytes [9]. However, it was latter suggested that the 2e$^-$ + 2e$^-$ pathway more reasonably applies to Pt-based catalysts [8]. Besides a chemical intuition, which would rather suggest a molecular than dissociative adsorption of O$_2$ in liquid environment (where strong adsorbates are generally present), rotating ring-disk (RRDE) measurements have also given experimental evidence of this assumption [8]. ORR on Pt-based catalysts involves several reaction steps, which include both chemical and electrochemical reactions. Among them, the rate determining step (rds) is considered the addition of the first electron to the adsorbed oxygen in a proton-coupled process [8, 10]. The number of electrons transferred in the rds can be determined by the Tafel slope obtained plotting the IR-corrected potential vs the logarithm of the kinetic current. In acidic environments, polycrystalline Pt catalysts exhibit a Tafel slope of about 60 mV dec^{-1} at potentials above ~0.85 V$_{RHE}$, while at more negative potentials a slope of 120 mV dec^{-1} is observed [11, 12]. The change in the Tafel slope has been ascribed to surface

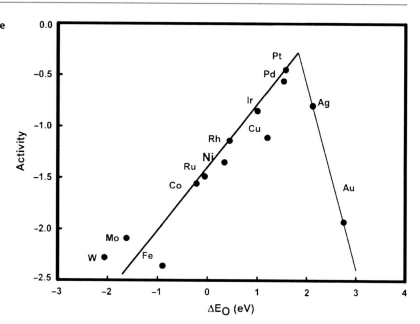

Platinum-Based Cathode Catalysts for Polymer Electrolyte Fuel Cells, Fig. 2 DFT activities as a function of the oxygen binding energy derived by density functional theory (DFT) calculations. The volcano is in good agreement with experiment, showing that Pt is the best catalysts for oxygen reduction (Ref. [16])

coverage by irreversible adsorption of HO_{ad}, which starts at 0.8–0.85 V and increases rapidly at more positive potentials. The Tafel slope of 60 mV dec^{-1} is consistent with the first electron charge transfer as the rds under Temkin conditions of adsorption of oxygen-containing species, while the slope of 120 mV dec^{-1} is characteristic for the same rds under Langmuirian conditions [13]. Adsorbed OH not only blocks active Pt site for the ORR but also alters the adsorption energy of ORR intermediates [14]. Given the first electron transfer to the adsorbed oxygen as rds and for a first-order dependence of the kinetics of ORR, the current density (i) has been proposed [8] to be determined either by the free Pt sites available for the adsorption of O_2, $(1 - \Theta_{ads})$, or by the change of Gibbs energy of adsorption of reactive intermediates (ΔG_{ads}):

$$i = nFKC_{O_2}(1 - \Theta_{ads})^x \exp(-\beta FE/RT) \times \exp(-\gamma \, r\Theta_{ads}/RT) \quad (4)$$

where F, x, β, γ, and R are constants, n is the number of electrons, c_{O2} is the oxygen concentration in the solution, K is the chemical rate constant, E is the applied potential, $r\Theta_{ads}$ is a parameter characterizing the rate of change of ΔG_{ads} with surface coverage, and Θ_{ads} is the fraction of electrode surface sites covered by all the adsorbed species [8]. The sensitivity of the Pt catalytic activity to other adsorbed species which block Pt-active sites also meant that great care must be taken to avoid trace impurities during measurements [15]. Modification of the Pt surface atomic or electronic structure leading to a reduced adsorption of hydroxyls (i.e., increasing the $(1 - \Theta_{ads})$ term in Eq. 4) without changing the overall mechanism will result in higher catalyst activity. Besides the influence of the adsorbated species, the ORR activity of Pt-based catalysts is also affected by the oxygen binding energy to the metal surface. Theoretically calculated oxygen binding energy as a function of the ORR activity for several metals shows a volcano plot (Fig. 2) indicating that too strong and too weak O_2 binding energy are both not favorable for ORR, according to the well-known Sabatier principle [16]. It has been later shown that the strength of the oxygen–metal bond interaction and the interaction of Pt with adsorbed species both strongly depend on the Pt d-band center and on the Pt–Pt interatomic distance [17].

As mentioned above, to minimize the amount of Pt required for a given level of activity, Pt catalyst is generally dispersed as small

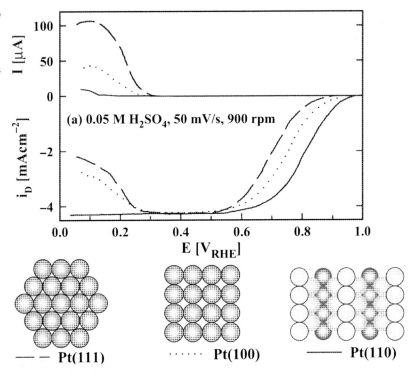

Platinum-Based Cathode Catalysts for Polymer Electrolyte Fuel Cells, Fig. 3 Oxygen reduction on Pt(hkl) in 0.05 M H_2SO_4 *lower part*, disk currents; *upper part*, ring currents. *Top view* of the surface structure of Pt(111), Pt(100), and Pt(110) (Ref. [8])

particles on high surface area carbon supports (Pt/C). It was proposed from the very beginning a modification of the Pt/C activity with the variation of the Pt particle size, i.e., for Pt particles in the range of 1–12 nm, the highest mass activity is obtained for Pt particles of about 3 nm [18]. Kinoshita [19] suggested that the maximum in mass activity observed at about 3 nm particle size is correlated with the large surface fraction of Pt atoms on the (100) and (110) crystal faces, while for smaller particles size the (111) crystal facets start to predominate and correlations to Pt single-crystal ORR activities have been drawn (Fig. 3). Newer results cannot necessarily confirm the maximum of mass activity at a specific particle size, showing a constant increase in mass activity with decreasing particle size, while the specific activity is constantly dropping in the same direction. This effect was correlated to the increase of potential of total zero charge (*pztc*) with increasing particle size towards large particles and bulk and as a consequence to the higher oxidation probability of small particles [20, 21].

Recent Advances in Pt-Based Catalyst Activity

In the last decades the fundamental understanding of the factors controlling the ORR has dramatically improved and is now guiding next-generation catalyst development. Although significant advances have been made in the area of non-Pt cathode catalysts, Pt-based catalysts are currently the most promising option to provide high catalytic activity towards ORR. Most emerging approaches focus on controlling the surface structure and composition of catalytic nanoparticles to achieve higher ORR activity with less amount of Pt. Improved ORR-specific activities have been reported alloying Pt with non-noble metals [22] or producing Pt core–shell nanoparticles (i.e., Pt layers supported on non-noble metal nanoparticles) [23, 24]. Typically, Pt-alloy catalysts consist of either binary or ternary systems containing different transition metals such as Co, Fe, or Ni. Compared to pure Pt, ORR activities of these catalysts are up to a factor of 4 higher [22], thus allowing the

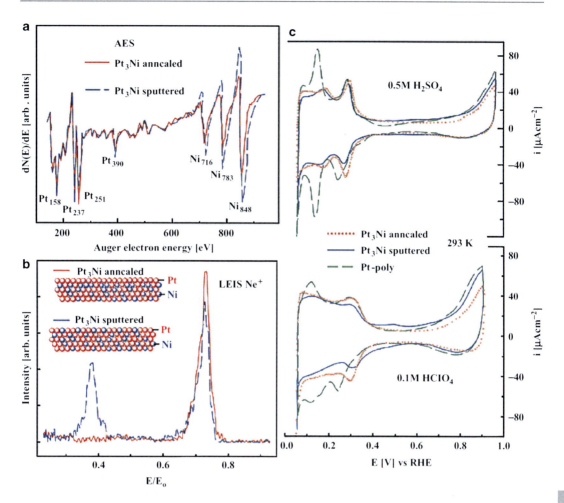

Platinum-Based Cathode Catalysts for Polymer Electrolyte Fuel Cells, Fig. 4 (a) Auger electron spectroscopy of as-sputtered and annealed Pt$_3$Ni. (b) Low-energy ion-scattering spectra of the same two surfaces, also shown is a schematic picture. (c) Cyclic voltammetry of Pt and two Pt$_3$Ni surfaces at 293 K in 0.5 M H$_2$SO$_4$ and 0.1 M HClO$_4$ (Ref. [28])

reduction of catalyst loading without loss of PEFC performance. The commercially available PtCo/C system provides MEA measurement values of specific and mass activity close to the targets set by the US Department of Energy for automotive applications, for example, 1.2 mA cm^{-2}$_{Pt}$ and 0.39 Am g^{-2}$_{Pt}$ at 0.9 V$_{SHE}$ [25]. Studies on the electrochemical stability of Pt alloys under PEFC operation show contradictory results. Some groups found an improved durability [26] whereas others reported accelerated catalyst degradation [27]. Contrasting results in the literature arise from the strong influence of several variable parameters (composition, particle size, and shape), which make it difficult to draw a definitive conclusion on the performance of Pt-alloy supported nanoparticles. As described above, thin-film single-crystal electrodes can serve as valuable model systems to understand the effect of structural aspects such as single-crystal facet orientation, size, and composition on the ORR activity. Studies on model electrodes have shown that while as-sputtered Pt-alloy catalysts present randomly distributed Pt atoms on the surface, annealed samples exhibit a coordinated pure "Pt skin" (see Fig. 4) [28, 29]. On the surface of these Pt-skin alloys, ORR proceeds via the same reaction mechanism described above for pure Pt, but the cyclic voltammetry in acid electrolytes shows a shift in the onset of oxide formation towards more positive potentials (up to 50 mV). This finding indicates that Pt-alloying leads to a fundamental change in the adsorption coverage of OH spectator species (i.e., the $(1 - \Theta_{ads})$ term in Eq. 4) [28].

Compared to pure Pt, the Pt-skin alloys show a shift in the d-band center which influences both the metal bond strength and the adsorption of the oxygenated spectators [17, 29]. For the Pt-skin alloys, the increase in the 5d vacancies results in an increase of the 2p electron donation from O_2 to the catalyst, leading to a weakening of the O–O bond compared to pure Pt. Given also the structure sensitivity of Pt to adsorbed anions, the synergy between the optimum in the surface crystal structure and electronic properties led to the highest catalytic activity ever reported to the best of our knowledge [29]. The ORR activity at 0.9 V_{RHE} for $Pt_3Ni(111)$ skin is about 1 order of magnitude higher than Pt(111) single crystal and about 90 times higher than a standard commercial dispersed Pt/C in comparative RDE measurements [29]. On the basis of these findings, a breakthrough in catalyst development would be the development of nanocatalysts able to reproduce the $Pt_3Ni(111)$ skin structure. Recently, truncated-octahedral Pt_3Ni particles with predominantly (111) facets were synthesized, showing activity values up to four times those of commercial Pt/C [30]. New results from Carpenter et al. demonstrated the synthesis and also high activity of PtNi-octaeder catalysts matching the aforementioned DOE targets by increasing the specific activity by a factor of 5 [31]. To mimic the Pt-alloy skin structure, electrochemical surface dealloying concept has been also proposed. The latter consists in a selective electrodissolution of non-noble metal atoms from bimetallic precursors in order to achieve a core–shell structure with a Pt-rich shell and a poor alloy particle core [24]. Increase in ORR activity (about three to sixfold) has been reported after Cu or Co dealloying from PtCu/C and PtCo/C, respectively, compared to Pt/C.

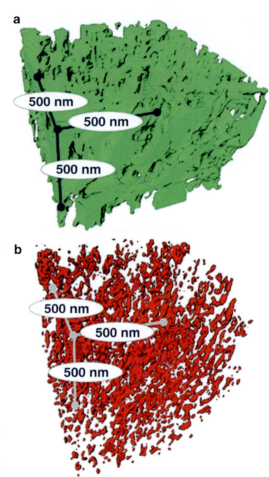

Platinum-Based Cathode Catalysts for Polymer Electrolyte Fuel Cells, Fig. 5 3D pore structure of the pristine cathode catalyst layer (**a**) and after 1,000 start/stop cycling (**b**). The *colored phase* represents the pore structure (Ref. [32])

Corrosion Issues

To reach the requirements of performance and durability for both automotive and stationary PEFC applications [20], catalyst durability has become an important issue both for of academic and industrial R&D. Cathode degradation in operating PEFCs mainly occurs under transient conditions, leading both to Pt dissolution/degradation and carbon-support corrosion. At typical operational pH conditions and above ~0.9 V, two main mechanisms lead to Pt degradation: (i) diffusion of dissolved Pt species

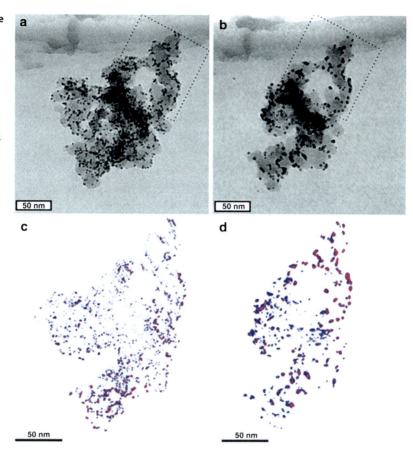

Platinum-Based Cathode Catalysts for Polymer Electrolyte Fuel Cells, Fig. 6 Standard TEM images (a, b) and tomography images (c, d) of the same catalyst location after 0 (a, c) and 3,600 (b, d) degradation cycles between 0.4 and 1.4 VRHE with a scan rate of 1 V s^{-1} (Ref. [33])

towards the electrolyte membrane where they are chemically reduced by crossover hydrogen and then precipitated, leading to a loss of electrically connected Pt particles, and (ii) Ostwald ripening of Pt inside the cathode electrode leading to a loss of Pt surface area due to nanoparticle growth. Regarding the durability of carbon support, even though carbon oxidation reaction can thermodynamically occur at potential as low as 0.206 V$_{SHE}$, due the sluggish kinetic of this reaction, carbon corrosion represents only a minor contribution to the cell degradation under steady-state operation. However, the corrosion becomes significantly accelerated during start/stop cycling due to the propagation of a hydrogen/air front through the anode compartment which leads to elevated potentials at the cathode (up to 1.5 V). Carbon-support corrosion leads to the so-called cathode thinning caused by loss of carbon void volume, which results in an increased oxygen diffusion resistance and loss of Pt surface area. Figure 5 shows morphology changes visualized by 3D-rendering SEM micrographs of Pt/C catalyst slices before and after 1,000 start/stop cycling [32]. The effect of start/stop cycling (3,600 cycles between 0.4 and 1.4 V$_{SHE}$) on the surface area of Pt catalyst is shown in Fig. 6 [33].

To improve the PEFC cathode durability, both materials and system approaches have been proposed in the last years. From the system point of view, improvements in cathode durability can be achieved by minimizing the residence time of the hydrogen–air front in the anode compartment. On the materials side, Pt alloys have shown higher durability compared relative to Pt-based electrodes [34]. However, since transition metal reduction potential is below that of hydrogen, once dissolved they will remain in the electrolyte membrane, leading to H⁺ exchange and reduction of the fuel cell performance. To reduce start/stop degradation of the catalyst support, graphitized carbon particles have been proposed as alternative to carbon-black, showing improved durability compared to the latter but also smaller surface area than carbon-black (Table 1) [4]. Furthermore, since the support is still based on carbon materials, the oxidation reaction is only delayed but not completely avoided. A more straightforward approach to completely prevent carbon corrosion consists in replacing carbon-based supports with more stable supports or in using unsupported Pt catalysts. In the last years, a growing number of publications report the use of carbides, nitrides, or conductive oxides as alternative Pt supports. Improved durability has been obtained for WC compared to Vulcan XC-72R over 100 cycles between 0 and 1.4 V [35]. However, carbides as well as nitrides compounds are still thermodynamically unstable under the oxidative conditions of a PEFC cathode. Therefore, their oxidation is expected after long cycling leading to a loss of their electronic conductivity and surface area. Differently, doping binary oxides and using advanced synthesis methods can lead to stable materials under the most relevant PEFC cathode conditions with high electronic conductivity and high surface area. Pt supported on $Mo_{0.3}Ti_{0.7}O_2$ have shown improved durability as well as activity towards ORR compared to commercial Pt/C [36]. After 5,000 potential cycling at room temperature between 0 and 1.1 V, Pt/$Mo_{0.3}Ti_{0.7}O_2$ showed a decrease of activity at 0.9 V of 8 %, while the activity of Pt/C and PtCo/C decreased of 50.6 % and 25.8 %, respectively, under in the same experimental conditions. Under long stability test (10,000 potential cycling between 0.3 and 1.3 V_{RHE}), Pt/SnO_2 have showed only a slight decrease of the electrochemical surface area, while strong degradation was observed for Pt/C catalyst prepared and tested with the same procedure [37]. Furthermore, under fuel cell tests Pt/SnO_2 showed similar performance than Pt/C catalyst. A different non-carbon catalyst support developed by 3M is based on an oriented array of organic whiskers, with a length below 1 mm, a diameter of about 50 nm, and a number density of about 30–40 whiskers mm^{-2} [25]. The complete catalyst is achieved by coating the whisker structure with the desired catalyst thin film by physical vapor deposition or magnetron sputtering (see Fig. 7) [38]. These nanostructured thin-film (NSTF) catalysts present for the same catalyst loading significantly lower electrochemical surface area compared to conventional Pt/C catalysts; but the high catalyst utilization of almost 100 % and negligible losses within the catalyst layer lead to ten times higher area-specific activity and 50 % higher mass activity

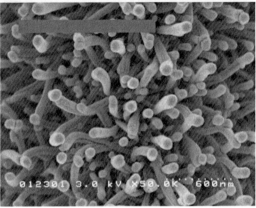

Platinum-Based Cathode Catalysts for Polymer Electrolyte Fuel Cells, Fig. 7 Scanning electron micrographs of typical NSTF catalysts as fabricated on a microstructured catalyst transfer substrate, seen (*top*) in cross section with original magnification of ×10,000, and (*bottom*) in plain view with original magnification of ×50,000 (Ref. [38])

than conventional Pt/C. Furthermore, NSTF catalysts are much more resistant to loss of surface area from high voltage cycling than Pt/C. It was reported that while Pt/C loses 90 % of the initial surface area in 2,000 cycles between, a NSTF-Pt catalyst asymptotically approached a maximum of 33 % surface area loss in 9,000 cycles between [38].

Conclusions and Future Directions

In the view of a sustainable energy system, more and more attention has been directed towards the development of efficient energy conversion systems such as PEFCs. However, in order to achieve a widespread commercialization, still cost and durability issues must be solved. In the last decades, system optimization and fundamental understanding of the ORR mechanism at the PEFC cathode have led to a drastic reduction of the Pt loading and thus of the cost system. On the basis of the findings achieved so far, a breakthrough in catalyst development would be the production of nanocatalysts able to reproduce the structure and electronic properties of the $Pt_3Ni(111)$ model electrodes. On the durability side, carbon-support replacement represents an urgent target to achieve. Promising results have been obtained using oxide-based supports; in addition this strategy can open new possibilities also in the search for more active catalysts because the establishment of a metal-support interaction may lead to a further activity enhancement.

Cross-References

▶ Electrocatalysis - Basic Concepts, Theoretical Treatments in Electrocatalysis via DFT-Based Simulations
▶ Oxygen Reduction Reaction in Acid Solution
▶ Platinum-Based Anode Catalysts for Polymer Electrolyte Fuel Cells
▶ Polymer Electrolyte Fuel Cells (PEFCs), Introduction

References

1. Srinivasan S, Manko DJ, Koch H, Enayetullah MA, Appleby AJ (1990) Recent advances in solid polymer electrolyte fuel cell technology with low platinum loading electrodes. J Power Sources 29:367–387
2. Raistrick ID (1986) Diaphragms, separators, and ion exchange membranes. In: Van Zee JW, White RE, Kinoshita K, Burney HS (eds) The Electrochemical Society softbound proceeding series, Pennington
3. Taylor EJ, Anderson EB, Vilambi NRK (1992) Preparation of high-platinum-utilization gas diffusion electrodes for proton-exchange-membrane fuel cells. J Electrochem Soc 139:L45–L46
4. Yu PT, Gu W, Zhang J, Makharia R, Wagner FT, Gasteiger HA (2009) Carbon support requirements for highly durable fuel cell operation. In: Büchi FN, Inaba M, Schmidt TJ (eds) Polymer electrolyte fuel cell durability. Springer, New York
5. Neyerlin KC, Gu WB, Jorne J, Gasteiger H (2007) A study of the exchange current density for the hydrogen oxidation and evolution reactions. J Electrochem Soc 154:B631–B635
6. Wagner FT, Lakshmanan B, Mathias M (2010) Electrochemistry and the future of the automobile. J Phys Chem Lett 1:2204–2209
7. Sepa DB, Vojnovic MV, Damjanovic A (1981) Reaction intermediates as a controlling factor in the kinetics and mechanism of oxygen reduction at platinum electrodes. Electrochim Acta 26:781–793
8. Markovic NM, Schmidt TJ, Stamenkovic V, Ross PN (2001) Oxygen reduction reaction on Pt and Pt bimetallic surfaces: a selective review. Fuel Cells 1:105–115
9. Yeager E (1984) Electrocatalysts for O_2 reduction. Electrochim Acta 29:1527–1537
10. Sidik RA, Anderson AB (2002) Density functional theory study of O_2 electroreduction when bonded to a Pt dual site. J Electroanal Chem 528:69–76
11. Paulus UA, Schmidt TJ, Gasteiger HA, Behm RJ (2001) Oxygen reduction on high-surface area Pt/Vulcan carbon catalyst: a thin-film rotating ring-disk electrode study. J Electroanal Chem 495:134–145
12. Subramanian NP, Greszler TA, Zhang J, Gu W, Makharia R (2012) Pt-oxide coverage-dependent oxygen reduction reaction (ORR) kinetics. J Electrochem Soc 159:B531–B540
13. Grgur BN, Markovic NM, Ross PN (1997) Temperature-dependent oxygen electrochemistry on platinum low-index single crystal surfaces in acid solutions. Can J Chem 75:1465–1471
14. Markovic NM, Gasteiger HA, Grgur BN, Ross PN (1999) Oxygen reduction reaction on Pt(111): effects of bromide. J Electroanal Chem 467:157–163
15. Paulus UA, Schmidt TJ, Gasteiger HA (2003) Poisons for the oxygen reduction. In: Vielstich W, Gasteiger HA, Lamm A (eds) Handbook of fuel cells, vol 2. Wiley, Chichester

16. Norskov JK et al (2004) Origin of the overpotential for oxygen reduction in fuel cell cathode. J Phys Chem B 108:17886–17892
17. Stamenkovic V et al (2006) Changing the activity of electrocatalysts for oxygen reduction by tuning the surface electronic structure. Angew Chem Int Ed 45:2897–2901
18. Peuckert M, Yoneda T, Dalla Betta RA, Boudart M (1986) Oxygen reduction on small supported platinum particles. J Electrochem Soc 133:944–947
19. Kinoshita K (1990) Particle size effects for oxygen reduction on highly dispersed platinum in acid electrolytes. J Electrochem Soc 137:845–848
20. Rabis A, Rodriguez P, Schmidt TJ (2012) Electrocatalysis for polymer electrolyte fuel cells: recent achievements and future challenges. ACS Catal 2:864–890
21. Mayrhofer KJJ, Blizanac BB, Arenz M, Stamenkovic V, Ross PN, Markovic NM (2005) The impact of geometric and surface electronic properties of Pt-catalysts on the particle size effect in electrocatalysis. J Phys Chem B 109:14433–14440
22. Gasteiger HA, Kocha SS, Sompalli B, Wagner FT (2005) Activity benchmarks and requirements for Pt, Pt-alloy, and non-Pt oxygen reduction catalysts for PEMFCs. Appl Catal Environ 56:9–35
23. Wei ZD et al (2008) Electrochemically synthesized Cu/Pt core-shell catalysts on a porous carbon electrode for polymer electrolyte membrane fuel cells. J Power Sources 180:84–91
24. Koh S, Strasser P (2007) Electrocatalysis on bimetallic surfaces: modifying catalytic reactivity for oxygen reduction by voltammetric surface dealloying. J Am Chem Soc 129:12624–12625
25. Debe MK (2012) Electrocatalyst approaches and challenges for automotive fuel cells. Nature 7:43–51
26. Yu P, Pemberton M, Plasse P (2005) PtCo/C cathode catalyst for improved durability in PEMFCs. J Power Sources 144:11–20
27. Zignani C, Antolini E, Gonzalez ER (2008) Evaluation of the stability and durability of Pt and Pt-Co/C catalysts for polymer electrolyte membrane fuel cells. J Power Sources 182:83–90
28. Stamenkovic V, Schmidt TJ, Ross PN, Markovic NM (2002) Surface composition effects in electrocatalysis: kinetics of oxygen reduction on well-defined Pt_3Ni and Pt_3Co alloy surfaces. J Phys Chem B 106:11970–11979
29. Stamenkovic VR et al (2007) Improved oxygen reduction activity on $Pt_3Ni(111)$ via increased surface site availability. Science 315:493–497
30. Wu JB et al (2010) Truncated octahedral Pt_3Ni ORR electrocatalysts. J Am Chem Soc 132:4984–4985
31. Carpenter MK, Moylan TE, Kukreja RS, Atwan MH, Tessema MM (2012) Solvothermal synthesis of platinum alloy nanoparticles for oxygen reduction electrocatalysis. J Am Chem Soc 134:8535–8542
32. Schulenburg H, Schwanitz B, Linse N, Scherer GG, Wokaun A (2011) 3D imaging of catalyst support corrosion in polymer electrolyte fuel cell. J Phys Chem C 115:14236–14243
33. Meier JC et al (2012) Degradation mechanisms of Pt/C fuel cell catalysts under simulated start–stop conditions. ACS Catal 2:832–843
34. Hartnig C, Schmidt TJ (2011) Simulated start–stop as a rapid aging tool for polymer electrolyte fuel cell electrodes. J Power Sources 196:5564–5572
35. Chhina H, Campbell S, Kesler O (2007) Thermal and electrochemical stability of tungsten carbide catalyst supports. J Power Sources 164:431–440
36. Ho VTT et al (2011) Nanostructured $Ti_{0.7}Mo_{0.3}O_2$ support enhances electron transfer to Pt: high-performance catalyst for oxygen reduction reaction. J Am Chem Soc 133:11716–11724
37. Masao A, Noda S, Takasaki F, Ito K, Sasaki K (2009) Carbon-Free Pt electrocatalysts supported on SnO_2 for polymer electrolyte fuel cells. Electrochem Solid State Lett 12:B119–B122
38. Debe MK, Schmoeckel AK, Vernstrom GD, Atanasoski R (2006) High voltage stability of nanostructured thin film catalysts for PEM fuel cells. J Power Sources 161:1002–1011

Polarography

Michael Heyrovsky
J. Heyrovsky Institute of Physical Chemistry of the ASCR, Prague, Czech Republic

In its primary meaning, polarography is the method of potential-controlled *electrolysis with dropping mercury electrode*, usually of drop time 3–5 s, as working electrode. The physicochemical properties of mercury are specially advantageous in electrochemistry. As liquid element it can be prepared in the highest purity by repeated distillation, its surface is homogeneous and isotropic on atomic scale. The dropping renews spontaneously the electrode surface irrespective of processes that could have occurred on it, which makes the results of measurements highly reproducible. Mercury has a high surface tension, and in contact with a solution, many dissolved species have the tendency to adsorb on its surface, depending on its potential. In electrochemistry mercury has of all metals the highest hydrogen overvoltage which gives mercury electrodes wide potential range for following electrode reactions

Polarography

in negative direction. On the other hand, in a relatively narrow range of positive potentials, metallic mercury dissolves into ions which may interact with components of the solution thus enabling their determination. In general, polarography is better suited for following cathodic than anodic processes. In order to keep the resistance of the polarographic cell as low as possible, the electrolyzed solutions should contain a supporting electrolyte, usually of 0.1–1 M/l concentration. In the simple *"classical" polarography* the mean current flowing through the cell with dropping mercury electrode is recorded as function of direct potential applied to the dropping electrode with respect to an electrode maintaining constant potential, the reference electrode. The presence of an electroactive substance shows on the polarographic curve usually in the form of a concentration-dependent current increase to a limiting value, or "wave," in a definite potential range.

Systematic experimental research resulted in distinction of several basic kinds of polarographic currents which theoreticians formulated by quantitative equations for their instantaneous time course and then for their mean time course during the drop life:

- *Charging current* given by charging electrical double layer of the dropping electrode by ions from the solution
- *Diffusion current* given by the rate of diffusion of the electroactive species from solution to the surface of the dropping electrode
- *Kinetic current* given by the rate of a chemical reaction producing the electroactive species in the vicinity of the electrode
- *Catalytic current* given by the rate of a catalytic process generating the electroactive species near or at the electrode surface
- *Adsorption current* given by the rate and extent of adsorption of electroactive and surface active species from the solution at the surface of the electrode.

Besides the mentioned basic currents on polarographic curves, *current maxima can occur*, according to experimental conditions either "of the first kind"(tapering shape) or "of the second kind"(rounded shape). "Classical" polarography

has been used for experimental research in electrochemistry or for qualitative and quantitative electroanalytical determinations, its advantage being simple operations and a relatively cheap apparatus. It allows to analyze electroactive species down to 10^{-6} M/l concentration, in case of catalytic currents even lower. In analytical practice after recording polarographic curves, the *"polarometric"* (or *"amperometric"*) *titrations* can be introduced with dropping mercury electrode maintained at constant potential of the limiting current, indicating actual concentration of the electroactive substance.

The "classical" polarography has been fully developed and internationally widely utilized by the nineteen forties. The technical progress in the middle of twentieth century led to various modifications of the primary simple method. In *"alternating current polarography"* on the direct voltage applied to the cell is superimposed alternating voltage of small amplitude, and instead of direct current the resulting alternating current is recorded, producing current peaks instead of "waves" on the current-voltage curves; the peaks apart of being the measure of concentration indicate the occurrence of possible adsorption/desorption processes at the electrode surface. In *"square-wave polarography"* the small alternating voltage superimposed on the polarizing voltage is of rectangular shape and the instantaneous current is measured at the end of each individual voltage pulse, i.e., with lowered charging current component – this considerably increases the sensitivity of the method. The fact that the instantaneous charging current at the dropping mercury electrode decreases with time of the drop life has been also utilized in *"tast polarography"* where the total polarographic current is recorded before the drop falls off, i.e., when the charging current component of the total current is minimized and the sensitivity for diffusion current is thus increased. In *"pulse polarography"* the individual drops of the dropping mercury electrode are polarized by gradually increasing pulses of constant voltage, while the current is measured at the end of the drop life when the charging current component is negligible, the result being again increased sensitivity. *"Differential pulse polarography"* records differences between instantaneous currents at the ends of subsequent drops which produces polarographic curves with current peaks in place of waves, also with high sensitivity. Besides the abovementioned methods, many other modifications of polarography have been described.

In order to speed up the method of classical polarography instead of recording current-voltage curves graphically, the cathode ray oscilloscope was used providing their optical display. Because in the *"oscillographic polarography"* the time scale is different than in the classical method, the shape of the current-voltage curves is different and hence the theory of classical polarography cannot be directly applied. It is more so in the *"alternating current oscillographic polarography,"* a method belonging to chronopotentiometry, where the dropping electrode is polarized by alternating current and the curves of potential versus time and of the derivative of potential/time dependence versus potential are followed; as the experimental conditions are different, the results considerably differ, qualitatively and quantitatively, from those of classical polarography.

When the dropping of the mercury electrode is stopped and the current-potential curve is recorded with one stable hanging mercury drop, the method formally belongs to *voltammetry* – according to accepted terminology polarography is voltammetry with dropping mercury electrode. However, as long as the hanging mercury drop is maintained as working electrode, that particular voltammetry can be considered as polarography in wider sense because of similar background conditions. If the potential scan is applied to the electrode in one direction, it is **linear voltammetry**, and if it is applied in one direction and back, it is **cyclic voltammetry**. Cyclic voltammetry offers a simple check of reversibility of an electrode process – a direct comparison of the cathodic and anodic reaction. In analytical applications the voltammetry with hanging mercury drop has the advantage over classical polarography in that an active species can be accumulated at the electrode under constant potential for some time and then electrochemically determined during a following voltage scan – by that *voltammetric "stripping"* method an electroactive species can be determined in nanomolar and lower concentrations.

If the cell with hanging mercury drop electrode is polarized by constant current and the time changes of the electrode potential are recorded, the thus introduced *chronopotentiometric method* provides effects which can be interpreted on the basis of experience with polarography and which extend polarographic results further. That is, for example, the case with catalytic evolution of hydrogen. In current-controlled electrolysis the electrode potential changes at the rate given by the intensity of the polarizing current – in this way at mercury electrode the negative potentials are reached in general faster in chronopotentiometry than in voltammetry and the catalysis of hydrogen evolution occurs at less negative potentials. A particularly well-developed signal of hydrogen catalysis, called *"peak H,"* is produced in *constant current derivative stripping chronopotentiometry* (or chronopotentiometric stripping analysis, CPSA). This "peak H" is used in bioelectrochemistry to study and determine biomacromolecules such as peptides, proteins, nucleic acids, or polysaccharides.

After decades of development in science and technology, polarography is being generally understood not as a single research method but rather as a field of electrochemistry based on the unique properties of mercury electrodes.

Cross-References

► Mercury Drop Electrodes
► Polarography
► Potentiometry
► Voltammetry of Adsorbed Proteins

References

1. Schmidt H, Stackelberg M v (1964) Modern polarographic methods. Academic, New York
2. Meites L (1965) Polarographic techniques, 2nd edn. Wiley, Interscience, New York
3. Kalvoda R (1965) Techniques of oscillographic polarography, 2nd edn. Elsevier, SNTL, Amsterdam

4. Heyrovský J, Kůta J (1966) Principles of polarography. Academic, New York
5. Heyrovský J, Zuman P (1968) Practical polarography. Academic, London/New York
6. Bond AM (1980) Modern polarographic methods in analytical chemistry. Marcel Dekker, New York
7. Baizer MM, Lund H (1983) Organic electrochemistry. Marcel Dekker, New York
8. Meites L, Zuman P, Rupp EB, Fenner TL, Narayanan A (1983) Handbook series in organic electrochemistry, vol I-VI. CRC, Boca Raton
9. Meites L, Zuman P, Rupp EB, Fenner TL, Narayanan A (1986) Handbook series in inorganic chemistry, vol I-VII. CRC, Boca Raton
10. Brainina KZ, Neyman E (1993) Electroanalytical stripping methods. Wiley-Interscience, New York
11. Crow DR (1994) Principles and applications of electrochemistry, 4th edn. Blackie Academic Professional, London
12. Galus Z (1994) Fundamentals of electroanalysis, 2nd edn. Elis Horwood/Polish Scientific Publishers PWN, New York/Warsaw
13. Kissinger PT, Heineman WR (1996) Laboratory techniques in electroanalytical chemistry. Marcel Dekker, New York
14. Vanýsek P (1996) Modern techniques in electroanalysis. Wiley, New York
15. Bard AJ, Faulkner LR (2001) Electrochemical methods, 2nd edn. Wiley, New York
16. Scholz F (2002) Electroanalytical methods. Springer, Berlin
17. Paleček E, Scheller F, Wang J (eds) (2005) Electrochemistry of nucleic acids and proteins. Elsevier, Amsterdam
18. Wang J (2006) Analytical electrochemistry, 3rd edn. Wiley, New York

Pollutants in Water - Electrochemical Remediation Using Ebonex Electrodes

Nigel J. Bunce and Dorin Bejan
Department of Chemistry, Electrochemical Technology Centre, University of Guelph, Guelph, ON, Canada

Introduction

Ebonex$^{\circledR}$ is the trade name for an electrically conductive ceramic suboxide of titanium, having the approximate composition Ti_4O_7 and conductivity comparable to that of carbon.

Early research on Ebonex, including its discovery in 1981, has been described by Hayfield, who noted that an impetus for its development was to find an electrode material that would be resistant to highly corrosive environments [1]. As a ceramic material, Ebonex powder can be sintered into a variety of shapes and also porosities [2].

Ti_4O_7 is one of a wide range of defined compositions TiO_x ($0.5 < x < 2$); those between that of Ebonex ($x = 1.75$) and $x = 1.9$ have a triclinic crystal structure and are known as Magnéli phases Ti_nO_{2n-1}. Titanium dioxide cannot be reduced to the metal with hydrogen, carbon, or carbon monoxide but is reduced to TiO_x suboxides at $\sim 1,200$–$1,300\ ^{\circ}C$, usually with hydrogen.

To date, the most frequent application of Ebonex (named for its black color) has been in cathodic protection, especially of the steel rods used to reinforce concrete [1]. There is an extensive literature on Ebonex coated with various other materials (noble metals, carbon, lead dioxide), whose deposition is favored by Ebonex's high overpotential for hydrogen and oxygen evolution. Applications for coated Ebonex include batteries, fuel cells, and dechlorination of chlorinated pollutants [2], but the focus of this article is the use of uncoated Ebonex as an electrode material for electrolysis of pollutant substances, most often for electrochemical oxidation.

Concepts

Ebonex has many of the properties of a non-active anode, at which reactive hydroxyl species are produced upon oxidation of water. Most non-active anodes are based on main group elements, such as PbO_2, SnO_2, and boron-doped diamond (BDD), whereas Ebonex is based on titanium, a first-row transition metal.

The mechanistic distinction between active and non-active anodes is considered elsewhere in this encyclopedia [3]. Both types of anodes act indirectly, by electrochemical oxidation of water to reactive anode-bound species that can bring about chemical oxidation of a substrate.

At an anode **A**, the relevant reactions are given by Eqs. (1) and (2):

$$\mathbf{A} + H_2O \rightarrow \mathbf{A} \sim \!\!^\bullet OH + H^+ + e^- \quad (1)$$

$$\mathbf{A} \sim \!\!^\bullet OH \rightarrow \mathbf{A} = O + H^+ + e^- \quad (2)$$

At a non-active anode, the sorbed hydroxyl radicals ("physisorbed active oxygen") from Reaction (1) can initiate free radical (one-electron) reactions of oxidizable substrates. Often, this leads to complete mineralization of organic materials (also called electrochemical combustion).

Active anodes are dimensionally stable anodes based on noble metal oxides such as RuO_2, IrO_2, and Pt (=PtO_x under anodic polarization), at which the initially formed $\mathbf{A} \sim \!\!^\bullet OH$ is readily oxidized further to give chemisorbed active oxygen (Eq. 2). Oxidations at active anodes rarely give efficient mineralization of the substrate. At both types of anode, the parasitic production of molecular O_2 lowers the current efficiency for substrate oxidation.

Ebonex has a wide range of potentials over which water is stable (approximately -1 V to $+2.7$ V vs. Ag/AgCl in aqueous sulfuric acid) [2], but shifted significantly in the expected negative direction in base [4]. Most relevant to the present perspective is the similarity of the anodic range to that of boron-doped diamond, for which values of the overpotential for the OER are ~ 1.3 V [3].

Typical evidence that Ebonex functions as a non-active anode is that when Ebonex was compared with different anodes for oxidations of acetaminophen (AP) and p-benzoquinone (BQ), Ebonex behaved much more like BDD (non-active) than like Ti/IrO_2-Ta_2O_5 (active) in terms of the low yield of BQ formed from AP, and in the rates of transformation versus complete mineralization [5, 6]. The oxidation of coumarin to the fluorescent product, 7-hydroxycoumarin, is also a useful qualitative probe for the presence of hydroxyl species in these oxidations [7, 8]. The activity of Ebonex is consistently less than that of BDD, suggesting a lower concentration of hydroxyl species. In competition experiments involving substrates that are resistant to direct oxidation but susceptible to hydroxyl radical attack, substrate loss was slower at Ebonex anodes than at BDD under similar conditions but with less discrimination between substrates, suggesting that at Ebonex the hydroxyl radicals are bound less tightly to the anode [7].

The behavior of Ebonex is more complicated than that of the conventional non-active anodes PbO_2, SnO_2-Sb_2O_5, and BDD. Passivation towards oxygen evolution occurs at high current density, because the surface becomes oxidized to less conductive Magnéli phases or even to TiO_2 [9]. Bejan et al. [6] found that when a pair of Ebonex electrodes was polarized for a period of time and then the current was interrupted, depolarization occurred with evolution of gases, indicating oxidation or reduction of the Ebonex surface upon anodic or cathodic polarization, respectively. This is consistent with the working paradigm that the non-active behavior of an Ebonex anode is accompanied by oxidation of the Magnéli phase towards the composition TiO_2. The gradual loss of anode activity occurs because the $\mathbf{A} = O$ species from Ebonex fails to oxidize added substrates, making the surface oxidation essentially irreversible, contrasting typical active anodes such as Ti/RuO_2. Recovery of the anode performance can be achieved by reducing the inactive deposit back to an active form. This paradigm may parallel observations by Beck and Gabriel [10] that Ti/TiO_2 ceramic materials can be made conducting under cathodic polarization, suggesting the possible electrochemical formation of the Magnéli phase. The resulting surface can catalyze the reduction of other substrates.

Consistent with this explanation for anode passivation, when an Ebonex anode was used for repeated oxidations of p-nitrosodimethylaniline, the rate constant for loss of the substrate gradually decreased; partial anodic activity was restored by cathodic polarization of the deactivated anode [6]. In the oxidation of sulfide ion, a constituent of geothermal sour brines, the gradual loss of activity of an Ebonex anode was overcome by employing periodic polarity reversal in order to reduce the over-oxidized surface layer back to the Magnéli phase [11]. Sulfide was

Results

Initial voltammetric studies had seemed to suggest that Ebonex is inert towards both oxidation and reduction of added substrates. The well-known ferrocyanide/ferricyanide couple shows smaller currents at Ebonex than at, e.g., platinum, with the oxidation and reduction currents shifted to more positive and more negative potentials, respectively. This evidence for slow electron transfer indicates that high overpotentials are needed to drive electrolysis at Ebonex [9]. Based on voltammetry, Grimm et al. [12] concluded that Ebonex was inert towards the oxidation of phenol, because the addition of 5 mM phenol to supporting electrolyte caused only a small current increase in the potential range of water stability. Likewise, Scott and Cheng [13] reported no voltammetric oxidation current for oxidation of oxalic acid at an Ebonex anode; however, they did observe an increase in the cell voltage upon prolonged polarization, consistent with gradual oxidative inactivation of the Ebonex anode.

Smith et al. [14] had reported that the porosities of porous and hardened Ebonex were 12–15 % and <2 %, respectively, and that the voltammetric behavior of Ebonex in terms, for example, of the potential range of water stability is highly dependent on porosity [9]. Recently, Kitada et al. [15] reported selective preparation of macroporous monoliths of Magnéli phases having widely different porosities and bulk densities.

The idea that porous Ebonex might be used as a flow-through permeable electrode, originally suggested by Chen et al. [16], was exploited by Zaky and Chaplin [17], who used a porous, cylindrical Ebonex anode as a reactive electrochemical membrane for the oxidation of several model compounds. These included oxalic acid and p-.methoxyphenol that might have been expected to be inert to oxidation based on the previous voltammetry studies [12, 13] – for example, well-defined oxidation currents were obtained with p-methoxyphenol at slow sweep rates [17]. Under electrolysis conditions, contaminated influent water was passed through the cylinder, while a current was passed between the cylinder (anode) and a central stainless steel rod (cathode). With oxalic acid as substrate, the concentration of substrate in the effluent from the cylinder and in the permeate both decreased with the applied current, essentially to zero in the permeate. p-Methoxyphenol could also be remediated, in terms of both the substrate concentration and the COD. At low current densities (=low anode potential), a significant proportion of the COD removed was associated with material adsorbed to the anode, which is typical of anode fouling by phenolic compounds upon electrochemical oxidation. Fouling was eliminated at an anode potential >2 V versus SHE, consistent with an indirect oxidation mechanism, corresponding to the production of hydroxyl radicals under these conditions.

Scialdone et al. [18] compared Ebonex with several other anodes for the mineralization of 1,2-dichloroethane at pH 2. In this application, Ebonex performed poorly in comparison with BDD (CE $\sim$12 % vs. $\sim$50 %) and no better than active anodes such as Pt and $Ti/IrO_2\text{-}Ta_2O_5$. Whether the current yield could be improved by periodic polarity reversal was not studied.

Nowack et al. [19] used Magnéli phase TiO_x for the oxidation of cyanide ion to cyanate (which hydrolyzed to ammonia and carbonate) at pH $\sim$13. From initial cyanide concentrations in the range 0.5–2.0 g L^{-1}, the specific energy ranged from 12 to 260 kWh per kg cyanide, depending on the extent of remediation of cyanide.

Chen et al. [16] achieved complete mineralization of trichloroethylene at an Ebonex anode, the chief products being CO_2 (and about 10 % CO), chloride, and chlorate. The authors determined the fractions of the anodic current carried out to form O_2, carbon oxides, and ClO_3^- and observed the expected relationship that the current efficiency for TCE oxidation decreased at the expense of that for O_2 with increasing anode potential. Spin trapping experiments were carried out to search for hydroxyl radical intermediates, although the identity of the spin adduct was later challenged as not definitive for $HO^{\bullet}$ [7].

Considering the activity of Ebonex as a cathode, it was found [20, 21] that the resistance of Ebonex was so great towards the reduction of sulfite ions that sulfite could be oxidized to sulfate using combinations of Ebonex cathode with various anodes even in undivided cells. Chen et al. [22] likewise reported that uncoated Ebonex was inactive towards the reductive dechlorination of trichloroethylene.

When Ebonex cathodes were used for the reduction of aqueous nitrate ion, comparable amounts of ammonia and elemental nitrogen were formed, as at many other cathodes [23]. Complete denitrification was achieved by coupling the electrochemical reduction to oxidation of ammonia in the presence of chloride ion, which was oxidized to hypochlorite at the anode. The chemical oxidation of ammonia to nitrogen by hypochlorite corresponds to breakpoint chlorination in water treatment. The advantage of Ebonex in this application is its ability to undergo periodic polarity reversal, in order to avoid fouling of the cathode by carbonate deposits in hard water. The optimum experimental arrangement comprised a pair of Ebonex electrodes whose polarity was periodically reversed, along with a "full time" anode (e.g., Ti/IrO_2-Ta_2O_5) to optimize the oxidation of chloride to hypochlorite. This procedure parallels an early application of Ebonex electrodes for swimming pool electrochlorinators, which also involve electrochemical hypochlorination, with polarity reversal to avoid the accumulation of hard water scale [1].

Advantages and Future Directions

Ebonex electrodes have been used for electrolysis of both organic and inorganic pollutants, as anodes or as cathodes. Ebonex is produced from inexpensive starting materials (titania and hydrogen) and has excellent corrosion resistance and the ability, as a conductive ceramic, to be fabricated into a variety of shapes. Mechanistically, Ebonex has characteristics of both active and non-active anodes – it furnishes reactive hydroxyl radicals but also undergoes surface oxidation. However, the "higher oxides" produced from Ebonex are more oxidized Magnéli phases than Ti_4O_7 and have little or no oxidizing power towards added substrates. In comparison with BDD as a non-active anode, Ebonex exhibits similar mineralization ability but with lower current efficiency, and its activity must be maintained by employing periodic polarity reversal with the combination Ebonex anode–Ebonex cathode in order to reduce the over-oxidized Ti_4O_7 phase back to its reactive form. Polarity reversal appears to be a promising application for the remediation of aqueous wastes that contain hardness cations. Unlike BDD, which tends to be degraded at high pH, Ebonex electrodes can be used in both acidic and alkaline solutions. However, there has been relatively little comparison of Ebonex with other non-active anodes, and its full range of applications is probably yet to be discovered.

References

1. Hayfield PCS (2002) Development of a new material – Monolithic Ti_4O_7 Ebonex® Ceramic. Royal Society of Chemistry/Metal Finishing Information Services, Cambridge, UK/Herts
2. Walsh FC, Wills RGA (2010) The continuing development of Magnéli phase titanium sub-oxides and Ebonex® electrodes. Electrochim Acta 55:6342–6351
3. Bunce NJ, Bejan D (2013) Oxidation of organic pollutants on active and non active anodes. In: Savinell RF, Ota K, Kreysa G (eds) Encyclopedia of applied electrochemistry. Springer
4. Pollock RJ, Houlihan JF, Bain AN, Coryea BS (1984) Electrochemical properties of a new electrode material Ti_4O_7. Mater Res Bull 19:17–24
5. Waterston K, Wang JW, Bejan D, Bunce NJ (2006) Electrochemical waste water treatment: electrooxidation of acetaminophen. J Appl Electrochem 36:227–232
6. Bejan D, Malcolm JD, Morrison L, Bunce NJ (2009) Mechanistic investigation of the conductive ceramic Ebonex as an anode material. Electrochim Acta 54:5548–5556
7. Bejan D, Guinea E, Bunce NJ (2012) On the nature of the hydroxyl radicals produced at boron-doped diamond and Ebonex anodes. Electrochim Acta 69:275–281

8. Kisacik I, Stefanova A, Ernst S, Baltruschat H (2013) Oxidation of carbon monoxide, hydrogen peroxide and water at a boron doped diamond electrode: the competition for hydroxyl radicals. Phys Chem Chem Phys 15:4616–4624
9. Smith JR, Walsh FC, Clarke RL (1998) Electrodes based on Magnéli phase titanium oxides: the properties and application of Ebonex® materials. J Appl Electrochem 28:1021–1033
10. Beck F, Gabriel W (1985) Heterogeneous redox catalysis at titanium/titanium dioxide cathodes. Reduction of nitrobenzene. Angew Chem 97:765–767
11. El-Sherif S, Bejan D, Bunce NJ (2010) Electrochemical oxidation of sulfide ion in synthetic sour brines using periodic polarity reversal at Ebonex electrodes. Can J Chem 88:928–936
12. Grimm J, Bessarabov D, Maier W, Storck S, Sanderson RD (1998) Sol–gel film-preparation of novel electrodes for the electrocatalytic oxidation of organic pollutants in water. Desalin 115:295–302
13. Scott K, Cheng H (2002) The anodic behaviour of Ebonex® in oxalic acid solutions. J Appl Electrochem 32:583–589
14. Smith JR, Nahle AH, Walsh FC (1997) Scanning probe microscopy studies of Ebonex® electrodes. J Appl Electrochem 27:815–820
15. Kitada A, Hasegawa G, Kobayashi Y, Kanamori K, Nakanishi K, Kageyama H (2012) Selective preparation of macroporous monoliths of conductive Titanium Oxides Ti_nO_{2n-1} (n = 2, 3, 4, 6). J Am Chem Soc 134:10894–10898
16. Chen G, Betterton EA, Arnold RG (1999) Electrolytic oxidation of trichloroethylene using a ceramic anode. J Appl Electrochem 29:961–970
17. Zaky AM, Chaplin BP (2013) Porous substoichiometric TiO_2 anodes as reactive electrochemical membranes for water treatment. Environ Sci Technol 47:6554–6563
18. Scialdone O, Galia A, Filardo G (2008) Electrochemical incineration of 1,2-dichloroethane: effect of the electrode material. Electrochim Acta 53:7220–7225
19. Nowack N, Heger K, Korneli D, Rheindorf A (2005) Oxidation of cyanides in wastewaters (low-concentration region) by fixed-bed electrolysis. Chem Ing Tech 77:1927–1936
20. Scott K, Taama WM (1997) Electrolysis of simulated flue gas solutions in an undivided cell. J Chem Technol Biotechnol 70:51–56
21. Scott K, Cheng H, Taama W (1999) Zirconium and Ebonex® as cathodes for sulphite ion oxidation in sulphuric acid. J Appl Electrochem 29:1329–1338
22. Chen G, Betterton EA, Arnold RG, Ela WP (2003) Electrolytic reduction of trichloroethylene and chloroform at a Pt- or Pd-coated ceramic cathode. J Appl Electrochem 33:161–169
23. Kearney D, Bejan D, Bunce NJ (2012) The use of Ebonex electrodes for the electrochemical removal of nitrate ion from water. Can J Chem 90:666–674

Polyelectrolytes, Films-Specific Ion Effects in Thin Films

Natascha Schelero and Regine von Klitzing
Stranski-Laboratorium, Institut für Chemie, Fakultät II, Technical University Berlin, Berlin, Germany

Introduction

This entry addresses specific ion effects in thin films with thicknesses in the nano- to micrometer range and focuses on the effect of monovalent cations and anions on the structure of thin films. First, thin organic adsorbed films, so-called polyelectrolyte multilayers (PEMs) which are prepared by sequential adsorption of polyanions and polycations on a charged surface [10], are presented. Second, thin liquid (aqueous) films are discussed. These are thin layers of a continuous phase through which the dispersed phase (bubbles, droplets, solid particles) of colloidal dispersions such as foams, emulsions, and suspensions interacts. Both PEMs and liquid films have one thing in common: The amount as well as the type of ions plays a central role in determining the properties of such thin films.

Specific Ion Effects on the Properties of Polyelectrolyte Multilayers

Polyelectrolyte multilayers (PEMs) are prepared by layer-by-layer (LbL) technique, where polycations and polyanions are alternately adsorbed from aqueous solutions [10]. The impact of the amount and type of salt on the structural properties, mobility, and the swelling behavior of polyelectrolyte multilayers has been investigated for more than 10 years [11, 12, 15, 18, 22–24, 28–30, 37, 40].

Dubas and Schlenoff [12] reported for the first time a tendency of increasing PEM thickness with increasing size of monovalent ions, which has been confirmed by several other studies in the meanwhile

Polyelectrolytes, Films-Specific Ion Effects in Thin Films,

Fig. 1 Scheme of the ion-specific effect on the complexation of polyanions and polycations

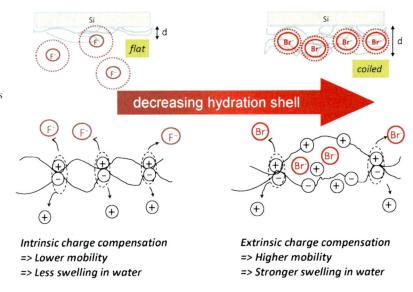

[11, 29, 30, 37, 40]. The observed ion-specific effects were associated with the hydrophobicity of the ions and the ions' affinity to the polyelectrolyte. Later, Klitzing et al. explained this effect by the difference between cosmotropic and chaotropic ions [37]. The influence of specific anions on certain properties of PEMs has been investigated extensively by Salomäki et al. [28–30]. The anions tested within this work followed the famous Hofmeister series and gave an increasing storage shear modulus with increasing ion size. The latter result is in contradiction to the increase in chain mobility found later by Nazaran et al. [24].

Wong et al. showed that the specific ion effects on building up polyelectrolyte multilayers become important above a certain ionic strength, namely, 0.1 M for anions and 0.25 M for cations [40].

Recently, Liu et al. have broadened the field of ion-specific effects on polyelectrolyte multilayers by investigation of the deposition of PSS/PDADMAC multilayers in mixed electrolyte solutions (PDADMAC: poly(diallyldimethylammonium chloride)) [22]. It appeared that the effects of anions of a mixed electrolyte solution on the deposition of PEMs are nonadditive. Moreover, in a mixed electrolyte solution containing both chaotropic and cosmotropic anions, the multilayer buildup is dominated by chaotropic anions.

The PSS/PDADMAC multilayer growth in a single electrolyte solution containing one type of chaotropic anions (e.g., Br^-, ClO_3^-) is nonlinear (increasing increment with increasing number of deposited layers). In contrast to this, in the presence of cosmotropic anions like F^-, CH_3COO^-, $H_2PO_4^-$, and SO_2^{4-}, a linear growth is observed: The multilayer buildup is determined by the conformation of PDADMAC chains [11, 30, 36, 37, 40]. An overview of the ion-specific effect is shown in Fig. 1.

The effects of monovalent anions and cations on the properties can be understood as following [37]: Cosmotropic ions have a well-ordered large hydration shell. Chaotropic ions are large with a significant polarizability, a weak electric field, and their hydration water can be easily removed. That makes it easier for chaotropic ions to adapt to the environment, e.g., the polyion. This might lead to a stronger attraction between them, which could be a kind of "bridging" and/or overlap of hydration shells of ions and the oppositely charged groups of the polyions. This leads to a stronger screening and chain coiling followed by increase in thickness and roughness [30, 37, 40], comparable to the effect of increasing ionic strength or decreasing charge density. The effect of anions is much larger than the effect of cations, since anions have a much larger difference in polarizability than typical cations due to their larger variety of their diameter [40]. Also theoretical calculations predict an effect of the ion type on their distribution around a polyelectrolyte [8]. Short-range dispersion forces have to be added to DLVO forces to describe this effect.

The stronger attraction between polyion and oppositely charged chaotropic ions explains the dominating effect of chaotropic ions in comparison to cosmotropic ones [22]. A further consequence is extrinsic compensation of the polyion charge. In contrast to this, in the presence of cosmotropic ions, the polyion charge is rather compensated by oppositely charged polyions, i.e., intrinsic charge compensation [36]. Since the mobility of polymer chains in the multilayer depends on the density of complexation sites between polyanions and polycations, extrinsic charge compensation leads to a higher chain mobility (liquefied system) than intrinsic one (glassy system) [24]. The degree of chain mobility determines also the type of growth. More mobile PEMs show a tendency for a nonlinear growth, while intrinsically charge compensated chains show a rather linear growth [11, 30, 36, 37, 40]. PEMs can be also liquefied after preparation, by adding chaotropic ions [24]. At high ion concentration, the PEMs can be even erased, which is more efficient with chaotropic ions than with cosmotropic ones: chaotropic ions can easily destroy the complexation sites. Another consequence of extrinsic charge compensation is that PEMs which are built up in the presence of chaotropic ions can swell stronger in water than PEMs which are formed in the presence of cosmotropic ions [11, 15, 28]. The lower density of complexation sites leads to a kind of sponge-like swelling of the PEM. This explains the higher permeability in case of chaotropic ions [15]. In contrast to this, the density of a dry PEM is higher, when they were prepared with chaotropic ions. This is due to the higher flexibility of the individual polymer chains caused by stronger electrostatic screening along the polyelectrolyte chains [11].

Specific Ion Effects on Thin Liquid Films

Interactions between the opposing film surfaces are important for the stability of a liquid film like foam films, emulsion films, etc. Related to this, the adsorption of ions at the film interfaces plays a decisive role.

Many experimental and theoretical results contradict each other concerning the ion adsorption at the air/water interface, since they are sensitive for different length scales. Some methods are only sensitive for the surface layer and not for subsurface layers, others are sensitive for both. For the stability of colloidal systems, interfacial forces over a range of several tens of nanometers are important and related to an average surface charge across the complete interfacial region.

This surface interaction causes an excess pressure normal to the film interfaces, called disjoining pressure, which is the sum of repulsive electrostatic (π_{el}), attractive van der Waals (π_{vdW}), and repulsive steric pressures (π_{st}). Adapted from these interactions, two different types of thin films can be distinguished: common black films (CBFs), stabilized by π_{el} and Newton black films (NBFs), stabilized by π_{st}.

Wetting Films

Solely, in an asymmetric film (e.g., wetting film), one can determine the sign and precise value of the overall charge of the free air/water surface. This technique has successfully been used by the authors to probe the existence of negative surface charges at the air/water interface [9, 16].

A simple system to study ion-specific effects is a water film on a substrate, e.g., Silicon. Schelero and Klitzing studied water films at different ionic strengths and in the presence of different types of salts [33] (see Fig. 2).

At a fixed concentration of 10^{-4} M sodium salts, the film thickness increases in the order of $F^- < Cl^- < I^-$, i.e., with increasing ion size (Fig. 2a). After the addition of iodide salts (qualitatively the same for other halides) to the water wetting film, the film thickness increases in the order of $Cs^+ < K^+ < Na^+$, i.e., with decreasing

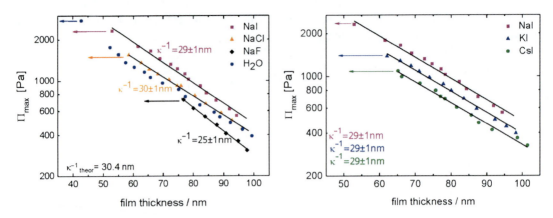

Polyelectrolytes, Films-Specific Ion Effects in Thin Films, Fig. 2 Disjoining pressure isotherms of aqueous wetting films on Si wafers

ion size (Fig. 2b). Under the assumption that the air/water interface is negatively charged, this result clearly shows that with increasing ion size, i.e., decreasing charge density, ions show a stronger tendency to adsorb at the surface. Enhancement of anions at the air/water interface leads to an increase of stability and film thickness due to an increase of the surface potential. The same adsorption effect is found for cations, but with the reversed effect on the surface potential and thus on the film thickness and stability. This is a clear further indication for an excess of negative charges at the air/water interface. It is worth to note that anions have such a drastic effect at the air/water interface although the surface is negatively charged. This demonstrates the dominating effect of the hydration shell of ions over electrostatics on their adsorption at the air/water interface.

Foam Films

Foams and emulsions are achieved due to adsorption of foam stabilizing agents like surfactants at the interface between the dispersed and continuous phases. The foam stability is often related to the stability of thin liquid films formed between two air bubbles. All considered foam films are stabilized by ionic surfactant.

Independent of the salt type, the addition of salt leads to several counteracting stabilization mechanisms in foam films [5–7, 13, 34]: The first mechanism originates from the electrostatic screening between the opposing equally charged film interfaces, leading to thinner films and often to a lower film stability. The second mechanism is due to lateral screening of the charges of adjacent surfactant head groups within the adsorption layer. This leads to the formation of more condensed films which promote foam film stability. Another mechanism affecting the stability of foam films is that electrolyte influences the surface activity of the surfactant. With the addition of salt, the dielectric constant of the solvent is reduced [27]. The dielectric properties of the medium where the head groups are located are directly linked with the degree of dissociation of the surfactant [5, 14, 19, 34]. Furthermore, the degree of dissociation changes due to different ion binding to the surfactant. The concentration of surfactant at the air/water interface can be either enhanced or reduced by adding electrolytes. For example, the adsorption of SDS at the air/water interface increases with increasing salt concentration due to electrostatic screening [20, 35].

Sentenac and Benattar were the first who observed ion-specific effects in foam films. They investigated the influence of LiCl and CsCl on aerosol-OT (AOT) films at a fixed surfactant concentration close to the cmc [35]. The addition of LiCl leads to about 10-nm thicker films than in case of CsCl. The disjoining pressure isotherms of the AOT/CsCl systems can be described by the classical DLVO approach

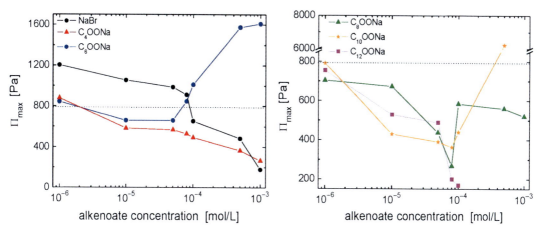

Polyelectrolytes, Films-Specific Ion Effects in Thin Films, Fig. 3 Disjoining pressure isotherms of LiDS (*filled circles*), SDS (*open squares*), and CsDS (*filled triangles*) at a surfactant concentration of 10–4 M. The solid line refers to the results from PB calculation

while it fails in case of LiCl. Apparently, an additional repulsion arises in the LiCl containing π(h) isotherm. The authors argue that it is a consequence of the reduced screening due to the large hydration shell of Li$^+$. Another explanation could be that the adsorption of Li$^+$ at the surface is weaker because of its hydrated radius.

Recently, a detailed study of specific ion effects on SDS foam films showed that chaotropic anions show a strong adsorption at the negatively charged surfaces of foam films [32] in analogy to the negatively charged air/water interface in wetting films. Again, especially the pronounced anion effects at a negatively charged surface imply that the adsorption of ions to the film surfaces is governed rather by ion-specific than electrostatic interactions. The order of the effect of halides and alkalis is the same as that detected for aqueous wetting films on Si of the pure salt solutions [33].

It is important to realize that the addition of ions which differ from the original counterion of the surfactant molecules leads to competitive ion adsorption. This could result in an exchange of the native counterion of the surfactant. Generally, ion-specific effects on foam films can be investigated in two ways: (i) due to the addition of salt as already discussed before and (ii) by variation of the counterion of the surfactant. The latter approach has the advantage of having only one type of counterion present in the solution.

A lot of work has been published concerning the effect and nature of counterions on the properties of surfactants (e.g., cmc, aggregation number, cloud point and ionization number) [1–4, 21, 31, 38, 39, 41], but only few publications deal with the counterion effect on the properties of foam or foam films. Interestingly, the counterion size has an opposite effect on foaming and foam stability. Pandey et al. found decreasing foamability of dodecyl sulfate surfactants with the order Li$^+$ > Na$^+$ > Cs$^+$ > Mg^{2+} as counterion. This has been explained by differences in micellar stability and diffusion of monomers. Due to the larger hydration shell of Li$^+$, Li$^+$ binds less strongly to the SD$^-$ headgroups which leads to stronger repulsion between the SD$^-$ headgroups than in case of CsDS. This leads to higher cmc of SDS and higher monomer to micelle ratio. Obviously, for the highly dynamic process of foaming, the fast adsorption of monomers to the film interfaces is much more efficient than the slow one of micelles followed by the required breakage of micelles.

While there is no significant ion effect on foam stability below the surfactant's cmc, above cmc, the foam stability in the presence of CsDS and Mg(DS)$_2$ is much larger than for LiDS and NaDS. This indicates that the presence of stable micelles is essential for high foam stability. Surface viscosity measurements correlated well with the foam stability trends and gave the following order

LiDS < NaDS < CsDS, indicating that the molecules of CsDS are more tightly packed at the air/water interface, which is also supported by surface tension measurements [26].

In opposite to macroscopic foams, the type of counterion affects the thickness and stability of single foam films, already below the cmc [32]. By keeping the headgroup constant, a decrease in film thickness and stability with increasing size of the counterion is found (Fig. 3). This result is independent of the charge of the headgroup and is explained by a decreasing surface charge due to decreasing degree of dissociation. It indicates that below surface saturation concentration, surface charges play an important role for stabilization and above the cmc (i.e., surface saturation), the packing density of the surfactant is dominating.

Future Directions

So far, for PEMs mainly the effects of monovalent ions are reported, which follow more or less the Hofmeister series. It is still unclear if this simple image as illustrated in Fig. 1 can be generalized to other ion types. Another open question is: If the polarizability [25] or rather the ion hydrophobicity [17] presents the decisive parameter.

Ion-specific effects on wetting films and foam films are mainly dominated by the affinity of ions to adsorb at the air/solution interface. This has a pronounced influence on the film thickness and stability. Additionally, foam film properties are affected by the propensity of ions to interact with the headgroups of the foam stabilizer. The open question is: If these findings can be generalized to other surfaces.

Cross-References

▶ Conductivity of Electrolytes
▶ Electrolytes, Classification
▶ Ion Properties
▶ Polyelectrolytes, Properties
▶ Polyelectrolytes, Simulation

References

1. Bales BL, Benrraou M, Zana R (2002) Characterization of micelles of quaternary ammonium surfactants as reaction media I: Dodeclytrimethylammonium bromide and chloride. J Phys Chem B 106:1926–1939
2. Bales BL, Benrraou M, Zana R (2002) Krafft temperature and micelle ionization of aqueous solutions of cesium dodecyl sulfate. J Phys Chem B 106:9033–9035
3. Bales BL, Shahin A, Lindblad C, Almgren MJ (2000). Timeresolved fluorescence quenching and electron paramagnetic resonance studies of the hydration of lithium dodecyl sulfate micelles. J Phys Chem B 104: 256–263 and references therein
4. Benrraou M, Bales BL, Zana R (2003) Effect of the nature of the counterion on the properties of anionic surfactants. 1. cmc, ionization degree at the cmc and aggregation number of micelles of sodium, cesium, tetramethylammonium, tetraethylammonium, tetrapropylammonium, and tetrabutylammonium dodecyl sulfates. J Phys Chem B 107:13432–13440
5. Bergeron V (1997) Disjoining pressures and film stability of alkyltrimethylammonium bromide foam films. Langmuir 13:3474–3482
6. Bergeron V, Radke C (1992) Equilibrium measurements of oscillatory disjoining pressures in aqueous foam films. Langmuir 8:3020–3026
7. Black IJ, Herrington TM (1995) Determination of thickness and rupture pressure for aqueous foam part 1.- films stabilized by anionic surfactantsfilms. J Chem Soc Faraday Trans 91:4251–4260
8. Boström M, Williams D, Ninham B (2002) The influence of ionic dispersion potentials on counterion condensation on polyelectrolytes. J Phys Chem B 106:7908–7912
9. Ciunel K, Armelin M, Findenegg GH, von Klitzing R (2005) Evidence of surface charge at the air/water interface from thin-film studies on polyelectrolyte-coated substrates. Langmuir 21:4790–4793
10. Decher G (1997) Fuzzy nanoassemblies: toward layered polymeric multicomposites. Science 277:1232–1237
11. Dodoo S, Steitz R, Laschewsky A, von Klitzing R (2011) Effect of ionic strength and type of ions on the structure of water swollen polyelectrolyte multilayers. Phys Chem Chem Phys 13:10318–10325
12. Dubas ST, Schlenoff JB (1999) Factors controlling the growth of polyelectrolyte multilayers. Macromolecules 32:8153–8160
13. Exerowa D, Kolarov T, Khristov K (1987) Direct measurement of disjoining pressure in black foam films. Colloids Surf 22:161–169
14. Fainerman V, Lucassen-Reynders E (2002) Adsorption of single and mixed ionic surfactants at fluid interfaces. Adv Colloid Interface Sci 96:295–323
15. Haitami AE, Martel D, Ball V, Nguyen HC, Gonthier E, Labbé P, Voegel JC, Schaaf P, Senger B, Boulmedais F (2009) Effect of the supporting

electrolyte anion on the thickness of pss/pah multi-layer films and on their permeability to an electroactive probe. Langmuir 25:2282–2289
16. Hänni-Ciunel K, Schelero N, von Klitzing R (2009) Negative charges at the air/water interface and their consequences for aqueous wetting films containing surfactants. Faraday Discuss 141:41–53
17. Horinek D, Serr A, Bonthius D, Bostroem M, Kunz W, Netz R (2008) Molecular hydrophobic attraction and ion-specific effects studied by molecular dynamics. Langmuir 24:1271–1283
18. Kharlampieva R, Pristinski D, Sukhishvili S (2007) Hydrogenbonded multilayers of poly(carboxybetaine)s. Macromolecules 40:6967–6972
19. Kralchevsky PA, Danov KD, Broze G, Mehreteab A (1999) Thermodynamics of ionic surfactant adsorption with account for the counterion binding: effect of salts of various valency. Langmuir 15:2351–2365
20. Kumar MK, Ghosh P (2006) Coalescence of air bubbles in aqueous solutions of ionic surfactants in presence of inorganic salt. Chem Eng Res Des 84:703–710
21. Kumar MK, Sharma D, Khan ZA (2001) Cloud point phenomenon in ionic micellar solutions: a SANS study. Langmuir 17:2549–2551
22. Liu G, Hou Y, Xiao X, Zhang G (2010) Specific anion effects on the growth of a polyelectrolyte multilayer in single and mixed electrolyte solutions investigated with quartz crystal microbalance. J Phys Chem B 114:9987–9993
23. Mermut O, Barret C (2003) Effects of charge density and counterions on the assembly of polyelectrolyte multilayers. J Phys Chem B 107:2525–2530
24. Nazaran P, Bosio V, Jaeger W, Anghel D, von Klitzing R (2007) Lateral mobility of polyelectrolyte chains in multilayers. J Phys Chem B 111:8572–8581
25. Ninham BW, Yaminsky V (1997) Ion binding and ion specificity: the hofmeister effect and onsager and lifshitz theories. Langmuir 13:2097–2108
26. Pandey S, Bagwe RP, Shah DO (2003) Effect of counterions on surface and foaming properties of dodecyl sulfate. J Colloid Interface Sci 267:160–166
27. Sack H (1927) Über die Dielektrizitätskonstanten von Elektrolytlösungen bei geringen Konzentrationen. Ph. D. thesis, Eidgenössischen Technischen Hochschule in Zürich
28. Salomäki M, Kankare J (2008) Specific anion effect in swelling of polyelectrolyte multilayers. Macromolecules 41:4423–4428
29. Salomäki M, Laiho M, Kankare J (2004) Counteranion-controlled properties of polyelectrolyte multilayers. Macromolecules 37:9585–9590
30. Salomäki M, Tervasmäki P, Areva S, Kankare J (2004) The hofmeister anion effect and the growth of polyelectrolyte multilayers. Langmuir 20:3679–3683
31. Satake I, Iwamatsu I, Hosokawa S, Matuura R (1963) The surface activities of bivalent metal alkyl sulfates. i. On the micelles of some metal alkyl sulfates. Bull Chem Soc Japan 36:204–209

32. Schelero N, Hedicke G, Linse P, Klitzing RV (2010) Effects of counterions and co-ions on foam films stabilized by anionic dodecyl sulfate. J Phys Chem B 114:15523–15529
33. Schelero N, Klitzing RV (2011) Ion specific effects on thin aqueous wetting films. Soft Matter 7:2936–2942
34. Schulze-Schlarmann J, Buchavzov N, Stubenrauch C (2006) A disjoining pressure study of foam films stabilized by tetradecyl trimethyl ammonium bromide c14tab. Soft Matter 2:584–594
35. Sentenac D, Benattar JJ (1998) Long range hydration effects in electrolytic free suspended black films. Phys Rev Lett 81:160–163
36. Klitzing RV (2006) Internal structure of polyelectrolyte multilayer assemblies. Chem Phys Phys Chem 8:5012–5033
37. Klitzing RV, Wong J, Jaeger W, Steitz R (2004) Short range interactions in polyelectrolyte multilayers. Curr Opin Colloid Interface Sci 9:158–162
38. Wen W-Y, Saito S (1964) Apparent and partial molal volumes of five symmetrical tetraalkylammonium bromides in aqueous solutions. J Phys Chem 68:2639–2644
39. Wirth HE, LoSurdo AJ (1972) Proton magnetic resonance in concentrated aqueous solutions of tetraalkylammonium bromides and inorganic halides at 25 and 65.deg. J Phys Chem 76:130–132
40. Wong J, Zastrow H, Jaeger W, Klitzing RV (2009) Specific ion versus electrostatic effects on the construction of polyelectrolyte multilayer. Langmuir 25:14061–14070
41. Yu Z-Y, Zhang X, Zhou Z, Xu G (1990) Physicochemical properties of aqueous mixtures of tetrabutylammonium bromide and anionic surfactants. Effects of surfactant chain length and salinity. J Phys Chem 94:3675–3681

Polyelectrolytes, Properties

Yoav D. Livney
Faculty of Biotechnology and Food Engineering, The Technion, Israel Institute of Technology, Haifa, Israel

Introduction

Polyelectrolytes are polymers with ionizable groups [1–9]. In polar solvents, such as water, these groups ionize by either dissociation (e.g., carboxyl: $-COOH<=>-COO^-+H^+$), or proton-association (e.g., amine: $-NH_2+H^+<=>-NH_3^+$), which also produces counterions in the surrounding solution. The ionization in aqueous solution, which is pH and ionic strength dependent, confers unique properties to the polyelectrolytes, dominated by strong long-range electrostatic interactions [10], and by entropic effects of counterions [6, 11–13]. Both living organisms and human technology make many various uses of the versatility of polyelectrolyte behavior [8]. Examples of polyelectrolytes include polyanions (or polyacids), like DNA, polyacrylic acid [6]; polycations (or polybases), like chitosan [14, 15], polylysine [16, 17]; and polyampholytes, i.e., polymers carrying both positive and negative charges [18], like most proteins.

The presence of charged groups leads to several important differences in solution behavior of polyelectrolytes, compared to uncharged polymers [6]: (i) the crossover concentration from dilute to semidilute is much lower; (ii) a pronounced peak (whose wavevector magnitude increases with $c^{1/2}$) appears in the scattering function of homogeneous polyelectrolyte solutions, not in that of nonionic polymer solutions; (iii) the osmotic pressure of polyelectrolyte solutions in pure water is several orders of magnitude higher than that of uncharged polymers (at same concentration), and it increases almost linearly with polymer concentration, independently of molecular weight. These facts, along with the strong dependence of the osmotic pressure of the polyelectrolyte solutions on added salt, show it is mainly due to counterion contribution; (iv) the viscosity of polyelectrolyte solutions varies linearly with $c^{1/2}$ (Fuoss' law [19]), while for uncharged polymers, it varies linearly with c; (v) compared to solutions of uncharged polymers, polyelectrolytes in semidilute solutions follow unentangled dynamics over a much wider concentration range [6]. While the vast majority of the studies on polyelectrolytes were focused on aqueous solutions, a few studies considered nonaqueous polyelectrolyte solutions [20, 21].

The properties of polyelectrolytes in solutions and at charged surfaces depend on the fraction of ionized groups, solvent quality for the polymer, dielectric constant of the medium, salt concentration, and polymer–substrate interactions [5–7].

Properties of Polyelectrolytes and Their Solutions

Structural Properties

Similarly to nonionic polymers, polyelectrolytes are macromolecules made of covalently linked repeating units (mers, or monomers), which may be identical, forming a homopolymer, or different, forming a heteropolymer or copolymer. The monomers of different types may be randomly distributed along the chain, grouped into blocks of same type, forming a block-copolymer (e.g., an A_nB_m diblock copolymer is made of two blocks, one of n A monomers, and the other of m B monomers), or specifically ordered in a particular sequence, as in DNA, RNA, and proteins. The linear sequence of monomers of different types has dramatic impact on the properties of the copolymer. In particular, being polyelectrolytes, the distribution of positive and negative charges along the chain defines the charge distribution and charge density (local and global), properties which intensely affect chain conformation in solution, and interactions between molecules of the same polymer, as well as with other species in solution [22]. Additionally, polyelectrolytes, like other polymers, may have various architectural structures. They may be linear, linear with side groups, branched, hyperbranched (i.e., with branched-branches), star-shaped, dendritic (i.e., formed in "generations" of branching points, from a central point outward), and more. Polyelectrolytes may also be cross-linked to form polyelectrolyte gels, notably polyelectrolyte hydrogels, some of which have super-absorbent capabilities (e.g., polyacrylic acid hydrogels used in diapers). The backbone configuration, which describes the stereochemical positioning of side groups, or tacticity, may either be isotactic (all consecutive side groups are oriented the same way), syndiotactic (alternating sides), or atactic (random orientations (e.g., as in polystyrene sulfonate [23]).

Typical anionic groups occurring in bio-polyelectrolytes include carboxyl ($-COO^-$, e.g., in pectin, alginate, and proteins), phosphate ($>PO_4^-$, e.g., in DNA), sulfate ($-SO_4^-$, e.g., in carrageenan). The typical cationic group in

Polyelectrolytes, Properties

bio-polyelectrolytes is amine ($-NH_3^+$, e.g., in chitosan and in proteins).

Electrochemical Properties

Aqueous polyelectrolyte solutions are electrically conductive, due to the presence of both the polyions and the counterions [24]. Strongly acidic groups (e.g., phosphate and sulfonate) ionize in solution over a very wide pH range. Strongly acidic groups have an acid-dissociation constant (K_a) for which pK_a is typically <2, while weakly acidic groups typically have a pK_a >2. Organic bases (e.g., amines) are weak bases and thus basic side groups have an association constant for protonation (K_b) for which pK_b is typically <10. This means that weak acidic and basic groups are only partially ionized at intermediate pH conditions, and so the fractional ionization and the consequent charge-related properties of weak polyelectrolytes are thus strongly pH dependent. Additionally, according to the theory of weak polyacids by Katchalsky (1954), due to interaction between the charged groups along a polymer chain, the dissociation constant will depend on both ionic strength in the solution (degree of shielding) and the distance between adjacent charged groups on the polymer. Katchalsky developed the following relationship: $pH = pK_0 - \log[(1-\alpha/)\alpha] + 0.4343e\Psi/kT$, where pK_0 is the intrinsic pK_a, α is the degree of dissociation, e is the elementary charge, and Ψ is the electrostatic potential around the polyelectrolyte. The last term is the contribution of the polyelectrolyte field to the standard free energy of ionization of a single group, above the free energy of ionization where no such field is acting (pK_0). According to Katchalsky, the dissociation constant, when half of the carboxylic groups are protonated, $pKa(\alpha = 0.5)$, may be determined by plotting pH as function of $\log[(1-\alpha/)\alpha]$. By plotting the apparent dissociation constant, $pK_a(app) \equiv pH-\log[(1-\alpha/)\alpha]$, as function of α, the intrinsic dissociation constant (the dissociation constant of a single acidic group when all neighboring acidic groups are uncharged) is obtainable by extrapolation to $\alpha = 0$. For example, the pK_0 and $pK_a(\alpha = 0.5)$ have been determined [25] by potentiometric titration using NMR spectroscopy (in D_2O) for hyaluronan (a linear polysaccharide abundant in mammalian tissues consisting of alternating monosaccharide units, N-acetyl glucosamine, and glucuronic acid, connected by β-(1-3) and β-(1-4) glycosidic bonds, respectively). As NaCl concentration increased (10, 50, 100 mM) and the consequent Debye-screening length decreased (3.04, 1.36, 0.96 nm), the pK_a ($\alpha = 0.5$) remained expectedly unchanged (3.14, 3.16, 3.15) but the pK_0 decreased (3.05, 2.82, 2.81) until the Debye-screening length approached the length of the repeating disaccharide unit of hyaluronic acid ($\sim$1 nm) [25].

The electrostatic attraction between polyelectrolyte chains and counterions in solutions can result in condensation of counterions on polyelectrolytes. The counterion condensation [26] appears to be due to a fine interplay between the electrostatic attraction of a counterion to a polymer chain and the loss of the translational entropy by counterions due to their localization in the vicinity of the polymer chain [6]. The attraction of added ions to fixed charges on the polyelectrolyte chain and the ability to reduce this attraction by screening of competing small ions is the basis for ion exchange technologies, including water purification and ion exchange chromatography [27].

Another important aspect of the screening effect is commonly known as the "electric double layer" (EDL), a concept which simplistically describes the fact that near a charged surface in solution, a layer of oppositely charged ions is formed, to maintain electroneutrality [28]. The region containing the strongly bound counterions is often called the Stern Layer, while the adjacent region where ions move freely influenced by the electric and thermal forces is called the diffuse layer, or the Gouy–Chapman layer [28, 29]. The effective thickness of the EDL in the linearized Poisson–Boltzmann theory may be described by the Debye length, k^{-1} [28, 30], $\kappa^{-1} = \sqrt{\frac{\varepsilon_r \varepsilon_0 k_B T}{2 N_A e^2 I}}$, where ε_r is the relative permittivity of the electrolyte solution, ε_0 is the electric permittivity of vacuum, k_B is the Boltzmann constant, T is the absolute temperature, N_A is Avogadro's number, e is the elementary charge, and I is the ionic strength.

Because the electrical potential at the surface of a colloidal particle in solution cannot be directly and unambiguously measured, the potential at the hydrodynamic slip-plane, called the zeta potential (ζ), is commonly studied, based on measurements of electrophoretic mobility, i.e., the terminal velocity of a colloidal particle, e.g., a polyelectrolyte molecule, under a constant electric field, divided by field intensity [29]. For polyampholytes, the pH where the electrophoretic mobility is zero is the isoelectric point, i.e., the pH where the numbers of positive and negative charges on the particle are equal [29]. There are several theories relating electrophoretic mobility to ζ, each based on specific assumptions; hence, the choice of a proper theory must be based on assuring the validity of these assumptions for the system studied. The most commonly cited in this respect is the Smoluchowski theory [31]. Smoluchowski's theory is valid for particles of any shape, provided the (local) curvature radius, a, largely exceeds the Debye length κ^{-1}, i.e., $\kappa a \gg 1$. According to Smoluchowski, the electrophoretic mobility, u_e, is related to ζ by $u_e = \varepsilon_r \varepsilon_0 \zeta / \eta$, known as the Helmholtz–Smoluchowski equation, where η is the dynamic viscosity of the liquid [29, 31]. For $\kappa a < 1$, the Hückel–Onsager equation applies: $u_e = 2\varepsilon_r \varepsilon_0 \zeta / 3\eta$ [29]. Additional situations and the relevant theories available are detailed by Delgado et al. [29].

Lyotropic Properties (Hofmeister Series: Ion-Specific Effects)

Ion-specific effects are universal in biology, biochemistry, chemistry, and chemical engineering [32]. They were first systematically studied by Franz Hofmeister [32–34], who observed the different protein precipitating effects ("salting-out") of various salts, and described them in terms of the "water withdrawing power of salts." Hofmeister effects, or series, refer to the relative effect of anions or cations on a wide range of phenomena, not only related to proteins or to polyelectrolytes [35–38]. Advances in experimental and computational methodologies have led to insights into the underlying molecular mechanisms, although a deeper molecular understanding still seems to be elusive. The principal

reason appears to be that the Hofmeister series emerges from a combination of a general effect of cosolutes (salts, etc.) on solvent structure, and of specific interactions between the cosolutes and the solute (protein or other biopolymer) [39]. Originally, it was thought that an ion's influence on macromolecular properties was caused at least in part by "making" or "breaking" bulk water structure. Recent time-resolved and thermodynamic studies of water molecules in salt solutions, however, demonstrate that bulk water structure is not central to the Hofmeister effect. Instead, models are being developed that depend upon direct ion–macromolecule interactions as well as interactions with water molecules in the first hydration shell of the macromolecule [40], and in the first couple of hydration layers of the ion [41, 42]. Small ions of high charge density (kosmotropes) bind water molecules strongly, whereas large monovalent ions of low charge density (chaotropes) bind water molecules weakly relative to the strength of water–water interactions in bulk solution [43, 44]. Kosmotropic anions tend to cause "salting-out" of polyelectrolytes, nonionic polymers, and other colloids and cosolutes. Chaotropic anions and kosmotropic cations tend to cause "salting in" (although at high concentrations most salts cause salting-out) [38, 45–49]. Positively charged macromolecular systems may show inverse Hofmeister behavior only at relatively low-salt concentrations, but revert to a direct Hofmeister series as the salt concentration is increased [50]. The topic of ion-specific effects is also dealt with in other sections of this encyclopedia.

Rheological Properties of Polyelectrolyte Solutions

Dobrynin and coworkers [6, 51] have extended and generalized the scaling theory of de Gennes [52] and Pfeuty [53] to both unentangled and entangled regimes of intrinsically flexible polyelectrolyte solutions. In semidilute solutions, the electrostatic persistence length of a polyelectrolyte is assumed to be proportional to the Debye-screening length. If the salt concentration is low, the unentangled semidilute concentration

Polyelectrolytes, Properties

regime spans three to four decades in polymer concentration. When comparing the rheological properties of polyelectrolytes to those of neutral polymers, it can be generalized that the viscosity of polyelectrolyte solutions is proportional to the square root of polymer concentration $\eta \sim c^{1/2}$ (Fuoss' law [54]), while for solutions of uncharged polymers at the same concentration, the viscosity is proportional to polymer concentration [6]. Moreover, there is no concentration regime where reduced viscosity η/c of solutions of neutral polymers decreases with polymer concentration ($\eta/c \sim c^{-1/2}$ for polyelectrolytes in the Fuoss regime) [6]. Polyelectrolytes should form entanglements at the same relative viscosity as neutral polymer solutions ($\eta \approx 50\eta(s)$) (s=solvent), and in the entangled regime of salt-free polyelectrolyte solutions $\eta \sim c^{3/2}$ [51]. While the viscosity of a hydrophilic polyelectrolyte (e.g., poly(acrylic acid) (PAA)) increases smoothly with increasing neutralization as pH is raised, the viscosity of a hydrophobic polyelectrolyte (e.g., poly(methacrylic acid) (PMA) remains almost constant at low pH, and increases abruptly as pH reaches a critical value, indicating a globule-to-coil transition upon charging, because water is a poor solvent for the uncharged PMA, but good for the charged polyelectrolyte [6, 55].

Intermolecular and Surface Properties

Intermolecular interactions between different polyelectrolytes have a plethora of implications and applications in science and technology. In principle, different polyelectrolytes may present either repulsive or attractive intermolecular interactions [56]. Similar charge sign favors repulsion, which, at low concentrations, tends to lead to co-solubility, but at high concentrations, may lead to segregative phase separation, due to thermodynamic incompatibility [57]. Opposite charge signs on polyelectrolytes may lead to association driven by attractive Coulombic interactions [58], resulting in electrostatic-complex formation, and possibly, to associative phase separation, also termed complex coacervation [56, 59, 60]. This phenomenon was first observed for proteins by Kossel in 1896 [61]

and for protein–polysaccharide coacervation in 1911 by Tiebackx [62] who observed the appearance of opacity or precipitation upon mixing of gelatin and gum arabic in acid conditions. It was first systematically studied for protein–polysaccharide complexes by Bungenberg de Jong and Kruyt [63], and for synthetic polyelectrolytes by Michaels & Miekka [64]. Protein–polysaccharide interactions have since been extensively studied and reviewed [59, 60, 65–70], and so were polyelectrolyte complexes in general [71–75]. Intermolecular polyelectrolyte interactions are mainly dependent on pH (which affects ionization), ionic strength (causing screening), and the stoichiometric ratio of the macroions involved and the total concentration. Insoluble complex coacervates become progressively more soluble with increasing salt concentration [75]. Other important factors are the molecular characteristics of the polyelectrolytes (molecular weight, charge density, conformation, etc.), solvent quality (including presence of cosolvents, and cosolutes, ion-specific effects), mixing procedures, and conditions (temperature, pressure, shear, etc.) [56, 59, 65, 67, 70]. Recently, computer simulations are helping in gaining new insights into polyelectrolyte complexation [74, 76–79]. The net charge of two macroions might be of same sign, and yet association may be favorable, thanks to local "patches" of high charge density, e.g., as in case of a protein slightly above its isoelectric pH (but with densely positive charge patches) associating with an anionic polyelectrolyte [80, 81]. By carefully controlling the ratio of the oppositely charged polyelectrolytes, either precipitated coacervates (mutual charge neutralization) or soluble complexes (excess charge due to surplus of one of the polyelectrolytes) may be formed, which may be applied, e.g., for the formation of nanocomplexes useful as protective vehicles for bioactive compounds in clear solutions [82, 83].

An important application of electrostatic complexation of oppositely charged polyelectrolytes is known as the "layer-by-layer deposition" or electrostatic self-assembly [84–87], which is the alternating deposition of positively and negatively charged polyelectrolytes onto a charged surface. The surface could be that of a solid or of another liquid phase, and may have any shape. Various materials have been used with this technique, e.g., synthetic polyelectrolytes, biopolymers (proteins, DNA, polysaccharides) and more, mainly for delivery and controlled release applications [88, 89]. A typical example is the nanocoating of emulsion droplets, emulsified with a low molecular weight ionic surfactant (e.g., SDS), by several alternating layers of oppositely charged polyelectrolytes (e.g., chitosan and pectin) [90], as a way to form a microcapsule. Sometimes the core particle is later dissolved to leave a hollow multilayered shell [91]. From a fundamental viewpoint, the adsorption of a polyelectrolyte on a surface is an intriguing process, which questionably reached equilibrium, and whose kinetics has only been scarcely studied and reviewed [92].

Block-copolymer polyelectrolytes, with at least one hydrophobic block (in the main chain or a grafted side chain), may demonstrate self-assembly behavior in aqueous solution [93], and may adsorb onto hydrophobic surfaces, which may be facilitated by increasing ionic strength [94]. Moreover, such block-copolymeric amphiphiles typically have a much lower critical micellization concentration (CMC) [95] and a much longer micelle life time (by several orders of magnitude) compared to low molecular weight surfactants [96]. Emulsions stabilized by block-copolymeric polyelectrolytes are generally much more stable than their counterparts stabilized by low molecular weight surfactants of similar hydrophilic–lipophilic balance (HLB) [97]. Micelle formation and micellar structure depends on various parameters like block lengths, salt concentration, pH, and solvent quality, and their stability depends on electro-steric stabilization forces [98]. Certain proteins, like β-casein, are natural examples of such "block-copolymeric amphiphiles" [99, 100]. These properties of amphiphilic polyelectrolytes are useful for various applications, including encapsulation and delivery of hydrophobic bioactives [101], like drugs [102, 103] and nutraceuticals [104, 105], formation of self-assembled nanoreactors [106], and more.

Similarly charged polyelectrolytes may interact attractively by bridging action of mediating counterions. The higher the charge density of the polyelectrolytes, the higher the required charge density of the mediating counterion. An example is the preferred potassium-induced gelation of k-carrageenan (one sulfate group per disaccharide repeating unit) vs. the preferred calcium-induced gelation of i-carrageenan (two sulfate groups per disaccharide repeating unit) [107]. An important additional example includes calcium-induced gelation of alginate, particularly block-wise guluronic rich type, for which the egg-box model for cooperative calcium-assisted cross-link zones was proposed [108]. This type of quick ion-induced gelation has found important applications, notably in gel-bead formation by dripping of alginate solution into calcium-ion containing solution, e.g., for encapsulation purposes [109].

Osmotic Pressure and Gel Swelling of Polyelectrolyte Systems

In dilute polyelectrolyte solutions without added salt, the Poisson–Boltzmann cylindrical cell model accounts fairly well for thermodynamic and some transport properties observed [110–112]. Accordingly, the osmotic pressure in such solutions may be expressed in a virial expansion as commonly used with only two terms [110]:

$$\Pi = CRT(1/M + A_2 C)$$

where M is the molecular weight, C the polymer concentration (g/mL), and A_2, the second virial coefficient, is:

$$A_2 = 4\pi^{3/2} \rho(z) N_A R_F / M^2$$

where $\rho(z)$ is the penetration function, which is constant (0.21) in good solvents [113], N_A is Avogadro's number, and R_F is the Flory radius (for a single chain in a good solvent; $R_F \sim M^{3/5}$) [110].

Above the overlap-onset concentration, c*, according to the Scaling Theory of de Gennes [52], the osmotic pressure (Π_{os}) of a semidilute neutral polymer solution is principally the thermal energy κT per correlation volume:

$$\Pi_{os} \approx kT/\xi^3 \quad c > c*$$

where ξ is the correlation length. In polyelectrolyte solutions, there is an important additional contribution of ions to the total osmotic pressure ($\Pi_{os} = \Pi_p + \Pi_i$) [51]. Even though ions may pass through the membrane (used for measuring the osmotic pressure) which separates the polyelectrolyte from the pure solvent, Donnan equilibrium requires that charge neutrality would be maintained on both sides of the membrane [114–117]. When salt ions concentration is much lower than polyelectrolyte counterions concentration ($c >> 2Ac_s$; A is the number of monomers between uncondensed charges, and c_s is the number-density salt concentration, assuming monovalent ions), then $\Pi_i \approx kTc/A$ (c is the monomer number-density concentration). On the other extreme, at high salt concentration, the counterions are distributed almost uniformly on both sides of the membrane, and salt redistributes to maintain charge neutrality, which also contributes to the osmotic pressure [51, 117]: $\Pi_i \approx kTc^2/4A^2c_s$ ($c << 2Ac_s$). Combining these two expressions to extrapolate the two extremes, one obtains: $\Pi_i \approx kTc^2/(4A^2c_s + Ac)$. Hence, the total osmotic pressure, which is the sum of all these polymer and ionic contributions, is [51]:

$$\Pi_i \approx kT\left[c^2/\left(4A^2c_s + Ac\right) + 1/\xi^3\right]$$

At low-salt concentrations, the ionic (c/A) contribution generally dominates over the semidilute polymer contribution ($1/\xi^3$), while at high salt concentrations, both the ionic ($c^2/4A^2c_s$) and polymer contributions to the osmotic pressure are much smaller than in low-salt solutions. For the vast majority of polyelectrolyte systems studied, the ionic contribution dominates both at low and at high salt concentrations, and the polymer contribution term may be negligible [51].

Polyelectrolyte gels are capable of swelling to much greater extents than their uncharged counterparts, because of high osmotic pressure due to dissociated counterions. Gel swelling, according to the widely accepted Flory-Rehner [118] conception, is governed by the additive contributions of the osmotic pressure, Π_{os}, acting to swell the gel, and the elastic pressure of the network, which at swelling equilibrium counteracts and balances the osmotic pressure, i.e., the swelling pressure Π_{sw} (which equals zero at swelling equilibrium), is:

$$\Pi_{sw} = \Pi_{os} - \Pi_{net}$$

Theories of rubber elasticity [119], such as the "affine network theory" [120] or the "phantom network theory" [121], provide expressions for the network pressure, depending on cross-link functionality and network topology. For a perfect tetrafunctional network without trapped entanglements, the elastic network pressure is given by [120]:

$$\Pi_{net} = k_BTA\frac{\phi R^2}{NR_0^2}$$

where N is the number of statistical segments between cross-links, and R and R_0 are the mean square end-to-end distances of a network strand at a concentration Φ and in its reference state, respectively. The prefactor, A, according to the phantom network theory, predicts $A = 0.5$, while the affine network theory predicts $A = 1$ [120, 122]. Additional theoretical treatments of this intriguing problem may be found, e.g., in [123–125], and in recent years, computer simulations [9, 74, 126–129] are contributing significantly to advance our understanding and gain new insights of polyelectrolytes and their networks. Polyelectrolyte gels have numerous applications, such as superabsorbents [130, 131] (in diapers, hygienic and wound dressing products, and in agriculture [132] – "water-holding" for soil improvement), in sensor technologies [133], in ion exchange resins [134, 135], in food technology [136] (gel-textured products), in biomedical applications [137, 138] environmentally responsive hydrogels [139] have applications in implants, in drug targeting, and many more.

Future Directions

While much progress has been made over the years in our understanding of polyelectrolytes, which enabled numerous important applications, much work remains to be done, particularly in terms of the effects of solvent and cosolutes, including Hofmeister series and water-structure effects on polyelectrolytes, dynamics of gel swelling, adsorption on surfaces, formation of multilayered films, and self-assembly of amphiphilic macroions. The advancements in computer capabilities will continue to facilitate simulations, which are becoming progressively more explicit in the ability to model atomistic structure and dynamics of macromolecules and surrounding solvent components. Improvement of organic synthesis capabilities, along with better thermodynamic understanding and more powerful computerization tools, will enable design of new polymeric architectures, including protein-like sequence-based polyampholytes for programmed conformational and functional behaviors, enabling novel nanotechnologies for advanced applications.

Cross-References

▶ Ion Properties
▶ Ions at Solid-Liquid Interfaces
▶ Polyelectrolytes, Films-Specific Ion Effects in Thin Films
▶ Polyelectrolytes, Simulation
▶ Specific Ion Effects, Evidences
▶ Specific Ion Effects, Theory

References

1. Barrat JL, Joanny JF (1996) Theory of polyelectrolyte solutions. J Adv Chem Phys 94:1–66
2. Forster S, Schmidt M, Antonietti M (1992) Experimental and theoretical investigation of the electrostatic persistence length of flexible polyelectrolytes at various ionic strengths. J Phys Chem 96:4008–4014
3. Forster S, Schmidt M (1995) Polyelectrolytes in solution. Adv Polym Sci 120:51–133
4. Tanford C (1961) Physical chemistry of macromolecules. Wiley, New York/London
5. Hara M (ed) (1993) Polyelectrolytes, science and technology. Marcel Dekker, New York
6. Dobrynin A, Rubinstein M (2005) Theory of polyelectrolytes in solutions and at surfaces. Prog Polym Sci 30:1049–1118. doi:10.1016/j.progpolymsci.2005.07.006
7. Mandel M (1988) Polyelectrolytes. In: Mark HF, Bikales N, Overberger CG, Menges G (eds) Polyelectrolytes, 2nd edn. Wiley, New York
8. Katchalsky A (1954) Problems in the physical chemistry of polyelectrolytes. J Polym Sci 12:159–184. doi:10.1002/pol.1954.120120114
9. Dobrynin AV (2008) Theory and simulations of charged polymers: from solution properties to polymeric nanomaterials. Curr Opin Colloid Int Sci 13:376–388. doi:10.1016/j.cocis.2008.03.006
10. Radeva T (ed) (2001) Physical chemistry of polyelectrolytes. Marcel Dekker, New York
11. Dobrynin AV, Rubinstein M (2001) Counterion condensation and phase separation in solutions of hydrophobic polyelectrolytes. Macromolecules 34:1964–1972. doi:10.1021/Ma001619o
12. Khokhlov AR (1980) Collapse of weakly charged poly-electrolytes. J Phys A-Math Gen 13:979–987
13. Khokhlov AR (1980) Swelling and collapse of polymer networks. Polymer 21:376–380
14. Muzzarelli RAA, Muzzarelli C (2005) Chitosan chemistry: relevance to the biomedical sciences. Adv Polym Sci 186:151–209
15. Makhlof A, Tozuka Y, Takeuchi H (2011) Design and evaluation of novel pH-sensitive chitosan nanoparticles for oral insulin delivery. Eur J Pharm Sci 42:445–451. doi:10.1016/j.ejps.2010.12.007
16. Shih I-L, Shen M-H, Van Y-T (2006) Microbial synthesis of poly(ε-lysine) and its various applications. Bioresour Technol 97:1148–1159. doi:10.1016/j.biortech.2004.08.012
17. Hiraki J, Ichikawa T, S-i N, Seki H, Uohama K, Seki H, Kimura S, Yanagimoto Y, Barnett JW Jr (2003) Use of ADME studies to confirm the safety of ε-polylysine as a preservative in food. Regul Toxicol Pharmacol 37:328–340. doi:10.1016/s0273-2300(03)00029-1
18. Dobrynin AV, Colby RH, Rubinstein M (2004) Polyampholytes. J Polym Sci, Part B: Polym Phys 42:3513–3538. doi:10.1002/polb.20207
19. Fuoss RM (1948) Viscosity function for polyelectrolytes. J Polymer Sci 3:603–604
20. Hara M (2001) Polyelectrolytes in nonaqueous solutions. In: Radeva T (ed) Polyelectrolytes in nonaqueous solutionsedn. Marcel Dekker, New York
21. Hara M (1993) Polyelectrolytes in nonaqueous solution. In: Hara M (ed) Polyelectrolytes in nonaqueous solutionedn. Marcel Dekker, New York
22. Sperber B, Schols HA, Stuart MAC, Norde W, Voragen AGJ (2009) Influence of the overall charge and local charge density of pectin on the complex formation between pectin and beta-lactoglobulin. Food Hydrocol 23:765–772. doi:10.1016/j.foodhyd.2008.04.008
23. Li C, Shen J, Peter C, van der Vegt NFA (2012) A chemically accurate implicit-solvent

24. Vink H (2001) Conductance of polyelectrolyte solutions, anisotropy, and other anomalies. In: Radeva T (ed) Conductance of polyelectrolyte solutions, anisotropy, and other anomaliesedn. Marcel Dekker, New York

25. Tømmeraas K, Wahlund P-O (2009) Poly-acid properties of biosynthetic hyaluronan studied by titration. Carbohydr Polym 77:194–200. doi:10.1016/j.carbpol.2008.12.021

26. Manning GS (1969) Limiting laws and counterion condensation in polyelectrolyte solutions I. Colligative Properties. J Chem Phys 51:924–933

27. HaddadPR JPE (1990) Ion chromatography: principles and applications. Elsevier, Amsterdam

28. DA RusselWB S, Schowalter WR (1989) Colloidal dispersions. Cambridge University Press, Cambridge, UK

29. Delgado AV, Gonzalez-Caballero F, Hunter RJ, Koopal LK, Lyklema J (2007) Measurement and interpretation of electrokinetic phenomena. J Colloid Interface Sci 309:194–224. doi:10.1016/j.jcis.2006.12.075

30. Bohinc K, Kralj-Iglic V, Iglic A (2001) Thickness of electrical double layer. Effect of ion size. Electrochim Acta 46:3033–3040

31. von Smoluchowski M (1921) Elektrische endosmose und stroemungsstroeme. In: Greatz L (ed) Elektrische endosmose und stroemungsstroemeedn. Barth J.A., Leizig

32. Kunz W, Henle J, Ninham BW (2004) 'Zur Lehre von der wirkung der Salze' (about the science of the effect of salts): Franz Hofmeister's historical papers. Curr Opin Colloid Int Sci 9:19–37. doi:10.1016/j.cocis.2004.05.005

33. Hofmeister F (1887) About regularities in the protein precipitating effects of salts and the relation of these effects with the physiological behaviour of salts. Arch exp Path Pharm 24:247–260

34. Hofmeister F (1888) Zur lehre von der wirkung der salze. Naunyn-Schmiedeberg's Arch Pharmacol 25:1–30

35. Kunz W, Lo Nostro P, Ninham BW (2004) The present state of affairs with Hoffffieister effects. Curr Opin Colloid Interface Sci 9:1–18. doi:10.1016/j.cocis.2004.05.004

36. Karlstrom G, Carlsson A, Lindman B (1990) Phase diagrams of nonionic polymer -water systems. Experimental and theoretical studies of the effects of surfactants and other cosolutes. J Phys Chem 94:5005–5015

37. Jungwirth P, Winter B (2008) Ions at aqueous interfaces: from water surface to hydrated proteins. Annu Rev Phys Chem 59:343–66

38. Livney YD, Portnaya I, Faupin B, Ramon O, Cohen Y, Cogan U, Mizrahi S (2003) Interactions between inorganic salts and polyacrylamide in aqueous solutions and gels. J Polym Sci, Part B: Polym Phys 41:508–519

39. Cacace MG, Landau EM, Ramsden JJ (1997) The hofmeister series: salt and solvent effects on interfacial phenomena. Q Rev Biophys 30:241–277. doi:10.1017/s0033583597003363

40. Zhang YJ, Cremer PS (2006) Interactions between macromolecules and ions: the hofmeister series. Curr Opin Chem Biol 10:658–663. doi:10.1016/j.cbpa.2006.09.020

41. Collins KD (2006) Ion hydration: implications for cellular function, polyelectrolytes, and protein crystallization. Biophys Chem 119:271–281. doi:10.1016/j.bpc.2005.08.010

42. Collins KD, Neilson GW, Enderby JE (2007) Ions in water: characterizing the forces that control chemical processes and biological structure. Biophys Chem 128:95–104. doi:10.1016/j.bpc.2007.03.009

43. Collins KD (1995) Sticky ions in biological systems. Proc Natl Acad Sci USA 92:5553–5557

44. Collins KD (1997) Charge density-dependent strength of hydration and biological structure. Biophys J 72:65–72

45. Collins KD (2004) Ions from the hofmeister series and osmolytes: effects on proteins in solution and in the crystallization process. Methods 34:300–311. doi:10.1016/j.ymeth.2004.03.021

46. Livney YD, Ramon O, Kesselman E, Cogan U, Mizrahi S, Cohen Y (2001) Swelling of dextran Gel and osmotic pressure of soluble dextran in the presence of salts. J Polym Sci, Part B: Polym Phys 39:2740–2750

47. Collins KD, Washabaugh MW (1985) The Hofmeister effect and the behaviour of water at interfaces. Quart Rev Biophys 18:323–422

48. von Hippel PH, Schleich T (1969) Ion effects on solution structure of biological macromolecules. Acc Chem Res 2:257–265

49. Hamabata A, von Hippel PH (1973) Model studies on the effects of neutral salts on the conformation stability of biological macromolecules. II. Effects of vicinal hydrophobic groups on the specificity of binding of ions to amide groups. Biochem J 7:1271

50. Zhang Y, Cremer PS (2009) The inverse and direct Hofmeister series for lysozyme. Proc Natl Acad Sci 106:15249–15253. doi:10.1073/pnas.0907616106

51. Dobrynin AV, Colby RH, Rubinstein M (1995) Scaling theory of polyelectrolyte solutions. Macromolecules 28:1859–1871. doi:10.1021/ma00110a021

52. de Gennes PG (1979) Scaling concepts in polymer physics. Cornell University Press, Ithaca, NY

53. Pfeuty P (1978) Conformation des polyelectrolytes ordre dans les solutions de polyelectrolytes. J Phys (Paris) 39:C2–149

54. Fuoss RM (1948) Viscosity function for polyelectrolytes. J Polym Sci 3:603–604

55. Katchalsky A, Eisenberg H (1951) Molecular weight of polyacrylic and polymethacrylic acid. J Polym Sci 6:145–154

56. Tolstoguzov V, Srinivasan D, Paraf A (1997) Protein-polysaccharide interactions. CRC Press, Boca Raton

57. Grinberg VY, Tolstoguzov VB (1997) Thermodynamic incompatibility of proteins and polysaccharides in solutions. Food Hydrocol 11:145–158

58. Dautzenberg H (2001) Polyelectrolyte complex formation in highly aggregating systems: methodical aspects and general tendencies. In: Radeva T (ed) Polyelectrolyte complex formation in highly aggregating systems: methodical aspects and general tendencies. Marcel Dekker, New York, NY

59. de Kruif CG, Weinbreck F, de Vries R (2004) Complex coacervation of proteins and anionic polysaccharides. Curr Opin Colloid Int Sci 9:340–349

60. Livney YD (2008) Complexes and conjugates of biopolymers for delivery of bioactive ingredients via food. In: Garti N (ed) Complexes and conjugates of biopolymers for delivery of bioactive ingredients via food, 1st edn. Woodhead, Cambridge, England

61. Kossel A (1896) Ueber die basischen stoffe des zellkerns. Hoppe-Seiler's Z Physiol Chem 22:176–187

62. Tiebackx FW (1911) Simultaneous coagulation of two colloids. Zeitschrift fuer Chemie und Industrie der Kolloide 8:198–201

63. Bungenberg de Jong HG, Kruyt HR (1929) Coacervation (partial miscibility in colloid systems). Proc Acad Sci Amsterdam 32:849–856

64. Michaels AS, Miekka RG (1961) Polycation–polyanion complexes: preparation and properties of poly(vinylbenzyltrimethylammonium) poly (styrenesulfonate). J Phys Chem 65:1765–1773

65. Schmitt C, Turgeon SL (2011) Protein/polysaccharide complexes and coacervates in food systems. Adv Colloid Interface Sci 167:63–70. doi:10.1016/j.cis.2010.10.001

66. McClements DJ (2006) Non-covalent interactions between proteins and polysaccharides. Biotechnol Adv 24:621–625

67. Turgeon SL, Beaulieu M, Schmitt C, Sanchez C (2003) Protein-polysaccharide interactions: phase-ordering kinetics, thermodynamic and structural aspects. Curr Opin Colloid Int Sci 8:401–414

68. Tolstoguzov V (2000) Compositions and phase diagrams for aqueous systems based on proteins and polysaccharides. Int Rev Cytol 192:3–31

69. Doublier JL, Garnier C, Renard D, Sanchez C (2000) Protein-polysaccharide interactions. Curr Opin Colloid Int Sci 5:202–214

70. Schmitt C, Sanchez C, Sobry-Banon S, Hardy J (1998) Structure and technofunctional properties of protein-polysaccharide complexes: a review. Crit Rev Food Sci Nutr 38:689–753

71. Thunemann AF, Muller M, Dautzenberg H, Joanny JFO, Lowen H (2004) Polyelectrolyte complexes. In: Schmidt M (ed) Polyelectrolytes with defined molecular architecture II. Springer, Berlin

72. Ulrich S, Seijo M, Stoll S (2006) The many facets of polyelectrolytes and oppositely charged macroions complex formation. Curr Opin Colloid Int Sci 11:268–272. doi:10.1016/j.cocis.2006.08.002

73. Sukhishvili SA, Kharlampieva E, Izumrudov V (2006) Where polyelectrolyte multilayers and polyelectrolyte complexes meet. Macromolecules 39:8873–8881. doi:10.1021/ma061617p

74. Hoda N, Larson RG (2009) Explicit- and implicit-solvent molecular dynamics simulations of complex formation between polycations and polyanions. Macromolecules 42:8851–8863. doi:10.1021/ma901632c

75. Jvd G, Spruijt E, Lemmers M, Cohen Stuart MA (2011) Polyelectrolyte complexes: bulk phases and colloidal systems. J Colloid Interface Sci 361:407–422. doi:10.1016/j.jcis.2011.05.080

76. da Silva FLB, Jonsson B (2009) Polyelectrolyte-protein complexation driven by charge regulation. Soft Matter 5:2862–2868. doi:10.1039/b902039j

77. de Vries R, Stuart MC (2006) Theory and simulations of macroion complexation. Curr Opin Colloid Int Sci 11:295–301

78. da Silva FL, Lund M, Joensson B, Aakesson T (2006) On the complexation of proteins and polyelectrolytes. J Phys Chem B 110:4459–4464

79. Cooper CL, Dubin PL, Kayitmazer AB, Turksen S (2005) Polyelectrolyte-protein complexes. Curr Opin Colloid Int Sci 10:52–78

80. Weinbreck F, de Kruif CG (2003) Complex coacervation of globular proteins and gum arabic. Roy Soc Chem 284:337–344, Special Publication

81. Harnsilawat T, Pongsawatmanit R, McClements DJ (2006) Characterization of b-lactoglobulin-sodium alginate interactions in aqueous solutions: a calorimetry, light scattering, electrophoretic mobility and solubility study. Food Hydrocol 20:577–585

82. Ron N, Zimet P, Bargarum J, Livney YD (2010) Beta-lactoglobulin-polysaccharide complexes as nanovehicles for hydrophobic nutraceuticals in non-fat foods and clear beverages. Int Dairy J 20:686–693

83. Zimet P, Livney YD (2009) Beta-lactoglobulin and its nanocomplexes with pectin as vehicles for omega-3 polyunsaturated fatty acids. Food Hydrocol 23:1120–1126. doi:10.1016/j.foodhyd.2008.10.008

84. Bertrand P, Jonas A, Laschewsky A, Legras R (2000) Ultrathin polymer coatings by complexation of polyelectrolytes at interfaces: suitable materials, structure and properties. Macromol Rapid Commun 21:319–348. doi:10.1002/(sici)1521-3927(20000401)21:7<319::aid-marc319>3.0.co;2-7

85. Decher G (1997) Fuzzy nanoassemblies: toward layered polymeric multicomposites. Science 277:1232–1237. doi:10.1126/science.277.5330.1232

86. Decher G, Hong JD, Schmitt J (1992) Buildup of ultrathin multilayer films by a self-assembly process.3. Consecutively alternating adsorption of anionic and cationic polyelectrolytes on charged surfaces. Thin Solid Films 210:831–835. doi:10.1016/0040-6090(92)90417-a
87. Sukhishvili SA (2005) Responsive polymer films and capsules via layer-by-layer assembly. Curr Opin Colloid Int Sci 10:37–44. doi:10.1016/j.cocis.2005.05.001
88. Sukhorukov GB (2001) Designed nano-engineered polymer films on colloidal particles and capsules. Stud Interface Sci 11:383–414
89. Ogawa S, Decker EA, McClements DJ (2004) Production and characterization of O/W emulsions containing droplets stabilized by lecithin-chitosan-pectin mutilayered membranes. J Agric Food Chem 52:3595–3600
90. Aoki T, Decker EA, McClements DJ (2005) Influence of environmental stresses on stability of O/W emulsions containing droplets stabilized by multi-layered membranes produced by a layer-by-layer electrostatic deposition technique. Food Hydrocol 19:209–220
91. Radtchenko IL, Sukhorukov GB, Leporatti S, Khomutov GB, Donath E, Mohwald H (2000) Assembly of alternated multivalent ion/polyelectrolyte layers on colloidal particles. Stability of the multilayers and encapsulation of macromolecules into polyelectrolyte capsules. J Colloid Interface Sci 230:272–280. doi:10.1006/jcis.2000.7068
92. Cohen Stuart MA, Kleijn JM (2001) Kinetics of polyelectrolyte adsorption. In: Radeva T (ed) Kinetics of polyelectrolyte adsorptionedn. Marcel Dekker, New York
93. Kotz J, Kosmella S, Beitz T (2001) Self-assembled polyelectrolyte systems. Prog Polym Sci 26:1199–1232. doi:10.1016/s0079-6700(01)00016-8
94. Amiel C, Sikka M, Schneider JW, Tsao YH, Tirrell M, Mays JW (1995) Adsorption of hydrophilic-hydrophobic block-copolymers on silica from aqueous-solutions. Macromolecules 28:3125–3134. doi:10.1021/ma00113a015
95. Torchilin VP (2001) Structure and design of polymeric surfactant-based drug delivery systems. J Control Release 73:137–172. doi:10.1016/s0168-3659(01)00299-1
96. Zana R (2005) Dynamics in micellar solutions of amphiphilic block copolymers. In: Zana R (ed) Dynamics in micellar solutions of amphiphilic block copolymersedn. CRC Press/Taylor & Francis, New York
97. Perrin P, Millet F, Charleux B (2001) Emulsions stabilized by polyelectrolytes. In: Radeva T (ed) Emulsions stabilized by polyelectrolytes. Marcel Dekker, New York
98. Forster S, Abetz V, Muller AHE (2004) Polyelectrolyte block copolymer micelles. In: Schmidt M (ed) Polyelectrolytes with defined molecular architecture II, vol 166. Springer, Berlin, pp 173–210. doi:10.1007/b10951
99. Horne DS (2002) Casein structure, self-assembly and gelation. Curr Opin Colloid Interface Sci 7:456–461
100. Livney YD, Schwan AL, Dalgleish DG (2004) A study of beta-casein tertiary structure by intramolecular crosslinking and mass spectrometry. J Dairy Sci 87:3638–3647
101. Livney YD (2010) Milk proteins as vehicles for bioactives. Curr Opin Colloid Interface Sci 15:73–83
102. Shapira A, Davidson I, Avni N, Assaraf YG, Livney YD (2012) β-casein nanoparticle-based oral drug delivery system for potential treatment of gastric carcinoma: stability, target-activated release and cytotoxicity. Eur J Pharm Biopharm 80:298–305. doi:10.1016/j.ejpb.2011.10.022
103. Kataoka K, Harada A, Nagasaki Y (2001) Block copolymer micelles for drug delivery: design, characterization and biological significance. Adv Drug Deliv Rev 47:113–131. doi:10.1016/s0169-409x(00)00124-1
104. Zimet P, Rosenberg D, Livney YD (2011) Re-assembled casein micelles and casein nanoparticles as nano-vehicles for [omega]-3 polyunsaturated fatty acids. Food Hydrocol 25:1270–1276. doi:10.1016/j.foodhyd.2010.11.025
105. Semo E, Kesselman E, Danino D, Livney YD (2007) Casein micelle as a natural nano-capsular vehicle for nutraceuticals. Food Hydrocol 21:936–942
106. Vriezema DM, Aragones MC, Elemans J, Cornelissen J, Rowan AE, Nolte RJM (2005) Self-assembled nanoreactors. Chem Rev 105:1445–1489. doi:10.1021/cr0300688
107. Michel AS, Mestdagh MM, Axelos MAV (1997) Physico-chemical properties of carrageenan gels in presence of various cations. Int J Biol Macromol 21:195–200. doi:10.1016/s0141-8130(97)00061-5
108. Grant GT, Morris ER, Rees DA, Smith PJC, Thom D (1973) Biological interactions between polysaccharides and divalent cations: the egg-box model. FEBS Lett 32:195–198. doi:10.1016/0014-5793(73)80770-7
109. Gombotz WR, Wee SF (1998) Protein release from alginate matrices. Adv Drug Deliv Rev 31:267–285. doi:10.1016/s0169-409x(97)00124-5
110. Wang LX, Bloomfield VA (1990) Osmotic-pressure of polyelectrolytes without added salt. Macromolecules 23:804–809. doi:10.1021/ma00205a018
111. Fuoss RM, Katchalsky A (1951) The potential of an infinite rod-like molecule and the distribution of the counter ions. Proc Natl Acad Sci USA 37:579–589
112. Katchalsky A (1971) Polyelectrolytes. Pure Appl Chem 26:327–373

113. Noda I, Kato N, Kitano T, Nagasawa M (1981) Thermodynamic properties of moderately concentrated solutions of linear polymers. Macromolecules 14:668–676. doi:10.1021/ma50004a042

114. Tombs MP, Peacocke AR (1974) The osmotic pressure of biological macromolecules. Clarendon, Oxford

115. Donnan FG (1933) Some considerations relating to membrane equilibria and the secondary swelling of protein gels. J Soc Leather Trade Chem 17:136–143

116. Hill TL (1956) A fundamental studies. On the theory of the donnan membrane equilibrium. Discussions of the Faraday Society 21:31–45. doi:10.1039/DF9562100031

117. Hill TL (1957) Electrolyte theory and the donnan membrane equilibrium. J Phys Chem 61:548–553

118. Flory PJ, Rehner J Jr (1943) Statistical mechanics of crosslinked polymer networks II. Swelling. J Chem Phys 11:521–526

119. Flory PJ (1979) Molecular theory of rubber elasticity. Polymer 20:1317–1320. doi:10.1016/0032-3861(79)90268-4

120. Flory PJ (1953) Principles of polymer chemistry. Cornell University Press, Ithaca

121. James HM, Guth E (1953) Statistical thermodynamics of rubber elasticity. J Chem Phys 21:1039–1049

122. Skouri R, Schosseler F, Munch JP, Candau SJ (1995) Swelling and elastic properties of polyelectrolyte gels. Macromolecules 28:197–210. doi:10.1021

123. Khokhlov AR, Starodubtzev SG, Vasilevskaya VV (1993) Conformational transitions in polymer gels – theory and experiment. Adv Polym Sci 109:123–175

124. Katchalsky A, Michaeli I (1955) Polyelectrolyte gels in salt solutions. J Polym Sci 15:69–86

125. Katchalsky A, Lifson S, Eisenberg H (1951) Equation of swelling for polyelectrolyte gels. J Polym Sci 7:571–574

126. Khokhlov AR, Khalatur PG (2005) Solution properties of charged hydrophobic/hydrophilic copolymers. Curr Opin Colloid Interface Sci 10:22–29. doi:10.1016/j.cocis.2005.04.003

127. Holm C, Joanny JF, Kremer K, Netz RR, Reineker P, Seidel C, Vilgis TA, Winkler RG (2004) Polyelectrolyte theory. In: Polyelectrolytes with defined molecular architecture II, vol 166. Springer, Berlin, pp 67–111. doi:10.1007/B11349

128. Liao Q, Dobrynin AV, Rubinstein M (2003) Molecular dynamics simulations of polyelectrolyte solutions: osmotic coefficient and counterion condensation. Macromolecules 36:3399–3410. doi:10.1021/ma0259968

129. Stevens MJ, Kremer K (1995) The nature of flexible linear polyelectrolytes in salt-free solution – a molecular-dynamics study. J Chem Phys 103:1669–1690. doi:10.1063/1.470698

130. Krul LP, Nareiko EI, Matusevich YI, Yakimtsova LB, Matusevich V, Seeber W (2000) Water super absorbents based on copolymers of acrylamide with sodium acrylate. Polym Bull 45:159–165. doi:10.1007/pl00006832

131. Zohuriaan-Mehr MJ, Kabiri K (2008) Superabsorbent polymer materials: a review. Iran Polym J 17:451–477

132. Kazanskii KS, Dubrovskii SA (1992) Chemistry and physics of agricultural hydrogels. Adv Polym Sci 104:97–133

133. Saunders JR, Abu-Salih S, Khaleque T, Hanula S, Moussa W (2008) Modeling theories of intelligent hydrogel polymers. J Comput Theor Nanosci 5:1942–1960. doi:10.1166/jctn.2008.1001

134. Sata T, Yang WK (2002) Studies on cation-exchange membranes having permselectivity between cations in electrodialysis. J Membr Sci 206:31–60. doi:10.1016/s0376-7388(01)00491-4

135. Mizutani Y (1990) Ion-exchange membranes with preferential permselectivity for monovalent ions. J Membr Sci 54:233–257. doi:10.1016/s0376-7388(00)80612-2

136. Morris ER, Nishinari K, Rinaudo M (2012) Gelation of gellan – a review. Food Hydrocol 28:373–411. doi:10.1016/j.foodhyd.2012.01.004

137. He CL, Kim SW, Lee DS (2008) In situ gelling stimuli-sensitive block copolymer hydrogels for drug delivery. J Control Release 127:189–207. doi:10.1016/j.jconrel.2008.01.005

138. Scranton AB, Rangarajan B, Klier J (1995) Biomedical applications of polyelectrolytes. In: Peppasand NA, Langer RS (eds) Biomedical applications of polyelectrolytesedn. Springer-Verlag Berlin, Berlin

139. Park TG, Hoffman AS (1992) Synthesis and characterization of Ph- and or temperature-sensitive hydrogels. J Appl Polym Sci 46:659–671. doi:10.1002/app.1992.070460413

Polyelectrolytes, Simulation

Barbara Hribar-Lee and Vojko Vlachy
Faculty of Chemistry and Chemical Technology, University of Ljubljana, Ljubljana, Slovenia

Introduction

Nature and synthetic chemistry have provided polyelectrolytes of different shapes: They can be rod-like as, for example, DNA, or flexible (chain-like) as are many of the synthetic polyelectrolytes. Moreover, they can change their conformation in solution and, under the influence of external conditions such as nature of the

Polyelectrolytes, Simulation

solvent, salt content, pH, or temperature, undergo the transition from globular to extended state. In the simplest case, the polyelectrolyte solutions contain long and often highly charged polyions and the related number of counterions to render the systems electroneutral. Thermodynamic, transport, and structural measurements indicate that their properties are governed by the Coulomb interaction, as also by the short-range, solvent mediated, specific ion effects. The role of the hydrophobic interaction, especially in the systems containing benzene groups, or in conjugated polyelectrolytes, is also important. Altogether we deal with a complicated system where we need to treat accurately both, the long-range Coulomb interaction (hundred angstroms), as well as the short-range (few angstroms only) forces.

Cylindrical Cell Model

First computer simulations of simple rod-like model of polyion were performed in the cell model approximation [1–4]. In [1] the polyions were treated as rigid cylinders of finite length. The counterions were charged hard spheres embedded in the dielectric continuum within the cell. The osmotic coefficient was found to decrease with the decreasing degree of polymerization. The concentration dependence of this quantity exhibited minimum, shifted toward lower concentrations when the degree of polymerization increased. Later, in [2, 3], the distributions of mono- and divalent counterions around very long cylindrical polyion were simulated. Important conclusion was that the Poisson-Boltzmann equation provided semiquantitatively correct results for the osmotic coefficient. Inclusion of the ion-ion correlations through the modified Poisson-Boltzmann theory led to considerably better agreement between theory and simulations, especially when divalent counterions were present in solution. The computer simulations were also used to calculate osmotic coefficient via the contact theorem [4, 5]. The latter two studies demonstrated that neglecting of the interionic correlations by the mean field

approach leads to a partial compensation of the shortcomings of the Poisson-Boltzmann osmotic equation. In another contribution, Le Bret and Zimm [6] used Monte Carlo method to investigate the ion size effects on distribution of ions in solution of rod-like polyelectrolyte mimicking DNA. Extensive Monte Carlo simulations and comparison with the other polyelectrolyte theories were performed by Mills et al. [7, 8]. The grand canonical simulations of the rod-like polyelectrolyte were performed in [9, 10]. The mean activity coefficient of simple electrolyte in solution was calculated and, very interestingly, the results for divalent counterions indicated the "charge inversion" [9]. Extensive simulations of the rod-like polyions in the cell model and comparisons with the modified PB theory were performed by Das et al. [11, 12] and Piñero et al. [13]. In the latter study, the emphasis was on exploration of the catalytic effect on ions, caused by presence of rod-like polyions.

Flexible Polyelectrolytes Interacting via the Screened Coulomb Potential

Polyions are not fully extended in solution and those, containing hydrophobic groups, may collapse into the globule under certain conditions. Valleau simulated flexible polyelectrolyte (a constrained "necklace" model) in ionic solution [14]. Average conformations of the polyion immersed in a primitive-model aqueous electrolyte were studied for several model parameters. Similar studies were performed by Woodward and Jönsson using a bead-spring polyelectrolyte model [15]; they found the screened Coulomb potential between charged species (polyion monomers or ions) to be an excellent approximation for systems with low and moderate added salt concentrations of 1:1 electrolyte. For divalent counterions, this is not true anymore. Seidel [16] studied the persistence length for the same model potential and found an evident influence of the flexibility of the underlying neutral chain on properties of the charged chain.

Modeling Flexible Polyelectrolytes with Explicit Ions

Kremer and coworkers [17–19] performed very complete molecular dynamics simulations for multichain polyelectrolyte systems. The Coulomb interaction between the monomers (bead-spring model polyelectrolyte) and counterions was treated explicitly. Experimental results for the osmotic pressure and the structure factor were reproduced. In addition, the persistence length and end-to-end distance of polyelectrolyte chains were calculated. The authors showed that the chains exhibit significant departures from the fully extended conformation even at low chain densities. Furthermore, the chains contracted significantly before they overlapped. At high polymer density and poor solvent conditions [18], equivalent to strongly screened electrostatic interaction, the chains were found to be extremely collapsed. The pearl necklace conformations were observed [19] and analyzed in detail, as was also the position of the first peak of the structure factor with respect to the monomer density. To study the shift in the apparent dissociation constant, Monte Carlo simulations of linear weak polyacids have been performed in the cell model approximation [20]. Widom's particle-insertion method was utilized for the purpose. Simulation results were compared with experiments. Thermodynamic properties of a model solution with chain-like polyions and hard sphere counterions were, for the purpose of comparison with the integral equation theory, published in [21]. There were other studies in this direction. For example, conformational characteristics of single flexible polyelectrolyte chain of 150 univalent and negatively charged beads, connected by a harmonic-like potential in the presence of an equal number of positive counterions, were studied in molecular dynamics simulations by Jesudason et al. [22]. Extensive computer simulations of polyelectrolyte solutions were performed by Dobrynin and coworkers [23–26] (for review see [24–26]). In [23], they evaluated osmotic coefficients and counterion distribution functions of rod-like and flexible polyelectrolyte chains with explicit counterions by using

molecular dynamics approach. Osmotic pressure was studied also by Chang and Yethiraj [27], and very recently by Carrillo and Dobrynin [28]. In this last paper, the polyelectrolyte solution was modeled as an ensemble of bead-spring chains of charged Lennard-Jones particles with explicit counterions and salt ions. The simulations showed that in dilute and semidilute polyelectrolyte solutions, the electrostatic-induced chain persistence length scaled with the solution ionic strength as $I^{-1/2}$. This dependence is due to the counterion condensation on polyions. The simulations confirmed that the peak position in the polymer scattering function scaled with the polymer concentration c_p in dilute polyelectrolyte solutions as $c_p^{1/3}$. In semidilute polyelectrolyte solutions, and for low concentration of added salt, the position of this peak shifts toward the large values of the wave number. The paper contains important citations of previous relevant studies. The rheology of dilute salt-free polyelectrolyte solutions was studied by Stoltz and coworkers [29] using Brownian dynamics simulations and coarse-grained bead-spring polyelectrolyte model with explicit counterions. An overview of the simulation studies and theories of polyelectrolyte solutions was provided by Yethiraj [30].

Explicit Water Simulations of Polyelectrolyte Solutions

All-atom simulations of polyelectrolyte systems are less frequent and they only became emerging in last 5 years. Among the first such studies was the paper of Molnar and Rieger [31]. These authors studied the "like-charge attraction" between polyanions observed in the presence of multivalent cations on a fully atomistic scale. As a relevant example, they examined the interaction of negatively charged carboxylic groups of sodium polyacrylate molecules with divalent calcium ions in explicit water. They showed that Ca^{2+} ions initially associate with single chains of polyacrylates; strongly shielded polyanions approach each other and eventually precipitate. Chialvo and Simonson [32] examined the solvation behavior of short-chain polystyrene sulfonate in aqueous electrolyte solutions by molecular dynamics simulation. The explicit atomistic picture of all species was used. The goal was to determine the hydration effects on the conformation of the polyelectrolyte with varying degree of sulfonation and varying the valence of counterions. Chang and Yethiraj [33] studied dilute salt-free solutions of charged flexible polymer molecules in poor solvents. The simulations suggest that the presence of explicit solvent molecules can be an important aspect of polyelectrolyte behavior in poor solvents. In another contribution, Ju and coworkers [34] simulated solution of poly-methacrylic acid in aqueous solutions at various degrees of charge density. They observed that water molecules may act as a bridging agent between two neighboring carboxylic groups. These bridged water molecules stabilize the rod-like chain conformation and display different dynamic properties from the bulk water. A similar study was recently published by Sulatha and Natarajan [35]. Druchok and coworkers [36–38] presented a molecular dynamics simulation in explicit water of a solution of aliphatic 3,3- and 6,6-ionene oligocations with sodium co-ions and fluorine, chlorine, bromine, and iodine counterions. The purpose of these studies was to investigate how the increasing number of methylene groups (increased hydrophobicity) affected the specific ion interaction between the counterions in solution and quaternary ammonium group on the ionene backbone. The results were able to explain some experimental observations in ionene solutions and weakly charged polyelectrolytes in general.

Simulations of Polyelectrolytes at Surfaces

The adsorption of polyelectrolytes to surfaces is a problem of growing interest stimulated by many industrial applications. Explicit and implicit solvent models were used in studying this problem via computer simulation [39, 40]. Using molecular and Brownian dynamics simulations and freely jointed models of polyelectrolytes, Reddy and Yethiraj [39] established that the solvent plays a dominant role in the adsorption of polyelectrolytes in poor solvents, and that the many-body effects qualitatively influence the adsorption characteristics and mechanism. The effects of surface charge density, charge distributions, solvent quality, and short-ranged interactions were studied by Carillo and Dobrynin by using the molecular dynamics simulations [25, 41].

Future Directions

The progress in this area of research is hampered by uncertainties in the current force fields, mixing rules, and other details of simulation protocols. In addition, some thermodynamic properties of solution like, the enthalpy or heat capacity of solution, cannot be simulated with sufficient accuracies to be tested against the experimental data for polyelectrolyte solutions. Further development of the force fields and methods to calculate solution thermodynamic parameters is needed to advance this area of science.

Cross-References

▶ Polyelectrolytes, Properties
▶ Specific Ion Effects, Theory

References

1. Vlachy V, Dolar D (1982) Monte Carlo studies of polyelectrolyte solution at low degrees of polymerization. J Chem Phys 76:2010–2014. doi:10.1063/1.443174
2. Bratko D, Vlachy V (1982) Distribution of counterions in the double layer around a cylindrical polyion. Chem Phys Lett 90:434–438. doi:10.1016/0009-2614(82)80250-9
3. Bratko D, Vlachy V (1985) Monte Carlo studies of polyelectrolyte solutions. Effect of polyelectrolyte charge density. Chem Phys Lett 115:294–298. doi:10.1016/0009-2614(85)80031-2
4. Vlachy V (1982) On the virial equation for the osmotic pressure of linear polyelectrolytes. J Chem Phys 77:5823–5825. doi:10.1063/1.443741

5. Piñero J, Bhuiyan LB, Reščič J, Vlachy V (2006) Coulomb correlation between counterions in the double layer around cylindrical polyions. Acta Chim Slov 53:316–323

6. Le Bret M, Zimm B (1984) Monte Carlo determination of the distribution of ions about a cylindrical polyelectrolyte. Biopolymers 23:271–285

7. Mills P, Anderson CF, Record MT Jr (1985) Monte-Carlo studies of counterion DNA interactions – comparison of the radial-distribution of counterions with predictions of other poly-electrolyte theories. J Phys Chem 89:3984–3994. doi:10.1021/j100265a012

8. Mills P, Paulsen MD, Anderson CF, Record MT Jr (1986) Monte-Carlo simulations of counterion accumulation near helical DNA. Chem Phys Lett 129:155–158. doi:10.1016/0009-2614(86)80188-9

9. Vlachy V, Haymet ADJ (1986) A grand canonical Monte Carlo simulation study of polyelectrolyte solutions. J Chem Phys 84:5874–5880. doi:10.1063/1.449898

10. Mills P, Anderson CF, Record MT Jr (1986) Grand canonical Monte-Carlo calculations of thermodynamic coefficients for a primitive model of DNA salt-solutions. J Phys Chem 90:6541–6548. doi:10.1021/j100282a025

11. Das T, Bratko D, Bhuiyan LB, Outwaite CW (1995) Modified Poisson-Boltzmann theory applied to polyelectrolyte solutions. J Phys Chem 99:410–418. doi:10.1021/j100001a061

12. Das T, Bratko D, Bhuiyan LB, Outwaite CW (1997) Polyelectrolyte solutions containing mixed valency ions in the cell model: a simulation and modified Poisson-Boltzmann study. J Chem Phys 107:9197–9207. doi:10.1063/1.475211

13. Piñero J, Bhuiyan LB, Reščič J, Vlachy V (2008) Ionic correlations in the inhomogeneous atmosphere surrounding cylindrical polyions. Catalytic effects of polyions. J Chem Phys 128:214904

14. Valleu JP (1989) Flexible polyelectrolyte in ionic solution: a Monte Carlo study. Chem Phys 129:163–174. doi:10.1016/0301-0104(89)80001-1

15. Woodward CE, Jönsson B (1991) Monte Carlo and mean field studies of a polyelectrolyte in salt solution. Chem Phys 155:207–219. doi:10.1016/0301-0104(91)87021-M

16. Seidel C (1996) Polyelectrolyte simulation. Ber Bunsenges-PCCP 100:757–763

17. Stevens MJ, Kremer K (1995) The nature of flexible linear polyelectrolytes in salt-free solution – a molecular-dynamics study. J Chem Phys 103:1669–1690. doi:10.1063/1.470698

18. Micka U, Holm C, Kremer K (1999) Strongly charged, flexible polyelectrolytes in poor solvents: molecular dynamics simulations. Langmuir 15:4033–4044. doi:10.1021/la981191a

19. Limbach HJ, Holm C, Kremer K (2002) Structure of polyelectrolytes in poor solvent. Europhys Lett 60:566–572. doi:10.1209/epl/i2002-00256-8

20. Ullner M, Woodward CE (2000) Simulations of the titration of linear polyelectrolytes with explicit simple ions: comparisons with screened coulomb models and experiments. Macromolecules 33:7144–7156

21. Bizjak A, Reščič J, Kalyuzhnyi YV, Vlachy V (2006) Theoretical aspects and computer simulations of flexible charged oligomers in salt-free solutions. J Chem Phys 125:214907. doi:10.1063/1.2401606

22. Jesudason CG, Lyubartsev AP, Laaksonen A (2009) Conformational characteristics of single flexible polyelectrolyte chain. Eur Phys J E 30:341–350. doi:10.1140/epje/i2009-10532-5

23. Liao Q, Dobrynin AV, Rubinstein M (2003) Molecular dynamics simulations of polyelectrolyte solutions: osmotic coefficient and counterion condensation. Macromolecules 36:3399–3410. doi:10.1021/ma0259968

24. Dobrynin AV (2004) Molecular simulations of charged polymers. In: Kotelyanskii M, Theodorou DN (eds) Simulation methods for polymers. Marcel Dekker, New York, pp 259–312

25. Dobrynin AV, Rubinstein M (2005) Theory of polyelectrolytes in solutions and at surfaces. Prog Polym Sci 30:1049–1118. doi:10.1016/j.progpolymsci.2005.07.006

26. Dobrynin AV (2008) Theory and simulations of charged polymers: from solution properties to polymeric nanomaterials. Curr Opin Coll Interface Sci 13:376–388. doi:10.1016/j.cocis.2008.03.006

27. Chang R, Yethiraj A (2005) Osmotic pressure of salt-free polyelectrolyte solutions: a Monte Carlo simulation study. Macromolecules 38:607–616. doi:10.1021/ma0486952

28. Carrillo YJ-M, Dobrynin AV (2011) Polyelectrolytes in salt solutions: molecular dynamics simulations. Macromolecules 44:5798–5816. doi:10.1021/ma2007943

29. Stoltz C, de Pablo JJ, Graham MD (2007) Simulation of nonlinear shear rheology of dilute salt-free polyelectrolyte solutions. J Chem Phys 126:124906. doi:10.1063/1.2712182

30. Yethiraj A (2009) Liquid state theory of polyelectrolyte solutions. J Phys Chem B 113:1539–1551. doi:10.1021/jp8069964

31. Molnar F, Rieger J (2005) "Like-charge attraction" between anionic polyelectrolytes: molecular dynamics simulations. Langmuir 21:786–789. doi:10.1021/la048057c

32. Chialvo AA, Simonson JM (2005) Solvation behavior of short-chain polystyrene sulfonate in aqueous electrolyte solutions: a molecular dynamics study. J Phys Chem B 109:23031–23042. doi:10.1021/jp053512e

33. Chang R, Yethiraj A (2006) Dilute solutions of strongly charged flexible polyelectrolytes in poor solvents: molecular dynamics simulations with explicit solvent. Macromolecules 39:821–828. doi:10.1021/ma051095y

34. Ju S-P, Lee W-J, Huang C-I, Cheng W-Z, Chung Y-T (2007) Structure and dynamics of water surrounding

the polymethacrylic acid. A molecular dynamics study. J Chem Phys 126:224901

35. Sulatha MS, Natarajan U (2011) Origin of the difference in structural behavior of poly(acrylic acid) and poly(methacrylic acid) in aqueous solution discerned by explicit-solvent explicit-ion MD simulations. Ind Eng Chem Res 50:11785–11796. doi:10.1021/ie2014845

36. Druchok M, Hribar–Lee B, Krienke H, Vlachy V (2008) A molecular dynamics study of short–chain polyelectrolytes in explicit water: toward the origin of ion–specific effects. Chem Phys Lett 450:281–285. doi:10.1016/j.cplett.2007.11.024

37. Druchok M, Vlachy V, Dill KA (2009) Computer simulations of ionenes, hydrophobic ions with unusual solution thermodynamic properties. The ion-specific effects. J Phys Chem B 113:14270–14276. doi:10.1021/jp906727h

38. Druchok M, Vlachy V, Dill KA (2009) Explicit – water molecular dynamics study of a short – chain 3,3 ionene in solution with sodium halides. J Chem Phys 130:134903-1–134903-8. doi:10.1063/1.3078268

39. Reddy G, Yethiraj A (2010) Solvent effects in polyelectrolyte adsorption: computer simulation with explicit and implicit solvent. J Chem Phys 132:074903. doi:10.1063/1.3319782

40. Reddy G, Chang R, Yethiraj A (2006) Adsorption and dynamics of a single polyelectrolyte chain near a planar charged surfaces: molecular dynamics simulation with explicit solvent. J Chem Theory Comput 2:630–636. doi:10.1021/ct050267u

41. Carillo J-MY, Dobrynin AV (2007) Molecular dynamics simulation of polyelectrolyte adsorption. Langmuir 23:2472–2482. doi:10.1021/la063079f

Polymer Electrolyte Fuel Cells (PEFCs), Introduction

Günther G. Scherer
Electrochemistry Laboratory, Paul Scherrer Institute, Villigen, Switzerland

Introduction

Novel highly efficient conversion technologies for mobility (electromobility) and combined heat and power systems (CHP) with independence on fossil fuels, in particular crude oil, are of utmost interest to face the energy challenges of the future [1, 2]. Fuel cell technology, based on the "cold" combustion of a fuel in an electrochemical cell, can fulfill these requirements. Another area of application for fuel cell technology is portable electric and electronic devices, where the argument of potentially higher energy density as compared to today's available battery technologies, hence, longer time of operation, is of prime interest [3].

The polymer electrolyte fuel cell concept, utilizing a thin ion-conducting polymer film, can fulfill the specifications of many of these applications. Typically operating at temperatures around 100 °C, this technology is the choice for many of these applications, due to their flexibility in start at ambient temperature, load following behavior, flexible power demand, etc. However, limitations exist for operation temperatures below 100 °C in the freedom of choice of fuel, limiting this essentially to hydrogen at high purity.

In this context, the installation of new supply infrastructures for alternative fuels, e.g., H_2, is an important additional economical and political factor, in particular for mobility applications. Dedicated well-to-wheel energy analysis has clearly shown that energy conversion in fuel cells has to be based on fuels derived from renewable sources, particularly when hydrogen is the fuel [4].

The Solid Polymer Electrolyte

The characteristics of any fuel cell technology are determined by specifics of the electrolyte chosen, e.g., low or high temperature, acidic or alkaline. However, general statements can be made, which hold for any electrolyte, due to its universal function it has to fulfill in an electrochemical cell.

This concerns the transport of charge in ionic form from one interface to the other within the cell. The ion conduction should be carried by an ion, which is produced at one electrode and consumed at the other to avoid losses caused by concentration gradients, hence minimize losses in the ionic circuit [5]. This majority ion should carry as much charge as possible through the

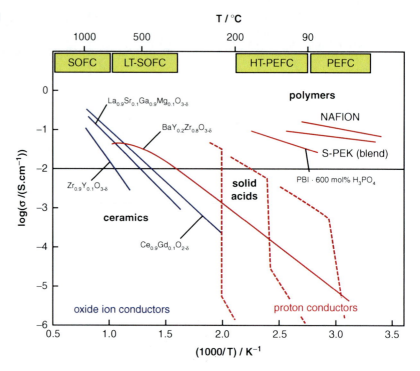

Polymer Electrolyte Fuel Cells (PEFCs), Introduction, Fig. 1 Specific conductivity versus temperature of different electrolytes interesting for fuel cell applications (Adopted from K. D. Kreuer)

electrolyte; as a consequence, its transport number (Hittorf number) should be as high as possible, ideally one. For high power applications with an aqueous electrolyte, the concentration of this respective anion or cation must be high and exclusively responsible for conduction. Typically, a specific ionic conductivity in the range of 100 mS/cm is required, which at a current density of 1 A/cm^2 and an electrolyte gap of 100 μm would yield an ohmic voltage loss in the range of 100 mV. As seen in Fig. 1, the conductivities of various electrolytes are strongly temperature dependent and cover a wide temperature range, depending on the ion-conducting species in its respective environment (material). Again, electrolytes are of interest, which offer the opportunity that the ionic species is participating in the fuel cell reaction.

Further, the electrolyte material has to act as separator to (1) avoid crossover and, as a consequence, "hot" combustion of the fuel and the oxidant at the respective counter electrode and (2) to avoid touching of the electrodes by short circuiting the electronic pathway. Thereby, the electrolyte/separator gap should be as narrow as possible to lower the ohmic contribution to the overall voltage loss. In case of a liquid electrolyte, it can be absorbed into an inert matrix, providing porosity high enough for an ionic conduction path with low tortuosity for the ion and at the same time a high enough bubble pressure to suppress gas crossover. An ideal concept is the one of a solid electrolyte, which fulfills the dual electrolyte and separator function at the same time, providing a transfer number of one for the current carrying ionic species, as in the case of solid polymer electrolytes (ion-exchange membranes), thereby excluding a contribution of electronic conduction.

One has to emphasize that the electrolyte has to sustain the potential window of the respective fuel cell reaction, given by the Gibbs free energy, at least over the device lifetime, specified for a certain application.

The idea to utilize a "solid ion-exchange membrane electrolyte" in a secondary cell employing metal electrodes was first published by Grubb in 1959 [6], reflecting the fact that a solid electrolyte acts in the dual function of ion conductor and separator. This concept was extended to H_2/O_2

$$-(CF_2-CF_2)_k-(CF_2-CF)_l-$$
$$|$$
$$O-(CF_2-CF-O)_m-CF_2-CF_2-SO_3H$$
$$|$$
$$CF_3$$

Polymer Electrolyte Fuel Cells (PEFCs), Introduction, Fig. 2 Generalized structure formula of Nafion®-type membranes (DuPont) with $m = 1, l = 1, k = 5-7$

fuel cells in 1960 [7]. The solid polymer electrolyte consisted of a heterogeneous membrane, where particles of sulfonated cross-linked polystyrene bonded into sheet form with an inert binder. Actually, the realization of this concept happened only a few years after the preparation of a free-standing synthetic polymeric ion-exchange membrane was described by Juda et al. for the first time [8].

Many different types of polymer membranes containing fixed ionic groups have been explored since. Up to today, the materials of choice are perfluoro sulfonic acid membranes of the Nafion-type (Fig. 2), for the first time described by Grot [9]. These materials, similar products have later been developed by other companies (Asahi Glass, Asahi Kasei, Solvay, 3M, Gore, etc.), have been continuously improved for their application in fuel cells. A comprehensive overview of membrane development can be found in [10].

The Concept of the Polymer Electrolyte Fuel Cells

The idea of using a thin ion-conducting polymer membrane, *solid polymer electrolyte*, as electrolyte and separator can lead to different concepts of cells. Firstly, when ionic charges, anions or cations, are chemically bound to the polymer (ionomer) network, the respective countercharge can move freely within the polymer volume, preconditioned a certain volumetric charge density within the polymer exists, which under uptake of, e.g., water, leads to phase separation into a hydrophobic polymer (backbone) and a hydrophilic, charge-containing phase [11]. As a consequence, the polymer morphology allows the continuous transport of this ion from one electrode interface to the other. One has to

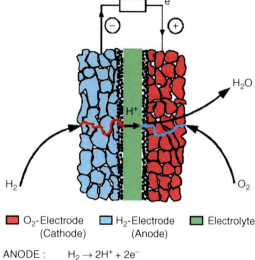

ANODE: $H_2 \to 2H^+ + 2e^-$
CATHODE: $2H^+ + 2e^- + 1/2\, O_2 \to H_2O$

Polymer Electrolyte Fuel Cells (PEFCs), Introduction, Fig. 3 Simplified scheme with an acidic solid polymer electrolyte, e.g., the polymer electrolyte fuel cell (*PEFC*). Fuel, H_2; Oxidant, O_2. Only porous gas diffusion electrodes and electrolyte are shown; cell housing is not shown [12]

emphasize that ionic conductivity increases with water content. The optimization of the in situ water content of the solid electrolyte is of paramount importance for an optimal performance, whereby the visco-mechanical properties of the polymer sheet have to be balanced.

A simplified scheme of the electrochemical heart of a PEFC is displayed in Fig. 3, where the "central solid electrolyte" is contacted by two porous gas diffusion electrodes (GDLs), which are in intimate contact to the membrane surface (see below, three phase boundary). At the interface to the membranes, the GDLs contain nanoparticles of platinum (black dots) as electrocatalyst.

Depending on the fuel, H_2, MeOH (gas or liquid), or other, the solid polymer electrolyte properties have to be tailored, in particular for their properties towards separation of fuel and oxidant.

Generally speaking, the ion conduction in the electrolyte contributes to the losses (ohmic losses, see Fig. 3) in the cell voltage.

Hence, one of the major tasks in fuel cell development is the reduction of these losses in the ionic circuit part, because the work available in the external electronic circuit should be as high as possible. This concerns the reduction of the membrane thickness to as low values as possible, today down to a few microns, and the optimization of the specific conductivity at low water content (see water management).

In aqueous acidic electrolytes the ionic conduction is provided by an H^+ ion, respectively H $(H_2O)_n^+$. As mentioned above, H^+ is created by the anodic oxidation of H_2 as a fuel, and the conducting species is transported to the cathode, where it is consumed in the direct (ideally four electrons) cathodic reduction of molecular oxygen. Hereby, water as the reaction product appears at the cathode side of the cell.

In an anion exchange membrane (not shown), the ionic conduction is provided by OH^--ions, created by the ORR at the cathode. Principally, OH^- ions are carriers for the O^{2-} ions, produced as intermediate by the cathodic reduction reaction of molecular O_2 and the follow-up reaction with a water molecule. After conduction, OH^- reacts with the H^+ created at the anode and yields water as a product at the anode side.

Solid polymer electrolytes on the basis of a polymer cation exchange membrane in H^+-form or an anion exchange polymer membrane in OH^--form can be considered as quasi-aqueous electrolytes, whereby the water is absorbed in the phase separated ionic nano morphology of the respective material. This nano morphology forms ionic pathways through the polymeric membrane connecting the two fuel cell electrodes.

In cells operated at temperatures below 100 °C, liquid water will be the reaction product. For operation temperatures above the boiling point of water, a cell with an aqueous or quasi-aqueous electrolyte can be operated, however, at the expense of pressurizing it to avoid loss of the water and, as a consequence, concentration and conductivity changes in the electrolyte. Hence, water management of a PEFC operated at temperatures below 100 °C is an engineering issue and of utmost importance for many high power applications. Novel methods have been introduced to detect liquid water in PEFCs and to help understanding product water removal water [13].

The Role of Electrocatalysis and the Electrode/Electrolyte Interface

Next to the voltage losses due to ionic conduction, voltage losses due to the activation overvoltages of the respective electrode reactions arise, as described by the Butler-Volmer, respectively the Tafel equation [14]. At the same value of current density the overvoltage for the HOR is much smaller than for the ORR, due to the simpler electron transfer kinetics. Overvoltage for an electrochemical reaction shows also a strong temperature dependency, as does activation energy, and decreases with increasing temperature.

Generally, for fuel cells with an acidic electrolyte, platinum is the electrocatalyst of choice at temperatures at these respective temperatures, due to stability requirements in this particular environment. High dispersion of the catalyst is required, taking into account its nature as a precious metal, to provide a high surface area to volume ratio. Platinum nanoparticles in the diameter range of a few nanometers, typically 2–5 nm, supported on carbon particles are utilized. Different carbons with different surface area (up to 800 m^2/g) are used. The amount of platinum in the catalyst powder is specified as weight % per g of carbon and further as mg loading per cm^2 geometric electrode area.

Platinum is prone to CO adsorption at temperatures of around 100 °C and below. For this reason, care has to be taken with respect to the purity of hydrogen, in particular for hydrogen liberated from a C-containing fuel by a reforming process (steam reforming, partial oxidation, autothermal reforming).

Alkaline electrolytes promise to allow cheaper, non-precious metal catalysts like nickel and its alloys.

There is a common problem to all material selections for the different fuel cell types,

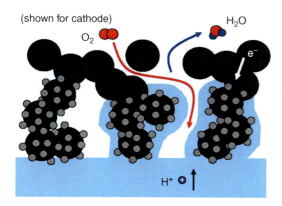

Polymer Electrolyte Fuel Cells (PEFCs), Introduction, Fig. 4 Electrode layer (interphase) with three phase boundary (schematic) of a polymer electrolyte fuel cell (cathode side). *Blue*, polymer electrolyte; *black*, carbon particles; *grey*, platinum nanoparticles (Adopted from L. Gubler)

namely, to generate an optimal electrolyte/electrode interface. One has to realize that gaseous molecular reactants are converted into ionic species, solvated/hydrated in their electrolyte medium, by exchanging electrons with the electrode material. This reaction occurs at the so-called triple phase boundary (triple point), where electron conducting, ion conducting, and gas phase (eventually dissolved in the electrolyte) join to each other. To allow a high surface area for current generation, the electrolyte/electrode interface is extended into a three-dimensional *interphase* with a thickness of a few μm. For a polymer electrolyte fuel cell, this *interphase* is drafted in Fig. 4 for the cathode (oxygen) side.

Types of PEFCs

Hydrogen is the preferred fuel for PEFCs. However, a PEFC can also be fed by liquid or gaseous methanol, called direct methanol fuel cell, DMFC. Other fuels based on alcohols, e.g., the direct ethanol fuel cell, DEFC, are subject to research.

In the past, most concepts for an alkaline fuel cell have been described with an aqueous alkaline electrolyte. Recently, also anion exchange membranes, alkaline solid polymer electrolytes, have been considered [15].

H$_2$-Fed PEFC

This type of H$_2$-fed fuel cell, due to its high achievable specific and volumetric power density (in the range of 1 kW/kg and 1 kW/l and above), its cold start behavior, and its fast load following properties, has found the interest of automotive industry and is under development by all major automotive companies.

A thin ion-conducting polymer sheet, typically in the range of 25 μm, is utilized as solid electrolyte, i.e., electrolyte and separator, in the PEFC (Fig. 3). For thermal stability reasons, only cation exchange membranes in the H$^+$-form have been considered up to today for technical applications (see also alkaline fuel cell). In comparison to cells with liquid electrolyte, this PEFC-concept offers the advantage that the "electrolyte" is chemically bound within the polymer matrix and only water as reactant, in addition to the gases, appears in the peripheral system components. Membranes are perfluorinated or partially fluorinated polymers, with side chains ending in pendant acid groups, e.g., the sulfonic acid group –SO$_3$H.

Under operation, the membrane has to contain some water, e.g., 15 water molecules per sulfonic acid group, to provide the necessary specific conductivity. This fact causes consequences for some of the system auxiliaries, as the gases, hydrogen and air, have to be humidified before flowing into the cell to sustain the hydration level in the membrane.

The acidic electrolyte asks for a precious metal catalyst; hence platinum supported on carbon particles serves as electrocatalyst, with typical Pt loadings of ca. 0.1 mg/cm^2 at the anode side and ca. 0.4 mg/cm^2 at the cathode side. Platinum nanoparticles, typically a few (3–5) nm in diameter, are deposited on various carbon substrates (e.g., 20–40 %Pt/C) by wet chemical processes and then further processed in combination with solubilized ionomer material and binder(s) (PTFE) to yield an ink, which then is applied either to the electrolyte membrane surface (CCM, catalyst coated membrane) or to the GDL to form the GDE.

Extensive characterization of these electrocatalysts for the fuel cell reactions has shown that the electrocatalytic activity of these nanoparticles is by a factor 10 lower than the activity of a polycrystalline platinum surface for the respective reactions, HOR or ORR [16]. This difference is not fully understood yet.

An interesting proprietary approach has been followed recently by the 3M Company, creating a continuous electrode area covered by nanoscale Pt-whiskers. Reactivity of these electrocatalytic layers is in the range of polycrystalline platinum surfaces [17].

Today, pure hydrogen is considered as the ideal fuel. Due to its CO-content, hydrogen derived from C-containing fuels (by steam reforming (SR), partial oxidation (POX), or autothermal reforming (ATR) followed by preferential oxidation (PROX)) would need specific measures in terms of electrocatalysis, e.g., PtRu as bifunctional electrocatalyst, to allow oxidation of CO. As an alternative, purification methods of the anode gas stream prior to entering the cell down to only a few ppm CO (depending on operation temperature) are necessary.

The electrocatalytic layers are contacted by gas diffusion layers (GDLs), allowing the gases, H_2 and O_2, as reactants to be passed to the interface and liquid H_2O as product removed from the interface. These GDLs consist of sheets of carbon cloth or paper with typically 100 μm thickness and 50 % porosity (Fig. 2). This arrangement of membrane, electrocatalytic layers, and gas diffusion layers is colloquially called membrane–electrode-assembly (MEA).

Most of the applications of this fuel cell type are developed for air operation, taking advantage of utilizing one reactant from the ambient environment. However, there exist also some applications in which pure oxygen is employed as oxidant. Due to the higher partial pressure of oxygen and the absence of electrochemically inert nitrogen as the majority component in the cathodic gas stream, humidification issues of the cell can be strongly simplified [18, 19].

Methanol-Fed Fuel Cell (PEFC-Type)

The same concept of combination of an acidic solid polymer electrolyte and acid stable precious metal electrocatalysts can also be applied to a methanol-fed fuel cell (Direct Methanol Fuel Cell, DMFC). Methanol can be fed in liquid or vapor form, mixed with water. The methanol molecule CH_3OH is electrochemically converted at the anode according to

$$\text{Anodic reaction}: CH_3OH + H_2O = 6\,H^+ 6e^- + CO_2 \quad (\text{Methanol Oxidation Reaction, MOR})$$

$$\text{Cathodic reaction}: O_2 + 4H^+ + 4e^- = 2H_2O$$
$$(\text{Oxygen Reduction Reaction, ORR})$$

$$\text{Overall reaction}: CH_3OH + H_2O + O_2$$
$$= CO_2 + 3H_2O$$

As seen from above, the MOR occurs under the consumption of one molecule of water, necessary to oxidize carbon in the CH_3OH molecule to CO_2, while potentially six protons are liberated. The ORR can be formulated the same way as in a hydrogen-fed cell.

The equilibrium potential for the anode reaction is at 0.02 V versus the Normal Hydrogen Electrode (NHE); hence, a theoretical cell voltage close to the H_2/O_2 fuel cell should be observed. In practice, the OCV is lower, for several reasons: Due to the similarity of methanol and water as solvents (e.g., solubility parameter), methanol also penetrates into the water-swollen polymer membrane and passes to the cathode (methanol crossover), causing a mixed potential at a lower value than the oxygen potential. Further, the anode reaction requires a binary or ternary bifunctional catalyst, containing next to platinum one metal component (or two) providing the splitting reaction of the water molecule involved to liberate oxygen for the carbon oxidation at lower potentials as compared to platinum. Examples would be Ru, Sn, Mo, others, or even combinations of two of these together with platinum as the hydrogen liberating catalyst part.

Intermediate oxidation species of the methanol molecule, e.g., –COH, may adsorb and poison the platinum catalyst surface, thereby impeding the full oxidation reaction.

These kinetic losses at the anode lead to a high anode overvoltage and, therefore, the cell voltage in the DMFC is lower at a respective value of current density as compared to a H_2-fed cell, as is the achievable power density.

Direct methanol fuel cell concepts with liquid electrolyte have also been considered in the past. One advantage of circulating liquid electrolytes is the option to cool the cell without additional cooling fluid.

Cross-References

▶ Fuel Cells, Principles and Thermodynamics
▶ Polymer Electrolyte Fuel Cells, Membrane-Electrode Assemblies
▶ Polymer Electrolyte Fuel Cells, Oxide-Based Cathode Catalysts

References

1. International Energy Agency (updated June 2011) Clean energy progress report, Paris
2. International Energy Agency (2004) Hydrogen and fuel cells, review of national R&D programs. OECD Publishing. ISBN 9789264108837
3. e.g., http://www.sfc.com/en/
4. http://www.roads2hy.com/pub_download.asp?PageIndex=1
5. Appleby AJ (1994) J Power Sources 49:15
6. Grubb WT (1959) J Electrochem Soc 106:275
7. Grubb WT (1960) J Electrochem Soc 107:131
8. Juda W, McRae WA (1953) US Patent 2,636,851
9. Grot WG (1972). Chem Ing Techn 44:163, 167
10. Scherer GG (ed) (2008) Fuel cells I. Adv Polym Sci 215. Springer, Berlin/Heidelberg. ISBN 978-3-540-69755-8; Fuel cells II (2008), Adv Polym Sci 216. Springer, Berlin/Heidelberg. ISBN 978-3-540-69763-3
11. Kreuer K-D (2007) Proton conduction in fuel cells. In: Limbach HH, Schowen RL, Hynes JT, Klinman JP (eds) Hydrogen-transfer reactions, vol 1. Wiley-VCH, Weinheim, pp 709–736, chap. 23
12. Scherer GG (1990) Ber Bunsenges Physik Chem 94:1008
13. Boillat P, Scherer GG (2012) Neutron imaging. In: Wang H, Yuan X-Z, Lui H (eds) PEM fuel cell diagnostic tools. CRC Press, Boca Raton, p 255
14. Schmickler W (1996) Grundlagen der Elektrochemie. Vieweg, Braunschweig
15. Merle G, Wessling M, Nijmeijer K (2011) J Membr Sci 377:1
16. Gasteiger HA, Kocha SS, Sompali B, Wagner FT (2005) Appl Catal B Environ 56:9
17. Debe M (2012) Nature 486:43
18. http://info.industry.siemens.com/data/presse/docs/m1-isfb07033403e.pdf
19. Büchi FN, Paganelli G, Dietrich P, Laurent D, Tsukada A, Varenne P, Delfino A, Kötz R, Freunberger SA, Magne P-A, Walser D, Olsommer D (2007) Fuel Cells 7:329

Polymer Electrolyte Fuel Cells, Mass Transport

Felix N. Büchi and Pierre Boillat
Paul Scherrer Institut, Villigen PSI, Switzerland

Introduction

Mass transport in electrochemical reactions is defined as the transport of reactants and products which does not include the transport of electric or ionic charges. In the case of polymer electrolyte fuel cells (PEFCs), mass transport includes the following phenomena:

1. The transport of gaseous reactants (hydrogen and oxygen) to the reaction sites
2. The transport of the product water away from the reaction sites.

Mass transport loss is defined as the loss in performance of the fuel cell due to limitations in mass transport processes. This performance loss is usually attributed to a reduction of the oxygen activity (associated to its partial pressure) at the electrode, in comparison to the oxygen partial pressure at the cell inlet. The accumulation of water in the transport pathways of the gaseous reactants can lead to increased mass transport losses and instability and can result in accelerated degradation.

The operation of fuel cells depends on mass transport on a variety of different scales, from the distribution of gas flow between different cells of a stack down to transport on the nanometer scale inside the electrode.

At high current densities (>1 A/cm^2) required for the automotive application, or high humidity conditions, the mass transport losses can account for more than 100 mV of cell voltage losses [1]. Detailed understanding of the processes is therefore required to increase the electrochemical cell efficiency. In order to gain accurate insight into the complex mass transport processes, specific experimental methods and modeling approaches are developed.

Mass Transport on the Cell and Stack Level

In most PEFC designs, the reactant gases are actively supplied to the stack by feeding a given gas flow. The gas mass flow exceeds the quantity strictly needed for the electrochemical reaction, so that a sufficient concentration is present in the outlet section of the cell. Ideally, the gas flow should be distributed homogeneously to the flow channels of a cell and between the cells of a stack, but various processes can lead to inhomogeneous distributions as described below.

Channel-to-Channel Distribution of Gas Flow

A typical flow field consists of a number of flow channels running parallel to each other (see Fig. 1, top). The distribution between the different channels is determined by the flow resistance of each channel, which can vary from channel to channel due to the following reasons:

- Structural differences of the gas flow channel sections. Such differences can stem from manufacturing imprecision or inhomogeneous distribution of the compression force.
- The presence of liquid water in the flow channels and/or in the manifolds. As the gas flow velocity is an important parameter for water removal (c.f. below), this can lead to a "latching" effect: the gas flow is reduced in channels where liquid water appears, which in turns leads to less water removal in these sections.

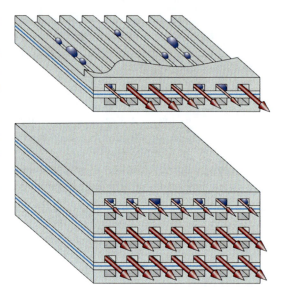

Polymer Electrolyte Fuel Cells, Mass Transport, Fig. 1 Accumulation of water in the flow channels can disturb the distribution of gas on the cell level (*channel to channel, see top*) and on the stack level (*cell to cell, see bottom*)

Different channels of the cell can be considered as placed in parallel, both concerning the gas flow and the electrical current flow. Thus, the complete absence of flow in one section of the cell will not lead to cell reversal, but to the absence of current production in this section. Nevertheless, long-term degradation effects can be expected, in particular in the absence of hydrogen.

Cell-to-Cell Distribution of Gas Flow

The distribution of gas flow between different cells of a stack can suffer from the same perturbations as described above between different channels (see Fig. 1 bottom). Again, both structural differences and accumulation of liquid water can be the cause of inhomogeneous distribution of gas flows.

The cells in a stack are connected in parallel with respect to the gas flow but in series with respect to the current. The implication is that in complete absence of gas flow in a cell, cell voltage reversal will occur which will induce a rapid damage (e.g., due to carbon corrosion). As a consequence, individual cell voltage

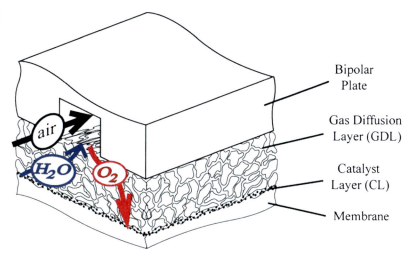

Polymer Electrolyte Fuel Cells, Mass Transport, Fig. 2 Mass transport through the cathode GDL, interconnecting the channel of the bipolar plate and the catalyst layer

monitoring and effective purging strategies are essential for the durable operation of a stack.

Water Removal

A limited quantity of water can be removed as water vapor. This quantity is defined by the magnitude and relative humidity of the incoming gas flows. Water produced in excess of this quantity needs to be removed as a liquid. In the flow channels, water forms droplets which grow as they are fed by produced water transiting through the GDL. When the shear force produced by the gas flow is sufficient to offset the adhesion force of the droplet, it will detach and be carried away by the gas flow (or in some cases adhere to the wall or GDL in a downstream section of the flow channel). The typical size a droplet can grow until it detaches depends primarily on the velocity of the gas flow and on the gas viscosity. If the shear force is low (typically, for the hydrogen channels and/or for low flows), the droplets can grow large enough to form a "slug," meaning a complete filling with water of one fraction of the channel.

Transport in the Gas Diffusion Layer

The gas diffusion layer (GDL) connects the millimeter-scaled gas channels in the bipolar plate with the catalyst layer, where the electrochemical reaction takes place on the surface of nanometer-sized catalyst particles (see Fig. 2).

Gas diffusion layers are used to collect current over the channels and to provide access for the gases under the flow field ribs. Mass, charge, and heat are transported in and through the GDL in the respective pressure, concentration, potential, and temperature fields. The GDLs need to fulfill two antagonistic tasks: they need to be highly permeable for the reactant transport and highly conductive for charge and heat. The porous materials consist typically of carbon fibers with a diameter of 6–8 μm, in the form of either carbon papers or clothes. The maximum of the resulting pore size distribution usually lies in the range of tens of micrometers, and due to the orientation of the fibers, the pore and solid structures are often anisotropic. Porosities typically range in the order of 0.75–0.85 and the thickness is typically between 200 and 300 μm [2].

The main transport mechanism for the gaseous reactants through the GDL is diffusion in the concentration gradient between the gas channel and the surface of the catalyst layer and is described by Fick's first law:

$$N_{eff} = -D_{eff} \frac{\partial c}{\partial x} \qquad (1)$$

where N_{eff} is the effective flux, D_{eff} the effective diffusion coefficient, c the species concentration, and x the distance.

In the porous gas diffusion layer, the effective diffusion coefficient is different from the intrinsic coefficient of the species in a gas mixture in free volume because (i) the solid structure of the GDL acts as a resistance which might be increased by (ii) the presence of liquid water in the pores of the GDL. The effect of multicomponent diffusion is described by the Stefan–Maxwell relation, which can be found in textbooks [3] and is not specific to fuel cells. This is therefore not discussed further here.

The effect of the solid obstructing the gas transport is determined by the porosity ε and tortuosity τ of the GDL. Tortuosity describes the elongation of the direct transport distance x by the solid structure of the GDL, and definition of the porosity is obvious. Liquid water may occupy part of the pore space, decreasing porosity and increasing tortuosity. This effect depends on the amount of liquid water present in the GDL, which is given as the saturation s, the fraction of the pore space occupied. The effective diffusion coefficient is therefore defined as:

$$D_{eff} = -D \frac{\varepsilon}{\tau}(1-s)^n \qquad (2)$$

The effect of the saturation is not linear and described by the factor n in Eq. 2.

In the past years advanced characterization techniques such as synchrotron-based X-ray tomography have allowed to accurately determine the structure of the GDL materials including the effects of compression in the cell [4]. In combination with electrochemical methods [5, 6], this has lead to accurate determination of the effective diffusion coefficients as defined in Eq. 2. D_{eff} in dry GDL, depending on the compression and the direction (in- or through-plane), is in the range between 0.6 and 0.2 times the diffusion coefficient in the free volume (see Fig. 3).

Recently, the effect of liquid water has been quantified using tomographic techniques [7]. With liquid saturation reaching values up to 0.3–0.4 in the cathode, the effective diffusion coefficient D_{eff} drops considerably below 0.1 of the free volume value. This illustrates well that gas transport through the GDL can become a limiting factor at high current densities.

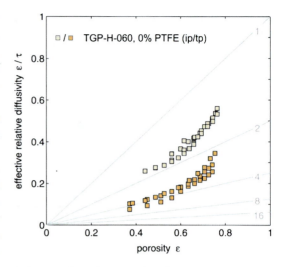

Polymer Electrolyte Fuel Cells, Mass Transport, Fig. 3 Effective relative diffusivity ε/τ as a function of the porosity of plain Toray 060 GDL (without PTFE) for different compressions (expressed in porosity). *ip* in-plane, *tp* through-plane. The *gray lines* indicate iso-tortuosity levels

The transport of liquid water through the GDL is not well understood at this point in time. Liquid water is transported by a pressure gradient between the catalyst layer and the gas channel, where it is removed from the cell. However, the buildup of the transportation paths depends on the local pore sizes and the surface properties (contact angle) of the GDL. Understanding has been gained in ex situ experiments by determining the liquid saturation as function of the capillary pressure [8, 9].

However, in operating cells the situation is more complicated, as the channel/rib structure of the flow field imposes inhomogeneous temperature and wetting boundary conditions. Further the local condensation/evaporation equilibrium strongly affects the liquid phase. The understanding of the transport of liquid water in the operating cell therefore also needs advanced in situ methods such as neutron radiography [10, 11] or X-ray tomography [12]; see section Experimental Methods for Mass Transport Analysis.

Transport of charge (as electrons) and heat in the GDL takes place in the solid part of the gas diffusion layer. As the fibers in usual materials are normally oriented in the in-plane direction,

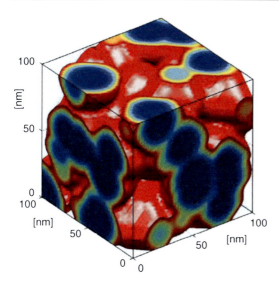

Polymer Electrolyte Fuel Cells, Mass Transport, Fig. 4 Modeled catalyst layer cutout with a side length of 100 nm. The shown random structure includes 64 carbon particles with a diameter of 28 nm colored in blue. The ionomer coating has a thickness of 10 nm and is shown in colors from *green* to *red*

both heat and electric conduction show a high anisotropy, with in-plane conductivities up to ten times higher than the through-plane values.

Transport in the Catalyst Layer

The catalyst layer (CL) is also a porous structure. However, the pore size is much smaller (between 10 and few hundreds of nanometers) as compared to tens of micrometers in the GDL. Further, because the electrochemical conversion occurs here, charge is transported by both protons and electrons and the electrochemically interlinked charge transport becomes more complicated than the one in the GDL. Figure 4 shows a small schematic cutout of the CL showing the tortuous arrangement of the electronically (carbon) and ion-conducting (ionomer) phases. Reactants are predominantly transported in the pore space (void).

In the catalyst layer educts and water are transported in the void and the ionomer phases. Again, because of the porous and tortuous structure, the effective transport parameters deviate substantially from the properties of the respective pure phases. However, in the case of the CL, due to the small scale and the thinness of the layer, it is experimentally difficult to accurately determine the geometry of the single phases. Also the compositions and structures of the CL, when using different preparation methods, have a higher variation than in the case of the GDL. Therefore, the effective transport properties are less well known. Generally, due to the relatively low porosity (typically 0.3–0.5) and high tortuosity of the CL, the effective transport parameters in all the phases are considerably lower (typically a factor of 0.05–0.2) than the one in the respective pure material.

In the catalyst layer, the presence of liquid water also complicates the mass transport processes. On the one hand, the liquid may fill part of the pore space, affecting the diffusional transport of the reactants in the pores. On the other hand, the relative humidity also strongly affects the ionic conductivity of the ionomer phase. As the humidity and/or presence of liquid water strongly varies with temperature and current density, the transport properties for gas and proton transport change significantly during operation.

Modeling

It has become obvious from the previous sections that transport of matter, heat, and charge on all scales of the cell is complex and interlinked and has a strong influence on the stationary and dynamic performance of the cell. As the experimental analysis is difficult and some of the parameters are poorly known, understanding of the important transport phenomena has also strongly relied on modeling of single components or cells.

Newman et al. [13] and De Levie [14] have developed the theoretical understanding of porous electrodes in the 1960s of the last century. Pioneering work for an entire cell (in 1D and isothermal conditions) has been done by Springer

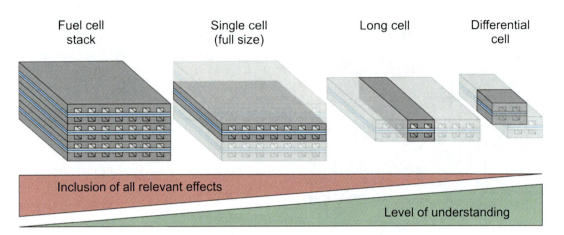

Polymer Electrolyte Fuel Cells, Mass Transport, Fig. 5 Illustration of the downscaling possibilities for experimental fuel cell studies

et al. [15] in the 1990s. Thereafter modeling efforts have seen an exponential growth till today with more and more sophisticated approaches being developed. With time models grew from one-dimensional to full three-dimensional treatment. The energy field and the complexity of two-phase flow (gaseous and liquid) in the different domains have been studied, and models for the time domain were developed, e.g., [16].

However, even with today's advanced computational capabilities, modeling is not able yet to capture the entire parameter space of a single cell without extensive experimental parameterization. In particular the two-phase transport in the catalyst layer and the GDL are still poorly understood. It will require a joint experimental and computational effort to close this important gap in the future.

Experimental Methods for Mass Transport Analysis

The most common analysis tool used for assessing PEFC performance is the recoding of current–voltage characteristics. However, the information obtained from such a characteristic is in most cases not sufficient to draw conclusions about mass transport processes. Therefore an insight into different possibilities of advanced analysis is given here.

Choices of Cell/Stack Design

As presented in the previous paragraphs, mass transport processes are relevant for different scales, from the processes inside the electrode up to the gas flow distribution in a complete stack. When performing experimental analysis of mass transport, the choice of cell/stack hardware will determine which transport processes will be included in the experiment. Some possible hardware designs (c.f. Fig. 5) are listed here:

- *Full-size stack*. The relevance for real operation is obvious, but analysis possibilities are limited.
- *Full-size single cell*. Some mass transport effect (cell-to-cell distribution) is not included but can be emulated by operating the cell with a constant differential pressure instead of a constant gas flow.
- *"Long cell"* (cell with a single channel or a few channels with technically relevant lengths). Channel-to-channel distribution effects may be reduced. On the contrary, the impact of a single channel blocking will be emphasized.
- *"Differential cell"* (cell with a few short parallel channels, operated with high stoichiometry to keep technically relevant gas velocities). The effect of gas consumption and water accumulation along the channels is not included.

The degree of understanding of the experimental results will usually increase when going

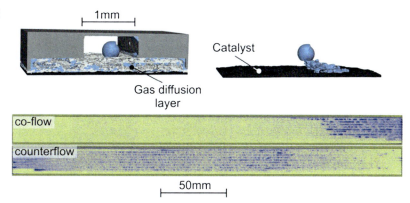

Polymer Electrolyte Fuel Cells, Mass Transport, Fig. 6 Illustration of in situ water visualization methods. *Top*: synchrotron 3D imaging of water in a differential cell. *Bottom*: neutron imaging of a long cell with an active area of 60 cm^2 for the H$_2$/air co- and counterflow arrangements

from a full technical stack to a differential cell, but the relevance for real stack operation becomes less obvious. Even on the scale of a differential cell, effects such as rib/channel distribution of reactants and current density make the interpretation of experimental results difficult, and experiments on further simplified cells are sometimes appropriate.

Visualization Methods

Because the occurrence of liquid water is a major factor for mass transport losses, visualization methods play an important role in the analysis of mass transport processes in PEFCs, in particular for the analysis of liquid water distribution.

- *Neutron imaging* [11] is used for the in situ analysis of the water distribution in the flow channels and GDLs (Fig. 6). Usual fuel cell construction materials including aluminum and, to some extent, steel are readily transparent for neutrons. Spatial resolution is limited, preventing the observation of single pores inside GDLs.
- *Synchrotron X-ray tomography* [12] is used for the in situ analysis of tridimensional water distribution in GDLs (Fig. 6). This method offers high spatial resolution. Limitations include the requisite of cell constructions transparent enough to X-rays and the issue of cell radiation damage. *Soft X-ray imaging* [17] has also been reported for the measurement of water in GDLs. Similarly to synchrotron X-ray imaging, strong limitations are set on the cell construction to make it transparent enough to X-rays.
- *Transparent fuel cells* [18] can be used to visualize the water inside flow channels. They have the advantage of allowing the use of inexpensive imaging equipment. The major drawback is the necessity of using transparent flow field and housing materials, which have different physical characteristics (contact angle, thermal properties) than commonly used materials.
- *Magnetic resonance imaging* [19] has been reported for observing the distribution of liquid water, in particular in the membrane. This method shares with neutron imaging the characteristic of being sensitive to different hydrogen isotopes, making labeling experiments possible. The drawbacks include a limited spatial resolution and the constraint of using nonconductive housing materials.

The visualization of gas flow or reactant distribution is scarcely used due to the limited possibilities of in situ measurements. For cells having a visual access to the flow channels, some optical methods have been reported:

- *Particle velocimetry* measurements for the analysis of the gas flow distribution [20]
- The use of *O$_2$-sensitive dyes* for the visualization of oxygen concentration distribution [21]

Electrochemical Analysis Methods

A few electrochemical analysis methods which can be used to characterize the impact of mass transport limitations on cell operation are listed here. These methods are sometimes used in combination with water visualization to draw correlations between water accumulation and mass transport losses.

- **Electrochemical Impedance Spectroscopy (EIS)** consists in measuring the voltage response to sinusoidal current perturbation with different frequencies. This method can yield important information about mass transport limitations, but the interpretation of the results must be done very carefully. In particular, it has been shown that the cell global impedance spectrum is strongly affected by gas consumption along the channels [22]. Even on the scale of a differential cell, local effects on the rib/channel scale can dominate the resulting spectrum [23].
- The measurement of the **limiting current density** [24] can be used to obtain information about the diffusive structure of the cathode GDL. In most designs, it is necessary to use diluted oxygen (typ. 1–5 %) to lower the limiting current density within the reachable range.
- **Multiple gas analysis** consists in comparing the performance of the cell operated with different gases. Usually, the cathode gas is changed from air to pure O_2 and/or to helox (a mixture of helium and oxygen). The differences in performance can point out mass transport limitations. The use of different gases can induce parasitic effects (e.g., dry out when using helox), but these can be avoided by using only short periods of operation with O_2/helox [11].

Conclusions and Outlook

Mass transport in PEFC is a multi-scale issue ranging from the nanometer scale in the catalyst layer to the meter scale in the flow field channels. A reasonable understanding of the different processes exists, mainly based on the insight from specialized experimental setups. However, further mitigation of cell voltage and degradation losses associated with the different transport phenomena is required. This will need a combination of even more advanced experimental tools and model-based comprehension of the different processes to improve structures and materials.

Acknowledgments Figures 3 and 4 are adapted from the PhD thesis of R. Flückiger (PSI/ETHZ) and part of Fig. 6 from the PhD thesis of J. Eller (PSI/ETHZ); the authors would like to thank for making the original figures available.

Cross-References

▶ Polymer Electrolyte Fuel Cells (PEFCs), Introduction
▶ Polymer Electrolyte Fuel Cells, Membrane-Electrode Assemblies

References

1. Oberholzer P, Boillat P, Kaestner A, Lehmann EH, Scherer GG, Schmidt TJ, Wokaun A (2013) Characterizing local O_2 diffusive losses in GDLs of PEFCs using simplified flow field patterns. J Electrochem Soc 160:F659–F669
2. El-kharouf A, Mason TJ, Brett DJL, Pollet BG (2012) Ex-situ characterisation of gas diffusion layers for proton exchange membrane fuel cells. J Power Sources 218:393–404
3. Bird RB, Steward WE, Lightfoot EN (1960) Transport phenomena. Wiley, New York
4. Becker J et al (2009) Determination of material properties of gas diffusion layers: experiments and simulations using phase contrast tomographic microscopy. J Electrochem Soc 156:B1175–B1181
5. Flückiger R et al (2008) Anisotropic, effective diffusivity of porous gas diffusion layer materials for PEFC. Electrochim Acta 54:551–559
6. Kramer D et al (2008) Electrochemical diffusimetry of fuel cell gas diffusion layers. J Electroanal Chem 612:63–77
7. Rosén T et al (2012) Saturation dependent effective transport properties of PEFC gas diffusion layers. J Electrochem Soc 159:F536–F544
8. Fairweather JD, Cheung P, St-Pierre J, Schwartz DT (2007) A microfluidic approach for measuring capillary pressure in PEMFC gas diffusion layers. Electrochem Comm 9:2340–2345
9. Gostick JT, Ioannidis MA, Fowler MW, Pritzker MD (2008) Direct measurement of the capillary pressure characteristics of water–air–gas diffusion layer systems for PEM fuel cells. Electrochem Comm 10:1520–1523

10. Kramer D et al (2005) In situ diagnostic of two-phase flow phenomena in polymer electrolyte fuel cells by neutron imaging: part A. Experimental, data treatment, and quantification. Electrochim Acta 50:2603–2614
11. Boillat P et al (2012) Impact of water on PEFC performance evaluated by neutron imaging combined with pulsed helox operation. J Electrochem Soc 159: F210–F218
12. Eller J et al (2011) Progress in in situ X-ray tomographic microscopy of liquid water in gas diffusion layers of PEFC. J Electrochem Soc 158:B963–B970
13. Newman JS, Tobias CW (1962) Theoretical analysis of current distribution in porous electrodes. J Electrochem Soc 109:1183–1191
14. de Levie R (1963) On porous electrodes in electrolyte solutions: I. Capacitance effects. Electrochim Acta 8:751
15. Springer TE, Zawodzinski TA, Gottesfeld S (1991) Polymer elektrolyte fuel cell model. J Electrochem Soc 138:2334–2341
16. Zaglio M, Wokaun A, Mantzaras J, Büchi FN (2011) 1d-modelling and experimental study of the Pefc dynamic behaviour at load increase. Fuel Cells 11:526–536
17. Sasabe T, Deevanhxay P, Tsushima S, Hirai S (2011) Investigation on the effect of microstructure of proton exchange membrane fuel cell porous layers on liquid water behavior by soft X-ray radiography. J Power Sources 196:8197–8206
18. Zhang FY, Yang XG, Wang CY (2006) Liquid water removal from a polymer electrolyte fuel cell. J Electrochem Soc 153:A225–A232
19. Minard KR, Viswanathan VV, Majors PD, Wang L-Q, Rieke PC (2006) Magnetic resonance imaging (MRI) of PEM dehydration and gas manifold flooding during continuous fuel cell operation. J Power Sources 161:856–863
20. Martin J, Oshkai P, Djilali N (2005) Flow structures in a U-shaped fuel cell flow channel: quantitative visualization using particle image velocimetry. J Fuel Cell Sci Technol 2:70–80
21. Takada K et al (2011) Simultaneous visualization of oxygen distribution and water blockages in an operating triple-serpentine polymer electrolyte fuel cell. J Power Sources 196:2635–2639
22. Schneider IA, Kramer D, Wokaun A, Scherer GG (2007) Oscillations in gas channels. J Electrochem Soc 154:B770–B782
23. Schneider IA, Bayer MH, von Dahlen S (2011) Locally resolved electrochemical impedance spectroscopy in channel and land areas of a differential polymer electrolyte fuel cell. J Electrochem Soc 158:B343–B348
24. Baker DR, Caulk DA, Neyerlin KC, Murphy MW (2009) Measurement of oxygen transport resistance in PEM fuel cells by limiting current methods. J Electrochem Soc 156:B991–B1003

Polymer Electrolyte Fuel Cells, Membrane-Electrode Assemblies

Makoto Uchida
Fuel Cell Nanomaterials Center,
Yamanashi University, 6-43 Miyamaecho,
Kofu, Yamanashi, Japan

The Electrode Structure and Electrochemical Reactions of MEAs (Membrane-Electrode Assemblies)

In general, a polymer electrolyte fuel cell (PEFC) consists of a polymer electrolyte membrane (proton exchange membrane) in contact with a porous anode and a porous cathode. The construction of a single cell of a PEFC is schematically shown in Fig. 1. Hydrogen and oxygen gases are supplied to the anode and cathode compartments, respectively, at which the hydrogen oxidation reaction (HOR) and oxygen reduction reaction (ORR) occur, thereby generating an electrical current:

$$\text{Anode} : H_2 \rightarrow 2H^+ + 2e^-$$

$$\text{Cathode} : 1/2O_2 + 2H^+ + 2e^- \rightarrow H_2O$$

$$\text{Cell reaction} : H_2 + 1/2O_2 \rightarrow H_2O$$

The MEA is composed of three main parts, e.g., polymer electrolyte membrane (PEM), gas diffusion medium, and catalyst layer (CL). The membrane, with hydrophilic proton-conducting channels embedded in a hydrophobic structural matrix, plays a key role in the operation of PEFCs. The PEMs for PEFCs commonly use perfluorosulfonic acid (PFSA) electrolytes such as Nafion®, with the chemical structure shown in Fig. 2, because of its high proton conductivity as well as chemical and thermal stability [1]. The gas diffusion medium (GDM), including both the microporous layer (MPL) and the gas diffusion layer (GDL), which typically is based on carbon fibers, is also an important component. The GDM is designed with three distinct

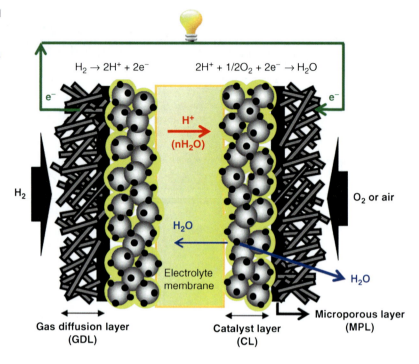

Polymer Electrolyte Fuel Cells, Membrane-Electrode Assemblies, **Fig. 1** Schematic diagram of an MEA for a PEFC

Polymer Electrolyte Fuel Cells, Membrane-Electrode Assemblies, Fig. 2 Chemical structure of perfluorosulfonic acid polymer

functions in mind: to transport reactant gases to the catalyst layers (CLs), to conduct electrons in or out of the CLs, and to exhaust product water from the cathode CL to the gas flow channel [2–7]. The CL consists of the electrocatalyst (Pt nanoparticles supported on carbon black, Pt/C) and a proton-conducting polymer electrolyte ionomer as a binder. In particular, the performance of the cathode CL controls the performance of the cell as a whole. The MEA, which consists of anode and cathode, each including their respective GDM and CL, is sandwiched on both sides of the membrane and placed between two bipolar plates, both containing gas flow channels, as well as coolant channels.

A typical transmission electron microscope image and schematic diagram of a Pt/C electrocatalyst are shown in Fig. 3a, b. The carbonaceous fine powder (carbon black), with a structure in which the 10–50 nm primary particles are connected together as in a string of beads, is generally used for the carbon support. The Pt nanoparticulates (2–5 nm in diameter) are supported on the surface of the carbon black. The Pt/C itself is coated with the PFSA ionomer. A proton generated by the HOR on the Pt/C in the anode CL is transported to the membrane through PFSA ionomer channels. A schematic depiction of the ORR on the Pt/C with an ideal PFSA film in the cathode CL is shown in Fig. 4. In the cathode CL, the proton from the membrane is transported to the Pt on the carbon support through the PFSA ionomer channels. The electrons generated simultaneously by the HOR in the anode are conducted to the gas diffusion medium connected with the bipolar plate through the carbon support network. The electrons in the cathode from the bipolar plate are conducted to the Pt on the carbon through the carbon support network and then react with oxygen (ORR) and the protons mediated via the PFSA. The gas transport, both of the

Polymer Electrolyte Fuel Cells, Membrane-Electrode Assemblies 1671

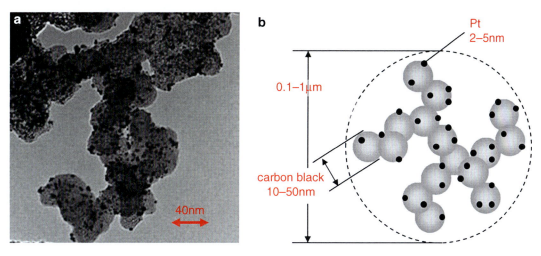

Polymer Electrolyte Fuel Cells, Membrane-Electrode Assemblies, Fig. 3 (a) Transmission electron microscope image. (b) Schematic diagram of the Pt/C

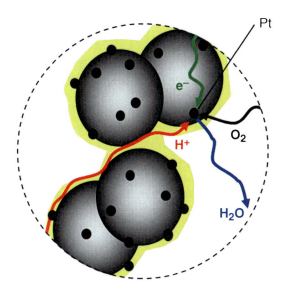

Polymer Electrolyte Fuel Cells, Membrane-Electrode Assemblies, Fig. 4 Schematic depiction of the ORR on the Pt/C with ideal PFSA film in the CL

hydrogen in the anode and the oxygen in the cathode, is mediated by the void channels existing in the carbon black network structure. If the catalyst layer needs to have more void volume for gas transport, extractable pore-forming materials can be used. The water generated in the cathode is exhausted from the MEA through the gas channels or is returned to the membrane side through the PFSA ionomer channels.

The key point of the MEA development for the PEFC is how to promote the reactions on the surface of the platinum catalyst in the CLs, particularly at the cathode. To increase the reaction area that can contribute to the catalyst reduction, the supply channels of oxygen and hydrogen to the reactive sites, the conductive channels for protons, and the exit paths for the generated water all become important. In the present essay, mainly the development history and the current design for the reduction of Pt in the CL will be described.

Microscopic regions in which solid, gas, and liquid meet at a so-called three-phase boundary exist within the CL and are the locations at which the ORR proceeds. The properties and dimensions of these components control the performance of the fuel cell. In the PEFC, the reactant gases are transported through ionomer films that cover the electrocatalyst before they react at the latter. Thus, the film thickness must be small enough to avoid slowing down the gas diffusion. On the other hand, the film must be thick enough to avoid slowing down the ionic conduction (Fig. 1). Similarly, the support material must provide sufficient volume for electronic conduction yet provide enough void volume for effective gas mass transport. Thus, it is clear that there will be dimensions at which gas transport, electronic conduction, and ionic conduction will be simultaneously optimized.

In the early stages of PEFC development, there was no concept of incorporating the polymer electrolyte in the interior of the CL. Although the PFSA membrane is used as the electrolyte, the membrane cannot make effective ionic contact deeply within the electrode as a liquid electrolyte can. The three-phase (gas/electrolyte/electrode) boundary as the reaction site therefore existed only at the interface between the membrane and the CL, with the result that the electrodes had low catalyst (platinum) utilization. Consequently, they required a high catalyst loading (4 mg/cm^2) [8]. Subsequently, PFSA solutions in alcoholic solvents were adopted as the most popular approach; these were soaked into the CLs to increase the contact area between the Pt particles and the PFSA. Thereby, the Pt loadings have so far been decreased to 0.35–0.5 mg/cm^2 [9–11]. Since the Pt utilization (U_{Pt}) values, defined as the percentage of possible Pt area that is electrochemically active, were still low (15–20 %), in spite of these approaches [9], many techniques were investigated to increase the interfacial area between Pt and PFSA. The PFSA solutions were impregnated into the CLs, thereby enabling the Pt loadings to decrease and at the same time U_{Pt} values to increase [12, 13]. It has been reported that the latter have reached 60 ~ 80 %, calculated from the electrochemically active surface area (ECSA) measured by cyclic voltammetry (CV). These values represent a quite high-utilized Pt ratio in the CL. However, in the present state, the performance of the PEFC remains relatively low due to combined problems in the CL, even though the U_{Pt} shows almost the maximum possible.

Furthermore, for cost reduction, the amount of Pt used in the PEFC industry will be required to be reduced to one-tenth. In more detail, a conventional, present-day automobile is using a precious metal catalyst (Pt) in an amount of about 10 g as an exhaust gas catalyst. On the other hand, the prototypes of current fuel cell vehicles, with fuel cell stacks of about 100 kW,

use ca. 100 g Pt per vehicle. Therefore, it is necessary to achieve a 0.1 g Pt/kW level of one-tenth, which is also the developmental goal for industry and the national project in order to achieve a cost close to the current vehicle level. However, the cell performance would be expected to decrease significantly, based on the reduction of the Pt amount in the CL. Therefore, the development of high-performance CLs is necessary to reduce the Pt amount without performance loss.

Design and Evaluation of Catalyst Layers with High Pt Effectiveness

An ideal CL requires the composition of the polymer electrolyte on the Pt/C surface to simultaneously optimize the gas diffusion and proton conductance, as shown in Fig. 4. However, the realization of an ideal CL is difficult; the polymer electrolyte in the actual CL does not distribute uniformly, as shown in Fig. 5. The Pt catalyst which is not covered by the polymer electrolyte cannot form a three-phase boundary and therefore does not contribute to the reaction. The Pt cannot be used effectively, and the utilization of Pt is low. Consequently, for the purpose of increasing the contact surface area between polymer electrolyte and Pt, many research and development efforts have been conducted, for example, the method of using organic solvents with high boiling point [12] or the use of a colloidal polymer electrolyte [13]. Watanabe and coworkers [14, 15] characterized the microstructure of the hydrophobic gas diffusion electrode for phosphoric acid fuel cells. In their works, the CL was claimed to have two distinctive pore distributions, with a boundary of ca. 0.1 μm. The smaller pores (primary pores) were identified as the spaces between the primary particles of the carbon support and those connected together (aggregates) in their agglomerates, and the larger ones (secondary pores) were identified as the spaces between the agglomerates. Moreover, the effects of the microstructure on the carbon support in the CL on the performance of PEFCs were investigated in detail; the primary

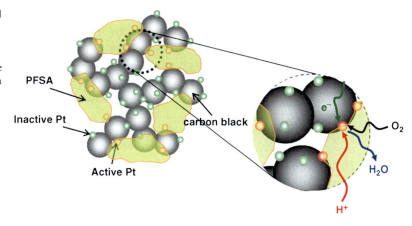

Polymer Electrolyte Fuel Cells, Membrane-Electrode Assemblies, Fig. 5 Schematic depiction of the ORR on the Pt/C with actual PFSA film in the CL

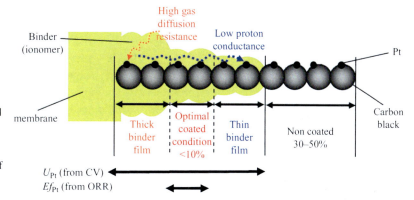

Polymer Electrolyte Fuel Cells, Membrane-Electrode Assemblies, Fig. 6 Schematic depiction of the variation of PFSA film thickness with depth within the CL

pores with distributed phosphoric acid mainly functioned as the reaction sites in the optimized CL structure due to the location of a large part of Pt particles (>80 %). The optimization of both the microstructure of the carbon support and the dispersion of the polymer electrolyte promoted the improvement of the effectiveness of Pt. Extending the design concept of CLs, the original design for the electrode model of the PEFC, as distinct from that of a liquid electrolyte system, was proposed by investigations of M. Uchida et al., e.g., reference [16]. Subsequently, various approaches for the optimization of the rheological properties of the CL ink by the use of improved solvent mixtures and associated procedures were examined. As a result, the U_{Pt} values calculated from the ECSA were improved to 70–97 % [17]. On the other hand, it has been shown that it is still necessary to reduce the Pt loading of MEAs, according to industry requirements, to about one-tenth, for cost reduction. It is clear that there is little opportunity for reducing the Pt loading based on the results of the U_{Pt} values alone. However, it was realized that there were no clear indices to evaluate "the practical utilization" under the various actual operating conditions for PEFCs.

In recent research, a new evaluation method for the utilization under actual operating conditions was proposed as an "effectiveness of platinum (Ef_{Pt})" [18], based on the comparison of ORR mass activities for the MEA with those measured with the channel flow technique at the same temperature. The Ef_{Pt} value indicates the extent to which Pt particles exist in the CL in an optimal condition of binder coverage. A schematic depiction of the variation of PFSA film thickness with depth within the CL is shown in Fig. 6. In the case of U_{Pt}, the concept of Pt utilization can be rationalized from Fig. 6, because the U_{Pt} values are not affected by either proton conductance or the gas diffusion resistance of the ionomer binder. The U_{Pt} value could in principle reach 100 %, if ionomer covers all of the Pt particles, based on the definition of U_{Pt}. In typical, conventional results, however, the U_{Pt} values of the MEA have indicated about 50–70 % under actual operating conditions.

On the other hand, in the case of Ef_{Pt} values evaluated for the important situation of air feed, or at higher current densities with O_2 feed, the optimal ionomer-coated condition was found to be the most effective for the ORR performance. Of course, both the thin and thick ionomer films also affected the ORR, but the contribution was relatively low, because the ionic path connectivity, proton conductance, and gas diffusion resistance all affected the cell performance loss. The values of Ef_{Pt} were found to be 10 % or less under actual operating conditions at the state-of-the-art CL structures. These values show that the ratio of catalyst existing in an effective reaction environment might be small during actual operation. Similar trends have also been observed from various other research efforts, including modeling studies [19]. Such results have indicated that we have much room for improvement for the design of the CL.

Future Directions

For the commercialization of PEFCs, in order to reduce the Pt loading without serious loss of cell performance and durability, it has been proposed that maximizing the Ef_{Pt} is necessary. In the relationship between the Pt catalyst powder and ionomer in a conventional CL, much of the Pt existing in mesopores of the carbon agglomerates cannot connect with the ionomer. Moreover, the primary particles of the carbon support contain many nanopores. The Pt in such nanopores also does not connect with the ionomer [16, 20]. Therefore, we must improve the Ef_{Pt} by improving the catalyst loading method, the support materials, and development of new ionomers. In addition, we should try to improve the ORR activity of the catalyst material itself. Of course, adequate durability of the MEA under practical conditions must also be secured.

The various challenges of the improvement in the ORR activity and the durability have been advancing by means of alloying, formation of Pt skin layers, and core-shell preparation methods. Approaches based on non-precious metals are proceeding as well. In addition, the further optimization of the thin CL structure by use of new, durable support materials (e.g., graphitized carbon blacks and conductive ceramic supports) should help to improve the effectiveness of Pt and the durability, as well as specific gas transport problems in the CL under high current density operating conditions.

Cross-References

▶ Hydrocarbon Membranes for Polymer Electrolyte Fuel Cells
▶ Platinum-Based Cathode Catalysts for Polymer Electrolyte Fuel Cells
▶ Polymer Electrolyte Fuel Cells, Oxide-Based Cathode Catalysts
▶ Polymer Electrolyte Fuel Cells (PEFCs), Introduction
▶ Polymer Electrolyte Fuel Cells, Perfluorinated Membranes

References

1. Yeager HL, Steck A (1981) Cation and water diffusion in Nafion ion exchange membranes: influence of polymer structure. J Electrochem Soc 128:1880–1884
2. Lim C, Wang CY (2004) Effects of hydrophobic polymer content in GDL on power performance of a PEM fuel cell. Electrochim Acta 49:4149–4156
3. Park GG, Sohn YJ, Yang TH, Yoon YG, Lee WY, Kim CS (2002) Effect of PTFE contents in the gas diffusion media on the performance of PEMFC. J Power Sources 131:182–187
4. Lin GY, Nguyen TV (2005) Effect of thickness and hydrophobic polymer content of the gas diffusion layer on electrode flooding level in a PEMFC. J Electrochem Soc 152:A1942–A1948
5. Park S, Lee JW, Popov BN (2008) Effect of PTFE content in microporous layer on water management in PEM fuel cells. J Power Sources 177:457–463
6. Wang ED, Shi PF, Du CY (2008) Treatment and characterization of gas diffusion layers by sucrose carbonization for PEMFC applications. Electrochem Commun 10:555–558

7. Gostick JT, Fowler MW, Ioannidis MA, Pritzker MD, Volfkovich YM, Sakars A (2006) Capillary pressure and hydrophilic porosity in gas diffusion layers for polymer electrolyte fuel cells. J Power Source 156:375–387
8. Appleby AJ, Yeager EB (1986) Solid polymer electrolyte fuel cells (SPEFCs). Energy 11:137–152
9. Ticianelli EA, Derouin CR, Srinivasan S (1988) Localization of platinum in low catalyst loading electrodes to attain high power densities in SPE fuel cells. J Electroanal Chem 251:275–295
10. Ticianelli EA, Derouin CR, Redondo A, Srinivasan S (1988) Methods to advance technology of proton exchange membrane fuel cells. J Electrochem Soc 135:2209–2214
11. Srinivasan S, Ticianelli EA, Derouin CR, Redondo A (1988) Advances in solid polymer electrolyte fuel cell technology with low platinum loading electrodes. J Power Sources 22:359–375
12. Wilson MS, Valerioand JA, Gottesfeld S (1995) Low platinum loading electrodes for polymer electrolyte fuel cells fabricated using thermoplastic ionomers. Electrochim Acta 40:355–363
13. Uchida M, Aoyama Y, Eda N, Ohta A (1995) New preparation method for polymer-electrolyte fuel cells. J Electrochem Soc 142:463–468
14. Watanabe M, Tomikawa M, Motoo S (1985) Experimental analysis of the reaction layer structure in a gas diffusion electrode. J Electroanal Chem 195:81–93
15. Watanabe M, Makita K, Usami H, Motoo S (1986) New preparation method of a high performance gas diffusion electrode working at 100 % utilization of catalyst clusters and analysis of the reaction layer. J Electroanal Chem 197:195–208
16. Uchida M, Fukuoka Y, Sugawara Y, Eda N, Ohta A (1996) Effects of microstructure of carbon support in the catalyst layer on the performance of polymer-electrolyte fuel cells. J Electrochem Soc 143:2245–2252
17. Gasteiger HA, Kocha SS, Sompalli B, Wagner FT (2005) Activity benchmarks and requirements for Pt, Pt-alloy, and non-Pt oxygen reduction catalysts for PEMFCs. Appl Catal Environ 56:9–35
18. Lee M, Uchida M, Yano H, Tryk DA, Uchida H, Watanabe M (2010) New evaluation method for the effectiveness of platinum/carbon electrocatalysts under operating conditions. Electrochim Acta 55:8504–8512
19. Xia Z, Wang Q, Eikerling M, Liu Z (2008) Effectiveness factor of Pt utilization in cathode catalyst layer of polymer electrolyte fuel cells. Can J Chem 86:657–667
20. Uchida M, Park YC, Kakinuma K, Yano H, Tryk DA, Kamino T, Uchida H, Watanabe M (2013) Effect of the state of distribution of supported Pt nanoparticles on effective Pt utilization in polymer electrolyte fuel cells. Phys Chem Chem Phys 15:11236–11247

Polymer Electrolyte Fuel Cells, Oxide-Based Cathode Catalysts

Akimitsu Ishihara
Yokohama National University, Hodogaya-ku, Yokohama, Japan

Introduction

Oxide-based cathode catalysts are entirely new non-precious metal cathode catalysts for low-temperature fuel cells such as polymer electrolyte fuel cells (PEFCs). These catalysts were developed from a viewpoint that high chemical stability was essentially required for the cathode for PEFCs. The cathode catalysts for PEFCs are exposed to an acidic and oxidative atmosphere, that is, a strong corrosive environment, therefore, even platinum nanoparticles dissolved during a long-time operation. This instability of electrocatalysts is one of the factors which hindered the wide commercialization of PEFCs.

Group 4 and 5 metal oxides, which are well known as valve metals, are expected to have high chemical stability even in acid electrolyte. Table 1 shows the solubility of platinum black powder and some oxide-based materials prepared by various methods in 0.1 M H_2SO_4 at 30 °C under atmospheric condition [1]. The solubility of the oxide-based materials was smaller than that of platinum black. Therefore, these oxide-based materials were chemically stable in acid electrolyte under atmospheric condition. However, these oxides were almost insulators. In addition, oxygen molecules hardly adsorb on a surface of the oxides with no defects. Therefore, the surface of group 4 and 5 metal oxides must be modified to have definite catalytic activity for oxygen reduction reaction (ORR). The surface modifications were classified into four [1]: (1) formation of complex oxide layer containing active sites, (2) substitutional doping of nitrogen, (3) creation of oxygen defects without using carbon and nitrogen, and (4) partial oxidation of compounds including carbon and nitrogen.

Polymer Electrolyte Fuel Cells, Oxide-Based Cathode Catalysts, Table 1 Solubility of platinum black powder and oxide-based materials prepared by various methods in 0.1 mol dm^{-3} H$_2$SO$_4$ at 30 °C under atmospheric condition

Catalysts	Preparation method	Solubility/μmol dm^{-3}
TaO$_x$N$_y$ (powder)	Nitridation of Ta$_2$O$_3$ with NH$_3$	0.33
TaO$_x$N$_y$ (thin film)	Reactive sputtering	0.20
ZrO$_x$N$_y$ (thin film)	Reactive sputtering	0.041
TiO$_{2-x}$ (plate)	Heat treatment	0.36 (50 °C)
Pt black (powder)	–	0.56

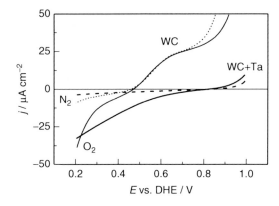

Polymer Electrolyte Fuel Cells, Oxide-Based Cathode Catalysts, Fig. 1 Potential–current curves for a Ta-added WC and a pure WC under N$_2$ and O$_2$ atmosphere in 0.1 M H$_2$SO$_4$ at 30 °C

Surface Modifications of Oxides

Formation of Complex Oxide Layer Containing Active Sites

Tungsten carbide (WC) and molybdenum carbide have the platinum-like electronic structure. Therefore, these carbides are expected to be active for oxygen reduction reaction. However, tungsten and other carbides were unstable at the high potential region in an acid electrolyte. Tantalum (Ta) was added to tungsten carbide to form thin protective film. The pure WC corroded to form WO$_3$ and CO$_2$ at high potential region. However, the tantalum addition to tungsten carbide catalyst showed high electrochemical stability even at high potential region. Tungsten and tantalum formed complex hydroxide films, which had high corrosion resistance. The surface of the WC with Ta addition was covered by the complex hydroxide film. Figure 1 shows the potential–current curves for a Ta-added WC and a pure WC in N$_2$ and O$_2$ atmosphere [2]. The current observed under N$_2$ reflected electrochemical stability of the catalysts. The current difference under O$_2$ and N$_2$ corresponded to the ORR. The WC with Ta addition was stable, and the onset potential for the ORR was observed at about 0.8 V versus dynamic hydrogen electrode (DHE). XPS suggested that the surface of the WC with Ta addition was mainly consisted of complex hydroxide and there existed WC near the surface. Because the thin hydroxide film of W and Ta was formed on the WC, the catalytic activity was remained and the stability increased. However, long-term durability should be investigated.

Substitutional Doping of Nitrogen

In order to obtain a narrower band-gap of these oxides, the substitutional doping of N might be ineffective since its p states contribute to the band-gap narrowing by mixing with O 2p states.

Figure 2a, b show the potential–current curves of tantalum nitride (Ta$_3$N$_5$) and tantalum oxynitride (TaO$_{0.92}$N$_{1.05}$) under N$_2$ or O$_2$ in 0.1 M H$_2$SO$_4$ [3]. Ta$_3$N$_5$ had poor catalytic activity for the ORR. On the other hand, the ORR current with TaO$_{0.92}$N$_{1.05}$ started to flow at 0.8 V versus Reversible Hydrogen Electrode (RHE), indicating that TaO$_{0.92}$N$_{1.05}$ had definite catalytic activity for the ORR. Other nitrogen-doped oxides such as TaO$_x$N$_y$, NbO$_x$N$_y$, HfO$_x$N$_y$-C, and Zr$_2$ON$_2$ also have some catalytic activity for the ORR [4–7]. Because pure oxynitrides also have wide band-gap, it is predicted that pure oxynitrides might have poor catalytic activity for the ORR. These results indicate that the surface of these oxynitrides might have some defects which create the donor levels close to the edge level of the conduction band, that is, close to the Fermi level. The electrons in the donor level due to the surface defects probably participate in the ORR. Therefore, not only

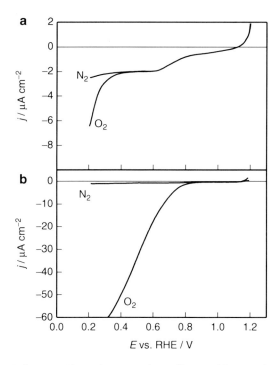

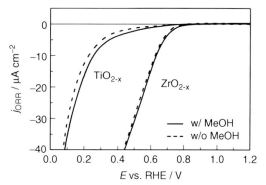

Polymer Electrolyte Fuel Cells, Oxide-Based Cathode Catalysts, Fig. 3 Potential–current curves of ZrO_{2-x} and TiO_{2-x} in 0.1 M H_2SO_4 under O_2 with or without 0.1 M methanol at 30 °C

Polymer Electrolyte Fuel Cells, Oxide-Based Cathode Catalysts, Fig. 2 Potential–current curves of tantalum nitride (Ta_3N_5) and tantalum oxynitride ($TaO_{0.92}N_{1.05}$) under N_2 or O_2 in 0.1 M H_2SO_4 at 30 °C

substitutional doping of N but also creating surface defects was required to enhance the catalytic activity for the ORR. Recently, the addition of carbon to TaON was found to be effective to increase the ORR current [4]. The additive carbon into oxynitride might form the electron conduction path. This result indicated that the suitable electron conduction path must be constructed to obtain larger ORR current.

Creation of Oxygen Defects Without Using Carbon and Nitrogen

Adsorption of oxygen molecules on the surface was required as the first step to proceed the ORR. The presence of the surface defects sites is required to adsorb the oxygen molecules on the surface of the oxides.

ZrO_{2-x} prepared by sputtering with ZrO_2 target under Ar atmosphere had clear catalytic activity for the ORR [8]. The onset potential of the sputtered ZrO_{2-x} for the ORR was observed below 0.9 V versus RHE in 0.1 mol dm^{-3} H_2SO_4 at 30 °C. Other metal oxides such as Co_3O_{4-x}, TiO_{2-x}, SnO_{2-x}, and Nb_2O_{5-x} prepared by sputtering also showed ORR activity to some extent. In addition, these oxides were inactive for methanol oxidation. Figure 3 shows the potential–current curves of ZrO_{2-x} and TiO_{2-x} in 0.1 M H_2SO_4 under O_2 with or without 0.1 M methanol [8]. The presence of methanol did not affect the ORR activity of the oxides. Therefore, these oxides could be a possible candidate substituting the platinum cathode for direct methanol fuel cells.

TiO_x/Ti, $Ti_{0.7}Zr_{0.3}O_x/Ti$, $Ti_{0.3}Ta_{0.7}O_x/Ti$, ZrO_x/Ti, and TaO_x/Ti prepared by dip-coating method at 450 °C [9] and highly dispersed tantalum oxide prepared by electrodeposition in a nonaqueous solution [10] also showed high ORR activity. These oxides might have some oxygen defects on the surface. Therefore, there are various methods in preparing oxides with oxygen defects.

Partial Oxidation of Compounds Including Carbon and Nitrogen

The surface oxidation state would be changed by the control of a partial oxidation of non-oxides. Carbonitrides of group 4 and 5 elements were used as precursors for a partial oxidation. Carbonitrides form a complete solid solution of carbide and nitride. A partial

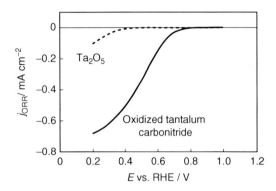

Polymer Electrolyte Fuel Cells, Oxide-Based Cathode Catalysts, Fig. 4 Potential–oxygen reduction current curves of oxidized tantalum carbonitride and commercial tantalum oxide in 0.1 M H_2SO_4 at 30 °C

oxidation of carbonitride, that is, the substitution for O atoms, brings an increase in ionic components among the chemical bonding, because the energy difference between the electron orbital of metal ion and O $2p$ orbital is higher than that between electron orbital of metal ion and C or N $2p$. The electronic state might be controlled by the control of the degree of the oxidation.

Figure 4 shows the potential–oxygen reduction current curves of oxidized tantalum carbonitride and commercial tantalum oxide [11]. Compared with commercial Ta_2O_5, the oxidized tantalum carbonitride had definite ORR activity. Oxidized tantalum carbonitride had Ta_2O_5 structure, and some deposited carbon was contained. A surface-sensitive conversion-electron-yield x-ray absorption spectroscopy analysis revealed that the oxidized tantalum carbonitride with high ORR activity contained oxygen-vacancy sites, providing both oxygen adsorption sites and local electron conduction paths [12]. The deposited carbon might form local micro electron conduction path and play an important role of the formation of the oxygen vacancies. In addition, it was found that nitrogen in precursors was essentially required to have high ORR activity. The nitrogen might concern the formation of the oxygen vacancies as active sites. Although it is necessary to clarify the role of carbon and nitrogen, it is presumed that the oxygen vacancies act as active sites.

Catalyst Design

The four surface modifications provided the principle of the catalyst design of oxide-based cathodes. In order to increase the ORR current density, the following three points are required: (1) increase of density of oxygen defects such as vacancies as active sites, (2) preparation of highly dispersed fine particles to increase surface area, and (3) optimization of electron conduction path. It is considered that there are various methods to prepare the catalysts which satisfy these conditions.

Future Directions

Group 4 and 5 oxide-based cathodes have basically high chemical stability in acid electrolyte. In order to obtain larger ORR current, high dispersion of nano-sized oxides is one solution. The increase in surface area, the creation of oxygen defects, and the improvement of electronic conductivity are expected by becoming the oxides to be nano-sized. However, the surface energy of nano-sized oxides increases, indicating that the solubility may increase. It would be important to investigate the optimal size of the oxides.

Because carbon deteriorates in high potential, both carbon support and deposited carbon might become a key factor of catalyst degradation. In case of oxide-based cathode, the active sites are not related to the carbon. Therefore, carbon-free cathode would be possible in the future.

Cross-References

▶ Fuel Cells, Non-Precious Metal Catalysts for Oxygen Reduction Reaction
▶ Platinum-Based Cathode Catalysts for Polymer Electrolyte Fuel Cells

References

1. Ishihara A, Ohgi Y, Matsuzawa K, Mitsushima S, Ota K (2010) Progress in non-precious metal oxide-based cathode for polymer electrolyte fuel cells. Electrochim Acta 55(27):8005–8012
2. Lee K, Ishihara A, Mitsushima S, Kamiya N, Ota K (2004) Stability and electrocatalytic activity for oxygen reduction in WC+Ta catalyst. Electrochim Acta 49(21):3479–3485
3. Ishihara A, Lee K, Doi S, Mitsushima S, Kamiya N, Hara M, Domen K, Fukuda K, Ota K (2005) Tantalum oxynitride for a novel cathode of PEFC. Electrochem Solid-State Lett 8(4):A201–A203
4. Kikuchi A, Ishihara A, Matsuzawa K, Mitsushima S, Ota K (2009) Tantalum-based compounds prepared by reactive sputtering as a new non-platinum cathode for PEFC. Chem Lett 38(12):1184–1185
5. Ohnishi R, Katayama M, Takanabe K, Kubota J, Domen K (2010) Niobium-based catalysts prepared by reactive radio-frequency magnetron sputtering and arc plasma methods as non-noble metal cathode catalysts for polymer electrolyte fuel cells. Electrochim Acta 55(19):5393–5400
6. Chisaka M, Suzuki Y, Iijima T, Sakurai Y (2011) Effect of synthesis route on oxygen reduction reaction activity of carbon-supported hafnium oxynitride in acid media. J Phys Chem C 115:20610
7. Maekawa Y, Ishihara A, Mitsushima S, Ota K (2008) Catalytic activity of zirconium oxynitride prepared by reactive sputtering for ORR in sulfuric acid. Electrochem Solid-State Lett 11(7): B109–B112
8. Liu Y, Ishihara A, Mitsushima S, Kamiya N, Ota K (2007) Transition metal oxides as DMFC cathodes without platinum. J Electrochem Soc 154(7): B664–B669
9. Takasu Y, Suzuki M, Yang H, Ohashi T, Sugimoto W (2010) Oxygen reduction characteristics of several valve metal oxide electrodes in $HClO_4$ solution. Electrochim Acta 55(27):8220–8229
10. Seo J, Cha D, Takanabe K, Kubota J, Domen K (2012) Highly-dispersed Ta-oxide catalysts prepared by electrode position in a non-aqueous plating bath for polymer electrolyte fuel cell cathodes. Chem Commun 48:9074–9076
11. Ishihara A, Tamura M, Matsuzawa K, Mitsushima S, Ota K (2010) Tantalum oxide-based compounds as new non-noble cathodes for polymer electrolyte fuel cell. Electrochim Acta 55(26):7581–7589
12. Imai H, Matsumoto M, Miyazaki T, Fujieda S, Ishihara A, Tamura M, Ota K (2010) Structural defects working as active oxygen-reduction sites in partially oxidized Ta-carbonitride core-shell particles probed by using surface-sensitive conversion-electron-yield x-ray absorption spectroscopy. Appl Phys Chem 96:191905

Polymer Electrolyte Fuel Cells, Perfluorinated Membranes

Kohta Yamada
Asahi Glass Co. Ltd., Research Center, Kanagawa-ku, Yokohama, Japan

Introduction

Polymer electrolyte fuel cells (PEFCs) have attracted great interest for applications to automotive and residential cogeneration systems, because of their high power density and low operational temperature. One of the key materials for PEFCs is proton exchange membranes, which are electrolyte assembled between anode and cathode. The performance and the lifetime of membrane electrode assemblies (MEAs) for PEFCs depend on the properties of the membranes, which are chemical or electrochemical stability, mechanical strength in the form of membrane, proton conductivity in the range of operational temperatures, water transport between electrodes, and gas-barrier property to avoid fuel/oxidant crossover between electrodes. In addition, from the viewpoint of fabrication and commercialization of MEAs, the membranes should be requested appropriate properties for manufacture handling in the form thin film, low-cost, material reliability, and quality control. Although many types membranes have been developed to realize the above properties, the present essay will focus on perfluorinated sulfonic acid membrane (PFSA membrane), because they are extensively used in MEAs for PEFC systems due to high chemical stability compared with hydrocarbon membranes.

Development of Membranes

The concept of the usage of an ion exchange membrane as electrolyte in fuel cells dates back to the 1940s [1, 2]. Firstly, the development of ion

Polymer Electrolyte Fuel Cells, Perfluorinated Membranes, Fig. 1 Chemical structure of PFSA polymers.

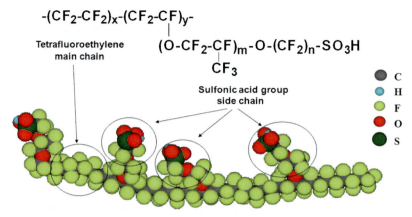

exchange membranes was eagerly performed using sulfonated poly(phenol-formaldehyde)- or sulfonated poly(styrene-divinylbenzene)-based polymers [3–6]. However these membranes were rapidly degraded by the hydroxyl radicals, which were thought to be by-products of the reaction on cathodes, oxygen reduction reaction [7, 8]. From the viewpoint of the chemical stability of electrolyte in MEA for a fuel cell, the PFSA membrane was firstly used by GE in 1966 for National Aeronautics and Space Administration (NASA) [9]. Although the stability problems of the electrolyte were alleviated in some extent by the use of the perfluorinated chemicals, there still were concerns about long-term stability of the electrolytes in electrochemical cells. Especially, long-term operations of fuel cells often require stable performance greater than 5,000 h for automobiles or 60,000 h for residential use under operating conditions. In the operations, the membranes are requested to be stable under severe conditions, simultaneously oxidizing, reducing and thermally circumstances [10].

Figure 1 describes the chemical structure of PFSA polymers, which was widely used as perfluorinated electrolytes for PEFCs. The structure is based on copolymers of tetrafluoroethylene (TFE) and perfluorovinyl ether monomers with a side chain of sulfonic acid group. While many acid groups have been examined in ion exchange systems for several decades, the sulfonic acid group is still dominant due to high stability and proton conductivity. The ion exchange capacity of the polymers (IEC meq/g = 1,000/EW, EW: equivalent weight) can be controlled by changing the constituents x and y in the figure. As examples of these polymers, Nafion® from DuPont, Flemion® Asahi Glass Co. Ltd., and Aciplex® Asahi Kasei Chemicals are well known. Dow's short side chain monomer structure is also known, even though this polymer is no longer supplied commercially. As for durability test of these PFSA membranes in a fuel cell, Nafion (Nafion120 membranes) was demonstrated for approximately 60,000 h MEA operation at a cell temperature of 60 °C and 100 % RH [11]. However, under low humidity conditions, it has been clarified that even PFSA membranes significantly deteriorate in a short period of time [12–14]. The fuel cell operation under low humidity conditions is required especially for automotive applications to increase energy density of the PEFC systems.

In order to develop durable membranes under low humidity conditions, it was of utmost importance that the degradation mechanism was understood. It was commonly believed that the hydrogen peroxide formed in the catalyst layer diffuses into the membrane and subsequently the

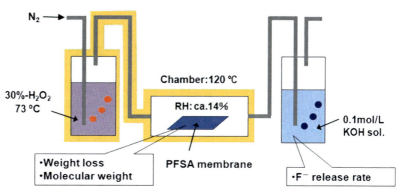

Polymer Electrolyte Fuel Cells, Perfluorinated Membranes, Fig. 2 Schematic representation of apparatus for the accelerated degradation test.

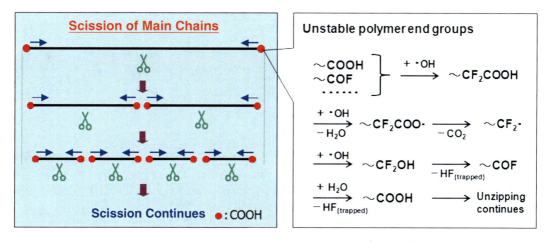

Polymer Electrolyte Fuel Cells, Perfluorinated Membranes, Fig. 3 Decomposition mechanism of PFSA polymer under low humidity conditions.

chemical degradation of the PFSA polymer proceeded via a peroxide radical attack. Therefore, to investigate the degradation mechanism of the PFSA membrane itself under low humidity conditions, an accelerated test method was developed, in which the membrane was exposed to a gas stream containing hydrogen peroxide [15]. The schematic representation of the apparatus for the accelerated test is shown in Fig. 2. As an indicator of the decomposition of the PFSA membrane, fluoride ion (F^-) in the effluent gas was trapped in a KOH solution and analyzed. As a result, membrane degradation was thought to proceed via the following two steps as shown in Fig. 3. Hydrogen peroxide or presumably the hydroxyl radical decomposes the polymer by:

1. An unzipping reaction at the unstable polymer end groups
2. Scission of the main chains of the PFSA polymer

An unzipping reaction at the unstable polymer end groups was initially discussed by Curtin et al. [16]. Also, the scission of the main chains occurs, which produces new unstable end groups and allows for the unzipping reaction to continue [15].

Based on the analysis of degradation mechanism above, some durable membranes have been investigated recently in order to protect from

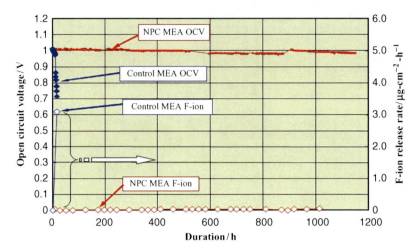

Polymer Electrolyte Fuel Cells, Perfluorinated Membranes, Fig. 4 Open circuit voltage durability of 25cm^2 MEAs at 120°C and 18%RH. Hydrogen and air was flowed at 50 and 200ml/min for anode and cathode, respectively. For NPC MEA, the thickness of NPC membrane was 40μm. For control MEA, Flemion®SH50 membrane of 50μm was used.

hydroxyl radical attack. As an example, cross-linking the polymer with a cationic radical quencher, which will protect the sulfonic acid group, was carried out [17]. The ionically cross-linking polymer membrane (new polymer composite membrane, NPC membrane) showed high chemical stability resistant to the degradation caused by the hydroxyl radical and has also improved mechanical properties at high temperatures and low humidity conditions. Figure 4 shows the OCV durability of MEAs with NPC membrane at 120 °C under low humidity. Open circuit voltage (OCV) durability tests are often used as an accelerated test of MEAs. The cell temperature was kept at 120 °C and the relative humidity of hydrogen and air was 18 %. The thickness of the NPC membrane was 40 μm. As a reference, MEA with Flemion SH 50 membrane was used. As shown in Fig. 4, the reference MEA failed within 10 h of operation, releasing a large amount of fluoride ion. In contrast, the MEA with NPC membrane showed excellent stability over 1,000 h, and the fluoride ion release rate was approximately 2×10^{-8} g (F$^-$) cm^{-2} h^{-1}, which was less than 1 % of the reference MEA and would predict a membrane-based MEA life of >6,000 h. This result demonstrates the exceptional chemical stability of the NPC membrane against the degradation caused by the hydroxyl radicals even at 120 °C and low humidity.

Future Directions

Although the highly durable PFSA membranes have been developed as mentioned above, significant cost reduction of MEA is also required for the widespread commercialization of PEFC systems for automotive and residential applications. Especially, the improvement of the MEA power density is thought to be indispensable for the purpose. Recently, in order to attain higher MEA performance, the PFSA ionomers for catalyst coating have been intensively investigated. Especially, to reduce overpotential for cathode reaction (oxygen reduction reaction), the improvement of conductivity by increasing IEC in the ionomers, the synthesis of high oxygen-permeable ionomers, and the control of hydrophobic properties in cathode layer by solvent-soluble perfluorinated polymers have been expected.

Cross-References

▶ Electrochemical Perfluorination
▶ Electrolytes, Classification
▶ Fuel Cells, Principles and Thermodynamics
▶ Polymer Electrolyte Fuel Cells, Membrane-Electrode Assemblies
▶ Polymer Electrolyte Fuel Cells (PEFCs), Introduction

References

1. Grubb WT (1957) Proceedings of the 11th annual battery research and development conference, p 5
2. Grubb WT (1959) USP 2,913,511
3. Adams B et al (1935) J Soc Chem Ind 54
4. Adams B et al (1936) Br Pat 450:308
5. D'Alelio G (1944) USP 2,366,007
6. Abrams IM (1956) Ind Eng Chem 48:1469
7. LaConti AB (1977) Proceedings of the symposium on electrode materials and process for energy conversion and storage, vol 77, p 354
8. Stek AE et al (1995) Proceedings of the 1st international symposium on new materials for fuel cell systems, vol 74
9. Grot WG (1994) Macromol Symp 82:161
10. LaConti (2003) Mechanisms of membrane degradation. In: Handbook of fuel cells, vol 3. Wiley, p 647, England
11. St-Pierre J et al (2000) J New Mater Electrochem Syst 3:99
12. Endoh E et al (2004) Electrochem Solid State Lett 7:A209
13. Liu W, Cleghorn S (2005) Electrochem Soc Trans 1:263
14. Inaba M et al (2006) Electrochim Acta 51:5746
15. Hommura S et al (2008) J Electrochem Soc 155:A29
16. Curtin DE et al (2004) J Power Sources 131:41
17. Endoh E (2008) Electrochem Soc Trans 12:41

Potentiometric pH Sensors at Ambient Temperature

Metini Janyasupab[1], Ying-Hui Lee[2] and Chung-Chiun Liu[1]
[1]Electronics Design Center, and Chemical Engineering Department, Case Western Reserve University, Cleveland, OH, USA
[2]Chemical Engineering Department, National Tsing Hua University, Hsinchu, Taiwan

Introduction

Monitoring pH is performed countless times on a daily basis. This is especially true in clinical analysis, environmental monitoring, and the food industry, where potentiometric pH sensors have become necessary analytically routine. In spite of the well-established technology, pH sensors still encounter problems with long-term stability and poor selectivity from certain reducing agents and alkaline ions. In general, potentiometric pH sensors can be categorized in three classes: (i) glass, (ii) metal/metal oxides, and (iii) polymers. In this entry, we will discuss the fundamentals, demonstrate the mechanism, and provide advantages and limitations of each class, including the trends that recent studies have demonstrated in potentiometric pH sensors.

Glass electrodes were established nearly 100 years ago and have been widely employed for pH measurement in aqueous solutions. Typical construction for the pH measurement is shown in Fig. 1 [1]. The wall of the glass tube is relatively thick and strong, and the crucial glass membrane bulb underneath is made as thin as possible, separating the internal solution and the external solution (the solution to be determined) for the pH measurement. The surface of the glass membrane is protonated by the internal and the external solution until the equilibrium is reached. This leads to the potential difference between both sides of the glass membrane. The Nernst equation of the potentials on both sides of the glass membrane is

$$E_1 = E_{10} + \frac{RT}{F} \ln[H^+]_{in}$$
$$= E_{10} + 0.0591 \log[H^+]_{in}$$

$$E_2 = E_{20} + \frac{RT}{F} \ln[H^+]_{out}$$
$$= E_{20} + 0.0591 \log[H^+]_{out}$$

The potential inside, E_1, and outside, E_2, also depend upon standard reduction potential. E_{10}, and E_{20}. E_1, and, outside, E_2 have a logarithmic relationship to concentration of hydrogen ion with a scale factor of temperature (T) in Kelvin, Faraday's constant (F), and ideal gas constant (R). At ambient temperature (~298 K), the scale factor of $(RT.F^{-1})$. ln 10 can be simplified to

Potentiometric pH Sensors at Ambient Temperature, Fig. 1 Typical construction of the combination design for the glass electrode [1]

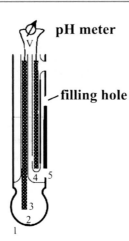

1: External solution
2: Glass elecctrode internal solution
3: Ag/AgCl reference electrode
4: Ag/AgCl reference electrode (internal solution)
5: Junction

0.0591 V/pH. Therefore, the potential difference across the glass membrane is

$$E = E_0 + 0.0591 \log \frac{[H^+]_{in}}{[H^+]_{out}}$$
$$= E_0 + 0.0591(pH_{out} - pH_{in})$$

Since the concentration of hydrogen ions is fixed, it can be incorporated in the constant

$$E = E_0' + 0.0591 pH_{out}$$

Accordingly, a small amount of H^+ can pass through liquid junction denoted by label (5) in Fig. 1. Consequently, the pH of the external solution can be determined through the potential difference. The constants in the above equations, E_{10}, E_{20}, and E_0, will change according to the composition of the glass membrane. The composition of the glass today is 21–33 % Li_2O, 2–4 % Cs_2O, 3–5 % La_2O_3 (Nd_2O_3, Er_2O_3), 2–4 % CaO (BaO), and SiO_2 (till 100 %) from most glass electrode manufacturers [2].

The glass electrodes are chosen due to their high selectivity, reliability, detection limit, long-term stability, independence of redox interferences, and ease of use [3]. However, longer response time is needed for the glass electrodes to reform a hydrated layer before use. Both acid error and alkali ion error exist in the glass electrode. The thickness of the hydrated layer will become thinner due to the acid stripping, causing the measured pH to remain higher than the real value. Alkali ions, especially sodium ions, can replace proton, subsequently suppressing the proton activity. As a result, the pH value decreases. Furthermore, major disadvantages of the glass electrode are (i) high cost, (ii) large size, (iii) inflexibility, and (iv) mechanical fragility. Many factors (e.g., solution temperature, glass composition, glass thickness, and the shape of the glass tip) contribute to the high impedance of the glass electrodes. Among these factors, the temperature is highly considered since it affects the pH measurement in two ways. The temperature changes the dissociation degree of ions in solution; mean while, the resistance across the glass membrane is reduced with higher temperatures. Consequently, temperature compensation is required for the pH measurement [4]. In addition to the temperature instability, the presence of internal solution limits the use of glass electrodes in a vertical position [3]. pH glass microelectrodes are dedicated and fragile with a limited lifetime. The trade-off between resistivity and leakage is unpreventable. Therefore, the glass electrodes are impractical for the miniaturization of pH-sensing devices. Instead, an all-solid-state electrode is desirable for many specialized applications [5–7].

Metal/Metal Oxide Electrode

To determine pH in food and physiological processes, a pH electrode must be non-fragile,

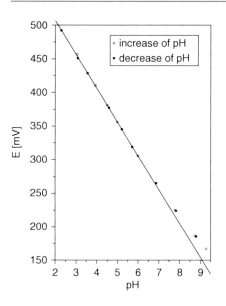

Drink	pH(RuO$_2$)	pH(glass)
Coca-Cola	2.52 ± 0.02	2.49 ± 0.01
Fanta (12 % of lemon juice)	2.59 ± 0.03	2.63 ± 0.02
Red wine	3.65 ± 0.10	3.34 ± 0.01
Orange juice	4.30 ± 0.15	3.86 ± 0.02
Milk	6.73 ± 0.03	6.69 ± 0.02

Potentiometric pH Sensors at Ambient Temperature, Fig. 2 (*Left*) Potentiometric calibration curve of screen-printed RuO$_2$ in universal buffer and (*right*) comparison to a commercial glass electrode in some drink and beverages [18] (Reprinted from Analytica Chimica Acta, 1997. **351**(1–3): p. 143–149 with permission from Elsevier)

adequately small, and able to penetrate the tissue of samples. Due to the intrinsic properties of glass, pH measurement in microscale in these specialized systems by conventional glass electrode is impractical. Therefore, a new alternative for pH measurement uses metal/metal oxide, classified as one of solid-state pH electrodes. There are several metal/metal oxides, e.g., molybdenum [8], lead [9], iridium [10–12], cobalt [13], palladium [14], manganese [15], or ruthenium [16] oxide, exhibiting Nerstian behaviors at ambient temperature. In principle, a reversible metal/metal oxide electrode utilizes a change of electronic charge at the electrode-electrolyte interface, describing by an oxygen intercalation mechanism to hydrogen ion response as the following [17]:

$$MO_x + 2yH^+ + 2ye^- \leftrightarrow MO_{x-y} + yH_2O$$

The electrode potential depends on the ion exchange on metal sites of the OH group on the surface which can be expressed as [16]

$$MO_x(OH)_z + yH^+ + ye^- \leftrightarrow MO_{x-y}(OH)_{y+z}$$

$$E = E_0' + 0.0591 \log\left[\frac{MO_x(OH)_z}{MO_{x-y}(OH)_{y+z}}\right] + 0.0591 pH_{out}$$

The metal oxide must be sparingly soluble in tested solution and sufficiently resistive to strong corrosion in acidity or alkalinity. The intercalation equilibrium reaction must be reach as quickly to provide electron transfer to the working electrode, and it is contributed only by H$^+$ ions. As shown in Fig. 2, Koncki and Mascini demonstrated [18] a RuO$_2$ screen-printed pH sensor, successfully fabricated on a plastic platform which is flexible, sufficiently strong, and safe to test in food processing. In this work, the pH values of several beverages such as Coca-Cola and red wine were examined by RuO$_2$ in comparison to a commercial glass sensor. Since RuO$_2$ (and other metal oxides) rely on the proton/ion exchange mechanism, the potentiometric responses are not necessarily contributed by pH change. Therefore, the linear range of the sensor is limited (up to pH = 8); ruthenium encounters dissolution from alkaline ions beyond this point in potentiometric pH measurement.

Potentiometric pH Sensors at Ambient Temperature,

Fig. 3 (*Left*) Comparison of IrO$_2$-TiO$_2$-30-70-mol% and a commercial pH glass electrode, (*right*) response time from pH change between 4 and 12 (inserts: acid and basic region from titration) [19] (Reprinted from Analytica Chimica Acta, 2008. **616**(1): p. 36–41 with permission from Elsevier)

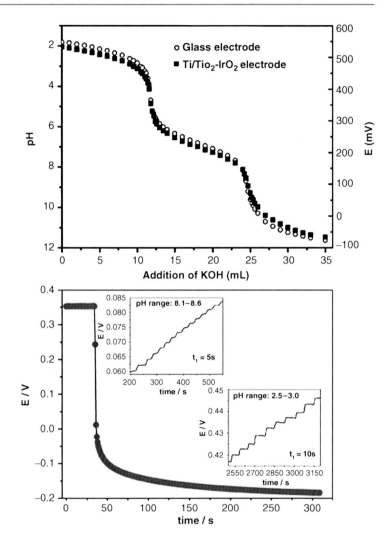

In clinical analysis, Marzouk and coworkers [12] also reported the use of metal/metal oxide (IrO$_2$) pH sensor by monitoring ischemic rabbit heart muscle. However, the sensor had low sensitivities in the presence of Na$^+$, K$^+$, Li$^+$, NH$_4^+$, Ca^{2+}, Mg^{2+}, dissolved oxygen, lactate, ascorbate, and urate. In 2008, Da Silva et al. [19] showed promising results of IrO$_2$ design, incorporating 30 % mol. IrO$_x$ and 70 % mol. TiO$_2$ in a binary system, which shows performance comparable to a commercial pH glass membrane electrode in Fig. 3 (linear range from pH 4 to 12). According to this experiment, the sensor does not exhibit hysteresis effect and is highly selective against alkaline ions: Li$^+$, Na$^+$, and K$^+$. The pH measurement by IrO$_2$ is nearly similar to the glass electrode; however, it still suffers interference by strong reducing agents such as ascorbic acid, H$_2$O$_2$, Fe^{2+}, I$^-$, or SO$_3^{2-}$ [1]. Recently, Shim et al. [20] reported an improvement of Ru oxide pH sensor by using 3D nanoporous Ru oxide film on gold substrate. Facilitated by a reverse micelle nonionic surfactant, L$_2$, Ru oxide can be electrochemically deposited as shown by Fig. 4a, b, and the surface roughness is modified by Triton-X. In Fig. 4c, the potentiometric response is determined from pH 2 to 11 by adding NaOH and reversely titrated by HCl in comparison to the commercial pH glass electrode. The L$_2$-eRuO shows a reliable near Nerstian behaviors (sensitivity of −60.5 mV/pH, r^2 = 0.9997) at response time of 6–8 s in the pH range of 2–11. Controlling the oxide thickness and

Potentiometric pH Sensors at Ambient Temperature, Fig. 4 (a) SEM and (b) TEM images of L$_2$-eRuO of 5 % wt. RuCl$_3$; comparison of pH response between a glass pH electrode and an L$_2$-eRuO toward (c) pH values and (d) calibration curve of NaOH and HCl titration in universal buffer solution [20] (Reprinted from Microchimica Acta, 2012. 177(1–2): p. 211–219 with permission from Springer Science and Business Media)

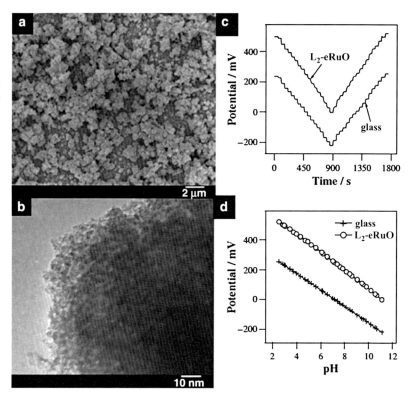

preventing parasite reactions involved in oxygen intercalation and ion exchange would overcome metal dissolution and provide better performance over the traditional pH glass electrode.

Polymeric pH Sensor

Because of the effect of severe interference by reducing agents and toxicity, metal/metal oxide pH sensors are less attractive in physiological measurements. Coexisting alkaline ions and long monitoring times are growing concerns for in vivo analysis using metal oxide. This leads to another class of pH sensor by selectively modifying pH electrode with polymers. In 1980, polymeric pH sensor, using poly (1,2-diaminobenzene) thick film coated on platinum electrode, was elaborated by Heineman et al. [21]. The sensor detected good potentiometric response (sensitivity of 57 mV/pH, $r^2 = 0.991$) from pH 4 to 10. In general, a pH-sensitive polymer network consists of a backbone polymer, carrying weak acid or basic groups in order to obtain charge from ion exchange or protonation [22]. For example, polypropylenimine (PPI) [23, 24], polyethylenimine (PEI) [25], poly (vinyl chloride) (PVC) [26], polypyrrole (PPy), and polyaniline (PANI) [27, 28] exhibit in general Nerstian behaviors toward H$^+$ ions changes. Lakard et al. [29] reported a comparison of PPy, PPPD, PEI, and PANI for potentiometric polymer-based pH sensors. In Fig. 5 (left), surface morphology of each membrane is characterized by SEM images. As shown in Fig. 5 (right), all polymers exhibit near Nerstian response between pH 2 and 10. Among these polymers, PEI and PPI show the fastest response time within 15 s, whereas the others take a few minutes to reach steady state. On the other hand, PANI exhibits the highest sensitivity of 52 mV/pH with an acceptable correlation coefficient ($r^2 = 0.957$). PPPD maintains the optimum r^2 value of 0.995, implying a highly correlated response toward H$^+$ ions change. Furthermore, flexible thin film PANI coated on a carbon

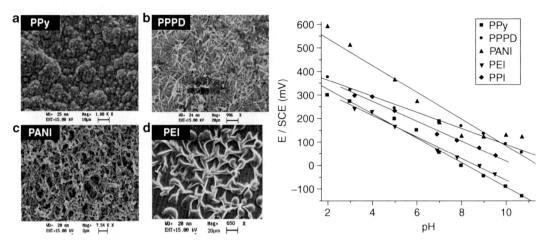

Potentiometric pH Sensors at Ambient Temperature, Fig. 5 SEM and calibration curves of PPy, PPPD, PANI, and PEI on Pt electrode [29] (Reprinted from Polymer, 2005. 46(26): p. 12233–12239 with permission from Elsevier)

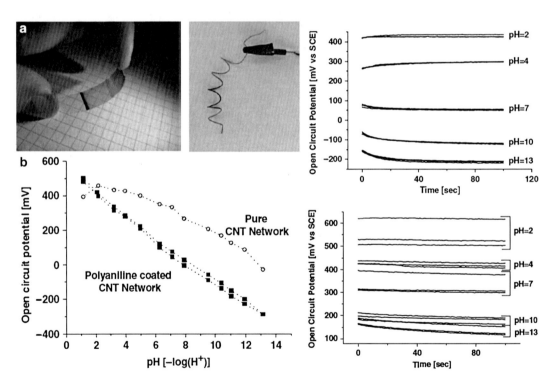

Potentiometric pH Sensors at Ambient Temperature, Fig. 6 Transparent and flexible CNT/PANI pH sensor (a), pH dependence of open circuit time of CNT/PANI in comparison to pure CNT network (b), and reproducibility (n = 3) of the sensor dipping in pH 2 → 7 → 13 → 4 → 10 in sequence [30]

nanotube (CNT) pH sensor was elaborated by Kaempgen and Roth [30]. The CNT network has a high mechanical respect ratio, highly conductive to ions, and chemically stable. Figure 6 illustrates features of a plastic PANI/CNT film and a simple wire for pH sensors. The novel electrode can detect from pH 1 to 13 and has a good linearity with excellent sensitivity of 58 mV/pH. The potential is stabilized within a few seconds and remains unchanged for a long

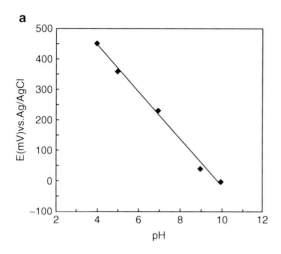

Potentiometric pH Sensors at Ambient Temperature, Fig. 7 (a) Potentiometric pH calibration curve, (b) disposable microelectrode lab chip sensor, and (c) SEM image of self-assembly polystyrene nanobead [32] (Reprinted from Talanta, 2010. **83**(1): p. 1–8 with permission from Elsevier)

period of time (5 min). Further stability testing was carried out and drift of 2 mV/pH was observed after 5 h and in extreme acidity and alkalinity (pH of 1 and 13, respectively). Nevertheless, PANI/CNT has shown highly reproducible results by measuring pH of 2 → 7 → 13 → 4 → 10, accordingly in separate trials (Fig. 6, right). In comparison to pure CNT network, PANI-coated CNT clearly exhibit both higher sensitivity and reproducibility, providing a promising advancement to polymeric solid-state pH sensors.

Functionalizations of metal/metal oxide or carbon nanotechnology with polymers are reported in potentiometric pH designs. Li et al. [31] demonstrated a simple fabricated polybisphenol A (PBPA), electropolymerized on indium thin oxide (ITO) for pH sensing. The PBPA/ITO exhibits a sensitivity of 58.6 ± 1.4 mV/pH, near theoretical value, and can be stored for up to 12 days. Possible interferents such as Na^+, K^+, Cl^-, and SO_4^{2-} show no significant response. Jang and coworkers developed self-assembly nanobead polystyrene, deposited on Au/Bi in lab chip sensor. As shown in Fig. 7, the microfluid chip has two ion-selective membranes to H^+ and NO_3^- species.

The potentiometric response of the nanobead polystyrene is measured as shown Fig. 7a. This sensor has the sensitivity of 77.031 mV/pH, exhibiting super Nerstian behavior from pH range 4–10. However, interference by alkaline ions and reducing agent interference of pH electrode have not been reported in this study.

Taouil et al. [33] reported functionalized polypyrroles, poly(11-N-pyrrolylundecanoic acid) (PUPA), which shows enhancement in conductivity from PPy. Similar alkyl chain PPy, poly (N-undecypyrrole) (PUP), was synthesized containing non-protonatable group. Figure 8 (left) shows the presence of carboxylic acid group in PUPA and other identical chemical composition to PUP film, characterized by XPS spectrum. PUPA exhibits near Nerstian behavior in the range of pH 4–9 shown in Fig. 8 (right). In comparison, PUPA has an excellent sensitivity of 51 mV/pH ($r^2 = 0.995$), higher than that of PUP (33 mV/pH, $r^2 = 0.994$) and PPy (48 mV/pH, $r^2 = 0.996$). After 30 days, the PUPA shows a slight change in performance (sensitivity of 48 mV/pH, $r^2 = 0.996$), whereas, the sensitivity of PPy decreases to 33 mV/pH ($r^2 = 0.992$). Based on the result, the behavior of PUPA can be explained by the presence of carboxylic acid

Potentiometric pH Sensors at Ambient Temperature, Fig. 8 XPS spectrum and potentiometric pH response of PUPA (a) and PPU (b) [33] (Reprinted from Synthetic Metals, 2010. 160(9–10): p. 1073–1080 with permission from Elsevier)

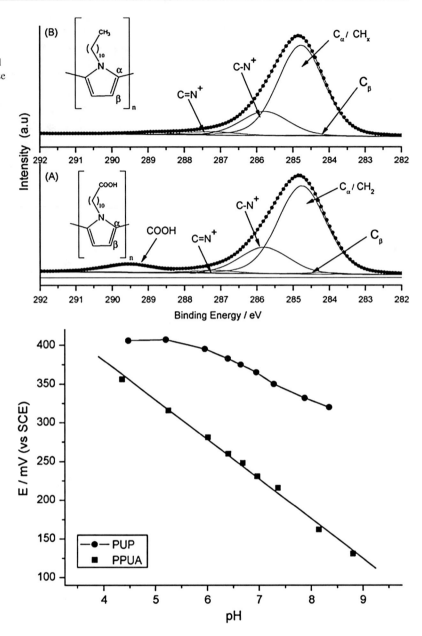

group that can be reversibly deprotonated according to the pH value. In contrast, PUP, which contains nitrogen from pyrrole, confirms that protonation is limited because these nitrogen atoms are sterically hindered [33]. For interference, a high concentration (~1 M) of several existing ions – Li^+, K^+, Na^+, Ca^{2+}, Cl^-, and SO_4^{2-} – shows no significant change in both PUPA and PUP. However, CO_3^{2-} severely affects the sensor showing an increase of 15 mV/pH on both films, indicating further improvement of polypyrrole potentiometric sensors.

This entry reviews the development of various pH sensors and provides assessment of these pH sensors for further development in order to meet the specific need. pH detection by metal/metal oxide sensors and polymer-based sensors involves in protonation or oxygen intercalation mechanism. Therefore, strong reducing agents or

Potentiometric pH Sensors at Ambient Temperature

alkaline ions are a major challenge to improve selectivity. Long-term stability from the film deterioration also needs to be addressed for practical use. Nevertheless, both classes can provide high mechanical strength, stiffness, and non-fragile materials, desirable for many practical uses. In view of replacing glass membrane pH sensor, hybridization in pH sensor designs of carbon nanotechnology, protonable group functionalization, or metal/metal oxide and polymer combination are expected to provide further improvements for potentiometric pH sensors.

Cross-References

▶ pH Electrodes - Industrial, Medical, and Other Applications
▶ Potentiometry

References

1. Kurzweil P (2009) Metal oxides and ion-exchanging surfaces as pH sensors in liquid: state-of-the-art and outlook. Sensors 9(6):4955–4985
2. Belyustin AA (2010) The centenary of glass electrode: from Max Cremer to F. G. K. Baucke. J Solid State Electrochem 15(1):47–65
3. Vonau W, Guth U (2006) pH Monitoring: a review. J Solid State Electrochem 10(9):746–752
4. Kohlmann FJ (2003) pH technical handbook. Hach Company
5. Zhang X, Ogorevc B, Wang J (2002) All solid-state pH nano-electrode based on polyaniline thin film electrodeposited onto ion-beam etched carbon fiber. Anal Chim Acta 452:1
6. Slim C et al (2008) Polyaniline films based ultramicroelectrodes sensitive to pH. J Electroanal Chem 612(1):53–62
7. Li J, Du Y, Fang C (2007) Developing an iridium oxide film modified microelectrode for microscale measurement of pH. Electroanalysis 19(5):608–611
8. Shuk P, Ramanujachary KV, Greenblatt M (1996) New metal-oxide-type pH sensors. Solid State Ionics 86–88, Part 2(0):1115–1120
9. Eftekhari A (2003) pH sensor based on deposited film of lead oxide on aluminum substrate electrode. Sens Actuators B Chem 88(3):234–238
10. Hitchman ML, Ramanathan S (1988) Evaluation of iridium oxide electrodes formed by potential cycling as pH probes. Analyst 113(1):35–39
11. VanHoudt P, Lewandowski Z, Little B (1992) Iridium oxide pH microelectrode. Biotechnol Bioeng 40(5):601–608
12. Marzouk SAM et al (1998) Electrodeposited iridium oxide pH electrode for measurement of extracellular myocardial acidosis during acute ischemia. Anal Chem 70(23):5054–5061
13. Meruva RK, Meyerhoff ME (1998) Catheter-type sensor for potentiometric monitoring of oxygen, pH and carbon dioxide. Biosens Bioelectron 13(2):201–212
14. Yun M et al (2004) Electrochemically grown wires for individually addressable sensor arrays. Nano Lett 4(3):419–422
15. Cherchour N et al (2011) pH sensing in aqueous solutions using a MnO2 thin film electrodeposited on a glassy carbon electrode. Electrochim Acta 56(27):9746–9755
16. Liao Y-H, Chou J-C (2008) Preparation and characteristics of ruthenium dioxide for pH array sensors with real-time measurement system. Sens Actuators B Chem 128(2):603–612
17. Głab S et al (1989) Metal-metal oxide and metal oxide electrodes as pH sensors. Crit Rev Anal Chem 21(1):29–47
18. Koncki R, Mascini M (1997) Screen-printed ruthenium dioxide electrodes for pH measurements. Anal Chim Acta 351(1–3):143–149
19. da Silva GM et al (2008) Development of low-cost metal oxide pH electrodes based on the polymeric precursor method. Anal Chim Acta 616(1):36–41
20. Shim JH et al (2012) A nanoporous ruthenium oxide framework for amperometric sensing of glucose and potentiometric sensing of pH. Microchimica Acta 177(1–2):211–219
21. Heineman WR, Wieck HJ, Yacynych AM (1980) Polymer film chemically modified electrode as a potentiometric sensor. Anal Chem 52(2):345–346
22. Richter A et al (2008) Review on hydrogel-based pH sensors and microsensors. Sensors 8(1):561–581
23. van Duijvenbode RC et al (1999) Synthesis and protonation behavior of carboxylate-functionalized poly (propyleneimine) dendrimers. Macromolecules 33(1):46–52
24. van Duijvenbode RC, Borkovec M, Koper GJM (1998) Acid–base properties of poly (propylene imine) dendrimers. Polymer 39(12):2657–2664
25. Herlem G et al (2001) pH sensing at Pt electrode surfaces coated with linear polyethylenimine from anodic polymerization of ethylenediamine. J Electrochem Soc 148(11):E435–E438
26. Liu X-J et al (2007) Potentiometric liquid membrane pH sensors based on calix[4]-aza-crowns. Sens Actuators B Chem 125(2):656–663
27. Gill EI et al. (2008) Novel conducting polymer composite pH sensors for medical applications. In: Katashev A, Dekhtyar Y, Spigulis J (eds) 14th Nordic-Baltic conference on biomedical engineering and medical physics. Springer, Berlin/Heidelberg, p 225–228

28. Ansari R et al (2012) Solid-state Cu (II) ion-selective electrode based on polyaniline-conducting polymer film doped with copper carmoisine dye complex. J Solid State Electrochem 16(3):869–875
29. Lakard B et al (2005) Potentiometric pH sensors based on electrodeposited polymers. Polymer 46(26):12233–12239
30. Kaempgen M, Roth S (2006) Transparent and flexible carbon nanotube/polyaniline pH sensors. J Electroanal Chem 586(1):72–76
31. Li Q et al (2011) A novel pH potentiometric sensor based on electrochemically synthesized polybisphenol A films at an ITO electrode. Sens Actuators B Chem 155(2):730–736
32. Jang A et al (2010) Potentiometric and voltammetric polymer lab chip sensors for determination of nitrate, pH and Cd(II) in water. Talanta 83(1):1–8
33. Taouil AE et al (2010) Effects of polypyrrole modified electrode functionalization on potentiometric pH responses. Synth Met 160(9–10):1073–1080

Potentiometry

Winfried Vonau
Kurt-Schwabe-Institut fuer Mess- und Sensortechnik e.V. Meinsberg, Waldhein, Germany

Introduction

Potentiometry is an electrochemical measurement technique. The term was introduced in connection with potential determinations of electrochemical measuring chains by W. Ostwald [1]. Often instead of potentiometric measurement the measurement of galvanic voltages is spoken about. Mainly, the measuring setup consists of two electrodes (galvanic cell). These half cells are called indicator (or measuring) and reference electrode. Additionally, a high-impedance potential measuring device is necessary. Figure 1 shows a scheme of such an arrangement also called potentiometric measuring cell or chain. An important feature of the method is the fact that it is measured under the conditions of no current flow. The measurand may be used to determine the analytical quantity of interest, i.e., the activity of some components of an analyte solution or the partial pressure of

gases (see also high-temperature electrochemistry). The potential that develops in the electrochemical cell is the result of the free energy change that would occur if the chemical phenomena were to proceed until the equilibrium condition has been fulfilled.

For the potentiometric measurement, according to Eq. 1, the potential of the cell can be expressed in terms of the potential developed by the indicator electrode (E_{ind}) and the reference electrode (E_{ref}) under consideration of the junction potential (E_j):

$$E = (E_{ind} - E_{ref}) + E_j \qquad (1)$$

An important mathematical basis of the potentiometry also of high interest for practical measurements is formed by the Nernst Eq. 2. It allows the calculation of required analytical values, e.g., ion activity (a_i) as function of concentration, from the voltage E produced by any electrochemical cell (see Eq. 3):

$$E = E^{\ominus} - (RT/nF)lnQ \qquad (2)$$

$E^{\ominus}=$ standard electrochemical cell potential (voltage).
$R =$ gas constant.
$T =$ temperature in Kelvin.
$n =$ number of moles of electrons transferred in the balanced equation or the charge/valency of the ion. In the case of the hydrogen ion, $n = 1$.
$F =$ Faraday constant $= 96.485$ C mol^{-1}.
$Q =$ mass-action expression (approximated by the equilibrium expression).

$$E = E^{\ominus} - (RT/nF)ln\, a_i \qquad (3)$$

A distinction is made between direct potentiometry and potentiometric titration. First mentioned is the method which makes use of the single measured electrode potential, for example, to determine ions using different ion-selective electrodes (ISE). While the potential of the reference electrode must be independent on the composition of the measurement medium, the indicator electrode should be fast responding and highly sensitive and selective for special ions in a large concentration range. Figure 2 shows in principle the course of the curves for direct potentiometric determinations of a cation (i) and an anion (ii) at $T = 298$ K with pC and pA as negative logarithms of the cation a_{C^+} and anion activities a_{A^-} which have to be recorded for the calibration of potentiometric chains before measuring.

Potentiometric titrations (neutralization titrations, oxidation–reduction titrations, precipitation titrations, complex formation titrations, differential titrations) provide information different from a direct potentiometric measurement. The equivalence point of the reaction (see Fig. 3) will be revealed by a sudden change in potential in the plot of potential readings against the volume of the titrating solution. One electrode has to be maintained at a constant (possibly unknown) reference potential. The other electrode must serve as an indicator of the changes in ionic concentration and has to respond rapidly. The solution should be stirred during the titration. Compared to direct potentiometric measurements, potentiometric titrations generally offer increased accuracy and precision because potentials are used to detect rapid changes in activity that occurs at the equivalence point of the titration. Furthermore, the influences of liquid junction potentials and activity coefficients are minimized.

In both, direct potentiometry and potentiometric titration redox reactions as well as equilibrium adjustments cause the function of the electrochemical cells.

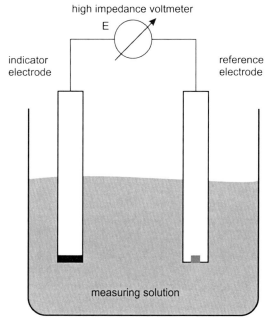

Potentiometry, Fig. 1 Scheme of a potentiometric measuring chain

Electrodes for Potentiometry

There are several possibilities to generate the reference potential in potentiometric measuring chains. Based on research results of Le Blanc [2] at first, the standard hydrogen electrode (SHE)

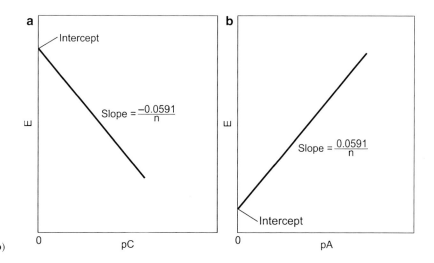

Potentiometry, Fig. 2 Potentiometric calibration curves for cations (**a**) and anions (**b**)

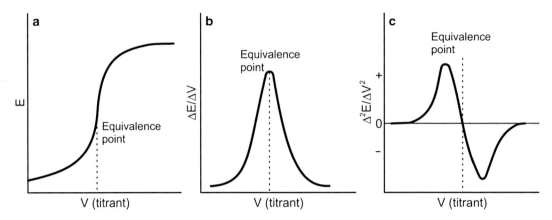

Potentiometry, Fig. 3 Potentiometric titration curves, (**a**) typical plot of E versus volume, (**b**) typical first derivative titration curve, (**c**) typical second derivative titration curve

was established for this purpose. Using the shorthand notation for electrochemical cells, the SHE can be described as follows:

$$Pt, H_2(g, 101.325\,kPa)/H^+$$
$$(aq, a = 1.00)//2H^+(aq) + 2e^- \rightleftharpoons H_2(g)$$

The potential of that electrode was defined as the origin of the electrochemical potential scale and was arbitrarily declared to be zero for all temperatures. Because of their complicated handling and some other disadvantages, the standard hydrogen electrode today is used only for very precise measurements, especially for the calibration and testing of pH buffer solutions. It is not trivial to realize exactly the experimental conditions that are necessary. Instead of the standard hydrogen electrode (SHE) more commonly the normal hydrogen electrode (NHE) is used, which contains an acid solution of the concentration 1 mol/L.

In practice, the so-called reference electrodes of the second kind, e.g., the Ag/AgCl, Cl^- electrode or the Hg/Hg_2Cl_2, Cl^- (calomel) electrode, are preferably applied because they are more convenient to use. Their potential is stable and well defined in relation to the potential of the SHE. Such electrodes are commercially available in a variety of chemical systems and embodiments for different applications [3]. A substantial drawback of conventional reference electrodes is their liquid electrolyte filling. The performance of such electrodes is position dependence, and they are mechanically fragile and suitable to only a limited extend for miniaturization or for applications at high pressures and temperatures. In recent years several solutions for all solid-state arrangements were introduced, such as electrodes with solid polymer electrolyte, reference electrodes based on bronzes from the transition metals molybdenum and tungsten, or electrodes containing essentially an Ag/AgCl-reference element and a solid crystalline KCl melt. A current overview is given in [4].

Concerning the indicator electrodes often it is distinguished between metallic and membrane electrodes. The potential of a metallic electrode is determined by a redox reaction at the interphase electrode/solution. Basically there are three different kinds of metal-based indicator electrodes (Table 1):

Membranes as location for the potential determining reaction in indicator electrodes as part of potentiometric measuring chains can be divided in crystalline and noncrystalline membranes.

Noncrystalline membranes	Crystalline membranes
Silicate (i) and chalcogenide glasses (ii)	Single crystals (v)
Liquid–liquid ion exchanger (iii)	Separate or mixed polycrystals (vi)
Immobilized liquid–liquid/PVC matrix (iv)	

(i). Oxidic glasses with special compositions can be part of potentiometric sensors to

Potentiometry

determine the activity of H^+, Na^+, K^+, NH_4^+, Cs^+, Rb^+, Li^+, Ag^+, and several other cations [5] as well as of the redox potential [6].

Examples. H^+ (more precise H_3O^+) activity 72 wt% SiO_2, 22 wt% Na_2O, 6 wt% CaO [7].

Redox potential 50 wt-% SiO_2, 39 wt% Fe_2O_3, 7 wt% Na_2O, 4 wt% Li_2O [8].

(ii). Using such materials leads to ISE for the determination of several heavy metals but also for anions such as F^- [9]. Besides compact sensors it is possible to fabricate chalcogenide glass electrodes in thin film technology [10].

Example. Fe^{3+} activity $Fe_2Se_{60}Ge_{28}Sb_{12}$ [11].

(iii). One of the most famous liquid membrane electrodes has been used for calcium determination. This electrode works by an ion-exchange process. The cation exchanger is an aliphatic diester of phosphoric acid, $(RO)_2PO_2^-$, where each R group is an aliphatic hydrocarbon chain containing between 8 and 16 carbons. The phosphate group can be protonated but has a strong affinity for Ca^{2+}. The cation exchanger is dissolved in an organic solvent and held in a porous compartment between the analyte solution and an internal reference calcium chloride solution. The ion exchanger uptakes Ca^{2+} into the membrane.

Example. $Ca^{2+}(aq) + 2(RO)_2PO_2^-$ $(organic) \rightleftharpoons [(RO)_2PO_2]_2Ca(organic)$.

(iv). An alternative to liquid membrane electrodes is to use a polymer matrix-based membrane, which is composed of a polymer such as polyvinylchloride (PVC), a plasticizer, and the ion carrier or exchanger. The response of these electrodes is highly selective and they have replaced many liquid membrane electrodes.

Polymer-based electrodes are available, e.g., for the determination of K^+, NH_4^+, Ba^{2+}, Ca^{2+}, Cl^-, and NO_3^-.

Example.

K^+ activity ... Valinomicin [12]

(v). It is also known to use membranes consisting of (doped) single crystals. At the inner phase boundary, the single crystals are in contact with other solids (all solid-state electrodes). In order to get reproducible and long-term stable potentials in these cases, it is necessary to create non-blocked (reversible) interfaces.

Example. Fluoride can be analyzed potentiometrically using the half cell $F^-/$ LaF_3 (doped with Eu)/AgF/Ag [13].

(vi). Besides the usage of metals in contact with one of its hardly soluble salts (electrodes of the 2nd kind) for the potentiometric ion determination (Ag/AgCl membrane to monitor chloride, see above), sometimes mixed compositions are used. As a result, a lot of electrode properties can be improved (mechanical and light stability, lower detection limit).

Example. Potentiometric chloride determination with heterogeneous membranes (Fig. 4).

Cross Selectivity

Full selectivity for exactly one type of ion is never given. ISE have only a particular sensitivity for a special type of ion but often show interference with ions with similar chemical properties or a similar structure. For this reason the cross sensitivity to other ions that may be present in the

Potentiometry, Table 1 Kinds of metal based indicator electrodes

Electrodes of the 1st kind	Electrodes of the 2nd kind	Redox electrodes
	A metal is coated with one of its salt precipitate	
A metal is in contact with a solution containing its cation, e.g., Zn/Zn^{2+}, Ag/Ag^+, Cu/Cu^{2+}	An example is the Ag/AgCl electrode to measure the activity of chloride ions in a solution	An inert metal is in contact with a solution containing soluble oxidized and reduced forms of redox half-reactions
E is a function of the cation activity in the system metal/metal ion	E is a function of the anion activity	Electrode materials are often Au, Pt, or Pd
Example:	Example:	E is determined by the ratio of reduced and oxidized species
$Ag^+ + e^- \rightleftharpoons Ag$	$Ag^+ + e^- \rightleftharpoons Ag$ $Ag^+ + Cl^- \rightleftharpoons AgCl(\Downarrow)$ $\overline{AgCl(\Downarrow) + e^- \rightleftharpoons Ag + Cl^-}$	Example: $Fe^{3+} + e^- \rightleftharpoons Fe^{2+}$

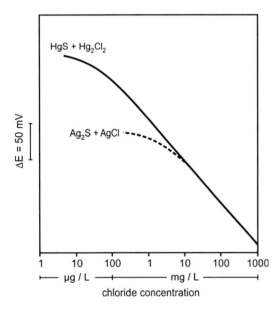

Potentiometry, Fig. 4 Measuring ranges for two different potentiometric chloride electrodes with heterogeneous membrane

sample solution must always be taken into consideration when selecting an ISE for a potentiometric determination. One of the best-known examples of such a cross sensitivity is the so-called alkali error of pH glass electrodes. Unfortunately, there are only very few ion-selective electrodes that have a linear range similar to that of pH glass electrodes. The use of an ISE is normally restricted to a concentration range of 4–5 powers. If an ISE is used for a measurement often the Nernst Eq. 3 must be extended by the contribution made by the particular interfering ion for the evaluation of the measured potential. This leads to the Nikolskij Eq. 4:

$$E = E - (RT/nF) \ln \left(a_i + K_{j/i}a_j\right) \quad (4)$$

$K_{j/i}$ is the selectivity coefficient of the ISE for an interfering ion S. This is a factor that describes the influence of the interfering ion in relationship to the ion to be measured (i). These selectivity coefficients are known for the most important interfering ions for an ISE, and therefore, a simple estimation can be made to estimate if an interfering ion present in the sample solution will influence the measured value or not.

Future Directions

Until today, there are still no satisfactory electrode membrane materials for the potentiometric determination of several important ions such as SO_4^{2-} and PO_4^{3-} so that here remains a need of research for the future.

For a lot of potentiometric indicator electrodes solid-state embodiments are available widely equivalent to the electrolyte containing systems concerning the electrode performance. Because this does not apply for all measuring electrodes (e.g., for pH glass electrodes) and for reference electrodes to the full extent, there is also still a need for development in this field.

Especially for medical applications it is often necessary to use miniaturized equipment. Due to this fact the engineers are faced with the task to develop new potentiometric electrodes by microtechnological methods, e.g., thin or thick film technology. This creates the opportunities to come to a number of miniaturized multi-sensors as part of lab on chip systems increasingly required in biotechnology and other areas.

Cross-References

▶ Potentiometric pH Sensors at Ambient Temperature
▶ Reference Electrodes

References

1. Drucker C (1943) Ostwald-Luther Hand- und Hilfsbuch zur Ausführung physikochemischer Messungen. Dover, New York
2. Le Blanc M (1893) Die elektromotorischen Kräfte der Polarisation II. Z Phys Chem 12:133–358
3. Lisdat F, Moritz W, Müller L (1990) Referenzelektroden für wässrige und geschmolzene Systeme. Zeitschrift für Chemie 30:427–433
4. Vonau W, Oelßner W, Guth U, Henze J (2010) An all-solid-state reference electrode. Sens Actuators B 144:368–373
5. Eisenman G (1967) The physical basis for the ionic specifity of the glass electrode. In: Eisenman G (ed) Glass electrodes for hydrogen and other cations, principles and practice. Marcel Dekker, New york
6. Nikolskij BP, Schulz MM (1971) US Patent US 3 773 642
7. MacInnes DA, Dole M (1930) The behavior of glass electrodes of different compositions. J Am Chem Soc 52:29–36
8. Nikolskij BP, Schulz MM, Pisarevskij AM, Beljustin AA, Bolchnzeva SK, Dolidse WA, Tarasova WM, Karatschenzeva JM, Dolmasowa LI (1973) German Patent DE 2 134 101
9. Vlassov Y, Bychkov E (1987) Ion-selective chalcogenide glass electrodes. Ion-Selective Rev 9:5–93
10. Kloock JP, Mourzina YG, Ermolenko Y, Doll T, Schubert J, Schöning MJ (2004) Inorganic thin-film sensor membranes with PLD-prepared chalcogenide glasses. Challenges and implementation. Sensors 4:156–162
11. König C (1993) Entwicklung und Charakterisierung einer Chalkogenidglaselektrode zur selektiven Detektion von Eisen(III)ionen. Dissertation. Goethe University Frankfurt/M., Germany
12. Pioda L, Stankova V, Simon W (1969) Highly selective potassium ion responsive liquid membrane electrode. Anal Lett 2:665–674
13. Fjedly TA, Nagy K (1980) Fluoride electrodes with reversible solid-state contacts. J Electrochem Soc 127:1299

Potentiostat

Manuel Lohrengel
University of Düsseldorf, Düsseldorf, Germany

Introduction

In the beginning, electrochemical experiments were reduced to potential measurements of systems, which were stationary or in equilibrium. But with time, scientists became interested in systems apart from equilibrium [1] or even in time-dependent reactions. This was realized by galvanostatic experiments (Potentiostat), which were by two reasons advantageous: They were easily realized and guaranteed a constant reaction rate, which was relevant in some cases. Moreover, time-dependent reactions could be monitored, if the potential was recorded vs. time. These charging curves were the main technique to follow electrode kinetics up to the sixties of the last century.

The potential, however, or more precise, the potential drop somewhere within the electrode interface, is said to be the cause of all electrochemical reactions. Therefore, scientists tried to control the potential of electrodes or cells.

Less effective was the use of a power supply and variable resistors, controlled by human intervention [2] or by slow electromechanical devices [3, 4]. More effective was the combination of a controlled power supply with an electronic control circuit, which worked fast and unattended. This was realized by *Hickling* [5], who called the instrument for the first time "potentiostat." The name is nowadays somewhat misleading as scientists do not want to keep the potential constant but wish to vary it in a predetermined way in most cases.

Definition

A potentiostat is a controlled electric power supply, which adjusts the potential drop across an active or passive dipole to a desired value. To do this, the device has to measure the actual potential drop and to force currents through the dipole, until actual value and desired value become equal. In electrochemical systems, the interest to keep the potential drop across a complete cell is limited to rare cases, e.g., charging or discharging of batteries. More interesting are kinetics of single electrodes, which means an electric control of the interface electrode/electrolyte. An additional device is needed to measure the electrolyte potential, the so-called reference electrode. As a result, a three-electrode setup is obtained (working, counter, and reference electrodes). Reference electrodes include at least one interface *ionic conductor/ electronic conductor* with an additional potential drop which must be taken into account.

The popularity of potentiostats is reflected by a large number of publications discussing concepts, function, and applications [6–13].

Basic Circuit

Perfect combinations of control circuit and power source are operational amplifiers, which were built from discrete components in the fifties and become much more popular as integrated circuits from 1960 onward.

Most potentiostats are based on operational amplifiers. Figure 1 shows the fundamental setup. The output voltage U_{out} of an operational amplifier depends on the difference of the input voltages U_+ and U_- according to

$$U_{out} = A \cdot (U_+ - U_-) \quad (1)$$

with an open-loop gain $A > 10^6$. A difference $U_+ - U_-$ causes an increasing output voltage

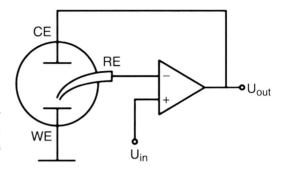

Potentiostat, Fig. 1 Fundamental potentiostat circuit with an operational amplifier

and, therefore, forces currents through the cell. This reduces the difference $U_+ - U_-$ close to zero due to the negative feedback of the output to U_-.

The limits of the output are controlled by the power supply and the particular circuit design. Typical values range from ± 5 V to ± 100 V and currents up to ± 100 mA. This value can be increased to some 10 A or even more by additional booster amplifiers. Input resistances from 10^9 Ω to 10^{12} Ω prevent notable current flow through the reference electrode.

The open-loop gain A decreases with increasing frequency, typically by a factor of 10 per decade of frequency and A comes close to 1 (unity gain) in the MHz region. This means the useful bandwidth is at least one or two orders of magnitude lower than at unity gain. This is especially important for pulse experiments or experiments at higher frequencies.

Another limitation is the slew rate, which describes the rate of change at the output in response to a large amplitude input step. Typical values such as 10^6 V/s are less informative as electrochemical cells mean capacitive loads. A potential step ΔU and an electrode capacitance ΔU yield a charge

$$\Delta Q = C \cdot \Delta U \quad (2)$$

which must be provided by the amplifier. This means at least a time interval Δt determined by the amplifier's maximum current i_{max}:

$$\Delta t = \frac{\Delta Q}{i_{max}} = \frac{C \cdot \Delta U}{i_{max}} \quad (3)$$

Potentiostat, Fig. 2 Potentiostat system with main amplifier (*MA*), impedance buffer (*IB*), and two independent optional current detection systems: current-to-voltage converter (*CVC*) and differential amplifier (*DA*)

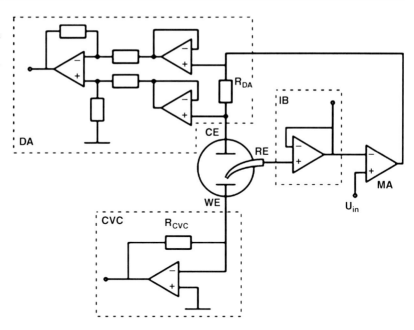

Accordingly, adjustment times of 30–100 μs have to be expected in real systems for a potential step of 1 V.

Potentiostat Systems

The basic setup controls the potential, but no information is returned from the experiment. A potentiostat system allows monitoring the actual potential, recording the cell current, and mixing different input signal such as potential steps, sweeps, sinusoidal signals, and dc levels. Figure 2 shows some concepts.

Potential Monitor

Reference electrodes are sensitive to current flow, due to their large source resistance and/or to small exchange currents of the redox system. Therefore, reference electrodes are usually connected via special high-impedance amplifiers (voltage followers, IB in Fig. 2).

In thermodynamics the electrolyte is thought to be at zero potential and the electrode potential is given. In experiments, however, the working electrode is at ground potential, and therefore, the sign is inverted. This can cause some confusion.

Current Monitor

Current-to-Voltage Converter

A current-to-voltage converter (CVC) is connected to the working electrode in Fig. 2. The cell current I flows also through R_{CVC} because the input resistances of the amplifier are extremely high. The negative feedback, however, keeps the potential of the inverting input to the same level as the non-inverting input, i.e., zero potential (virtual ground). Therefore, the output voltage of the CVC is given by

$$U_{CVC} = -R_{CVC} \cdot I \qquad (4)$$

and can be used to record the current. This concept is also used for IR-drop elimination (▶ iR-Drop Elimination).

This circuit is most often used, but some disadvantages must be respected. The CVC must be able to deliver at least the same (large) currents as the main amplifier, and virtual ground is not necessarily guaranteed for fast changing or noisy signals.

Differential Amplifier

The potential drop across a resistor R_{DA} between counter electrode and main amplifier output can also be used for current monitoring.

The signal is buffered by a more complex differential amplifier [14] (preferentially an instrumentation amplifier, DA in Fig. 2). The cell current I is given by

$$I = -\frac{U_{DA}}{R_{DA}} \quad (5)$$

The amplifiers maximum output current can be small as only recording devices must be supported. Disadvantageous is the limitation of the main amplifier output current by R_{DA}, especially if high values of R_{DA} are chosen to detect small currents.

It was useful in some cases to use both systems (CVC and DA) in parallel. The resistor R_{DA} was short-circuited in one direction to allow high currents, recorded by the current-to-voltage amplifier, and very small currents in the other direction, recorded by the differential amplifier. Moreover, the differential amplifier could be modified into an auto-ranging system to record current transients after large potential steps [15].

Alternative Concepts
The current-to-voltage amplifier is replaced in some concepts by other systems such as simple resistors between working electrode and ground potential [16, 17], negative resistances (gyrator circuits, [18, 19]), logarithmic current amplifiers, and current-photon-converters [20], often in context with IR-drop elimination (▶ iR-Drop Elimination).

Composition of Input Signals
An adder circuit is presented in Fig. 3. The output voltage U_{add} is the sum of all input signals according to

$$U_{add} = -(U_1 + U_2 + U_3 + \ldots + U_n) \quad (6)$$

if all resistors are equal. The output signal is connected to the non-inverting input of the main amplifier (MA) in Fig. 2. The advantage of this concept is the control of the composed signal at the adder output.

A modified popular concept is the adder potentiostat (Fig. 4) which combines summation

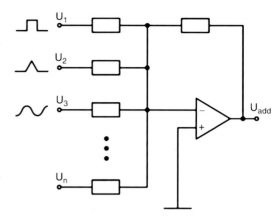

Potentiostat, Fig. 3 Adder circuit to compose an input signal from different sources U_1 to U_n

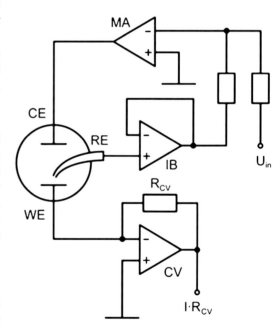

Potentiostat, Fig. 4 Adder potentiostat with main amplifier (MA), impedance buffer (IB), and current-to-voltage converter (CV)

and potential control in one amplifier. Due to resistor network, the circuit would draw currents from the reference electrode. To avoid this, an additional impedance buffer as in Fig. 4 must be added. The composed input signal cannot be monitored.

Digital Concepts

Modern potentiostats are equipped with computer interfaces (e.g., USB), at least to transfer experimental data such as potential and current. Optional is a complete control by the computer, e.g., current ranges (eventually auto-ranging), potential programs, and frequency response characteristics.

The main amplifier in Fig. 2 is substituted in completely digitized concepts by an analog-to-digital converter, a processor, and a digital-to-analog converter. The transfer function, especially the frequency-dependent amplification A, can be easily modified. Such concepts are more complex and not applicable for extremely fast applications.

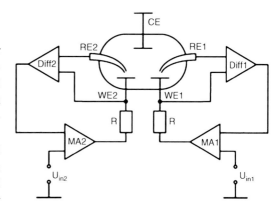

Potentiostat, Fig. 5 Concept of multipotentiostat with counter electrode (CE) connected to ground, 2 working electrodes (WE), 2 reference electrodes (RE), 2 main amplifiers (MA), and 2 differential amplifiers (Diff)

Multi-electrode Concepts

Most electrochemical cells consist of three electrodes: working electrode, counter electrode, and reference electrode. In some applications, however, the working electrode is split into several separate electrodes, e.g., in rotating disc/ring systems (Rotating Disc Electrode). This requires an independent potential control of the working electrodes.

An early concept of a bipotentiostat was presented in [21] but suffered from interactions between the working electrodes. This was overcome by some modifications shown in [22]. More popular is a bipotentiostat based on the adder concept [12], which was presented for the first time in [23]. The circuit is very complex and cannot be extended to more than two working electrodes.

More flexible concepts are based on a counter electrode connected to ground (Fig. 5, [24, 25]). Some can be extended to more than two working electrodes, and every electrode can be equipped with an individual ▶ reference electrode to avoid interactions due to current flow in the electrolyte and to minimize IR drops (▶ iR-Drop Elimination).

Galvanostats

The introduction of operational amplifiers, starting after 1950, made galvanostatic setups much less complex but potentiostats as well. Potentiostats seemed to be more versatile, and thus, galvanostatic mode was simply added as an option in many cases. Every potentiostat, however, can be changed into a galvanostat by simple modifications, see entry ▶ Galvanostat.

Future Directions

The quality of commercial potentiostat systems is nowadays almost perfect. The electrical performance, computer, and software support allow effective and convenient investigations of main-stream problems with minor efforts. But this is a risk, as "...*most students and many professional scientists adopt the "black box" approach to electrochemical instrumentation, and fail to appreciate the operation and limitations of various circuits employed.*" [10].

Future challenges result from less common components, e.g., electrodes (microelectrodes, porous electrodes, scanning probes, reference electrodes with extremely high source resistances), cell geometry (thin film cells, capillary cells, industrial reactors), and special requirements such as noise reduction, high frequencies, high voltages, and extremely small or large currents.

Cross-References

▶ Potentiometry
▶ Reference Electrodes

References

1. Tafel J (1905) Polarisation during cathodic hydrogen development. Z Phys Chem 50:641–712
2. Lingane JJ, Kolthoff IM (1939) Fundamental studies with the dropping mercury electrode. I. The Ilkovic equation of polarographic diffusion currents. J Am Chem Soc 61:825–834. doi:10.1021/ja01873a016
3. Caldwell CW, Parker RC, Diehl H (1944) Apparatus for automatic control of electrodeposition with graded electrode potential. Ind Eng Chem 16:532–535
4. Lingane JJ (1945) Automatic apparatus for electrolysis at controlled potential. Ind Eng Chem 17:332–333
5. Hickling A (1942) Studies in electrode polarisation Part IV- The automatic control of the potential of a working electrode. Trans Faraday Soc 38:27–33
6. von Fraunhofer JA, Banks CH (1972) Potentiostat and its applications. Butterworths, London
7. Schroeden RR (1972) Operational amplifier instruments for electrochemistry. In: Mattson JS, Mark HB, MacDonald HC (eds) Computers in chemistry and instrumentation, vol 2. Marcel Dekker, New York
8. Gabe DR (1972) Bibliography of potentiostat design. Br Corros J 7:236–238
9. Harrar JE (1975) Techniques, apparatus, and analytical applications of controlled-potential coulometry. In: Bard AJ (ed) Electroanalytical chemistry, vol 8. Marcel Dekker, New York
10. Macdonald DD (1977) Transient techniques in electrochemistry. Plenum, New York
11. Greef R (1978) Instruments for use in electrode process research. J Phys E 11:1–15. doi:10.1088/0022-3735/11/1/001
12. Bard AJ, Faulkner LR (1980) Electrochemical methods: fundamentals and applications. Wiley, New York
13. McKubre MCH, MacDonald DD (1984) Electronic instrumentation for electrochemical studies. In: White RE, O'M Bockris J, Yeager E (eds) Comprehensive treatise of electrochemistry, vol 8, Experimental Methods in Electrochemistry. Plenum, New York
14. Mumby JE, Perone SP (1971) Potentiostat and cell design for the study of rapid electrochemical systems. Chem Instrum 3:191–227
15. Lohrengel MM, Schultze JW, Speckmann HD, Strehblow HH (1987) Growth, corrosion and capacity of copper oxide films investigated by pulse techniques. Electrochim Acta 32:733–742
16. Booman GL, Holbrook WB (1965) Electroanalytical controlled-potential instrumentation. Anal Chem 37:1793–1809
17. Brown ER, McCord TG, Smith DE, DeFord DD (1966) Some investigations on instrumental compensation of nonfaradaic effects in voltammetric techniques. Anal Chem 38:1119–1129. doi:10.1021/ac60241a004
18. Gabrielli C, Keddam M (1974) Progres recents dans la mesure des impedances electrochimiques en regime sinusoidal. Electrochim Acta 19:355–362
19. Lamy C, Herrmann CC (1975) A new method for ohmic-drop compensation in potentiostatic circuits: stability and bandpass analysis, including the effect of faradaic impedance. J Electroanal Chem 59:113–135
20. Faulkner LR (1987) In: Fleischmann MS, Pons S, Rolison DR, Schmidt PP (eds) Ultramicroelectrodes. Datatech Systems, Morganton
21. Anderson LB, Reilley CN (1965) Thin-layer electrochemistry: steady-state methods of studying rate processes. J Electroanal Chem 10:295–305
22. Gutwillinger P, Schade G (1979) Simplified bipotentiostat. Electrochim Acta 24:1135–1136
23. Napp DT, Johnson DC, Bruckenstein S (1967) Simultaneous and independent potentiostatic control of two indicator electrodes. Application to the copper (II)/copper(I)/copper system in 0.5M potassium chloride at the rotating ring-disk electrode. Anal Chem 39:481–485. doi:10.1021/ac60248a01
24. Kinza H, Lohse H (1975) Electrocatalytic reduction of the nitrate ion by a positive electrode potential. Z Phys Chem Leipzig 256:233–249
25. Burshtein RKH, Vilinskaya VS, Knots LL, Kushnev VV, Lentser BI, Tarasevich MR (1972) Apparatus for automatic control of ring-disk electrode potentials. Elektrokhimiya 8:1183–1187

Predominance-Zone Diagrams for Chemical Species

Ignacio Gonzalez and Alberto Rojas-Hernández
Department of Chemistry, Universidad Autónoma Metropolitana-Iztapalapa, México, Mexico

Determination of the Thermodynamic Equilibrium State of a Multicomponent and Multi-Reacting System

The study of processes involving several chemical species in solution (e.g., electrochemistry, geochemistry, biochemistry, hydrometallurgy, chemical analysis, and environmental chemistry)

Predominance-Zone Diagrams for Chemical Species

generally requires a precise knowledge of the stability of the different species present in the system in each phase, as well as their coexistence.

The thermodynamic equilibrium of the components of the system provides a useful model to determine the stability of its species and phases. The law of mass action allows calculating the chemical composition of a given system at equilibrium [1].

For a large number of chemical species or equilibria, it is necessary to use iterative calculations in order to determine the thermodynamic equilibrium of the system, due to the set of nonlinear equations (e.g., mass or charge balance of each phase and the law of mass action of each of the independent equilibria).

Perrin [2] and Ingri et al. [3] proposed computational programs to calculate the chemical composition of a system with a limited number of components in thermodynamic equilibrium. Based on these, others have been proposed that apply different algorithms to solve the nonlinear set of equations involved in multicomponent and multi-reacting systems [2–8]. On the other hand, the algorithms based on Gibbs energy minimization of the system have been improved [9–12]. It has been demonstrated that the two types of algorithms are equivalent [13].

Graphic Representations of Systems in Thermodynamic Equilibrium

The study of multicomponent and multi-reacting systems involves many calculations, and it is common to give excessive importance to the mathematical problem, disregarding the chemical problem to be solved. For this reason, the application of several graphic representations and methods has been proposed; they help to consider a given set of thermodynamic equilibrium conditions with a better perspective.

The first type of graphic representation is that of distribution and logarithmic diagrams, representing species fractions (in linear or logarithmic form) as a function of composition variables of the system [14, 15]. The second class is that of the reaction prediction diagrams, and to

this class belong the Latimer [16, 17], Frost [18, 19], and Ellingham [20–22] diagrams. The third kind is that of the predominance-zone diagrams (PZD), and to this type belong Pourbaix [23] diagrams.

From this point to the end of the article, more information will be given for the PZD concerning algorithms, methods of construction, and applications.

Predominance-Zone Diagrams and their Construction Algorithms

PZD establish zones in a multidimensional space where generally one species has higher quantity or concentration with respect to others of a previously defined set (the condition of predominance). The mathematical problem is to find the boundaries from which this predominance is defined in that multidimensional space. The dimensions of the space are generally chemical composition variables, like pH, concentration or amount of species (in linear or logarithmic form), and electrode potential.

The Pourbaix diagrams (also known as potential-pH (Eh-pH) diagrams since this is the simplest representation proposed by Marcel Pourbaix) are the more famous PZD in electrochemistry. There are many papers in the scientific literature that describe their construction algorithms (Fig. 1).

The majority of these articles establish equations involving the Gibbs energy of the systems or the chemical potentials of species. Some of those papers have been compiled in recent works [24, 25] that furthermore describe construction for multicomponent and multi-reacting systems, where one of the axes of the diagram represents the potential of one electrode, while the others represent other chemical composition variables.

There are also few algorithms that use the law of mass action for independent equilibria instead of the chemical potentials of the species [26].

The Pourbaix diagrams are generally achieved as a final representation that is applied to describe or interpret several electrochemical problems;

Predominance-Zone Diagrams for Chemical Species, Fig. 1 Pourbaix diagram of americium species for $-\log[\text{Am}]_{\text{Total}} = 10.0$ [24]. *Dashed lines* represent the oxidation (*upper*) and reduction (*lower*) of water

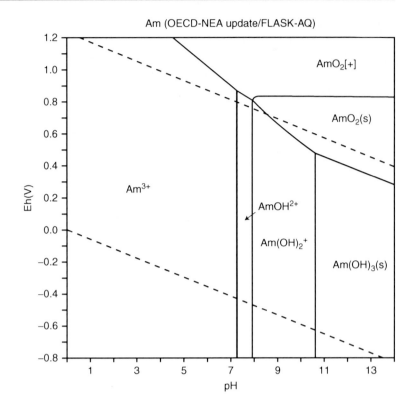

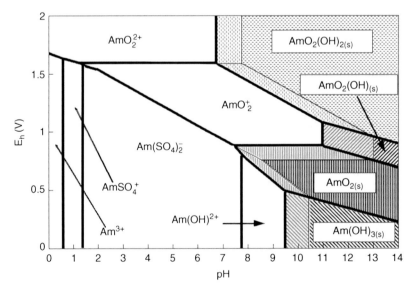

Predominance-Zone Diagrams for Chemical Species, Fig. 2 Pourbaix-type diagram of americium species at $-\log[\text{SO}_4'] = 0.523$ in the Eh/pH space, constructed with the MGSE. $\text{SO}_4' = \text{HSO}_4^- + \text{SO}_4^{2-}$. *Dashed black* zones represent the coexistence of condensed phases for a $-\log[\text{Am}]_{\text{Total}} = 8.0$, while *dashed black* and *green* zones represent the coexistence of condensed phases for a $-\log[\text{Am}]_{\text{Total}} = 6.0$ [29]. In this diagram, unlike that of Fig. 1, the formation of complexes of americium with sulfates in solution is considered

Predominance-Zone Diagrams for Chemical Species, Fig. 3 Different kinds of PZD for the Mg(II)-phosphates-ammonia-water system, constructed with the MGSE [29]. $PO_4'' = PO_4' = H_3PO_4 + H_2PO_4^- + HPO_4^{2-} + PO_4^{3-}$ and $NH_3' = NH_3 + NH_4^+$. (**a**) Predominance-zone diagram for Mg(II)-soluble species at $-\log[NH_3'] = -0.3$ and $-\log[PO_4''] = 1.0$ in the $-\log[Mg]_{Total}/pH$ space. (**b**) Condensed phase diagram (CPD) for Mg(II)-insoluble species at $-\log[NH_3'] = -0.3$ in the $-\log[PO_4'']_{Total}/pH$ space. (**c**) Predominance-existence diagram (PED) for all Mg(II) species at $-\log[NH_3'] = -0.3$ and $-\log[PO_4''] = 1.0$ in the $-\log[Mg]_{Total}/pH$ space

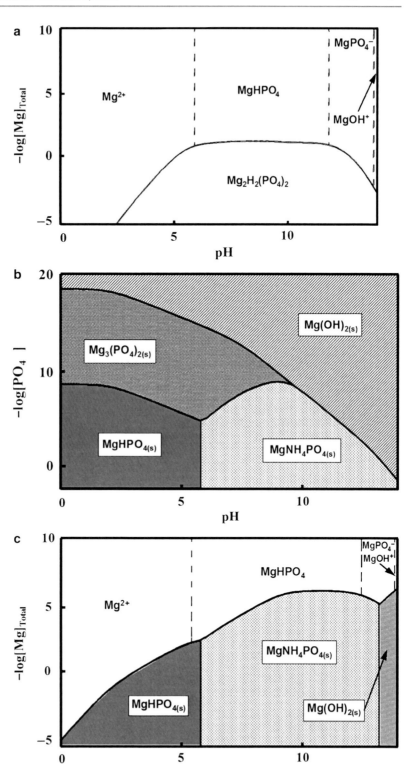

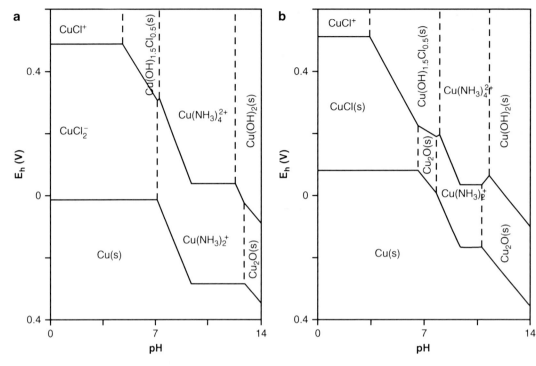

Predominance-Zone Diagrams for Chemical Species, Fig. 4 Pourbaix-type diagrams for copper species [49, 50] at $-\log[NH_3''] = -\log[Cl'] = 0$, constructed with the MGSE [26, 29]. $NH_3'' = NH_3' = NH_3 + NH_4^+$ and $Cl' = Cl^-$. (**a**) $-\log[Cu]_{Total} = 0.001$ M. (**b**) $-\log[Cu]_{Total} = 0.1$ M

nevertheless there are some papers that use them as an intermediate way of obtaining another graphic representation where there is no axis of electrode potential. One example of this is the paper of Osseo-Asare and Brown, used to describe some extraction processes for systems in hydrometallurgy [27].

It is true that there are many systems whose graphic representations do not require an electrode potential axis because the involved processes do not exchange electrons. It is thus unnecessary to use the Pourbaix diagram as an intermediate step because this type of diagram may be constructed directly, following construction algorithms completely analogous to those used to achieve the Pourbaix diagrams. Some examples of these algorithms can be found in the scientific literature [10, 28] (Fig. 2).

Remarkable inside the last kind of algorithms is the **method of generalized species and equilibria** (MGSE) [26, 29–36]. The MGSE extends the method of Gaston Charlot for the study of chemical equilibrium in solution [15, 37] through a definition of **conditional constant** (similar to a former concept proposed by Gerold Schwarzenbach [38] and developed further by Anders Ringbom [39, 40]). Its algorithms for PZD allows for the selection of the graphic representation spaces in terms of chemical composition variables of **generalized species** in order to relate the predominance boundaries in these spaces with **conditional constants of generalized equilibria**. The last feature gives the MGSE a great capability of prediction and interpretation of multicomponent and multi-reacting systems in thermodynamic equilibrium (Fig. 3).

Computational Methods for the Construction of Predominance-Zone Diagrams

There are many computational programs available nowadays to construct PZD. Some of these can be used online by Internet, covering the costs of services, like the Thermo-Calc [41] or

Predominance-Zone Diagrams for Chemical Species, Fig. 5 PZD for zinc-soluble species in the $-\log[NH_3']$/pH space, constructed with the MGSE [26, 29]. $NH_3' = NH_3 + NH_4^+$. (**a**) Without considering the ternary complexes $Zn(NH_3)_2(OH)_2$ and $Zn(NH_3)(OH)_3^-$ [52]. (**b**) Considering these ternary complexes [29], which model could better represent the reality?

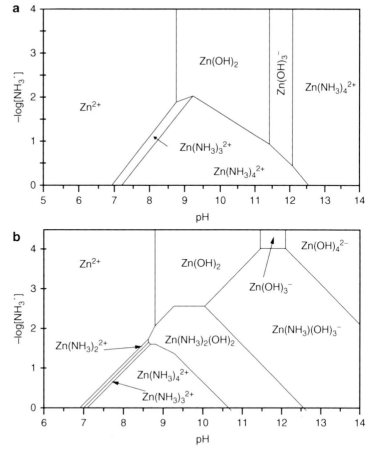

CSIRO Thermochemistry System [42]. Others, like STABCAL, have appeared in scientific literature references [43]. Nevertheless, possibly the more powerful computational set of programs available to construct PZD is the suite MEDUSA [44] that uses algorithms of the computational program SOLGASWATER [10].

Some Applications of Predominance-Zone Diagrams

One of the main applications of PZD is to evaluate the relative importance of species on systems in thermodynamic equilibrium for a set of given initial or imposed conditions. For this reason, some references that have applied PZD in different fields of electrochemistry are compiled in the final paragraphs of this essay.

Corrosion
Pourbaix diagrams have been proposed initially in order to explain corrosion phenomena, and for this reason the application of the PZD to study problems in this field may yet be found in electrochemical literature [45]. Nevertheless, other kinds of PZD have also been used to explain corrosion phenomena [46].

Electrocrystallization and Electrodeposition
There are many references that describe different applications of PZD to predict and interpret electrocrystallization and electrodeposition processes in multicomponent systems [47–65] (Fig. 4).

Pourbaix diagrams are used in many of these cases [47, 50, 63], but the application of other kinds of PZD is also common when the process requires it [47, 49, 52, 56, 60] (Fig. 5).

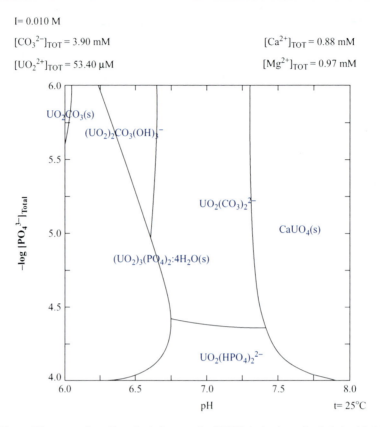

Predominance-Zone Diagrams for Chemical Species, Fig. 6 Predominance-zone diagram of uranyl species for a soil water in the high plains of Mexico, constructed with MEDUSA software [44] in the $-\log[PO_4^{3-}]_{Total}/pH$ space. The formation equilibrium constants of all considered species (45 soluble and 26 insoluble, from 6 soluble components) were taken from the HYDRA database (included with MEDUSA), only slightly modified for UO_2CO_3 and $(UO_2)_2CO_3(OH)_3^-$ species and substituting $UO_2(H_2PO_4)_2$ for $UO_2(HPO_4)_2^{2-}$ in the input data file [70]. The selected concentration conditions for the different species take into account the experimental measurements as well as the temperature (T) and the ionic strength (I)

Environmental Chemistry and Electrochemistry

PZD have a great application potential to develop studies related to environmental chemistry and electrochemistry. In the first place, we want to illustrate this with some studies for the research and remediation of Cr(VI) [66–68]. Some other cases of environmental studies that apply PZD are metal recovery from wastes [69], mobility of ionic and neutral species in several systems [70], chemical and biochemical leaching processes [71], and characterization of species in natural systems [72] (Fig. 6).

Coagulation and Flocculation

Coagulation and flocculation processes are used to separate dissolved and suspended solids from waters, mainly to purify the waters [73, 74]. Coagulants with opposite electrical charges to those of the suspended solids are added to the water to neutralize the negative charges on dispersed non-settleable solids such as clay and color-producing organic substances. Aluminum and iron ions are commonly used as inorganic coagulants [74]. These ions are directly added from sacrificial electrodes for electrocoagulation and electroflocculation. The PZD should be used

Predominance-Zone Diagrams for Chemical Species 1709

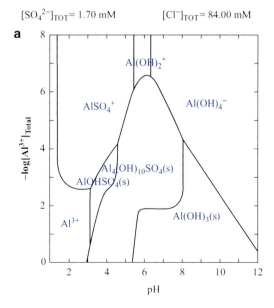
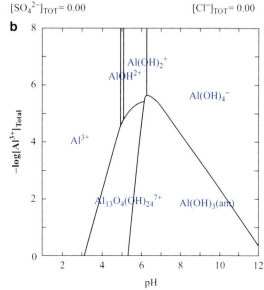

Predominance-Zone Diagrams for Chemical Species, Fig. 7 Predominance-zone diagrams of aluminum species constructed with MEDUSA software [44], in the $-\log[Al^{3+}]_{Total}$/pH space. The formation equilibrium constants of all considered species (13 soluble and 5 insoluble, from 4 soluble components) were taken from the HYDRA database (included with MEDUSA). (**a**) PZD obtained considering the mixed Al(III)-Sulfate-OH$^-$ complexes and the insoluble amorphous hydroxide, Al(OH)$_{3(am)}$, at specific conditions of total sulfate and chloride concentrations. It may be that the coexistence zones of mixed insoluble complexes are overestimated due to the use of formation equilibrium constants of crystalline phases. (**b**) PZD obtained in the absence of sulfates and chlorides

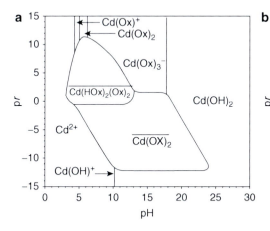
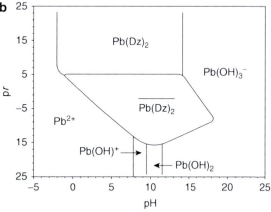

Predominance-Zone Diagrams for Chemical Species, Fig. 8 Predominance-zone diagrams in extraction (PZDE) [36], constructed with the MGSE [29], in the pr/pH space. The nonaqueous phase is trichloromethane. pr = $-\log r$ where r represents the nonaqueous volume/aqueous volume ratio, the bar over the species takes into account the species dissolved in nonaqueous phase, HOx represents oxine, and HDz represents dithizone. (**a**) PZD of Cd(II) species at a [oxine noncomplexed] = 0.1 M; this concentration considers the aqueous phase volume. (**b**) PZD of Pb(II) species at a [dithizone noncomplexed] = 0.1 M; this concentration considers the aqueous phase volume

Future Directions

to establish the optimal experimental conditions (i.e., concentration, pH) for the coagulation process. Soluble metallic hydroxyl complexes are required to separate suspended solids; meanwhile, insoluble metallic hydroxides are required for the separation of soluble chemical species (Fig. 7).

Future Directions

Pourbaix diagrams have been extensively used in corrosion studies, as well as for stability diagrams in geochemistry.

But the application of these and other PZD has practically not included other chemistry and electrochemistry fields (e.g., synthesis, separations, and characterizations), in spite of their great power to interpret results and predict operational conditions for several processes. Therefore, the application of PZD should be promoted as the powerful tools that they are.

Some fundamental developments should be undertaken to construct and interpret PZD in systems with a distribution of species between several phases (e.g., ionic exchange, adsorption, and spreading and growing of species on surfaces), as has been done for liquid-liquid extraction systems [36] (Fig. 8).

Finally, these kinds of representations should be developed to study chemical equilibria for chemical reactions taking place in soft matter (pastes, amorphous systems and polymers, among others).

Cross-References

- ▶ Activity Coefficients
- ▶ Anodic Reactions in Electrocatalysis - Oxidation of Carbon Monoxide
- ▶ Cell, Electrochemical
- ▶ Degradation of Organics, Use of Combined Electrochemical-Ultrasound
- ▶ Disinfection of Water, Electrochemical
- ▶ Electrolytes, Classification
- ▶ Electrolytes, Thermodynamics

- ▶ Environmentally Accepted Processes for Substitution and Reduction of Cr(VI)
- ▶ Metal Ion Removal by Cathodic Reduction
- ▶ Solvents and Solutions

References

1. Perrin DD, Sayce IG (1967) Computer calculation of equilibrium concentrations in mixtures of metal ions and complexing species. Talanta 14:833–842. doi:10.1016/0039-9140(67)80105-X
2. Perrin DD (1965) Multiple equilibria in assemblages of metal ions and complexing species: a model for biological systems. Nature 206:170–171. doi:10.1038/206170a0
3. Ingri N, Kakolowicz W, Sillén LG, Warnqvist B (1967) High-speed computers as a supplement to graphical methods – V: Haltafall, a general program for calculating the composition of equilibrium mixtures. Talanta 14:1261–1286. doi:10.1016/0039-9140(67)80203-0
4. Zeleznik FJ, Gordon S (1968) Calculation of complex equilibria. Ind Eng Chem 60:27–57. doi:10.1021/ie50702a006
5. Helgeson HC (1968) Evaluation of irreversible reactions in geochemical processes involving minerals and aqueous solutions – I. Thermodynamic relations. Geochim Cosmochim Acta 32:853–877. doi:10.1016/0016-7037(68)90100-2
6. Ting-Po I, Nancollas GH (1972) EQUIL. General computational method for the calculation of solution equilibriums. Anal Chem 44:1940–1950. doi:10.1021/ac60320a007
7. Dunsmore HS, Midgley D (1972) Computer calculatiodoi:10.1016/0039-9140(88)80166-8n of the composition of equilibrium mixtures in solution. Anal Chim Acta 72:121–126. doi:10.1016/S0003-2670(01)82955-X
8. Leung VW-H, Darvell VW, Chan AP-C (1988) A rapid algorithm for solution of the equations of multiple equilibrium systems – RAMESES. Talanta 35:713–718. doi:10.1016/0039-9140(88)80166-8
9. Karpov IK, Chudnenko KV, Kulik DA (1997) Modeling chemical mass transfer in geochemical processes: thermodynamic relations, conditions of equilibria, and numerical algorithms. Am J Sci 297:767–806. doi:10.2475/ajs.297.8.767
10. Eriksson G (1979) An algorithm for the computation of aqueous multi-component, multiphase equilibria. Anal Chim Acta 112:375–383. doi:10.1016/S0003-2670(01)85035-2
11. Gautam R, Seider WD (1979) Computation of phase and chemical equilibrium: Part I. Local and constrained minima in Gibbs free energy. AICHE J 25:991–999. doi:10.1002/aic.690250610

12. Smith WR (1980) The computation of chemical equilibria in complex systems. Ind Eng Chem Fundam 19:1–10. doi:10.1021/i160073a001
13. Smith WR, Missen RW (1982) Chemical reaction equilibrium analysis: theory and algorithms. Wiley, New York
14. Högfeldt E (1983) Graphic presentation of equilibrium data. In: Kolthoff IM, Elving PJ (eds) Treatise on analytical chemistry, Part I, Vol 2, Sect D, 2nd edn. Wiley, New York
15. Rojas-Hernández A, Ramírez MT, González I, Ibanez JG (1995) Predominance zone diagrams in solution chemistry: dismutation processes in two component systems (M-L). J Chem Educ 72:1099–1105. doi:10.1021/ed072p1099
16. Latimer WM (1952) The oxidation state of the elements and their potentials in aqueous solutions. Prentice-Hall, New York
17. Freiser H (1994) Enhanced Latimer potential diagrams via spreadsheets. J Chem Educ 71:786–788. doi:10.1021/ed071p786
18. Frost AA (1951) Oxidation potential-free energy diagrams. J Am Chem Soc 73:2680–2682. doi:10.1021/ja01150a074
19. Martínez-de-Ilarduya JM, Villafañe F (1994) A warning for Frost diagrams users. J Chem Educ 71:480–482. doi:10.1021/ed071p480
20. Ellingham HJT (1944) Reducibility of oxides and sulfides in metallurgical processes. J Soc Chem Ind 63:125–130
21. Gaskell DR (2001) Metal production: Ellingham diagrams. In: Buschow KHJ, Cahn R, Flemings MC, Ilschner B, Kramer EJ, Mahajan S, Veyssiere P (eds) Encyclopedia of materials: science and technology, vol 10. Elsevier, Oxford, pp 5481–5486. ISBN:0-08-0431526
22. Kishimoto H, Yamaji K, Brito ME, Horita T, Yokokawa H (2008) Generalized Ellingham diagrams for utilization in solid oxide fuel cells. J Min Metall Sect B 44B:39–48. doi:10.2298/JMMB0801039K
23. Pourbaix M (1966) Atlas of electrochemical equilibria in aqueous solutions (English edition). Pergamon, Oxford
24. Takeno N (2005) Atlas of Eh-pH diagrams: intercomparison of thermodynamic databases. Open File Report No.419. AIST Geological Survey of Japan, Tokyo. http://www.gsj.jp/GDB/openfile/files/no0419/openfile419e.pdf. Accessed 1 Jan 2012
25. Winston R (ed) (2000) Uhlig's corrosion handbook. 2nd edn. Chaps 6 and 7. Wiley, New York
26. Rojas-Hernández A, Ramírez MT, Ibáñez JG, González I (1991) Construction of multicomponent Pourbaix diagrams using generalized species. J Electrochem Soc 138:365–371. doi:10.1149/1.2085590
27. Osseo-Asare K, Brown TH (1979) A numerical method for computing hydrometallurgical activity-activity diagrams. Hydrometallurgy 4:217–232. doi:10.1016/0304-386X(79)90013-6
28. Garrels RM, Christ CL (1965) Solutions, minerals and equilibria. Freeman, San Francisco
29. Rojas-Hernández A (1996) The method of generalized species and equilibria for the study of chemical systems in equilibria with buffering conditions: theory and algorithms of predominance-zones diagrams (Spanish). PhD thesis. Universidad Autónoma Metropolitana, Unidad Iztapalapa, Mexico
30. Rojas A, González I (1986) Relationship of two-dimensional predominance-zone diagrams with conditional constants for complexation equilibria. Anal Chim Acta 187:279–285. doi:10.1016/S0003-2670(00)82919-0
31. Rojas-Hernández A, Ramírez MT, González I, Ibanez JG (1991) Relationship of multidimensional predominance-zone diagrams with multiconditional constants for complexation equilibria. Anal Chim Acta 246:435–442. doi:10.1016/S0003-2670(00)80983-6
32. Rojas-Hernández A, Ramírez MT, González I (1992) Multi-dimensional predominance-zone diagrams for polynuclear chemical species. Anal Chim Acta 259:95–104. doi:10.1016/0003-2670(92)85080-P
33. Rojas-Hernández A, Ramírez MT, González I (1993) Equilibria among condensed phases and a multi-component solution using the concept of generalized species. Part I. Systems with mixed complexes. Anal Chim Acta 278:321–333. doi:10.1016/0003-2670(93)85116-2
34. Rojas-Hernández A, Ramírez MT, González I (1993) Equilibria among condensed phases and a multi-component solution using the concept of generalized species. Part II. Systems with polynuclear species. Anal Chim Acta 278:335–347. doi:10.1016/0003-2670(93)85117-3
35. Rojas-Hernández A, Ramírez MT, González I (1996) Distribution of mononuclear chemical species in two-phase multicomponent systems using generalized species and equilibria. Quim Anal 15(Suppl.1):S4–S8. http://www.seqa.es/site/index.php?option=com_content&task=view&id=113&Itemid=112. Accessed 1 Jan 2012
36. Páez-Hernández ME, Ramírez MT, Rojas-Hernández A (2000) Predominance-zone diagrams and their application to solvent extraction techniques. Talanta 51:107–121. doi:10.1016/S0039-9140(99)00276-3
37. Charlot G (1969) Traité de chimie analytique générale, vol 1. Masson, Paris
38. Schwarzenbach G (1957) Complexometric titrations. Methuen, London
39. Ringbom A (1958) The analyst and the inconstant constants. J Chem Educ 35:282–288. doi:10.1021/ed035p282
40. Ringbom A (1963) Complexation in analytical chemistry. Wiley-Interscience, New York
41. http://www.calphad.com/thermo_calc.html. Accessed 1 Jan 2012
42. Zhang L, Sun S, Jahanshahi S, Chen C, Bourke B, Wright S, Somerville M (2002) CSIRO's multiphase

43. Wang HH, Young CA (1996) Mass-balanced calculations of Eh-pH diagrams using STABCAL. In: Woods R, Richardson P, Doyle FM (eds) Electrochemistry in minerals and metal processing IV, vol 96-6. The Electrochemical Society, Pennington

44. Puigdomenech I (2011) Making Equilibrium diagrams using sophisticated algorithms (MEDUSA). KTH Royal Institute of Technology. Department of Chemistry. http://www.kemi.kth.se/medusa/ Accessed 1 Jan 2012

45. Mouanga M, Ricq L, Douglade J, Berçot P (2007) Effects of some additives on the corrosion behaviour and preferred orientations of zinc obtained by continuous current deposition. J Appl Electrochem 37:283–289. doi:10.1007/s10800-006-9255-3

46. Arzola S, Palomar-Pardave ME, Genesca J (2003) Effect of resistivity on the corrosion mechanism of mild steel in sodium sulfate solutions. J Appl Electrochem 33:1233–1237. doi:10.1023/B:JACH.0000003855.95788.12

47. Becerril-Vilchis A, Meas Y, Rojas-Hernández A (1994) Electrodeposition of americium and physicochemical behavior of the solution. Radiochim Acta 64:99–105

48. Soto AB, Arce EM, Palomar-Pardavé M, González I (1996) Electrochemical nucleation of cobalt onto glassy carbon electrode from ammonium chloride solutions. Electrochim Acta 41:2647–2655. doi:10.1016/0013-4686(96)00088-6

49. Nila C, González I (1996) Thermodynamics of Cu-H2SO4-Cl–H2O and Cu-NH4Cl-H2O based on predominance-existence diagrams and Pourbaix-type diagrams. Hydrometallurgy 42:63–82. doi:10.1016/0304-386X(95)00073-P

50. Nila C, González I (1996) The role of pH and Cu(II) concentration in the electrodeposition of Cu(II) in NH$_4$Cl solutions. J Electroanal Chem 401:171–182. doi:10.1016/0022-0728(95)04278-4

51. Trejo G, Ortega R, Meas Y, Ozil P, Chainet E, Nguyen B (1998) Nucleation and growth of zinc from chloride concentrated solutions. J Electrochem Soc 145:4090–4097. doi:10.1149/1.1838919

52. Rodríguez-Torres I, Valentin G, Lapicque F (1999) Electrodeposition of zinc-nickel alloys from ammonia-containing baths. J Appl Electrochem 29:1035–1044. doi:10.1023/A:1003610617785

53. Ibrahim MAM (2000) Copper electrodeposition from non-polluting aqueous ammonia baths. Plat Surf Finish 87:67–72

54. Rodríguez-Torres I, Valentin G, Chanel S, Lapicque F (2000) Recovery of zinc and nickel from electrogalvanisation sludges using glycine solutions. Electrochim Acta 46:279–287. doi:10.1016/S0013-4686(00)00583-1

55. Trejo G, Ruiz H, Borges RO, Meas Y (2001) Influence of polyethoxylated additives on zinc electrodeposition from acidic solutions. J Appl Electrochem 31:685–692. doi:10.1023/A:1017580025961

56. Mendoza-Huizar LH, Robles J, Palomar-Pardavé M (2002) Nucleation and growth of cobalt onto different substrates Part I. Underpotential deposition onto a gold electrode. J Electroanal Chem 521:95–106. doi:10.1016/S0022-0728(02)00659-9

57. Díaz-Arista P, Antaño-López R, Meas Y, Ortega R, Chainet E, Ozil P, Trejo G (2005) EQCM study of the electrodeposition of manganese in the presence of ammonium thiocyanate in chloride-based acidic solutions. Electrochim Acta 51:4393–4404. doi:10.1016/j.electacta.2005.12.019

58. Ortiz-Aparicio JL, Meas Y, Trejo G, Ortega R, Chapman TW, Chainet E, Ozil P (2007) Electrodeposition of zinc-cobalt alloy from a complexing alkaline glycinate bath. Electrochim Acta 52:4742–4751. doi:10.1016/j.electacta.2007.01.010

59. Ballesteros JC, Diaz-Arista P, Meas Y, Ortega R, Trejo G (2007) Zinc electrodeposition in the presence of polyethylene glycol 20000. Electrochim Acta 52:3686–3696. doi:10.1016/j.electacta.2006.10.042

60. Alonso Alejandro R, Lapidus Gretchen T, González I (2007) A strategy to determine the potential interval for selective silver electrodeposition from ammoniacal thiosulfate solutions. Hydrometallurgy 85:144–153. doi:10.1016/j.hydromet.2006.08.009

61. Ortiz-Aparicio JL, Meas Y, Trejo G, Ortega R, Chapman TW, Chainet E, Ozil P (2008) Effect of quaternary ammonium compounds on the electrodeposition of ZnCo alloys from alkaline gluconate baths. J Electrochem Soc 155:D167–D175. doi:10.1149/1.2823000

62. Poisot-Diaz ME, González I, Lapidus GT (2008) Electrodeposition of a silver-gold alloy (DORE) from thiourea solutions in the presence of other metallic ion impurities. Hydrometallurgy 93:23–29. doi:10.1016/j.hydromet.2008.02.015

63. Alonso Alejandro R, Lapidus GT, González I (2008) Selective silver electroseparation from ammoniacal thiosulfate leaching solutions using a rotating cylinder electrode reactor (RCE). Hydrometallurgy 92:115–123. doi:10.1016/j.hydromet.2008.02.001

64. Rios-Reyes CH, Granados-Neri M, Mendoza-Huizar LH (2009) Kinetic study of the cobalt electrodeposition onto glassy carbon electrode from ammonium sulfate solutions. Quim Nova 32:2382–2386. doi:10.1590/S0100-40422009000900028

65. Lin ZB, Xie BG, Chen JS, Sun JJ, Chen GN (2009) Nucleation mechanism of silver during electrodeposition on a glassy carbon electrode from a cyanide-free bath with 2-hydroxypyridine as a complexing agent. J Electroanal Chem 633:207–211. doi:10.1016/j.jelechem.2009.05.015

66. Barrera-Díaz C, Palomar-Pardavé M, Romero-Romo M, Martínez S (2001) A comparison between chemical and electrochemical methods for the reduction of hexavalent chromium in aqueous solution. Proceedings of the electrochemical society 2001. 23(Energy and electrochemical processes for a cleaner environment):486–495

67. Barrera-Díaz C, Palomar-Pardavé M, Romero-Romo M, Martínez S (2003) Chemical and electrochemical considerations on the removal process of hexavalent chromium from aqueous media. J Appl Electrochem 33:61–71. doi:10.1023/A:1022983919644
68. Lugo-Lugo V, Barrera-Díaz C, Bilyeu B, Balderas-Hernández P, Ureña-Nuñez F, Sánchez-Mendieta V (2010) Cr(VI) reduction in wastewater using a bimetallic galvanic reactor. J Hazard Mater 176:418–425. doi:10.1016/j.jhazmat.2009.11.046
69. Rodríguez-Torres I, Valentin G, Lapicque F (1999) Recovery of zinc and nickel species from electro-galvanization sludges. Hung J Ind Chem 1-(Conference Proceedings, Novel Chemical Reaction Engineering for Cleaner Technologies):62–67. Accessed 1 Jan 2012
70. Guzmán ETR, Regil EO, Gutiérrez LRR, Alberich MVE, Hernández AR, Regil EO (2006) Contamination of corn growing areas due to intensive fertilization in the high plane of Mexico. Water Air Soil Poll 175:77–98. doi:10.1007/s11270-006-9114-1
71. Núñez-López RA, Meas Y, Gama SC, Borges RO, Olguín EJ (2008) Leaching of lead by ammonium salts and EDTA from *Salvinia minima* biomass produced during aquatic phytoremediation. J Hazard Mater 154:623–632. doi:10.1016/j.jhazmat.2007.10.101
72. Bautista-Flores AN, Rodríguez de San Miguel E, de Gyves J, Jönsson JÅ (2010) Optimization, evaluation, and characterization of a hollow fiber supported liquid membrane for sampling and speciation of lead(II) from aqueous solutions. J Membrane Sci 363:180–187. doi:10.1016/j.memsci.2010.07.028
73. Bratby J (2006) Coagulation and flocculation in water and wastewater treatment. IWA Publishing, London. Accessed 1 Jan 2012
74. http://iwawaterwiki.org/xwiki/bin/view/Articles/CoagulationandFlocculationinWaterand Wastewater Treatment. Accessed 1 Jan 2012

Primary Batteries for Medical Applications

Prabhakar A. Tamirisa, Kaimin Chen and Gaurav Jain
MECC, Medtronic Inc., Minneapolis, MN, USA

Introduction

Primary or non-rechargeable batteries are used to power several major types of implanted medical devices such as pacemakers, implanted cardioverter defibrillators (ICDs), cardiac and hemodynamic monitors, cardiac resynchronization therapy (CRT) devices, neurostimulators (electrical stimulation in regions such as spinal cord, deep brain, sacral nerve, vagal nerve, cochlea, gastric nerve, phrenic nerve), and drug pumps. Examples of non-implanted medical devices that can be powered by primary batteries include hearing aids, external defibrillators, infusion pumps, transcutaneous electrical nerve stimulators (TENs), and monitors such as Holter monitors, pulse oximeters, heart rate monitors, blood pressure monitors, and blood glucose sensors. Batteries for implanted medical applications are subject to regulations of national agencies such as FDA in the United States.

Energy and power density (ratio of energy and power deliverable by a battery to its volume) of batteries determine their suitability for most electrical devices. Other important criteria for use in medical devices are safety and reliability in operation during lifetime of the device (up to 10 years or longer), predictability of battery performance characteristics, and provision of device replacement indicators based on battery depletion and mechanical design features such as shape and size to allow efficient packaging in the device and for patient comfort.

High energy density is an important feature for primary batteries since this minimizes battery size and maximizes service life. Consequently, lithium metal anode-based batteries, which possess high energy density due to the high electrochemical capacity (3,860 mAh/g) and oxidation potential (3.04 V vs. SHE) of lithium, constitute the vast majority of primary batteries used in medical applications, especially implanted medical devices. Non-implanted medical devices use commercially available batteries, including non-lithium-based primary batteries. A prominent example is the use of Zn/air batteries in hearing aids. Primary batteries used in implanted medical devices are often classified in terms of their power capabilities as low (10–100 μW), medium (100 μW–10 mW), and high-rate batteries (up to 10 W). Power capability of a battery is a function of the electrochemical characteristics of electrode

Primary Batteries for Medical Applications, Fig. 1 Ragone plots of medium (MR) and high-rate (HR) hybrid chemistries compared to SVO and I_2-based battery systems; ERI = elective replacement indicator, DSVO = decomposition SVO, CSVO = combination SVO (Reprinted from Ref. [8], Copyright (2006), with permission from Elsevier)

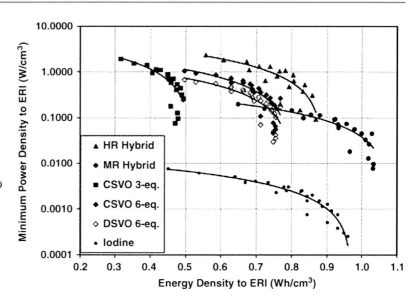

and electrolyte materials as well as mechanical design such as electrode and separator design. In the past decades, several battery chemistries and mechanical designs have been developed to meet the often stringent criteria for medical devices. This article will highlight the characteristics of these chemistries and mechanical designs.

Evolution of Primary Battery Chemistries in Implanted Medical Applications

The earliest primary battery used in implanted medical devices based on Zn/HgO [1] suffered from disadvantages such as hydrogen evolution during operation, sudden approach to depletion, and poor longevity due to high self-discharge. Alternative energy systems such as the rechargeable nickel-cadmium batteries and nuclear batteries were able to overcome the longevity issue but posed other challenges. In the case of rechargeable batteries, patients were faced with the burden of recharging the batteries, while nuclear batteries came with a regulatory burden to track the radioisotopes used in the batteries, especially after explant of the devices. Most of these challenges were overcome with the introduction of the lithium metal-based batteries in the early 1970s. The first lithium anode battery (Li/I_2)-based pacemaker was implanted in 1972 [2–5]; the battery did not evolve gases during its operation, thus allowing hermetic encapsulation inside a device. As peak power demands increased in the 1980s and 1990s, due to the introduction of new medical devices such as neurostimulators, implanted drug pumps, and ICDs, new chemistries such as $Li/SOCl_2$, Li/MnO_2, and Li/SVO were developed to meet the demand [3, 6, 7]. With the introduction of devices with greater microprocessor power, memory, and communication systems, complex therapy algorithms and long-distance telemetry were introduced in medical devices. In order to maintain peak power characteristics of the Li/SVO system, but improve energy density to minimize device volume, hybrid chemistries based on CF_x-SVO cathode systems have been introduced recently [8]. Figure 1 showcases the improvements that are achieved with the hybrid chemistry in comparison to batteries with a non-hybrid cathode chemistry such as Li/SVO or Li/I_2.

Lithium Primary Battery Characteristics

Chemical Characteristics

Lithium batteries can be classified on the basis of the chemical identity of electrochemical couples and separators. Li/I_2 battery is based on a cathode

Primary Batteries for Medical Applications

consisting of a charge-transfer complex of poly (vinyl pyridine) and iodine; the electrolyte and separator, LiI is formed in-situ upon contact between lithium metal and the cathode mixture, $Li/SOCl_2$ battery is based on a liquid cathode (containing a lithium salt for ionic conduction) coupled with a porous carbon support to deposit the solid discharge reaction products. Other commonly used battery chemistries consist of solid cathodes in combination with lithium anode and liquid electrolyte; anode and cathode are separated using a porous insulator. Lithium batteries for implanted medical devices have a well-defined operating temperature: the body temperature at 37 °C.

I. Li/I_2

Li/I_2 battery, first introduced as a power source for pacemakers in 1972, is a solid electrolyte battery with I_2 as the cathode [2]. To enhance cathode conductivity, I_2 is mixed with poly(vinyl pyridine) (PVP) to create an electronically conductive charge transfer complex. A thin layer of PVP is also applied on the lithium anode to improve cell resistance. One unique characteristic of Li/I_2 is that the discharge product, LiI, serves as a separator as well as the electrolyte. When I_2 is in contact with lithium during the construction of the battery, a layer of LiI is formed and serves as the separator. Being an ionic conductor, LiI also serves as the electrolyte. However, as discharge progresses, more LiI is formed and eventually leads to a large cell resistance increase and rapid decrease in cell operating voltage. Li/I_2 has an excellent volumetric energy density at 1 Wh/cc and has been a reliable power source for low-power applications such as implantable pacemakers for the past 40 years.

II. $Li/SOCl_2$

$Li/SOCl_2$ uses liquid $SOCl_2$ as the cathode; $LiAlCl_4$ which is used as the electrolyte salt is dissolved in the cathode. A porous carbon electrode serves as the cathode current collector. As discharge proceeds, nonconductive LiCl and S deposit inside the porous carbon electrode resulting in a significant increase in cell resistance and rapid voltage drop near the end of service. $Li/SOCl_2$ batteries have been used to power various implantable neurological devices such as deep brain stimulators (DBS) for reducing symptoms associated with Parkinson's disease, sacral nerve stimulators for bladder control issues, spinal cord stimulators for chronic back and leg pain, and gastric stimulators for relieving chronic nausea and vomiting symptoms associated with gastroparesis and for treating obesity.

III. Li/MnO_2

Li/MnO_2 uses heat-treated MnO_2 as the cathode and $LiClO_4$ in a mixed solvent system of propylene carbonate (PC) and dimethoxyethane (DME) as the electrolyte. The applicable capacity of MnO_2 corresponds to 1 electron equivalent per MnO_2. Li/MnO_2 was developed as a power source in cardio-myostimulators and has also been developed for use in pacemakers and ICDs [4].

IV. Li/SVO

Li/SVO was developed for powering implantable cardioverter defibrillators (ICDs) in 1980s [3, 7]. The majority of today's ICDs are powered by Li/SVO. The cathode in Li/SVO is a silver vanadium oxide with a stoichiometry of $Ag_2V_4O_{11}$ and the electrolyte is typically 1.0 M $LiAsF_6$ in PC/DME. The applicable capacity of SVO corresponds to six equivalent electrons per $Ag_2V_4O_{11}$. Two methods have been used to synthesize SVO: decomposition method ($2AgNO_3 + 2V_2O_5 = Ag_2V_4O_{11} + 2 NO_2 + ½ O_2$) and combination method ($Ag_2O + 2V_2O_5 = Ag_2V_4O_{11}$) [9, 10]. The combination reaction produces a crystalline and phase pure SVO, which provides stable power capability in the second portion of the discharge curve over time [11]. Li/SVO has also been adopted to power implantable neurological devices such as deep brain, sacral nerve, and spinal cord stimulators.

Primary Batteries for Medical Applications, Table 1 Summary of primary lithium battery characteristics and their applications in implanted medical devices

Battery chemistry	Overall cell reaction	Operating voltage range	Common electrolyte	Common implantable applications
Li/I_2	$Li + I_2 = LiI$	2.8–2.2 V	LiI (the product)	Pacemakers
$Li/SOCl_2$	$4Li + 2SOCl_2 = 4LiCl + S + SO_2$	3.6–3.0 V	$SOCl_2/LiAlCl_4$	Neurostimulators
Li/MnO_2	$Li + MnO_2 = LiMnO_2$	3.0–2.2 V	$LiClO_4/PC/DME$	Pacemakers, ICDs
Li/SVO	$6 Li + Ag_2V_4O_{11} = 2Ag + Li_6V_4O_{11}$	3.2–2.2 V	$LiAsF_6/PC/DME$	ICDs, neurostimulators
Li/CF_x	$Li + CF = C + LiF$	3.0–2.2 V	$LiBF_4/GBL$ $LiBF_4/GBL/DME$	Pacemakers, drug pump, and neurostimulators
Li/CF_x-SVO	$Li + CF = C + LiF$ $6 Li + Ag_2V_4O_{11} = 2Ag + Li_6V_4O_{11}$	3.2–2.2 V	$LiBF_4/GBL/DME$ $LiAsF_6/PC/DME$	Pacemakers, drug pump, ICDs, and neurostimulators

V. Li/CF_x

Li/CF_x uses CF_x (typically, $x \sim 1.0$) as the cathode and $LiBF_4$ in GBL-based solvents as the electrolyte. CF_x is obtained through fluorination of carbon at elevated temperatures with F_2 gas diluted with N_2. The morphology of CF_x often takes the shape of its carbon precursor. The deliverable capacity of CF_x is affected by the degree of fluorination [12]. The discharge products of CF_x are carbon and LiF. Since Li/CF_x batteries have a rapid decrease in voltage near the end of service, a second cathode component with a slightly lower operating voltage may be added to CF_x to provide a voltage-based end of service indicator [13]. Li/CF_x batteries have found uses in pacemakers, drug pumps, and neurostimulators such as cochlear implants.

VI. Li/CF_x-SVO

Li/CF_x-SVO uses both CF_x and SVO ($Ag_2V_4O_{11}$) as the cathode [8]. The cathode can be prepared as a mixture of CF_x and SVO or as discrete CF_x and SVO layers bonded to the current collector [13, 14]. The ratio of CF_x to SVO can be tailored to suit specific applications. Electrolytes used in Li/CF_x-SVO are typically $LiBF_4$ in GBL-based electrolytes or $LiAsF_6$ in PC/DME-based electrolyte. The choice of electrolyte depends on the power requirement of the applications. Li/CF_x-SVO combines the energy density advantage of CF_x and the power capability of SVO to produce a unique cathode that is suitable for implanted applications with a wide range of power requirements. Li/CF_x-SVO has been developed to power pacemakers, ICDs, implantable drug pump, neurostimulator for treating chronic back and leg pain, and deep brain stimulator (DBS) for reducing symptoms associated with Parkinson's disease (Table 1).

Mechanical Design

External shape and size constraints for batteries are typically dictated by the shape and the allocated internal volume of the medical device. Design of the internal components of batteries is guided by energy, power, and safety requirements. Energy and power densities of batteries are inversely correlated. Increased electrode area minimizes battery resistance and hence maximizes power density but at the expense of energy density since large electrode area is achieved through a larger fraction of inert materials in the battery such as current collectors, separators, spacers, and other insulators. Batteries for high-power applications such as ICDs use stacked or coiled designs wherein thin layers (100–300 μm) of electrode materials are applied to current collectors and either stacked or coiled to achieve

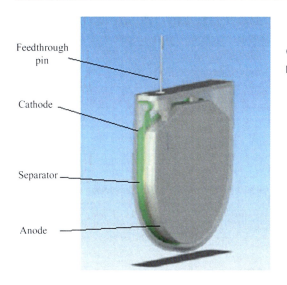

Primary Batteries for Medical Applications, Fig. 2 D-shaped monolithic battery with anode, separator, and cathode. Feedthrough pin connected to anode; case connected to cathode

high electrode area (40–100 cm^2) and low resistance. Conversely, energy density is maximized by increasing the fraction of active materials relative to the inerts, by lowering electrode area (2–15 cm^2) and increasing electrode thickness (up to 8 mm). Low/medium rate batteries use thicker plate like electrodes and smaller areas seen in monolithic or single-plate designs. In general coiled electrodes are enclosed in prismatic cases (Fig. 2), while monolithic electrodes are used most commonly in D-shaped cases (see Fig. 3). Stacked electrodes are attractive from the standpoint of being able to provide additional shape flexibility that is generally unavailable to coiled electrodes. Cylindrical and coin shaped batteries are typically not common in implanted medical devices since they do not pack efficiently in the devices. However, cylindrical miniature batteries have been developed recently for use in injectable medical devices (see Fig. 4) [15].

Unlike batteries used in commercial and non-implanted medical devices, implanted medical device batteries are hermetically sealed in welded metal cases with glass-to-metal feedthroughs based on lithium-resistant glasses such as TA-23

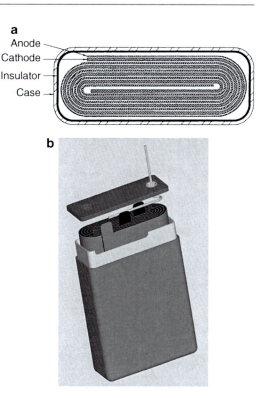

Primary Batteries for Medical Applications, Fig. 3 (a) Cross section view of a coiled electrode configuration used in high-rate designs; (b) assembly view of a coiled electrode configuration in insulator and battery case, showing attachment in the headspace (Reprinted from Ref. [7], Copyright (2001), with permission from Elsevier)

and CABAL-12 to make connections to electrodes. To guarantee reliable containment of the electrodes and electrolyte and functionality of the feedthrough pins, materials for metal cases, covers, and feedthrough pins are selected by screening for materials degradation through pathways such as corrosion and chemical attack. Battery enclosures are designed to be compatible with volume changes due to expansion of electrodes and gas generation during service; typically gas generation is minimized or eliminated by design. Furthermore, medical device batteries are designed to possess high safety margins, often with redundant safety features such as multiple layers of separators that can shutdown in the event of an internal short circuit.

Primary Batteries for Medical Applications, Fig. 4 Photographs of cylindrical, pin-shaped batteries: (**a**) lithium-ion rechargeable battery, QL0003I, and (**b**) the lithium/CFx primary battery, QC0025B (Reprinted from Ref. [15], Copyright (2006), with permission from Elsevier)

Modeling for Design and Performance

Model-based predictions of battery performance and longevity are required to provide depletion indicators for batteries such as RRT (recommended replacement time), typically based on discharge voltage or resistance, to allow patients to schedule a replacement surgery before batteries reach EOS (end of service). Electrical characteristics of primary batteries used in medical applications have been modeled in a variety of ways including empirically or thermodynamics-based voltage curve models [16, 17], equivalent circuit models [18, 19], resistance models [11, 20], and more recently physics-based comprehensive models [17, 21–23] that can be used to predict pulse performance [24], resistance characteristics, and, furthermore, medical device performance characteristics such as capacitor charge time in ICDs [25]. Statistical models developed by considering the distributions of the battery model parameters and device use conditions can be used to develop longevity predictions, which can be validated using real-time clinical/field performance data from the devices.

Design generation occurs early in the development cycle of batteries when energy/power requirements and shape/dimensional constraints are known and can also be based on models. Sizing of electrodes and consequently of the battery is guided by a balance of materials approach [6, 26], various design rules formulated for safety/reliability considerations (e.g., minimum values of separator thickness), and the intricacies of the chemistry such as incidence of parasitic reactions [26, 27] and expectations of energy/power density based on empirical performance data gathered from early prototypes.

Future Directions

Primary battery chemistries and mechanical designs for medical applications evolved to meet increasing power demands while attempting to maintain or exceed the high energy density achieved in the Li/I$_2$ battery. This trend is likely to continue due to interest in developing medical devices with long-range telemetry capabilities and newer therapy modalities that require higher currents. Simultaneously, the search for new cathode chemistries capable of higher energy density may be expected to continue. Platform technologies such as the CF$_x$-SVO hybrid cathodes [8], which achieve good performance over a much broader region of the energy-power density space in comparison to non-hybrid technologies, may become more common in the future resulting in reduced development cycle times and increased performance reliability. Depending on the use conditions of the devices, other types of hybrid technologies such as combinations of primary and rechargeable batteries may be developed. In the case of implanted medical devices, materials selection for battery components may be guided by the need to make the devices MRI (magnetic resonance imaging) safe [28, 29].

Development of ultralow power electronics and advanced packaging technologies such as wafer-level packaging has begun to enable creation of leadless miniature devices that can be implanted near the site of therapy via minimally invasive procedures such as injection through a needle or a catheter [30, 31]. Miniature batteries with high energy densities and yet adequate power capabilities are required for such applications in order to achieve parity with the traditional implanted devices, especially in longevity and features that require high peak currents such as telemetry. Miniature batteries are being developed in new shapes, such as pin style or bobbin style [15], to enable integration with miniature, leadless devices. Thin-film solid-state batteries are attractive from the standpoint of integrating a minimally packaged battery into the device at the wafer or board level [32]. However, current designs of thin-film batteries which are produced using thin electrode and electrolyte layers (<5 μm) to minimize internal resistances are not sufficiently energy dense for many typical medical devices [33], which has prompted interest in three-dimensional battery electrode architectures to enhance energy and power density in solid-state and miniature batteries [34–37]. Since miniaturization results in increased inert material relative to the electrochemically active materials, energy and power density values are smaller in comparison to traditional medical device batteries. Therefore, innovations in packaging, mechanical design of electrodes, and development of new chemistries (electrodes and electrolyte materials) are required to achieve miniature batteries that exceed the current standard of 1 Wh/cc.

Use of predictive modeling can be expected to play an increasing role in the design and development phases of batteries for medical applications. Development of accelerated testing protocols and use of accelerated degradation models may become more prevalent to ensure highest levels of reliability in design and operation of batteries.

Cross-References

▶ Primary Batteries, Comparative Performance Characteristics
▶ Primary Batteries, Selection and Application

References

1. Ruben S (1947) Alkaline dry cell. US Patent 2422045
2. Holmes CF (2007) The lithium/iodine-polyvinylpyridine battery – 35 years of successful clinical use. ECS Trans 6(5):1–7. doi:10.1149/1.2790382
3. Holmes CF (2003) Electrochemical power sources and the treatment of human illness. Interface, vol 12. The Electrochemical Society, Pennington, NJ, USA
4. Root MJ (2008) Implantable cardiac rhythm device batteries. J Cardiovasc Transl Res 1(4):254–257. doi:10.1007/s12265-008-9054-9
5. Mallela VS, Ilankumaran V, Rao NS (2004) Trends in cardiac pacemaker batteries. Indian Pack Electrophysiol J 4(4):201–212
6. Skarstad PM (2004) Battery and capacitor technology for uniform charge time in implantable cardioverter-defibrillators. J Power Sources 136(2):263–267. doi:10.1016/j.jpowsour.2004.03.010
7. Crespi AM, Somdahl SK, Schmidt CL, Skarstad PM (2001) Evolution of power sources for implantable cardioverter defibrillators. J Power Sources 96:33–38
8. Chen K, Merritt DR, Howard WG, Schmidt CL, Skarstad PM (2006) Hybrid cathode lithium batteries for implantable medical applications. J Power Sources 162(2):837–840. doi:10.1016/j.jpowsour.2005.07.018
9. Liang CC, Bolster ME, Murphy RM (1982) Metal oxide composite cathode material for high energy density batteries. US Patent 4310609
10. Crespi AM (1993) Silver vanadium oxide cathode material and method of preparation. US Patent 5221453
11. Crespi A, Schmidt C, Norton J, Chen K, Skarstad P (2001) Modeling and characterization of the resistance of lithium/SVO batteries for implantable cardioverter defibrillators. J Electrochem Soc 148(1):A30–A37
12. Nakajima T, Groult H (2005) Fluorinated materials for energy conversion. Elsevier Science, p 592
13. Weiss DJ, Cretzmeyer JW, Crespi AM, Howard WG, Skarstad PM (1993) Electrochemical cells with end-of-service indicator. US Patent 5180642
14. Gan H, Takeuchi ES (2005) SVO/CFx parallel cell design within the same casing. US Patent 6926991
15. Nagata M, Saraswat A, Nakahara H, Yumoto H, Skinlo DM, Takeya K, Tsukamoto H (2005) Miniature pin-type lithium batteries for medical

applications. J Power Sources 146(1–2):762–765. doi:10.1016/j.jpowsour.2005.03.156

16. Root MJ (2010) Lithium–manganese dioxide cells for implantable defibrillator devices—discharge voltage models. J Power Sources 195(15):5089–5093. doi:10.1016/j.jpowsour.2009.12.083

17. Strange DA, Rayman S, Shaffer JS, White RE (2011) Physics-based lithium ion silver vanadium oxide cathode model. J Power Sources 196(22):9708–9718. doi:10.1016/j.jpowsour.2011.07.057

18. Schmidt CL, Skarstad PM (1997) Development of an equivalent-circuit model for the lithium/iodine battery. J Power Sources 65:121–128

19. Skarstad PM, Schmidt CL (1993) Modeling the discharge behavior of the lithium/iodine battery. J Power Sources 43–44:111–118

20. Root MJ (2011) Resistance model for lithium-silver vanadium oxide cells. J Electrochem Soc 158(12): A1347. doi:10.1149/2.049112jes

21. Davis S, Takeuchi ES, Tiedemann W, Newman J (2007) Simulation of the Li–CF[sub x] system. J Electrochem Soc 154(5):A477. doi:10.1149/1.2714323

22. Gomadam PM, Merritt DR, Scott ER, Schmidt CL, Skarstad PM, Weidner JW (2007) Modeling lithium/hybrid-cathode batteries. J Power Sources 174(2):872–876. doi:10.1016/j.jpowsour.2007.06.164

23. Gomadam PM, Merritt DR, Scott ER, Schmidt CL, Skarstad PM, Weidner JW (2007) Modeling Li/CF [sub x]-SVO hybrid-cathode batteries. J Electrochem Soc 154(11):A1058–A1064. doi:10.1149/1.2779963

24. Davis S, Takeuchi ES, Tiedemann W, Newman J (2008) Simulation of pulse discharge of the Li-CFx system. J Electrochem Soc 155(1):A24–A28. doi:10.1149/1.2800168

25. Gomadam PM, Brown JR, Scott ER, Schmidt CL (2007) Predicting charge-times of implantable cardioverter defibrillators. ECS Trans 6(5):15–23

26. Crespi AM, Skarstad PM (1993) Parasitic reactions and the balance of materials in lithium batteries for implantable medical devices. J Power Sources 43–44:119–125

27. Untereker DF (1978) The use of a microcalorimeter for analysis of load-dependent processes occurring in a primary battery. J Electrochem Soc 125(12):1907–1912

28. Levine GN, Gomes AS, Arai AE, Bluemke DA, Flamm SD, Kanal E, Manning WJ, Martin ET, Smith JM, Wilke N, Shellock FS (2007) Safety of magnetic resonance imaging in patients with cardiovascular devices. Circulation 116(24):2878–2891. doi:10.1161/CIRCULATIONAHA.107.187256

29. Faris OP, Shein M (2006) Food and drug administration perspective: magnetic resonance imaging of pacemaker and implantable cardioverter-defibrillator patients. Circulation 114(12):1232–1233. doi:10.1161/CIRCULATIONAHA.106.647800

30. McLaughlin BL, Smith B, Lachapelle J, Traviglia D, Sriram TS, O'Dowd D (2011) Ultra-high density packaging technology for injectable medical devices. In: 2011 annual international conference of the IEEE engineering in medicine and biology society, EMBC, Boston. IEEE, pp 2897–2900. doi:10.1109/IEMBS.2011.6090798

31. Miniaturization: Smaller, more efficient pacemakers, Medtronic Inc. http://www.medtronic.com/innovation/smarter-miniaturization.html. Accessed Sep. 17, 2013

32. Chen G, Fojtik M, Kim D, Fick D, Park J, Seok M, Chen M-T, Foo Z, Sylvester D, Blaauw D (2010) Millimeter-scale nearly perpetual sensor system with stacked battery and solar cells. Paper presented at the 2010 IEEE international solid-state circuits conference digest of technical papers (ISSCC)

33. Patil A, Patil V, Wook Shin D, Choi J-W, Paik D-S, Yoon S-J (2008) Issue and challenges facing rechargeable thin film lithium batteries. Mater Res Bull 43(8–9):1913–1942. doi:10.1016/j.materresbull.2007.08.031

34. Nathan M, Golodnitsky D, Yufit V, Strauss E, Ripenbein T, Shechtman I, Menkin S, Peled E (2005) Three-dimensional thin-fim Li-ion microbatteries for autonomous MEMS. J Microelectromechan S 14(5):879–885

35. Long JW, Dunn B, Rolison DR, White HS (2004) Three-dimensional battery architectures. Chem Rev 104:4463–4492. doi:10.1021/cr020740l

36. Zhang H, Yu X, Braun PV (2011) Three-dimensional bicontinuous ultra-fast and-discharge bulk battery electrodes. Nat Nanotechnol 6:277–281. doi:10.1038/nnano.2011.38

37. Baggetto L, Oudenhoven JFM, van Dongen T, Klootwijk JH, Mulder M, Niessen RAH, de Croon MHJM, Notten PHL (2009) On the electrochemistry of an anode stack for all-solid-state 3D-integrated batteries. J Power Sources 189(1):402–410. doi:10.1016/j.jpowsour.2008.07.076

Primary Batteries for Military Applications

George Blomgren
Blomgren Consulting Services, Lakewood, OH, USA

Introduction

There are many battery types that are in use for military applications. One of the most common is

Primary Batteries for Military Applications

Primary Batteries for Military Applications, Table 1 Lithium sulfur dioxide batteries in military applications

Designation	Nominal voltage (V)	Energy (Wh)	Weight (g)	Applications
BA-5093/U	23.4	77.3	635	Respirators
BA-5557A/U	16/13	54	410	Digital message devices
BA-5588A/U	13	35	290	Radios, respirators
BA-5590A/U	26/13	185	1,021	SINCGARS radios, chemical detectors, satellite radios, jammers, loudspeakers, range finders, countermeasures
BA-5598A/U	13	87	650	PRC-77 radios, direction finders, sensors
BA-5599A/U	7.8	50	450	Test sets, sensors

the standard batteries of commerce often found in consumer applications. This is an example of "dual use" which requires designers of military devices to accommodate batteries already in commerce to save costs. Since there are no unique features to the batteries used in this way, they will not be further discussed. Three other types of batteries are also used in military applications. Active batteries are designed for special military applications, but do not have special activation requirements. The second type is reserve batteries that require a physical or chemical step for activation. This type assures the military of very long shelf life since the activation usually involves introducing the electrolyte to the dry cell. Thus, the electrodes are maintained in a dry state prior to activation, and corrosion of the active materials does not occur in the shelf period. The third type of battery for military applications is thermally activated batteries. Again, a long shelf life is guaranteed since the electrolyte is present as a solid until thermal activation. The high temperature required to melt the electrolyte also allows very high currents since the kinetics and mass transport are greatly reduced at high temperature. These three types of applications and the battery types used will be discussed in turn.

Active Batteries

The most prominent among active batteries for military applications are lithium sulfur dioxide batteries, lithium manganese dioxide batteries, and lithium thionyl chloride batteries.

Lithium sulfur dioxide batteries have taken over from magnesium batteries as the main power source for communications batteries. Most of the applications involve multicell battery packs where the individual cells are larger than typical consumer cells, often of the F or DD size and have about 3 Ah capacity and are designed in a spirally wound configuration. Often a dual voltage is required for the application for different functions, which is accomplished by different wiring for the cells with concomitant switching. A list of typical battery sizes and designations is given in Table 1.

Lithium manganese dioxide batteries have become very prominent in recent years for military applications which are now widespread. Like sulfur dioxide batteries these are usually found in multicell packs often with dual voltage and cell sizes. The voltage of the system is similar to sulfur dioxide, so no major device changes are required. Typical batteries with their applications are given in Table 2 [1].

As is clear from Tables 1 and 2, the BA-5590 and BA-5390 batteries are the most widely used and deployed. All of the designations are from the US Department of Defense.

Lithium thionyl chloride batteries are not as numerous in production, but the applications are very widespread. This is mainly because of the very high specific energy and energy density of this battery type. Many different designs and form factors are used in the cells. The most

Primary Batteries for Military Applications, Table 2 Military applications for lithium manganese dioxide batteries

Designation	Nominal voltage (V)	Energy (Wh)	Weight (g)	Applications
BA-5312/U	10.8	41	275	PRC-112G survival radios
BA-5347/U	5	40	290	Thermal weapon sights, test sets
BA-5360/U	8.1	65	320	Digital communication devices
BA-5367/U	2.7	3.25	20	Night vision devices
BA-5368/U	10.8	12	140	Memory hold function, decoding devices
BA-5372/U	5.4	2.3	20	Memory hold function, encoding devices
BA-5380/U	5.4	45	230	Ground navigation sets, chemical agent monitors, respirators
BA-5390/U	27/13.5	250	1,350	SINCGARS radios, chemical agent detectors, satellite radios, jammers, loudspeakers, range finders, countermeasures

common type is the bobbin type in which the outer electrode is a thin sheet of lithium pressed against the stainless steel can. The next layer is a separator against the lithium foil. Next is the bobbin of highly porous carbon bonded into a bobbin by polytetrafluoroethylene fibers and pressed into a metal collector which is either a cylinder (for larger cells) or a pin for smaller cells. This arrangement gives very high energy to the system (among the highest energy density and specific energy for any system), although the current capability is only moderate. For high current cells, the electrodes are spirally wound in a similar fashion to the Li/SO_2 or Li/MnO_2 cells. The energy of this type is not as high as for the bobbin construction. A third geometry is the disc type which resembles a coin-type consumer battery in construction. Finally, a fourth type of construction is the large box cell, which holds many stacked electrodes and can have capacities as high as 16,500 Ah. This latter type is no longer in use, but was the main power supply for intercontinental ballistic missile silos for many years in the USA.

Reserve Batteries

This type of battery is noted for very long shelf life, often greater than 20 years. In the construction, the electrolyte is physically separated from the active materials of the electrodes until the cell is ready for use and is stored in a reservoir prior to activation [2]. The battery type is generally separated by the mode of activation. One type is activated by a force used to break a glass ampoule containing the electrolyte solution or forcibly breaking a diaphragm. This force may be handled manually or electrically. Single cells are generally activated by opening the ampoule, while multiple cells are usually activated by breaking a diaphragm and forcing the electrolyte into the cells, often with a bellows arrangement. This type of reserve cell is usually has a lithium anode with either a solid cathode or liquid cathode. The second type of reserve cell is generally spin activated. The spin is provided by the rotation of a projectile. The centripetal force on the spinning cell is used to open the electrolyte container as well as rapidly force the electrolyte into the interstices of the cells to complete the circuit. The chemistry of the cells may involve zinc alkaline battery chemistry, lithium battery chemistry, or in older versions lead fluoroborate cell chemistry. The latter involves a chemistry similar to rechargeable lead acid batteries, but utilizes a different electrolyte, fluoboric acid, rather than sulfuric acid. This along with the electrode design allows very high current capability to the cells. Typical applications for reserve batteries are fuzing of various kinds of munitions and projectiles as well as missile applications and even self-destruct mechanisms. The proximity fuze of World War II was made possible by the development of the lead fluoboric acid cell.

Thermal Reserve Batteries

This novel battery type utilizes solid inorganic salt electrolytes in a design which are present in the preactivated state as a simple pellet between electrodes [3]. There are also interspersed between electrodes a heat pellet which can be activated in a multielectrode design by a heat tape or bridge wire. Sometimes the activation is provided by a mechanical striking device or by an inertial device similar to the spin-activated reserve cells above. Thus when activation is called for, the ignitor is activated which in turn ignites the heat pellet which then melts the salt electrolyte which flows into the electrodes to complete the circuit. Again, they have multiple chemistries and may be single-cell- or multiple-cell construction. The salt pellet is usually a eutectic mixture of alkali metal salts, usually chlorides. The anodes may have lithium (as an alloy) or calcium metal. The lithium cells are most common and usually used FeS_2 cathodes, while calcium cells usually use calcium chromate cathodes. Typical applications are in missiles, bombs (munitions of all types), emergency escape systems, etc. They have very long shelf life (up to 25 years) and very fast activation. The activated stand time after activation may be very short, however, as little as 5 min or as long as an hour or so.

Cross-References

▶ Primary Batteries, Comparative Performance Characteristics
▶ Primary Batteries, Selection and Application

References

1. Reddy T (2011) Lithium primary batteries. In: Reddy T, Linden D (eds) Linden's handbook of batteries, 4th edn. McGraw Hill, New York
2. Chua D, Meyer B, Epply W, Swank J, Ding M (2011) Reserve military batteries. In: Reddy T, Linden D (eds) Linden's handbook of batteries, 4th edn. McGraw Hill, New York
3. Lamb C (2011) Thermal batteries. In: Reddy T, Linden D (eds) Linden's handbook of batteries, 4th edn. McGraw Hill, New York

Primary Batteries, Comparative Performance Characteristics

George Blomgren
Blomgren Consulting Services,
Lakewood, OH, USA

Background

An electrochemical cell is a device which converts chemical energy into electrical energy. It has several requirements, namely, each chemically active material is confined to a single electrode; each electrode must have a means of conducting the electrical current generated to an external circuit, with one electrode acting as the positive pole and the other as the negative pole; the electrodes must be separated from each other by an electronically insulating but ionically conductive phase; and the entire assembly of electrodes, current leads, and insulating phase should be contained in a body which provides isolation from deleterious atmospheric effects (in general, air is harmful to metallic electrodes, and evaporation of liquid conductive phases will inactivate the cell). Such electrochemical cells are commonly called batteries, although strictly speaking, only an assemblage of cells should be called a battery or battery pack.

The electrochemical cells of interest here are primary batteries. They differ from rechargeable batteries in that they may only be used for one discharge and cannot be recharged (of course they may be partially discharged in a pulse mode may times until they are exhausted). They are of particular interest for applications that are infrequent or involve very small currents if they are continuous or frequent. They also must have adequate shelf life to fit the application, have reasonable energy content and sufficient power capability for the application, have a cost which is acceptable to the consumer, and have high reliability.

Primary batteries have a long history. The dominant cell type in use until World War II was the carbon-zinc cell also called the

Leclanche cell, with a zinc anode and a manganese dioxide cathode and a mildly acidic electrolyte. The chemistry dates back to the nineteenth century, but it was not fully developed for portable power until the dawn of the twentieth century. The war brought the alkaline battery onto the scene with a zinc anode, mercuric oxide cathode, and a concentrated alkaline electrolyte (KOH). In the 1950s the manganese oxide cathode was developed for the alkaline battery, and this system, also called the alkaline cell, began to take over from carbon-zinc cells in popularity among consumers because of its greater rate capability as well as its low leakage characteristics. The next major development in primary batteries was with lithium anode cells beginning in the 1970s. Li/CF_x cells, Li/MnO_2 cells, and Li/FeS_2 cells made their appearance at this time followed by liquid cathode cells starting with Li/SO_2 cells, then by $Li/SOCl_2$ and Li/SO_2Cl_2 cells. The development of lithium cells required the use of nonaqueous electrolytes. This development of lithium cells relied on methods of low moisture exposure and led to the use of dry rooms for development and manufacturing facilities. They also required a high level of cleanliness because of the sensitivity of the chemistry to small amounts of impurities. The major gain in energy density and specific energy were the main drivers for these advances. This gain was enabled by the very negative potential of the lithium anode giving rise to high-voltage cells and the concomitant increase in energy. Another feature of lithium cells was the need for tight sealing of the container to prevent the ingress of oxygen and moisture, but that was balanced by the low rate of gassing of the lithium anode in the presence of the specially chosen organic and inorganic solvents, in contrast to zinc anode cells in aqueous media. These details are discussed in depth in the articles describing the individual cells. Mention should be made that the use of liquid inorganic cathode electrolytes is only possible because of the thin, passivating layer of insoluble product which protects the lithium from aggressive reaction with the solvents. This type of protective layer is also developed on the lithium surface due to reaction with the organic solvents, such as propylene carbonate and dimethoxyethane, which are commonly used with the solid cathode materials.

Definitions

Convenient ways to compare primary batteries are the specific energy of the finished cell in watt hours per kilogram (Wh/kg), the energy density in watt hours per liter (Wh/l), and the voltage of the cell under open circuit or during a low current drain. The voltage is an intensive property and therefore is independent of the size of the battery as are the Wh/kg and Wh/l. In many batteries, the voltage is defined as sloping; that is, it declines steadily as the battery is discharged. In cases like this, the nominal voltage is usually selected as the starting voltage under low current drain or at open circuit. Because the voltage may also be affected by the corrosion/passivation properties of the anode, it may be most useful to measure the voltage under a low current drain. In selecting a battery type for an application, it is important to consider that if it is especially weight sensitive, the Wh/kg is most important and if it is especially volume sensitive, the Wh/l is most important. If the electronics within the device can be damaged by too high a voltage, then this property is most important. Of course, the rate capability (or current capability) is also quite important to the particular application. This is a more subtle property since the voltage behavior of a battery with increasing current is often highly nonlinear. Furthermore, the behavior of the voltage with time at a given current is often highly nonlinear. Therefore, a much more detailed consideration is usually appropriate in determining the suitability of a battery for a higher current application (at low currents, most batteries have a linear voltage current relationship defined by the internal resistance of the cell). Nevertheless, a property called the specific power (W/kg) and power density (W/l) is generally tabulated for primary batteries. Unfortunately, these properties are not well defined as noted above, so they are not of great use except as a very general guide to power capability. It would be better if the industry would standardize on some definite definitions for these properties so that comparisons would be consistent, but they would not in any case be useful for a broad range of applications. A common designation of present-day aqueous batteries is "dry cell batteries." This term derives from much earlier technology in which many batteries were constructed with an excess quantity of electrolyte and the entire cell was contained in a box type of container. This type of battery was called a "wet cell battery." Because of the loss of energy density due to excess electrolyte, wet cells are no longer in common use as primary batteries.

Many battery sizes and dimensions have been standardized by electrical multinational organizations. The main organizations in this field are ANSI (American Standards Institute), IEC (International Electrotechnical Commission), and UL (Underwriters Laboratory). These organizations have also developed standardized tests for safety and performance in accordance with the most common applications for consumer batteries. For example, heavy and light flashlight tests, radio tests, and toy tests have been devised and are generally reported by the battery producers. For a comprehensive review of standardization of primary batteries, see [1].

Characteristics and Applications of Different Battery Systems

The common battery systems for consumers often have overlapping uses due to the design of the devices. For example, a flashlight with a tungsten bulb may be powered equally by carbon-zinc, alkaline, and in some cases lithium iron disulfide batteries. The circuit is simply a resistor (the tungsten filament). The resistance, however, is temperature dependent and thus it is sensitive to the voltage which controls the current of the battery – a higher voltage gives a higher current due to Ohm's Law – which results in a higher temperature of the filament as a quasi steady state is approached. If the current is too high, the temperature of the filament may exceed the stability limit of the filament, and the filament will separate or even melt if the temperature rise is too

fast. Therefore, if the filament is designed to accommodate the higher voltage of the lithium iron disulfide cell, it may also be used. In a LED flashlight, the color depends on the temperature which in turn depends on the voltage, while the light output depends on the power (voltage x current). For white light, the voltage needed is about 3.4 V, which may be obtained from almost any battery source, e.g., 2 alkaline cells, when a DC to DC converter is installed. It is important to use the recommended battery with the particular flashlight. For example, some flashlights recommend 2 or 3 AA alkaline batteries, while some recommend 2 or 3 lithium iron disulfide AA batteries. Some LED flashlights now operate on lithium manganese dioxide batteries, for example, CR123, while many are now designed for lithium ion or nickel-metal hydride rechargeable cells.

Many electronic devices such as wrist watches and clocks, remote controls, and electronic keys are operated by primary batteries. For small devices, alkaline or lithium button/coin cells are often utilized. These common miniature batteries are also used in very low drain devices such as electronic scales, clocks, and watches. The larger carbon-zinc cells are utilized in many clock devices because of the low drain rate needed for clock escapements. Children's toys are common applications for primary batteries. They are safe for children to handle, although parental supervision should be exercised with miniature batteries so that small children do not swallow them. Carbon-zinc and alkaline batteries are most commonly used in these applications. Lithium batteries are commonly used in photo applications – cameras and separate flash units. Their portability and easy replacement are advantages compared to some rechargeable batteries in these uses. In general, the highest drain devices can benefit from the use of lithium primaries, intermediate drain devices can benefit from alkaline batteries, while the lowest drain devices may use carbon-zinc batteries unless leakage and shelf life are problematic for the application. Smoke detectors and carbon monoxide detectors commonly use 9 V or AA batteries, and alkaline batteries are often the battery of choice because the alarms

Primary Batteries, Comparative Performance Characteristics, Table 1 Typical characteristics and applications of the most common types of primary batteries

System	Characteristics	Applications
Carbon-zinc (Leclanche or zinc chloride)	Common, low cost, widely available in different sizes	Flashlights, portable audio devices, toys, instruments
Alkaline (zinc-manganese dioxide – Zn/MnO_2)	Most popular general purpose battery with good low-temperature performance, medium rate capability, and low cost. Also found in many sizes	Many types of electronic instruments as well as toys and mechanical devices of all types
Mercury (Zn/HgO)	High capacity, flat discharge, good shelf life, but disposal problems due to mercury	Limited to devices that have controlled use and disposal. Has been widely employed in electronics in the past
Silver zinc (Zn/Ag_2O)	High capacity, flat discharge, good shelf life, but expensive. Usually made in miniature form	Hearing aids, watches, photography in small sizes. Military and aerospace in large sizes
Zinc air (Zn/O_2)	Very high energy density, low cost, very sensitive to environmental conditions and duty cycle	Hearing aids, pagers, medical devices, military electronics, and other special applications
Lithium liquid cathode	High to very high energy density, long shelf life, very good performance over wide temperature range but expensive	Utility meter readers, military electronics, aerospace uses. Toxic liquids inside require special protection
Lithium solid cathode	High to very high energy density, long shelf life, very good performance over wide temperature range and duty cycle. Lower cost than liquid cathode cells	Replacement for alkaline button cells and cylindrical applications. Different volt cathodes suit different applications
Lithium solid electrolyte	Very long shelf life but used only for low power applications	Medical electronics and implantable devices
Magnesium (Mg/MnO_2)	High capacity with long shelf life unless activated by discharge. Must be consumed in short period of time after activation	Formerly used in many military applications such as communications devices. Now mostly replaced by lithium batteries
Mercad (Cd/HgO)	Long shelf life and stable voltage. Low energy density and disposal problems	Limited to controlled applications requiring long shelf life and stable voltage

require substantial current over a fairly long time to ensure alerting everyone in the vicinity of the alarm. It is interesting that hearing aids are often best served by zinc air batteries because they are used in a continuous or near-continuous mode (e.g., for all awake hours) and have very high energy content for their size, thus giving good value. A disadvantage for this system is that the cells have a short shelf life after they are activated, so intermittent use is not recommended. Another factor to consider in selecting a battery type is how much damage to the device would be sustained if the battery leaks electrolyte. This is especially important in carbon-zinc cells which have the greatest tendency to leak, followed by alkaline cells, which leak very corrosive alkaline electrolyte when leakage does occur (e.g., when the device is accidentally left in the on condition

for a long time) and finally for lithium cells which are nearly hermetically sealed and very seldom display leakage. In summary, the best use of a given battery type is the one that best satisfies the use profile at the least cost and the least potential for damage [2].

Table 1 gives typical characteristics and applications of the most common types of primary batteries.

Carbon-Zinc Battery

This battery system has existed for over 100 years and still is a major battery particularly in the consumer market. It is still widely used for flashlights and portable radios as well as low drain devices such as digital clocks. It is widely available all over the world in most standard cell sizes. It has been supplanted in many consumer electronic applications because of its low rate capability and leakage/reliability problems; however, it is still widely used in the developing world for a great many applications.

Alkaline Zinc-Manganese Dioxide Battery

Most of the industrialized world now uses the alkaline battery as the battery of choice for consumer electronics as well as many other electrical devices. The advantage over carbon-zinc cells is the higher rate capability, low leakage rate, long shelf life, and better reliability and low-temperature operation. It is intermediate in cost between carbon-zinc and lithium battery types. Like carbon-zinc, it is available in many standard sizes and types. It is commonly used in 9 V devices such as smoke detectors and carbon monoxide detectors, wherein six small cells are arranged in a rectangular metal container and leakage is minimal. This is important for expensive and safety-related devices because of the corrosive nature of the leakage liquid (KOH in most cases) which can ruin the device without the knowledge of the user. It has been available since the 1950s.

Mercury Battery

This battery type has been phased out in most applications because of the problems of disposal. The product of the reaction is liquid mercury which has a tendency to pool and can create a toxic hazard in landfills. Most countries have banned the use of mercury batteries in consumer applications for this reason. Other battery systems such as lithium primary batteries, zinc air, and alkaline batteries have to a large extent replaced mercury batteries.

Silver Zinc Battery

This battery is mainly produced in small-size (button) cells because of the cost of the cathode material (silver oxide). The battery has been produced in two forms, one using argentous oxide and the other argentic oxide or a mixture of argentous and argentic oxide. It has also been used in several military applications where the performance attributes are more important than the cost factor.

Zinc Air Battery

The fact that the cathode does not carry the weight of the active material (oxygen from the air) has given this battery outstanding energy per unit of weight and volume. It has lifetime issues, however, due to the difficulty of maintaining the correct moisture balance (high humidity results in battery flooding, while low humidity results in cathode drying) since it must be open to the air during operation. Schemes have been devised to limit the access of air during off periods, but these are cumbersome and are not in general use. The batteries are generally sealed with a pull-off tab that is to be removed on activation, and then the battery is used directly. Generally the lifetime is only a matter of a few weeks or months depending on the battery design, but for applications such as hearing aids, the much higher energy than other small cells makes this the battery of choice. The battery was used extensively at one time in railroad signaling applications but has now generally been replaced by sealed cells.

Lithium Liquid Cathode Battery

This battery type mainly involves the use of thionyl chloride or sulfuryl chloride liquid cathode electrolyte system. Because the liquid cathode electrolyte fills all the space usually taken up by the electrolyte (in the separator as well as in the pores of the cathode), the battery has very high energy density per unit weight and volume. The liquid cathodes are all toxic, so the cells are always hermetically sealed to prevent the escape of the toxic, corrosive vapors or liquids. Because of the construction, the cells are expensive. The main applications have been in industrial uses such as meter readers, counters of all kinds, in aerospace, and medical and military applications. The combination of high energy, long shelf life, and wide temperature range has allowed the field to develop substantially. The cells were even seen as having a wide service in downhole drilling rigs, where sensors need to be battery powered, at temperatures even exceeding the melting point of lithium. The voltage ranges from 3.5 to almost 4 V for the different liquid cathodes.

Lithium Solid Cathode Battery

This battery type has been designed with several cathode materials over the years, but the most common cathodes now are manganese dioxide (MnO_2), carbon monofluoride (CF_x), and iron disulfide (FeS_2). The first two operate around 3 V, while the last operates at 1.6 V and can therefore be used in place of carbon-zinc or alkaline zinc-manganese dioxide batteries. The first two types have been widely used in aerospace and medical and military applications in addition to consumer batteries. Because of the different voltages of the first two systems from aqueous batteries, they are made in special sizes so they cannot be used in place of a nominal 1.5 V battery and damage the electronic device. This is not true of the iron disulfide cathode cell, so it is made in conventional AA and AAA sizes. Because of restrictions on the amount of lithium in a single cell, larger sizes of iron disulfide batteries are not made for consumer applications. As in the liquid cathode batteries, the high energy (and power) as well as the long shelf life and wide temperature range of use have allowed many applications for these batteries. The use of silver vanadium oxide has seen wide use in medical applications because of its much higher rate capability and is widely implanted for cardiac defibrillator devices.

Lithium Solid Electrolyte Battery

A special lithium cell uses an iodine cathode with an in situ formed solid electrolyte which is very stable and used in implantable devices in medicine. These devices have been widely used in pacemakers.

Magnesium Manganese Dioxide Battery

This battery has seen applications in many military devices. Like the zinc air battery, the shelf life is very long in the unused state, in this case because magnesium is heavily passivated by the aqueous electrolyte. When the cell is used, however, the magnesium metal is exposed to fresh electrolyte and a strong corrosion reaction is initiated. This uses up much of the magnesium in wasteful reaction. For some military applications, the long shelf life is more important than the wasteful generation of hydrogen due to corrosion. Modern weapons generally use a variation of lithium batteries for most deployments now.

Mercad Battery

This system utilizes two toxic materials, mercuric oxide and cadmium metal. Because of its low voltage, however, it is thermodynamically stable, and if it is well sealed, it has a very long lifetime. It also has a very stable voltage and can be used as a voltage reference in electronic circuits. It is seeing less and less usage because of disposal issues and the availability of stable lithium cells with equivalent lifetime and voltage stability.

Other battery types which have been studied in some detail but are not in common use include aluminum air cells, magnesium air cells, and iron air cells.

Cross-References

▶ Primary Batteries, Selection and Application

References

1. Wicelinski S (2011) Battery standardization. In: Reddy T, Linden D (eds) Linden's handbook of batteries, 4th edn. McGraw Hill, New York
2. Linden D, Reddy T (2011) An introduction to primary batteries. In: Reddy T, Linden D (eds) Linden's handbook of batteries, 4th edn. McGraw Hill, New York

---

Primary Batteries, Selection and Application

George Blomgren
Blomgren Consulting Services, Lakewood, OH, USA

The selection of particular batteries for any given application is a complex decision based on operating and shelf life, power capability, sensitivity to hazards, and cost. Battery characteristics are tabulated in many sources [1]. It is important after

Primary Batteries, Selection and Application, Table 1 Qualitative comparison of primary battery types (1 to 7 where 1 is best)

System	Voltage	Wh.kg	W/kg	Profile flat?	Low temp.	High temp.	Shelf life	Cost
Carbon zinc	7	7	5	7	6	7	7	1
Alkaline	6	4	3	6	4	5	6	2
Zinc mercuric oxide	4	3	5	2	5	3	4	5
Zinc silver oxide	2	3	2	2	4	3	4	6
Zinc air	5	2	3	2	5	5	–	3
Lithium liquid cathode	1	1	1	1	1	2	1	5
Lithium solid cathode	1	1	1	2	2	3	2	3

examining these issues that the user or application developer makes measurements of the selected battery under the conditions of use of the application. The profile of current and voltage during operation and rest should be carefully documented, so the range of these properties during use under different conditions can be ascertained. Among these properties are ambient temperature and battery/device temperature during operation, maximum and minimum currents likely to be encountered, range of input impedance of the device as a function of frequency, as well as the frequency dependent impedance of the selected battery so that the impedance can be matched for optimum utilization of the battery. Part of the burden of the utilization of the battery rests with the application designer, who should be aware of battery characteristics so that optimum designs can be arrived at in parallel with battery selection. The device design often dictates the volume available for the battery as well as the voltage requirement. Sometimes the weight of the battery is very important for portability concerns, for example, with portable, handheld tools. In some cases, the criteria will favor choosing a secondary (rechargeable) battery over a primary one or primary over secondary. This is one of the most important decisions to be made. Cost of the battery is an important issue because cost may affect the acceptance of the ultimate consumer of the device. Users have much less freedom in making these selections since the voltage and current characteristics of the design greatly affect the suitability of the battery. It is often, but not always, desirable to accept the device manufacturer's recommendation for battery selection. For example, when a new version of a battery appears such as the lithium-iron sulfide battery or the nickel oxide additive to the zinc-manganese alkaline battery, it may be better to switch from a recommended standard alkaline cell to power the device.

Table 1 gives some general characteristic rankings of primary battery types that should be of help in narrowing the selection of a battery.

Obviously, these rankings are only semiquantitative and do not imply a linear relationship. However, for a first look at battery types, these can be useful. A more detailed study is necessary to make final selections as discussed above. Sometimes several battery types are selected and tested for optimum characteristics for the particular application under study.

An important area to be considered in battery selection is safety of the battery in combination with the device. Part of the safety consideration is the design of the battery compartment. In some cases, the battery compartment should be designed for maximum heat dissipation – in some cases, they should be designed for maximum electrical isolation. Electrical isolation can be important for devices which are not in waterproof containers and can result in battery short circuiting if the device is immersed in water. Small batteries such as miniature alkaline or lithium coin cells should be enclosed in the device in a way that is difficult for small children to access the battery. Many of these small cells have been swallowed by children with the result of serious medical complications. Considerations of battery design are complex, and the size, electrode configuration, and format of the cell are critical to the energy content and power capability of the cell and should be part of the overall selection criteria.

Often, the consumer is confronted with the need to decide which battery is best to replace an expired battery. It is generally advisable to follow the device manufacturer's recommendation for the chemistry of the battery. For example, if a carbon zinc battery is chosen rather than an alkaline cell because of lower cost, the result may be very poor utilization of the battery if the application is not appropriate and opposes the recommendation of the manufacturer of the device. If a comparison among different battery manufacturers for a given type of battery is desired, there are many independent studies of different batteries available to the consumer.

Cross-References

▶ Primary Batteries, Comparative Performance Characteristics

References

1. Linden D, Reddy T (2011) An introduction to primary batteries. In: Reddy T, Linden D (eds) Linden's handbook of batteries, 4th edn. McGraw Hill, New York

---

Primary Battery Design

Jack Marple
Research and Development, Energizer Battery Manufacturing Inc, Westlake, OH, USA

Introduction

Effective battery design is a balance between the intended application, the required operating voltage, rate requirements, shelf life, operating temperature range, reliability, safety, and product costs. Each of these considerations must then be applied to the anode, cathode, separator, electrolyte, and other features such as pressure vents, safety devices, and tolerance to abuse. Commercially, a few electrochemical systems have won out in the competitive field of primary batteries for general consumer use. These include Leclanche'(also known as carbon zinc) and its variant zinc chloride, alkaline zinc manganese dioxide, zinc nickel oxyhydroxide, zinc air, lithium iron disulfide, lithium manganese dioxide, lithium carbon mono-fluoride, lithium thionyl chloride, and lithium sulfur dioxide. Interestingly, in some cases, batteries may be referred to by the chemistry of the anode and cathode and in other cases, by the inventor, the appearance, the choice of electrolyte, or even the product size.

There are many references which provide details about the chemistry of each of these types of batteries [1]. But detail relative to product design considerations is more limited. It is the intent of this article to focus on material properties and how they impact product design, cell format, and application.

Battery Format

Primary batteries are generally available in two basic form factors, cylindrical and coin. Within these form factors, the arrangement of the working electrodes can vary considerably depending on the volumetric differences in anode and cathode materials, changes in the volume of these materials during electrochemical discharge, the application current, and the necessary interfacial surface area need to support the current. Additionally, factors such as material electronic conductivity, electrolyte ionic conductivity, separator requirements, and safety features of the battery need to be considered.

The following cross-sectional drawings provide a quick comparison as to how these considerations influence the final battery design and electrode configurations. Figures 1, 2, and 3 show the cross sections of the three most common cylindrical cell designs. Figure 1 is referred to as a "carbon zinc" construction where the center carbon electrode, which is used as a cathode current collector and gas vent, is most evident, and

Primary Battery Design, Fig. 1 (Carbon zinc) Zn can/anode with MnO_2 cathode and a porous carbon rod current collector and vent. Low rate – oldest construction

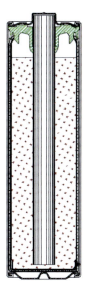

Primary Battery Design, Fig. 3 (Lithium) Wound layers of cathode coated onto a metal current collector with a solid Li foil anode which also serves as a current collector. High rate – newest construction

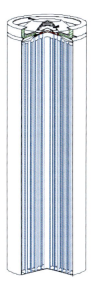

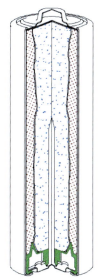

Primary Battery Design, Fig. 2 (Alkaline) Suspended Zn particle anode gel with a brass current collector and outer MnO_2 cathode with the container as current collector. Moderate rate

the zinc cell container also serves as the anode. Note that, volumetrically, the center of the cell is mostly cathode which is manganese dioxide.

Figure 2 is a typical alkaline cell construction which uses the same electrochemical couple as the "carbon zinc" cell. However, in this case the zinc is a powder, in the center of the cell, and the container is made of an inactive material, such as steel. Like carbon zinc cells, the cathode is manganese dioxide. The electrolyte is based on potassium hydroxide (or in some coin cells sodium hydroxide), which explains the use of the term "alkaline" to describe these cells.

Figure 3 represents a jelly roll construction most often used for lithium metal anode cells. The cathode can be made of manganese dioxide, carbon mono-fluoride, or iron disulfide. The use of lithium as an anode rather than zinc necessitates that the electrolyte be organic due to the extreme reactivity of lithium metal with water.

These three constructions represent products designed for increasing rate capability (Fig. 3 > Fig. 2 > Fig. 1). They also represent advances over time in terms of commercial production, with Fig. 1 being older than Fig. 2 which is itself older than Fig. 3.

To better understand the differences in rate capability, application, and performance, it is necessary to look at each product in terms of its key components of anode, cathode, separator, and electrolyte.

Anode

In Figs. 1 and 2, each cell makes use of zinc metal as the anode half couple. However, when the anode also serves as the outer container it must be solid which limits the reaction surface area. For a carbon zinc cell, the surface area of the anode is limited to the interfacial area, creating a much longer diffusion path through the cathode than in Fig. 2 or 3. In the alkaline cell construction shown in Fig. 2, the zinc surface area has both an interfacial component and a contribution from particle size. However, limitations to increasing surface area, based on particle size, must be considered. These include zinc particle to particle contact, zinc oxidation, the need for cathode overbalance to prevent water decomposition and hydrogen gas generation during deep discharge, and stability of the zinc in this alkaline environment, which particularly in today's zero mercury cell design can contribute to gas generation and reduced shelf life.

Another means of achieving high interfacial surface area of the anode is shown in Fig. 3. In this case the anode is once again a solid metal, but a high surface area is achieved by winding together long, thin layers of cathode and anode. Using this configuration, electrode interfacial surface areas of approximately 200 cm^2 can be achieved in AA cells, which is at least 20 times higher than that of the other AA cell configurations.

These variations in cell and anode designs also increase the challenge of collecting the current from each electrode. In the case of carbon zinc cells shown in Fig. 1, the zinc container serves as its own current collector. In the case of Fig. 2, a brass nail often is positioned within the center of the zinc particle suspension thereby relying upon particle to particle contact for current collection. In Fig. 3, the lithium metal anode often serves as its own current collector with the aid of a small metal tab which connects the lithium metal foil to the outer cell container.

Anode Alloys

While the anode is referenced relative to its base metal composition, anodes are seldom pure metal. In the case of carbon zinc cells, lead is often added at about 0.5 % to help with the metal drawing and can manufacturing process. Alkaline cells today, with the removal of mercury from the gel formulations, rely upon a number of different elements to minimize hydrogen generation resulting from water decomposition. Common elements include In, Al, Bi, Sn, and Pb at about the 500 ppm level. Lithium cells often contain Al, at about 0.5 % level to increase the tensile strength and to control the formation of a passivation layer from the reaction of the lithium metal with the organic electrolyte components [2].

Cathode

Interestingly, commercial cells using MnO_2 as the cathode material are available in all three configurations. However, while the cathode has the same base composition, changes in purity and crystal structure are required for each system. Carbon zinc cells are normally based on raw ore, while alkaline cells require higher purity electrolytically plated MnO_2. Lithium cells require even higher purity material based on thermal treatment of the electrolytic MnO_2 in order to create a crystal structure which is more stable in organic electrolytes. In addition to MnO_2, lithium can also be coupled with other cathodes such as iron disulfide, carbon monofluoride, thionyl chloride, and SO_2. With the exception of iron disulfide, which is considered a 1.5 V system, all others are considered 3 V systems. To avoid damage to devices, these

Primary Battery Design, Fig. 4 Storage maintenance based on rating drain test at 20 C, to 0.9 V

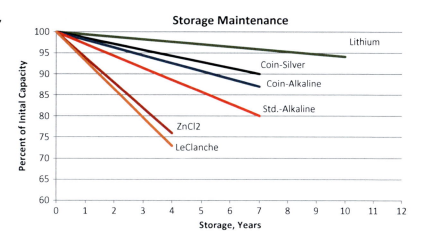

3 V systems are not available in standard 1.5 V formats such as AA cells.

Cathode design considerations include the conductivity of the active material; its stability relative to various forms of carbon conductors; volumetric changes of the reaction products; particle size influenced by electrode thicknesses and rate requirements; the use of processing aids such as polymers, surfactants, and rheology aids; and how the active materials will interface with a current collector. Additionally, in the case of lithium cells, it is critical to condition the cathode components prior to cell assembly to remove moisture and volatile process aids which can leach into the organic electrolyte and then react with lithium metal. Such reactions are almost always detrimental to cell impedance and product shelf life.

Figure 4 shows a comparison of storage maintenance as a function of product age for common primary systems. Actual shelf life varies by manufacturer and the cell's chemistry. Material purities, passivation additives, seal integrity, and temperature may further influence shelf life.

When discussing cathodes for primary cells, it is important to make a distinction between solid materials, mentioned above, and those in which the cathode electrochemical component is not in the physical electrode structure, but rather exist in a liquid or gas state. Examples include zinc air cells, lithium thionyl chloride cells, and lithium sulfur dioxide cells. In each case, the physical cathode structure serves as a catalytic reaction site for a gas or liquid which is reduced most often on a high surface area carbon matrix. The polymeric binder in these structures is usually Teflon, and the carbon may be coated with a catalyst (e.g., MnO_2 in an air electrode). An important design consideration in each case is the stability of the carbon structure and the mobility of the cathode gas or liquid within this structure throughout the life of the battery.

Separators

For the most part, battery separators can be grouped into two categories: nonwovens and microporous membranes. The choice of a separator for each battery system must take into consideration such factors as cost, thickness, processibility, formability, pore size, tortuosity of pores, tensile strength, melt point, range of operating temperature stability, puncture resistance, dielectric strength, particle sizes of the anode and cathode materials, and stresses developed during product discharge.

Typically, carbon zinc and alkaline cells use nonwoven separators. In these cases, cost is an important design consideration, and the high conductivity of the electrolyte enables thicker separator materials. However, the thickness of the separator also defines the gap or distance between electrodes, which can impact the cell's ability to support high current densities. Pores are not well defined and are typically on the order of several microns, while the thicknesses are often several hundreds of microns.

In contrast, lithium cells require thin microporous separators based on the low conductivity of organic electrolytes. These microporous separators more closely resemble membranes with a thickness of 25 μm or less with pore diameters in the hundredths of microns. They most often are polyolefins consisting of polyethylene or polypropylene. Despite their thinness, up to three layers of separately extruded films may be used.

In selecting a separator for a particular cell design, it is also important to consider its stability in the electrolyte, ability to wet out quickly by the electrolyte, ability to maintain electrolyte within its pore structure throughout the cell's life, and the impact it has on ionic mobility of the electrolyte salt. Since separator pores are seldom channels straight through the separator, the difference in true ion path versus the separator thickness is referred to as tortuosity. A certain amount of tortuosity is desired to prevent shorting, due to particle penetration and soft shorts from dendritic bridging between electrodes, while maintaining low tortuosity is often required to achieve the best high rate performance. Therefore, striking the right balance of pore size and structure is an important design consideration [3].

Electrolytes

The selection of the electrolyte formulation for a cell design is a critical decision, which impacts a wide range of cell characteristics, including operating temperature range, shelf life, rate capability, product safety, product application voltage range, solubility of impurities, and solubility of active components and reaction products. Optimized electrolytes are often more than just a solute in a solvent; they may contain other additives such as surfactants, voltage control agents, film formation components, and Le Chatelier's principle additives to control chemical equilibriums and solubility. In many battery systems, the anode and electrolyte need to form a stable passivation film on the surface of the anode. These films are referred to as the SEI or solid electrolyte interfaces. SEI layers are highly important to controlling wasteful corrosion reactions while still allowing for unhindered ionic mobility through these passivation films.

Electrolyte conductivity provides the greatest impact of electrolyte on product design. As a reference point, the radically different product designs shown in Figs. 2 and 3 are, in fact, driven by the conductivity of the electrolytes. Alkaline cells have high concentrations of KOH in water, yielding conductivities of approximately 500 S/cm [4]. However, lithium cells must utilize organic solvents with low ionic salt concentrations because of the incompatibility of lithium metal with water-based electrolytes. Such organic electrolytes are more typically 10 S/cm [5], a factor of 50 times less conductive. To counter this large difference in conductivity, a lithium cell designer must dramatically increase the interfacial surface area of the electrodes and additionally minimize the gap or distance between the electrodes. As a result, an alkaline AA size battery typically has an interfacial electrode area of 10 cm^2, while an AA lithium cell is approximately 200 cm^2. Additionally, the distance between electrodes may be 200 μm in an alkaline cell but only 20 μm for a lithium cell. By increasing the interfacial electrode area by a factor of 20 and decreasing the distance between electrodes by a factor of 10, the disadvantage of lower electrolyte conductivity by a factor of 50 can be offset such that the lithium cell significantly outperforms the alkaline cell on high rate applications.

Figure 5 shows a comparison of service and operating voltage on a typical application drain of 200 mA for $ZnCl_2$, alkaline, and $LiFes_2$, AA cells. The differences in both rate capability and energy delivered are evident for these product design evolutions.

Total Product Design: Container and Format

Battery systems are often compared based on energy per weight or energy per volume. A plot of these two energy densities is referred to as

Primary Battery Design, Fig. 5 Comparison of operating voltage and service on a 200 mA, AA battery rating drain

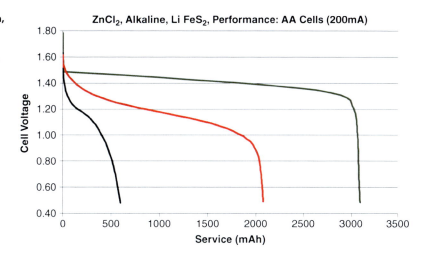

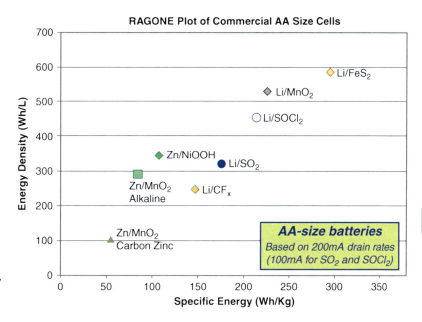

Primary Battery Design, Fig. 6 Ragone plot of primary AA cells

a Ragone plot. Figure 6 is a Ragone plot for commercial, primary, AA cells based on a common drain rate of 200 mA, consistent with many of today's device requirements. Figure 6 clearly shows the energy differences for the three products shown in Figs. 1, 2, and 3. While other systems have been developed, these systems have not found wide acceptance outside of the military or niche markets such as photo applications, high-end flashlights, remote meter sensors, and the medical industry.

From a battery design standpoint, the challenge is to maximize active components and their utilization while minimizing non-active components such as the cell container, current collectors, seals, head space, separators, conductors, and void required for reaction products. The percent of both volume and weight that these non-active components contribute to any given battery system is dependent on format and size; the larger the battery, the greater the percentage of active components. Electrolyte is of course needed for any system but can take on a more important role when it is also involved in the discharge reaction. This is particularly true for alkaline cells.

For many consumer applications, the focus is more on Wh/cc versus Wh/Kg. On a volume basis, the anode and cathode volumes account for 18 %, 34 %, and 38 % of the total AA cell volume for carbon zinc, alkaline, and lithium iron disulfide, respectively. These values change considerably when electrolyte is included: 56 %, 43 %, and 56 % respectively. In general, these systems possess less than 40 % active volume percentages without electrolyte and less than 60 % with electrolyte. These active material utilization design limitations still exist despite having the benefit of more than 100, 50, and 20 years in product optimization respectively. Thus, when considering the potential energy of an electrochemical couple, it is important to realize that these values need to be reduced by as much as 60 % for defining the practical energy.

In addition to volumetric optimization, weight optimization is also important in today's ever decreasing in size portable devices. Figure 7 shows a breakdown of the components in percentage of total cell weight for one of the highest gravimetric systems, lithium iron disulfide. Note that close to 50 % of the cell's weight consist of non-electrochemically active materials with the majority of this weight associated with the battery container.

Coin Cells

Most of the above discussion on primary battery design has focused on cylindrical cells. While the design considerations are, for the most part, similar for coin cells, the increase in container

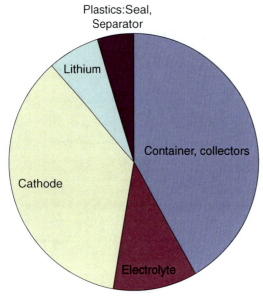

Primary Battery Design, Fig. 7 Contributions of materials to total weight of a LiFeS$_2$ cell

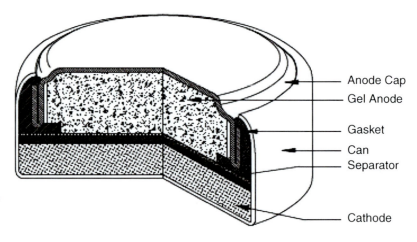

Primary Battery Design, Fig. 8 Cross section of a silver oxide cell

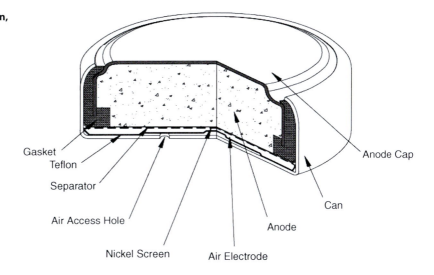

Primary Battery Design, Fig. 9 Cross section of a zinc air cell

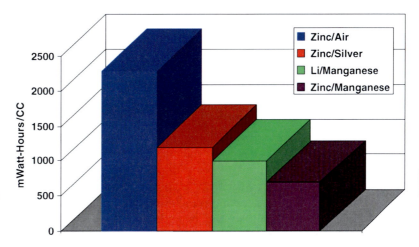

Primary Battery Design, Fig. 10 Volumetric energy densities for common coin cells

interfaces, container volume, and decreased container strength place increased demand on volumetric or internal void considerations. Figures 8 and 9 show the cross-sectional profiles of two common coin cell chemistries.

The zinc air cell design is unique in that the cathode active material is supplied by oxygen in the air and, other than a thin reactive electrode sites for absorption and reduction of oxygen, the cells internal volume is primarily occupied by zinc anode. However, void volume must be reserved to account for the volumetric expansion of zinc to zinc oxide upon discharge.

Figures 10 and 11 provide a comparison of the volumetric energy densities in the various commercial coin cell chemistries. While zinc air cells by far exceed the other systems, on a watt-hour per cc basis, its application has been mostly limited to hearing aid applications. This application limitation is in part because the path way for oxygen into the cell also allows moisture in and out of the cell, as well as carbon dioxide which may cause wasteful carbonation of the anode components. As a result, the zinc air chemistry has been limited to applications with duration of a month or less.

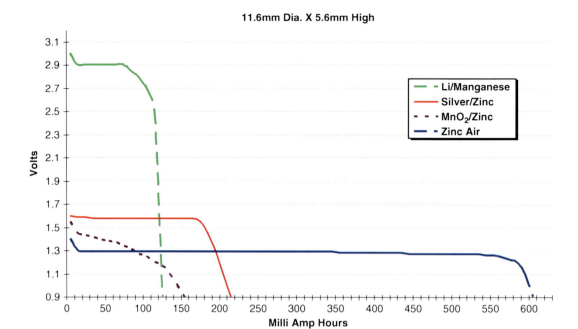

Primary Battery Design, Fig. 11 Discharge profiles of various coin cell chemistries

Future Direction

Future designs of primary batteries will be strongly influenced by both shifts in applications and material science. A review of device applications by NPD [6] shows the following growth areas from 2005 to 2010 for primary batteries: wireless game controllers, wireless sensors, garage door openers, remote controls, and games/toys, while emerging market segments include LED lights, remote sensors, 3-D glasses, medical devices, ultralow power sensors, and military applications like night vision, GPS, and laser sights.

Rapid changes in material science such as nanomaterials, graphene, conductive polymers (both ionic and electronic), and ionic liquids will open battery designs to new formats such as thin printed and battery on board, non-replaceable primary batteries [7, 8]. These trends, further coupled with the rise in popularity of low-power applications, may transition some of today's secondary electrochemistries into primary applications without the added cost of charge control systems required for secondary batteries.

Cross-References

▶ Primary Batteries, Comparative Performance Characteristics
▶ Primary Batteries, Selection and Application

References

1. Linden D (2002) Handbook of batteries, 3rd edn. McGraw-Hill, New York
2. Marple WJ (2005) Nonaqueous electrochemical cell with improved energy density. US Patent No 6849360
3. Marple WJ (1994) Separator/electrolyte combination foe a nonaqueous cell. US Patent No. 5290414
4. Dow Website (2012) Form No 609-02127-804, Conductivity of solutions. http://msdssearch.dow.com/PublishedLiteratureDOWCOM/dh_003c/0901b8038003ccb2.pdf?filepath=liquidseps/pdfs/noreg/609-02127.pdf&fromPage=GetDoc
5. Blomgren G (1983) Properties of electrolytes. In: Gabano J-P (ed) Lithium batteries. Academic, New York
6. NPD (2010) Consumer battery usage. US Market, Commissioned by Energizer Battery Company
7. Sapru V (2011) Opportunities in the primary lithium battery market, battery power. Frost & Sullivan Webcom Communications, Greenwood Village, CO 80111
8. Svastano D (2011) Advancements in printed battery technology are driving growth, printed electronics now. http://www.printedelectronicsnow.com/articles/2011/01/advancements-in-printed-battery-technology-are-dri

Probe Beam Deflection Method

Cesar Alfredo Barbero
Department of Chemistry, National University of Rio Cuarto, Rio Cuarto, Cordoba, Argentina

Introduction

Different techniques can be used to study electrochemical phenomena. Classic electrochemical studies are based on the study of electron fluxes at the electrode/electrolyte interface [1]. Using different ways of system perturbation and/or to measure the current-potential response of electrochemical systems, together with studies on the effect of electrolyte media (e.g., pH), it has been possible to obtain some information about electrochemical mechanisms. However, it has become clear that purely electrochemical techniques have significant limitations. To overcome those deficiencies, a plethora of in situ spectroscopic techniques has been developed in recent years. In those techniques, spectroscopies have been combined with electrochemical perturbations to help to understand complex electrochemical phenomena. Most of those techniques have been applied to study the electrode/electrolyte interface [2], by measuring changes occurring at the interface itself [3] or on the whole electrode [4]. On the other hand, the reaction occurring at the electrode necessarily created fluxes of solution species (ions, neutrals, solvent) from/to the electrode/electrolyte interface. The study of such fluxes has been less common, inasmuch that fluxes could control the electrochemical systems, such as batteries, sensors, fuel cells, or supercapacitors. One reason is that few techniques are able to measure concentration gradients coupled to electrochemical reactions. The classical technique is optical interferometry. It has low sensitivity and slow response and both complex experimental setup and analysis. It has been mainly used to study electroplating cells and similar systems [5]. Some other techniques are able to detect ion fluxes in solution: radiotracer

detectors, pH sensors, fluorescence or absorption measurement on grid electrodes, scanning electrochemical microscopy, ring-disk voltammetry, surface plasmon resonance, confocal microscopy, and diffraction spectroelectrochemistry. They have been seldom used to study electrochemical systems. Probe beam deflection is a technique which allows fast monitoring of concentration gradients related to electrochemical phenomena and has been used to study a wide range of electrochemical phenomena [6]. The mechanism of detection in PBD involves deflection (refraction) of the probe beam by the refractive index gradient. While PBD techniques could be used to measure any electrochemical phenomena, it is difficult to relate the observed signal with electrochemical reactions when there are reactions in solution or several ions are exchanged. On the other hand, the ion exchange coupled to electrochemical reactions, between an electrode surface and a film confined at it, and the electrolyte solution is quite straightforward to be studied by PBD techniques.

Basic Principles

The usual scheme of a probe beam deflection experiment is depicted in Fig. 1. A solid/electrolyte is set up and a laser beam is set up to travel parallel to the surface. The interface is usually planar; however, spherical or cylindrical solids can be used. The only difference is the geometry of the interaction between probe beam and the interface. The deflection of a probe beam, which travels close to an electrode along a path length l, in an electrolyte of refractive index n (Fig. 1) could be understood as a distortion in the wave front of the beam. The wave speed increases according to $v = c/n$. In conditions of small deflection, the geometric optic approximation could be used [6]:

$$\theta(x,t) = \left(\frac{L}{n}\frac{\partial n}{\partial C}\right)\left(\frac{\partial C(x,t)}{\partial x}\right) \qquad (1)$$

where C is the concentration, l is the interaction path length, n is the refractive index of the bulk, and dn/dC is the variation of refractive index with

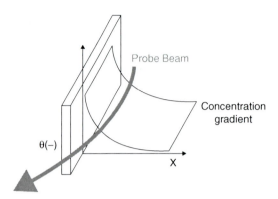

Probe Beam Deflection Method, Fig. 1 Probe beam deflection in a concentration gradient

concentration. The sign of the deflection depends on the change of refractive index with concentration (dn/dc).

The dn/dc constant is positive for liquids or solids dissolved in a solvent because the refractive index increases with concentration. Therefore, an increase of concentration at the electrode surface will be accompanied by a negative deflection (Fig. 1). On the other hand, when gases are dissolved in a liquid phase, the refractive index decreases; therefore, dn/dc has negative values. This has to be taken into account when the PBD data of electrochemical systems with gaseous reactants/products (e.g., hydrogen fuel cell) is analyzed. One way to make both signals proportional is to make the beam-electrode distance close to zero. However, the beam waist cannot be zero. From laser beam optics [7], it is known that a beam can be focused to a small spot, but there is a relationship between the waist ($w(x)$) and the path length (b) of the beam in front of the electrode. Since the interaction between the probe beam and the concentration gradient assumes that the beam is nearly cylindrical with a fixed waist, the electrode width has to be smaller than b. Otherwise, the waist at the extremes will be very different from the waist at the center of the electrode and the probe beam will sample different concentration gradients.

The possible perturbations (controlled potential or current) used in PBD could be all the usually used in electrochemistry. To obtain the equations governing the PBD signal for different electrochemical techniques, the concentration gradient has to be obtained. An electrochemical system could be modelled as a planar electrode of width w in contact with a semi-infinite fluid layer where diffusion occurs (being the diffusion layer several times thinner than w). The latter condition implies that border effects could be neglected and that interaction path length is $l = w$.

Using the mass transport equations and the boundary conditions, the variation of concentration gradient with time and distance could be obtained. Therefore, the calculation could use the methods developed previously in electrochemistry to calculate the concentration gradients in front of the electrode. In general, to obtain the equations to simulate the PBD response, the following steps have to be carried out: (i) obtain the concentration profile at distance x, for the technique under study, $C(x, t)$; (ii) differentiate with respect to x, to obtain $dC(x,t)/dx$; (iii) combine with Eq. 1 to obtain the dependence of deflection with time and space $\theta(x, t)$.

One characteristic of the technique, shared with purely electrochemical techniques where only electron fluxes are monitored, is its lack of specificity. That is, not individual concentration gradients but the sum of all concentration gradients is detected. This could be an advantage because it allows measuring all possible ion exchange not only of some ions (e.g., massive ions for EQCM) like other techniques. In the case of surface and film species, only concentration gradients due to ion exchange will be detected. Moreover, if the measurements are carried out in the presence of binary electrolytes that is electrolyte containing only one anion and one cation (e.g., ClNa), the diffusion of ions and its migration are necessarily coupled [8]. In that way, the diffusion of one ion is coupled to the other and a single gradient will be measured. A different situation exists when redox reactions in solution are monitored (e.g., $Fe(CN)_6^{-4}$ oxidation). There, the gradient of reactants, products, and supporting electrolyte (if present) has to be considered. Therefore, the interpretation of the PBD signal in terms of electrode reactions becomes difficult or impossible.

Probe Beam Deflection Method

Following such procedure, the PBD signal ($q(x, t)$) could be calculated, when a potential pulse (from a potential where no reaction occurs to a potential where all the reactant is converted) is applied to the electrode and an electrode reaction occurs during the whole measurement. The profile follows Eq. 2:

$$\theta(x,t) = \left(\frac{l}{n}\frac{\partial n}{\partial C}\right)\frac{C_o}{\sqrt{\pi Dt}}e^{-x^2/4Dt} \qquad (2)$$

where D is the diffusion coefficient and C_0 is the concentration of the redox species. As it can be seen, the PBD signal depends both on the time (t) and the beam-electrode distance (x). The PBD signal is zero at $t = 0$ (unlike the current which has an infinite value) and reaches a maximum value at a finite time (t_{max}):

$$\sqrt{t_{max}} = \frac{(x - x_0)}{\sqrt{2D}} \qquad (3)$$

Equation 3 could be used to measure the diffusion coefficient of the species because x could be set up experimentally. Equation 2 can be used to simulate the PBD signal of a continuous electrochemical reaction. That is, when the redox species are transformed during the whole time span of the measurement. This is the case of redox reactions in solutions (e.g., Fe^{+2} electrochemically oxidized to Fe^{+3}), but in such cases, several concentration gradients have to be considered and interpretation is difficult. A simpler system involves metal deposition by reduction of a metal ion (e.g., Ag^+) to metal (Ag^0), where only one concentration gradient have to be considered.

A more interesting case deals with discontinuous reactions, where the reaction takes place in a short time span which is negligible compared with the duration of the measurement. An example is a monolayer of metal (e.g., Ag) which dissolves by oxidation giving short pulse of metal ions. The PBD signal profile follows Eq. 4:

$$\theta(x,t) = \left(\frac{l}{n}\frac{\partial n}{\partial C}\right)\frac{C_s}{\sqrt{\pi Dt}}\frac{x}{2Dt}e^{-x^2/4Dt} \qquad (4)$$

where D is the diffusion coefficient of the mobile species and C_s is the surface concentration of the redox species. Equation 4 is similar to Eq. 2 but the time profiles are sharper and the time for the maximum PBD signal (t_{max}) obeys Eq. 5:

$$\sqrt{t_{max}} = (x - x_0)/\sqrt{6D} \qquad (5)$$

The discontinuous reaction with a binary electrolyte is completely described by Eq. 4. Discontinuous reactions occur in electroactive films, adsorbed redox monolayers, ion desorption in double layers, stripping of underpotential metal deposit, stripping of metals in Hg amalgams, etc.

The experimental setup allows measuring the PBD signal along with the current and potential; therefore, the data analysis involves comparison of both kinds of data, since the absolute value of beam-electrode distance (x) cannot be set experimentally and could be measured by fitting the experimental measurement with simulation. In that sense, experimental PBD data can be post-processed to extrapolate the data to $x = 0$ (electrode surface), therefore making it comparable with current. Another way to perform such comparison involves the so-called sampled deflection voltammetry. If the potential is pulsed between two oxidation states, C_0 (or C_s) will follow a potential-concentration relationship (e.g., Nernst equation). Therefore, the PBD signal at each extreme potential measures the effect of such relationship on the ion exchange. Since the related sampled current voltammetry technique gives the changes of electronic charge with potential, both data can be compared and the relationship between electron and ion exchange ascertained. Analytical equations similar to Eqs. 2 and 4 have been obtained for different electrochemical techniques, including constant current and alternating current or potential. On the other hand the important case when the potential is cyclically scanned at constant rate between different potentials (cyclic voltadeflectometry, CVD), which is directly related to cyclic voltammetry, cannot be solved analytically. Therefore, numerical calculation techniques have been devised, mainly based

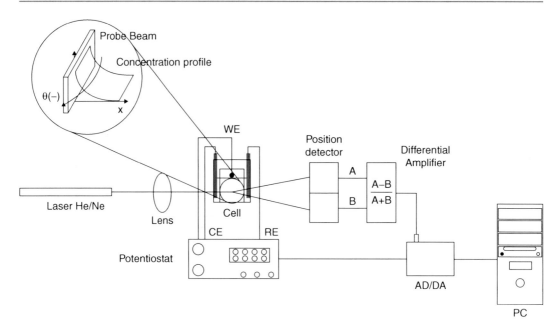

Probe Beam Deflection Method, Fig. 2 Experimental setup for probe beam deflection techniques

on finite differences [6]. Additionally, the semi-integration numerical procedure has been used both to simulate the CVD signal and to post-process the experimental data [6].

Experimental Details

The application of PBD techniques to study electrochemical phenomena requires integrating an optical setup to the electrochemical cell (Fig. 2).

A laser (usually a high pointing stability He-Ne laser) is loosely collimated by a lens to pass parallel to planar working electrode. The electrode is placed inside a rectangular optical cell where counter and reference electrodes are placed outside the beam path. The refracted laser beam impinges onto a lateral position detector, usually a spliced photodiode. The signal of each half of the photodiode is fed into a differential amplifier which ratios the difference of signal to its sum. In that way, intensity fluctuations in the laser are cancelled out. The differential signal is then digitized by an A/D converter and fed into a computer. The computer, trough a D/A converter, also provides the signal used to control the potentiostat and input the actual current and potential signals. The measurement speed is determined by the amplifier bandwidth, which is in the tens of kHz range. Therefore, millisecond time resolution is easily achieved. The cell is mounted onto an X-Y stage which allows adjusting the cell tilt to make the electrode surface parallel to the laser beam. The X-Y stage has an additional micrometric screw which allows moving the surface in the direction perpendicular to the beam direction. In that way, the beam-electrode distance (x) could be adjusted in relative way with micrometric resolution. The usual electrode surface is a flat conductive material (e.g., glassy carbon). An electroactive film can be deposited onto the surface as a thin film. In the case of porous electrodes, the powdered materials can be mixed with a binder (e.g., Nafion) into an ink and dip coated onto a flat surface.

Applications

The mainstream application of PBD techniques has been the study of redox coupled ion exchange of electroactive films [9]. The electroactive material could be redox polymers (e.g., polyvinylferrocene [10]), conductive polymers (e.g., polyaniline [11]), redox oxides (e.g., cobalt oxide [12]), or inorganic solid complexes (e.g., Prussian blue [13]). Using binary electrolytes (e.g., HCl), the observed PBD signal can be directly related to the actual ion fluxes occurring between the film and solution. The signal/noise ratio could be always improved by forming thicker films (increasing C_s in Eq. 4). The method was used in aqueous and nonaqueous electrolytes [9]. The techniques allowed detecting ion exchange not predicted by simple models of charge electroneutrality. An interesting film was built by layer-by-layer self-assembly. In this case, the film is anisotropic and PBD gives unique information on the influence of the outer layer charge on the ion exchange kinetics [14].

A related system involves thin (one or few monolayers) of surface species either covalently bonded or adsorbed onto the solid substrate. Since the effective surface concentration (C_s) is quite smaller than for thicker films, the PBD signal can be too weak to measure. One way to overcome this problem involves increasing the active area of the electrode while maintaining the geometrical area, by using porous electrodes. In that way, the adsorption of redox intermediates (e.g., CO) of methanol fuel cells on mesoporous metal electrodes have been studied successfully [15]. Double/layer charging of conductive surfaces involves the exchange of ions with the electrolyte solution. Highly porous electrode materials are applied in supercapacitors and deionizers. The large active area makes them easy to study the specific ion adsorption by PBD [16] and the potential of zero charge to be determined [17]. PBD techniques have also been used to study other electrochemical systems, such as metal deposition/dissolution, hydrogen adsorption, underpotential deposition, silicon dissolution, graphite intercalation, and oscillating oxidation of formic acid [6].

Advantages and Disadvantages

PBD techniques' biggest advantage is to be a simple and fast way to ascertain the ion exchange of surface confined species, including electroactive films, surface redox species, and double layer on porous electrodes. Cyclic voltadeflectometry techniques could be used in an exploratory way, equivalent to cyclic voltammetry, to study the ion exchange. Then, chonodeflectometry could be used to obtain quantitative data on the ion exchange.

The main disadvantage of PBD techniques is its lack of specificity. The deflection signal is related with the sum of all concentration gradients present in the experiment. Therefore, the understanding of a PBD measurement of a soluble redox system, where several concentration gradients (due to reduced and oxidized species, summed to those of the supporting electrolyte) could be present, is quite difficult.

Future Directions

A future application of PBD techniques would be the study of ion exchange across liquid/liquid interfaces like those found in ITIES (interface between two immiscible electrolyte solutions) systems. Similarly, the effect of potential on the ion transfer across artificial or natural membranes could be studied by PBD.

The PBD techniques' usefulness could be significantly extended by making it more specific. This can be achieved by measuring the photothermal signal due to the optical absorption of colored ions. Using modulated laser or LED sources and phase-sensitive detectors, it is possible to measure the individual ion concentration gradients.

Cross-References

▶ Electrochemical Quartz Crystal Microbalance
▶ Ion Mobilities
▶ Radiotracer Methods
▶ Scanning Electrochemical Microscopy (SECM)
▶ UV–Vis Spectroelectrochemistry

References

1. Bard AJ, Faulkner L (1984) Electrochemical methods. Wiley, New York
2. Abruña HD (1991) Electrochemical Interfaces. Modern techniques for in-situ interface characterization. VCH, Weinheim
3. Gale RJ (1988) Spectroelectrochemistry. Theory and practice. Plenum Press, New York
4. Buttry DA (1991) Applications of the quartz crystal microbalance in electrochemistry. In: Bard AJ (ed) Electroanalytical chemistry. M. Dekker, New York, p 17
5. Shaposhnik VA, Vasileva VI, Praslov DB (1995) Concentration fields of solutions under electrodialysis with ion-exchange membranes. J Membr Sci 101:23–30
6. Lang GG, Barbero CA (2012) Laser techniques for the study of electrode processes (Monographs in electrochemistry). Springer, Berlin
7. Alda J (2003) Laser and gaussian beams propagation and transformation. In: Encyclopedia of optical engineering. Marcel Dekker, New York
8. Newman JS, Thomas-Alyea KE (2004) Electrochemical systems. Wiley-IEEE, New York
9. Barbero CA (2005) Ion exchange at the electrode/electrolyte interface studied by probe beam deflection techniques. Phys Chem Chem Phys 7:1875–1884
10. Barbero C, Calvo EJ, Miras MC, Koetz R, Haas O (2002) A probe beam deflection study of ion exchange at poly(vinylferrocene) films in aqueous and nonaqueous electrolytes. Langmuir 18:2756–2764
11. Barbero C, Miras MC, Haas O, Kötz R (1991) Direct in situ evidence for proton/anion exchange in polyaniline films by means of probe beam deflection. J Electrochem Soc 138:669–672
12. Barbero C, Planes GA, Miras MC (2001) Redox coupled ion exchange in cobalt oxide films. Electrochem Comm 3:113–116
13. Plichon V, Besbes S (1990) Mirage detection of counter-ion flux between Prussian blue films and electrolyte solutions. J Electroanal Chem 284:141–153
14. Grumelli DE, Wolosiuk A, Forzani E, Planes GA, Barbero C, Calvo EJ (2003) Probe beam deflection study of ion exchange in self-assembled redox polyelectrolyte thin films. Chem Commun 3014–3015
15. García G, Bruno MM, Planes GA, Rodriguez JL, Barbero CA, Pastor E (2008) Probe beam deflection studies of nanostructured catalyst materials for fuel cells. Phys Chem Chem Phys 10:6677–6685
16. Bruno MM, Cotella NG, Miras MC, Barbero CA (2005) Porous carbon–carbon composite replicated from a natural fibre. Chem Commun 5896–5898
17. Planes GA, Miras MC, Barbero CA (2005) Double layer properties of carbon aerogel electrodes measured by probe beam deflection and AC impedance techniques. Chem Commun 2146–2148

Protein Engineering for Electrochemical Applications

Anna Joëlle Ruff and Ulrich Schwaneberg
Lehrstuhl für Biotechnologie, RWTH Aachen
University, Aachen, Germany

Introduction

Enzymes are molecular machines to sustain life in organisms through chemical reactions. Enzymes are by natural design not engineered to be driven efficiently by electrical current since redox-active centers, which catalyze chemical reaction, are often deeply buried inside of an enzyme. The protein shell that surrounds a catalytic center functions often as an isolator and prevents "loss of electrons." The latter would lead to formation of reactive radical species which could harm organisms and futile consumption of "energy" by wasting reduction equivalents such as NADH or NADPH. Enzymes have therefore usually sophisticated electron transfer pathways and control mechanism for oxidation/reduction and reduction reactions. Furthermore, the optimal reaction conditions of enzymes in their natural environment (low substrate/product concentrations, neutral pH, viscosity, salt concentration, ambient temperature) differ from application conditions in bioelectrocatalysis, for instance, in diagnostics (e.g., glucose determination [1]; enzymatic biofuel cells [2]), fine chemical production [3–5], or applications in bleaching or biofuel production [6]. Especially attractive are applications in which cost-effective electrical current from renewable sources can directly be used to drive enzymes.

Protein engineers have developed methods for rational and evolutive protein engineering which empowers them to design enzymes that are tailored for electrochemical applications ("bioelectrozymes"). Bioelectrozymes are usually optimized in specific properties such as stability (e.g., oxidative, thermal) and electron transfer rates (higher activity, mediator acceptance/specificity). An emphasis is often

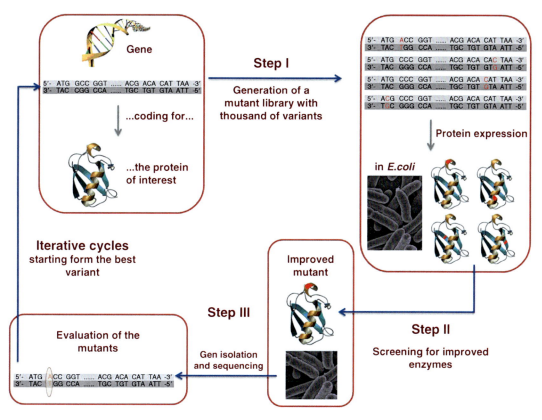

Protein Engineering for Electrochemical Applications, Fig. 1 Directed evolution scheme. Step I: Diversity generation on DNA level. Step II: Screening for improved variants on protein level. Step III: Isolating gene(s) that encode for improved variant(s) and used for the next round of evolution, starting with Step I [5]

given to the design of the interface between the electrode and enzyme to ensure an efficient electrical communication. These design principles are described in the following two paragraphs: Protein Engineering Principles and Design of Bioelectrozymes.

Protein Engineering Principles

Protein engineers can redesign enzymes by rational design and/or directed evolution. Rational design strategies are much less time consuming compared to directed evolution campaigns which require the screening of thousands of variants in iterative round of diversity generation and screening. Rational design requires however a deep molecular and structural understanding based on crystal structures or a reliable homology model to rationally reengineer properties of bioelectrozymes. Focused enzyme properties such as activity, selectivity, thermal resistance, shift of pH profiles [7, 8]. Modifications for oriented immobilization are more successfully improved by using rational design strategies and designed enzyme variants can simply be ordered as synthetic genes. Rationally not understood properties are generally tackled by directed evolution which does not require any structural information. Enzyme properties improved by directed evolution comprise stabilities toward pH, temperature, salt, organic solvent, tolerance toward high concentrations of substrate or product, and acceptance of alternative cofactor systems like mediators that shuttle electrons between electrode and enzyme. Figure 1 shows the standard directed evolution scheme in three steps (Step I: Diversity generation on the gene level,

Step II: Screening for improved variants on the protein level, and Step III: Isolating gene(s) that encode for improved variant(s) followed by subsequent iterative rounds to improve enzyme properties). Iterative rounds are important for the success of a directed evolution experiments since enzyme properties are gradually improved, for instance, activities are usually improved by a factor of 1.5–2.5 per round of evolution [3]. A standard directed evolution campaign takes usually 12–24 months and allows within physical constrains to reengineer all kind of enzyme properties that can be reflected in a screening systems. Library sizes of 1,200–2,000 variants proved often to be sufficient to find improved variants [9] despite the size of the protein sequence space. For instance, 64 Mio different peptides are available with a peptide length of six amino acids since nature provides 20 amino acid building blocks per amino acid position (20^6 different variants). Limitations in diversity generation and high throughput screening technologies that arise from the combinatorial complexity of the protein sequence space are discussed in recent reviews [3, 10]. Time requirements of 12–24 months for tailored enzymes are often still too long to fit into timelines of process developments. Notably directed enzyme evolution and natural evolution (Charles Darwin in the seventeenth century [11]) are very different in their nature: (A) Direct evolution selects on the enzyme level, natural evolution on the organism level, and (B) Directed evolution mutates *one* gene with a mutation rate of 1–5 mutations per gene and selects usually for improved variants; natural evolution can exchange and recombine genes or accumulate mutations which are not beneficial (natural drift); additionally natural mutation rates are less than 1 per 10^{6-7} bp.

Recently unifying protein engineering concepts with combined evolutive and rational design approaches are emerging in which directed evolution is used to identify hot spots or regions that are responsible for the targeted property and subsequent computational studies are used to identify and direct focus mutagenesis studies on residues in close proximity [12, 23] with a high likelihood to target residues that influence the targeted property.

Prominent oxidoreductases in bioelectrocatalysis are glucose oxidase and glucose dehydrogenase (e.g., applications in diabetes diagnostics), mono-oxygenases (e.g., important roles in synthesis of steroids and secondary metabolites), laccases (e.g., applications in textile bleaching and lignin treatment), bilirubin oxidase (e.g., biofuel cells), and peroxidases (e.g., general analytic applications). Applications of oxidoreductases in enzymatic biofuel cells are summarized comprehensively in reviews [5, 13].

Design of Bioelectrozymes

The main challenge in bioelectrozymes design is to ensure an efficient electrical communication between the electrode and the catalytic redox center of an enzyme. The glucosylation shell and/or amino acids surrounding an active-site can act as an isolator and effectively prevent an electrical communication between an electrode and a catalytic redox center. Strategies for efficient electrical communication comprise direct wiring of electrodes to the catalytic redox center through a conducting polymer [14], employment of mediators (small molecules) as electron shuttles which can come in close proximity to redox centers or electron pathways in enzymes [15, 16], and protein engineering strategies that are governed by the principle to reduce the distance between an electrode and the catalytic center or an electron transfer pathway within an enzyme [2]. Discovered engineering principles for efficient electrical communication comprise: I. Truncating enzymes at the N-, C-terminus or shortening of a loop in a protein to expose redox-active catalytic sites so that a conducting support can be positioned in close proximity (examples are a microperoxidase and a laccase [5]. II. Amino acid substitutions in close proximity to the active-site affect interactions that alter specificities (e.g., through sterical demands) or electrical communication with a mediator (potentials, charged interactions [23]). An alternative are specific binding sites that are generated at the protein surface in close proximity

Protein Engineering for Electrochemical Applications

Protein Engineering for Electrochemical Applications, Fig. 2 Example of a reaction scheme of an electrochemical-driven conversion in microtiter plate format [18] of the model substrate 12-pNCA [21] by P450 BM3 with CoSep as mediator. Catalytic activity is monitored through a yellow color development due to para-nitrophenolate formation at 410 nm

to electron pathways Mediator [15]. III. Amino acid exchanges at exposed surface regions by genetic engineering for oriented immobilization of enzymes with one or multiple contact points that are located opposite of a catalytic center [17]. Oriented immobilization on electrodes surface ensures often an efficient communication resulting in high power outputs [2, 5]. Especially the reengineering of the mediator-enzyme interface ofers promising avenues for identifying in mutant libraries improved bioelectrozymes and to match application demands, for instance, in drug metabolite detection by using an electrochemical microtiter plates ([18, 19]; see Fig. 2) or in fine chemical production [20].

Structural and molecular understanding of mediated electron transfer enables a paradigm shift from a mediator acceptance screening to a rational mediator design which considers only stability and electron transfer performance parameters. The rational mediator design would employ a subsequent enzyme engineering step to ensure an efficient electron transfer between mediator and enzyme which would open novel and exciting opportunities for enzymes in bioelectrocatalysis.

electrochemical devices), fine chemical production using cost-efficient electrochemical reactors with low down-stream processing costs, and switchable interactive materials [22]. Fundamental design principles in reengineering enzymes (see section Design of Bioelectrozymes) and designing the interface between electrode and enzyme for efficient electrical communication have been discovered in the last decade. A molecular understanding of the interface between mediator and enzyme will likely allow in the near future to develop ideal enzyme/mediator couples and overcome current strategies in which a variety of mediators are synthesize from a mediator "lead structure" that was found experimentally to drive the wild-type enzyme. A combined approach to rationally construct ideal mediators (from an application point of view) and to reengineer efficiently the interface to the interacting enzyme will ensure a rapid development of the field due to the economic benefit that tailor-made bioelectrozymes will offer in the above-mentioned application. The combinatorial complexity of the protein sequence combined with the sophisticated genetic engineering methods renders it very likely that exciting bioelectrozymes wait to be discovered in the upcoming decade.

Outlook/Future Perspective

Bioelectrozymes will very likely have a bright future in diagnostics (employed in miniaturized

Cross-References

▶ Biofilms, Electroactive

- ▶ Electrobioremediation of Organic Contaminants
- ▶ Environmentally Accepted Processes for Substitution and Reduction of Cr(VI)

References

1. Yu EH, Prodanovic R, Güven G, Ostafe R, Schwaneberg U (2011) Electrochemical oxidation of glucose using mutant glucose oxidase from directed protein evolution for biosensor and biofuel cell applications. Appl Biochem Biotechnol 165:1448–1457
2. Wong TS, Schwaneberg U (2003) Protein engineering in bioelectrocatalysis. Curr Opin Biotechnol 14:590–596
3. Ruff AJ, Dennig A, Schwaneberg U (2013) To get what we aim for: progress in diversity generation methods. FEBS J 280(13):2961–2978
4. Hollmann F, Arends IWCE, Buehler K, Schallmey A, Buehler B (2011) Enzyme-mediated oxidations for the chemist. Green Chem 13:226–265
5. Güven G, Prodanovic R, Schwaneberg U (2010) Protein engineering – an option for enzymatic biofuel cell design. Electroanalysis 22:765–775
6. Liu H, Zhu L, Bocola M, Chen N, Spiess AC, Schwaneberg U (2013) Directed laccase evolution for improved ionic liquid resistance. Green Chem 15:1348–1355
7. Jakob F, Martinez R, Mandawe J, Hellmuth H, Siegert P, Maurer K-H, Schwaneberg U (2013) Surface charge engineering of a Bacillus gibsonii subtilisin protease. Appl Microbiol Biotechnol 97(15):6793–6802
8. Martinez R, Jakob F, Tu R, Siegert P, Maurer K-H, Schwaneberg U (2013) Increasing activity and thermal resistance of *Bacillus gibsonii* alkaline protease (BgAP) by directed evolution. Biotechnol Bioeng 110:711–720
9. Tee KL, Schwaneberg U (2007) Directed evolution of oxygenases: screening systems, success stories and challenges. Comb Chem High Throughput Screen 10:197–217
10. Agresti JJ, Antipov E, Abate AR, Ahn K, Rowat AC, Baret J-C, Marquez M, Klibanov AM, Griffiths AD, Weitz DA (2010) Ultrahigh-throughput screening in drop-based microfluidics for directed evolution. Proc Natl Acad Sci USA 107:4004–4009
11. Campbell NA, Reece JB (2006) Biologie 6. überarbeitete Auflage. Pearson Studium 503–520
12. Vojcic L, Despotovic D, Maurer K-H, Zacharias M, Bocola M, Martinez R, Schwaneberg U (2013) Reengineering of subtilisin Carlsberg for oxidative resistance. Biol Chem 394:79–87
13. Zhu Z, Momeu C, Zakhartsev M, Schwaneberg U (2006) Making glucose oxidase fit for biofuel cell applications by directed protein evolution. Biosens Bioelectron 21:2046–2051
14. Katz E, Willner I (2003) A biofuel cell with electrochemically switchable and tunable power output. J Am Chem Soc 125:6803–6813
15. Shehzad A, Panneerselvam S, Linow M, Bocola M, Roccatano D, Mueller-Dieckmann J, Wilmanns M, Schwaneberg U (2013) P450 BM3 crystal structures reveal the role of the charged surface residue Lys/Arg184 in inversion of enantioselective styrene epoxidation. Chem Commun (Camb Engl) 49:4694–4696
16. Ströhle FW, Cekic SZ, Magnusson AO, Schwaneberg U, Roccatano D, Schrader J, Holtmann D (2013) A computational protocol to predict suitable redox mediators for substitution of NAD(P)H in P450 monooxygenases. J Mol Catal B: Enzym 88:47–51
17. Abian O, Grazú V, Hermoso J, González R, García JL, Fernández-Lafuente R, Guisán JM (2004) Stabilization of penicillin G acylase from *Escherichia coli*: site-directed mutagenesis of the protein surface to increase multipoint covalent attachment. Appl Environ Microbiol 70:1249–1251
18. Ley C, Schewe H, Ströhle FW, Ruff AJ, Schwaneberg U, Schrader J, Holtmann D (2013) Coupling of electrochemical and optical measurements in a microtiter plate for the fast development of electro enzymatic processes with P450s. J Mol Catal B: Enzym 92:71–78
19. Ley C, Zengin Çekiç S, Kochius S, Mangold K-M, Schwaneberg U, Schrader J, Holtmann D (2013) An electrochemical microtiter plate for parallel spectroelectrochemical measurements. Electrochim Acta 89:98–105
20. Holtmann D, Mangold K-M, Schrader J (2009) Entrapment of cytochrome P450 BM-3 in polypyrrole for electrochemically-driven biocatalysis. Biotechnol Lett 31:765–770
21. Schwaneberg U, Schmidt-Dannert C, Schmitt J, Schmid RD (1999) A continuous spectrophotometric assay for P450 BM-3, a fatty acid hydroxylating enzyme, and its mutant F87A. Anal Biochem 269:359–366
22. Van Rijn P, Tutus M, Kathrein C, Zhu L, Wessling M, Schwaneberg U, Böker A (2013) Challenges and advances in the field of self-assembled membranes. Chem Soc Rev 42:6578–6592
23. Arango Gutierrez E, Mundhada H, Meier T, Duefel H, Bocola M, Schwaneberg U (2013) Reengineered glucose oxidase for amperometric glucose determination in diabetes analytics. Biosensors and Bioelectronics 50:84–90

Pulse and Step Methods

Manuel Lohrengel
University of Düsseldorf, Düsseldorf, Germany

Introduction

If one physical parameter is suddenly changed in an electrochemical system, which is stationary or

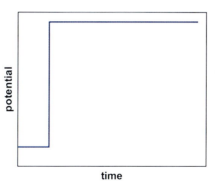 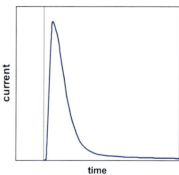

Pulse and Step Methods, Fig. 1 Schematic representation of a potential step (*left*) and the corresponding current transient (*right*). The increasing part of the current transient is dominated by the electronic setup (typically <<100 µs)

in equilibrium, the system will respond with a relaxation process to return to more stable conditions. The physical parameter may be random, such as pressure, temperature, current density, magnetic field, light intensity, electrolyte concentration, or electrode area, but in most cases the potential will be stepped because interface processes are fundamentally controlled by the potential.

In a double-step experiment the second step is often of same amplitude but inverse direction and can compensate, at least in part, effects of the first one. Such experiments are usually called pulse experiments. The intention behind step or pulse experiments will focus on two aspects:

- Analysis of the electrolyte, such as determination of components and their concentration, of concentration gradients, and, especially, of their changes with time [1–4]. Most techniques were developed for and used in polarography.
- Analysis of material and charge transport through the interface electrode/electrolyte, such as dissolution and deposition, especially adsorption, corrosion, passivation, galvanics, shaping, and polishing [4].

The response to a potential step is a current transient, a time-dependent current density (Fig. 1). These experiments are classified as chronoamperometry. Potential transient as a response to current step experiments, formerly also called charging curves [5], is related to chronopotentiometry. These transients are recorded and interpreted. Step experiments are large signal, time domain experiments and investigate processes far from equilibrium; they are not limited to special conditions, e.g., a linear response as in frequency domain techniques (impedance spectroscopy). Most restrictions come from the limits of the experimental setup, such as maximum output values and slew rates of potentiostats or galvanostats or dynamics of the recording systems.

Analysis in the Electrolyte

Chronoamperometry

The current density of a redox reaction will change, if the potential is changed. Chronoamperometry usually starts at potentials without faradaic processes. The response to a potential step, the current transient, contains the re-arrangement of the electrode interface, mainly double-layer charging, which is indicated by a current peak and can be reduced to some 10 µs in a suitable potentiostatic setup. Assuming a simple reaction

$$O + ne^- \rightarrow R \qquad (1)$$

no product R in the beginning and a planar electrode with the area A, a time-dependent faradaic current I will be observed after the peak, which can be expressed by the Cottrell equation [6]

$$I = \frac{nFAc_\infty \sqrt{D_O}}{\sqrt{\pi \cdot t}} \qquad (2)$$

with the initial concentration c_∞ and the diffusion coefficient D_O of O.

Deviations will occur for long times due to convection, for especially shaped electrodes

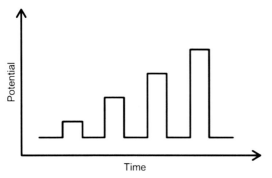

Pulse and Step Methods, Fig. 2 Schematic pulse program of normal-pulse voltammetry

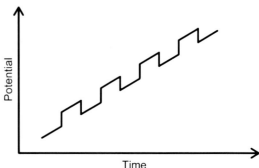

Pulse and Step Methods, Fig. 3 Schematic pulse program of differential-pulse voltammetry

(spherical or microelectrodes) and, of course, for more complex reactions.

The electrode reaction can be inverted by a second potential step of different sign and same amplitude at $t = \tau$ (double potential step or pulse chronoamperometry). The first transient follows Eq. 2 and the second one, however, is given by [7, 8]:

$$I = nFAc_\infty \sqrt{D_O} \left[\frac{1}{\sqrt{\pi(t-\tau)}} - \frac{1}{\sqrt{\pi \cdot t}} \right] \quad (3)$$

Pulse Voltammetry

Pulse voltammetry is a subset of chronoamperometry with special pulse sequences. Pulse voltammetry was introduced as *square wave polarography* [9] to increase sensitivity in polarography.

Normal-pulse voltammetry consists of a series of rectangular pulses with increasing amplitude (Fig. 2). The potential between the pulses is chosen so that no reaction occurs. The currents are measured just before the end of the pulses to suppress effects of interface charging. The analysis follows the Cottrell equation.

Series of rectangular pulses (typical length some 10 ms) with constant amplitude ΔE are superimposed to linear potential ramp in *differential-pulse voltammetry* (Fig. 3). The currents are detected twice per pulse, just before the pulse and before the end of the pulse. The differences are plotted versus potential of the linear sweep. Current peaks will be observed for reversible systems with heights:

$$I = \frac{nFAc\sqrt{D}}{\sqrt{\pi \cdot t}} \left[\frac{1 - \exp\left(\frac{nF\Delta E}{2RT}\right)}{1 + \exp\left(\frac{nF\Delta E}{2RT}\right)} \right] \quad (4)$$

Staircase voltammetry is similar to linear sweep techniques. The current is detected at the end of each step and, thus, suppresses interface charging effects, just as in the other techniques. The resulting current-potential curve is similar to the corresponding sweep but excludes charging currents.

Chronocoulometry

Chronocoulometry is very similar to chronoamperometry. The charge transient is recorded, which is yielded by electronic integration in analog setups or by numeric integration in digital systems. Advantages are noise reduction, as integration suppresses higher frequencies, and a more obvious separation of double-layer charging and faradaic current.

Chronopotentiometry

In chronopotentiometry, the potential responses (potential transient) to current steps are recorded. If the system is initially stationary or close to equilibrium, the potential will change rapidly due to charging of the electrode interface. This fast step corresponds to the current peak in chronoamperometry. After electrode charging, the potential becomes dominated by the reaction of electroactive species, the faradaic part.

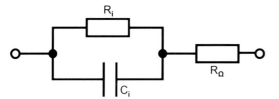

Pulse and Step Methods, Fig. 4 Simplified impedance between reference electrode and working electrode: electrolyte resistance R_Ω, interface resistance R_i, and interface capacitance C_i

Analysis of the Interface Electrode/Electrolyte

Diffusion-controlled processes in the electrolyte were investigated and analyzed by solving Fick's laws of diffusion. Effects of the impedance of the interface, e.g., charging of the electrode interface, were eliminated.

A simplified equivalent of this impedance is shown in Fig. 4. Investigations are usually carried out by impedance spectroscopy [10]. Step experiments, however, are advantageous in some cases: they are much faster (μs to ms instead of some 100 s) and are not limited to linear response (equilibrium or stationary conditions, small amplitudes).

Determination of Electrolyte Resistance

The electrolyte resistance R_Ω (Fig. 4) between reference and working electrode can be determined by short current pulses [5]. This is necessary for some concepts of iR compensation (▶ iR-Drop Elimination).

Determination of Interface Capacitance

The interface capacitance can be measured by short pulses [11–13]. The current transient of a potential step ΔU is dominated by capacitive charging of the electrode interface for short times. The interface capacitance C_i in Fig. 5 has to be charged via R_Ω with a time constant $R_\Omega \cdot C_i$. This charge Δq_i and the interface capacitance C_i are obtained from simultaneous integration of the current density i_i according to

$$C_i = \frac{\Delta q_i}{\Delta U} = \frac{\int i_i \partial t}{\Delta U} \qquad (5)$$

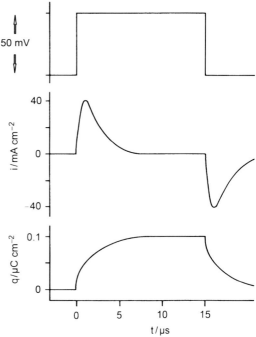

Pulse and Step Methods, Fig. 5 Measurement of interface capacitance by short pulses. From *top*: potential, current density and charge density

Typical time constants and, thus, pulse lengths are around 20 μs (Fig. 5).

An interpretation of the potential dependent capacitance (Mott-Schottky analysis [14]) yields information about semiconducting properties of the electrode. The corresponding experiment requires a modified technique in non-stationary systems, because the complete function must be measured in extremely short times to avoid fundamental system changes. This was done by superimposing small pulse to each step of a potential staircase (Fig. 6). With this technique, the complete analysis requires less than 500 μs [15].

Determination of Random Reactions

Surface reactions on the electrode are extremely complex. The current transient after a potential step can include:
- Dielectric phenomena (double-layer charging, dielectric relaxation [16])
- Adsorption (physisorption, chemisorption, ordered sublayers)

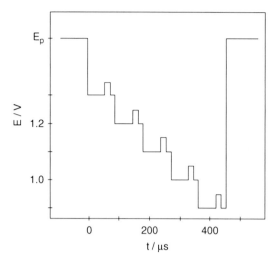

Pulse and Step Methods, Fig. 6 Example of pulse measurements of the potential dependent capacitance for Mott-Schottky analysis. The electrode was pre-polarized to E_p

- Formation of new phases (passivation, galvanic deposition, including kinetic phenomena such as nucleation [17], and charge and material transport such as solid-state diffusion, migration, tunneling of electrons or holes)
- Electrode dissolution (corrosion, electropolishing, shaping by anodic dissolution [18], often accompanied by formation of supersaturated product films).

The current density i will often follow a "universal law" [16] (well known as Curie-von Schweidler law in dielectric systems [19])

$$i \propto t^{-1} \qquad (6)$$

Less common is a current increasing with time, which is observed in autocatalytic systems, e.g., if a kinetic hindrance is overcome by the potential pulse. A description in these cases is often possible by a modified law

$$i \propto t^n \qquad (7)$$

An analysis of current transients according to Eq. 7 becomes easy in a double-logarithmic plot. The slope of linear parts yields the exponent n, as

$$\frac{\partial \log i}{\partial \log t} = n \qquad (8)$$

The size of n classifies the process: self-inhibiting processes for $n < 0$, stationary processes for $n = 0$, and autocatalytic processes for $n > 0$. Typical examples of n and corresponding reactions are [20]:

$n = -1$	High-field oxide growth; dielectric relaxation
$n = -0.5$	Diffusion
$n = 0$	Stationary corrosion; 1-dim. growth of nuclei (whisker) at a constant number of nuclei
$n = 1$	1-dim. progressive nucleation (whisker) or 2-dim. at constant number
$n = 2$	2-dim. progressive nucleation (hemisphere, dendrites) or 3-dim. at constant number
$n = 3$	3-dim. progressive nucleation

Several mechanisms may be superimposed in real systems. In this case, Eq. 2 becomes more complex:

$$i = A + B \cdot t^n + C \cdot t^m + \ldots \qquad (9)$$

Identification of single processes according to Eq. 9 requires data acquisition over one or two decades in time and current density. Therefore, the complete transient has to cover many orders of magnitude, e.g., current densities from 1 A/cm^2 to 10 nA/cm^2 and a time scale from 1 μs to 10^3 s. This requires especially designed electronic devices and strategies to handle the numerous data [21].

Future Directions

Step and pulse techniques for investigations of redox reactions in the electrolyte are well developed for many years. Future research will focus on faster and more complex reactions.

Investigations of random processes by pulse experiments, which are not limited by diffusion in the electrolyte, are much less common. The increasing interest of industry to introduce pulse techniques, such as pulse plating [22], formation of gradient layers, or pulse electrochemical machining (PECM) [23], will help making these techniques more popular.

Cross-Reference

▶ iR-Drop Elimination

References

1. Wang J (2006) Analytical Electrochemistry. Wiley-VCH, New York
2. Brett CMA, Brett AMO (1998) Electrochemistry principles methods and applications. Oxford Science, Oxford
3. Bard AJ, Faulkner LR (1980) Electrochemical methods: fundamentals and applications. Wiley, New York
4. Macdonald DD (1977) Transient techniques in electrochemistry. Plenum, New York
5. Vetter KJ (1967) Electrochemical kinetics theoretical and experimental aspects. Academic, New York
6. Cottrell FG (1903) Residual current in galvanic polarization regarded as a diffusion problem. Z Phys Chem 42:385–431
7. Smit NM, Wijnen MD (1960) Square-wave electrolysis I cyclic potential-step method. Rec Trav Chim 79:5–21
8. Kimmerle F, Chevalet J (1969) The determination of heterogeneous rate constants by double-step potentiostatic method. J Electroanal Chem 21:237–255
9. Barker GC, Jenkin IL (1952) Square-wave polarography. Analyst 77:685–696
10. Macdonald JR (1987) Impedance spectroscopy. Wiley, New York
11. Wagner C (1950) Determination of the concentrations of cation and anion vacancies in solid potassium chloride. J Electrochem Soc 97:72–74. doi:10.1063/1.1747460
12. Gilman S (1964) Electrochemical surface oxidation of platinum. Electrochim Acta 9:1025–1046. doi:10.1016/0013-4686(64)85049-0
13. Krischer CC, Osteryoung RA (1965) A method for the dynamic measurement of capacity at electrode interfaces. J Electrochem Soc 97:735–739. doi:10.1149/1.2423677
14. Stimming U, Schultze JW (1976) The capacity of passivated iron electrodes and the band structure of the passive layer. Ber Bunsenges Physik Chem 80:1297–1302
15. König U, Lohrengel MM, Schultze JW (1987) Computer supported pulse measurements of current and capacity during fast oxide formation on iron. Ber Bunsenges Physik Chem 91:426–431
16. Jonscher AK (1983) Dielectric relaxation in solids. Chelsea Dielectric Press, London
17. Schultze JW, Lohrengel MM, Roß D (1983) Nucleation and growth of anodic oxide films. Electrochim Acta 28:973–984
18. Olsson COA, Landolt D (2003) Passive films on stainless steels- chemistry structure and growth. Electrochim Acta 48:1093–1104. doi:10.1016/S0013-4686(02)00841-1
19. von Schweidler E (1907) Studien über die Anomalien im Verhalten der Dielektrika. Ann Phys 24:711–770
20. Lohrengel MM (1993) Pulse measurements for the investigation of fast electronic and ionic processes at the electrode/electrolyte interface. Ber Bunsenges Physik Chem 112:440–447
21. Lohrengel MM (1993) Thin anodic oxide layers on aluminum and other valve metals: high field regime. Mater Sci Eng R11:243–294
22. Chandrasekar MS, Pushpavanam M (2008) Pulse and pulse reverse plating- conceptual advantages and applications. Electrochim Acta 53:3313–3322. doi:10.1016/j.electacta.2007.11.054
23. Lohrengel MM (2005) Pulsed electrochemical machining of iron in $NaNO_3$ fundamentals and new aspects. Mater Manufact Proc 20:1–9

Q

Quantum Dot Sensitization

Taro Toyoda
Department of Engineering Science, The
University of Electro-Communications, Chofu,
Tokyo, Japan

Introduction

The increasing demand for renewable and low-cost energy has engendered some outstanding research in the field of next-generation solar cells. Conventional Si solar cells rely on high-quality materials, since the photo-generated carriers in the device remain in the same materials until they are extracted at the selective contacts. There is a great deal of interest in the technological applications of titanium dioxide (TiO_2) to dye-sensitized solar cells (DSCs) made with nanostructured TiO_2 electrodes, because of their high photovoltaic conversion efficiency, which exceeds 10 % [1]. In DSCs, the applications of organic dye molecules as a photosensitizer, nanostructured TiO_2 as an electron transport layer, and an iodine redox couple for hole transport dramatically improve the light harvesting efficiency. Photosensitizers can quickly separate the photo-generated carriers into two different media. In the case of a DSC, an organic dye acts as a photosensitizer on a cell formed by complementary nanostructured electron and hole transport materials. With Ru-based organic dyes adsorbed on nanostructured TiO_2 electrodes, the large surface area enables more efficient absorption of the solar light energy. Many different dye compounds have been designed and synthesized to achieve high photovoltaic conversion efficiency. It is likely that ideal photosensitizer for DSCs will be realized by co-adsorption of a few different dyes for optical absorption of an extended near infrared region [2]. However, attempts to sensitize electrodes with multiple dyes have achieved only limited success to date. Further effort is needed to improve DSCs and, in addition, the production cost should be lowered in order to replace conventional Si-based solar cells in practical applications. The main undertaking for those developing next-generation solar cells is to improve the photovoltaic conversion efficiency, together with the long time stability. One of the promising approaches is to replace the organic dyes by inorganic substances with high optical absorption characteristics and longer stability over time. Recently, as an alternative to organic dyes, semiconductor quantum dots (QDs) have been studied for their light harvesting capability compared to other sensitizers [3–8].

Semiconductor Quantum Dots as Photoelectrochemical Sensitizers

When semiconductor materials are translated to the nanoscale, new physical and chemical properties are realized for useful applications. One emerging area of nanoscience being at the

G. Kreysa et al. (eds.), *Encyclopedia of Applied Electrochemistry*, DOI 10.1007/978-1-4419-6996-5,
© Springer Science+Business Media New York 2014

interface of chemistry, physics, biology, and materials science is the field of semiconductor QDs, whose unique properties have attracted great attention by researchers during the last two decades. Different strategies for the synthesis of semiconductor QDs have been developed, so that their composition, size, shape, and surface protection can be controlled nowadays to an exceptionally high degree. Semiconductor QDs exhibit attractive characteristics as sensitizers due to their tunable bandgap (or HOMO-LUMO gap) by size control, which can be used to match the absorption spectrum to the spectral distribution of solar light. Moreover, semiconductor QDs possess higher extinction coefficients than conventional metal-organic dyes, and greater intrinsic dipole moments, leading to rapid charge separation [9]. The demonstration of multiple exciton generation (MEG) by impact ionization has fostered an interest in colloidal semiconductor QDs [10–12]. The efficient formation of more than one photo-induced electron–hole pair (exciton) upon the absorption of a single photon is a process, not only of a great current scientific interest, but is potentially important for optoelectronic devices that directly convert solar radiant energy into electricity. The demonstration of MEG by impact ionization in colloidal semiconductor QDs could push the thermodynamic photovoltaic conversion efficiency limit of solar cells by up to 44 % [13] from the current 31 % of the Shockley-Queisser detailed balance limit [14]. Accordingly, one of the most attractive configurations to exploit these fascinating properties of semiconductor QDs is the quantum dot-sensitized solar cell (QDSC) [15, 16] due to its high photoactivity, process realization, and low cost of production [17].

The general scheme of a QDSC device is similar to the DSC concept. Light excites electron–hole pairs in the QDs. The electrons are injected into nanostructured TiO_2 electrode (photosensitization), and they are transported to the transparent conductive oxide electrode. The holes are injected into the liquid or solid electrolyte which acts as a hole-transporting medium. The resulting holes are transported to the counterelectrode, where the oxidized

counterpart of the redox system is reduced [8]. QDSCs offer an undoubted potential in obtaining high photovoltaic conversion efficiency and low-cost production. The use of semiconductor QDs as sensitizers goes back to 1990s [18–23]. However, it is in the recent 10 years that semiconductor QDSCs have attracted much attention. The photovoltaic conversion efficiencies of QDSCs lags behind those of DSCs, while the use of semiconductor QDs as light absorbers requires the development of new strategies in order to push the performance of QDSCs. The poor performance of QDSCs may be ascribed to the difficulty of assembling the semiconductor QDs into a mesoporous TiO_2 matrix to obtain a well-covered QD layer on the TiO_2 crystalline surface. The other problem one would encounter is the selection of an efficient electrolyte in which the metal chalcogenide can run stably without serious degradation. Even though QDSCs still have low photovoltaic conversion efficiency, they have attracted significant attention among researchers as promising third-generation photovoltaic devices due to a rapid increase of photovoltaic conversion efficiencies, around 4–5 % at 1 sun illumination (AM 1.5, 100 mW/cm^2) [24, 25]. Moreover, the signature of MEG has also been observed in PbSe QDSCs recently. Therefore, fundamental studies on the mechanism and preparation of QDSCs are necessary and important for the boosting of photovoltaic performance of QDSCs in the field of third-generation solar cells.

The morphology of TiO_2 electrodes including crystal structure is important for satisfactory assembly of QDSCs for improving the photovoltaic conversion efficiency. On the other hand, one of the limiting factors for the overall photovoltaic performance of QDSCs is the transport of photogenerated electrons through nanostructured TiO_2 network. TiO_2 is the recipient of injected electrons from optically excited QDs and provides the conductive pathway from the site of electron injection to the transparent back-contact. Thus, TiO_2 electrodes with a higher degree of order than those conventionally made from a disordered assembly of nanoparticles are desirable for achieving high photovoltaic efficiency of

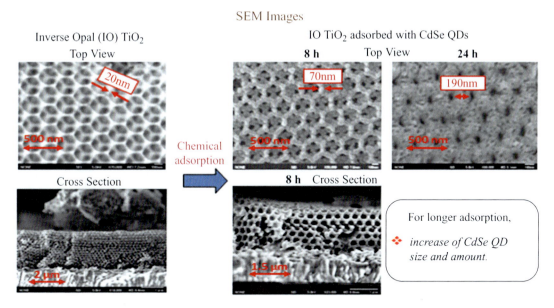

Quantum Dot Sensitization, Fig. 1 SEM images of inverse opal (IO) TiO2 electrode (left) and that adsorbed with CdSe quantum dots (right)

QDSCs through improvements in the electron transfer rate and the effective QD assembly, as well as good hole electrolyte penetration [4].

Future Perspectives

Starting from quite low photovoltaic conversion efficiencies, QDSCs have grown very rapidly to values around 4~5 %. Breakthroughs will come from:
(1) materials,
(2) surface treatments, and
(3) nanocomposite absorbers. Also, it is necessary to develop suitable combinations of QDs and nanostructured TiO2 electrodes with different morphologies to enhance photovoltaic conversion efficiencies of QDSCs (Fig. 1).

References

1. Chiba Y, Islam A, Watanabe KR, Koide N, Han L (2006) Dye-sensitized solar cells with conversion efficiency of 11.1%. Jpn J Appl Phys 43:L638–L640
2. Park B-W, Inoue T, Ogomi Y, Miyamoto A, Fujita S, Pandy SS, Hayase S (2011) Electron injection from linearly linked two dyes molecules to metal oxide nanoparticles for Dye-sensitized solar cells covering wavelength range from 400 to 950 nm. Appl Phys Express 4:012301
3. Niitsoo O, Sarker SK, Pejoux P, Rühle S, Cahen D, Hodes G (2006) Chemical bath deposited CdS/CdSe-sensitized porous TiO2 solar cells. J Photochem Photobiol A182:306–311
4. Diguna LJ, Shen Q, Kobayashi J, Toyoda T (2007) High efficiency of CdSe quantum-dot-sensitized TiO2 inverse opal solar cells. Appl Phys Lett 91:023116
5. Mora-Seró I, Bisquert J (2010) Breakthroughs in the development of semiconductor-sensitized solar cells. J Phys Chem Lett 1:3046–3052
6. Ruhle S, Shalom M, Zaban A (2010) Quantum-dot-sensitized solar cells. Chem Phys Chem 11:2290–2304
7. Emin S, Singh SP, Han L, Satoh N, Islam A (2011) Colloidal quantum dot solar cells. Solar Energy 85:1264–1282
8. Hetsch F, Xu X, Wang H, Kershaw SV, Rogach AL (2011) Semiconductor nanocrystal quantum dots as solar cell components and photosensitizers: material, charge transfer, and separation aspects of some topologies. J Phys Chem Lett 2:1879–1887
9. Underwood DF, Kippeny T, Rosenthal SJ (2001) Charge carrier dynamics in CdSe nanocrystals: implications for the use of quantum dots in novel photovoltaics. Eur Phys J D 16:241–244
10. Schaller RD, Sykora M, Pietryga JM, Klimov VI (2006) Seven excitons at a cost of one: redefining the limits for conversion efficiency of photons into charge carriers. Nano Lett 6:424–429

11. Trinh MT, Houtepen AJ, Schins JM, Hanrath T, Piris J, Knulst W, Goossens APLM, Siebbeles LDA (2008) In spite of recent doubts carrier multiplication does occur in PbSe nanocrystals. Nano Lett 8:1713–1718

12. Nozik AJ (2008) Multiple exciton generation in semiconductor quantum dots. Chem Phys Lett 457:3–11

13. Hanna MC, Nozik AJ (2006) Solar conversion efficiency of photovoltaic and photoelectrolysis cells with carrier multiplication absorbers. J Appl Phys 100:074510

14. Shockley W, Queisser HJ (1961) Detailed balance limit of efficiency of p-n junction solar cells. J Appl Phys 32:510–519

15. Nozik AJ (2002) Quantum dot solar cells. Physica E 14:115–120

16. Klimov VI (2006) Mechanism for photogeneration and recombination of multiexcitons in semiconductor nanocrystals: implications for lasing and solar energy conversion. J Phys Chem B 110:16827–16845

17. Gur I, Fromer NA, Geier ML, Alivisatos AP (2005) Air-stable, all-inorganic nanocrystal solar cells processed from solution. Science 310:462–465

18. Gopidas KR, Bohorquez M, Kamat PV (1990) Photophysical and photochemical aspects of coupled semiconductors. Charge transfer processes in colloidal CdS-TiO_2 and CdS-AgI systems. J Phys Chem 9:6435–6440

19. Vogel R, Pohl K, Weller H (1990) Sensitization of highly porous, polycrystallne TiO_2 electrodes by quantum sized CdS. Chem Phys Lett 174:241–246

20. Liu D, Kamat PV (1993) Photoelectrochemical behavior of thin CdSe and coupled TiO_2/CdSe semiconductor films. J Phys Chem 97:10769–10773

21. Vogel R, Hoyer P, Weller H (1994) Quantum-sized PbS, Ag_2S, Sb_2S_3, and Bi_2S_3 particles as sensitizers for various nanoporous wide- bandgap semiconductors. J Phys Chem 98:3138–3188

22. Zaban A, Micic OI, Gregg BA, Nozik AJ (1998) Photosensitization of nanoporous TiO_2 electrodes with InP quantum dots. Langmuir 14:3153–3156

23. Toyoda T, Saikusa K, Shen Q (1999) Photoacoustic and photocurrent studies of highly porous TiO_2 electrodes sensitized by quantum-sized CdS. Jpn J Appl Phys 38:3185–3186

24. Gónzalez-Pedro V, Xu X, Mora-Seró I, Bisquert J (2010) Modeling high-efficiency quantum dot sensitized solar cells. ACS Nano 10:5783–5790

25. Chang JA, Rhee JH, Im SH, Lee YH, Kim H-J, Seok SI, Nazeeruddin MK, Grätzel M (2010) High performance nanostructured inorganic–organic heterojunction solar cells. Nano Lett 10:2609–2612

R

Radiotracer Methods

György Inzelt
Department of Physical Chemistry, Eötvös
Loránd University, Budapest, Hungary

Introduction

The application of radiotracer methods in electrochemistry dates back to the pioneering works by György Hevesy in 1914. The aim of these studies was to demonstrate that isotopic elements can replace each other in both electrodeposition and equilibrium processes (Nernst law). Nevertheless, Joliot's fundamental work in 1930 is considered by electrochemists as a landmark in the application of radiochemical (nuclear) methods in electrochemistry.

Radiochemical methods, such as tracer methods [1–3], Mössbauer spectroscopy [4], neutron activation [5], thin layer activation (TLA) [5], ultrathin layer activation (UTLA) [5], and positron lifetime spectroscopy [6], are applied for the study of a wide range of electrochemical surface processes. The most important areas are as follows: adsorption and electrosorption occurring on the surface of electrodes; the role of electrosorption in electrocatalysis; deposition and dissolution of metals; corrosion processes; the formation of surface layers, films on electrodes (e.g., polymer films), and investigation of migration processes in these films; study of the dynamics of electrosorption and electrode processes under steady-state and equilibrium conditions (exchange and mobility of surface species); and electroanalytical methods (e.g., radiopolarography).

Application of Radiotracers in Electrochemistry

The main advantages of using radiotracers are as follows:
- The radiation emitted by radiotracers is generally easy to detect and measure with high precision.
- The radiation emitted is independent of pressure, temperature, and chemical and physical state.
- Radiotracers do not affect the system and can be used in nondestructive techniques.
- The radiation intensities measured furnish direct information concerning the amount of the labeled species, and no special models are required to draw quantitative conclusions.

Metal Dissolution

In the case of metal dissolution studies the principle of the methods used is based on the labeling of a component of the metal phase by one of its radioactive isotopes and calculating the dissolution rate of the metal specimen by measuring either the increase in radiation coming from the solution phase or the decrease in radiation coming from the solid phase.

G. Kreysa et al. (eds.), *Encyclopedia of Applied Electrochemistry*, DOI 10.1007/978-1-4419-6996-5,
© Springer Science+Business Media New York 2014

The main steps here are as follows:

- Introduction of the radioisotope into the specimen. The task can be achieved at least by three methods: (a) through melting, (b) by electrolytic deposition of the radioactive metal, and (c) by subjecting the metal specimen to neutron irradiation in a nuclear reactor.
- Measurement of the changes in radiation intensity caused by the dissolution process.
- A variation is the application of backscattering for the study of the electrochemical formation and dissolution of thin metal layers.

Electrosorption

Various methods have been developed for in situ radiotracer adsorption studies depending on the requirements of the problems to be studied. In case of in situ studies, the central problem is how to separate the signal (radiation) to be measured from the background radiation and how to attain the optimal ratio of these quantities.

From this point of view methods can be divided into two main groups:

- Radiation of the solution background is governed and minimized by self-absorption of the radiation, i.e., by the attenuation of the radiation intensity by the radioactive medium itself (thin-foil method).
- Background radiation intensity is minimized by mechanical means (thin-gap method, electrode lowering technique).

As to the role of the labeled species in the radiotracer study of adsorption phenomena, two different versions of the method may be distinguished. In the first, the direct method, the species to be studied is labeled and the radiation measured gives direct information on the adsorption of this species. However, this method cannot be used in several cases owing to technical restrictions related to the very nature of the radiotracer method (the available concentration range is limited; no distinction can be made between the adsorption of the labeled compound studied and that of a product formed from it; the number of commercially available labeled compounds is restricted).

In the case of indirect radiotracer methods, instead of labeling the species to be studied another adequately chosen labeled species (the indicator species) is added to the system, and the adsorption of this component is followed by the usual radiotracer measuring technique. The adsorption of the indicator species should be related to that of the species studied. The nature of this link could be different in different systems.

Radiotracer technique offers a unique possibility of demonstrating the occurrence of specific adsorption of an ion by labeling it and studying its adsorption in the presence of a great excess of other ions, electrolytes (supporting electrolyte). For these studies at least a difference of one to two orders of magnitude in concentrations should be considered. Under such conditions with nonspecific adsorption, determined by coulombic interactions, no significant adsorption of the labeled species, present in low concentration, could be observed. In contrast to this, the observation of a measurable adsorption can be considered as a proof of the occurrence of the specific adsorption. Using isotopes emitting soft β^--radiation, (e.g., ^{14}C, ^{35}S, ^{36}Cl), the self-absorption of the radiation in the solution phase is so high that the thickness of the solution layer effective in the measured solution background radiation is very low. In this case the surface concentration, Γ, can be easily calculated from the radiation intensity measured (I_T):

$$\Gamma = \frac{I_a}{I_s} \cdot \frac{c}{\mu\rho\gamma} = \frac{I_T - I_s}{I_s} \frac{c}{\mu\rho\gamma}$$
$$= \left(\frac{I_T}{I_s} - 1\right) \frac{c}{\mu\rho\gamma} \quad (1)$$

where I_s and I_a are the intensities of the radiation coming from the solution phase and from the adsorbed layer, respectively, c is the concentration of the labeled species present in the solution phase, μ is the mass absorption coefficient of the radiation, γ is the roughness factor of the electrode, and ρ is the density of the solution (Fig. 1).

Various versions of tracer methods were applied for studies of electrodes with polycrystalline smooth or rough surfaces and well-defined surfaces [3].

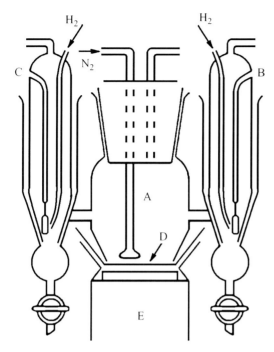

Radiotracer Methods, Fig. 1 Three-electrode cell used for the radiotracer studies. (*A*) Is the main compartment that contains the electrolyte and the labeled compound (ions). (*B* and *C*) Are the compartments of the reference and counter electrodes, respectively. (*D*) Is the working electrode, usually a gold-covered plastic foil (on the gold other metals and/or films can be deposited in advance). (*E*) Is the scintillation detector

Future Directions

It is a well-established in situ technique in electrochemistry. As described the novel radiochemical methods are continuously built in its arsenal widening the opportunities of the application of radiotracer methods.

References

1. Horányi G (1999) Radiotracer studies of adsorption/sorption phenomena at electrode surfaces. In: Wieckowski A (ed) Interfacial electrochemistry, theory, experiment, and applications. Marcel Dekker, New York, pp 477–491
2. Horányi G (ed) (2004) Radiotracer studies of interfaces. In: Interface science and technology, vol 3. Elsevier, Amsterdam
3. Krauskopf EK, Chan K, Wieckowski A (1987) In situ radiochemical characterization of adsorbates at smooth electrode surfaces. J Phys Chem 91:2327–2332
4. Kálmán E, Lakatos M, Kármán FH, Nagy F, Klencsár Z, Vértes A (2005) Mössbauer spectroscopy for characterization of corrosion products and electrochemically formed layers. In: Freund HE, Zewi I (eds) Corrosion reviews. Freund Publishing House, Tel Aviv, pp 1–106
5. Horányi G, Kálmán E (2005) Recent developments in the application of radiotracer methods in corrosion studies. In: Marcus PH, Mansfeld F (eds) Analytical methods in corrosion science and engineering. CRC, Boca Raton, pp 283–333
6. Süvegh K, Horányi TS, Vértes A (1988) Characterization of the β-Ni(OH)2/β-NiOOH system by positron lifetime spectroscopy. Electrochim Acta 33:1061–1066

Raman Spectroelectrochemistry

Salma Bilal
National Centre of Excellence in Physical Chemistry, University of Peshawar, Peshawar, Pakistan

Introduction

Raman Spectroscopy

Raman spectroscopy (named after C. V. Raman) is a spectroscopic technique used to study vibrational, rotational, and other low-frequency modes in a system. Generally Raman spectroscopy relies on inelastic or Raman scattering of monochromatic light, usually from a laser in the visible, near-infrared, or near ultraviolet range. The Raman effect occurs when light impinges upon a molecule and interacts with the electron cloud of the bonds of that molecule. Typically, a sample is illuminated with a laser beam. Scattered light from the illuminated spot is collected with a lens and sent through a monochromator. The laser light interacts with phonons or other excitations in the system, resulting in the energy of the laser photons being shifted up or down. The shift in energy gives information about the phonon modes in the system.

**Raman Spectroelectrochemistry,
Fig. 1** Schematic diagram of Raman scattering

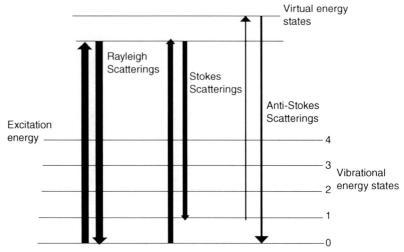

For a molecule it is necessary that the molecular vibration or rotation must cause a change in a component of molecular polarizability in order to be Raman active. During the experiment radiation scattered by the sample is detected. Wavelengths close to the laser line (due to elastic Rayleigh scattering) are filtered out, and those in a certain spectral window away from the laser line are dispersed onto a CCD detector. The Raman scattering depends upon the polarizability (the amount of deformation of electron cloud) of the molecules. The amount of the polarizability and its change of the bond will determine the intensity and frequency of the Raman shift. The photon excites one of the electrons into a virtual state. For a molecule it is necessary that the molecular vibration or rotation must cause a change in a component of molecular polarizability. During the experiment radiation scattered by the sample is detected. When the photon is released the molecule relaxes back into a vibrational energy state. The molecule will typically relax into the first vibrational energy state and this generates Stokes Raman scattering. If the molecule is excited from an elevated vibrational energy state, it may relax into the vibrational ground energy state and the Raman scattering is then called anti-Stokes Raman scattering. Since most molecules are in their vibrational ground state at ambient temperature, the intensities of Stokes lines are higher than those of the anti-Stokes lines. This is the reason why only the Stokes lines are recorded as the Raman spectrum (Fig. 1).

The development of modern Raman spectrometers has led to a more efficient registration of weak Raman signals. Spectra that in the old equipments took several hours to register can often be recorded within few minutes, using a modern spectrometer. In the early spectrometers, the separation of Rayleigh and Raman scattering was accomplished by double or triple monochromators, but after this complex process with many components, only a small amount reaches the detector. The development of efficient notch filters allows simple separation of Raman and Rayleigh light and usually more than 30 % of the scattered light can be detected. Since the scattered light is detected more efficiently, low-power lasers (about 20 mW) can be used. The advantages are that these lasers are relatively cheap, do not require external cooling, and are less likely to burn the sample. The wide range of accessible wavelengths (100–4,000 cm^{-1}), the possible resolution (1 cm^{-1} or better), fast response of the detection setup, and the lack of interference by water (the most commonly employed electrolyte in electrochemistry) because of its weak Raman scattering have further added plus points to this development.

Raman Spectroelectrochemistry

Raman spectroscopy can be coupled with an electrochemical setup, since vibrational information is very specific for the chemical bonds in molecules and between molecules and electrode surfaces.

This technique can also be used for the study of more complex systems such as complicated carbon nanostructures or electrodes modified with a film of an intrinsically conducting polymer. The dependence of the observed features (in the Raman spectra of polyconjugated molecules) on the excitation line wavelength is an intrinsic property of these macromolecular systems [1]. This phenomenon significantly complicates vibrational investigations of conducting polymers based on Raman spectroscopy. Raman analysis of conducting polymers requires the registration of the spectra not only for different oxidation states of the polymer but also using different excitation wavelengths. Figure 2a shows development of Raman spectra during charging of a polyaniline (PANI) film in 0.5 M H_2SO_4, deposited on a gold disk electrode. Sample was illuminated with 514.5 Ar^+-ion laser Coherent Innova 70. When recorded at different applied potentials, the Raman spectra of the conducting polymer film provide useful information about the changes occurring in the polymer structure upon oxidation or reduction. At lower potential the spectra show only two bands, but as the potential increases the intensity of the bands increases and new bands appear showing the change in the structure of polymer with changing potential up to $E_{SCE} = 0.2$ V and then decreases with further potential increase. Normally the intensity of a particular band increases with increasing potential if it is related to the redox process of the polymer. At potentials more positive than the formal redox potential, these bands show a decreasing trend, while bands corresponding to the reduced form of the polymeric material show a decreasing trend with increasing potential. The appearance of strong bands around 1,620 cm^{-1} at lower potentials corresponds to the benzenoid structure in the polymer. Appearance of the strong band at 1,169 cm^{-1} and around 1,510 and 1,579 cm^{-1} is caused by modes of the quinoid structure, characteristics of the oxidized form of the polymer. With red excitation line ($l_0 = 647.1$ nm, Fig. 2b) the spectra show features different from the one in Fig. 2. Numerous bands that were not observed in Fig. 2 can clearly be seen here. Hence, the combination of recording Raman spectra at different excitation wavelengths can provide useful information about the changes in the structure of the sample upon changing the potential.

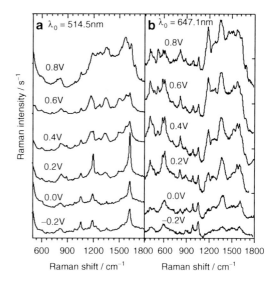

Raman Spectroelectrochemistry, Fig. 2 In situ Raman spectra of polyaniline in 0.5 M H_2SO_4 at different electrode potentials and different excitation wavelengths, as indicated

Surface-Enhanced Raman Spectroscopy (SERS)

A great disadvantage in any application of Raman spectroscopy is the extremely small cross-section of Raman scattering, between 10^{-13} and 10^{-26} cm^2/molecule. The larger value is obtained under resonance Raman conditions. With the discovery of a particular enhancement effect (upto 10^6) by Fleischmann et al. [2], that affects only species in close contact with the metal electrode surface and slightly later by Jeanmaire et al. [3] demonstrated surprisingly the feasibility of vibrational studies of electrochemical interfaces with Raman spectroscopy (for an overview, see [4]). Because of the large enhancement, small sample volumes can be used and current detections are in picomole to femtomole range [5]. Surface-enhanced Raman spectra can only be obtained if the molecules to be investigated are adsorbed to the so-called SERS-active substrate. The most common SERS-active substrates are coinage metals (i.e., copper, silver, and gold), island films consisting of small metal particles deposited on glass surface, cold-deposited films, lithographically produced metal spheroids, etc. [6].

If the laser excitation wavelength matches with the one of the band maxima in the UV–vis spectra of the sample, the intensity of the scattered light is strongly increased. This is called "resonance enhancement" [7]. The problem with low sensitivity of Raman signals can be overcome when surface-enhanced resonance Raman (SERRS) is used. An enhancement factor of about 10^{10} and 10^{11} has been claimed for a dye molecule in SERRS experiments. In SERRS it is generally agreed that different effects must contribute to the enhancement of the Raman signals.

Preparation of the electrode is a main step in electrochemical SERS studies. In general, the surface of the electrode is roughened to provide the desired enhancement and an artificially increased surface area. As a common practice, roughening of the electrode is done by applying oxidation–reduction electrode potential cycling in a suitable electrolyte solution. Chloride-containing electrolyte solutions have frequently been used for coinage metals. Although chloride strongly adsorbs on these surfaces, however, studies have shown that the adsorbed chlorides can be washed off completely, leaving no spectroscopic evidence in the subsequently recorded spectra. Surface roughening can also be done by keeping the electrode potential at the upper limit for some time. Further information on electrochemical surface treatment procedure can be found elsewhere [8]. The enhancement mechanisms can be divided into electromagnetic field enhancement and chemical enhancement. In the former mechanism the electromagnetic field of the light at the surface

is enhanced under conditions of surface plasmon resonance. If the surface plasmons are in resonance with the incident light, then both the resonance laser light and scattered Raman light are amplified through their interaction with the surface. Coinage metals are dominant as SERRS substrate as the resonance condition is satisfied at the visible frequencies, commonly used for Raman spectroscopy. The chemical enhancement mechanism is related to specific interactions, i.e., electronic coupling between metal and molecule. The interaction results in an increased cross-section of the Raman signal of the adsorbed molecule in SERRS experiment, compared to the cross-section of the free molecule in a normal Raman measurement. The chemical enhancement mechanism is restricted to molecules in the first layer of metal, since electronic transitions are only possible when the metal and molecules are in close contact [9, 10]. It has been known since early SERS observations that the relative intensities of SERS bands are different from those of free molecules in solution, and it has been speculated that this may be connected with the effect of orientation of the adsorbed molecule on the surface [11]. The electromagnetic enhancement theory explains the dependence of the Raman signal intensity on molecular orientation by describing how the electric field at the surface of metal particles couples with the polarizability tensors [12]. The greatest surface enhancement is observed for vibration that involves changes in the molecular polarizability perpendicular to the metal surface. Comparison of the relative band intensities in SERS and normal Raman spectra allow the average orientation and distance from the surface of the adsorbate functional groups to be determined [13]. Some other terminologies related to SERS are as follows:

- Surface-Unenhanced Raman Spectroscopy (SUERS)
- This terminology is used when non-coinage metals like platinum, illuminating wavelengths that are known not to support surface enhancement (i.e., 514.5 nm for gold electrode) or smooth surfaces, are used.

- Deenhanced Surface Raman Spectroscopy (DESERS)
- Spectroscopy performed with surfaces where enhancement has been quenched by deposition/adsorption of foreign metals atoms has been designated as deenhanced SERS (DESERS), and for a complete overview and further details see [14–16].

Applications of Raman Spectroelectrochemistry

Raman spectroelectrochemistry is the method of choice to analyze the structure of all kinds of materials, e.g.:

1. Carbon materials like graphite, graphene, carbon nanotubes, and fullerenes.
2. Information about crystalline or amorphous structure and crystallite size can be obtained from shift and shape of vibration modes.
3. Useful information can be obtained about the changes occurring in the structure of the sample material upon charging.
4. The state and quality of nanostructured materials for battery research as well as lithium content can be concluded by their characteristic vibration modes.
5. Worthy information can be obtained for the orientation of molecules on the surface of electrode in case of self-assembled monolayers (SERS).

References

1. Louarn G, Lapkowski M, Quillard S, Pron A, Buisson PJ, Lefrant S (1996) Vibrational properties of polyaniline-isotope effect. J Phys Chem 100:6998
2. Fleischmann M, Hendra PJ, McQuillan AJ (1974) Raman spectra of pyridine adsorbed at a silver electrode. Chem Phys Lett 26:163
3. Jeanmair DL, van Duyne RP (1977) Surface Raman spectroelectrochemistry: part I. Heterocyclic, aromatic, and aliphatic amines adsorbed on the anodized silver electrode. J Electroanal Chem 84:1
4. Haynes CL, MacFarland AD, van RP D (2005) Surface-enhanced Raman spectroscopy. Anal Chem 77:338A
5. Kneipp K, Kneipp H, Itzkan I, Dasari RR, Feld MS, Zander C, Enderlein J, Keller RA (2002) Single molecule detection in solution: methods and applications. Wiley-VCH Verlag, Berlin

6. Moskovits M (1985) Surface-enhanced spectroscopy. Rev Mod Phys 57:783
7. Sathayanarayana DN (2004) Theory and application. In: Vibrational spectroscopy. New Age International, New Delhi
8. Holze R (1994) Spectroscopic methods in electrochemistry: new tools for old problems. Bull Electrochem 10:45
9. Campion A, Kambhampati P (1998) Surface-enhanced Raman scattering. Chem Soc Rev 27:241
10. Ruperez A, Laserrna JJ (1996) Modern techniques in Raman spectroscopy. Wiley, Chichester
11. Creighton JA (1983) Surface raman electromagnetic enhancement factors for molecules at the surface of small isolated metal spheres: the determination of adsorbate orientation from sers relative intensities. Surf Sci 124:209
12. Grabbe ES, Buck RP (1989) Surface-enhanced Raman spectroscopic investigation of human immunoglobulin G adsorbed on a silver electrode. J Am Chem Soc 111:8362
13. Herne TM, Ahern AM, Garell RL (1991) Surface-enhanced Raman spectroscopy of tripeptides adsorbed on colloidal silver. Anal Chim Acta 246:75
14. Holze R (2009) Surface and interface analysis: an electrochemists tool box. Springer, Berlin/Heidelberg
15. Mayer P, Holze R (2003) Pyridine as a probe molecule for surface enhanced Raman spectroscopy of the silver-modified glassy carbon/solution interface. Surf Sci 522:55
16. Kania S, Holze R (1998) Surface enhanced Raman spectroscopy of anions adsorbed on foreign metal modified gold electrodes. Surf Sci 408:252

Rare Earth Metal Production by Molten Salt Electrolysis

Hongmin Zhu
School of Metallurgical and Ecological Engineering, University of Science and Technology Beijing, Beijing, China

Introduction

Rare earth metals (REM) include 15 lanthanides elements (Ln) from atomic number 57–71 located in the IIIB group of the periodic table and scandium (Sc) and yttrium (Y) elements, as listed in Table 1. The most common oxidation state for the rare earth metal elements is RE^{3+}, but some other

Rare Earth Metal Production by Molten Salt Electrolysis, Table 1 Classification of the rare earth metals

Element	Symbol	Atomic number	Melting point/°C	Valence
Scandium	Sc	21	1,539	3
Yttrium	Y	39	1,509	3
Lanthanum	La	57	920	3
Cerium	Ce	58	795	3,4
Praseodymium	Pr	59	935	3,4
Neodymium	Nd	60	1,024	3
Promethium	Pm	61		3
Samarium	Sm	62	1,072	2,3
Europium	Eu	63	826	2,3
Gadolinium	Gd	64	1,312	3
Terbium	Tb	65	1,356	3,4
Dysprosium	Dy	66	1,407	3
Holmium	Ho	67	1,470	
Erbium	Er	68	1,522	
Thulium	Tm	69	1,545	
Ytterbium	Yb	70	824	
Lutetium	Lu	71	1,656	

oxidation states, RE^{2+} and RE^{4+}, are occasionally found. For example, cerium (Ce), and to a much lesser extent Pr and Tb, can form RE^{4+} ions, and Sm, Eu, and to a lesser extent Yb can form RE^{2+} ions. These deviations from "normal" behavior (formation of only RE^{3+}) are sometimes attributed to the special stability of empty, half-filled, or filled shells: $Ce^{4+}(4f^0)$, $Eu^{2+}(4f^7)$, $Yb^{2+}(4f^{14})$, but $Pr^{4+}(4f^1)$ and $Sm^{2+}(4f^6)$ do not fully satisfy this criterion.

The high activity of REM makes it impossible to be obtained from an aqueous solution; therefore, currently most of the REM have been produced by molten salt electrolysis and some by thermal reduction method due to the existence of multivalent ions in the molten salts as shown in Table 1. Compared to the thermal reduction method, molten salt electrolysis is a relatively economical one because the process is continuous and easy to control. And it has been widely used in the industry to produce a single rare earth metal such as La, Ce, Pr, Nd and mixed rare earth metal alloys.

Usually two types of electrolyte systems in the industry have been applied, one is chloride system composing of $RECl_3$–KCl and another is fluoride–oxide system of RE_2O_3–REF_3–LiF [1]. Currently, in the USA, Japan, and Kazakhstan, fluoride–oxide system is employed to produce the mixed rare earth metals with a 20–25 kA capacity. While Germany applies chloride system to produce rare earth metals and the current capacity reaches 50 kA. Before 1997 most of the rare earth metals in China were produced from chloride system, only a small amount of rare earth metals from fluoride–oxide system. Since 2000, China has completely employed fluoride–oxide system to produce rare earth metals. Usually, the current efficiency is dependent on the electrolyte system, and less than 50 % for chloride system and less than 87 % for fluoride–oxide system are shown in Table 2. This is caused by the multiple oxidation states of the rare earth metal elements in the molten salts.

The most common raw materials for the REM molten salt electrolysis are in the RE^{3+} state, such as RE_2O_3, $RECl_3$. But RE^{2+} still exists to a certain extent in the molten salts, especially in the chloride melts, some rare earth metal elements have presented a higher level of divalent oxidation states, such as neodymium, samarium, europium, dysprosium, thulium, and ytterbium, which result in a lower current efficiency. For Sm and Eu molten salt electrolysis processes, even no metals can be obtained at the cathodes due to a cyclic transformation of $Sm^{2+}/Sm^{3+}(Eu^{2+}/Eu^{3+})$ and $Sm^{3+}/Sm^{2+}(Eu^{3+}/Eu^{2+})$ on the electrodes during electrolysis. And some of the rare earth metal elements show tetravalent oxidation states at the chlorine pressure far in excess of atmospheric pressure, such as Ce^{4+}. Most of the rare earth metal elements in oxidation state of +4 are not stable in chloride melts, because the reaction occurs according to the following equation: $RE^{4+} + Cl^- = RE^{3+} + 1/2Cl_2$.

However, the multiple oxidation states are also dependent on the electrolyte composition and temperature. In general, the RE^{2+} is more stable in the chloride melts than in the fluoride melts.

Electrolyte System

Usually, binary or ternary systems are used as the supporting electrolyte to satisfy the

Rare Earth Metal Production by Molten Salt Electrolysis, Table 2 The parameters for REM and REM alloy production by molten salt electrolysis

	Chloride system		Fluoride–oxide system	
Rare earth metals	Electrolyte composition	Current efficiency	Electrolyte composition	Current efficiency
La	$LaCl_3$–KCl	30–50 %	LaF_3–LiF–La_2O_3	Max 87 %
Ce	$CeCl_3$–KCl	30–50 %	CcF_3–LiF–Ce_2O_3	
Pr	$PrCl_3$–KCl	35–40 %	PrF_3–LiF–Pr_2O_3	Max 80 % $\sim$ 82 %, 10kA, max 24kA
Nd	$NdCl_3$–KCl	Very low	NdF_3–LiF–Nd_2O_3	80 % $\sim$ 82 %,10kA
Gd			GaF_3–LiF–Ga_2O_3 Ga–Fe alloy	
Dy			D_Y–Fe alloy	Max 80 %
Yb			YbF_3–LiF–Yb_2O_3	
Mixed rare earth metals	$ReCl_3$–KCl	35–50 %	RE_2O_3–REF_3–LiF	65 %

The current efficiency is obtained from the industrial data

requirements for viscosity, density, conductivity, and melting point of the molten salts. When $RECl_3$ is selected as the raw material, according to the theoretical decomposition voltages of chlorides in Table 3, Li (Na, KCl, Rb, Cs)Cl or Be (Mg, Ca)Cl_2 are suitable for the supporting electrolytes, usually KCl–$RECl_3$ is applied in the industry due to the comprehensive properties. Most of the rare earth metal fluorides have presented higher theoretical decomposition voltages than their chlorides, and some even are close to the alkaline and alkaline earth fluorides. Only LiF, CaF_2, SrF_2, and BaF_2 can satisfy the requirements for the supporting electrolyte for the active rare earth metal oxide components such as Nd_2O_3 and La_2O_3 as given in Table 4. The electrolyte is generally composed of Re_2O_3–LiF–ReF_3.

Electrode Reactions

Chloride Melts

The typical electrode reactions for the chlorides molten salt electrolysis are as follows:

The cathode reaction:

$$RE^{3+} + 3e = RE \qquad (1)$$

Rare Earth Metal Production by Molten Salt Electrolysis, Table 3 The theoretical decomposition voltages for chlorides

T/°C E/V	400	600	800	1,000	1,200	1,400
LiCl	3.66	3.49	3.37	3.26	3.16	3.06
NaCl	3.61	3.42	3.24	3.14	3.00	2.89
KCl	3.86	3.66	3.48	3.35	3.23	3.11
RbCl	3.84	3.65	3.48	3.35	3.22	3.10
CsCl	3.90	3.71	3.56	3.43	3.31	3.20
$BeCl_2$	2.04	1.92	1.81	1.72	1.63	1.55
$MgCl_2$	2.77	2.61	2.47	2.36	2.25	2.14
$CaCl_2$	3.59	3.44	3.29	3.17	3.05	2.93
$SrCl_2$	3.74	3.59	3.44	3.31	3.19	3.07
$BaCl_2$	3.90	3.74	3.58	3.43	3.31	3.19
YCl_3	2.90	2.75	2.60	2.48	2.36	2.24
$LaCl_3$	3.12	2.96	2.81	2.68	2.57	2.46
$CeCl_3$	3.06	2.90	2.75	2.63	2.51	2.41
$PrCl_3$	3.07	2.91	2.76	2.64	2.52	2.41
$NdCl_3$	3.03	2.87	2.72	2.60	2.49	2.38
$PmCl_3$	2.90	2.66	2.41	2.14	1.86	1.57
$SmCl_3$	2.98	2.81	2.68	2.55	2.43	2.31
$EuCl_3$	2.63	2.46	2.33	2.19	2.06	1.94
$GdCl_3$	2.90	2.74	2.61	2.50	2.39	2.28
$TbCl_3$	2.88	2.73	2.60	2.49	2.38	2.27
$DyCl_3$	2.80	2.62	2.45	2.30	2.16	2.02
$HoCl_3$	2.91	2.75	2.60	2.47	2.35	2.23
$ErCl_3$	2.84	2.68	2.52	2.38	2.25	2.13
$TmCl_3$	2.81	2.64	2.48	2.33	2.20	2.07
$YbCl_3$	2.75	2.59	2.43	2.29	2.15	2.02
$LuCl_3$	2.81	2.59	2.36	2.12	1.86	1.60

Rare Earth Metal Production by Molten Salt Electrolysis, Table 4 The theoretical decomposition voltages for fluorides

T/°C E/V	400	600	800	1,000	1,200	1,400
LiF	5.72	5.52	5.32	5.16	5.02	4.89
NaF	5.24	5.02	4.80	4.60	4.45	4.31
KF	5.18	4.97	4.76	4.60	4.45	4.31
RbF	5.08	4.88	4.68	4.54	4.40	4.27
CsF	5.10	4.91	4.75	4.63	4.51	4.39
BeF_2	4.78	4.62	4.48	4.34	4.20	4.06
MgF_2	5.21	5.04	4.86	4.68	4.50	4.35
CaF_2	5.76	5.59	5.42	5.25	5.09	4.93
SrF_2	5.71	5.54	5.37	5.20	5.04	4.88
BaF_2	5.69	5.52	5.36	5.20	5.05	4.90
YF_3	5.37	5.21	5.06	4.90	4.77	4.65
LaF_3	5.29	5.13	4.96	4.80	4.65	4.49
CeF_3	5.25	5.08	4.92	4.75	4.59	4.43
PrF_3	5.25	5.09	4.93	4.77	4.61	4.45
NdF_3	5.22	5.06	4.90	4.74	4.58	4.42
PmF_3						
SmF_3	5.18	5.01	4.85	4.69	4.53	4.38
EuF_3	4.84	4.67	4.50	4.34	4.19	4.04
GdF_3	5.28	5.11	4.95	4.79	4.63	4.50
TbF_3	5.31	5.14	4.98	4.82	4.67	4.54
DyF_3	5.25	5.09	4.93	4.77	4.62	4.50
HoF_3	5.27	5.10	4.94	4.78	4.63	4.49
ErF_3	5.26	5.10	4.93	4.78	4.63	4.51
TmF_3	4.60	4.44	4.27	4.11	3.97	3.85
YbF_3	4.86	4.70	4.55	4.40	4.26	4.14
LuF_3	5.21	5.04	4.88	4.72	4.57	4.45

On the anode reaction:

$$2Cl^- = Cl_2(g) + 2e \qquad (2)$$

Therefore, the total reaction is

$$2RECl_3 = 2RE + 3Cl_2(g) \qquad (3)$$

Chlorides system is suitable to produce the rare earth metals or intermediate alloys with low melting point. The REM including La, Ce, Pr, and some mixed rare earth metal alloys have been produced from the chloride system. The disadvantages for the chlorides electrolysis lie in high hygroscopicity and volatility of electrolyte and low current efficiency.

Fluoride–Oxide Melts

In the fluoride–oxide system, the corresponding electrode reactions are the following:

The cathode reaction

$$RE^{3+} + 3e = RE \qquad (4)$$

The anode reaction

$$3O^{2-} + 2C = CO + CO_2 + 6e \qquad (5)$$

The total reaction

$$RE_2O_3 + 2C = 2RE + CO + CO_2 \qquad (6)$$

Fluoride–oxide system is suitable to produce high melting point rare earth metals and has been successfully used in production in La, Ce, Pr, Nb, and mixed rare earth alloys with a higher current efficiency than chloride system. But the electricity energy consumption is as high as 10 kWh/kg (REM), while the theoretical value for Nd electrolysis from Nd_2O_3–LiF–NdF_3 melts is 1.334 kWh/kg-Nd, showing utilization efficiency of the electricity energy of less than 16 %. Currently the widely used cell capacity in industry is 5 kA though the highest single cell capacity has already reached 25 kA.

Chlorides Electrolysis

Current Efficiency

It is reported that the maximum current efficiency (CE) for chlorides electrolysis is 70 %, usually in the range of 30–50 % in the industry depending on the operational conditions such as the water content of the raw materials, current density, temperature, and impurities. Especially, the water in the raw materials greatly reduces the current efficiency [2]. Under the same operation conditions, the CE value is mainly related to RE^{2+} stability in the melts. The existence of RE^{2+} in the melts will make the electrode reaction Eq. 7 occur prior to RE^{3+} reduction to metal RE, and the reduction product RE will react with RE^{3+} to form RE^{2+} according to Eq. 8, resulting in the reduction product RE transfer to RE^{2+}, and therefore current efficiency is reduced by reducing the amount of the RE.

$$RE^{3+} + e \rightarrow RE^{2+} \qquad (7)$$

$$RE^{3+} + RE \rightarrow RE2^{+} \qquad (8)$$

The higher the stability of RE^{2+} is, the lower the current efficiency is. The stability of RE^{2+} is dependent on the rare earth metal element itself and electrolyte composition. For example, Sm, Eu, and Yb form highly stable RE^{2+} ions in the chloride melts and are unable to obtain corresponding metals by molten salt electrolysis due to a very low current efficiency. If Sm, Eu, and/or Yb ions exist in the chloride system as trace impurities, the current efficiency will be lowered significantly. This is explained later by the electrochemical analysis results. Besides Sm, Eu, and Yb, nearly all the rare earth metal elements form RE^{2+} ions in the chloride melts. For example, in alkaline chloride melts, electrochemical investigation has showed that Nd^{3+} is reduced on an inert cathode through two steps, according to Eqs. 9 and 10. The presence of Nd^{2+} will cause a lower current efficiency. So the chloride electrolytic process is not subject to Nd production in industry. In addition to Nd, the other REM also have similar trends to Nd.

$$Nd^{3+} + e = Nd^{2+} \qquad (9)$$

$$Nd^{2+} + 2e = Nd^{0} \qquad (10)$$

The possibility of formation and existence of the different oxidation states in molten chlorides is confirmed by direct measurements of redox potentials of lanthanides that were performed by a potentiometric method. The ratio between their concentrations ($[Ln^{3+}]/[Ln^{2+}]$) is related to the redox potentials through the Eq. 11 [3].

$$E_{Ln3+/Ln2+} = E_{Ln3+/Ln2+}{}^{*} + RT/nF \ln Ln^{3+}/Ln^{2+} \qquad (11)$$

where $E_{Ln3+/Ln2+}{}^{*}$ is a conditional standard redox potentials.

Sm^{3+} Electrochemical Behaviors in Chloride Melts

Sm^{3+}/Sm^{2+} Transformation in Chloride Melts

Stability of samarium ions (Sm^{3+}, Sm^{2+}) in the alkaline chloride melts changes as functions of the solvent salt cations and temperature [4]. Sm^{3+} exhibits a higher stability for a larger solvent salt cation and lower temperature. Electrochemical reduction of Sm^{3+} into Sm^{0} in KCl–NaCl–CsCl melt at an inert cathode has been found to occur in two steps as shown in Eqs. 12 and 13. And the reduction of Sm^{2+} to Sm^{0} takes place at near the decomposition potential of the supporting electrolyte. In addition, Sm^{2+} losing one electron to form Sm^{3+} takes place at the anode in terms of reaction Eq. 14, making $Sm^{2+} \rightarrow Sm^{3+}$/transformation at the electrodes; therefore, this process can circulate in the cathode and anode, and therefore nearly no Sm metal can be obtained at the cathode, resulting an extremely low current efficiency. This is the reason why samarium cannot be produced from the chloride melts by molten salt electrolysis. It is reported that when the concentration of Sm^{3+} ions reach 0.1 wt% in the chloride melts, the current efficiency will be substantially decreased. Eu^{3+} behaves in nearly the same manner as Sm^{3+} in the chloride melts.

$$Sm^{3+} + e = Sm^{2+} \qquad (12)$$

$$Sm^{2+} + e = Sm^{0} \qquad (13)$$

$$Sm^{2+} - e = Sm^{3+} \qquad (14)$$

The Redox Potentials of Sm^{3+}/Sm^{2+}

The conditional standard potentials $E^{0}{}_{(Ln3+/Ln2+)}$ on the inert electrode [5] are strongly affected by the salt solvents. For example, at 1,073 K, these potentials were -1.880 and -0.742 V in molten equimolar KCl–CsCl mixture [6], -1.966 and -0.844 V in molten KCl [7], and -2.087 and -0.981 V in molten CsCl, for samarium and europium, respectively. The dependence of the conditional standard potentials Sm^{3+}/Sm^{2+} changes inversely with the cationic radii of the solvent.

While samarium ions are electrochemically reduced on an active electrode, such as nickel, silver, and aluminum electrodes, Sm alloy is formed with the electrode material according to Eqs. 15 and 16:

$$Sm^{2+} + 2e \ (Al) = Sm_xAl_y(alloy) \qquad (15)$$

$$Sm^{2+} + 2e \ (Ag) = Sm_xAg_y(alloy) \qquad (16)$$

For example, electroreduction of samarium ions in KCl–NaCl–CsCl eutectic melts on Pt electrode involves reduction of Sm^{3+} into Sm^{2+} at the potentials of -1.2 to -1.4 V, and the reduction at more negative potentials about -2.3 to -2.4 V corresponds to that of Sm^{2+} to Sm^0 and formation of Pt_xSm_y intermetallide confirmed by the voltammetry [4].

The Stability of RE²⁺ in the Chloride Melts

The stability of the divalent and trivalent rare earth metals ions may be judged upon by comparing the standard Gibbs energies of the $RECl_2$ and $RECl_3$ formation in a given condition. The relative stability of the two valences in the molten salts can be expressed by the equilibrium constant for the disproportionation reaction according to Eq. 17. Obviously, the more constant the equilibrium of the reaction to the right hand is, the more instable for the divalence is, and vice versa. The stability of the ions is dependent on the temperature, cation, and anion types, for example, divalence of the rare earth metals is more stable in the chloride system than in the fluoride–oxide system:

$$3RE^{2+} = 2RE^{3+} + RE \qquad (17)$$

Fluoride–Oxide Electrolysis

Current Efficiency

Currently, molten fluoride–oxide electrolysis has been widely applied to produce a single rare earth metal such as La, Ce, Pr, and Nd and mixed rare earth metals such as Gd–Fe, Dy–Fe Ho–Fe, and Er–Fe alloys in industry. The maximum cell capacity has reached 25 kA, and the current efficiency is usually above 80 % with an energy consumption of about 10.6 kWh/kg. Clearly, fluoride–oxide electrolysis has shown a higher current efficiency than chloride electrolysis. This is attributed to the positive effects of F^- on the stability of the higher oxidation state

Nd³⁺ Cathodic Processes

LiF–NdF₃

According to references [8, 9], the reduction process involves one step Eq. 18 in mixed chloride–fluoride melts, while two steps in pure chloride melts as shown in Eqs. 8 and 9.

$$Nd^{3+} + 3e \rightarrow Nd \qquad (18)$$

The addition of fluoride ions in chloride melts can stabilize the higher valences of cationic species, and thus, the intermediate step Eq. 9 is not pronounced in the fluoride–chloride melts [8]. Cyclic voltammetry [10] on a molybdenum electrode in $LiF–NdF_3$ melts at 810 °C has proved that reduction of Nd^{3+} ions into Nd in a single step; this is also confirmed by C. Hamela [11]. It is proved that the electrochemical reduction process is controlled by the diffusion of neodymium ions in the melts.

LiF–NdF₃–Nd₂O₃ System

The voltammetric characteristics of $LiF–NdF_3$ and $LiF–NdF_3–Nd_2O_3$ on tungsten electrodes at 900 °C are shown in Figs. 1 and 2 [12]. Due to much higher concentration of $[NdF_6]^{3-}$ species in the melts, the voltammetric characteristics for the two melts with or without Nd_2O_3 are similar.

The structure of the trivalent neodymium fluoride species in the melt is the octahedral $[NdF_6]^{3-}$ complex anion identified by Raman spectroscopy [13]. The dissolution of Nd_2O_3 results in the formation of neodymium oxyfluorides in the melts. It also has been found that the $[NdOF_5]^{4-}$ is present [10]. Actually, $[NdF_6]^{3-}$ and $[NdOF_5]^{4-}$ coexist in the melt since the concentration of NdF_3 is much

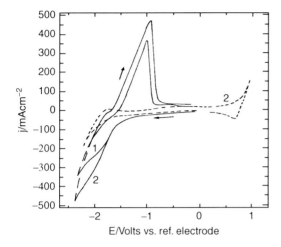

Rare Earth Metal Production by Molten Salt Electrolysis, Fig. 1 CV of fused LiF without (– – –) and with (—) NdF$_3$ (1 mol%) on tungsten electrode at 900 °C. dE/dt (mV s^{-1}) : (1) 20, (2) 100 [10]

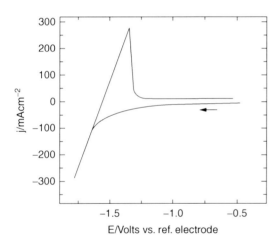

Rare Earth Metal Production by Molten Salt Electrolysis, Fig. 2 CV of LiF–NdF$_3$ eutectic composition melt with Nd$_2$O$_3$ on tungsten electrode at 900 °C; dE/dt = 100 mV s^{-1} [10]

higher than that of Nd$_2$O$_3$. While at the anode the oxidation of neodymium oxyfluorides [NdOF$_5$]$^{4-}$ generates oxygen according to the following Eqs. 19 and 20. This has been confirmed by the Ref. [14] that oxygen generation at the anode during the electrolytic production of neodymium from oxide–fluoride melts. The produced oxygen subsequently reacts with the carbon anode and then CO and CO$_2$ gases are produced.

$$2[\text{NdF}_6]^{3-} + 6e = 2\text{Nd}(s) + 12\,\text{F}^- \quad (19)$$

$$3[\text{NdOF}_5]^{4-} + 3\,\text{F}^- = 3/2\,\text{O}_2(g) + 3[\text{NdF}_6]^{3-} + 6e \quad (20)$$

However, in LiF–NdF$_3$–Nd$_2$O$_3$ melts, whether NdF$_3$ or Nd$_2$O$_3$ is consumed at the electrode is still controversial. Tamamura et al. [15] claim that NdF$_3$ is the raw material in the electrolytic production of neodymium in oxide–fluoride melts. According to Kaneko et al. [14], Stefanidaki [13], and Keller [16], only Nd$_2$O$_3$ is consumed by controlling the cell voltage in a low region. More work is needed to prove the above points.

Conclusions

In this chapter, the two electrolyte systems are introduced to produce rare earth metals by molten salt electrolysis. The involved electrode processes, current efficiency, and the oxidation states of the rare earth metal ions and their stability have been discussed. The lower current efficiency for chloride melts is caused by the higher stability of divalent ions of rare earth metals in the melts. Fluoride ions have lowered the stability of divalent ions; therefore a higher current efficiency is reached in the fluoride–oxide system. As an example, the electrochemical reduction process for NdF$_3$ and Nd$_2$O$_3$ has been discussed.

Future Directions

In the last two decades the world production of rare earth has increased remarkably. However, compared to the modern aluminium electrolysis Héroult-Hall cell, the electricity energy consumption for rare earth electrolysis is still very high due to the low current efficiency ($\leq 80\%$) and very high cell voltage (≥ 10V). Therefore, reducing the energy consumption for rare earth electrolysis will be the major future direction. It is also worth to mention that the anode gas management for the rare earth electrolysis should be considered to eliminate the emission of perfluorocarbons (PFCs, CF_4 and $C2F_6$).

Cross-References

▶ Refractory Metal Production by Molten Salt Electrolysis

References

1. Hecheng Lin (2000) World nonferrous metal. 3:9–13
2. Zhiqiang Zhao, Haijun Jia, Bingyi Yan (2005) Science and Technology of Baotou Steel Corporation, vol. 31, Supplement. Nov 2005
3. Khokhlov VA, Novoselova AV, Nikolaeva EV, Tkacheva OY, Salyulev AB (2007) Russ J Electrochem 43(8):961–967
4. Kushkhov KB, Vindizheva MK, Karashaeva RA, Tlenkopachevz MR (2010) Russ J Electrochem 46(6):691–701
5. Novoselova AV, Khokhlov VA, Shishkin VY (2001) Russ J Appl Chem 74(10):1672–1677
6. Novoselova AV, Shishkin VY, Khokhlov VA (1999) Rasplavy 6:32339
7. La–Lu Sc Y (1982) Gmelin handbook of inorganic chemistry: Rare Earth Elements, Part C 4b. Springer, Berlin
8. Mottot Y (1986) Thesis, University of Pierre et Marie Curie
9. Castrillejo Y, Bermejo MR, Barrado E, Martinez AM, Diaz Arocas P (2003) J Electroanal Chem 545:141
10. Stefanidaki E, Hasiotis C, Kontoyannis C (2001) Electrochim Acta 46:2665–2670
11. Hamela C, Chamelot P, Taxil P (2004) Electrochim Acta 49:4467–4476
12. Thoma RE (1966) Progress in science technology of the rare earths, vol 2. Pergamon Press, New York, p 90
13. Stefanidaki E, Photiadis GM, Hasiotis C, Kontoyannis C (2000) In: Berg R, Hjuler HA (eds) Progress in molten salt chemistry, vol 1. Elsevier, Paris, p 505
14. Kaneko A, Yamamoto Y, Okada C (1993) J Alloys Compd 193:44
15. Tamamura H, Shimooka T, Utsunomiya M (1990) Proc Seventh Int Symp Molten Salts 90–17:611
16. Keller R, Larimer KT (1997) Rare earths: science, technology and applications III. The Minerals, Metals & Materials Society, 175

Reactive Metal Electrode

Hirofumi Maekawa
Department of Materials Science and Technology, Nagaoka University of Technology, Nagaoka, Japan

Definition, Characteristics, and Advantages

In electrochemistry, a typical reaction cell is mainly composed of anode, cathode, reference electrode, and diaphragm. In most of the cathodic reduction, a diaphragm is required, which sometimes demands continuous electrolysis under high voltage. Some kinds of metal are used as an anode to decrease voltage between electrodes in practical electrochemical synthesis. Such electrodes are called sacrificial electrode, sacrificial anode, or sacrificial metal anode, and it is a nice tool in organic electrochemical synthesis [1–4]. Ionization of the metal material at the anode means the supply of one electron to the anode. The generated electron is used for passing current. Recently, sacrificial metal anode is positively applied to electrochemical synthesis, and the metal ion derived from anode sometimes plays important roles in chemical reaction and gives remarkable influences to regiochemistry, stereochemistry, selectivity of the products by coordination, catalytic effects, formation of a new reagent in situ, etc. [5, 6]. In these cases, the electrode is especially called reactive electrode, reactive anode, or reactive metal anode, but "reactive electrode" may be used as the same technical term instead of sacrificial electrode. Magnesium, aluminum, and zinc are generally available for reactive metal anodes, and an aprotic polar solvent such as DMF or NMP is frequently used as the solvent. There are also some reports on mixed electrode of carbon and sulfur [7] and on the Reformatsky reaction by behavior of the anode surface as activated zinc metal [8].

Application of reactive metal electrode gives several advantages in electrochemical synthesis.

Reactive Metal Electrode, Scheme 1 Stereoselective synthesis of pinocols

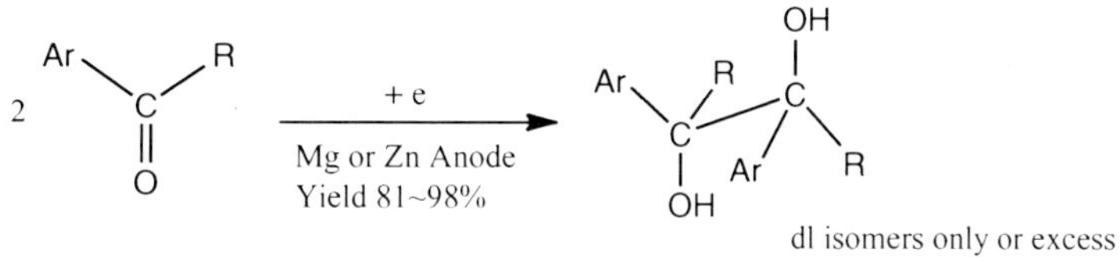

dl isomers only or excess

Reactive Metal Electrode, Scheme 2 Diacylation of Styrenes

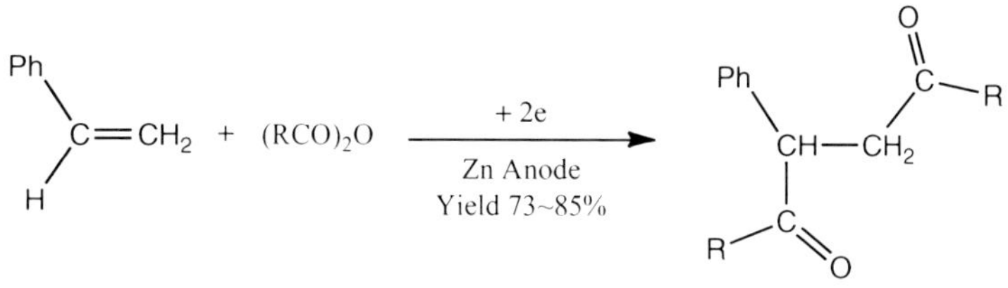

First one is electrolysis under low voltage without diaphragm. Second one is decreasing the amount of supporting electrolyte because the generated metal ion will continuously dissolve in the solvent and help carrying the current. Third one is appearance of different selectivity by behavior of the metal ion under electrolysis. In some cases, good choice of reactive metal anode material sometimes gets an organic compound with the metal ion to form an organometallic reagent and improves the yield of the reaction. Furthermore, suppression of oxidizing organic compounds at the anode is also a positive effect of reactive metal anode. A suitable reactive metal anode has to be chosen with consideration on redox potential of reagents and products.

Examples and Synthetic Application

In the early years after development of sacrificial electrode (reactive metal anode), some reviews were reported [1, 2, 4]. Many experimental results are also reported in recent years and typical examples of the reactions are shown below:

1. Diastereoselective Synthesis of Pinacols [9]

   Electrochemical reduction of benzaldehydes and aromatic ketones with zinc or magnesium anode gave the corresponding homo-coupling compounds, 1,2-diol, with high diastereoselectivity in high yield. Stereochemistry of the product is controlled by oxygen-metal-chelate complex. The metal ion is derived from the anode material in the electrochemical reaction.

2. Fixation of Carbon Dioxide [10]

   Electrochemical carboxylation of organic halides with aluminum, zinc, or magnesium anode brought about the selective formation of carboxylic acids in high yield. Carboxylation of organic halides suffered from esterification of carboxylate anion with unreacted organic halide. This method enabled us to synthesize carboxylic acids by trapping carboxylate anion with metal ion.

3. Synthesis of Drugs [11]

   Electrochemical carboxylation of a-halogenated arenes under the carbon dioxide atmosphere with zinc anode afforded the effective formation of phenylpropionic acid derivatives in high yield. This method is application of carboxylation of aromatic halides with reactive metal anode to synthesis of *anti*-inflammatory pharmaceutical drugs.

4. Nickel-Catalyzed Electroreductive Coupling [12]

   Nickel-catalyzed electroreductive homo-coupling reaction of organic halides and coupling reaction between organic halide and activated olefins could be carried out with iron anode in ionic liquid. With addition of small amount of DMF, coupling compounds were obtained at room temperature in good to high yield. This method is characterized by use of aromatic halide and nickel salt.

5. Electrochemical Synthesis of Polysilanes [13]

Electrochemical reduction of 1,1-dichlorodialkylsilanes and 1,2-dichlorotetraalkyldisilanes with magnesium anode in an undivided cell gave the corresponding polysilanes. Magnesium electrode was only effective for this synthesis of polysilanes, and platinum, zinc, nickel, and copper were less effective. This method is an application of this reactive metal anode to attractive polymer synthesis.

6. 1,2-Diacylation of Styrenes and Activated Olefins [14]

Electroreduction of styrenes or alkyl methacrylates in the presence of aliphatic acid anhydrides or N-acylimidazoles with zinc anode brought about novel one-pot vicinal double C-acylation to afford the corresponding 1,4-diketones in good to high yields. This is an example of cross-coupling reaction with no use of halogenated compounds.

Future Directions

It is quite interesting to know what the reactive metal anode plays a role in, but it is not easy to analyze the effects because the role of the reactive metal anode may be different, depending on a kind of the metal or each reaction. However, development of reactive metal anode exploited a new field of electrochemical synthesis in this 30 years. Typical examples are selective synthesis of carboxylated compounds, cross-coupling reaction of activated halogenated compounds and carbonyl compounds, and cross-coupling reaction between non-halogenated compounds. In recent years, many types of transition metal-catalyzed reactions have been reported in the field of organic synthesis, and some professors received the Nobel Prize on cross-coupling reactions. From the view of electrochemical synthesis and green sustainable chemistry, novel methods of reactive metal anode will be developed with involving application of transition metal-catalyzed reactions, ionic liquids, recovery of anode material, elemental strategy, etc.

Cross-References

▶ Electrochemical Fixation of Carbon Dioxide (Cathodic Reduction in the Presence of Carbon Dioxide)
▶ Electrosynthesis in Ionic Liquid
▶ Electrosynthesis of Polysilane

References

1. Chaussard J, Folest JC, Nedelec JY, Perichon J, Sibille S, Troupel M (1990) Use of sacrificial anodes in electrochemical functionalization of organic halides. Synthesis 5:369–381. doi:10.1055/s-1990-26880
2. Silvestri G, Gambino S, Filardo G (1991) Use of sacrificial anodes in synthetic electrochemistry. Processes involving carbon dioxide. Acta Chem Scand 45:987–992. doi:10.3891/acta.chem.scand.45-0987
3. Lund H, Hammerich O (Eds.) (2001) Organic electrochemistry 4th ed. Marcel Dekker, New York
4. Tokuda M (2006) Efficient fixation of carbon dioxide by electrolysis – facile synthesis of useful carboxylic acids – J Natur Gas Chem 15:275–281. doi:10.1016/S1003-9953(07)60006-1
5. Lehmkuhl H (1973) Preparative scope of organometallic electrochemistry. Synthesis 7:377–396. doi:10.1055/s-1973-22221
6. Tuck DG (1979) Direct electrochemical synthesis of inorganic and organometallic compounds. Pure Appl Chem 51:2005–2018. doi:10.1351/pac197951102005
7. Guillanton GL Do QT Simonet J (1990) Electrogenerated electrophilic reagents from sulphur: a new access to bis-methoxyaryl sulphides. J. C. S. Chem Commun: 1990:393–394. doi:10.1039/C39900000393
8. Schwarz KH, Kleiner K, Ludwig R, Schick H (1992) Electrochemically supported Reformatskii reaction: a convenient preparation of 2-substituted 1-ethyl 3-oxoalkanedioates. J Org Chem 57:4013–4015. doi:10.1021/jo00040a051
9. Thomas HG, Littmann K (1990) The use of sacrificial anodes for diastereoselective formation of pinacols. Synlett 12:757–758. doi:10.1055/s-1990-21241
10. Silvestri G, Gambino S, Filardo G, Gulotta A (1984) Sacrificial anodes in the electrocarboxylation of organic chlorides. Angew Chem Int Ed Engl 23:979–980. doi:10.1002/anie.198409791
11. Yamauchi Y, Hara S, Senboku H (2010) Synthesis of 2-aryl-3,3,3-trifluoropropanoic acids using electrochemical carboxylation of (1-bromo-2,2,2-trifluoroethyl)arenes and its application to the

synthesis of β, β, β-trifluorinated non-steroidal *anti*-inflammatory drug. Tetrahedron 65:473–479. doi:10.1016/j.tet.2009.11.053

12. Barhdadi R Courtinard C Nédélec JY Troupel M (2003) Room-temperature ionic liquids as new solvents for organic electrosynthesis. The first examples of direct or nickel-catalysed electroreductive coupling involving organic halides. Chem Comm 1434–1435. doi:10.1039/B302944A

13. Shono T Kashimura S Ishifune M Nishida R (1990) Electroreductive formation of polysilanes. J Chem Soc Chem Comm: 1160–1161. doi:10.1039/C39900001160

14. Yamamoto Y, Goda S, Maekawa H, Nishiguchi I (2003) Novel one-pot vicinal double C-acylation of styrenes and methacrylates by electroreduction. Org Lett 15:2755–2758. doi:10.1021/ol035020v

Reconstituted Redox Proteins on Surfaces for Bioelectronic Applications

Claudia Ley[1] and Dirk Holtmann[2]
[1]DECHEMA Research Institute, Frankfurt am Main, Germany
[2]DECHEMA Research Institute of Biochemical Engineering, Frankfurt am Main, Germany

Introduction

Many redox proteins contain a cofactor or more precisely a prosthetic group as a nondiffusible organic or inorganic compound located at the enzyme's active site. The cofactor plays an essential role for the enzyme's catalytic activity [1]. It is linked firmly to the protein backbone, and the linkage may be of non-covalent or covalent nature and is often accompanied by additional interactions between the cofactor and its protein surrounding (e.g., ionic or hydrophobic). The most prominent examples for cofactors of organic origin are heme and flavin adenine dinucleotide (FAD) which can be found in myoglobin and hemoglobin or in case of FAD in glucose oxidase (Fig. 1). Furthermore, pyrroloquinoline quinone (PQQ), the cofactor of, e.g., certain alcohol dehydrogenases, is of interest since it functions not only as cofactor but also as redox shuttle [1].

Inorganic cofactors like certain metal clusters can be found, e.g., in nitrogenases and hydrogenases. A further example is photosystem I, a protein complex participating in photosynthesis, which displays both, a quinone cofactor as well as several iron-sulfur clusters [2].

Extraction of the Prosthetic Group from the Holoprotein

The prosthetic group can be removed from the holoprotein, which is the complete functional protein unit, to yield the cofactor-free apoprotein (Fig. 2). This can be done in several ways which include dialysis, extraction with an organic solvent, or chromatographic steps. Most prominent is the method developed by Teale for the extraction of non-covalently bound heme [3]. After acidification of the solution, resulting in partial denaturation of the protein, the heme cofactor is extracted with methyl ethyl ketone yielding an organic heme-containing and an aqueous apoprotein-containing phase. The method has been applied, e.g., for myoglobin, hemoglobin, horseradish peroxidase, and catalase [3]. The preparation of the apoprotein as well as the reconstitution depends strongly on protein structure and the position of the cofactor in the protein [1]. For example, the heme of myoglobin is located close to the surface of the protein, and preparation of functional myoglobin apoprotein is well described. In contrast for P450 enzymes, apoprotein preparation is not easy since the protein has to be denatured to a high extent to access the deeply buried heme. Nevertheless, some interesting approaches have been developed to overcome these problems like heme transfer from the P450 to apomyoglobin [4]. When the prosthetic group is linked covalently to the protein, one option is also the direct expression of the apoprotein [5].

Apoprotein Reconstitution

The apoprotein can be reconstituted with its natural cofactor or an artificial one to obtain

Reconstituted Redox Proteins on Surfaces for Bioelectronic Applications,
Fig. 1 Structures of three prominent cofactors: (**a**) heme b (or iron protoporphyrin IX), (**b**) FAD, and (**c**) PQQ

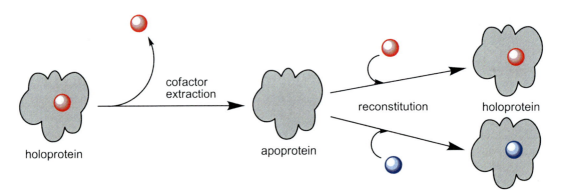

Reconstituted Redox Proteins on Surfaces for Bioelectronic Applications, Fig. 2 Extraction of the cofactor (*red*) from the holoprotein (*left*) yielding the apoprotein (*center*). The apoprotein can be reconstituted with the natural (*red*) or an artificial cofactor (*blue*) to yield again a functional holoprotein (*right*)

a functional holoenzyme (Fig. 2). The holoenzyme often displays a changed catalytic behavior when reconstituted with a nonnatural cofactor. In this way, it has been possible to evaluate catalytic mechanisms and gain insight into enzyme kinetics. The introduction of a modified cofactor which lacks the covalent binding site opens up the possibility to get information about the nature of the covalent bonding and if this bonding is necessary for catalysis. Furthermore, by changing substituents at the cofactor, the relation between its structure and electron transfer properties yields important information [6].

Reconstituted Redox Proteins at Electrode Surfaces

Several studies deal with direct electrochemistry of reconstituted myoglobin at hydrophilic electrode surfaces with nonnatural cofactors where, for example, substituents at the heme porphyrin ring or the central metal cation have been changed [6]. Thereby incorporation of an artificial heme resulted in shifts of the redox potential, alteration in electrode transfer kinetics, as well as different ligand-binding properties. In bioelectronic applications, redox proteins can normally not be attenuated directly via the electrode since they possess an insulating protein shell. One way to establish an electrical contact between the electrode and the protein's active site is the reconstitution with a cofactor to which an electroactive group has been bound so that the accessibility of the active site for electron transfer is improved. This has been done, for example, with myoglobin, to whose heme cofactor a photoactivable ruthenium group has been coupled [7]. As a result for the ruthenium reconstituted myoglobin electron transfer could be induced by light irradiation which is not seen for native myoglobin [7]. Another example for a photochemical approach is the reconstitution of the flavoprotein glucose oxidase at a photoisomerizable monolayer to which FAD was bound. In this study, the assembly could be converted from an insulating to a charge-transporting state by light irradiation [8]. Furthermore, glucose oxidase and D-amino acid oxidase have been reconstituted with ferrocene-modified FAD [9]. The artificial holoenzymes show direct electrical communication with the electrode and respond well to different substrate concentrations implying a possible use for amperometric biosensors. Also coupling of a ferrocene unit to hemin and reconstitution with horseradish peroxidase yielded a catalytically active enzyme which displays electroactivity [10]. The redox active groups introduced work as an electron relay or wire and enable a long-range electron transfer. The attachment of the cofactor to gold nanoparticles, or other nanomaterials like carbon nanotubes, facilitates electron transfer to the active site as it has been shown for glucose oxidase or PQQ (Fig. 3) [11, 12]. This is done in combination with immobilization of the nanomaterial-bound cofactor at an electrode surface where the nanoparticle may serve as an electron relay to the macroelectrode. Besides the direct access to the active site, the advantage of immobilizing the cofactor at an electrode surfaces is also the short distance between the electrode and the enzyme and the defined orientation of the reconstituted protein. Electrode surface modifications include mainly self-assembled monolayer systems (Fig. 3) or functionalized polymers. In addition to nanoparticles, the immobilization of PQQ at modified electrodes should be mentioned since it is used for the production of enzyme electrodes due to its electron shuttling function (Fig. 3). For example, apo-glucose dehydrogenase was reconstituted at a polyaniline/polyacrylic acid polymer modified with PQQ [13], and cholesterol oxidase was reconstituted with FAD which has been complexed to a PQQ-containing SAM [14]. Other interesting approaches include the reconstitution with DNA-modified cofactors, which are characterized by their good molecular recognition properties [15]. DNA oligomers were bound to hemin and reconstitution with apomyoglobin and apo-horseradish peroxidase to give catalytically active enzymes. The reconstituted proteins were immobilized at a surface by hybridization of the heme-bound DNA oligomer with its surface-bound counterpart [15]. Also in vivo reconstitution of apo-glucose dehydrogenase in *E.coli* cells was observed electrochemically at carbon paste electrodes [16].

Future Directions

Reconstituted proteins are of special interest for applications in biofuel cells, biosensors, bioelectronics, and nanobiotechnology since direct electron transfer between an electrode and

Reconstituted Redox Proteins on Surfaces for Bioelectronic Applications, Fig. 3 Immobilization of cofactors at gold electrodes (*left*) modified with self-assembled monolayers (SAM). (**a**) Immobilization of hemin at a SAM. This assembly was subsequently used for the reconstitution of horseradish peroxidase [17]. (**b**) Immobilization of the cofactor PQQ at an Au nanoparticle, which in turn is immobilized at the electrode via a SAM, for the reconstitution of glucose dehydrogenase [12]. (**c**) Immobilization of the cofactor FAD for reconstitution of glucose oxidase [18]. As electron relay, PQQ has been immobilized via a SAM in between the electrode and the cofactor

the cofactor at the enzymes active site is established. The reconstitution method enables a direct manipulation of the active site and may be used for production of tailor-made biocatalysts converting substrates of interest. Due to the unique orientation of reconstituted proteins at surfaces, they serve as molecular scaffolds for nanostructuring of surfaces.

Cross-References

- ▶ Biosensors, Electrochemical
- ▶ Cofactor Regeneration, Electrochemical
- ▶ Cofactor Substitution, Mediated Electron Transfer to Enzymes
- ▶ Direct Electron Transfer to Enzymes

References

1. Fruk L, Kuo C-H, Torres E, Niemeyer CM (2009) Rekonstitution von Apoenzymen als chemisches Werkzeug für die strukturelle Enzymologie und Biotechnologie. Angew Chem 121:1578–1603
2. Iwaki M, Itoh S (1989) Electron transfer in spinach photosystem I reaction center containing benzo-, naphtho- and anthraquinones in place of phylloquinone. FEBS Lett 256:11–16
3. Teale FWJ (1959) Cleavage of the heam-protein link by acid methylethylketone. Biochim Biophys Acta 35:543
4. Muller-Eberhard U, Liem HH, Yu CA, Gunsalus IC (1969) Removal of Heme from cytochrome P-450$_{CAM}$ by hemopexin and apomyoglobin associated with loss of P-450 hydroxylase activity. Biochem Biophys Res Commun 35:229–235
5. Kim J, Fuller JH, Kuusk V, Cunane L, Z-w C, Mathews FS, McIntire WS (1995) The cytochrome

subunit is necessary for covalent FAD attachment to the flavoprotein subunit of p-cresol methylhydroxylase. J Biol Chem 270:31202–31209

6. Mie Y, Sonoda K, Neya S, Funasaki N, Taniguchi I (1998) Electrochemistry of myoglobins reconstituted with azahemes and mesohemes. Bioelectrochem Bioenerg 46:175–184

7. Hamachi I, Tanaka S, Shinkai S (1993) Light-driven activation of reconstituted myoglobin with a ruthenium tris(2,2'-bipyridine) pendant. J Am Chem Soc 115:10458–10459

8. Yehezkeli O, Moshe M, Tel-Vered R, Feng Y, Li Y, Tian H, Willner I (2010) Switchable photochemical/electrochemical wiring of glucose oxidase with electrodes. Analyst 135:474–476

9. Riklin A, Katz E, Willner I, Stocker A, Bückmann AF (1995) Improving enzyme-electrode contacts by redox modification of cofactors. Nature 376:672–675

10. Ryabov AD, Goral VN, Gorton L, Csöregi E (1999) Electrochemically and catalytically active reconstituted horseradish peroxidase with ferrocene-modified hemin and an artificial binding site. Chem Eur J 5:961–967

11. Xiao Y, Patolsky F, Katz E, Hainfeld JF, Willner I (2003) "Plugging into enzymes": nanowiring of redox enzymes by a gold nanoparticle. Science 299:1877–1881

12. Zayats M, Katz E, Baron R, Willner I (2005) Reconstitution of apo-glucose dehydrogenase on pyrroloquinoline quinone-functionalized au nanoparticles yields an electrically contacted biocatalyst. J Am Chem Soc 127:12400–12406

13. Laurinavicius V, Kurtinaitiene B, Liauksminas V, Ramanavicius A, Meskys R, Rudomanskis R, Skotheim T, Boguslavsky L (1999) Oxygen insensitive glucose biosensor based on PQQ-dependent glucose dehydrogenase. Anal Lett 32:299–316

14. Vidal J-C, Espuelas J, Castillo J-R (2004) Amperometric cholesterol biosensor based on in situ reconstituted cholesterol oxidase on an immobilized monolayer of flavin adenine dinucleotide cofactor. Anal Biochem 333:88–98

15. Fruk L, Niemeyer CM (2005) Covalent hemin–DNA adducts for generating a novel class of artificial heme enzymes. Angew Chem Int Ed 44:2603–2606

16. Iswantini D, Kano K, Ikeda T (2000) Kinetics and thermodynamics of activation of quinoprotein glucose dehydrogenase apoenzyme in vivo and catalytic activity of the activated enzyme in Escherichia coli cells. Biochem J 350:917–923

17. Zimmermann H, Lindgren A, Schuhmann W, Gorton L (2000) Anisotropic orientation of horseradish peroxidase by reconstitution on a thiol-modified gold electrode. Chem Eur J 6:592–599

18. Willner I, Heleg-Shabtai V, Blonder R, Katz E, Tao G (1996) Electrical wiring of glucose oxidase by reconstitution of FAD-modified monolayers assembled onto au-electrodes. J Am Chem Soc 118:10321–10322

Redox Capacitor

Shigeaki Yamazaki and Masashi Ishikawa
Kansai University, Suita, Osaka, Japan

Introduction

Today, electrochemical capacitors (ECs), often called "electric double-layer capacitors," "supercapacitors," "ultracapacitors," and so on, have attracted worldwide research interest because of their potential applications as energy storage devices in many fields. The drawback of ECs is certainly their limited energy density, which restricts applications to power density over only few seconds. According to the situation, many research efforts have focused on designing new materials to improve energy and power density [1].

The energy capacity of ECs arises from either double-layer capacitance for electric double-layer capacitors (EDLCs) or pseudocapacitance for redox capacitors [2, 3]. The energy storage mechanism of EDLCs is based on non-faradic phenomena in electric double layer formed at an electrode/electrolyte interface. In regard to electrode active materials for EDLCs, carbon materials such as activated carbons have been most widely used [4] because of their reasonable cost, good electrical conductivity, and high specific surface area. However, there is a limitation in their specific capacitance; the gravimetric capacitance of most carbon materials does not linearly increase with an increase in the specific surface area above $\sim$1,200 $m^2 g^{-1}$ [5].

On the other hand, pseudocapacitance arises from a faradic reaction at an electrode surface as an oxidation or reduction process, which can provide higher capacitance and/or higher power capability than carbon electrodes utilized in conventional EDLCs. Electrode materials that exhibit such pseudocapacitive storage are typically organic materials such as conducting polymers and transition metal oxides [2, 4].

Organic Materials

Many investigators have reported promising research on ECs based on polymer electrodes, because some polymer materials not only are generally cheap and light and have suitable morphology and fast doping-undoping process but also can be relatively easily manufactured into ECs. There are many organic materials such as polypyrrole, polyaniline, polythiophene, and its derivatives for redox capacitors as shown in Scheme 1. These organic molecules can be operated in nonaqueous or aqueous electrolytes. It has been reported that polyaniline (Scheme 1b) [6], poly-3-methylthiophene (Scheme 1d) [7], and polyfluorophenylthiophene (Scheme 1g) [8] can be operated in a nonaqueous electrolyte, such as propylene carbonate containing tetraethylammonium tetrafluoroborate. Among them, poly-1,5-diaminoanthraquinone (Scheme 1j) [9] shows a high specific capacitance of 200–300 Fg^{-1} but only in an acidic aqueous electrolyte. Generally, conducting polymers do not have very good cycleability and capacity retention because swelling and shrinking of electroactive polymers lead to degradation during cycling.

To improve cycleability, many researchers proposed composite electrodes and new organic materials. The polymer modified by conductive materials such as carbon materials has been proposed. For example, Xia et al. presented a poly(2,2,6,6,-tetramethylpiperidinyloxy methacrylate)nitroxide polyradical/activated carbon composite as a negative electrode material [10]. The capacity of the composite electrode was 30 % larger than that of the pure activated carbon electrode. This composite electrode could be operated for over 1,000 cycles with only slight capacity loss. For another material, the network of cyclic indole trimer (CIT) is one of the most durable organic for ECs. Naoi et al. reported that the CIT electrode showed electrochemical redox activity and excellent cycleability in an aqueous H_2SO_4 electrolyte [11] as well as a nonaqueous electrolyte of $LiBF_4$ dissolved in EC + DMC [12]. In the aqueous system, the CIT electrode maintained its high capacity of 52 Ah kg^{-1} (187 F g^{-1}) after 100,000 cycling, while the capacity decreased down to 5 Ah kg^{-1} for a polyindole electrode. In the case of above nonaqueous electrolyte system, CIT maintained 70 % (55 Ah kg^{-1}) of its initial capacity after 50,000 cycles (Scheme 1).

Electric energy can be stored and delivered in conducting polymers as delocalized π-electrons are accepted and released during electrochemical doping-undoping, respectively. There are two types of doping process as follows:

p-doping (anion doping)
$$[\text{polymer}] + xA^- \leftrightarrow [(\text{polymer})x^+xA^-] + xe^-$$
n-doping (cation doping)
$$[\text{polymer}] + yC^+ + ye^- \leftrightarrow [(\text{polymer})y^-yC^+]$$

Conducting polymer ECs are classified into three types (Fig. 1). Type I capacitors utilize p-doping conducting polymers such as polypyrrole, polythiophene, and polyaniline for both positive and negative electrodes. The type I capacitors show low potential ($< 1V$), and only half of the total capacity can be utilized. In the type II capacitors, two different p-dope polymers are used and have different potential ranges of doping and undoing. The proton polymer battery is a practical application of this type. The type II capacitors provide higher capacity and higher working voltage than type I system. Type III devices utilize p- and n-doping conducting polymers such as polythiophene derivatives and polyacene, for both positive and negative electrodes, respectively. Type III capacitors can provide the widest operating voltage (2.5–3.0V), which is about two times higher than type II, and the highest energy density among these three types. The type III capacitors have similar discharge characteristics to batteries, where operating voltage drops very rapidly at the end of discharge.

Transition Metal Oxides

Transition metal oxides for ECs can generally provide higher energy density than conventional

Redox Capacitor

Redox Capacitor, Scheme 1 Organic materials for ECs include polypyrrole (**a**), polyaniline (**b**), polythiophene, its derivatives (**c**)–(**h**), *p*-benzoquinone (**i**), poly-1,5-diaminoanthraquinone (**j**), and cyclic indole trimer (**k**)

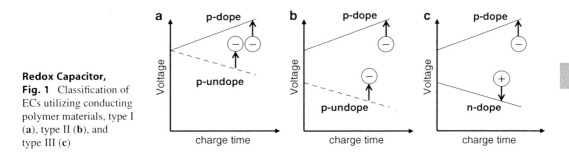

Redox Capacitor, Fig. 1 Classification of ECs utilizing conducting polymer materials, type I (**a**), type II (**b**), and type III (**c**)

carbon materials and better electrochemical stability than polymer materials. Among transition metal oxides, hydrous ruthenium oxide is the most promising electrode material because of its high theoretical specific capacitance (900–1,400 F g^{-1}) [13–17] depending on the hydration number, high conductivity, long cycle life, and good electrochemical reversibility as well as its high-rate capability. In 1995, Zheng et al. [13] reported that amorphous hydrous RuO$_2$ prepared by a sol–gel method exhibited a specific capacitance of 720 F g^{-1}. Such a high capacitance is attributed to hydrous surface layers that enable facile transport of electrons and protons. However, the capacitance decreased rapidly at high rates due to proton depletion and oversaturation in the electrolyte during charge–discharge cycling. A two-dimensionally controlled RuO_2 nanosheet was fabricated by Sugimoto et al. [18] for better electron transport. Moreover, in order to improve the rate capability, many researchers have attempted to prepare nano-sized particles of hydrous RuO_2 with carbon materials such as activated carbons [14–16, 19], carbon black [17, 20–22], and CNTs [23]. For example, Hu et al. [16] reported that nano-sized (3 nm) hydrous RuO_2/carbon composites exhibited high specific capacitance of 800–1,200 F g^{-1} (per RuO_2). Moreover, Naoi et al. [22] reported that a superhighly dispersed nano-sized (0.5–2.0 nm) hydrous RuO_2/ketjen black composite could be prepared by an in situ sol–gel process induced by ultracentrifugal mechanical agitation. After annealing at 150 °C, the composite showed a high specific capacitance of 821 F g^{-1} (per composite).

The scarcity and cost of the precious metals are major disadvantages for massive devices of ECs. Due to cost consideration, RuO_2, MnO_2 [24], CoOx [25], VOx [26], Fe_3O_4 [27, 28], Ni$(OH)_2$ [29], WC [30], and Mo_2C [30] have been studied as alternative materials. The research efforts have focused on compounds providing high-rate capability and capacitance.

Manganese oxides, characterized by low cost, abundance, and environmental compatibility, serve as a low cost replacement for precious metals. The first study reporting the capacitive behavior of manganese dioxide was published in 1999 by Lee and Goodenough [24]. This was followed by many studies that dealt with the pseudocapacitive properties of manganese dioxide materials prepared by dip coating, electrodeposition, etc. The capacitive properties of manganese oxides are sensitive to morphologies, crystal structures, cation valence, and defect chemistry. The relationship between the crystallographic structures and their pseudocapacitive properties has been investigated exhaustively by Brousse et al. [31] and Devaraj and Munichandraiah [32]. These and other groups reported that birnessite-type manganese oxides could provide the highest capacitance [33, 34].

Several groups have reported that MnO_2 is formed as an ultrathin film on a planar current collector, and anomalously high gravimetric capacitances can be observed [35, 36]. The implication of this finding has been translated to 3-dimensional electrode designs in which nanoscopic MnO_2 deposits are incorporated directly onto surface nanostructured carbon such as carbon nanotubes [36–39], templated mesoporous carbons [40], and carbon aerogel/nanoforms [41]. Such materials on the nanostructured carbon substrates serve a high specific surface area on a three-dimensional current collector, which facilitates the infiltration and rapid transport of an electrolyte into the nanoscopic MnO_2 phase (Table 1).

Redox Capacitor, Table 1 Summary of the various classes of transition metal oxides

Electrode material	Electrolyte	Working voltage (V)	Capacitance (F g^{-1})	References
$RuO_2 \cdot nH_2O$ powder	H_2SO_4	1.0	720 (n = 0.5)	[13]
$RuO_2 \cdot nH_2O$/activated carbon	H_2SO_4	1.0	720	[14]
Ruthenic acid nanosheet	H_2SO_4	1.0	658	[18]
$RuO_2 \cdot nH_2O$/activated carbon	H_2SO_4	1.0	1,200	[16]
$RuO_2 \cdot nH_2O$/ketjen black	H_2SO_4	0.8	821	[22]
$CoOx \cdot nH_2O$	NaOH	0.7	230	[25]
$VOx \cdot nH_2O$	KCl	1.0	167	[26]
Nanoporous Ni$(OH)_2$ film	KOH	0.65	578	[29]
Mn_3O_4	Na_2SO_4 adding Na_2HPO_4	1.0	230	[33]
MnO_2 thin film	Na_2SO_4	0.9	698	[35]
MnO_2/CNT/carbon paper	K_2SO_4	1.0	322	[36]
$\alpha MnO_2 \cdot nH_2O$/CNT	Na_2SO_4	0.6	277	[38]
MnO_2/mesoporous carbon	KCl	1.0	600	[40]

Novel Concept for Charge Storage Mechanism Involving "Electrolyte"

Recently, novel striking redox capacitors utilizing an electrolyte charge storage system have been attempted by few researchers [42–45]. Before them, such a system had been considered impossible for an electrochemical capacitor because of an incidental shuttle reaction or charge migration of active redox species in an electrolyte between electrodes. In 2007, Ishikawa et al. reported that EC containing an aqueous NaBr solution with a potentiostatic treatment at 70 °C exhibits much enhanced capacitance when compared to EC without the treatment [46]. In their later study, the mechanism of enhanced capacitance was found to be based on some redox processes involving bromine species (iodine species is also possible) in the electrolyte. This mechanism provides the reversible cycling of ECs, which can be stabilized especially by a pretreatment of an activated carbon electrode with bromine species. Utilizing this strategy, Yamazaki and Ishikawa et al. attained excellent capacitance and a high practical cell voltage of 1.8 V in spite of an aqueous electrolyte system (Fig. 2a and b).

Furthermore, the aqueous EC utilizing the bromide system shows outstanding performance (Fig. 3); its energy density at a low power density corresponding to a current density of 100 mA g^{-1} is about 1.2 times higher than a 2.7V class nonaqueous EC [triethylmethylammonium tetrafluoroborate (TEMABF$_4$) dissolved in propylene carbonate (PC)] in spite of an aqueous EC system. Moreover, the EC system has not only higher energy density but also much higher power density than the reference nonaqueous EC because of its high ionic conductivity (Fig. 3).

To achieve higher energy density and higher voltage operation than aqueous ECs utilizing the bromide system, nonaqueous electrolyte systems have also been proposed such as an ionic liquid electrolyte-based EC system [47] and lithium-ion capacitor system [48].

Future Advanced Developments for Redox Capacitors

Despite high capacitance of redox materials such as transition metal oxide, the cell voltage of conventional symmetric devices as introduced in the above chapter is limited to ~1V, and subsequently both power and energy densities remain unsatisfactory for industrial applications. Recently, therefore, asymmetric design often called a hybrid capacitor and nonaqueous EC system with the above materials such as polymer materials and transition metal oxides have been studied. The term "asymmetric design" means a capacitor using different materials with different operating potentials as negative and positive electrode materials, which can increase overall cell potential, resulting in enhanced energy and

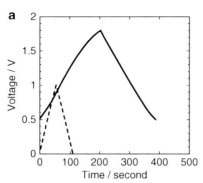 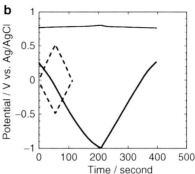

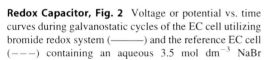

Redox Capacitor, Fig. 2 Voltage or potential vs. time curves during galvanostatic cycles of the EC cell utilizing bromide redox system (———) and the reference EC cell (– – –) containing an aqueous 3.5 mol dm^{-3} NaBr electrolyte at 1,000 mA g^{-1}; (**a**) voltage between positive and negative electrodes, (**b**) potential of negative and positive electrodes versus Ag/AgCl reference

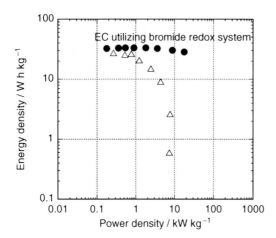

Redox Capacitor, Fig. 3 Ragone plots (relationship between energy density and power density) for the EC cell utilizing bromide redox reactions in the aqueous 3.5 mol dm^{-3} NaBr electrolyte (●) and the conventional nonaqueous EC containing 1.96 mol dm^{-3} TEMABF$_4$/PC(△)

power densities. Asymmetric hybrid capacitors include various combinations of positive and negative electrode materials: metal oxide/metal oxide [27], conducting polymer/metal oxide [49], metal oxide/carbon [50, 51], and conducting polymer/carbon materials [52].

Cross-References

▶ Electrical Double-Layer Capacitors (EDLC)
▶ Hybrid Li-Ion Based Supercapacitor Systems in Organic Media
▶ Lithium-Ion Batteries

References

1. Miller JR, Burke AF (2008) Electrochemical capacitors: challenges and opportunities for real-world applications. ECS Interface 17:53–57
2. Conway BE (1999) Electrochemical supercapacitors: scientific fundamentals and technological applications. Kluwer Academic/Plenum, New York
3. Miller JR, Simon P (2008) Electrochemical capacitors for energy management. Science 321:651–652
4. Beguin F, Frackowiak E (2009) Carbons for electrochemical energy storage and conversion systems. CRC Press, Boca Raton
5. Barbieri O, Hahn M, Herzog A, Kotz R (2005) Capacitance limits of high surface area activated carbons for double layer capacitors. Carbon 43:1303–1310
6. Prasad KR, Munichandraiah N (2002) Electrochemical studies of polyaniline in a gel polymer electrolyte. Electrochem Solid-State Lett 5:A271–A274
7. Mastragostino M, Paraventi R, Zanelli A (2000) Supercapacitors based on composite polymer electrodes. J Electrochem Soc 147:3167–3170
8. Villers D, Jobin D, Soucy C, Cossement D, Chahine R, Breau L, Belanger D (2003) The influence of the range of electroactivity and capacitance of conducting polymers on the performance of carbon conducting polymer hybrid supercapacitor. J Electrochem Soc 150:A747–A752
9. Hashmi SA, Suematsu S, Naoi K (2004) All solid-state redox supercapacitors based on supramolecular 1,5-diaminoanthraquinone oligomeric electrode and polymeric electrolytes. J Power Sources 137:145–151
10. Li H, Zou Y, Xia Y (2007) A study of nitroxide polyradical/activated carbon composite as the positive electrode material for electrochemical hybrid capacitor. Electrochim Acta 52:2153–2157
11. Machida K, Takenouchi H, Hiraki R, Naoi K (2005) Redox capacitor properties of indole derivatives-electrochemical characteristics and long-term cycleability of cyclic indole trimer. Electrochemistry 73:489–495
12. Machida K, Nakagawa Y, Ogihara N, Naoi K (2005) Redox capacitor properties of indole derivatives III-electrochemical characteristics of 5-carboxy cyclic indole trimers in non-aqueous electrolyte. Electrochemistry 73:1035–1041
13. Zheng J-P, Cygan P-J, Jow T-R (1995) Hydrous ruthenium oxide as an electrode material for electrochemical capacitors. J Electrochem Soc 142:2699–2703
14. Zheng J-P (1999) Ruthenium oxide-carbon composite electrodes for electrochemical capacitors. Electrochem Solid-State Lett 2:359–361
15. Ramani M, Haran BS, White RE, Popov BN, Arsov L (2001) Studies on activated carbon capacitor materials loaded with different amounts of ruthenium oxide. J Power Sources 93:209–214
16. Hu C-C, Chen W-C, Chang K-H (2004) How to achieve maximum utilization of hydrous ruthenium oxide for supercapacitors. J Electrochem Soc 151:A281–A290
17. Lee Y-H, Oh J-G, Oh H-S, Kim H (2008) Novel method for the preparation of carbon supported nano-sized amorphous ruthenium oxides for super capacitors. Electrochem Commun 10:1035–1037
18. Sugimoto W, Iwata H, Yasunaga Y, Murakami Y, Takasu Y (2003) Preparation of ruthenic acid nanosheets and utilization of its interlayer surface for electrochemical energy storage. Angew Chem Int Ed Engl 42:4092–4096
19. Zhang J, Jiang D, Chen B, Zhu J, Jiang L, Fang H (2001) Preparation and electrochemistry of hydrous ruthenium oxide/active carbon electrode materials for supercapacitor. J Electrochem Soc 148:A1362–A1367
20. Kim H, Popov BN (2002) Characterization of hydrous ruthenium oxide/carbon nanocomposite

supercapacitors prepared by a colloidal method. J Power Sources 104:52–61

21. Min M, Machida K, Jang J-H, Naoi K (2006) Hydrous RuO_2/carbon black nanocomposite with 3D porous structure by novel incipient wetness method for supercapacitors. J Electrochem Soc 153:A334–A338

22. Naoi K, Ishimoto S, Ogihara N, Nakagawa Y, Hatta S (2009) Encapsulation of nanodot ruthenium oxide into KB for electrochemical capacitors. J Electrochem Soc 156:A52–A59

23. Park J-H, Ko J-M, Park O-O (2003) Carbon nanotube/ RuO_2 nanocomposite electrodes for supercapacitors. J Electrochem Soc 150:A864–A867

24. Lee H-Y, Goodenough J-B (1999) Supercapacitor behavior with KCl electrolyte. J Solid State Chem 144:220–223

25. Hu C-C, Hsu T-Y (2008) Effects of complex agents on the anodic deposition and electrochemical characteristics of cobalt oxides. Electrochim Acta 53:2386–2395

26. Hu C-C, Huang C-M, Chang K-H (2008) Anodic deposition of porous vanadium oxide network with high power characteristics for pseudocapacitors. J Power Sources 185:1594–1597

27. Brousse T, Belanger D (2003) A hybrid Fe_3O_4-MnO_2 Capacitor in mild aqueous electrolyte. Electrochem Solid-State Lett 6:A244–A248

28. Wang S-Y, Wu N-L (2003) Operating characteristics of aqueous magnetite electrochemical capacitors. J Appl Electrochem 33:345–348

29. Zhao D-D, Bao S-J, Zhou W-J, Li H-L (2007) Preparation of hexagonal nanoporous nickel hydroxide film and its application for electrochemical capacitor. Electrochem Commun 9:867–874

30. Morishita T, Soneda Y, Hatori H, Inagaki M (2007) Carbon-coated tungsten and molybdenum carbides for electrode of electrochemical capacitor. Electrochim Acta 52:2478–2484

31. Brousse T, Toupin M, Dugas R, Athouel L, Crosnier O, Belanger D (2006) Crystalline MnO_2 as possible alternatives to amorphous compounds in electrochemical supercapacitors. J Electrochem Soc 153:A2171–A2180

32. Devaraj S, Munichandraiah N (2008) Effect of crystallographic structure of MnO_2 on its electrochemical capacitance properties. J Phys Chem C 112:4406–4417

33. Komaba S, Ogata A, Tsuchikawa T (2008) Enhanced supercapacitive behaviors of birnessite. Electrochem Commun 10:1435–1437

34. Inoue R, Nakayama M (2009) Pseudocapacitive properties of vertically aligned multilayered manganese oxide. Electrochem Solid-State Lett 12:A203–A206

35. Pang S-C, Anderson MA, Chapman TW (2000) Novel electrode materials for thin-film ultracapacitors: comparison of electrochemical properties of sol–gel-derived and electrodeposited manganese dioxide. J Electrochem Soc 147:444–450

36. Bordjiba T, Belanger D (2009) Direct redox deposition of manganese oxide on multiscaled carbon nanotube/ microfiber carbon electrode for electrochemical capacitor. J Electrochm Soc 156:A378–A384

37. Hu C-C, Wu Y-T (2004) Effects of electrochemical activation and multiwall carbon nanotubes on the capacitive characteristics of thick MnO_2 deposits. J Electrochem Soc 151:A2060–A2066

38. Raymundo-Pinero E, Khomenko V, Frackwiak E, Beguin F (2005) Performance of manganese oxide/ CNTs composites as electrode materials for electrochemical capacitors. J Electrochm Soc 152: A229–A235

39. Xie X, Gao L (2007) Characterization of a manganese dioxide/carbon nanotube composite fabricated using an in situ coating method. Carbon 45:2365–2373

40. Dong X, Shen W, Gu J, Xiong L, Zhu Y, Li H, Shi J (2006) MnO_2-embedded-in-mesoporous-carbon-wall structure for use as electrochemical capacitors. J Phys Chem B 110:6015–6019

41. Fischer AE, Pettigrew KA, Rolison DR, Stround RM, Long JW (2007) Incorporation of homogeneous carbon structures via self-limiting electroless deposition: implications for electrochemical capacitors. Nano Lett 7:281–286

42. Naitou M, Yamazaki S, Yamagata M, Ishikawa M (2008) Effects of bromide ion for electric double layer capacitors. In: The third Asian conference on electrochemical power sources (ACEPS-3) Abstract Pc-03, p.291, Soul

43. Lota G, Frackowiak E (2008) Striking capacitance of carbon/iodide interface. Electrochem Commun 11:87–90

44. Yamazaki S, Ito T, Yamagata M, Ishikawa M (2010) Performance of electrochemical capacitor utilizing bromide ion as redox species. In: 218th Electrochem. Soc. Meeting abstract #312, Las Vegas; Yamazaki S, Ito T, Murakumo Y, Naitou M, Shimooka T, Yamagata M, Ishikawa M, J Electrochem Soc (in contribution)

45. Lota G, Fic K, Frackwiak E (2011) Alkali metal iodide/carbon interface as a source of pseudocapacitance. Electrochem Commun 12:38–41

46. Shimooka T, Yamazaki S, Sugimoto T, Jyozuka T, Teraishi H, Nagao Y, Oda H, Matsuda Y, Ishikawa M (2007) Capacitance enhancement of aqueous EDLC Systems by electrochemical treatment. Electrochemistry 75:273–279

47. Yamazaki S, Ito T, Yamagata M, Ishikawa M, Non-aqueous electrochemical capacitor utilizing electrolytic redox reactions of bromide species in ionic liquid. Electrochim Acta (in press)

48. Ishikawa M, Yamazaki S, Yamagata M (2011) Novel designs of electrode-electrolyte interface for supercapacitors. In: 2nd International symposium on enhanced electrochemical capacitors (ISEECap'11) abstract p. 30, Poznan

49. Pasquier AD, Laforgue A, Simon P, Amatucci GG, Fauvarque J-F (2002) A novel asymmetric hybrid

Li₄Ti₅O₁₂/poly(fluorophenylthiophene) energy storage device. J Electrochem Soc 149:A302–A306

50. Brousse T, Marchand R, Taberna P-L, Simon P (2006) TiO₂(B)/activated carbon non-aqueous hybrid system for energy storage. J Power Sources 158:571–577

51. Demarconny L, Raymundo-Pinero E, Beguin F (2011) Adjustment of electrodes potential window in an asymmetric carbon/MnO₂ supercapacitor. J Power Sources 196:580–586

52. Balducci A, Henderson WA, Mastragostino M, Passerini S, Simon P, Soavi F (2005) Cycling stability of a hybrid activated carbon/poly(3-methylthiophene) supercapacitor with N-butyl-N-methylpyrrolidinium bis(trifluoromethanesulfonyl)imide ionic liquid as electrolyte. Electrochim Acta 50:2233–2237

Redox Processes at Semiconductors-Gerischer Model and Beyond

Frank Willig[1] and Lars Gundlach[2]
[1]Fritz-Haber-Institut der Max-Planck-Gesellschaft, Berlin, Germany
[2]Department of Chemistry and Biochemistry and Department of Physics and Astronomy, University of Delaware, Newark, DE, USA

Introduction

In this article we describe ideas and experimental results that are fundamental to electron transfer between molecules (redox ions) and the surface of semiconductor (SC) electrodes. We do not make any attempt here of covering the extensive literature on electrochemistry at semiconductor electrodes. Rather, experimental data are shown to illustrate relevant results. We consider only the transfer of one electron between a molecular monomer (redox ion) and the electrode. We do not consider electron transfer from dimers and higher aggregates and also not the more complicated processes like corrosion, etching, and tunneling through barriers. In the case of ultrafast injection from an excited dye molecule, we show results where the system is exposed to ultrahigh vacuum since the solvent environment would obscure the most interesting results obtained from time-resolved measurements of electron transfer. The effects arising from the addition of a solvent environment are mentioned.

Electron Transfer Between a SC Electrode and Redox Ions (Molecules) in Solution

The free energy of charge carriers in the semiconductor electrode (SC) is characterized by the Fermi energy and that of the charge exchanging molecules (redox ions) in solution by their redox potential. When the two subsystems are brought into contact with their free energy levels not too far apart, the two subsystems will exchange charges until a common electrochemical potential is established throughout the whole system [1] (Fig. 1).

The corresponding equilibrium situation is established via setting up a space-charge region with a corresponding band bending in the near-surface region of the semiconductor. Figure 2 illustrates the band bending for the lower edge of the conduction band and the upper edge of the valence band versus distance. Note that the applied potential (η) drops over the space-charge layer in the semiconductor and only a negligible fraction of the voltage drop occurs at the electrode surface. Thus, the rate constant of electron transfer remains virtually unchanged at the surface of the semiconductor when the applied voltage is changed. This is very different from a metal electrode. The arrow in Fig. 2 indicates the reduction in the electron concentration at the surface due to applied voltage because the latter enhances the barrier height for electrons moving from the bulk of the semiconductor to the surface.

Gerischer [1, 3] postulates a density of states function in the form of a Gaussian distribution for the reduced species in solution and a corresponding Gaussian distribution shifted toward the vacuum level for the oxidized species. The two distributions are labeled occupied and unoccupied in Fig. 1. Solvent configurations with the highest probability give rise to the two peaks, and the Gaussian distribution arises from different solvent configurations formed around the

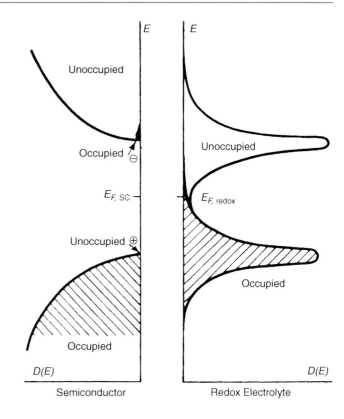

Redox Processes at Semiconductors-Gerischer Model and Beyond, Fig. 1 Scheme illustrating equilibrium between a semiconductor electrode and a redox electrolyte. The density of states D(E) in the valence band and conduction band on the left (labeled occupied and unoccupied, respectively) and one Gaussians each for the density of states D(E) of the reduced redox ions and the oxidized redox ions on the right (labeled occupied and unoccupied, respectively). The free energy is the same throughout the system, with the Fermi energy $E_{F,SC}$ at the same energy as the redox potential $E_{F,redox}$ (Reproduced from Fig. 23, Ref. [1])

reduced and the oxidized redox ions, respectively. The two peaks and the corresponding distributions for reduced and oxidized redox ions are shifted against each other on the energy axis due to the fact that the system has a slow and a fast polarization response to a sudden change in the charge of the redox ion. The polarization response of the electronic subsystem to a change in the charge is fast, instantaneous for the time scales considered here, and the polarization response from the change in the spatial coordinates of the solvent molecules is slow to a sudden change in the charge on the redox ion. Removing an electron from the reduced redox ion has to overcome the attraction of a more positively charged environment than is present when the electron is returned to the oxidized redox ion after the slow polarization response is already completed and the system is completely relaxed. Completing the slow polarization response after removal of the electron means that the effective positive charge in the environment has decreased around the now oxidized redox ion that carries less negative charge than the reduced redox ion. Thus, less energy is gained from the return of the electron to the oxidized redox ion after the environment has relaxed in response to the lowered negative charge than had to be spent in separating the electron from the reduced redox ion where the environment forms a higher positive charge around the redox ion with a higher negative charge, i.e., the reduced redox ion. Marcus [4] has presented a quantitative measure for the change in polarization energy for a system where the redox ions are conducting spheres carrying a different charge and the environment is a dielectric medium with two different dielectric constants, one for the fast response and the other for the slow response. Other authors (e.g., [5]) have added an additional slow polarization energy that arises from a change in the equilibrium coordinates of the atoms making up the molecule (redox ion) when the charge is changed on the molecule. The polarization energy has thus an outer (solvent) and an inner (atoms of the molecule) contribution. Classical Marcus theory

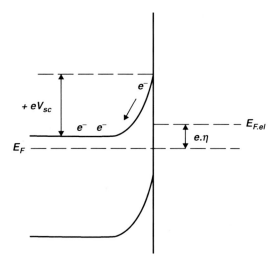

Redox Processes at Semiconductors-Gerischer Model and Beyond, Fig. 2 Band bending due to the space-charge layer near the surface of the semiconductor. The ordinate is energy and the abscissa distance. The Fermi level (E_F) is downshifted against the redox potential ($E_{F,el}$) by the applied potential η (multiplied by the elementary charge e). The SC is n-doped (Fermi level lies close to the conduction band edge in the bulk). The applied potential enhances the barrier for electrons moving from the bulk of the semiconductor to the surface and thus reduces the electron concentration at the surface (2) (Reproduced from Fig. 2.4, Ref. [2])

of the rate constant of electron transfer of a redox ion in a polarizable medium [4] predicts also a Gaussian energy dependence of the rate constant. As long as the rate constant of electron transfer is described with the tools of only classical physics assuming a fast and a slow polarization response, the predictions of classical Marcus theory for the rate constant of electron transfer are identical with those derived from the assumption of a density of states for the redox ions in the form of two Gaussian energy distributions, one for the reduced and the other for the oxidized species as illustrated in Fig. 1. The peaks of the two distributions are separated on the energy axis in Fig 1 by twice the amount of the above polarization energy. Classical Marcus theory of electron transfer and the Gerischer scenario (Fig. 1) both fail when quantum effects become important. The latter arises in the dynamics of electron transfer when strong coupling between the electronic states and high-energy vibrational modes in the molecules (redox ions) are incorporated into the model [5, 6]. This strong coupling is a common feature of different classes of molecules and redox ions [7–9]. The additional qualitative effect arising from high-energy (quantum) vibrational modes in the molecule (redox ion) is faster electron transfer compared to the classical model when electron transfer occurs to a low-lying acceptor, i.e., electron transfer becomes faster in the downhill high-energy wing of the energy distribution than is predicted by the classical calculation of the rate constant, means the shape deviates here from a Gaussian. Such quantum effects are automatic ingredients if quantum theory is employed for calculating rates of electron transfer [5, 6]. For a long period of time, quantum theory was used in the form of perturbation theory, where the maximum permissible strength of electronic interaction had an upper limit and thus the electron transfer time had a lower limit. Recently, fully quantum mechanical calculations of electron transfer have been presented without the restrictions of perturbation theory that can address ultrafast electron transfer at semiconductor electrodes, as will be described below. Experimental results will be shown below where the peak of the distribution curves shown in Fig. 1 corresponds to electron transfer in the range of a few femtoseconds. Conventional experimental measurements in electrochemistry can only access the time window of nanoseconds, at the most picoseconds. The corresponding rate constants correspond to the tails of the distribution curves, many order of magnitude smaller than at the peak. Rate constants in the wings can be visualized if a logarithmic plot of the rate constant versus energy is used instead of the linear plot shown in Fig. 1. In the wings the rate constants decrease about exponentially with increasing energy difference. Therefore, the energy range with sufficient overlap between electronic donor and acceptor states making a significant contribution to charge exchange across the interface is usually comprising only an energy interval twice the mean thermal energy ($2k_BT$ with T = temperature and k_B = Boltzmann's constant) above the respective band edge of the semiconductor.

A systematic variation in the type of redox ion and measurements of the corresponding exchange currents have been performed at metal electrodes [10] and at insulator electrodes, where charge can be injected by redox ions only into the valence band [11]. Electron transfer reactions at semiconductor/liquid interfaces were studied by the Marcus group with a Fermi golden rule approach [12, 13] and agreed reasonable well with experimental results by the Lewis group [14, 15]. Such data displays considerable uncertainty in the value of the reorganization energy for a specific redox ion even if the measurements were carried out in an identical ionic environment. Shifts in the redox energy at the electrode surface compared to the value measured against a reference electrode can also occur. Using such compiled data of the reorganization energies of redox ions, one can arrive at rough qualitative prediction concerning the value of the rate constant with a logarithmic plot instead of Fig. 1. Quantum effects arising in the downhill energy wing can make the prediction even less reliable. The uncertainty margin for the thus estimated rate constant should be expected in the range of a factor of 10–100. Depending on the type of measurement, the interfacial rate constant can be obtained with different dimensions, i.e., s^{-1}, cms^{-1}, cm^3s^{-1}, and cm^4s^{-1}. Making plausible assumptions about the reaction distance and reaction volume such values can be converted with an uncertainty margin.

Photocurrent Transient Due to Light Absorption in the Bulk of the SC and Interfacial Electron Transfer

The photocurrent due to the photo-generation of minority carriers is controlled by the discharge of the minority carriers from the surface of the electrode into the electrolyte and by competing recombination reactions of the minority charge carriers with the majority charge carriers. Of course, there are also competing side reactions of the minority carriers at the crystal surface, e.g., those leading to the corrosion of the electrode

surface. Several attempts can be found in the literature of obtaining the rate constant of interfacial electron transfer in the SC/redox electrolyte system from the time-resolved photocurrent response to optical bulk excitation of the SC electrode. This is not possible, however, since the photocurrent transient contains the influence of electron transfer to the redox ions in the electrolyte only indirectly as a reaction channel competing with recombination between minority and majority carriers near the surface of the electrode. Figure 3 illustrates the different processes in the energy versus distance diagram along with an equivalent circuit augmented by two current sources where the relevant physical processes can be introduced in the form of appropriate equations describing transport and reactions of the charge carriers [16]. Superimposed on the actual dynamics is the response of the system to a change in voltage arising from the passive elements like capacitors and resistors in the circuit as illustrated on top of Fig. 3.

The electrical photo response to excitation with a weak laser pulse of 10 ps duration absorbed in the bulk of Si is measured as time-dependent voltage drop across the external resistor R_M. If the time elapsed after the laser pulse is short compared to the RC constant of the circuit, the measured signal can be interpreted as photovoltage. It can be interpreted as photocurrent if the elapsed time is long compared to the RC constant. Figure 4 shows the response of an n-Si electrode in the ns time window to the absorption of a laser pulse of 10 ps duration. The black shaded area in Fig. 3 illustrates the generation of electron–hole pairs by the incident light inside of the Si material.

The apparent instantaneous initial rise of the signal shown in Fig. 4 arises from the separation of photo-generated electron–hole pairs that are generated inside the depletion layer. The ensuing slower rise is the diffusional flux of screened minority carriers arriving from the bulk at the edge of the depletion layer, convoluted with the RC response of the circuit. At the edge of the depletion layer, the holes are separated from the screening charge. This process is described by

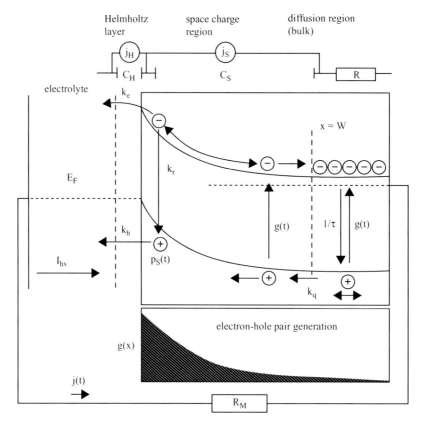

Redox Processes at Semiconductors-Gerischer Model and Beyond, Fig. 3 Light is impinging from the left and is absorbed in the SC electrode generating electron–hole pairs with function g(t). Holes are the minority carriers in the n-Si electrode. Screened minority carriers generated in the bulk (*right-hand side*) diffuse toward the depletion layer where they are separated from the screening charge (rate constant kq) and move from there to the electrode surface driven by the electric field of the depletion layer. The time-dependent concentration of holes at the electrode surface is p$_S$(t). Bimolecular recombination of holes and electrons occurs with rate constant k$_r$. A reduced redox ion transfers an electron to the hole at the surface with rate constant k$_h$; an oxidized redox ion accepts an electron from the surface of the SC electrode with rate constant k$_e$. The simplest equivalent circuit with two current sources for the equations describing reactions and transport of charge carriers is shown on *top* of the diagram. The electric response is measured as time-dependent voltage across the external resistor R$_M$ (Reproduced from Fig. 1, Ref. [16])

the phenomenological rate constant k$_q$. The signal is convoluted with the RC response, and the latter controls in particular the decay. The smooth solid curve is the total calculated response which includes the current sources describing transport and reactions of the charge carriers. The dashed curves are calculated with the current sources omitted. They represent the so-called RC response arising from the passive elements in the equivalent circuit shown on top of Fig. 3. The parameter in the inset is the band bending. Calculations have shown that the measured photovoltage is due to the displacement current flowing through the capacitor C$_S$ formed by the depletion layer. The measured photovoltage is not sensitive to the faradaic current at the interface because the capacitor C$_H$ is much larger than C$_S$ making the contribution from the current source in C$_H$ very small [16, 17]. Thus, the measured signal is not sensitive to interfacial electron transfer, specifically not to the rate constant k$_h$ in Fig. 3 for electron transfer from reduced redox ions in solution to the holes at the surface of the n-Si electrode.

Redox Processes at Semiconductors-Gerischer Model and Beyond, Fig. 4 Calculated and measured photovoltage across the external resistor R_M. A weak laser pulse (10^7 photons/mm^2) of 10 ps duration with 590 nm central wavelength impinged on the n-Si electrode. The thin curve is the calculated response including the current sources, whereas the dashed curve is the response without the current sources (so-called RC response). Time resolved is the diffusional flux of screened minority carriers arriving at the edge of the depletion layer convoluted with the RC response. Holes are separated from the screening charge and driven to the surface by the electric field in the depletion layer (Reproduced from Fig. 3, Ref. [16])

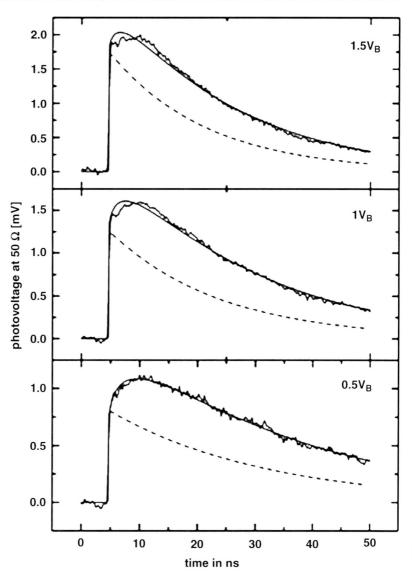

Recombination between the photo-generated holes (minority carriers) and electrons (majority carriers in the n-Si electrode) near the surface gives rise to a time-dependent dip in the photocurrent measured in response to a rectangular illumination of 1 ms duration (Fig. 5). The shape of the photocurrent signal in dependence on the bias voltage can be simulated by introducing a corresponding rate for bimolecular recombination between electrons and holes into the current source for the depletion layer (upper part of Fig. 3). Negative bias voltage lowers the band bending depicted in Fig. 3 and increases the dip, i.e., decreases the stationary photocurrent. The flat-band situation is reached at a bias of − 0.95 V as indicated by the disappearance of the initial photocurrent peak, labeled peak (filled squares in Fig. 5). Both the stationary photocurrent, labeled final (filled circles in Fig. 5), and the corresponding stationary recombination loss are linked to the stationary accumulation of holes at the interface. The stationary photocurrent disappears already at finite band bending of about 0.3 V with respect to the flat-band potential at − 0.95 V (Fig. 5).

The concentration of electrons in the depletion layer increases exponentially with the bias voltage going into negative direction. This leads to an increased recombination and a deeper dip in the photocurrent transient. Both the rate of electron transfer from the reduced redox ions to the holes at the electrode surface and the concentration of reduced redox ions must have an influence on the measured photocurrent of the holes since discharge into the electrolyte reduces the concentration of holes at the surface of the electrode and competes with recombination. A high concentration of the reduced redox ions combined with a reasonably high value of the rate constant k_h will keep the concentration of photo-generated holes fairly small at the surface of the n-Si electrode and thus reduce the recombination loss. On the other hand, the concentration of holes will increase at the surface if the concentration of the reduced redox ions is made very small slowing down discharge into the electrolyte. With a higher concentration of holes, there will be enhanced bimolecular recombination with the electrons. Bimolecular recombination of the holes with the electrons will dominate once a sufficient concentration of holes has accumulated at the surface. With only a slow discharge in the equations for the current source compared to bimolecular recombination, the photocurrent transient is predicted to develop an asymmetric shape with respect to switching on and off the illumination (Fig. 6). Moreover, the recombination dip should appear now as S-shaped time dependence of the initial photocurrent transient. Both features are very different when the photocurrent transient is measured in the presence of a high concentration of reduced redox ions with a fast discharge of the holes (Fig. 5). There is almost perfect agreement of the experimental data (noisy curve in Fig. 6) with the predicted

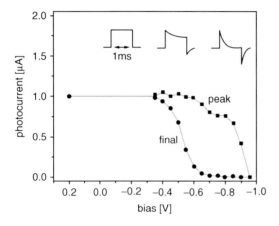

Redox Processes at Semiconductors-Gerischer Model and Beyond, Fig. 5 Peak value (*filled square*) and stationary plateau value (*filled circle*) of the photocurrent transient in response to a rectangular illumination pulse of 1 ms duration in dependence on the voltage bias. The inbuilt depletion layer voltage is decreased with the bias shifting in negative direction. The flat-band potential is at − 0.95 V. The reduced redox ions that can discharge the holes at the surface are 0.005 M 1,1′-dimethylferrocene (Reproduced from upper part of Fig. 4, Ref. [16])

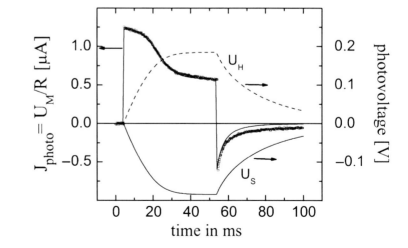

Redox Processes at Semiconductors-Gerischer Model and Beyond, Fig. 6 (Reproduced from upper part of Fig. 5, Ref. [16])

time dependence of the photocurrent transient (thin solid curve in Fig. 6). In summary, the photocurrent transient does not show the rate constant of electron transfer at the interface. From its shape one can obtain qualitative information on whether recombination or discharge is dominating the fate of the photo-generated minority carriers at the surface.

Photocurrent transient (voltage across R_M) in response to a rectangular illumination pulse. The signal is strongly asymmetric with respect to beginning and end of the illumination. The concentration of reduced redox ions is negligible, and recombination of the holes with electrons dominates their fate. The calculated thin curve virtually merges with the measured noisy curve except for some deviation at the right-hand side. The voltage calculated for the Helmholtz layer U_H and for the depletion layer U_S cannot be measured.

Electron Injection into a SC Electrode from Photoexcited Adsorbed Molecules

Several interesting effects can arise in dye-sensitized charge injection, for example, spin-dependent recombination kinetics that has been studied with organic insulator crystals functioning as electrodes [18]. There are several ways of light-induced charge injection, e.g., injection from the locally excited electronic dye molecule into a semiconductor or direct optical electron transfer from the ground state of an adsorbed molecule to states in the empty conduction band of a semiconductor. Experiments have shown that the latter process is less efficient than the first one [19]. The most impressive progress has been made in this field with the recent feasibility of measurements on the femtosecond time scale. The most important tool for studying the latter is a frequency tunable laser generating pulses of a few femtosecond duration. The most recent progress is the direct measurement of the energy distribution of the injected electron in the electronic acceptor states of the semiconductor [20, 21]. The data is collected as femtosecond two-photon photoemission signal [20–22].

The energy distribution of the corresponding excited molecular donor state has been obtained from the stationary absorption spectrum of the adsorbed molecule [22, 23]. Time-dependent interfacial electron transfer in the above system is probed either by femtosecond transient absorption spectroscopy [24–26], mostly applied in the case of a nanostructured electrode, or by femtosecond two-photon photoemission (fs-2PPE) spectroscopy [20, 21] which is more sensitive and able to time-resolve the reaction on a well-prepared planar surface in ultrahigh vacuum. Since the solvent environment of a traditional electrochemical system has been omitted from these systems, the investigations focus on the role of the high-energy (quantum) molecular vibrations in the dynamics. We note here that adding a solvent environment would cause a downward energy shift of the electronic levels in the adsorbed molecule, introduce additional inhomogeneous broadening of the electronic levels, increase the reorganization energy of the reaction, and would make transient absorption measurement extremely difficult in the most relevant time window shorter than 100 fs. The solvent when excited by an ultrashort laser pulse generates the so-called coherent artifact which obscures the actual signal. The ET reaction would still occur on the same time scale as in the absence of the solvent if the downshift of the electronic level due to the solvent environment is compensated and the molecular donor level still positioned high enough above the lower conduction band edge. Vibrational peaks would be broadened or completely obscured due to additional inhomogeneous broadening caused by the solvent environment. To avoid these difficulties and for employing fs-2PPE, the experimental data shown below was collected in ultrahigh vacuum.

The availability of time-resolved [20, 26] and frequency-resolved experimental data [22, 23] for the same systems has motivated several different theory groups to model ultrafast heterogeneous electron transfer with advanced theoretical tools [23, 27–34]. Recent quantum mechanical calculations of the injection dynamics [23, 27–30, 34] are not any more based on a perturbation approach as was customary for earlier model calculations but permit an arbitrarily high value for the electronic interaction energy between the excited state of the molecular donor and the electronic acceptor states of the solid. Thus, the new theoretical calculations can address the experimentally observed ultrashort electron transfer times of a few femtoseconds. The simplest assumption about the electronic coupling is a constant matrix element across the whole conduction band of the semiconductor. By choosing perylene as the molecular donor with its excited donor level located high above the lower conduction band edge of the wide-gap semiconductor TiO$_2$ [20], the most general case of a heterogeneous electron transfer reaction can be realized. This case is referred to as the wide-band limit [35], where the complete electron transfer spectrum is mapped as energy distribution of the injected electron onto the continuum of empty electronic states of the electrode [20]. Another advantage of using perylene as the donor is vibrational structure in the optical spectra being dominated by just one high-energy vibrational mode. Therefore, low-resolution spectra can be fitted by considering only this 0.17 eV skeletal stretch mode [23]. The injection dynamics predicted by the fully quantum mechanical calculations for the perylene/TiO$_2$ system is illustrated in Fig. 7.

The dynamics [27, 28, 37] is characterized by two different energy distributions, i.e., that of the donor state and that of the injected electron, as illustrated in Fig. 7. The corresponding electron injection from an excited donor state high above the lower edge of the conduction band occurs in the wide-band limit [35] and gives rise to an exponential decay in the population of the donor state due to electron transfer [37]. This general case is realized by the perylene dye/TiO$_2$ systems [20, 26]. Strong coupling of the electronic states to high-energy vibrational modes [7–9] is a characteristic feature of all the molecules that have been used as visible light sensitizers of wide bandgap semiconductors. This strong coupling leads to the wide energy spread and vibrational structure in the energy distribution of the injected electron. Just one high-energy vibrational mode

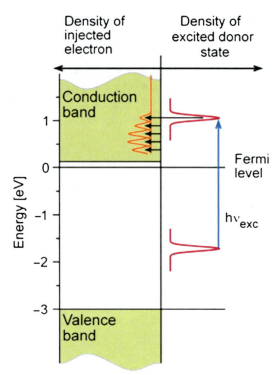

Redox Processes at Semiconductors-Gerischer Model and Beyond, Fig. 7 Scheme illustrating laser pulse induced ultrafast nonadiabatic heterogeneous electron transfer according to fully quantum mechanical model calculations [27, 28, 30, 31, 34]. The energy distribution of the excited molecular donor state (*upper red curve*) is much narrower than the energy distribution of the injected electron. The latter spreads from the donor state to lower energies in the electronic acceptor states and shows vibrational structure. For details see text (Reproduced from Fig. 1 Ref. [36])

is considered in Fig. 7 since this is sufficient for describing low-resolution spectra of perylene dye/TiO$_2$ systems [23, 38] used in the experiments. The energetic position of the donor state with respect to the conduction band of TiO$_2$ is known from UPS data combined with optical absorption spectra and independently from two-photon photoemission experiments [20]. Excitation by an ultrashort laser pulse creates the excited donor state at the respective photon energy above the ground state [20, 28]. When generated by an ultrashort laser pulse, the energy distribution of the donor state consists firstly of the width of this laser pulse, secondly of a Lorentzian which is controlled by the ET time of the system [29, 35], and thirdly by any inhomogeneous distribution of the states involved in the electronic transition [20]. The excited donor state shifts with higher photon energy to a higher vibrationally excited state (not shown in Fig. 1). Redistribution of vibrational excitation energy to other modes is much slower than ultrafast electron injection from the initially excited vibrational state in this system [39]. The dynamics in this system is very different from earlier injection scenarios where it was assumed that redistribution of vibrational excitation energy occurs either faster than electron transfer or at least on a comparable time scale. Figure 7 shows the energy distribution of the injected electron to spread from the excited donor state to acceptor states at much lower energies. The energy lost by the injected electron is used for exciting a high-energy (quantum) vibrational mode in the ionized molecule. Specific advantages of using the perylene/TiO_2 system for testing the theoretical predictions illustrated in Fig. 7 are firstly the high-lying excited singlet state which functions as the donor state and realizes the wide-band limit [35], secondly slow redistribution of vibrational excitation energy on the picosecond time scale [39] and much slower singlet–triplet conversion, and thirdly the fact that low-resolution spectra [23, 37] are dominated by just one high-energy vibrational mode, i.e., the 0.17 eV skeletal stretch mode.

The electron transfer spectrum is obtained as a cross section along the energy axis [20, 22] through the complete set of the fs-2PPE data [23, 39], in this case for the perylene/TiO_2 system with the –COOH anchor–bridge group (Fig. 8a). The injected electron spreads over a wide energy range reaching electronic acceptor levels more than 0.5 eV below the donor state. Convoluting a spectrum structured by the dominant 0.17 eV stretch mode of perylene with a Gaussian of 80 meV width (FWHM), which accounts for the instrumental response of the used time-of-flight (TOF) detector [20], results in the fit to the data points. Fully quantum mechanical model calculations have shown that the femtosecond two-photon photoemission (fs-2PPE) signal preserves both energy spread and vibrational

structure of the energy distribution of the injected electron [39]. The energetic width of the donor state has been extracted from the stationary absorption spectrum of the adsorbed molecules [22]. The scheme in Fig. 7 considers also additional broadening due to the short laser pulse. The energy distribution of the injected electron covers a much wider range than the excited donor state. The injected electron retains its energy distribution unchanged for a sufficiently long time in this system, here 200 fs [20] allowing for the measurement of the initial distribution. The electron is injected into surface states, probably created by the anchor groups where energy relaxation is much slower than the 40 fs time scale measured for energy relaxation in bulk states of TiO_2 [40]. By inserting different anchor–bridge groups between the perylene chromophore and the surface atoms of TiO_2 the injection time has been varied from 10 fs to one picosecond [20]. With perylene attached via the anchor–bridge group –CH_2–SH to an Ag electrode electron transfer from the excited state of the molecule occurs on the same time scale of a few fs [21] as for perylene attached via the –COOH group to the TiO_2 electrode. At the metal electrode it would be extremely difficult, however, to measure the initial energy distribution of the injected electron. The latter is immediately distorted by inelastic scattering processes occurring with the high density of thermal electrons in the metal electrode. From Figs. 7 and 8a, it is clear that strong coupling of the electronic states to high-energy vibrational modes controls the dynamics of electron transfer from a typical molecular donor like perylene to the semiconductor TiO_2. Figure 8b explains the choice of the delay time td = 40 fs at which the cross section along the energy axis has been taken to obtain the data points of Fig. 8a. Figure 8b shows a cross section along the time axis through the complete set of fs-2PPE data [20]. The transient contains an early peak arising from photoemission from the excited singlet state of perylene and a second peak due to photoemission from the injected electrons at the surface of TiO_2. The latter population is generated from the first population. Thus, the first peak decays with

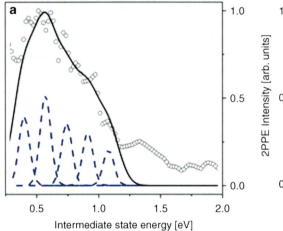 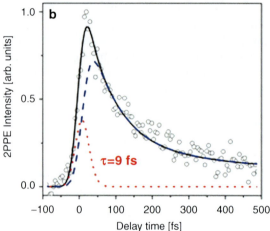

Redox Processes at Semiconductors-Gerischer Model and Beyond, Fig. 8 (a) Experimental data points are obtained from a cross section along the energy axis of the complete set of femtosecond two-photon photoemission data [20]. The data points are collected at a delay time of 40 fs determined from the data shown in (b). The energy distribution of the data points can be fitted using a structured spectrum predicted by a theoretical model for the perylene/TiO$_2$ system [39] and by convoluting with the response function of the used time-of-flight detector. (b) Experimental data points are obtained from a cross section along the time axis of the complete set of femtosecond two-photon photoemission data [20]. The transient contains two contributions; photoemission from the excited molecular donor state is followed by photoemission from the electronic acceptor states filled via electron transfer from the donor state. The time constant for the decay of the first peak is identical with that for the rise of the second peak, i.e., it is the electron transfer time, here 9 fs (Reproduced from Fig. 2 Ref. [36])

the same time constant that is controlling the rise of the second peak; it is the electron transfer time, i.e., 9 fs for the –COOH anchor–bridge group [20, 26]. The delay time td = 40 fs is chosen such that the peak due to photoemission from the injected electron is probed after the peak due to photoemission from the excited singlet state has decayed. The ultrafast injection time of 9 fs and the excitation of the high-energy vibrational mode in the ionized molecule demonstrate ultrafast nonadiabatic electron transfer as the dominant ET mechanism. Recent experimental results (Fig. 8) and recent quantum mechanical model calculations are in agreement, but both differ strongly from early quantum mechanical perturbation theory models where it was automatically assumed that nonadiabatic electron transfer is slow. The fully quantum mechanical model of the injection dynamics illustrated in Fig. 7 has the virtue of incorporating strong coupling of the electronic states to high-energy vibrational modes and allowing at the same time for ultrafast ET [27, 28, 30, 31, 34]. Theoretical model calculations of the above type need the input of parameter values which are taken either from experimental data or from specific theoretical calculations.

It is important to note here that the ultrafast injection dynamics are very different from Gerischer's much earlier intuitive injection scenario [41, 42]. The latter assumes strong thermal broadening of the excited donor state which is not borne out by the experimental data [22, 26]. Moreover, a thermally relaxed donor state is assumed in Gerischer's scenario which would imply for the perylene/TiO$_2$ system that electron transfer must be slowed down by a factor of 1,000 compared to the measured electron transfer times [20, 26] because intermolecular relaxation of vibrational energy has been measured for perylene in the time range of several ps [39]. A correspondingly slow electron injection in the picosecond time domain is realized with an excited donor state located below the conduction band edge rendering electron transfer thermally activated. An experimental example for this case has been found with electron injection from a dye J-aggregate into AgBr [43]. The latter represents a rather special situation, and the general case shown in Fig. 7 is not described by the Gerischer's scenario for dye-sensitized injection.

The recent fully quantum mechanical model of dye-sensitized electron injection has addressed also the specific border case where the excited donor state lies above but fairly close to the lower edge of the conduction band. For the latter case a slowdown in the electron transfer time is predicted compared to the case of the wide-band limit. Moreover, if the excited donor state is generated with a sufficiently short laser pulse, the population is predicted to show time-dependent oscillations in this border case. Time-dependent oscillations are also predicted for the populations of the injected electron and of vibrational states in the ionized molecule [37, 44]. The electronic level of a donor molecule can be shifted through a chemical modification of the chromophore; however, the chemical change might bring about additional changes in the system. A better defined energy shift of the donor level can be realized with a semiconductor quantum dot where the electronic levels shift with the size of the particle [45]. Recently, a low-lying excited donor state has been realized in the form of an exciton in a PbSe nanocrystal attached to the surface of TiO_2, and strong oscillations have been reported for a second harmonic signal probing this interface [46]. The latter signal could indeed indicate that the theoretically predicted electron transfer dynamics for the border case of a low-lying donor state [37, 44] has been realized in the system, but this has to be checked in more detail.

References

1. Gerischer H (1961) Semiconductor electrode reactions. In: Delahay P, Tobias CW (eds) Advances in electrochemistry and electrochemical engineering, vol 1. Interscience, New York, 4:139–232
2. Notten PHL, van den Meerakker JEAM, Kelly JJ (1991) Etching of III-V semiconductors an electrochemical approach. Elsevier Advanced Technology, Oxfordshire
3. Gerischer H (1960) Über den Ablauf von Redoxreaktionen an Metallen und an Halbleitern. I. Allgemeines zum Elektronenübergang zwischen einem Festkörper und einem Redoxelektrolyten. Z. phys. Chem. NF 26:223–247
4. Marcus RA (1965) J Chem Phys 43:679
5. Ulstrup J, Jortner J (1975) J Chem Phys 63:4358
6. Schmickler W (1976) Electrochim Acta 21:161
7. Halasinski TM, Weissman JL, Ruiterkamp R, Lee TJ, Salama F, Head-Gordon M (2003) J Phys Chem A 107:3660
8. Eastwood D, Gouterman M (1970) J Mol Spectrosc 35:359
9. Zakeeruddin SM, Nazeeruddin MK, Pechy P, Rotzinger FP, Humphry-Baker R, Kalyanasundaram K, Grätzel M (1997) Inorg Chem 36:5937
10. Hale JM (1971) The rates of reactions involving only electron transfer at metal electrodes. In: Hush NS (ed) Reactions of molecules at electrodes. Wiley 229, and references therein
11. Willig F (1981) Electrochemistry a the organic molecular crystal/aqueous electrolyte interface. In: Gerischer H, Tobias CW (eds) Advances in electrochemistry and electrochemical engineering, vol 12. Wiley 1:1, and references therein
12. Gao YQ, Marcus RA (2000) J Chem Phys 113:6351
13. Gao YQ, Georgievskii Y, Marcus RA (2000) J Chem Phys 112:3358
14. Fajardo AM, Lewis NS (1997) J Phys Chem B 101:11136
15. Pomykal KE, Lewis NS (1997) J Phys Chem B 101:2476
16. Schwarzburg K, Willig F (1997) J Phys Chem B 101:2451
17. Willig F (1988) Ber Bunsenges Phys Chem 92:1312–1319
18. Charle KP, Willig F (1988) Spin-dependent kinetics in dye-sensitized charge carrier injection into organic crystal electrodes. In: Conway BE, Bockris JOM, White RE (eds) Modern aspects of electrochemistry. Plenum, New York, pp 359–389
19. Wang L, Willig F, May V (2007) J Chem Phys 126:134110
20. Gundlach L, Ernstorfer R, Willig F (2007) Prog Surf Sci 82:355
21. Gundlach L, Willig F (2007) Chem Phys Lett 449:82
22. Gundlach F, Letzig T, Willig F (2009) J Chem Sci 121:561
23. Wang L, May V, Ernstorfer R, Willig F (2005) J Phys Chem B 109:9589
24. Rehm JM, McLendon GL, Nagasawa Y, Yoshihara K, Moser J, Grätzel M (1996) J Phys Chem 100:9577
25. Burfeindt B, Hannappel T, Storck W, Willig F (1996) J Phys Chem 100:16463
26. Ernstorfer R, Gundlach L, Felber S, Storck W, Eichberger R, Willig F (2006) J Phys Chem B 110:25383
27. Ramakrishna S, Willig F, May V (2000) Phys Rev B 62, R16330

28. Wang L, Willig F, May V (2006) Mol Simul 3:765
29. Persson P, Lundqvist MJ, Ernstorfer R, Goddard WA III, Willig F (2006) J Chem Theory Comput 2:441
30. Sebastian KL, Tachya M (2006) J Chem Phys 124:064713
31. Kondov L, Thoss M, Wang H (2006) J Phys Chem A 110:1364
32. Prezhdo OV, Duncan WR, Prezhdo VV (2009) Prog Surf Sci 84:30
33. Abuabara SG, Rego LGC, Batista VS (2005) J Am Chem Soc 127:1823
34. Mohr J, Schmickler W, Badiali JP (2006) Chem Phys 324:140
35. Muscat JP, Newns DM (1978) Prog Surf Sci 9:1
36. Gundlach L, Willig F (2012) Ultrafast photoinduced electron transfer at electrodes: the general case of a heterogeneous electron transfer reaction, Chem Phys Chem 13 doi:10.1002/cphc.201200151 p. 2877–2881
37. Ramakrishna S, Willig F, May V, Knorr A (2003) J Phys Chem B 107:607
38. Tsivlin DV, Willig F, May V (2008) Phys Rev B 77:035319
39. Zimmermann C, Willig F, Ramakrishna S, Burfeindt B, Pettinger B, Eichberger R, Storck W (2001) J Phys Chem B 105:9245
40. Gundlach L (2005) Dissertation, Freie Universität Berlin, http://wwwdissfu-berlinde/diss/receive/FUDISS_thesis_000000001801
41. Gerischer H (1972) Photochem Photobiol 16:243
42. Gerischer H, Willig F (1976) Reactions of excited dye molecules at electrodes. In: Boschke FL (ed) Topics in current chemistry. Springer, Berlin, pp 31–84
43. Troesken B, Willig F, Schwarzburg K, Ehret A, Spitler M (1995) J Phys Chem 99:5152
44. Ramakrishna S, Seideman T, Willig F, May V (2009) J Chem Sci 121:589
45. Brus LE (1984) J Chem Phys 80:4403
46. Tisdale WA, Williams KJ, Timp BA, Norris DJ, Aydil ES, Zhu XY (2010) Science 328:1543

Reference Electrodes

Heike Kahlert
Institut für Biochemie, Universität Greifswald, Greifswald, Germany

Introduction

Reference electrodes are necessary to control the potential of a working electrode (e.g., during voltammetric measurements) or to measure the potential of an indicator electrode in potentiometric measurements, since the Galvani potential difference of a single electrode is not measurable [1]. An ideal reference electrode would have the following characteristics:

(i) It is chemically and electrochemically reversible, i.e., its potential is governed by the Nernst equation, (ii) the potential should remain practically constant, when current flows through the electrochemical cell (ideally nonpolarizable electrode), and (iii) the thermal coefficient should be small. Whereas there is no practical reference electrode that offers all these properties to the same extent, a variety of reference systems very close to that ideal behavior exist [2, 3]. The choice and construction of the reference electrode depend on the experimental conditions such as temperature, pressure, size, nature and composition of the electrolyte, and the electrochemical method applied in the measurement. With voltammetric or amperometric techniques, the tolerated uncertainty in the potential is still relatively large since the analyte concentration is related primarily to the measured current, which is quite constant in a certain potential window. With potentiometric sensors, however, the zero-current potential is directly related to the activity of the analyte ion. The potential change at the reference electrode as a function of the sample composition must therefore be kept reliably small. In the two-electrode configuration in voltammetric measurements, one of the electrodes may also play the role of a reference electrode. The same metal can be used for both electrodes (the working and the counter/reference electrode) if the surface area of the reference electrode is much higher than that of the working electrode, since the electrode potential is varied by the current density, i.e., in that case the current density of the working electrode is much higher than that on the reference electrode and hence the potential change at the reference electrode is small. A classical example is the large nonpolarizable mercury pool electrode in Heyrovský dc polarography. The two-electrode configuration is also widely used in arrangements with microelectrodes as working electrodes. The primary standard in electrochemistry is the standard hydrogen electrode, the potential of which is zero by definition [4]. Because of the difficulties in handling with the standard hydrogen electrode, reference electrodes of the second kind or quasireference electrodes are widely used. If the reference electrode consists of compartments including electrolyte solutions, it has to be separated from the test solution. For this purpose, a diaphragm is used in order to prevent the mixing of both solutions while providing the conductivity between them. Common materials are porous like sintered glass and porcelain. Different arrangements of diaphragms are given in [3]. To prevent a contact between the analyte solution and the reference electrolyte, a second salt bridge (double salt bridge) is recommendable.

In the following, a short overview of the most widespread and most important reference electrodes will be given.

Quasireference Electrodes

Quasireference electrodes are simple noble metal wires (silver, gold, platinum) or plane layers of these metals maintaining a given but not well-defined potential during the course of an electrochemical measurement. They have to be calibrated with an inner standard or with respect to a conventional reference electrode. A reference redox couple like complexes between large ligands and a transition metal, e.g., the redox couple ferrocene/ferrocenium, can be used as an inner standard, especially in nonaqueous solutions [5–7]. The calibration with respect to a conventional reference electrode is only reliable when one can be sure that the potential of the quasireference electrode is the same in the calibration and in the application experiment. The main advantage of a quasireference is that no additional reference electrolyte solution is necessary, and hence, a contamination of the test solution by solvent or ions of a reference electrolyte is excluded.

The Hydrogen Electrode

The hydrogen electrode consists of a platinum wire or a platinum sheet covered with platinum black (i.e., platinized) and an acidic aqueous electrolyte solution. The hydrogen gas is usually continuously supplied. The reversible electrochemical equilibrium can be formulated as follows:

$$H_2 + 2H_2O \leftrightarrows 2H_3O^+ + 2e^- \tag{1}$$

The Galvani potential difference of such electrode is:

$$\Delta\phi = \Delta\phi^\ominus + \frac{RT}{2F} \ln a_{H_3O^+}^2 + \frac{RT}{2F} \ln \frac{1}{f_{H_2}}$$

$$= \Delta\phi^\ominus + \frac{RT}{F} \ln \frac{a_{H_3O^+}}{\sqrt{f_{H_2}}} \tag{2}$$

By definition, the potential of the standard hydrogen electrode (SHE) is zero at all temperatures. The standard conditions are defined as:

$$a_{H_3O^+} = 1$$

$$\left(\text{i.e., } a_{H_3O^+} = \gamma_{H_3O^+} + \frac{m_{H_3O^+}}{m^\ominus} \text{ with } \gamma_{H_3O^-} = 1, \right.$$

$$\left. m_{H_3O^+} = 1 \, \text{mol} \, \text{kg}^{-1}, \text{ and } m^\ominus = 1 \, \text{mol} \, \text{kg}^{-1} \right)$$

and $f_{H_2} = 1$ (i.e., the fugacity of hydrogen is calculated according to $f_{H_2} = \gamma_{H_2} \frac{p_{H_2}}{p^\ominus}$ with $\gamma_{H_2} = 1$, $p_{H_2} = 1 \, \text{bar} = 10^5 \, \text{Pa}$, and $p^\ominus = 1 \, \text{bar} = 10^5 \, \text{Pa}$). Earlier, $p^\ominus = 1$ standard atmosphere $= 101.325 \, \text{kPa}$ was used. It causes a difference in the potential of SHE of $+0.169 \, \text{mV}$, and this value has to be subtracted from the standard electrode potentials given previously in different tables. However, because of the uncertainty of the $E^\ominus$ values of about 1 mV, this difference can be neglected. As it is difficult to adjust the activity of hydronium ions to 1, for practical purpose, relative hydrogen electrodes (RHE) in acidic solutions are widely used. The nature and the concentration of the acid is the same in the reference and in the test solution. The potential of the electrode can be calculated by using well-known activity coefficients of the hydrochloric acid. The standard hydrogen electrode allows very precise measurements to be made; however, the demanding handling restricts its use [8]. For instance, the hydrogen gas must be of highest purity, esp. with respect to oxygen, H_2S, AsH_3, SO_2, CO, and HCN because these gases poison the platinum electrode. In solution, volatile substances, e. g., HCl, can be purged from the solution by the hydrogen gas, metals can be reduced at the electrode, redox systems may influence the electrode potential, etc.

Practical Reference Electrodes of the Second Kind

Electrodes of the second kind consist of a metal, a sparingly soluble salt of this metal (or an oxide or hydroxide of this metal), and an electrolyte containing the anions of the sparingly soluble compound to establish an equilibrium with the precipitate. The potential of the electrode depends on the fixed activity of the anion in the solution.

The silver/silver chloride electrode is an example of this kind of electrodes, and beside the calomel electrode most commonly used. The net reaction can be described by the following equation:

$$Ag^0 + Cl^- \leftrightarrows AgCl \downarrow + e^- \qquad (3)$$

The activity of the metal ions in the solution depends on the solubility equilibrium and can be described by the solubility product K_s according to:

$$a_{Ag^+} = \frac{K_s}{a_{Cl^-}} \qquad (4)$$

Thus, the electrode potential is proportional to the logarithm of the activity of the chloride in the electrolyte solution:

$$
\begin{aligned}
E &= E^{\ominus}(Ag, Ag^+) + \frac{RT}{F} \ln a_{Ag^+} \\
&= E^{\ominus}(Ag, Ag^+) + \frac{RT}{F} \ln K_s - \frac{RT}{F} \ln a_{Cl^-}
\end{aligned} \qquad (5)
$$

or

$$E = E_c^{\ominus\prime}(Ag, AgCl) - \frac{RT}{F} \ln a_{Cl^-} \qquad (6)$$

$E_c^{\ominus\prime}(Ag, AgCl)$ is the formal potential of the silver/silver chloride electrode including the solubility product.

The common arrangement of the silver/silver chloride electrode is that a silver wire is covered with silver chloride, which can be achieved chemically or electrochemically. The electrolyte solution in these reference systems is normally a potassium chloride solution (mostly saturated or 3M) and only seldom sodium or lithium chloride. Compared to the calomel electrode, the silver/silver chloride reference system has the great advantage that measurements at elevated temperatures are possible. Special devices have been developed based on the silver/silver chloride reference systems for measurements in high temperature aqueous solutions [9] and under changing pressure conditions [10]. Because reference systems based on silver/silver chloride can be produced in a very small size and also as planar layers, they are often used in microsystems.

The calomel electrode consists of mercury and mercury (I) chloride (calomel Hg_2Cl_2) in contact with a potassium chloride solution of constant activity. In case that the supporting electrolyte in the cell contains perchlorate anions, it is advisable to use NaCl instead of KCl since $KClO_4$ is sparingly soluble and could precipitate in the diaphragm. In most cases, saturated KCl solution is used; however, in such solution already at temperatures above 35 °C, a disproportion reaction takes place. The back reaction by cooling down the electrode is very slow so that a hysteresis of the electrode potential occurs. This is the reason why it is recommended that the calomel electrode only be used at temperatures in maximum up to 70 °C. In Table 1 electrode potentials of the silver/silver chloride electrode and for the calomel electrode at different temperatures and different concentrations of KCl are given. Note that the electrode potentials differ when other salts than potassium chloride (e.g., NaCl) are used because of the different solubility products.

The trend in the development of modern electroanalytical techniques is a miniaturization of

Reference Electrodes, Table 1 Electrode potentials of the calomel electrode and of the silver/silver chloride electrode at different temperatures and different concentrations of KCl (the concentrations are related to 25 °C) [3]

| | E/mV | | | |
| | Calomel | | Silver/silver chloride | |
$T/°C$	c_{KCl} 1 mol L^{-1}	Saturated	c_{KCl} 3 mol L^{-1}	Saturated
0			224.2	220.5
10	285.4	260.2	217.4	211.5
20	283.9	254.1	210.5	201.9
25	281.5	247.7	207.0	197.0
30	280.1	244.4	203.4	191.9
40	278.6	241.1	196.1	181.4
50	275.3	234.3	188.4	170.7
60	271.6	227.2	180.3	159.8
70	267.3	219.9	172.1	148.8
90	262.2	212.4	153.3	126.9

the measuring devices, including the reference electrodes. Although the development of new technological approaches such as the thick film and thin film techniques supported arrangements with planar reference electrodes, special problems with miniaturized reference electrodes still exist, and there is no general role for the conception of an integrated micro reference electrode. The lifetime of miniaturized reference electrodes is limited because of the non-negligible solubility of AgCl, the reduced electrolyte volume (rapid contamination as well as exhaustion and drying out of the reference electrolyte), the miniaturization of the diaphragm (small pores can easily be blocked), etc.

References

1. Bard AJ, Faulkner LR (2001) Electrochemical methods, 2nd edn. Wiley, New York, pp 24–28
2. Ives DJG, Janz GJ (1961) Reference electrodes. Academic, New York
3. Kahlert H (2010) Reference electrodes. In: Scholz F (ed) Electroanalytical methods, 2nd edn. Springer, Berlin, pp 291–308
4. Inzelt G (2006) Standard potentials. In: Bard AJ, Stratman M, Scholz F, Pickett CJ (eds) Inorganic electrochemistry, vol 7A, Encyclopedia of electrochemistry. Wiley-VCH, Weinheim, p 1

5. Gritzner G, Kuta J (1982) Recommendations on reporting electrode potentials in nonaqueous solvents. Pure Appl Chem 54:1527–1532
6. Gritzner G, Kuta J (1984) Electrode potentials in nonaqueous solvents (recommendations 1983). Pure Appl Chem 56:461–466
7. Lund H (1983) Reference electrodes. In: Baizer MM, Lund H (eds) Organic electrochemistry. Marcel Dekker, New York
8. Buck RP, Rondinini S, Covington AK, Baucke FGK, Brett CMA, Camoes MF, Milton MJT, Mussini T, Naumann R, Pratt KW, Spitzer P, Wilson GS (2002) Measurement of pH. Definition, standards, and procedures (IUPAC Recommendations 2002). Pure Appl Chem 74:2169–2200
9. Lvov SN, Gao H, Macdonald DD (1998) Advanced flow-through external pressure-balanced reference electrode for potentiometric and pH studies in high temperature aqueous solutions. J Electroanal Chem 443:186–194
10. Peters G (1997) A reference electrode with free-diffusion liquid junction for electrochemical measurements under changing pressure conditions. Anal Chem 69:2362–2366

Refractory Metal Production by Molten Salt Electrolysis

Pierre Taxil
Laboratoire de Génie Chimique, Université de Toulouse, Toulouse, France

General Comments

Certain elements of the middle part of the periodic table (columns IVB, VB, and VIB), devoted to transition metals, are so-called refractory metals because of their high fusion temperature (above 2,000 °C), which have relevant consequences in particular on their thermoelastic properties, microhardness, and a correlative industrial use as tools for metal working at high temperature [1, 2] and formally their use as resistance in light bulbs. The most relevant elements of this category are tantalum, niobium (col. IVB), hafnium and zirconium (VB), and molybdenum and tungsten (VIB). Another important common property of them is the thermodynamic stability of their oxide coatings which offers them high corrosion resistance in oxidizing conditions comparable to noble metals whereas they are considered intrinsically among the most reactive metals [3]. Their particularity of shutting oxidation currents and opening reduction currents puts them in the category of valve metals, with other well-known metals such as titanium [4]. These metals coated with their oxide layer in surface find a lot of applications: supercapacitors with Ta or Nb, with high dielectric constant and potential breakdown [5] for high-tech electronics (mobile phones, camcorders, and computers); microhardness for tools and surgery; and superconductivity of pure Nb and Nb alloys [5] for supermagnets. Their relative low abundance in the earth's crust makes them strategic substances, sensitive to the market price, recyclables and explains the trend to use them, for applications calling on their surface properties as a thin coating over usual metals.

The elaboration of refractory metals consists of reducing their salts or oxide phases by oxidant compounds. Molten salt electrolysis is appropriate for this application, for various reasons: First is their extended electrochemical window (more than 3 V) that allows the ions of most of the reactive metals to be discharged before the solvent and then lower cost, lower pollution, and easiness of processing, compared to metallothermic processes, are to be noticed [6]. Nevertheless, refractory metals being multivalent elements, the chemical composition of the electrolyte is critical for avoiding the formation of stable intermediate species instead of the metal during the electroreduction process.

This chapter aims to describe the elaboration process and refining of refractory metals by techniques involving molten salt media. Tungsten and molybdenum preparation will not be mentioned, as far for these elements, molten salt electrolysis is little reported and moreover controversial.

Niobium and Tantalum

These metals have similar properties, and they are together in one of their natural origins, the so-called columbite ore, containing columbium (former name of niobium) and tantalum. Nevertheless, their respective production and use rates are quite different: 1,000–1,500 t/year for Ta and about 16,000 t/year for Nb, far more present in the earth's crust than Ta, that explain their respective costs, in inverse ratio of their production level.

The main production sites of Ta are Australia (54 %), Brazil (13 %), Canada, China, South West Asia, and Europe, while Nb provides almost exclusively from Brazil (93 %).

Molten salt route is one relevant way for producing these metals; extractive metallurgy can provide the oxides (Nb_2O_5, Ta_2O_5) or the fluoride compounds (K_2NbF_7; K_2TaF_7).

Chloride and Fluoride Melts

Niobium oxides, provided by the ore treatment, are dissolved in chloride melts (NaCl-KCl or LiCl-KCl) in the form of $NbCl_5$ [7], and Nb metal is produced by electroreduction in the molten chloride solution. A French company, Cezus, developed this process at the industrial scale in the 1990s [8] for Nb and other refractory metals. Nevertheless, the cathodic process in pure chloride melts was proved to be too complex to be industrially valid, with a series of intermediate steps [8, 9], and provides nonadherent or powder metal layers with a low current efficiency. Better results are obtained in chloride melts containing fluoride ions because of the complexation of Nb in NbF_7^{2-} ions which are reduced in only two steps: $Nb^V \rightarrow Nb^{IV} \rightarrow Nb$ [10, 11], with current efficiencies less than 100 % since Nb^V reacts readily with Nb cathodic product (proportionation reaction) for giving the intermediate species, Nb^{IV}.

The complexing role of fluoride ions in the reduction of refractory metals is now well known for the metal recovery [12]. The technology of extracting tantalum in molten salts is based on the formation of K_2TaF_7, obtained by reaction of HF on the oxide Ta_2O_5 extracted from raw materials [14] before the reduction of this compound by sodium in the liquid phase:

$$K_2TaF_7 + 5 \ Na \rightarrow \ Ta + 5 \ NaF + 2 \ KF$$

Other salt compounds are added in the reactor for lowering the temperature between 400 °C and 900 °C below the fusion point of pure K_2TaF_7 [13]. Another reduction mode involves the electrochemical reduction of K_2TaF_7 in molten fluoride salts [5]. It is well stated now that the electroreduction of Ta^V in fluorides proceeds in a five-electron single step directly leading to Ta metal [14], and thus current efficiency of the preparation of Ta by the electrochemical route is close to 100 %.

Anode Reaction

Classically the anode material is mostly carbon, cheap, and an appropriate material. In pure fluoride melts, the reaction would be the discharge of fluoride ions, reacting with carbon to give perfluorocarbons, highly polluting gas, CF_4 or C_2F_6, at high overpotentials. For this reason, pure fluoride melts must be avoided for electrowinning refractory metals. Chlorides must be always present in the electrolyte in order to develop the discharge of Cl^- ions in Cl_2 at the anode.

Niobium and Tantalum Coatings

Taking into account on one hand the low resources of Nb and Ta on the earth and on the other hand that a great part of their properties concerns their surface only, it seems to be really reasonable to use them as surface coatings on more usual metals or carbon. The anodic metal is Ta or Nb, and the electrolysis allows the anodic metal to be transferred from the anode to the cathode surface, according to an electrorefining operation. During these last decades, successful coatings were realized in pure fluoride melts, with Ta and Nb [15, 16], exhibiting such characteristics (compacity, adherence on the substrate, and purity) that they are directly available for applications of the metal coating [16].

Niobium and Tantalum Alloys and Compounds

Superconducting niobium alloys with nickel, titanium, and germanium are prepared by coelectrodeposition in molten salts (Nb-Ge [17], Ti-Nb [18]); or metalliding: Ni-Nb [19].

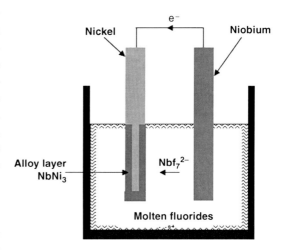

Refractory Metal Production by Molten Salt Electrolysis, Fig. 1 Principle scheme of metalliding of nickel with niobium

This process works as a galvanic cell where the anodic metal (Nb) dissolves in the bath. Nb ions react with the cathodic metal (Ni) to give a defined compound $NbNi_3$ (Fig. 1). Likewise, tantalum and niobium carbides are also prepared by coelectrodeposition of Ta and C [20] or metalliding [21].

Hafnium and Zirconium

Hafnium and zirconium are in the same column of the classification in the periodic table and have hence similar properties except in the nuclear field. Both are associated in the same ore, called zircon, containing only 2–3 % Hf compared to Zr. The production of zircon is abundant: 800,000 t/year in Australia (57 %), South Africa (25 %), and other countries. The treatment of zircon leads to the oxide ZrO_2, zirconia (with 2–3 % of HfO_2), which has its own and well-known applications as refractory ceramics. Electrowinning of pure Zr and pure Hf needs the dissolution of the oxides and their exhaustive separation in advance, Zr being neutron transparent and used as fuel sheath, while Hf is impervious to neutrons and involved in the control systems of nuclear reactors. Carbochlorination in molten chlorides of the oxides leads to $ZrCl_4$ with 2–3 % of $HfCl_4$.

Distillation of ZrCl$_4$-HfCl$_4$ Mixtures

This technology is based on the circulation, at 350 °C and under the atmospheric pressure, of the gaseous ascending ZrCl$_4$-HfCl$_4$ mixture going in contact with a solution AlCl$_3$-KCl, descending in the column. The distillation process leads to the collection of ZrCl$_4$, more soluble in the liquid phase in the bottom of the column, while pure HfCl$_4$ is recovered in the gas phase from the top of the column. The process, finalized by Cezus in 1981, offers the advantage of a significantly lower cost of chemicals than liquid-liquid extraction and more efficiency and more productivity than fractionated crystallization [22].

Zirconium and Hafnium Electrowinning

Zirconium is prepared from ZrCl$_4$ by the Kroll process (chemical reduction of the chloride by liquid magnesium at 900 °C), but the electrochemical route in molten salts is considered as a good alternative. As for Ta and Nb, pure chloride melts are not appropriate for the electrowinning to yield the metal because of too complex cathode process. The addition of fluoride ion source (KF) allows to stabilize Zr ions in the form of K$_2$ZrF$_6$ which is reduced in one step to Zr [23].

For hafnium electrowinning, electrolysis process is readily preferred to Kroll process; Cezus developed the process in pure chloride media from HfCl$_4$, reduced in Hf in a one single step [24]. However, the mechanism is controversial, other authors proposing two steps in pure chloride and a single step in chloride-fluoride melts [25].

Electrowinning by Direct Reduction of Metal Oxides (FFC Process)

Discovered in 1997 at the Cambridge University by **F**arthing, **F**ray, and **C**hen (FFC) and developed in the industry from 2001, this recent process involves a molten salt electrolyzer with a cathodic basket containing metal oxides and a carbon anode [26]. The cell polarization promotes the deoxidation of metal oxides by

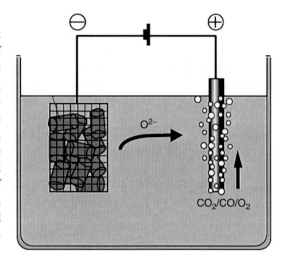

Refractory Metal Production by Molten Salt Electrolysis, Fig. 2 Principle scheme of the FFC process

a cathode reaction releasing O^{2-} ions in the bath, the ionic transfer of these ions towards the anode and the anodic formation of CO$_2$ (Fig. 2). The process is based on the ability of the solvent cation to gain oxide ions at the cathode interface.

Calcium and lithium having a high affinity for oxygen in molten salts are appropriate for this role. The solvent and temperature mostly selected are LiCl-CaCl$_2$ and 850–900 °C, respectively. The patent of the process claims that a host of metals such as Ti, Ge, Si, Zr, Hf, Sm, U, Ta, and Nb or their alloys can be produced with starting materials as oxides [27]. The most cited metal produced by this technology is now titanium (from TiO$_2$) but the English company Metalysis developed a complete device for the production of tantalum from Ta$_2$O$_5$ [28]. Alloys such as Nb$_3$Sn are also prepared by mixing the oxides of parent metals in the appropriate ratio for the alloy.

FFC process exhibits advantages compared to the other metallothermic processes such as Kroll process, in particular a lower cost and far less polluting, but it has major drawbacks:

- The low conductivity of the oxides and of the insufficient contact with the metal basket at the origin of low current efficiencies (less than 50 %).
- The anode reaction consumes carbon leading to the greenhouse gases CO and CO$_2$; moreover

CO_2 is partially soluble in molten solutions to react with O^{2-} to form carbonate ions which are reduced at the cathode, which causes further decrease in the current efficiency. Inert anodes are now being developed for discharging O^{2-} directly in the form of O_2, but a higher anodic voltage by 1 V is necessary.

Recycling of Refractory Metals

Recycling of refractory metals from wastes is an important issue today since they are in relatively low amount in the earth's crust and it should really be a substantial economy of expensive raw materials and also of energy. Molten salt electrolysis is proved to be appropriate to this function which can be assimilated in the case of metallic wastes to electrorefining. Today, waste treatment of other transition metals like actinides or lanthanides is a reality in the nuclear field, while other strategic elements such as silicon are expected to be recovered with a molten salt technology.

Cross-References

▶ Rare Earth Metal Production by Molten Salt Electrolysis

References

1. Schmid K (2006) Manufacturing engineering and technology. Pearson Prentice Hall, Upper Saddle River, pp 86–93
2. Veronski A, Hejwowski T (1991) Creep-resisting materials, Thermal fatigue of Metals. CRC Press, p 81–93
3. Pourbaix M (1963) Corrosion, Atlas d'équilibres thermodynamiques. Gauthier-Villard, Paris, pp 70–83
4. Young L (1961) Anodic oxide films. Acad. Press, New York; Young L (1954) Anodic oxide films on tantalum electrodes. Trans Faraday Soc, 50:153–171
5. Polak C (2009) Metallurgie et Recyclage du niobium et du tantale. Techniques de l'Ingénieur, M2 365v2 (2–16):1–3
6. Kermarrec M (2004) Applications Industrielles des Sels Fondus, Agence Rhône-Alpes Pour la Maîtrise des Matériaux, Savoie Technolac

7. Gonzales J et al (1998) Chlorination of niobium and tantalum ores. Thermochim Acta 311:61–69
8. Lamaze AP, Paillère P (1990) Fused salts for continuous production of multivalent metal from halide. European patent n°394154A
9. Ting G, Fung KW, Mamantov G (1976) Voltammetric studies of $NbCl_5$. J Electrochem Soc 123:624–631
10. Sienerth KD, Mamantov G Meeting ECS 1993, Proceedings of the symposium on molten salts, pp 365–372
11. Khalidi A, Bouteillon J (1993) Comportement électrochimique de $NbCl_5$. J Appl Electrochem 23:801
12. Chamelot P, Taxil P, Lafage B (1994) Voltammetric studies of Tantalum electrodeposition baths. Electrochem Acta 17:2571–2574
13. Agulyanski A (2003) Potassium fluorotantalate in solid, dissolved and molten conditions. J Fluorine chem 123:155–161
14. Taxil P, Mahenc J (1987) Formation of corrosion resistant layers by electrodeposition of refractory metals or by alloy electrowinning in molten fluoride. JAppl Electrochem 17:261–269; Polyakova LP et al (1992) Secondary processes during electrodeposition of tantalum in molten salts. J Appl Electrochem 22:628–637
15. Senderoff S, Mellors GW (1966) Coherent coatings of refractory metals. Science 153:1475–1481
16. Taxil P (1986) Dépôts de tantale et de niobium dans les sels fondus et leurs applications. Dissertation, Toulouse (France)
17. Cohen U (1983) Electrodeposition of niobium-germanium alloys. J Electrochem Soc 130:1480
18. Polyakova LP, Taxil P, Polyakov EG (2003) Electrochemical behaviour and codeposition of titanium and niobium in chloride-fluoride melts. J Alloys Comp 359:244–255
19. Taxil P, Qiao ZY (1985) Electrochemical alloying of nickel with niobium in molten fluorides. J Appl Electrochem 15:947–952
20. Stern KH, Gadomski ST (1983) Electrodeposition of tantalum carbides from molten salts. J Electrochem Soc 130:300–305; Massot L, Chamelot P, Taxil P (2006) Preparation of tantalum carbide films by reaction of electrolytic carbon coating with the tantalum coating. J Alloys Comp 424(1–2):199–203
21. Chamelot Massot P, Taxil P Preparation of tantalum carbide films by reaction of electrolytic carbon coating with the tantalum coating. J Alloys Comp
22. Tricot R (1994) Techniques de l'Ingénieur. Métallurgie Extractive et recyclage des métaux de transition, Zirconium et Hafnium, M2360
23. Steinberg MA, Sibert ME, Wainer E (1954) Process development of the electrolytic production of zirconium from K_2ZrF_6. J Electrochem Soc 101(2):63–78
24. Lamaze AP, Charquet D (1990) Development of hafnium tetrachloride electrolysis. In: Liddell KC

(ed) Refractory metals. The Minerals Metals and Materials Society

25. Serrano K, Boïko O, Chamelot P, Lafage B, Taxil P (1997) Electrochemical reduction of HfCl4 in molten salts. In: 5th symposium on molten salts chemistry and Technology, Dechema and V., Dresden (Germany)

26. Chen GZ, Fray DJ, Farthing TW, Chen GZ, Fray DJ, Farthing TW (2000) Direct electrochemical reduction of titanium dioxide to titanium in molten calcium chloride. Nature 407(6802):361–364

27. World patent WO 99/64638

28. Jeong SM et al (2007) Characteristics of an electrochemical reduction of Ta_2O_5 for the preparation of metallic tantalum in a $LiCl$-Li_2O molten salts. J Alloys comp 400:210–215

Regenerative Fuel Cells

Tsutomu Ioroi
AIST, Ikeda, Japan

Introduction

A regenerative fuel cell (RFC) is an electrochemical device that can store electrical energy using hydrogen as an energy medium. While in fuel cell mode, electrical power can be extracted from an RFC system by reacting stored hydrogen and oxygen (air). On the other hand, in electrolysis mode, the storage tank can be refilled with hydrogen (and oxygen) by operating the RFC as a water/steam electrolyzer using electric power and water. Therefore, RFCs are expected to be useful for electrical energy storage (EES) especially for medium- to large-scale systems. As an EES device, a secondary battery can provide a similar function to RFCs. However, one of the advantages of RFCs is that they are free from self-discharge because the active reactants are stored separately from the reactor. An RFC consists of two separate subunit: a fuel cell/electrolyzer subunit and an energy storage subunit. This separation also enables the independent design of the rated output/input power and energy capacity of the RFC system. Another advantage of RFCs over secondary batteries is their higher energy density. The theoretical energy density of the hydrogen/oxygen reaction

$$H_2 + 1/2O_2 \Rightarrow H_2O \qquad (1)$$

is 237.3 kJ/mol-H_2O (25 °C, HHV basis), which is equal to 3,660 Wh/kg-H_2O. While this value is simply calculated from the weight of a stoichiometric reactant, a more realistic value is expected to be 400–1,000 Wh/kg-system [1–3]. This is still higher than the value for the current Li-ion batteries ($C_6Li/LiCoO_2$, 387 Wh/kg) [4].

Materials and Structure of an RFC

The materials and structure of an RFC depend on the operating temperature (polymer electrolyte type or solid oxide type), cell design (unitized or separate fuel cell/electrolyzer units), and reactants (H_2/O_2, H_2/Cl_2, H_2/Br_2, etc.).

High-temperature RFCs based on solid oxide electrolyte technology are operated at 700–1,000 °C with H_2/O_2. The materials used for high-temperature RFCs are basically similar to those in solid oxide fuel cells. As an electrolyte material, 8 mol% Y_2O_3 stabilized ZrO_2 (YSZ) is typically used because of its relatively high ionic conductivity and chemical/mechanical stability. A porous mixture of Ni and YSZ (Ni/YSZ cermet) and perovskite-type oxides such as strontium-doped lanthanum manganite (LSM) are used as hydrogen and oxygen electrode materials, respectively [5]. The advantage of high-temperature RFCs is higher overall (round-trip) energy conversion efficiency (electric $\Rightarrow$ hydrogen $\Rightarrow$ electric). This is mainly due to electrode kinetic and thermodynamic factors. Under high-temperature conditions, the kinetics of the electrode reaction are accelerated so that electrode polarization, especially for oxygen redox reactions, is significantly mitigated compared to that in low-temperature RFCs. In addition, the $\Delta G/T\Delta S$ ratio of the water-splitting reaction decreases with an increase in temperature, which means that the demand for electric energy is reduced and a significant portion of the required energy can be provided as thermal

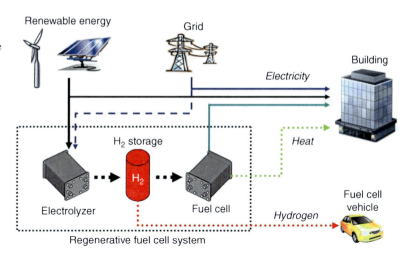

Regenerative Fuel Cells, Fig. 1 Renewable energy storage using a regenerative fuel cell system

energy [6]. The disadvantages are vulnerability to heat shock and long-term stability (lifetime), which limits the number of times the system can start-up/shutdown.

Low-temperature RFCs based on polymer electrolyte technology are operated at 60–80 °C and typically use H_2/O_2. The materials used in low-temperature RFCs are also similar to those in polymer electrolyte fuel cells and polymer electrolyte water electrolyzers. Perfluorinated membranes such as Nafion (DuPont) are typically used as an electrolyte. As catalyst materials, Pt is used for the hydrogen electrode, and Pt or a mixture of Pt/Ir/Ru metals or oxides is used for the oxygen electrode [7]. One of the advantages of low-temperature RFCs is the rapid start-up/shutdown of the system. The RFC system is operated alternately as a fuel cell and an electrolyzer, so that, in low-temperature RFCs, the restart time and standby power of the subunit that is not in operation can be reduced. The disadvantage is that low-temperature system are less efficient than high-temperature systems, which is mainly due to the higher polarization of oxygen electrode reactions and thermodynamic requirements. The overall efficiency of low-temperature RFCs can be significantly improved through the use of halogens, such as H_2/Cl_2 and H_2/Br_2, due to the greater reversibility of the redox reactions of halogens [8, 9]. However, the highly corrosive nature of halogen compounds is a major drawback. The overall efficiency can also be improved by using the waste heat from an RFC system, such as to supply hot water, by installing a cogeneration system, as shown in Fig. 1.

System Design and Potential Applications of RFC

In the basic design of an RFC, the system includes a dedicated (separate) fuel cell and electrolyzer, especially for low-temperature RFCs. Another possible design is the integration of the functions of a fuel cell and electrolyzer into a single electrochemical cell (unitized regenerative fuel cell, URFC). The advantages of a URFC are lower weight and volume, so that URFCs have been considered for space or military applications. In addition, URFCs have the potential advantage of a lower cost provided that electrochemical cells account for a significant portion of the system cost. On the other hand, the catalyst materials and electrode design must be balanced in a URFC. For example, the electrode should be flooded (hydrophilic) during electrolyzer mode but should be dry (hydrophobic) during fuel cell operation. The cathode material should be suitable for both anodic and cathodic reactions. This is an important issue especially for the oxygen electrode because redox reactions of oxygen are highly irreversible, which causes larger polarizations, as described above.

Examples of potential applications of RFCs are as follows:

- Energy storage for renewable energy sources [10]
- Energy storage for remote off-grid power sources [11]
- Energy storage for spacecraft [12]
- Stratospheric platform airship [13]
- High-altitude, long-endurance, solar-rechargeable aircraft [1].

Future Directions

RFCs may be able to offer high-density electrical energy storage. Therefore, in applications in which energy density is the first priority, advanced RFC system should be considered. However, the efficiency of energy conversion in RFCs is still inadequate compared to competing technologies such as secondary battery systems. To realize the commercialization of RFC energy storage system, low-cost, durable, and efficient materials for electrochemical cells (electrocatalysts, electrolyte, bipolar plate) and a lightweight system for hydrogen storage should be developed.

Cross-References

► Polymer Electrolyte Fuel Cells (PEFCs), Introduction
► Solid Oxide Fuel Cells, Introduction

References

1. Mitlitsky F, Myers B, Weisberg AH (1998) Regenerative fuel cell systems. Energy Fuel 12:56–71. doi:10.1021/ef970151w
2. Smith W (2000) The role of fuel cells in energy storage. J Power Sources 86:74–83. doi:10.1016/S0378-7753(99)00485-1
3. Barbir F, Molter T, Dalton L (2005) Efficiency and weight trade-off analysis of regenerative fuel cells as energy storage for aerospace applications. Int J Hydrogen Energ 30:351–357. doi:10.1016/j.ijhydene.2004.08.004
4. Bruce PG, Freunberger SA, Hardwick LJ, Tarascon JM (2012) Li-O$_2$ and Li-S batteries with high energy storage. Nat Mater 11:19–29. doi:10.1038/nmat3191

5. Hauch A, Ebbesen SD, Jensen SH, Mogensen M (2008) Highly efficient high temperature electrolysis. J Mater Chem 18:2331–2340. doi:10.1039/B718822F
6. Laguna-Bercero MA, Kilner JA, Skinner SJ (2011) Development of oxygen electrodes for reversible solid oxide fuel cells with scandia stabilized zirconia electrolytes. Solid State Ionics 192:501–504. doi:10.1016/j.ssi.2010.01.003
7. Ioroi T, Kitazawa N, Yasuda K, Yamamoto Y, Takenaka H (2000) Iridium oxide/platinum electrocatalysts for unitized regenerative polymer electrolyte fuel cells. J Electrochem Soc 147:2018–2022. doi:10.1149/1.1393478
8. Rugolo J, Huskinson B, Aziz MJ (2012) Model of performance of a regenerative hydrogen chlorine fuel cell for grid-scale electrical energy storage. J Electrochem Soc 159:B133–B144. doi:10.1149/2.030202jes
9. Kreutzer H, Yarlagadda V, Nguyen TV (2012) Performance evaluation of a regenerative hydrogen-bromine fuel cell. J Electrochem Soc 159:F331–F337. doi:10.1149/2.086207jes
10. Maclay JD, Brouwer J, Samuelsen GS (2006) Dynamic analysis of regenerative fuel cell power for potential use in renewable residential applications. Int J Hydrogen Energ 31:994–1009. doi:10.1016/j.ijhydene.2005.10.008
11. Agbossou K, Chahine R, Hamelin J, Laurencelle F, Anouar A, St-Arnaud JM, Bose TK (2001) Renewable energy systems based on hydrogen for remote applications. J Power Sources 96:168–172. doi:10.1016/S0378-7753(01)00495-5
12. Sone Y (2011) A 100-W class regenerative fuel cell system for lunar and planetary missions. J Power Sources 196:9076–9080. doi:10.1016/j.jpowsour.2011.01.085
13. Eguchi K, Fujihara T (2003) Research progress in solar powered technology for SPF airship. NAL Res Prog 2002–2003:6–9

Role of Separators in Batteries

Daniel Steingart
Department of Mechanical and Aerospace Engineering, Andlinger Center for Energy, the Environment Princeton University, Princeton, NJ, USA

Introduction

Within a battery the active components, that is, the electrodes, electrolytes, and current collectors, receive the majority of attention in the literature. The inert components of a battery, that is the casing but more importantly the separator, however, play a critical role what they allow and what they do not allow. Without binder particulate electrodes would not stay together; without a case the battery would have no protection from its environment; and without a mechanical separation between anode and cathode, there would be no impetus for current to be created through an electrochemical reaction as the active material materials would simply undergo a mutually passivating redox reaction (Fig. 1).

Unsung players, these materials must prevent short circuits while maximizing the cross section for ionic transport [1–5]:

1. Provide reliable and high porosity through its bulk to transport current in the form of ions between electrodes
2. Have sufficient, unchanging surface porosity at electrodes to allow even distribution of ion and therefore reaction current at the electrode
3. Physically separate the electrodes for the lifetime of the battery
4. Be an electronic insulator
5. Be wetted by an electrolyte without swelling
6. Be as uniform as possible
7. Prevent sediment (e.g., precipitated electrolyte salts, fractured electrode materials) from migrating across the cell and/or blocking electrode reaction sites.

The separator must serve this purpose in a dynamic environment. Electrodes can change shape and/or volume as a function of age and cycle number; electrolytes can age triggering precipitation. The separator must be considered and matched to its system so that it can provide the aforementioned functionality for the design life of the system. Figure 2 indicates the workhorse design of perhaps the most ubiquitous battery and separator material, the zinc-alkaline primary system.

The separator region should be as thin as possible without risking short circuits in a practical cell. The minimum electrolyte distance is limited by the unevenness of the electrode surfaces. Because electrode surfaces will change and deform over time, the electrolyte must be engineered to properly accept such change if the cell is to be safe and maximize its cycle life.

An electrolyte can be "too thick" as well. Again, thicker electrolytes are linearly more resistive, so for high-power cells the electrolyte resistance can lead to significant heating of a cell. In older lithium ion cells, this can trigger "thermal runaway." Thus, an electrolyte should

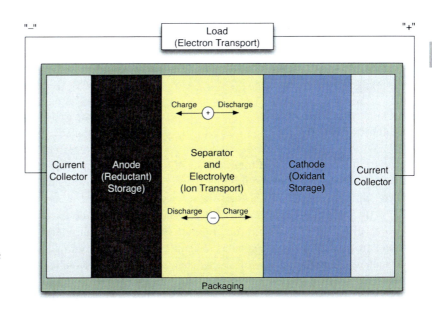

Role of Separators in Batteries, Fig. 1 Basic schematic of a closed electrochemical energy storage cell

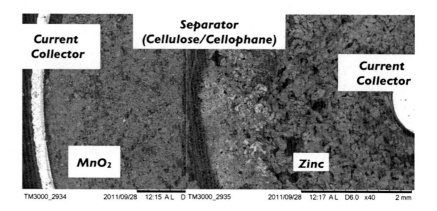

Role of Separators in Batteries, Fig. 2 Cross section of an AA Zinc-MnO$_2$ battery

be as thin as possible to maximize potential efficiency and minimize overheating.

This brief chapter will consider some material challenges across aqueous, aprotic, and high-temperature systems as a start for the reader's consideration of separators. This is by no means an attempt at an exhaustive approach, and so long as there are new battery electrodes and electrolytes, there will be new challenges for separator materials to meet. We will begin with a brief overview of battery separator materials and then consider design challenges for the lead acid, alkaline, lithium ion and molten metal battery systems. Not considered here are in any depth beyond this sentence are "overloaded" ion-conducting membranes such as proton exchange membranes (PEM), as well as O^{2+} conductors for high-temperature fuel cells: these systems combine the mechanical requirements of the separator with the ion transport requirements of the electrolyte. These systems typically have pore sizes on the order of 2 nm [6].

Separator Materials

Separator materials, first and foremost, need to be completely nonreactive and minimally interactive with the electrolyte they host. By minimally interactive, this is the razor's edge of not swelling significantly in the presence of an electrolyte (swelling indicates eventual mechanical degradation) while also being completely wetting within the electrolyte [4, 5, 7].

Lead acid battery separator materials have progressed significantly over the history of this workhorse chemistry and is a good indicator of the arrow of progress of the entire field. The first lead acid separators were natural rubbers that had moderate porosity (~55–65 %) with more sizes on the order of 1–10 μm. These separators were on the order of 500 μm thick. These systems suffered aging and embrittlement problems, and the separator was often the point of failure for these batteries. The next generation of separators for lead acid batteries were engineered cellulosic materials. These systems were actually worse mechanically than the original rubber systems, but had significantly better porosity and lower thicknesses, effectively halving the ohmic resistance with an equivalent electrolyte composition. This system was optimized for higher current density at lower cost; the aging disadvantage was an acceptable trade-off at the time.

The advent of engineered polymer brought a step change to the lead acid separator. With polyvinyl chloride systems, pore sizes decreased slightly compared to the cellulose-based systems, but the thickness deceased as much, and the mechanical and chemical stability of these systems improved markedly as well. As a result separators were no longer the age-limiting mechanisms for lead acid batteries, and conductivity effectively doubled again. Polyethylene systems improved the overall porosity to levels previously realized by natural rubber systems while maintaining the mechanical advantages of PVC. The next and final step change to lead acid separators was a move from engineered polymers

Role of Separators in Batteries

to engineered glasses. In these systems the porosity skyrocketed to up to 95 % and the conductivity of the electrolyte doubled again, all while maintaining the mechanical advantages of the engineered polymers. What this meant for the industry now is that a lead acid separator could be 10 times as thick as the original rubber separators (a condition which could allow for mass transport and mixing parallel to the electrode faces, while still maintaining a lower electrolyte conductivity.

In the current era lithium ion separators have gone through a similar engineering trajectory, and most systems now are either PE, PP, PvDF, polyolefin, or composites thereof with [5, 8, 9], engineered to be resistant to chemical attack by aprotic electrolytes. For alkaline systems, PP and PVA are analogous systems. Additionally, the low cost and abundance of both cellophane and cellulose make it an attractive choice of separator for low-cost zinc-alkaline primary cells.

Finally, high-temperature molten salt electrolyte batteries (NaS, Zebra) require completely inorganic separators capable of withstanding liquid metal temperature and chemical attack, effectively acidic conditions at temperatures >200 °C. Beta-Al_2O_3 has been significantly engineered to serve this role [10].

Structure of Separators

Beyond the materials development, the morphology induced by processing a given separator is a critical aspect of the separators performance. What follows is a brief summary of separators structure as enumerate by Arora [5]. Microporous separators can be manufactured from any of the above materials. Specifically, nonwoven separators (colloquially referred to as "nonwovens") are fibers of a given materials that are laid and processed in a dry method (perhaps a sintered mat) or a wet bonding process very similar to that of paper. The processing of the fibers can be exploited to provide or enhance mechanical or wetting properties [8].

Separators can also be made directly from solution-processed and/or cross-linked methods. Polyvinylidene difluoride (PVdF) is a popular choice for this approach because it can, when processed directly on an electrode containing a PVdF binder, form an excellent electrode/electrolyte interface (Fig. 2). These systems can also be cast and processed with ionic conductors "built in" in the form of gels or ion-conducting backbones. Such a system is shown in Fig. 3 [8, 11].

Given the constraints of a given materials system, the choice between cast or nonwovens, and methods of processing within these systems, is driven by the many, and often diametrically, opposed needs of a given system.

Design Considerations for Individual Systems

The devil is the in the details, and the choice of materials compatibility is insufficient to match a separator system to a battery chemistry. In this section we will briefly outline design issues and their solutions to a few different chemistries. This is intended to serve as guiding examples for the reader's own selection process of a separator material and system design.

Lead Acid Batteries

Lead acid batteries pose the following challenges to a separator. Both anode and cathode are subject to shape change and possible embrittlement, so the separator must be compliant enough to accommodate this type of change while also preventing material crossover. Electrolyte stratification is a significant issue for certain types of lead acid systems, so the pores must be big enough to accommodate natural or forced convection to "refresh" the electrolyte. In other lead acid systems, the generation and recombination of gas is critical to the operation of the cell, so the separator must allow these gases to recombine if need be.

The astute reader will note that the above constraints are at cross-purposes: flow must be maximized but particle transport must be minimized. As a result the following general practices take place. Depending on the specific lead acid system, either a PE or microglass separator will

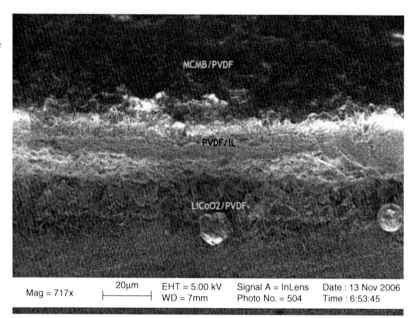

Role of Separators in Batteries, Fig. 3 A fully printed PVdF IL electrolyte gel from [11]

be used to minimize crossover or maximize flow. In certain system, composites of the two are employed where small porosity layers are used near the electrode to catch embrittled, cracked electrodes, and larger pores are used to catch allow for electrolyte convection.

Zinc-Alkaline Batteries

With zinc-alkaline batteries the separator must accommodate yet suppress the ramification of zinc upon cycling, while also preventing the formation of thick zinc oxide layers on the zinc electrode. This has led to the use of cellophane laminated to something like PE, to achieve a similar effect to the shape change prevention in the aforementioned lead acid cells.

Lithium Ion Batteries

As of this writing no commercial separator is reliable enough to prevent a lithium metal dendrite or ramification from short circuiting a cell upon charge. Due to the danger of this condition, lithium metal is never, or never in the author's knowledge, used as a secondary battery. Within lithium ion cells, however, the effect of shape change of an electrode is still an issue with graphitic carbon undergoing volume changes up to 30 % and silicon electrodes up to 400 % [3].

Beyond shape change, lithium ion cells have be designed to retard the inherent flammability of most high-performance aprotic electrolytes as well as the ignition danger of high-performance anodes and cathodes in overcharge or over-current and over-temperature conditions. Commercial separators [5] have been designed with a specific melting point (typically just above 130 °C), which when reached will cause porosity within the cell to diminish significantly. This porosity decrease has been shown to very quickly increase the resistance of a given electrolyte up to four orders of magnitude, effectively stopping any further reaction. This mechanism has been tested successfully against overcharge, over-temperature, mechanical intrusion (the "nail test"), and over-current conditions [5]. While not a guarantee against all fires, this separator shutdown design significantly reduces the threat of explosion in an off condition.

Conclusions and Further Readings

While not a formally *required* aspect of electrochemical energy storage devices or fuel cells, separators are an enabling technology that has shown above can greatly improve the power performance, cycle lifetime, and safety aspects of a battery. The students interested in this field should know that a separator can never be "too good" and that the industry will always appreciate further improvements in porosity, stability, safety, and costs. The reader is directed to the references following this review for deeper analysis.

Cross-References

▶ Primary Battery Design

References

1. Peabody C, Arnold CB (2011) The role of mechanically induced separator creep in lithium-ion battery capacity fade. J Power Sources 196:8147–8153. doi:10.1016/j.jpowsour.2011.05.023
2. Brodd RJ (2012) Batteries for sustainability. Springer, New York
3. Besenhard JO (1999) Handbook of battery materials. In: Besenhard JO (ed) Handbook of battery materials. Wiley-VCH, New York, p 648. ISBN 3-527-29469-4
4. Linden D, Reddy TB (2002) Handbook of batteries. McGraw-Hill Professional, New York
5. Arora P, Zhang ZJ (2004) Battery separators. Chem Rev 104:4419–4462. doi:10.1021/cr020738u
6. Agrawal RC, Pandey GP (2008) Solid polymer electrolytes: materials designing and all-solid-state battery applications: an overview. J Phys D: Appl Phys 41:223001. doi:10.1088/0022-3727/41/22/223001
7. Leung P, Li X, de León CP et al (2012) Progress in redox flow batteries, remaining challenges and their applications in energy storage. RSC Adv 2:10125–10156
8. Mehta V, Cooper JS (2003) Review and analysis of PEM fuel cell design and manufacturing. J Power Sources 114:32–53. doi:10.1016/S0378-7753(02)00542-6
9. Huang Z-M, Zhang YZ, Kotaki M, Ramakrishna S (2003) A review on polymer nanofibers by electrospinning and their applications in nanocomposites. Compos Sci Technol 63:2223–2253. doi:10.1016/S0266-3538(03)00178-7
10. Lu X, Xia G, Lemmon JP, Yang Z (2010) Advanced materials for sodium-beta alumina batteries: status, challenges and perspectives. J Power Sources 195:2431–2442. doi:10.1016/j.jpowsour.2009.11.120
11. Steingart D, Ho CC, Salminen J et al (2007) Dispenser printing of solid polymer-ionic liquid electrolytes for lithium ion cells. IEEE Polytronics, Vol. 1, pp 1:261–264

Ruthenium Oxides as Supercapacitor Electrodes

Wataru Sugimoto
Faculty of Textile Science and Technology, Shinshu University, Nagano, Japan

Background

Since the pioneering work by Trasatti and Buzzanca [1], the first to recognize that the "rectangular"-shaped cyclic voltammogram of a RuO_2 film (Fig. 1) resembled that of the carbon-based electric a double-layer capacitor, RuO_2 has attracted interest as a model system for the fundamental understanding of pseudocapacitive behavior of oxide electrodes as well as practical application towards high rate, lightweight, and small energy harvesting devices. The difference to an ideally polarizable electrode system is that there is some surface oxidation/reduction occurring in addition to the non-Faradaic electrical double-layer charging, which, in contrast to reactions occurring in batteries, is limited to a monolayer or a few layers on the electrode surfaces. For these systems, the electrode potential varies almost linearly with surface coverage, that is, with the charge passed during the reaction (similar to an electrical double-layer capacitor). The kinetics of such surface and near surface redox reactions are extremely fast and exhibits reversible or pseudo-reversible behavior. Consequently, the device behaves like an electrochemical double-layer capacitor with redox contribution, hence the termination and concept "pseudocapacitor."

Hydrous ruthenium oxide ($RuO_2 \cdot xH_2O$) nanoparticles represent one of the best-known electrode materials for aqueous supercapacitors providing high specific capacitance ranging from a few hundred to $\sim 1,000$ F g^{-1}. The high gravimetric and volumetric capacitance of RuO_2-based electrodes is appealing, especially where size and weight is taken more seriously than cost issues such as low-voltage miniaturized devices and other high value-added devices [2–4].

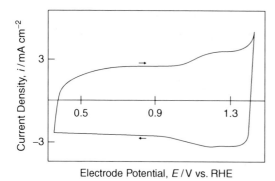

Ruthenium Oxides as Supercapacitor Electrodes, Fig. 1 A typical cyclic voltammogram of RuO$_2$ electrode in 1 M HClO$_4$ at 40 mV s^{-1} [1]

The weakness of RuO$_2$-based electrodes, besides its lack of abundance and cost, is that these electrodes perform better in aqueous compared to nonaqueous electrolytes, limiting the voltage window to about 1.0–1.2 V.

The study of hydrous RuO$_2$ nanoparticles prepared by a sol–gel method instituted a major progress in terms of gravimetric capacitance [5, 6], activating a worldwide surge in the study of RuO$_2$-based electrodes for electrochemical capacitor application. Much effort has since been devoted to research and development with particular emphasis on structure–property relations. Higher capacitance and enhanced power capability has been achieved by material design, and sophisticated characterization methods have contributed to the understanding of the Faradaic behavior of RuO$_2$-based electrodes. Owing to such extensive studies, the fundamental charge storage properties of RuO$_2$ are now much better understood.

The theoretical redox capacitance of RuO$_2 \cdot$ xH$_2$O is often calculated from the equation,

$$C_{\text{redox}} = \frac{nF}{m\Delta V} \quad (1)$$

where m is the molar mass of RuO$_2 \cdot x$H$_2$O, F is the Faraday constant 96,485 C mol^{-1}, and n is the number of electrons involved in the reaction within the potential window, ΔV [5–12]. A C_{redox} value of 1,358 F g^{-1} can be deduced for RuO$_2 \cdot$ 0.5H$_2$O assuming that all of the Ru atoms of the material change its oxidation state from Ru^{2+} to Ru^{4+} within a 1.0 V potential window [7, 13]. The C_{redox} is 970 F g^{-1} if the potential range is 1.4 V [5]. If a 4-electron reaction is assumed within a 1.35 V window, then the value exceeds 2,000 F g^{-1} [9, 14]. Note that in the above calculations, it is assumed that all of the Ru atoms in the crystallite are involved in the reaction (bulk reaction) and that Ru can take a number of oxidation states, which seems to be an overestimation based on various studies. It should also be kept in mind that this calculation neglects the contribution from the electric a double-layer capacitance C_{dl}.

The theoretical value of the electric a double-layer capacitance C_{dl} of RuO$_2$ is somewhat a controversial matter. Most of the C_{dl} values reported in literature for RuO$_2$-based electrodes are larger than the value for a mercury electrode in a diluted aqueous electrolyte solution, which is $C_{\text{dl}} \sim 20$ μF cm^{-2}. It is quite challenging to experimentally differentiate between the contribution from the electric a double-layer capacitance C_{dl} and redox-related Faradaic capacitance C_{redox}, as the (surface) redox-related pseudocapacitance for RuO$_2$ is a fast and reversible process [15]. The most widely reported value is 80 μF cm^{-2} [15–18]. Using this value as a probe, one can deduce the theoretical C_{dl} for a 1 nm RuO$_2$ particle as 680 F g^{-1}. In the case of hydrous RuO$_2$ with gravimetric capacitance $\sim$700 F g^{-1}, area-specific capacitance as high as 1,000 μF cm^{-2} has been reported using BET surface area [5, 6]. Such unusually high area-specific capacitance may be a result of a mixture of the non-Faradaic electric a double-layer capacitance and Faradaic redox-related processes. The surface area of hydrous oxides may also be underestimated by N$_2$ adsorption measurements depending on the pretreatment.

Synthesis, Chemical, and Physical Properties

The specific capacitance, as well as the cycle and rate performance, is affected by the various

Ruthenium Oxides as Supercapacitor Electrodes

parameters including particle size and porosity (electrochemically accessible surface area), crystal structure, water content, etc., which will vary as a result of the different preparation procedures and measurement conditions reported by numerous groups [1, 5, 6, 19–33]. Porous RuO_2 films can be prepared by pyrolysis of ruthenium precursors coated on a conducting substrate by dip coating or spraying, which are the processes adopted for the well-known dimensionally stable electrodes used in industrial electrolysis. The films are typically heat treated at moderately high temperature (450–550 °C) to prepare well-crystalline material that can withstand tortuous gas evolution with crystallite size in the range of 20–50 nm in diameter. Smaller, high-surface area RuO_2 nanoparticles can be obtained by soft-solution processing methods including chemical reduction of RuO_4 [34], precipitation from various precursors [35], sol–gel reactions [5, 6, 36], etc. Electrochemical methods such as electrooxidation of Ru metal [23, 24, 37–48] and electrodeposition [49–57] have also been applied. Heat treatment at relatively low temperatures (150 $\sim$ 200 °C) is often necessary to convert all Ru ions to the electrochemically stable Ru^{4+} state while maintaining small particle size. Mesoporous RuO_2 via soft and hard template synthesis has also been conducted [58].

Synthesis of RuO_2-MO_x composites and A_xRuO_y complex oxides is a classical approach to increase the utilization of RuO_2. Binary rutile-type $(Ru,M)O_2$ (M $=$ Sn^{4+}, Ti^{4+}, V^{4+}, Mo^{4+} etc.), pyrochlore-type $A_2Ru_2O_7$, and perovskite-type $ARuO_3$ are typical examples [59–67]. Many studies have focused on preparing composites with carbonaceous materials, including activated carbon [7–9], carbon black [10, 11, 68, 69], carbon nanotubes [11, 70–81], and more recently reduced graphite oxide [82–87]. Carbon-supported RuO_2 materials can be prepared by impregnation or colloidal methods [10, 88, 89]. Even a simple physical mixture of acetylene black and sol–gel-derived hydrous RuO_2 can drastically improve the rate performance, where the carbon black helps to increase porosity to supply extra pores for electrolyte permeation [19].

High capacitance can also be achieved by using a water-swellable layered structure or by exfoliation of the layered structure into elemental nanosheets. Specific capacitance of $\sim$400 F g^{-1} has been accomplished by using layered RuO_2 as the electrode material [90, 91]. The capacitance is exceptionally large considering the large micrometer size of the particles, which has been attributed to the ability of the interlayer to swell thereby allowing electrolyte permeation into the two-dimensional structure. RuO_2 nanosheet electrodes provide even higher capacitance, reaching $\sim$700 F g^{-1}. These two-dimensional materials exhibit unique electrochemical properties with remarkably large contribution from pseudocapacitance (as much as 50 %) [90, 91].

Charge Storage Mechanism

The cyclic voltammograms of hydrous RuO_2 heat treated at suitable temperatures are featureless and principally rectangular in shape in comparison to as-prepared or well-annealed material (Fig. 2). At heat-treatment temperatures below 150 °C, poor i-E response is observed particularly in the low potential region, which has been associated to the presence of Ru^{3+}. Considerable capacitance fading occurs for such non-optimized conditions, most likely due to the presence of soluble Ru species [35].

The fast and slow charges in thermally prepared RuO_2/Ti DSA-type electrodes have been attributed to the utilization of more accessible, mesoporous surfaces and less accessible, microporous inner surfaces, respectively [92]. Based on ellipsometry and ac impedance measurements on anodically and thermally prepared RuO_2, the fast- and slow-charging modes have been ascribed to the charging of the grain surfaces and incorporation of protons into the oxide grains, respectively [93]. Similar conclusions were derived by CV and CA measurements of anhydrous RuO_2 nanoparticles [15].

Various models have been proposed to explain the reasoning behind the observed maximum capacitance and power capability of sol–gel-derived hydrous RuO_2 as a function of

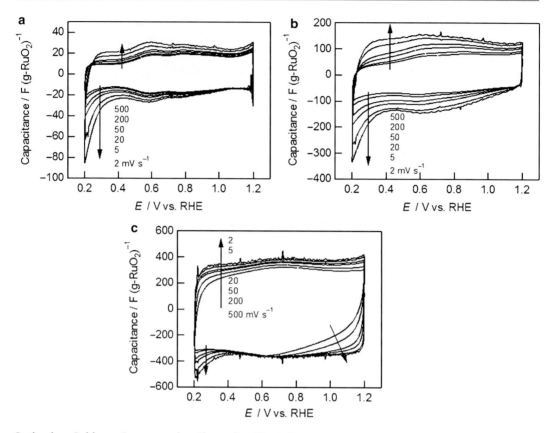

Ruthenium Oxides as Supercapacitor Electrodes, Fig. 2 Typical cyclic voltammograms of $RuO_2 \cdot nH_2O$ in 0.5 M H_2SO_4 with $n =$ (**a**) 0, (**b**) 0.3, and (**c**) 0.5 [33]

the heat-treatment temperature. The cluster-sized particles for low-temperature annealed hydrous RuO_2 afford high capacitance due to large surface area at the expense of power performance. Zheng et al. studied the interaction between the structural water and the rutile crystals by measuring the proton dynamics of hydrous RuO_2 using solid-state proton nuclear magnetic resonance (H-NMR) spectroscopy [28]. A maximum in proton activation energy was observed when the hydrous RuO_2 exhibited the maximum capacitance revealing the correlation between the proton mobility and the charge storage (Fig. 3) [28]. An optimum mixed percolation conduction mechanism has also been proposed to explain the volcano-plot behavior [31]. In addition to the relation between the maximum capacitance and water content in hydrous RuO_2, Sugimoto et al. derived a similar model based on electrochemical impedance spectroscopy to explain the frequency response [33]. It was suggested that the frequency response (power capability) was dominated by proton conduction within the hydrated micropores between the RuO_2 nanocrystallites. Kötz et al. reached similar conclusions based on X-ray photoelectron spectroscopy (XPS) studies [94] revealing two types of water in hydrous RuO_2: weakly bound physically adsorbed water and strongly bound chemically bound water. The chemically bound water was correlated to the heat-treatment temperature and increase in particle size. It has also been shown that the initial hydrous nature is not a prerequisite for the high capacitance [27]. X-ray absorption near-edge structure (XANES) and atomic pair-density function (PDF) analysis of synchrotron X-ray scattering measurements on a series of hydrous RuO_2 treated at different temperatures have revealed that hydrous RuO_2 is composed of disordered rutile-like RuO_2 nanocrystals

Ruthenium Oxides as Supercapacitor Electrodes, Fig. 3 (a) Specific capacitance and (b) apparent proton activation energy determined from H-NMR of the $RuO_2 \cdot H_2O$ as a function of the sample annealing temperature [28]

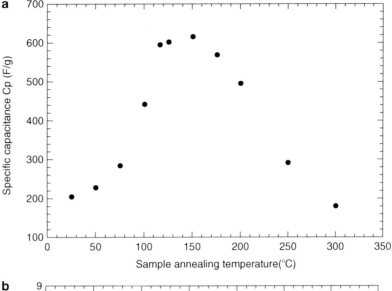

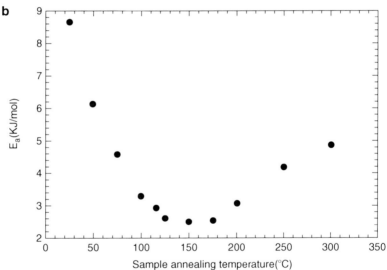

dispersed by boundaries of structural water, in contrast to the apparent amorphousness measured with conventional XRD [22, 31]. A parallel electron-proton conduction model, where the rutile-like RuO_2 nanocrystals support electronic conduction and structural water as the boundaries transport protons, was proposed to account for the maximum capacitance of $RuO_2 \cdot xH_2O$ at $x \sim 0.5$ (Fig. 4) [31].

In situ techniques have also been exploited to realize the effect of the heat-treatment temperature on the charge storage capability of ruthenium oxide. In situ XPS and XRD data gives evidence that no bulk reduction of anhydrous RuO_2 occurs until the electrode is polarized at potentials where hydrogen evolution occurs [95–97]. In situ resistivity measurements of anhydrous RuO_2 have shown that the resistivity is independent of potential within the hydrogen and oxygen evolution region, complementing the in situ XPS and XRD data showing that bulk reduction of Ru^{4+} does not take place in the potential region relevant to electrochemical capacitor applications [98]. Ru^{4+} reduction to Ru^{3+} has been shown to occur only below $E < 0.4$ V by in situ EXAFS and XANES [46, 99–101]. The reduction of Ru^{4+}

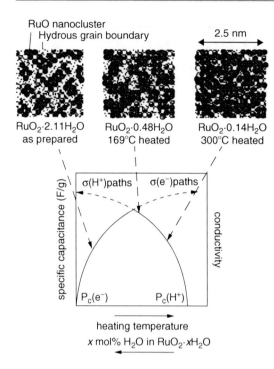

Ruthenium Oxides as Supercapacitor Electrodes, Fig. 4 Schematic illustration of the variation of the percolation volumes of the RuO$_2$ nanocrystals and the hydrous grain boundaries with changes in water content [31]

for the case of hydrous RuO$_2$ leads to a prominent change in resistivity as a function of the potential, in contrast to the case of anhydrous RuO$_2$ [98].

Cross-References

▶ Electrical Double-Layer Capacitors (EDLC)

References

1. Trasatti S, Buzzanca G (1971) Ruthenium dioxide: a new interesting electrode material. Solid state structure and electrochemical behaviour. J Electroanal Chem Interfacial Electrochem 29:A1–A5
2. Arnold CB, Wartena RC, Swider-Lyons KE, Pique A (2003) Direct-write planar microultracapacitors by laser engineering. J Electrochem Soc 150:A571
3. Sugimoto W, Yokoshima K, Ohuchi K, Murakami Y, Takasu Y (2006) Fabrication of thin-film, flexible, and transparent electrodes composed of ruthenic acid nanosheets by electrophoretic deposition and application to electrochemical capacitors. J Electrochem Soc 153:A255
4. Liu C-C, Tsai D-S, Susanti D, Yeh W-C, Huang Y-S, Liu F-J (2010) Planar ultracapacitors of miniature interdigital electrode loaded with hydrous RuO$_2$ and RuO$_2$ nanorods. Electrochim Acta 55:5768–5774
5. Zheng JP, Cygan PJ, Jow TR (1995) Hydrous ruthenium oxide as an electrode material for electrochemical capacitors. J Electrochem Soc 142:2699–2703
6. Zheng JP, Jow TR (1995) A new charge storage mechanism for electrochemical capacitors. J Electrochem Soc 142:L6–L8
7. Zhang J, Jiang D, Chen B, Zhu J, Jiang L, Fang H (2001) Preparation and electrochemistry of hydrous ruthenium oxide/active carbon electrode materials for supercapacitor. J Electrochem Soc 148:A1362
8. Ramani M, Haran BS, White RE, Popov BN, Arsov L (2001) Studies on activated carbon capacitor materials loaded with different amounts of ruthenium oxide. J Power Sources 93:209–214
9. Hu C-C, Chen W-C, Chang K-H (2004) How to achieve maximum utilization of hydrous ruthenium oxide for supercapacitors. J Electrochem Soc 151: A281
10. Kim H, Popov BN (2002) Characterization of hydrous ruthenium oxide/carbon nanocomposite supercapacitors prepared by a colloidal method. J Power Sources 104:52–61
11. Park JH, Ko JM, Park O (2003) Carbon nanotube/RuO$_2$ nanocomposite electrodes for supercapacitors. J Electrochem Soc 150:864
12. Min M, Machida K, Jang JH, Naoi K (2006) Hydrous RuO$_2$/carbon black nanocomposites with 3D porous structure by novel incipient wetness method for supercapacitors. J Electrochem Soc 153:A334
13. Lin C, Ritter JA, Popov BN (1999) Development of carbon-metal oxide supercapacitors from Sol-Gel derived carbon-ruthenium xerogels. J Electrochem Soc 146:3155
14. Hu C, Chen W (2004) Effects of substrates on the capacitive performance of RuOxnH$_2$O and activated carbon-RuOx electrodes for supercapacitors. Electrochim Acta 49:3469–3477
15. Sugimoto W, Kizaki T, Yokoshima K, Murakami Y, Takasu Y (2004) Evaluation of the pseudocapacitance in RuO$_2$ with a RuO$_2$/GC thin film electrode. Electrochim Acta 49:313–320
16. Siviglia P, Daghetti A, Trasatti S (1983) Influence of the preparation temperature of Ruthenium dioxide on its point of zero charge. Colloids Surf 7:15–27
17. Burke LD, Murphy OJ (1979) Cyclic voltammetry as a technique for determining the surface area of RuO$_2$ electrodes. J Electroanal Chem Interfacial Electrochem 96:19–27
18. Burke LD, Murphy OJ (1980) Surface area—Voltammetric charge correlation for RuO$_2$/TiO$_2$-based anodes. J Electroanal Chem Interfacial Electrochem 112:39–50

19. Zheng, J. P. Ruthenium Oxide-Carbon Composite Electrodes for Electrochemical Capacitors. Electrochem. Solid-State Lett. 1999, 2, 359.
20. Zheng JP, Jow TR (1996) High energy and high power density electrochemical capacitors. J Power Sources 62:155–159
21. Jow TR, Zheng JP (1998) Electrochemical capacitors using hydrous ruthenium oxide and hydrogen inserted ruthenium oxide. J Electrochem Soc 145:49
22. McKeown DA, Hagans PL, Carette LPL, Russell AE, Swider KE, Rolison DR (1999) Structure of hydrous ruthenium oxides: implications for charge storage. J Phys Chem B 103:4825–4832
23. Vukovic M, Cukman D (1999) Electrochemical quartz crystal microbalance study of electrodeposited ruthenium. J Electroanal Chem 474:167–173
24. Hu C-C, Huang Y-H (1999) Cyclic voltammetric deposition of hydrous ruthenium oxide for electrochemical capacitors. J Electrochem Soc 146:2465
25. Long JW, Swider KE, Merzbacher CI, Rolison DR (1999) Voltammetric characterization of ruthenium oxide-based aerogels and other RuO_2 solids: the nature of capacitance in nanostructured materials. Langmuir 15:780–785
26. Fang QL, Evans DA, Roberson SL, Zheng JP (2001) Ruthenium oxide film electrodes prepared at Low temperatures for electrochemical capacitors. J Electrochem Soc 148:A833
27. Kim I-H, Kim K-B (2001) Ruthenium oxide thin film electrodes for supercapacitors. Electrochem Solid-State Lett 4:A62
28. Fu R, Ma Z, Zheng JP (2002) Proton NMR and dynamic studies of hydrous ruthenium oxide. J Phys Chem B 106:3592–3596
29. Zheng JP, Xin Y (2002) Characterization of $RuO_2 \cdot H_2O$ with various water contents. J Power Sources 110:86–90
30. Long JW, Ayers KE, Rolison DR (2002) Electrochemical characterization of high-surface-area catalysts and other nanoscale electroactive materials at sticky-carbon electrodes. J Electroanal Chem 522:58–65
31. Dmowski W, Egami T, Swider-Lyons KE, Love CT, Rolison DR (2002) Local atomic structure and conduction mechanism of nanocrystalline hydrous RuO_2 from X-ray scattering. J Phys Chem B 106:12677–12683
32. Sugimoto W, Iwata H, Murakami Y, Takasu Y (2004) Electrochemical capacitor behavior of layered ruthenic acid hydrate. J Electrochem Soc 151:A1181
33. Sugimoto W, Iwata H, Yokoshima K, Murakami Y, Takasu Y (2005) Proton and electron conductivity in hydrous ruthenium oxides evaluated by electrochemical impedance spectroscopy: the origin of large capacitance. J Phys Chem B 109:7330–7338
34. Swider-Lyons KE, Love CT, Rolison DR (2005) Selective Vapor Deposition of Hydrous RuO2 Thin Films. J Electrochem Soc 152:C158
35. Chang K-H, Hu C-C (2004) Oxidative Synthesis of RuOx·nH2O with Ideal Capacitive Characteristics for Supercapacitors. J Electrochem Soc 151:A958
36. Murakami Y, Tsuchiya S, Yahikozawa K, Takasu Y (1994) Preparation of ultrafine RuO2 and IrO2 particles by a sol-gel process. J Mater Sci Lett 13:1773–1774
37. Hadži-Jordanov S, Angerstein-Kozlowska H, Conway BE (1975) Surface oxidation and H deposition at ruthenium electrodes: Resolution of component processes in potential-sweep experiments. J Electroanal Chem Interfacial Electrochem 60:359–362
38. Hadži-Jordanov, S.; Angerstein-Kozlowska, H.; Vukovic, M.; Conway, B. E. Reversibility and Growth Behavior of Surface Oxide Films at Ruthenium Electrodes. J. Electrochem. Soc. 1978, 125, 1471.
39. Birss V, Myers R, Angerstein-Kozlowska H, Conway BE (1984) Electron microscopy study of formation of thick oxide films on Ir and Ru electrodes. J Electrochem Soc 131:1502
40. Hadzi-Jordanov S, Angerstein-Kozlowska H, Vukovic M, Conway BE (1977) The state of electrodeposited hydrogen at ruthenium electrodes. J Phys Chem 81:2271–2279
41. Vuković M, Angerstein-Kozlowska H, Conway BE (1982) Electrocatalytic activation of ruthenium electrodes for the Cl_2 and O_2 evolution reactions by anodic/cathodic cycling. J Appl Electrochem 12:193–204
42. Vuković M, Valla T, Milun M (1993) Electron spectroscopy characterization of an activated ruthenium electrode. J Electroanal Chem 356:81–91
43. Marijan D, Čukman D, Vuković M, Milun M (1995) Anodic stability of electrodeposited ruthenium: galvanostatic, thermogravimetric and X-ray photoelectron spectroscopy studies. J Mater Sci 30:3045–3049
44. Liu T, Pell WG, Conway BE (1997) Self-discharge and potential recovery phenomena at thermally and electrochemically prepared RuO_2 supercapacitor electrodes. Electrochim Acta 42:3541–3552
45. Blouin M, Guay D (1997) Activation of ruthenium oxide, iridium oxide, and mixed Ru[sub x]Ir[sub 1−x] oxide electrodes during cathodic polarization and hydrogen evolution. J Electrochem Soc 144:573
46. Mo Y, Antonio MR, Scherson DA (2000) In situ Ru K-edge X-Ray absorption fine structure studies of electroprecipitated ruthenium dioxide films with relevance to supercapacitor applications. J Phys Chem B 104:9777–9779
47. Horvat-Radosevic V, Kvastek K, Vukovic M, Cukman D (2000) Electrochemical properties of ruthenised electrodes in the oxide layer region. J Electroanal Chem 482:188–201

48. Hu C-C, Wang C-C (2002) Improving the utilization of ruthenium oxide within thick carbon–ruthenium oxide composites by annealing and anodizing for electrochemical supercapacitors. Electrochem Commun 4:554–559

49. Hu C-C, Chiang H-R, Wang C-C (2003) Electrochemical and structural investigations of oxide films anodically formed on ruthenium-plated titanium electrodes in sulfuric acid. J Solid State Electrochem 7:477–484

50. Hu C, Liu M, Chang K (2008) Anodic deposition of hydrous ruthenium oxide for supercapaciors: effects of the AcO$^-$ concentration, plating temperature, and oxide loading. Electrochim Acta 53:2679–2687

51. Hu C, Liu M, Chang K (2007) Anodic deposition of hydrous ruthenium oxide for supercapacitors. J Power Sources 163:1126–1131

52. Park B, Lokhande CD, Park H, Jung K, Joo O (2004) Cathodic electrodeposition of RuO_2 thin films from $Ru(III)Cl_3$ solution. Mater Chem Phys 87:59–66

53. Wu H-M, Hsu P-F, Hung W-T (2009) Investigation of redox reaction of Ru on carbon nanotubes by pulse potential electrochemical deposition. Diam Relat Mater 18:337–340

54. Patake V, Lokhande C, Joo O (2009) Electrodeposited ruthenium oxide thin films for supercapacitor: effect of surface treatments. Appl Surf Sci 255:4192–4196

55. Zheng Y, Ding H, Zhang M (2008) Hydrous–ruthenium–oxide thin film electrodes prepared by cathodic electrodeposition for supercapacitors. Thin Solid Films 516:7381–7385

56. Ahn YR, Song MY, Jo SM, Park CR, Kim DY (2006) Electrochemical capacitors based on electrodeposited ruthenium oxide on nanofibre substrates. Nanotechnology 17:2865–2869

57. Zhitomirsky I, Gal-Or L (1997) Ruthenium oxide deposits prepared by cathodic electrosynthesis. Mater Lett 31:155–159

58. Sugimoto, W.; Makino, S.; Mukai, R.; Tatsumi, Y.; Katsutoshi; Fukuda; Takasu, Y.; Yamauchi, Y. Synthesis of Ordered Mesoporous Ruthenium by Lyotropic Liquid Crystals and Its Electrochemical Conversion to Mesoporous Ruthenium Oxide with High Surface Area. J. Power Sources 2012, in press.

59. Takasu Y, Murakami Y (2000) Design of oxide electrodes with large surface area. Electrochim Acta 45:4135–4141

60. Sugimoto W, Shibutani T, Murakami Y, Takasu Y (2002) Charge storage capabilities of rutile-type RuO[sub 2]-VO[sub 2] solid solution for electrochemical supercapacitors. Electrochem Solid-State Lett 5:A170

61. Yokoshima K, Shibutani T, Hirota M, Sugimoto W, Murakami Y, Takasu Y (2006) Electrochemical supercapacitor behavior of nanoparticulate rutile-type Ru1−xVxO2. J Power Sources 160:1480–1486

62. Jeong Y, Manthiram A (2000) Amorphous ruthenium-chromium oxides for electrochemical capacitors. Electrochem Solid-State Lett 3:205

63. Jeong YU, Manthiram A (2001) Amorphous tungsten oxide/ruthenium oxide composites for electrochemical capacitors. J Electrochem Soc 148:A189

64. Wang C-C, Hu C-C (2005) Electrochemical and textural characteristics of (Ru-Sn)Ox·nH$_2$O for supercapacitors. J Electrochem Soc 152:A370

65. Cao F, Prakash J (2001) Performance investigations of $Pb_2Ru_2O_6.5$ oxide based pseudocapacitors. J Power Sources 92:40–44

66. Schenk J, Wilde PM, Abdelmula E, Axmann P, Garche J (2002) New materials for supercapacitors. J Power Sources 105:182–188

67. Park B, Lokhande CD, Park H, Jung K, Joo O (2004) Preparation of lead ruthenium oxide and its use in electrochemical capacitor. Mater Chem Phys 86:239–242

68. Lee Y, Oh J, Oh H, Kim H (2008) Novel method for the preparation of carbon supported nano-sized amorphous ruthenium oxides for supercapacitors. Electrochem Commun 10:1035–1037

69. Naoi K, Ishimoto S, Ogihara N, Nakagawa Y, Hatta S (2009) Encapsulation of nanodot ruthenium oxide into KB for electrochemical capacitors. J Electrochem Soc 156:A52–A59

70. Fang W, Cyyan O, Sun C, Wu C, Chen C, Chen K, Chen L, Huang J (2007) Arrayed CNx NT–RuO2 nanocomposites directly grown on Ti-buffered Si substrate for supercapacitor applications. Electrochem Commun 9:239–244

71. Kim I-H, Kim J-H, Kim K-B (2005) Electrochemical characterization of electrochemically prepared ruthenium oxide/carbon nanotube electrode for supercapacitor application. Electrochem Solid-State Lett 8:A369

72. Qin X, Durbach S, Wu GT (2004) Electrochemical characterization on RuO_2• xH_2O/carbon nanotubes composite electrodes for high energy density supercapacitors. Carbon 42:451–453

73. Fang W-C, Leu M-S, Chen K-H, Chen L-C (2008) Ultrafast charging-discharging capacitive property of RuO[sub 2] nanoparticles on carbon nanotubes using nitrogen incorporation. J Electrochem Soc 155:K15

74. Deng GH, Xiao X, Chen JH, Zeng XB, He DL, Kuang YF (2005) A new method to prepare RuO2•xH2O/carbon nanotube composite for electrochemical capacitors. Carbon 43:1566–1569

75. Arabale G, Wagh D, Kulkarni M, Mulla IS, Vernekar SP, Vijayamohanan K, Rao AM (2003) Enhanced supercapacitance of multiwalled carbon nanotubes functionalized with ruthenium oxide. Chem Phys Lett 376:207–213

76. Kim Y-T, Tadai K, Mitani T (2005) Highly dispersed ruthenium oxide nanoparticles on carboxylated carbon nanotubes for supercapacitor electrode materials. J Mater Chem 15:4914

77. Lee J-K, Pathan HM, Jung K-D, Joo O-S (2006) Electrochemical capacitance of nanocomposite films formed by loading carbon nanotubes with ruthenium oxide. J Power Sources 159:1527–1531

78. Yuan C, Chen L, Gao B, Su L, Zhang X (2009) Synthesis and utilization of $RuO2•xH2O$ nanodots well dispersed on poly(sodium 4-styrene sulfonate functionalized multi-walled carbon nanotubes for supercapacitors. J Mater Chem 19:246

79. Kim I-H, Kim J-H, Lee Y-H, Kim K-B (2005) Synthesis and characterization of electrochemically prepared ruthenium oxide on carbon nanotube film substrate for supercapacitor applications. J Electrochem Soc 152:A2170

80. Reddy ALM, Ramaprabhu S (2007) Nanocrystalline metal oxides dispersed multiwalled carbon nanotubes as supercapacitor electrodes. J Phys Chem C 111:7727–7734

81. Kim JD, Kang BS, Noh TW, Yoon J-G, Baik SI, Kim Y-W (2005) Controlling the nanostructure of RuO_2/carbon nanotube composites by Gas annealing. J Electrochem Soc 152:D23

82. Wu Z-S, Wang D-W, Ren W, Zhao J, Zhou G, Li F, Cheng H-M (2010) Anchoring hydrous RuO_2 on graphene sheets for high-performance electrochemical capacitors. Adv Funct Mater 20:3595–3602

83. Wang, H.; Liang, Y.; Mirfakhrai, T.; Chen, Z.; Casalongue, H. S.; Dai, H. Advanced asymmetrical supercapacitors based on graphene hybrid materials. Nano Research 2011, 3.

84. Mishra AK, Ramaprabhu S (2011) Functionalized graphene-based nanocomposites for supercapacitor application. J Phys Chem C 115:14006–14013

85. Zhang J, Jiang J, Li H, Zhao XS (2011) A high-performance asymmetric supercapacitor fabricated with graphene-based electrodes. Energy Environ Sci 4:4009

86. Chen Y, Zhang X, Zhang D, Ma Y (2012) One-pot hydrothermal synthesis of ruthenium oxide nanodots on reduced graphene oxide sheets for supercapacitors. J Alloys Comp 511:251–256

87. Rakhi RB, Chen W, Cha D, Alshareef HN (2011) High performance supercapacitors using metal oxide anchored graphene nanosheet electrodes. J Mater Chem 21:16197

88. Sato Y, Yomogida K, Nanaumi T, Kobayakawa K, Ohsawa Y, Kawai M (1999) Electrochemical behavior of activated-carbon capacitor materials loaded with ruthenium oxide. Electrochem Solid-State Lett 3:113

89. Nanaumi T, Ohsawa Y, Kobayakawa K, Sato Y (2002) High energy electrochemical capacitor materials prepared by loading ruthenium oxide on activated carbon. Electrochemistry 70:681

90. Sugimoto W, Iwata H, Yasunaga Y, Murakami Y, Takasu Y (2003) Preparation of ruthenic acid nanosheets and utilization of its interlayer surface for electrochemical energy storage. Angew Chem Int Ed Engl 42:4092–4096

91. Fukuda K, Saida T, Sato J, Yonezawa M, Takasu Y, Sugimoto W (2010) Synthesis of nanosheet crystallites of ruthenate with an alpha-$NaFeO_2$-related structure and its electrochemical supercapacitor property. Inorg Chem 49:4391–4393

92. Ardizzone S, Fregonara G, Trasatti S (1990) "Inner" and "outer" active surface of RuO_2 electrodes. Electrochim Acta 35:263–267

93. Rishpon J, Gottesfeld S (1960) Resolution of fast and slow charging processes in ruthenium oxide films: an AC impedance and optical investigation. J Electrochem Soc 1984:131

94. Foelske A, Barbieri O, Hahn M, Kötz R (2006) An X-Ray photoelectron spectroscopy study of hydrous ruthenium oxide powders with various water contents for supercapacitors. Electrochem Solid-State Lett 9:A268

95. Rochefort D, Dabo P, Guay D, Sherwood P (2003) XPS investigations of thermally prepared RuO_2 electrodes in reductive conditions. Electrochim Acta 48:4245–4252

96. Chabanier C, Irissou E, Guay D, Pelletier JF, Sutton M, Lurio LB (2002) Hydrogen absorption in thermally prepared RuO_2 electrode. Electrochem Solid-State Lett 5:E40

97. Chabanier C, Guay D (2004) Activation and hydrogen absorption in thermally prepared RuO_2 and IrO_2. J Electroanal Chem 570:13–27

98. Barbieri O, Hahn M, Foelske A, Kotz R (2006) Effect of electronic resistance and water content on the performance of RuO_2 for supercapacitors. J Electrochem Soc 153:A2049

99. Stefan IC, Mo Y, Antonio MR, Scherson DA (2002) In situ Ru L II and L III edge X-ray absorption near edge structure of electrodeposited ruthenium dioxide films. J Phys Chem B 106:12373–12375

100. Mo Y, Bae IT, Sarangapani S, Scherson DA (2003) In situ Ru K-edge X-ray absorption spectroscopy of a high-area ruthenium dioxide electrode in a Nafion-based supercapacitor environment. J Solid State Electrochem 7:572–575

101. Mo Y, Cai W-B, Dong J, Carey PR, Scherson DA (2001) In situ surface enhanced Raman scattering of ruthenium dioxide films in acid electrolytes. Electrochem Solid-State Lett 4:E37

S

Salting-In and Salting-Out

Yizhak Marcus
Department of Inorganic and Analytical Chemistry,
The Hebrew University, Jerusalem, Israel

Description

Ions, being hydrated, bind water in their solvation shells, so that there remains less "free" water to accommodate other solutes. A general result of the presence of ions in a solution, disregarding any further interactions, is then an elevation of the activity coefficient y_N of a nonelectrolyte solute, marked by subscript $_N$, in a c_E molar solution of an electrolyte, marked by subscript $_E$, to $1/[1 - (V_W/1{,}000)h_E c_E]$ in molar units. Here V_W is the molar volume of the water and h_E is the sum of the hydration numbers of the ions constituting the electrolyte.

A further result of this is a diminished solubility of the nonelectrolyte solute, to maintain a constant activity of it at equilibrium with the pure solute. This phenomenon is called "*salting-out*" and is described up to fairly high concentrations of the electrolyte, c_E, by the Setchenov expression:

$$\log(s_N{}^*/s_N) = k_{NE} c_E \qquad (1)$$

Here $s_N{}^*$ is the solubility of the nonelectrolyte in the solvent (water) in the absence of the electrolyte and s_N is that in its presence. The coefficient k_{NE} is called the *Setchenov salting-out*

constant and depends on the natures of the nonelectrolyte and of the electrolyte as well as on the temperature at ambient pressures. There are some systems where the solubility is enhanced by the presence of the electrolyte so that $k_{NE} < 0$ and *salting-in* then occurs, but as a rule, $k_{NE} > 0$ and the solubility is diminished. Equation 1 is taken as the limiting expression, valid for small solubilities, where self-interactions of the nonelectrolyte can be disregarded. Therefore, Eq. 1 pertains also to infinite dilution of the electrolyte and is additive with respect to the contribution of each ion, k_{NI}, to the total $k_{NE} = \Sigma_I \, v_I k_{NI}$, the index $_I$ pertaining to individual ions of the salting agent (cations and anions, including mixed electrolytes) and where v_I is the stoichiometric coefficient.

The salting-out and salting-in phenomena pertain not only to the solubilities of the nonelectrolyte solutes, but also to the extractability of nonelectrolytes by solvents non-miscible with water, to phase-transfer catalysis, and other phenomena. The salting is not confined to aqueous solutions, where it was primarily studied and applied, but is found in all kinds of solutions of electrolytes, whatever may be the solvent.

The magnitude of k_{NE} generally increases with the molar volume of the nonelectrolyte, V_N, and with the intensity of hydration of the electrolyte. The latter can be described by the electrostriction that the ion causes [1] and this leads to the McDevit and Long [2] formulation for the Setchenov constant:

$$k_{NE} = -V_N V_{E \, elec}/(\ln 10) RT \kappa_{TW} \qquad (2)$$

G. Kreysa et al. (eds.), *Encyclopedia of Applied Electrochemistry*, DOI 10.1007/978-1-4419-6996-5,
© Springer Science+Business Media New York 2014

Here $V_{E\,elec}$ is the molar electrostriction by the electrolyte, and κ_{TW} is the isothermal compressibility of the water. The direct interactions of the ions of the electrolyte with molecules of the nonelectrolyte are ignored in this approach. Actually, values of k_{NE} predicted by Eq. 2 are several-fold larger than the experimental values [2, 3].

Salting-in cannot be described by this formulation and occurs with poorly hydrated ions and/or with nonelectrolytes that are more polar than the solvent (have a higher permittivity). An earlier theory, that of Debye and McAulay [4], related the salting to the electrical work of charging the ions in water compared with that in the presence of the nonelectrolyte. This work is proportional to the difference $(1/\varepsilon - 1/\varepsilon_W)$, where ε is the relative permittivity of the solution and $_W$ pertains to pure water. When this difference is negative, that is, for highly polar solutes that increase the permittivity, salting-in is predicted. Other early theories of salting-out and -in were reviewed in the book by Marcus and Kertes [5] and by Conway [6] among others.

The *scaled particle theory* has been applied more recently to the salting-out and -in phenomena. Shoor and Gubbins [7] applied it to the solubilities of nonpolar gases in electrolyte solutions, and Masterton and Lee [8] derived the Setchenov constants from it. Their expression is $k_{NE} = k_{\alpha N} + k_{\beta NE} + k_{\gamma E}$, where $k_{\alpha N}$ pertains to the work required for forming a cavity of the size of the solute, $k_{\beta NE}$ pertains to the interactions of the solute in the cavity with its surroundings, and $k_{\gamma E}$ converts from molar to mole-fraction units. Fairly complicated expressions result from this theory for the quantities $k_{\alpha N}$, $k_{\beta NE}$, and $k_{\gamma E}$ and for the calculation of which for a particular system, it is necessary to know V_E^{∞}, the standard partial molar volume of the electrolyte, the diameters σ_I and polarizabilities α_I of the ions, and the diameter σ_N, polarizability α_N, and Lennard-Jones energy parameter e_N/k_B of the solute. The values of $k_{\alpha N}$ are positive, those of $k_{\beta NE}$ are negative, and of $k_{\gamma E}$ may have either sign, depending on V_E^{∞} but is much smaller than the other terms, as found for the salting-out of some gases [8].

Ruckenstein and Shulgin [9] used the *Kirkwood-Buff fluctuation theory* to obtain an expression for the salting-out (of gases, but applicable to any nonelectrolyte solute) in electrolyte solutions. The resulting expression can be rewritten as:

$$k_{NE} = -(2.303 \times 2000) \times [G_{WW} - G_{EE} - 2(G_{WN} - G_{WE})] \quad (3)$$

where $G_{\alpha\beta} = {}_0\!\int^{\infty}(g_{\alpha\beta} - 1)4\pi r^2 dr$ is the Kirkwood-Buff integral, $g_{\alpha\beta}$ is the pair correlation function for species α and β (being water W, electrolyte E, or nonelectrolyte N) as a function of the distance r between the centers of their molecules. These integrals are related to the partial molar volumes and Eq. 3 can be transformed in dilute solutions to:

$$k_{NE} = -(2.303 \times 2000) \times [(-(\partial V_N/\partial c_E)/V_E^{\infty} + V_E^{\infty} - V_W^{*}] \quad (4)$$

The first term in the square brackets is generally small compared with the other two and may be neglected. Then salting-in occurs in cases where $V_E^{\infty} > V_W^{*}$, i.e., for bulky ions having a molar volume larger than that of pure water, V_W^{*}, but otherwise salting-out occurs. Mazo [10] in an equivalent derivation used the measured k_{NE} to evaluate the Kirkwood-Buff integral G_{NE} not otherwise accessible.

On the practical level, the salting-out properties of organic solutes, mainly hydrocarbons, by many salts were compiled by Xie et al. [11]. The dependence on the molar volume of the solute, predicted by Eq. 2, was confirmed. Conventional ionic salting constants $k_{NI}/dm^3\ mol^{-1}$ for benzene are shown in Table 1, compared with the earlier values derived from McDevit and Long [2]. On the basis $k_N(H^+) = 0$, the value for HCl yields the Cl^- value, that for NaCl then the Na^+ value, and these served as secondary reference ions for the conventional values. Instances of salting-in ($k_{NI} < 0$)

Salting-In and Salting-Out, Table 1 Conventional ionic salting-out parameters $k_I/dm^3\ mol^{-1}$ of benzene according to Eq. 1 from [2] and [11]

Cation	k_{NI} [2]	k_{NI} [11]	Anion	k_{NI} [2]	k_{NI} [11]
H^+	0	0	F^-	0.107	0.115
Li^+	0.093	0.094	Cl^-	0.048	0.048
Na^+	0.147	0.137	Br^-	0.004	0.018
K^+	0.118	0.108	I^-	−0.052	−0.042
Rb^+	0.092	0.092	OH^-	0.109	
Cs^+	0.040	0.040	SCN^-		−0.032
NH_4^+	0.045	0.055	NO_3^-	−0.028	−0.018
Mg^{2+}		0.174	ClO_3^-		−0.016
Sr^{2+}		0.176	ClO_4^-	−0.041	−0.041
Ba^{2+}	0.238	0.238	HCO_2^-		0.029
$(CH_3)_4N^+$	−0.15[a]	−0.303	$CH_3CO_2^-$		0.028
$(C_2H_5)_4N^+$	−0.25[a]	−0.625	$C_2H_5CO_2^-$		0.021
$(C_3H_7)_4N^+$	−0.41[a]		SO_4^{2-}	0.254	0.274

[a]From Deno and Spink [3]

by large ions are noted. In such cases, hydrophobic interactions are invoked as an explanation.

In the case of nonreactive gases, denoted by subscript $_G$, an empirical manner of looking at the salting-out has been proposed by Weisenberger and Schumpe [12] by means of the expression:

$$k_{GI} = \Sigma_I \nu_I [k_I + k_{G25} + h_G((t/^\circ C) - 25)] \quad (5)$$

k_I is an ion specific parameter that is relatively independent of the temperature, and k_{G25} and h_G are gas-specific parameters. The conventional numerical values for the k_I are shown in Table 2, on the basis $k_I(H^+) = 0$. To a good approximation $k_I/z_I = 0.080 \pm 0.002\ dm^3\ mol^{-1}$ for the cations of charge z_I, with a small increase in the alkaline earth and divalent transition metal series with the ionic radius, but for the alkali metal ions, the opposite trend is observed, except for Li^+. The values of $k_I = 0.0648$ for Cr^{3+} and $k_I = 0.1161$ for Fe^{3+} appear to be outliers. For the anions, less clear charge and size dependencies are seen, with $k_I/|z_I| = 0.072 \pm 0.007$, and for the halides, k_I decreases with increasing size. The gas parameters are normalized to $k(O_2) = 0$, and values of values of k_{G25} are listed in Table 3, for most of

Salting-In and Salting-Out, Table 2 Conventional ionic salting-out parameters $k_I/dm^3\ mol^{-1}$ of gases according to Eq. 5 from [12]

Cation	k_I(gases)	Anion	k_I(gases)
H^+	0	F^-	0.0920
Li^+	0.0754	Cl^-	0.0318
Na^+	0.1143	Br^-	0.0269
K^+	0.0922	I^-	0.0039
Rb^+	0.0839	OH^-	0.0839
Cs^+	0.0759	HS^-	0.0851
NH_4^+	0.0556	CN^-	0.0679
Mg^{2+}	0.1694	SCN^-	0.0627
Ca^{2+}	0.1762	NO_2^-	0.0795
Sr^{2+}	0.1881	NO_3^-	0.0128
Ba^{2+}	0.2168	ClO_3^-	0.1348
Mn^{2+}	0.1463	BrO_3^-	0.1116
Fe^{2+}	0.1523	IO_3^-	0.0913
Co^{2+}	0.1680	ClO_4^-	0.0492
Ni^{2+}	0.1654	IO_4^-	0.1464
Cu^{2+}	0.1675	HCO_3^-	0.0967
Zn^{2+}	0.1537	HSO_3^-	0.0549
Cd^{2+}	0.1869	$H_2PO_4^-$	0.0906
Al^{3+}	0.2174	CO_3^{2-}	0.1423
La^{3+}	0.2297		
Ce^{3+}	0.2406		
Th^{4+}	0.2709		

Salting-In and Salting-Out, Table 3 Conventional ionic salting-out parameters $k_{G25}/dm^3\ mol^{-1}$ of gases according to Eq. 5 from [12]

Gas	k_{G25}	Gas	k_{G25}	Gas	k_{G25}
H_2	−0.0218	O_2	0	C_3H_8	0.0240
He	−0.0353	NO	0.0060	n-C_4H_{10}	0.0297
Ne	−0.0080	N_2O	−0.0085	H_2S	−0.0333
Ar	0.0057	CH_4	0.0022	SF_6	0.0100
Kr	0.0071	C_2H_2	−0.0159		
Xe	0.0133	C_2H_4	0.0037		
N_2	−0.0010	C_2H_6	0.0120		

which $h_G < 0$ for $273 \leq t/°C \leq 353$. There is good qualitative agreement between the salting by the ions of the gases and the salting of benzene.

Future Directions

There is at present no general theory that permits the estimation of the salting properties, even of the limiting k_{NE} values, for given salts and non-electrolytes that relates them to readily available measurable properties of the salt and the solute. The electrostriction, scaled particle, and the Kirkwood-Buff theories do not answer this requirement, and research to meet this need should be forthcoming. In particular, direct attractive interactions between the ions of the salt and the solute that could lead to salting-in are not well understood so far. Extension of any theory to higher salt and solute concentrations should also be attempted.

Cross-References

▶ Gas Solubility of Electrolytes
▶ Ion Properties

References

1. Marcus Y (2011) Electrostriction in electrolyte solutions. Chem Rev 111:2761
2. McDevit WF, Long FA (1952) The activity coefficient of benzene in aqueous salt solutions. J Am Chem Soc 74:1773
3. Deno NC, Spink CH (1963) The McDevit-Long equation for salt effects on nonelectrolytes. J Phys Chem 67:1347
4. Debye P, McAulay J (1925) Theory of salting out. Z Phys 26:22
5. Marcus Y, Kertes AS (1969) Ion exchange and solvent extraction of metal complexes. Wiley, London, p 74 ff
6. Conway BE (1985) Local changes of solubility induced by electrolytes: salting-out and ionic hydration. Pure Appl Chem 57:263
7. Shoor SK, Gubbins KE (1969) Solubility of nonpolar gases in concentrated electrolyte solutions. J Phys Chem 78:498
8. Masterton WL, Lee TP (1970) Salting coefficients from scaled particle theory. J Phys Chem 74:1776
9. Ruckenstein E, Shulgin I (2002) Salting-out or –in by fluctuation theory. Ind Eng Chem Res 41:4674
10. Mazo RM (2006) A fluctuation theory analysis of the salting-out effect. J Phys Chem B 110:24077
11. Xie W-H, Shiu W-Y, Mackay D (1997) A review of the effect of salts on the solubility of organic compounds in seawater. Marine Environ Res 44:429
12. Weisenberger S, Schumpe A (1996) Estimation of gas solubilities in salt solutions at temperatures from 273 to 363 K. AIChE J 42:298

Scanning Electrochemical Microscopy (SECM)

Daniel Mandler
Institute of Chemistry, The Hebrew University, Jerusalem, Israel

Introduction

Scanning electrochemical microscopy (SECM) belongs to the family of techniques termed scanning probe microscopy (SPM). The latter was the consequence of the development of scanning tunneling microscopy in the mid 1980s by Binnig and Rohr [1]. SECM, similarly to other SPM techniques, is used for scanning interfaces with high resolution as a means of obtaining topographic as well as chemical or physical local characterization. The early studies of SECM were reported in the late 1980s by R. C. Engstrom [2, 3] and by A. J. Bard [4–6] who significantly advanced the application of SECM. Since its birth until 2012, more than 2,000 studies

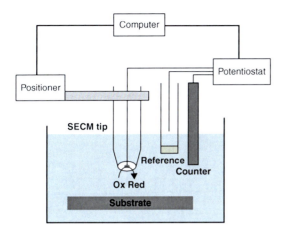

Scanning Electrochemical Microscopy (SECM), Fig. 1 Schematics of a SECM setup

have been published dealing with SECM, among them more than 30 reviews [7–14]. Seven international workshops were conducted on SECM, and the community of SECM users increases steadily and involves already more than 50 laboratories around the world. A few commercial instrumentations exists, and applications of SECM spans from imaging biological and medical samples to studying heterogeneous interfacial processes, patterning surfaces with enzymes and nanoparticles, and imaging fingerprints. A comprehensive book about SECM was published in 2001 [15] which is currently being updated and should appear soon in its second edition.

Basic Concepts

The essence of SECM is bringing a small electrode, i.e., micro- or nanoelectrode, close to an interface while recording its faradic current. In most cases the solution consists of one state of a redox couple that is being oxidized or reduced at the microelectrode. A typical setup of SECM is shown in Fig. 1. Once the interface enters the diffusion layer of the microelectrode, the latter can laterally scan the interface to obtain topographical and electrochemical information. Several modes of operation of SECM have been developed; the most common are the feedback and the direct modes. Additional modes include the generation–collection, transient, surface interrogation, and alternating current, which differ in the perturbation applied to the microelectrode or the method of measurement.

The feedback mode [5] is based on measuring changes in the steady-state current that flows across the SECM tip, which is the result of oxidizing or reducing electroactive species in the solution. While this faradic current is constant as the tip–surface distance is a few times the diameter of the tip, it can either increase (so-called positive feedback current) or decrease (negative feedback current) as the tip approaches the surface. A positive feedback current is observed upon approaching a conducting surface in which the electrogenerated species at the tip undergoes a reverse redox reaction on the surface. As the distance between the tip and the interface decreases, the diffusion is enhanced, and therefore the steady-state current increases. On the other hand, approaching a nonconducting surface causes the steady-state current to decrease because of the hindrance of the diffusion of electroactive species to the tip by the surface. The approach curves are usually normalized by dividing the current and distance by the steady-state current at infinity and the tip radius, respectively. Moreover, theoretical curves of the positive and negative feedback currents were constructed [5, 11, 16] (Fig. 2) from which the experimental distance is often estimated.

The shape of SECM tip is crucial and must be embedded in a flat insulator to assure the formation of a thin layer cell between the tip and the interface. The ratio between the radius of the entire microelectrode (including the insulating sheath) and that of the conducting part is termed RG and affects the feedback current. Interestingly, a positive feedback current is obtained even in the case where the surface is not attached to a power supply and is due to lateral electron transfer driven by a concentration cell that is formed between the tip–surface volume and the rest of the solution (Fig. 3).

Most of SECM studies were performed with micrometer size disk-shaped tips, which are

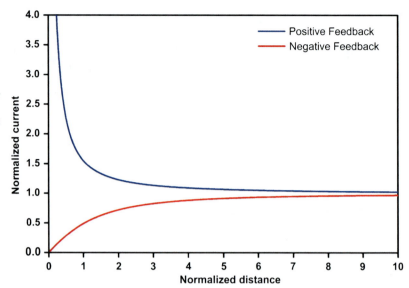

Scanning Electrochemical Microscopy (SECM), Fig. 2 The positive and negative theoretical feedback current obtained by approaching a conductor and insulator, respectively, with a microelectrode while generating electroactive species at the tip

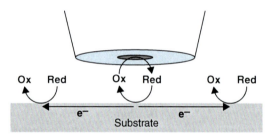

Scanning Electrochemical Microscopy (SECM), Fig. 3 Mechanism of lateral electron transfer that is responsible for a positive feedback current in cases where the conductive surface is unbiased

either commercially available or can be fabricated fairly easily. Yet, smaller submicrometer and nanometer tips have also been produced by different approaches such as pulling and through lithography [17–22]. Moreover, other types of microelectrodes, e.g., microband, cylindrical and conical shaped, have also been applied, and their steady-state current approaching curves are theoretically modeled [23–27]. Finally, numerous liquid–liquid electrochemistry studies have been conducted using micropipettes filled with an electrolyte solution that was immiscible with that in which the substrate was held.

The direct mode is, on the other hand, less popular and is based on focusing the electrical field between the surface, i.e., the working electrode and the tip that is now used as the counter electrode. The principal disadvantages of the direct mode are the fact that the faradic current is not sensitive to the tip–surface distance and that only conducting surfaces can be employed. Most studies using the direct mode aimed at local patterning of solid surfaces.

As mentioned above, other modes of operation have also been developed; among them the generation–collection mode [28–30] is presumably the most often used. Two approaches exist: a tip generation–substrate collection (TG/SC) where species are generated electrochemically at the tip and collected by the surface. The collection efficiency, which is defined as the ratio between the tip to surface currents, usually reaches unity (unless the species are unstable) due to the large area of the surface. On the other hand, the species can be generated at the surface either electrochemically or chemically and collected by the tip (surface generation–tip collection). The species can be generated electrochemically or chemically such as by enzymes that are attached onto the surface. This mode is particularly interesting and allows imaging of the surface, where the generated sites at the surface are localized and separated from each other so that their diffusion layers do not overlap.

Significant research has been devoted to studying homogeneous and heterogeneous kinetics by transient methods, mostly,

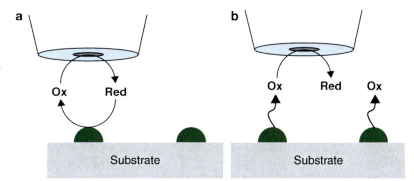

Scanning Electrochemical Microscopy (SECM), Fig. 4 Schematics of the feedback mode (a) and the generation–collection mode (b) of SECM for imaging and characterization of surface active sites

chronoamperometry, using SECM [31, 32]. In this case the transient current at the tip is measured at a fixed distance from the surface.

Applications

SECM has been utilized for a large number of applications in various fields. Applications can be categorized into three main activities: imaging, studying charge transfer processes, and patterning. Many of these activities have been reviewed [9, 10, 13, 14, 33, 34] and will only be briefly described.

Imaging

By scanning the SECM tip across a surface and measuring its faradic current, a 3D current image is obtained, which reflects the topography of the substrate and electrochemical activity. In most cases two imaging modes have been used: the feedback current, which is sensitive to the regeneration of the electroactive species generated at the tip on the substrate, and the generation–collection mode, which can map the substrate based on its catalytic activity. An additional difference between the two modes of imaging is that while in the generation–collection mode, the SECM tip is passive and used only to probe concentration and fluxes of electroactive species generated by the substrate, in the feedback mode, the species generated at the tip can react with or penetrate into the substrate affecting it.

The resolution of the image depends on the tip radius and the distance from the surface. By decreasing the size of the microelectrodes (using nanoelectrodes) and the tip–substrate distance, high-resolution imaging of various substrates has been achieved [19, 21, 35–37]. Moreover, controlling the tip–substrate distance is problematic on heterogeneous substrates because of difficulty to differentiate between the contribution of topography and changes in rate of electron transfer (due to heterogeneity) to the total current. Therefore, different approaches have been developed, such as the introduction of the shear-force mode, tip-position modulation, tip impedance, and the combination of the SECM and AFM, for monitoring the tip–substrate distance independent of the faradic current [11, 38].

During the years, imaging by SECM has been primarily directed towards biological systems. The generation–collection mode of SECM has been used for studying respiration and photosynthesis of living tissues and in particularly individual cells. The feedback mode has been very often used for studying the penetration of redox species across the cell membrane and their redox reaction inside. The aim of some of these studies has been to correlate between the cellular redox activity and biological activity.

Enzyme activity has been the subject of numerous studies by SECM [39–42]. The enzymes were usually attached onto a nonconductive surface, and either the feedback or the generation–collection modes were used (Fig. 4). Yet, there are a few studies that used a potentiometric mode in which the enzymatic activity affected local changes in pH that were measured with a pH microelectrode. DNA, proteins, and antibodies have also been targeted recently by SECM. Imaging DNA arrays and studying the hybridization of DNA strands including the detection of a single DNA mismatch have been reported [43–47]. The interaction between antibody and antigen by SECM has also been reported a label-free assay [48]. All these studies are based on monitoring the changes in the electrochemical activity of redox couples on the substrate as a result of the biological activity. Finally, imaging of biological membranes and the transport of mostly electroactive species across membranes, including bilayer lipid membranes, liposomes, and dentin, have been also examined by SECM [49–56].

Studying Charge Transfer

The major activity in SECM focused on studying charge transfer processes at different interfaces including solid–liquid, liquid–liquid, and liquid–air interfaces. The advantage of SECM lies in its ability to bring the electrochemical probe (tip) very close to the interface. As SECM tips have been miniaturized, and better positioners developed, the distance between the tip and the substrate decreased and reached, in some cases, that of a few nanometer. This means that very fast kinetics could be probed and nano-objects, such as catalysts, could be studied and characterized.

Heterogeneous electron transfer at the electrode–electrolyte interface has been the most common topic and numerous systems including thin polymeric films (conducting polymer, polyelectrolytes, and others), self-assembled monolayers, minerals and single crystals, nanoparticles, catalysts, membranes, and biological systems [57–63]. These studies have been supported by simulations and appropriate approximations, which enabled quantitative information of mass transport and kinetic parameters. For example, arrays of binary and tertiary metallic combinations of various metals were studied as a means of improving the design catalysts for the oxygen reduction reaction (ORR) as well as for hydrogen oxidation [58, 64, 65].

Lateral charge propagation (Fig. 3) in thin films has been also the subject of intensive research [66–71]. This research was not limited to thin films on solid supports but was extended to monolayers and polymeric films at the liquid–air interface. Steady-state (feedback approaching curves) and transient (chronoamperometry) techniques have been widely used in combination with the Langmuir technique.

Substituting the solid tip by a micropipette has been used for studying the liquid–liquid interface between two immiscible electrolyte solutions (ITIES) [16, 72–79]. Ion and electron transfer processes have been studied by bringing a micropipette (mostly filled with an aqueous electrolyte) close to another interface (either solid–liquid or liquid–liquid). The ITIES potential was either poised by the concentration of potential-determining ion, which caused a constant driving force for electron or ion transfer, or a potential was applied to the micropipette enabling to control the transfer of charge across the interface. This activity was expanded to many different and interesting systems such as water-ionic liquids and nanoparticles located at the interface [60, 66, 80].

A large number of surface reactions such as etching and dissolution, adsorption, protonation, and doping have also been studied by SECM [55, 62, 81, 82]. The approach was quite similar; the tip was either used to generate the species that diffuse to the substrate and drove the reaction, such as etching, or the tip perturbed the equilibrium concentration of species, e.g., protons, close to the substrate, which initiated the surface reaction.

Local Patterning

One of the major advantages of SECM originates from the fact that charge transfer at the surface below the tip is highly localized. This opened an elegant and straightforward approach for patterning a wide range of different surfaces with lateral resolution governed by the tip size and shape [83]. First studies focused on metal deposition and etching [84, 85], which were followed by semiconductor etching, conducting polymer deposition, and localized enzyme attachment [40, 86–88]. Most of these studies involved the feedback mode, in which highly reactive species generated at the tip reacted irreversibly with the surface.

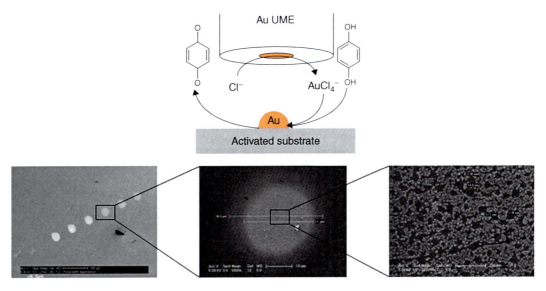

Scanning Electrochemical Microscopy (SECM), Fig. 5 Local deposition of Au nanoparticles by the local generation of a flux of gold ions and surface-catalyzed reduction by hydroquinone

The direct mode has also been used primarily for deposition of metals and conducting polymers [89–91]. Different approaches have been developed, such as the use of sacrificial microelectrode made of the metals to be deposited, which were anodically dissolved under well-controlled conditions [92, 93]. To increase the resolution of the deposit, a "chemical lens" mode was developed [94]. Micropipettes have also been used in an ITIES configuration as a means to generate a flux of silver ions that were locally deposited [95]. Recently, the local biocatalyzed deposition of nanoparticles has also been reported [96]. An example of local deposition of nanoparticles is shown in Fig. 5.

Different organic local modification, e.g., diazonium attachment, polymer reduction, and thiol desorption, have been demonstrated and studied [82, 97–106]. Local oxidation or desorption of self-assembled monolayers was followed by local deposition or specific attachment onto the modified patterns [105, 106]. The reduction of polymer, such as PTFE, made it possible to locally metalize or further polymerize the reducing sites [101–104]. More recently, a few approaches have been developed for the formation of either an anisotropic structure using a non-disk-shaped microelectrode and imprinting its structure [107] or by constructing a multiscaled electrochemical probe for SECM [108]. The probe was fabricated by wet chemical etching of an optical fiber bundle comprising 6,000 individually cladded optical cores, which was sputter-coated by a thin gold layer resulting in ca. 6,000 nanotips. These new SECM probes were then used to pattern a Teflon surface.

Future Directions

Evidently, SECM is becoming a common tool, primarily for the electrochemists, in a variety of scientific and applied fields, such as corrosion inhibition, single cell analysis, catalysis, medical diagnostic, and forensic science. There are two major technical challenges that the SECM community faces: the production of nanoelectrodes, which will be commercially available for a reasonable price, and integration of a reliable mechanism for controlling the tip–surface distance. It might well be that the combination of SECM and AFM, providing that the tips are mass-produced, will enable many more electrochemists and non-electrochemists to employ this valuable tool. If this happens, we will witness blooming of the SECM and its application in imaging of biological and other surfaces, studying very fast charge transfer processes at different interfaces, and studying single molecule and nano-object transformations and many more nano-modifications of different materials.

Cross-References

▶ Scanning Probe Methods
▶ Scanning Tunneling Microscopy Studies of Immobilized Biomolecules

References

1. Binnig G, Rohrer H (1982) Scanning tunneling microscopy. Helv Phys Acta 55(6):726–735
2. Engstrom RC, Weber M, Wunder DJ, Burgess R, Winquist S (1986) Measurements within the diffusion layer using a microelectrode probe. Anal Chem 58(4):844–848. doi:10.1021/ac00295a044
3. Engstrom RC, Meaney T, Tople R, Wightman RM (1987) Spatiotemporal description of the diffusion layer with a microelectrode probe. Anal Chem 59 (15):2005–2010. doi:10.1021/ac00142a024
4. Bard AJ, Fan FRF, Kwak J, Lev O (1989) Scanning electrochemical microscopy – introduction and principles. Anal Chem 61(2):132–138. doi:10.1021/ac00177a011
5. Kwak J, Bard AJ (1989) Scanning electrochemical microscopy – theory of the feedback mode. Anal Chem 61(11):1221–1227. doi:10.1021/ac00186a009
6. Bard AJ, Denuault G, Lee C, Mandler D, Wipf DO (1990) Scanning electrochemical microscopy – a new technique for the characterization and modification of surfaces. Acc Chem Res 23(11):357–363. doi:10.1021/ar00179a002
7. Mirkin MV, Horrocks BR (2000) Electroanalytical measurements using the scanning electrochemical microscope. Anal Chim Acta 406(2):119–146. doi:10.1016/s0003-2670(99)00630-3
8. Nagy G, Nagy L (2000) Scanning electrochemical microscopy: a new way of making electrochemical experiments. Fresenius J Anal Chem 366(6–7):735–744. doi:10.1007/s002160051567
9. Lu X, Wang Q, Liu X (2007) Review: recent applications of scanning electrochemical microscopy to the study of charge transfer kinetics. Anal Chim Acta 601(1):10–25. doi:10.1016/j.aca.2007.08.021
10. Amemiya S, Bard AJ, Fan F-RF, Mirkin MV, Unwin PR (2008) Scanning electrochemical microscopy. Annu Rev Anal Chem 1:95–131
11. Sun P, Laforge FO, Mirkin MV (2007) Scanning electrochemical microscopy in the 21st century. Phys Chem Chem Phys 9(7):802–823. doi:10.1039/b612259k
12. Stoica L, Neugebauer S, Schuhmann W (2008) Scanning electrochemical microscopy (SECM) as a tool in biosensor research. In: Renneberg RLF et al (eds) Biosensing for the 21st Century. Springer, Berlin, pp 455–492
13. Bertoncello P (2010) Advances on scanning electrochemical microscopy (SECM) for energy. Energy Environ Sci 3(11):1620–1633. doi:10.1039/c0ee00046a
14. Schulte A, Nebel M, Schuhmann W (2010) Scanning electrochemical microscopy in neuroscience. Annu Rev Anal Chem 3:299–318
15. Bard AJ, Mirkin MV (2001) Scanning electrochemical microscopy, 1st edn. Marcel Dekker, New York
16. Shao YH, Mirkin MV (1998) Probing ion transfer at the liquid/liquid interface by scanning electrochemical microscopy (SECM). J Phys Chem B 102 (49):9915–9921. doi:10.1021/jp9828282
17. Sun P, Zhang ZQ, Guo JD, Shao YH (2001) Fabrication of nanometer-sized electrodes and tips for scanning electrochemical microscopy. Anal Chem 73(21):5346–5351. doi:10.1021/ac010474w
18. Katemann BB, Schuhmann T (2002) Fabrication and characterization of needle-type Pt-disk nanoelectrodes. Electroanalysis 14(1):22–28. doi:10.1002/1521-4109 (200201)14:1<22::aid-elan22>3.0.co;2-f
19. Lugstein A, Bertagnolli E, Kranz C, Mizaikoff B (2002) Fabrication of a ring nanoelectrode in an AFM tip: novel approach towards simultaneous electrochemical and topographical imaging. Surf Interface Anal 33(2):146–150. doi:10.1002/sia.1178
20. Zoski CG (2002) Ultramicroelectrodes: design, fabrication, and characterization. Electroanalysis 14 (15–16):1041–1051. doi:10.1002/1521-4109 (200208)14:15/16<1041::aid-elan1041>3.0.co;2-8
21. Burt DP, Wilson NR, Weaver JMR, Dobson PS, Macpherson JV (2005) Nanowire probes for high resolution combined scanning electrochemical microscopy – atomic force microscopy. Nano Lett 5(4):639–643. doi:10.1021/nl050018d
22. Velmurugan J, Sun P, Mirkin MV (2009) Scanning electrochemical microscopy with gold nanotips: the effect of electrode material on electron transfer rates. J Phys Chem C 113(1):459–464. doi:10.1021/jp808632w
23. Selzer Y, Mandler D (2000) Scanning electrochemical microscopy. Theory of the feedback mode for hemispherical ultramicroelectrodes: steady-state and transient behavior. Anal Chem 72(11):2383–2390. doi:10.1021/ac991061m
24. Zoski CG, Mirkin MV (2002) Steady-state limiting currents at finite conical microelectrodes. Anal Chem 74(9):1986–1992. doi:10.1021/ac015669i
25. Combellas C, Fuchs A, Kanoufi F (2004) Scanning electrochemical microscopy with a band microelectrode: theory and application. Anal Chem 76 (13):3612–3618. doi:10.1021/ac049752s
26. Zoski CG, Liu B, Bard AJ (2004) Scanning electrochemical microscopy: theory and characterization of electrodes of finite conical geometry. Anal Chem 76(13):3646–3654. doi:10.1021/ac049938

27. Lindsey G, Abercrombie S, Denuault G, Daniele S, De Faveri E (2007) Scanning electrochemical microscopy: approach curves for sphere-cap scanning electrochemical microscopy tips. Anal Chem 79(7):2952–2956. doi:10.1021/ac061427c

28. Lee C, Kwak JY, Anson FC (1991) Application of scanning electrochemical microscopy to generation/collection experiments with high collection efficiency. Anal Chem 63(14):1501–1504. doi:10.1021/ac00014a030

29. Martin RD, Unwin PR (1998) Theory and experiment for the substrate generation tip collection mode of the scanning electrochemical microscope: application as an approach for measuring the diffusion coefficient ratio of a redox couple. Anal Chem 70 (2):276–284. doi:10.1021/ac970681p

30. Fernandez JL, Bard AJ (2004) Scanning electrochemical microscopy 50. Kinetic study of electrode reactions by the tip generation-substrate collection mode. Anal Chem 76(8):2281–2289. doi:10.1021/ac035518a

31. Bard AJ, Denault G, Friesner RA, Dornblaser BC, Tuckerman LS (1991) Scanning electrochemical microscopy – theory and application of the transient (chronoamperometric) SECM response. Anal Chem 63(13):1282–1288. doi:10.1021/ac00013a019

32. Slevin CJ, Macpherson JV, Unwin PR (1997) Measurement of local reactivity at liquid/solid, liquid/liquid, and liquid/gas interfaces with the scanning electrochemical microscope: principles, theory, and applications of the double potential step chronoamperometric mode. J Phys Chem B 101(50):10851–10859. doi:10.1021/jp972587i

33. Niu L, Yin Y, Guo W, Lu M, Qin R, Chen S (2009) Application of scanning electrochemical microscope in the study of corrosion of metals. J Mater Sci 44 (17):4511–4521. doi:10.1007/s10853-009-3654-x

34. Casero E, Vazquez L, Maria Parra-Alfambra A, Lorenzo E (2010) AFM, SECM and QCM as useful analytical tools in the characterization of enzyme-based bioanalytical platforms. Analyst 135 (8):1878–1903. doi:10.1039/c0an00120a

35. Hirata Y, Mizutani F, Inoue T, Yokoyama H (2003) Scanning electrochemical microscopy with high resolution. Biophys J 84(2):467A-A

36. Amemiya S, Guo J, Xiong H, Gross DA (2006) Biological applications of scanning electrochemical microscopy: chemical imaging of single living cells and beyond. Anal Bioanal Chem 386(3):458–471. doi:10.1007/s00216-006-0510-6

37. Wain AJ, Cox D, Zhou S, Turnbull A (2011) High-aspect ratio needle probes for combined scanning electrochemical microscopy-Atomic force microscopy. Electrochem Commun 13(1):78–81. doi:10.1016/j.elecom.2010.11.018

38. Mirkin MV, Nogala W, Velmurugan J, Wang Y (2011) Scanning electrochemical microscopy in the 21st century. Update 1: five years after. Phys Chem Chem Phys 13(48):21196–21212

39. Kranz C, Wittstock G, Wohlschlager H, Schuhmann W (1997) Imaging of microstructured biochemically active surfaces by means of scanning electrochemical microscopy. Electrochim Acta 42(20–22):3105–3111. doi:10.1016/s0013-4686(97)00158-8

40. Turyan I, Matsue T, Mandler D (2000) Patterning and characterization of surfaces with organic and biological molecules by the scanning electrochemical microscope. Anal Chem 72(15):3431–3435. doi:10.1021/ac000046a

41. Wittstock G (2001) Modification and characterization of artificially patterned enzymatically active surfaces by scanning electrochemical microscopy. Fresenius J Anal Chem 370(4):303–315. doi:10.1007/s002160100795

42. Gyurcsanyi RE, Jagerszki G, Kiss G, Toth K (2004) Chemical imaging of biological systems with the scanning electrochemical microscope. Bioelectrochemistry 63(1–2):207–215. doi:10.1016/j.bioelechem.2003.12.011

43. Yamashita K, Takagi M, Uchida K, Kondo H, Takenaka S (2001) Visualization of DNA microarrays by scanning electrochemical microscopy (SECM). Analyst 126(8):1210–1211. doi:10.1039/b105097b

44. Wang J, Zhou FM (2002) Scanning electrochemical microscopic imaging of surface-confined DNA probes and their hybridization via guanine oxidation. J Electroanal Chem 537(1–2):95–102. doi:10.1016/s0022-0728(02)01254-8, Pii s0022-0728(02)01254-8

45. Komatsu M, Yamashita K, Uchida K, Kondo H, Takenaka S (2006) Imaging of DNA microarray with scanning electrochemical microscopy. Electrochim Acta 51(10):2023–2029. doi:10.1016/j.electacta.2005.07.007

46. Diakowski PM, Kraatz H-B (2009) Detection of single-nucleotide mismatches using scanning electrochemical microscopy. Chem Commun 10:1189–1191. doi:10.1039/b819876d

47. Zhang Z, Zhou J, Tang A, Wu Z, Shen G, Yu R (2010) Scanning electrochemical microscopy assay of DNA based on hairpin probe and enzymatic amplification biosensor. Biosens Bioelectron 25 (8):1953–1957. doi:10.1016/j.bios.2010.01.013

48. Holmes JL, Davis F, Collyer SD, Higson SPJ (2011) A new application of scanning electrochemical microscopy for the label-free interrogation of antibody-antigen interactions. Anal Chim Acta 689 (2):206–211. doi:10.1016/j.aca.2011.01.033

49. Scott ER, White HS, Phipps JB (1991) Scanning electrochemical microscopy of a porous membrane. J Membr Sci 58(1):71–87. doi:10.1016/s0376-7388 (00)80638-9

50. Nugues S, Denuault G (1996) Scanning electrochemical microscopy: amperometric probing of diffusional ion fluxes through porous membranes and human dentine. J Electroanal Chem 408(1–2):125–140. doi:10.1016/0022-0728(96)04523-8

51. Bath BD, Lee RD, White HS, Scott ER (1998) Imaging molecular transport in porous membranes. Observation and analysis of electroosmotic flow in individual pores using the scanning electrochemical microscope. Anal Chem 70(6):1047–1058. doi:10.1021/ac971213i

52. Tsionsky M, Zhou JF, Amemiya S, Fan FRF, Bard AJ, Dryfe RAW (1999) Scanning electrochemical microscopy. 38. Application of SECM to the study of charge transfer through bilayer lipid membranes. Anal Chem 71(19):4300–4305

53. Amemiya S, Bard AJ (2000) Scanning electrochemical microscopy. 40. Voltammetric ion-selective micropipet electrodes for probing ion transfer at bilayer lipid membranes. Anal Chem 72(20): 4940–4948

54. Uitto OD, White HS, Aoki K (2002) Diffusive-convective transport into a porous membrane. A comparison of theory and experiment using scanning electrochemical microscopy operated in reverse imaging mode. Anal Chem 74(17):4577–4582. doi:10.1021/ac0256538

55. Slevin CJ, Liljeroth P, Kontturi K (2003) Measurement of the adsorption of drug ions at model membranes by scanning electrochemical microscopy. Langmuir 19(7):2851–2858. doi:10.1021/la0265641e

56. Ishimatsu R, Kim J, Jing P, Striemer CC, Fang DZ, Fauchet PM et al (2010) Ion-selective permeability of an ultrathin nanoporous silicon membrane as probed by scanning electrochemical microscopy using micropipet-supported ITIES tips. Anal Chem 82(17):7127–7134. doi:10.1021/ac1005052

57. Mirkin MV (1999) High resolution studies of heterogeneous processes with the scanning electrochemical microscope. Mikrochim Acta 130 (3):127–153. doi:10.1007/bf01244921

58. Shah BC, Hillier AC (2000) Imaging the reactivity of electro-oxidation catalysts with the scanning electrochemical microscope. J Electrochem Soc 147 (8):3043–3048. doi:10.1149/1.1393645

59. Quinn BM, Liljeroth P, Kontturi K (2002) Interfacial reactivity of monolayer-protected clusters studied by scanning electrochemical microscopy. J Am Chem Soc 124(43):12915–12921. doi:10.1021/ja0282137

60. Georganopoulou DG, Mirkin MV, Murray RW (2004) SECM measurement of the fast electron transfer dynamics between Au-38(1+) nanoparticles and aqueous redox species at a liquid/liquid interface. Nano Lett 4(9):1763–1767. doi:10.1021/nl049196h

61. Macpherson JV, Unwin PR (1996) Scanning electrochemical microscope-induced dissolution: theory and experiment for silver chloride dissolution kinetics in aqueous solution without supporting electrolyte. J Phys Chem 100(50):19475–19483. doi:10.1021/jp9614862

62. McGeouch C-A, Edwards MA, Mbogoro MM, Parkinson C, Unwin PR (2010) Scanning electrochemical microscopy as a quantitative probe of acid-induced dissolution: theory and application to dental enamel. Anal Chem 82(22):9322–9328. doi:10.1021/ac101662h

63. Souto RM, Gonzalez-Garcia Y, Battistel D, Daniele S (2012) In situ scanning electrochemical microscopy (SECM) detection of metal dissolution during zinc corrosion by means of mercury sphere-cap microelectrode tips. Chemistry Eur J 18(1):230–236. doi:10.1002/chem.201102325

64. Fernandez JL, Walsh DA, Bard AJ (2005) Thermodynamic guidelines for the design of bimetallic catalysts for oxygen electroreduction and rapid screening by scanning electrochemical microscopy. M-Co (M : Pd, Ag, Au). J Am Chem Soc 127(1): 357–365

65. Weng Y-C, Hsieh C-T (2011) Scanning electrochemical microscopy characterization of bimetallic Pt-M (M = Pd, Ru, Ir) catalysts for hydrogen oxidation. Electrochim Acta 56(5):1932–1940. doi:10.1016/j.electacta.2010.12.029

66. Liljeroth P, Quinn BM, Ruiz V, Kontturi K (2003) Charge injection and lateral conductivity in monolayers of metallic nanoparticles. Chem Commun 13:1570–1571. doi:10.1039/b302958a

67. Mandler D, Unwin PR (2003) Measurement of lateral charge propagation in polyaniline layers with the scanning electrochemical microscope. J Phys Chem B 107(2):407–410. doi:10.1021/jp021623x

68. O'Mullane AP, Macpherson JV, Unwin PR, Cervera-Montesinos J, Manzanares JA, Frehill F et al (2004) Measurement of lateral charge propagation in Os(bpy)(2)(PVP)(n)Cl Cl thin films: a scanning electrochemical microscopy approach. J Phys Chem B 108(22):7219–7227. doi:10.1021/jp049500v

69. Whitworth AL, Mandler D, Unwin PR (2005) Theory of scanning electrochemical microscopy (SECM) as a probe of surface conductivity. Phys Chem Chem Phys 7(2):356–365. doi:10.1039/b407397e

70. Sheffer M, Mandler D (2008) Why is copper locally etched by scanning electrochemical microscopy? J Electroanal Chem 622(1):115–120. doi:10.1016/j.jelechem.2008.05.005

71. Zhang J, Burt DP, Whitworth AL, Mandler D, Unwin PR (2009) Polyaniline Langmuir-Blodgett films: formation and properties. Phys Chem Chem Phys 11(18):3490–3496. doi:10.1039/b819809h

72. Wei C, Bard AJ, Mirkin MV (1995) Scanning electrochemical microscopy.31. Application of SECM to the study of charge-transfer processes at the liquid-liquid interface. J Phys Chem 99 (43):16033–16042

73. Tsionsky M, Bard AJ, Mirkin MV (1996) Scanning electrochemical microscopy.34. Potential dependence of the electron-transfer rate and film formation at the liquid/liquid interface. J Phys Chem 100(45):17881–17888

74. Shao YH, Mirkin MV (1997) Scanning electrochemical microscopy (SECM) of facilitated ion transfer at

the liquid/liquid interface. J Electroanal Chem 439 (1):137–143. doi:10.1016/s0022-0728(97)00378-1

75. Barker AL, Macpherson JV, Slevin CJ, Unwin PR (1998) Scanning electrochemical microscopy (SECM) as a probe of transfer processes in two-phase systems: theory and experimental applications of SECM-induced transfer with arbitrary partition coefficients, diffusion coefficients, and interfacial kinetics. J Phys Chem B 102(9):1586–1598. doi:10.1021/jp973370r

76. Shao YH, Mirkin MV (1998) Voltammetry at micropipet electrodes. Anal Chem 70(15):3155–3161. doi:10.1021/ac980244q

77. Barker AL, Gonsalves M, Macpherson JV, Slevin CJ, Unwin PR (1999) Scanning electrochemical microscopy: beyond the solid/liquid interface. Anal Chim Acta 385(1–3):223–240. doi:10.1016/s0003-2670(98)00588-1

78. Barker AL, Unwin PR, Amemiya S, Zhou JF, Bard AJ (1999) Scanning electrochemistry microscopy (SECM) in the study of electron transfer kinetics at liquid/liquid interfaces: beyond the constant composition approximation. J Phys Chem B 103(34):7260–7269. doi:10.1021/jp991414l

79. Ding ZF, Quinn BM, Bard AJ (2001) Kinetics of heterogeneous electron transfer at liquid/liquid interfaces as studied by SECM. J Phys Chem B 105 (27):6367–6374. doi:10.1021/jp0100598

79. Ding ZF, Quinn BM, Bard AJ (2001) Kinetics of heterogeneous electron transfer at liquid/liquid interfaces as studied by SECM. J Phys Chem B 105 (27):6367–6374. doi:10.1021/jp0100598

80. Lee S, Zhang YH, White HS, Harrell CC, Martin CR (2004) Electrophoretic capture and detection of nanoparticles at the opening of a membrane pore using scanning electrochemical microscopy. Anal Chem 76(20):6108–6115. doi:10.1021/ac049147p

81. Unwin PR, Bard AJ (1992) Scanning electrochemical microscopy.14. Scanning electrochemical microscope induced desorption – a new technique for the measurement of adsorption desorption-kinetics and surface-diffusion rates at the solid liquid interface. J Phys Chem 96(12):5035–5045

82. Forouzan F, Bard AJ, Mirkin MV (1997) Voltammetric and scanning electrochemical microscopic studies of the adsorption kinetics and self-assembly of n-alkanethiol monolayers on gold. Isr J Chem 37(2–3):155–163

83. Mandler D, Meltzer S, Shohat I (1996) Microelectrochemistry on surfaces with the scanning electrochemical microscope (SECM). Isr J Chem 36(1): 73–80

84. Mandler D, Bard AJ (1989) Scanning electrochemical microscopy – the application of the feedback mode for high-resolution copper etching. J Electrochem Soc 136(10):3143–3144. doi:10.1149/1.2096416

85. Mandler D, Bard AJ (1990) A new approach to the high-resolution electrodeposition of metals via the

feedback mode of the scanning electrochemical microscope. J Electrochem Soc 137(4):1079–1086. doi:10.1149/1.2086606

86. Mandler D, Bard AJ (1990) Hole injection and etching studies of GAAS using the scanning electrochemical microscope. Langmuir 6(9):1489–1494. doi:10.1021/la00099a010

87. Mandler D, Bard AJ (1990) High-resolution etching of semiconductors by the feedback mode of the scanning electrochemical microscope. J Electrochem Soc 137(8):2468–2472. doi:10.1149/1.2086965

88. Meltzer S, Mandler D (1995) Study of silicon etching in HBR solutions using a scanning electrochemical microscope. J Chem Soc Faraday Trans 91(6):1019–1024. doi:10.1039/ft9959101019

89. Sugimura H, Shimo N, Kitamura N, Masuhara H, Itaya K (1993) Topographical imaging of Prussian blue surfaces by direct-mode scanning electrochemical microscopy. J Electroanal Chem 346(1–2):147–160. doi:10.1016/0022-0728(93)85009-6

90. Kranz C, Ludwig M, Gaub HE, Schuhmann W (1995) Lateral deposition of polypyrrole lines by means of the scanning electrochemical microscope. Adv Mater 7(1):38–40. doi:10.1002/adma.19950070106

91. Kranz C, Gaub HE, Schuhmann W (1996) Polypyrrole towers grown with the scanning electrochemical microscope. Adv Mater 8(8):634. doi:10.1002/adma.19960080805

92. Meltzer S, Mandler D (1995) Microwriting of gold patterns with the scanning electrochemical microscope. J Electrochem Soc 142(6):L82–L84. doi:10.1149/1.2044252

93. Ammann E, Mandler D (2001) Local deposition of gold on silicon by the scanning electrochemical microscope. J Electrochem Soc 148(8): C533–C539. doi:10.1149/1.1381390

94. Borgwarth K, Heinze J (1999) Increasing the resolution of the scanning electrochemical microscope using a chemical lens: application to silver deposition. J Electrochem Soc 146(9):3285–3289. doi:10.1149/1.1392468

95. Yatziv Y, Turyan I, Mandler D (2002) A new approach to micropatterning: application of potential-assisted ion transfer at the liquid-liquid interface for the local metal deposition. J Am Chem Soc 124 (20):5618–5619. doi:10.1021/ja0257826

96. Malel E, Ludwig R, Gorton L, Mandler D (2010) Localized deposition of Au nanoparticles by direct electron transfer through cellobiose dehydrogenase. Chemistry-a European Journal 16(38):11697–11706. doi:10.1002/chem.201000453

97. Santos LM, Ghilane J, Fave C, Lacaze P-C, Randriamahazaka H, Abrantes LM et al (2008) Electrografting polyaniline on carbon through the electroreduction of diazonium salts and the electrochemical polymerization of aniline. J Phys Chem C 112(41):16103–16109. doi:10.1021/jp8042818

98. Cougnon C, Gohier F, Belanger D, Mauzeroll J (2009) In situ formation of diazonium salts from nitro precursors for scanning electrochemical microscopy patterning of surfaces. Angew Chem Int Ed 48(22):4006–4008. doi:10.1002/anie.200900498

99. Hauquier F, Matrab T, Kanoufi F, Combellas C (2009) Local direct and indirect reduction of electrografted aryldiazonium/gold surfaces for polymer brushes patterning. Electrochim Acta 54 (22):5127–5136. doi:10.1016/j.electacta.2009.01.059

100. Coates M, Cabet E, Griveau S, Nyokong T, Bedioui F (2011) Microelectrochemical patterning of gold surfaces using 4-azidobenzenediazonium and scanning electrochemical microscopy. Electrochem Commun 13(2):150–153. doi:10.1016/j.elecom.2010.11.037

101. Combellas C, Kanoufi F, Mazouzi D, Thiebault A (2003) Surface modification of halogenated polymers 5. Localized electroless deposition of metals on poly(tetrafluoroethylene) surfaces. J Electroanal Chem 556:43–52

102. Combellas C, Fuchs A, Kanoufi F, Mazouzi D, Nunige S (2004) Surface modification of halogenated polymers. 6. Graft copolymerization of poly (tetrafluoroethylene) surfaces by polyacrylic acid. Polymer 45(14):4669–4675

103. Combellas C, Kanoufi F, Mazouzi D (2004) Surface modification of halogenated polymers. 8. Local reduction of poly(tetrafluoroethylene) by the scanning electrochemical microscope – transient investigation. J Phys Chem B 108(50): 19260–19268

104. Combellas C, Kanoufi F, Nunige S (2007) Surface modification of halogenated polymers. 10. Redox catalysis induction of the polymerization of vinylic monomers. Application to the localized graft copolymerization of poly(tetrafluoroethylene) surfaces by vinylic monomers. Chem Mater 19(15): 3830–3839

105. Wittstock G, Hesse R, Schuhmann W (1997) Patterned self-assembled alkanethiolate monolayers on gold. Patterning and imaging by means of scanning electrochemical microscopy. Electroanalysis 9(10):746–750. doi:10.1002/elan.1140091003

106. Wittstock G, Schuhmann W (1997) Formation and imaging of microscopic enzymatically active spots on an alkanethiolate-covered gold electrode by scanning electrochemical microscopy. Anal Chem 69(24):5059–5066. doi:10.1021/ac970504o

107. Sheffer M, Mandler D (2008) Scanning electrochemical imprinting microscopy: a tool for surface patterning. J Electrochem Soc 155(3):D203–D208. doi:10.1149/1.2830543

108. Deiss F, Combellas C, Fretigny C, Sojic N, Kanoufi F (2010) Lithography by scanning electrochemical microscopy with a multiscaled electrode. Anal Chem 82(12):5169–5175. doi:10.1021/ac100399q

Scanning Probe Methods

Alexander Wiek and Rudolf Holze
AG Elektrochemie, Institut für Chemie,
Technische Universität Chemnitz,
Chemnitz, Germany

Introduction

The advent of micropositioners capable of moving a probe (e.g., a tiny metal tip) in closest proximity to a solid surface to be investigated (i.e., in nanometer distance) with high spatial resolution based predominantly on piezocrystal-driven actuators has made a variety of scanning probe methods (or scanning probe microscopies SPM) possible. Depending on the principle of measurement and the type of probe methods are named. Several methods can be used in different modes of operation (e.g., the scanning tunneling microscope can be run in the constant distance and the constant current mode); in addition, some methods have been developed into further variations. The following overview of established methods starts with a general description of the most often used variant of a method, and variations are included wherever it seemed appropriate. A classification of scanning probe microscopies has been provided [1, 2].

Scanning Tunneling Microscopy

An STM uses an atomically sharp probe tip of an electronically conducting material in close proximity ($\cong$1 nm) to the surface under investigation. The tip is rastered (scanned) relative to the surface using piezoelectric devices. Thus an STM can be used to monitor directly the local density of electronic surface states with atomic resolution. The current flowing between the tip and the surface when a small voltage is applied depends on exponential dependency on the tip-surface distance characteristic of an electronic tunneling process. This results in the remarkable vertical resolution of the apparatus. In a simple approximation [3] the tunneling current I can be

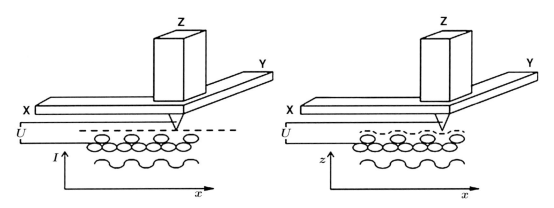

Scanning Probe Methods, Scheme 1 Scheme of scanning tunneling microscopy in the constant height mode (*left*) and the constant current mode (*right*); X, Y, and Z designate the respective piezodrive

given in terms of the local density of states (LDOS) $\rho_s(E_F)$ at the Fermi level, distance d, applied voltage U, and decay constant κ:

$$I \approx U\rho_s(E_F) \exp(-2\kappa d)$$

Using average values for a semiclassical square potential barrier and effective barrier height of 4 eV yields a value of $\kappa \cong 1$ Å^{-1}. Further details are provided elsewhere [4]. This causes the tunneling current to drop by about an order of magnitude per Å of increased distance finally resulting in the indicated vertical resolution. The very pronounced distance-current relationship contributes also in a very specific way to the high resolution: Although tips may be very sharp, they nevertheless show even in the perfect case a curvature on an atomic scale. Instead in the real world the tip is generally composed of several atoms because it is prepared, e.g., by cutting a wire at an oblique angle or by etching a thin wire [5]. Nevertheless, there will be most likely one atom protruding somewhat beyond the neighbors. As the tunneling currents drop by about an order of magnitude when the distance increases by 1 Å (100 pm), the dominant fraction of the tunneling current will flow across this particular atom. In case of extremely blunt tips, the image of the tip instead of the investigated surface will be recorded.

Probing the LDOS can be done in two fundamentally different ways: The tip can be scanned at constant height (distance) over the surface. This is possible only when the surface is smooth on an atomic scale; otherwise, the tip might crash into surface features. Alternatively the tip can be scanned in a constant tunneling current mode. In this case the actually measured current is fed into an electronic regulation circuit, which adjusts the actual tip-surface distance to a value resulting in a constant value of the tunneling current at all probed places. The signal fed into the z-axis piezodrive provides information about the local elevation. This mode works well even with strongly structured surfaces; because of the employed feedback circuitry, it is inherently slower. At first glance both methods seem to be equivalent, i.e., the results should be the same. Upon closer inspection differences appear. The first method operating at constant height yields results wherein the observed tunneling current taken as a measure of the tip-surface distance is influenced by the distance-current relationship mentioned above. Interpreting the observed pictures requires a sophisticated understanding of this relationship for the system under investigation. This problem is absent in the second mode. In order to get a correct topographic picture, a precise knowledge of the relationship between LDOS and actual atomic surface structure on an atomic level is needed. Both modes of operation are shown schematically below (Scheme 1).

Both designs were initially developed and applied under vacuum conditions yielding microscope pictures with atomic resolution [6, 7]. Very soon it was found that this design was also suitable for measurements at ambient temperature or even in the presence of an electrolyte solution [8]. The need to maintain the tip at a certain potential (bias) with respect to the surface (i.e., the electrode) under investigation and to keep this electrode itself at a selected potential adds to the complexity of the experiment. The tip acts as a probe for tunneling microscopy and as an ultramicroelectrode in the electrolyte solution. Attention has to be paid to conceivable Faradaic processes occurring at the tip. This current is usually minimized by coating the tip with an insulating material with only its apex exposed to the solution. A typical current of the STM of about 1–10 nA corresponds to 10^6 A cm^{-2} flowing between the tip and the probed section of the surface under investigation (typically 10^{-14} cm^2). Any Faradaic current at the tip caused by an electrochemical process would flow across the whole exposed surface area of the tip (about 10^{-8}–10^{-10} cm^2) (In addition, the tip is covered with an insulating material (wax, resin) leaving only the apex of the tip exposed to the electrolyte solution); thus, 10 nA would correspond to about 10–100 A cm^{-2}. This Faradaic current density is much smaller than the tunneling current; no distortion of the tunneling current has to be expected.

Chemical modification of the microscope tip most frequently prepared from metals like tungsten or gold results in surface properties useful for transferring "chemical sensitivity" to the tip. Chemical modification of the tip (by coating with polypyrrole or with self-assembled monolayers (SAM) [9]) resulting in, e.g., enhanced hydrogen bond or coordination bond interaction with species on the scanned surface results in enhanced tunneling electron transfer and increased brightness of the observed surface location or in higher contrast [10].

In order to operate an STM under in situ conditions, i.e., in the presence of an electrolyte solution, some conditions have to be fulfilled. The design of the STM must allow investigation of a horizontal surface at the bottom of the microscope. The tip has to be coated as

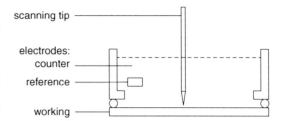

Scanning Probe Methods, Scheme 2 Scheme of an electrochemical cell for in situ investigations with an STM

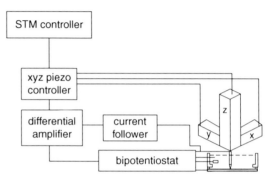

Scanning Probe Methods, Scheme 3 Scheme of a setup for electrochemical in situ investigations with an STM

completely as possible in order to minimize the Faradaic current. As the potential of the electrode surface under investigation has to be maintained at a fixed, controlled potential with respect to a reference electrode, a four-electrode arrangement requiring a corresponding bipotentiostat is necessary. The schematic drawing of the electrochemical cell as depicted below shows the major components (Scheme 2).

The peripheral components necessary for this experiment are indicated in the following scheme (Scheme 3).

An experimental setup specifically dedicated to electrochemical in situ investigations has been described in detail [11]. The preparation of suitable tips preferably prepared from metals like Pt, Pt–Ir, Ir, and W has been reviewed elsewhere [3]. The importance of operation under a controlled gas atmosphere, in particular the influence of dioxygen, has been examined extensively [12]. An experimental setup for measurements at elevated temperatures [13] has been described.

Fast scan measurements, i.e., for investigations of dynamics of surface diffusion or reconstruction, are done preferably in constant height instead of constant current mode because in this mode no electronic feedback circuit limiting response time and scan speed is involved. Obviously this works only with very smooth electrode surfaces. An electronic setup (bipotentiostat) allowing fast transient methods combined with scanning probe microscopies has been reported [14].

General overviews of STM studies of metal electrodes have been provided [3, 15–17]; an extensive review focused on ordered anion monolayers on metal electrode surfaces has been published [18] in addition. Charge-induced surface phase transitions on ordered Au(111) caused by increasing iodide adsorption from an aqueous electrolyte solutions have been observed [19]. The formation of copper sulfide nanostripe patterns on a Au(111) electrode surface formed by exposure of a single copper monolayer on this electrode exposed to bisulfide ions in the electrolyte solution has been studied with in situ STM [20]. Correlations between STM and SEM pictures have been discussed [21], and implications for reversible oxidative roughening have been pointed out [22]. Relationships between observed topography, electrochemical roughening parameters, and Raman spectroscopic features have been discussed [23]. Metal deposition including underpotential deposition processes on metal substrates [24–26] and on highly ordered pyrolytic graphite [27] has been studied frequently; the influence of solution additives (both organic molecules and inorganic anions) has been investigated [28–30]; organic adsorbate layers have been reviewed elsewhere [31]. Applications of STM to disordered (polycrystalline) materials including metallic glasses have been described [32]. Topography and local barrier height of metallic glass surfaces were measured. Step and island dynamics at liquid/solution interfaces have been studied, comparison with the solid/gas interface was drawn [33], and step and kink energies on Au (100) electrode surfaces could be derived from island studies with an STM [34]. Real-time observations of surface reactivity and mobility studied initially *ex situ* only [35] have been extended to in situ observations (There seems to be some debate over the limit between static and dynamic measurements related to the question, at which frequency of image registration (frames per second) a true real-time observation is done. Obviously this judgement – if necessary at all – has to be made with reference to the rate of change at the surface. In this text, no attempt is made to enter this discussion. Consequently in this text, no attempt is made to enter into this debate: "video STM" will not be separated from other "real-time STM.") recently. Reported examples included CO adsorption on Au(111) electrodes [36], adsorption of alkanethiols on Au(111) [37], and reductive desorption of self-assembled monolayers of hexanethiol from Au (111) surfaces [38, 39]. In the latter study, it was observed that desorption initiates at defects in the SAM, at missing rows and edges of vacancy islands. Both formation and final structure of self-assembling osmium-bipyridine complexes were monitored [40]. Adsorption and subsequent monolayer film formation of various protoporphyrins on a HOPG surface have been studied with both STM and AFM [41]. Studies of active metal dissolution [42] have been reported. In an investigation of electrodeposition of bismuth on a graphite surface, initially formation of small particles of about 10 nm diameter was observed [43]. Upon making contact with neighbors, these particles coalesce. Coarsening of platinum island deposits in the electrochemical double-layer potential region has been studied with STM [44]. Underpotential deposition of lead on a Cu(100) electrode surface revealed initially a high surface mobility [45]. Subsequent deposition of lead caused numerous changes of both structure and dynamics at the interface. Copper deposition and stripping at a gold electrode has been investigated [46]. Dissolution of highly polished copper surfaces showed roughening and formation of facets [47]. During selective dissolution of copper from Cu-Au alloys, pit formation and finally porosity were observed. AuCN adlayers were found during adsorption from aqueous solutions of $KAu(CN)_2$ on Au(111) surfaces [48]. Adsorption of NO from a KNO_2-containing electrolyte solution on a Rh(111) surface and the subsequent reduction were monitored in real time [49]. The reaction proceeds preferably at atomically flat terraces, not at surface defects. Initial reaction fronts were spatially concentrated, not randomly distributed.

A computational tool for analyzing observations obtained with STM based on a self-consistent semiempirical molecular orbital model has been described [50]. The electrodeposition of aluminum and titanium-aluminum alloys potentially useful as corrosion protection layers from room-temperature molten salt electrolytes has been studied with STM [51]. Underpotential deposition of cadmium on Ag(111) surfaces from ionic liquids has been monitored in situ with STM [52], and spinodal decomposition and surface alloying were observed. The use of STM beyond simple surface imaging including molecular identification, investigation of molecular reactivity, electron transfer kinetics, and nanofabrication has been reviewed elsewhere [53, 54]. Investigations of the semiconductor/solution interface beyond topographic ones [55] with varied tunnel gap distances and tip potentials allowed the separation of the effects of the tunneling barrier and the Schottky barrier at this interface [56].

An STM can be used to study the effects of illumination of a semiconductor surface by measuring local photovoltages and photocurrents [57]; surprisingly this approach has been employed so far only in *ex situ* studies. Gold/n-Si(111) nanocontacts (interface area about 10^{-12} cm^2) prepared by electrodeposition of gold onto n-Si(111):H substrates have been studied with an STM [58] in the presence of an aqueous solution of 0.02 M $HClO_4$. By varying tip voltage and electrode potential of the silicon substrate, a Schottky diode behavior of the nanocontact was verified.

An emerging application of the STM is the structuring of an electrode surface on a nanometer scale (nanostructuring). The use of an STM to deposit silver inside a polymer electrolyte film (Nafion$^{®}$) has been reported [59].

A combination of an STM with an SECM (see also below for this method) has been described [60]. The specifically adapted tip can be operated both in tunneling mode and as a probe electrode for scanning the surface in a large distance in the feedback mode measuring diffusion-controlled oxidation of a mediator. This way the topography and local reactivity can be studied with a single instrument and with high spatial reproducibility. The influence of the geometry of a nanometer-sized electrode has been discussed [61].

Atomic Force Microscopy

Beyond the tunneling current flowing between the tip and the surface, further interactions are effective between the tip and the surface. Spence et al. [62] have observed strain fringes on a graphite surface interacting with an STM tip extending 200 nm from the tunnel junction. This observation led to the development of the atomic force microscopy AFM (The terms SFM (scanning force microscopy) and AFM (atomic force microscopy) are used synonymously; the former term is used less frequently. Only in the initial stages of development the latter term was exclusively used for setups providing atomic (or better) resolution.) by Binnig et al. [63]. Depending on the design (including surface coating) of the tip van der Waals forces, electrostatic or magnetic forces can be monitored [64]. Generally forces between 10^{-9} and 10^{-6} N are measured; there have been reports describing measurements down to $3 \cdot 10^{-13}$ N [65]. They can be attractive or repulsive. When considering interatomic interactions the forces reach a minimum at the mechanical point contact; at smaller distances the repulsive interactions measured in the contact mode dominate; at higher distances the attractive interactions observed in the noncontact mode dominate. In the contact mode, the tip actually touches the surface. Obviously the electronic conductivity of the surface and the tip play no role in the operation; thus, nonconducting surfaces not suitable for the use of STM can be studied. The presence of a liquid between tip and surface provides no fundamental problem; of

Scanning Probe Methods, Scheme 4 Scheme of an atomic force microscope

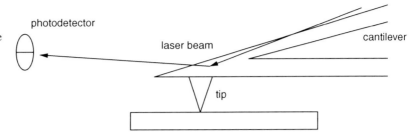

course the mechanical properties of the whole setup are changed. Nevertheless, electrochemical measurements in situ are possible.

Chemical sensitivity can be conferred to AFM by coating the tip with covalently linked monolayers which affect the tip-surface interaction; the method is called chemical force microscopy [66]. Additional modulation of the piezo actuator operating in z-direction and evaluation of the force signal can be used to measure the adhesion force between a surface and a chemically modified AFM tip [67]. Metal-coated AFM tips can be used in a scanning electrochemical microscopy (SECM, see below) mode [68] in studies of crystal dissolution or growth where surface processes are associated with considerable fluxes of species.

A cantilever with a sharp tip interacting with the surface under investigation is used. The actual bending of the cantilever is measured with a laser beam deflected from a mirrorlike surface spot on the back of the cantilever towards a position-sensitive photodetector. The measured signal is used to control the piezo actuators. A constant force mode in which the cantilever-surface distance is kept at a preset interaction force and a constant height mode of scanning operation are possible. The principle of operation is schematically outlined below (Scheme 4).

The mechanical properties of the tip-cantilever assembly are of central importance. Caused by the forces effective between the surface and tip, the cantilever is deformed. This deformation controls the overall performance of the microscope. The spring constant k and the resonance frequency ω_0 of the cantilever are particularly important. In order to be insensitive to mechanical noise from the environment, a high resonance frequency is desirable. A small spring constant in turn is required to detect weak forces. To obtain high resonance frequencies, stiff materials (silicon, silicon oxide, or silicon nitride) are used for the cantilever. A small spring constant can be maintained by limiting the mass of the device to a minimum by, e.g., microfabrication techniques. A typical cantilever has a length of 0.1 mm, a thickness of 1 mm, a spring constant around $0.1–1$ N m^{-1}, and a resonance frequency around 10–100 kHz. The development of single-wall carbon nanotubes (SWNT) as tips for AFM to be used also in electrochemical investigations has been described [69]. An alternative mode of operation without optical detection as described above employs a tip attached to a vibrating fork-like assembly. This approach has resulted in very high resolution; unfortunately it cannot be employed in an electrochemical environment because of the dampening effect of the electrolyte solution. The integration of an ultramicroelectrode into a tip for atomic force microscopy has been accomplished [70]. The electrochemically active area is located as a ring around the tip. It has been used in SECM measurements; simultaneously an AFM picture was obtained.

In actual operation in the contact mode, the tip touches the surface like the stylus of a record player. In the noncontact mode, the cantilever is oscillated at a frequency close to the resonance frequency with a large amplitude. In this mode, vertical long-range forces are probed, whereas lateral forces (friction-like forces in the plane of the sample surface) are almost noneffective. These forces have been employed in lateral force microscopy LFM.

Investigations published so far include metal dissolution (relevant to corrosion and corrosion inhibition) [71], underpotential metal deposition (upd) [72], and overpotential deposition (opd) [73]. Structural features of deposits, influence of electrolyte composition, electrode potential, etc. were reported. In a study employing both AFM and LFM, specific adsorption and phase changes at the polycrystalline silver/halide-containing electrolyte solution were investigated [74]. Whereas AFM provided topological imaging, LFM enabled detailed studies of adsorption and chemical reactions in adsorbate layers. Specifically adsorbed halide anions with their hydration shell stripped off provided higher friction values probed with LFM; hydrated anions in the outer Helmholtz layer not adsorbed to specific sites and with their hydration shell intact caused lower friction values. Using a colloidal probe (a silica particle attached to the cantilever of an AFM), the diffuse layer properties of a thiol-modified gold electrode have been investigated [75].

With chemically modified AFM tips, adhesion forces between the tip and a two-component self-assembled monolayer on a gold electrode have been studied [76]. Utilizing the different strengths of interaction between the modified tip (methyl and carboxyl terminating group functionalized) and the probed surface, SAM areas with methyl and carboxyl end groups could be distinguished.

Several reviews dealing with the fundamentals, experimental aspects, and applications have been published [77–79]. Operated in the constant force mode, the AFM can monitor changes in the thickness of a film (e.g., a metal hydroxide, which shows swelling/shrinking during redox processes) [80]. Dimensional changes of highly oriented pyrolytic graphite (HOPG) during lithium ion intercalation/deintercalation have been studied with an AFM [81]. During the first intercalation cycle, an irreversible increase of layer spacing was found. In following cycles, a reversible change of 17 % of the layer spacing was measured. Roughness effects caused by the formation of a solid electrolyte interface were taken into account by statistical analysis of the data. Electrochemical deposition and dissolution of molecular crystals of organic conductors

have been studied [82]. Morphological changes occurring during electropolishing of stainless steel in an ionic liquid have been identified with AFM [83].

AFM has been combined with nano-indentation and nano-scratching studies [84]. The hardness (and to a similar extent the friction coefficient) of a passivated titanium was three to four times higher under in situ conditions; this was assigned to a much faster repassivation process in the presence of the passivating electrolyte solution. Nanotribology, in particular surface friction force measurements of electrode surfaces modified with submonolayer foreign metal (upd deposits) with AFM, has been reported [85]. An AFM operated in the contact mode was used to scratch a surface of the aluminum alloy AA2024-T3 in contact with electrolyte solutions of different compositions (with/without chloride, dichromate) and under varying experimental conditions (stagnant/flowing solution) to gain insights into corrosion, protection, and breakdown [86].

Scanning Kelvin Probe Force Microscopy SKPFM

An AFM can also be used to probe the local Volta potential. Using a metal-coated silicon tip (e.g., Co-Cr40), first the topography of the surface under investigation is mapped using the tapping mode. In a second scan, the tip is moved along and kept at a constant distance of, e.g., 50 nm above every point on the surface. An AC voltage is applied to the tip generating an oscillating dipole. In the presence of an external field, this will in turn create a mechanical oscillation of the cantilever which can be detected using the standard features of the AFM. At every point of the scan, a DC ramp is added to the AC modulation. At the DC voltage where the oscillation of the cantilever vanishes, the potential on the tip and on the surface are the same. Thus, a map of the surface Volta potentials with respect to the tip is created. Because the potential of the tip might be unstable and could vary from experiment to experiment, calibration is necessary.

Scanning Probe Methods

A particularly reproducible reference is a nickel surface exposed to deionized water before the measurement [87]. In the absence of further calibration, this is the point of reference. The method cannot be applied in the presence of an electrolyte solution because of the large voltages applied to the tip; this would cause Faradaic reactions. Data from measurements of Volta potentials at corroding surfaces could be related to corrosion potentials of the same surface in contact with a solution because the linear correlation has been established before [88]. Nevertheless, studies at air or in the presence of ultrathin electrolyte films (i.e., under conditions frequently encountered in atmospheric corrosion) are possible. The general advantages of SKPFM, in particular the greatly enhanced spatial resolution, in comparison with SKP have been discussed in detail [88]. A critical review of applications of SKPFM focused on corrosion science with particular attention to possible artifacts and a comparison with SKP has been provided [89].

An AFM equipped with a suitably coated tip is employed; the experimental procedure has been outlined above. In a study of an aluminum alloy AA2024-T3, intermetallic particles and the matrix phase could be separated clearly [87]. The different surface films on these phases could be associated with their corrosion behavior. Inclusions and their corrosive behavior have been studied with a combination of SKPFM and AFM [90]. The effect of chloride-containing solution on corrosion at the matrix and the intermetallic particles was studied with SKPFM; in addition, light scratching with the AFM in the contact mode was applied to study the effect of the mechanical destabilization [91]. The intermetallic particles dissolved immediately after the film on their surface had been destabilized by mechanical abrasion. A general study of the influence of experimental parameters applied during emersion of the electrode, distance between tip and surface, influence of oxide coverages, etc. on the observed Volta potentials has been reported [92]; relationships to previous studies at emersed electrodes (e.g., [93]) and the topic of the adherence of the electrochemical double layer on an emersed electrode have been discussed. The influence of aluminum in magnesium alloys on atmospheric corrosion (in the absence/presence of CO_2) was studied with SKPFM [94]; a corrosion mechanism was suggested. Applications as well as limitations of SKPFM in studies of the surface of aluminum alloys have been reviewed thoroughly [95]. The surface of cast AlSi(Cu) alloys has been characterized with SKPFM [96]. Numerous particles of different compositions were detected, and they showed a positive Volta potential difference relative to the matrix with the actual value depending on the matrix composition. Filiform corrosion on epoxy-coated 1045 carbon steel was investigated with SKPFM [97]. Under coatings of 150 and 300 nm at 93 %, relative humidity samples were studied under air. Separation of active anode and cathode locations in the head of the filament could be identified. Microscopic and even submicroscopic aspects of electrochemical delamination have been studied with SKPFM [88].

Scanning Reference Electrode Technique

Localized very small variations of the electrode potential caused by current flow across the metal/solution interface over the surface of an electrochemically active material (e.g., a corroding metal) can be measured with a scanning reference electrode [98]. The local variations are picked up by a pair of very fine tips about 10 mm above the surface. The response of a twin platinum electrode has been modeled; results could be matched satisfactorily with real scans across localized events [99]. Instead of real reference electrodes, pseudo-reference electrodes like platinum or iridium tips or wires may be used. The tips pick up potential gradients normal to the current flux lines caused by the current flowing across the interface which are subsequently amplified and displayed. By scanning the tips across the surface, a map of local potential variations emerges. In more recent versions, a single tip is used. It vibrates at a frequency of 80 Hz using a piezocrystal. The potential change is picked up and amplified using a lock-in amplifier. A spatial resolution of 0.5 mm is possible [100]. At high electrolyte concentrations, the effect of ion flux caused by localized corrosion is swamped out, and the sensitivity of the method is diminished.

A commercial setup has been reported [101]. Operation with a fixed reference electrode and a rotating sample or a flat, fixed sample and a moving electrode have been described; their particular advantages and limitations have been reviewed [102]. In most applications, the rotation electrode systems appeared to be superior. Reported studies include investigations of corrosion protection coatings [100, 103], weld metal corrosion [104], pit initiation [105] (including hydrogen-promoted pitting [106]), and localized corrosion [107–109]. The method has been identified as being highly sensitive, compared to other methods in short time results can be obtained [110]. For inherent limitations and a comparison with SVET see below. Measurements of electrode potentials at higher resolution with a scanning probe setup as suited for STM and AFM have been reported [111].

Scanning Vibrating Electrode Technique SVET

A small electrode (typically a microelectrode of about 20 μm diameter) is scanned across the surface under investigation in a distance of about 100 μm. (This method, SVET, is also called current density probe CDP.) Any current flow across the sample/solution interface causes a potential drop in the solution which is probed by the microelectrode. Using previously established calibration data (with known current densities from a point source electrode), the measured voltage vectors are converted into current vectors [112]. Magnitude and direction of currents above the interface can thus be mapped. Cathodic and anodic processes can be localized; overlaying the obtained maps onto optical micrographs allows detection of visible surface features related to localized corrosion

phenomena [113, 114]. The technique is similar to the scanning reference electrode technique SRET (see above). Beyond the fact, that in comparison to the initially used two microreference electrodes with SRET, which have been later replaced by a single vibrating microelectrode, the actual way of vibrating the scanning electrode is different in SVET as compared to SRET [115]. This results in a limitation to measurement of DC currents only, which in turn have restricted the use of SRET to bare metals or coated metals with defects [116, 117].

Various possibilities to manufacture the required microelectrodes and their piezoelectric driver as well as the associated electronics have been described [114, 118]. Reported studies deal mainly with corrosion processes. With chromate-containing epoxy coatings on both steel and aluminum surfaces, a significant delay in the onset of anodic corrosion currents at a defect site was observed, whereas chromate-free epoxy coatings did not show this delay [114]. With the steel sample the cathodic current was observed at the defect site only with the chromate-free coating; on a chromate-containing epoxy, a cathodic current was also observed on this coating. With coatings of an intrinsically conducting polymer (poly(3-octyl pyrrole)), a further onset of any detectable current was observed both with coated steel and aluminum alloy [119]. Current density maps with the coated steel sample showed reduction currents on the polymer surface; oxidation was confined to the defect site. With the coated aluminum alloy sample, no significant oxidation was observed at the defect. Instead reduction was observed both on the polymer coating and the defect site; concomitant oxidation was observed locally under the coating. The localization of this processes seemed to be associated with specific interactions between the polymer and locally enriched copper (an alloy constituent). With pure aluminum, the oxidation current was more distributed over the coated surface, but still as far away from the defect as possible. The metal surface showed no pitting after removal of the coating. The influence of the application of the ICP coating (direct deposition by anodic electropolymerization onto the sample or casting from a polymer solution) was apparent in the study of polypyrrole-coated aluminum and aluminum alloys [120]. The former method yielded better corrosion protection because of stronger electronic (conductive) coupling. The action of dissolved cerium ions as corrosion inhibitors on steel has been investigated [113]. Contrary to previously published assumptions, these ions act as anodic inhibitor. Corrosion studies at metal matrix composites (MMC, reinforced 6092 aluminum composites) with SVET revealed corrosion initiation at localized anodic regions [121]. In a study of the corrosion at cut edges of galvanized steel, zinc oxide, zinc carbonate, and zinc hydroxide were suggested as reaction products of the anodic current observed with SVET over the zinc surface [122]. The cathodic current observed over the steel surface showed a behavior typical of a diffusion-limited oxygen reduction current; consequently with a pH microelectrode localized pH shifts were observed. In the presence of $SrCrO_4$ in the electrolyte solution, an increase of the Tafel slope of the anodic current indicative of the passivating effect of this inhibitor was found. In a solution of NaCl, the spatial patterns of deposition of corrosion products and anodic/cathodic currents could be matched with the SVET [123]. With an aqueous solution of $(NH_4)_2SO_4$, no such localized behavior was found.

Localized impedance measurements with a SVE have been described [124]; limitations and artifacts observed when using an SVE under specific corrosion conditions were discussed in detail [125].

Scanning Kelvin Probe SKP

The surface (or contact) potential of a solid- or a liquid-film-covered solid can be measured with a Kelvin probe [126, 127]. Essentially the Volta potential difference $\Delta\Psi$ between the two employed surfaces as described below is measured. In common abbreviation, this is also called measurement of a Volta potential Ψ (e.g., in [128]). As depicted below, a probe tip is brought

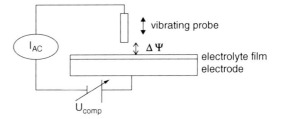

Scanning Probe Methods, Scheme 5 Scheme of Kelvin probe measurement

close to the surface under investigation (Scheme 5).

The tip and the adjacent surface form a condenser. When the distance between tip and surface is changed by vibrating the probe, an AC current flows; its magnitude depends on the existing potential difference. By adjusting the external bias voltage U_{comp}, this potential difference can be compensated; as a consequence the AC current vanishes. In most cases, relative changes of the local surface potential are of interest. In order to remove any unwanted influence of the probe surface potential, a material with constant surface potential is used (typically an etched Ni/Cr wire tip or a cylindrical probe of this material); thus, the measured local Volta potential depends only on the surface potential of the sample. A calibration of the probe is accomplished by measuring the corrosion potential with a conventional reference electrode touching the electrolyte film-covered surface under investigation with a Luggin capillary [126]. Simultaneous measurements with a SKP yield the desired Volta potential difference, which differs from the measured corrosion potential only by a constant difference typical of the experimental setup. The Volta potential is thus closely and directly related to the local corrosion potential [129–132]. Thus, spatially resolved measurements are useful in studies of localized processes like corrosion on heterogeneous surfaces. The required resolution can be obtained with piezodrives. The required compensation voltages are high enough to cause Faradaic processes in aqueous solution in the gap between the tip and the surface; obviously this method will work only in the absence of bulk solution.

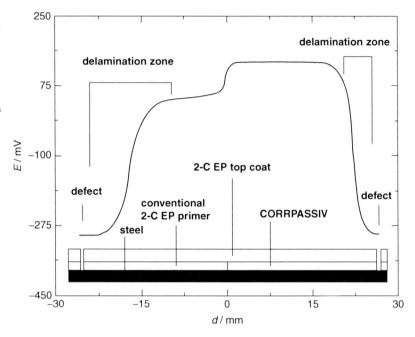

Scanning Probe Methods, Fig. 1 Potential profile measured with a SKP at a steel surface coated alternatively with standard primer and a new primer (see text for details) and top coat (Based on data in [127])

It works well with thin electrolyte solution films coating the corroding surface under investigation as frequently encountered in atmospheric corrosion [133, 134]. In its described setup, the distance between the tip (needle) and the surface is kept constant on a macroscopic level. In case of very rough, bended, or otherwise deformed surfaces, this might prove insufficient. A mode of operation with a height-regulated probe has been proposed (HR-SKP) [135].

Reported examples are mostly related to corrosion studies; the particular problems in relating Volta potential differences as measured with a SKP and local corrosion potentials have been treated in detail [131]. The effect of barrier layers and metal-surface pretreatment has been investigated [136, 137]. In a study of the effect of a corrosion protection primer containing the intrinsically conductive polyaniline (CORRPASSIV™) [128], the positive shift of the potential of the steel surface coated with this primer in comparison to the surface coated with a standard primer is evident, and the positive effect of the primer on the extent of the (much narrower) delamination zone is obvious (see below). With the standard primer, the delamination zone is larger by a factor of two (Fig. 1).

In a study of zinc-coated steel covered with a polymer topcoat, the mechanism of topcoat delamination was elucidated with high spatial resolution [138]. Depending on the details of the defect and the composition of the corroding atmosphere, the rate and type of delamination could be described. A similar study with a coated iron surface has been reported [139]. A comparison of results obtained with SKP, electrochemical impedance measurements, and cyclic voltammetry with respect to validity as corrosion prediction tool has been reported [140].

Differences in detected Volta potentials between pristine and corroded Al-Mg alloy surfaces could be related to the factors influencing thickness and conductivity of the corrosion product layers [141]. Corrosion layers developed in the presence of ion containing solutions yielded lower Volta potentials and showed higher conductivity. Cathodic delamination of polyaniline-based organic coatings on iron has been studied with SKP [142]. The roles of dioxygen reduction and of the polyaniline

Scanning Probe Methods, Fig. 2 Height and potential profile measured with a HR-SKP at an iron surface coated partially with zinc (Based on data in [134])

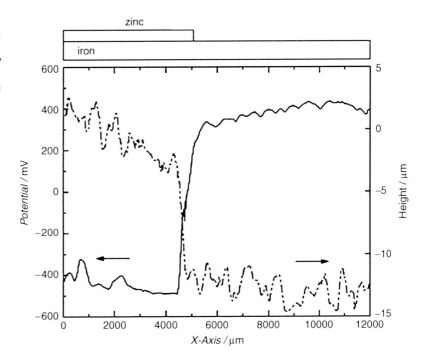

fraction in the coating were included in a corrosion mechanism.

The surface topography (i.e., the distance between the actual surface and the needle when the latter is kept at a constant distance from the sample itself) can be mapped simultaneously with the local potentials with a Kelvin probe equipped with an additional modulation setup [135]. A typical example of a zinc-coated iron surface with a thin polymer top coat shows the change of height (about 8 μm based on the deposition data) at the edge of the zinc coating (see figure below). The expected change of the Volta potential upon changing from zinc to iron is also observed. The roughness of the metal surface is visible in the plot (Fig. 2).

An iron surface coated with layers of latex of varying thickness yields a considerably different topographic picture (see below). The change of height is registered when an edge in the coating is passed. The changes of height (5 μm for every step) are well defined; the potential remains unchanged because the polymer coating has no influence (Fig. 3).

In a setup similar to the Evans drop experiment, the height and the potential of an iron surface with a drop of an aqueous solution of 0.5 **M** NaCl were scanned; results are displayed below (Fig. 4):

The lowest potential is measured in the center where corrosion (i.e., anodic dissolution of iron) attacks most aggressively. At the edges the potential increases somewhat; in this zone oxygen reduction proceeds. The potential changes around the drop imply the presence of an ultrathin electrolyte film because the potential reaches values of the bare iron surface only at a considerable distance from the edges of the macroscopically observed drop [135]. Filiform corrosion of automotive aluminum alloy AA6016 has been studied with SKP [143].

In earlier studies using a fixed Kelvin probe, corrosion kinetics and mechanisms were studied without the spatial resolution possible with the SKP [144–146]. The use of a Kelvin probe as a reference electrode in corrosion studies with very thin electrolyte films (2 μm) has been described [147]. The use of Kelvin probes to control and to monitor the potential has been reviewed [148].

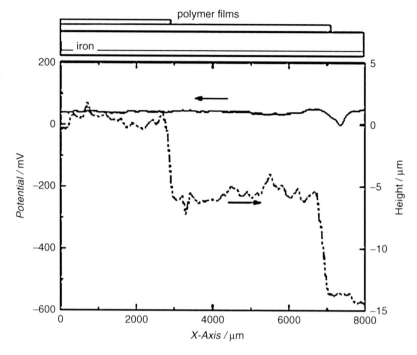

Scanning Probe Methods, Fig. 3 Height and potential profile measured with a HR-SKP at an iron surface coated with layers of latex (Based on data in [134])

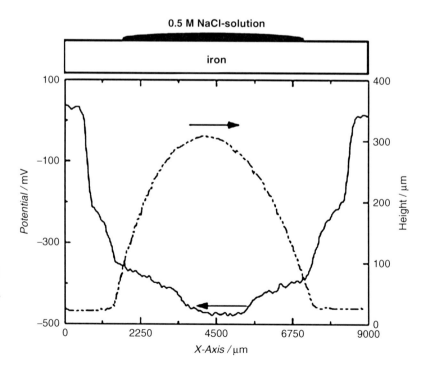

Scanning Probe Methods, Fig. 4 Height and potential profile measured with a HR-SKP at an iron surface with a drop of an aqueous solution of 0.5 M NaCl (Based on data in [134])

Future Directions

Still falling prices for the instruments described above will support even wider use. Combination with other surface-analytical methods will help in more careful interpretation of results.

Cross-References

▶ Scanning Electrochemical Microscopy (SECM)
▶ Scanning Tunneling Microscopy Studies of Immobilized Biomolecules

References

1. Friedbacher G, Fuchs H (1999) Pure Appl Chem 71:1337
2. Friedbacher G, Fuchs H (2003) Angew Chem 115:5804
3. Moffat TP (1999) In: Bard AJ, Rubinstein I (eds) Electroanalytical chemistry 21. Marcel Dekker, New York, p 211
4. Bonnel DA (ed) (1993) Tunneling microscopy and spectroscopy. VCH, New York
5. Legett G (1997) In: Vickerman JC (ed) Surface analysis – the principal techniques. Wiley, Chichester, p 393
6. Binnig G, Rohrer H, Gerber C, Weibel E (1982) Phys Rev Lett 49:57
7. Binnig G, Rohrer H (1983) Phys Bl 39:176
8. Lustenberger P, Rohrer H, Christoph R, Siegenthaler H (1988) J Electroanal Chem 243:225
9. Ito T, Bühlmann P, Umezawa Y (1999) Anal Chem 71:1699
10. Nishino T, Ito T, Umezawa Y (2003) J Electroanal Chem 550–551:125
11. Wilms M, Kruft M, Bermes G, Wandelt K (1999) Rev Sci Instrum 70:3641
12. Zhang J, Ulstrup J (2007) J Electroanal Chem 599:213
13. Stamenkowvic V, Lucas C, Tripkovic D, Strmenik D, Markovic NM (2006) 210 Electrochemical Society meeting, Cancun, 29 Oct–03 Nov 2006, Ext Abstr #1913
14. Schindler W, Bucharsky E, Behrend C (2003) Extended abstracts of the 203rd meeting of the Electrochemical Society, Paris, 27 Apr–02 May 2003, Ext Abstr 2314
15. Gewirth AA, Hanson KJ (1993) Interface 2:37
16. Sonnenfeld R, Schneir J, Hansma PK (1990) In: White RE, Bockris JO'M, Conway BE (eds) Modern aspects of electrochemistry 21. Plenum Press, New York, p 1
17. Arvia AJ (1990) In: Gutiérrez C, Melendres C (eds) Spectroscopic and diffraction techniques in interfacial electrochemistry, vol 320, NATO ASI Series C Vol. Kluwer, Dordrecht, p 449
18. Magnussen OM (2002) Chem Rev 102:679
19. Gao XP, Edens GJ, Weaver MJ (1994) J Phys Chem 98:8074
20. Friebel D, Schlaup C, Broekmann P, Wandelt K (2007) Phys Chem Chem Phys 9:2142
21. Gómez J, Vázquez L, Baró AM, Alonso C, González E, Gónzalez-Velasco J, Arvía AJ (1988) J Electroanal Chem 240:77
22. Holland-Moritz E, Gordon J II, Borges G, Sonnenfeld R (1991) Langmuir 7:301
23. Siperko LM (1990) J Electrochem Soc 137:2791
24. Carnal D, Müller U, Siegenthaler H (1994) J Phys IV (4):297
25. Obretenov W, Schmidt U, Lorenz WJ, Staikov G, Budevski E, Carnal D, Müller U, Siegenthaler H, Schmidt E (1993) J Electrochem Soc 140:692
26. Nichols RJ, Schröer D, Meyer H (1993) Scanning 15:266
27. Chen ZF, Li J, Wang EK (1994) J Electroanal Chem 373:83
28. Haiss W, Lackey D, Sass JK, Meyer H, Nichols RJ (1992) Chem Phys Lett 200:343
29. Nichols RJ, Bach CE, Meyer H (1993) Ber Bunsenges Phys Chem 97:1012
30. Wu ZL, Zang ZH, Yau SL (2000) Langmuir 16:3522
31. Itaya K (1998) Prog Surf Sci 58:121
32. Wiesendanger R, Ringger M, Rosenthaler L, Hidber HR, Oelhafen P, Rudin H, Güntherodt HJ (1987) Surf Sci 181:46
33. Giesen M (2001) Prog Surf Sci 68:1
34. Dieluweit S, Giesen M (2002) J Electroanal Chem 524:194
35. Guo XC, Madix RJ (2003) Acc Chem Res 36:471
36. Shue C-H, Yang L-YO, Yau S-L, Itaya K (2005) Langmuir 21:1942
37. Yamada R, Uosaki K (1998) Langmuir 14:855
38. Wano H, Uosaki K (2001) Langmuir 17:8224
39. Wano H, Uosaki K (2005) Langmuir 21:4024
40. Hudson JE, Abruna HD (1996) J Phys Chem 100:1036
41. Tao NJ, Cardenas G, Cunha F, Shi Z (1995) Langmuir 11:4445
42. Robinson RS (1988) J Electrochem Soc 135:143C
43. Nishitani R, Kasuya A, Nishina Y (1993) Z Phys D 26:S42
44. Xu Q, He T, Wipf DO (2007) Langmuir 23:9098
45. Wu HC, Yau SL (2001) J Phys Chem B 105:6965
46. Andersen JET, Møller P (1995) J Electrochem Soc 142:2225
47. Wu YC, Pickering HW, Gregory DS, Geh S, Sakurai T (1991) Surf Sci 246:468
48. Sawaguchi T, Yamada T, Okinaka Y, Itaya K (1995) J Phys Chem 99:14149
49. Zang ZH, Wu ZL, Yau SL (1999) Langmuir 15:8750
50. Kawamoto T (2004) Phys Chem Chem Phys 6:4913

51. Aravinda CL, Mukhopadhyay I, Freyland W (2004) Phys Chem Chem Phys 6:5225
52. Pan G-B, Freyland W (2007) Phys Chem Chem Phys 9:3286
53. Tao NJ, Li CZ, He HX (2000) J Electroanal Chem 492:81
54. Gonzalez-Martin A, Bhardwaj RC, O'M Bockris J (1993) J Appl Electrochem 23:531
55. Uosaki K, Koinuma M (1992) Faraday Discuss 94:361
56. Hiesgen R, Krause M, Meissner D (2000) Electrochim Acta 45:3213
57. Hiesgen R, Meissner D (1998) Adv Mater 10:619
58. Hugelmann M, Schindler W (2004) Appl Phys Lett 85:3608
59. Craston DH, Lin CW, Bard AJ (1988) J Electrochem Soc 135:785
60. Treutler TH, Wittstock G (2003) Electrochim Acta 48:2923
61. Sklyar O, Treutler TH, Vlachopoulos N, Wittstock G (2005) Surf Sci 597:181
62. Spence JCH, Lo W, Kuwabara M (1990) Ultramicroscopy 33:69
63. Binnig G, Quate CF, Gerber C (1989) Phys Rev Lett 63:2669
64. Burnham NA, Colton RJ (1993) In: Bonnell DA (ed) Scanning tunneling microscopy and spectroscopy. VCH, New York, p 191
65. Martin Y, Williams CC, Wickramsinghe Y (1987) J Appl Phys 61:4723
66. For a review see Noy A, Vezenov DV, Lieber CM (1997) Annu Rev Mater Sci 27:381
67. Rosazeiser A, Weilandt E, Hild S, Marti O (1997) Meas Sci Technol 8:1333
68. Holder MN, Gardner CE, Macpherson JV, Unwin PR (2005) J Electroanal Chem 585:8
69. Wilson NR, Cobden DH, Macpherson JV (2002) J Phys Chem B 106:13102
70. Kranz C, Friedbacher G, Mizaikoff B, Lugstein A, Smoliner J, Bertagnolli E (2001) Anal Chem 73:2491
71. Cruickshank BJ, Gewirth AA, Rynders RM, Alkire RC (1992) J Electrochem Soc 139:2829
72. Chen CH, Vesecky SM, Gewirth AA (1992) J Am Chem Soc 114:451
73. Manne S, Hansma PK, Massie J, Elings VB, Gewirth AA (1991) Science 251:183
74. Kautek W, Dieluweit S, Sahre M (1997) J Phys Chem B 101:2709
75. Rentsch S, Siegenthaler H, Papastavrou G (2007) Langmuir 23:9083
76. Balss KM, Fried GA, Bohn PW (2002) J Electrochem Soc 149:C450
77. Sonnenfeld R, Schneir J, Hansma (1990) In: White RE, Bockris, JO'M, Conway BE (eds) Modern aspects of electrochemistry 21. Plenum, New York, p 1
78. Cataldi TRI, Blackham IG, Briggs GAD, Pethica JB, Hill HAO (1990) J Electroanal Chem 290:1
79. Takano H, Kenseth JR, Wong S-S, O'Brien JC, Porter MD (1999) Chem Rev 99:2845
80. Häring P, Kötz R (1995) J Electroanal Chem 385:273
81. Campana FP, Kötz R, Vetter J, Novak P, Siegenthaler H (2005) Electrochem Commun 7:107
82. Hillier AC, Ward MD (1994) Science 263:1261
83. Abbott AP, Capper G, McKenzie KJ, Glidle A, Ryder KS (2006) Phys Chem Chem Phys 8:4214
84. Seo M, Kurata Y (2003) Electrochim Acta 48:3221
85. Nielinger M, Baltruschat H (2007) Phys Chem Chem Phys 9:3965
86. Schmutz P, Frankel GS (1999) J Electrochem Soc 146:4461
87. Schmutz P, Frankel GS (1998) J Electrochem Soc 145:2285
88. Rohwerder M, Hornung E, Stratmann M (2003) Electrochim Acta 48:1235
89. Rohwerder M, Turcu F (2007) Electrochim Acta 53:290
90. Leblanc P, Frankel GS (2002) J Electrochem Soc 149:B239
91. Schmutz P, Frankel GS (1998) J Electrochem Soc 145:2295
92. Guillaumin V, Schmutz P, Frankel GS (2001) J Electrochem Soc 148:B163
93. Kolb DM, Rath DL, Wille R, Hansen WN (1983) Ber Bunsenges Phys Chem 87: 1108; Hansen WN, Kolb DM (1979) J Electroanal Chem 100:493; Stuve EM, Krasnopoler A, Sauer DE (1995) Surf Sci 335:177; Rath DL, Kolb DM (1981) Surf Sci 109:641
94. Bengtsson Blücher D, Svensson J-E, Johansson L-G, Rohwerder M, Stratmann M (2004) J Electrochem Soc 151:B621
95. Muster TH, Hughes AE (2006) J Electrochem Soc 153:B474
96. Fratila-Apachitei LE, Apachitei I, Duszczyk J (2006) Electrochim Acta 51:5892
97. Leblanc PP, Frankel J (2004) Electrochem Soc 151:B105
98. Trethewey KR, Sargeant DA (1992) Metal and Materials 8:378
99. Badger SJ, Lyon SB, Turgoose S (1998) J Electrochem Soc 145:4074
100. Kinlen PJ, Menon V, Ding YW (1999) J Electrochem Soc 146:3690
101. EG&G Instruments, Princeton Applied Research, P. O. Box 2565, Princeton, NJ 08543, USA
102. Sargeant DA (1997) Corros Prevent Control 44:91
103. Gasparac R, Martin CR (2001) J Electrochem Soc 148:B138
104. Voruganti VS, Luft HB, DeGeer D, Bradford SA (1991) Corrosion 47:343
105. Sargeant DA, Ford C, Corderoy J (1991) Corros Prev Control 38:12
106. Yu JG, Luo JL, Norton PR (2002) Langmuir 18:6637
107. Trethewey KR, Sargeant DA, Marsh DJ, Tamimi AA (1993) Corros Sci 35:127

108. Bellanger G, Rameau JJ (1995) Electrochim Acta 40:2519
109. Sargeant DA, Hainse JGC, Bates S (1989) Mat Sci Tech 5:487
110. Voruganti VS, Luft HB, De Geer D, Bradferd SA (1991) Corrosion 47:343
111. VEECO data sheet DS57 (2003) Rev A0
112. Isaacs HS (1991) J Electrochem Soc 138:722
113. Isaacs HS, Davenport AJ, Shipley A (1991) J Electrochem Soc 138:390
114. He J, Gelling VJ, Tallman DE, Bierwagen GP (2000) J Electrochem Soc 147:3661
115. Sargeant DA (1997) Corros Prevent Control 44:91
116. Sekine I, Yuasa M, Tanaka K, Tsutsumi Tl, Koizumi F, Oda N, Tanabe H, Nagai M (1994) Shikizai 67:424
117. Sekine I (1997) Prog Org Coat 31:73
118. Scheffey C (1988) Rev Sci Instrum 59:787
119. He J, Gelling VJ, Tallman DE, Bierwagen GP, Wallace GG (2000) J Electrochem Soc 147:3667
120. He J, Tallmann DE, Bierwagen GP (2000) J Electrochem Soc 151:B644
121. Ding H, Hihara LH (2005) J Electrochem Soc 152:B161
122. Ogle K, Baudu V, Garrigues L, Philippe X (2000) J Electrochem Soc 147:3654
123. Ogle K, Morel S, Jacquet D (2006) J Electrochem Soc 153:B1
124. Bayet E, Huet F, Keddam M, Ogle K, Takenouti H (1997) J Electrochem Soc 144:L87
125. Bayet E, Huet F, Keddam M, Ogle K, Takenouti H (1999) Electrochim Acta 44:4117
126. Kelvin L (1898) Philos Mag 46:82
127. Adamson AW (1990) Physical chemistry of surfaces. Wiley, New York
128. Wessling B, Posdorfer J (1999) Electrochim Acta 44:2139
129. Stratmann M, Streckel H (1990) Corros Sci 30:681
130. Stratmann M, Wolper M, Streckel H, Feser R (1991) Ber Bunsenges Phys Chem 95:1365
131. Yee S, Oriani RA, Stratmann M (1991) J Electrochem Soc 138:55
132. Stratmann M, Wolpers M, Streckel H, Feser R (1991) Ber Bunsenges Phys Chem 95:1365
133. Grundmeier G, Schmidt W, Stratmann M (2000) Electrochim Acta 45:2515
134. Stratmann M, Streckel H, Feser R (1991) Farbe und Lack 97:9
135. Wapner K, Schoenberger B, Stratmann M, Grundmeier G (2005) J Electrochem Soc 152:E114
136. Grundmeier G, Stratmann M (1999) Appl Surf Sci 141:43
137. Barranco V, Thiemann P, Yasuda HK, Stratmann M, Grundmeier G (2004) Appl Surf Sci 229:87
138. Fürbeth W, Stratmann M (1995) Fresenius J Anal Chem 353:337
139. Stratmann M, Feser R, Leng A (1994) Electrochim Acta 39:1207
140. Posdorfera J, Wessling B (2000) Fresenius J Anal Chem 367:343
141. Juzeliunas E, Sudavicius A, Jüttner K, Fürbeth W (2003) Electrochem Commun 5:154
142. Williams G, Holness RJ, Worsley DA, McMurray HN (2004) Electrochem Commun 6:549
143. McMurray HN, Coleman AJ, Williams G, Afseth A, Scamans GM (2007) J Electrochem Soc 154:C339
144. Stratmann M (1987) Corros Sci 27:869
145. Stratmann M, Streckel H (1990) Corros Sci 30:697
146. Leng A, Stratmann M (1993) Corros Sci 34:1657
147. Stratmann M, Streckel H, Kim KT, Crockett S (1990) Corros Sci 30:715
148. Frankel GS, Stratmann M, Rohwerder M, Michalik A, Maier B, Dora J, Wicinski M (2007) Corr Sci 49:2021

Scanning Tunneling Microscopy Studies of Immobilized Biomolecules

Pau Gorostiza[1] and Juan Manuel Artés[2,3]
[1]Institució Catalana de Recerca i Estudis Avançats (ICREA), Barcelona, Spain
[2]Institució Catalana de Recerca i Estudis Avançats (ICREA), Institute for Bioengineering of Catalonia (IBEC), Barcelona, Spain
[3]Electrical and Computer Engineering, University of California Davis, Davis, CA, USA

Electron Transfer in Biology

Electron transfer (ET) is the process involving the movement of an electron from an atom or a molecule to another. ET is essential in many chemical and biological processes [1], such as cellular respiration, photosynthesis, and most of the enzyme-catalyzed reactions [2]. In particular, charges are required to move over long distances through proteins in most biological energy transduction pathways [2]. Many theoretical and experimental approaches [2–10] have been reported since Marcus [1] introduced the theoretical basis for ET. Briefly, the rate for a nonadiabatic ET process at fixed distance between an electron donor (A) and an acceptor (B) can be described by the semiclassical Marcus theory using Eq. 1 [1–5].

$$k_{ET} = \sqrt{\frac{4\pi^3}{h^2 \lambda k_b T}} H_{AB}^2 \exp\left\{-\frac{(\Delta G + \lambda)}{4\lambda k_b T}\right\} \quad (1)$$

The rate for ET (k_{ET}) depends on the driving force (ΔG^0) for the ET reaction, the reorganization energy (λ), and the electronic coupling (H_{AB}) between A and B at the transition state. k_B and h are the Boltzmann and Planck constants, respectively, and T is the temperature. For electron tunneling reactions, H_{AB} has an exponential dependence on the distance between donor and acceptor (d_{AB}) [1]. Thus, when the driving force equals the reorganization energy, the rate for ET has an exponential dependence on d_{AB} (Eq. 2). The key parameter in this relationship is the distance decay factor β, characteristic of the ET mechanism, and the medium between donor and acceptor.

$$k_{ET} \propto \exp(-\beta d_{AB}) \quad (2)$$

Theoretical efforts established models for the electronic coupling and the ET pathway along biomolecular structures [11–13] which predict that distance decay factors of biomolecules depend on their structure [2, 11]. In the case of proteins, $\beta \approx 1 \text{ A}^{-1}$ was expected for a beta sheet structure [3, 14] and 1.26 A^{-1} for an alpha helix [14, 15] (see Fig. 1b). Other theoretical reports focused on fluctuations and interferences between ET pathways that account for the coherence of the ET mechanism [9, 13, 16]. Quantum coherence has become an intensive area of research in biology [16–21]. Coherent quantum processes have been found to play important roles in nature, with examples as diverse as photosynthesis [19] or the way birds are able to navigate by sensing Earth's magnetic field [22].

This review focuses on experimental advances to study ET in individual biomolecules, with special emphasis on intra- and intermolecular ET process in redox proteins studied by electrochemical scanning tunneling microscopy (ECSTM) and spectroscopy (ECTS). Implications for the development of electronic devices based on single molecules are also discussed.

Blue Copper Proteins as Model Systems for Biological ET

Metalloproteins are convenient systems to study biological ET, and copper-containing enzymes are important among them [23]. Blue copper proteins include mononuclear proteins involved in intermolecular ET such as plastocyanins and azurins as well as multicopper enzymes [23, 24]. These copper centers exhibit an extremely intense light absorption band, which is responsible for their deep blue color, and a very small Cu parallel hyperfine splitting in their electron paramagnetic spectra [25, 26].

Pseudomonas aeruginosa azurin (Az) is a 14.6 KDa globular protein that has a blue copper center located 8 Å below the Az surface and a solvent-exposed disulfide bridge between Cys 3 and Cys 26 [27] (see Fig. 1a). Azurin has represented the benchmark for the study of ET in proteins. The disulfide bridge in azurin was used in pulse radiolysis experiments to investigate intraprotein ET pathways [3]. These native cysteines have been also used to immobilize the protein on gold surfaces to study ET using electrochemical [28] and scanning probe microscopy techniques [26, 29–32]. In an insightful experimental approach, the intraprotein ET distance decay constant was obtained by measuring the ET rate between $Cu^{I/II}$ center and redox probes ($Ru^{II/III}$) attached to histidine residues mutated at different points on the surface of the protein [14]. These studies yielded $\beta \approx 1 \text{ Å}^{-1}$, in agreement with previous calculations [11]. This result was in the same range than the one found for ET through saturated alkane chains, suggesting that the ET mechanism was the same in both systems [2, 4], namely, superexchange tunneling. Figure 1b shows the experimental results for different redox proteins in the context of ET tunneling reactions in different media for comparison [33].

An alternative to bulk electrochemical approaches is the study of redox proteins confined to the surface of an electrode. In order to perform electrochemical experiments, azurin can be bound covalently through native or introduced cysteines to a gold electrode. Another strategy is to use non-covalent interactions with a linker self-assembled monolayer (SAM) covering the working electrode. This has been used to study the distance dependence of the ET reaction involving azurin using variable-length alkanethiol SAMs [6, 34]. By these means, the previously reported distance decay factor ($\beta = 0.9$–1 Å^{-1}) was confirmed [35–37]. These electrochemical approaches revealed detailed thermodynamic and kinetic properties [38–41].

Although the native environment of proteins is a buffered aqueous solution, the ET of protein

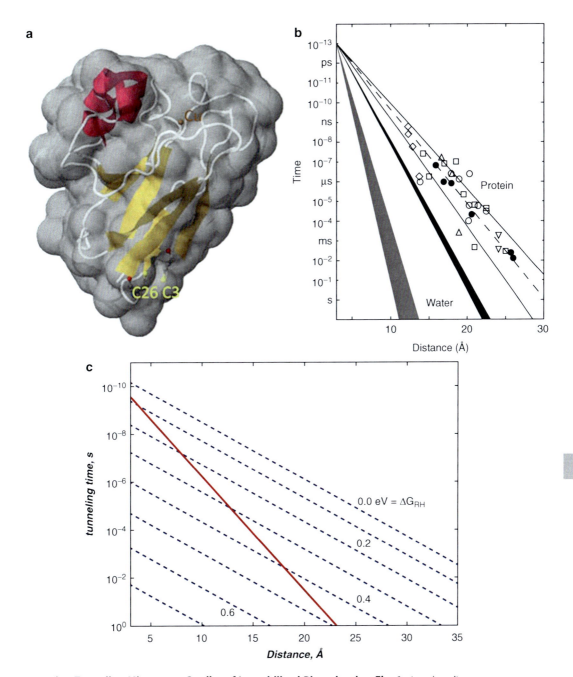

Scanning Tunneling Microscopy Studies of Immobilized Biomolecules, Fig. 1 (continued)

films on electrodes has been also studied in air or "solid state" configuration, in order to explore their application to electronic devices [42–47]. In the case of azurin and bacteriorhodopsin, temperature-dependent changes in ET mechanism in air conditions have been recently reported [44, 47]. However, from a fundamental perspective, it is difficult to compare results obtained in air and under electrochemical control [42].

Mechanistic Studies of Biological ET

The study of ET in azurin under electrochemical control has revealed that ET occurs in the nanosecond to microsecond scale (see Fig. 1b). Biological redox machines need tunneling times in the microseconds range to function properly [2]. This sets a maximum donor–acceptor tunneling distance of about 2 nm, which is far below the actual distances found between many redox enzyme centers [2, 4, 48]. ET between distant redox centers could occur by a multistep tunneling mechanism, which displays smoother distance dependence (compare blue and red lines respectively in Fig. 1c). It has been reported that multistep tunneling can be faster than a single tunneling event at certain distances [4, 49, 50]. Oxidizable protein residues like tryptophan and tyrosine are suspected to support these long-range ET pathways, acting as semiconductor relay elements to facilitate charge transfer [51].

Recently, the redox properties of azurin have been modulated by engineering the protein residues surrounding the redox center [52–54]. These interesting results represent a step forward in understanding the protein ET process and offer interesting perspectives for technological applications. Azurin has proven useful to build biomemory devices [55, 56] and has also been proposed for other novel electronic devices [32, 57–59] (see section "Conductance and Spectroscopic Measurements and the Birth of Molecular Electronics").

The electrochemical methods discussed above have been used to study ET in a variety of other metalloproteins [2, 3, 15, 60, 61] including cytochrome [15], stellacyanin [15], and multicopper enzymes [3]. DNA bases have also been found to participate and mediate ET [62], in analogy to the results found for proteins. Charge transfer in DNA is thought to be fundamental in DNA oxidative processes, like oxidative damage and its repair [63]. Thus, ET in DNA offers interesting opportunities for fundamental research and technological applications. The field has experienced intensive research in the last years and has been recently reviewed in [63, 64].

Photolysis and electrochemical methods have pioneered the study of ET in biomolecules. However, in bulk measurements the properties of individual molecules are obscured by ensemble averages [39, 40, 65]. Recently, some reports show the importance of the variability in the thermodynamic and kinetic properties of molecules [39, 40]. Thus, efforts are still needed in order to understand ET processes at the level of single molecules.

Scanning Tunneling Microscopy Studies of Immobilized Biomolecules, Fig. 1 (a) Azurin 3-D representation (PDB accession code: 1AZU[27]). α-helix and β-sheet structures are represented as *pink* and *yellow ribbons*, respectively. The Cu center and the two cysteines establishing the disulfide bridge (C3–C26) are represented as ball and stick models. The solvent accessible protein surface is overlayed. (**b**) Tunneling ET time versus donor–acceptor separation for ET in Ru-modified proteins: azurin (*solid circle*); cytochrome C (*cyt c, empty circle*); myoglobin (*empty triangle*); cyt b562 (*empty square*). The *solid lines* illustrate the tunneling pathway predictions for coupling along β-strands ($\beta = 1$ Å^{-1}) and α-helices ($b = 1.3$ Å^{-1}); the *dashed line* illustrates a 1.1 Å^{-1} distance decay. Distance decay for electron tunneling through water is shown as a *black* wedge. Estimated distance dependence for tunneling through vacuum is shown as a *grey* wedge for comparison. From 4. Reprinted with permission. (**c**) Distance dependences of the rates of single-step and two-step electron tunneling reactions. *Red* solid line indicates theoretical distance dependence for a single-step, ergoneutral ($\Delta G°_{RP} = 0$) tunneling process ($\beta = 1.1$ Å^{-1}). *Blue dashed lines* indicate distance dependence calculated for two-step ergoneutral tunneling with the indicated standard free-energy changes for the first step. From 2. Reprinted with permission

Biological ET Studied by Electrochemical Tunneling Microscopy and Spectroscopy

High-Resolution Imaging in Liquids

Since its invention in 1982 [66], scanning tunneling microscopy (STM) has allowed high-resolution studies of surfaces. STM takes advantage of the quantum mechanical tunneling effect. It is based on an atomically sharp conductive tip that scans a conductive sample recording the tunneling current established between them as a consequence of a difference of potential (U_{bias}) applied between the electrodes. In the most commonly used constant current mode, this tunneling current is used as feedback signal that allows a 3-D mapping of the sample surface. Limited only by the ability to record the tunneling current, STM can operate in different environments, from ultrahigh vacuum (UHV) to liquid environments including aqueous conductive solutions. When STM is performed in liquid environment under control of the potential applied to the electrodes, the technique is known as electrochemical STM (ECSTM) (see Fig. 2a). In order to operate STM in an electrochemical environment, tip faradaic currents must be reduced to allow a reliable measurement of the tunneling current. To this end, the tip area exposed to the solution has to be minimized [67]. Due to its high resolution in current measurements and sub-nanometric positioning, ECSTM is a unique tool to perform single-molecule experiments. It is also a very advantageous technique for biomolecules because it allows keeping the sample in nearly physiological conditions and full electrochemical control while offering the possibility of single-molecule resolution [26, 31, 35, 68, 69].

Conductance and Spectroscopic Measurements and the Birth of Molecular Electronics

STM can be also used to perform spectroscopic experiments at fixed points of the sample surface with sub-nanometric resolution. Tunneling spectroscopic measurements include current–distance and current–potential recordings and can also be performed in solution and under electrochemical control (electrochemical tunneling spectroscopy, ECTS). In current–voltage (I(V)) ECTS, the absolute tip and sample potentials, and the potential drop in the tunneling gap can be scanned while the tunneling current is recorded, providing information about the energy and bias dependence of the conductance. In current–distance (I(z)) ECTS, the tip to sample distance is scanned while the current is recorded, which allows studying the distance dependence of the ET process at the single-molecule level, and directly testing the models of Eqs. (1) and (2). In this context, current–distance recordings by ECTS can be compared to electrochemical measurements with redox probes introduced at different distances along a well-defined axis of the protein (see section "Blue Copper Proteins as Model Systems for Biological ET"). Thus, the ECSTM tip constitutes an electrochemical probe whose Fermi energy can be adjusted at any desired value within the solvent window and whose position with respect to an individual protein can be fixed or scanned along any direction with sub-nanometer accuracy [31, 70].

Advances in tunneling microscopy and spectroscopy led to breakthroughs in the field of molecular electronics, a discipline that deals with the conductance measurement of single molecules and the use of their electronic properties for the development of devices [71]. Molecular electronics has grown considerably since Aviram and Ratner made the first theoretical predictions about the conductance of a single molecule between two electrodes [72]. Several experimental approaches have been devised in order to measure the conductance of molecules [73, 74]. These techniques can be divided in two broad groups: approaches based on thin molecular films and single-molecule measurement techniques [73]. Scanning probe microscopies have proven useful for single-molecule experiments, and approaches based on AFM and STM have result in powerful techniques used to determine single-molecule conductance. In the STM-break junction approach (see Fig. 2b) [75], the STM tip is brought to contact and retracted from the sample repetitively while the current is recorded.

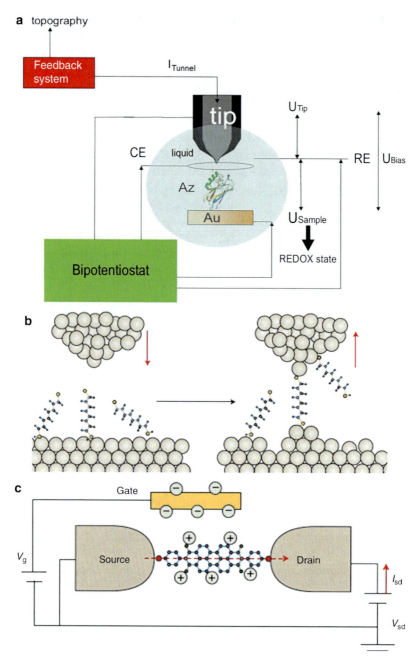

Scanning Tunneling Microscopy Studies of Immobilized Biomolecules, Fig. 2 (a) Electrochemical scanning tunneling microscopy (ECSTM) setup scheme. Electrochemical potentials are indicated: tip potential (U_{TIP}), sample potential (U_{SAMPLE}), and bias voltage (U_{BIAS}). I_{Tunnel} indicates the recorded tunneling current, CE counter or auxiliary electrode, RE reference electrode, and Az is azurin bound to the gold (Au) working electrode. (b) Measurement of single-molecule conductance using a scanning tunneling microscopy (STM) break junction. An STM tip is pushed into contact with molecules adsorbed on an electrode, during which the molecules have a chance to bridge the tip and substrate electrodes. As the tip electrode is pulled away again, the bridging molecules break contact with one of the two electrodes. Each breakage is revealed as a step in the conductance trace. Statistical analysis of the individual traces allows determination of the conductance of the molecule. Reproduced from 71 with permission. (c) A redox molecule is covalently linked to two electrodes

An exponential current distance plot is usually obtained, described by Eq. 3:

$$G = \frac{1}{U_{bias}} = G_0 \exp(-\beta d) \qquad (3)$$

where G is the conductance, I the tunneling current, U_{bias} is the bias voltage applied between the electrodes, G_0 is the quantum conductance ($G_0 = 2e^2/h \approx 77.4\ \mu S$), β is the distance decay factor, and d the distance between the electrodes. During approach and retract cycles, a molecule can be covalently bound between the electrodes, and this causes plateaus or steps in the current–distance recording. The current of the step corresponds to the conductance of the system in a moderate bias voltage range (Eq. 3). This technique allows obtaining thousands of single-molecule conductance measurements that are used to construct conductance histograms [75].

In a complementary method, the STM tip is brought to a fixed distance from the sample surface and the current is recorded while molecules diffuse in the tunneling gap [76]. When a molecule binds to both electrodes, a sudden jump is observed in the current trace that can be used to determine single-molecule conductance using Eq. 3.

In order to reduce the size of integrated circuits, the technologies used to fabricate them have been continuously changing. As circuit features approach the nanometer scale, it has been proposed that devices (diodes, transistors, resistors, capacitors, and conductive tracks) could be made from molecules with suitable properties, instead of patterned doped silicon layers [65]. For that purpose, it is essential to characterize their electronic properties at the level of single molecules, and the molecular junction methods described above offer many advantages to do it.

The Fermi level of the electrodes in a molecular junction can be adjusted by tuning the potentials in the system [77]. When the Fermi levels of the electrodes align with energetic levels in the molecule, the charge transfer probability is maximized, corresponding to a high conductance situation. In contrast, when electrode levels are far away from the energetic level of the molecule, a low conductance situation is normally found. The conductance of the junction can thus be changed as a function of potentials. Most of the energy levels in molecules lie outside of the experimentally accessible window, which is limited by solvent hydrolysis, and thus often the overlap with HOMO or LUMO levels is only partial [77]. In the case of a redox molecule, the redox level becomes accessible, and the situation turns out to be similar to the one found in the classical field effect transistors (FETs), with the junction electrodes (tip and sample) acting as source and drain, and the redox level being adjusted by means of electrochemical potentials, which act as the gate electrode [78]. In this way the molecular conductance can be varied with the electrochemical potentials (Fig. 2c) [77].

Application to Organic Molecules

Scanning probe microscopy and spectroscopy techniques have been used to determine the conductance of hydrocarbon chains [75, 76] and small organic molecules [79–82]. In these studies, it is important to emphasize that although the lateral size of molecules observed by ECSTM matches the crystallographic size, the vertical dimension or apparent height is a complex convolution between topography and electronic properties of the molecule [26, 29, 31, 68, 83]. It was first observed using ECSTM that Fe-protoporphyrin molecules on graphite display a resonant-like behavior of the tunneling ET [68] (Fig. 3).

Scanning Tunneling Microscopy Studies of Immobilized Biomolecules, Fig. 2 (continued) in an electrochemical cell. (Counter electrode is not shown here for clarity.) In this configuration, the electrochemical gate voltage (Vg) shifts the molecular energy levels up and down relative to the Fermi energy levels of the source and drain electrodes, and the current is driven by a finite bias voltage (Vsd) between the two electrodes. Single-molecule conductance is modulated by electrochemical potentials by these means. Reproduced from 71 with permission

Scanning Tunneling Microscopy Studies of Immobilized Biomolecules,
Fig. 3 STM image of an FePP molecule embedded in an ordered array of PP molecules when the substrate was held at −0.15 (A), −0.30 (B), −0.42 (C), −0.55 (D), and −0.65 V (E), respectively. (F)–(J) are the corresponding plots of the cross sections along the *white line* indicated in (A). The data were symmetrized with respect to the center. Reproduced from 68 with permission

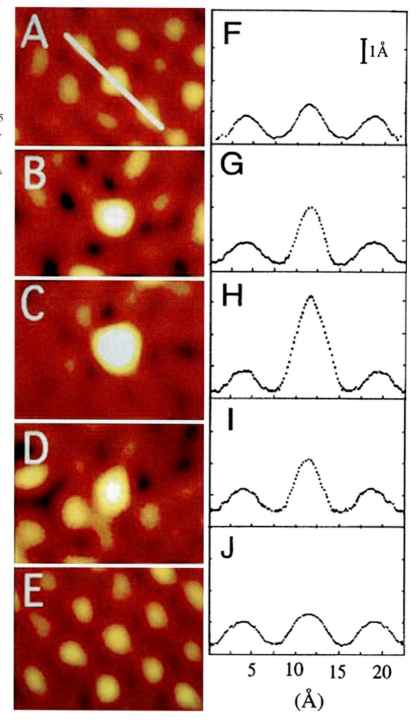

The contrast in ECSTM images was potential dependent. Figure 3 displays a series of images showing single Fe-protoporphyrin molecules at different substrate potentials (Fig. 3a–e), with the corresponding profiles (Fig. 3f–j). The apparent height of the molecule was changing with the potential applied and the maximum contrast was obtained close to the redox level of the molecule.

Comparing these measurements with embedded reference molecules consisting of protoporphyrin molecules without metal ion, it was demonstrated that ET is mediated by the Fe ion in this case. These results demonstrated resonant tunneling as the mechanism for ET in this system [83].

Other interesting electronic effects were evidenced in later single-molecule junction studies. The conductance in the junction can be modulated depending on different factors including the anchoring group [84–86], mechanical interactions [87–89], temperature effects [90], and electrochemical gating [77, 79, 80]. This has been used to demonstrate and characterize different single-molecule devices like switches [91], diodes [92], or transistors [32, 80].

Application to Biomolecules

Scanning probe microscopy and spectroscopy techniques have also been used to determine the conductance of DNA [93–95], peptides [96–98], and proteins [32, 99]. Several experimental approaches have demonstrated the gating behavior shown in Fig. 3 for different redox molecules, including redox proteins [26, 35, 100–103]. The first ECSTM images of azurin were reported in a combined ECSTM and AFM approach [29]. It was demonstrated that ECSTM imaging was potential dependent, and two-step tunneling was suggested as the ET mechanism in azurin [69]. Approaches were based either on direct binding of azurin to a gold electrode [100] or through hydrophobic interactions with alkanethiols SAMs [35]. Spectroscopic-like ECSTM experiments were performed by imaging with different applied potentials. Figure 4 shows a series of ECSTM images of azurin bound to gold through an octanethiol SAM at different substrate potentials. Comparing the images in Fig. 4 A to E, a contrast dependence on potentials again arises. These reports demonstrated the role of the Cu ion in the ET process involving azurin [35, 100]. Figure 4F shows the dependence of the contrast (the apparent height measured by ECSTM) on overpotential (defined as redox potential of the molecule minus substrate potential applied). In analogy to the early Fe-protoporphyrin results, the maximum apparent height for azurin is obtained at zero overpotential (corresponding to the protein redox potential).

Based on these results, theoretical models have been refined [8] and two-step tunneling has been established as the ET mechanism involving azurin [26, 35]. These conclusions are based on ECSTM imaging at different overpotential and bias voltage using ECSTM and are supported by spectroscopic measurements [31, 32, 58] (see below). Besides, azurin has been studied using AFM approaches [30] demonstrating conductance modulation as function of mechanical properties and interesting negative differential resistance features in current–voltage results. However, most AFM studies were performed in air without electrochemical control. Mechanical studies on single-azurin proteins in liquid under electrochemical control should provide more details about the ET process. Overall, the interesting electronic features displayed by azurin suggest that it is a good candidate for building electronic nanodevices and encourage further investigation [32, 58].

Other biomolecules have been imaged using STM [101, 102, 104, 105]. Figure 5a shows an ECSTM image of engineered cytochrome molecules bound to a gold substrate [101]. The interfacial ET process was studied in this protein and led to similar conclusions as azurin [101]. In that study, the cytochrome conductance was estimated (in the order of 1 nS, $10^{-5}G_0$) and in a later report, the STM-BJ approach was used in air to find conductance characteristics of cytochrome mutants, finding an orientation-dependent conductance ranging from 1 to 3.6 $10^{-5}G_0$ [99]. In a recent work, ECSTM was used to study laccase (a multicopper oxidase) immobilized on gold electrodes in electrochemical environment [105].

Figure 5b shows an ECSTM image of hydrogenase molecules immobilized on a gold electrode through a linker SAM [102]. ECSTM imaging combined with a complete electrochemical study was used to estimate the enzyme catalytic turnover rate at the single-molecule level [102].

STM has also been used to study DNA [106–108] and has proven useful to resolve structural and electronic properties [109].

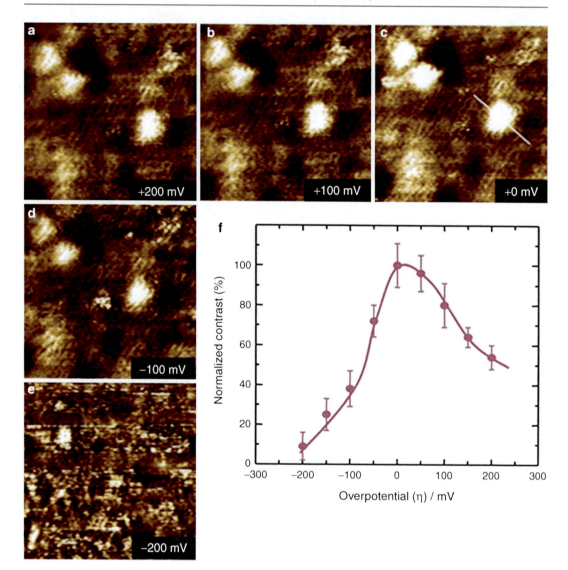

Scanning Tunneling Microscopy Studies of Immobilized Biomolecules, Fig. 4 A series of STM images showing in situ observations of redox-gated electron-tunneling resonance arising from single-azurin molecules. The images were obtained by using the azurin/octanethiol/Au(111) system in NH$_4$Ac buffer (pH 4.6) with a fixed bias voltage (defined as $V_{bias} = E_T - E_S$) of -0.2 V but variable substrate overpotentials (vs. the redox potential of azurin, +100 mV vs. SCE): +200 (A), +100 (B), 0 (C), -100 (D), and -200 mV (E). Scan area is 35 × 35 nm. (F). A correlation between the normalized contrast and the overpotential, showing an asymmetric dependence with a maximum on/off ratio of ≈ 9. Reproduced with permission from 35

Conductance characteristics of DNA bases in solution and in DNA oligomers have been obtained using molecular electronics techniques [95, 110–112]. These methods have been proposed as an alternative approach to DNA sequencing [113] in a setup involving nanopores made using carbon nanotubes [114].

Azurin as a Model System for Spectroscopic Studies and Molecular Electronics Devices

The exponential dependence of ET rate with the donor–acceptor distance (Eq. 2) was first verified in bulk photolysis studies by Gray et al. [12]

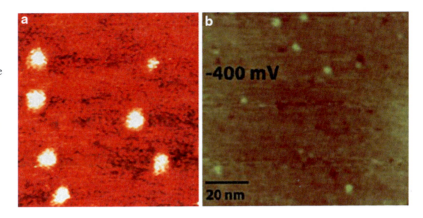

Scanning Tunneling Microscopy Studies of Immobilized Biomolecules, Fig. 5 (a) ECSTM image of cytochrome molecules on a gold substrate. Reproduced with permission From 101. (b) ECSTM image of hydrogenase molecules. Reproduced with permission from 102

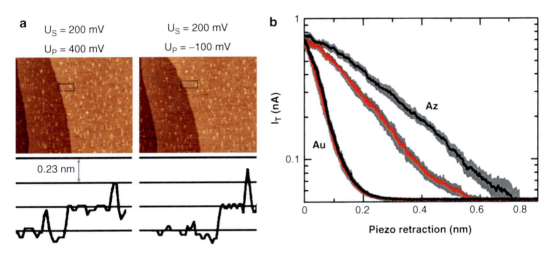

Scanning Tunneling Microscopy Studies of Immobilized Biomolecules, Fig. 6 (a) Azurin proteins on an atomically flat gold surface imaged by ECSTM in 50 mM ammonium acetate solution. Representative images (100 nm × 100 nm and 1 nm colored height scale) obtained at the indicated probe potentials (U_P) and sample potential U_S = 200 mV/SSC (corresponding to oxidized azurin). Squares in the image indicate the region used to measure the height profiles shown below each image. The *horizontal lines* in height profiles are visual guides corresponding to Au < 111 > monatomic steps (0.23 nm high independently of potentials). The average height over N = 30 proteins is 0.22 ± 0.01 nm at U_P = 400 mV and 0.29 ± 0.01 nm at U_P = −100 mV. Reprinted with permission From 31. (b) Current–distance plots on bare gold (labeled "Au") and on azurin-coated gold (labeled "Az") averaged at U_S = 200 mV (*black traces*) and U_S = −300 mV (*red traces*) at constant bias of 200 mV. Reprinted with permission from 31

(Fig. 1b), but its direct demonstration at the single-protein level was only reported recently [31].

Figure 6 shows ECSTM images of a gold sample where azurin is covalently bound to the surface. As discussed in section "Application to Organic Molecules," it was previously reported that the apparent height of azurin molecules studied by ECSTM changes with the potentials applied in the system [26, 35]. The ECSTM images in Fig. 6a show this effect, where profiles corresponding to the same zone of the sample are placed below each image. While the height of monatomic gold step remains constant as function of the tip potential, changes in the apparent height of the protein indicate dependence with the redox activity of the protein.

In order to obtain a deeper insight into ET mechanism in azurin, current–distance measurements under electrochemical control provided the distance decay constants of the ET process [31].

These results were in agreement with previous reports [29] but spanned a broader potential range and conditions. Figure 6b shows current–distance averages of azurin bound to gold at two different potentials (labeled "Az"). They can be compared to the control experiment performed on a clean gold sample (lines labeled 'Au' in Fig. 6b). The behavior of azurin samples is again potential dependent, while results for clean gold remain potential independent. The distance-decay factor obtained for the gold plots is 1 Å^{-1}, as expected for a superexchange tunneling process between two metal atoms [115](see section "Blue Copper Proteins as Model Systems for Biological ET"). In contrast, betas obtained for azurin sample are 0.4 Å^{-1} and 0.5 Å^{-1} for oxidized and reduced azurin, respectively. These distance-decay factors obtained for azurin are below the one corresponding to a single tunneling step. In fact, the azurin plots of Fig. 6b display a similar slope to the discontinuous lines in Fig. 1c (0.5 Å^{-1}, corresponding to calculations of a multistep ET), which indicates that the situation in the ECSTM setup is similar to the one expected in an interprotein ET involving azurin. The ECSTM probe would thus be acting as a partner protein with adjustable electrochemical potential. The conclusion emerging from these results is that ET in azurin occurs by multistep tunneling in the ECSTM configuration and that this mechanism is probably relevant to interprotein ET processes of the protein [31].

As introduced in section "Conductance and Spectroscopic Measurements and the Birth of Molecular Electronics," the field of molecular electronics aims to understand electronic phenomena in individual molecules and use them as the fundamental components of nanoscale devices for electronic applications. To achieve this goal, fundamental questions about the electronic conductance of a single-molecule must be answered first [73, 74], mainly regarding single-molecule conductance and its dependence on different factors (see section "Conductance and Spectroscopic Measurements and the Birth of Molecular Electronics"). Then, in the case of synthetic molecules, they must be designed such that they display the desired electronic properties

to perform specific functions in electronic circuits. On the other hand, a large number and variety of proteins have evolved to carry out many different functions with an optimal performance, and they are excellent candidates to form nanodevices, provided that their electronic performance can be properly characterized. In the previous section, we have introduced powerful methods based on STM and AFM to determine and modulate single-molecule conductance. These methods have been applied to proteins recently [32, 99]. Using ECSTM in a STM-BJ approach [75], azurin immobilized on gold was studied under electrochemical control in buffer solution [32]. Current–distance plots shown in Fig. 6b present current steps that resemble those reported for small organic compounds using the same setup. These steps were absent in control experiments performed on a clean gold surface (compare red traces of Fig. 7b with black traces in the inset of Fig. 7b), and thus they were interpreted as transient formations and ruptures of molecular junctions with azurin bridging the two electrodes (the Au substrate and ECSTM probe, see Fig. 7a). Collecting hundreds of current steps during one experiment allowed building conductance histograms and determining single-azurin conductance using Eq. 3. By performing histograms at different sample potentials, conductance modulation with overpotential ("redox gating") was demonstrated. In Fig. 7c, we find a plot showing conductance of azurin (red circles) as function of electrochemical gate potential. It can be noticed that azurin conductance depends on potentials applied, while the conductance obtained in control experiments on non-redox Zn–azurin junctions (black squares) remains potential independent. This demonstrates again that the Cu center is fundamental for the ET process and that the conductance can be modulated with electrochemical potentials. Such conductance modulation is analogous to the modulation of current in a field effect transistor (see section "Conductance and Spectroscopic Measurements and the Birth of Molecular Electronics" above) and yields an on/off ratio of 20 in conductance modulation (see Fig. 7b). In order to test in situ this behavior on a single azurin, the

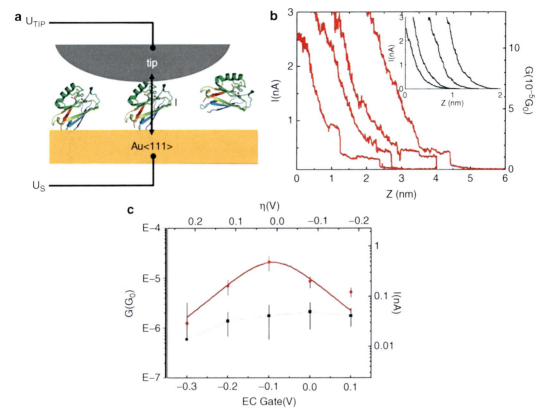

Scanning Tunneling Microscopy Studies of Immobilized Biomolecules, Fig. 7 (a) Scheme showing the ECSTM-BJ approach applied to the study of azurin in electrochemical environment. Reprinted with permission from 32. (b) Raw data examples of I(z) plots exhibiting current plateaus or steps corresponding to single-azurin junction formation. The conductance values are represented in the right axis (G0 = $2e2/h \approx 77.4$ μS). The inset shows curves obtained on a clean gold surface. A horizontal offset was applied in all of them for clarity. Reprinted with permission from 32. (c) Conductance values obtained from the center of the Gaussian fit of the conductance histograms peak as a function of EC gate potential at constant 0.3 V bias for azurin (*red circles*) and non-redox Zn–azurin junctions (*black squares*). Upper axis represents overpotential (η = U_S- U_{AZ}). Error bars indicate the full width half-maximum (fwhm) of the Gaussian fit on each conductance peak. The red plot shows a fit of the numerical version[97] corresponding to the formalism for a two-step electron transfer process[35]. Control measurements with non-redox Zn–Az are indicated with a *dashed black line* as visual guidance. All experiments were performed in 50 mM ammonium acetate solution (pH 4.55). Reprinted with permission from 32

current–time approach introduced in section "Conductance and Spectroscopic Measurements and the Birth of Molecular Electronics" was used [76]. Briefly, having the ECSTM tip at a controlled distance from the surface, the current was recorded in order to detect spontaneous formation of single-azurin junctions. Upon the detection of such single-azurin junctions, the reference electrode potential (acting as gate electrode) was swept. Conductance of a single azurin determined in situ was found to be modulated with the electrochemical gate. These results constituted a proof of concept of a single-wired protein transistor [32]. More recently, the current-voltage characteristics of single azurin were recorded both in tunneling and wired configurations, which yield the lowest transition voltage reported to date (0.4 V) [116].

Conclusions and Future Directions

In summary, STM has enabled single-biomolecule imaging in quasi-physiological conditions and helped elucidating the corresponding ET mechanisms. Many biomolecules can be imaged at single-molecule level using ECSTM, once a suitable strategy for its immobilization on an atomically flat electrode is found. In addition, conductance characteristics of individual biomolecules have demonstrated the proof of principle of biomolecule-based nanodevices. However, certain factors influencing the tunneling gap remain to be understood, and there are several experimental challenges ahead. For example, the role of ionic strength in ET must be assessed in future ECSTM studies and could lead to improvements in the performance of biomolecule-based devices. Furthermore, techniques providing mechanical information at the single-molecule level like AFM should be complemented with full electrochemical control in order to enhance the biological interpretation and relevance of ET studies. Although the basis for single-biomolecule ET studies is well established, more powerful conductance measurements and imaging techniques are needed to quantitatively study ET in multi-center redox proteins and protein complexes, even in live cells. New experimental tools will enable a deeper understanding of ET pathways of complex biomolecules and of cellular bioenergetics. Together with computer simulations, such experimental advances will provide new concepts and applications of nanobiodevices.

Cross-References

▶ AFM Studies of Biomolecules
▶ Biomolecules in Electric Fields
▶ Biosensors, Electrochemical
▶ Direct Electron Transfer to Enzymes
▶ Electrodeposition of Electronic Materials for Applications in Macroelectronic- and Nanotechnology-Based Devices
▶ Enzymatic Electrochemical Biosensors
▶ Interfaces Modified with Electroactive Biological Species
▶ Micro- and Nanoelectrodes
▶ Protein Engineering for Electrochemical Applications

▶ Reconstituted Redox Proteins on Surfaces for Bioelectronic Applications
▶ Scanning Electrochemical Microscopy (SECM)
▶ Scanning Probe Methods
▶ Voltammetry of Adsorbed Proteins

References

1. Marcus RA, Sutin N (1985) Electron transfers in chemistry and biology. Biochim Biophys Acta 811:265–322
2. Gray HB, Winkler JR (2010) Electron flow through metalloproteins. Biochim Biophys Acta - Bioener 1797:1563–1572
3. Farver O, Pecht I (2011) Electron transfer in blue copper proteins. Coordin Chem Rev 255:757–773
4. Gray HB, Winkler JR (2003) Electron tunneling through proteins. Q Rev Biophys 36:341–372
5. Reece SY, Nocera DG (2009) Proton-coupled electron transfer in biology: results from synergistic studies in natural and model systems. Ann Rev Biochem 78:673–699
6. Zhang JD, Kuznetsov AM, Medvedev IG, Chi QJ, Albrecht T, Jensen PS, Ulstrup J (2008) Single-molecule electron transfer in electrochemical environments. Chem Rev 108:2737–2791
7. Kuznetsov AM, Ulstrup J (2010) Proton and proton-coupled electron transfer with paradigms towards single-molecule systems. J Phys Org Chem 23:647–659
8. Kuznetsov AM, Ulstrup J (2000) Theory of electron transfer at electrified interfaces. Electrochim Acta 45:2339–2361
9. Page CC, Moser CC, Chen X, Dutton PL (1999) Natural engineering principles of electron tunnelling in biological oxidation-reduction. Nature 402:47–52
10. Vuilleumier R, Tay K, Jeanmairet G, Borgis D, Boutin A (2012) Extension of Marcus picture for electron transfer reactions with large solvation changes. J Am Chem Soc 134:2067–2074
11. Beratan DN, Betts JN, Onuchic JN (1991) Protein electron-transfer rates set by the bridging secondary and tertiary structure. Science 252:1285–1288
12. Beratan DN, Onuchic JN, Winkler JR, Gray HB (1992) Electron-tunneling pathways in proteins. Science 258:1740–1741
13. Beratan DN, Skourtis SS, Balabin IA, Balaeff A, Keinan S, Venkatramani R, Xiao D (2009) Steering electrons on moving pathways. Accounts Chem Res 42:1669–1678
14. Langen R, Chang IJ, Germanas JP, Richards JH, Winkler JR, Gray HB (1995) Electron-tunneling in proteins - coupling through a beta-strand. Science 268:1733–1735

15. Winkler JR, Di Bilio AJ, Farrow NA, Richards JH, Gray HB (1999) Electron tunneling in biological molecules. Pure App Chem 71:1753–1764
16. Prytkova TR, Kurnikov IV, Beratan DN (2007) Coupling coherence distinguishes structure sensitivity in protein electron transfer. Science 315:622–625
17. Engel GS, Calhoun TR, Read EL, Ahn T-K, Mancal T, Cheng Y-C, Blankenship RE, Fleming GR (2007) Evidence for wavelike energy transfer through quantum coherence in photosynthetic systems. Nature 446:782–786
18. Lee H, Cheng Y-C, Fleming GR (2007) Coherence dynamics in photosynthesis: protein protection of excitonic coherence. Science 316:1462–1465
19. Panitchayangkoon G, Hayes D, Fransted KA, Caram JR, Harel E, Wen J, Blankenship RE, Engel GS (2010) Long-lived quantum coherence in photosynthetic complexes at physiological temperature. Proc Natl Acad Sci USA 107:12766–12770
20. Collini E, Wong CY, Wilk KE, Curmi PMG, Brumer P, Scholes GD (2010) Coherently wired light-harvesting in photosynthetic marine algae at ambient temperature. Nature 463:644–647
21. Nick EM (2011) Quantum coherence in (Brain) microtubules and efficient energy and information transport. J Phys Conf Ser 329:012026
22. Gauger EM, Rieper E, Morton JJL, Benjamin SC, Vedral V (2011) Sustained quantum coherence and entanglement in the avian compass. Phys Rev Lett 106:040503
23. Solomon EI, Sundaram UM, Machonkin TE (1996) Multicopper oxidases and oxygenases. Chem Rev 96:2563–2605
24. Lowery MD, Solomon EI (1992) Axial ligand bonding in blue copper proteins. Inorg Chim Acta 198:233–243
25. Solomon EI, Hare JW, Dooley DM, Dawson JH, Stephens PJ, Gray HB (1980) Spectroscopic studies of stellacyanin, plastocyanin, and azurin - electronic-structure of the blue copper sites. J Am Chem Soc 102:168–178
26. Alessandrini A, Corni S, Facci P (2006) Unravelling single metalloprotein electron transfer by scanning probe techniques. Phys Chem Chem Phys 8:4383–4397
27. Adman ET, Jensen LH (1981) Structural features of azurin at 2.7 a-resolution. Isr J Chem 21:8–12
28. Chi QJ, Zhang JD, Friis EP, Andersen JET, Ulstrup J (1999) Electrochemistry of self-assembled monolayers of the blue copper protein Pseudomonas aeruginosa azurin on Au(111). Electrochem Comm 1:91–96
29. Friis EP, Andersen JET, Madsen LL, Moller P, Ulstrup J (1997) In situ STM and AFM of the copper protein Pseudomonas aeruginosa azurin. J Electroanal Chem 431:35–38
30. Davis JJ, Wang N, Morgan A, Tiantian Z, Jianwei Z (2005) Metalloprotein tunnel junctions: compressional modulation of barrier height and transport mechanism. Faraday Discuss 131:167–179
31. Artes JM, Diez-Perez I, Sanz F, Gorostiza P (2011) Direct measurement of electron transfer distance decay constants of single redox proteins by electrochemical tunneling spectroscopy. ACS Nano 5:2060–2066
32. Artés JM, Díez-Pérez I, Gorostiza P (2011) Transistor-like behavior of single metalloprotein junctions. Nano Lett. doi:10.1021/nl2028969
33. Wenger OS, Leigh BS, Villahermosa RM, Gray HB, Winkler JR (2005) Electron tunneling through organic molecules in frozen glasses. Science 307:99–102
34. Zhang J, Welinder AC, Chi Q, Ulstrup J (2011) Electrochemically controlled self-assembled monolayers characterized with molecular and submolecular resolution. Phys Chem Chem Phys 13:5526–5545
35. Chi QJ, Farver O, Ulstrup J (2005) Long-range protein electron transfer observed at the single-molecule level: In situ mapping of redox-gated tunneling resonance. Proc Natl Acad Sci USA 102:16203–16208
36. Chi Q, Zhang J, Jensen PS, Christensen HEM, Ulstrup J (2005) Long-range interfacial electron transfer of metalloproteins based on molecular wiring assemblies. Faraday Discuss 131:181–195
37. Fujita K, Nakamura N, Ohno H, Leigh BS, Niki K, Gray HB, Richards JH (2004) Mimicking protein-protein electron transfer: voltammetry of Pseudomonas aeruginosa azurin and the thermus thermophilus Cu-A domain at omega-derivatized self-assembled-monolayer gold electrodes. J Am Chem Soc 126:13954–13961
38. Khoshtariya DE, Dolidze TD, Shushanyan M, Davis KL, Waldeck DH, van Eldik R (2010) Fundamental signatures of short- and long-range electron transfer for the blue copper protein azurin at Au/SAM junctions. Proc Natl Acad Sci USA 107:2757–2762
39. Salverda JM, Patil AV, Mizzon G, Kuznetsova S, Zauner G, Akkilic N, Canters GW, Davis JJ, Heering HA, Aartsma TJ (2010) Fluorescent cyclic voltammetry of immobilized azurin: direct observation of thermodynamic and kinetic heterogeneity. Angew Chem 122:5912–5915
40. Patil AV, Davis JJ (2010) Visualizing and tuning thermodynamic dispersion in metalloprotein monolayers. J Am Chem Soc 132:16938–16944
41. Monari S, Battistuzzi G, Dennison C, Borsari M, Ranieri A, Siwek MJ, Sola M (2010) Factors affecting the electron transfer properties of an immobilized cupredoxin. J Phys Chem C 114:22322–22329
42. Ron I, Pecht I, Sheves M, Cahen D (2010) Proteins as solid-state electronic conductors. Accounts Chem Res 43:945–953
43. Ron I, Sepunaru L, Itzhakov S, Belenkova T, Friedman N, Pecht I, Sheves M, Cahen D (2010) Proteins as electronic materials: electron transport through

solid-state protein monolayer junctions. J Am Chem Soc 132:4131–4140

44. Sepunaru L, Pecht I, Sheves M, Cahen D (2011) Solid-state electron transport across azurin: from a temperature-independent to a temperature-activated mechanism. J Am Chem Soc 133:2421–2423

45. Mentovich ED, Belgorodsky B, Richter S (2011) Resolving the mystery of the elusive peak: negative differential resistance in redox proteins. J Phys Chem Lett 2:1125–1128

46. Maruccio G, Marzo P, Krahne R, Passaseo A, Cingolani R, Rinaldi R (2007) Protein conduction and negative differential resistance in large-scale nanojunction arrays. Small 3:1184–1188

47. Sepunaru L, Friedman N, Pecht I, Sheves M, Cahen D (2012) Temperature-dependent solid-state electron transport through bacteriorhodopsin: experimental evidence for multiple transport paths through proteins. J Am Chem Soc. doi:10.1021/ja2097139

48. Bennati M, Robblee JH, Mugnaini V, Stubbe J, Freed JH, Borbat P (2005) EPR distance measurements support a model for long-range radical initiation in E. coli ribonucleotide reductase. J Am Chem Soc 127:15014–15015

49. Gray HB, Winkler JR (2005) Long-range electron transfer. Proc Natl Acad Sci USA 102:3534–3539

50. Lin J, Balabin IA, Beratan DN (2005) The nature of aqueous tunneling pathways between electron-transfer proteins. Science 310:1311–1313

51. Shih C, Museth AK, Abrahamsson M, Blanco-Rodriguez AM, Di Bilio AJ, Sudhamsu J, Crane BR, Ronayne KL, Towrie M, Vlcek A, Richards JH, Winkler JR, Gray HB (2008) Tryptophan-accelerated electron flow through proteins. Science 320:1760–1762

52. Lancaster KM, George SD, Yokoyama K, Richards JH, Gray HB (2009) Type-zero copper proteins. Nat Chem 1:711–715

53. Marshall NM, Garner DK, Wilson TD, Gao Y-G, Robinson H, Nilges MJ, Lu Y (2009) Rationally tuning the reduction potential of a single cupredoxin beyond the natural range. Nature 462:113–116

54. Lancaster KM, Farver O, Wherland S, Crane EJ, Richards JH, Pecht I, Gray HB (2011) Electron transfer reactivity of type zero *Pseudomonas aeruginosa* azurin. J Am Chem Soc 133:4865–4873

55. Lee T, Kim S-U, Min J, Choi J-W (2010) Multilevel biomemory device consisting of recombinant azurin/cytochrome c. Adv Mater 22:510

56. Kim SU, Yagati AK, Min JH, Choi JW (2010) Nanoscale protein-based memory device composed of recombinant azurin. Biomaterials 31:1293–1298

57. Yau ST, Qian G (2005) A prototype protein field-effect transistor. Appl Phys Lett 86:103508–103511

58. Alessandrini A, Salerno M, Frabboni S, Facci P (2005) Single-metalloprotein wet biotransistor. Appl Phys Lett 86:133902–133905

59. Maruccio G, Biasco A, Visconti P, Bramanti A, Pompa PP, Calabi F, Cingolani R, Rinaldi R, Corni S, Di Felice R, Molinari E, Verbeet MP, Canters GW (2005) Towards protein field-effect transistors: report and model of a prototype. Adv Mater 17:816–822

60. Winkler JR, Nocera DG, Yocom KM, Bordignon E, Gray HB (1982) Electron-transfer kinetics of pentaammineruthenium(III)(histidine-33)-ferricytochrome c. Measurement of the rate of intramolecular electron transfer between redox centers separated by 15.ANG. in a protein. J Am Chem Soc 104:5798–5800

61. Bortolotti CA, Siwko ME, Castellini E, Ranieri A, Sola M, Corni S (2011) The reorganization energy in cytochrome c is controlled by the accessibility of the heme to the solvent. J Phys Chem Lett 2:1761–1765

62. Kelley SO, Barton JK (1999) Electron transfer between bases in double helical DNA. Science 283:375–381

63. Barton JK, Olmon ED, Sontz PA (2011) Metal complexes for DNA-mediated charge transport. Coordin Chem Rev 255:619–634

64. Venkatramani R, Keinan S, Balaeff A, Beratan DN (2011) Nucleic acid charge transfer: black, white and gray. Coordin Chem Rev 255:635–648

65. Lindsay S (2009) Introduction to nanoscience. Oxford University Press, New York

66. Binnig G, Rohrer H (1982) Scanning tunneling microscopy. Helv Phys Acta 55:726–735

67. Guell AG, Diez-Perez I, Gorostiza P, Sanz F (2004) Preparation of reliable probes for electrochemical tunneling spectroscopy. Anal Chem 76:5218–5222

68. Tao NJ (1996) Probing potential-tuned resonant tunneling through redox molecules with scanning tunneling microscopy. Phys Rev Lett 76:4066–4069

69. Friis EP, Andersen JET, Kharkats YI, Kuznetsov AM, Nichols RJ, Zhang JD, Ulstrup J (1999) An approach to long-range electron transfer mechanisms in metalloproteins: in situ scanning tunneling microscopy with submolecular resolution. Proc Natl Acad Sci USA 96:1379–1384

70. Diez-Perez I, Guell AG, Sanz F, Gorostiza P (2006) Conductance maps by electrochemical tunneling spectroscopy to fingerprint the electrode electronic structure. Anal Chem 78:7325–7329

71. Tao NJ (2006) Electron transport in molecular junctions. Nat Nanotechnol 1:173–181

72. Aviram A, Ratner MA (1974) Molecular rectifiers. Chem Phys Lett 29:277–283

73. Chen F, Hihath J, Huang ZF, Li XL, Tao NJ (2007) Measurement of single-molecule conductance. Ann Rev Phys Chem 58:535–564

74. Nichols RJ, Haiss W, Higgins SJ, Leary E, Martin S, Bethell D (2010) The experimental determination of the conductance of single molecules. Phys Chem Chem Phys 12:2801–2815

75. Xu BQ, Tao NJJ (2003) Measurement of single-molecule resistance by repeated formation of molecular junctions. Science 301:1221–1223
76. Haiss W, Nichols RJ, van Zalinge H, Higgins SJ, Bethell D, Schiffrin DJ (2004) Measurement of single molecule conductivity using the spontaneous formation of molecular wires. Phys Chem Chem Phys 6:4330–4337
77. Li X, Xu B, Xiao X, Yang X, Zang L, Tao N (2006) Controlling charge transport in single molecules using electrochemical gate. Faraday Discuss 131:111–120
78. Pobelov IV, Li ZH, Wandlowski T (2008) Electrolyte gating in redox-active tunneling junctions-an electrochemical STM approach. J Am Chem Soc 130:16045–16054
79. Xu BQQ, Li XLL, Xiao XYY, Sakaguchi H, Tao NJJ (2005) Electromechanical and conductance switching properties of single oligothiophene molecules. Nano Lett 5:1491–1495
80. Albrecht T, Guckian A, Ulstrup J, Vos JG (2005) Transistor-like behavior of transition metal complexes. Nano Lett 5:1451–1455
81. Hong W, Manrique DZ, Moreno-Garcia P, Gulcur M, Mishchenko A, Lambert CJ, Bryce MR, Wandlowski T (2011) Single molecular conductance of tolanes: an experimental and theoretical study on the junction evolution in dependence on the anchoring group. J Am Chem Soc 134:2292–2304
82. Visoly-Fisher I, Daie K, Terazono Y, Herrero C, Fungo F, Otero L, Durantini E, Silber JJ, Sereno L, Gust D, Moore TA, Moore AL, Lindsay SM (2006) Conductance of a biomolecular wire. Proc Natl Acad Sci USA 103:8686–8690
83. Schmickler W, Tao N (1997) Measuring the inverted region of an electron transfer reaction with a scanning tunneling microscope. Electrochim Acta 42:2809–2815
84. Cheng ZL, Skouta R, Vazquez H, Widawsky JR, Schneebeli S, Chen W, Hybertsen MS, Breslow R, Venkataraman L (2011) In situ formation of highly conducting covalent Au-C contacts for single-molecule junctions. Nat Nanotechnol 6:353–357
85. Kamenetska M, Koentopp M, Whalley AC, Park YS, Steigerwald ML, Nuckolls C, Hybertsen MS, Venkataraman L (2009) Formation and evolution of single-molecule junctions. Phys Rev Lett 102:126803–126806
86. Venkataraman L, Park YS, Whalley AC, Nuckolls C, Hybertsen MS, Steigerwald ML (2007) Electronics and chemistry: varying single-molecule junction conductance using chemical substituents. Nano Lett 7:502–506
87. Joachim C, Gimzewski JK, Aviram A (2000) Electronics using hybrid-molecular and mono-molecular devices. Nature 408:541–548
88. Diez-Perez I, Hihath J, Hines T, Wang Z-S, Zhou G, Mullen K, Tao N (2011) Controlling single-molecule conductance through lateral coupling of [pi] orbitals. Nat Nanotechnol 6:226–231
89. Bruot C, Hihath J, Tao N (2011) Mechanically controlled molecular orbital alignment in single molecule junctions. Nat Nanotechnol 7:35–40
90. Hines T, Diez-Perez I, Hihath J, Liu H, Wang Z-S, Zhao J, Zhou G, Muellen K, Tao N (2010) Transition from tunneling to hopping in single molecular junctions by measuring length and temperature dependence. J Am Chem Soc 132:11658–11664
91. Auwarter W, Seufert K, Bischoff F, Ecija D, Vijayaraghavan S, Joshi S, Klappenberger F, Samudrala N, Barth JV (2011) A surface-anchored molecular four-level conductance switch based on single proton transfer. Nat Nanotechnol 7:41–46
92. Diez-Perez I, Hihath J, Lee Y, Yu LP, Adamska L, Kozhushner MA, Oleynik II, Tao NJ (2009) Rectification and stability of a single molecular diode with controlled orientation. Nature Chem 1:635–641
93. Xu BQ, Zhang PM, Li XL, Tao NJ (2004) Direct conductance measurement of single DNA molecules in aqueous solution. Nano Lett 4:1105–1108
94. Hihath J, Xu BQ, Zhang PM, Tao NJ (2005) Study of single-nucleotide polymorphisms by means of electrical conductance measurements. Proc Natl Acad Sci USA 102:16979–16983
95. Chang S, Huang S, He J, Liang F, Zhang P, Li S, Chen X, Sankey O, Lindsay S (2010) Electronic signatures of all four DNA nucleosides in a tunneling gap. Nano Lett 10:1070–1075
96. Cardamone DM, Kirczenow G (2010) Electrochemically gated oligopeptide nanowires bridging gold electrodes: novel bio-nanoelectronic switches operating in aqueous electrolytic environments. Nano Lett 10:1158–1162
97. Xiao XY, Xu BQ, Tao NJ (2004) Conductance titration of single-peptide molecules. J Am Chem Soc 126:5370–5371
98. Scullion L, Doneux T, Bouffier L, Fernig DG, Higgins SJ, Bethell D, Nichols RJ (2011) Large conductance changes in peptide single molecule junctions controlled by pH. J Phys Chem C 115:8361–8368
99. Della Pia EA, Elliott M, Jones DD, Macdonald JE (2011) Orientation-dependent electron transport in a single redox protein. ACS Nano 6:355–361
100. Alessandrini A, Gerunda M, Canters GW, Verbeet MP, Facci P (2003) Electron tunnelling through azurin is mediated by the active site Cu ion. Chem Phys Lett 376:625–630
101. Della Pia EA, Chi QJ, Jones DD, Macdonald JE, Ulstrup J, Elliott M (2011) Single-molecule mapping of long-range electron transport for a cytochrome b(562) variant. Nano Lett 11:176–182
102. Madden C, Vaughn MD, Díez-Pérez I, Brown KA, King PW, Gust D, Moore AL, Moore TA (2011) Catalytic turnover of [FeFe]-hydrogenase based on single-molecule imaging. J Am Chem Soc 143:1577–1582

103. Davis JJ, Hill HAO, Bond AM (2000) The application of electrochemical scanning probe microscopy to the interpretation of metalloprotein voltammetry. Coordin Chem Rev 200–202:411–442
104. Rakshit T, Banerjee S, Mukhopadhyay R (2010) Near-metallic behavior of warm holoferritin molecules on a Gold(111) surface. Langmuir 26: 16005–16012
105. Climent V, Zhang J, Friis EP, Østergaard LH, Ulstrup J (2011) Voltammetry and single-molecule in situ scanning tunneling microscopy of laccases and bilirubin oxidase in electrocatalytic dioxygen reduction on Au(111) single-crystal electrodes. J Phys Chem C 116:1232–1243
106. Lindsay SM, Thundat T, Nagahara L (1988) Adsorbate deformation as a contrast mechanism in stm images of bio-polymers in an aqueous environment - images of the unstained, hydrated dna double helix. J Microsc-Oxford 152:213–220
107. Nagahara LA, Lindsay SM, Barris BJ, Thundat TG, Knipping U (1988) Imaging bio-polymers using scanning tunneling microscopy in solution. Biophy J 53:A396–A396
108. Lindsay SM, Thundat T, Nagahara L, Knipping U, Rill RL (1989) Images of the DNA double helix in water. Science 244:1063–1064
109. Shapir E, Cohen H, Calzolari A, Cavazzoni C, Ryndyk DA, Cuniberti G, Kotlyar A, Di Felice R, Porath D (2008) Electronic structure of single DNA molecules resolved by transverse scanning tunnelling spectroscopy. Nat Mater 7:68–74
110. He J, Lin L, Zhang P, Lindsay S (2007) Identification of DNA basepairing via tunnel-current decay. Nano Lett 7:3854–3858
111. Chang S, He J, Kibel A, Lee M, Sankey O, Zhang P, Lindsay S (2009) Tunnelling readout of hydrogen-bonding-based recognition. Nat Nanotechnol 4:297–301
112. Huang S, He J, Chang S, Zhang P, Liang F, Li S, Tuchband M, Fuhrmann A, Ros R, Lindsay S (2010) Identifying single bases in a DNA oligomer with electron tunnelling. Nat Nanotechnol 5: 868–873
113. Lindsay S, He J, Sankey O, Hapala P, Jelinek P, Zhang P, Chang S, Huang S (2010) Recognition tunneling. Nanotechnology 21
114. Liu H, He J, Tang J, Liu H, Pang P, Cao D, Krstic P, Joseph S, Lindsay S, Nuckolls C (2010) Translocation of single-stranded DNA through single-walled carbon nanotubes. Science 327:64–67
115. Nagy G, Wandlowski T (2003) Double layer properties of Au(111)/H2SO4 (Cl) + Cu2+ from distance tunneling spectroscopy. Langmuir 19: 10271–10280
116. Artés JM, López-Martínez M, Giraudet A, Díez-Pérez I, Sanz F, Gorostiza P (2012) Current-voltage characteristics and transition voltage spectroscopy of individual redox proteins. J Am Chem Soc 134:20218–20221

Selective Electrochemical Fluorination

Toshio Fuchigami
Department of Electrochemistry,
Tokyo Institute of Technology, Midori-ku,
Yokohama, Japan

Introduction

Organofluorine compounds are classified into two groups: perfluoro compounds and partially fluorinated compounds. The compounds in the former class are widely utilized as functional materials, while those in the latter family find biological uses as pharmaceuticals and agrochemicals. Perfluoro compounds are manufactured by converting all C-H bonds to C-F bonds using electrochemical fluorination in anhydrous liquid HF as a solvent with a nickel anode (see electrochemical perfluorination).

In contrast, selective electrochemical fluorination can be commonly achieved in aprotic solvents (acetonitrile, dichloromethane, dimethoxyethane, nitromethane, sulfolane, etc.) containing fluoride ions to provide mostly mon- and/or difluorinated products [1–4]. Electrolyses are conducted at constant potentials slightly higher than the first oxidation potential of the substrate by using a platinum or graphite anode. Constant current electrolysis is also effective for selective fluorination in many cases. Choice of the combination of a supporting fluoride salt and an electrolytic solvent is most important to accomplish efficient selective fluorination because competitive anode passivation (the formation of a nonconducting polymer film on the anode surface that suppresses faradaic current) takes place very often during the electrolysis. Pulse electrolysis is in many cases effective in order to avoid such passivation. Therefore, difficult to oxidize fluoride salts, which do not cause the passivation of the anode and have strongly nucleophilic F-, are generally recommended as the supporting fluoride salts. Thus, room temperature molten salts such as R_3N-nHF ($n = 3$–5),

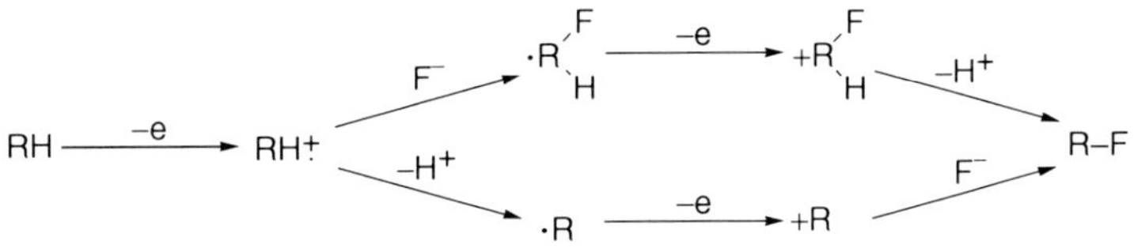

Selective Electrochemical Fluorination 1869

Selective Electrochemical Fluorination, Scheme 1 General pathway for electrochemical fluorination

Selective Electrochemical Fluorination, Scheme 2 Electrochemical fluorination of naphthalene

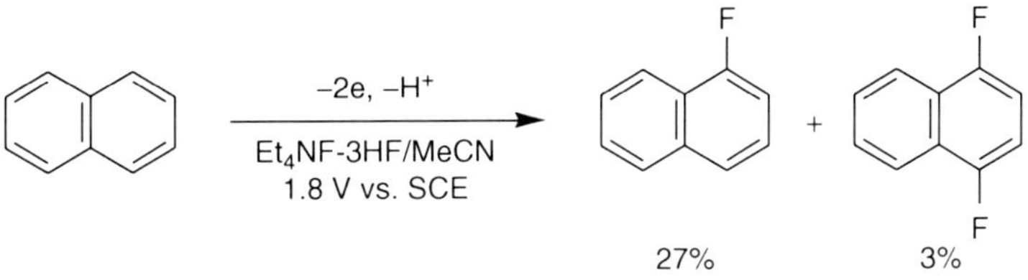

$-2e, -H^+$
$Et_4NF-3HF/MeCN$
1.8 V vs. SCE

27% 3%

$R_4NF\text{-}nHF$ ($n = 3$–5), and pyridine polyhydrogen fluoride salt (Py-nHF) are most often used, and even R_4NBF_4 and R_4NPF_6 salts are effective in some cases [1–4]. Particularly when HF supporting salts and low hydrogen overvoltage cathodes such as platinum are used, the reduction of protons (hydrogen evolution) occurs predominantly at the cathode during the electrolysis. Therefore, a divided cell is not always necessary for the fluorination under such conditions.

In aprotic solvent, F^- becomes more nucleophilic; however, the reactivity of F^- is quite sensitive to the water content of the electrolysis system because a hydrated F^- is a weak nucleophile. Drying of both the solvent and electrolyte is therefore necessary to optimize the formation of fluorinated products.

Since the discharge potential of the fluoride ion is extremely high ($> + 2.9$ V versus SCE at Pt anode in MeCN), the fluorination proceeds via a (radical) cation intermediate as shown in Scheme 1, which is the general pathway for anodic nucleophilic substitutions.

Selective Electrochemical Fluorination in Organic Solvents

Some typical examples are shown below, although various examples have been reported so far in the literature.

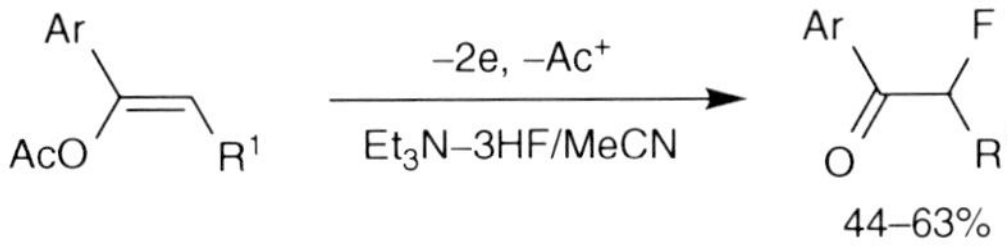

$-2e, -Ac^+$
$Et_3N-3HF/MeCN$
44–63%

Selective Electrochemical Fluorination, Scheme 3 Electrochemical fluorination of olefin

Aromatic compounds such as benzene, substituted benzenes, and naphthalene are selectively fluorinated by constant potential anodic oxidation in Et_4N-3HF/MeCN [5, 6] (Scheme 2).

Electrochemical fluorination of olefins provides mono- and/or difluorinated products. In the cases of α-acetoxystyrene and 1-acetoxy-3,4-dihydronaphthalene, the corresponding α-fluoroketones are formed as shown in Scheme 3 [7].

Although anodic benzylic substitution reactions take place readily, anodic benzylic fluorination does not always occur. The major competitive reaction is acetamidation when MeCN is used as a solvent. For example, electrochemical benzylic fluorination in MeCN proceeded selectively when the benzylic position was substituted by an electron-donating group, as shown in Scheme 4 [8]. In contrast, α-acetoamidation became the major reaction when the phenyl group had no electron-donating substituent.

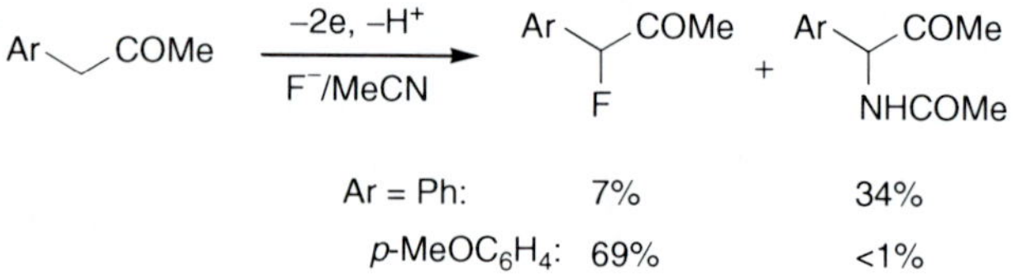

Selective Electrochemical Fluorination, Scheme 4 Electrochemical benzylic fluorination

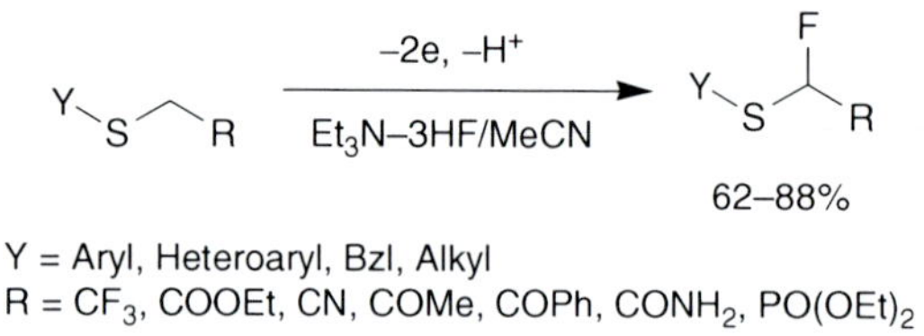

$$Ar \diagdown COMe \xrightarrow[F^-/MeCN]{-2e, -H^+} Ar \diagdown COMe \over F + Ar \diagdown COMe \over NHCOMe$$

Ar = Ph: 7% 34%
p-MeOC$_6$H$_4$: 69% <1%

$$Y \diagdown S \diagdown R \xrightarrow[Et_3N-3HF/MeCN]{-2e, -H^+} Y \diagdown S \diagup R \over F$$

62–88%

Y = Aryl, Heteroaryl, Bzl, Alkyl
R = CF$_3$, COOEt, CN, COMe, COPh, CONH$_2$, PO(OEt)$_2$

Selective Electrochemical Fluorination, Scheme 5 Electrochemical fluorination of various sulfides

Anodic fluorination of sulfides having α-electron-withdrawing groups proceeds quite well to provide the corresponding α-fluorinated products in good yields as shown in Scheme 5 [9, 10]. The fluorination proceeded by way of a Pummerer-type mechanism via the fluorosulfonium cation (**A**), as shown in Scheme 6 [10, 11]. Thus, when R is an electron-withdrawing group, the deprotonation of **A** is significantly facilitated, and consequently, the fluorination proceeds efficiently.

Various sulfur-containing heterocyclic compounds and heterocycles having a phenysulfenyl group are also anodically fluorinated as shown in Schemes 7 and 8 [12, 13]. In these cases, electron transfer is initiated from the sulfur atom. As shown in Scheme 6, the regioselectivity is not thermodynamically controlled but kinetically controlled.

The selectivity of fluorination was strongly influenced by supporting fluoride salts (Scheme 9) [14]. Since Et$_3$N-3HF contains the free base Et$_3$N, the difluorinated product, once formed, was dehydrofluorinated to the monofluoro product.

As mentioned, anode passivation takes place very often, which results in poor yield and low current efficiency. In order to avoid such passivation, various mediators such as halide ions, triarylamines, and iodoarenes can be used as shown in Schemes 10 and 11 [15, 16].

Moreover, ethereal solvents such as 1,2-dimethoxyethane (DME) are much more suitable than MeCN for the anodic fluorination of various heterocyclic sulfides (Scheme 12) [17]. The pronounced solvent effect of DME could be explained in terms of the significant enhanced nucleophilicity of fluoride ions as well as the suppression of anode passivation and overoxidation of fluorinated products.

Electrochemical Fluorination in Room Temperature Poly HF Molten Salt

Solvent-free electrochemical fluorination is an alternative method for preventing anode passivation and acetoamidation [18, 19]. As already mentioned, handling extremely corrosive and poisonous anhydrous HF in a laboratory setting is accompanied by serious hazards and experimental difficulties. Molten salts such as 70 % HF/pyridine (Olah's reagent) and commercially available Et$_3$N-3HF [20] are often used to replace anhydrous HF. Other molten salts with the general formula R$_4$NF-nHF (n > 3.5, R = Me, Et, and n-Pr) are useful in selective electrochemical fluorination. These electrolytes

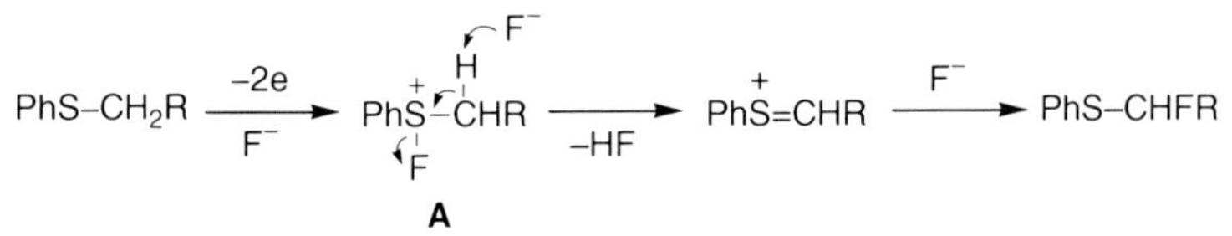

Selective Electrochemical Fluorination

Selective Electrochemical Fluorination, Scheme 6 Reaction mechanism for electrochemical fluorination of sulfides

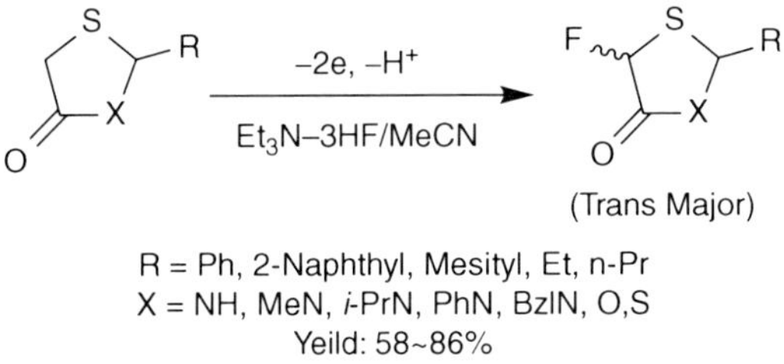

Selective Electrochemical Fluorination, Scheme 7 Electrochemical fluorination of sulfur-containing heterocyclic compounds

R = Ph, 2-Naphthyl, Mesityl, Et, n-Pr
X = NH, MeN, i-PrN, PhN, BzlN, O,S
Yeild: 58~86%

are nonviscous liquids that have high conductivity and anodic stability.

Electrochemical fluorination of various aromatic compounds such as benzene, substituted benzenes' toluene, and quinolines is achieved at high current densities using these molten fluoride salts in the absence of organic solvent with good to high current efficiencies (66–90 %) (Schemes 13 and 14) [21–23].

Selective electrochemical fluorination of cyclic ketones and cyclic unsaturated esters in Et_3N-5HF is also accomplished to provide ring-opening and ring-expansion fluorinated products respectively, as shown in Schemes 15 and 16 [24, 25]. Selective electrochemical formyl hydrogen-exchange fluorination of aliphatic aldehydes affords acyl fluorides using Et_3N-5HF [26].

The fluorination of cyclic ethers, esters, lactones, and cyclic and acyclic carbonates can be achieved by anodic oxidation of a large amount of the liquid substrates and a small amount of Et_4NF-4HF (only 1.5–1.7 equiv. of F^- to the ether) at a high current density (150 mA cm^{-2}) (Schemes 17 and 18) [27].

Electrochemical fluorination of various phenol and α-naphthol derivatives can be achieved in Et_3N-5HF using carbon fiber cloth as an anode to provide 4,4-difluorocyclohexadienone derivatives in good yields (Scheme 19) [28].

Electrochemical fluorination of adamantanes is also possible. Mono-, di-, tri-, and tetrafluoroadamantanes can be selectively prepared from adamantanes by controlling oxidation potentials, and the fluorine atoms are introduced selectively at the tertiary carbons, as shown in Scheme 20 [29]. Adamantanes bearing functional groups such as ester, cyano, and acetoxymethyl moieties were also selectively fluorinated.

Future Directions

Highly efficient and selective fluorination of organic compounds using green sustainable chemistry is one of the most important goals of modern organofluorine chemistry. Selective electrochemical fluorination had been unexplored until about 20 years ago. Great progress has been made in this area and various new electrochemical methodologies have been developed for fluorination using various room temperature molten fluoride salts with and without organic solvents. Particularly, volatile organic solvent-free selective electrochemical fluorination in such molten salts seems to be promising as an environmentally friendly electrochemical process.

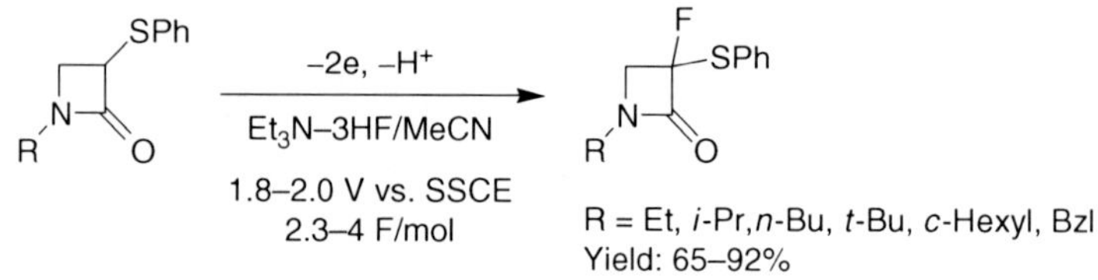

Selective Electrochemical Fluorination, Scheme 8 Electrochemical fluorination of β-lactams

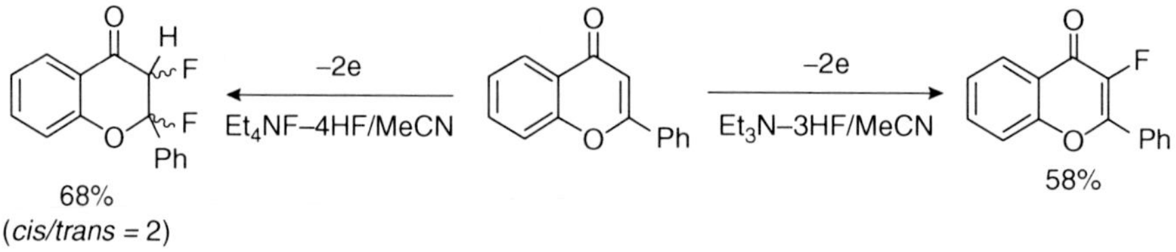

R = Et, *i*-Pr, *n*-Bu, *t*-Bu, *c*-Hexyl, Bzl
Yield: 65–92%

68%
(*cis/trans* = 2)

58%

Selective Electrochemical Fluorination, Scheme 9 Effects of supporting fluoride salts on fluorinated product selectivity

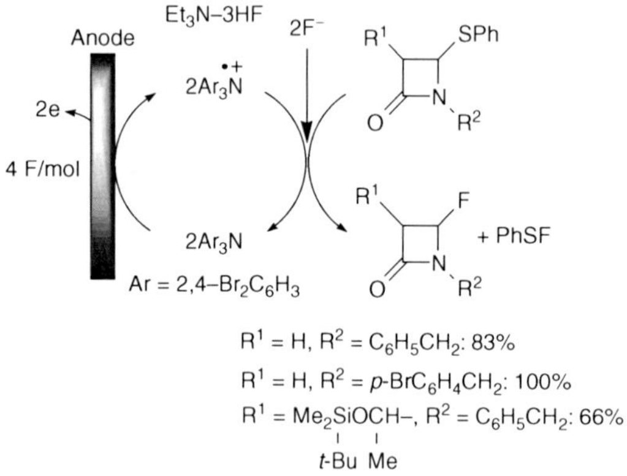

R^1 = H, R^2 = $C_6H_5CH_2$: 83%
R^1 = H, R^2 = *p*-$BrC_6H_4CH_2$: 100%
R^1 = $Me_2SiOCH-$, R^2 = $C_6H_5CH_2$: 66%
      t-Bu Me

Selective Electrochemical Fluorination, Scheme 10 Electrochemical fluorodesulfurization of β-lactams using triarylamine mediator

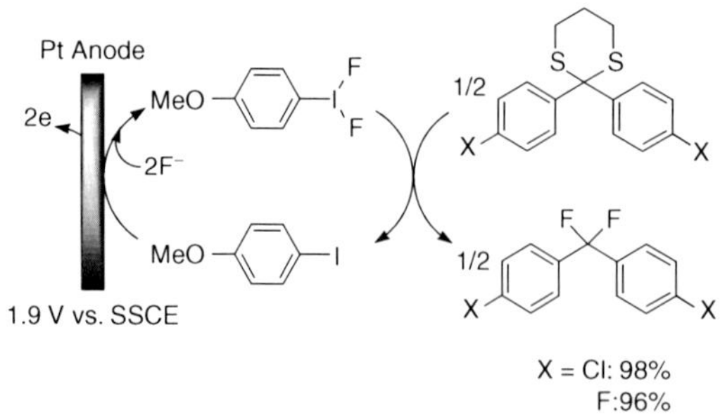

in MeCN: 0%
DME: 92%

Selective Electrochemical Fluorination, Scheme 12 Solvent effect on electrochemical fluorination

X = Cl: 98%
F: 96%

Selective Electrochemical Fluorination, Scheme 11 Electrochemical fluorodesulfurization of dithioacetals using iodoarene mediator

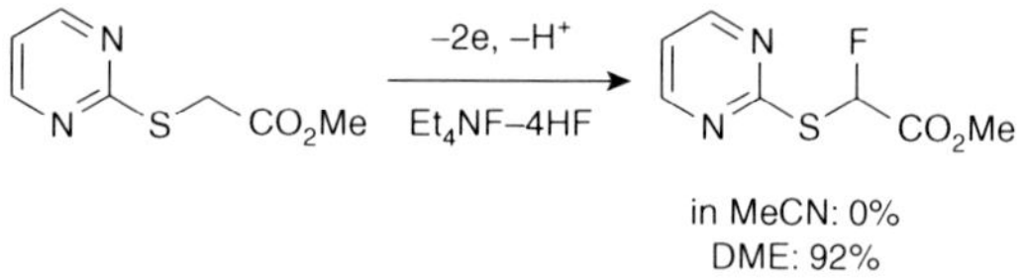

90%

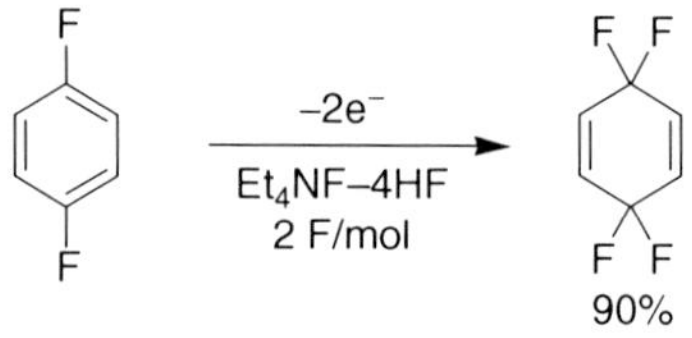

Selective Electrochemical Fluorination, Scheme 13 Solvent-free electrochemical fluorination of *p*-difluorobenzene in poly HF salt

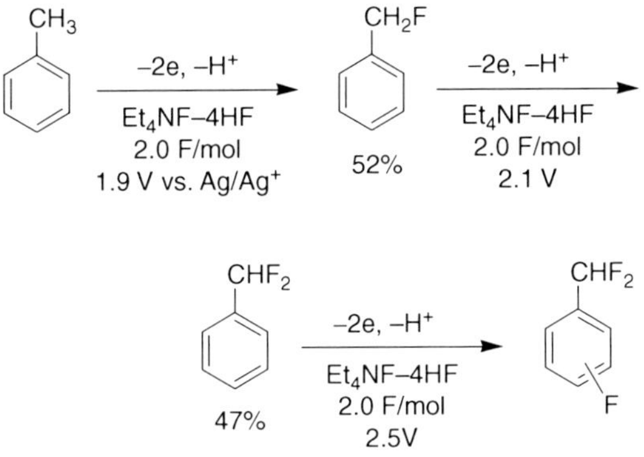

Selective Electrochemical Fluorination

1873

Selective Electrochemical Fluorination, Scheme 14 Solvent-free electrochemical fluorination of toluene in poly HF salt

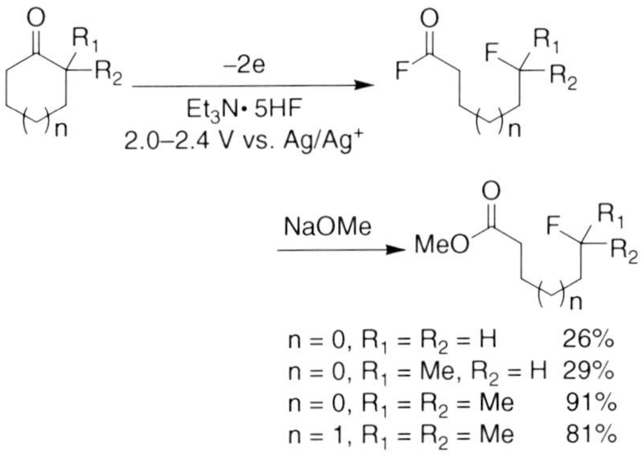

Selective Electrochemical Fluorination, Scheme 15 Solvent-free electrochemical fluorination of cyclic ketones in poly HF salt

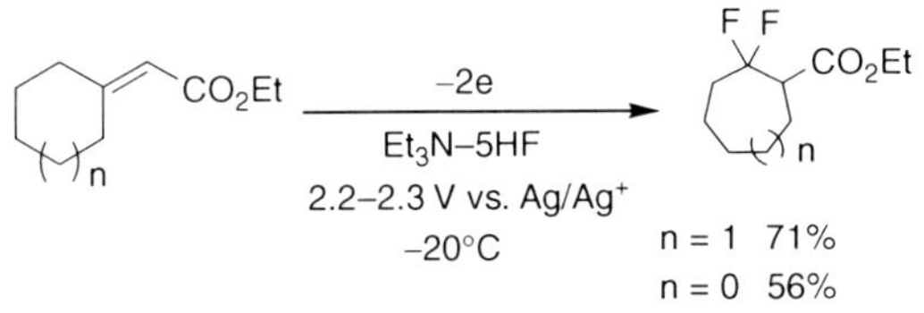

Selective Electrochemical Fluorination, Scheme 16 Solvent-free electrochemical fluorination of cyclic unsaturated esters in poly HF salt

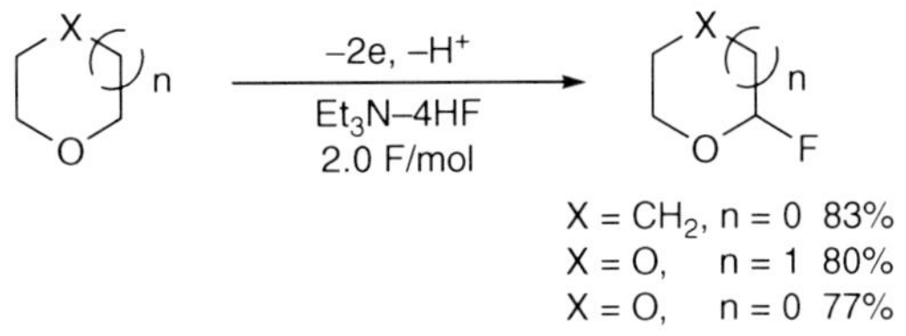

Selective Electrochemical Fluorination, Scheme 17 Solvent-free electrochemical fluorination of cyclic ethers in poly HF molten salt

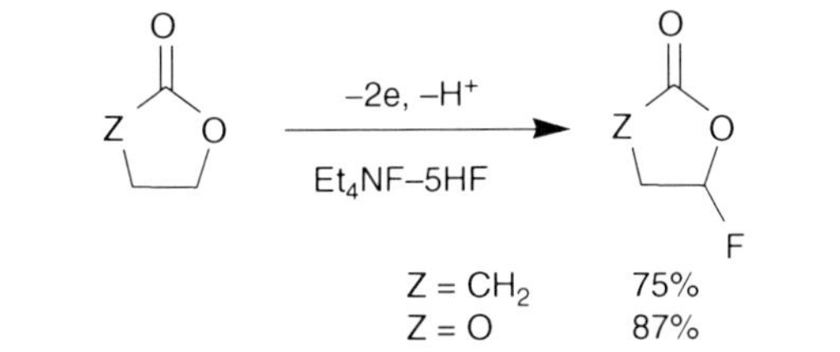

Selective Electrochemical Fluorination, Scheme 18 Solvent-free electrochemical fluorination of lactone and cyclic carbonate in poly HF salt

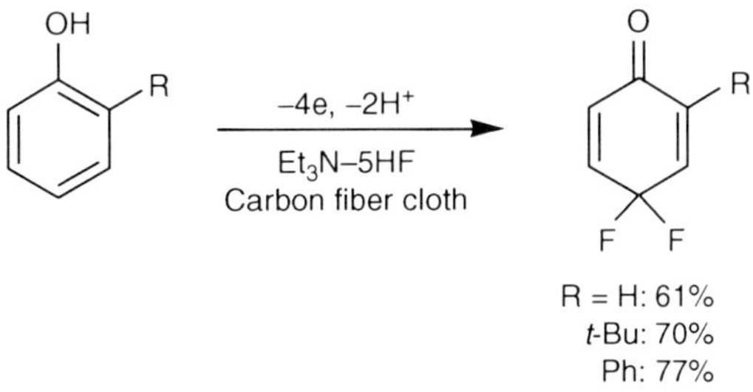

Selective Electrochemical Fluorination, Scheme 19 Solvent-free electrochemical fluorination of phenols in poly HF salt

Selective Electrochemical Fluorination, Scheme 20 Solvent-free electrochemical fluorination of adamantanes in poly HF salt

Cross-References

▶ Anodic Substitutions
▶ Electroauxiliary
▶ Electrosynthesis in Ionic Liquid

References

1. Ronzhkov IN (1983) Anodic fluorination. In: Baizer MM, Lund H (eds) Organic electrochemistry, 2nd edn. Dekker, New York
2. Childs WV, Christensen L, Klink FW, Kolpin CF (1990) Anodic fluorination. In: Lund H, Baizer MM (eds) Organic electrochemistry, 3rd edn. Dekker, New York
3. Fuchigami T (2000) Electrochemical partial fluorination. In: Lund H, Hammerich O (eds) Organic electrochemistry, 4th edn. Dekker, New York
4. Fuchigami T (1999) Electrochemistry applied to the synthsis of fluorinated organic substances. In: Mariano PS (ed) Advances in electron transfer chemistry, vol 6. JAI Press, Greenwich CT
5. Knunyants IL,Rozhkov IN, Bukhtiarov AV, Goldin MM, Kudryavtseu RV (1970) New method for introducing fluorine into the aromatic ring. Izv Akad Nauk SSSR, Ser Khim 1207–1208.
6. Rozhkov IN (1976) Radical-cation mechanism of the anodic fluorination of organic compounds. Russ Chem Rev 45:615–629
7. Laurent EG, Tardivel R, Thiebault H (1983) Un nouveau moded'accès aux α-fluorocétones : l'oxydation anodique des acétates d'énol en présence d'ions fluorures. Tetrahedron Lett 24:903–906
8. Laurent EG, Marquet B, Tardivel R, Thiebault H (1987) Nouvelle methode de preparation de cetones, ester et nitrile benzyliques α-fluoride OU α, α-difluores. Tetrahedron Lett 28:2359–2362
9. Fuchigami T, Shimojo M, Konno A, Nakagawa K (1990) Electrolytic partial fluorination of organic compounds. 1. Regioselective anodic monofluorination of organosulfur compounds. J Org Chem 55:6074–6075
10. Konno A, Nakagawa K, Fuchigami T (1991) New mechanistic aspects of anodic monofluorination of haloalkyl and alkyl phenyl sulfides. J Chem Soc Chem Commun 1027–1029
11. Fuchigami T, Konno A, Nakagawa K, Shimojo M (1994) Electrolytic partial fluorination of organic compounds. 12. Selective anodic monofluorination of fluoroalkyl and alkyl sulfides. J Org Chem 59:5937–5941
12. Fuchigami T, Narizuka S, Konno A (1992) Electrolytic partial fluorination of organic compounds. 4. Regioselective anodic monofluorinatiora of 4-thiazolidinones and its application to the synthesis of monofluoro β-lactams. J Org Chem 57:3755–3757
13. Narizuka S, Fuchigami T (1993) Electrolytic partial fluorination of organic compounds. 8. Highly regioselective anodic monofluorination of β-lactams. J Org Chem 58:4200–4201
14. Hou Y, Higashiya S, Fuchigami T (1999) Electrolytic partial fluorination of organic compounds. 32. Regioselective anodic mono- and difluorination of flavones. J Org Chem 64:3346–3349
15. Fuchigami T, Mitomo K, Ishii H, Konno A (2001) Electrolytic partial fluorination of organic compounds: part 44. Anodic gem-difluorodesulfurization using triarylamine mediators. J Electroanal Chem 507:30–33
16. Fuchigami T, Fujita T (1994) Electrolytic partial fluorination of organic compounds. 14. The first electrosynthesis of hypervalent iodobenzene difluoride derivatives and its application to indirect anodic gem-difluorination. J Org Chem 59:7190–7192
17. Hou Y, Fuchigami T (2000) Electrolytic partial fluorination of organic compounds XL. Solvent effects on anodic fluorination of heterocyclic sulfides. J Electrochem Soc 147:4567–4572

18. Momota K, Morita M, Matsuda Y (1993) Electrochemical fluorination of aromatic compounds in liquid $R_4NF \cdot mHF$-part I. Basic properties of $R_4NF \cdot mHF$ and the fluorination of benzene, fluorobenzene, and 1,4-difluorobenzene. Electrochim Acta 38:1123–1130
19. Fuchigami T, Inagi S (2011) Selective electrochemical fluorination of organic molecules and macromolecules in ionic liquids. Chem Commun 47: 10211–10223
20. Franz F (1980) Ueber trishydrofluoride tertiaerer amine und ihren einsatz als fluorierungsmittel. J Fluorine Chem 15:423–434
21. Momota K, Horio H, Kato K, Morita M, Matsuda Y (1995) Electrochemical fluorination of aromatic compounds in liquid $R_4NF \cdot mHF$-Part IV. Fluorination of chlorobenzene. Electrochim Acta 40:233–240
22. Momota K, Yonezawa T, Hayakawa Y, Kato K, M M, Matsuda Y (1995) Synthesis of fluorocyclohexadienes by the electrochemical fluorination of p-difluorobenzenes on a preparative scale. J Appl Electrochem 25:651–658
23. Momota K, Mukai K, Kato K, Morita M (1998) Electrochemical fluorination of aromatic compounds in liquid $R_4NF \cdot mHF$. Part VI. The fluorination of toluene, monofluoromethylbenzene and difluoromethylbenzene. Electrochim Acta 43: 2503–2514
24. Chen S-Q, Hatakeyama T, Fukuhara T, Hara S, Yoneda N (1997) Electrochemical fluorination of aliphatic aldehydes and cyclic ketones using Et_3N-5HF electrolyte. Electrochim Acta 42: 1951–1960
25. Hara S, Chen S-Q, Hoshio T, Fukuhara T, Yoneda N (1996) Electrochemically induced fluorinative ring expansion of cycloalkylideneacetates. Tetrahedron Lett 37:8511–8514
26. Yoneda N, Chen S.-Q, Hatakeyama T, Hara S, Fukuhara T (1994) Selective electrochemical formyl hydrogen-exchange fluorination of aliphatic aldehydes to prepare acyl fluorides using HF-base solutions. Chem Lett 849–850
27. Hasegawa M, Ishii H, Cao Y, Fuchigami T (2006) Regioselective anodic monofluorination of ethers, lactones, carbonates, and esters using ionic liquid fluoride salts. J Electrochem Soc 153: D162–D166
28. Fukuhara T, Akiyama Y, Yoneda N, Tada T, Hara S (2002) Effective synthesis of difluorocyclohexadienones by electrochemical oxidation of phenols. Tetrahedron Lett 43: 6583–6585
29. Aoyama M, Fukuhara T, Hara S (2008) Selective fluorination of adamantanes by an electrochemical method. J Org Chem 73:4186–4189

Semiconductor Electrode

Kohei Uosaki
International Center for Materials Nanoarchitectonics (WPI-MANA), National Institute for Materials Science (NIMS), Tsukuba, Japan

Band Structures of Metal, Semiconductor, and Insulator

Electrochemistry of semiconductor electrode is quite different from that of metal electrode as their electronic structures are different, and it is important to understand band structures of metal, semiconductor, and insulator as described below [1].

When a large number of atoms are bound together, bands are formed. Figure 1 shows band structures of (a) metal, (b) insulator, and (c) semiconductor. Metal has high electronic conductivity as a band partially filled by electrons is formed. For example, Li, an alkaline metal, has a fully filled 1 s band and a half-filled 2 s band, which contributes to metallic conduction. Insulator and semiconductor have essentially the same band structures with fully filled bands and totally vacant bands separated by forbidden bands where no electronic states are present. The difference between insulator and semiconductor is simply the difference in the width of the forbidden band separating fully filled band and totally vacant band, i.e., energy gap, E_g. Electrons are thermally excited from the fully filled band, i.e., valence band, to the totally vacant band, i.e., conduction band, leaving holes in the valence band. Both electrons in the conduction band and holes in the valence band contribute to electronic conduction. It is easy to imagine that a number of excited electrons and, therefore, conductivity are larger as the energy gap is smaller. The conductivity is known to be proportional to $\exp(-E_g/2kT)$, where k is the Boltzmann constant and T is the

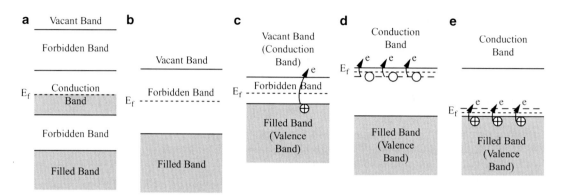

Semiconductor Electrode, Fig. 1 Band diagrams of (**a**) metal, (**b**) insulator, (**c**) intrinsic semiconductor, (**d**) n-type semiconductor, and (**e**) p-type semiconductor

absolute temperature. Thus, solids with large energy gap are insulator, and those with relatively small energy gap are semiconductor. C, Si, and Ge are elements in group 14 of periodic table and form solids with diamond structure. The energy gaps of C (diamond), Si, and Ge at 300 K are 5.5, 1.11, and 0.67 eV, respectively. The conductivity at 293 K increases as energy gap decreases as $\sim 10^{-13}$, 1.56×10^{-3}, and 2.17 S/m for C (diamond), Si, and Ge, respectively. Thus, C (diamond) is an insulator and Si and Ge are semiconductors. Since the electronic properties of these semiconductors are intrinsically determined, they are specifically called "intrinsic" semiconductor. The electronic properties of semiconductors, however, can be controlled by doping other elements. For example, the electronic conductivity of the solids of 14 group elements is significantly increased if they are doped with 13 group elements such as B or with 15 group elements such as P, as 13 and 15 group elements are ionized to donate electron to the conduction band (Fig. 1d) and accept electron from and leaving hole to the valence band (Fig. 1e), respectively. They are called "extrinsic" or "impurity" semiconductors, and the former and the latter are called n- and p-type semiconductors, respectively, as negatively charged carrier, electron, and positively charged carrier, hole, are the majority carrier of the former and the latter, respectively. Actually, "extrinsic" not "intrinsic" semiconductors are used in most of the applications.

Occupancy of electronic states at a given energy, $f(E)$, is determined by Fermi distribution function:

$$f(E) = 1/\left[1 + \exp\{(E - E_f)/kT\}\right] \quad (1)$$

where E_f is the Fermi energy representing the Fermi level, which is also shown in Fig. 1 by dotted line. At $T = 0$, $f(E) = 1$ for energies less than E_f and zero for energies greater than E_f. When $T > 0$, $f(E_f) = 1/2$ and occupancy decreases quickly as E increases. The Fermi level of metal is practically equal to the highest energy level of electrons in the conduction band, and those of insulator and intrinsic semiconductor are at the middle of forbidden band between filled and vacant bands. Those of n- and p-type semiconductors are just below the bottom of the conduction band and just above the top of the valence band, respectively, although they would be within the band if the doping density is very high.

Historical Background of Semiconductor Electrochemistry

The first report on the electrochemistry of semiconductors was published in 1955 by Brattain and Garrett [2], who examined the electrochemical properties of germanium, although Bardeen described in his Nobel lecture [3] that he and Brattain used p-Si/electrolyte interface to create an inversion layer by applying the field to the interface. Actually this experiment led to the point-contact transistor. The history of semiconductor photoelectrochemistry, however, started over 100 years ago when Becquerel [4] found that an electric current flowed if one of the electrodes immersed in a dilute acid was illuminated by light. Although the concept of semiconductor did not exist at that time, it is now clear that the electrodes he used had semiconductor properties. Since the publication of Brattain and Garrett, the knowledge of semiconductor electrochemistry has grown steadily, and several reviews and books on this subject appeared by the early 1970s [5–12]. Semiconductor electrochemistry has significantly progressed since the 1970s triggered by the publication of Fujishima and Honda [13], who reported that water can be split to hydrogen and oxygen under illumination using a TiO_2/electrolyte solution/metal cell and the first oil crisis that took place just after their publication. Photoelectrochemical energy conversion using semiconductor electrodes attracted many research groups as one of the possible means of converting solar energy to chemical energy, hydrogen, and/or electricity. Although not only the photoelectrochemical conversion efficiency increased and the application of semiconductor electrodes widened but also the understanding of the electrochemical properties of semiconductors progressed quite significantly as a result of intensive research in 1970s and 1980s [14–17], interest in semiconductor electrochemistry declined in 1990s because photoenergy conversion efficiency and stability of semiconductor electrodes were still too low to be practical for energy conversion and dye-sensitized solar cell [18], in which dye not semiconductor is excited, has attracted more attention. Recently, photoelectrochemistry of semiconductor seems to attract interests again.

Electronic Structure of Semiconductor-Electrolyte Interfaces

When a semiconductor is in contact with an electrolyte solution and they are in equilibrium, the electrochemical potentials of the two phases should be equal. The electrochemical potentials of the semiconductor and the electrolyte solution are represented by the Fermi level and the redox potential, respectively. Since the Fermi levels of n- and p-type semiconductors are situated just below the bottom of the conduction band and just above the top of the valence band, respectively, as mentioned above and the redox potentials of most of the commonly used electrolytes are found between the bottom of the conduction band and above the top of the valence band, i.e., within the energy gap, electrons should be transferred from the n-type semiconductor to the solution and from the solution to the p-type semiconductor upon contact for equilibrium to be established between the two phases. Because of the relatively low free carrier density in semiconductors, the excess charge should be compensated by the ionization of donors for n-type or acceptors for p-type semiconductors. Thus, the energy bands of semiconductor are bent up and down at n- and p-type semiconductor-electrolyte interfaces, respectively. Figure 2 shows the energy band diagram of an n-type semiconductor-electrolyte interface in equilibrium (a), under forward bias (b), and under reverse bias (c) in an ideal situation [19]. At equilibrium, as a result of electron transfer from the n-type semiconductor to the solution, the energy bands are bent up as mentioned above. Within the band-bending region, free (majority) carriers, electrons for an n-type semiconductor, and holes of a p-type semiconductor are depleted, and the region is called the space charge (depletion) region or layer. The existence of the space charge region is the origin of many interesting characteristics of semiconductor electrodes. When an external potential is applied, the position of the Fermi level of the semiconductor relative to the solution redox potential changes, but the band edge energy stays constant and only the amount of band bending is affected. When the potential of the n-type semiconductor is made negative relative to the equilibrium potential, electrons are transferred from the semiconductor to the solution and cathodic current flows continuously (Fig. 2b). If the rate of electrochemical

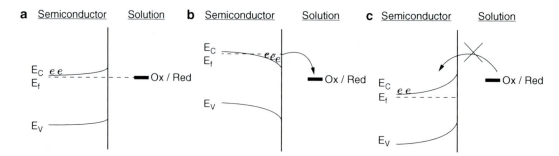

Semiconductor Electrode, Fig. 2 Energy diagrams at an n-type semiconductor-electrolyte solution interface (**a**) at equilibrium, (**b**) under forward bias, and (**c**) under reverse bias. E_C (, or ; or : is required). Energy of the bottom of the conduction band, E_V (, or ; or : is required). Energy of the top of the valence band, E_f (, or ; or : is required). Fermi level, and Ox/Red (, or ; or : is required). Redox potential [19]

reaction at the surface is slow, electrons are accumulated near the surface and an accumulation layer is formed. On the other hand, when the potential of the n-type semiconductor is made positive relative to equilibrium potential, electrons cannot be transferred from the solution to the semiconductor due to the existence of the space charge region, i.e., potential barrier, and, therefore, no anodic current flows (Fig. 2c). At a p-type semiconductor-electrolyte interface, similar arguments are valid and anodic current flows under positive polarization, but no cathodic current flows under negative polarization. Thus, rectification is expected at the semiconductor-electrolyte interfaces.

The potential distribution within the semiconductor can be calculated by solving the Poisson equation [16]:

$$d^2V(x)/dx^2 = eN/\varepsilon_0\varepsilon \quad (2)$$

with the boundary conditions of $V(0) = V_{fb}$ and $V(\infty) = V$, where $V(x)$, V_{fb}, and V are the potential of the semiconductor at x from the semiconductor surface, the flat band potential at which the band bending is 0, and the electrode potential of the semiconductor electrode, respectively, with respect to a common reference electrode, ε_0 is the permittivity of the vacuum, ε is the dielectric constant of the semiconductor, N is the concentration of the donor (N_D) for an n-type or the acceptor (N_A) for a p-type semiconductor, and e is the elementary charge. The thickness of the space charge region, W, is given approximately by

$$W = (2\varepsilon_0\varepsilon\Delta\phi_{SC}/eN)^{1/2} \quad (3)$$

where $\Delta\phi_{SC}$ is the potential drop within the semiconductor and is given by $V_{fb} - V$. If the reverse bias of an n-type semiconductor is increased significantly, the Fermi level becomes closer to the valence band than to the conduction band at the surface, and the surface concentration of holes (the minority carrier) exceeds that of electrons (the majority carrier) at the surface. This situation is called inversion.

One must note that the situation shown in Fig. 2 is of an ideal case where the surface state density is low and the carrier density is not too high. If either the surface state density [20] or the carrier density [21] is very high, an appreciable portion of the potential drop occurs within the Helmholtz layer, i.e., the electrolyte side of the interface. In an extreme case, all of the potential drop occurs within the Helmholtz layer. This situation is often called Fermi level pinning [20].

Figure 3 shows experimentally observed current-potential relations of the (a) p- [22] and (b) n-GaAs [23] electrodes in 10 mM HCl in dark. As described above, while a large cathodic current and a negligible anodic current were observed at n-GaAs electrode, a large anodic current was observed but only a small cathodic current flowed at p-GaAs electrode. The cathodic current corresponds to hydrogen evolution and the anodic current to GaAs dissolution.

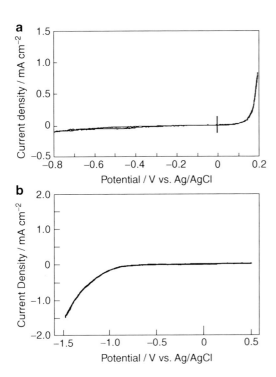

Semiconductor Electrode, Fig. 3 Experimentally observed current-potential relations at (**a**) p- [22] and (**b**) n-GaAs [23] electrodes in 10 mM HCl solution in dark

Effect of Light Illumination on Electrochemical Behavior

When semiconductor electrode in contact with a solution containing a redox couple of Red and Ox is Illuminated by light, the energy of which is larger than the energy gap of the semiconductor, electrons are excited from the valence band to the conduction band, leaving holes in the valence band as shown in Fig. 4a for p-type semiconductor electrode of which the majority carrier is hole. Even when the potential of the electrode, which is defined by the Fermi level, is more positive than the redox potential, Ox/Red, excited electrons in the conduction band have higher electronic energy than the reversible redox potential and cathodic current should flow. In the case of n-type semiconductor of which the majority carrier is electron, photoinduced oxidation takes place at more negative potential than the redox potential as hole generated by photoexcitation has much more positive potential than the redox potential. Excited electrons in p-type and holes in n-type semiconductor may attack semiconductor itself rather than reduce Ox and oxidize Red, respectively. The most important application of photoelectrochemistry of semiconductor electrode is water splitting to generate hydrogen, i.e., solar-fuel conversion, which requires multi-electron transfer reactions, i.e., hydrogen evolution reaction (HER) at cathode and oxygen evolution (OER) at anode. For these

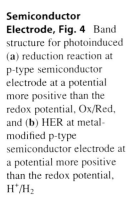

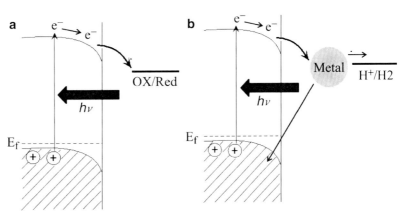

Semiconductor Electrode, Fig. 4 Band structure for photoinduced (**a**) reduction reaction at p-type semiconductor electrode at a potential more positive than the redox potential, Ox/Red, and (**b**) HER at metal-modified p-type semiconductor electrode at a potential more positive than the redox potential, H^+/H_2

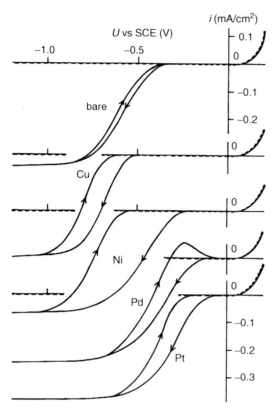

Semiconductor Electrode, Fig. 5 Current-potential relations at p-GaP electrode modified by various metals measured in a borate-buffered 0.5 M K_2SO_4 (pH 9.5) solution under illumination [24]

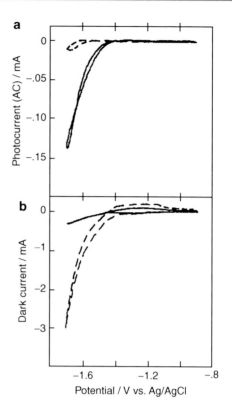

Semiconductor Electrode, Fig. 6 Current-potential relations at a bare (*solid line*) and Pt-modified (*broken line*) p-GaAs electrodes measured in a 1 M NaOH solution (**a**) in dark and (**b**) under illumination [25]

multi-electron transfer reactions, adsorption of intermediate species is a key process and electrochemical reaction rate strongly depends on the electrocatalytic nature of the electrode. Generally, semiconductor surface is not electrocatalytically active and the surface is often modified by catalytic metal to improve the reaction rate as shown in Fig. 4b for HER at p-type semiconductor electrode. Figure 5 shows the effect of metal modification for HER at p-GaP [24]. Improvement of HER and OER rates also improve the stability of the electrode as otherwise excited electrons and holes may reduce and oxidize, respectively, semiconductor itself as mentioned before. Actually, only a very limited number of n-type semiconductors, particularly of low energy gap, which can absorb more solar energy, can be used as photoanode for OER. Modification of semiconductor surface by metal does not necessary improve efficiency but sometimes leads to the efficiency decrease as metal acts as recombination center or forms Schottky junction. Figure 6 shows that currents at p-GaAs in dark and under illumination increased and decreased, respectively, by Pt treatment, and it was proved by impedance and photoluminescence measurements that high-density surface states, which act as a recombination center for electron and hole and a mediator for dark current, are introduced on the surface by the Pt treatment [25]. Various attempts have been made to avoid the direct contact of metal to the semiconductor surface so that metal acts as catalyst but not as recombination center. For example, we have recently reported that p-Si(111) electrode modified by a molecular layer with viologen moiety, which acts as electron relay, and Pt, which

Semiconductor Electrode

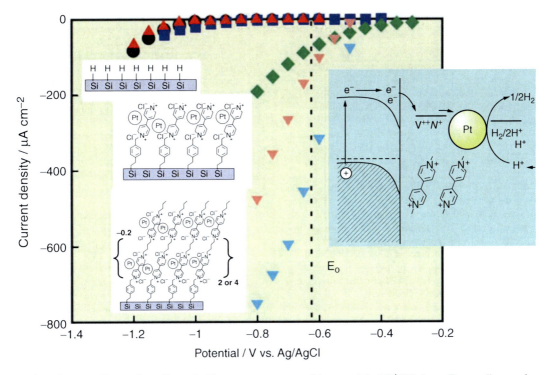

Semiconductor Electrode, Fig. 7 Photocurrent-potential relations at hydrogen-terminated and viologen layer-Pt-modified p-Si(111) electrode in 0.1 M Na$_2$SO$_4$ aqueous solution [26]. The dotted line shows the reversible potential of H$^+$/H2. Inset: Energy diagram for the hydrogen evolution reaction at the p-Si(111) modified with Pt/viologen layers. The *dotted line* shows the reversible potential of H$^+$/H$_2$

acts as HER catalyst, showed high photoelectrochemical HER rate as shown in Fig. 7 [26].

Conclusion and Future Prospects

After band structures of metal, insulator, and semiconductors are described and historical background of semiconductor electrochemistry is presented, electronic structure of semiconductor/electrolyte solution interface is discussed in relation to the unique electrochemical behavior of semiconductor electrode. Finally, effect of illumination as well as the surface modification on the electrochemical behavior of semiconductor electrode are described. Fundamental knowledge of semiconductor electrode presented here should be very important for the future development of photoelectrochemical and photocatalytic energy conversions including water splitting, CO$_2$ reduction, and dye sensitized solar cell.

Cross-References

▶ Electrocatalysis - Basic Concepts, Theoretical Treatments in Electrocatalysis via DFT-Based Simulations
▶ Electrocatalysis, Fundamentals - Electron Transfer Process; Current-Potential Relationship; Volcano Plots
▶ Electrode

References

1. Cox PA (1987) The electronic structure and chemistry of solids. Oxford University Press, Oxford

2. Brattain WH, Garrett CGB (1955) Experiments on the Interface between Germanium and an Electrolyte. Bell Syst Tech J 34:129
3. Bardeen J (1964) Semiconductor research leading to the point contact transistor. Nobel lectures, physics 1942–1962. Elsevier, Amsterdam, p 318
4. Becquerel E (1839) Mémoire sur les effets électriques produits sous l'influence des rayons solaires. C R Acad Sci 9:561
5. Green M (1959) Electrochemistry of the semiconductor-electrolyte interface. In: Bockris J O'M (ed) Modern aspects of electrochemistry, vol 2. Butterworths, London, pp 343–407
6. Dewald JF (1959) Semiconductor electrodes. In: Hannay NB (ed) Semiconductors, vol 140, ACS monograph. Reinhold, New York, pp 727–752
7. Gerischer H (1961) Semiconductor electrode reactions. In: Delahay P (ed) Advances in electrochemistry and electrochemical engineering, vol 1. Interscience, New York, pp 139–232
8. Holmes PJ (ed) (1962) The electrochemistry of semiconductors. Academic, London
9. Myamlim VA, Pleskov YV (1967) Electrochemistry of semiconductors. Plenum Press, New York
10. Gerischer H (1970) Physical chemistry: an advanced treatise, vol IXA. Academic, New York
11. Morrison SR (1971) Surface phenomena associated with the semiconductor/electrolyte interface. Prog Surf Sci 1:105
12. Pleskov YV (1973) Electric double layer of the semiconductor-electrolyte interface. In: Danielli JF, Rosenberg MD, Cadenhead DA (eds) Progress in Surface and Membrane Science, vol. 7. Academic Press, New York, p 57
13. Fujishima A, Honda K (1972) Electrochemical photolysis of water at a semiconductor electrode. Nature 238:37
14. Morrison SR (1980) Electrochemistry at semiconductors and oxidized metal electrodes. Plenum Press, New York
15. Pleskov YV, Gurevich YY (1986) Semiconductor photoelectrochemistry. Consultants Bureau, New York
16. Uosaki K, Kita H (1986) Theoretical aspects of semiconductor electrochemistry. In: White RE, Bockris J O'M, Conway BE (eds) Modern aspects of electrochemistry, vol. 18. Plenum, New York, pp 1–60
17. Hamnett A (1987) Semiconductor electrochemistry. In: Compton RG (ed) Comprehensive chemical kinetics, vol 27. Elsevier, Amsterdam, pp 61–246
18. O'Regan B, Grätzel M (1991) A low-cost, high-efficiency solar cell based on dye-sensitized colloidal TiO_2 films. Nature 353(6346):737–740
19. Uosaki K, Koinuma M, Ye S (1999) Electronic and morphological structures of semiconductor electrodes. In: Wieckowski A (ed) Interfacial electrochemistry – theory, experiment and applications. Marcel Dekker, New York, pp 737–755
20. Bard AJ, Bocarsly AB, Fan F-RF, Walton EG, Wrighton MS (1980) The concept of Fermi level pinning at semiconductor/liquid junctions. Consequences for energy conversion efficiency and selection of useful solution redox couples in solar devices. J Am Chem Soc 102:3671
21. Uosaki K, Kita H (1983) Effect of the Helmholtz layer capacitance on the potential distribution at semiconductor/electrolyte interface and the linearity of the Mott-Schottky plot. J Electrochem Soc 130:895
22. Koinuma M, Uosaki K (1994) In situ observations of atomic resolution image and anodic dissolution process of p-GaAs in HCl solution by electrochemical atomic force microscope. Surf Sci Lett 311:L737
23. Koinuma M, Uosaki K (1994) In situ observation of anodic dissolution process of n-GaAs in HCl solution by electrochemical atomic force microscope. J Vac Sci Technol B 12:1543
24. Nakato Y, Tonomura S, Tsubomura H (1976) The catalytic effect of electrodeposited metals on the photo-reduction of water at p-type semiconductors. Ber Bunsenges Phys Chem 80:1289
25. Uosaki K, Shigematsu Y, Kaneko S, Kita H (1989) Photoluminescence and impedance study of p-GaAs/electrolyte interfaces under cathodic bias: Evidence for flat-band potential shift during illumination and introduction of high-density surface states by Pt treatments. J Phys Chem 93:6521
26. Masuda T, Shimazu K, Uosaki K (2008) Construction of mono- and multimolecular layers with electron transfer mediation function and catalytic activity for hydrogen evolution on a hydrogen-terminated Si(111) surface via Si-C Bond. J Phys Chem C 112:10923

Semiconductor Junctions, Solid-Solid Junctions

Vidhya Chakrapani
Department of Chemical and Biological Engineering, Rensselaer Polytechnic Institute, Troy, NY, USA

Metal–Semiconductor Junction

A metal-semiconductor junction is an integral part of any semiconductor device, and hence, it is crucial to understand their nature. Properties of a semiconductor–metal junction often closely resembles that of a semiconductor-electrolyte junction, as both can be rectifying in nature. However, the space-charge region in the metal is usually neglected because the high density of states causes very little penetration of the electric field beyond the surface.

Semiconductor Junctions, Solid-Solid Junctions, Fig. 1 Energy band diagram of a metal and an n-type semiconductor junction (**a**) before contact, (**b**) after contact and thermal equilibrium has been attained

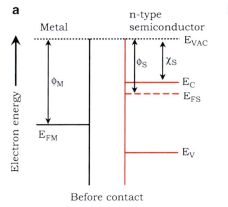
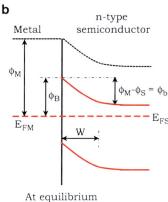

Consider the case of a metal and an n-type semiconductor contact, such that the Fermi level of an n-doped semiconductor, E_{FS}, is above the Fermi level of the metal, E_{FM}. The energy band diagram of the two materials before contact is shown in Fig. 1a. When the semiconductor and the metal are contacted (Fig. 1b), electrons from the conduction band of the semiconductor start to flow into the metal until both Fermi levels equilibrate. This results in an electron depletion region in the semiconductor, consisting of fixed positively charged donor atoms. This creates a negative field that lowers the band edges of the semiconductor in the bulk with respect to the energy at the interface.

Electrons flow into the metal until equilibrium is reached between the diffusion of electrons from the semiconductor into the metal and the drift of electrons caused by the electric field created by the ionized impurity atoms. The positive charge of the donor atoms in the semiconductor is compensated by a negative charge in the metal, thus maintaining the electroneutrality of the junction. Typically, the density of states in metal are higher than in a semiconductor. As a result, the electron accumulation region in the metal is thin, and is primarily restricted to the surface. The built-in field, ϕ_{bi}, at the interface occurs entirely within the semiconductor. Since the electron affinity and bandgap are invariant in the semiconductor, the bending of band edges results in the concomitant change in the energy of the vacuum level, E_{VAC}, as shown in Fig. 1b.

The Schottky barrier height, ϕ_B, is defined as the energy difference between the Fermi level of the metal and the band edge of the semiconductor where the majority carriers reside. The barrier height of the junction consisting of an n-type semiconductor and a metal is given by the equation:

$$\phi_B = \phi_M - \chi_S \qquad (1)$$

where ϕ_M is the work function of the metal and χ_S is the electron affinity of the semiconductor. For p-type semiconductor, the barrier height is given by the difference between the valence band edge and the Fermi energy in the metal:

$$\phi_B = E_V - \phi_M = \frac{E_g}{q} + \chi_S - \phi_M \qquad (2)$$

where E_V and E_g are the valence band energy and bandgap of the semiconductor respectively.

ϕ_{bi} is the built-in potential across the depletion layer is given by the difference between the Fermi energy of the metal and that of the semiconductor.

$$\phi_{bi} = \phi_B - \left(\frac{E_C - E_F}{q}\right); \text{n-type} \qquad (3)$$

A metal–semiconductor junction will therefore form a barrier for electrons and holes if the Fermi energy of the metal is somewhere between the conduction and valence band edge. Such a contact that is rectifying in nature is called a Schottky contact.

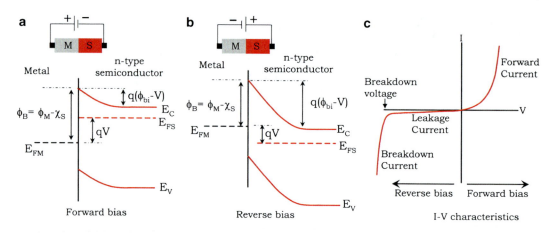

Semiconductor Junctions, Solid-Solid Junctions, Fig. 2 Energy band diagram of a metal–semiconductor junction under (**a**) forward and (**b**) reverse bias. (**c**) I–V characteristics of the junction

The bending of band edges of the semiconductor creates a space-charge region of thickness, W, which is determined by the doping density and dielectric constant of the semiconductor and is given by the equation:

$$W = \sqrt{\frac{2\varepsilon_S(\phi_{bi} - V)}{qN_d}} \quad (4)$$

where ε_S and N_d are the permittivity and donor density in the semiconductor. V is the applied voltage.

Forward and Reverse Biasing

For a Schottky junction at zero bias, the net current is zero because equal numbers of electrons on the metal side and on the semiconductor side cross the energy barrier and move to the other side. When a positive bias is applied to the metal with respect to the n-type semiconductor (Fig. 2a), the Fermi energy of the metal is lowered with respect to the Fermi energy in the semiconductor. This reduces the built-in field and the depletion region within the semiconductor. Hence, higher number of electrons diffuse towards the metal than the number drifting into the semiconductor. This is called forward biasing. When the applied voltage is comparable or greater than the built-in potential, a positive current, termed "forward current" flows through the junction.

Conversely, when a negative voltage is applied to the metal, its Fermi energy increases with respect to the Fermi energy in the semiconductor. The potential barrier for the electrons in the semiconductor increases, yielding a larger depletion region and built-in electric field at the interface. See Fig. 2b. The Schottky barrier height, however, remains unchanged so that the small leakage current that flows is independent of the applied voltage. This is true until a critical reverse bias is reached, when the junction breaks down, and large amount of current, termed "breakdown current," flows through the external circuit. Unless this current is limited by the external circuitry, it can cause overheating that can permanently damage the junction. Figure 2c shows the current–voltage characteristics of the junction under both forward and reverse biasing.

Under forward biasing, the current across a metal–semiconductor junction is mainly due to the flow of majority carriers. Different mechanisms of charge transport exist that contribute to the overall current:

1. Drift-Diffusion: The flux is limited by the rate of diffusion of charge carriers into the metal as a result of concentration gradient across the depletion layer.
2. Thermionic emission: The flow of energetic charge carriers that have thermal energies large enough to overcome the energy barrier, ϕ_B.

3. Quantum-mechanical tunneling: For barrier thickness of $\sim$1-4 nm, the wave-nature of the electrons enables penetration through thin barriers. This is especially true for heavily doped semiconductors.

In a given junction, all the above mechanisms contribute to the overall current, but one process may be dominant over the others depending on device conditions.

Ohmic Contacts

A metal–semiconductor contact is said to be ohmic if its resistance is independent of the applied voltage and is small enough to allow free flow of charge carriers in or out of the semiconductor. Low-resistance ohmic contacts are critical to the performance of any high-current device. According to Eq. 1, any metal whose work function is similar to that of the semiconductor will have zero Schottky barrier height. For an n-type semiconductor, this means that the work function of the metal must be close to or smaller than the electron affinity of the semiconductor. In case of a p-type semiconductor, the work function of the metal must be close to or larger than the sum of the electron affinity and the bandgap energy of the semiconductor for the contact to be ohmic. Although the experimentally determined barrier heights often differ from those calculated from the known values of work functions, general trends predicted by Eq. 1 can still be observed for many common semiconductor such as Si.

Surface States and Fermi Level Pinning

A surface can be considered as a defect, because it abruptly terminates a three-dimensional crystal and introduces electronic states, called "surface states," within the band structure. Since these states are in general are different from bulk lattice states, their presence significantly alter electronic and optical properties of materials. Consider, for example, a freshly cleaved semiconductor surface in vacuum. The cleaving process results in host of dangling bonds that have incompletely filled atomic orbitals. Dangling bonds are high energy site, and they have strong tendency to interact with neighboring atoms in the crystal lattice (intrinsic surface states) or other atoms or molecules from the ambient (extrinsic states) in order to lower their energy. The resulting rearrangement of the surface atoms compared to the atomic arrangement in the bulk is termed as "surface reconstruction." Surface reconstruction can be observed experimentally by various techniques such as low energy electron diffraction (LEED) and scanning tunneling microscopy (STM) and is specific to the material under consideration. Intrinsic surface states are categorized into two types: (1) Ionic states (also called Tamm states) result from heteronuclear splitting of the atomic bonds and typically give rise to surface states that lie close to the band edges. Such states are commonly found in wide bandgap semiconductors such as transition metal oxide and chalcogenides semiconductor where bonding between atoms have high ionic component. (2) Schottky states are found in pure covalent solids such as Ge and Si. Here atomic overlap of the dangling bonds with each other or with the orbitals of the bulk lattice cause splitting of the molecular orbitals into bonding and antibonding states that lie relatively close to the band edges.

The effect of surface states on the bulk electronic properties depends on the relative position of the states with respect to the intrinsic band edges of the material. Surface states in some material, such as a freshly cleaved GaAs surface in vacuum, form bands that partly overlap with the intrinsic band edges, such that the overlapping states become indistinguishable from the bulk states. Since the surface states lie outside the band gap, they do not induce any band bending at the surface, i.e., the position of the band edges with respect to the Fermi level is same throughout the bulk and surface of the semiconductor. If the surface state lies within the bandgap of the material and if the density of such states is high, then it leads to the "pinning" of Fermi level in the manner shown in Fig. 3b and c for an n- and p-type semiconductors. Density of states can be in general as high as $10^{15}/\text{eVcm}^2$ for compound semiconductors. In comparison, a nominally doped semiconductor with 10^{18} dopant atoms/cm^3 has a carrier density of $\sim 10^{10}$ electrons/cm^2

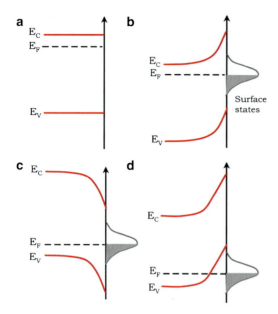

Semiconductor Junctions, Solid-Solid Junctions, Fig. 3 Effect of surface states on band structure. (**a**) Semiconductor without any surface states and hence no Fermi-level pinning. (**b**) n-type semiconductor with a depletion region and upward band bending. (**c**) p-type semiconductor with a hole depletion layer and resulting downward band bending. (**d**) p-type semiconductor with a hole accumulation region and upward band bending

at the surface. The nearly five orders of magnitude higher density of surface states then dictate the equilibrium Fermi level position.

If surface states can trap electrons, then it acts an acceptor. For an n-type semiconductor, the pinning of Fermi level at the surface trap states leads to depletion of majority charge carriers near the surface. See Fig. 3b. The resulting upward bending of the bands near the surface lowers the surface conductivity. Conversely, donor-type trap states in a p-doped material creates positive charges at the surface that bends the band downwards. See Fig. 3c. This type of band bending is seen in materials with extrinsic surface states formed as result of adsorption of highly electronegative species such as oxygen from the ambient. Moist oxygen in the atmosphere is known to have electron affinity high enough to oxidize many common semiconductors. For example, oxygen adsorption on n-type Ge surface leads to depletion of charge of majority charge carriers and an upward band bending similar to that shown in Fig. 3b. These surface states on a p-type Ge lie below the valence band [1]. The electron transfer from Ge into the surface states leads to a hole accumulation region on the surface with an upward band bending (Fig. 3d).

When a semiconductor with surface states is in contact with a metal surface, it results in a Schottky contact whose band bending is similar to that shown in Fig. 1b except that now there is an additional band bending induced by the surface states. According to Eq. 1, the Schottky barrier height, ϕ_B, for an n-type semiconductor increases with increase in work function, ϕ_M, of the metal. However, for semiconductors whose Fermi level is pinned by surface states, this dependence of ϕ_B on ϕ_M is weak. This is due to the electron exchange process that takes place almost exclusively between the surface states and the metal and has negligible influence on the built-in field and the Fermi level in the semiconductor. As a consequence, ϕ_B is mostly determined by the surface properties of the semiconductor rather than the work function of the metal. This leads to difficulty in forming ohmic contacts on many large band gap compound semiconductors such as GaAs, GaN. For in-depth analysis of the surface charging process, see Bardeen theory on surface states [2].

Furthermore, materials such as p-type GaN or SiC have a large bandgap together with high electron affinity, such that the position of their valence band edge is very low in the electron energy scale. Unfortunately, there exists no metal whose work function is high enough to form ohmic contacts to these materials. This leaves us with finding alternative ways of making ohmic contacts.

Two frequently used techniques for making ohmic contacts are tunnel and annealed/alloyed contacts.

Tunnel Contacts

Tunnel junctions are also metal–semiconductor junctions with positive Schottky barrier height. But contacts are fabricated on high dopant concentrations region, such that the depletion region in the semiconductor is extremely thin, and carriers can readily tunnel across the barrier.

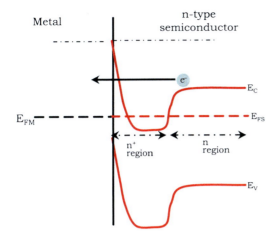

Semiconductor Junctions, Solid-Solid Junctions, Fig. 4 Band diagram showing the ohmic behavior of a tunnel contact. Thin depletion width formed as a result of heavy doping in the n⁺ region enables electron tunneling

See Fig. 4. The typical doping density for such contact is 10^{19} cm^{-3} or higher, which results in the depletion region of the order of 10 nm or less. For low contact resistance, the position of conduction band edge of the semiconductor should be low enough to permit electron tunneling into empty energy states on the other side of the barrier.

Tunnel junctions are also used as diodes, as they allow electron tunneling only at certain voltages and are capable of fast operation. It finds applications as oscillators and high-frequency signal generators. In multijunction photovoltaic (PV) cells, tunnel junctions are used for optically low-loss, series connections between consecutive solar cells. These junctions are made of a highly doped, wide bandgap semiconductor to minimize light absorption or scattering at these contacts.

Annealed and Alloyed Contacts

Annealed contacts are formed by deposition of metal with low Schottky barrier height, such as aluminum on n-type Si. Postdeposition annealing at high temperature, in reducing environment such as forming gas, helps to passivate surface states and lowers the barrier height and contact resistivity.

Alloyed contacts to silicon are usually silicides of transition metal, such as titanium or platinum silicides. Such contacts are often made by depositing the transition metal and subsequently annealing at temperature past the metal/Si eutectic point to help form the appropriate stoichiometric metal silicide. Alternatively, these contacts can also be formed by direct sputtering or ion implantation of the metal silicide, followed by annealing to reduce junction resistance.

Semiconductor–Semiconductor Junction

Semiconductor–semiconductor junctions can be broadly classified into homo- and heterojunctions depending on the materials that form the junction. In homojunctions, the contacting materials are compositionally the same except for the type of the dopant atoms, and the charge segregation occurs as a result of Fermi level offset created by the change in the type of dopant, typically from n- to p-type, across the junction. In heterojunctions, potential offset are created as a result of the change in both dopant as well as composition, that both the bandgap and band edge positions of the semiconductors.

Homojunctions: p–n Diode

A p–n homojunction device is the building block for all transistors, light emitting diodes, and photovoltaic solar cells. For simplicity, let us first consider p–n junction in which the p and n regions are uniformly doped with acceptor and donor concentrations denoted by N_a and N_d, respectively (Fig. 5a). At equilibrium, the Fermi level or the electron chemical potential is the same throughout the materials, and the resulting band bending is depicted in Fig. 5b. The nature of the band bending is similar to that occurring in a metal-semiconductor interface except that now there are two space-charge or depletion regions in a p–n diode. The built-in potential, ϕ_{bi}, is given by

$$\phi_{bi} = \frac{kT}{q} \ln \frac{N_d N_a}{n_i^2} \quad (5)$$

where n_i is the intrinsic carrier concentration of the material. The built-in potential for homojunction p–n diode is independent of the

Semiconductor Junctions, Solid-Solid Junctions,
Fig. 5 Distribution of charge carriers (**a**) and band diagram (**b**) of a p–n junction diode at equilibrium. (**c**) I–V characteristics under dark conditions during forward and reverse biasing and under illumination

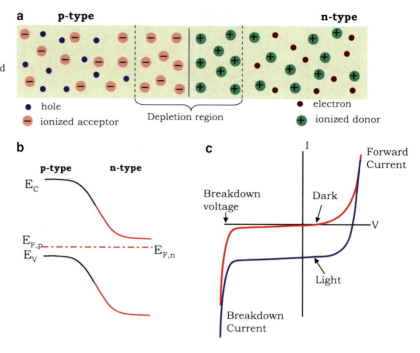

electron affinity and depends only on the dopant concentration, i.e., ϕ_{bi} increases with increase in dopant concentrations (Fermi level is closer to band edges). Note that even though the concentration of the dopant change abruptly across the junction, the concentration of holes and electron change gradually and depends on the extent of band bending at the interface.

The depletion width in this case is the sum of the depth width of both n- and p- doped region and is given by

$$W = W_N + W_P = \sqrt{\frac{2\varepsilon_s(\phi_{bi} - V)}{q}\left(\frac{1}{N_a} + \frac{1}{N_d}\right)} \quad (6)$$

So far we have only considered the case of "abrupt" junctions where dopant concentration changes abruptly across the interface. Example of this type is when the doping material is "implanted" into the substrate by means of a high-energy ion source and then annealed at high temperatures over a relatively short period. On the other hand, heating for long period of time tends to drive the dopant atoms into the material and thus gives rise to a diffused junction that has a "graded" transition profile. Such junctions are called graded p–n junctions. Often the junctions are created with highly asymmetrical doping concentration, such as a junction created by implantation or diffusion of high concentration of n-dopant into a lightly doped p-type substrate or vice versa. A highly asymmetrical junction is called a one-sided junction and can be an n^+p or a p^+n junction, where n^+ and p^+ denote the heavily doped sides. From Eq. 6, we see that the width of the space-charge region is inversely proportional to the dopant concentration; the depletion width is smaller in the heavily doped side. The band bending for a heavily doped one-sided junction then looks similar to that of a metal–semiconductor junction, with heavily doped side behaving as a metal, and its depletion layer can often be neglected.

When $N_d > N_a$, the total depletion width can be approximated as

$$W \approx W_P = \sqrt{\frac{2\varepsilon_s(\phi_{bi} - V)}{qN_a}} \quad (7)$$

Conversely, when $N_a > N_d$, then

$$W \approx W_N = \sqrt{\frac{2\varepsilon_s(\phi_{bi} - V)}{qN_d}} \qquad (8)$$

Forward and Reverse Biasing: Schottky Diode Versus p–n Junction

The current–voltage characteristic of a p–n diode looks similar to that of a metal–semiconductor junction and is shown in Fig. 5c. Forward biasing of the p-n junction (−ve voltage to n-side) reduces the barrier height. Electrons now diffuse from the n-doped side into the p-doped region. This is called minority carrier injection. Similarly, holes are injected from the p-doped region into the n-doped side. During reverse bias, there are few electrons (minority carriers) on the p-side and few holes on the n-side. Therefore, the current is negligibly small.

Although Schottky and p–n diodes follow the same I–V expression, significant differences exist between the two in their conduction mechanism. Operation of a Schottky diode involves only the majority carriers. Whereas, the forward current in a p–n junction is due to minority carrier injection from both the n- and p-sides. Further, there is negligible minority carrier injection (and storage) at the Schottky junction. Therefore, they can operate at higher frequencies than a p–n diode where switching speed is limited by the recombination of injected minority carriers.

Let's examine the I–V characteristics of the diode under illumination. Light is a driving force that can produce a diode current without any external voltage. Illumination of a p–n diode with a light of frequency greater than the bandgap of the material produces e–h pairs across the illuminated volume. The generated minority carriers are then swept across the junction by the built-in field, and cause a current to flow from the p-doped region to n-doped region through the external circuit. See Fig. 5c. Note that the current is negative in the forward direction because its direction is opposite to that generated using the an external bias. This is the operating principle of a p–n junction solar cell.

Junction Capacitance

For a p–n junction in reverse bias, the width of the depletion region is proportional to the square root of the applied voltage, and becomes wider at higher reverse bias voltages. As long as the junction is kept in reverse bias, there will be no significant flow of current. As shown in Fig. 5a, the electroneutral n and p regions are separated by the space-charge region that is depleted of both the charge carriers. This situation is similar to a parallel plate capacitor that is separated by a dielectric or an insulator. Therefore, a p-n junction can be modeled as a parallel plate capacitor. The thickness of the depletion region in a p-n junction is small, usually on the order of microns. Hence, they have considerable capacitance. Although, it results in a capacitive load that is undesirable for many integrated circuits, it finds use as a varactor, where its voltage variable capacitance is beneficial in devices such as parametric amplifiers, oscillators, and tuning devices.

The depletion-layer capacitance, C_{dep}, is given by

$$C_{dep} = A\frac{\varepsilon_s}{W} \qquad (9)$$

Inserting Eq. 9 into Eq. 6, we obtain:

$$\frac{1}{C_{dep}^2} = \frac{2(\phi_{bi} - V)}{qN\varepsilon_s A^2} \qquad (10)$$

A plot of $1/C_{dep}^2$ and applied voltage, V gives a straight line. The dopant concentration of a one-sided junction can be estimated from the slope of the curve. Extrapolation of the straight line to $1/C_{dep}^2 \rightarrow 0$ gives the built-in or flat band potential. From Eq. 10, we see that one way to reduce depletion layer capacitance in circuits is to increase the thickness of the space-charge region, W, which can be achieved by reducing the dopant concentration.

Junction Breakdown

Previously, we saw that reverse biasing a Schottky diode at large voltages can cause the junction to break down and conduct current. See Fig. 2c. The same is true for a p–n diode.

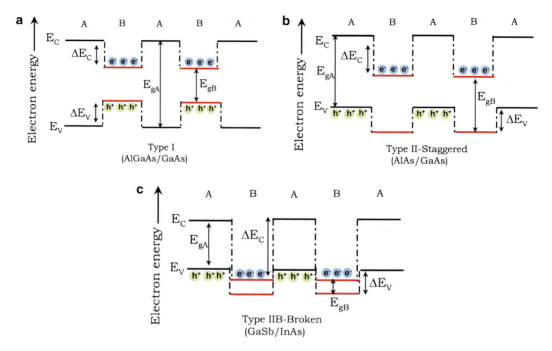

Semiconductor Junctions, Solid-Solid Junctions, Fig. 6 Schematic diagrams showing electron and hole confinement in the three type of multiheterojunctions formed between two semiconductors A and B

Concept of junction breakdown is exploited in many practical devices such as Zener diodes, where external circuits are designed to prevent large currents and overheating of the device so that diode is undamaged by the breakdown.

Zener diodes are essentially heavily doped p–n junction diodes that operate in the breakdown mode with a breakdown voltage that is tightly controlled by the manufacturer. The breakdown voltage, called Zener voltage, can be greatly reduced or modulated by controlling the extent of doping process. The electric field induced by the high dopant concentration result in reduced barrier heights such that tunneling of electrons from the valence band of the p-type material into the empty conduction band states of the n-type material is possible. Zener diodes are ideal for applications that require the generation of a reference voltage or as a voltage stabilizer for low-current applications.

Heterojunctions

Junctions formed between dissimilar materials are termed heterojunctions. Heterojunction devices made from compound semiconductors such as group III arsenides and nitrides form the basis for today's high-speed optoelectronics.

Ideal Junctions

In an ideal heterostructure, junctions considered have atomically sharp interfaces, such that defects and strain resulting from lattice mismatch do not influence the electrical and optical properties. The band structure and the resulting properties depend mainly on two factors: first, the bandgaps of the contacting materials and second, the electron affinity and work function of the materials, in other words their band edge positions with respect to the vacuum level. Depending upon these values, different scenarios arise for the band lineup, which are illustrated in the Fig. 6.

Type I: Straddled Structure

The bandgap of one semiconductor is completely contained within the bandgap of the other, i.e., $E_{C,A} > E_{C,B}$ and $E_{V,A} < E_{V,B}$ as shown in Fig. 6a. The discontinuities of the band edges are such that both types of carriers, electrons and holes, need energy (ΔE_C and ΔE_V, resp.) to move from the material B with the smaller bandgap to material A with a larger bandgap. On the other hand, both carriers in A have a driving force of ΔE_C and ΔE_V, to inject into material B. This results in accumulation of both types of charge carriers in the smaller bandgap semiconductor.

An example of type I structure is GaAs/AlGaAs heterostructure. The energy gap of gallium arsenide (GaAs) is 1.43 eV and that of aluminum gallium arsenide (AlGaAs) (e.g., $Al_{0.3}Ga_{0.7}As$) is 1.86 eV. The bandgap of AlGaAs is direct as long as the Al content is less than 40 %. In multiheterojunctions, where periodic structure consisting of alternate layers of AlGaAs and GaAs with thickness of each layer in nanometers, electrons and holes accumulation in thin GaAs layer by the potential barrier on either sides result in strong quantum confinement effects with a high probability for radiative recombination. This effect is exploited in heterojunction laser diodes [3].

In general, electron mobilities in doped semiconductors are limited by scattering at ionized impurities. In the case of a AlGaAs/GaAs/AlGaAs heterojunction, increasing the n-doping of AlGaAs increases the number of electrons injected into GaAs, where it forms a two-dimensional electron gas. However, these electrons are now physically separated from the ionized donors in AlGaAs, and hence, their mobilities are not limited by impurity scattering. Carrier mobilities larger than 10^7 cm^2/Vs have already been demonstrated in GaAs, which is close to the theoretical limit [4]. This novel way of doping a material is called modulation doping.

Type II: Staggered Structure

Here the band lineups of the contacting materials overlap and give rise to a type II structure that is staggered (Fig. 6b). The resulting energy scheme is sometime referred to as "spatially indirect bandgap," because electrons and holes are spatially confined in different materials. Examples of type II structures are GaAs and aluminum arsenide (AlAs), whose bandgaps are 1.43 and 2.16 eV,

respectively, and indium phosphide (InP, $E_g = 1.35$ eV) and indium antimonide (InSb, $E_g = 0.17$ eV).

Type IIB: Broken

Here, there is no overlap between the band lineups of the materials. Type IIB structure are in fact special case of type II structure, where material B is either a small bandgap semiconductor or a semimetal. This results in little or no barrier for the holes in material A and the electrons in the material B. See Fig. 6c. The superlattices of InAs ($E_g = 0.36$ eV) and GaSb ($E_g = 0.7$ eV) are an examples of type III structure.

Dopant Modulation: Iso- and Anisotype Heterojunctions

So far we have seen charge segregation in heterojunctions resulting from compositional differences across the interface without considering any dopant changes. In other words, we were able to produce junctions with specific properties between materials of the same doping type and or even those having identical carrier concentrations. This type of heterojunction is referred to as an "isotype" heterojunction junction. When the contacting materials of the junction is oppositely doped, such as a p–n diode, they form what is referred to as an "anisotype" heterojunction. It is common practice to denote the doping type of the wide band gap material with a upper case letter and that of the smaller gap semiconductor with a lower case letter. We can then have Pn, pN, Np, and nP junctions of the diode type and Nn, nN, Pp, and pP junctions of the isotype.

Figure 7a–c shows the band profiles of an isotype Nn and anisotype Np and Pn heterojunctions, respectively. As we can see, all junctions have band discontinuities at the interface. The nature and effect of such discontinuities on the device properties depends on its application. Let us consider the case of Np heterojunction in detail. Example of this would be an n-type AlGaAs grown epitaxially on p-type GaAs. As shown in Fig. 7b, the band lineup at

Semiconductor Junctions, Solid-Solid Junctions, Fig. 7 Band diagrams of heterojunction formed between AlGaAs and GaAs for various doping types. (**a**) Nn isotype heterojunction (**b**) Np, and (**c**) Pn anisotype heterojunctions

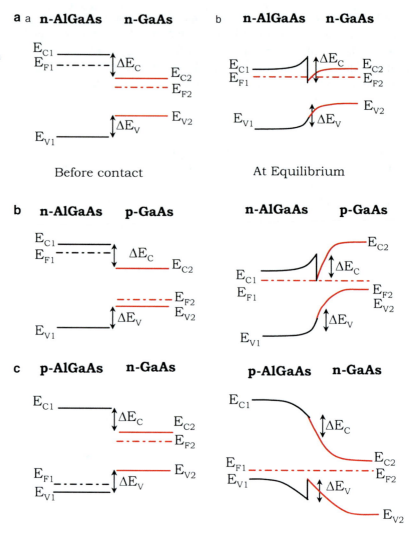

equilibrium has a "spike and notch" in the conduction band profile, with a potential barrier for electrons in GaAs (ΔE_C). If the doping concentration of the wider gap semiconductor is lower than the narrower gap material, which is usually the case in heterojunction bipolar transistors, then most of the electrostatic potential drop (or band bending) occurs in the mildly doped side. The resulting potential spike at the interface represents the potential barrier for the majority carrier electrons in AlGaAs. This barrier is, however, easily overcome during the forward biasing of the device. The presence of ΔE_C driving force results in electron injection into GaAs with substantial kinetic velocity ($\sim 10^8$ cm/s) and near-ballistic electron transport through it [5]. A potential notch in the GaAs will lead to accumulation of injected electrons and serve as high recombination centers. Although notches such as these can be detrimental for the transistor operation, it is highly advantageous for devices that require high efficiency of radiative recombination, such as in double heterojunction lasers. In bipolar transistors, notch effects are reduced by heavy doping of the small bandgap semiconductor. The valence band offset (ΔE_V) ensures that there is negligible hole injection from GaAs to AlGaAs, which is the key to increasing transistor performance. The extent and type of band edge offsets can be further widened by using quaternary alloys such as GaInAsP, which gives us additional knobs for compositional changes.

Real Heterojunctions

So far we only considered ideal heterojunctions. Often times in real heterojunction, the lattice constants of the contacting materials are never precisely the same. The lattice mismatch between the crystals gives rise to a network of so-called misfit dislocations, which can extend several nanometers into the bulk lattice. Although the details of such defects are beyond the scope of this book, the interface states introduced by dislocations and defects, similar to the surfaces states at the metal–semiconductor junction, induce additional band bending at the surface.

Future Direction

With the advent of sophisticated techniques such as molecular beam epitaxy (MBE) and metal organic chemical vapor deposition (MOCVD), synthesis of heterostructure such as multiple quantum wells or superlattices with precise interface layer down to one monolayer have been routinely possible. This not only allows modulation of electronic properties such as carrier confinement and concentration profile, but also optical confinement and wave guiding properties with appropriate choice of refractive indices of the materials. Such precise controls over the growth and material properties have opened the field of "band gap engineering".

Cross-References

▶ Semiconductor–Liquid Junction: From Fundamentals to Solar Fuel Generating Structures
▶ Semiconductors, Principles

References

1. Bardeen J, Morrison SR (1954) Surface barriers and surface conductance. Physica 20:873–884. doi:10.1016/S0031-8914(54)80199-6

2. Bardeen J (1947) Surface states and rectification at a metal semi-conductor contact. Phys Rev 71:717–727. doi:10.1103/PhysRev.71.717
3. Esaki L (1985) Compositional Superlattices. In: Parker EHC (ed) The Technology and Physics of Molecular beam epitaxy, p185 Plenum Press, New York
4. Pfeiffer L, West KW, Stormer HL, Baldwin KW (1989) Electron mobilities exceeding 10^7 cm^2/Vs in modulation doped GaAs. Appl Phys Lett 55:1888–1890. doi: 10.1063/1.102162
5. Picraux ST, Doyle BL, Tsao JY (1990) Structure and characterization of strained layer superlattices. In: Pearsall T P (ed) Strained-layer superlattices: materials science and technology, vol 33 Academic Press, New York

Semiconductor–Liquid Junction: From Fundamentals to Solar Fuel Generating Structures

Hans J. Lewerenz
Joint Center for Artificial Photosynthesis, California Institute of Technology, Pasadena, CA, USA

Introduction

Historically, the investigation of the solid–liquid interface has seen four major breakthroughs: van Troostwijk and Deiman reported the first splitting of water in 1789 using a spark discharge source [1]. Becquerel observed the photoelectric effect at the solid–liquid interface in 1839 [2] and, 183 years after the first water splitting, Fujishima and Honda reported the light-induced dissociation of water at a TiO_2 rutile electrode [3]. Three years later, Gerischer published an article demonstrating that a rectifying contact can be realized at the semiconductor–redox electrolyte junction upon judicious choice of the semiconductor–electrolyte pairing [4]. This latter work laid the basis for all present energy-converting electrochemical devices for the conversion of sunlight into electricity or fuels at the solid–liquid interface [5]. Numerous reports followed after this inception of photoelectrochemical energy conversion [6–16] which included the development of regenerative photoelectrochemical solar cells [17–21], water splitting half-cells [22, 23], excitonic solar cells [24], and the dye sensitization cell of Graetzel [25, 26] which represents the first photoelectrochemical solar cell that has been realized as a technical device.

The fundamentals of the operational principles of photoelectrochemical solar cells have so far only been outlined for a single electron transfer reaction as an outer sphere charge transfer process according to the Marcus-Gerischer theory [27, 28]. Many interesting catalytic reactions, however, are characterized by multi-electron transfer reactions. This is not only true for the tri-iodide/iodide redox couple used in many photovoltaic electrochemical systems but, also, for photocatalytic reactions such as hydrogen evolution, oxygen evolution, and carbon dioxide reduction [29, 30] which are 2-, 4-, and 6-electron transfer reactions (the latter for CO_2 reduction to methanol). The present situation of the field is therefore presently less than satisfying, particularly in the light of theoretical attempts to describe water dissociation based on free energy calculations that neglect entropy factors such as solvation shell formation where reorganization energies can easily reach values of 1 eV [31–33]. Solvation is an ultrafast process and the reaction of charged intermediates will have to be taken into account in a detailed description of a photoelectrocatalytic reaction. In addition, the development of theoretical concepts that describe multi-electron charge transfer events including their kinetics would be highly desirable.

In this work, the present status of the field of semiconductor electrochemistry with regard to energy conversion processes is reviewed. Naturally, this includes the derivation of the basic concepts and a (selected) overview of systems that operate either in the photovoltaic or the photocatalytic mode. Due to their importance for storable renewable energy, the principles of solar fuel generation are treated in some detail with emphasis on photoelectrochemical water splitting. In the outlook (part 6), selected advanced concepts will be described.

Classical Concept of the Semiconductor–Electrolyte Junction

Gerischer has demonstrated that the potential of a redox electrolyte in solution represents an electrochemical potential [34, 35]. The Fermi level of a semiconductor can also be considered as electrochemical potential. Equilibrium formation demands that the electrochemical potentials equilibrate. The schematic in Fig. 1 shows the respective density of states (DOS) of a series of semiconductors and of a redox solution before contact formation. The solution DOS is given by the Marcus-Gerischer theory [27, 36] and is characterized by Gaussian distributions for the oxidized (unoccupied) and reduced (occupied) component of the redox couple (see Eq. 1).

$$D_{ox,red} = w_{ox,red} \, (E) \, c_{ox,red}$$
$$= c_{ox,red}(E) \, (4\pi\lambda kT)^{-\frac{1}{2}}$$
$$\times \exp\left[-\frac{(E)_{ox,red} - (E)^2}{4\lambda kT}\right] \quad (1)$$

The widths of the distributions are directly related to the (entropic) reorganization energy, λ, of the solvent that results from the change in the ion solvation shell upon a change of the charge of the central ion. The redox level is defined as the energy at which the DOS of the oxidized and that of the reduced species is equal, as shown in Fig. 1 for the $Cu^{+/2+}$ redox couple. Figure 1 shows also the hydrogen redox potential and the thermodynamic value for oxygen evolution from water (at pH 0) in comparison to the band edges and energy gaps of semiconductors that have been employed for photoelectrochemical energy conversion [18, 20, 37–39].

The work function of the saturated hydrogen electrode has been taken to be 4.6 eV [40]. The potential scale can be easily converted to the energy scale by multiplication with –q (elementary charge). One sees that the Gaussian distribution of the redox DOS does not overlap well with the band edges of most of the shown semiconductors. Figure 2 shows the behavior in more detail, indicating that although the occupation

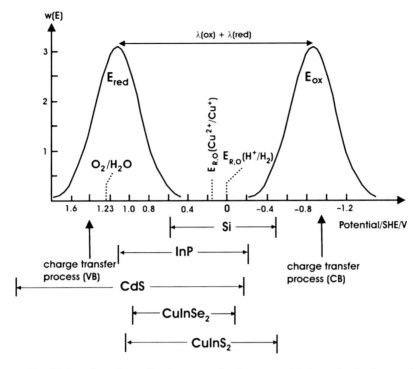

Semiconductor–Liquid Junction: From Fundamentals to Solar Fuel Generating Structures, Fig. 1 Marcus-Gerischer distribution of the density of states for a one-electron charge transfer redox couple (Cu(II)/Cu(I)) in comparison to the band edges of selected semiconductors assuming a sum of the reorganization energies, λ, of 2 eV; also shown are the redox potentials for the water oxidation and reduction reactions assuming a work function of the normal hydrogen electrode of 4.6 eV [40]; E_{ox}, E_{red} denote the respective maxima of the DOS for the oxidized and reduced states of the redox couple, the band edge positions have been determined from electron affinities and flatband potential measurements

probability w(E) appears to vanish, its value at the redox energy is about 2×10^{-4} of the maximum value, which, for concentrated solutions (~0.1 M), still provides sufficient ions for charge transfer from or to semiconductor surfaces.

Although the redox energy in solution is sometimes referred to as the Fermi level of a solution, it should be noted that, contrary to the situation in a metal or semiconductor, the electrons on the ions in solution do not form a free electron gas type of ensemble, characterized by an occupation limit. Also, in solution, the DOS at the redox energy can be rather small despite the rather large value of the reorganization energies in water of ~1 eV. For charge transfer between semiconductor and redox couple (see below), the energetic overlap of the oxidized states in solution with the semiconductor conduction band (electrons) and, analogously, that of the reduced states with the valence band holes is one important parameter in the process.

Figure 3 shows the situation upon contact formation between an n-type semiconductor and a redox electrolyte that forms a rectifying junction. In Fig. 3a, the charge distribution is shown for concentrated electrolytes where the counter-charge to that of the semiconductor space charge is concentrated at the outer Helmholtz plane. In the ideal case which is only rarely realized as, for example, for hydrogen- or methyl-terminated (111)-oriented silicon [41–44], the contribution from surface states can be neglected as their density is in the range of 10^{10} states/eVcm^2 [45, 46]. In that case, the potential drop in the Helmholtz layer is small (negligible) and the dominant part of the contact potential difference (CPD) drops across the semiconductor

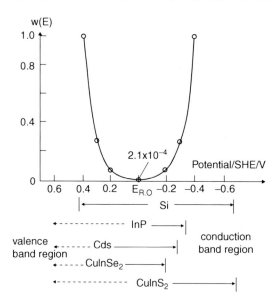

Semiconductor–Liquid Junction: From Fundamentals to Solar Fuel Generating Structures, Fig. 2 Plot of the probability function $w(E)$, normalized in the shown potential interval, near the redox energy, $E_{R.O}$, for the semiconductors and their band edges as presented in Fig. 1 (see text)

space charge region that forms, according to the differences in the respective capacitances. In the presence of surface states with an appreciable density of states, however, the junction energetics change compared to the ideal situation. The total CPD is then divided between the potential drop across the Helmholtz layer and that in the semiconductor space charge region. This results in only a partial increase of the band bending upon changing the redox potential, resulting in a slope smaller than 1 (see Fig. 4 below).

The derivation of the dark current–voltage characteristic at the ideal semiconductor–electrolyte contact is based on the assumption that the voltage dependence of the current results from the voltage dependence of the semiconductor majority carriers at the surface, n_s. Figure 3 shows an energy scheme of the contact formation. The characteristic parameters are the CPD, the barrier height, Φ_{bh}, the band bending, eV_{bb}, and the (almost negligible) potential drop across the Helmholtz layer, $\Delta\Phi_{HH}$. The solution redox potential and the Fermi level of the semiconductor are electrochemical potentials,

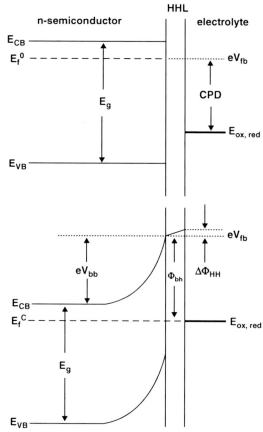

Semiconductor–Liquid Junction: From Fundamentals to Solar Fuel Generating Structures, Fig. 3 Energy schematic for junction formation at an ideal semiconductor–redox electrolyte contact; *upper* part: before contact; *lower* part: in contact, including a small potential drop across the Helmholtz double layer; E_{CB} conduction band edge, E_{VB} valence band edge, E_g energy gap, eV_{bb} band bending, CPD contact potential difference, Φ_{bh} barrier height at the semiconductor surface, E_F^0 fm level before contact formation, eV_{fb} flat band potential (energy) which equals E_F^0, E_F^C Fermi level after contact formation, $\Delta\Phi_{HH}$ potential drop across Helmholtz layer

i.e., they can be described by a chemical potential term and a Galvani potential term that can be divided into the Volta potential (the actual electrostatic potential) and a surface dipole term. The equilibration of both electrochemical potentials in the dark results in charge exchange, and an according electrostatic field. It has been assumed that no changes of the surface dipoles occur upon contact formation which is

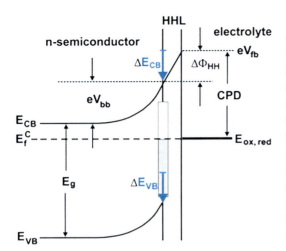

Semiconductor–Liquid Junction: From Fundamentals to Solar Fuel Generating Structures, Fig. 4 Energy relationships at the semiconductor–electrolyte interface for semiconductors with non-negligible surface state density; note the band edge shift $\Delta E_{CB,VB}$ (blue arrows) and the corresponding potential drop in the Helmholtz layer, $\Delta \Phi_{HH}$, due to surface states. Partial Fermi level pinning is observed, indicating that the CPD is now distributed between semiconductor band bending and Helmholtz layer potential as indicated

a simplification made for intelligibility. In the lower part of Fig. 3, a small potential drop in the Helmholtz layer has been included. This results in a corresponding shift of the band edges at the semiconductor surface. The barrier height at the semiconductor surface, Φ_{bh}, is lower than before contact formation by $\Delta \Phi_{HH}$. Accordingly, the energy difference $eV_{fb} - E_{ox,\,red}$ is reduced to eV_{bb}. The concentration of electrons at the surface is typically given by the Boltzmann approximation of the Fermi Dirac integral and valid for non-degenerate semiconductors:

$$n_S = n_0 \cdot e^{-\frac{E_{CB} - E_F^S}{kT}} \quad (2)$$

The energetic difference between the conduction band edge at the surface and Fermi level, given in the nominator in the exponent of Eq. 2 equals the band bending eV_{bb}. n_0 denotes the majority electron concentration in the bulk, given by the doping level (room temperature assumed).

As a consequence, the electron surface concentration is drastically reduced for larger band bendings. The electrochemical current from the semiconductor to the solution is usually described by a tunneling process from the semiconductor surface to the oxidized species of the redox couple. This current is hence dependent on the tunneling probability for outer sphere charge transfer to fully solvated ions that form the outer Helmholtz layer, on the surface concentration of electrons and on vibrational frequency factors that are related to nuclear coordinates. In the Marcus-Gerischer approximation, the current is written as

$$j_D = const \cdot f(v_{CB,VB}, T) n_s c_{R,O} \sqrt{\frac{1}{\lambda}}$$
$$\times \exp\left[-\frac{(E_{CB,VB} - E_{R,O})^2}{4\lambda kT}\right], \quad (3)$$

where the subscripts R, O refer to reduced and oxidized species and v is a frequency factor [47]. Taking Eq. 2 into account, the voltage dependence of the dark current at a semiconductor–electrolyte junction is given (for a reduction reaction) by

$$j_D(V=0) = const \cdot f(v,T) \cdot c_O \cdot w(E) \cdot n_0$$
$$\times \exp\left[-\frac{E_{CB}(x=0) - E_F}{kT}\right]. \quad (4)$$

The dark current–voltage characteristic shows the typical diode behavior with the applied voltage lowering the band bending, thus increasing exponentially the electron concentration at the surface and can be expressed as

$$j_D(V) = j_0\left(e^{\frac{eV}{kT}} - 1\right), \quad (5)$$

with

$$j_0 = const \cdot f(v,T) \cdot c_O \cdot w(E) \cdot n_S(V=0) \quad (6)$$

Despite the formal similarity of Eq. 5 with rectifying solid state junctions, it should be noted that although the current j_0 is related to

Schottky barrier thermionic emission theory [48], contained in the term n_S, solution parameters are dominating Eq. 6. The potential dependence, however, results from the thermal population increase of the surface electron concentration due to the upward shift of the Fermi level upon forward biasing.

Often, the exponential dependence of the dark current at semiconductor–electrolyte contacts is interpreted as Tafel behavior [49], since the Tafel approximation of the Butler–Volmer equation [50] also shows an exponential increase of the current with applied potential. One should, however, be aware of the fundamental differences of the situation at the metal–electrolyte versus the semiconductor–electrolyte contact. In the former, applied potentials result in an energetic change of the activated complex [51] that resides between the metal surface and the outer Helmholtz plane. The supply of electrons from the Fermi level of the metal is not the limiting factor; rather, the exponential behavior results from the Arrhenius-type voltage dependence of the reaction rate that contains the Gibbs free energy in the exponent. It is therefore somewhat misleading to refer to Tafel behavior at semiconductor–electrolyte contacts.

In most realistic situations, however, deviations from the ideal behavior, described in Eqs. 2–5, exist. Typically, semiconductors exhibit surface states of varying density and, for increased doping levels, the total CPD can be distributed more evenly between the semiconductor band bending and the potential drop in the Helmholtz layer. In that case, the simple relation between current density and voltage (Eq. 5) does not hold anymore. A schematic for Fermi level pinning at the electrolyte interface is shown in Fig. 4.

In analogy to the solid state situation, the simplest approach to obtain a photocurrent–voltage characteristic is to add a light-induced current component of opposite sign using for the light-induced current $j_L = e \cdot n_{Ph}(E_g)(1 - R)$ where $n_{Ph}(E_g)$ denotes the number of photons $(s^{-1} \, cm^{-2})$ with energy above the band gap of the absorber material

and R is the reflectivity. The photocurrent–voltage characteristic then reads as:

$$j_{ph} = j_D + j_L = j_0 \left(e^{\frac{eV}{kT}} - 1 \right) - j_L \qquad (7)$$

Figure 5 shows energy band diagrams for the dark and photocurrent dependence on applied voltage. The current j_L as described in Eq. 7 is only dependent on the energy gap, its nature (direct or indirect), and on the reflectivity of the material. To include other material-related parameters, Gaertner developed an idealized model [52] that contains the absorption coefficient, the space charge layer width, given by the doping, the built-in voltage, and the minority carrier diffusion length. The model does not consider surface recombination and assumes complete collection at the back ohmic contact. Also, recombination within the space charge region has been excluded. The light-induced (Gaertner)-current is now also a function of potential via the square root dependence of the extension of the space charge region on the bias voltage:

$$j_L(V) = -eI_0 \left(1 - \frac{e^{-\alpha W(V)}}{1 + \alpha L_h} \right) - \frac{ep_0 D_p}{L_h} \qquad (8)$$

The Gaertner model, used for solid state devices, can be used to determine minority carrier diffusion lengths and the flatband potential at semiconductor–electrolyte junctions [53]. With the advent of photoelectrochemical energy conversion in the 1970s, models have been developed that were specifically addressed to the semiconductor–electrolyte boundary [54–59], taking into account the specific situation at the reactive boundary by introducing the charge transfer rate and the surface recombination velocity as parameters.

In a model that considers the situation at the semiconductor–electrolyte contact, the charge transfer rate and the surface recombination velocity have to be included [59]. In such a model, the currents related to charge transfer and to surface recombination are expressed by the excess carrier concentration at the surface $(x = 0)$ according to

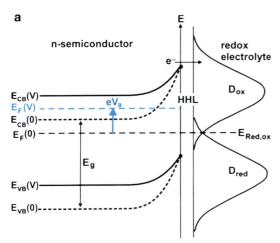
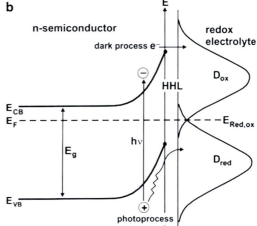

Semiconductor–Liquid Junction: From Fundamentals to Solar Fuel Generating Structures, Fig. 5 Energy band schematic for forward biasing at semiconductor–electrolyte junctions for n-type absorbers; (**a**) *dark*, eV$_a$, applied negative voltage (*blue arrow*), energy bands and Fermi level for applied voltage are labeled as a function of V; *dashed lines*: valence and conduction band in equilibrium at $V = 0$; *full lines*: after applied voltage; (**b**) illuminated; note the sign inversion of the light-induced current in Fig. 5b compared to Fig. 5a

$$j_{Ph} = qk_r \Delta_p(0), \qquad (9)$$

and for the current due to surface recombination one uses

$$j_{SR} = qS_r \Delta_p(0). \qquad (10)$$

The expression for the excess hole concentration at the surface is obtained from superposition of the solutions of the differential equations for the space charge region and the quasi-neutral diffusion region ($x > W$)

$$\Delta p(0) = \frac{P_0\left(1 - \frac{(1+\alpha L_p R_j) e^{-\alpha W}}{1+\alpha L_p}\right)}{(1+S_j)\frac{D_p}{L_p} e^{-\frac{W^2}{2L_d^2}} + k_r + S_r}, \qquad (11)$$

where L_D denotes the Debye length, P_0 the photon flux (unit cm^{-2} s^{-1}) and the parameter R_j is a function of absorption coefficient (α), thickness (d), space charge layer extension (W), and minority carrier diffusion length (L_p):

$$R_j = \frac{2\left[e^{-(\alpha - 1/L)\cdot(d-W)} - e^{-2(d-W)/L}\right]}{(1 - L\alpha)\cdot(1 - e^{-2(d-W)/L})}. \qquad (12)$$

S_j is given by

$$S_j = \frac{2e^{-2k}}{1 - 2e^{-2k}} \text{ with } k = \sqrt{\frac{2}{\pi}\frac{L_d}{D_p}P_0 e^{-\alpha W}}. \qquad (13)$$

For comparison with experimental parameters, the voltage-dependent surface recombination velocity is defined such that with increasing band bending, the effective surface recombination velocity decreases due to an exponential reduction of majority carriers at the surface

$$Sr(V) = S_{r0}\cdot e^{-fV}, \qquad (14)$$

where f is a parameter that describes the influence of the decrease of the majority carrier concentration at the surface on the surface recombination velocity. With increasing bias in the reverse direction of the junction, the surface recombination velocity decreases and the quantum yield increases. Since, in the presence of surface states, partial Fermi level pinning exists [60], the applied voltage does not change the semiconductor band bending with a slope of 1. Therefore, the exponent f is usually chosen smaller than the inverse Boltzmann factor 1/38.6 V^{-1} at room

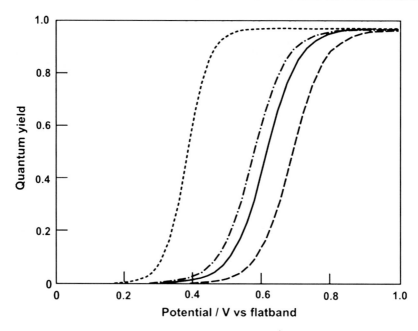

Semiconductor–Liquid Junction: From Fundamentals to Solar Fuel Generating Structures, Fig. 6 Dependence of the quantum yield Q on applied potential and on effective surface recombination velocity and charge transfer rates; material parameters: $P_0 = 6.25 \cdot 10^{16}$ cm^{-2} s^{-1}; $L_p = 500$ μm; $D_p = 11.65$ cm^2 s^{-1}; $\alpha = 780$ cm^{-1}; S_{rb} (back contact recombination velocity) 107 cm s^{-1}; $d = 500$ μm; *full line*: $k_r = 1$ cm s^{-1}, $S_r = 10^6$ (S_{r0}) exp(-20 V); *dashed line*: same data but $k_r = 0.5$ cm s^{-1}; *dotted line*: $k_r = 1$ cm s^{-1}, $S_r = 10^5$ exp(-20 V); *dashed-dotted line*: same data but exponent in S_r is -30 V (see text)

temperature. For f = 38.6, an applied voltage equivalent to the thermal voltage of 0.026 V yields S_r $(kT/q) = S_{r0}/e$. For f smaller than 38.6, a less steep decrease of $S_r(V)$ with the applied potential is obtained.

Figure 6 shows the quantum yield Q ($Q = j_{Ph}/qP_0$), plotted for several combinations of surface recombination velocities, charge transfer rate, and the voltage factor f. Material parameters are those of crystalline Si. With the assumptions made, one sees that the quantum yield shows the typical S-shape behavior of samples that exhibit surface recombination and a relatively steep increase of Q is noted for more anodic potentials. Saturation is reached, depending on the respective parameters, at an applied voltages ranging between +0.5 V and close to 1 V positive of the flatband situation ($V = 0$).

The smallest losses are observed for the lowest S_{r0} value (10^5 cm s^{-1}) and for the larger value of f (30 V) (dashed-dotted line in Fig. 6). The onset of the quantum yield occurs at ~ 0.15 V and saturation is reached at 0.5 V. If the exponential is smaller (20 V), the reduction of the surface recombination velocity is smaller, resulting in a lowered quantum yield (dotted line). The photocurrent onset is now at about 0.3 V positive from flatband and saturation is reached at +0.8 V. For $S_{r0} = 10^6$ cm s^{-1}, the voltage dependence shows further reduction of Q despite the higher value for f (30 V). The influence of a change in the charge transfer rate is shown when comparing the full and the dashed line.

The excess minority carrier concentration can be viewed as an additional stationary concentration term that enters the chemical potential although this terminology is strictly only valid in equilibrium. Extending the consideration to a stationary illumination situation where an excess carrier concentration profile exists for a given minority carrier lifetime and for defined boundary conditions, the concept of quasi-Fermi levels can be introduced. Using the Boltzmann approximation of the Fermi Dirac integral [61] and extending the relationship for the carrier concentration in the dark to the situation under illumination, one obtains

$$p^*(x) = N_{VB} \cdot e^{-\frac{pE_F^*(x) - E_{VB}}{kT}} = p_0 + \Delta p(x), \quad (15)$$

$$n^*(x) = N_{CB} \cdot e^{-\frac{E_{CB} - nE_F^*(x)}{kT}} = n_0 + \Delta_n(x), \quad (16)$$

where $E_F^*(x)$ denotes the quasi-Fermi level under illumination. The subscripts p and n refer to the considered carriers: p holes in n-type semiconductors; n: electrons (majority carriers) in n-type semiconductors. The quasi-Fermi level is given by the ratio of the excess carrier concentration and the dark carrier concentration. Solving for $E_F^*(x)$ shows that the light-induced effect is predominantly acting on the minority carriers, as the dark concentration of minority carriers in doped semiconductors is several orders of magnitude smaller than that of the majority carriers (mass action law of semiconductors) and the excess carrier concentration Δp, Δn, is typically higher than or similar to the dopant concentration.

$$_pE_F^*(x) = E_F - kT \ln \frac{p^*(x)}{p}$$

$$= E_F - kT \ln \left(1 + \frac{\Delta p(x)}{p} \right) \quad (17)$$

$$_nE_F^*(x) = E_F + kT \ln \frac{n^*(x)}{n}$$

$$= E_F + kT \ln \left(1 + \frac{\Delta n(x)}{n} \right) \quad (18)$$

It has been shown that the maximum attainable photovoltage is given by the energetic difference of the quasi-Fermi levels at the surface [62]:

$$eV_{Ph}^{max} = \left| {_nE_F^*(0)} - {_pE_F^*(0)} \right| \quad (19)$$

Quasi-Fermi levels are also considered for the description of the stability of semiconductors against photocorrosion, as will be described below.

In Fig. 7, a compilation of examples of the output characteristic of photovoltaic photoelectrochemical solar cells, adapted from the literature, is displayed. This figure shows the need for benchmarking of (photo)electrochemical devices, absorbers, and catalysts. It can be seen that the photovoltaic solar conversion efficiency of each of the cells has been determined with different light intensity and spectral behavior ranging from natural sunlight to solar simulation to halogen lamps.

The half-cell photoelectrochemical systems shown in Fig. 7 operate at reactive interfaces. Therefore, stability and strategies to achieve long-term stable operations are a mandatory condition for any technical realization. Stability or robustness is measured in chronoamperometric experiments or by determination of the turnover number (TON). The observation of rectification of n-type GaAs with the negative polyselenide redox couple points to a pronounced change of the surface/interface dipole of GaAs in contact with this redox electrolyte that also contains OH^- ions. Both aspects, the suppression of photocorrosion and the energetic alignment of the absorber to the redox solution, indicate the importance of controlling the surface chemistry and surface kinetics. They will be treated in the subsequent sections.

Stability at the Reactive Electrolyte Interface

The sustained dissolution or the surface transformation into passivating films of semiconductors in contact with electrolytes is limiting the lifetime of energy-converting devices and considerable efforts have been made to overcome this deficiency [65–69]. Rather early, criteria have been developed and recently been addressed again that allow to determine whether a semiconductor is *thermodynamically* stable [70, 71]. The method relates the position of the quasi-Fermi levels at the surface (see Eqs. 17 and 18) with those of the anodic or cathodic decomposition energies. The calculation of the decomposition levels is based on the respective corrosion reaction and a few

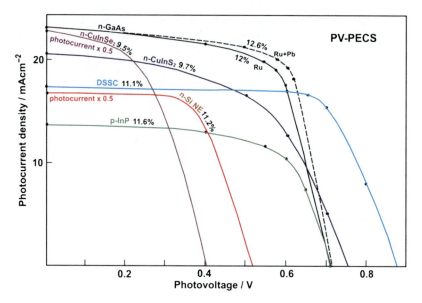

Semiconductor–Liquid Junction: From Fundamentals to Solar Fuel Generating Structures, Fig. 7 Overview of selected output power characteristics of photoelectrochemical solar cells operating in the photovoltaic mode; note that here, also 2-electron transfer redox couples have been used, for which the Marcus-Gerischer theory does not apply; the output power characteristics have been normalized and the respective efficiencies are given at each characteristic; the conditions (illumination intensity and source) are as follows: n-GaAs: 95mWcm^{-2} (sunlight) [15, 17]; p-InP: 55 mWcm^{-2} (sunlight) [37], note that the photocurrent is cathodic but has been plotted here as positive (sign inversion); n-CuInSe$_2$: 99 mWcm^{-2}, simulated AM 1 [19], note that the photocurrent is twice as large as shown; n-CuInS$_2$: 64 mWcm^{-2} natural sunlight [21]; nanoemitter (NE) Si: 100 mWcm^{-2}, W-I lamp [63, 64], note that the photocurrent is twice as large as shown; DSSC dye-sensitized solar cell: AM 1.5 global, solar simulator [25]; electrolytes: GaAs: alkaline polyselenide solution; InP: acidic V(II)/V(III); CuInSe$_2$, CuInS$_2$, and Si nanoemitter (NE) cell: acidic I$^-$/I$_3^-$; DSSC: I$^-$/I$_3^-$ in organic solution

typical decomposition reactions for n- and p-type compound semiconductors are given here:

$$p - InP + 6h^+_{VB,SS} + nH_2O + Cl$$
$$\Rightarrow In^{3+}(H_2O)_{n-y}Cl + PO_y^{3-y} + 2yH^+_{aq} \quad (20)$$

$$p - InP + 3e^-_{CB}(h\nu) + 3H^+_{aq} \Rightarrow I_n^0 + PH_3 \quad (21)$$

In Eq. 20, the anodic corrosion of p-type InP upon biasing the semiconductor positive from its flatband potential leads to the formation of an accumulation layer with a high concentration of holes at the surface. In the presence of hydrochloric acid, this results in decomposition where indium hydroxo-chlorides and phosphates or phosphites are formed. Reaction (21) describes the cathodic decomposition of InP under illumination. Another typical example is the anodic photocorrosion of n-CdS that results in the formation of a passivating sulfur film on the surface:

$$n - CdS + 2h^+_{VB}(h\nu) + aq \Rightarrow Cd^{2+}_{aq} + S^0 \quad (22)$$

Photo- and dark corrosion, however, can also be employed to judiciously modify semiconductor surfaces. The method is rather empirical but allows the in situ formation of interphases and of interfacial films that protect the photoelectrode while simultaneously allowing for efficient charge transport across the films. Examples are p-InP solar cells (photovoltaic and photoelectrocatalytic) [18, 22, 72, 73], n-CuInSe$_2$ [19, 74] and nanoemitter structures with n- and p-Si nanoemitter structures with n- and p-Si [75, 76]. Examples will be presented in section 5 below.

The thermodynamic stability considerations are based on decomposition levels introduced

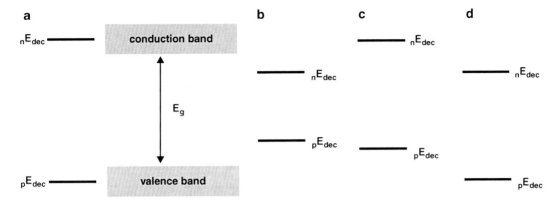

Semiconductor–Liquid Junction: From Fundamentals to Solar Fuel Generating Structures, Fig. 8 Thermodynamic stability criteria for semiconductor–electrolyte contacts; (**a**) stable material, (**b**) instable, (**c**) cathodically stable, (**d**) anodically stable (see text)

by Gerischer [71, 77]. The calculation is done by combining the respective decomposition reaction with the redox reaction of the reversible hydrogen reference electrode. In aqueous electrolytes, the anodic and cathodic decomposition reactions of a compound semiconductor MX are written as:

$$MX + zh^+ + aq \rightarrow M^{z+}_{aq} + X \quad (23)$$

$$MX + ze^- + aq \rightarrow M + X^{z-}_{aq} \quad (24)$$

In the calculation of the decomposition levels, the hydrogen electrode is used as source or sink of electrons according to:

$$zH^+_{aq} + ze^- \Longleftrightarrow \frac{z}{2}H_2 \quad (25)$$

Combining Eqs. 23 and 24 with Eq. 25 yields

$$MX + zH^+_{aq} \Longleftrightarrow M^{z+}_{aq} + X + \frac{z}{2}H_2 \quad (26)$$

(anodic decomposition), and for the cathodic decomposition reaction one gets

$$MX + \frac{z}{2}H_2 + aq \Longleftrightarrow M + X^{z-}_{aq} + H^{z+}_{aq} \quad (27)$$

With the knowledge of the Gibbs reaction enthalpy ΔG°, the standard decomposition potentials in relation to the electrochemical potential scale can be calculated:

$$\frac{\Delta G^0}{zF} = {_p}E_{dec} - \frac{\Delta G^0}{zF} = {_n}E_{dec} \quad (28)$$

In Eq. 28, ${_p}E_{dec}$ and ${_n}E_{dec}$ denote the anodic and cathodic decomposition energy, respectively. Two concepts are used to describe the stability of semiconductors at the electrolyte contact. First, the general thermodynamic stability against corrosion is determined form the position of the semiconductor band edges with respect to the decomposition levels, as shown in Fig. 8. Situations shown encompass complete stability, complete instability, and stability against anodic or cathodic corrosion.

Under illumination, stability against photocorrosion is expressed by comparing the energetic position of the quasi-Fermi level at the surface with the respective decomposition level. Accordingly, this stability definition depends on the illumination level and on surface recombination, as can be seen in Fig. 9. If the quasi-Fermi level for excess holes (photoanodes) is located energetically below the respective decomposition potential, corrosion occurs. Analogously, if ${_n}E_F^*(x=0)$ is located above ${_n}E_{dec}$, cathodic photocorrosion occurs. Stability is obtained if ${_p}E_F^*(0) > {_p}E_{dec}$ (photoanode) and for ${_n}E_F^*(0) < {_n}E_{dec}$ (photocathode), as shown in Fig. 9. Also shown are changes due to different

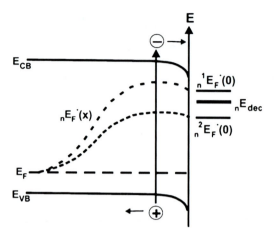
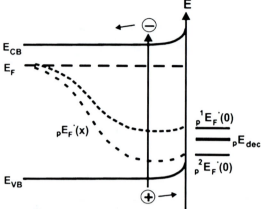

Semiconductor–Liquid Junction: From Fundamentals to Solar Fuel Generating Structures, Fig. 9 Stability assessment against photocorrosion by relative alignment of quasi-Fermi levels and anodic and cathodic decomposition levels (see text); *left*: p-semiconductor; note that the decomposition level is located above $_n^2E_F^*(0)$, indicating stability but below $_n^1E_F^*(0)$, indicating photocorrosion; for photoanodes, the *right-hand* image shows the according relations for $_p^2E_F^*(0)$ and $_p^1E_F^*(0)$ where stable operation is observed for $_p^1E_F^*(0)$ located energetically above $_pE_{dec}$

illumination intensity and surface recombination velocities. In situations, where the absorber is characterized by a large surface area compared to the geometric one (roughness, nano- and microstructures), the increase in the current density due to an increased area is compensated by a reduction of the excess carrier concentration per unit surface area which results in a smaller splitting of the minority and majority quasi-Fermi levels. This might result in improved stability but also decreases the photovoltage.

Generally, the situation is more complex. Several photo- and dark corrosion reactions, each with different decomposition levels, are possible. For long-term stability of photoelectrochemical systems, minute contributions from corrosion reactions to the overall current can result in poor stability and it is difficult to identify such "hidden" parallel reactions.

As a consequence, several concepts and strategies have been developed for improved stability. They encompass a chemical stability approach, such as surface termination of Si by hydrogen or CH_3 [41–46]. However, since even highly ordered single crystal surfaces are characterized by atomic terraces, step edges, and kink sites, solvolytic splitting of backbonds can occur that results in either terrace dissolution or in initiation of oxidative processes that continue [78, 79] despite an almost complete saturation of surface dangling bonds [80]. Kinetic stabilization, as claimed for the dye sensitization cell [81], depends on the turnover numbers (TONs) reached and how the TON relates to lifetime.

A rather surprising stability concept resulted from investigations on group VIb transition metal dichalcogenides such as WSe_2 and MoS_2. Initially, stability was explained by the d-band character of the valence band as well as of the conduction band. In this case, the photoexcitation would not result in the removal of electrons from bonding orbitals because the transition would be quasi-intrametallic [82]. This argumentation neglects, however, that optical range d-d transitions are forbidden, and it thus lacks an explanation for the high absorption. The latter is actually attributed to A and B excitons that are energetically close to the fundamental absorption edge [83] and to the fact that the photogenerated electron hole pairs relax to the band edges where their lifetime is enhanced due to the slow back transition (compare Fig. 10). The stability is explained by the nature of the topmost valence band which consists of nonbonding M – X (M: W, Mo; X: S, Se) orbitals. Hence, the interatomic bonds remain unaffected by the photoexcitation and, indeed,

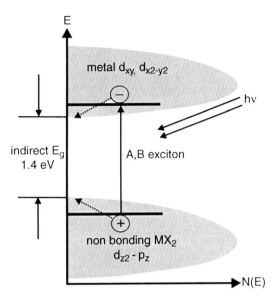

Semiconductor–Liquid Junction: From Fundamentals to Solar Fuel Generating Structures, Fig. 10 Schematic of the energy relations of a MX$_2$ layered group VI transition metal dichalcogenide near the materials fundamental absorption edge; the band gap energy of 1.4 eV refers to WSe$_2$, (see text)

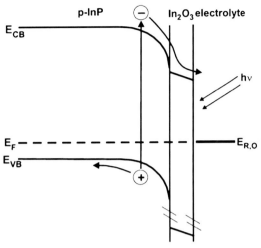

Semiconductor–Liquid Junction: From Fundamentals to Solar Fuel Generating Structures, Fig. 11 Energy band scheme for a p-InP/indium oxide film/electrolyte junction; note that the oxide conduction band is energetically located below the conduction band edge of InP

high stability has been observed on these electrodes. As the materials from this class are mostly earth-abundant, they are promising candidates for photoelectrochemical solar energy conversion. The drawback has been that a large-scale thin film preparation of electronic quality materials has not yet been achieved, despite several efforts [84–86].

A rather straightforward approach is the physical protection of electrodes by interfacial films. Some of the most stable systems have employed in situ grown films that have protected the photoelectrodes, both, cathodically and anodically [18, 19, 22, 37, 63, 74]. Examples are InP photocathodes, operating in the photovoltaic and in the photoelectrocatalytic mode, ternary chalcopyrite materials with thicker interphases and systems where the deposition of metals that are then anodized to form protective oxide films [87]. If the protective coatings are characterized by a substantial larger band gap (oxides, nitrides) than the absorber, the excess carriers have to tunnel through these films or an efficient hopping mechanism must exist. The latter is difficult to design, and films of tunneling thickness ($d \leq 1$ nm) need to be highly conformal and well attached to the surface which often is difficult to realize. An alternative protection by interfacial films would involve thicker films that are energetically aligned such that the excess minority carriers are transported across these films via conduction (photocathodes) or valence (photoanodes) band transport. A schematic for the film formation on InP photocathodes is shown in Fig. 11.

Upon illumination, the photogenerated excess electrons diffuse or drift to the surface. Even for a thick interfacial oxide film, the carrier transport is uninhibited if a cliff exists in the conduction band between the semiconductor and the oxide. As a consequence, the oxide phase can be considerably thicker than for tunnel films without losses in the carrier collection efficiency at the solution interface. Although this type of protection has already shown exceptional stability as can be seen in Fig. 12, the application of this principle is not straightforward as the detailed surface chemistry that influences interfacial dipoles and hence the band alignment has to be known and controlled. For InP, a cyclic

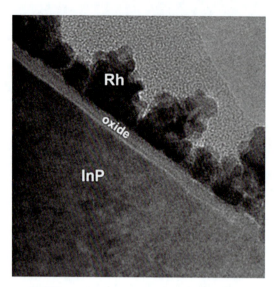

Semiconductor–Liquid Junction: From Fundamentals to Solar Fuel Generating Structures, Fig. 12 TEM side view of the InP/oxide/Rh structure after a charge of 45 C cm^{-2} has been passed; note that the interface between InP and the oxide phase remains completely unaffected whereas a dissolution of more than 10 mm of InP would have occurred at a dissolution rate of 1

voltammetry procedure under illumination resulted in the formation of an approximately 5-nm thick oxidic corrosion layer whose thickness increased to 7 nm upon photoelectrodeposition of Rh [72].

The oxidic film serves also as substrate for the photoelectrodeposition of Rh. The nature of the last step of the charge transport at the junction InP/Film/Rh/electrolyte has not yet fully been elucidated. Whether the excess electrons move ballistically in the 5-nm large Rh particles, being injected as hot electrons into the solution, or whether they partially or fully thermalize to the Fermi level, is not yet fully understood. The application of this stability principle for photoanodes encounters the difficulty that the valence band maximum of interfacial oxides is located below the valence band edge of the absorber. Using sulfide films would be an alternative if their band gap energy allows transmission of the incoming light. Another possibility is the use of interfacial dipole layers that alter the energy band alignment. This approach will be outlined below.

Interfacial Aspects: Recombination, Charge Transfer, and Dipoles

Besides the stability issues encountered at the solid–liquid interface, its electronic quality defines the efficiency of the system. This encompasses the control of surface recombination but, also, the control of the energy band alignments between light absorbers, interfacial films, and electrocatalysts. Accordingly, two avenues have to be followed: (i) the reduction of the surface recombination velocity (compare Eqs. 10, 11, 14) and (ii) the use of surface- or interface dipoles for band alignment tuning purposes.

Examples of the former include ion adsorption experiments that have been performed in the initial phase of the development of photoelectrochemical solar cells [14, 15, 17], the directed termination of surface dangling bonds by adatoms, and the participation of surface states in the carrier exchange with the solution. Figure 13a–d shows a summarizing schematic of these principles. Figure 13a shows the influence of the formation of a chemical bond between surface atoms that are electronically acting as surface state radicals (E_S) with unpaired electrons for strong electronic interaction with species (E_R) that have been introduced by external processes by solution chemistry [15, 17, 88] or using gas phase treatments. The induced split into bonding and antibonding states (E_{bd}, E_{ab}) removes these states from the energy gap, thereby drastically reducing the surface recombination velocity. Figure 13b displays the passivation of dangling bonds by selective termination of these bonds with a species that rather ideally terminates the lattice without leading to sterical constraints. H-termination of Si, for example, is very efficiently done on (111) surfaces, but on Si (100), the two dangling bonds are too close to allow for ideal termination. The figure shows how the etching in concentrated NH$_4$F solutions attacks the kink sites at the step edges of atomic terraces, removing terraces according to the fluorination of the surface site and dissolution as a tetravalent species. The process involves two steps where solvolytic splitting of Si backbonds by a water molecule occurs. Si atoms that exhibit two

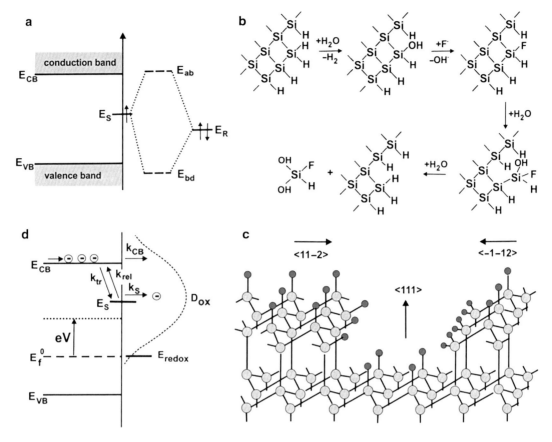

Semiconductor–Liquid Junction: From Fundamentals to Solar Fuel Generating Structures, Fig. 13 Surface phenomena at the semiconductor–electrolyte phase boundary; (**a**) removal of surface states by strong interaction with an ambient species E_R, E_{ab}, E_{bd} antibonding and bonding levels, respectively (solution, gas phase); (**b**) chemical dissolution scheme of Si in concentrated ammonium fluoride solution in a two-dimensional representation; (**c**) development of stable atomic terraces and terrace edges by etching in ammonium fluoride; *light gray* spheres: Si, *small dark gray* spheres: hydrogen; (**d**) advantageous effect of surface states in mediating charge transfer; k_{tr} trapping rate, k_{rel}: carrier release rate, k_S: surface charge transfer rate, k_{CB}: charge transfer rate of conduction band electrons, E_S energetic position of the surface state (see text)

backbonds to the lattice, as shown on the right-hand side of Fig. 13c, are hence susceptible to dissolution and finally, only terraces with single-bond H-terminated Si remain, as experimentally observed in AFM experiments where the atomic step edges and their crystallographic direction can be observed [89]. The role of surface states in the charge transfer process is schematically displayed in Fig. 13d. A surface state (E_S) mediated oxidation reaction through hole capture can be written as:

$$E_S + h_{VB}^+ \iff E_S^+ \qquad (29)$$

$$E_S^+ + R_{Red} \Rightarrow E_S + R_{Ox}$$

The photocurrent resulting from this process is a function of the surface state density, the capture cross section of the state(s) and of the occupation level that depends on the position of the Fermi level which changes with applied voltage and is indicated in the figure [90, 91].

Surface and interface dipoles are known to shift energetic alignments [92]. The energetic shifts can rather easily reach values in the eV range and, therefore, the assessment and, in particular, the control of interface dipoles is of

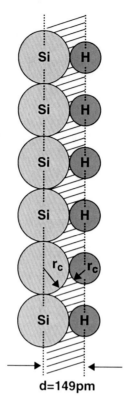

Semiconductor–Liquid Junction: From Fundamentals to Solar Fuel Generating Structures, Fig. 14 Idealized microcapacitor arrangement for (111): (1 × 1) H-terminated Si; r_c, covalent radii (see text)

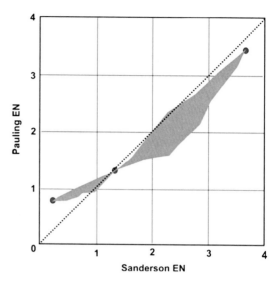

Semiconductor–Liquid Junction: From Fundamentals to Solar Fuel Generating Structures, Fig. 15 Plot of Pauling versus Sanderson electronegativities for 37 elements; the data spread is shown by the *gray area*

considerable and general importance for the functioning of devices. As an example, the hydrogen termination of a Si (111) surface is considered. It is (somewhat unrealistically) assumed that the H-termination is perfect. In reality, kink sites and some surface defects can be terminated differently, as has been shown by HREELS experiments [80]. On Si (111), the potential drop due to the occurrence of a molecular dipole between the topmost Si atom and H is determined in what one could call a microcapacitor model [93] by $V = Q/C$, Q is the total charge and C the capacitance. In this approach, Q is determined by the partial charge shift, $\Delta \rho$, between Si and H due to the differences in electronegativity, the number of Si–H bonds at the surface ($n_B = 8 \times 10^{14}$ cm^{-2} for Si (111)) and the elementary charge, q. The capacitance is determined by the static dielectric constant of the structure and the distance between the Si and H planes, d, as shown in Fig. 14. A principal challenge is the determination of the dielectric constant and in first approximations for the structure, a value between 6 and 10 has been taken deduced from that of hydrogenated amorphous silicon [94, 95]. The electronegativity is calculated using Sanderson's data that are based on the concepts of "compactness" of an atom and on electronegativity equilibration in a compound [96], allowing simple calculation of the electronegativity (EN) of molecules. The difference between Pauling and Sanderson electronegativities is presented in Fig. 15. Marked differences occur for smaller values of EN. With the EN equilibration concept, one can calculate, for example, the EN of a methyl group and determine the partial charge transfer also for methylated Si surfaces that are known to terminate Si (111) surfaces in an almost ideal manner [97]. The ENs of Si and H are 2.138 and 2.592, respectively. The partial charge shift, $\Delta \rho$, is determined by

$$\Delta \rho = \frac{(S_m - S_i)}{\Delta S_i}, \tag{30}$$

with ΔSi (Si) $= 3.04$, $S_H = 2.592$ and $S_{Si} = 2.138$, one obtains $\Delta \rho = 0.149$. For the covalent radii, one uses $r_H = 32$ pm and $r_{Si} = 116.9$ pm, hence $d = 148.9$ pm. The potential drop is then calculated according to

$$\Delta V = \frac{qn_B\Delta\rho d}{\varepsilon\varepsilon_0 A}, \tag{31}$$

obtaining $\Delta V = 0.5$ V using $\varepsilon_M = 6$. For $\varepsilon = 10$, the value reduces to 0.3 V, which is much closer to the observed value in the synchrotron radiation photoelectron spectroscopy (SRPES) experiment where a value between 0.25 and 0.3 eV is observed [98]. Although the experimental result is described considerably better by the larger value for e, at real surfaces, the Si-H-H scissor vibrational mode, Si-OH groups, and some Si hydrocarbon bonding are observed in HREELS [80]. These surface terminations as well as the lattice termination at kink sites are not contained in the simple microcapacitor model and the calculated values can only be considered an estimate. At step edges and kink sites, for example, the dipole projection perpendicular to the surface is substantially reduced (compare Fig. 13c). The resulting ambiguity not only relates to the determination of surface core level shifts but, more generally, also to the band offsets in heterojunctions for solid state devices.

The methyl group electronegativity is calculated according to

$$S_m = \left[\prod_{i=1}^{N} S_i\right]^{1/n}, \tag{32}$$

yielding $S(CH_3) = 1.749$ which indicates that the dipole with the Si surface atoms is orientated opposite to the situation for H-termination. Since the main contribution to this dipole occurs perpendicular to the surface, the distance is approximated by the covalent radii of Si and C which yields $r_C = 194$ pm. The calculation of the potential drop at methyl-terminated Si(111) gives $\Delta\rho = 0.31$ and $\Delta V \sim 1$ V with the dipole pointing toward the surface. This should be visible as a lowering of the Si electron affinity and UPS analysis on this surface is planned where the shift in the secondary electron cut-off allows the determination of $\chi = hv - E_g - S_W$ (photon energy, energy gap, and spectral width of the UP spectrum).

Long-term stability via suited band alignment of light absorbers with thicker passivating films (see Fig. 11) will need the incorporation of interfacial dipole layers that are stable enough to withstand subsequent passive film deposition. Recently, organic films have been prepared that are stable up to 300°C and that show a clear dipole effect, measured by work function changes of Ru metal deposited onto the films [99]. Two organic films, APTES (Aminopropyl-triethoxysilane) and UDTS (undecenyltrichlorosilane) have been used. They were deposited onto SiO_2 and on top of these films, metallic Ru was deposited. The work function of Ru was measured for three situations: for deposition onto the films and for the SiO_2–Ru system without interphase. Flatband voltage measurements of the Si/SiO_2/film/Ru structures show that compared to the situation without film, APTES induces a flatband shift of +0.2 V whereas UDTS leads to a negative shift of -0.5 V. The according dipole orientations are summarized in Fig. 16.

The application of the selective application of interfacial films for energy band alignment in photoelectrochemical energy-converting systems is yet in its infancy, but the concept promises a very significant step toward the realization of stable and robust devices such as monolithically integrated water splitting structures. Such systems will be reviewed below.

Photoelectrocatalysis Systems: Water Splitting Approaches

The light-assisted dissociation of water is presently in the focus of a series of research initiatives around the globe. The background is the attempt to produce fuels, in particular hydrogen, in a renewable, carbon neutral manner. Since biomimetic approaches are relatively far from robustness and sufficient efficiency, the development of mostly inorganic material-based systems and structures appears more favorable presently. Also, the adaption of the concepts of nature that uses Earth-abundant components, albeit with low efficiency and stability, neglects that natural

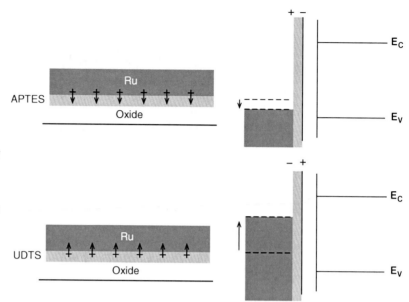

Semiconductor–Liquid Junction: From Fundamentals to Solar Fuel Generating Structures, Fig. 16 Influence of interfacial organic films on the energy alignment in MIS type metal–semiconductor structures; as substrate, Si was used, indicated by the position of its valence and conduction band, E_V, E_C, respectively; also indicated is the orientation of the dipoles in the films (see text)

systems are characterized by entropic losses due to the maintenance of life, reproduction, and generation of highly ordered structures, thereby reducing the available free energy. Although the excitation energy to the reactive site of the water oxidase is transferred by multichromophoric Foerster transfer [100] and the enzyme is thus not directly illuminated, it has to be replaced every 20–30 min [101]. Therefore it is rather unlikely to derive stable and efficient systems from an approach that is very closely adapted from nature. On the other hand, nature provides a series of concepts, such as the Z-scheme, insights into excitation energy transfer and oxidative cycles in oxygen evolution [102] that are of tremendous importance for the further development of the field.

To consider the photolysis of water based on the inorganic equivalent of the Z-scheme in nature, semiconductor tandem-type structures, such as shown in Fig. 17, can be analyzed. It should be noted that the structure shown in Fig. 17 represents a solid state analogon of two half-cells for the photoelectrolysis of water in a joint scheme. Therefore, compared to solid state devices with two pn junctions, the band bending of the structure shown develops at the respective electrolyte contact or by forming Schottky-type junctions with the electrocatalysts at the surfaces which then form electrolyte contacts. The ohmic contacts at the semiconductor–semiconductor rear interface show a short-circuit condition. The contacts are (typically) produced by highly doped degenerate p^+ and n^+ regions, as shown in the figure or by metallic contacts. The latter, however, would have to be thin enough to be transparent to the impinging light. It has been assumed that the excess minority carriers on each side only react to oxidize water or to reduce hydronium ions which results in electron and hole accumulation at the surface (inhibition of charge transfer and low surface recombination assumed) until the corresponding quasi-Fermi levels exceed the energy necessary to induce the reaction. The resulting photovoltage then provides the free energy to split water at a potential of $V = 1.23$ V $+ V_W$ (working potential of the anode) $+ V_W$ (cathode). The working potentials depend on the activity of the electrocatalysts and on the total current density. Figure 18 visualizes the voltage and current matching and, in addition, implicitly contains the overall losses in these structures.

The material with the smaller energy gap defines the current matching condition. Therefore, in Fig. 18, the current at the maximum power point of the photoanode is matched with the current of opposite sign from the photocathode. Since the

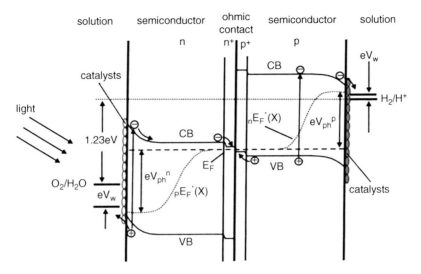

Semiconductor–Liquid Junction: From Fundamentals to Solar Fuel Generating Structures, Fig. 17 Energy schematic of a Z-scheme analogon light absorber/catalyst structure using a tandem-type cell design; n- and p-type semiconductors act as photoanode and photocathode, respectively (see text); VB, CB: valence and conduction band edges, respectively; n$^+$, p$^+$ denote the highly doped n- and p-type regions that form the ohmic contact between the semiconductors, V_W is the voltage at which the device operates (working voltage), V_{ph}^n, V_{ph}^p are the photovoltages in the respective semiconductors at the maximum power point; $_nE_F^*(x)$, $_pE_F^*(x)$ denote the quasi-Fermi levels for excess electrons and holes, respectively

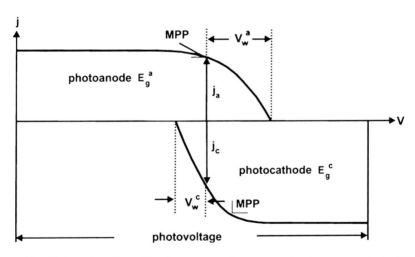

Semiconductor–Liquid Junction: From Fundamentals to Solar Fuel Generating Structures, Fig. 18 Photocurrent and photovoltage relationships in a tandem structure for water splitting; MPP, maximum power point; E_g^a, E_g^c, energy gap of photoanode (a) and photocathode (c), assuming that $E_g^a > E_g^c$ as typical for such structures; the photocurrent obtained from the anode is thus smaller than that from the cathode as shown; V_w^a, V_w^c are the working potentials for anode and cathode, respectively; j_a, j_c denote anodic and cathodic photocurrents

energy gap of the photocathode has been assumed to be smaller, the material absorbs more photons from the solar spectrum, resulting in a larger short-circuit current.

Hence, the current matching occurs at a potential where this cell component does not have its optimum performance, e.g., at potentials larger than that at the maximum power point.

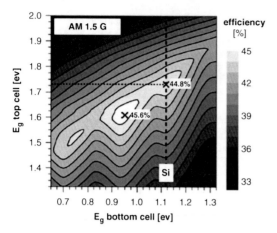

Semiconductor–Liquid Junction: From Fundamentals to Solar Fuel Generating Structures, Fig. 19 Theoretical efficiency for tandem structures and air mass 1.5 global; *dashed* and *dotted lines* show structures that would be based on Si as the bottom cell (see text)

The losses that one encounters relative to the thermodynamic value of 1.23 V are rather large. They can be deduced from the schematics in Figs. 17 and 18. The reaction overvoltages η for the anode and cathode reaction are about 0.4 and 0.1 V for noble metal catalysts and the residual band bending to drive the reaction is given by $e(V_w^a + V_w^c)$ which amounts at least to 0.3 eV. The energetic difference between the band edges and the bulk Fermi level can be assumed to be 0.3–0.4 eV. The total losses are then in the range of 1.1–1.2 eV. Accordingly, the sum of the energy gaps for the semiconductor pair should reach about 2.4 eV. The estimate made here neglects, however, solution resistance losses as well as losses due to hydronium transport in membranes. For a practical system, the assumption of a sum of energy gaps that reaches 2.5–2.6 eV appears more realistic. If one uses non-noble metal electrocatalysts, their kinetics are generally poorer and the needed voltage will most likely further increase even if one uses large surface area catalyst materials to reach a given exchange current density. Figure 19 shows theoretical efficiency calculations for tandem structures in dependence of the component energy gaps. With Si as photocathode base material, one would have to find a photoanode material with an energy gap between 1.6 and 1.8 eV. Besides this demand on the energy gap, further requirements include (i) the absolute position of the valence band with respect to the water oxidation potential, (ii) earth-abundant components, (iii) high quality optoelectronic properties, and (iv), possibly, stability. The latter might be circumvented by novel approaches of physical protection of surfaces with conformal corrosion protective layers either prepared (photo)electrochemically or by ALD [103] (atomic layer deposition).

Realizing a solar fuel generator has to take into consideration the separation of gaseous products and ion exchange through membranes. Such systems could be planar with embedded small photoactive structures connected and mechanically held by a membrane as seen in Fig. 20.

A vertical design solar fuel generator would consist of nano- or microwire absorber material with either radial or axial p-n junctions and of an ohmic contact between them, according to the scheme in Fig. 17. Due to reflection between the wires, metallic electrocatalysts at the side walls of the structures and metallic ohmic contacts absorb too much of the incoming light. Therefore, research efforts are needed for the preparation and anchoring of photoactive electrocatalysts where the photonic excess energy is not lost but is transferred to the water splitting reaction if the energy band alignments are appropriate (see Figs. 14 and 16). Already known materials are RuS_2 compounds, possibly layered group VIb transition metal dichalcogenides and perovskites [104]. Figure 21 shows a schematic of a part of a rod-type vertical solar fuel generator. Only the principle of the structure, consisting of a Si microwire as the core element, is shown. The structure is embedded in a proton-conducting membrane. In an advanced design [105], a p^+-n Si structure is formed and the upper part of the microwire consists of an n-type large gap photoanode material, as indicated in Fig. 21. The cathode part of the structure (lower part of the microwire) uses electrons as majority carriers.

A possible operational principle of a microwire structure could consist of a (radial) p^+-n Si junction that connects via an ohmic contact to a photoanode material (upper part of the

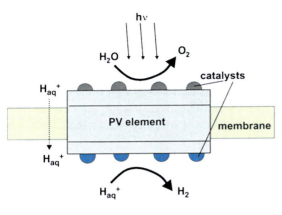
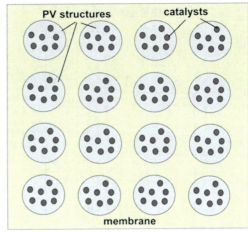

Semiconductor–Liquid Junction: From Fundamentals to Solar Fuel Generating Structures, Fig. 20 Schematic of a planar arrangement for a solar fuel generator, consisting of photovoltaically active disks embedded into an ion-conducting, gas-separating membrane; besides proton (hydronium) reduction, CO_2 reduction can also be carried out at the cathode side; the size of the PV elements and their distance has to be adjusted to the ionic conductivity of the membrane and the solution

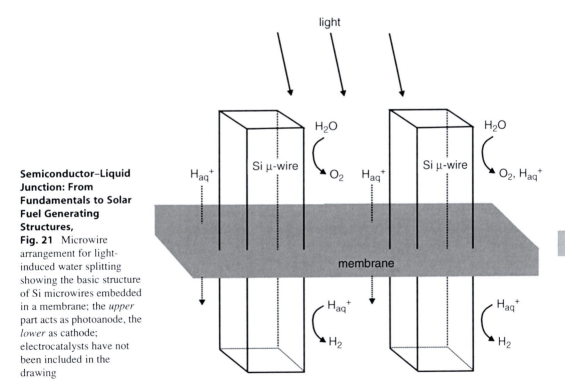

Semiconductor–Liquid Junction: From Fundamentals to Solar Fuel Generating Structures, Fig. 21 Microwire arrangement for light-induced water splitting showing the basic structure of Si microwires embedded in a membrane; the *upper* part acts as photoanode, the *lower* as cathode; electrocatalysts have not been included in the drawing

structure in Fig. 21). In this structural arrangement, the light-generated excess holes in n-Si move to the interface with the photoanode where the current flow is maintained by reaction with the anode majority carriers (electrons) and the majority electrons from n-Si are consumed in the reduction reaction (lower part of the structure) The total current is controlled by that of the

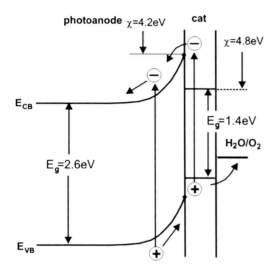

Semiconductor–Liquid Junction: From Fundamentals to Solar Fuel Generating Structures, Fig. 22 Idealized energy scheme $E(x)$ assuming that instead of a metallic electrocatalyst, a photoactive electrocatalyst labeled *cat* has been deposited onto a photoanode; data (electron affinities and energy gaps) have been taken for WO_3 (photoanode) and RuS_2; band labeling as in above figures; χ denotes the electron affinity

photoanode because of its larger energy gap. The total free energy separation of the cathode electrons and of the anode holes must exceed the voltage necessary to split water including overpotentials. If the microwire structure of Fig. 21 is coated with electrocatalysts that reside on the side and top areas of the absorbers, reflection losses of the incoming light are substantial, particularly if the aspect ratio is large. Typical structures grown have roughly 80-μm length and 2–3-μm width of the wires. To reduce these substantial losses, one can envisage geometries that suppress these losses or, alternatively, one could use *photoactive* electrocatalysts. In Fig. 22, an energy scheme for a realistic structure has been drawn, consisting of a RuS_2 photoelectrocatalyst on top of the photoanode material WO_3. RuS_2 is known to be electrocatalytically active in the dark, but the material exhibits an energy gap of 1.4 eV and is photoactive. The scheme uses the known values for the electron affinity of the two materials but neglects band alignment shifts due to interfacial dipoles which could be substantial as discussed above.

Remaining in the ideal picture, the schematic in Fig. 22 shows that losses due to thermalization occur only for photons with energy between 1.4 and 2 eV since, for this range, the photogenerated electrons cannot overcome the barrier to the conduction band of WO_3, resulting in recombination in RuS_2.

In a radial tandem design with a photocathode in contact with the anode, most of the photons with the intermediate energy will, however, reach the low gap tandem partner where they can contribute to the photocurrent. Two more aspects are worth mentioning: The absorption within a thin layer of RuS_2 will be considerable less than in a metal and the band alignment drawn here did not include any interfacial interactions between absorber and photocatalyst, and the actual band alignment is likely to be different. As shown in Fig. 16, this alignment can be controlled to some extent.

Presently, a few efficient photocathodes exist. They encompass p-WSe_2, Si photovoltaic p-n junctions, and p-InP. Figure 23 shows the output characteristic of the InP half-cell [72] and possible improvements if a larger photovoltage with less surface/interface recombination could be realized.

The experimental data (open circles in Fig. 23) show a variation at high photocurrents due to the formation of hydrogen bubbles and their release. One also sees that the S-shaped behavior for voltages larger than that at the maximum power point indicates surface recombination, indicating that in this complex structure, despite the high photocurrents, substantial current losses occur at increased voltage where the absorber is nearing its flatband situation. The output characteristic for negligible surface would result in a relative efficiency increase by ∼50 %. If the total contact potential difference between the H^+/H_2 redox couple could be exploited, the half-cell would show excellent and unequaled efficiency. Despite the already good output, the system is characterized by losses, presumably at the several interfaces (see Fig. 12). An advantage of using custom-made thin (3 μm) film homoepitaxially

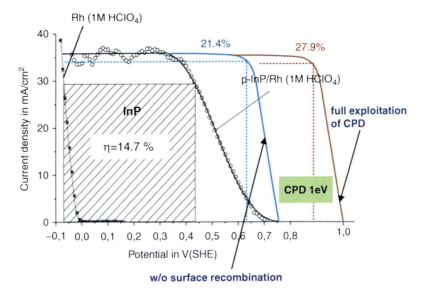

Semiconductor–Liquid Junction: From Fundamentals to Solar Fuel Generating Structures, Fig. 23 In P epitaxial thin film photocathode performance and development potential for absence of surface states ($\eta = 21.4\ \%$) and for additional full use of the contact potential difference (CPD) ($\eta = 27.9\ \%$); the original data were obtained in a 3 electrode standard potentiostatic arrangement at 100 mWcm^{-2} illumination with a W-I lamp; the crossed signal shows the I-V characteristic of a Rh wire

grown InP is the control of thickness, doping level, and the possibility of reusable substrates if a suitable wafer bonding method is developed. Earth-abundant photocathode materials such as Si or group VIb layered transition metal dichalcogenides show smaller photovoltages and, in case of the transition metal dichalcogenides, a method for growth of planar large area samples is yet to be developed.

Modifications of dye-sensitized solar cell architectures can also be employed for water photolysis. A structure that consists of titania, the typically used dyes and an attached oxygen evolution catalyst such as IrO$_2$, is shown in Fig. 24.

In such experiments, the time scale for Reaction (3) is typically faster than that of the energy-converting process 4 by a factor of about 4–5, with process 4 having a time constant in the ms range. Therefore, an interfacial film, band aligned with the collector conduction band and the excited dye state S^* (see Fig. 24), could reduce the back-reaction 3, similar to the approach advocated in Fig. 11 and the according text. With the indicated

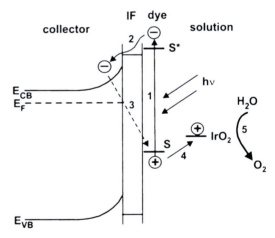

Semiconductor–Liquid Junction: From Fundamentals to Solar Fuel Generating Structures, Fig. 24 Schematic on sensitization based water splitting half-cell (photoanode); processes are labeled according to: *1* photoexcitation of the dye S to S^*, *2* charge injection into collector (TiO$_2$) conduction band via an interfacial film (IF) with large energy gap, *3* reduction of excited dye ground state, *4* hole transfer to the electrocatalyst, *5* electrochemical interface reaction (water oxidation)

band alignment, this film could be thicker than necessary for a tunnel process. It will then strongly suppress/quench the back-reaction (3), increasing substantially the efficiency. A system with an interfacial film as drawn here is not yet been realized but is rather straightforward when using ALD techniques allowing to prepare large gap oxide materials, for example.

From the efficiency scheme in Fig. 19, the use of photoanode materials with energy gaps between 1.8 eV (for InP) and $\sim$ 1.7 eV is desirable. In addition, the absolute positions of the band edges have to match the energetic restraints for water splitting. Presently, absorbers such as WO_3, Fe_2O_3, $BiVO_4$, TaON, Ta_3N_5, perovskites, and doped/alloyed TiO_2 are considered as candidates (see also above) for photoanodes but, so far, their energy gaps, valence band position and optoelectronic properties are not fulfilling the criteria for the desired materials. As an alternative, efforts in combinatorial research for the development of novel absorbers and catalysts have been introduced, using predominantly ink jet printing [106] and sputter methods. However, even if new promising materials are identified, they have to be optimized electronically and their interface behavior must be controlled. This type of research demands new efforts in interdisciplinary collaboration between chemists, materials scientists, device physicists, and photoelectrochemists.

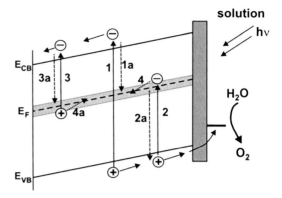

Semiconductor–Liquid Junction: From Fundamentals to Solar Fuel Generating Structures, Fig. 25 Concept and energy schematic for an intermediate band photoanode; *full lines*: excitation processes, *dashed numbered lines*: recombination processes; the intermediate band (*light gray*) is assumed to be partially filled, therefore, E_F is drawn in the middle of this band; a passivation layer that also serves as contact (*darker gray*) is indicated; processes 1–3 represent optical excitation, processes 1a–3a the respective recombination, and processes 4, 4a show the motion of the light-generated excess carriers in the intermediate band (see text)

Beyond Classical Photoelectrochemistry

The nontraditional approaches in photoelectrochemistry encompass systems and effects where the conversion efficiency exceeds that of the single junction Shockley-Queisser limit (defect level absorbers, multiple exciton generation, singlet fission). In a second approach, the use of hot electrons for initiating electrochemical reactions which otherwise would necessitate large overvoltages is attempted. Besides the direct use of non-thermalized hot electrons from the absorber surface [107], recently, contributions from the decay of surface plasmons have been investigated, too [108, 109]. The third approach is based on processes observed in nature, such as Förster and Dexter excitation energy transfer [110, 111]. These topics will be presented below briefly.

In intermediate band solar cell designs, a partially filled band within the gap of a larger energy gap material provides an additional initial state for photo-transitions of electrons (see Fig. 25). The additional excitation routes besides those from the valence band to the conduction band, e.g., from the valence band into the unoccupied states of the intermediate band and from the intermediate band to the conduction band, allow to harvest considerably more photons than possible for a single junction device. Since the states of the intermediate band can also act as recombination sites, it is advantageous to form a depletion layer in the material throughout the range of light absorption. Historically, this concept has been successfully implemented in amorphous Si solar cells. It allows a favorable competition of the drift process with respect to recombination. Intrinsic semiconductors have extended space charge regions and are particularly suited for realization of this concept. Therefore, besides allowing the use of longer wavelength photons for energy conversion, the concept is also applicable to low mobility materials that are prepared in the search for novel photoanode and photocathode materials in photoelectrochemistry. Instead of migration via hopping, drift along the band edges becomes possible which is a much faster process. It is assumed that the mobility in the intermediate band is small and that the processes 4 and 4a in Fig. 25 are slow compared to the drift along the valence and conduction band edges.

By contacting the material with two high band gap semiconductors, it can be shown that, under certain restrictions regarding the properties of the states of the intermediate band, the output voltage is still given by the total energy gap between the valence and the conduction bands [112]. The theoretical efficiency for this type of cell is close to the limiting efficiency of a triple junction cell, e.g., $\sim$63 % [112].

In a water splitting application, electrocatalysts would be anchored onto the surface and the quasi-Fermi level separation for the valence and conduction bands under illumination has to exceed the total voltage needed for water splitting. Since novel materials, developed for photoelectro-catalysis, are often alloyed to some extent, it would be important to accurately determine the formation of intermediate bands in these semiconductors. A method to analyze electronic defects in semiconductors at room temperature with high detection sensitivity is Brewster angle spectroscopy [113, 114]. The method allows to detect wavelength sensitive the Brewster angle and the reflectivity in this angle and has been used to identify defect levels in GaAs and in the ternary chalcopyrite $CuInS_2$ which, together with low temperature photoluminescence, provided an unprecedented energy level scheme for point defects in this solar material [114].

The quantum yield of single absorbers can also be increased by multiple exciton generation (MEG) [115] and singlet fission [116]. The former process, also known as impact ionization, uses photons of higher energy to generate two or more electron hole pairs and has been observed in semiconductor quantum dots [117]. Recently, the corresponding charge injection, resulting in

a current measurement related to this effect, has been communicated [118]. Singlet fission into triplets results in energetically lower excitations. For both processes, MEG and singlet fission, it will be difficult to meet the requirements of providing a photovoltage in excess of $\sim$1.8 V and, although of possible importance for photovoltaic applications, their value for photoelectrochemical water splitting or carbon dioxide reduction has to be doubted.

The use of higher energy electrons from single particle excitations of surface plasmons is presently investigated. Earlier work of photoemission into electrolytes of delocalized surface plasmons showed that electrons are emitted with energies of the plasmon resonance energy [119]. The surface plasmon resonance frequency, ω_s, for rough Ag films, treated here as nearly free electron metal, is given by

$$\omega_s = \frac{\omega_p}{\sqrt{1 + \varepsilon_M}}, \qquad (33)$$

where ε_M denotes the dielectric constant of the adjacent medium which was water in the photoemission experiment, and ω_p is the volume plasmon resonance frequency given by the free electron term (neglecting damping), modified by the effective mass approximation

$$\omega_p = \sqrt{\frac{ne^2}{\varepsilon_0 m*}}. \qquad (34)$$

Here, n is the metal electron density and $m*$ the effective mass. At the liquid interface, $w_s \sim$ 3.5 eV, and Fig. 26 shows the result of a photoemission experiment that also indicates the quantum yield reached in the experiment. The experiment clearly states that high energy (hot) electrons can be emitted from a surface plasmon resonance.

Although the electron emission yield is rather low, it should be considered that in photoemission yield experiments, the work function threshold has to be surmounted. Since the work function in these experiments is about 3 eV [120], the (excess) kinetic energy is only 0.5 eV

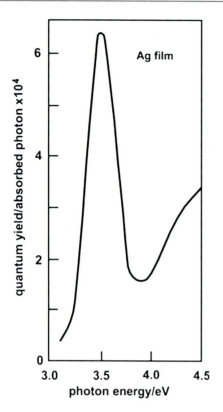

Semiconductor–Liquid Junction: From Fundamentals to Solar Fuel Generating Structures, Fig. 26 Yield photoemission in the spectral range of the (delocalized) Ag surface plasmon [119]; potential −0.2 V (NHE), 1 N H$_2$SO$_4$; note that the quantum yield is below 10^{-3} (see text)

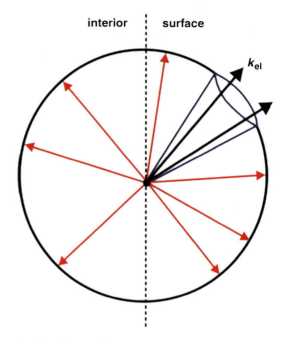

Semiconductor–Liquid Junction: From Fundamentals to Solar Fuel Generating Structures, Fig. 27 Visualization of the (assumed isotropic) electron momentum distribution from a surface plasmon decay; *blue lines* represent the emission cone in a photoemission yield experiment, *black arrows* indicate two examples for emitted electrons with wave vector k_{el}; *red arrows* indicate non-escaping electrons and the *dotted line* indicates the momentum distribution toward the surface or interior of the sample (see also text)

and only those electrons can leave the solid that have sufficient momentum perpendicular to the surface. This condition is typically described by the escape cone, given by

$$\left|k_f^\perp\right| = \frac{1}{\hbar}\sqrt{2m*(E_f - \Phi)}. \quad (35)$$

Assuming translational symmetry parallel to the surface of crystalline or polycrystalline material, the electron wave vector of the final state perpendicular to the surface, $k_f^\perp$, is smaller due to the difference of the final state energy, E_f, and the work function, Φ, of the sample. The emission cone is given by the angular condition, $\theta = \cos^{-1}\left(\frac{|k^\perp|}{|k|}\right)$. Thus, the total yield is considerably reduced since only a reduced amount of excited electrons can leave the solid at this low excess energy. This is indicated in Fig. 27. The actual quantum yield can be higher by at least an order of magnitude. This could be estimated using spherical energy surfaces, but this approximation does not hold well for a d-band metals such as Ag.

Whereas the resonance frequency of delocalized surface plasmons is given by the adjacent medium, which for Ag at the electrolyte interface is located in the near-UV range (350 nm), localized excitations from noble metal nanoparticles occur in the visible spectral range [121] and their resonance frequency can be tuned by their size and shape [122].

The photon scattering cross section is given by the particle polarizability α

$$\sigma_{SC} \propto \left\{\frac{2\pi}{\lambda_{Ph}}\right\}^4 |\alpha|^2, \quad (36)$$

and α depends on the particle size, V_{NP}, and the dielectric functions of the nanoparticle and the surrounding medium, ε_{NP}, ε_M:

$$\alpha \propto V_{NP} \frac{\varepsilon_{NP} - \varepsilon_M}{\varepsilon_{NP} + 2\varepsilon_M} \qquad (37)$$

The surface plasmon resonance for this case is given by $\varepsilon_{NP} = -2\varepsilon_M$. In addition, the excitation of localized plasmons (LSPs), mediated by particle interactions, for instance, on nanostructured metallic surfaces, can result in field enhancements of the order of 10^3. This can influence the temperature profile at and near plasmonic nanoparticles and result in a rate increase of electrochemical reactions. Although the hot electron yield form LSP is presently not known, it can be speculated that it is larger than that of the rough films (Fig. 26) and one might be able to induce electrochemical reactions by overcoming, for example, the initial barrier for the first step in CO_2 reduction [123].

Compared to the rather unlikely application of MEG or singlet fission in light-induced water splitting, the future use of carrier-less excitation energy transfer appears more realistic. This will depend, however, on the control in the preparation of nanoparticles and quantum dots that could act as donor or acceptor in a Foerster transfer chain. Although recognized rather early as a process in natural systems [124], Foerster transfer has more recently been studied in detail in natural photosynthesis where the light-harvesting complexes are spatially separated from the oxidase and hydrogenase enzymes that contain the catalytically active $CaMn_4O_4$ and Fe–Fe or Fe–Ni sites [125].

The first observation of non-radiative excitation energy transfer was made by Cario and Franck [126]. They investigated Hg and Th vapor and illuminated with the resonance line of mercury and found emission spectra from both atoms although Thallium did not absorb the light from the mercury resonance line. Since radiation by re-absorption was not possible, only a non-radiative energy transfer could have been operative with the Hg atoms as donor (sensitizer) and the Th atoms as acceptors.

The description of the effect dates back to Perrin [127] who investigated fluorescence polarization, i.e., the angular correlation between excited and emitted photons from fluorescence molecules in solutions. To explain the observed rather long distance effect in the depolarization, Perrin proposed that excitation energy is transferred between the molecules via a dipole–dipole interaction. He further supposed that the interaction takes place between identical molecules at a fixed frequency. This resulted in an error in the calculation of the intermolecular distance by a factor of about 100, yielding a distance of $R \sim 1.2 \lambda$ (the excitation wavelength). For green light ($\lambda = 500$ nm), the value would be 600 nm instead of the typical values between 1 and 10 nm. Spectral broadening in liquids occurs over a wide spectral range and the probability that donor and acceptor molecules will have the same frequency at the same time is considerably smaller than one as assumed by Perrin. Foerster calculated the probability of a resonance between two narrow-band oscillators. Due to the spread of the donor and acceptor frequencies, the resulting low probability leads to a shortened time for transfer and a corresponding decrease of the distance over which the interaction occurs efficiently [110]. The principle interaction is displayed schematically in the Jablonski diagram, Fig. 28. The radiation-less transfer of the excitation energy is indicated by the dashed lines. Since the transfer occurs after vibrational relaxation, indicated by the blue arrows, the process is relatively slow in agreement with the assumed weak coupling between the transition dipole moments of the interacting molecules. Accordingly, the process is described quantum mechanically by first order perturbation theory and includes vibrational processes. The Foerster transfer rate, k_{FT}, is described in the framework of Fermi's golden rule:

$$k_T = 4\pi^2 |V_{mn}|^2 \delta(v_n - v_m) \qquad (38)$$

$$k_{FT} = 4\pi^2 |V_{mn}|^2 \delta(v_n - v_m)$$

It is given by the interaction matrix element V_{mn} that contains the dipole–dipole term and the energy conservation (energies v_n, v_m) where the donor and acceptor states are $|m\rangle$ and $|n\rangle$, respectively,

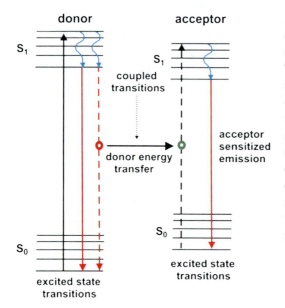

Semiconductor–Liquid Junction: From Fundamentals to Solar Fuel Generating Structures, Fig. 28 Jablonski diagram of Foerster resonance energy transfer; S_0, S_1, ground and excited states of donor and acceptor molecules; color code: *blue*: vibronic relaxation, *black*: excitations, *red*: full lines – fluorescence, *dashed line* coupled Foerster transition (see text)

written in Dirac's bracket notation [128]. The delta function in Eq. 38 defines the resonance condition (compare Fig. 28).

The interaction matrix element V_{mn} is given by the product of the donor and acceptor transition dipole moments, $\boldsymbol{\mu}_m$ and $\boldsymbol{\mu}_n$, respectively. It is modulated by a relative orientation factor, κ, of the transition dipoles with respect to each other. The matrix element in Eq. 38 then becomes

$$V_{mn} = \frac{l^2}{4\pi\varepsilon_0\varepsilon \cdot r^3} \vec{\mu}_m \vec{\mu}_n \kappa, \quad (39)$$

where l is a factor that describes the difference of interaction between vacuum and the considered medium (dielectric constant ε) and the individual transition dipole moments are given by the expression:

$$\vec{\mu}_{fi} = -e \sum_k \langle E_f | \vec{r}_k | E_i \rangle \langle v_f | v_i \rangle$$

$$= \vec{\mu}_\varepsilon^{f,i} S(v_f, v_i) \quad (40)$$

The first term in the sum on the right-hand side describes the electronic interaction and the second Dirac bracket denotes the vibrational terms. The right-hand side of Eq. 40 condenses these terms into an electronic matrix element and an overlap integral (the brackets are used to describe integrals in Dirac's notation) between vibrational states in the initial (subscript i) and final states (subscript f). The Foerster transition moments can then be expressed by the respective transition dipole moments of donor (subscript D) and acceptor (subscript A) molecules, e.g., $\vec{\mu}_m = S_m \vec{\mu}_D$; $\vec{\mu}_n = S_n \vec{\mu}_A$ and the Foerster transfer rate is then given by:

$$k_{FT} = \frac{l^4 \kappa^2}{4\varepsilon_0^2 n^4 r^6} \sum_{n,m} \mu_D^2 \mu_A^2 S_m^2 S_n^2 \delta(v_m - v_n) \quad (41)$$

This expression contains the well-known distance dependence ($r^{-1/6}$) and the spectral overlap integral:

$$J = \int \frac{A(v)F(v)}{v^4} dv \quad (42)$$

The Franck-Condon factors S in Eq. 41 describe the vibrational overlap. The excitation energy transfer rate can thus be described by the product of the fluorescence spectrum of the donor and the absorption spectrum of the acceptor molecule. The total rate is given by:

$$k_{FT} = 8.8 \times 10^{17} \frac{\kappa^2}{n^2 \tau_D r^6} \int \frac{A(v)F(v)}{v^4} dv \quad (43)$$

For an efficiency of energy transfer of 0.5, one obtains the Foerster radius, i.e., the distance for that energy transfer between donor and acceptor part in the interaction. This distance is given by

$$R_0^6 = 8.8 \times 10^{17} \frac{\kappa^2}{n^4} J, \quad (44)$$

and the rate can be written in terms of the distance and the donor fluorescence decay rate (inverse lifetime) k_D

$$k_T = k_D \left(\frac{R_0}{R}\right)^6. \quad (45)$$

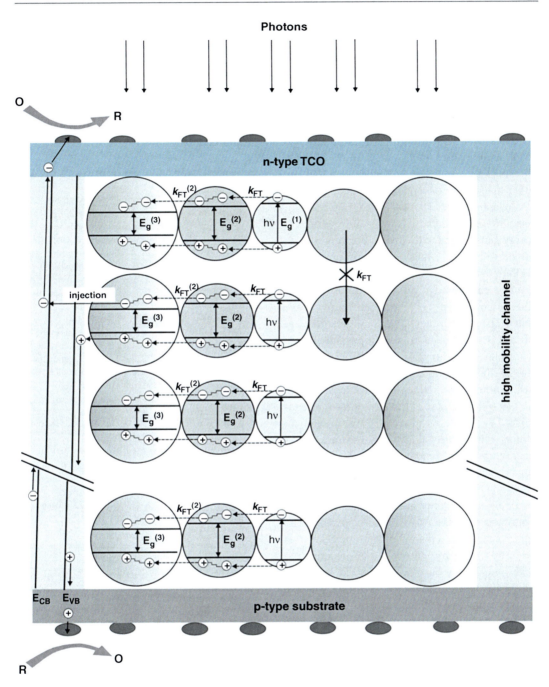

Semiconductor–Liquid Junction: From Fundamentals to Solar Fuel Generating Structures, Fig. 29 Schematic of a photoelectrocatalytic structure based on Foerster transfer in quantum size nanoparticles; size-dependent energy gaps ($E_g^{(x)}$, $x = 2, 3$) and Foerster transfer rates, k_{FT}, are indicated; for the smallest particle, E_g is largest (not shown); ellipsoids: catalyst particles; TCO: transparent conductive oxide; the phase labeled high mobility channel is identical to the semiconductor phase on the left-hand side (see text)

Typically, R_0 is in the order of 5–10 nm and this mechanism provides a different mode for efficient excitation energy transfer without radiation or actual charge carrier (electron) transport.

One might hypothesize about solar energy transfer devices that are built using more robust materials than those used in photosynthesis and about architectures that would allow efficient photon energy transfer over distances that are larger than those for tunneling. Such a structure is shown in Fig. 29. The structure is composed of quantum size nanoparticles that have different energy gaps related to their size. For neighboring particles, Foerster transfer will occur only in the horizontal direction, as indicated, but interaction between identical particles has a low probability, as indicated by the crossed arrow. If the system is compartmentalized by junctions to intrinsic semiconductors with an extended space charge layer as shown in the left-hand side of Fig. 29, the excitation energy is transferred into electron hole pairs that become separated in the electric field across the bordering semiconductor. The respective charges reach the catalytic centers that are drawn as nanoparticles where they induce oxidation and reduction reactions. Not shown in Fig. 29 are direct photon excitation processes in particles 2 and 3 (labeled regarding their energy gap) which will strongly contribute to the total photocurrent. The main achievable photovoltage, however, is given by the energy gap of the semiconducting phase, the energy band alignment with the contact phases (TCO and p-type substrate) and the energetic alignment of the electrolyte with the catalyst particles and the TCO or p-substrate semiconductor.

Acknowledgments Valuable discussions with S. Fiechter, K. Skorupska, (Helmholtz Zentrum Berlin), H. Atwater, N.S. Lewis, P. Narang, A. Leenher, M. Shaner (California Institute of Technology), G. Parsons (North Carolina State University), and T. Hannappel (Technical University Ilmenau, Germany) are gratefully acknowledged. JCAP acknowledgement: This material is based upon work performed by the Joint Center for Artificial Photosynthesis, a DOE Energy Innovation Hub, supported through the Office of Science of the U.S. Department of Energy under Award No. DE-SC0004993.

Cross-References

▶ Semiconductor Electrode
▶ Semiconductor Junctions, Solid-Solid Junctions
▶ Semiconductors, Principles

References

1. Van Troostwijk AP, Deiman JR (1789) Obs Phys 35:369
2. Becquerel AE (1839) Compt Rend Acad Sci 9:145
3. Fujishima A, Honda K (1972) Nature 238:37
4. Gerischer H (1975) J Electroanal Chem 58:263
5. Lewerenz HJ (2010) Tailoring of interfaces for the photoelectrochemical conversion of solar energy. In: Kolb DM, Alkire RC, Ross PN, Lipkowski J (eds) Advances in electrochemistry and electrochemical engineering. Wiley-VCH, Weinheim, Germany, pp 61–181
6. Gerischer H (1966) J Electrochem Soc 113:1174
7. Nozik AJ (1976) Appl Phys Lett 29:150
8. Tributsch H, Bennett JC (1977) J Electroanal Chem 81:97
9. Hodes G, Manassen J, Cahen D (1976) Nature 261:403
10. Bocarsly AB, Bookbinder DC, Dominey RN, Lewis NS, Wrighton MS (1980) J Am Chem Soc 102:3683
11. Lewerenz HJ, Gerischer H, Lübke M (1984) J Electrochem Soc 131:100
12. Kautek W, Gerischer H, Tributsch H (1980) J Electrochem Soc 127:2471
13. Bard AJ, Bocarsly AB, Fan FRF, Walton EG, Wrighton MS (1980) J Am Chem Soc 102:3671
14. Miller B, Heller A, Menezes S, Lewerenz HJ (1981) Faraday Discuss Chem Soc 70:223
15. Heller A, Lewerenz HJ, Miller B (1980) Ber Bunsenges Phys Chem 84:592
16. Licht S, Tenne R, Dagan G, Hodes G, Manassen J, Cahen D, Triboulet R, Rioux J, Levy-Clement C (1985) Appl Phys Lett 46:608
17. Parkinson BA, Heller A, Miller B (1978) Appl Phys Lett 33:521
18. Heller A, Miller B, Lewerenz HJ, Bachmann KJ (1980) J Am Chem Soc 102:6555
19. Menezes S, Lewerenz HJ, Bachmann KJ (1983) Nature 305:615
20. Tenne R, Wold A (1987) Appl Phys Lett 47:707
21. Lewerenz HJ, Goslowsky H, Husemann K-D, Fiechter S (1986) Nature 321:687
22. Heller A, Vadimsky RG (1981) Phys Rev Lett 46:1153
23. Bard AJ, Fox MA (1995) Acc Chem Res 28:141
24. Gregg BA (2003) J Phys Chem B 107:4688
25. O'Reagan B, Grätzel M (1991) Nature 353:737
26. Grätzel M (2001) Nature 414:338
27. Marcus RA (1956) J Chem Phys 24:966
28. Gerischer H (1961) Z Phys Chem N F 27:48
29. Tributsch H (2007) Electrochim Acta 52:2302

30. Tanaka K, Ooyama D (2002) Coord Chem Rev 226:211
31. Norskov JK, Bligaard T, Rossmeisl J, Christensen CH (2009) Nat Chem 1:37
32. Taylor CD, Wasiliski SA, Filhol J-S, Neurock M (2006) Phys Rev B 73:165402
33. Pliego JR Jr, Riveros JM (2001) J Phys Chem A 105:7241
34. Gerischer H, Ekardt W (1983) Appl Phys Lett 43:393
35. Khan SUM, Kainthla RC, Bockris JO'M (1987) J Phys Chem 91:5974
36. Gerischer H (1991) J Phys Chem 95:1356
37. Schulte K, Lewerenz HJ (2002) Electrochim Acta 47:2633
38. Lewerenz HJ, Lübke M, Menezes S, Bachmann KJ (1981) Appl Phys Lett 39:798
39. Miller B, Heller A (1976) Nature 262:680
40. There is some uncertainty regarding the vacuum work function of the normal hydrogen electrode. While calculations indicate values of 4.4 eV to 4.5 eV, experimental data have consistently yielded higher values between 4.7 eV and 4.8 eV. Taking into account this discrepancy, a value of 4.6 eV has been chosen here as a mean value
41. Lewerenz HJ, Bitzer T (1992) J Electrochem Soc 139:L21
42. Lewerenz HJ, Bitzer T, Gruyters M, Jacobi K (1993) J Electrochem Soc 140:L44
43. Jacobi K, Gruyters M, Geng P, Bitzer T, Aggour M, Rauscher S, Lewerenz HJ (1995) Phys Rev B 51:5437
44. Bansal A, Li X, Yi SI, Weinberg WH, Lewis NS (2001) J Phys Chem B 105:10266
45. Rauscher S, Dittrich T, Aggour M, Rappich J, Flietner H, Lewerenz HJ (1995) Appl Phys Lett 66:3018
46. Webb LJ, Lewis NS (2003) J Phys Chem B 107:5404
47. Marcus RA (1981) Int J Chem Kinet 13:865
48. Crowell CR, Rideout VL (1969) Solid-State Electron 12:89
49. Petrii OA, Nazmutdinov RR, Bronshtein MD, Tsirlina GA (2007) Electrochim Acta 52:3493
50. Bockris JO'M, Reddy AKN (1970) Modern electrochemistry. Plenum, New York, p 880
51. Barbara PF, Meyer TJ, Ratnerm MA (1996) J Phys Chem 100:13148
52. Gärtner WW (1959) Phys Rev 116:84
53. Lewerenz HJ, Jungblut H (1995) Photovoltaik; Grundlagen und Anwendungen. Springer, Heidelberg/New York
54. Reiss H (1978) J Electrochem Soc 125:937
55. Gobrecht J, Gerischer H (1979) Sol Energy Mater 2:131
56. Wilson R (1977) J Appl Phys 48:4297
57. Reichman J (1981) Appl Phys Lett 35:251
58. Khan SUM, Bockris JO'M (1984) J Phys Chem 88:2504
59. Schlichthörl G, Lewerenz HJ (1998) J Electroanal Chem 443:9
60. Lewerenz HJ (1993) J Electroanal Chem 356:121
61. Aymerich-Humet X, Serra-Mestres F, Millàn J (1981) Solid-State Electron 24:981
62. Memming R (2000) In: Semiconductor electrochemistry, applications. Wiley-VCH, Weinheim (Chap 11)
63. Stempel T, Aggour M, Skorupska K, Munoz AG, Lewerenz HJ (2008) Electrochem Commun 10:1184–1186
64. Lewerenz HJ (2011) J Electroanal Chem 662:184
65. Tributsch H (1977) Ber Bunsenges Phys Chem 81:361
66. Cooper G, Noufi R, Frank AJ, Nozik AJ (1982) Nature 295:578
67. Lewerenz HJ, Skorupska K, Aggour M, Stempel-Pereira T, Grzanna J (2009) J Solid State Electrochem 13:185
68. Wang P, Zakeeruddin SM, Moser JE, Nazeeruddin MK, Sekiguchi T, Graetzel M (2003) Nat Mater 2:402
69. Lewis JS, Hoertz PG, Glass JT, Parsons GN (2012) J Vac Sci Technol A 30:010803
70. Chen S, Wang L-W (2012) arXiv:1203.1970v1
71. Gerischer H (1977) J Electroanal Chem 82:133
72. Lewerenz HJ, Heine C, Skorupska K, Szabo N, Hannappel T, Vo-Dinh T, Campbell SA, Klemm HW, Munoz AG (2010) Energy Environ Sci 3:748
73. Munoz AG, Lewerenz HJ (2010) Chem Phys Chem 11:1603
74. Menezes S, Lewerenz HJ, Betz G, Bachmann KJ, Kötz R (1984) J Electrochem Soc 131:3030
75. Lewerenz HJ, Skorupska K, Munoz AG, Nüsse N, Lublow M, Vo-Dinh T, Kulesza P (2011) Electrochim Acta 56:10726
76. Lewerenz HJ (2011) Electrochim Acta 56:10713
77. Gerischer H (1978) J Vac Sci Technol 15:1422
78. Allongue P, Kieling V, Gerischer H (1995) Electrochim Acta 40:1353
79. Lewerenz HJ, Aggour M, Murrell C, Kanis M, Jungblut H, Jakubowicz J, Cox PA, Campbell SA, Hoffmann P, Schmeißer D (2003) J Electrochem Soc 150:E 185
80. Bitzer T, Gruyters M, Lewerenz HJ, Jacobi K (1993) Appl Phys Lett 63:397
81. Benkö G, Kallioinen J, Myllyperkiö P, Trif F, Korppi-Tommola JEI, Yartsev AP, Sundström V (2004) J Phys Chem B 108:2862
82. Gobrecht J, Gerischer H, Tributsch H (1978) Ber Bunsenges Phys Chem 82:1331
83. Goldberg AM, Bea AR, Lévy FA, Davis EA (1975) Philos Mag 32:367
84. Hoffmann WK, Lewerenz HJ (1988) Sol Energy Mater 17:369
85. Tsirlina T, Cohen S, Cohen H, Sapir L, Peisach M, Tenne R, Matthaeus A, Tiefenbacher S, Jaegermann W, Ponomarev EA, Lévy-Clément C (1996) Sol Energy Mater Sol Cells 44:457
86. Delphine SM, Jayachandran M, Sanjeeviraja C (2011) Crystallogr Rev 17:281
87. Cahen D (1984) Appl Phys Lett 45:746
88. Gerischer H (1969) Surf Sci 13:265

89. Jakubowicz J, Jungblut H, Lewerenz HJ (2003) Electrochim Acta 49:137
90. Vanmaekelbergh D (1997) Electrochim Acta 42:1121
91. Leng WH, Zhang Z, Zhang JQ, Cao CN (2005) J Phys Chem B 109:15008
92. Christensen NE (1988) Phys Rev B 37:4528
93. Mönch W (1994) Surf Sci 299(300):928
94. Wang W-L, Zunger A (1994) Phys Rev Lett 73:1039
95. Lu JJ, Chen J, He YL, Shen WZ (2007) J Appl Phys 102:063701
96. Sanderson RT (1983) J Am Chem Soc 105:2259
97. Hunger R, Fritsche R, Jaeckel B, Jaegermann W, Webb LJ, Lewis NS (2005) Phys Rev B 72:045317
98. Lewerenz HJ, Aggour M, Murrell C, Kanis M, Hoffmann P, Jungblut H, Schmeißer D (2002) J Non Cryst Solids 303:1
99. Park KJ, Parsons GN (2012) J Vac Sci Technol A 30:01A162
100. Jang S, Newton MD, Silbey RJ (2004) Phys Rev Lett 92:218304
101. W. Lubitz, Private communication
102. Lubitz W, Reijerse EJ, Messinger J (2008) Energy Environ Sci 1:15–31
103. Peng Q, Lewis JS, Hoertz PG, Glass JT, Parsons GN (2012) J Vac Sci Technol A 30:010803
104. Osterloh FE (2008) Chem Mater 20:35
105. H.A. Atwater, K. Fountaine, M. Shaner, Private communication
106. Woodhouse M, Herman GS, Parkinson BA (2005) Chem Mater 17:4318
107. Ross RT, Nozik AJ (1982) J Appl Phys 53:3813
108. Lublow M, Skorupska K, Zoladek S, Kulesza PJ, Vo-Dinh T, Lewerenz HJ (2010) Electrochem Commun 12:1298
109. Munoz AG, Skorupska K, Lewerenz HJ (2010) Chem Phys Chem 11:2919
110. Förster T v (1948) Ann Phys 2:55
111. Dexter DL (1953) J Chem Phys 21:836
112. Luque A, Martí A (1997) Phys Rev Lett 78:5014
113. Lewerenz HJ, Dietz N (1991) Appl Phys Lett 59:1470
114. Lewerenz HJ, Dietz N (1993) J Appl Phys 73:4975
115. Nozik AJ (2001) Ann Rev Phys Chem 52:193
116. Hanna MC, Nozik AJ (2006) J Appl Phys 100:074510
117. Schaller RD, Klimov VI (2004) Phys Rev Lett 92:186601
118. Sambur JB, Novet T, Parkinson BA (2010) Science 330:63
119. Neff H, Sass JK, Lewerenz HJ (1984) Surf Sci 143:L356
120. Sass JK, Lewerenz HJ (1977) Journ de Phys 38(C5):277
121. Stuart HR, Hall DG (1996) Appl Phys Lett 69:2327
122. Catchpole KR, Polman A (2008) Appl Phys Lett 93:191113
123. Le M, Ren M, Zhang Z, Sprunger PT, Kurtz RL, Flake JC (2011) J Electrochem Soc 158:E 45 (and references therein)
124. Oppenheimer JR (1941) Phys Rev 60:158
125. Scholes GD (2003) Annu Rev Phys Chem 54:57
126. Cario G, Franck J (1923) Z Phys 17:202
127. Perrin J, Hebd CR (1927) Seances Acad Sci 184:1097
128. Dirac PAM (1958) The principles of quantum mechanics. Clarendon, Oxford, p 180

Semiconductors Group IV, Electrochemical Decomposition

Paul A. Kohl
Georgia Institute of Technology, School of Chemical and Biomolecular Engineering, Atlanta, GA, USA

Introduction

Silicon is the dominant semiconductor for integrated circuits. As a result of its use in integrated circuits and other unique attributes, the electrochemical oxidation and/or dissolution of silicon (decomposition) is one of the most studied topics in applied electrochemistry. First, the oxidation of silicon forming silicon dioxide, an essential part of modern integrated circuits, is an electrochemical process. Silicon dioxide is used to form the gate electrode in field-effect transistors. Silicon dioxide is also used to isolate transistors from each other and is used as a masking material during the doping of the source and drain regions of the transistors. Second, the complexation and dissolution of silicon dioxide in an aqueous electrolyte is the most reliable way to clean a silicon surface. Surface impurities become trapped in or on the silicon dioxide surface and are swept away during dissolution of the layer. Third, luminescence from porous silicon (PS) was observed in 1990, which sparked worldwide curiosity of the origin and commercialization of the material [1]. A silicon-based electroluminescent source could revolutionize optical communications on integrated circuits and in other applications. However, this has not come to pass. Finally, the electrochemical or photoelectrochemical oxidation of silicon can be used to shape silicon into structures of interest for microelectromechanical devices and sensors.

Oxidation of Silicon

The tetravalent state is the stable form of a silicon when oxidized. The mechanism of oxidation and dissolution of silicon and germanium was first studied by Turner et al. in an interest to understand the etching and cleaning of silicon and germanium surfaces [2]. The mechanism of the oxidation and dissolution of semiconductor is like that of a metal except that (1) two types of charge carriers can be involved, valence band holes and conduction band electrons, and (2) the density of charge carriers at the solid-liquid interface is much smaller for the semiconductor than for a metal. The electrochemical reaction of silicon in an aqueous solution is given by Eq. 1:

$$Si + 3H_2O + 4ah^+ \rightarrow H_2SiO_3 + 4H^+ + 4 \times (1 - a)e^- \quad (1)$$

where h^+ represents a hole in the valence band, e^- is an electron in the conduction band, and "a" is the fraction of the total charge passed belonging to holes. The value of "a" can be determined by performing electrochemical experiments using photoexcited carriers in n-type and p-type silicon and properly biasing the silicon electrode, such as by Propst [3]. Nitric acid can be used as the oxidizing agent in the etching (e.g., cleaning) of silicon [4]. The reaction for silicon etching in nitric acid solutions is given by Eq. 2 [2]:

$$3Si + 4HNO_3 + 9H2O + 4ah^+ \rightarrow 3H_2SiO_3 + NO + 4(1 - a)e^- \quad (2)$$

The species H_2SiO_3 (or a hydrated form of SiO_2, $SiO_2 + H_2O$) is not soluble in water. A complexing agent for Si(IV) can be used to dissolve the insoluble oxide from the surface. There are very few complexing agents for Si (IV) making silicon dioxide stable in a wide variety of solvents, which is highly valuable in integrated circuit manufacturing. HF is commonly added to aqueous and nonaqueous solutions to dissolve the oxide, as shown in Eq. 3:

$$H_2SiO_3 + 6HF \; SiF_6^{2-} + 2H^+ + 3H_2O \quad (3)$$

Electropolishing and Formation of Porous Silicon (PS)

The direct electrochemical etching of silicon in an HF or buffered HF solution (mixture of HF and KF to stabilize the pH) is a complex function of current density and HF concentration. A dynamic process takes place where silicon dioxide is created and dissolved. Thus, the rate-limiting reaction is a function of current density (creation of Si (IV)), HF concentration (dissolution of Si(IV)), agitation, and other factors. A typical current-voltage curve for a moderately doped p-type silicon in 1% HF is shown in Fig. 1 [5].

As the potential is scanned from negative to positive values in Fig. 1, the anodic current initially rises exponentially. Eventually, the current deviates from its exponential rise and a peak, J_1, is observed. At potentials positive of J_1, the current becomes relatively independent of potential. In this potential region, the rate of oxidation is high compared to the rate of dissolution of the oxide, and the overall rate is limited by the dissolution of the surface oxide. The silicon enters into a traditional electropolishing mode. In electropolishing, the surface is covered by oxide and dissolution is the rate-determining step. A rough surface can be made smooth by preferential oxidation of the surface peaks (rough areas have a high primary current distribution) compared to the remainder of the surface resulting in an oxide-covered surface. The overall rate of the reaction is then determined by the rate of dissolution of the oxide across the surface, which is relatively uniform, resulting isn a smoothened surface.

In the exponential region in Fig. 1, hydrogen gas evolution is observed and a PS layer is observed in the silicon. In the transition region between the exponential region and the electropolishing region, PS may also be formed.

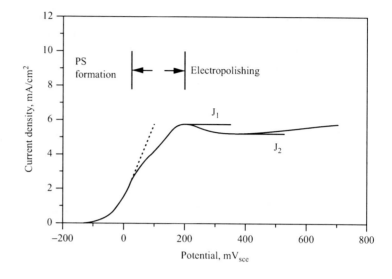

Semiconductors Group IV, Electrochemical Decomposition, Fig. 1 Current versus voltage for moderately doped p-Si in 1% HF solution

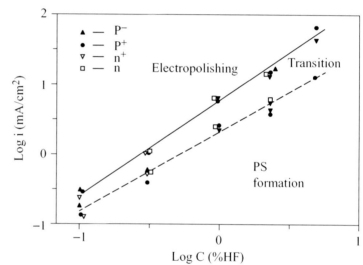

Semiconductors Group IV, Electrochemical Decomposition, Fig. 2 Formation conditions for porous silicon (PS). The solid line is the peak current density, and the dashed line is the maximum slope from Fig. 1 (above)

No hydrogen evolution is observed in the electropolishing region. A wide variety of pore structures can be created during etching silicon in the PS region, as identified in Fig. 1. The pores created in PS can range from nanometer size and highly branched to many micrometers in size and straight walled. The pore structure depends on the doping type and doping level of the silicon, HF concentration, and illumination level and direction (front or back illumination).

Most theories concerning PS have dealt with only isolated aspects of the creation of PS. No comprehensive theory has yet been presented to account for the full range of effects and conditions. Figure 2 summarizes the current and HF concentration ranges for the electropolishing, PS, and transition region. The different regions can be better understood by dividing the voltage drop in the electrochemical path into different phases. The phases include the silicon substrate, the space-charge region within the semiconductor, the Helmholtz region in solution, the surface oxide on the silicon, and the electrolyte. The dominate phase in the current path can be changed creating the different surface structures. For example, in heavily doped silicon oxidized in the exponential region of Fig. 1, the Helmholtz region in solution dominates the potential drop. In lightly doped silicon, the space-charge region dominates.

Mechanism of Silicon Oxidation and Dissolution

There are two competing pathways in the oxidation and dissolution of silicon which involve either HF or H_2O. An aqueous solution spontaneously reacts with silicon to form a passivating oxide. The presence of HF leads to the dissolution of the oxide-forming H_2SiF_6. Silicon can be oxidized to the tetravalent state consuming valence band holes or injecting electrons in the conduction band, as mentioned above. The dissolution of a silicon ion results in a hydrogen-terminated surface (Si-H). Oxidation of the hydrogen-terminated surface silicon results in formation of a fluoride-terminated surface which weakens the Si-Si back bonds (those bonds between the surface silicon species and the remainder of the silicon crystal). The number of holes can vary between two and four depending on the orientation of the silicon, electrolyte, and other factors. When the back bonds are weakened, electrons can be directly injected into the conduction band, thus changing the value of "a" in Eq. 1. Reaction of the partially oxidized silicon intermediate with HF can result in formation of an F-Si bond and a hydrogen atom, which can lead to evolution of hydrogen gas.

Future Directions

Silicon remains the dominant semiconductor in integrated circuits and will remain so for many years to come. The enormous infrastructure of the integrated circuit industry ensures that plentiful quantities of low-cost silicon will be available for other uses, such as solar cells, sensors, and microelectromechanical devices. The oxidation, dissolution, cleaning, and shaping of silicon will remain a critical aspect of those technologies. Future insights into controlling and understanding the oxidation and dissolution of silicon can be expected. Issues remain to be discovered concerning PS; however, its use is currently limited to niche applications such as sensors. The emission of visible light from PS remains essentially a curiosity because of its irreversible nature.

Cross-References

▶ Semiconductor Electrode
▶ Semiconductor Junctions, Solid-Solid Junctions
▶ Semiconductor–Liquid Junction: From Fundamentals to Solar Fuel Generating Structures
▶ Semiconductors, Principles

References

1. Canham LT (1990) Silicon quantum wire array fabrication by electrochemical and chemical dissolution of wafers. Appl Phys Lett 57:1046
2. Turner DR (1960) On the Mechanism of Chemically Etching Germanium and Silicon. J Electrochem Soc 107:810
3. Propst E, Kohl PA (1993) The Photoelectrochemical Oxidation of n-Si in Anhydrous HF – Acetonitrile. J Electrochem Soc 140:L78
4. Gator HC, Lavine MC (1965) Semiconductor Electrochemistry. In: Gibson AF, Burgess RE (eds) Progress in semiconductors, vol 9. Hayward, London, p 1
5. Zhang XG (2004) Morphology and Formation Mechanisms of Porous Silicon. J Electrochem Soc 151:C69

Semiconductors Groups II-IV and III-V, Electrochemical Deposition

I-Wen Sun[1] and Po-Yu Chen[2]
[1]Department of Chemistry, National Cheng kung University, Tainan, Taiwan
[2]Department of Medicinal and Applied Chemistry, Kaohsiung Medical University, Kaohsiung, Taiwan

Introduction

First introduced in the late 1970s [1–5], electrodeposition is a low-cost, large-area semiconductor growth technique of thin films for applications in macroelectronic devices such as solar panels and large-area displays. Chalcogenide compounds (sulfides, selenides, and tellurides) afford larger deviations from stoichiometry without altering their transport properties (self-compensation effects), which may be due to their increased ionicity, favoring the use of electrodeposition [6]. Several review papers and books on the electrodeposition of semiconductors are available [6–12].

The electrodeposition of compound semiconductors has been performed using several approaches. In two-stage methods, layers of the component elements are first electrodeposited sequentially and then annealed to convert them into compounds. Alternatively, the electrodeposition of compound semiconductors can be achieved by direct codeposition from a single bath containing all the precursors needed for creating the compound. In electrochemical atomic layer epitaxy (EC-ALE), the semiconductor compound is formed by successive deposition of atomic layers of the component elements by a surface-limited electrochemical reaction such as underpotential deposition (UPD). Each layer is formed one at a time for each component, and layer-by-layer growth is performed by running the depositions in cycles. Usually, EC-ALE cycles involve switching deposition solutions and the potential to form each atomic layer of each component. The number of cycles determines the thickness of the deposited film.

This entry focuses on the electrodeposition of binary II-VI and III-V compounds. The electrodeposition of metal oxide semiconductors such as ZnO is covered in another entry.

The electrodeposition of III-V compounds is complicated by the very negative reduction potential of III elements relative to that of V elements. Consequently, electrodeposited III-V compounds are mostly limited to GaAs and InSb, though AlSb, GaSb, InAs, and some ternary compounds have also been reported.

Compared with the electrodeposition of III-V semiconductors, that of II-VI semiconductors is relatively easy. Therefore, many reports are available on the electrodeposition of II-VI semiconductors from aqueous baths, nonaqueous baths, molten salts, and ionic liquids. II-VI semiconductors are formed by metallic element Zn, Cd, or Hg and the semimetals S, Se, or Te.

II-VI compounds

CdS

From Acidic Aqueous Solutions

Acidic aqueous solutions used for the cathodic electrodeposition of CdS semiconductors usually contain Cd^{2+} and $S_2O_3^{2-}$ (thiosulfate) ions as the sources of Cd and S, respectively [13–17]. Sometimes, complexing agents such as EDTA are used to coordinate Cd^{2+} ions [14]. Gelatin has been added to stabilize small sulfur particles (colloidal sulfur in the form of S_8) to provide better conditions for the electrodeposition [16]. According to Denisson [18], two overall processes explain the formation of CdS by electrodeposition:

$$Cd^{2+} + S_2O_3^{2-} + 2e^- \rightarrow CdS + SO_3^{2-} \quad (1)$$

$$2Cd^{2+} + S_2O_3^{2-} + 6H^+ + 8e^- \rightarrow 2CdS + 3H_2O \quad (2)$$

However, Zarębska and Skompska [15] believe that (Eq. 1) and (Eq. 2) do not reflect all possible reactions. In addition, Takahashi et al. proposed another mechanism [16]:

$$Cd^{2+}_{(ads)} + S_{8(ads)} + 2e^- \rightarrow CdS + S_{7(ads)} \quad (3)$$

S^{2-} (from Na_2S) and SCN^- have also been used as the source of S in the electrodeposition of CdS. S^{2-} (from Na_2S) has been employed for the atom-by-atom growth of CdS thin films via UPD on Au(1 1 1) [19] and SCN^- has been used for the atom-by-atom growth of CdS thin film via the electrochemical reduction of Cd^{2+}-SCN^- complex [20].

From Nonaqueous Solutions

DMSO (dimethylsulfoxide) has been used as an organic solvent for the cathodic electrodeposition of CdS because elemental S is soluble in DMSO. $Cd(ClO_4)_2$ and $CdCl_2$ have been used as the sources of Cd [21, 22]. It was supposed that S^{2-} species is produced during cathodic electrolysis and reacts with Cd^{2+} to form CdS.

From Molten Salts or Ionic Liquids

CdS has been cathodically electrodeposited from ionic liquid methyltributylphosphonium ($P_{1,4,4,4}$) tosylate containing Cd^{2+} (from $CdCl_2$) and $S_2O_3{}^{2-}$(from $Na_2S_2O_3$) at 130–150 $^{\circ}$C [23]. The following mechanisms were supposed for the formation of CdS in this ionic liquid [24].

$$S_2O_3{}^{2-} + 2e^- \leftrightarrow S^{2-} + SO_3{}^{2-} \qquad (4)$$

$$S^{2-} + Cd^{2+} \rightarrow CdS \qquad (5)$$

Other Electrochemical Approaches

CdS has also been prepared from aqueous solutions by other methods, with electrodeposition as an important step. CdS nanotubes have been formed by the self-assembly of CdS nanowires that were prepared by electrodepositing Cd into an AAO (anodic aluminum oxide) template followed by sulfurization [25]. A hybrid electrochemical/chemical (E/C) method was employed to synthesize epitaxially oriented CdS on graphite surfaces [26], which involves the electrodeposition of Cd, the conversion of Cd to $Cd(OH)_2$, and the formation of CdS via the displacement of OH^- by S^{2-}. Another E/C method in which H_2S was used as the sulfur source and reacted with electrodeposited Cd at 300 $^{\circ}$C was reported in the preparation of CdS/S core-shell nanoparticles [27].

EC-ALE has been used for the formation of CdS on Au(1 1 1) surfaces. The process involves the layer-by-layer electrodeposition of S and Cd in sequence [28, 29].

CdSe

From Acidic Aqueous Solutions

The cathodic electrochemical codeposition of Cd and Se is usually accomplished in acidic aqueous solutions containing Cd^{2+} and $SeO_3{}^{2-}$ [30–43]. $SeO_3{}^{2-}$ ions are usually introduced by the addition of SeO_2 or H_2SeO_3. In the acidic media, SeO_2 dissolved in H_2O to be H_2SeO_3. The formation of CdSe can be simply expressed as the following reaction [44]:

$$H_2SeO_3 + Cd^{2+} + 6e^- + 4H^+ \rightarrow CdSe + 3H_2O \qquad (6)$$

Many substrates have been employed for the electrodeposition of CdSe. Those that induce special morphologies of CdSe are mentioned here. AAO has been employed as a template for the dc (direct current) electrochemical deposition of CdSe nanorod arrays [31]. Epitaxial CdSe layers have been electrodeposited onto a (1 1 1) InP single crystal [32]. HOPG (highly oriented pyrolytic graphite) has been used as a substrate for the electrodeposition of CdSe nanowires [33] via the cyclic electrodeposition/stripping technique [35] in which excess Cd and Se in the nanowires are selectively stripped. CdSe nanowires have been prepared via electrodeposition using photoresist templates [34]. CdSe quantum-dot arrays have been prepared via electrodeposition on a graphene basal plane using mesoporous silica thin-film templates [39].

Pulse electrodeposition (pulsed potential or pulsed current) has been also employed for CdSe formation [30]. In addition, EC-ALE has been employed for the electrochemical synthesis of CdSe on Au substrates through the layer-by-layer UPD of Se and Cd in sequence [45, 46].

A special example of CdSe electrodeposition has been carried out on Ti substrates in acidic medium containing $CdSO_4$ and SeO_2 with no external current source [47]. The potential different between the anode Cd rod and the cathode Ti rod is the driving force of the electrodeposition.

A simple two-step method for the preparation of CdSe has been developed which consists of the electrodeposition of Cd on a nickel substrate and the subsequent dipping of the Cd film in an acidic SeO_2 solution [48]. In an acidic medium, Cd tends to dissolve and therefore, CdSe forms through the reaction of Cd and H_2SeO_3.

The electrodeposition of CdSe has also been used for the photoelectrochemical detection of Cd^{2+} ions in aqueous solutions [36].

From Basic Aqueous Solutions

Basic aqueous solutions used for the electrodeposition of CdSe semiconductor usually contain

Cd^{2+} (from various Cd salts) and $SeSO_3^{2-}$ (selenosulfate, selenosulfite, or selenosulfite; from Na_2SeSO_3 or elemental Se dissolved in excess SO_3^{2-}) as the sources of Cd and Se, respectively. Cd^{2+} ions are usually complexed with organic ligands such as citrate [49], NTA (nitrilotriacetate) [50–52], or EDTA [53]. Complexing agents are introduced to prevent the precipitation of $Cd(OH)_2$ in the alkaline solutions.

Several mechanisms have been reported to explain the formation of CdSe by electrodeposition. Tena-Zaera et al. [51] supposed that CdSe is electrodeposited via the following reaction:

$$Cd^{2+} + Se + 2e^- \rightarrow CdSe \qquad (7)$$

It must be emphasized that $SeSO_3^{2-}$ ions in the electrodeposition baths are formed by the dissolution of elemental Se in SO_3^{2-} solution. The formation of CdSe can also be explained by the codeposition of Cd and Se, or the initial formation of Se^{2-} followed by CdSe precipitation onto the electrode surface [54]:

$$SeSO_3^{2-} + 2e^- \rightarrow Se^{2-} + SO_3^{2-} \qquad (8)$$

$$Cd^{2+} + Se^{2-} \rightarrow CdSe \qquad (9)$$

However, the mechanism [53] where EDTA is used as the complexing agent in an ammonia buffer may be closest to the real reaction:

$$(NH_4)_2[Cd(EDTA)] + (NH_4)_2SeSO_3 + 2e^- \rightarrow$$
$$CdSe + (NH_4)_2EDTA + (NH_4)_2SO_3 \qquad (10)$$

Equation 10 can be viewed as the complete form of Eqs. 8 and 9.

EC-ALE has also been employed for the electrodeposition of CdSe on Ag(1 1 1) surfaces via the layer-by-layer UPD of Se and Cd in sequence [55]. A pH 8.5 ammonia buffer prepared from $HClO_4$ and NH_3 has been used as the electrolyte in which $CdSO_4$ and Na_2SeO_3 are introduced as the sources of Cd and Se, respectively.

A method related to the electrochemical anodization of a Cd anode in basic Se^{2-} solution

was reported. The CdSe formed via the reaction $Cd^{2+} + Se^{2-} \rightarrow CdSe$ [56].

From Nonaqueous Solutions

Many reports related to the electrodeposition of CdSe from nonaqueous solutions used DMSO as the solvent because elemental Se is soluble in it [57–61]. $CdCl_2$ or $Cd(ClO_4)_2$ is used as the source of Cd. Sometimes, supporting electrolytes such as LiCl are added. By electrodeposition on AAO templates, CdSe nanowire arrays can be obtained [57]. The epitaxial electrodeposition of CdSe has been achieved on Au(1 1 1) surfaces [58]. One report mentioned that the size and bandgap of the electrodeposited CdSe quantum dots can be varied by applying mechanical strain to the Au substrate (evaporated on mica) during the electrodeposition. The authors attributed this behavior to a change in the lattice spacing of the strained Au(1 1 1) and consequently to the lattice mismatch between Au and CdSe [59]. According to a report [61], the formation of CdSe electrodeposited from DMSO containing Cd^{2+} and elemental Se is as follows:

$$Se_{(sol)} + 2e^- \rightarrow Se^{2-}_{(sol)} \qquad (11)$$

$$Cd^{2+}_{(sol)} + Se^{2-}_{(sol)} \rightarrow CdSe_{(s)} \qquad (12)$$

CdSe has also been electrodeposited from various DMSO and H_2O mixtures containing $CdSO_4$ and SeO_2 [62].

Other organic solvents used for the electrodeposition of CdSe include ethylene glycol [63] and diethylene glycol [64] containing Cd^{2+} and SeO_2. Sometimes, EDTA is employed to complex with Cd^{2+}. Both constant-potential electrodeposition and pulse electrodeposition were used.

Other Electrochemical Approaches

A two-step method has been developed for the growth of CdSe films. A Se film is first electrodeposited onto a polycrystalline Au surface from an aqueous solution containing SeO_3^{2-} and then the Se-modified Au electrode is immersed into another aqueous solution containing Cd^{2+} where the Se layer is cathodically stripped to generate CdSe film in situ

(Se $+ 2e^- \rightarrow$ Se^{2-}; Cd^{2+} + Se$^{2-} \rightarrow$ CdSe) [65]. CdSe nanoparticles have been electrochemically prepared on Au electrodes by using a similar two-step technique. The difference is that the Au electrode is modified by molecular templates formed by self-assembled monolayers (SAMs) before the electrodeposition of Se is carried out [66].

CdTe

CdTe is the most popular semiconductor prepared by electrodeposition due to its application in thin-film solar cells. A large number of papers thus discuss this process. Some examples are given below. They include all the electrolytes used for the electrodeposition of CdTe.

From Acidic Aqueous Solutions

The electrodeposition of CdTe from acidic aqueous solutions has been carried out on various substrates, including the single crystals of Au(1 1 1) or Au(1 0 0) [67], n-Si(1 0 0) [68], and Si(1 1 1) [69]. CdTe has also been electrodeposited on CdS to form thin-film CdS/CdTe solar cells [70].

The acidic baths usually contain Cd^{2+} (from various Cd salts) and TeO$_2$ as the sources of Cd and Te, respectively [67–81]. In general, HTeO$_2^+$ is the stable and active species of TeO$_2$ in acidic electrodepositing baths. The formation of CdTe via electrodeposition from acidic aqueous baths has been explained by several proposed reactions. One of these reactions is:

$$3H^+ + Cd^{2+} + HTeO_2^+ + 6e^- \rightarrow CdTe + 2H_2O \tag{13}$$

The above reaction can be separated into two individual reactions related to the reduction of Cd (II) and Te(IV), respectively [6]:

$$Cd^{2+} + 2e^- \rightarrow Cd \tag{14}$$

$$3H^+ + HTeO_2^+ + 4e^- \rightarrow Te + 2H_2O \tag{15}$$

Because the reaction shown below is spontaneous [6], CdTe formation takes place via solid-state reaction once the precursors are reduced to elemental Cd and Te:

$$Cd + Te \rightarrow CdTe \quad \Delta G^0_{CdTe} = -106.7 \text{ kJ/mol} \tag{16}$$

Because reaction (Eq. 16) is spontaneous, the UPD of Cd can occur on a Te surface. The formation of CdTe, therefore, may occur through the following two steps [76]:

$$HTeO_2^+ + 3H^+ + 4e^- \rightarrow Te + 2H_2O \tag{17}$$

$$Te + Cd^{2+} + 2e^- \rightarrow CdTe \tag{18}$$

This mechanism explains why CdTe can be obtained at an electrodeposition potential that is more positive than the reduction potential of Cd^{2+} in the electrodeposition baths. Regardless of which mechanism is adopted, the overall process is a six-electron reductive reaction, as shown in Eq. 13.

A much older theory [82] assumes that HTeO$_2^+$ ions are reduced in a six-electron step to H$_2$Te which then reacts in a chemical step to precipitate CdTe:

$$H_2Te + Cd^{2+} \rightarrow CdTe + 2H^+ \tag{19}$$

Various electrochemical techniques have been employed to deposit CdTe from acidic aqueous baths. Constant-potential electrodeposition is frequently employed, but pulse-current electrodeposition is also used [72, 73]. Continuous cycling of the potential has also been used to grow CdTe layer-by-layer on Si substrates [69]. EC-ALE has also been employed for the growth of CdTe, in which a flow cell is used to conveniently change the deposition baths for Cd or Te electrodeposition [77]; the procedures include the constant-potential electrodeposition of Cd and Te in sequence. Hybrid organic-inorganic semiconducting systems have also been developed [74]. The inorganic part is the electrodeposited CdTe and N-methyl[60] fulleropyrrolidine is the organic layer.

From Basic Aqueous Solutions

Most basic aqueous electrolytes used for the electrodeposition of CdTe are ammoniacal buffer solutions (NH_3/NH_4^+ buffer, with NH_4^+ coming from various NH_4^+ salts) that contain Cd^{2+} and TeO_3^{2-} [83–92]. Cd^{2+} and TeO_3^{2-} are introduced by the addition of various Cd salts and TeO_2, respectively. TeO_2 reacts with H_2O to become H_2TeO_3, which dissociates into $HTeO_3^-$ or TeO_3^{2-} in alkaline solutions.

Murase et al. supposed that the growth of CdTe via electrodeposition from alkaline ammoniacal buffers followed the following steps [84]:

$$TeO_3^{2-} + 6H^+ + 4e^- \rightarrow Te + 3H_2O \quad (20)$$

$$Cd(NH_3)_4^{2+} + Te + 2e^- \rightarrow CdTe + 4NH_3 \quad (21)$$

The second step (Reaction 21) is believed to be through the UPD of Cd on Te sites. The overall process of reactions (Eq. 20) and (Eq. 21) can be represented as:

$$Cd(NH_3)_4^{2+} + TeO_3^{2-} + 6H^+ + 6e^- \rightarrow$$
$$CdTe + 4NH_3 + 3H_2O \quad (22)$$

The same group also studied the effect of visible-light irradiation on the growth of CdTe films. It was found that both deposition current density and current efficiency for the CdTe electrodeposition were enhanced by photoirradiation and, as a result, the time required for a given amount of deposition was dramatically reduced. They called this technique photo-assisted electrodeposition [91].

Arai et al. studied the effect of chloride ions on CdTe electrodeposition under white-light irradiation [86]. They found that the increase in the deposition rate of CdTe under irradiation of visible light was depressed with increasing concentration of chloride ions. The as-deposited CdTe layer prepared from a chloride electrolyte has n-type conduction, even though the CdTe layer obtained from the sulfate electrolyte is p-type.

EC-ALE has also been applied to the electrodeposition of CdTe on Ag(1 1 1) via alternating UPD of Te and Cd [93].

Some complexing agents such as 2,2′-bypyridine and ethylenediamine (en) are introduced to facilitate the electrochemical growth of CdTe. Dergacheva et al. indicated that the introduction of 2,2′-bypyridine makes the reduction potentials of Te(IV) anions and Cd (II) ions closer [94]. Murase et al. used ethylenediamine instead of ammonia in the electrodeposition baths. This allowed the temperature of the electrodeposition baths to be increased to 363 K, resulting in highly crystalline CdTe deposits without any post-treatment [95].

From Molten Salts or Ionic Liquids

Reports on the electrodeposition of CdTe from molten salts or ionic liquids are relatively rare. However, the electrodeposition of CdTe from the Lewis basic ionic liquid 1-ethyl-3-methylimidazolium chloride/tetrafluoroborate ($EMICl\text{-}BF_4$) has been achieved [96]. $CdCl_2$ and $TeCl_4$, used as the sources of Cd and Te, respectively, can dissolve in the Lewis basic ionic liquid via a complexing reaction with excess chloride ions in the ionic liquid. Te(IV) can be reduced to Te through a four-electron step and CdTe electrodeposits can be obtained via the UPD of Cd on the deposited Te.

Other Electrochemical Approaches

EC-ALE has also been employed for the formation of CdTe via the UPD of Te and Cd in sequence. The UPD of Cd has been performed in an acidic solution containing $CdSO_4$ and the UPD of Te has been carried out in a basic solution in which TeO_2 was dissolved [97]. EC-STM (electrochemical scanning tunneling microscopy) has been used to study the EC-ALE of CdTe growth. A two-step EC-ALE process consisting of ex situ Te UPD from a basic Te solution, followed by in situ Cd UPD from an acidic solution, has been reported [98].

CdTe nanowires have been electrodeposited onto AAO templates [99].

ZnS

The atomic epitaxial electrodeposition of ZnS on Au(1 1 1) surfaces has been reported [100]. The process includes the UPD of S carried out from a basic aqueous solution of Na_2S. Then, the S-modified electrode is deposited with Zn via Zn UPD from an acidic $ZnSO_4$ solution.

ZnSe

From Acidic Aqueous Solutions

Most electrodeposited ZnSe is prepared by cathodic deposition from acidic aqueous baths containing Cd^{2+} (from various Cd salts) and SeO_2 [38, 101–116]. In acidic conditions, SeO_2 exists in the form of H_2SeO_3. There are two possible pathways for the electrochemical formation of ZnSe [117, 118]:

The first pathway is:

$$H_2SeO_3 + 4H^+ + 4e^- \rightarrow Se + 3H_2O \quad (23)$$

$$Zn^{2+} + Se + 2e^- \rightarrow ZnSe \quad (24)$$

The overall reaction can be represented as:

$$H_2SeO_3 + Zn^{2+} + 4H^+ + 6e^- \rightarrow ZnSe + 3H_2O \quad (25)$$

The second pathway is:

$$Se + 2e^- + 2H^+ \rightarrow H_2Se \quad (26)$$

$$Zn^{2+} + H_2Se \rightarrow ZnSe + 2H^+ \quad (27)$$

Which pathway takes place depends on the electrodeposition conditions, such as the applied electrode potentials.

The electrodeposition of three-dimensional (3D) macroporous ZnSe semiconductors has also been reported [101]. Macroporous structures of ZnSe have been fabricated because highly ordered macroporous structures consisting of semiconductors with high refractive indexes are well-known to be photonic crystals. A template consisting of 3D-ordered arrays of silica spheres is employed. Galvanostatic and potentiostatic electrodeposition is then carried out to deposit ZnSe into the interstitial spaces between the silica

spheres. After the template is removed, a 3D-ordered macroporous ZnSe is obtained.

By introducing appropriate species (As or Ga) to the electrodepositing baths, n-type or p-type ZnSe thin films can be prepared [106].

From Basic Aqueous Solutions

ZnSe can be formed by electrodeposition from an alkaline solution containing Zn^{2+} and $SeSO_3^{2-}$ ions [119, 120]. The formation of ZnSe is supposed to follow the reaction:

$$Zn^{2+} + SeSO_3^{2-} + 2e^- \rightarrow ZnSe + SO_3^{2-} \quad (28)$$

However, the chemical precipitation of ZnSe may take place via the direct hydrolysis of $SeSO_3^{2-}$ ions in basic solutions [121]:

$$Zn^{2+} + SeSO_3^{2-} + 2OH^- \rightarrow ZnSe + SO_4^{2-} + H_2O \quad (29)$$

In order to prevent the chemical precipitation of ZnSe, complexing agents such as EDTA and NTA are used to form strong complex ions with Zn^{2+} [119, 120]. In addition, the solution pH is controlled to limit the concentration of hydroxide ions.

EC-ALE has been employed for the formation of ZnSe from basic aqueous solutions on Ag(1 1 1) [122] or In P(1 1 1) and GaAs(1 0 0) wafers [120].

Most reports study the electrodeposition of ZnSe in either acidic or basic aqueous solutions. The electrodeposition of ZnSe from both types of aqueous solution was discussed in one report [123] in which Zn^{2+} and SeO_2 were used in acidic solutions, whereas EDTA or NTA complex ions of Zn^{2+}, and $SeSO_3^{2-}$ ions were used in basic solutions.

From Nonaqueous Solutions

DMSO is frequently used as the solvent to prepare nonaqueous solutions for the electrodeposition of ZnSe because elemental Se is soluble. Supporting electrolytes such as LiCl and $LiClO_4$ have been used [124, 125]. The growth of ZnSe films is achieved via the reaction [124]:

$$Se + Zn^{2+} + 2e^- \rightarrow ZnSe \quad (30)$$

The epitaxial growth of ZnSe films on (1 1 1) or (1 0 0) n-type InP single-crystalline substrates can thus be achieved.

Gal et al. reported the electrochemical deposition of ZnSe and (Zn,Cd)Ze from a DMSO solution containing elemental Se, $Cd(ClO_4)_2$, and supporting electrolyte $LiClO_4$ [126].

From Molten Salts or Ionic Liquids

Sanchez et al. reported the electrodeposition of ZnSe from $CdCl_2$-NaCl molten salt at 550 °C [127]. $ZnCl_2$ and SeO_2 are introduced to produce Zn(II) and Se(IV) ions. Electrodeposition is then carried out on SnO_2-covered glass sheets. Yellow, transparent, and adherent thin films containing up to 90 % well-crystallized ZnSe are obtained. The high crystallinity is due to the high working temperature of the molten salt.

ZnTe

From Acidic Aqueous Solutions

Many reports on the electrodeposition of ZnTe from acidic aqueous solutions have been published. Most of these reports used Zn^{2+} (from various Zn salts) and TeO_2 as the sources of Zn and Te, respectively [128–134]. In acidic solutions, $HTeO_2^+$ is the stable and active species of TeO_2 in which Te(IV) can be reduced to Te or Te(-II). Neumann-Spallart et al. reported that the formation of ZnTe from aqueous solutions occurs via the following reactions [131]:

$$HTeO_2^+ + 4e^- + 3H^+ \rightarrow Te + 2H_2O \quad (31)$$

$$Zn^{2+} + Te + 2e^- \rightarrow ZnTe \quad (32)$$

Because the formation of ZnTe is spontaneous (negative free-energy change), reaction (32) is possibly a UPD of Zn on Te sites. If the electrodeposition potentials are shifted to more negative values, the formation of ZnTe may be:

$$Te + H^+ + 2e^- \rightarrow HTe^- \text{ or } Te + 2H^+ + 2e^- \rightarrow$$
$$H_2Te \text{ (at lower pH)}$$
$$(33)$$

$$HTe^- (\text{or } H_2Te) + Zn^{2+} \rightarrow ZnTe + H^+ (\text{or } 2H^+)$$
$$(34)$$

Citrate buffers are usually used as electrolytes for the electrodeposition of ZnTe [135–140]. Citrate can form complex ions with Zn^{2+}. Therefore, Zn^{2+} ions exist in the form of $Zn(Cit)_2^{4-}$ in citrate buffer solutions and $Zn(Cit)_2^{4-}$ ions are the species involved in reactions (31)–(34). The cathodic electrodeposition of ZnTe has been carried out using three kinds of citric acid electrolyte, each with a different counter anion (i.e., sulfate, chloride, and nitrate) to investigate the anion effect [139]. The authors reported that the type of anions affects the morphology, composition, and structure of the obtained ZnTe films.

ZnTe nanowires have been prepared by pulsed electrodeposition using an AAO membrane as the template [135, 136]. In addition to common Zn salts, ZnO has also been used for electrodeposition of ZnTe in acidic baths [141, 142].

From Nonaqueous Solutions

Ethylene glycol has been employed as a solvent to prepare the electrodeposition bath for ZnTe deposition [143]. $ZnCl_2$, $TeCl_4$, and KI were dissolved for the electrodeposition. Heo et al. reported the electrodeposition of ZnTe from methanol, acetonitrile, and propylene carbonate solutions containing $ZnCl_2$, $TeCl_4$, and $NaClO_4$ [144]. The authors claimed that propylene carbonate is the best solvent because it is aprotic and allows higher working temperatures. A higher working temperature improves the crystallinity of the ZnTe films. ZnTe nanowires have been prepared by pulse-reverse electrodeposition onto TiO_2 nanotubular arrays from propylene carbonate solution containing $ZnCl_2$, $TeCl_4$, and $NaClO_4$ [145].

From Molten Salts or Ionic Liquids

The electrodeposition of ZnTe has been achieved on nickel substrates from a 40–60 mol% zinc chloride-1-ethyl-3-methylimidazolium chloride ionic liquid containing propylene carbonate (PC) as a cosolvent at 40 °C [146]. PC is added in order to reduce the viscosity and the melting point of the ionic liquid. $ZnCl_2$ and $TeCl_4$ can dissolve in this ionic liquid to provide Zn(II) and

Te(IV) ions, respectively. ZnTe may form via the following two reactions:

$$Zn^{2+} + Te + 2e^- \rightarrow ZnTe \qquad (35)$$

$$Te + 2e^- \rightarrow Te^{2-}, \ Zn^{2+} + Te^{2-} \rightarrow ZnTe \qquad (36)$$

Reaction (35) is the UPD of Zn on Te deposits. With the appropriate electrodeposition potentials and ratio of Zn(II)/Te(IV), the stoichiometric growth of ZnTe can be achieved.

An unusual method (not electrodeposition) of ZnTe growth on Zn substrates was investigated at 640 K by using the following ion exchange and chemical reactions in LiCl-KCl molten salt [147]:

$$2Zn_{(substrate)} + Te^{4+}_{(in \ molten \ salt)} \rightarrow \\ 2Zn^{2+} + Te_{(on \ substrate)} \qquad (37)$$

$$Zn_{(substrate)} + Te_{(on \ substrate)} \rightarrow \\ ZnTe_{(on \ substrate)} \qquad (38)$$

Other Electrochemical Approaches

A two-step method has been reported to form ZnTe films on Au electrodes [148]. In the first step, Te is electrodeposited from an acidic aqueous solution in which TeO_2 is dissolved. The ZnTe film is then prepared via the potentiostatic or potentiodynamic reduction of the Te film in a solution containing Zn^{2+} ions. During the reduction step, Te^{2-} is produced and reacts with Zn^{2+} to form ZnTe.

Bi_2S_3

From Acidic Aqueous Solutions

Acidic aqueous solutions consisting of Bi^{3+} (from $Bi(NO_3)_3$) and $S_2O_3^{2-}$ (from $Na_2S_2O_3$) ions have been used for the electrodeposition of Bi_2S_3 semiconductors [149–151]. EDTA and citrate have been used to form complex ions with Bi^{3+}. Yesugade et al. proposed that Bi_2S_3 films form via the following reactions [151]:

$$S_2O_3^{2-} + 2H^+ \rightarrow S + SO_2 + H_2O \qquad (39)$$

$$3S + 6e^- \rightarrow 3S^{2-} \qquad (40)$$

$$2Bi^{3+} + 3S^{2-} \rightarrow Bi_2S_3 \qquad (41)$$

Saitou et al. let the S colloids produced from the decomposition of $S_2O_3^{2-}$ ions to completely precipitate and then used the top clear solution for electrodeposition. The authors indicated that no S existed in the solution and that S was produced from the following reaction [150]:

$$SO_3^{2-} + 2H_2O + 4e^- \rightarrow S + 6OH^- \qquad (42)$$

Yan et al. proposed that S colloids can be produced from the electrochemical reduction of $S_2O_3^{2-}$ ions [149]:

$$S_2O_3^{2-} + 6H^+ + 4e^- \rightarrow 2S + 3H_2O \qquad (43)$$

Bi_2S_3 films are then produced according to the reactions of (40) and (41).

From Basic Aqueous Solutions

Bi_2S_3 films have been prepared by anodically oxidizing a Bi rod anode [152] or a Bi-electrodeposited Pt electrode (Pt/Bi) [153] in basic aqueous solutions containing Na_2S. The electrodeposition of Bi onto Pt electrode is carried out in an aqueous solution containing Bi $(NO_3)_3$. Bi_2S_3 formed by this method is believed to follow the reaction [152]:

$$2Bi + 3HS^- \rightarrow Bi_2S_3 + 3H^+ + 6e^- \qquad (44)$$

The authors pointed out that HS^- ions are the predominant species at pH 13. However, another group reported that Bi_2S_3 is electrodeposited through the following reaction [153]:

$$2Bi + 3S^{2-} \rightarrow Bi_2S_3 + 6e^- \qquad (45)$$

The hydrogen photoproduction efficiency of the Bi_2S_3 semiconductor was examined in a hydrogen sulfide solution. H_2O is reduced by accepting the photogenerated electrons to produce H_2, and S^{2-} is oxidized by holes in the valence band of the semiconductor.

From Nonaqueous Solutions

A mixture containing acetic acid and formaldehyde has been employed for the electrodeposition of Bi_2S_3 semiconductor films on stainless steel and fluoride-doped tin oxide (FTO)-coated glass substrates [154]. $Bi(NO_3)_3 \cdot 5H_2O$ and thiourea $(CS(NH_2)_2)$ were respectively dissolved in acetic acid and in formaldehyde as the sources of Bi and S, respectively, and then the two solutions were mixed in various volumetric proportions.

DMSO, in which elemental S is soluble, has been used as solvent for the electrodeposition of Bi_2S_3. $BiCl_3$ was used to provide Bi^{3+} ions and $LiClO_4$ was employed as the supporting electrolyte [155].

Other Electrochemical Approaches

EC-ALE has been employed for the electrochemical formation of Bi_2S_3 semiconducting films on $Au(1\ 1\ 1)$ surfaces [156]. Bi_2S_3 is formed by the electrodeposition of S from alkaline Na_2S solution and the electrodeposition of Bi from acidic $Bi(NO_3)_3$ solution in the layer-by-layer method.

Cu_2S

From Acidic Aqueous Solutions

Cu_2S semiconducting films have been prepared by electrodeposition from acidic aqueous solutions containing $CuSO_4$, $Na_2S_2O_3$, and EDTA [157, 158]. The solution pH was adjusted by HCl. The presence of EDTA in aqueous solutions was found to improve the life time of the deposition bath as well as the adhesion of the electrodeposits. The formation of Cu_2S was proposed to be via the following reactions [157, 158]:

$$Cu^{2+} + e^- \rightarrow Cu^+ \qquad (46)$$

$$2Cu^+ + S^{2-} \rightarrow Cu_2S \qquad (47)$$

S^{2-} is produced from the electrochemical reduction of $S_2O_3^{2-}$.

From Basic Aqueous Solution

Cu_2S has been formed by anodically electrolyzing copper sheets in alkaline Na_2S solution.

The growth of Cu_2S can be represented as the following reaction [159]:

$$2Cu + S^{2-} \rightarrow Cu_2S + 2e^- \qquad (48)$$

Cd-Bi-S and Cd-Zn-S

The low efficiency of CdS/Cu_2S solar cells might be due to the mismatch of lattice constants and electron affinities of the two compounds. Zn has been introduced into CdS to match lattice constants and electron affinities. Cd-Bi-S films have been prepared for the same purpose. Cd-Bi-S has been prepared by electrodeposition from aqueous solution containing $CdSO_4$, $Bi(NO_3)_3$, and $Na_2S_2O_3$. Cd-Zn-S has been formed by electrodeposition from aqueous solution containing $CdSO_4$, $ZnSO4$, and $Na_2S_2O_3$ [160].

$ZnIn_2S_4$

$ZnIn_2S_4$ semiconductor films have been prepared by electrodeposition on Ti substrates from acidic aqueous solution containing $ZnCl_2$, $InCl_3$, and $Na_2S_2O_3$. As for many other semiconductors prepared by electrodeposition, annealing in a nitrogen flow is necessary to obtain hexagonal phase $ZnIn_2S_4$. The photocatalytic ability of $ZnIn_2S_4$ was investigated using the photoelectrocatalytic inactivation of *Escherichia coli* (*E. coli*). More than 3 logs of *E. coli* were killed within 60 min with the $ZnIn_2S_4$ film under visible light and a 0.6 V positive potential [161].

Cu_2ZnSnS_4

High performance Cu_2ZnSnS_4 (CZTS) solar cells have been prepared by co-sputtering and vapor phase sulfurization. Schurr et al. developed an alternative method in which Cu, Zn, and Sn are electrodeposited in one step and subsequent sulfurization (thermal evaporation of S) is carried out to form the final compound semiconductor [162]. A single-step electrodeposition of Cu_2ZnSnS_4 was developed by Jeon et al. [163]. Both potentiostatic electrodeposition and pulsed deposition are carried out in aqueous solutions containing $CuSO_4$, $ZnSO_4$, $SnSO_4$, and $Na_2S_2O_3$. Trisodium citrate is used as a complexing agent, and tartaric acid is used as

Semiconductors Groups II-IV and III-V, Electrochemical Deposition 1937

a pH control solution. Near-stoichiometric CZTS thin films were obtained by potentiostatic electrodeposition. Nanowire structures, however, were obtained with pulsed electrodeposition.

III-V compounds

AlSb

The electrodeposition of AlSb cannot be achieved from aqueous solution because the reduction potential is far beyond the aqueous electrolyte potential limit. Furthermore, the high volatility of Sb complicates the electrodeposition of AlSb from high-temperature molten salt electrolytes. Consequently, the electrodeposition of AlSb is limited to be in the room-temperature molten salts which are also termed ionic liquids. Freyland and coworkers [164, 165] first explored the nanoscale electrodeposition of AlSb on Au(111) using in situ scanning probe techniques such as STM and STS from $AlCl_3$-1-butyl-3-methylimidzolium chloride (1:1) ionic liquid containing SbC_3. At a potential positive to 0.0 V (vs. an Al/Al(III) quasireference electrode), only Sb was deposited. The codeposition of AlSb occurred at more negative potentials. The deposition obtained at -0.9 V was Sb-rich whereas that at -1.5 V was Al-rich. Homogeneous distributed stoichiometric AlSb with a band gap of 2.0 ± 0.2 eV was obtained at -1.1 V.

Gandhi et al. [166] reported the use of a pulse-reverse potential technique to control the stoichiometry of the AlSb deposit from a Lewis acidic $AlCl_3$-1-ethyl-3-methylimidzolium chloride (1.5:1) ionic liquid containing 4 mM $SbCl_3$. Typically, the cathodic pulse was set at -0.6 V for 0.5–1 s and the anodic pulse was set at 0.7 V for 1–2 s. The deposition was performed with 300–1,000 pulse cycles. A longer cathodic pulse increased the Al content and a larger anodic pulse increased the Sb content. The as-deposited AlSb was amorphous and transformed into a crystalline phase after being annealed at 350 °C for 2 h in argon.

GaAs

Although the reduction potentials of As and Ga are separated by about 0.6 V, the negative Gibbs free-energy of formation of GaAs, ΔG^o (298K) $=$ -83.67 kJ/mol, indicates that a mixture of Ga and As is thermodynamically unstable and will convert to GaAs with either excess As or Ga.

Villegas and Stickney [167] studied the EC-ALE deposition of GaAs on polycrystalline and single-crystal Au electrodes. Reductive UPD from $HAsO_2$ and oxidative UPD from AsH_3 resulted in low-coverage ordered As layers on the (100) and (110) surfaces of Au. Higher-coverage ordered structures were obtained by the reductive deposition of As on low-index surfaces. The reductive UPD of Ga from $Ga_2(SO_4)_3$ on low-index surfaces of Au has been achieved. The successive UPD of As and Ga atomic layers results in ordered GaAs structures on Au(100) and (110) surfaces.

Perrault [168] performed thermodynamic calculations and some potentiostatic experiments on the electrodeposition of GaAs in aqueous solutions made by the dissolution of Ga_2O_3 and As_2O_3. He concluded that the electrodeposition of GaAs cannot be achieved from a simple aqueous solution.

Chandra and Khare [169] investigated the codeposition of GaAs from solutions prepared by mixing Ga and As_2O_3 in concentrated HCl followed by dilution and adjustment of pH with KOH. The study showed that the Ga content in the deposits increased with pH. The Ga content increased and then decreased with increasing current density and electrolysis time. Increasing temperature decreased the Ga content.

Yang et al. [170] investigated the electrodeposition of GaAs using Ga_2O_3 and As_2O_3 in alkaline (KOH, pH $>$ 12) and $GaCl_3$ and As_2O_3 in acid (HCl, pH $<$ 3) aqueous electrolytes on Ti substrates for its high hydrogen overpotential. In alkaline solutions, GaAs was deposited with potentiostatic electrolysis. The Ga content in the deposits increased at more negative potentials. The deposits obtained between -1.65 and -1.73 V (vs. Hg/HgO) had a relatively constant composition of about 62% Ga. The composition of the deposits obtained at room temperature was rather uniform, but that at high temperatures was nonuniform. In acid solutions, a galvanostatic technique is more suitable for the deposition of thin, more compact, and better adhesion deposits than is a potentiostatic method. A higher current produced higher Ga content in the deposit. The best deposits were obtained at pH 2.5 and a current density of between 10 and 50 A/cm^2. The as-deposited GaAs films obtained from both alkaline and acid solutions were amorphous but transformed into a crystalline phase with some loss of As after being annealed in N$_2$ at 250–300 $^{\circ}$C.

Gao et al. [171] prepared n-type polycrystalline stoichiometric Ga$_{0.93}$As$_{1.07}$ on Ti and Ga$_{0.91}$As$_{1.09}$ on SiO$_2$-coated glass, respectively, using galvanostatic deposition in HCl solutions. The parameters of the deposition are: a Ga/AsO concentration of 7×10^{-4}/5.6×10^{-3} M, a pH of 0.93, and a current density of 5.56 mA/cm^2 on SnO$_2$-coated glass and a Ga/AsO of 1.4×10^{-1}/1.7×10^{-3} M, a pH of 0.64, and a current density of 2.50 mA/cm^2 on Ti.

The electrodeposition of 3D microporous GaAs, using self-assembled silica sphere arrays as a template, was attempted by Lee et al. [101] from aqueous solutions of 7.5×10^{-1} M Ga (NO$_3$)$_3$ and 5.0×10^{-2} M As$_2$O$_3$ at 25 $^{\circ}$C with a current density of 2 mA/cm^2. The quality of the porous GaAs was strongly determined by the Ga (NO$_3$)$_3$/As$_2$O$_3$ ratio; a ratio lower than 1 resulted in the formation of As oxide and As and poor quality in terms of uniformity and crystallinity.

Mahalingam et al. [172] electrodeposited GaAs thin film from an aqueous solution of GaCl$_3$ and As$_2$O$_3$. A high current density or a very negative potential was required at pH values lower than 1.0. Porous films were obtained at pH > 3.0. The Ga content increased with more negative deposition potential, and near-stoichiometric GaAs films were obtained between –0.6 and –0.8 V (vs. SCE). Increasing the temperature increased the current density and lowered the Ga content. The optimum pH, deposition potential, and temperature were found to be pH 3.0, –0.8 V, and 60 $^{\circ}$C, respectively. Annealing at 350 $^{\circ}$C for 2 h further improved the crystallinity of the GaAs films.

Gheorghies et al. [173, 174] studied the effect of pH, deposition potential, current density, and temperature on the composition of electrodeposited GaAs from both acid and alkaline solutions. In both cases, electrodeposited films were amorphous and annealing was required to obtain polycrystalline GaAs.

Lajnef et al. [175] electrodeposited GaAs on a porous silicon substrate. The electrolyte was prepared by dissolving Ga and As$_2$O$_3$ in concentrated HCl and adjusting the pH to 1. The deposition was carried out with constant-current electrolysis at 273 K with agitation. After annealing at 300 $^{\circ}$C under N$_2$ for 1 h, GaAs films with a mixed phase of orthorhombic and cubic GaAs were identified with XRD.

Kozlov et al. [176] prepared polycrystalline GaAs by the sequential galvanostatic electrodeposition of layers of Ga from alkaline and As from acidic solutions. The two-layer electrodeposit was annealed to form GaAs by reaction-diffusion. The Ga diffusion coefficient was determined to be in the temperature range of 80–190 $^{\circ}$C.

The pulse plating method was adopted by Murali and Trivedi [177] to prepare 1.0–2.0-mm thick polycrystalline single-phase GaAs films with a direct band gap of 1.40 eV on Ti and SnO$_2$-coated glass substrates from an aqueous solution containing 0.20 M GaCl$_3$ and 0.15 M As$_2$O$_3$ at a pH of 2 and at room temperature. The current density was 50 mA/cm^2. The ON time was varied between 3 s and 15 s and the corresponding OFF time was in the range 27–15 s. This resulted in a duty cycle in the range of 10–50 %.

The electrodeposition of GaAs at 300 $^{\circ}$C from GaCl$_3$-KCl molten salt under galvanostatic conditions was reported by Tremillon and coworkers [178]. Wicelinsski and Gale [179] reported the electrodeposition of GaAs from a Lewis acidic AlCl$_3$-1-butylpyridiium chloride or AlCl$_3$-1-methyl-3-ethylimidazolium chloride melts at 40 $^{\circ}$C. To avoid possible contamination from Al, Carpenter and Verbrugge [180] investigated the electrodeposition of GaAs from a Lewis basic GaCl$_3$-1-methyl-3-ethylimidazolium chloride (40/60 mol%) melt containing AsCl$_3$. The deposition of Ga results from the reduction of GaCl$_4^-$ whereas the deposition of As is due to the

Semiconductors Groups II-IV and III-V, Electrochemical Deposition

reduction of $AsCl_4^-$. Their data showed that As is more noble than Ga in the melt, and that Ga and As can be codeposited between 0 and -1 V vs. an Al reference electrode. The deposits were, however, not homogenous in composition. The deposition of stoichiometric GaAs films from molten salts deserves further study.

GaSb

The electrochemical codeposition of GaSb from aqueous solution is difficult due to the large deposition potential difference between the two components. Paolucci et al. [181] formed GaSb thin films via the sequential electrodeposition of Sb and Ga films and a mild thermal annealing at 100 °C. Acid $SbCl_3$ and alkaline GaO_3^{3-} introduced as $GaCl_3$ solutions was employed to minimize the hydrogen evolution and the formation of SbH_3. A similar approach was adopted by Kozlov and Bicelli [182].

InSb

Mengoli et al. [183] synthesized InSb thin film using a two-stage approach; successive electrodepositions of individual layers in the order Sb and In from separate acidic chloride baths, followed by subsequent thermal annealing. Kozlov et al. [184] studied the two-stage synthesis of InSb by electrodepositing In over an electrodeposited Sb substrate with annealing. They found that the In diffusion coefficient was higher for amorphous Sb than for a crystalline Sb substrate.

The widely separated standard potentials of uncomplexed In^{3+} and Sb^{3+} (-0.340V and 0.205 V, respectively) make the stoichiometric codeposition of InSb difficult. It is necessary to add a complexing agent to bring the deposition potentials of the two elements closer to facilitate codeposition. Ortega and Herrero[5] reported that InSb free of In or Sb metal on a Ti substrate could be electrodeposited from a citric acid solution, but detailed experimental information was not given. Recently, Fulop et al. [185] reported the deposition of μm-thick polycrystalline InSb from acid aqueous electrolytes at room temperature using sodium citrate-citric acid as the complexing agent. The composition of the deposited layers depends primarily on the ratio of $InCl_3$ and $SbCl_3$ in the electrolyte, and is less sensitive to the deposition potential and pH. XRD data revealed that the surface diffusion of atoms led to post-deposition crystallite grain growth and phase separation into zinc blende InSb and excess metals.

The possibility of using EC-ALE to form InSb compounds was evaluated by Wade et al. [186] using $In_2(SO_4)_3$ and Sb_2O_3 as precursors in acid aqueous baths at room temperature. They showed that the potential must be adjusted carefully before a buffer layer formed. Metallic gray InSb films were obtained with 200 cycles.

The anodic alumina-template-assisted electrodeposition of InSb nanowire arrays was first attempted by Zhang et al. [187] using aqueous baths containing $InCl_3$, $SbCl_3$, $C_6H_8O_7 \cdot H_2O$, and $K_3C_6H_5O_7 \cdot H_2O$ at pH 2.2 adjusted by 7.0 M HCl. The citrate ions were used as the complexing agent. Near-stoichiometric polycrystalline InSb nanowire arrays with a diameter of 50 nm were obtained between –0.85 and –1.5 V (vs. Ag/AgCl) at 16 °C inside the nanochannels of the anodic alumina membrane and subsequently annealed in Ar at 425 K for 6 h, then at 720 K for 4 h. The InSb nanowires were characterized by XRD, SEM, TEM, and Raman spectroscopy. Anodic alumina membranes were also adopted for the templated deposition of InSb nanowires by Khan et al. [188, 189], who employed solutions containing 0.1 M $SbCl_3$, 0.15 M $InCl_3$, 0.36 M citric acid, and 0.17 M potassium citrate at pH 1.8. They claimed that single-crystalline InSb nanowires were obtained at -1.5 V vs. Ag/AgCl at room temperature without subsequent annealing.

Li et al. [190] reported the electrodeposition of nanoropes and terraced micropyramids of In-Sb at room temperature using galvanostatic electrolysis from aqueous electrolytes containing 0.01 M tartaric acid (TA) and 0.01 M sodium dodecyl sulfate (SDS) in addition to $InCl_3$ and $SbCl_3$. Nanoropes were obtained from solutions of 0.134 mM $SbCl_3$ and 0.066 mM $InCl_3$ with a current density of 1 mA/cm^2, whereas micropyramids were obtained from solution of 0.2 mM $SbCl_3$ and 0.4 mM $InCl_3$ with a current density of 4.0 mA/cm^2. Although the detailed mechanism for the formation of these nanostructures is still unclear, it is believed that SDS plays a role by selectively adsorbing to certain crystallographic facets.

The electrodeposition of InSb from a 1-ethyl-3-methylimidazolium chloride/tetrafluoroborate room-temperature ionic liquid (molten salt) was investigated by Yang et al. [191] using $SbCl_3$ and $InCl_3$ as precursors. The composition of the In-Sb codeposits can be varied by the deposition potential and bath concentrations. At a potential at which the deposition of Sb and In is mass-transport limited, InSb compound can be obtained from solutions containing equal moles of Sb(III) and In(III). The electrodeposited InSb was a p-type semiconductor which exhibited a direct optical transition with an optical bandgap of 0.20 eV.

InAs

Amorphous InAs was electrodeposited by Ortega and Herrero [5] from solutions that contained $InCl_3$, $AsCl_3$, and citric acid at −1.40V (vs. SCE). It converted to fcc InAs after being annealed at 400 °C. The use of EC-ALE for the electrodeposition of InAs on a Si substrate was reported by Wade et al. [192], who obtained InSb films up to 160-nm thick by alternating the electrodeposition of atomic layers of As and In by UPD at −0.25 V (vs. Ag/AgCl) from 2.5 mM As_2O_3 solution buffered at pH 4.8 with 50.0 mM CH_3COONa and at −0.3 V (vs. Ag/AgCl) from 0.3 mM In_2O_3 solution at pH 3.0, respectively. The film bandgap of 0.45 eV is higher than that (0.36 eV) for single-crystal InSb.

Kozlov et al. [193] reported the two-stage electrodeposition of InAs from aqueous solutions. First, a 3-μm-thick As film was electrodeposited on an Fe substrate from a HCl solution containing As_2O_3 solution at 50 °C with a current density of 1.5 A dm^{-2}. A 1-mm-thick layer was subsequently electrodeposited at 25 °C on the As surface using a pulsed current from a pH 1.8 aqueous $InCl_3$ bath. After electrodeposition, the In-As samples were annealed to form the InAs compound via reaction-diffusion at the In-As interface. The values of the In diffusion coefficient in the temperature range 60–140 °C were evaluated from XRD data.

Ternary Compounds

Paolucci and coworkers reported the formation of several ternary compounds by sequential electrodeposition and annealing. Layers of Sb, In, and Ga were electrodeposited successively and annealed at 400 °C to produce $In_xGa_{1-x}Sb$. The films with $x \geqq 0.91$ consisted of a single phase whereas films with $x \leqq 0.84$ consisted of two phases, one rich in In and the other in Ga [183]. The electrodeposition of In on an AsSb layer produced $InAs_{1-x}Sb_x$ that consisted of two phases after annealing [194]. Films of $GaAs_{1-x}Sb_x$ were prepared by the electrodeposition of Ga on As-Sb alloy and annealing [195].

Tsuda and Hussey [196] examined the galvanostatic electrodeposition of $Al_xIn_ySb_{100-x-y}$ in a Lewis acidic $AlCl_3$-1-ethyl-3-methylimidazolium chloride (2:1) ionic liquid containing In(I) and Sb(III). The deposits obtained at a high applied current density are most likely mixtures of $Al_xIn_ySb_{100-x-y}$ and InSb.

Future Directions

There has been a significant progress in the understanding of electrodeposition of II-VI and III-V semiconductors. Low cost, simplicity, and scalability remain to be the interest in this field. Selecting the deposition conditions including the type of precursor, electrolyte, pH, current density, potential, and annealing condition is important for optimizing the properties of the electrodeposited materials. Atomic scale deposition and epitaxial growth of the materials are possible.

Although in many cases, the electrodeposited thin films show properties comparable to those prepared by classical methods in practical applications, still many of the as-grown films do not exhibit the desired semiconducting properties. Improving the electrodeposited materials properties is still required. Advanced shape and morphology control and new functionalities are of particular interest because these may lead to new practical applications that are not yet anticipated. While electrodeposition of III-V semiconductors is less investigated than II-VI semiconductors due to the difficulty in finding the proper solvents, the recent appearing of new ambient temperature ionic liquids that have large electrochemical window and wide working temperature range may provide a solution for this problem. Electrodeposition is certainly a promising route for preparing new semiconducting materials in the future.

Cross-References

▶ Compound Semiconductors, Electrochemical Decomposition
▶ Electrodeposition of Electronic Materials for Applications in Macroelectronic- and Nanotechnology-Based Devices
▶ Electrolytes, Classification
▶ High-Temperature Molten Salts
▶ Ionic Liquids
▶ Non-Aqueous Electrolyte Solutions
▶ Semiconductor Electrode
▶ Semiconductors Group IV, Electrochemical Decomposition
▶ Semiconductors, Electrochemical Atomic Layer Deposition (E-ALD)
▶ Semiconductors, Principles

References

1. Danaher WJ, Lyons LE (1978) Photoelectro-chemical cell with cadmium telluride film. Nature (London) 271:139
2. Kröger FA (1978) Cathodic deposition and characterization of metallic or semiconducting binary alloys or compounds. J Electrochem Soc 125:2028
3. Panicker MPR, Knaster M, Kröger FA (1978) Cathodic deposition of cadmium telluride from aqueous electrolytes. J Electrochem Soc 125:566
4. Fulop GF, Taylor RM (1985) Electrodeposition of semiconductors. Annu Rev Mater Sci 15:197
5. Ortega J, Herrero J (1989) Preparation of In X (X = P, As, Sb) thin films by electrochemical methods. J Electrochem Soc 136:3388
6. Lincot D (2005) Electrodeposition of semiconductors. Thin Solid Films 487:40

7. Zein El Abedin S, Endres F (2006) Electrodeposition of metals and semiconductors in air- and water-stable ionic liquids. Chem Phys Chem 7:58
8. Endres F (2002) Ionic liquids: solvents for the electrodeposition of metals and semiconductors. Chem Phys Chem 3:144
9. Dharmadasaz IM, Haigh J (2006) Strengths and advantages of electrodeposition as a semiconductor growth technique for applications in macroelectronic devices. J Electrochem Soc 153:G47
10. Pandey RK, Sahu SN, Chandra S (1996) Handbook of semiconductor electrodeposition. Marcel Dekker, New York
11. Hodes G (1995) In: Rubinstein I (ed) Physical electrochemistry. Electrodeposition of II-VI semiconductors. Marcel Dekker, New York, p 515
12. Schlesinger M (2000) In: Schlesinger M, Paunovic M (eds) Modern electroplating. Electrodeposition of semiconductors, Wiley, New York, p 585
13. Murali KR, Matheline M, John R (2009) Characteristics of brush electrodeposited CdS films. Chalcogenide Lett 6:483
14. Lade SJ, Uplane MD, Lokhande CD (1998) Studies on the electrodeposition of CdS films. Mater Chem Phys 53:239
15. Zarębska K, Skompska M (2011) Electrodeposition of CdS from acidic aqueous thiosulfate solution—invesitigationof the mechanism by electrochemical quartz microbalance technique. Electrochim Acta 56:5731
16. Takahashi M, Hasegawa S, Watanabe M, Miyuki T, Ikeda S, Iida K (2002) Preparation of CdS thin films by electrodeposition: effect of colloidal sulfur particle stability on film composition. J Appl Electrochem 32:359
17. Chi YJ, Fu HG, Qi LH, Shi KY, Zhang HB, Yu HT (2008) Preparation and photoelectric performance of ITO/TiO$_2$/CdS composite thin films. J Photochem Photobiol A 195:357
18. Dennison S (1993) Studies of the cathodic electrodeposition of cadmium sulfide from aqueous solution. Electrochim Acta 38:2395
19. Şişman İ, Alanyalıoğlu M, Demir Ü (2007) Atom-by-atom growth of CdS thin films by an electrochemical co-deposition method: effects of pH on the growth mechanism and structure. J Phys Chem C 111:2670
20. Yoshida T, Yamaguchi K, Kazitani T, Sugiura T (1999) Atom-by-atom growth of cadmium sulfide thin films by electroreduction of aqueous Cd^{2+}–SCN^- complex. J Electroanal Chem 473:209
21. Behar D, Rubinstein I, Hodes G, Cohen S, Cohen H (1999) Electrodeposition of CdS quantum dots and their optoelectronic characterization by photoelectrochemical and scanning probe spectroscopies. Superlattices Microstruct 25:601
22. Aguilera A, Jayaraman V, Sanagapalli S, Suresh Singh R, Jayaraman V, Sampson K, Singh VP (2006) Porous alumina templates and nanostructured

CdS for thin film solar cell applications. Sol Energy Mater Sol Cells 90:713

23. Izgorodin A, Winther-Jensen O, Winther-Jensen B, MacFarlane DR (2009) CdS thin-film electrodeposition from a phosphonium ionic liquid. Phys Chem Chem Phys 11:8532

24. Rami M, Benamar E, Fahoume M, Ennaoui A (1999) Growth analysis of electrodeposited CdS on ITO coated glass using atomic force microscopy. Phys Status Solid A172:137

25. Zhou SM, Feng YS, Zhang LD (2003) A two-step route to self-assembly of CdS nanotubes via electrodeposition and dissolution. Eur J Inorg Chem 1794

26. Anderson MA, Gorer S, Penner RM (1997) A hybrid electrochemical/chemical synthesis of supported, luminescent cadmium sulfide nanocrystals. J Phys Chem B 101:5895

27. Gorer S, Ganske JA, Hemminger JC, Penner RM (1998) Size-selective and epitaxial electrochemical/chemical synthesis of sulfur-passivated cadmium sulfide nanocrystals on graphite. J Am Chem Soc 120:9584

28. Demir U, Shannon C (1994) A scanning tunneling microscopy study of electrochemically grown cadmium sulfide monolayers on Au(111). Langmuir 10:2794

29. Foresti ML, Loglio F, Innocenti M, Bellassai S, Carlà F, Lastraioli E, Pezzatini G, Bianchini C, Vizza F (2010) Confined electrodeposition of CdS in the holes left by the selective desorption of 3-mercapto-1-propionic acid from a binary self-assembled monolayer formed with 1-octanethiol. Langmuir 26:1802

30. Swaminathan V, Subramanian V, Murali KR (2000) Characteristics of CdSe films electrodeposited with microprocessor based pulse plating unit. Thin Solid Films 359:113

31. Shen CM, Zhang XG, Li HL (2001) DC electrochemical deposition of CdSe nanorods array using porous anodic aluminum oxide template. Mater Sci Eng A303:19

32. Cachet H, Cortes R, Froment M, Maurin G (1997) Epitaxial electrodeposition of cadmium selenide thin films on indium phosphide single crystal. J Solid State Electrochem 1:100

33. Li Q, Brown MA, Hemminger JC, Penner RM (2006) Luminescent polycrystalline cadmium selenide nanowires synthesized by cyclic electrodeposition/stripping coupled with step edge decoration. Chem Mater 18:3432

34. Erenturk B, Gurbuz S, Corbett RE, Claiborne SAM, Krizan J, Venkataraman D, Carter KR (2011) Formation of crystalline cadmium selenide nanowires. Chem Mater 23:3371

35. Kressin AM, Doan VV, Klein JD, Sailor MJ (1991) Synthesis of stoichiometric cadmium selenide films via sequential monolayer electrodeposition. Chem Mater 3:1015

36. Liang Y, Kong B, Zhu A, Wang Z, Tian Y (2012) A facile and efficient strategy for photoelectrochemical detection of cadmium ions based on in situ electrodeposition of CdSe clusters on TiO_2 nanotubes. Chem Commun 48:245

37. Sarangi SN, Sahu SN (2004) CdSe nanocrystalline thin films: composition, structure and optical properties. Physica E 23:159

38. Chubenko EB, Klyshko AA, Petrovich VA, Bondarenko VP (2009) Electrochemical deposition of zinc selenide and cadmium selenide onto porous silicon from aqueous acidic solutions. Thin Solid Films 517:5981

39. Kim YT, Han JH, Hong BH, Kwon YU (2010) Electrochemical synthesis of CdSe quantum-dot arrays on a graphene basal plane using mesoporous silica thin-film templates. Adv Mater 22:515

40. Rashwan SM, Abd El-Wahab SM, Mohamed MM (2007) Electrodeposition and characterization of CdSe semiconductor thin films. J Mater Sci Mater Electron 18:575

41. Pawar SM, Moholkar AV, Rajpure KY, Bhosale CH (2006) Electrosynthesis and characterization of CdSe thin films: optimization of preparative parameters by photoelectrochemical technique. J Phys Chem Solids 67:2386

42. Bieńkowski K, Strawski M, Maranowski B, Szklarczyk M (2010) Studies of stoichiometry of electrochemically grown CdSe deposits. Electrochim Acta 55:8908

43. Koh JL, Teh LK, Romanato F, Wong CC (2007) Temperature dependence of electrochemical deposition of CdSe. J Electrochem Soc 154:D300

44. Shpaisman N, Givan U, Patolsky F (2010) Electrochemical synthesis of morphology-controlled segmented CdSe nanowires. ACS Nano 4:1901

45. Lister TE, Stickney JL (1996) Formation of the first monolayer of CdSe on Au(111) by electrochemical ALE. Appl Surf Sci 107:153

46. Mathe MK, Cox SM, Flowers BH Jr, Vaidyanathan R, Pham L, Srisook N, Happek U, Stickney JL (2004) Deposition of CdSe by EC-ALE. J Cryst Growth 271:55

47. Eriksson S, Gruszecki T, Carlsson P, Holmström B (1995) Electrochemically deposited polycrystalline n-CdSe thin films studied with laterally resolved photoelectrochemistry and SEM. Thin Solid Films 269:14

48. Lokhande CD, Lee EH, Jung KD, Joo OS (2005) A simple two-step method for preparation of cadmium selenide film on nickel substrate. Mater Chem Phys 93:399

49. Lokhande CD, Lee EH, Jung KD, Joo OS (2005) Electrosynthesis of cadmium selenide films from sodium citrate–selenosulphite bath. Mater Chem Phys 91:399

50. Majidi H, Baxter JB (2011) Electrodeposition of CdSe coatings on ZnO nanowire arrays for extremely thin absorber solar cells. Electrochim Acta 56:2703

51. Tena-Zaera R, Ryan MA, Katty A, Hodes G, Bastide S, Lévy-Clément C (2006) Fabrication and characterization of ZnO nanowires/CdSe/CuSCN *eta*-solar cell. C R Chimie 9:717

52. Hodes G, Grunbaum E, Feldman Y, Bastide S, Lévy-Clément C (2005) Variable optical properties and effective porosity of CdSe nanocrystalline films electrodeposited from selenosulfate solutions. J Electrochem Soc 152:G917

53. St K, Láng G, Heusler KE (2001) The electrodeposition of CdSe from alkaline electrolytes. Electrochim Acta 47:955

54. Skyllas-Kayacos M, Miller B (1980) Studies in selenious acid reduction and cadmium selenide film deposition. J Electrochem Soc 127:869

55. Loglio F, Innocenti M, Acapito FD, Felici R, Pezzatini G, Salvietti E, Foresti ML (2005) Cadmium selenide electrodeposited by ECALE: electrochemical characterization and preliminary results by EXAFS. J Electroanal Chem 575:161

56. Ham D, Mishra KK, Rajeshwa K (1991) Anodic electrosynthesis of cadmium selenide thin films. J Electrochem Soc 138:100

57. Xu D, Shi X, Guo G, Gui L, Tang Y (2000) Electrochemical preparation of CdSe nanowire arrays. J Phys Chem B 104:5061

58. Golan Y, Margulis L, Rubinstein I, Hodes G (1992) Epitaxial electrodeposition of CdSe nanocrystals on gold. Langmuir 8:749

59. Ruach-Nir I, Daniel Wagner H, Rubinstein I, Hodes G (2003) Structural effects in the electrodeposition of CdSe quantum dots on mechanically strained gold. Adv Funct Mater 13:159

60. Xu D, Guo G, Guo Y, Zhang Y, Gui L (2003) Nanocrystal size control by bath temperature in electrodeposited CdSe thin films. J Mater Chem 13:360

61. Henríquez R, Badán A, Grez P, Muñoz E, Vera J, Dalchiele EA, Marottib RE, Gómez H (2011) Electrodeposition of nanocrystalline CdSe thin films from dimethyl sulfoxide solution: nucleation and growth mechanism, structural and optical studies. Electrochim Acta 56:4895

62. Singh K, Mishra SSD (2000) Photoelectrochemical studies on galvanostatically formed cadmium selenide films using mixed solvents. Sol Energy Mater Sol Cells 63:275

63. Kokate AV, Suryavanshi UB, Bhosale CH (2006) Structural, compositional, and optical properties of electrochemically deposited stoichiometric CdSe thin films from non-aqueous bath. Sol Energy 80:156

64. Murali KR, Chitra K, Elango P (2008) Properties of CdSe films pulse electrodeposited from non-aqueous bath. Chalcogenide Lett 5:273

65. Myung N, de Tacconi NR, Rajeshwar K (1999) Electrosynthesis of cadmium selenide films on a selenium-modified gold surface. Electrochem Commun 1:42

66. Choi SJ, Woo DH, Myung N, Kang H, Park SM (2001) Electrochemical preparation of cadmium selenide nanoparticles by the use of molecular templates. J Electrochem Soc 148:C569

67. Vidu R, Ku JR, Stroeve P (2006) Growth of ultrathin films of cadmium telluride and tellurium as studied by electrochemical atomic force microscopy. J Colloid Interface Sci 300:404

68. Gómez H, Henríquez R, Schrebler R, Córdova R, Ramírez D, Riveros G, Dalchiele EA (2005) Electrodeposition of CdTe thin films onto n-Si (1 0 0): nucleation and growth mechanisms. Electrochim Acta 50:1299

69. Jackson F, Berlouis LEA, Rocabois P (1996) Layer-by-layer electrodeposition of cadmium telluride onto silicon. J Cryst Growth 159:200

70. Duffy NW, Peter LM, Wang RL, Lane DW, Rogers KD (2000) Electrodeposition and characterisation of CdTe films for solar cell applications. Electrochim Acta 45:3355

71. Calixto ME, McClure JC, Singh VP, Bronson A, Sebastian PJ, Mathew X (2000) Electrodeposition and characterization of CdTe thin films on Mo foils using a two voltage technique. Sol Energy Mater Sol Cells 63:325

72. Rastogi AC, Sharma RK (2009) Properties and mechanism of solar absorber CdTe thin film synthesis by unipolar galvanic pulsed electrodeposition. J Appl Electrochem 39:167

73. Sharma RK, Singh G, Rastogi AC (2004) Pulsed electrodeposition of CdTe thin films: effect of pulse parameters over structure, stoichiometry and optical absorption. Sol Energy Mater Sol Cells 82:201

74. Mitzithra C, Kaniaris V, Hamilakis S, Kordatos K, Kollia C, Loizos Z (2011) Development and study of new hybrid semiconducting systems involving Cd chalcogenide thin films coated by a fullerene derivative. Mater Lett 65:1651

75. Lepiller C, Cowache P, Guillemoles JF, Gibson N, Özsan E, Lincot D (2000) Fast electrodeposition route for cadmium telluride solar cells. Thin Solid Films 361–362:118

76. Yang SY, Chou JC, Ueng HY (2010) Influence of electrodeposition potential and heat treatment on structural properties of CdTe films. Thin Solid Films 518:4197

77. Venkatasamy V, Jayaraju N, Cox SM, Thambidurai C, Happek U, Stickney JL (2006) Optimization of CdTe nanofilm formation by electrochemical atomic layer epitaxy (EC-ALE). J Appl Electrochem 36:1223

78. Lepiller C, Lincot D (2004) New Facets of CdTe Electrodeposition in acidic solutions with higher tellurium concentrations. J Electrochem Soc 151:C348

79. Gómez H, Henríquez R, Schrebler R, Riveros G, Córdova R (2001) Nucleation and growth mechanism of CdTe at polycrystalline gold surfaces

analysed through $\Delta m/t$ simulation transients. Electrochim Acta 46:821

80. Saraby-Reintjes A, Peter LM, Özsan ME, Dennison S, Webster S (1993) On the mechanism of the cathodic electrodeposition of cadmium telluride. J Electrochem Soc 140:2880

81. Soliman M, Kashyout AB, Shabana M, Elgamal M (2001) Preparation and characterization of thin films of electrodeposited CdTe semiconductors. Renew Energy 23:471

82. Danaher WJ, Lyons LE (1983) Galvanostatic deposition of thin films of cadmium and tellurium. Aust J Chem 36:1011

83. Miyake M, Inui H, Murase K, Hirato T, Awakura Y (2004) Analytical TEM study of CdTe layer electrodeposited from basic ammoniacal aqueous electrolyte. J Electrochem Soc 151:C712

84. Murase K, Watanabe H, Mori S, Hirato T, Awakura Y (1999) Control of composition and conduction type of CdTe film electrodeposited from ammonia alkaline aqueous solutions. J Electrochem Soc 146:4477

85. Dergacheva MB, Statsyuk VN, Fogel LA (2004) Electrodeposition of CdTe films from ammonia-chloride buffer electrolyte. Russ J Appl Chem 77:226

86. Arai K, Murase K, Hirato T, Awakura Y (2006) Effect of chloride ions on electrodeposition of CdTe from ammoniacal basic electrolytes. J Electrochem Soc 153:C121

87. Murase K, Tanaka Y, Hirato T, Awakura Y (2005) Electrochemical QCM studies of CdTe formation and dissolution in ammoniacal basic aqueous electrolytes. J Electrochem Soc 152:C304

88. Murase K, Uchida H, Hirato T, Awakura Y (1999) Electrodeposition of CdTe films from ammoniacal alkaline aqueous solution at low cathodic overpotentials. J Electrochem Soc 146:531

89. Arai K, Hagiwara S, Takayama S, Murase K, Hirato T, Awakura Y (2006) Galvanic contact deposition of CdTe from ammoniacal basic electrolytes at elevated temperatures using an autoclave-type electrolysis vessel. Electrochem Commun 8:605

90. Arai K, Hagiwara S, Murase K, Hirato T, Awakura Y (2005) Galvanic contact deposition of CdTe layers using ammoniacal basic aqueous solution. J Electrochem Soc 152:C237

91. Murase K, Matsui M, Miyake M, Hirato T, Awakura Y (2003) Photoassisted electrodeposition of CdTe layer from ammoniacal basic aqueous solutions. J Electrochem Soc 150:C44

92. Murase K, Watanabe H, Hirato T, Awakura Y (1999) Potential-pH diagram of the Cd-Te-NH_3-H_2O system and electrodeposition behavior of CdTe from ammoniacal alkaline baths. J Electrochem Soc 146:1798

93. Forni F, Innocenti M, Pezzatini G, Foresti ML (2000) Electrochemical aspects of CdTe growth on the face (111) of silver by ECALE. Electrochim Acta 45:3225

94. Dergacheva MB, Statsyuk VN, Fogel LA (2007) Electrochemical preparation of semiconducting CdTe films from ammonia-chloride solutions containing 2,2'-bipyridine. Russ J Appl Chem 80:66

95. Murase K, Honda T, Yamamoto M, Hirato T, Awakura Y (2001) Electrodeposition of CdTe from basic aqueous solutions containing ethylenediamine. J Electrochem Soc 148:C203

96. Hsiu SI, Sun IW (2004) Electrodeposition behaviour of cadmium telluride from 1-ethyl-3-methylimidazolium chloride tetrafluoroborate ionic liquid. J Appl Electrochem 34:1057

97. Colletti LP, Stickney JL (1998) Optimization of the growth of CdTe thin films formed by electrochemical atomic layer epitaxy in an automated deposition system. J Electrochem Soc 145:3594

98. Lay MD, Stickney JL (2004) EC-STM studies of Te and CdTe atomic layer formation from a basic Te solution. J Electrochem Soc 151:C431

99. Hackney Z, Mair L, Skinner K, Washburn S (2010) Photoconductive and polarization properties of individual CdTe nanowires. Mater Lett 64:2016

100. Gichuhi A, Shannon C (1999) A scanning tunneling microscopy and X-ray photoelectron spectroscopy study of electrochemically grown ZnS monolayers on Au(111). Langmuir 15:5654

101. Lee YC, Kuo TJ, Hsu CJ, Su YW, Chen CC (2002) Fabrication of 3D macroporous structures of II-VI and III-V semiconductors using electrochemical deposition. Langmuir 18:9942

102. Manzoli A, Santos MC, Machado SAS (2007) A voltammetric and nanogravimetric study of ZnSe electrodeposition from an acid bath containing Zn (II) and Se(IV). Thin Solid Films 515:6860

103. Mahalingam T, Kathalingam A, Velumani S, Lee S, Sun MH, Deak KY (2006) Electrochemical synthesis and characterization of zinc selenide thin films. J Mater Sci 41:3553

104. Manzoli A, Eguiluz KIB, Salazar-Banda GR, Machado SAS (2010) Electrodeposition and characterization of undoped and nitrogen-doped ZnSe films. Mater Chem Phys 121:58

105. Riveros G, Gómez H, Henríquez R, Schrebler R, Marotti RE, Dalchiele EA (2001) Electrodeposition and characterization of ZnSe semiconductor thin films. Sol Energy Mater Sol Cells 70:255

106. Samantilleke AP, Boyle MH, Young J, Dharmadasa IM (1998) Electrodeposition of n-type and p-type ZnSe thin films for applications in large area optoelectronic devices. J Mater Sci Mater Electron 9:289

107. Soundeswaran S, Senthil Kumar O, Dhanasekaran R, Ramasamy P, Kumaresen R, Ichimura M (2003) Growth of ZnSe thin films by electrocrystallization technique. Mater Chem Phys 82:268

108. Moses Ezhil Raj A, Mary Delphine S, Sanjeeviraja C, Jayachandran M (2010) Growth of ZnSe thin layers on different substrates and their structural consequences with bath temperature. Physica B 405:2485

109. Kathalingam A, Mahalingam T, Sanjeeviraja C (2007) Optical and structural study of electrodeposited zinc selenide thin films. Mater Chem Phys 106:215
110. Bouroushian M, Kosanovic T, Spyrellis N (2005) Oriented [1 1 1] ZnSe electrodeposits grown on polycrystalline CdSe substrates. J Cryst Growth 277:335
111. Chandramohan R, Sanjeeviraja C, Mahalingam T (1997) Preparation of zinc selenide thin films by electrodeposition technique for solar cell applications. Phys Stat Sol (A) 163:R11
112. Murali KR, Dhanapandiyana S, Manoharana C (2009) Pulse electrodeposited zinc selenide films and their characteristics. Chalcogenide Lett 6:51
113. Bouroushian M, Kosanovic T, Spyrellis N (2006) A pulse plating method for the electrosynthesis of ZnSe. J Appl Electrochem 36:821
114. Gudage YG, Deshpande NG, Sagade AA, Sharma R (2009) Room temperature electrosynthesis of ZnSe thin films. J Alloys Compd 488:157
115. Kowalik R, Fitzner K (2009) Analysis of the mechanism for electrodeposition of the ZnSe phase on Cu substrate. J Electroanal Chem 633:78
116. Kowalik R, Żabiński P, Fitzner K (2008) Electrodeposition of ZnSe. Electrochim Acta 53:6184
117. Shibata M, Kobayashi T, Furuya N (1997) Studies of oxidation of As ad-atoms at Pt and Au electrodes using the EQCM technique. J Electroanal Chem 436:103
118. Natarajan C, Sharon M, Lévy-Clément C, Neumann-Spallart M (1994) Electrodeposition of zinc selenide. Thin Solid Films 237:118
119. Kosanovic T, Bouroushian M, Spyrellis N (2005) Soft growth of the ZnSe compound from alkaline selenosulfite solutions. Mater Chem Phys 90:148
120. Riveros G, Guillemoles JF, Lincot D, Meier HG, Froment M, Bernard MC, Cortes R (2002) Electrodeposition of epitaxial ZnSe films on InP and GaAs from an aqueous zinc sulfate-selenosulfate solution. Adv Mater 14:1286
121. Chaudhari GN, Sardesai SN, Sathaye SD, Rao VJ (1992) Structural properties of ZnS_xSe_{1-x} thin films on gallium arsenide (110) substrate. J Mater Sci 27:4647
122. Pezzatini G, Caporali S, Innocenti M, Foresti ML (1999) Formation of ZnSe on Ag(111) by electrochemical atomic layer epitaxy. J Electroanal Chem 475:164
123. Bouroushian M, Kosanovic T, Spyrellis N (2005) Aspects of ZnSe electrosynthesis from selenite and selenosulfite aqueous solutions. J Solid State Electrochem 9:55
124. Henríquez R, Gómez H, Riveros G, Guillemoles JF, Froment M, Lincot D (2004) A novel approach for the electrodeposition of epitaxial films of ZnSe on (111) and (100) InP using dimethylsulfoxide as a solvent. Electrochem Solid-State Lett 7:C75
125. Henríquez R, Gómez H, Riveros G, Guillemoles JF, Froment M, Lincot D (2004) Electrochemical deposition of ZnSe from dimethyl sulfoxide solution

and characterization of epitaxial growth. J Phys Chem B 108:13191
126. Gal D, Hodes G (2000) Electrochemical deposition of ZnSe and (Zn, Cd)Se films from nonaqueous solutions. J Electrochem Soc 147:1825
127. Sanchez S, Lucas C, Picard GS, Bermejo MR, Castrillejo Y (2000) Molten salt route for ZnSe high-temperature electrosynthesis. Thin Solid Films 361–362:107
128. Mahalingam T, John VS, Rajendran S, Ravi G, Sebastian PJ (2002) Annealing studies of electrodeposited zinc telluride thin films. Surf Coat Technol 155:245
129. Murali KR, Rajkumar PR (2006) Characteristics of pulse plated ZnTe films. J Mater Sci Mater Electron 17(5):393
130. Mahalingam T, John VS, Rajendran S, Sebastian PJ (2002) Electrochemical deposition of ZnTe thin films. Semicond Sci Technol 17:465
131. Neumann-Spallart M, Königstein C (1995) Electrodeposition of zinc telluride. Thin Solid Films 265:33
132. Kashyout AB, Aricò AS, Antonucci PL, Mohamed FA, Antonucci V (1997) Influence of annealing temperature on the opto-electronic characteristics of ZnTe electrodeposited semiconductors. Mater Chem Phys 51:130
133. Königstein C, Neumann-Spallart M (1998) Mechanistic studies on the electrodeposition of ZnTe. J Electrochem Soc 145:337
134. Rakhshani AE, Pradeep B (2004) Thin films of ZnTe electrodeposited on stainless steel. Appl Phys A 79:2021
135. Yang YW, Li L, Ye M, Wu YC, Xie T, Li GH (2007) Electrochemical deposition and properties of ZnTe nanowire array. Chin Phys Lett 24:2973
136. Li L, Yang Y, Huang X, Li G, Zhang L (2005) Fabrication and characterization of single-crystalline ZnTe nanowire arrays. J Phys Chem B 109:12394
137. Ishizaki T, Ohtomo T, Fuwa A (2004) Electrodeposition of ZnTe film with high current efficiency at low overpotential from a citric acid bath. J Electrochem Soc 151:C161
138. Ishizaki T, Ishizaki T, Fuwa A (2004) Structural, optical and electrical properties of ZnTe thin films electrochemically deposited from a citric acid aqueous solution. J Phys D: Appl Phys 37:255
139. Ishizaki T, Saito N, Takai O, Asakura S, Goto K, Fuwa A (2005) An investigation into the effect of ionic species on the formation of ZnTe from a citric acid electrolyte. Electrochim Acta 50:3509
140. Bouroushian M, Kosanovic T, Karoussos D, Spyrellis N (2009) Electrodeposition of polycrystalline ZnTe from simple and citrate-complexed acidic aqueous solutions. Electrochim Acta 54:2522
141. Bozzini B, Lenardi C, Lovergine N (2000) Electrodeposition of stoichiometric polycrystalline, ZnTe on n^+-GaAs and Ni–P. Mater Chem Phys 66:219

142. Bozzini B, Baker MA, Cavallotti PL, Cerri E, Lenardi C (2000) Electrodeposition of ZnTe for photovoltaic cells. Thin Solid Films 361–362:388

143. Chaure NB, Jayakrishnan R, Nai JP, Pandey RK (1997) Electrodeposition of ZnTe films from a nonaqueous bath. Semicond Sci Technol 12:1171

144. Heo P, Ichino R, Okido M (2006) ZnTe electrodeposition from organic solvents. Electrochim Acta 51:6325

145. Gandhi T, Raja KS, Misra M (2009) Synthesis of ZnTe nanowires onto TiO_2 nanotubular arrays by pulse-reverse electrodeposition. Thin Solid Films 517:4527

146. Lin MC, Chen PY, Sun IW (2001) Electrodeposition of zinc telluride from a zinc chloride-1-ethyl-3-methylimidazolium chloride molten salt. J Electrochem Soc 148:C653

147. Kuroda K, Kobayashi T, Sakamoto T, Ichino R, Okido M (2005) Formation of ZnTe compounds by using the electrochemical ion exchange reaction in molten chloride. Thin Solid Films 478:223

148. Han DH, Choi SJ, Park SM (2003) Electrochemical preparation of zinc telluride films on gold electrodes. J Electrochem Soc 150:C342

149. Wang Y, Huang JF, Cao LY, Zhu H, He HY, Wu JP (2009) Preparation of Bi_2S_3 thin films with a nanoleaf structure by electrodeposition method. Appl Surf Sci 255:7749

150. Saitou M, Yamaguchi R, Oshikawa W (2002) Novel process for electrodeposition of Bi_2S_3 thin films. Mater Chem Phys 73:306

151. Yesugade NS, Lokhande CD, Bhosale CH (1995) Structural and optical properties of electrodeposited Bi_2S_3, Sb_2S_3, and As_2S_3 thin films. Thin Solid Films 263:145

152. Grubač Z, Metikoš-Huković M (2002) Electrodeposition of thin sulfide films: nucleation and growth observed for Bi_2S_3. Thin Solid Films 413:248

153. Bessekhouad Y, Mohammedi M, Trari M (2002) Hydrogen photoproduction from hydrogen sulfide on Bi_2S_3 catalyst. Sol Energy Mater Sol Cells 73:339

154. Killedar VV, Katore SN, Bhosale CH (2000) Preparation and characterization of electrodeposited Bi_2S_3 thin films prepared from non-aqueous media. Mater Chem Phys 64:166

155. Georges C, Tena-Zaera R, Bastide S, Rouchaud JC, Larramona G, Lévy-Clément C (2007) Electrochemical deposition of Bi_2S_3 thin films using dimethylsulfoxide as a Solvent. J Electrochem Soc 154:D669

156. Öznülüer T, Demir Ü (2002) Formation of Bi_2S_3 thin films on Au(111) by electrochemical atomic layer epitaxy: kinetics of structural changes in the initial monolayers. J Electroanal Chem 529:34

157. Anuar K, Zainal Z, Hussein MZ, Saravanan N, Haslina I (2002) Cathodic electrodeposition of Cu_2S thin film for solar energy conversion. Sol Energy Mater Sol Cells 73:351

158. Anuar K, Zainal Z, Hussein MZ, Ismail H (2001) Electrodeposition and characterization of Cu_2S thin films from aqueous solution. J Mater Sci Mater Electro 12:147

159. Schimmel MI, Bottechia OL, Wendt H (1998) Anodic formation of binary and ternary compound semiconductor films for photovoltaic cells. J Appl Electrochem 28:299

160. Lokhande CD, Yermune VS, Pawar SH (1991) Electrodeposition of Cd-Bi-S and Cd-Zn-S films. J Electrochem Soc 138:624

161. Yu H, Quan X, Zhang Y, Ma N, Chen S, Zhao H (2008) Electrochemically assisted photocatalytic inactivation of *Escherichia coli* under visible light using a $ZnIn_2S_4$ film electrode. Langmuir 24:7599

162. Schurr R, Hölzing A, Jost S, Hock R, Voβ T, Schulze J, Kirbs A, Ennaoui A, Lux-Steiner M, Weber A, Kötschau I, Schock HW (2009) The crystallisation of Cu_2ZnSnS_4 thin film solar cell absorbers from co-electroplated Cu–Zn–Sn precursors. Thin Solid Films 517:2465

163. Jeon M, Shimizu T, Shingubara S (2011) Cu_2ZnSnS_4 thin films and nanowires prepared by different single-step electrodeposition method in quaternary electrolyte. Mater Lett 65:2364

164. Aravinda CL, Freyland W (2006) Nanoscale electrocrystallisation of Sb and the compound semiconductor AlSb from an ionic liquid. Chem Commun 16:1703

165. Mann O, Aravinda CL, Freyland W (2006) Microscopic and electronic structure of semimetallic Sb and semiconducting AlSb fabricated by nanoscale electrodeposition: an in situ scanning probe investigation. J Phys Chem B 110:21521

166. Gandhi T, Raja KS, Misra M (2008) Room temperature electrodeposition of aluminum antimonide compound semiconductor. Electrochim Acta 53:7331

167. Villegas I, Stickney JL (1992) Preliminary studies of GaAs on Au(100), (110), and (111) surfaces by electrochemical atomic layer epitaxy. J Electrochem Soc 139:686

168. Perrault GG (1989) Thermodynamics of gallium arsenide electrodeposition. J Electrochem Soc 136:2845

169. Chandra S, Khare N (1987) Electrodeposition gallium arsenide film: I. Preparation, structural, optical ad electrical studies. Semicond Ci. Technol 2:214

170. Yang MC, Landau U, Angus JC (1992) Electrodeposition of GaAs from aqueous electrolytes. J Electrochem Soc 139:3480

171. Gao Y, Han A, Lin Y, Zhao Y, Zhang J (1994) Electrodeposition and characterization of GaAs polystallie thi films. J Appl Phys 75:549

172. Mahalingam T, Lee S, Lim H, Moon H, Kim YD (2006) Electrosynthesis and characterization of GaAs in acid solutions by potentiostatic method. Sol Energy Mater Sol Cells 90:2456

173. Gheorghies C, Gheorghies L (2002) Preparation of GaAs thin films from acid aqueous solution. J Optoelectron Adv Mater 4:97912

174. Gheorghies C, Gheorghies L, Fetecau G (2007) Electrodeposition of GaAs thin films from alkaline aqueous solution. J Optoelectron Adv Mater 9:2795
175. Lajnef M, Chtourou R, Ezzaouia H (2010) Electric characterization of GaAs deposited on porous silicon by electrodeposition technique. Appl Surf Sci 252:3058
176. Kozlov VM, Bozzini B, Bicelli LP (2004) Formation of GaAs by annealing of two-layer Ga-As electrodeposits. J Alloys Compd 379:209
177. Murali KR, Trivedi DC (2006) Preparation of pulse pated GaAs films. J Phys Chem Solid 67:1432
178. Dioum IG, Vedel J, Tremillon B (1982) Properties of arsenic in molten potassium tetrachlorogallate at 300°C: Formation of gallium arsenide. J Electroanal Chem 139:329
179. Wicelinski SP, Gal RJ (1986) In: Saboungo L, Newman DS, Johson K, Inman D (eds) Fifth international symposium on molten salts (PV86-1). The Electrochemical Society, Pennington, GaAs film formation from low temperature chloroaluminate melts. NJ, p. 144
180. Carpenter MK, Verbrugge MW (1990) Electrochemical codeposition of gallium and arsenic from a room temperature chlorogallate melt. J Electrochem Soc 137:123
181. Paolucci F, Mengoli G, Musiani MM (1990) An electrochemical route to GaSb thin films. J Appl Electrochem 20:868
182. Kozlov VM, Bicelli LP (2000) Influence of temperature and of structure of antimony substrate on gallium diffusion into the GaSb semiconductor compound. J Alloys Compd 313:161
183. Mengoli G, Musiani MM, Paolucci F (1991) Synthesis of indium antimonide (InSb) and indium gallium antimonide (In$_x$Ga$_{1-x}$Sb) thin films from electrodeposited elemental layers. J Appl Electrochem 21:863
184. Kozlov VM, Agrigento V, Mussati G, Bicelli LP (1999) Influence of the structure of the electrodeposited antimony substrate on indium diffusion. J Alloys Compd 288:255
185. Fulop T, Bekele C, Landau U, Angus J, Kash K (2004) Electrodeposition of polycrystalline InSb from aqueous electrolytes. Thin Solid Films 449:1
186. Wade TL, Vaidyanathan R, Happek U, StickneyJL JL (2001) Electrochemical formation of a III–V compound semiconductor superlattice: InAs/InSb. J Electroanal Chem 500:332
187. Zhang X, Hao Y, Meng G, Zhang L (2005) Fabrication of highly ordered InSb nanowire arrays by electrodeposition in porous anodic aluminum membranes. J Electrochem Soc 152:C664
188. Khan MI, Wang X, Bozhilov KN, Ozkan CS (2008) Templated fabrication of InSb nanowires for nanoelectronics. J Nanomater. doi:1155/2008/678759
189. Khan MI, Wang X, Jing X, Bozhilov KN, Ozkan CS (2009) Study of a single InSb nanowire fabricated via DC electrodeposition in porous templates. J Nanosci Nanotechnol 9:26391
190. Li GR, Lu XH, Tong YX (2008) Electrochemical reduction synthesis of In–Sb nanoropes and terraced micropyramids. Electrochem Commun 10:127
191. Yang MH, Yang MC, Sun IW (2003) Electrodeposition of indium antimonide from the water-stable 1-ethyl-3-methylimidazolium chloride/tetrafluoroborate ionic liquid. J Electrochem Soc 150:C544
192. Wade TL, Ward LC, Maddox CB, Happek U, Stickney JL (1999) Electrodeposition of InAs. Electrochem Solid-State Lett 2:616
193. Kozlov VM, Bozzini B, Bicelli LP (2004) Preparation of InAs by annealing of two-layer In–As electrodeposits. J Alloys Compd 366:152
194. Mengoli G, Musiani MM, Paolucci F (1992) Synthesis of InAs and InAs$_{1-x}$Sb$_x$ from electrodeposited layers of indium, arsenic and As-Sb alloy. J Electroanal Chem 332:199
195. Andreoli P, Cattarin S, Musiani MM, Paolucci F (1995) Electrochemical approaches to GaAs$_{1-x}$Sb$_x$ thin films. J Electroanal Chem 385:265
196. Tsuda T, Hussey CL (2008) Electrodeposition of photocatalytic AlInSb semiconductor alloys in the Lewis acidic aluminum chloride-1-ethyl-3-methylimidazolium chloride room temperature ionic liquid. Thin Solid Film 516:622

Semiconductors, Electrochemical Atomic Layer Deposition (E-ALD)

John Stickney
University of Georgia, Athens, GA, USA

Introduction

Electrodeposition has been a thin film formation method since the nineteenth century, though it is thought by some to be more prone to contamination and less precise than corresponding vacuum processes. With the introduction of the Cu Damascene process for the formation of interconnects in ultra-large-scale integration (ULSI), it is now clear that electrodeposition can be as clean as needed. The development of E-ALD signals that electrodeposition can also be precise.

Molecular beam epitaxy (MBE) is a thin film deposition method with exceptional control over the deposit growth. It is the author's contention that there are no fundamental reasons why electrodeposition cannot achieve a similar degree of control.

This entry describes the electrodeposition of nanofilms using atomic layer deposition (ALD) [1]. ALD is the formation of nanofilms of materials one atomic layer at a time using surface limited reactions (SLRs). SLRs occur only at the surface, and once the surface is covered, the reaction stops. Electrochemical atomic layer deposition (E-ALD) is the electrochemical form of ALD, though it has also been referred to as electrochemical atomic layer epitaxy (EC-ALE) [2, 3], EC-ALD, ECALE, and ECALD.

Surface Limited Reactions (SLR)

Electrochemical SLRs are referred to as underpotential deposition (UPD) [4], where an atomic layer of a first element deposits on a second, at a potential prior to (under) that needed to deposit the first element on itself. The term "atomic layer" refers to a deposit no more than one atom thick, with a coverage less than a monolayer (ML). A ML corresponds to the deposition of one atom for each surface atom. UPD results from a larger interaction energy between two elements than between the element and itself, resulting in a surface compound or alloy.

In electrodeposition, metal ion reduction to the element,

$$M^{2+} + 2e^- = M,$$

is fundamentally controlled by the applied potential, which dictates the activity ratio of products to reactants at equilibrium, according to the Nernst equation:

$$E = E^o_{M^{2+}} - \frac{RT}{2F} \ln \frac{\text{Products}}{\text{Reactants}}$$

In conventional bulk metal electrodeposition, the metal has an activity of 1. In UPD, however, the product is an atomic layer of the element, with a different electronic structure than the bulk resulting from its interaction with the substrate. The activity is then less than 1, shifting the equilibrium potential positive and accounting

for the underpotential. By definition, UPD does not result in a bulk deposit, regardless of the deposition time.

E-ALD Cycle Chemistry

Each pair of elements has their own chemistry, and some result in UPD and some do not. The size of the underpotential varies with the chemistry as well. UPD varies with structure and coverage, which are a function of the electrolyte, additives, and substrate orientation. To develop an E-ALD cycle, the electrochemistry between the pair must be investigated. A cycle is the sequence of steps used to deposit one stoichiometric layer of the desired material: an atomic layer for a pure element, a bilayer for a I:I compound, etc. An E-ALD cycle for CdS (Fig. 1) is illustrative: An atomic layer of S is deposited on one of Cd, and an atomic layer of Cd is deposited on one of S.

$$S^{2-} = S_{upd} + 2e^-$$

$$Cd^{2+} + 2e^- = Cd_{upd}$$

This cycle is composed of four steps: oxidative UPD of sulfur from a S^{2-} ion solution, a blank rinse, reductive UPD of cadmium from a Cd^{2+} ion solution, and a second blank rinse. Separate solutions and potentials are used for each reactant and the blank, providing extensive control over deposit growth, composition, and morphology. A nanofilm is formed by repeating the cycle, with the deposit thickness determined by the number of cycles. This linear relationship is a good indication of an ALD process and a layer by layer growth mechanism.

A cartoon of E-ALD suggests that at "the" UPD potential, a ML deposits on the surface and there is no more deposition until the formal potential ($E^{o'}$) for bulk deposition is reached. The truth is that different pairs of elements result in different coverages at different potentials. Polycrystalline substrates result in polycrystalline deposits. Each substrate crystallographic

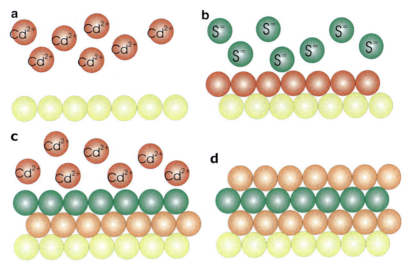

Semiconductors, Electrochemical Atomic Layer Deposition (E-ALD), Fig. 1 Schematic of a cycle for the deposition of CdS. (**a**) Cd^{2+} solution, reductive UPD forming Cd atoms; (**b**) $S^=$ solution, oxidative UPD forming S atoms; (**c**) Cd^{2+} solution, reductive UPD forming Cd atoms; (**d**) 1.5 ML of CdS on Au substrate

orientation may result in a different deposit orientation, with a different atomic density. In addition, different elements result in different structures with different coverages at different potentials. Grain boundaries, steps, and kinks can have different energetics, and result in different atomic densities and coordination, all a function of potential. On a given single crystal substrate, a sequence of different structures with different coverages, all less than an ML, can form as a function of potential [5].

For the alternated deposition of two elements, formation of a compound is a thermodynamically preferred result. The amount of the first element on the surface helps determine the amount of the next. For a binary compound like CdS, the elemental atomic layer coverages should be identical, beyond the first few cycles where the substrate may have an influence. The same compound should result over a range of underpotentials. The more of the first element deposited, the more of the second that will deposit, as the stoichiometry of the deposit controls the relative coverages. For CdS, the larger the underpotential used to deposit S, the less that will deposit. Consequently, in the subsequent step, less Cd will deposit, maintaining the CdS stoichiometry. CdS will still grow, though at a lower rate (nm/cycle). To form a conformal stoichiometric deposit, the applied potentials and solution compositions for the E-ALD cycle should be optimized.

There are a number of cycle chemistries that have been investigated so far, depending on the material being formed. The CdS cycle (Fig. 1) is oxidative UPD followed by reductive UPD (O-R) [6–8]. In practice, few compounds can be formed using the O-R cycle because of a lack of suitable negative oxidation state precursors (like S^{2-}), for which oxidative UPD is practicable. Other cycle chemistries include: reductive UPD followed by reductive UPD (R-R), R-R with a reductive strip (R-R-R), R-R with an oxidative strip (R-R-O), bait and switch (B&S) [9], and surface limited redox replacement (SLRR) [10–13].

Ideally, E-ALD could be used to form deposits on any substrate; however, the need for some conductivity limits substrates to metals and some semiconductors, though as the charges correspond only to the deposition of an AL, very thin or low conductivity substrates can be used. In addition, substrates are generally limited to those materials where UPD can be performed, in order that the first AL is homogeneous and conformal. An understanding of the surface chemistry of the substrate in the reactant solutions will help in identifying which substrates are practicable. For example, Si is difficult to use as a substrate because of its oxide layer, while the oxide on Ge can be reduced electrochemically, suggesting it would make a better substrate [9]. Note also that the conditions needed to deposit the first atomic layer should differ from the steady-state conditions used to deposit the majority of the nanofilm, as the chemistry on the substrate will differ from that on the deposit. There are examples where the conditions used to form a nanofilm are systematically adjusted for the first 10 or more cycles before steady state. Why adjustments were need for 10 cycles is a active topic of study, and appear to be related to interfacial stress.

It may appear from the above discussion that each cycle results in a conformal compound layer over the whole substrate surface. However, the rules for thin film formation still apply: Better deposits form on better substrates and a lattice mismatch still results in strain. Strain will increase with the number of cycles until the critical thickness is achieved, whereupon threading dislocations and other defects appear. The substrate may be powder, foil, a polycrystalline film, a single crystal, or a wafer. For a given substrate, each low index plane will show a different affinity for the depositing element and result in a different coverage and interface structure. Where facets intersect, grain boundaries will form and the cycle conditions chosen will determine the resulting deposit morphology. Use of lower underpotentials for a given material will generally result in more deposition, which can lead to roughening at grain boundaries, while larger underpotentials could minimize deposition at the boundaries (Fig. 2). Even when a single crystal is used as a substrate, steps and defects could nucleate structures which grow cycle by cycle, roughening the surface.

UPD can also result in more than an ML in some cases, where small underpotentials are used [5]. If the amount of an element deposited in a cycle is in excess of that needed to deposit a ML of the compound, a second compound layer could be initiated, possibly nucleating asperities and promoting roughing. To prevent roughening, it is advisable to choose underpotentials which will only result in the amount needed to form a compound monolayer or less. It is much better to deposit too little each cycle then too much.

Exchange Current (i_{ex})

One of the major differences between E-ALD and various vacuum-based deposition methodologies, besides use of a condensed phase, is temperature. Temperature is used in vacuum-based methods to control reactivity and provide surface diffusion to limit roughening. E-ALD is performed at room temperature, or below the solvent boiling point, which is considered "low temperature" for thin film formation. Despite not heating, E-ALD deposits are formed under equilibrium conditions. That is, a reactant is introduced to the deposit at a specific potential, and left until there is no net deposition current: equilibrium. For fast electrochemical deposition reactions at equilibrium, the process is dynamic: Atoms are both depositing and dissolving at the same time, with no net current. The rates of depositing and dissolving are equal though abosite and referred to as the exchange current (i_{ex}). Ideally, atoms in high energy sites dissolve and other ions deposit into more stable sites. i_{ex} takes the place of surface diffusion, as the net product of this repeated deposition and dissolution is to create a surface where only the most stable atoms remain, promoting optimization of the surface structure. If overpotentials are used, i_{ex} decreases, leading to rougher deposits.

Materials Deposited

An increasing number of groups around the world have been using E-ALD to form materials. Compounds formed include: most of the II-VI compounds, including ZnSe [14], CdTe [2, 3] and CdS [7, 15, 16], III-V compounds such as InAs [17, 18], IR detector materials InSb [19] and HgCdTe [20], thermoelectric materials such as the IV-VI compounds PbS [21, 22], PbSe [23, 24, 25], and PbTe [24, 25], as well as Sb_2Te_3 [26], and Bi_2Te_3 [26, 27]. In addition, phase change materials such as GeSbTe [28], and the photovoltaic materials CdTe [29], Ge [9, 30], CIS [31], and CIGS have been grown. More recently, elemental deposits of metals have been formed using a cycle referred to as surface

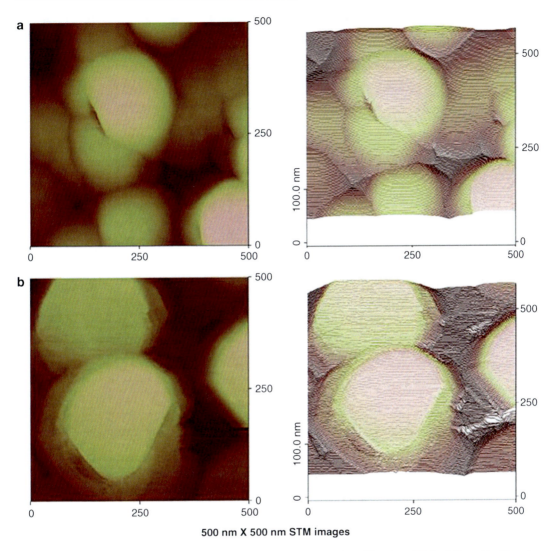

Semiconductors, Electrochemical Atomic Layer Deposition (E-ALD), Fig. 2 STM images: (a) Au vapor deposited on glass; (b) 30 superlattice periods, each consisting of 3 cycles of PbSe, followed by 15 cycles of PbTe, on Au vapor deposited on glass. Note the nearly atomically flat terraces resulting from the E-ALD deposit

limited redox replacement (SLRR) [10, 11]. Metals deposited using SLRR include: Pt [10, 11, 32], Ag, Cu [33, 34], Pd, Rh and Ru [35].

E-ALD Hardware

Most E-ALD deposits are formed using some type of electrochemical flow cell that allows for the rapid exchange of solutions, in combination with automation (Electrochemical ALD L.C., Athens, GA), where the cycle can be programmed, and allowed to run for as many cycles as desired. Graduate students manually performing cycles usually tire and make mistakes somewhere after the 10th cycle. The more elements in the deposit, the more solution lines required. More complex sequences of potentials and solution exchanges can be performed as well, such as the formation of a superlattice [19, 24, 36, 37], where two or more materials or compounds are alternated to form materials with unique lattice constants. The optical and electronic properties of a superlattice can be a function of the thicknesses of the constituent nanofilms, or their repeat distance or period (Fig. 2).

Acknowledgment Support of the National Science Foundation, Division of Materials Research, Award #1006747, is duly acknowledged.

Cross-References

▶ Compound Semiconductors, Electrochemical Decomposition
▶ Semiconductors Groups II-IV and III-V, Electrochemical Deposition
▶ Semiconductors, Principles

References

1. Suntola T, Antson J (1976) Apparatus and method for growing thin compound films. 75-2553048, 2553048, 19751126
2. Gregory BW, Norton ML, Stickney JL (1990) Thin-layer electrochemical studies of the underpotential deposition of cadmium and tellurium on polycrystalline Au, Pt and Cu electrodes. J Electroanal Chem 293(1–2):85–101
3. Gregory BW, Stickney JL (1991) Electrochemical atomic layer epitaxy (ECALE). J Electroanal Chem 300(1–2):543–561
4. Kolb DM (1978) Physical and electrochemcial properties of metal monolayers on metallic substrates. In: Gerischer H, Tobias CW (eds) Advances in electrochemistry and electrochemical engineering, vol 11. Wiley, New York, p 125
5. Stickney JL, Rosasco SD, Song D, Soriaga MP, Hubbard AT (1983) Superlattices formed by electrodeposition of silver on iodine-pretreated Pt(111); studies by LEED, Auger spectroscopy and electrochemistry. Surf Sci 130(2):326–347
6. Gu CK, Xu H, Park M, Shannon C (2009) Synthesis of metal-semiconductor core-shell nanoparticles using electrochemical surface-limited reactions. Langmuir 25(1):410–414
7. Foresti ML, Pezzatini G, Cavallini M, Aloisi G, Innocenti M, Guidelli R (1998) Electrochemical atomic layer epitaxy deposition of CdS on Ag(111): an electrochemical and STM investigation. J Phys Chem B 102(38):7413–7420
8. Colletti LP, Teklay D, Stickney JL (1994) Thin-layer electrochemical studies of the oxidative underpotential depostiion of sulfur and its application to the electrochemical atomic layer eptiaxy deposition of CdS. J Electroanal Chem 369(1–2):145–152
9. Liang X, Zhang Q, Lay MD, Stickney JL (2011) Growth of Ge nanofilms using electrochemical atomic layer deposition, with a "Bait and Switch" Surface-limited reaction. J Am Chem Soc 133(21):8199–8204
10. Brankovic SR, Wang JX, Adzic RR (2001) Metal monolayer deposition by replacement of metal adlayers on electrode surfaces. Surf Sci 474(1–3):L173–L179
11. Mrozek MF, Xie Y, Weaver MJ (2001) Surface-enhanced Raman scattering on uniform platinum-group overlayers: preparation by redox replacement of underpotential-deposited metals on gold. Anal Chem 73(24):5953–5960
12. Kim Y-G, Kim JY, Vairavapandian D, Stickney JL (2006) Platinum nanofilm formation by EC-ALE via redox replacement of UPD copper: studies using in-situ scanning tunneling microscopy. J Phys Chem B 110(36):17998–18006
13. Vasilic R, Dimitrov N (2005) Epitaxial growth by monolayer-restricted galvanic displacement. Electrochem Solid-State Lett 8(11):C173–C176
14. Pezzatini G, Caporali S, Innocenti M, Foresti ML (1999) Formation of ZnSe on Ag(111) by electrochemical atomic layer epitaxy. J Electroanal Chem 475(2):164–170
15. Demir U, Shannon C (1994) A scanning tunneling microscopy study of electrochemically grown cadmium sulfide monolayers on Au(111). Langmuir 10(8):2794–9
16. Streltsov ES, Labarevich II, Talapin DV (1994) Electrochemical formation of monolayer films of cadmium-sulfide on the Au surface. Dokl Akad Nauk Bel 38(5):64
17. Wade TL, Ward LC, Maddox CB, Happek U, Stickney JL (1999) Electrodeposition of InAs. Electrochem Sol State Lett 2(12):616–618
18. Innocenti M, Forni F, Pezzatini G, Raiteri R, Loglio F, Foresti ML (2001) Electrochemical behavior of As on silver single crystals and experimental conditions for InAs growth by ECALE. J Electroanal Chem 514(1–2):75–82
19. Wade TL, Vaidyanathan R, Happek U, Stickney JL (2001) Electrochemical formation of a III-V compound semiconductor superlattice: InAs/InSb. J Electroanal Chem 500(1–2):322–332
20. Venkatasamy V, Jayaraju N, Cox SM, Thambidurai C, Stickney JL (2007) Studies of Hg(1-x)CdxTe formation by electrochemical atomic layer deposition and investigations into bandgap engineering. J Electrochem Soc 154(8):H720–H725
21. Torimoto T, Takabayashi S, Mori H, Kuwabata S (2002) Photoelectrochemical activities of ultrathin lead sulfide films prepared by electrochemical atomic layer epitaxy. J Electroanal Chem 522(1):33–39
22. Oeznuelueer T, Erdogan I, Sisman I, Demir U (2005) Electrochemical atom-by-atom growth of PbS by modified ECALE method. Chem Mater 17(5):935–937

23. Vaidyanathan R, Stickney JL, Happek U (2004) Quantum confinement in PbSe thin films electrodeposited by electrochemical atomic layer epitaxy (EC-ALE). Electrochim Acta 49(8): 1321–1326
24. Banga D, Stickney JL (2007) PbTe – PbSe superlattice formation by electrochemical atomic layer deposition. ECS Trans 6 (2, State-of-the-Art Program on Compound Semiconductors 46 (SOTAPOCS 46) and Processes at the Semiconductor/Solution Interface 2): 439–449
25. Vaidyanathan R, Cox SM, Happek U, Banga D, Mathe MK, Stickney JL (2006) Preliminary studies in the electrodeposition of PbSe/PbTe superlattice thin films via electrochemical atomic layer deposition (ALD). Langmuir 22(25):10590–10595
26. Zhu W, Yang JY, Zhou DX, Xiao CJ, Duan XK (2008) Development of growth cycle for antimony telluride film on Au(111) disk by electrochemical atomic layer epitaxy. Electrochim Acta 53(10): 3579–3586
27. Zhu W, Yang JY, Gao XH, Bao SQ, Fan XA, Zhang TJ, Cui K (2005) Effect of potential on bismuth telluride thin film growth by electrochemical atomic layer epitaxy. Electrochim Acta 50(20): 4041–4047
28. Liang X, Jayaraju N, Thambidurai C, Zhang Q, Stickney JL (2011) Controlled electrochemical formation of GexSbyTez using Atomic Layer Deposition (ALD). Chem Mater 23(7):1742–1752
29. Huang BM, Colletti LP, Gregory BW, Anderson JL, Stickney JL (1995) Preliminary studies of the use of an automated flow-cell electrodeposition system for the formation of CdTe thin-films by electrochemical atomic layer epitaxy. J Electrochem Soc 142(9): 3007–3016
30. Liang X, Kim Y-G, Gebergziabiher DK, Stickney JL (2010) Aqueous electrodeposition of Ge monolayers. Langmuir 26(4):2877–2884
31. Banga D, Jarayaju N, Sheridan L, Kim Y-G, Perdue B, Zhang X, Zhang Q, Stickney J (2012) Electrodeposition of CuInSe2 (CIS) via Electrochemical Atomic Layer Deposition (E-ALD). Langmuir 28(5):3024–3031
32. Dimitrov N, Vasilic R, Vasiljevic N (2007) A kinetic model for redox replacement of UPD layers. Electrochem Solid State Lett 10(7):D79–D83
33. Kim JY, Kim YG, Stickney JL (2008) Cu nanofilm formation by electrochemical atomic layer deposition (ALD) in the presence of chloride ions. J Electroanal Chem 621(2):205–213
34. Viyannalage LT, Vasilic R, Dimitrov N (2007) Epitaxial growth of Cu on Au(111) and Ag(111) by surface limited redox replacement-An electrochemical and STM study. J Phys Chem C 111(10): 4036–4041
35. Thambidurai C, Kim Y-G, Stickney JL (2008) Electrodeposition of Ru by atomic layer deposition (ALD). Electrochim Acta 53(21):6157–6164
36. Zou S, Weaver MJ (1999) Surface-enhanced Raman spectroscopy of cadmium sulfide/cadmium selenide superlattices formed on gold by electrochemical atomic-layer epitaxy. Chem Phys Lett 312(2–4): 101–107
37. Banga D, Kim Y-G, Stikney J (2011) PbSe/PbTe Superlattice Formation via Electrochemical Atomic Layer Depostion (E-ALD). J. Electrochem. Soc. 153: D99–D106.

Semiconductors, Principles

Michael A. Filler
School of Chemical and Biomolecular Engineering, Georgia Institute of Technology Atlanta, GA, USA

Introduction

Semiconductors are a class of materials whose behavior is intermediate between that of insulators and metals. Their properties largely result from (1) an electronic structure with a moderately sized region (0.5–3.5 eV) devoid of energy levels known as the bandgap and (2) the ability to tune their properties via the intentional incorporation of impurity atoms. This chapter will outline the basic physics and chemistry that dictates semiconductor properties. We begin with a description of important semiconductor crystal structures and discuss how their unique electronic properties arise from the arrangement of, and bonding between, atoms in a lattice. The equilibrium concentration of charge carriers is subsequently considered and connected to changes in electrical conductivity. Here, we outline only the most important concepts as they relate to semiconductor electrochemistry and note that fully rigorous treatments are available in most solid-state physics textbooks [1, 2]. Although inorganic semiconductors are the focus of this chapter, the study of organic semiconductors has burgeoned in recent years, and the reader is referred to any number of excellent discussions of the subject [3].

Crystal Structure

The crystal structures of all extended solids are comprised of a Bravais lattice, which is a periodic arrangement of points in space, and a basis, which is the atom or atoms associated with each lattice point [1]. While a single-atom basis is sufficient for some crystal structures, including body-centered cubic (BCC) or face-centered cubic (FCC), many materials have multiple-atom bases. Most semiconductors, including those illustrated in Fig. 1, fall into the latter class of materials and are briefly summarized below. The structure of bulk crystals are described here, but note that synthesis method or nanoscale confinement can yield alternatives [4–6].

A significant fraction of semiconductors exhibit diamond cubic (DC), zinc blende (ZB), or wurtzite (WZ) crystal structures. The DC crystal structure observed for C, Si, and Ge consists of a FCC lattice and a two-atom basis. Alternatively, one can also think of the DC crystal structure as two interpenetrating FCC lattices with monoatomic bases. The zinc blende (ZB) crystal structure is identical to DC except that the two atoms comprising the basis are different. Many group III–V and II–VI semiconductors, including GaAs, GaP, and CdTe, exhibit a ZB crystal structure. When the atomic radii and/or electronegativities of the atoms comprising a ZB crystal are sufficiently different, the wurtzite (WZ) structure is often observed [7]. ZnO, CdSe, and CdS are common WZ semiconductors. Some semiconductors, including ZnS, exhibit intermediate bonding characteristics and can therefore be prepared as bulk materials with a ZB or a WZ crystal structure.

The transition metal oxides, such as Fe_2O_3 or TiO_2, can exhibit a variety of crystal structures. For example, the rutile, anatase, and brookite phases of TiO_2 naturally occur in nature, but a myriad of others are known [8]. Transition metal chalcogenides, another class of semiconductors which include MoS_2, WS_2, and WSe_2, crystallize with a complex non-covalent layered structure where the transition atom is situated between two hexagonally closed packed chalcogenide layers [9].

Electronic Structure

Despite the complexity of describing the interaction between $\sim 10^{22}$ atoms/cm^3 in a bulk semiconductor, many of the most important features of its electronic structure can be easily understood within the context of molecular orbital (MO) theory. We begin here and then adapt the results to account for the behavior of solid-state materials. Consider how the atomic orbitals (AOs) of isolated oxygen atoms combine to yield MOs in molecular oxygen (i.e., O_2). As shown in Fig. 2, the valence electrons of isolated oxygen atoms occupy 2s and 2p AOs. Upon formation of a covalent bond, bonding and antibonding MOs with energies below and above the original 2s AOs, respectively, are formed. The electrons previously occupying the 2s AO of each atom are now shared between the two oxygen atoms and completely fill these MOs. A similar situation occurs for the four 2p valence electrons. Importantly, the O_2 2p bonding MOs are filled with electrons and the antibonding MOs are not. The energy difference between the highest occupied molecular orbital (HOMO) and lowest unoccupied molecular orbital (LUMO) is known as the HOMO-LUMO gap [10].

Figure 3 illustrates the key features of semiconductor electronic structure. The situation in semiconductors is complicated by lattice symmetry and spin-orbit coupling but is analogous to that described above for the O_2 molecule. When many atoms are arranged on a lattice, bonding and antibonding MOs are derived from the original AOs. Since the interaction between any two atoms in a lattice is similar, it is not too surprising that the bonding and antibonding MOs cluster around similar energies. Although a finite energy interval separates all energy levels within the solid, the spacing becomes so small for large ensembles of atoms that they can be treated continuously. Semiconductors have two important regions of tightly space energy levels, known as the "valence band" and "conduction band." The energy levels at the top of the valence band and bottom of the conduction band are known as the "valence band edge" (E_V) and the

Semiconductors, Principles

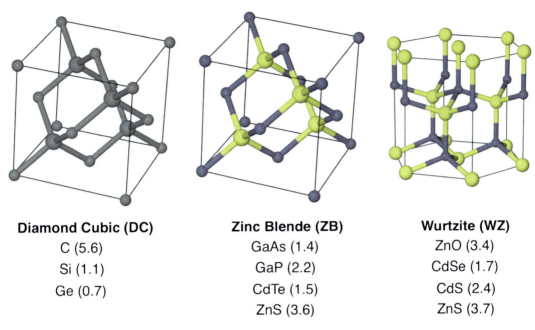

Semiconductors, Principles, Fig. 1 Schematic illustrations of the diamond cubic (*DC*), zinc blende (*ZB*), and wurtzite (*WZ*) crystal structures. Example semiconductors for each structure are listed with the bandgap at 300 K (in eV)

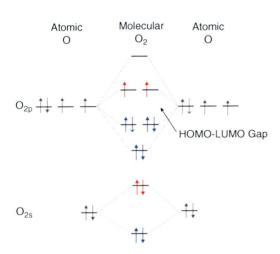

Semiconductors, Principles, Fig. 2 Molecular orbital (*MO*) diagram for O_2. Electrons from atomic orbitals (*AOs*), bonding MOs, and antibonding MOs are shown in *grey*, *blue*, and *red*, respectively

"conduction band edge" (E_C), respectively. The "bandgap" (E_g) is the region between the band edges where no electronic states exist and is analogous to the HOMO-LUMO gap of a molecule.

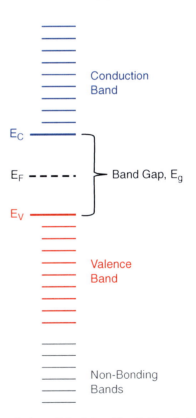

Semiconductors, Principles, Fig. 3 Key features of semiconductor electronic structure

Carrier Statistics

We now describe the behavior of charge carriers in an intrinsic semiconductor (i.e., pure) at equilibrium. The electrical properties of any extended solid depend on the position of the Fermi level, defined as the highest occupied state at $T = 0$ K. An alternative definition, stemming from the Fermi-Dirac statistics that govern the distribution of electrons, the Fermi level is the energy at which the probability of finding an electron is ½. If the Fermi level falls within a band, the band is partially filled and the material behaves as a conductor. As shown in Fig. 3, the valence and conduction band edges of an intrinsic semiconductor straddle the Fermi level. At $T = 0$ K, no conduction is possible since all of the states in the valence band are completely filled with electrons while all of the states in the conduction band are empty.

Thermal energy at $T > 0$ K can excite electrons from the valence to conduction band. The excited conduction band electron leaves behind an empty state in the valence band often termed a "hole." When such a transition occurs, both bands are now partially filled and conduction can take place. It is important to note that charge carrier motion in a semiconductor is strongly influenced by scattering events at atomic centers and by the electric fields that exist between those points. To simplify the mathematical description of these phenomena, the masses of the electrons and holes are often described within the effective mass approximation where m_o, the free electron mass, is replaced with the effective mass (m_e^* for electrons or m_h^* for holes).

The concentration of electrons in the conduction band (n) and holes in the valence band (p) can be determined by summing the charge carriers that occupy each electronic state:

$$n = \int_{E_C}^{E_{top}} Z(E)F(E)dE \tag{1}$$

$$p = \int_{E_{bottom}}^{E_V} Z(E)[1 - F(E)]dE \tag{2}$$

where $Z(E)$ is the density of states and $F(E)$ is the Fermi-Dirac distribution. For most common situations, the Fermi-Dirac distribution can be approximated by an exponential function, and integration of Eqs. 1 and 2 with E_{top} replaced by $+\infty$ and E_{bottom} replaced by $-\infty$ produces the following closed-form solutions:

$$n = 2\left(\frac{m_e^* k_B T}{2\pi\hbar^2}\right)^{\frac{3}{2}} \exp\left(\frac{E_F - E_C}{k_B T}\right) \tag{3}$$

$$p = 2\left(\frac{m_h^* k_B T}{2\pi\hbar^2}\right)^{\frac{3}{2}} \exp\left(\frac{E_V - E_F}{k_B T}\right) \tag{4}$$

where m_e^* is the effective mass of an electron, m_h^* is the effective mass of a hole, E_F is the Fermi level, E_C is the conduction band edge, E_V is the valence band edge, k_B is Boltzmann's constant, $\hbar$ is Planck's constant divided by 2π, and T is the absolute temperature.

Since a hole is created in the valence band for each electron that is excited to the conduction band, equal numbers of electrons and holes are present in an intrinsic semiconductor at equilibrium:

$$n = p = n_i \tag{5}$$

where n_i is the intrinsic carrier concentration. Combining Eqs. 3, 4, and 5 yields

$$n_i = 2.51 \times 10^{19} \left(\frac{m_e^*}{m_o}\frac{m_h^*}{m_o}\right)^{\frac{3}{4}} \left(\frac{T}{300}\right)^{\frac{3}{2}} \exp\left[-\left(\frac{E_g}{2k_B T}\right)\right] \tag{6}$$

where n_i has units cm^{-3}, m_o is the mass of a free electron, $E_g = E_c - E_v$ is the bandgap, and the remaining variables have the same meaning as above.

As can be determined for Si at 300 K from Eq. 6, only about 1 electron per 10^{13} contributes to conduction! While higher temperatures can further increase conductivity, via additional interband transitions, this is not practical for many applications. Thus, the intentional addition of substitutional impurity atoms to a semiconductor lattice – a process known as doping – is used to increase the number of electrons in the conduction band or holes in the

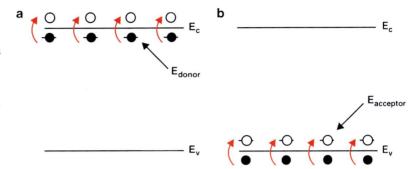

Semiconductors, Principles, Fig. 4 Schematic illustration of donor atoms adding electrons to the conduction band and (b) acceptor atoms removing electrons from the valence band

valence band. A doped semiconductor is called extrinsic. Dopant atoms are classified as donors or acceptors depending on whether they donate or accept an electron from atoms comprising the crystal, respectively. For the case of Si, group V atoms (e.g., P) are used as donors whereas III atoms (e.g., B) are used as acceptors. Whether a dopant atom behaves as a donor or acceptor depends on the position of its energy level relative to the band edges. Donor/acceptor energy levels are situated a few meV below/above the conduction/valence band edge as shown in Fig. 4. While these atoms do not contribute charge carriers at 0 K, a small amount of thermal energy can excite an electron from the donor level to the conduction band or from the valence band into the acceptor level. The proximity of dopant energy levels to either band edge ensures that the dopant atoms are completely ionized under most conditions. An extrinsic semiconductor where the majority of charge carriers are electrons or holes is referred to as n-type (negative) or p-type (positive), respectively.

Figure 5 illustrates that the position of the Fermi level will move away from the middle of the bandgap as the doping concentration increases. When donors or acceptors are added, the Fermi level will shift toward the conduction or valence band edge, respectively. The Fermi level position (E_F) can be determined for totally ionized donors and acceptors, respectively, as:

$$E_F = E_C - kT\ln\left(\frac{n}{N_D}\right) \quad (7)$$

$$E_F = E_V + kT\ln\left(\frac{p}{N_A}\right) \quad (8)$$

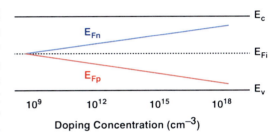

Semiconductors, Principles, Fig. 5 The Fermi level (E_F) shifts toward the conduction band and valence band with the addition of donors (E_{Fn}) and acceptors (E_{Fp}), respectively

where N_D is the donor concentration, N_A is the acceptor concentration, and the remaining variables have the same meaning as above.

With this information, the conductivity (σ) of an extrinsic semiconductor can be calculated via

$$\sigma = ne\mu_e + pe\mu_h \quad (9)$$

where μ_e is the electron mobility, μ_h is the hole mobility, e is the charge of an electron, and the remaining variables have the same meaning as above.

Cross-References

▶ Compound Semiconductors, Electrochemical Decomposition
▶ Semiconductor Electrode
▶ Semiconductor Junctions, Solid-Solid Junctions
▶ Semiconductor–Liquid Junction: From Fundamentals to Solar Fuel Generating Structures

References

1. Ashcroft NW, Mermin N (1976) Solid state physics. Harcourt Brace College, Fort Worth
2. Kittel C (1995) Introduction to solid state physics, 7th edn. Wiley, New York
3. Brutting W (2005) Physics of organic semiconductors. Wiley-VCH, Weinheim
4. Yeh CY, Lu ZW, Froyen S, Zunger A (1992) Zincblende-Wurtzite polytypism in semiconductors. Phys Rev B 46:10086–10097
5. Huang J, Kovalenko MV, Talapin DV (2010) Alkyl chains of surface ligands affect polytypism of CdSe nanocrystals and play an important role in the synthesis of anisotropic nanoheterostructures. J Am Chem Soc 132:15866–15868
6. Dick KA, Thelander C, Samuelson L, Caroff P (2010) Crystal phase engineering in single InAs nanowires. Nano Lett 10:3494–3499
7. Tan MX, Laibinis PE, Nguyen ST, Kesselman JM, Stanton CE, Lewis NS (1994) Principles and applications of semiconductor photoelectrochemistry. In: Karlin KD (ed) Progress in inorganic chemistry. Wiley, New York
8. Samsonov GV (1982) The oxide handbook. IFI/Plenum Press, New York
9. Bouroushian M (2010) Electrochemistry of metal chalcogenides. In: Scholz F (ed) Monographs in electrochemistry, 1st edn. Springer, New York
10. Atkins PW (1994) Physical chemistry, 5th edn. W.H. Freeman, New York

Sensors

Winfried Vonau and Manfred Decker
Kurt-Schwabe-Institut fuer Mess- und
Sensortechnik e.V. Meinsberg, Waldheim,
Germany

Introduction

In general a sensor is an artificial organ for the recognition and quantification of physical parameters or chemical measurands. A widespread survey can be found in a sensor series edited by Göpel, Hesse, and Zemel in the beginning of the 1990s [1] describing in the different volumes physical as well as chemical sensors in detail.

A chemical sensor – which is in the focus of this entry – enables the user to gain quantitative information about the composition of the investigated media. The resulting sensor signal has to represent the concentration, respectively, the activity of the measurands in the analyzed matrix just in real time without the use of extensive or even time-consuming preparation processes. In direct contact with the investigated media, the sensor interacts with the analyte and the extent of this process is the measure for the quantitative determination. The sensory information can be extracted from chemical reactions of the analyte taking place at the sensing element or by the registration of analyte-depending changes of physical properties. Basic statements concerning definition and classifications for chemical sensors [2] and electrochemical biosensors [3] have been published by the International Union of Pure and Applied Chemistry IUPAC.

In general a sensor is built up of two main parts – a *receptor* and a *transducer*. In some cases a third part – a *separator* – is introduced to increase the selectivity of the device, e.g., by the use of an analyte permeable membrane. The essential receptor in a chemical sensor has to convert the chemical information into a kind of energy which can be determined by the transducer. This component itself has to transform the information in a favorable signal suitable for instrumental analysis, mostly an electrical domain measuring voltage or current. Often the transducer is supplemented with an amplifying element and last but not least connected with an evaluation unit. The following scheme shows the principle design of a sensor and equipment (Fig. 1).

For the establishment of the receptor selectivity, different chemical interactions can be used. For example, the output of potentiometric sensors is strongly influenced by the equilibrium of the analyte with the sensitive layer of the chemosensor. Polymeric matrix membrane-based ion-selective electrodes are utilizing the concentration-dependent extraction of the analyte in the organic layer, while the analyte-dependent shift of potential of ion-selective sensors based on electrodes of the second kind can be described by the solubility product of the hardly soluble salt and the resulting Nernst equation of the electrochemical base reaction. Further equilibria can be predicated on ion exchanges, complexation, or adsorption effects. The interplay of analyte and receptor is determined by

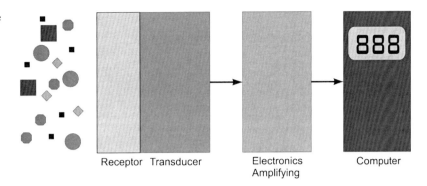

Sensors, Fig. 1 Scheme of the assembly of a chemical sensor

interactions between, e.g., dipole-dipole, ion-ion, donor-acceptor, and on hydrogen bonds. Especially biosensors benefit from these effects for their often distinguished selectivity. These forces are the base for the establishment of the required geometry of the biological component as well as the driving power for the specific reaction between the sensoric part and the targeted molecules. The shape recognition for the analyte uses the so-called key-and-lock principle.

Further discrimination between analyte and other interfering components can be obtained by kinetic selectivity. This is mainly applied in sensor devices working according to the amperometric (especially in biosensors) or thermal mode.

Sensor Classification

For the classification of the different chemical receptor principles, the following list is proposed by IUPAC [2]:

a. Optical sensors use the following principles for their measurement:
   - Absorbance of the analyte itself or after a reaction/coordination with an appropriate indicator substance
   - Reflectance by an immobilized indicator
   - Luminescence generated by the reaction of an analyte and the receptors placed on the sensor
   - Fluorescence using the measurement of emitted light or of quenching effects
   - Refractive index as measurand for the composition of the analyte solution
   - Optothermal effect caused by concentration-dependent light absorption on the sensor surface
   - Light scattering induced by particles in the analyte solution.

b. Electrochemical sensors can be divided in the following groups:
   - Voltammetric, respectively, amperometric sensors using the concentration-dependent current which is generated at the indicator electrode by the electrochemical transformation of the analyte itself or of an electrochemical active product after an appropriate reaction chain. The working electrode can be an inert metal as well as a chemically modified electrochemical active surface.
   - Potentiometric sensors establishing in combination with a reference electrode a potential dependent on the concentration/activity of the analyte.
   - ChemFETs (chemical field effect transistors) combined with a reference electrode. Although the transducer principle varies from the potentiometric sensors, the result is comparable.
   - Potentiometric solid electrolyte sensors for the gas measurement at high temperatures.

c. Electrical sensors using changes of physical properties of the sensor caused by the analyte:
   - Metal oxide semiconductor sensors for gas measurements tracking changes of reversible redox processes at the sensor surface caused by the analyte

- Sensors based on organic semiconductors using charge-transfer interactions between analyte and surface
- Conductivity sensors
- Sensors benefiting from changes of permittivity.

d. Sensors utilizing analyte-dependent mass changes:
  - Quartz microbalances using the piezoelectric principle for the detection of the adsorbed analyte load
  - Sensors based on surface acoustic wave for the determination of deposited mass on the sensor surface.

e. Magnetic devices which detect changes of paramagnetic properties of gases for the detection of analytes like oxygen.

f. Thermometric or calorimetric devices using the heat effect evolving by a chemical reaction or an adsorption process for quantification.

This listing may be incomplete and the classification of the sensors can slightly differ depending on the personal view of the users or on the transducing principle.

A special group of sensors are the so-called biosensors benefiting for the determination and quantification of the analyte from the same measuring principles like the chemosensors. But this kind of sensor is in addition equipped with one or even more additional biochemical or biological components. These substances generate a biological transformation or complex reaction cascade which can be tracked by the basic chemical sensor device; the resulting signal of the transducer represents the concentration of the analyte. The recognition system of the sensor can utilize enzymes, antibody/antigens, DNA sequences, proteins, membrane receptors, and even whole cells. A comprehensive overview concerning different receptor types and corresponding transducer devices for electrochemical biosensors is, e.g., given in [4].

Sensor Characterization

The essential and desirable claims for chemical (and physical) sensors can be described as follows:

- *Ruggedness.* This term comprises the ability of a sensor device to show the same performance under different operation conditions. On one hand sensors should withstand mechanical or thermal stress without damage or loss of sensitivity. This demand is fulfilled for most physical sensors. Chemical sensors on the other hand must also be rugged in different chemical environments. Their selectivity should not be affected by the composition of the matrix and by chemical changes during the measurement process – instead of the targeted analyte composition. This is the crucial criterion for the evaluation of the sensor quality.

- *Reversibility.* This describes the feature of a chemical sensor to react continuously on increasing as well as decreasing concentrations of the analyte. This has to be valid for processes in thermodynamically reversible cases like they are predominant in ion-selective electrodes as well as for sensors which use thermodynamically irreversible detection steps predominant in amperometric sensor devices [5].

- *Dynamic range.* This sensor parameter describes the concentration range between the detection limit and the saturation range of the sensor device and is illustrated in Fig. 2. Under defined conditions the detection limit is determined by sensor interactions with, in the best case, the solvent matrix alone or, in common measurements, by further present interferants. At the upper detection limit – called the saturation range – the sensing sites of the sensor are completely occupied by the analyte or the interfering substances. This context for different amounts of interferants – plots, e.g., obtained by measurements with ion-selective electrodes – is shown in Fig. 2. Still the best dynamic feature is presented by the pH glass electrode covering a measuring range for more than 14 decades of H_3O^+ activities.

- *Limit of detection* (LOD). A brief description of the limit of detection is given, e.g., by the IUPAC in a statistical overview and results in the following general definition: *"The limit of detection, expressed as a concentration or*

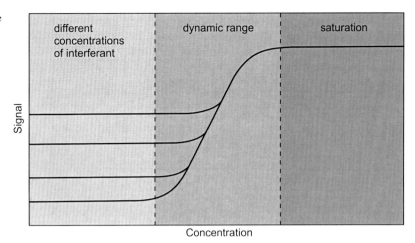

Sensors, Fig. 2 Response curve of sensors for different interferant concentrations according to [5]

quantity, is derived from the smallest measure that can be detected with reasonable certainty for a given analytical procedure" [6]. LOD and further analytical parameters like the limit of quantification (LOQ) are important for the evaluation of the sensor performance.

- *Selectivity*. This parameter describes the ability of a sensor device to respond selectively to a group of analytes or in the best case of a specific receptor interaction only on a single analyte. The procedures for the quantitative determination of the selectivity coefficients vary often – unfortunately also between the different producers of the same kind of sensor. The user must inform herself or himself thoroughly about the circumstances of sensor qualification and the advised measurement conditions to come to a well-founded decision for the suitable probe.
- *Response time*. The dynamic performance is an important factor for the selection of an appropriate sensor type. For example, a response time t_{95} describes the period necessary for the achievement of 95 % of the final sensor signal after a change of the sample matrix. The *time constant* of a sensor is reflecting the time needed to reach 63 % of the final value. A response curve representing the response times t_{90} and t_{98} is shown in Fig. 3 [7].
- *Lifetime*. During this period the sensor should remain his guaranteed functionality. This parameter can be divided in the *shelf time* – representing the period of storage of the device without sensory drawbacks – and the *operating time* the sensor can be used for analysis.

Further requirements in sensor devices are dictated mostly by economic and technological needs and comprehend the following parameters:

- Low acquisition costs
- Easy handling
- Long calibration intervals
- Low maintenance requirements
- Further miniaturization
- Mechanical robustness

Future Directions

Further progress in sensor developments in this context is mainly focused on the improvement of sensor characteristics, miniaturization of the sensing devices, the introduction of technologies suitable for mass production, and the combination of different sensing probes and principles in sensor arrays on a single device [8]. This allows to gain space- and time-resolved information of different chemical and physical parameters, e.g., in chemical and biotechnological processes. Finally this leads to the utilization of microsystem technologies in the preparation of sensory surfaces by thin film and thick film processes. Additionally the design should include

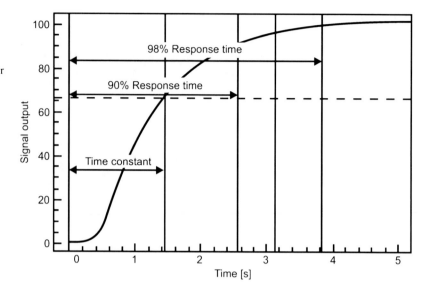

Sensors, Fig. 3 Schematic illustration of the time constant and different response times of a sensor according to [7]

the required components for signal transformation and signal transport – ideally combined with telemetric equipment. Together with the integration of microelectromechanical systems (MEMS), complex micro-total analytical systems (μTAS) have been constructed with several interesting application fields in life sciences, biotechnology, and medicine [9–11]. The application of nanocomposites and nanotubes [12, 13] offers furthermore the possibility to create sensor surfaces with new interesting electrochemical features and sensitivities.

In medical sciences research efforts are, e.g., concentrated on long-term working implantable or bloodless measuring sensor probes for rapid measurement of important parameters like glucose and related derivatives [14]. The development of new and tailor-made biosensor components still enables the researcher to broaden the application field of sensors on further analytes and sample matrices [15, 16].

Cross-References

- Amperometry
- Biosensors, Electrochemical
- Conductometry
- Cyclic Voltammetry
- pH Electrodes - Industrial, Medical, and Other Applications
- Potentiometry

References

1. Göpel W, Hesse J, Zemel JN (1989) Sensors a comprehensive survey, vol 1 fundamentals and general aspects. VCH Verlagsgesellschaft mbH, Weinheim
2. Hulanicki A, Glab S, Ingman F (1991) Chemical sensors: definitions and classifications. Pure Appl Chem 63:1247–1250
3. Thévenot DR, Toth K, Durst RA, Wilson GS (1999) Electrochemical biosensors: recommended definitions and classifications. Pure Appl Chem 71:2333–2348
4. Aizawa M (1991) Principles and applications of electrochemical and optical biosensors. Anal Chim Acta 250:249–256
5. Janata J (2009) Principles of chemical sensors. Springer Dordrecht, Heidelberg/London/New York, pp 4–10
6. Mocak J, Bond AM, Mitchell S, Scollary G (1997) A statistical overview of standard (IUPAC and ACS) and new procedures for determining the limits of detection and quantification: application to voltammetric and stripping techniques. Pure Appl Chem 69:297–328
7. Göpel W, Hess J, Zemel JN Göpel W, Hesse J, Zemel JN (1989) Sensors a comprehensive survey, vol 1, Fundamentals and general aspects. VCH Verlagsgesellschaft mbH, Weinheim, p 25

8. Bratov A, Abramova N, Ipatov A (2010) Recent trends in potentiometric sensor arrays – a review. Anal Chim Acta 678:149–159
9. Sadik OA, Aluoch AOA, Zhou A (2009) Status of biomolecular recognition using electrochemical techniques. Biosens Bioelectron 24:2749–2765
10. Lieberzeit PA, Dickert FL (2008) Rapid bioanalysis with chemical sensors: novel strategies for devices and artificial recognition membranes. Anal Bioanal Chem 391:1629–1639
11. Bashir R (2004) BioMEMS: state-of-the-art in detection, opportunities and prospects. Adv Drug Deliv Rev 56:1565–1586
12. Arya SK, Saha S, Ramirez-Vick JE, Gupta V, Bhansali S, Singh SP (2012) Recent advances in ZnO nanostructures and thin films for biosensor applications: review. Anal Chim Acta 737:1–21
13. Siangproh W, Dungchai W, Rattanarat P, Chailapakul O (2011) Nanoparticle-based electrochemical detection in conventional and miniaturized systems and their bioanalytical applications: a review. Anal Chim Acta 690:10–25
14. Kotanen CN, Moussy FG, Carrara S, Guiseppi-Elie A (2012) Implantable enzyme amperometric biosensors. Biosens Bioelectron 35:14–26
15. Su L, Jia W, Hou C, Lei Y (2011) Microbial biosensors: a review. Biosens Bioelectron 26:1788–1799
16. Holford TRJ, Davis F, Higson SPJ (2012) Recent trends in antibody based sensors. Biosens Bioelectron 34:12–24

Silicon Production by Molten Salt Electrolysis

Toshiyuki Nohira
Graduate School of Energy Science, Kyoto University, Kyoto, Japan

Introduction

For industrial use, silicon can be classified into three categories according to their purity: metallurgical grade silicon (MG-Si, 2N), solar grade silicon (SOG-Si, 6N–7N), and semiconductor grade silicon (SEG-Si, 11N–12N). Currently MG-Si is produced by carbothermic reduction of silica. The world production of MG-Si, excluding ferrosilicon, was ca. 1.5 million metric ton in 2011 [1]. SOG-Si is mainly produced from MG-Si by Siemens process which involves the combined reaction of hydrogen reduction and thermal decomposition of $SiHCl_3$. The world production of SOG-Si was estimated to be approximately 0.15 million metric ton in 2011. SEG-Si is also produced chiefly by Siemens process, and its world production was estimated to be about 0.03 million metric ton in 2011. Until now, silicon has not been practically produced by molten salt electrolysis. However, the attempt to produce silicon by molten salt electrolysis has a long history. It has been reported [2] that, as early as in 1865, Ullik prepared pure silicon and metallic potassium by the electrolysis of a molten $KF-K_2SiF_6$ mixture. Since then, many researchers have been tackling with the production of silicon by molten salt electrolysis. Due to the rapid growth of photovoltaic market in the last decade, the research and development to produce SOG-Si by the energy-efficient molten electrolysis instead of the energy-intensive Siemens process has been actively pushed forward. This entry summarizes principal methods to produce silicon by molten salt electrolysis according to the types of molten salts.

Fluoride Melts

The production of silicon by molten salt electrolysis is classified as either electrowinning of silicon from silicon compound or electrorefining of MG-Si. One of the most important molten salt systems for both electrowinning and electrorefining is LiF-KF or LiF-NaF-KF containing K_2SiF_6. Many important works regarding this molten salt were undertaken at Stanford University in 1970s and 1980s [2–4]. Although several fluoride melts were studied, it was reported that only LiF-KF and LiF-NaF-KF melts gave deposits with good quality [4]. These fluoride melts have the advantages of high conductivity, low viscosity, wide electrochemical window, and high solubility for many compounds. On the other hand, the disadvantages are high corrosiveness and difficulty in removing the solidified salt from the deposits. K_2SiF_6 was selected as silicon compound for electrowinning owing to the abundance of H_2SiF_6 as a by-product of phosphate fertilizer production. Cohen reported [5] that both epitaxial and polycrystalline continuous films of dense, coherent, and well-adherent silicon were obtained from LiF-KF containing K_2SiF_6 at 750 °C. As the substrate material, silver gave better deposits in terms of uniformity and adherence compared with tungsten and niobium [5]. Relatively inexpensive graphite substrate was also shown to be suitable for the electrodeposition of silicon [6, 7]. It was reported that more than 99.98 % purity was achieved for the electrodeposited silicon by using starting materials with 99–99.5 % purity [8].

Silica (SiO_2) has also been used as the starting material for electrowinning of silicon in fluoride melts because SiO_2 is extremely abundant and relatively easy to be purified. Cryolite (NaF_3AlF_6) melt was used to obtain silicon or Al-Si alloy in the Hall-Heroult cell which is commercially used for aluminum production [9–12]. However, this process seems to have the problem of slow deposition rate in the case of solid silicon deposition. In the case of liquid Al-Si alloy production, on the other hand, other purification processes are required to obtain high-purity silicon. Mixtures of alkali and alkali earth fluorides were also used for electrowinning of silicon from SiO_2 [13]. Powdery silicon was reported to be deposited from NaF-CaF_2 and KF-CaF_2 containing SiO_2 at 1,000 °C. Electrolysis at the temperature above the melting point of silicon is desirable from the viewpoint of productivity. Elwell et al. obtained molten silicon from BaO-SiO_2-BaF_2 melt at 1,465 °C [14].

$CaCl_2$-based molten salts at 500–950 °C [16–26]. The unique reduction mechanism of insulating and solid SiO_2 has been proposed and verified [16, 18, 21, 22, 25]. It depends on the concentration (more precisely, the activity) of O^{2-} ion in the molten salt whether the direct reduction route or the dissolution-deposition route is dominant [22, 25].

Since $CaCl_2$-based molten salt containing sufficient amount of CaO dissolves SiO_2 to produce Si(IV) ions, this system can be used also for electrorefining of MG-Si. Cai et al. reported electrorefining of MG-Si conducted in molten $CaCl_2$-$NaCl$-CaO-SiO_2 (80:10:5:5 mol%) at 850–950 °C [27].

Derivative molten salts are the mixtures of chloride and fluoride like KCl-KF and NaCl-KCl-NaF. In these molten salts, both K_2SiF_6 and SiO_2 can be used as the silicon source [28, 29]. Electrorefining of MG-Si was also reported in molten KCl-NaF at 800 °C [30].

Another type of electrodeposition of silicon was reported, in which $SiCl_4$ gas was used as the silicon source [31, 32]. $SiCl_4$ is produced either by chlorination of MG-Si or carbochlorination of SiO_2 and easily purified by distillation. Although it was reported that electrodeposition occurred in molten LiCl-KCl at 450 °C [31], high vapor pressure of $SiCl_4$ made the operation difficult at high temperature. At a lower temperature of 150 °C, electrodeposition of silicon from $SiCl_4$ was also reported for equimolar $EMPyrCl$-$ZnCl_2$ (EMPyr = N-ethyl-N-methylpyrrolidinium) [32].

Chloride Melts

Alkali and alkali earth chlorides have also been studied for both electrowinning and electrorefining of silicon. Molten $CaCl_2$ and $CaCl_2$-based molten salts are ones of the most frequently employed melts. Since the direct electrochemical reduction of solid TiO_2 in molten $CaCl_2$, which is called FFC Cambridge process, was reported by Fray et al. [15], many research groups have been working on the electrochemical reduction of SiO_2 in molten $CaCl_2$ and

Future Directions

Molten salt electrolysis should be developed for the production of MG-Si and SOG-Si as energy-efficient and environmentally sound processes. Since the required purity and cost are different between Mg-Si and SOG-Si, a proper approach should be taken for each purpose. For SOG-Si, special attention should be paid for the impurities of boron, phosphorous, and carbon, because they are very difficult to be removed by after treatment like unidirectional solidification process.

Cross-References

▶ Electrolytes, Classification
▶ High-Temperature Molten Salts
▶ Silicon, Electrochemical Deposition

References

1. U.S. Geological Survey. http://minerals.usgs.gov/minerals/pubs/commodity/silicon/mcs-2013-simet.pdf
2. Elwell D, Feigelson RS (1982) Electrodeposition of solar silicon. Sol Energy Mater 6:123–145
3. Cohen U, Huggins RA (1976) Silicon epitaxial growth by electrodeposition from molten fluorides. J Electrochem Soc 123:381–383
4. Elwell D, Rao GM (1988) Electrolytic production of silicon. J Appl Electrochem 18:15–22
5. Cohen U (1977) Some prospective applications of silicon electrodeposition from molten fluorides to solar cell fabrication. J Electron Mater 6:607–643
6. Rao GM, Elwell D, Feigelson RS (1981) Electrodeposition of silicon onto graphite. J Electrochem Soc 128:1708–1711
7. Elwell D, Feigelson RS, Rao GM (1983) The morphology of silicon electrodeposits on graphite substrates. J Electrochem Soc 130:1021–1025
8. Rao G, Elwell D, Feigelson RS (1982) Characterization of electrodeposited silicon on graphite. Sol Energy Mater 7:15–21
9. Grjotheim K, Matiasovsky K, Fellner P, Silny A (1971) Electrolytic deposition of silicon and of silicon alloys. I. Physicochemical properties of the sodium hexafluoroaluminate-aluminum oxide-silica mixtures. Can Met Quart 10:79–82
10. Boe G, Grjotheim K, Matiasovsky K, Fellner P (1971) Electrolytic deposition of silicon and of silicon alloys. II. Decomposition voltages of components and current efficiency in the electrolysis of the Na_3AlF_6-Al_2O_3-SiO_2 mixtures. Can Met Quart 10:179–183
11. Boe G, Grjotheim K, Matiasovsky K, Fellner P (1971) Electrolytic deposition of silicon and of silicon alloys. III. Deposition of silicon and aluminum using a copper cathode. Can Met Quart 10:281–285
12. Oishi T, Watanabe M, Koyama K, Tanaka M, Saegusa K (2011) Process for solar grade silicon production by molten salt electrolysis using aluminum-silicon liquid alloy. J Electrochem Soc 158: E93–E99
13. Elwell D (1981) Electrowinning of silicon from solutions of silica in alkali metal fluoride/alkaline earth fluoride eutectics. Sol Energy Mater 5:205–210
14. DeMattei RC, Elwell D, Feigelson RS (1981) Electrodeposition of silicon at temperatures above its melting point. J Electrochem Soc 128: 1712–1714
15. Chen GZ, Fray DJ, Farthing TW (2000) Direct electrochemical reduction of titanium dioxide to titanium in molten calcium chloride. Nature 407: 361–364
16. Nohira T, Yasuda K, Ito Y (2003) Pinpoint and bulk electrochemical reduction of insulating silicon dioxide to silicon. Nat Mater 2:397–401
17. Jin X, Gao P, Wang D, Hu X, Chen GZ (2004) Electrochemical preparation of silicon and its alloys from solid oxides in molten calcium chloride. Angewandte Chemie 116:751–754
18. Yasuda K, Nohira T, Amezawa K, Ogata YH, Ito Y (2005) Mechanism of direct electrolytic reduction of solid SiO_2 to Si in molten $CaCl_2$. J Electrochem Soc 152:D69–D74
19. Yasuda K, Nohira T, Ogata YH, Ito Y (2005) Direct electrolytic reduction of solid silicon dioxide in molten LiCl-KCl-$CaCl_2$ at 773 K. J Electrochem Soc 152: D208–D212
20. Pistorius PC, Fray DJ (2006) Formation of silicon by electro-deoxidation, and implications for titanium metal production. J South Afr Inst Min Metall 106:31–41
21. Xiao W, Jin X, Deng Y, Wang D, Hu X et al (2006) Electrochemically driven three-phase interlines into insulator compounds: electroreduction of solid SiO_2 in molten $CaCl_2$. Chemphyschem 7:1750–1758
22. Yasuda K, Nohira T, Hagiwara R, Ogata YH (2007) Diagrammatic representation of direct electrolytic reduction of SiO_2 in molten $CaCl_2$. J Electrochem Soc 154:E95–E101
23. Wang D, Jin X, Chen GZ (2008) Solid state reactions: an electrochemical approach in molten salts. Annu Rep Sect C Phys Chem 104:189
24. Cho SK, Fan F-RF, Bard AJ (2012) Formation of a silicon layer by electroreduction of SiO_2 nanoparticles in $CaCl_2$ molten salt. Electrochim Acta 65:57–63
25. Xiao W, Wang X, Yin H, Zhu H, Mao X et al (2012) Verification and implications of the dissolution-electrodeposition process during the electroreduction of solid silica in molten $CaCl_2$. RSC Adv 2:7588–7593
26. Yasuda K, Nohira T, Kobayashi K, Kani N, Tsuda T et al (2013) Improving purity and process volume during direct electrolytic reduction of solid SiO_2 in molten $CaCl_2$ toward production of solar-grade silicon. Energy Technology 1:245–252
27. Cai J, Luo X, Haarberg GM, Kongstein OE, Wang S (2012) Electrorefining of metallurgical grade silicon in molten $Cacl_2$ based salts. J Electrochem Soc 159:D155–D158
28. Andriiko AA, Panov EV, Boiko OI, Yakovlev BV, Borovik OY (1997) Dependence of the K_2SiF_6 content in the cathodic deposit on the melt composition during electrodeposition of powder-like silicon from the KCl-KF-K_2SiF_6 melt containing silicon dioxide. Russ J Electrochem 33:1343–1345

29. Cai Z, Li Y, He X, Liang J (2010) Electrochemical behavior of Silicon in the (NaCl-KCl-NaF-SiO$_2$) molten salt. Metall Mater TransB 41:1033–1037
30. Zou XY, Xie HW, Zhai YC, Lang XC, Zhang J (2011) Electrolysis process for preparation of solar grade silicon. Adv Mater Res 391–392:697–702
31. Matsuda T, Nakamura S, Ide K-i, Nyudo K, Yae S et al (1996) Oscillatory behavior in electrochemical deposition reaction of polycrystalline silicon thin films through reduction of silicon tetrachloride in a molten salt electrolyte. Chem Lett 569–570
32. Nishimura Y, Nohira T, Morioka T, Hagiwara R (2009) Electrochemical reduction of silicon tetrachloride in an intermediate-temperature ionic liquid. Electrochemistry 77:683–686

Silicon, Electrochemical Deposition

Tetsuya Osaka and Hiroki Nara
Faculty of Science and Engineering, Waseda University, Okubo, Shinjuku-ku, Tokyo, Japan

Introduction

Environmental issues such as global warming have become increasingly important in recent years, and the development of low-emission vehicles such as hybrid electric vehicles (HEVs) and battery-operated electric vehicles (BEVs) and the utilization of renewable energies have been attracting much attention [1–3]. The progress in portable electronic devices has also been significant. The development of these applications mainly depends on the advancement in the technology of energy storage, for which lithium ion battery (LIB) is most because of its high energy density and high power output.

Si-Based Anodes for Lithium Secondary Battery

The silicon anode system, along with Sn anode system [4–10], is promising for the anode of LIBs because of the high theoretical capacity of 4,200 mAh per gram of Si [11, 12], which is higher than that of graphite (372 mAh g^{-1}).

There are many papers on investigations of silicon-based anodes. A long-life n-Si anode with 2,000 mAh g^{-1} of capacity sustainable for more than 3,000 cycles has been reported when the anode thickness was as thin as 50 nm [13]. In order to increase the capacity of the Si battery system, electrodes of thicker Si films were examined, but such electrodes did not hold the structure during charge–discharge cycles because of a large change in volume of the active material during the cycles [14].

To realize a long-life anode using Si, some attempts have been proposed, such as the introduction of core-shell structures [15–18], a mesoporous structure [19–21], and microstructures [17, 22–26], to prevent a decrease in paths of electricity during the operation. Recently, an electrodeposited Li–Si metallic anode was reported to show a capacity greater than 500 mAh cm^{-2}, which was retained for 50 cycles [27]. On the other hand, it has been reported that composites of SiO and carbon prepared by mechanical grinding showed a high capacity exceeding 3,000 mAh g^{-1} [28].

Electrodeposited Si Anode for Lithium Secondary Battery

A novel Si-O-C composite for use as a Li battery anode was recently proposed by Osaka group. It was prepared by the electrodeposition from an organic solution [29, 33]. The electrodeposition method was introduced to form a Si-containing anode from an organic solution, based on the assumption that the composite of Si with an organic/inorganic compound withstands the stress during the anode operation. Organic/inorganic compounds formed by the reduction of the organic solvent, as in the formation of a resultant solid electrolyte interphase (SEI) [30] layer on anodes in Li ion batteries, are well known to exhibit permeability to Li$^+$ ion as well as chemical/electrochemical stability [26, 30–32]. Based on these considerations, to realize a Si and organic/inorganic compound which would buffer the stress and exhibit Li$^+$ permeability, the electrochemical co-reduction

Silicon, Electrochemical Deposition 1967

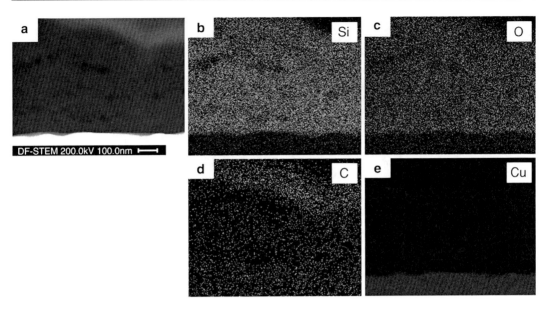

Silicon, Electrochemical Deposition, Fig. 1 (a) Cross-sectional dark-field image, and (b–e) the elemental mapping of Si, O, C, and Cu of the deposit, respectively

of Si and the solvent was performed directly on the current collector. While there is a report on Si-related material formed by electrodeposition for the anode of Li battery, their reversible capacity and cycle-life are lower than those of the conventional carbon anode [24]. Nevertheless, the Si-O-C composite anode showed an outstanding cycle durability [29].

The Si-O-C composite anode was prepared by electrodeposition from the bath containing a PC solution containing SiCl$_4$ and a supporting electrolyte. Figure 1 indicates the cross-sectional dark-field image and the elemental mappings of deposit obtained with scanning transmission electron microscope with energy dispersive X-ray analyzer (STEM-EDX). The elements of Si, O, and C confirmed to be dispersed in nanometric scale in the deposit. Figure 2 indicates the Z-contrast image obtained by high-angle annular dark-field scanning transmission electron microscopy (HAADF-STEM) and the transmission electron diffraction (TED) pattern. No rings were observed in the TED pattern obtained from the point in the Z-contrast image, which suggests that the deposit was amorphous. Consequently, a homogeneous Si-O-C composite in the nanometer scale confirmed to be successfully formed directly on a Cu substrate by electrodeposition in the organic solvent containing silicon source.

Figure 3 indicates the results of charge discharge test on the Si-O-C composite anode. In the first cycle, the charge (lithiation) and discharge (delithiation) capacities were 4,777 and 1,404 mAh g^{-1} of Si. The large irreversible capacity at first cycle could attribute to a conversion reaction of SiO$_x$ to SiO$_y$ (x > y) and Li$_2$O and a SEI formation. Although the irreversible capacity was observed in several tens of initial cycles, after that, the cycle efficiency reached 98–99 %. The Si-O-C composite anode exhibited an outstanding durability against charge–discharge cycles and delivered the discharge capacity of 1,045 mAh g^{-1} at the 2,000th cycle and 842 mAh g^{-1} of Si even at the 7,200th cycle. These superior results were achieved by the uniform dispersion of SiO$_x$ in the organic/inorganic compound synthesized by the electrodeposition of silicon in an organic solvent, which were confirmed by STEM-EDX and X-ray photoelectron microscopy (XPS). The reasons why the Si-O-C composite anode withstood thousands of cycles were discussed from the standpoint of the crack formation

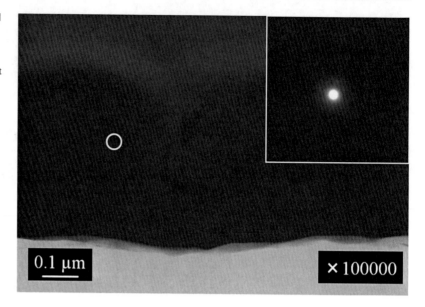

Silicon, Electrochemical Deposition, Fig. 2 Z-contrast image obtained by HAADF-STEM and TED pattern at the point indicated by the circle in the Z-contrast image

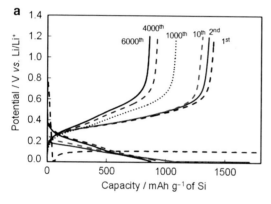
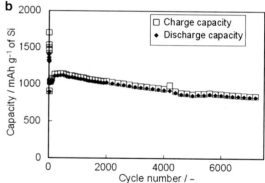

Silicon, Electrochemical Deposition, Fig. 3 (**a**) Charge–discharge curves at 1st, 2nd, 10th, 1,000th, 4,000th, and 6,000th cycle and (**b**) charge and discharge capacity against the number of cycles, of the Si-O-C composite anode measured by beaker-type three electrodes cell with lithium foils as counterelectrodes and reference electrodes. The capacity value was normalized with respect to the calculated amount of Si in the deposit. The applied current density was 250 μA cm^{-2} (1.0 C-rate)

attributed to the reaction of silicon with lithium during charge and discharge cycles.

Future Directions

Silicon-based anodes for lithium secondary batteries have been put into practical use by some manufacturers. However, the increase of the capacity is limited so far, that is to say, the capacity of Si-based anode is ca. 120 % of that of a conventional graphite anode although the theoretical capacity of Si anode is about 10 times as high as that of a conventional graphite anode. Therefore, the development of silicon-based anodes is expected to last for over 10 years. There are various synthesis methods of silicon material, which are dry process such as chemical vapor deposition and sputterings and wet process such as electrodeposition and

Silicon, Electrochemical Deposition

electroless-deposition. Among them, the electro-deposition could be beneficial for industry, because the electrodeposition is an established process which is easy and cheap.

Cross-References

► Lithium-Ion Batteries
► Lithium-Sulfur Batteries

References

1. Goodenough JB, Kim Y (2010) Challenges for rechargeable Li batteries. Chem Mater 22:587–603. doi:10.1021/cm901452z
2. Liu C, Li F, Ma LP, Cheng HM (2010) Advanced materials for energy storage. Adv Mater 22:E28–E62. doi:10.1002/adma.200903328
3. Scrosati B, Garche J (2010) Lithium batteries: status, prospects and future. J Power Sources 195:2419–2430. doi:10.1016/j.jpowsour.2009.11.048
4. Aifantis KE, Brutti S, Hackney SA, Sarakonsri T, Scrosati B (2010) SnO(2)/C nanocomposites as anodes in secondary Li-ion batteries. Electrochim Acta 55:5071–5076. doi:10.1016/j.electacta.2010.03.083
5. Hassoun J, Elia GA, Panero S, Scrosati B (2011) A high capacity, template-electroplated Ni-Sn intermetallic electrode for lithium ion battery. J Power Sources 196:7767–7770. doi:10.1016/j.jpowsour.2011.04.028
6. Mukaibo H, Momma T, Mohamedi M, Osaka T (2005) Structural and morphological moditications of a nanosized 62 atom percent Sn-Ni thin film anode during reaction with lithium. J Electrochem Soc 152:A560–A565. doi:10.1149/1.1856913
7. Mukaibo H, Momma T, Osaka T (2005) Changes of electro-deposited Sn-Ni alloy thin film for lithium ion battery anodes during charge discharge cycling. J Power Sources 146:457–463. doi:10.1016/j.jpowsour.2005.03.043
8. Mukaibo H, Osaka T, Reale P, Panero S, Scrosati B, Wachtler M (2004) Optimized Sn/SnSb lithium storage materials. J Power Sources 132:225–228. doi:10.1016/j.jpowsour.2003.12.042
9. Mukaibo H, Yoshizawa A, Momma T, Osaka T (2003) Particle size and performance of SnS2 anodes for rechargeable lithium batteries. J Power Sources 119:60–63. doi:10.1016/s0378-7753(03)00125-3
10. Nara H, Fukuhara Y, Takai A, Komatsu M, Mukaibo H, Yamauchi Y, Momma T, Kuroda K, Osaka T (2008) Cycle and rate properties of mesoporous tin anode for lithium ion secondary batteries. Chem Lett 37:142–143. doi:10.1246/cl.2008.142
11. Bourderau S, Brousse T, Schleich DM (1999) Amorphous silicon as a possible anode material for Li-ion batteries. J Power Sources 81:233–236. doi:10.1016/s0378-7753(99)00194-9
12. Weydanz WJ, Wohlfahrt-Mehrens M, Huggins RA (1999) A room temperature study of the binary lithium-silicon and the ternary lithium-chromium-silicon system for use in rechargeable lithium batteries. J Power Sources 81:237–242. doi:10.1016/s0378-7753(99)00139-1
13. Takamura T, Ohara S, Uehara M, Suzuki J, Sekine K (2004) A vacuum deposited Si film having a Li extraction capacity over 2000 mAh/g with a long cycle life. J Power Sources 129:96–100
14. Kasavajjula U, Wang C, Appleby AJ (2007) Nano- and bulk-silicon-based insertion anodes for lithium-ion secondary cells. J Power Sources 163:1003–1039
15. Cui LF, Ruffo R, Chan CK, Peng HL, Cui Y (2009) Crystalline-amorphous core-shell silicon nanowires for high capacity and high current battery electrodes. Nano Lett 9:491–495. doi:10.1021/nl8036323
16. Yen YC, Chao SC, Wu HC, Wu NL (2009) Study on solid-electrolyte-interphase of Si and C-coated Si electrodes in lithium cells. J Electrochem Soc 156:A95–A102. doi:10.1149/1.3032230
17. Ji L, Lin Z, Alcoutlabi M, Zhang X (2011) Recent developments in nanostructured anode materials for rechargeable lithium-ion batteries. Energy Environ Sci 4:2682–2699
18. Scrosati B, Hassoun J, Sun Y-K (2011) Lithium-ion batteries. A look into the future. Energy Environ Sci 4:3287–3295
19. Chen W, Fan ZL, Dhanabalan A, Chen CH, Wang CL (2011) Mesoporous silicon anodes prepared by magnesiothermic reduction for lithium ion batteries. J Electrochem Soc 158:A1055–A1059. doi:10.1149/1.3611433
20. Cho J (2010) Porous Si anode materials for lithium rechargeable batteries. J Mater Chem 20:4009–4014. doi:10.1039/b923002e
21. Kim H, Cho J (2008) Superior lithium electroactive mesoporous Si@carbon core-shell nanowires for lithium battery anode material. Nano Lett 8:3688–3691. doi:10.1021/nl801853x
22. Kim H, Han B, Choo J, Cho J (2008) Three-dimensional porous silicon particles for use in high-performance lithium secondary batteries. Angew Chem Int Ed 47:10151–10154
23. Magasinski A, Dixon P, Hertzberg B, Kvit A, Ayala J, Yushin G (2010) High-performance lithium-ion anodes using a hierarchical bottom-up approach. Nat Mater 9:353–358
24. Schmuck M, Balducci A, Rupp B, Kern W, Passerini S, Winter M (2010) Alloying of electrodeposited silicon with lithium-a principal study of applicability as anode material for lithium ion batteries. J Solid State Electrochem 14:2203–2207

25. Ji L, Zhang X (2009) Evaluation of Si/carbon composite nanofiber-based insertion anodes for new-generation rechargeable lithium-ion batteries. Energy Environ Sci 3:124–129
26. Szczech JR, Jin S (2011) Nanostructured silicon for high capacity lithium battery anodes. Energy Environ Sci 4:56–72
27. Lv R, Yang J, Wang J, Nuli Y (2011) Electrodeposited porous-microspheres Li-Si films as negative electrodes in lithium-ion batteries. J Power Sources 196:3868–3873
28. Yamada M, Ueda A, Matsumoto K, Ohzuku T (2011) Silicon-based negative electrode for high-capacity lithium-ion batteries: "SiO"-carbon composite. J Electrochem Soc 158:A417–A421
29. Momma T, Aoki S, Nara H, Yokoshima T, Osaka T (2011) Electrodeposited novel highly durable SiOC composite anode for Li battery above several thousands of cycles. Electrochem Commun 13:969–972. doi:10.1016/j.elecom.2011.06.014
30. Xu K (2010) Electrolytes and interphasial chemistry in Li ion devices. Energies 3:135–154. doi:10.3390/en3010135
31. Verma P, Maire P, Novak P (2010) A review of the features and analyses of the solid electrolyte interphase in Li-ion batteries. Electrochim Acta 55:6332–6341
32. Winter M, Appel WK, Evers B, Hodal T, Moller KC, Schneider I, Wachtler M, Wagner MR, Wrodnigg GH, Besenhard JO (2001) Studies on the anode/electrolyte interface in lithium ion batteries. Monatsh Chem 132:473–486. doi:10.1007/s007060170110
33. Nara H, Yokoshima T, Momma T, Osaka T (2012) Highly durable SiOC composite anode prepared by electrodeposition for lithium secondary batteries. Energy Environ Sci 5:6500–6505. doi:10.1039/c2ee03278c

Single Chamber Solid Oxide Fuel Cell

Takashi Hibino
Graduate School of Environmental Studies,
Nagoya University, Nagoya, Japan

Introduction

Single-chamber solid oxide fuel cells (SC-SOFCs) have attracted considerable attention in recent years as they have the potential to reduce the material and manufacturing costs of SOFCs [1, 2]. SC-SOFCs are operated with a feed mixture of fuel and oxidant gas, which no longer require the use of expensive and voluminous gas separators and high-temperature gas sealing materials. In addition, a simplified gas manifold structure is expected, due to gas feed to only one side of the fuel cell stack. SC-SOFCs have several additional benefits:

- The mechanical and thermal resistance of SOFCs can be enhanced, due to the simplified cell structure.
- Carbon deposition is less of a problem when hydrocarbons are used as fuels, due to the presence of a large amount of oxygen in the mixture.
- Various types of SOFC stacks can be designed, due to high degrees of freedom in the electrode arrangement.

The operation principle of SC-SOFCs is based on high selectivity of each electrode toward the target electrode reaction: one electrode (anode) must be electrochemically active for oxidation of the fuel but should be inert to the reduction of oxygen, while the other electrode (cathode) must behave vice versa. A theoretical interpretation of this selectivity has been reported by Riess et al. [3]. However, it is in practice difficult to realize such ideal selectivity in high-temperature conditions, because (1) in the case of hydrogen or carbon monoxide fuel, most of them are subject to the non-electrochemical oxidation before arriving at the triple-phase boundaries, and (2) in the case of hydrocarbon fuels, the target electrode reaction cannot overwhelm the counter electrode reaction in rate at the electrode. For this reason, experimental demonstration of SC-SOFCs that can generate high power output has not been accomplished until 1999.

Improvement of SC-SOFCs

One breakthrough in SC-SOFCs was accomplished by Hibino et al. [4]. An SC-SOFC was operated using yttria-stabilized zirconia (YSZ), Ni and $La_{0.8}Sr_{0.2}MnO_3$ (LSM) as the electrolyte, and anode and cathode, respectively. When the fuel cell was supplied with a methane-air mixture (methane/oxygen mole ratio = 2:1) at 950 °C, it generated an open-circuit voltage (OCV) of 795 mV and a peak power density of 121 mW cm^{-2}. Half-cell measurements were carried out for these electrodes. Partial oxidation of methane by oxygen proceeded at a very fast rate on the Ni anode, while no significant reaction between methane and oxygen was observed on the LSM cathode. As a result, the potentials of the Ni anode and the LSM cathode were -900 and -0 mV, respectively, against a Pt reference electrode exposed to air. They interpreted that the observed OCV was due to fast electrode reaction of the produced hydrogen and carbon monoxide (syngas) on the Ni anode and to selective electrode reaction of unreacted oxygen on the LSM cathode.

$$\text{At the anode}: \quad CH_4 + 1/2O_2 \rightarrow 2H_2 + CO \quad (1)$$

$$2H_2 + O^{2-} \rightarrow H_2O + 2e^- \quad (2)$$

$$CO + O^{2-} \rightarrow CO_2 + 2e^- \quad (3)$$

$$\text{At the cathode}: \quad O_2 + 4e^- \rightarrow 2O^{2-} \quad (4)$$

The performance of SC-SOFCs could be further improved by the simultaneous addition of 25 wt% gadolinia-doped ceria (GDC) and 15 wt% MnO_2, respectively, to the anode and cathode [5]. The resulting fuel cell generated an OCV of 833 mV and a peak power density of 162 mW cm^{-2} in a methane-air mixture feed at 950 °C. They explained that the improved performance was due to high electro-catalytic activity of each electrode for the corresponding target reaction.

Hibino et al. also attempted to reduce the operating temperature of the SC-SOFCs by using ceria-based electrolytes [6]. An SC-SOFC consisted of 0.15-mm thick $Ce_{0.8}Sm_{0.2}O_{1.9}$ (SDC) electrolyte, Ni-SDC cermet anode, and $Sm_{0.5}Sr_{0.5}CoO_3$ (SSC) cathode. Using a ethane-air mixture feed (ethane/oxygen mole ratio $= 1{:}1$), an OCV of about 900 mV and a peak power density of about 400 mW cm^{-2} were obtained at 500 °C. Hibino et al. also demonstrated that this fuel cell could operate not only when the two electrodes are placed on opposite sides of the electrolyte but also when the two electrodes are placed on the same side of the electrolyte [7]. Interestingly, this type of fuel

cell showed an increased power density with decreasing gas between the two electrodes, suggesting that this electrode arrangement allows for a reduction in the ohmic resistance of the cell without the use of a thin electrolyte film.

Hibino et al. also found the following important results in the above-related works. The first key factor for successful operation of SC-SOFCs at reduced temperatures is the use of higher hydrocarbons, including ethane, propane, and butane, as fuels [8].

- The secondary key factor for stable operation of SC-SOFCs at reduced temperatures is the use of the heating effect from reaction (1) to increase the cell temperature [9].
- The third key factor for high performance SC-SOFCs at reduced temperatures is the improvement of the anode by the addition of various noble metals such as Pd and Ru [10, 11].

As a result of these efforts, the power density increased to 644 mW cm^{-2}. The temperature required for operating their SC-SOFCs was also reduced to 200 °C. It appears that their SC-SOFCs finally outperform conventional dual-chamber SOFCs not only in terms of a simpler cell structure but also on performance. However, some serious challenges still remain for their SC-SOFCs as described below:

- A part of the fuel is subjected to deep oxidation on the anode and cathode.
- The selectivity of the anode and cathode becomes lower as the fuel utilization increases.
- All the fuel passed on the cathode side is in principle exhausted from the SOFC.
- Therefore, their SC-SOFCs inherently show the low efficiency of the overall fuel cell system.

Recent Progress in SC-SOFCs

Currently, attempts are under way by many researchers for improving SC-SOFCs using various innovative techniques for addressing the above challenges. In this section, recent advances in SC-SOFC development are briefly summarized with respect to the electrode material, stack design, and application.

Anode Materials for SC-SOFCs

In SC-SOFCs, the electro-catalytic activity of the anode is considerably influenced by the quantity of syngas produced through reaction (1). In 2005, Zhang et al. promoted the propane oxidation reaction by depositing a Ru + CeO_2 catalyst, where 7 wt% Ru was physically mixed with CeO_2, onto a Ni-SDC anode [12]. An increase of the quantity of syngas enhanced the performance of an SC-SOFC with a propane-air mixture feed, especially at low temperatures. They also succeeded the promotion of the methane oxidation reaction on the anode in an SC-SOFC with a methane-air mixture feed at high temperatures by depositing a 10 wt% Ru-impregnated CeO_2 catalyst onto a $Ni-(Sc_2O_3)_{0.1}(ZrO_2)_{0.9}$ anode surface [13].

Hao et al. conducted modeling of the catalytic reactions inside a porous anode in SC-SOFCs, identifying three distinct reaction zones at different depths within the anode: a methane deep oxidation region in the outer layer; a steam reforming region in the middle layer; a water-gas shift region in the bottom layer [14]. In other words, syngas required for reactions (2) and (3) is mainly produced by the steam reforming and water-gas shift reactions inside the electrode. Based on the above proposal, Gaudillère et al. inserted a 3.9 wt% Pt-impregnated CeO_2 catalyst inside a Ni-GDC cermet anode, enhancing the hydrogen production from the water-gas shift reaction and consequently improving the anode properties [15].

Cathode Materials for SC-SOFCs

In SC-SOFCs, the cathode must be as much as inert to reaction (1) and only selective to reaction (4). While LSM is the most promising cathode candidate for high-temperature SC-SOFCs, the position of SSC as the cathode preferred for intermediate-temperature SC-SOFCs still remains uncertain. In 2004, Suzuki et al. examined several composite mixtures of SDC and $La_{0.8}Sr_{0.2}Co_{0.2}Fe_{0.8}O_3$ (LSCF) as cathodes for SC-SOFCs in a propane-air mixture feeds at intermediate temperatures [16]. The cathodic overpotential decreased with increasing SDC content in the SDC-LSCF cathode. They explained that the addition of SDC to the cathode decreased the number of catalytically active sites of LSCF for fuel oxidation, improving the selectivity of the cathode toward the oxygen reduction reaction. A similar result was reported by Morales et al., who employed a mixture of $La_{0.5}Sr_{0.5}CO_{3-\delta}$, GDC, and AgO as a cathode for SC-SOFCs [17].

In 2008, Shao et al. reported a $SDC-Ba_{0.5}Sr_{0.5}Co_{0.8}Fe_{0.2}O_{3-\delta}$(BSCF) cathode for SC-SOFCs at reduced temperatures [18]. In a methane-air mixture feed, an SC-SOFC using this cathode exhibited peak power densities of about 570 mW cm^{-2} at 600 °C, which is comparable to that obtained when using the SDC-SSC cathode in similar conditions. However, Yan et al. pointed out the low chemical stabilities of BSCF to CO_2 under fuel cell conditions [19]. This material thermodynamically reacts with CO_2 to form carbonates.

Stack Designs

Although SC-SOFCs have the possibility to provide various types of stack designs, their cell structure may not allow for a compact stack connecting tightly several cells without a gas separator, due to the direct chemical reaction between syngas produced on the anode in one cell and oxygen remaining on the cathode in another cell. Indeed, Liu et al. investigated the effect of the distance between the two same cells, Ni-YSZ|YSZ|LSM, in a methane-air mixture feed at 700 °C [20]. Both the OCV and power output of their two-cell stack were significantly reduced by decreasing the distance. The influence of the distance was negligible at a distance of 4 mm or longer.

To avoid the above problem, the following two types of stack designs were proposed.

In 2009, Wei et al. reported a star-shaped stack comprised of four Ni-YSZ|YSZ|LSM cells [21]. This stack generated an OCV of approximately 3.5 V and a peak power output of 421 mW in a methane-air mixture feed at 700 °C. They concluded that the symmetric design can ensure the identical operation of each cell. On the other hand, as described in the previous section, SC-SOFCs allows for a coplanar electrode arrangement, where the two electrodes reside on the same surface of the electrolyte. Also in 2009, Kuhn et al. fabricated cells with one, two, three, four, five and ten pairs of electrodes with a width of 260 μm, thickness of 17 μm, and interelectrode gap of 114 μm [22]. The OCV roughly increased with increasing number of the electrode lines, although the maximum OCV value was as low as 0.8 V.

Promising Applications of SC-SOFCs

In spite of the above efforts, the energy conversion efficiency of SC-SOFCs is still not sufficiently high to use them as practical electric generators. Alternatively, SC-SOFCs can also be applied for the energy recovery from engine exhaust and for the cogeneration of chemical and electricity. In the former, unburnable hydrocarbons are no longer needed. Yano et al. investigated the feasibility of applying an SC-SOFC stack to power generators for exhaust energy recovery [23]. Twelve-cell stack exhibited a power output of 1 W or higher in the exhaust from a motorcycle. In the latter, this reactor works as a pre-reformer for conventional SOFCs. Shao et al. examined the cogeneration of syngas and electricity in a methane-oxygen mixture feed for an SC-SOFC integrated with a GdNi/Al_2O_3 catalyst [24]. High power output, high methane conversion, and high H_2 and CO selectivity could be achieved.

Future Directions

A number of research groups proposed selective electrode materials and effective cell designs. As a consequence, SC-SOFCs could provide fuel cell performance comparable to that of conventional SOFCs. Moreover, some new avenues of SC-SOFCs for various applications were proposed on the basis of cell designs, thermal properties, and fabrication processes. However, challenges still remain for further improvements in SC-SOFCs, especially in terms of the energy conversion efficiency. At present, ideally selective anode and cathode materials for fuel oxidation and oxygen reduction, respectively, are not yet available for SC-SOFCs, and further efforts are needed to realize these.

Cross-References

▶ Solid Oxide Fuel Cells, Introduction
▶ Solid Oxide Fuel Cells, Thermodynamics

References

1. Yano M, Tomita A, Sano M, Hibino T (2007) Recent advances in single-chamber solid oxide fuel cells: a review. Solid State Ionics 177:3351–3359. doi:10.1016/j.ssi.2006.10.014
2. Kuhn M, Napporn TW (2010) Single-chamber solid oxide fuel cell technology-from its origins to today's state of the art. Energies 3:57–134. doi:10.3390/en3010057
3. Riess I, Vanderput PJ, Schoonman J (1995) Solid oxide fuel-cells on uniform mixtures of fuel and air. Solid State Ionics 82:1–4. doi:10.1016/0167-2738(95)00210-W
4. Hibino T, Wang S, Kakimoto S, Sano M (1999) Single chamber solid oxide fuel cell constructed from an yttria-stabilized zirconia electrolyte. Electrochem Solid State Lett 2:317–319. doi:10.1149/1.1390822
5. Hibino T, Wang S, Kakimoto S, Sano M (2000) One-chamber solid oxide fuel cell constructed from a YSZ electrolyte with a Ni anode and LSM cathode. Solid State Ionics 127:89–98. doi:10.1016/S0167-2738(99)00253-2
6. Hibino T, Hashimoto A, Inoue T, Tokuno J, Yoshida S, Sano M (2000) A low-operating-temperature solid oxide fuel cell in hydrocarbon-air mixtures. Science 288:2031–2033. doi:10.1126/science.288.5473.2031
7. Hibino T, Hashimoto A, Suzuki M, Yano M, Yoshida S, Sano M (2002) A solid oxide fuel cell with a novel geometry that eliminates the need for preparing a thin electrolyte film. J Electrochem Soc 149:A195–A200. doi:10.1149/1.1431573
8. Hibino T, Hashimoto A, Inoue T, Tokuno J, Yoshida S, Sano M (2000) Single-chamber solid oxide fuel cells at intermediate temperatures with various hydrocarbon-air mixtures. J Electrochem Soc 147:2888–2892. doi:10.1149/1.1393621

9. Hibino T, Hashimoto A, Inoue T, Tokuno J, Yoshida S, Sano M (2001) A solid oxide fuel cell using an exothermic reaction as the heat source. J Electrochem Soc 148:A544–A549. doi:10.1149/1.1368098

10. Hibino T, Hashimoto A, Yano M, Suzuki M, Yoshida S, Sano M (2002) High performance anodes for SOFCs operating in methane-air mixture at reduced temperatures. J Electrochem Soc 149:A133–A136. doi:10.1149/1.1430226

11. Tomita A, Hirabayashi D, Hibino T, Nagao N, Sano M (2005) Single-chamber SOFCs with a Ce0.9Gd0.1O1.95 electrolyte film for low-temperature operation. Electrochem Solid State Lett 8:A63–A65. doi:10.1149/1.1836120

12. Shao Z, Haile SM, Ahn J, Ronney PD, Zhan Z, Barnett SA (2005) A thermally self-sustained micro solid-oxide fuel-cell stack with high power density. Nature 435:795–798. doi:10.1038/nature03673

13. Zhang CM, Sun LL, Ran R, Shao ZP (2009) Activation of a single-chamber solid oxide fuel cell by a simple catalyst-assisted in-situ process. Electrochem Commun 11:1563–1566. doi:10.1016/j.elecom.2009.05.048

14. Hao Y, Goodwin DG (2008) Efficiency and fuel utilization of methane-powered single-chamber solid oxide fuel cells. J Electrochem Soc 155:B666–B674. doi:10.1016/j.jpowsour.2008.04.072

15. Gaudillère C, Vernoux P, Farrusseng D (2010) Impact of reforming catalyst on the anodic polarisation resistance in single-chamber SOFC fed by methane. Electrochem Commun 12:1332–1335. doi:10.1016/j.elecom.2010.07.010

16. Suzuki T, Jasinski P, Anderson HU, Dogan F (2004) Role of composite cathodes in single chamber SOFC. J Electrochem Soc 151:A1678–A1682. doi:10.1149/1.1786071

17. Morales M, Piñol S, Segarra M (2009) Intermediate temperature single-chamber methane fed SOFC based on Gd doped ceria electrolyte and La(0.5)Sr(0.5)CoO(3-delta) as cathode. J Power Sources 194:961–966. doi:10.1016/j.jpowsour.2009.05.027

18. Zhang CM, Zheng Y, Ran R, Shao ZP, Jin WP, Xu NP, Ahn J (2008) Initialization of a methane-fueled single-chamber solid-oxide fuel cell with NiO plus SDC anode and BSCF plus SDC cathode. J Power Sources 179:640–648. doi:10.1016/j.jpowsour.2008.01.030

19. Yan A, Maragou V, Arico A, Cheng M, Tsiakaras P (2007) Investigation of a Ba0.5Sr0.5Co0.8Fe0.2O3-delta based cathode SOFCII. The effect of CO2 on the chemical stability. Appl Catal B 76:320–327. doi:10.1016/j.apeatb.2007.06.010

20. Liu ML, Lü Z, Wei B, Huang XQ, Zhang YH, Su WH (2010) Effects of the single chamber SOFC stack configuration on the performance of the single cells. Solid State Ionics 181:939–942. doi:10.1016/j.ssi.2010.05.028

21. Wei B, Lü Z, Huang XP, Liu ML, Jia D, Su WH (2009) A novel design of single-chamber SOFC micro-stack operated in methane-oxygen mixture. Electrochem Commun 11:347–350. doi:10.1016/j.elecom.2008.11.037

22. Kuhn M, Napporn TW, Meunier Vengallatore MS, Therriault D (2009) Miniaturization limits for single-chamber micro-solid oxide fuel cells with coplanar electrodes. J Power Sources 194:941–949. doi:10.1016/j.jpowsour.2009.05.034

23. Yano M, Nagao M, Okamoto K, Tomita A, Uchiyama Y, Uchiyama N, Hibino T (2008) A single-chamber SOFC stack operating in engine exhaust. Electrochem Solid State Lett 11:B29–B33. doi:10.1149/1.2824502

24. Shao ZP, Zhang CM, Wang W, Su C, Zhou W, Zhu ZH, Park HJ, Kwak C (2011) Electric power and synthesis Gas Co-generation from methane with zero waste gas emission. Angew Chem Int Ed 50:1792–1797. doi:10.1002/anie.201006855

SnO_2 Gas Sensor

Yasuhiro Shimizu
Graduate School of Engineering,
Nagasaki University, Nagasaki, Japan

Introduction

In 1962, two important research results, which are the very beginning of the present progress and development of semiconductor gas sensors, were reported independently: one was a ZnO thin film gas-sensing device reported by Seiyama et al. [1] and the other porous SnO_2 ceramic one applied for a Japanese patent by Taguchi [2]. This type of gas sensors utilizes changes in their resistance or conductance in detecting a target gas originating from electronic interactions between the semiconductor sensor materials and the target gas. After the numeral efforts to improve gas-sensing performance, SnO_2-based semiconductor gas sensors were put on the market as gas leakage monitors for town gas and liquid petroleum gas in 1968. Since then, SnO_2 has been the most attractive semiconductor gas sensor material among various kinds of semiconductor metal oxides so far developed and/or demonstrated to date as sensor materials.

This article explains basic properties of SnO_2 as a semiconductor gas sensor material and then discusses several factors affecting its gas-sensing properties. New approaches and challenges directed to further improving the gas-sensing performance are also introduced briefly.

Basic Properties of SnO_2

SnO_2 crystallizes in the rutile structure *($P4_2/mnm$ symmetry)* at ambient pressure [3], wherein Sn^{4+} ions are six coordinate and O^{2-} ions three coordinate, with a lattice constant of a = b = 0.4737 nm and c = 3.186 nm [4]. The appearance of SnO_2 powder is white in color and it is really a wide bandgap (E_g = 3.6 eV) semiconductor. But, owning to the formation of intrinsic oxygen vacancies in the lattice under the ambient oxygen partial pressure [5], SnO_2 shows n-type semiconductivity (see Eqs. 1 and 2 written on the basis of the Kröger-Vink notation). Donor electrons are likely excited to the conduction band at elevated temperatures and are available for electronic conduction.

$$2Sn_{Sn}{}^x + O_o{}^x \rightleftharpoons 2Sn_{Sn}{}' + V_o + 1/2O_2(g) \quad (1)$$

$$2Sn_{Sn}{}' \rightleftharpoons 2Sn_{Sn}{}^x + 2e' \quad (2)$$

Conventional Fabrication Procedure of SnO_2 Gas Sensors

SnO_2 powder can be prepared conventionally by pyrolysis around 400 °C and subsequent calcination at suitable temperature of tin oxalate precipitated by the reaction between $SnCl_2 \cdot 2H_2O$ and $H_2C_2O_4 \cdot H_2O$ [6]. Crystallite sizes as well as particle sizes of the resultant SnO_2 powder can be controlled by the final calcination conditions. Usually one oxide particle consists of several crystallites, and therefore one should note that the crystallite size calculated by the Scherrer equation based on X-ray diffraction data of the oxide powder is smaller than the particle size

observed by SEM or TEM. Of course, single crystal SnO_2 fine particles can be fabricated in some cases: e.g., neutralization of a cold solution of $SnCl_4$ with an aqueous ammonia solution, followed by drying and calcination [7]. In addition, another kind of single crystal, i.e., SnO_2 whiskers, can be fabricated by the oxidation of metallic tin at 1,100 °C under low O_2 partial pressure [8]. But, such single crystal SnO_2 materials are very special raw materials from the crystallographic point of view.

Pure SnO_2 powder is rarely used as a gas sensor material. To enhance the gas-sensing performance, i.e., the magnitude of gas response, selectivity, and long-term stability, SnO_2 powders doped with or mixed with different foreign metal oxide(s) and those loaded with different noble metal(s), namely, SnO_2-based sensor materials, are used depending upon the kind of target gases and their operating conditions. SnO_2-based gas sensors are fabricated conventionally in the shape of thick films on the surface of flat substrates equipped with a pair of interdigitated electrodes or ceramic tubes equipped with a pair of belt-type electrodes. In fabricating the thick film sensor, a SnO_2-based sensor material is first mixed with an organic or inorganic binder and a suitable solvent to make a paste, and then the resultant paste is applied or screen-printed on the substrate. Then, the thick film sensor is subjected to firing to burn out the organic binder and/or organic solvent if necessary, subsequently calcination to promote partial sintering among the SnO_2-based particles to ensure the mechanical strength of the thick film sensor. If calcination at low temperature would be adopted, one can observe relatively high gas response at initial operation of the sensor. But, the response tends to decrease with increasing the operation cycles, probably due to a decrease in active sites for gas detection induced by the progress of partial sintering and/or rearrangement of surface ions on the sensor material. Thus, the calcination conditions, i.e., temperature and period, should be selected to be more severe ones than those for the normal operation conditions of the sensor.

Measurement of Gas-Response Properties

As for commercially available SnO_2-based gas sensors, heaters are set up adjacent the sensor materials, e.g., on the back side of the substrate or inside the ceramic tubes, so as to keep the sensor materials at the most suitable operating temperature. In the case of laboratory tests, electronic furnaces are used alternatively as heating devices. In this case, the sensor is set in a small test chamber, which is in the line of a gas flow system, located inside an electronic furnace. Different heating methods of the sensor material may lead to certain changes in both gas response at certain temperature and its operating temperature dependence, owing to different preheating conditions of the target gas before leaching the most active sites in the sensor and possible temperature slope inside the sensor films. This point should be considered in comparing the gas-sensing performance among the sensors fabricated with the same sensor material but operated by different heating methods, especially for referring to the data reported in literatures.

Variations in electronic resistance of SnO_2-based gas sensors can be measured directly by a digital multimeter or indirectly by a fall-of-potential method. In the latter method, the sensor resistance can be calculated from the voltage drop across a load resistor connected in series with the sensor under a given applied voltage value. When the gas-response properties are evaluated at temperatures higher than 100 °C, we can use a DC power source. But, if the gas-response properties would be studied at room temperature or temperatures less than 100 °C by employing the DC power source, we have to pay special attention on the value of relative humidity (water contents) in the ambient atmosphere and then polarization effect of physisorbed water on the surface of porous sensors. Gas-sensing properties should be measured under completely dry conditions, without any chemisorbed and physisorbed water on the sensor surface, at room temperature, and then effect of relative humidity on the gas-sensing properties should be evaluated additionally. Otherwise, a porous

gas sensor may work also as an ionic-type humidity sensor at room temperature, and its response to an actual target gas may be negligible small in comparison with that to relative humidity. Such a gas sensor cannot be used at room temperature in a practical application fields. In addition, from the viewpoints of quick response and recovery, the operation of semiconductor gas sensors at room temperature is a big challenge, except for very special cases.

Typical Gas-Response Properties

Gas-Sensing Mechanism

As stated above, semiconductor gas sensors utilize changes in their resistance or conductance in detecting a target gas originating from electronic interactions between the semiconductor sensor materials and the target gas. When the target gas is an inflammable gas and the sensor is operated in air at elevated temperatures, the resistance change arises mainly from the consumption of negatively charged oxygen adsorbates on the surface of the sensor material by the reaction with the inflammable gas. Namely, when n-type semiconductive SnO_2 grains are used as a sensor material in air, the surface of SnO_2 grains has already been covered with negatively charged oxygen adsorbates, such as O^- in the temperature range of 300–400 °C in air, for example. The formation of such oxygen adsorbates extracts electrons from the SnO_2 bulk according to the following Eq. 3, leading to the formation of the space-charge region at the surface of SnO_2 grains and a potential barrier at the grain boundaries (see Fig. 1a, b). Then the sensor is in high resistance level in air.

$$1/2O_2(g) + 2e^- \rightleftharpoons 2O^- \qquad (3)$$

When the sensor is exposed to a certain concentration of H_2 in air at elevated temperature, for example, the oxygen adsorbates are consumed by the reaction with H_2 and a new, but a lower, surface coverage of oxygen adsorbates is achieved at the steady state. By this reaction, a certain number of electrons trapped

SnO₂ Gas Sensor

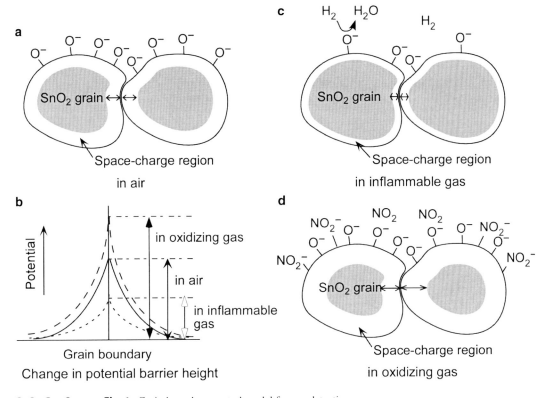

SnO₂ Gas Sensor, Fig. 1 Grain boundary control model for gas detection

by oxygen adsorbates return to the SnO₂ bulk, leading to a decrease in potential barrier and then a lower resistance level (see Fig. 1c). With increasing the target gas concentration, the surface coverage of oxygen adsorbates decreases and the sensor resistance decreases as well. Thus, this resistance change as a function of the target gas concentration can be used in detecting the existence of the target gas. The kind of oxygen adsorbates to be reacted with the inflammable gas depends on the operating temperature of the sensors: O_2^- (in a low temperature range less than 150 °C), O^- (in the middle temperature range less than 400 or 500 °C), and O^{2-} (in the high temperature range higher than 400 or 500 °C) [9]. On the other hand, in the case of an oxidizing gas, its chemisorption as a negatively charged species makes the potential barrier higher, leading to a higher resistance level in comparison with the original air level (see Fig. 1d).

Definition of Gas Response

Response of an n-type semiconductor gas sensor to an inflammable gas is usually defined as a ratio (R_a/R_g) of the sensor resistance in air (R_a) to that in a target gas (R_g), while the reverse ratio (R_g/R_a) is defined for the case of the detection of an oxidizing gas. The same definition had been used as a measure of gas "sensitivity" by many researchers including the author of the present article until quite recently. But, gas "sensitivity" is the slope of the calibration curve, that is, sensor response versus target gas concentration according to the IUPAC definition. If a straight line, which passes through the origin, could be expected between the response and the target gas concentration over the wide gas concentration range, the response defined only at the specific target gas concentration is closely related to the "sensitivity", although the units of these two measures are different. Actually, however, such a linear relationship is hard to be obtained and

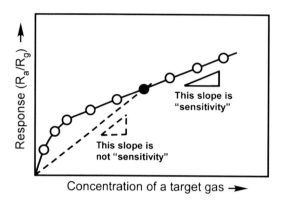

SnO₂ Gas Sensor, Fig. 2 Typical relationship between gas response and target gas concentration

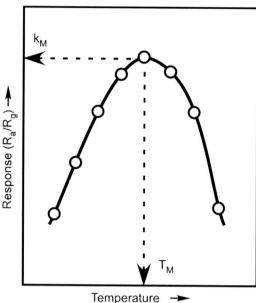

SnO₂ Gas Sensor, Fig. 3 Temperature dependence of gas response

different slope values are likely observed at different concentration ranges (see Fig. 2). For most cases, a larger slope value, i.e., a higher sensitivity value, is likely achieved in a lower gas concentration range [10–12], while exceptional results are also reported [13]. At present, therefore, the use of the word of "response" is recommended, provided that R_a/R_g defined only at one concentration value of the target gas is employed as a measure of the sensor performance.

Temperature-Dependent Gas-Response Properties

Gas response of n-type semiconductor gas sensors to inflammable gases in air usually exhibits volcano-type dependence against operating temperature (see Fig. 3), i.e., the response increases with a rise in operating temperature, reaches a maximum value (k_M) at certain temperature (T_M), and then decreases with further increasing temperature [14, 15]. The increase in response below T_M can be explained by the acceleration of the reaction of oxygen adsorbates with the inflammable gas. But, the decrease in response above T_M can be ascribed to several reasons. Both the decrease in the amount of oxygen adsorbates due to their desorption at higher temperature and the increase in charge carrier concentration of n-type semiconductor by the thermal excitation lead to lower sensor resistance in air and in turn lower response even if the same amount of oxygen adsorbates will be consumed by the reaction with the inflammable gas. Furthermore, the re-adsorption speed of oxygen adsorbates is expected to increase with a rise in operating temperature, leading to a lower change in the surface coverage of oxygen adsorbates upon exposure to the inflammable gas. Therefore, the control of the reactivity of oxygen adsorbates with the inflammable gas by any methods is of primary importance in achieving higher response at lower operating temperature. In the case of oxidizing gases, volcano-type-dependent response properties are also observed [16, 17]. In this case, one can consider the following scenario: the need for much thermal activation for a larger amount of chemisorption below T_M, whereas desorption at higher temperature and the increase in charge carrier concentration of n-type semiconductor by the thermal excitation above T_M. Thus, the increase in the number of adsorption sites is important in improving the response to oxidizing gases.

Important Factors Affecting Gas-Response Properties

The factors discussed in this section are mainly effective for improving the response to inflammable gases. Therefore, some arguments described below are not valid for the enhancement of response to oxidizing gases.

Particle Size Effect

The increase in a surface to volume ratio of n-type semiconductive sensor materials, i.e., the use of smaller metal oxide grains, is an effective method, since the number of electrons participating the reaction and/or interaction with a target gas at the grain surface becomes significant against the total number of electrons available for conduction inside the grain. If the diameter (D) of each grain would be less than two times of the depth (L) of the space-charge region in air at certain conditions, the space-charge region occupies the entire grain and the sensor is in a very high resistance level. Under such conditions, i.e., a grain control region [18, 19], the surface effect controls thoroughly the total resistance of each grain and therefore the whole sensor consisting of such fine grain, leading to amazingly high response to a target gas. For a thin sputtered SnO_2 film, the L value is calculated to be 3 nm at 250 °C [20].

Mixing of Foreign Metal Oxides

The gas response of SnO_2 can be enhanced by the mixing of appropriate foreign metal oxide(s) during its fabrication. Solid solution among the metal oxides is excluded from the present discussion here. This phenomenon can be explained by one or combined effects of the following reasons. First, the sintering and/or grain growth of SnO_2 particles may be restrained by the foreign metal oxide added. Second, the foreign metal oxide particle may act as a new adsorption and/or activation site for a target gas. Third, the formation of p-n junctions, if the added foreign metal oxide is p-type semiconductor [21]. This leads to a wider space-charge region and/or different energy states of electrons at grain boundaries. The CuO-SnO_2 sensor material developed for a H_2S sensor contains p-n junctions and shows high resistance in air, but the formation of metallic CuS from CuO in H_2S atmosphere is considered to be the main reason for its extremely high H_2S response [22].

Doping of Foreign Metal Oxides

The gas response of SnO_2 can be improved by doping with a small amount of Al_2O_3 [23]. The doping of such oxides also results in a decrease in electrons and therefore an increase in sensor resistance in air, by so-called a valency control method as expressed by the following equations. Therefore, such response enhancement can be considered to arise from the decrease in the number of electrons available for conduction in air. Namely, when the number of electrons in the SnO_2 particle is reduced by the valency control in air, the ratio of electrons trapped and released by oxygen adsorbates to that in the particle becomes bigger, even if the same numbers of oxygen adsorbates would react with a target gas.

$$Al_2O_3 \rightleftharpoons 2Al_{Sn}{}' + 3O_o{}^x + V_o \qquad (4)$$

$$1/2O_2 + V_o \rightleftharpoons O_o{}^x + 2h \qquad (5)$$

$$e' + h \rightleftharpoons null \qquad (6)$$

The addition of foreign metal oxides beyond their solubility limits may hinder the SnO_2 grain growth and then contribute to enhance the gas response also by the grain size effect. To make clear the role of dopants (additives), detailed investigation is, of course, indispensable.

Loading of Noble Metals

Please note that the present paper is written based on the following criterion: the words "noble metal loading" and "noble metal loaded" should be used, if there is no evidence for the solid solution of the noble metal into the SnO_2 lattice. Loading of a small amount of a noble metal on SnO_2 particles has been demonstrated to be effective for improving gas response. This promoting effect can be explained mainly by either one of the following two roles of the noble metal loaded: chemical and electronic sensitization [24]. The chemical sensitization is the ability of the loaded metal, such as Pt [25], for activating an

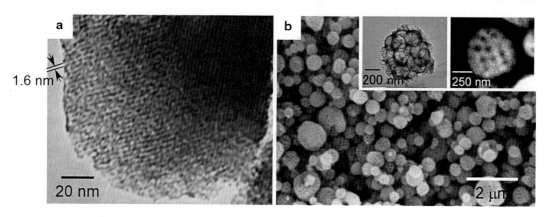

SnO₂ Gas Sensor, Fig. 4 (a) Meso-porous and (b) macro-porous SnO₂ powder

inflammable gas and then promoting the reaction with oxygen adsorbates, leading to lower surface coverage of oxygen adsorbates at the steady state. The electronic sensitization is based on the ability of the loaded metal, such as Pd [26], for extracting electrons from SnO_2 particles in air and partial releasing to SnO_2 particles upon exposure to inflammable gases, due to the change in oxidation state of the loaded metal.

Gas Diffusion Effect

Porous thick film structure itself acts a promoter for enhancing gas response or anti-promoter, depending upon the kind of target gases, oxidation activity of sensor materials used, and then the thickness of the films, provided that a pair of electrodes is located at the innermost region of the film [27]. If the oxidation activity of the sensor material is mild, H_2 response can be enhanced by employing a porous thicker film, due to the limitation of oxygen diffusion, in comparison with H_2 diffusion, into the most sensitive region. But if the oxidation activity is high, H_2 response decreased markedly owing to the consumption of H_2 at the surface region of the thick film and then lower H_2 concentration in the most sensitive region.

Thus, total design of semiconductor gas sensors by taking accounts of possible response-determining factors is of primary importance in promoting gas-sensing properties.

Future Direction

The use of surfactants and polymeric microspheres has been proven to be effective for controlling meso- and macroporous structures of SnO_2 sensor materials (see Fig. 4, [28, 29]) and then for improving gas-sensing properties [30, 31]. Another new trend is fabrication and application of SnO_2 nanowires and nanofiber mats to sensor materials [16, 32]. These approaches are aiming at enhancing the surface effect to the total sensor response. Micro-hot plate substrates fabricated by MEMS technology are very effective for lowering the power consumption and their potential has been demonstrated by many researchers [33, 34]. To achieve further improvement of gas-sensing properties of SnO_2-based semiconductor gas sensors, besides strict and tailored design of surface chemistry of SnO_2-based sensor materials, total design and control of physicochemical processes inside porous SnO_2-based sensors is absolutely necessary.

Cross-References

▶ Electrochemical Sensor of Gaseous Contaminants
▶ Electrochemical Sensors for Environmental Analysis
▶ Sensors

References

1. Seiyama T, Kato A, Fujiishi K, Nagatani N (1962) New detection for gaseous components using semiconductive thin film. Anal Chem 34:1502–1503
2. Taguchi N (1962) Gas alarm devices. JPB_1970038200
3. Wüller W, Kearley GJ, Ling CD (2012) Ab initio parametrized polarizable force for rutile-type SnO_2. Theor Chem Acc 131:1216–1223. doi:10.1007/s00214-012-1216-9
4. Wyckoff RG (1982) Crystal structure, 1: 250, Krieger, USA
5. Lantto V, Rantala TS, Rantala TT (2000) Experimental and theoretical studies on the receptor and transducer functions of SnO_2 Gas sensors. Electron Technol 33(1/2):22–30
6. Shimizu Y, Kai S, Takao Y, Hyodo T, Egashira M (2000) Correlation between methylmercaptan Gas-sensing properties and its surface chemistry of SnO_2-based sensor materials. Sens Actuators B 65:349–357
7. Xu X, Tamaki J, Miura N, Yamazoe N (1990) Relationship between gas sensitivity and microstructure of porous SnO_2. Denki Kagaku 58(12):1143–1148
8. Egashira M, Matsumoto T, Shimizu Y, Iwanaga H (1988) Effect of surface treatments on gas-sensing characteristics of tin oxide whiskers. J Mater Sci Lett 7:1122–1123
9. Yamazoe N, Fuchigami J, Kishikawa M, Seiyama T (1979) Interactions of tin oxide surface with O_2, H_2O and H_2. Surf Sci 86:335–344
10. Wang L, Lou Z, Zhang T, Fan H, Xu X (2011) Facile synthesis of hierarchical SnO_2 semiconductor microspheres for gas sensor application. Sens Actuators B 155:285–289
11. Hwang I, Lee E, Kim S, Choi J, Cha J, Lee H, Ju B, Lee J (2011) Gas sensing properties of SnO_2 nanowires on micro-heater. Sens Actuators B 154:295–300
12. Zhan Z, Wang W, Zhu L, An W, Biswas P (2010) Flame aerosol reactor synthesis of nanostructures SnO_2 thin films: high gas-sensing properties by control of morphology. Sens Actuators B 150:609–615
13. Firooz AA, Mahjoub AR, Khodadadi AA (2009) Highly sensitive CO and ethanol nanoflower-like SnO_2 sensor among various morphologies obtained by using single and mixed ionic surfactant templates. Sens Actuators B 141:89–96
14. Yu Q, Wang K, Luan C, Geng Y, Lian G, Cui D (2011) A dual-functional highly responsive gas sensors fabricated from SnO_2 porous nanosolid. Sens Actuators B 159:271–276
15. Hackner A, Habauzit A, Müller G (2010) Resistive and surface ionisation of SnO_2 gas sensing layer. Sens Actuators B 146:433–439
16. Hwang I, Kim S, Choi J, Jung J, Yoo D, Dong K, Ju B, Lee J (2012) Large-scale fabrication of highly sensitive SnO_2 nanowire network gas sensors by single step vapor phase growth. Sens Actuators B 165:97–103
17. Shaalan NM, Yamazaki T, Kikuta T (2011) Influence of morphology and structure geometry on NO gas-sensing characteristics of SnO_2 nanostructures synthesized via a thermal evaporation method. Sens Actuators B 153:11–16
18. Xu C, Tamaki J, Miura N, Yamazoe N (1991) Grain size effects on gas sensitivity of porous SnO_2-based elements. Sens Actuators B 3:147–155
19. Ippommatsu M, Ohnisi H, Sasaki H, Matsumoto T (1991) Study on the sensing mechanism of tin oxide flammable gas sensors using the hall effect. J Appl Phys 69:8368–8374
20. Ogawa H, Nishikawa M, Abe A (1982) Hall measurement studies and an electrical conductive model of tin oxide ultrafine particle films. J Appl Phys 53:4448–4454
21. Chen Y, Yu L, Feng D, Zhuo M, Zhang M, Zhang E, Xu Z, Li Q (2012) Superior ethanol-sensing properties based on Ni-doped SnO_2 p-n heterojunction hollow spheres. Sens Actuators B 166–167:61–67
22. Maekawa T, Tamaki J, Miura N, Yamazoe N (1991) Sensing behavior of CuO-loaded SnO_2 element for H_2S detection. Chem Lett 1991:575–578
23. Jain K, Pant RP, Lakshmikumar ST (2006) Effect of Ni doping on thick film SnO_2 gas sensor. Sens Actuators B 113:823–829
24. Yamazoe N, Miura N (1992) Some basic aspects of semiconductor gas sensors. In: Yamauchi S (ed) Chemical sensor technology, vol 4. Kodansha-Elsevier, New York, pp 19–42
25. Cricenti A, Generosi R, Scarselli MA, Perfetti P, Siciliano P, Serra A, Tepore A, Almeida J, Coluzza C, Margaritondo G (1996) Pt:SnO_2 thin films for gas sensor characterized by atomic force microscopy and X-ray photoemission spectromicroscopy. J Vac Sci Technol B 14:1527–1530
26. Matsushima S, Teraoka Y, Miura N, Yamazoe N (1988) Electronic interaction between metal additives and tin dioxide in tin dioxide-based gas sensors. Jpn J Appl Phys 27:1798–1802
27. Shimizu Y, Egashira M (1999) Basic aspects and challenges of semiconductor gas sensors. MRS Bull 24:18–24
28. Hashimoto M, Inoue H, Hyodo T, Shimizu Y, Egashira M (2008) Preparation and gas sensor application of ceramic particles with submicron-size spherical macropores. Sens Lett 6:1–4
29. Hyodo T, Nishida N, Shimizu Y, Egashira M (2002) Preparation and gas-sensing properties of thermally stable mesoporous SnO_2. Sens Actuators 83:209–215
30. Shimizu Y, Hyodo T, Egashira M (2004) Mesoporous semiconducting oxides for gas sensor application. J Eur Ceram Soc 24:1389–1398
31. Shimizu Y, Hyodo T, Egashira M (2004) Meso- to macro-porous oxides as semiconductor gas sensors. Catal Surveys from Asia 8:127–135

32. Kim I, Jeon E, Choi S, Choi D, Tuller HL (2010) Electrospun SnO_2 nanofiber mats with thermo-compression step for gas sensing applications. J Electroceram 2–4:159–167. doi:10.1007/s10832-010-9607-6
33. Andio MA, Browning PN, Morris PA, Akbar SA (2012) Comparison of gas sensor performance of SnO_2 nano-structures on microhotplate platforms. Sens Actuators B 165:13–18
34. Berger F, Sanchez J, Heintz O (2009) Detection of hydrogen fluoride using SnO_2-based gas sensors: understanding of the reactional mechanism. Sens Actuators B 143:152–157

Soil Remediation, Use of Combined (Coupled) Technologies

M. Carmen Lobo, Araceli Pérez-Sanz,
M. Mar Gil-Diaz and Antonio Plaza
IMIDRA, Alcalá de Henares, Madrid, Spain

Introduction

Nowadays, soil contamination is a world problem. It is driven or exacerbated by human activity such as inadequate agricultural and forestry practices, industrial activities, tourism, and industrial and urban expansion. These activities have a negative impact, preventing the soil from performing its functions and services to humans and ecosystems. The European Commission estimates that nearly 3.2 million sites across Europe could have contaminated soil. Soil contamination can be caused by fuel leaking from tanks in factories, use of heavy machinery in farming, and building of roads and other infrastructure, in general due to the industrial activities. Moreover, agricultural activities can lead to contamination processes due to the excessive use of pesticides and organic residues. Soil degradation due to contamination represents an increasing hazard that requires the development of improved methods for soil remediation with two main objectives: (1) remove the contaminants from the soil and (2) allow soil to recover its functionality [1].

Over decades, different technologies have been applied to soil remediation, e.g., incineration, inertization, soil washing, and soil venting, in order to remove pollutants according to economical criteria and time. In relation to the variability of soil types and the degree of disturbance of soil functions, there are considerable differences among the remediation methods.

Electrokinetic remediation (EK) is a developing technique that has a significant potential for in situ remediation of low permeability soils contaminated with heavy metals and/or organic compounds. The technique involves applying a low direct current or a low potential gradient to electrodes that are inserted into the ground [2, 3].

Currently, the state of the art suggests the use of combined technologies which add benefits and overlook the limitations of each principle/strategy of remediation. In this sense, the number of articles evaluating the possibility of technologies coupled to electrokinetics is increasing. Below, some applications for the treatment of contaminated soils are discussed.

Coupled Electrokinetic-Bioremediation

The use of bioremediation is receiving great interest due to the low cost and energy required by this process and its sustainable character. In order to optimize this application, it often needs to be combined with other strategies. This requirement is driven by the need to remediate over a short timescale, in the case of ex situ bioremediation, or the need to promote long-term effectiveness under a regime such as monitored natural attenuation or permeable reactive barriers [4]. While the ability of microorganisms to biodegrade a number of organic contaminants is understood, in practice, physical, chemical, and microbial limitations decrease the effectiveness of the technique and lead to a slow or incomplete biodegradation of the contaminants. The application of an electric field (EK process) leads to controlling the flow of ions and water through the soil. This fact makes it easy to manipulate the pH and redox, besides increasing the bioavailability of nutrients and electron donors and acceptors in the field.

According to Lohner [4], the potential benefits of electrokinetic and electrochemical processes coupled with bioremediation include enhancement of pollutant bioavailability by means of electrokinetic mobilization, increase of restricted soil bacteria mobility by electrokinetic transport processes, electrokinetic-induced mass transfer and transport of ionic electron acceptors and nutrients, and electrochemical production of limited electron donors (H_2) and acceptors (O_2).

In recent years, several studies have evaluated the benefits of the abovementioned, applying an electric field to enhance the biodegradation process in polluted soil samples with different compounds, aromatic compounds, herbicides, gasoline hydrocarbons, or trichloroethylene [5–8]. With the aim to increase the solubility of the organic-forming complexes and/or increase the electroosmotic flow, the use of chemical or biological surfactants has been assayed in some studies [9, 10].

Wick et al. [8], after a review on the fundamental interactions in electrobioremediation, pointed out that, although empirical electrobioremediation appears to be safe, effective, and economically interesting compared to other remediation techniques, there is still a need for mechanistic understanding of the molecular processes affecting the release and transport of organic pollutants in soil. The dispersion and physiology of pollutant degrading microorganisms should be well understood before the technology can be fully exploited. In this sense, Acuña et al. [11] showed the increase in the biodegradation of hydrocarbons by applying direct current to a contaminated soil previously treated by land farming.

It is worth mentioning that the experiments shown in the various references have been tested on a laboratory scale. Further studies in the field would be needed to validate the use of this technology.

A particular case of bioremediation is known as phytoremediation, where plants are used to remove metals from the soils.

Coupled Electrokinetic-Phytoremediation

Phytoremediation is defined as the use of living plants and their associated microorganisms to remove, degrade, or sequester inorganic and organic pollutants from in situ remediation of contaminated soils, sludges, sediments, and groundwater. It has been reported to be an effective, nonintrusive, inexpensive, aesthetically pleasing, and socially accepted technology [12]. Different authors have reviewed the advantages of this strategy [13–22]. Originally, the technique was applicable to heavy metal contaminants, but it was also suitable for organic pollutants [15]. Early estimates of the cost indicate that phytoremediation is cheaper than conventional remediation methods. In contrast, phytoremediation does have certain disadvantages and limitations. This technology is limited by depth (roots) and also by the solubility and the availability of pollutants. Although it is faster than natural attenuation, phytoremediation requires longtime periods and is restricted to sites with low contaminant concentration. The plant biomass obtained from phytoextraction requires proper disposal as a hazardous waste. Contaminants may be transferred to the environment and/or the food chain. In the last decades, the development of phytoremediation techniques has shown several limitations associated with the low rate of growth and biomass production of hyperaccumulators. Although a scarce number of investigations have been reported on phytoremediation assisted by the electromobilization of heavy metals, the combined technology seems to be an ameliorated implement for soil remediation [1]. Electrokinetic- and phytoremediation have been proven effective in the remediation of sites where the contamination level is low or moderate. The fact that the technique can be performed in situ without complex devices allows the simultaneous growth of crops in the contaminated site to be treated. Lim et al. [23] optimized the time and electric field applied to each Indian mustard (Brassica juncea) plant in a combined remediation experiment. In another experimental study [24], a constant voltage of

30 V was continuously applied to a contaminated soil that supported the growth of perennial ryegrass (Lolium perenne cv Elka). Other authors have assayed the influence of the type of electric current applied. In this sense, Bi et al. [25, 26] have shown the best results in the combined technology application when alternative current is applied, in comparison with direct current.

In light of the results obtained from these preliminary approaches to the combined technology, several aspects still need to be adjusted. The use of changes in the polarity of electrodes during the process could avoid fixed redistribution of heavy metals and pH values in soils which are associated to different rates of plant biomass and phytoextractions, depending on the proximity to electrodes. Thereby, the combined technique is homogenously applied to the contaminated site. As a consequence, further studies should be necessary to evaluate the impact of the coupled technology on a field scale.

Electrokinetic-Thermal Desorption

In situ thermal methods have been used for site remediation since 1980s, beginning in the form of steam injection and thermal conductive heating with in situ electrical resistance heating (ERH) being patented in 1996 and commercialized shortly thereafter [27]. Studies from Davis [28, 29] evaluated how heat or steam injection can enhance in situ soil remediation. Different principles govern the thermal desorption process, including thermal, physicochemical, fluid and energy transport, and hydraulics.

Thermal principles include heat transport, thermal conduction and diffusivity, heat capacity, and heat of vaporization. Besides, many pollutants can be removed as a vapor at temperatures below that of the boiling point of water or at steam temperatures at a boiling point of water greater than 100 °C. Under these conditions, gas mixtures are present during the process besides carbon dioxide, which is at saturation in natural groundwater [30] and other gases dissolved in the groundwater as nitrogen or argon [31]. With the removal of the organic pollutants to the subsurface, additional gases are generated, including nitrogen and ammonia from the reduction of nitrates and nitrites in the groundwater and hydrogen sulfide from the reduction of water and bicarbonate [27].

Steam stripping can be defined as a process where contaminants partition from soil, water, or NAPL phases into the vapor phase (steam) and are removed (stripped) from their source areas by vapor flow [27].

Physical and chemical principles include such properties as aqueous solubility and density and are relevant from the point of view of pollutant migration.

Other properties related to fluid and energy transport are viscosity, sorption, vapor transport, and hydrolysis. Gas viscosities will increase about 30 % with an increase of the temperature around 100 °C [28], facilitating transport through porous media.

The soil-water sorption coefficient affects only the aqueous phase transport of pollutants; this coefficient will decrease in relation to the temperature. In this way, the ability of the hot fluids to remove pollutants from the soil increases significantly. The effect of the temperature is specific to the soil type, water content, and type of pollutants.

Another factor to consider is the vapor transport, which controls the volatilization of a chemical compound. Vapor pressure increases with temperature, and the ebullitions occur when the sum of the partial pressures of the chemical compound, water, and associated dissolved gases exceeds hydrostatic plus capillary pressures [31].

Hydrolysis involves the chemical reaction with water, without regard to redox conditions, dissolved minerals, or the presence of soil microbes. Pollutants need to be dissolved in water to produce the reaction that can be modified by pH and/or temperature [27].

The number and characteristics of the process involved in the thermal desorption technology and its influence in the soil according to its properties make the implementation of the coupled technology difficult. In general, thermal desorption coupled with electrokinetic treatment has not been well studied.

Electrokinetic-Fenton Process

During the electrokinetic process, the electroosmotic flow can play the most important role in the removal of pollutants. To enhance the development of the treatment, an integration of the EK process with another treatment, such as the use of the Fenton process, could be adequate.

Fenton and Fenton-like processes have been tested as an effective technology for the degradation of different organics in soil and wastewater [32–34]. The destruction of organic pollutants is conducted through the free radicals generated from the H_2O_2 and Fe_2^+. The use of zerovalent iron or iron minerals instead of Fe^+ in a Fenton-like reaction can be effective in the degradation of organic pollutants [35]. The effectiveness of the process depends on different reaction parameters, such as pH, type of inorganic slats, natural organic matter, dosage of the reactive iron species, and the molecular structure of the pollutant.

Different authors have conducted research using the combination of electrokinetics with a Fenton process. First results were obtained by Long [36] where soils contaminated with phenol and 4-chlorophenol were treated. 1 vol/cm as the potential gradient and water as the processing fluid were the EK conditions used. Iron powder and $FeSO_4$ were used as a catalyst in the Fenton process. During treatment, the removal and destruction of the pollutants were observed. The degree of efficiency was dependent on the soil characteristics, the type of pollutants, and the catalyst.

Yang and Liu [37] evaluated the remediation of a soil contaminated with trichloroethylene by an in situ EK-Fenton process. They found that organic matter and mineral particle size affect the process feasibility. Likewise, the type of catalyst and its dosage could affect the reaction mechanisms and consequently the efficiency of the remediation.

An ultrasonic oxidation process has been proven for the removal of hazardous pollutants from water [38, 39]. The chemical effects of high-intensity ultrasound led to a variety of energy species (H_2O_2 and HO^-) in solution. This phenomenon is known as cavitation [40] and leads to subsequent chemical reactions in three phases [34]. Ince et al. [41] demonstrated the use of ultrasound as a catalyst of aqueous reaction systems in environmental applications.

Coupled Electrokinetic-Electrochemical Process

The combination of EK and electrooxidation (EK-EO) was developed by Wang et al. [42]. In this coupled process, electrooxidation on the anode surface and the EK process of anionic impurities under the electric field occur and, as a consequence, an enhancement of the organic pollutant is removed in the electrolysis process. The authors suggested that both electrooxidation-electroreduction and electromigration pathways coexisted in EK-EO process.

In a similar study, Cong et al. [43] found that system pH affected the migration of pollutants in the soil, the removal of weak organic acids being easier at higher pH values. After 140 min of treatment, the removal efficiency of chlorophenol was around 85 %.

Pazos et al. [44] applied the technology to removal of a dye from a kaolinite. Experimental results showed that EK treatment did not result in any significant removal of the pollutant; however, the use of $NaHPO_4$ in the processing fluid increased the EO flux, enhancing the desorption of the dye from the kaolinite and preventing the acidification process. These authors further applied a two-stage process combining EK remediation and liquid electrochemical oxidation for the removal of organic pollutants [45, 46]. In their experiments, the liquid collected during electrokinetic remediation was oxidized by electrochemical treatment. This oxidation was accomplished using graphite electrodes, and the pollutants were totally degraded in less than 2 days.

Considering that some of these experiments have been carried out using kaolinite, it should be necessary to know the feasibility of the technique when the sample is a real soil.

Coupled Electrokinetic-Permeable Reactive Barriers

The permeable reactive barrier (PRB) is a process that contains a reactive zone for the destruction of organic compounds and for the removal of specific inorganic compounds. This barrier is applied in situ for groundwater remediation in three different ways: continuous trench, funnel and gate, and reactive vessel [47]. Different materials can be used as fill material in the barrier; limestone, activated carbon, zeolites, or hydroxyapatite can act as chemical precipitators or adsorbents. Even microorganisms can be used to act as a biological reactor to biodegrade the pollutants.

Fe^0 is one of the most popular materials used in the barriers [48–51]. The process involves the corrosion of Fe^0, which provides the electrons necessary for the reduction of compounds such as trichloroethylene and chromium (VI) [47].

Considering the benefits of the use of electrokinetic technology and the advantages of the barriers, different authors [52–55] have considered the application of the electrokinetic process coupled with permeable reactive barriers (PRBs). Depending on the nature of the pollutants, different fillings will be selected for use in the barrier. Another factor to be considered to optimize the process is the thickness of the barrier. Cang et al. [56] have used this coupled technology using a permeable barrier of zero valence for treating a chromium contaminated soil. In this assay, the authors showed the feasibility of this technique to remove 72 % chromium after 384 h of treatment. Due to the experiments carried out on a lab scale, scaled-up research and field application would be necessary to demonstrate the efficiency of this coupled technology in different types of soils.

Future Directions

Given the complexity of the soil matrix, its different physico-chemical and biological properties, as well as the climate of the area, each situation of contamination in soil should be considered as a particular case. Currently the combination of technologies opens new possibilities for the implementation of decontamination strategies adding benefits and overlooking the limitations of each principle/strategy. The state of the art in the application of electrokinetic technology shows interesting applications to soil remediation, combining its use with other physico-chemical and biological strategies. Howewer most of the experiments are conducted in lab scale, then, the scaled up research and field application would be necessary to demonstrate the efficiency of these coupled technologies in different types of soils.

Cross-References

▶ Cell, Electrochemical
▶ Degradation of Organics, Use of Combined Electrochemical-Ultrasound
▶ Electrobioremediation of Organic Contaminants
▶ Electrochemical Bioremediation
▶ Electrochemical Cells, Current and Potential Distributions
▶ Electrochemical Treatment of Landfill Leachates
▶ Electro-Fenton Process for the Degradation of Organic Pollutants in Water
▶ Electrokinetic Barriers for Preventing Groundwater Pollution
▶ Electrokinetic Remediation, Cost Estimation
▶ Electrokinetics in the Removal of Chlorinated Organics from Soils
▶ Electrokinetics in the Removal of Hydrocarbons from Soils
▶ Electrokinetics in the Removal of Metal Ions from Soils
▶ Organic Pollutants, Direct Electrochemical Oxidation

References

1. Lobo MC, Pérez-Sanz A, Martínez-Iñigo MJ, Plaza A (2009) Influence of coupled electrokinetic phytoremediation on soil remediation. In: Reddy J, Cameselle C (eds) Electrochemical remediation technologies or polluted soils, sediments and groundwater. Wiley, Hoboken

2. Acar YB, Alshawabkeh AN (1993) Principles of electrokinetic remediation. Environ Sci Technol 27:2638–47
3. Ugaz A, Puppala S, Gale RJ, Acar YB (1994) Complicating features of electrokinetic remediation of soils and slurries: saturation effects and the role of the cathode electrolysis. Commun Chem Eng 129:183–200
4. Lohner ST, Tiehm A, Jackman SA, Carter P (2009) Coupled electrokinetic-bioremediation: applied aspects. In: Reddy J, Cameselle C (eds) Electrochemical remediation technologies or polluted soils, sediments and groundwater. Wiley, Hoboken, pp 389–416
5. Lee HS, Lee K (2001) Bioremediation of diesel-contaminated soil by bacterial cells transported by electrokinetic. J Microbiol Biotechnol 11(6):1038–1045
6. Jackman SA, Maini G, Sharman AK, Sunderland G, Knowles CJ (2001) Electrokinetic movement and biodegradation of 2,4-dichloropheoxyacetic acid in silt soil. Biotechnol Bioeng 74(1):40–48
7. Niqui-Arroyo JL, Bueno-Montes M, Posada-Baquero R, Ortega-Calvo JJ (2006) Electrokinetic enhancement of phenanthrene biodegradation in creosote-polluted clay soil. Environ Pollut 142(2): 326–332
8. Wick LY, Shi L, Harms H (2007) Electrobioremediation of hydrophobic organic soil contaminants. A review of fundamental interactions. Electrochim Acta 52(10):3441–3448
9. Alcántara MT, Gómez J, Pazos M, Sanromán MA (2008) Combined treatment of PAHs contaminated soils using the sequence extraction with surfactant–electrochemical degradation. Chemosphere 70: 1438–1444
10. Gonzini O, Plaza A, Di Palma C, Lobo MC (2010) Electrokinetic remediation of gasoil contaminated soil enhanced by rhamnolipid. J Appl Electrochem 40:1239–1248
11. Acuña AJ, Tonin N, Pucci GN, Wick L, Pucci OH (2010) Electrobioremediation of an unsaturated soil contaminated with hydrocarbon after landfarming treatment. Portugaliae Electrochim Acta 28(4): 253–263. doi:10.4152/pea.201004253
12. Alkorta I, Hernandez-Allica J, Becerril JM, Amezaga I, Albizu I, Garbisu C (2004) Recent findings on the phytoremediation of soils contaminated with environmentally toxic heavy metals and metalloids such as zinc, cadmium, lead and arsenic. Rev Environ Sci Biotechnol 3:71–90
13. Chaney RL, Malik M, Li YM, Brown SL, Brewer EP, Angle JS, Baker AJ (1997) Phytoremediation of metals. Curr Opin Biotechnol 8:279–284
14. Wenzel WW, Adriano DC, Salt D, Smith R (1999) Phytoremediation: a plant-microbe-based remediation system. In: SSSA (ed) Bioremediation of contaminated soils, vol 37, Agronomy monograph. SSSA, Madison, pp 457–508
15. Meagher RB (2000) Phytoremediation of toxic elemental and organic pollutants. Curr Opin Plant Biol 3:153–162
16. NavariIzzo F, Quartacci MF (2001) Phytoremediation of metals. Tolerance mechanisms against oxidative stress. Minerva Biotechnol 13:73–83
17. Salt DE, Baker AJM (2001) Phytoremediation of metals. In: Rehm HJ, Reed G (eds) Biotechnology, vol 11b, Environmental processes II, soil decontamination. Wiley VCH, New York, pp 386–397
18. Lasat MM (2002) Phytoextraction of toxic metals – a review of biological mechanisms. J Environ Qual 31:109–120
19. McGrath SP, Zhao FJ, Lombi E (2002) Phytoremediation of metals, metalloids, and radionuclides. Adv Agron 75:1–56
20. Wolfe AK, Bjornstad DJ (2002) Why would anyone object? An exploration of social aspects of phytoremediation acceptability. Crit Rev Plant Sci 21(5):429–438
21. McGrath SP, Zhao FJ (2003) Phytoextraction of metals and metalloids. Curr Opin Biotechnol 14:277–282
22. Padmavathiamma PK, Li LY (2007) Phytoremediation technology: hyperaccumulation metals in plants. Water Air Soil Pollut 184:105–126
23. Lim J, Salido AL, Butcher DJ (2004) Phytoremediation of lead using Indian mustard (Brassica juncea) with EDTA and electrodics. Microchem J 76(2004):3–9
24. O'Connor CS, Lepp NW, Edwards SR, Sunderland G (2003) The combined use of electrokinetic remediation and phytoremediation to decontaminate metal-polluted soils: a laboratory-scale feasibility study. Environ Monit Assess 84(1–2): 141–148
25. Bi R, Schlack M, Siefert E, Lord R, Conelly H (2010) Alternating current electrical field effects on lettuce (lactuca sativa) growing in hydroponic culture with and without cadmium contamination. J Appl Electrochem 40(6):1217–1233
26. Bi R, Schlack M, Siefert E, Lord R, Conelly H (2011) Influence of electrical field (AC and DC) on phytoremediation of metal polluted soils with rapeseed (Brassica napus) and tobacco (Nicotiana tabacum). Chemosphere 83(3):318–326
27. Smith GJ (2009) Coupled electrokinetic-thermal desorption. In: Reddy KR, Cameselle C (eds) Electrochemical remediation technologies for polluted soils, sediments and groundwater. Wiley, Hoboken
28. Davis EL (1997) How heat can enhance in situ soil and aquifer remediation: important chemical properties and guidance on choosing the appropriate technique, EPA/540/S-97/502
29. Davis EL (1998) Steam injection for soil and aquifer remediation, EPA/540/S-97/505

30. Stumm W, Morgan JJ (1981) Aquatic chemistry. An introduction emphasizing chemical equilibria in natural waters, 2nd edn. Wiley, New York
31. Amos RT, Mayer U, Belkins BA, Delin GN, Wiliams RC (2005) Use of dissolved and vapor-phase gases to investigate methanogenic degradation of petroleum hydrocarbon contamination in the subsurface. Water Resour Res 41(1):1–15
32. Bowers AR, Eckenfelder WW, Gaddipati P, Mosaen RM (1989) Treatment of toxic or refractory wastewater with hydrogen peroxide. Water Sci Technol 21:447–486
33. Watts RJ, Udell MD, Rauch PA, Leung SW (1990) Treatment of pentachlorophenol-contaminated soil using Fenton´s reagent. Hazard Waste Hazard Mater 7(4):335–345
34. Yang GCC (2009) Electrokinetic-chemical oxidation/reduction. In: Reddy KR, Cameselle C (eds) Electrochemical remediation technologies for polluted soils, sediments and groundwater. Wiley, Hoboken, pp 439–471
35. Greenberg RS, Andrews T, Kakaarla PKC, Watts RJ (1998) In-situ Fenton-like oxidation of volatile organics: laboratory, pilot and full-scale demonstration. Remediation 8:2,29–42
36. Long YW (1997) A study on treatment of phenol and 4-chlorophenol contaminated soil by electrokinetic-Fenton process. M.S. thesis, National Sun Yat-Sen University Kaohsiung, Taiwan (in chinese)
37. Yang GC, Liu CY (2001) Remediation of TCE contaminated soils by in situ EK-Fenton process. J Hazard Mater 85(3):317–331
38. Vinodgopal K, Peller K, Makogon O, Kamta PV (1998) Ultrasonic mineralization of reactive, textile azo dye, remazol black B. Water Res 32:3646–3650
39. Kidak R, Ince NH (2006) Effects of operating parameters on sonochemical decomposition of phenol. J Hazard Mater 137:1453–1457
40. Suslick KS, Goodale JW, Wang HH, Schubert PF (1983) Sonochemistry and sonocatalysis of metal carbonyls. J Am Chem Soc 105:5781–5785
41. Ince NH, Tezcandi G, Belen RK, Apikyan IG (2001) Ultrasound as a catalyzer of aqueous reaction system-the state of the arte and environmental applications. Appl Catal Environ 29:167–176
42. Wang A, Qu J, Liu H, Ge J (2004) Degradation of azo dye Acid Red 14 in aqueous solution by electrokinetic and electrooxidation process. Chemosphere 55(9):1189–1196
43. Cong Y, Ye Q, Wu Z (2005) Electrokinetic behaviour of chlorinated phenols in soil and their electrochemical degradation. Process Safety Environ Prot 83:178–183
44. Pazos M, Cameselle C, Sanroman MA (2008) Remediation of dye-polluted kaolinite by combination of electrokinetic remediation and electrochemical treatment. Environ Eng Sci 25:419–428
45. Alcantara MT, Pazos M, Cameselle C, Sanroman MA (2008) Electrochemical remediation of phenanthrene from contaminated kaolinite. Environ Geochem Health 30:89–94
46. Gomez J, Alcantara MT, Pazos M, Sanroman MA (2009) A two-stage process using electrokinetic remediation and electrochemical degradation for treating benzo[a]pyrene spiked kaolin. Chemosphere 74:1516–1521
47. Weng CH (2009) Coupled electrokinetic permeable reactive-barriers. In: Reddy J, Cameselle C (eds) Electrochemical remediation technologies for polluted soils, sediments and groundwater. Wiley, Hoboken, pp 483–503
48. Blowes DW, Ptacek CJ, Benner SG, Mc Re CWT, Bennet TA, Puls RW (2000) Treatment of inorganic contaminants using permeable reactive barriers. J Contam Hydrol 45:123–137
49. Phillips DH, Watson DB, Roh Y, Gu B (2003) Mineralogical characteristics and transformations during long term operation for a zero valent iron reactive barrier. J Environ Qual 32:2033–2045
50. Ebert M, Kober R, Parbs A, Plagentz V, Schafer D, Dahmke A (2006) Assessing degradation rates of chlorinated ethylenes in column experiments with commercial iron materials used in permeable reactive barrier. Enviorn Sci Technol 40:2004–2010
51. Lai KCK, Lo IC (2008) Removal of chromium (VI) by acid washed zero valent iron under various groundwater geochemistry conditions. Environ Sci Technol 42:1238–1244
52. Weng CH, Lin YT, Lin TK, Kao CM (2007) Enhancement of electrokinetic remediation of hyper Cr (VI) contaminated clay by zero valent iron. J Hazard Mater 149(2):292–302
53. Yuan C, Chiang TS (2007) The mechanisms of arsenic removal form soil by electrokinetic process coupled with iron permeable reaction barrier. Chemosphere 67:1533–1542
54. Chung HI, Lee MH (2007) A new method for remedial treatment of contaminated clayey soils by electrokinetics coupled with permeable reactive barrier. Electrochim Acta 52:3427–3431
55. Ruiz D, Anaya JM, Ramirez V, Alba GI, García MG, Carrillo-Chávez A, Teutli MM, Bustos E (2011) Soil Arsenic removal by a permeable reactive barrier of iron coupled to an electrochemical process. Int J Electrochem Sci 6:548–560
56. Cang L, Zhou D, Wu D, Alshawabkeh AN (2010) Coupling electrokinetics with permeable reactive barriers of zero-valent iron for treating a chromium contaminated soil. Sep Sci Technol 44:2188–2202

Solid Electrolytes

Ulrich Guth
Kurt-Schwabe-Institut für Mess- und
Sensortechnik e.V. Meinsberg,
Waldheim, Germany
FB Chemie und Lebensmittelchemie, Technische
Universität Dresden, Dresden, Germany

Definition

Solid ionic conductors that can be used in electrochemical cells as an electrolyte are called solid electrolytes. In such compounds only one ion is mobile (see entry ► Solid State Electrochemistry, Electrochemistry Using Solid Electrolytes). Generally, any conductor with a high ionic transference number can serve as an electrolyte. Often, the definition after Patterson is used who described solids with a transference number > 0.99 as solid electrolytes [1]. The transference number is not a fixed value. It depends on the temperature and the partial pressure of the gas involved in the chemical reaction with the mobile ion. Therefore, all solids are more or less conductors with a mixed ionic and electronic conductivity, so-called mixed conductors. For the application in sensors and fuel cells, only a window concerning temperature and partial pressure is suitable. This is also called as electrolytic domain. The phenomenon that solids exhibit a high ionic conductivity is also designed as fast ion transport.

Classification of Solid Electrolytes

Solid electrolytes can be classified with regard to:
- Mobile ions as
  Cation conductors such as alkali ion conductors (Li^+, Na^+, K^+), Ag^+, Cu^+, and H^+ conductors, respectively
  Anion conductors with those ions as O^{2-}, F^-, Cl^-
- The substances (independent on the mobile species)

Oxidic (oxide), carbonatic, sulfatic, and phosphatic solid electrolytes
- The defect chemistry
  Solid electrolytes based on chemical disorder
  Solid electrolytes based on structural disorder (super-ionic conductors)
- The dimension in which the mobile ions can move
  Two-dimensional solid electrolytes, e.g., β-alumina ($Na_2O·Al_2O_3$)
  Three-dimensional solid electrolytes, e.g., Nasicon $Na_{1+x}Zr_2Si_xP_{3-x}O_{12}$ ($0 \leq x \leq 3$).

For applications in industry the oxide ion conductors based on doped zirconia are the most important. They are used in millions of elements in SOFC (see entry ► Solid Oxide Fuel Cells, Introduction) and oxygen sensors for see entry ► Combustion Control Sensors, Electrochemical, *lambda sensors* for catalyst control in car engines, sensors for *heat treatments of steel* (steel hardening), and sensors for measurement of see entry ► Molten Steel, Measurement of Dissolved Oxygen (steel production). For application in fuel cells and batteries, also H^+ conductors as well as Na^+ and Li^+ conductors have been investigated intensively. Surveys on different types of solid electrolytes were given in more detail by Chandra [2], Rickert [3], Möbius [4], Shuk et al. [5], and Kudo and Fueki [6].

Fast Ion Conductors (Super Ionic Conductors)

If there are more places in the lattice than mobile ions, the ions can occupy different lattice sites. Such solids are called super-ionic conductors. In this classification, in turn, it is distinguished between those solid electrolytes which exhibit that structure above the temperature of a phase transition (sometimes called as type I) and those which show that behavior over the whole temperature range (type II). As examples for the Type I, α-AgI and δ-Bi_2O_3 are mentioned, which are formed after the phase transitions:

$$\gamma - AgI(\text{cubic, ZnS-type}) \xrightarrow{147\,°C} \alpha$$
$$- AgI(\text{cubic, body- centred})$$

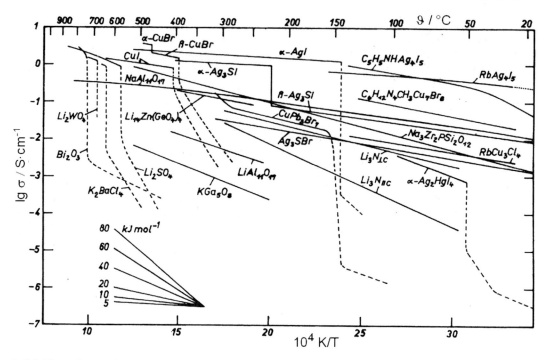

Solid Electrolytes, Fig. 1 Temperature dependence on the conductivity of ionic conductors with structural disorder [4]

$$\alpha - Bi_2O_3 \text{(monoclin)} \xrightarrow{729\,°C} \delta - Bi_2O_3 \text{(cubic)}$$

The entropy for the phase transition in AgI $\Delta_{tr}S = 14.5\,\text{J/(molK)}$ is higher than the melting entropy $\Delta_f S = 11.3\,\text{J/(molK)}$ [7]. That means the change in disorder in the solid phase transition is higher as in the solid–liquid transformation.

RbAg$_4$I$_5$ is the inorganic silver ion conductor having the highest conductivity over a broad temperature range, slightly lower than that in α-AgI. Its conductivity $\sigma = 0.26$ S/cm at 25 °C is comparable to that of H$_2$SO$_4$ in aqueous solutions [6] (Fig. 1).

Oxide Ion Conductors

Some oxides which are cubic (ThO$_2$, CeO$_2$) or polymorph (ZrO$_2$) can dissolve a remarkable amount of aliovalent ions and form solid solutions. As a result the conductivity increases and in the case of ZrO$_2$ the cubic high temperature is stabilized down to normal temperature (see entry ▶ Defects in Solids).

Such solid solutions are solids with chemical disorder because the ionic defects are generated by chemical composition and distributed statistically. Their concentration is independent on temperature and oxygen partial pressure. The activation energy for ion conduction is much higher ($\approx$100 kJ/mol) as compared to that for solids with structural disorder ($\approx$20–40 kJ/mol).

Zirconia Based Solid Electrolytes

The substitution of Zr^{4+} by Ca^{2+} in zirconia can be expressed by following equation:

$$CaO \xrightarrow{ZrO_2} Ca''_{Zr} + V^{\circ\circ}_O + O^x_O \qquad (1)$$

The concentration of cation defects is equal to that of oxide ion vacancies $[Ca''_{Zr}] = [V^{\circ\circ}_O]$. The conductivity increases with increasing amount of CaO because the defects are statistically distributed and have no inference with each other. The maximum in conductivity is reached at about 12 mol-% CaO. More CaO leads to a decrease due to forming of neutral aggregates

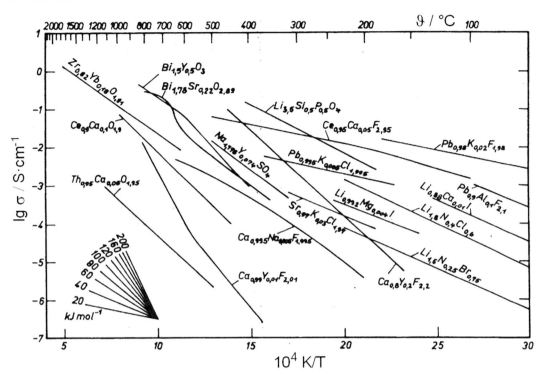

Solid Electrolytes, Fig. 2 Temperature dependence of the conductivity of solid electrolyte with chemical disorder [4]

$[Ca''_{Zr} - V_O^{\circ\circ}]^x$ which are not able to transport electric current. In the case of yttria-stabilized zirconia (YSZ), the substitution with 8 mol-% Y_2O_3 leads complete stabilization with the maximum conductivity and is preferred for most electrochemical applications. For that composition the following formula is obtained:

$$(ZrO_2)_{0.84}(Y_2O_3)_{0.08} = Zr_{0.84}Y_{0.16}O_{1.92}.$$

8 mol-% oxide ion vacancies are formed. Here the concentration of yttrium defects is

$$\left[Y'_{Zr}\right] = 2\left[V_O^{\circ\circ}\right] \text{ and } \left[V_O^{\circ\circ}\right] = \tfrac{1}{2}\left[Y'_{Zr}\right] \quad (2)$$

As early as 1899 Nernst suggested that zirconia is a conductor for oxide ions [8]. Not before 1943 Carl Wagner explained the mechanism of the oxygen ion transport via defects [9]. Since that time much work has been done to investigate the doping with different aliovalent ions, to determine their phase parameter, and to measure the conductivity and transference numbers [7, 10].

The temperature dependence of the conductivity can be described by a simple hopping model by means of an Arrhenius-like equation:

$$\sigma = \sigma_0 \times \exp\left[\frac{E_a}{RT}\right] \quad (3)$$

where E_a is the activation energy and σ_0 the conductivity for $T \to \infty$ which can be obtained by means of a log σ versus $1/T$ plot. Sometimes (depending on the conductivity model) also the relation

$$\sigma T = \sigma'_0 \times \exp\left[\frac{\Delta H^{\neq}}{RT}\right] \quad (4)$$

is used where $\Delta H^{\neq}$ is the activation enthalpy. The relation between both activation terms is given by $E_a - \Delta H^{\neq} = RT$. Concerning the linearity of plots only a little difference can be recognized (Fig. 2).

At high temperatures and extremely low or high oxygen partial pressures, oxygen can be removed from the lattice or built into the lattice.

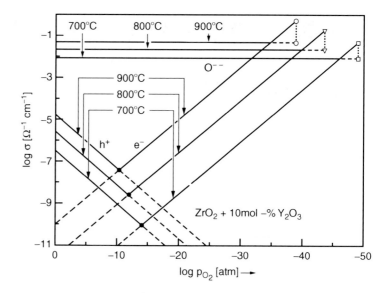

Solid Electrolytes, Fig. 3 Ionic and electronic conductivity in dependence on oxygen partial pressure for YSZ [3]

This leads to an electronic conductivity via excess electrons σ_e or holes σ_h which can be calculated by [11]

$$\sigma_e \left[\frac{S}{cm}\right] = 1.31 \cdot 10^7 \cdot \exp(-3.88 eV/kT) \cdot [p(O_2)/\text{atm}]^{-1/4} \quad (5)$$

$$\sigma_h \left[\frac{S}{cm}\right] = 2.35 \cdot 10^2 \cdot \exp(-1.67 eV/kT) \cdot [p(O_2)/\text{atm}]^{1/4} \quad (6)$$

In Fig. 3 the partial conductivities of ions and electronic carriers are plotted logarithmically vs. oxygen partial pressure. For the application as a solid electrolyte, the difference between ionic and hole conductivity should be at least 2–3 orders of magnitude. The transference number for electronic conductivity is between 0.01 and 0.001.

Other Oxide Electrolyte Systems

Ceria-Based Electrolytes Ceria-based solid electrolytes have received an increasing attention for so-called IT-SOFC (solid oxide fuel cells for intermediate temperature) because of their high conductivity. But its electrolytic domain is relatively small so that the working temperature must not be higher than 700 °C without a remarkable loss in efficiency due to the electronic conductivity. By doping with aliovalent stable oxides, a remarkable increase of conductivity can be obtained. The optimal composition was found to be $(CeO_2)_{0.8}(GdO_{1.5})_{0.2} = Ce_{0.8}Gd_{0.2}O_{1.9}$ (CGO):

$$2GdO_{1.5} \xrightarrow{CeO_2} 2Gd'_{Ce} + V_O^{\circ\circ} + 3O_O$$

An overview on ceria-based electrolytes is given in [12]. By adding of Tb_4O_7 or Pr_6O_{11} (rare earth ions occur in different valence states), mixed conductors with an enhanced electronic conductivity are obtained, which may be used as electrode materials for IT-SOFC.

Bismuth Oxide Based Electrolytes The cubic high temperature phase of bismuth oxide δ-Bi_2O_3 can be stabilized by adding of two- and three-valent ions as Sr^{2+} or Y^{3+} [5]. This is accompanied by a decrease in conductivity as compared to the pure δ-phase. Bismuth oxide-based electrolytes exhibit the highest ionic conductivity from all oxide systems, but they are not stable under reducing conditions and therefore not suitable for the most applications of oxide ion electrolytes. They can be used in cells for the oxygen enrichment in air.

Perovskite Based Electrolytes In mixed oxides of the perovskite-type oxide, general formula ABO_3, ion vacancies can be generated by substitution of both A and B cations in valence stable oxides, e.g., in gallates as $LaGaO_3$. In this oxide the electro neutrality is given with $z_+ \, \upsilon_+ \, (La^{3+}, Ga^{3+}) = +6$ and negative charge $z_- |\upsilon_-| = -6(O^{2-})$. When less positive than negative charges are present by doping with lower valent ions and Ga^{3+} is stable in its valence state, the system can only respond by forming of oxygen vacancies:

$$LaGaO_3 \xrightarrow{SrO,MgO} La_{0.8}Sr_{0.2}Mg_xGa_{1-x}O_{3-\delta}$$

The conductivity of $La_{0.8}Sr_{0.2}Mg_{0.15}Ga_{0.85}O_{3-\delta}$ as the composition with the highest conductivity is about one order of magnitude higher than that for YSZ. Those compounds were firstly described by Ishihara [13].

Solid electrolytes are produced by ceramic procedures using powders as precursors and organic binder. The slip or plastic mass can be formed by slip casting, green tape technology, hot and cold pressing, and extruding. In a thermal process below the melting temperature of the substance, the sintering process, the small particles grow together to a more or less dense solid. Ceramic technologies are described in detail in [14].

Cross-References

▶ Solid Electrolytes Cells, Electrochemical Cells with Solid Electrolytes in Equilibrium
▶ Solid Oxide Fuel Cells, History
▶ Solid Oxide Fuel Cells, Introduction
▶ Solid Oxide Fuel Cells, Thermodynamics

References

1. Patterson JW (1971) Conduction domains for solid electrolytes. J Electrochem Soc 118:1033–1039
2. Chandra S (1981) Super ionic solids. North Holland, Amsterdam/New York/Oxford
3. Rickert H (1982) Electrochemistry of solids. Springer, Berlin/Heidelberg/New York
4. Möbius HH (2011) Inheritance, personal communication 232
5. Shuk P, Wiemhöfer HD, Guth U, Göpel W, Grennblatt M (1996) Oxide ion conducting solid electrolytes based on Bi_2O_3. Solid State Ionics 89:179–196
6. Kudo T, Fueki K (1990) Solid State Ionics. VCH, New York

7. Kudo T (1997) Survey of types of solid electrolytes. In: Gellings PJ, Bouwmeester HJM (eds) Solid state chemistry CRC handbook. CRC Press, Boca Raton/New York/London/Tokyo
8. Nernst W (1899) On the electrolytic conductivity of solids at very high temperatures (in German). Z Elektrochem 6:41–43
9. Wagner C (1943) On the mechanism of the electrical current conductivity in the Nernst glower (in German). Naturwiss 31:265–268
10. Möbius HH (1965) The Nernst mass, its history and present importance (in German). Naturwissen 52:529–536
11. Park J-H, Blumenthal RN (1989) Electronic transport in 8 mole percent Y_2O_3-ZrO_2. J Electrochem Soc 136:2867–2876
12. Inaba H, Tagawa H (1996) Ceria-based solid electrolytes. Solid State Ionics 83:1–16
13. Ishihara T, Matssuda H, Takita Y (1994) Doped $LaGaO_3$ perovskite type oxide as a new oxide ionic conductor. J Am Chem Soc 116:3801–03
14. Kingery WD, Bowen HK, Uhlmann DR (1976) - Introduction to ceramic. Wiley, New York

Solid Electrolytes Cells, Electrochemical Cells with Solid Electrolytes in Equilibrium

Ulrich Guth
Kurt-Schwabe-Institut für Mess- und Sensortechnik e.V. Meinsberg, Waldheim, Germany
FB Chemie und Lebensmittelchemie, Technische Universität Dresden, Dresden, Germany

Solid Electrolyte Cells

The electrochemical cell, as shown schematically in Fig. 1, consists of gas-tight ceramic electrolyte in the form of tubes, discs, planar substrates, or thick films which is sandwiched by precious metal like platinum or silver as electrodes. Both electrodes are in close contact with two separate gas compartments. Such cells often modified by other electrode materials like electric conducting oxides (chromites, manganites) are commonly used as gas sensors, fuel cells, and electrolysis cells. Yttria-stabilized zirconia (YSZ: $Zr_{0.84}Y_{0.16}O_{1.92}$) as oxide ionic conductor is widely used as an electrolyte.

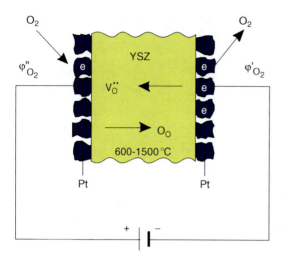

Solid Electrolytes Cells, Electrochemical Cells with Solid Electrolytes in Equilibrium, Fig. 1 General setup of an electrochemical cell with solid electrolytes yttria-stabilized zirconia (YSZ)

Electrochemical Equilibrium

The generation of an emf in a gas concentration cell can be explained as follows: Oxygen will move from the side with higher partial pressure toward the side with lower partial pressure. That is impossible if the solid electrolyte is gas tight. In the tendency oxygen can only move through the electrolyte as an oxide ion (O^{2-}). Oxygen takes up four electrons from electronic conducting material (here platinum) and moves through the electrolyte as O^{2-} (cathodic process). On the side with lower oxygen partial pressure, the reverse process takes place (anodic process). The electrode reactions and the cell reaction can be described by the following equations:

Cathode	$\frac{1}{2}O_2(g) + V_O^{\cdot\cdot}(YSZ) + 2e^-(Pt) \rightleftarrows O_O(YSZ)$
Anode	$O_O(YSZ) \rightleftarrows \frac{1}{2}O_2(g) + V_O^{\cdot\cdot}(YSZ) + 2e^-(Pt)$
Cell reaction	$\frac{1}{2}O_2(g) \rightleftarrows \frac{1}{2}O_2(g)$

The equilibrium arrows stand for an electrochemical equilibrium. In this case the turnover of the cathode and anode reactions is very small and equal.

Such cells symbolized by

$$O_2(\varphi'_{O_2}), Pt|(ZrO_2)_{0.84}(Y_2O_3)_{0.08}|Pt, O_2(\varphi''_{O_2})$$

can be regarded as oxygen concentration cells. The Kröger-Vink notation (see entry ▶ Kröger-Vinks Notation of Point Defects") was used to show how the formed defects are involved in the electrochemical reaction equations. By this relative notation, the deviation in position and charge can be expressed as compared to the perfect lattice sides ($V_O^{\cdot\cdot}$ means an oxide ion vacation double positively charged as compared to O^{2-}). The O^{2-} ion can be described according to the used Kröger-Vink notation as O_O^x. x means there is no deviation from the perfect lattice and therefore neutral charged, and the index O symbolizes the normal occupied position.

From the thermodynamic point of view, the change in the Gibbs energies can be expressed in terms of the stoichiometric sum of electrochemical potentials. For one phase I the electrochemical potential is the sum of chemical potential of oxygen μ_{O_2} and the electrical work $zF\varphi^I$ where z is the charge number, F the Faraday's constant, and φ the electrical potential:

$$\tilde{\mu}^I_{O_2} = \mu_{O_2} + zF\varphi^I \quad (1)$$

In the case of an electrochemical equilibrium, the electrochemical potentials of oxygen on both sides are equal:

$$\Delta_r G = \sum v_i \tilde{\mu}_i = 0 \quad (2)$$

For simplicity we assume that only two phases the electronic conducting I and the ionic conducting II are electrochemically involved in the reaction. The chemical work on the left side is equal to the electric work on the right side:

$$\sum v_i \mu_i^{I,II} = zF(\varphi^I - \varphi^{II})$$
$$\Delta\varphi_{eq}^{I,II} = (\varphi^I - \varphi^{II})_{eq} = \frac{\sum v_i \mu_i}{zF} \quad (3)$$

With the partial pressure dependence of the chemical potential

$$\mu_{O_2} = \mu_{O_2}^o + \ln p_{O_2} \qquad (4)$$

the Nernst's equation results:

$$\begin{aligned}
\Delta\varphi_{eq}^{I,II} &= \frac{\sum v_i\left(\mu_i^o + RT\ln p_i\right)}{zF} \\
&= \frac{\sum v_i\mu_i^o}{zF} + \frac{RT}{zF}\ln\sum v_i\ln p_i \qquad (5)
\end{aligned}$$

$$\Delta\varphi = \Delta\varphi^O + \frac{RT}{zF}\ln\prod p_i^{v_i} \qquad (6)$$

For the *isothermal oxygen concentration cell*, the cell voltage U_{eq} or E (emf), the Nernst's equation is expressed by

$$-E = U_{eq} = \frac{RT}{4F}\ln\frac{p_{O_2}''}{p_{O_2}'} \qquad (7)$$

With $R = 8.321$ VAs/(mol K) and $F = 96{,}483$ As/mol, on one side air with 50 % r. h. ($p_{O_2}'' = 20923.6\,Pa$) and T = 800 °C, one yields

$$U/mV = -230.3 + 53.23\lg p_{O_2}/Pa \qquad (8)$$

More general with dry air as a reference for the volume concentration φ_{O_2},

$$\varphi(O_2)/vol\% = \exp\left(3.042 - \frac{U_{eq}/mV}{0.0215\,T/K}\right) \qquad (9)$$

is obtained. Oxygen concentration cells were firstly investigated by Peters and Möbius (1958) [1]. The analytical method with thermodynamic gas cells based on the Nernst's equation is called as gas potentiometry.

In reducing gases, which are in chemical equilibrium (e.g., H_2, H_2O; CO, CO_2; water gas), the electrode reactions of such *reaction cells* are expressed by

$$\begin{aligned}
&H_2O(g) + V_O^{\cdot\cdot}(YSZ) \\
&\quad + 2e^-(Pt) \rightleftarrows Oo\ (YSZ) + H_2(g) \qquad (10)
\end{aligned}$$

$$\begin{aligned}
&CO_2(g) + V_O^{\cdot\cdot}(YSZ) \\
&\quad + 2e^-(Pt) \rightleftarrows Oo\ (YSZ) + CO(g). \qquad (11)
\end{aligned}$$

The oxygen partial pressure is determined by the mass law constant K_p depending on the temperature. In the case of H_2, H_2O mixtures, the emf is obtained by the insertion of the temperature function of log K_p into the Nernst's equation [2, 3]:

$$\log K_p = 2.947 - 13008\,K/T \qquad (12)$$

$$\begin{aligned}
-E/mV &= U_{eq}/mV \\
&= 0.0496\frac{T}{K}\lg\left[K_p\left(\frac{p_{H_2O}}{p_{H_2}}\right)\right]^2 \qquad (13)
\end{aligned}$$

The Nernst's equation for reaction cells is obtained:

$$\begin{aligned}
-E/mV &= U_{eq}/mV = -1290.6 \\
&\quad + \left[0.2924 - 0.0992\log\left(\frac{p_{H_2O}}{p_{H_2}}\right)\right]T/K
\end{aligned}$$
$$(14)$$

For an easy handling of the equations, the oxygen partial pressure at the reference electrode is assumed to be 1 bar (or nearly 1 atm). Similar equations can be derived for other gas equilibria like CO, CO_2; water gas [1, 2].

These calculations are the basic of potentiometric sensors which are applied in broad field of industry and traffic. Main fields of application are the fast measuring of oxygen concentration in gases and liquid metals such as flue gases of combustion in steam boilers, in glass and ceramic making industries. By combination of sensor signals with stoichiometric and thermodynamic relations, a complex characterization of gas phases under equilibrium conditions is possible.

Moreover, dissolved oxygen in molten metals or glasses or oxygen generated by heterogeneous equilibrium can take part in the electrochemical reaction:

$$\frac{1}{2} O_2(dissolv.Fe, l) + V_O^{\cdot\cdot} (YSZ)$$
$$+ 2e^-(Fe) \rightleftarrows O_O (YSZ) \tag{15}$$

This reaction Eq. 15 is the basic principle of measurement of dissolved oxygen during the steel production [4]:

$$FeO(s) + V_O^{\cdot\cdot} (YSZ)$$
$$+ 2e^- (Pt) \rightleftarrows O_O (YSZ) + Fe(s) \tag{16}$$

Cell consisting of metal, metal oxide mixtures electrode, and a reference electrode can be used to determine thermodynamic values of Gibbs energy, enthalpy, and entropy firstly described by Kiukkola and Wagner [5]:

$$Fe, FeO, Pt|(ZrO_2)_{0.84}(Y_2O_3)_{0.08}|Pt, O_2(\varphi_{o_2}'') \tag{17}$$

For such measurements the oxygen partial pressure of the reference electrode should be low in order to avoid the oxygen permeation through the solid electrolyte.

Furthermore, also such reaction can be electrochemically active in which at least one partner is involved in other equilibrium reactions with oxygen, e.g.,

$$CO_2 + H_2 \rightleftarrows CO + H_2O \tag{18}$$

$$C + CO_2 \rightleftarrows 2\, CO \tag{19}$$

$$2\, NH_3 \rightleftarrows N_2 + 3H_2 \tag{20}$$

$$CH_3OH \rightleftarrows CO + 2\, H_2 \tag{21}$$

$$2\, NO \rightleftarrows N_2 + O_2 \tag{22}$$

Whether or not an electrochemical equilibrium can be established depends mainly on the temperature and the catalytic activity of the electrode material. Porous platinum is an active material and promotes the chemisorption of molecular oxygen on its surface as the initial step for the electrochemical reduction of oxygen:

$\frac{1}{2} O_2(g) \rightarrow O_{ads} (Pt)$	Chemisorption
$O_{ads} (Pt) \rightarrow O_{ads,tpb}$	Surface diffusion to triple phase boundary
$O_{ads} (Pt) + V_O^{\cdot\cdot} (YSZ)$ $+2e^- (Pt) \rightleftarrows Oo(YSZ)$	Charge transfer

On gold surface this reaction does not occur. This electrode material is preferred for so called mixed potential sensors which can be applied for the measurements of hydrocarbons and nitric oxides. Depending on electrode material, the thermodynamic window can be opened in a broad range of temperatures (400–1,500 °C) and oxygen partial pressures (10–10^{-20} bar). The thermodynamic expressions for different gas and solid state electrodes are given in detail by Möbius [2, 3].

The thermodynamic relations can be illustrated by the diagram shown in Fig. 2. The Nernst voltage (emf) is plotted vs. the temperature. In this diagram the thermodynamic existence of oxides (e.g., Fe, FeO) and the application fields for *potentiometric measurements* (sensors) are seen.

The measured emf can be easily expressed in change of the Gibbs free energy by

$$\Delta_R G = -z F U_{eq} = z F E$$

The mean applications for potentiometric (emf) measurements are:

- Analysis of trace gases in pure inert gases, e.g., oxygen or hydrogen in Ar or N_2 [1, 6]
- Potentiometric water gas analysis [2, 7]
- Measurements of oxygen in industrial flue gases, e.g., (see entry ► Combustion Control Sensors, Electrochemical) [8]
- Lambda probe for automotive catalysis [9]
- Gas analysis in flames, determination of burn-off, and optimization of burners [2, 10, 11] (see entry ► Gas Titration with Solid Electrolytes)
- Measurements of oxygen in high temperature processes, e.g., porcelain and ceramic fabrication [12, 13]

Solid Electrolytes Cells, Electrochemical Cells with Solid Electrolytes in Equilibrium,

Fig. 2 Nernst cell voltage (emf) versus temperature for galvanic cells using solid electrolyte [2]

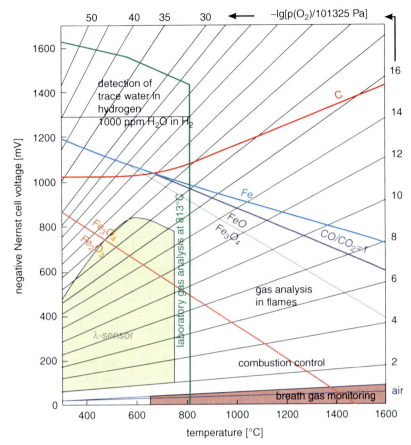

- In situ measurement in nitriding and nitrocarburizing gas mixtures for steel hardening [14, 15]
- Measurements of dissolved oxygen in molten metals [4, 16] (see entry ▶ Molten Steel, Measurement of Dissolved Oxygen).

Cross-References

▶ Solid Electrolytes
▶ Solid Oxide Fuel Cells, Introduction
▶ Solid Oxide Fuel Cells, Thermodynamics

References

1. Peters H, Möbius H-H (1958) Electrochemical investigation of equilibria $CO + 1/2\ O_2 \rightleftharpoons CO_2$ and $C + CO \rightleftharpoons 2CO$ (in German). Z phys Chem (Leipzig) 209:298–309
2. Möbius HH (1991) Solid state electrochemical potentiometric sensors for gas analysis. In: Göpel W, Hesse J, Zemel JN (eds) Sensors. VCH, Weinheim, p 1118
3. Möbius HH (1965) Oxide ion conducting solid electrolyte and their possibilities of application. Basic principles of gas potentiometry (in German). Z physik Chem (Leipzig) 230:396–412
4. Fischer WA, Janke D (1975) Metallurgic electrochemistry (in German). Stahleisen, Düsseldorf
5. Kiukkola K, Wagner C (1957) Measurement on galvanic cells involving solid electrolytes. J Electrochem Soc 104:379
6. Weissbart J, Ruka R (1961) Oxygen gauge. Rev Sci Instrum 32:593–595
7. Hartung R (1996) On the analysis of CO_2, H_2 and CO, H_2O-mixtures by water-gas potentiometry with solid electrolyte cells. Fresenius J Anal Chem 356:228–232
8. Shuk P, Bailey E, Guth U (2008) Zirconia oxygen sensor for the process application: state-of-the-art. Mod Sens Technol 90:174–184
9. Riegel J, Neuman H, Wiedenmann HM (2002) Exhaust gas sensors for automotive emission control. Solid State Ion 783:152–153
10. Harbeck W, Guth U (1990) Determination of burn-off of gas flames by means of gas potentiometric method (in German). Gas-Wärme-International (Essen) 39:10–24
11. Lorenz H, Tittmann K, Sitzki L, Trippler S, Rau H (1996) Gas-potentiometric method with solid

electrolyte oxygen sensors for the investigation of combustion. Fresenius J Anal Chem 356:215–220
12. Kämpfer K, Prescher E, Möbius HH (1991) The use of solid-electrolyte probes in porcelain production kilns. Ceramic forum International 68:126–131
13. Möbius HH, Handler T (1998) Control of gas composition enduring burning process to guarantee the colouring of ceramics (in German). Keram Z 50:918–933
14. Hartung R, Möbius HH, Teske K, Berg HJ, Böhmer S (1984) Potentiometric determination of H_2, H_2O partial pressure ratio in NH_3, H_2, H_2O, N_2 mixtures by means of solid electrolyte cells (in German). Z Chem (Leipzig) 24:447–448
15. Teske K, Berg HJ, Worm V, Böhmer S, Hartung R, Möbius HH (1985) Potentiometric determination of NH_3, H_2 partial pressure ratio in NH_3, H_2, H_2O, N_2 mixtures by means of solid electrolytes (in German). Z Chem (Leipzig) 25:95–96
16. Ullmann H (1993) Ceramic gas sensors, Akademie Verlag Berlin, pp 135–146

Solid Oxide Fuel Cells, Direct Hydrocarbon Type

Scott Barnett
Department of Materials Science and Engineering, Northwestern University, Evanston, IL, USA

Introduction

Solid oxide fuel cells (SOFCs) are highly efficient electrical generation devices [1]. An advantage of SOFCs is their ability to utilize abundant fuels such as natural gas with less fuel processing than other fuel cell types [2–4]. In particular, SOFCs utilize the CO present after hydrocarbon reforming, whereas most low-temperature fuel cells are poisoned by CO. This makes it possible to simplify the fuel processor by eliminating CO cleanup steps [4]. Furthermore, various studies have shown that SOFCs can operate effectively with hydrocarbons present in the fuel stream, further easing fuel-processing requirements. Thus, a key motivation for direct-hydrocarbon SOFCs is to achieve a lower-cost balance of plant. In addition, chemical processing applications require the SOFC to work in the presence of hydrocarbons. Examples include conversion of natural gas into syngas while at the same time producing electricity [5, 6] and ethylene synthesis from ethane [7]. There is also interest in electricity storage, where it may be desirable for the SOFC to intentionally produce methane [8].

In this review, a fairly broad definition of a "direct-hydrocarbon" fuel cell will be used – a SOFC in which a hydrocarbon-containing fuel stream is inlet directly into the anode compartment. Such SOFCs must have anodes that are able to catalytically or electro-chemically convert the hydrocarbon and operate stably without coking. This definition encompasses the fuel-processing strategies partial external reforming, internal reforming, and direct-hydrocarbon feed operation. The various direct-hydrocarbon SOFC strategies are discussed in section "Strategies For Hydrocarbon-Fueled SOFCs."

There is a major distinction between SOFCs operating with methane and those operating with higher hydrocarbons ranging from ethane to diesel fuel. Methane, discussed in section "Direct-Methane SOFCs," is a quite stable molecule and hence can be used in undiluted form in SOFCs without substantial coking. On the other hand, higher hydrocarbons, discussed in section "Higher Hydrocarbon Fuels," often cause substantial coke or tar deposits on the anode or fuel-compartment surfaces, even if they are balanced with substantial amounts of steam or carbon dioxide. Results on the performance of SOFC anodes with these various fuels will be described below, and the electrochemical and catalytic mechanisms by which SOFCs utilize different hydrocarbon fuels will be discussed. While most of this discussion will focus on the widely used Ni-based anodes, alternative anode materials will also be described, particularly their potential to work with hydrocarbons without coking. Finally, most hydrocarbons contain some level of impurities such as H_2S, which tends to poison anodes. Impurity effects are beyond the scope of the present review, and the interested reader is referred to excellent reviews on this topic [9, 10].

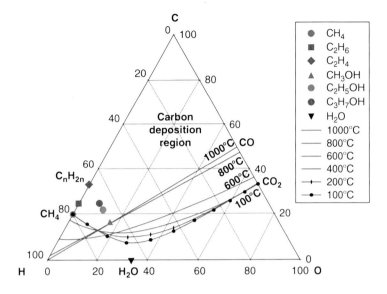

Solid Oxide Fuel Cells, Direct Hydrocarbon Type, Fig. 1 Coking/non-coking boundaries plotted on a C-H-O composition triangle, along with the positions of several relevant compounds (From Ref. [21])

Strategies for Hydrocarbon-Fueled SOFCs

The hydrocarbon fuel-processing method – ranging from external reforming to direct-hydrocarbon feed into the cell – is usually decided based on the hydrocarbon type and the power output desired [11]. SOFC stacks demonstrated to date have primarily utilized external reforming. Here, a separate fuel processor converts the hydrocarbon fuel – mixed with one or more of H_2O, CO_2, or air – into a H_2- and CO-rich mixture. The high-temperature heat from the SOFC is sufficient to maintain the reformer at temperature, an advantage over low-temperature fuel cells [3]. In some cases, the reforming process is intentionally incomplete. This leaves residual methane in the fuel, which is then reformed on the SOFC anode; note that the widely used Ni-based anodes exhibit quite good catalytic activity [12, 13]. This approach is advantageous because the endothermic reforming reaction helps to cool the stack [14–16]. Another possibility is direct on-anode internal reforming, where the fuel mixture is introduced into the anode compartment without prior reforming. A variant on this is to add a separate reforming catalyst in the anode compartment or directly on the anode [17–19]. If this reforming catalyst is physically separated from the anodes, the ability to transport heat from the stack to the reformer may be a significant limitation [20]. Finally, in direct-hydrocarbon utilization, a hydrocarbon-rich fuel is introduced directly into the anode compartment.

The most common strategy for maintaining stable coke-free operation is to maintain the fuel within the equilibrium non-coking range. The C-O-H composition triangle in Fig. 1 [21] shows that the coking/non-coking boundary is near a 1:1 C/O ratio at $\geq 800\ °C$ but varies considerably at lower temperature. Pure hydrocarbons lie well within the coking range and can only be made non-coking by addition of sufficient H_2O, CO_2, and/or O_2. On the other hand, coking may be suppressed kinetically in some cases.

The above hydrocarbon utilization methods have various advantages and disadvantages. External partial oxidation reforming has the advantages that the feed gas (air) does not have to be stored, the reformer is compact, the exothermic reaction helps to heat the incoming fuel, and it can be used with a wide range of hydrocarbons. Unfortunately, the achievable fuel efficiency is relatively low and the cell performance may be compromised by N_2 dilution of the fuel [22]. This is the method of choice for small-scale systems where efficiency is not critical and the heat is needed to maintain stack

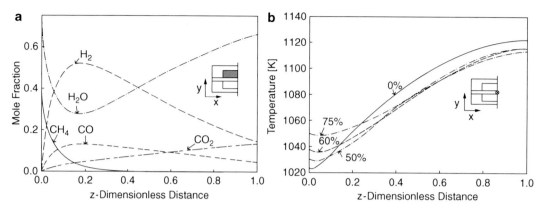

Solid Oxide Fuel Cells, Direct Hydrocarbon Type, Fig. 2 SOFC stack simulation results showing (**a**) fuel constitution during internal steam reforming and (**b**) temperature gradients for internal reforming with steam (0 %) or varying percent levels of exhaust gas recycle [16]

temperature. On-anode internal partial oxidation reforming has also been demonstrated [23].

For larger-scale systems where efficiency is important, H_2O and CO_2 reforming are utilized. Considerable fuel-processing hardware is required in the external reforming case, including a reformer, a heat exchanger to maintain temperature during endothermic reforming, and either a water tank and evaporator or a system for recycling the H_2O-CO_2-rich exhaust into the fuel stream [3, 4]. Note that the latter is a difficult proposition given the lack of high-temperature gas pumps. Internal reforming has the advantage of eliminating the external reformer. Although it has not been demonstrated at the stack level, considerable interest remains in more aggressive approaches where more or less pure hydrocarbons are introduced directly into the stack [24–27]. The motivation for this approach is the potential for achieving high efficiency combined with minimal fuel-processing balance of plant.

Direct-Methane SOFCs

Methane Internal Reforming

Internal reforming of methane and natural gas in SOFCs has been demonstrated, both with and without partial pre-reforming [12, 28–30]. External pre-reforming of natural gas is used to eliminate higher hydrocarbons that could cause coking. Reformer conditions (e.g., temperature) are controlled in order to produce a methane-rich gas that is then on-anode internally reformed [28]. This approach helped to enable a remarkable 60 % electrical efficiency in a 1.5 kW SOFC system [31].

Although internal reforming improves efficiency by helping to cool the stack, it can be challenging to maintain a reasonably uniform temperature and hydrogen content throughout the stack; severe cooling can occur near the fuel inlet if the reforming reaction is too fast [14, 16, 30]. Internal reforming stack modeling plays an important role in choosing pre-reformer and stack operating conditions [15, 16]. Simulation results show that fuel composition and temperature gradients can be reduced by increasing the degree of pre-reforming (but at the expense of efficiency) or by decreasing the reforming rate by, e.g., decreasing the cell operating temperature [12].

Figure 2 illustrates a 3D model prediction of the fuel-gas composition and cell temperature gradients predicted in the fuel-flow direction for a planar SOFC stack in a co-flow configuration [16]. Figure 2a shows that the H_2 content is lowest at the fuel inlet; this yields a low local cell current that, combined with the fast reforming reaction, leads to the low inlet temperature shown in Fig. 2b. The temperature and

Direct-Methane Utilization

An alternative to internal reforming is pure or nearly pure methane feed. This approach is sometimes termed "gradual internal reforming," which occurs via H_2O and CO_2 cell reaction products [32]. Conventional Ni-YSZ anode-supported SOFCs can be utilized with some modification. While the experimental work to date has been done with single cells, stack-level simulations have been reported [6, 26, 33].

An important measure to prevent coking on Ni-based anodes is the insertion of an inert porous "barrier layer," between the fuel flow and the anode. The barrier, composed of an oxide material with little tendency for coking, helps eliminate coking by reducing the methane partial pressure and trapping H_2O and CO_2 reaction products in the anode [6, 26, 33, 35]. Two designs have been utilized: a separate barrier layer inserted between the fuel flow and the SOFC anode and an integral barrier provided by an oxide cell support.

Separate barrier layers allow the use of unmodified SOFCs but raise concerns about increased concentration polarization [12]. However, it has been shown that the barrier layer is not needed beyond the first $\sim$30 % of the fuel-flow channel [6], beyond which the CH_4 content is low and the product H_2O and CO_2 content is sufficient that coking is not a problem. SOFCs with integral barrier layers include segmented-in-series designs [36] and anode-supported designs utilizing a conducting oxide, typically doped $SrTiO_3$, instead of Ni-YSZ [34, 37–39]. With no separate barrier, there is no concentration polarization penalty.

Figure 3 illustrates how the barrier layer, in this case the La-doped $SrTiO_3$ (SLT) support of the SOFC, helps to eliminate coking. The compositional gradients arise due to various processes: H_2 electrochemical oxidation to H_2O near the anode-electrolyte interface, reforming reactions that consume CH_4, CO_2, and H_2O while producing H_2 and CO, shift reactions ($CO + H_2O = CO_2 + H_2$), and gas diffusion. For example, CO_2 and H_2O are produced near the electrolyte-anode interface and diffuse out of the anode into the fuel stream. For Ni-YSZ supports (Fig. 3a), the reforming reactions reduce CO_2 and H_2O to very low mole fractions on the anode-compartment side of the anode, where the CH_4 content is still relatively high. Coking can readily occur here in the presence of Ni. For SLT supports (Fig. 3b), relatively small amounts of H_2 and CO are produced, and H_2O and CO_2 mole fractions are relatively high, because reforming is limited to the thin Ni-cermet layer. This, along with the relatively low CH_4 fraction within the Ni-YSZ layer, leads to non-coking conditions. While the SLT support is exposed to coking conditions (higher CH_4 and lower H_2O/CO_2 partial pressures), it is difficult for solid carbon to nucleate on oxide surfaces [40, 41].

Figure 4 shows stability tests of Ni-YSZ and SLT-supported cells. For the SLT-supported cell (Fig. 4a), stable operation is observed even for a current density as low as 0.1 A/cm^2, because a relatively small amount of H_2O/CO_2 production is sufficient to avoid coking. For the Ni-YSZ-supported cell (Fig. 4b), a current density >1.4 A/cm^2 is required to maintain stable operation [42]. The SLT-supported cells also showed good stability in direct natural gas utilization [39]. The ability to redox cycle these cells without degradation [39] is likely important to allow periodic oxidative removal of carbon and contaminants after long-term hydrocarbon operation.

Since stable coke-free operation results because of the cell current, a cessation of cell current will cause coking that could permanently damage the stack. Figure 5 shows the results of different length current interruptions for a SOFC operated in pure dry methane at 1.8 A/cm^2 at 750 °C. For the 1.5 and 6 min interruptions, the voltage exceeded the pre-interruption value immediately after current resumption, before

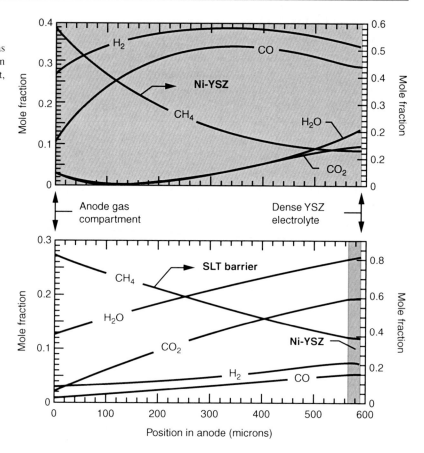

Solid Oxide Fuel Cells, Direct Hydrocarbon Type, Fig. 3 Predicted gas constitution versus position in the SOFC anode support, for a current density of 1 A/cm². The top shows a cell with a conventional NiYSZ support and the bottom a $Sr_{0.8}La_{0.2}TiO_3$ support. Both cell types had a NiYSZ anode functional layer, YSZ electrolyte, and LSMYSZ ($LSM = La_{0.8}Sr_{0.2}MnO_3$) cathode (From Ref. [34])

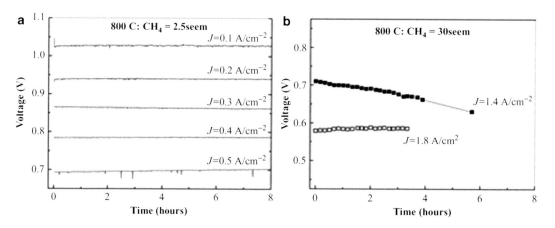

Solid Oxide Fuel Cells, Direct Hydrocarbon Type, Fig. 4 Cell voltage versus time at fixed current densities from SLT-supported. (a) and NiYSZ-supported (b) SOFCs tested at 800 °C (From Ref. [34])

gradually relaxing to the pre-interruption value. This voltage increase was explained as an effect of minor carbon deposition on the anode [43, 44], and the subsequent voltage decrease was explained by carbon removal in the non-coking gas composition achieved during high-current operation [44]. After the 10 min interruption, the voltage remained low and never recovered

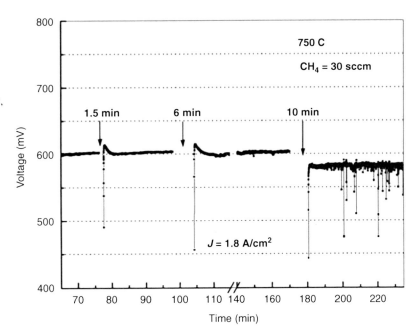

Solid Oxide Fuel Cells, Direct Hydrocarbon Type, Fig. 5 Voltage versus time for a NiYSZ anode-supported SOFC with barrier layer, during current interruptions of 1.5, 6, and 10 min durations (From Ref [45])

to the pre-interruption value, suggesting that there was sufficient carbon deposition, presumably causing Ni-YSZ micro-cracking [42], to permanently degrade the cell. These results suggest that anode damage can be avoided in direct-methane stacks by designing the SOFC system to allow flushing of the methane from the anode compartment within in a few minutes of a current interruption.

In practice, SOFC systems will operate on natural gas fuel (containing several percent of higher hydrocarbons) [46]. If these higher hydrocarbons are not removed by a pre-reforming step, they exacerbate coking compared to pure methane [12, 29]. In recent studies, single Ni-YSZ anode-supported SOFCs with barrier layers [29] and SLT-supported SOFCs [39] were tested using surrogate natural gas (5 % ethane and 2.5 % propane). Similar to the pure methane case, stable operation was achieved in both cases with sufficiently high cell current density, with the SLT-supported cell being stable to <0.2 A/cm^2. No coking was observed on the cells, but carbonaceous deposits were observed in the anode compartment due to the higher hydrocarbons. The addition of ~33 % air to the natural gas eliminated these deposits [29].

Such air additions as well as CO_2 additions were shown to also reduce the cell current density required to maintain stable operation [45].

The above results were for SOFC button cells, and more work is needed to demonstrate direct-methane utilization in stacks. One step towards this goal is simulation of stack operation. Such results indicate that in a tubular stack with a barrier layer over the first 30 % of the fuel-flow field, the anode-gas composition is maintained in the thermodynamically non-coking regime. Furthermore, the direct-methane feed case yielded overall stack performance nearly as good as the anode recycle case [33].

Electrochemical Methane Partial Oxidation

The ability to work stably with methane or natural gas fuel allows the application of SOFCs for electrochemical partial oxidation, where both electricity and a chemical product – syngas – are produced [5, 47]. Measured outlet gas constitution matches well with that predicted at equilibrium, due to the good catalytic activity of the Ni anode or an added catalyst. For a SOFC operated at 750 °C with current and methane flow rate set to yield an O^{2-}/CH_4 ratio of ~1.2, the methane conversion is nearly 100 % and the

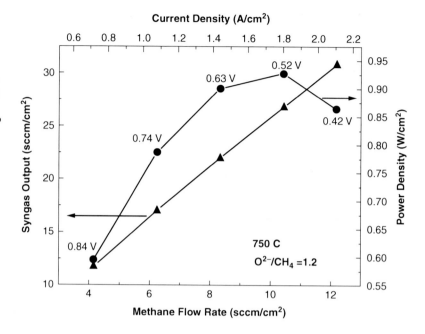

Solid Oxide Fuel Cells, Direct Hydrocarbon Type, Fig. 6 Syngas output and power density versus methane flow rate (and cell current) with the ratio O^{2-}/CH_4 maintained constant at 1.2. Cell operating voltages are also given (From Ref. [5])

outlet gas is rich in H_2 and CO. As shown in Fig. 6, the power density is competitive with conventional SOFCs, and the syngas output is competitive with mixed-conducting ceramic membrane reactors [48]. Modeling of an EPOx stack utilizing a barrier layer near the fuel inlet showed that coking conditions are avoided and that current density and temperature gradients are relatively small [6].

Oxide-Based Anodes

Besides the Ni-based anodes discussed above, a number of anode materials have been investigated. A review of this field is beyond the scope of this article, but the interested reader is referred to recent review articles [49, 50]. Results on direct-methane operation of SOFCs with oxide anodes generally show little tendency for coking, but performance results are mixed. On the one hand, some anodes have shown quite high polarization resistance with CH_4 fuel [51]. This result is not particularly surprising, given the relative stability of CH_4 and the poor catalytic activity of many oxides. Since the oxide does not provide a pathway for direct-methane oxidation, reforming, or coking, the fuel is essentially inert. On the other hand, a number of oxides have shown reasonable polarization resistance in CH_4, although higher than in H_2 fuel. The exact electrochemical oxidation mechanisms have not been elucidated. A number of reviews of anode materials are available [11, 49, 50, 52–54].

Higher Hydrocarbon Fuels

Three approaches have been utilized for "direct" utilization of higher hydrocarbon fuels in SOFCs [11]. First, the higher hydrocarbon is partially externally reformed, yielding a methane-rich mixture [55] that is then internally reformed in the SOFC [22, 56–58]. For the SOFC, this is similar to methane internal reforming as discussed above and is not discussed further below. Second, the hydrocarbon is mixed with H_2O, CO_2, and/or air in order to allow for internal reforming. Third, the pure hydrocarbon is introduced directly into the anode compartment, relying on a non-coking anode composition to avoid deleterious carbonaceous deposits.

Internal Reforming

Internal reforming of hydrocarbons ranging from propane to dodecane has been successfully

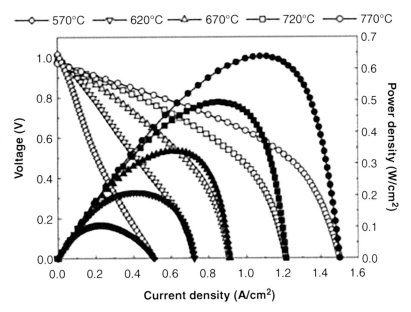

Solid Oxide Fuel Cells, Direct Hydrocarbon Type, Fig. 7 Voltage and power density versus current density for a Ni-YSZ | YSZ | LSCF-GDC cell, tested in 5 % isooctane/9 % air/86 % CO_2 at the anode and ambient air at the cathode at different cell temperatures (From Ref. [19])

utilized in SOFCs. An example of on-anode reforming is propane internal partial oxidation using air as the oxygen source. Partial oxidation produces substantial hydrogen via reforming on either a Ni-YSZ anode [59] or a Ru-based catalyst layer on the fuel-flow side of the anode [18]. The catalyst layer approach was shown to work in a tubular SOFC stack with partial oxidation of various fuels including propane [60]. Single cells also worked well using internal partial oxidation of *iso*-octane fuel, although in this case a Ru-based catalyst, supported on a separate porous oxide layer placed against the cell's anode side, was necessary to avoid coking on the Ni-based anode [17]. Disadvantages of the above partial oxidation reforming approach are the relatively low fuel efficiency and increased anode polarization resistance due to dilution of the fuel with nitrogen. H_2O and CO_2 reforming can yield higher efficiency [22] with no fuel dilution, with the disadvantage being the additional system requirements as discussed in section "Strategies For Hydrocarbon-Fueled SOFCs." Furthermore, in implementing internal reforming of *iso*-octane with H_2O and CO_2, it is necessary to add a small amount of air to the reformate mixture in order to maintain stable operation without coking, even though the fuel composition was outside the thermodynamic coking range in Fig. 1 [17, 19, 61].

This result may be explained by the observation that tar deposits from hydrocarbons are only removed by oxygen, not H_2O, at SOFC operating temperatures [62]. Figure 7 shows the performance of a typical anode-supported cell [19]. Because of the relatively low hydrogen content in the reformed fuel, open-circuit voltages are ~1.0 V and concentration polarization is pronounced, but power densities are only slightly lower than when the cells were run with pure hydrogen fuel.

Direct-Hydrocarbon Operation

In the above results, stable coke-free SOFC operation was enabled by balancing the hydrocarbon fuel with H_2O, CO_2, or O_2 in order to achieve a composition where solid carbon is not stable (Fig. 1). It has also been demonstrated that SOFCs can be operated with reasonable stability in a range of pure hydrocarbons. With Ni-based anodes, it is well known that carbon fibers form that can damage the anode [24, 62, 63]. Furthermore, the presence of steam may not prevent this process, even in amounts beyond those normally expected to prevent coking [24]. Thus, direct operation with higher hydrocarbons requires replacement of Ni with other electronically conducting oxides or a metal such as Cu. Indeed, when operated with higher hydrocarbons such as *n*-butane, polyaromatic compounds (tars) formed

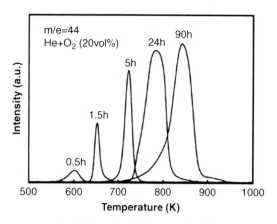

Solid Oxide Fuel Cells, Direct Hydrocarbon Type, Fig. 8 Temperature-programmed oxidation curves in 20 vol % O_2 + He with samples of porous YSZ/CeO$_2$ exposed to butane at 973 K for different times (From Ref. [62])

in vapor phase reactions are deposited in the anode; the formation rate is negligible below 973 K but becomes large above. For cells operated at 973 K, the moderate amounts that form over short times can actually improve performance by providing additional electronic conductivity [63]. However, for long enough times, increasing amounts of tar deposits and the compounds become more stable, as indicated by the increased area and peak temperature of the temperature-programmed oxidation peaks after varying hydrocarbon exposure times in Fig. 8 [62]. Tar deposition will ultimately block active sites on the anode and degrade performance. The deposits are readily removed at SOFC operating temperatures by oxygen exposure (Fig. 8), but are not removed by H_2O until much higher temperatures. Thus, such SOFC would have to be regenerated periodically by a redox cycle. Also, such cells have not yet yielded power densities comparable to those with Ni-based anodes, in part because the operating temperature must be maintained $\leq$973 K.

Future Directions

The search for new anode materials continues, with composites including nanoscale structures [53, 64, 65] constituting a promising avenue – it is challenging to find a single material that provides all of the properties desired. An approach that seems likely to bear fruit is the oxide anode-supported cell; besides its functionality for non-coking hydrocarbon operation and redox cycling, there is potential to improve electrocatalytic properties with new anode functional layer materials and to introduce materials with desired catalytic properties in a fuel-flow-side catalyst layer.

The direct internal reforming of pre-reformed natural gas has been demonstrated in SOFC systems, but the other approaches described above have mostly been implemented only in button cells. Thus, there is a need for stack-level simulations and demonstrations to more fully vet these approaches.

Acknowledgement The author gratefully acknowledges financial support from the Department of Energy Basic Energy Sciences DE-FG02-05ER46255.

Cross-References

▶ Solid Oxide Fuel Cells, History
▶ Solid Oxide Fuel Cells, Introduction
▶ Solid Oxide Fuel Cells, Thermodynamics

References

1. Minh NQ (1993) Ceramic fuel-cells. J Am Ceram Soc 76:563
2. Atkinson A et al (2004) Advanced anodes for high-temperature fuel cells. Nat Mater 3:17
3. Song CS (2002) Fuel processing for low-temperature and high-temperature fuel cells - challenges, and opportunities for sustainable development in the 21st century. Catal Today 77:17
4. van den Oosterkamp PF (2006) Critical issues in heat transfer for fuel cell systems. Energy Conversion and Management 47:3552
5. Pillai MR, Bierschenk DM, Barnett SA (2008) Electrochemical partial oxidation of methane in solid oxide fuel cells: effect of anode reforming activity. Catal Lett 121:19
6. Zhu H, Kee RJ, Pillai MR, Barnett SA (2008) Modeling electrochemical partial oxidation of methane for cogeneration of electricity and syngas in solid-oxide fuel cells. J Power Sources 183:143
7. Fu XZ et al (2011) CO2 Emission free co-generation of energy and ethylene in hydrocarbon SOFC reactors with a dehydrogenation anode. Physical Chemistry Chemical Physics : PCCP 13:19615

8. Bierschenk DM, Wilson JR, Barnett SA (2011) High efficiency electrical energy storage using a methane–oxygen solid oxide cell. Energy Environ Sci 4:944

9. Zha S, Cheng Z, Liu M (2007) Sulfur poisoning and regeneration of Ni-based anodes in solid oxide fuel cells. J Electrochem Soc 154:B201

10. Cheng Z et al (2011) From Ni-YSZ to sulfur-tolerant anode materials for SOFCs: electrochemical behavior, in situ characterization, modeling, and future perspectives. Energy Environ Sci 4:4380

11. Cimenti M, Hill JM (2009) Direct utilization of liquid fuels in SOFC for portable applications: challenges for the selection of alternative anodes. Energies 2:377

12. Mogensen D, Grunwaldt JD, Hendriksen PV, Dam-Johansen K, Nielsen JU (2011) Internal steam reforming in solid oxide fuel cells: status and opportunities of kinetic studies and their impact on modelling. J Power Sources 196:25

13. Hecht ES et al (2005) Methane reforming kinetics within a Ni-YSZ SOFC anode. Appl Catal A General 295:40

14. Janardhanan VM, Deutschmann O (2007) Modeling of solid-oxide fuel cells. Zeitschrift für Physikalische Chemie 221:443

15. Aguiar P, Adjiman CS, Brandon NP (2005) Anode-supported intermediate-temperature direct internal reforming solid oxide fuel cell. J Power Sources 147:136

16. Nikooyeh K, Jeje AA, Hill JM (2007) 3D modeling of anode-supported planar SOFC with internal reforming of methane. J Power Sources 171:601

17. Zhan Z, Barnett SA (2006) Operation of ceria-electrolyte solid oxide fuel cells on iso-octane–air fuel mixtures. J Power Sources 157:422

18. Zhan Z, Barnett SA (2005) Use of a catalyst layer for propane partial oxidation in solid oxide fuel cells. Solid State Ionics 176:871

19. Zhan Z, Barnett SA (2005) An Iso-octane fueled solid oxide fuel cell. Science 308:844

20. Brus G, Szmyd JS (2008) Numerical modelling of radiative heat transfer in an internal indirect reforming-type SOFC. J Power Sources 181:8

21. Sasaki K, Teraoka Y (2003) Equilibria in fuel cell gases - II. The C-H-O ternary diagrams. J Electrochem Soc 150:A885

22. Botti JJ (2003) Paper presented at the SOFC VIII, Paris

23. Shao Z et al (2005) A thermally self-sustained micro solid-oxide fuel-cell stack with high power density. Nature 435:795

24. McIntosh S, Gorte RJ (2004) Direct hydrocarbon solid oxide fuel cells. Chem Rev 104:4845

25. Park SD, Vohs JM, Gorte RJ (2000) Direct oxidation of hydrocarbons in a solid-oxide fuel cell. Nature 404:265

26. Zhu HY, Colclasure AM, Kee RJ, Lin YB, Barnett SA (2006) Anode barrier layers for tubular solid-oxide fuel cells with methane fuel streams. J Power Sources 161:413

27. Lin YB, Zhan ZL, Liu J, Barnett SA (2005) Direct operation of solid oxide fuel cells with methane fuel. Solid State Ionics 176:1827

28. Ahmed K, Seshadri P, Ramprakash Y, Jiang SP, Foger K (1997) Internal steam reforming of partially pre-reformed natural gas in SOFC stacks. Proceedings of the fifth international symposium on solid oxide fuel cells (SOFC-V), vol **97**, p 228

29. Biershcenk D, Pillai M, Lin Y, Barnett S (2010) Effect of ethane and propane in simulated natural Gas on the operation of Ni-YSZ anode supported solid oxide fuel cells. Fuel Cells 10:1129

30. Peters R, Riensche E, Cremer P (2000) Pre-reforming of natural gas in solid oxide fuel-cell systems. J Power Sources 86:432

31. Payne R, Love J, Kah M (2009) Generating electricity at 60% electrical efficiency from 1–2 kW SOFC products. SOFC-XI, ECS Trans 25(231)

32. Klein J-M, Hénault M, Roux C, Bultel Y, Georges S (2009) Direct methane solid oxide fuel cell working by gradual internal steam reforming: analysis of operation. J Power Sources 193:331

33. Zhu H, Colclasure, AM, Kee RJ, Lin Y, Barnett SA. (2006) Paper presented at the seventh european SOFC forum, Lucerne, Switzerland

34. Pillai MR et al (2008) Solid oxide fuel cell with oxide anode-side support. Electrochem Solid State Lett 11: B174

35. Lin Y, Zhan Z, Barnett SA (2006) Improving the stability of direct-methane solid oxide fuel cells using anode barrier layers. J Power Sources 158:1313

36. Pillai MR, Gostovic D, Kim I, Barnett SA (2007) Short-period segmented-in-series solid oxide fuel cells on flattened tube supports. J Power Sources 163:960

37. Ma Q, Tietz F, Leonide A, Ivers-Tiffée E (2010) Anode-supported planar SOFC with high performance and redox stability. Electrochem Commun 12:1326

38. Ma Q, Tietz F, Sebold D, Stöver D (2010) Y-substituted SrTiO3–YSZ composites as anode materials for solid oxide fuel cells: interaction between SYT and YSZ. J Power Sources 195:1920

39. Pillai MR, Kim I, Bierschenk DM, Barnett SA (2008) Fuel-flexible operation of a solid oxide fuel cell with Sr0.8La0.2TiO3 support. J Power Sources 185:1086

40. Marina OA, Canfield NL, Stevenson JW (2002) Thermal, electrical, and electrocatalytical properties of lanthanum-doped strontium titanate. Solid State Ionics 149:21

41. Marina OA, Pederson LR (2002). In: Huijismans J (ed) The fifth european solid oxide fuel cell forum (European Fuel Cell Forum, Lucerne, Switzerland, 2002), vol 1, pp 481–489

42. Lin Y, Zhan Z, Liu J, Barnett SA (2005) Direct operation of solid oxide fuel cells with methane fuel. Solid State Ionics 176:1827

43. Liu J, Barnett SA (2003) Operation of anode supported solid oxide fuel cells on methane and natural gas. Solid State Ionics 158:11
44. Mallon C, Kendall K (2005) Sensitivity of nickel cermet anodes to reduction conditions. J Power Sources 145:154
45. Pillai M, Lin Y, Zhu H, Kee RJ, Barnett SA (2010) Stability and coking of direct-methane solid oxide fuel cells: effect of CO2 and air additions. J Power Sources 195:271
46. Blazek C, Grimes J, Freeman P (1994). In: Preprints of the 207th ACS annual meeting Div of fuel chem. (San Diego, California, 1994), vol 39, pp 476
47. Zhu HY, Kee RJ, Pillai MR, Barnett SA (2008) Modeling electrochemical partial oxidation of methane for cogeneration of electricity and syngas in solid-oxide fuel cells. J Power Sources 183:143
48. Bouwmeester HJM (2003) Dense ceramic membranes for methane conversion. Catal Today 82:141
49. Cowin PI, Petit CTG, Lan R, Irvine JTS, Tao SW (2011) Recent progress in the development of anode materials for solid oxide fuel cells. Adv Energy Mater 1:314
50. Tsipis EV, Kharton VV (2011) Electrode materials and reaction mechanisms in solid oxide fuel cells: a brief review. III. Recent trends and selected methodological aspects. J Solid State Electrochem 15:1007
51. Haag J, Bierschenk DM, Barnett SA, Poppelmeier KR (2011) Structural, chemical, and electrochemical characteristics of LaSr2Fe2CrO9-<delta>−based solid oxide fuel cell anodes. Solid state ionics (in press)
52. Tao S, Irvine JTS (2004) Discovery and characterization of novel oxide anodes for solid oxide fuel cells. Chem Rec 4:83
53. Gorte RJ, Vohs JM (2009) Nanostructured anodes for solid oxide fuel cells. Curr Opin Colloid Interface Sci 14:236
54. Jiang SP, Chan SH (2004) A review of anode materials development in solid oxide fuel cells. J Mater Sci 39:4405
55. Timmermann H, Sawady W, Reimert R, Ivers-Tiffée E (2010) Kinetics of (reversible) internal reforming of methane in solid oxide fuel cells under stationary and APU conditions. J Power Sources 195:214
56. Ahmed K, Gamman J, Foger K (2002) Demonstration of LPG-fueled solid oxide fuel cell systems. Solid State Ionics 152:485
57. Ahmed K, Foger K (2003) Thermodynamic analysis of diesel reforming options for SOFC systems. Elec Soc S 2003:1240
58. Ahmed K, Foger K (2010) Fuel processing for high-temperature high-efficiency fuel cells. Ind Eng Chem Res 49:7239
59. Zhan Z, Jiang L, Scott AB (2004) Operation of anode supported solid oxide fuel cells on propane-Air fuel mixtures. Appl Catal A Gen 262:255
60. Cheekatamarla P, Finnerty C, Cai J (2008) Internal reforming of hydrocarbon fuels in tubular solid oxide fuel cells. Int J Hydrogen Energy 33:1853
61. Zhan ZL, Barnett SA (2006) Solid oxide fuel cells operated by internal partial oxidation reforming of iso-octane. J Power Sources 155:353
62. He H, Vohs JM, Gorte RJ (2005) Carbonaceous deposits in direct utilization hydrocarbon SOFC anode. J Power Sources 144:135
63. Kim T et al (2006) A study of carbon formation and prevention in hydrocarbon-fueled SOFC. J Power Sources 155:231
64. Kobsiriphat W et al (2010) Nickel- and ruthenium-doped lanthanum chromite anodes: effects of nano-scale metal precipitation on solid oxide fuel cell performance. J Electrochem Soc 157:B279
65. Kurokawa H, Yang L, Jacobson CP, De Jonghe LC, Visco SJ (2007) Y-doped SrTiO3 based sulfur tolerant anode for solid oxide fuel cells. J Power Sources 164:510

Solid Oxide Fuel Cells, History

Subhash C. Singhal
Pacific Northwest National Laboratory,
Richland, WA, USA

Introduction

A solid oxide fuel cell (SOFC) converts chemical energy of a fuel (such as hydrogen or methane) to electricity through a series of electrochemical reactions; no combustion process is involved. As a result, fuel cell efficiencies are not limited by Carnot efficiencies since the electrical work is directly converted from a substantial fraction of the enthalpy associated with the electrochemical oxidation of the fuel to water and/or carbon dioxide. Hence, fuel cells can deliver higher electrical conversion efficiencies when compared with traditional technologies such as coal-fired power plants and electrical generators based on internal combustion engines. In addition to high efficiency, SOFC allows the use of a variety of fuels ranging from hydrogen to hydrocarbons, and secondly, SOFC produces significant amount of exhaust heat, which can be used in combined heat and power systems (CHP). These fuel cells

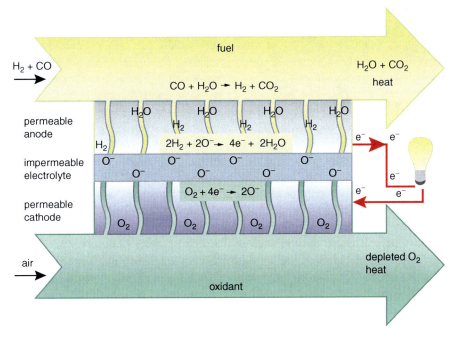

Solid Oxide Fuel Cells, History, Fig. 1 Operating principle of a solid oxide fuel cell

also produce no or very low levels of SO_x and NO_x emissions.

A typical SOFC is shown in Fig. 1. It essentially consists of two porous electrodes, separated by a dense, oxide ion-conducting electrolyte. Oxygen gas molecules on the cathode side react with incoming electrons coming from the external circuit to form oxygen ions, which migrate through the oxide ion-conducting electrolyte to the anode. At the anode, oxide ions react with H_2 or other fuels to form H_2O (and/or CO_2), liberating electrons, which flow from the anode through the external circuit to the cathode to produce electricity. Provided that both fuels and oxygen are supplied constantly, the continuous electrochemical reactions can steadily generate electricity.

Historical Developments

History of solid oxide electrolytes and of SOFC development in early years has been nicely summarized by Möbius [1]. Interest in solid ion conductors first arose in connection with the development of electric lighting devices. In 1897, Walther Nernst suggested [2, 3] that a solid electrolyte in the form of a thin rod could be made electrically conducting by means of an auxiliary heating appliance and then kept glowing by the passage of an electric current. At first Nernst mentioned only "lime, magnesia, and those sorts of substances" as appropriate conductors; later investigations led to his observation "that the conductivity of pure oxides rises very slowly with temperature and remains relatively low, whereas mixtures possess an enormously much greater conductivity." Subsequently, many mixed oxides exhibiting high conductivity at elevated temperatures, including the particularly favorable composition 85 % zirconia and 15 % yttria [4], the so-called Nernst mass, were identified. Nernst was convinced that his filaments were ionic conductors, and he assumed that in yttria stabilized zirconia (YSZ), the yttria provided the necessary charge carriers [3]. However, it was not until 1943 that Carl Wagner [5] recognized the existence of vacancies in the anion sublattice of mixed oxide solid solutions and explained the conduction mechanism of the Nernst glowers. We now know that Nernst lamp filaments were indeed oxide ion conductors and the platinum contacts behaved as air electrodes. Nernst lamps could thus be thought of as the first solid electrolyte cells.

Solid oxide fuel cells (SOFCs) have come a long way to become practical power generation devices since the initial experiments by Nernst. The first conceptual SOFC is believed to have been demonstrated in 1937 by Bauer and Preis [6]. However, the more focused studies on solid oxide electrolytes and SOFCs only began after the pioneering 1943 work by Carl Wagner who attributed the electrical conductivity in mixed oxides such as doped ZrO_2 to the presence of oxygen vacancies [6]. In 1957, Kiukkola and Wagner reported thermodynamic investigations on metal/metal oxide systems, for the first time using CaO-stabilized ZrO_2 (especially $Zr_{0.85}Ca_{0.15}O_{1.85}$) as solid electrolyte [7]. This paper of Kiukkola and Wagner [7] stimulated many activities in various parts of the world in the field of solid-state electrochemistry using solid oxide electrolytes. For example, Weissbart and Ruka used $Zr_{0.85}Ca_{0.15}O_{1.85}$ for the measurement of oxygen concentration in a gas phase using a high-temperature galvanic cell [8]. After 1960, a rapidly increasing number of applications for patents and of papers concerning SOFCs appeared in several countries. In the USA, in 1962, Ruka and other scientists at Westinghouse Electric Corporation published a paper titled simply as "A Solid Electrolyte Fuel Cell" [9]. This initial effort became the foundation of Westinghouse's cathode-supported tubular SOFCs; based on this cell design, Westinghouse successfully produced and tested several 5–250 kW size SOFC power systems from mid-1980s to early 2000s.

In Europe, in 1958, Palguyev and Volchenkova published conductivity measurements on $3ZrO_2.2CeO_2$ + 10 wt% CaO and other systems [10]. From 1960 onwards, under the leadership of Karpachov, results of a broadly based research program on cells with solid oxide electrolytes appeared from the Ural branch of the Academy of Sciences of the USSR [11]. Tannenberger et al. at the Battelle Institute in Geneva presented a thin film fuel cell concept in a 1962 patent [12]. Also in 1962, Sandstede from the Battelle Institute in Frankfurt gave the first report on the use of hydrocarbons as fuel in solid oxide cells [13]. At about the same time, fuel cell work was started in France by Kleitz [14], and in Britain, a patent was filed in 1963 [15] to form fuel cells by depositing layers on a porous metallic carrier. In Japan, Takahashi published in 1964 his first results obtained on fuel cells with solid oxide electrolytes [16].

One focal point of work on solid oxide fuel cells during this early period was the development of electrode and interconnection materials. An early problem was poor adhesion of the anode layers to the electrolyte. Spacil as early as 1964 found the now well-known solution of using nickel closely mixed with solid electrolyte material as the anode [17]. It was considerably more difficult to find a suitable inexpensive cathode material. Uranium oxide and indium oxide with different additives were proposed as cathode materials in 1960s, but electronically conducting perovskites soon began to dominate the development for both cathode and interconnect [18]. The use of $La_{1-x}Sr_xCoO_3$ for the air electrode of solid oxide fuel cells marked the beginning, followed in 1967 by recommendations of $PrCoO_3$ and of mixtures of the oxides of Pr, Cr, Ni, and Co. Strontium-doped lanthanum chromite was proposed by Meadowcroft in 1969 [19], which was later used as the interconnection material in Westinghouse SOFCs. These materials were intensively studied in the research laboratories of Brown Boveri, where under the leadership of Rohr from 1964, solid electrolyte fuel cells and oxygen sensors were investigated. Between 1969 and 1973, more than 100 oxide substances were synthesized and tested as electrode materials for SOFCs [20]. $LaNiO_3$ doped with Bi_2O_3 and $LaMnO_3$ doped with SrO proved to be particularly suitable, and since 1973, Sr-doped $LaMnO_3$ has been extensively used in SOFCs by commercial developers worldwide.

Since the mid-1980s, several SOFC materials and designs have been explored [21, 22]; in particular, anode-supported planar SOFCs have become quite popular because of performance

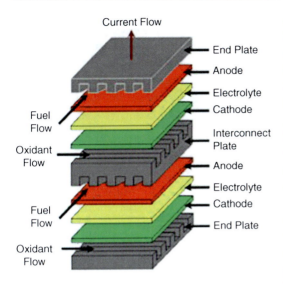

Solid Oxide Fuel Cells, History, Fig. 2 Planar solid oxide fuel cell design

and cost considerations. Today, SOFCs in many different designs and containing different cell materials are produced for power generation in small (few watts) to large (several hundred kWs) sizes. These are reviewed in the following sections.

SOFC Designs

The three active components of the cell, the electrolyte and the two electrodes, can be configured in a number of different geometric ways. The two most common designs are planar and tubular and their many variants.

Planar SOFC Design

In the planar design, a series of cell components are configured as thin, flat plates, then electrically connected to build up desirable electrochemical performance. A schematic of a generic planar SOFC design is shown in Fig. 2. The planar cells can be either electrolyte supported, electrode supported, or metal supported. Each of these designs can also have a number of interesting variants; for example, the planar SOFC may be in the form of a circular disk fed with fuel from the central axis, or it may be in the form of a square or rectangular plate fed from the edges. Planar designs offer several potential advantages, including simpler and less expensive manufacturing processes and higher power densities than tubular cells described in the next section. However, planar designs need high-temperature gastight seals between the components in the SOFC stack; such seals are not necessary with tubular cells. Seal development remains as the most challenging area in successfully commercializing planar SOFCs.

The initial planar SOFC configurations employed a thick electrolyte as the support, which required an operating temperature often higher than 900 °C. Advances in ceramic processing have allowed fabrication of thin electrolytes, 10 μm or thinner, by low-cost conventional ceramic processing techniques such as tape casting, tape calendaring, slurry sintering, screen printing, or by plasma spraying. As a consequence, anode-supported planar cell stacks have been extensively fabricated and tested by a number of developers for long-term operation.

Tubular SOFC Design

Tubular SOFCs may be of a large diameter (>15 mm), or of much smaller diameter (<5 mm), the so-called microtubular cells. Also, the tubes may be flat and joined together to give higher power density and easily printable surfaces for depositing the electrode layers. In a typical tubular SOFC, as illustrated by the Siemens Westinghouse design in Fig. 3, the cell tube is porous doped lanthanum manganite fabricated by extrusion/sintering and is closed at one end. The cell components, dense YSZ electrolyte, porous Ni-YSZ anode, and doped lanthanum chromite interconnect in the form of thin layers, were initially deposited by an electrochemical vapor deposition process [23] but later by a more cost-effective and automation-capable atmospheric plasma spraying process.

The single biggest advantage of tubular cells over planar cells is that they do not require any high-temperature seals to isolate oxidant from the fuel, and this makes performance of tubular cell stacks very stable over long periods of times

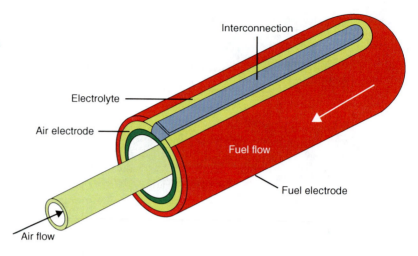

Solid Oxide Fuel Cells, History, Fig. 3 Tubular solid oxide fuel cell design

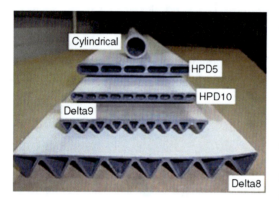

Solid Oxide Fuel Cells, History, Fig. 4 Alternate geometry tubular cells investigated by Siemens Westinghouse

(several years). However, their areal power density is much lower (about 0.2 W/cm^2) compared to planar cells (up to 2 W/cm^2 for single cells and at least 0.5 W/cm^2 for stacks), and manufacturing costs higher. The volumetric power density is also lower for tubular cells than for planar cells. To increase the power density and reduce the physical size and cost of tubular SOFC stacks, alternate tubular geometry cells have been investigated. Such alternate geometry cells combine all the advantages of the tubular SOFCs, such as not requiring high-temperature seals, while providing higher areal and volumetric power densities. Figure 4 shows several of such designs investigated by Siemens Westinghouse; the performance of these cells is higher than that of cylindrical tubular cells but still lower than that of anode-supported planar cells.

SOFC Applications and Power Systems

Because of their superior electrical efficiency and fuel flexibility, SOFC-based power systems enable numerous applications at various power levels, from a few watt to megawatt size. Following sections describe various applications of SOFC-based power systems that have been demonstrated or are under development.

Small SOFC Systems for Residential CHP Applications

A major application for SOFCs is at 1–5 kW level to supply combined heat and power (CHP) to residential buildings utilizing natural gas as the fuel. Early SOFC CHP units were designed, produced, and tested by Hexis (Switzerland) utilizing planar SOFCs. A system, named "Galileo 1000," has been developed to supply an electrical power of 1 kW and a thermal capacity of about 2.5 kW; over 50 such units are presently under demonstration. Ceres Power (UK) has developed a metal-supported planar SOFC system, utilizing ceria-based electrolyte, for generating electricity and central heating requirement (including hot water) for a typical home. Metallic-supported SOFC stack operates at relatively low temperatures ($\sim 550\,°C$) and tends to be lightweight; as a result, Ceres

Solid Oxide Fuel Cells, History, Fig. 5 A 1 kW SOFC CHP unit (*left*, SOFC system; *right*, hot water tank) utilizing Kyocera's anode-supported flat tubular cells

Power's unit is compact and wall-mountable, which may provide access to both the boiler replacement and new residential applications. Ceramic Fuel Cells Ltd. (Australia) utilizes planar SOFCs for residential CHP units; over 150 of its 1 kW units have been or are being demonstrated in various parts of the world.

Tubular SOFC design is also attractive for residential units because of its reliable and stable performance over long periods. Kyocera (Japan) has developed anode-supported flat tubular cells, which are being utilized by several organizations in Japan (Osaka Gas, Tokyo Gas, etc.) to fabricate and test SOFC-based residential CHP systems. The first trial operation of such a 1 kW unit in a residential house was conducted in 2005–2006; the average electric efficiency and hot water heat recovery efficiency were 44.1 % (LHV) and 34 % (LHV), respectively. Several hundred such units, illustrated in Fig. 5, are now installed and operating in homes throughout Japan and are currently being commercialized.

Toto Ltd. (Japan) is producing and testing 2 kW size CHP units using cathode-supported tubular cells. Siemens Westinghouse (USA) working with Fuel Cell Technologies (Canada) also produced and tested about a dozen 3–5 kW size CHP units using cathode-supported tubular cells; these units performed well with very stable performance for up to 1 year; unfortunately, both of these organizations are now out of the SOFC business.

Large SOFC Systems for Distributed Power Generation

Starting in 1986, Westinghouse (whose fossil energy business including SOFCs was acquired by Siemens in 1998) successfully fabricated about 15 integrated SOFC systems with a power range from 0.4 kW to 220 kW and tested them on customer sites. The longest test was conducted on a 100 kW atmospheric power generation system, shown in Fig. 6, using tubular SOFCs. The stack consisted of 1,152 cells (2.2 cm diameter and 150 cm active length) in 48 cell bundles of 24 cells each. This system operated for over 36,750 h in the USA, the Netherlands, Germany, and Italy on desulfurized natural gas without any detectable performance degradation at efficiency of ~46 %. This was the first successful demonstration of the solid oxide fuel cells for large-scale power generation.

Siemens Westinghouse also produced a 220 kW pressurized SOFC/gas turbine hybrid system, which was installed and tested at the National Fuel Cell Research Center on the campus of the University of California-Irvine (USA). This system was the world's first demonstration of a pressurized SOFC generator coupled with a microturbine generator. The system accumulated nearly 3,400 h of run-time while operating at a calculated net AC electrical efficiency of 53 %. Analysis indicates that with such systems in large MW sizes, an electrical efficiency of ~ 70 % is achievable.

As mentioned previously, Siemens/Westinghouse is no longer in the SOFC business. However, other organizations are continuing work on large-scale SOFC systems. Chief among these are Versa Power Systems

Solid Oxide Fuel Cells, History, Fig. 6 Siemens Westinghouse's 100 kW SOFC cogeneration system

Solid Oxide Fuel Cells, History, Fig. 7 Five 100 kW size Bloom Energy SOFC systems installed at eBay Headquarters (USA)

(in collaboration with FuelCell Energy, USA), United Technologies (in collaboration with Delphi Corporation, USA), and Rolls Royce Fuel Cell Systems (UK/USA); these organizations are supported by the US Department of Energy's Solid State Energy Conversion Alliance (SECA) program that was initiated in the year 2000 to reduce the cost of SOFC power generation systems. Mitsubishi Heavy Industries in Japan is also developing large size systems. Another company, Bloom Energy (USA) that started in 2001, has announced the manufacture and installation of several hundred 100 kW size SOFC systems, shown in Fig. 7, to commercial customers such as Adobe Systems, Bank of America, Coca Cola Company, eBay, FedEx, Google, Staples, Wal-Mart, Bank of America, and Safeway.

Portable SOFC Power Systems

The portable applications generally require power in the range from milliwatts to a few hundred watts. Proton exchange membrane fuel cells (including direct methanol fuel cells) are fuel cells of choice for such portable applications due to their light weight and low operating temperature; however, SOFC-based units are also being developed, primarily for certain military, leisure, emergency, and transportation applications because of their superior fuel flexibility, not requiring hydrogen. SOFCs can be successfully operated on fuels such as propane, gasoline, diesel, kerosene, JP-8 military fuel, ethanol, and other biofuels. Challenges arising for SOFCs in portable applications are that stacks must be light, have a short start up time, and be thermally

Solid Oxide Fuel Cells, History, Fig. 8 Ultra Electronics AMI's 50 W (*left*) and 250 W (*right*) portable SOFC systems

sustaining. Quick startup time is particularly crucial for portable applications; however, it is very difficult to achieve because of relatively low thermal shock resistance of ceramic components. One approach is to use microtubular SOFCs, which are capable of tolerating thermal shock resistance. Moreover, microtubular design can give reasonable volumetric power densities because the power density increases inversely proportional to the tube diameter.

Microtubular SOFCs have been successfully integrated into portable power units by Ultra Electronics AMI (USA). Figure 8 shows its 50 W and 250 W systems that use propane as the fuel to produce continuous power. The 50 W systems, running on propane, provide power for ground sensors, unmanned aerial vehicles, and robots. The 250 W systems are fueled by propane or LPG and are used to extend military mission durations and deliver off-grid power for electronics, radios, and computers. The use of globally available fuels in such portable SOFC systems eliminates complicated logistics.

SOFC-Based Transportation Auxiliary Power Units (APUs)

Another application of SOFC systems is in the transportation sector. As for portable applications, the proton exchange membrane fuel cell is generally regarded as the fuel cell of choice for transportation applications, particularly for propulsion to replace the internal combustion engine. Proton exchange membrane fuel cells require pure hydrogen, with no carbon monoxide, as the fuel to operate successfully. However, presently no hydrogen infrastructure exists, and onboard reformer systems to produce hydrogen from existing fuel base (gasoline, diesel) are technically challenging, complex, and expensive. Furthermore, it is difficult to eliminate all carbon monoxide from the reformate stream. In contrast, SOFCs can use carbon monoxide along with hydrogen as fuel, and their higher operating temperature and availability of water on the anode side make on-cell or in-stack reformation of hydrocarbon fuels feasible. Although not practical for propulsion, the application of SOFCs in the transportation sector will be for onboard auxiliary power units (APUs). Such APUs, operating on existing fuel base, will supply the ever-increasing electrical power demands of luxury automobiles, recreational vehicles, and heavy-duty trucks for various comfort items, such as refrigerators, televisions, stereos, even computers, and microwaves.

Delphi Corporation has developed a 3–5 kW size SOFC APU system using anode-supported planar cells. This unit is intended to operate on gasoline or diesel, which is reformed through partial oxidation within the APU unit. In 2008, Delphi Corporation (USA) and Peterbilt Motors Co. (USA) successfully demonstrated the operation of a Delphi's SOFC APU in powering a Peterbilt Model 386 truck's "hotel" loads (Fig. 9). The Delphi SOFC APU provided power for Model 386's electrical system and air conditioning and maintained the truck's batteries, all while the truck's diesel engine was turned off. Delphi hopes to commercialize such SOFC APUs in the next few years.

Solid Oxide Fuel Cells, History, Fig. 9 Delphi's SOFC APU mounted underneath a Peterbilt truck cabin

Summary and Future Directions

Solid oxide fuel cells utilize an electrolyte material that was first investigated by Walther Nernst in Germany in late 1890 s. Research, development, and demonstration of SOFCs accelerated in the last 30 years, and they are now on the verge of commercialization for certain applications. SOFCs offer potential for various applications with a wide power range from milliwatts to megawatts because of their fuel flexibility and combined high-quality heat and power. Extensive research is being conducted to decrease the cost of these power systems, particularly in the areas of cell and stack materials, dc to ac power conditioning systems, and other balance-of-plant components. With decreased cost, SOFC power systems are expected to provide a widespread low- or no-pollution technology to produce electricity. The first SOFC commercialization is expected to be for residential CHP for which many successful demonstrations by several companies have been accomplished. Large-scale distributed power generation also got a recent boost when Bloom Energy (USA) manufactured, delivered, and installed many 100 kW size SOFC systems for commercial customers.

The progress in SOFC technology and its commercialization status has been detailed in various scientific publications, in books [21, 22], and in proceedings of two international conferences devoted solely to solid oxide fuel cells: first, the biennial international symposia on solid oxide fuel cells starting in 1989 [24–35], and second, the SOFC Forums of the European Fuel Cell Forum [36–44]. These two series of proceedings give an excellent snapshot in time and progress achieved in SOFC technology over the last 25 years.

Cross-References

▶ Solid Electrolytes
▶ Solid Oxide Fuel Cells, Direct Hydrocarbon Type
▶ Solid Oxide Fuel Cells, Introduction
▶ Solid Oxide Fuel Cells, Thermodynamics

References

1. Hans-Heinrich M (2003) In: Singhal SC, Kendall K (eds) High temperature solid oxide fuel cells: fundamentals, design and applications. Elsevier, Oxford, UK, pp 23–51
2. Nernst W (1899) Electrical glow-light. US Patent No. 623,811
3. Nernst W (1899) Uber die elektrolytische Leitung Fester Korper bei sehr hohen Temperaturen. Z Elektrochem 6:41–43

4. Nernst W (1901) Material for electric-lamp glowers. US Patent No. 685,730
5. Wagner C (1943) Über den mechanismus der elektrischen Stromleitung im Nernststift. Naturwissenschaften 31:265–268
6. Bauer E, Preis H (1937) Über Brennstoff-Ketten mit Festleitern. Z Elektrochem 43:727–732
7. Kiukkola K, Wagner C (1957) Measurements on galvanic cells involving solid electrolytes. J Electrochem Soc 104:379–387
8. Weissbart J, Ruka R (1961) Oxygen gauge. Rev Sci Instrum 32:593–595
9. Weissbart J, Ruka R (1962) A solid electrolyte fuel cell. J Electrochem Soc 109:723–726
10. Palguev SF, Volchenkova ZS Das Problem der festen Elektrolyte für Brennstoffzellen. Tr. Inst. Khim. Akad. Nauk SSSR, Ural Filial, 2 (1958) 183–200; C. A. 54 (1960) 9542 i
11. Karpachov SV, Palguyev SF, Chebotin WN, Neuimin AD, Filyayev AT, Perfilyev MV et al. Tr. Inst. Elektrokhim. Akad. Nauk SSSR, Ural Filial, 1 (1960 foll.)
12. Tannenberger H, Schachner H, Simm W Festelektrolytbrennstoffelement. DE-Patent 1,471,768, filed May 22, 1963; Swiss priority May 23,1962
13. Binder H, Köhling A, Krupp H, Richter K, Sandstede G (1963) Elektrochemische Oxydation von Kohlenwasserstoffen in einer Festelektrolyt-Brennstoffzelle bei Temperaturen von 900-1000 °C. Electrochim Acta 8:781–793
14. Besson J, Deportes C, Kleitz M: 1er Colloque sur les piles à combustible. Bellevue, 6 déc 1962. Contrat de recherches 61 FR 136. "Étude des électrolytes solides pour piles à combustible à haute température" and "Utilisation des électrolytes dans les piles à combustible à haute temperature." In: Les piles à combustible. Éditions Technip, Paris 1965, 87–102 and 303–323
15. Williams KR, Smith JG (1963) Fuel cell. Great Britain Patent No. 1,049, 428, filed August 15, 1963
16. Takahashi T (1966) Research and development of fuel cells in Japan. J Electrochem Soc 34:60–69
17. Spacil HS (1964) Electrical device including nickel-containing stabilized zirconia electrode. US Patent No. 3,503,809, filed October 30,1964
18. Tedmon CS Jr, Spacil HS, Mitoff SP (1969) Cathode material and performance in high-temperature zirconia electrolyte fuel cells. J Electrochem Soc 116:1170–1175
19. Meadowcroft DB (1969) Some properties of strontium-doped lanthanum chromite. Brit J Appl Phys 2 (J. Phys. D) 2 :1225–1233
20. Fischer W, Kleinschmager H, Rohr FJ, Steiner R, Eysel HH (1972) Hochtemperatur-Brennstoffzellen mit keramischen Elektrolyten zum Umsatz billiger Brennstoffe. Chem Ing Tech 44:726–732
21. Kendall K, Singhal SC (2003) High temperature solid oxide fuel cells: fundamentals, design and applications. Elsevier, Oxford, UK

22. Huang K, Goodenough JB (2009) Solid oxide fuel cell technology. Woodhead Publishing, Cambridge, UK
23. Pal UB, Singhal SC (1990) Electrochemical vapor deposition of Yttria-Stabilized Zirconia films. J Electrochem Soc 137:2937–2941
24. Singhal SC (ed) (1989) SOFC-I, PV89-11, the electrochemical society proceedings series, Pennington, NJ
25. Grosz F, Zegers P, Singhal SC, Yamamoto O (eds) (1991) SOFC-II, Commission of the European Communities, Luxembourg
26. Singhal SC, Iwahara H (eds) (1993) SOFC-III, PV93-4, the electrochemical society proceedings series, Pennington, NJ
27. Dokiya M, Yamamoto O, Tagawa H, Singhal SC (eds) (1995) SOFC-IV, PV95-1, the electrochemical society proceedings series, Pennington, NJ
28. Stimming U, Singhal SC, Tagawa H, Lehnert W (eds) (1997) SOFC-V, PV97-40, the electrochemical society proceedings series, Pennington, NJ
29. Singhal SC, Dokiya M (eds) (1999) SOFC-VI, PV99-19, the electrochemical society proceedings series, Pennington, NJ
30. Yokokawa H, Singhal SC (eds) (2001) SOFC-VII, PV2001-16, the electrochemical society proceedings series, Pennington, NJ
31. Singhal SC, Dokiya M (eds) (2003) SOFC-VIII, PV2003-07, the electrochemical society proceedings series, Pennington, NJ
32. Singhal SC, Mizusaki J (eds) (2005) SOFC-IX, PV2005-07, the electrochemical society proceedings series, Pennington, NJ
33. Eguchi K, Singhal SC, Yokokawa H, Mizusaki J (eds) (2007) SOFC-X, the electrochemical society transactions, vol 7
34. Singhal SC, Yokokawa H (eds) (2009) SOFC-XI, the electrochemical society transactions, vol 25
35. Singhal SC, Eguchi K (eds) (2011) SOFC-XII, the electrochemical society transactions, vol 35
36. Bossel U (ed) (1994) First European solid oxide fuel cell forum proceedings. European Fuel Cell Forum, Oberrohrdorf, Switzerland
37. Thorstensen B (ed) (1996) Second European solid oxide fuel cell forum proceedings. European Fuel Cell Forum, Oberrohrdorf, Switzerland
38. Stevens P (ed) (1998) Third European solid oxide fuel cell forum proceedings. European Fuel Cell Forum, Oberrohrdorf, Switzerland
39. McEvoy AJ (ed) (2000) Fourth European solid oxide fuel cell forum proceedings. European Fuel Cell Forum, Oberrohrdorf, Switzerland
40. Huijsmans J (ed) (2002) Fifth European solid oxide fuel cell forum proceedings. European Fuel Cell Forum, Oberrohrdorf, Switzerland
41. Mogensen M (ed) (2004) Sixth European solid oxide fuel cell forum proceedings. European Fuel Cell Forum, Oberrohrdorf, Switzerland
42. Kilner J (ed) (2006) Seventh European solid oxide fuel cell forum proceedings. European Fuel Cell Forum, Oberrohrdorf, Switzerland

43. Steinberger-Wilckens R (ed) (2008) Eighth European solid oxide fuel cell forum proceedings. European Fuel Cell Forum, Oberrohrdorf, Switzerland
44. Irvine JTS (ed) (2010) Ninth European solid oxide fuel cell forum proceedings. European Fuel Cell Forum, Oberrohrdorf, Switzerland

Solid Oxide Fuel Cells, Introduction

Jennifer L. M. Rupp
Electrochemical Materials, ETH Zurich,
Zurich, Switzerland

Fuel cells are electrochemical conversion devices that convert chemical energy into electrical energy and heat. Conversion efficiency is purely based on an electrochemical reaction involving thermodynamics and kinetics, without involving the process of combustion. A simplistic view of a fuel cell can be envisaged as a cross between a battery (chemical energy converted directly into electrical energy) and a heat engine (a continuously fueled, air breathing device); therefore, fuel cells are sometimes referred to as *electrochemical engines* [1]. Currently, there exist various types of fuel cell technologies that differ in their operation temperature, material constituents, fuel tolerance, ionic carriers, and *efficiencies*. Among these, solid oxide fuel cells (SOFCs) offer the highest efficiencies, fuel flexibility (hydro carbons, i.e., natural gas), quiet operation, and low to zero emission compared to heat engines. State-of-the-art SOFCs consist of three major constituents, namely, a dense oxygen ion (O^{2-})-conducting electrolyte separating two porous mixed conducting electrodes (cathode and anode), (Fig. 1). The electrodes are exposed to two different gas compartments, namely, ait for the cathode and fuel (i.e., methane of hydrogen) for the anode. At the cathode, oxygen is reduced and ions move through the electrolyte to the anode where they recombine with an oxidized gaseous fuel. Aimed electrons are formed, as well as water, heat, and carbon oxides (in the case of hydrocarbons). The driving force for the electrochemical reaction is the difference in chemical potential and its gradient across the fuel cell. It is important to note that SOFCs are unique in their potential efficiency and ability to oxidize almost any fuel. Over the last decades, fuel cells have been most commonly discussed as part of a *broader hydrogen economy*. Indeed this holds true for proton-conducting fuel cells such as proton exchange membrane fuel cells (PEMFCs) at low temperatures, where small amounts of carbon in the fuel lead to irreversible damage [2]. In the case of SOFCs, oxygen ions are conducted to equilibrate the partial pressure gradient across the electrolyte. Essentially any hydrocarbon can be oxidized at the anode. Hence, SOFCs, with their ability to operate at high efficiencies and reduced CO_2 emission, function as important alternatives to our current hydrocarbon fuel infrastructure. The efficiency to convert chemical to electrical energy is 45–60 % for a state-of-the-art SOFC.

In SOFC systems several cells are connected through series or parallel interconnections to form a so-called stack. A single SOFC exhibits voltages of 1 V and an average power density of 1 W/cm^2; note that up to 2 W/cm^2 has also been reported for lab cells [3, 4]. Cell stacks allow accommodation for a wide range of power needs, system weights, and volumes.

On a large scale, power is generated at several hundreds of kilowatts *(kW)* to megawatts *(MW)* range for power plants, military or household applications. For such systems, typified by developers such as Bloom Energy, Siemens Westinghouse, Kyocera, Hexis, and Rolls-Royce, high operation temperatures around 800–1000 °C are targeted in order to reach high efficiencies and fast transport kinetics. Cogeneration of heat can be used for powering gas turbines to further increase efficiency in the "high-temperature" SOFC systems. On a smaller scale, SOFCs are excellent power generation systems for the *MW to kW range* in the form of micro-SOFCs integrated onto Si-chip technology and as auxiliary power units (APUs) typified by developers such as SiEnergy, Lilliputian, Ceres Power, Versa Power, eZelleron, and Topsoe Fuel Cell. These offer attractive alternatives to current battery-based technologies for the portable

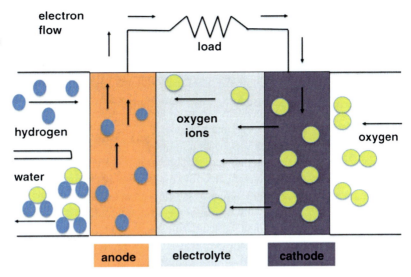

Solid Oxide Fuel Cells, Introduction, Fig. 1 Operating principle of a solid oxide fuel cell

electronic and transportation sectors, due to their high volumetric and specific energy (see also Ragone plots in Refs. [3, 5, 6] for details). In such micro fuel cell systems the chemical to electrical energy conversion is not restricted by the materials' storage capacity, as is the case in a battery. Instead, an external fuel compartment allows for net grid independence and constant conversion. Lowering the operation temperatures is especially important for the MW to kW range in portable applications to assure high SOFC stack performance at low volumetric packaging density. Here, intermediate (600–800 °C) or even low-temperature (200–600 °C) SOFC operation range is targeted. This allows the use of a broader set of materials; it is less demanding on the system's seals and support structures and also on new processing techniques such as microstructuring on Si-wafer technologies. In addition to faster start-up and cooldown times, reduced degradation of cell components and long-term stability are realized. Challenges related to the reduced operation temperature include overcoming sluggish transport kinetics through improved materials and designs. These reduce the thermodynamic theoretic cell voltage and conversion efficiency. Main losses occur due to polarization losses (overpotentials) at the electrode-electrolyte interfaces and ohmic resistances of the electrolyte. Typically, losses occur as a result of the interplay of design and materials property of the electrolyte. Its ohmic resistance scales inversely proportional with its thickness (distance between electrodes) and depends on the electrolyte's resistivity, as well as its geometry, as a material characteristic. In accordance to the Nernst-Einstein relation, general lowering of the operation temperature leads to systematic lowering of the oxygen ion diffusivity and an increase in resistivity of the electrolyte. Strategies to realize high oxygen ion diffusivities comprise determining suitable metal oxides with increased ionic conductivity compared to the state of the art, reduction of electrolyte thickness from thick-to-thin film electrolyte processing, and design change from electrolyte-based to electrode-based fuel cell support structures.

In the following, SOFC materials are detailed in view of the state of the art and the lowering of their operation temperatures at high fuel cell conversion efficiencies:

Electrolytes require a high oxygen ionic conductivity and a dense microstructure to allow for gas separation of fuel and air compartment. Electronic conductivity is to be avoided since this would potentially short-circuit the SOFC. Chemical and thermomechanical stability of electrolytes with electrode materials is also an important requirement to assure long-term stability. Cubic 8 mol% yttria-doped zirconia is, due to its sufficiently and predominantly ionic conductivity, the state-of-the-art material used as the electrolyte in SOFCs [40]. Among zirconia solid solutions, scandia doping contributes the highest ionic conductivity and would allow cost reduction compared to yttria solutes. However, close radii mismatch of dopant Sc^{3+} and host Zr^{4+} leads to cluster formation of vacancy-dopant cation with a reduced and unstable ion mobility over time. Therefore, system development of scandia-doped zirconia was not further pursued. A general drawback of zirconia-based electrolytes is their reactivity towards lanthanum-based cathodes (i.e., $LaCoO_3$ or $LaMnO_3$) forming insulating lanthanum zirconates phases at the interface [41]. This was one aspect that motivated investigating alternative materials such as ceria-based electrolytes. These offer increased ionic conductivity by up to five times and high chemical stability towards cathodes. Various trivalent cationic dopants have been tested for ceria, (e.g., Gd_2O_3, Sm_2O_3, Y_2O_3) to introduce oxygen vacancies for high ionic conduction [7–9]. Intermediate-to low-temperature regime gadolinia-doped ceria, in particular, is pursued as an interesting electrolyte alternative, due to its lowered resistivity compared to zirconia-based ones [42]. Stabilized bismuth oxides, such as $Er_{0.4}Bi_{1.6}O_3$ or $Dy_{0.08}W_{0.04}Bi_{0.88}O_{1.56}$, reveal increased ionic conduction up to two orders of magnitude [3]. However, long-term degradation issues occur due to decomposition under reducing conditions at the anode site. Nevertheless bilayer electrolytes, such as coupling ceria- and bismuth-based electrolytes or ceria- and zirconia-based electrolytes [43], continue to be researched to guarantee highest ionic conduction and long-term stability at low temperatures.

Anodes perform electrooxidation of fuel by catalyzing the reaction and facilitating fuel access and product removal [1]. These require sufficient reaction sites for the fuel oxidation and electronic conductivity to transfer electrons from the oxidant to the cell components, i.e., electrolyte and current collectors. In high-temperature SOFCs, porous cermets, made from a percolating metal phase and a ceramic such as nickel-yttria-stabilized zirconia, are well established. Nickel exhibits a high redox stability and catalytic activity. However, grain growth and the different thermal expansion coefficient towards electrolyte interface lead to stresses and poor long-term stability when used alone. If combined with a ceramic phase such as zirconia (or ceria), a percolating network of metallic nickel-ceramic pores can be formed, which fulfills the requirements of an anode. Full ceramic anodes such as $(La, Sr)TiO_3$ or $(La, Sr, Cr)MnO_3$ are also subjects of investigations [10, 11]. One of their advantages over metal-binary oxide cermets is their high thermal stability. For instance, metals of the cermets show a tendency towards unfavorable grain growth and even dewetting towards metal oxide components at higher temperatures. Lowering the operation temperature raises the opportunity for noble metal electrodes such as platinum, ruthenium, palladium, or silver as anodes for SOFC [12–16]. These are good electrocatalysts, which are currently being investigated as micro-SOFC system anodes at temperatures below 600 °C. Their application is restricted to low temperatures since low vapor pressure of suboxides prevents stability at temperatures above 600–800 °C.

Cathodes reduce oxygen from air, transport the oxygen ion carriers to the electrolyte, and distribute the necessary electronic carriers for the oxygen reduction reaction. The state-of-the-art cathode materials are perovskites based on transition metal ions, specifically $(La, Sr)MnO_3$ [17] and $(La, Sr)(Co, Fe)O_3$ [18, 19]. Both cathode materials differ in their volume participating in the oxygen reduction reaction and in their carrier conductivity [20]. In the case of $(La, Sr)MnO_3$, a predominant electronic conductivity is characteristic, which restricts the active oxygen reduction reaction zone to the so-called triple-phase boundaries where cathode, electrolyte, and air meet. A mixed ionic-electronic conduction takes place for $(La, Sr)(Co, Fe)O_3$ cathodes. Here, oxygen is reduced at the cathode surface and its reduced ions are distributed through the volume of the cathode towards the electrolyte. Mixed conductivity and competing diffusion pathways of the surface vs. volume are an important factor in increasing the oxygen exchange rate. Currently, the greatest losses in power of SOFC are related to overpotentials developing at the cathode-electrolyte interface and sluggish oxygen exchange rates [21–23]. Besides kinetics, thermal expansion mismatch and phase formation of the cathode-electrolyte interface are also important in material considerations: the electronically conductive $(La, Sr)MnO_3$ reveals the lowest thermal expansion mismatch towards standard zirconia-based electrolytes. However, phase formation of lanthanum zirconates continues to pose a challenge for long-term operations. Besides these common cathode materials, further perovskite compositions such as $(Ba, Sr)(Co, Fe)O_3$, and $(Pr, Sr)MnO_3$ [24–27]; pyrochlore ruthenates such as $(Bi, Pb, Y)_2Ru_2O_7$ [28–30]; and Ruddlesden Popper phase lanthanum nickelates [31, 32] are discussed. For low to intermediate temperatures, catalytic noble metals such as platinum or platinum-based alloys are also suitable cathodes for SOFC. However, agglomeration and thermal stability require further attention towards long-term stability.

Summary and Future Directions

Solid oxide fuel cells (SOFC) are promising electrochemical energy conversion systems to produce power for portable, mobile, and stationary applications ranging from several W to MW. The advantage that SOFCs have over other fuel cell systems is their direct operation on hydrocarbons and air (without being restricted to a hydrogen distribution net). Operation principle, temperature regime, and materials were introduced and are detailed in further articles.

As a new future trend, ongoing research is being undertaken on lowering the operation temperature of SOFCs and new device prototypes, such as micro-SOFCs, in order to use them as replacements for today's batteries in portable electronics. Advantages of this technology include grid independency and external chemical energy storage in a compartment. In order to achieve lower operation temperatures, new material compositions with higher carrier transfer and increased catalytic activities (electrodes) are being researched [1, 3, 10, 33]. In addition thin film SOFC components are being investigated to reduce ohmic resistances of electrolytes and realize metastable (strained, oriented, or interface modulated) microstructures with increased carrier concentrations and mobilities [34–39].

Cross-References

▶ Solid Electrolytes
▶ Solid Oxide Fuel Cells, Direct Hydrocarbon Type
▶ Solid Oxide Fuel Cells, History
▶ Solid Oxide Fuel Cells, Thermodynamics

References

1. Brett DJL, Atkinson A, Brandon NP, Skinner SJ (2008) Intermediate temperature solid oxide fuel cells. Chem Soc Rev 37:1568
2. Haile SM (2003) Fuel cell materials and components. Acta Mater 52:5981
3. Wachsman ED, Lee KT (2011) Lowering the temperature of solid oxide fuel cells. Science 344:935
4. Virkar AV, Chen J, Tanner CW, Kim JW (2000) The role of electrode microstructure on activation and concentration polarizations in solid oxide fuel cells. Solid State Ionics 131:2000
5. Evans A, Bieberle-Hütter A, Galinski H, Rupp JLM, Ryll T, Barbara S, Tölke R, Gauckler LJ (2009) Micro-solid oxide fuel cells: status, challenges, and chances. Chemical Monthly 140:975
6. Evans A, Bieberle-Hütter A, Rupp JLM, Gauckler LJ (2009) Review on microfabricated micro-solid oxide fuel cell membranes. J Power Sources 194:119
7. Mogensen M, Sammes NM, Tompsett GA (2000) Physical, chemical and electrochemical properties of pure and doped ceria. Solid State Ionics 192:63
8. Tuller HL (2000) Ionic conduction in nanocrystalline materials. Solid State Ionics 131:143
9. Rupp JLM, Gauckler LJ (2006) Microstrain and self-limited grain growth in nanocrystalline ceria ceramics. Solid State Ionics 54:2513

10. Cowin PI, Petit CTG, Lan R, Irvine JTS, Tao SW (2011) Recent progress in the development of anode materials for solid oxide fuel cells. Adv Energy Mater 1:314

11. Ruiz-Morales JC, Canales-Vazquez J, Savaniu C, Marrero-Lopez D, Zhou WZ, Irvine JTS (2006) Disruption of extended defects in solid oxide fuel cell anodes for methane oxidation. Nature 439:568

12. Galinski H, Ryll T, Elser P, Rupp JLM, Bieberle-Hutter A, Gauckler LJ (2010) Agglomeration of Pt thin films on dielectric substrates. Phys Rev B B82:235415-1–235415-11

13. Ryll T, Galinski H, Schlagenhauf L, Elser P, Rupp JLM, Bieberle-Hutter A, Gauckler LJ (2011) Microscopic and nanoscopic three-phase-boundaries of platinum thin-film electrodes on YSZ electrolyte. Adv Funct Mater 21:565

14. Ryll T, Galinski H, Schlagenhauf L, Rechberger F, Ying S, Gauckler LJ, Mornaghini FCF, Ries Y, Spolenak R, Dobeli M (2011) Dealloying of platinum-aluminum thin films: electrode performance. Phys Rev B 84:125408-1–125408-8

15. Popke H, Mutoro E, Luerssen B, Janek J (2012) Oxidation of platinum in the epitaxial model system Pt (111)/YSZ(111): quantitative analysis of an electrochemically driven PtOx formation. J Phys Chem C 116:1912

16. Kim YB, Holme TP, Gur TM, Prinz FB (2011) Surface-modified low-temperature solid oxide fuel cell. Adv Funct Mater 21:4684

17. Laguna-Bercero MA, Kilner JA, Skinner SJ (2010) Performance and characterization of (La, Sr)MnO3/YSZ and La0.6Sr0.4Co0.2Fe0.8O3 electrodes for solid oxide electrolysis cells. Chem Mater 22:1134

18. Cai ZH, Kubicek M, Fleig J, Yildiz B (2012) Chemical heterogeneities on La0.6Sr0.4CoO3-delta thin films-correlations to cathode surface activity and stability. Chem Mater 24:1116

19. Berenov AV, Atkinson A, Kilner JA, Bucher E, Sitte W (2010) Oxygen tracer diffusion and surface exchange kinetics in La0.6Sr0.4CoO3-delta. Solid State Ionics 181:819

20. Fleig J, Maier J (2004) The polarization of mixed conducting SOFC cathodes: Effects of surface reaction coefficient, ionic conductivity and geometry. J Eur Ceram Soc 24:1343

21. Chroneos A, Yildiz B, Tarancon A, Parfitt D, Kilner JA (2011) Oxygen diffusion in solid oxide fuel cell cathode and electrolyte materials: mechanistic insights from atomistic simulations. Energ Environ Sci 4:2774

22. Han JW, Yildiz B (2012) Mechanism for enhanced oxygen reduction kinetics at the (La, Sr)CoO3-delta/(La, Sr)(2)CoO4+delta hetero-interface. Energ Environ Sci 5:8598

23. Mutoro E, Crumlin EJ, Biegalski MD, Christen HM, Shao-Horn Y (2011) Enhanced oxygen reduction activity on surface-decorated perovskite thin films for solid oxide fuel cells. Energ Environ Sci 4:3689

24. Wang L, Merkle R, Baumann FS, Fleig J, Maier J (2007) (Ba(x)Sr(1-x))(Co(y)Fe(1-y))O(3-delta) Perovskites as SOFC cathode material: electrode-electrolyte reactions and electrochemical characterisation. Solid Oxide Fuel Cells 10 (Sofc-X), Pts 1 and 2 7:1015

25. Shao ZP, Haile SM (2004) A high-performance cathode for the next generation of solid-oxide fuel cells. Nature 431:170

26. Yang Z, Harvey AS, Infortuna A, Schoonman J, Gauckler LJ (2011) Electrical conductivity and defect chemistry of Ba(x)Sr(1-x)CoyFe(1-y)O(3-delta) perovskites. J Solid State Electr 15:277

27. Harvey AS, Litterst FJ, Yang Z, Rupp JLM, Infortuna A, Gauckler LJ (2009) Oxidation states of Co and Fe in Ba1-xSrxCo1-yFeyO3-delta (x, y=0.2-0.8) and oxygen desorption in the temperature range 300-1273 K. Phys Chem Chem Phys 11:3090

28. Ryll T, Brunner A, Ellenbroeck S, Bieberle-Huetter A, Rupp JLM, Gauckler LJ (2010) Crystallization and electrical conductivity of amorphous bismuth ruthenate thin films deposited by spray pyrolysis. Phys Chem Chem Phys 12:13933

29. Gauckler LJ, Beckel D, Buergler BE, Jud E, Mücke UP, Prestat M, Rupp JLM, Richter J (2004) Solid oxide fuel cells: Systems and materials. Chimia 58:837

30. Vannier RN, Chater RJ, Skinner SJ, Kilner JA, Mairesse G (2003) Oxygen transfer in BIMEVOX materials. Solid State Ionics 160:327

31. Chroneos A, Parfitt D, Kilner JA, Grimes RW (2010) Anisotropic oxygen diffusion in tetragonal La2NiO4 +delta: molecular dynamics calculations. J Mater Chem 20:266

32. Ryll T, Reibisch P, Schlagenhauf L, Bieberle-Huetter A, Dobeli M, Rupp JLM, Gauckler LJ (2012) Lanthanum nickelate thin films deposited by spray pyrolysis: Crystallization, microstructure and electrochemical properties. J Eur Ceram Soc 32:1701

33. Snyder GJ, Haile SM (2004) Fuel cell materials and components. Advanced Materials for Energy Conversion Ii 34:33

34. Guo XX, Maier J (2009) Ionically conducting two-dimensional heterostructures. Adv Mater 21:2619

35. Rupp JLM (2012) Ionic diffusion as a matter of lattice-strain for electroceramic thin films. Solid State Ionics 207:1

36. Santiso J, Burriel M (2011) Deposition and characterization of epitaxial oxide thin films for SOFCs. J Solid State Electrochem 15:985

37. Kushima A, Yildiz B (2010) Oxygen ion diffusivity in strained yttria stabilized zirconia: where is the fastest strain? J Mater Chem 20:4809

38. Schichtel N, Korte C, Hesse D, Zakharov N, Butz B, Gerthsen D Janek J (2010) On the influence of strain on ion transport: microstructure and ionic conductivity of nanoscale YSZ|Sc2O3 multilayers. Phys Chem Chem Phys 12:14596

39. Rupp JLM, Scherrer B, Schauble N, Gauckler LJ (2010) Time-Temperature-Transformation (TTT) Diagrams for crystallization of metal oxide thin films. Adv Funct Mater 20:2807

40. Guo X, Waser R (2013) Electrical properties of the grain boundaries of oxygen ion conductors: acceptor-doped airconia and ceria. Prog Mater Sci 51(2):151–210
41. Weber SB, Lein HL, Grande T et al. (2013) Lanthanum zirconate thermal barrier coatings deposited by spray pyrolysis. Surface & Coatings Technology 227:10–14
42. Rupp JLM Infortuna A, Gauckler LJ. Thermodynamic stability of gadolinia-doped ceria thin film electrolytes for micro-solid oxide fuel cells. Journal of the American Ceramic Society 90(6):1792–1797
43. Muecke UP, Beckel D, Bernard A et al. Micro solid oxide fuel cells on glass ceramic substrates. Adv Funct Mater 18(20):3158–3168

Solid Oxide Fuel Cells, Thermodynamics

Harumi Yokokawa
The University of Tokyo, Institute of Industrial Science, Tokyo, Japan

Description and Characteristics

Solid oxide fuel cells are energy convertor from chemical energy (fuel) to electricity by using oxide ion conductors as electrolyte [1–3]. Associated cell components and mass flows for charge carriers are schematically presented in Fig. 1. The most important processes in SOFCs are the oxygen incorporation to produce O^{2-}, transportation of oxide ions in electrolyte, and the electrochemical oxidation of fuels using thus transported oxide ions [1]. In this article, roles of thermodynamics in RD&D of SOFC will be examined in two different categories; one is from the thermodynamic functions and their utilization, the other being technological issues in stack/system developments.

The following thermodynamic functions or thermodynamic variables are quite important in understanding SOFC systems and materials behaviors in SOFCs:

1. Electrochemical potentials of charged species: Behaviors of charged particles are described in terms of the electrochemical potential of oxide ions as major carrier, $\eta(O^{2-})$, or protons as minor carrier, $\eta(H^+)$. The flux of such species can be determined as

$$J(O^{2-})\mathrm{d}x = -\sigma(O^{2-}) * \mathrm{d}\eta(O^{2-}) \qquad (1)$$

where $\mathrm{d}\eta(O^{2-})$ is a slope of their electrochemical potentials, $\sigma(O^{2-})$ being their conductivity. Under the local thermodynamic equilibrium (LTE) approximation, the electrochemical potentials of charged species and chemical potentials of neutral species can be defined at any places in a system as follows:

$$0.5\mu(O_2) = \eta(O^{2-}) - 2\eta(e^-) \qquad (2)$$

$$\mu(H_2O) = \eta(O^{2-}) + 2\eta(H^+) \qquad (3)$$

Distribution of such chemical potentials will be determined so as to provide the given flows of oxide ions or electrons. Figure 2 indicates how those (electro)chemical potentials of oxide ions, electrons, and oxygen gas behave inside electrolyte under cell operation. Since $\sigma(O^{2-})$ is high, the slope of $\eta(O^{2-})$ is small even at a high current density. On the other hand, $J(e^-)$ is limited although a steep gradient appears in $\eta(e^-)$. A typical example of oxygen potential distribution across the cell components is given in Fig. 1 for intermediate-temperature (second-generation) SOFCs.

2. Overpotential in terms of oxygen potential: When a current is passed, overpotentials appear at electrode/electrolyte interfaces as shown in Fig. 2. These overpotentials, ΔV_C and ΔV_A, can be also characterized in terms of the oxygen potential differences as follows.

This information on the oxygen potential provides a basis of examining the electrode chemical behaviors. Particularly, the air electrode containing transition metal oxides in the perovskite lattice exhibit oxygen potential-dependent phase relations, leading to considerations on stability/reactivity of interfaces with electrolyte in terms of operation conditions (direction and magnitude of current) [2].

$$nF\Delta V_C = \Delta\mu(O_2)^C \qquad (4)$$

Solid Oxide Fuel Cells, Thermodynamics, Fig. 1 Schematic view of cell components for the second-generation SOFC and associated mass flows or oxygen potential distribution across the cell

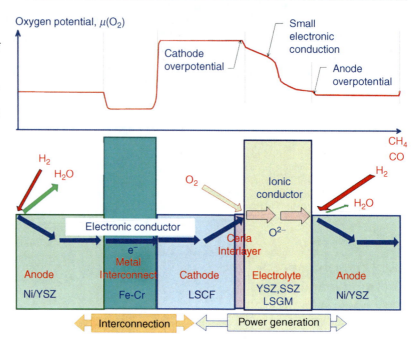

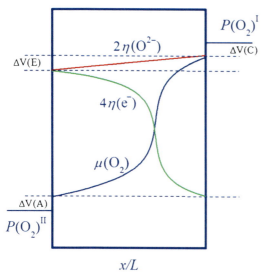

Solid Oxide Fuel Cells, Thermodynamics, Fig. 2 The distribution of electrochemical potentials of oxide ions and electrons together with the oxygen potential inside electrolyte placed between air and fuel. $\Delta V(E)$ is the ohmic loss inside electrolyte and $\Delta V(C)$, $\Delta V(AE)$ correspond to the overpotentials of cathode and anode

3. Elemental chemical potentials: Any interfaces among the cell components can be examined thermodynamically in terms of the chemical potentials of elements involved. For this purpose, the generalized chemical potential diagrams [4] are quite useful.

4. Valence stability of transition metal binary oxides and stabilization energy of double oxides: In order to understand the physicochemical meaning of chemical processes at interfaces or reactions with gases, it is essential to write down a chemical reaction and to examine it in terms of enthalpy and entropy changes. Since the entropy does not depend on elements but on the chemical reaction types [5], the fundamental properties of chemical reactions are determined predominantly by the enthalpy change for reactions. Since those properties can be well interpreted in terms of the valence stability of transition metal ions as well as the stabilization energy of double oxides from their constituent oxides [2], this provides a basis of interpreting observed chemical processes in terms of the valence state of the transition metal oxides in targeted materials. This makes it possible also to understand differences among the transition metal oxides in selecting electrode materials or their relations to chemical compatibility with electrolyte materials.

5. Volume changes: Defect chemistry provides the fundamental information to predict the volume changes as functions of temperature and oxygen potential. This in turn provides a basis of analyzing the mechanical compatibility among the cell components.
6. Exergy analyses: Since heats evolved from SOFC can be well treated in terms of exergy, it becomes important to analyze the SOFC systems with the exergy analyses [6].

These considerations based on thermodynamic functions can be made by utilizing thermodynamic database including advanced software. Thermodynamic database MALT [7] or FACTSage [8] are used widely. Evaluations of thermodynamic data have been also progressed on SOFC-related materials [9].

The roles of the thermodynamics can be also discussed in terms of technological issues in development of SOFC stacks/systems as follows [10]:

1. *Fuel flexibility:* One of the merits of SOFCs is fuel flexibility. Since the charged species associated with oxidants is transported in SOFCs, any kinds of fuels can be utilized in principles. Since nickel anodes are weak against carbon deposition or sulfur poisoning, fuels are used in SOFCs after fuel processing is made. Thus, fuel chemistry based on the thermodynamics is one of the major components in understanding SOFC systems.
2. *High conversion efficiency:* In addition to the Gibbs energy-based constrains concerning the theoretical efficiency, lowering of conversion efficiency come from Joule losses, electrode overpotential term, and oxygen permeation/leakage [3].
3. *Expecting low cost:* Since SOFC does not utilize precious metals, materials cost must be low. Even so, it becomes essential to achieve high functionalities at high temperatures. This requires a wide range of the thermodynamic properties for high-temperature materials including ceramics and alloys.
4. *High durability/reliability:* Materials thermodynamics are particularly important in SOFC because not only fabrication processes but also operation processes are fully understood on a basis of thermodynamic considerations on compatibility among dissimilar materials in cells and on interactions with gaseous impurities [11].

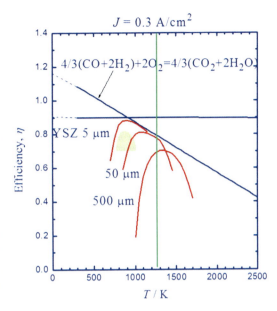

Solid Oxide Fuel Cells, Thermodynamics, Fig. 3 Thermodynamic theoretical conversion efficiency for reforming SOFC and further lowering in efficiency due to the Joule loss and electrochemical oxygen permeation through the YSZ electrolyte for difference thickness

In what follows, further discussions are made on selected topics:

1. Conversion Efficiency

One of the most important features in SOFCs is an increase of conversion efficiency by utilizing heats evolved as a result of the electrochemical reaction to the reforming process which will be made at a lower temperature than the SOFC operation temperature but at a temperature high enough for the reforming process. In Fig. 3, this is expressed by using HHV(High Heating Value)-based efficiency for the direct methane SOFC and reforming SOFCs as given in the following equation:

$$\text{Electrochemical}: CH_4 + 2O_2 = CO_2 + 2H_2O \quad (5)$$

$$\begin{aligned}\text{Reforming}: CH_4 &+ 1/3(CO_2 + 2H_2O) \\ &= 4/3(CO + 2H_2)\end{aligned} \quad (6)$$

Electrochemical :
$$4/3(CO + 2H_2) = 4/3(CO_2 + 2H_2O) \qquad (7)$$

Those lowering effects of conversion efficiency which are originated from electrolytes can be calculated from the conductivity characteristics. A typical example is given in Fig. 3, where the theoretical efficiency based on the reforming SOFCs is combined with the contributions of lowering due to the Joule loss and the oxygen permeation. At lower temperatures, the efficiency is limited by the Joule losses, whereas the electronic shorting (in other words, the electrochemical oxygen permeation) limits the efficiency at high temperatures.

Another important contribution to lowering efficiency comes from the overpotential of electrodes. To analyze the efficiency of SOFC systems, it is essential to evaluate the amount and temperature of evolved heats and to design how to use these heats in the system. An attempt was made to derive differences of electrolytes to be used and of fuels to be oxidized in SOFC systems within the framework of exergetic analyses [6].

2. Materials Compatibilities

SOFC is fabricated at high temperatures and is operated around 750–900 °C. This makes it necessary to consider materials compatibility of two different kinds; one is the chemical stability at interfaces among cell components, the other being the mechanical stability among stacked components.

Interface stabilities are well thermodynamically treated by using the generalized chemical potential diagrams [4] and are well interpreted in terms of the valence stability of transition metal oxides and the stabilization energy of perovskite oxides from the constituent oxides [2]. Examples are given in Fig. 4, where the lanthanum manganite $LaMnO_3$ and the lanthanum cobaltite $LaCoO_3$ are compared. Since the stabilization energy for $LaMnO_3$ and $LaCoO_3$, $\delta(LaMO_3(M=Mn,Co))$, is about the same, difference in reactivity is originated from the valence stability of transition metal oxides, $\Delta[M^{3+}:M^{2+}](M=Mn,Co)$. This leads to

a situation that the stable LSM cathode is utilized directly with YSZ, whereas active cathodes such as LSC, LSF, and LSCF are used with an interlayer of doped ceria between cathode and YSZ electrolyte.

Since those cathodes include the transition metal oxide, their valence state may change under polarization. This means that the interface stability depends on temperature and oxygen potential so that operation conditions should affect the interface stability. A most typical example is the LSM/YSZ interfaces. Since the tetravalent Mn ions are relatively stable in the perovskite lattice, this causes the formation of A-site deficiency in $La_{1-y}MnO_3$. Since this is related with the valence stability, behaviors associated with the A-site deficiency is sensitively dependent on the Sr content in LSM, polarization directed and its magnitude, and temperature.

3. Degradation Mechanism

About degradations, it is generally expected that with decreasing temperature, degradation may decrease and also that severer reactions with impurities will lead to severer degradation. However, observed degradation behaviors are quite different from this expectation in the following aspects [11]:

(a) Since the active cathodes are used for the intermediate or low temperature SOFCs, degradation of such SOFCs is large compared with the high-temperature SOFCs which use only stable materials for respective cell components.

(b) The most stable LSM is weakest against the chromium attack. In other words, active cathodes such as LSCF do not exhibit serious poisoning effects due to reactions with Cr-containing species.

In view of these examples available, the proper considerations have to be made on degradation mechanism by separating effects into those for gaseous channel, electrical path, and electrochemically active sites.

In many cases, gaseous channels, electrical paths, and electrochemically active sites consist of the same materials. In this sense, chemical

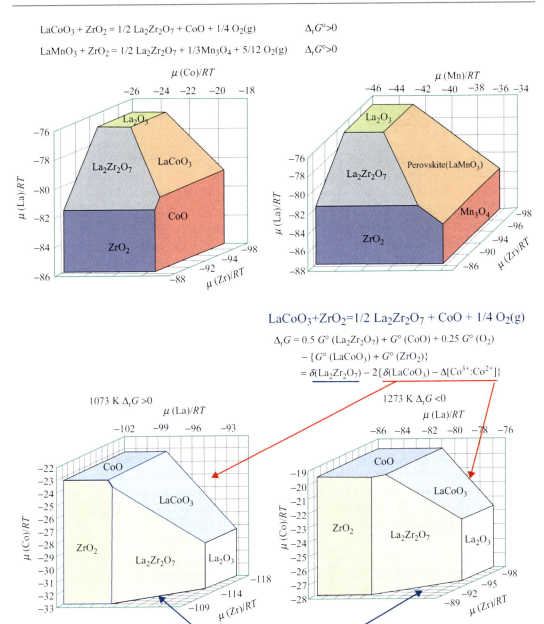

Solid Oxide Fuel Cells, Thermodynamics, Fig. 4 Chemical potential diagrams for the La-Mn-Zr-O and the La-Co-Zr-O system and analyses in terms of the stabilization energy and the valence stability of transition metal oxides

reactions of electrode components with gaseous impurities are important.

For cathode materials, thermodynamic features of reactions with gaseous impurities are essentially the same as those reactions with YSZ electrolyte. Important poisoning elements are Cr and S. As described briefly above, the stable LSM does not react with Cr- or S-containing species. However, the Cr poisoning is severe, while SO_2 poisoning is not severe. This is mainly because the electrochemically deposition reaction is available for LSM-Cr system, but not available for LSM-S system. The Cr poisoning of LSM is caused by Cr_2O_3 deposited at the electrochemically active sites, and volatilization of this Cr_2O_3 is not well promoted so that Cr_2O_3 tends to be accumulated, leading to destructive degradation. For the LSCF cathodes, behaviors against Cr and S are different, although both give rise to the $SrCrO_4$/$SrSO_4$ formation, respectively. When $SrCrO_4$ is formed preferentially at the surface of the LSCF layer, this formation sites can act as the trapping sites. If trapping is sufficient, no degradation is caused from the $SrCrO_4$ formation unless the amount of formed $SrCrO_4$ is small to block the gas permeation or the electrical path. For the $SrSO_4$ formation, this takes place near to the electrochemically active sites, leading immediate lowering in performance. This difference in the formation reaction sites seems to be correlated with difference in reaction; that is, for $SrSO_4$ formation from cathode materials and SO_2, oxygen is needed. Therefore, the electrochemically active sites are convenient for the $SrSO_4$ formation from the kinetic reasons.

For anode materials, sulfur is the most important poisoning impurity. This is not only come from fuels but also from cell components like metal interconnects. From the chemical point of view, the interaction of Ni and S is not so strong concerning compound formation, but it exhibits interesting features that the adsorption of sulfur quickly took place on nickel; in this aspect, interaction is rather strong. In addition, sulfur on nickel is competing with oxygen atoms on nickel to promote dissociation of hydrogen molecules, leading to a slightly lowering of performance but not destructive. With increasing partial pressure of sulfur, degradation behavior turns to be destructive. This is mainly because of promote diffusion of nickel due to the adsorption/dissolution of impurities on/in nickel.

Future Directions

1. Materials thermodynamic and kinetic properties

It has been found that the thermodynamic features can provide useful information on materials behaviors in SOFCs. As a next step, it is highly suggested that the kinetic properties such as nucleation or diffusion should be connected with those thermodynamic features. Particularly, reactions and diffusion can be well consistently considered in terms of chemical potentials of metals or cations. By combining with diffusion and nucleation data, this framework can provide a wider basis of understanding materials behavior. Defect chemistry that describes defects in oxides system is found to be very important to understand mechanical instability of SOFC stacks. Thus, the above combined approach will also be helpful to considerations on mechanical instability.

2. Thermodynamic description of minor impurities in gases or minor carrier in electrolyte.

It has been found that poisoning effects, if any, are originated from the impurity in a quite low partial pressure in order of 10^{-12}–10^{-6} atm. In order to treat those gaseous species, the accurate thermodynamic data for plausible gaseous species are needed.

For protons in electrolytes, it has been found that effects of minor carrier appear significantly in electrode reactions rather than in transport properties in electrolyte (12). Further considerations will be needed to treat protons within the local thermodynamic equilibrium approximation.

Cross-References

▶ Solid Oxide Fuel Cells, History

References

1. Kawada T, Yokokawa H (1997) Materials and characterization of solid oxide fuel cells. In: Nowotny J, Sorrell CC (eds) Electrical properties of ionic solids. Trans Tech Publications, Uetikon-Zuerich, pp 187–248
2. Yokokawa H (2003) Understanding materials compatibility. Annu Rev Mater Res 33:581–610
3. Yokokawa H, Sakai N, Horita T, Yamaji K, Brito ME (2005) Solid oxide electrolytes for high temperature fuel cells. Electrochemistry 73:20–30

4. Yokokawa H (1999) Generalized chemical potential diagram and its applications to chemical reactions at interfaces between dissimilar materials. J Phase Equilib 20(3):258–287
5. Spear KE (1976) High temperature reactivity. In: Treatise on solid state chemistry, vol 4, Plenum, New York, pp 115–192
6. Williams M, Yamaji K, Horita T, Sakai N, Yokokawa H (2009) Exergetic studies of intermediate-temperature, solid oxide fuel cell electrolytes. J Electrochem Soc 156(4):B546–B551
7. Thermodynamic Database MALT for Windows. http://www.kagaku.com/malt/index.html
8. F*A*C*T (Facility for the Analysis of Chemical Thermodynamics). http://www.crct.polymtl.ca/fact/index.php
9. Nicholas Grundy A, Hallstedt B, Gauckler LJ (2004) Assessment of the La-Sr-Mn-O system. Calphad 28:191–201
10. Yokokawa H, Horita T (2012) Solid oxide fuel cell materials: durability, reliability and cost. In: Meyers RA (ed) Encyclopedia of sustainability science and technology. Springer, New York/London, pp 9934–9968
11. Yokokawa H, Yamaji K, Brito ME, Kishimoto H, Horita T (2011) General considerations on degradation of SOFC anodes and cathodes to impurities in gases. J Power Sources 196:7070–7075
12. Yokokawa H (2012), Thermodynamic and SIMS analyses on the role of proton/water in oxygen reduction process and related degradations in solid oxide fuel cells, Solid State Ionics 225: 6–17

Solid State Dye-Sensitized Solar Cell

Henry Snaith and Pablo Docampo
Clarendon Laboratory, Department of Physics, Oxford University, Oxford, UK

In this entry, we outline the basic operating principles of solid-state dye-sensitized solar cells and identify the aspects of the technology which have been a focus of intensive research effort over the last few years. We specifically look at the influence of different mesoporous metal oxides, charge collection, and more strongly absorbing sensitizers.

Introduction

The world's current energy consumption rate has been rising steadily, nearly doubling in the last 30 years. At present, the majority of the energy is produced from combusting coal, natural gas, and oil, which is leading to dangerous climate change [1, 2]. In an effort to meet this demand, renewable energy sources such as wind, solar, geothermal, or tidal sources have been harnessed to produce around 6 % of the total energy consumed, from which sunlight is the most abundant natural source of energy among them. If we were able to harvest the entire sunlight incident on the planet's surface for just 1 h, the whole planet's yearly energy requirements could be met. Completely covering earth's surface with solar cells is not an option, but with below 0.7 % coverage of the land mass available, and with devices with a power conversion efficiency of only 8 %, we would be able to satisfy the whole world's energy needs. With higher efficiencies, an even smaller fraction of the Earth's surface is required.

Most of the technological development effort in photovoltaics has been directed at inorganic semiconductor photovoltaics, which can achieve power conversion efficiencies for single junction devices, for example, GaAs, of up to 29 % and over 37 % in multijunction devices [3]. These cells, albeit being very efficient, tend to require complex and expensive fabrication protocols, which make them expensive to produce, and environmentally costly, in terms of the energy input and waste produced during manufacture.

Some "low-cost" alternative concepts have been introduced over the last 20 years, such as organic, quantum dot-based, or other solution processed inorganic semiconductor-based solar cells [4–7]. Manufacture of all of these concepts is compatible with existing printing and coating equipment, making them ideal candidates for high-throughput, mass-producible solar technology [8]. This latter aspect is important because if our goal is to supply the global energy demand and to be able to substitute other sources of energy that contribute to climate change, a high potential production rate is critical in order to cover the required hundreds of thousands of squared kilometers.

Dye-sensitized solar cells (DSCs), as introduced by O'Regan and Grätzel in 1991, offer

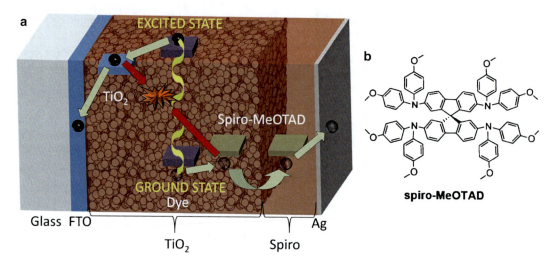

Solid State Dye-Sensitized Solar Cell, Fig. 1 (a) ssDSSC structure schematic and charge transport mechanism. *Green arrows* represent the charge transport paths while the *red arrows* represent the recombination path. (b) Chemical structure of 2,2′,7,7′-tetrakis(N,N-di-p-methoxyphenylamine)-9,9′-spirobifluorene (spiro-MeOTAD)

a low-cost and mass-producible system [9]. These devices are composed of a mesoporous electron-transporting metal oxide, usually TiO_2, sensitized with a monolayer of light-absorbing dye adsorbed to the surface. Absorbed photons excite electrons from the ground state to the excited state of the dye and the electron is subsequently transferred into, and transported through, the TiO_2 mesostructure to the collection electrode, usually fluorine-doped tin oxide (FTO). The dye is also contacted by a redox-active electrolyte, usually iodide/tri-iodide based, which regenerates the oxidized dye molecules and transports the holes to the counterelectrode, typically platinum-coated FTO glass.

The use of liquid electrolyte leads to significant technological challenges, as the devices must be carefully sealed to avoid leaks. Furthermore, the archetypal redox couple iodide tri-iodide is corrosive. In order to address these issues, in 1997, Yanigida and coworkers and, in 1998, Bach et al. replaced the liquid electrolyte with an organic solid-state alternatives, Polypyrrole [10], and 2,2′,7,7′-tetrakis(N,N-di*p*-methoxyphenylamine)-9,9′-spirobifluorene (spiro-MeOTAD) [11]. Spiro-MeOTAD has emerged to be an extremely effective HTM and is still employed in the best-performing current devices. A schematic of the device structure is shown in Fig. 1. Since the ionization potential of the iodide/tri-iodide couple is shallower than that of spiro-MeOTAD, theoretically, higher maximum voltages should be attainable with solid-state devices and would result in more efficient devices [12, 13]. It has been shown that this loss limits the liquid-electrolyte-based devices to around 15 % efficiency, while 20 % power conversion efficiencies should be possible with a solid-state HTM [13]. At present, the best reported solid-state dye-sensitized solar cells, which incorporate spiro-MeOTAD as the HTM, achieve above 7 % efficiencies [14] (Fig. 2).

Discussion

In order to achieve the predicted efficiency values in real operating solid-state dye-sensitized solar cells, the limiting factors must be first identified. The ssDSSC is largely limited by the fact that the optimized device thickness is only 2 μm, which is significantly thinner than the optical depth of the composite. This in turn limits the choice of dyes that can be used to efficiently harvest light, since

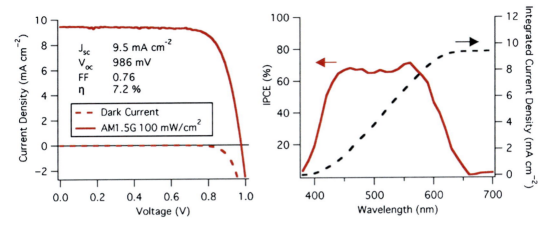

Solid State Dye-Sensitized Solar Cell, Fig. 2 (*left*) JV characteristics of the highest efficiency reported SDSSC and (*right*) its corresponding incident photon-to-electron conversion efficiency spectrum as a function of monochromatic wavelength. The right axis of the graph indicates the current density expected to be generated under standard solar conditions (This figure has been reprinted in full with permission from [14]. Copyright 2011 American Chemical Society)

they must absorb light very strongly [15, 16]. Infiltration of the mesoporous TiO$_2$ with the solid-state HTM is generally identified as the cause for these thickness limitations in the form of poor diffusion lengths, or light-harvesting capability of the dye sensitizers [17, 18]. However, it has been shown that sufficient (>60 %) pore-filling with the current state-of-the-art materials is feasible for thick electrodes (up to 5 μm) [19, 20]. Despite this, there have been no reports of thick films outperforming 2–3-μm thick devices. In fact, short-circuit currents are generally found to be lower with the majority of dyes once the film thickness exceeds 2.5 μm, suggesting that pore-filling is not the main reason for insufficient charge collection. Other factors that limit the electron diffusion length must be identified in order to maximize the collection of photo-generated charges.

Alternative Metal Oxide Photoanodes

Charge transport in solid-state DSSCs based on nanocrystalline TiO$_2$ appears to be "electron-limited" at short-circuit, meaning that the transport of electrons through the mesoporous TiO$_2$ is slower than the transport of holes through the hole transporter [21–23]. The electron transport is generally described by a charge multiple-trapping model: [24, 25] After dye excitation, electrons are transferred into the mesoporous oxide where they relax and become localized in sub-bandgap states (electronic states localized within the band gap). Then, they are thermally released into the conduction band, where they are free to move until they become trapped again [26–29]. This iterative process continues on until charges reach the electrodes and are extracted into the external circuit.

Since charges are only mobile when they are released into the conduction band, it is possible that employing materials with higher mobilities than TiO$_2$ may result in faster transport of charges through the photoanodes. Alternative materials with suitable energy levels for the conduction and valence bands and faster electron mobilities are ZnO [30–32] and SnO$_2$ [33, 34]. The bulk mobility for these two materials is around 200 cm^2 V^{-1} s^{-1} [35, 36], nearly two orders of magnitude above the value for TiO$_2$ which is around 1 cm^2 V^{-1} s^{-1} [37].

When applied in devices, ZnO nanoparticle-based photoanodes incorporating spiro-MeOTAD as the hole transporter can achieve up to 0.5 % efficiency [38, 39], while it is possible to reach over 1.5 % with copper thiocyanate (CuSCN) as the hole transporter (CuSCN is an inorganic solution processable p-type semiconductor) [40]. SnO_2 photoelectrodes fare little better, with maximum efficiencies of well under 0.5 % efficiency for pristine photoanodes [41]. However, when a blocking layer of TiO_2 or MgO is deposited over the pristine SnO_2 mesostructure, the performance can easily reach over 3 % power conversion efficiencies [41–44], and over 1 % power conversion efficiencies for poly(3-hexylthiophene-2,5-diyl) (P3HT), a polymeric organic semiconducting hole transporter [45]. This approach of growing a "shell" of a different material on top of the mesoporous metal oxide is often referred to as a "core-shell" approach. The properties of the "core" material (in this case SnO_2) can be optimized for highly efficient long range charge transport, and the properties of the "shell" material (in this case TiO_2) can be optimized to interface correctly with the dye-sensitizer and maximize charge generation, while inhibiting recombination between electrons in the porous oxide and holes in the HTM (Fig. 3).

Nevertheless, regardless of the choice of metal oxide, be it TiO_2, SnO_2, or ZnO, the electron transport dynamics are similar for all systems [47–50]. In fact, different diffusion coefficients are correlated to changes in the crystal size of the mesoporous oxide and not the constituting materials [49, 51–55]. A similar effect is also found when comparing photoanodes with very different morphologies, such as one dimensional nanotubes against random nanoparticle networks, where comparable diffusion coefficients have been reported due to the similar crystal sizes [56, 57]. The fraction of trapped charges compared to the free charges in the conduction band determines the average diffusion transport through the mesostructure, but this parameter should also be influenced by the long range mobility of the free charges. The fact that such different materials exhibit similar diffusion properties suggests that "grain boundaries" and mesoscale morphology have a significant influence on the long range mobility of free charges [58].

Nanostructured Photoanodes

Factors that determine the power conversion efficiency of complete devices are light absorption, charge generation, and charge collection efficiencies. Charge generation efficiency is wholly dependent on the intrinsic electron transfer properties of a particular dye and its regeneration by the hole transporter. Light absorption efficiency is high if the active layer of a device is thick enough to absorb most of the light over the bandwidth of the sensitizer. Charge collection efficiency depends on the balance of diffusion and recombination in a device. Combining these two together leads to the definition of the electron diffusion length (D_L) $D_L = \sqrt{D_e \tau}$, where D_e is the diffusion coefficient and τ the recombination lifetime.

Charges are only collected efficiently when they are generated within an electron diffusion length of the electrode. To achieve high charge collection efficiency of over 99 %, the film must be three times thinner than the diffusion length. For this reason, a requirement for high power conversion efficiency devices is that the diffusion length is long compared to the light absorption depth, guaranteeing both high charge collection and high light absorption efficiencies. A common approach to improve the transport through the films, and hence improve diffusion lengths and consequently allow thicker and better light-absorbing films, incorporates one-dimensional nanostructures, i.e., nanowires, in the photoanode design [59–62]. The most salient feature of devices fabricated from highly crystalline nanowires is much faster electron transport as compared to standard nanoparticle-based photoanodes [60].

Application of this concept in solar cells has led to the development of ZnO nanowire-based photoanodes, which achieve competitive power conversion efficiencies with the addition of thin insulator layers. In terms of devices, pristine ZnO nanowires achieve substantially lower than 1 % power conversion efficiencies [63, 64], while surface coatings, such as ZrO_2 [63], MgO [63, 65], or TiO_2 [62], have been reported to significantly improve the performance of the fabricated

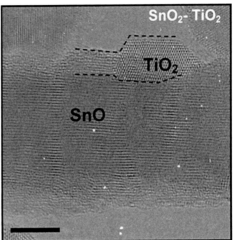
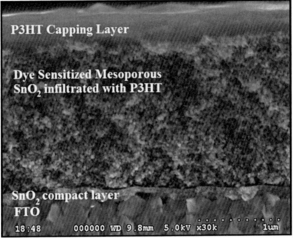

Solid State Dye-Sensitized Solar Cell, Fig. 3 (*left*) TiO$_2$ nanoparticle mesostructure (Reproduced with permission from [46]). (*right*) HRTEM image of a SnO$_2$ nanocrystal within the mesoporous films fabricated from sintered nanoparticle paste and coated with TiO$_2$ coating from a 20 mM TiCl$_4$ solution (Reprinted with permission from [43]. Copyright 2010 American Chemical Society). (*bottom*) Cross-section of a completed device based on a SnO$_2$ mesostructure and P3HT as the hole transporter [45] (Reproduced by permission of the PCCP Owner Societies)

devices, with the best reported efficiency of 5.65 % for TiO$_2$-coated ZnO nanowires [62]. Photoanodes incorporating TiO$_2$ nanowires have also been developed, and this system can achieve a substantial 4.9 % power conversion efficiency [66].

Electron transport through single crystalline nanowire photoanodes is considerably faster than in their nanoparticle-based counterparts, to the point that it can match hole transport through the small molecule spiro-MeOTAD at short-circuit. For this reason, devices incorporating alternative hole transporters with higher mobilities have been introduced over the years, such as P3HT [67] and CuSCN, and significant power conversion efficiencies of over 1.7 % can be achieved in conjunction with dye-sensitized ZnO nanowire photoanodes [68, 69].

While maximizing the charge diffusion length should enable much thicker devices to be fabricated, which still exhibit extremely efficient charge collection efficiency, there is another basic issue with the solid-state DSSC which may well prove to impose the most significant thickness limitation. The HTM is assumed to be, and indeed chosen to be, "transparent" where the dye absorbs, in order to allow the sun light to be absorbed solely in the dye-sensitizer. However, in order to achieve good conductivity in the HTM, it must be partially oxidized (p-doped), which results in new "sub-bandgap" optical transitions, and hence light absorption in the visible and near infrared [70]. The light absorbed in the doped hole-conductor does not result in efficiently generated charge, but is a major loss. This absorption is hence termed "parasitic" absorption and is not advantageous [71, 72]. If the dye absorbs very strongly, then the parasitic absorption losses are low. However, the conventional dyes which work extremely well in the thicker electrolyte-based cells absorb light very weakly in the near-IR. This important region of the spectrum is hence "unobtainable" for the present generation of solid-state DSSCs due to a larger fraction of light absorbed in the HTM than in the dye [71]. Other approaches to maximizing the device performance include the design of high surface area photoanodes in order to achieve more optically dense layers, and hence maximize light absorption in a thin layer, and therefore maintaining a high charge collection efficiency and fractional absorbance in the dye.

High surface area photoanodes can be achieved, for instance, by the use of block copolymers [74–77], which direct the formation of mesoporous structures, capable of delivering higher surface areas than standard nanoparticle devices, while still maintaining high enough charge transport through the films, with over 5 % power conversion efficiencies demonstrated [55]. Other approaches utilize preformed nanostructured particles which are then deposited on the conductive substrate, rather than assembling the structure in situ, usually in the form of hollow-sphere like particles [78]. As far as charge transport is concerned, the optimum photoanode would be a single crystalline TiO_2 film, which is highly porous. As a significant step in this direction, mesoporous single crystals (MSCs) of anatase TiO_2 have been demonstrated, with crystal sizes ranging from a few hundred nm up to 3 μm,

and independently tunable pore sizes varying from 20 nm to 250 nm [73] (Fig. 4).

Thin films of mesoporous single crystal assemblies exhibit "nanowire-like" transport while still maintaining high enough surface area for dye adsorption. A technological advantage is that when processed into films, the mesoporous single crystal assemblies do not require thermal sintering to ensure good electrical connectivity between the crystals, and efficiencies of over 3 % in solid-state DSSCs have been demonstrated for devices processed entirely below 150 °C [73]. This low temperature processing greatly simplifies manufacture, enables a broad choice of substrates, including plastic foil for low-cost, flexible solar cells [79], and will enable multiple junction ssDSSC fabrication, which by itself should enable a steep increase in efficiency in the near future.

Inorganic Absorbers

One of the disadvantages of using dyes as the light-absorbing material is that devices achieve optimum performance when a monolayer is deposited [80, 81], and hence, it is quite challenging to achieve high enough surface area for dye loading while maintaining the fast transport characteristics of larger crystals. Inorganic absorbers can significantly bypass this issue, since they generally exhibit good intrinsic charge transport characteristics, and can therefore be used in comparatively "thick" absorber layers of between 2 and 10 nm in thickness, (as opposed to the molecular nm layer of dye), enabling much higher light absorption in micron-thick mesoporous films.

One of the leading candidates for this class of materials is antimony sulfide [82], which can be used with CuSCN [83, 84] to achieve over 3.7 % efficiencies [84]. However, the most efficient devices are fabricated with hole-transporting polymers, such as P3HT, with efficiencies of over 6 % demonstrated [85, 86]. Other inorganic absorbers have also been developed, such as CdSe, which can achieve over 1.5 % efficiency with P3HT as the hole transporter [87] and over

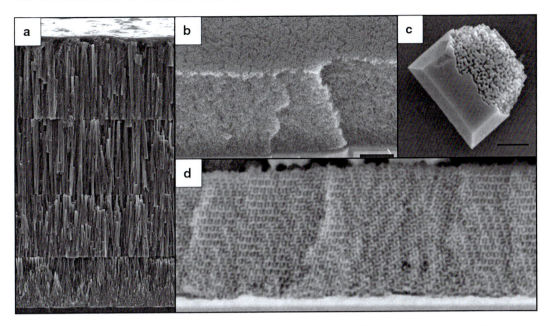

Solid State Dye-Sensitized Solar Cell, Fig. 4 Cross-sectional SEMs of: (**a**) ZnO nanowires (Reprinted with permission from [62]. Copyright 2012 American Chemical Society). (**b**) Triblock copolymer templated photoanodes (Reprinted with permission from [55]. Copyright 2012 WILEY-VCH Verlag GmbH & Co. KGaA, Weinheim). (**c**) Mesoporous single crystals (Reproduced with permission from [73]. Copyright 2013, Rights Managed by Nature Publishing Group). (**d**) Gyroidal assembly (Reproduced with permission from [74]. Copyright 2008 American Chemical Society)

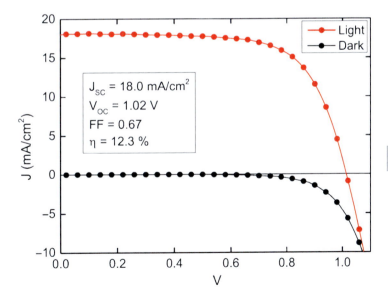

Solid State Dye-Sensitized Solar Cell, Fig. 5 JV curve characteristics of the best-performing solid-state perovskite-based solar cell (Reproduced with permission of The Royal Society of Chemistry from [92])

2 % for the CuSCN alternative [88, 89], or PbS [90], which can achieve over 1.3 % efficient cells with spiro-MeOTAD as the hole transporter [91] (Fig. 5).

Recently, a new family of materials has emerged with superior photovoltaic properties than both the dye-sensitized or inorganic sensitized absorbers. These are organolead halide perovskites which were first introduced by Miyasaka and coworkers as a sensitizer for electrolyte-based cells [93]. Over the last year, these perovskite absorbers have proved themselves superior for use in solid-state sensitized devices, in the first instance due to extremely strong and broad light absorption in films as thin as 500 nm [94, 95]. It has also become apparent that in addition to absorbing light and transferring charge a short distance into mesoporous TiO_2, the perovskite absorber itself can sustain long range and extremely fast transport of both electrons and holes simultaneously [92, 94]. With this realization, the mesoporous TiO_2 has become redundant, since it is no longer required for long range electron transport, and the perovskite absorber layer can assume the roles of light absorption and electron transport, and even long range hole transport at the same time [92, 94]. The maximum efficiency of the perovskite-based solar cells has now exceeded 12 %, as can be seen in Fig. 5, and is set to rise considerably further in the next few years.

Conclusions

Solid-state sensitized solar cells are a very promising candidate for low-cost solar energy generation, and represent an extremely rich research platform for investigating optoelectronic processes in organic, inorganic, and hybrid semiconductor materials. The current limitation to achieve predicted theoretical efficiencies appears to be thickness constraints, and hence insufficient light absorption with the current best-performing dyes. One proposed solution consists of utilizing one-dimensional or long range crystalline materials to improve the diffusion lengths of electrons in the mesostructure, and in this way, charge can be collected efficiently in thicker films. While the system has been thoroughly developed and proven, the current state-of-the-art hole transporters are not completely transparent, and hence, problems arise from absorption of photons in these materials. An alternative solution consists of utilizing inorganic absorbers which can be processed in relatively "thick" films of several

nanometers, rather than a single monolayer, as is required for efficient dye-sensitized devices. In this way, the optical density of the film is much improved, requiring much thinner mesostructures to operate efficiently. Among these, organolead halide perovskites have been shown to achieve very high power conversion efficiencies of over 12 % and offer a very promising alternative to the currently established photovoltaic industry.

References

1. Oreskes N (2004) The scientific consensus on climate change. Science 306:1686
2. Anderson K, Bows A (2011) Beyond 'dangerous' climate change: emission scenarios for a new world. Philos Trans R Soc A Math Phys Eng Sci 369:20–44
3. Green MA, Emery K, Hishikawa Y, Warta W, Dunlop ED (2013) Solar cell efficiency tables (version 41). Progr Photovolt Res Appl 21:1–11
4. Li G, Shrotriya V, Huang J, Yao Y, Moriarty T, Emery K, Yang Y (2005) High-efficiency solution processable polymer photovoltaic cells by self-organization of polymer blends. Nat Mater 4:864–868
5. McDonald SA, Konstantatos G, Zhang S, Cyr PW, Klem EJD, Levina L, Sargent EH (2005) Solution-processed PbS quantum dot infrared photodetectors and photovoltaics. Nat Mater 4:138–142
6. Katagiri H, Jimbo K, Maw WS, Oishi K, Yamazaki M, Araki H, Takeuchi A (2009) Development of CZTS-based thin film solar cells. Thin Solid Films 517:2455–2460
7. Jeong S, Lee B-S, Ahn S, Yoon K, Seo Y-H, Choi Y, Ryu B-H (2012) An 8.2% efficient solution-processed $CuInSe_2$ solar cell based on multiphase $CuInSe_2$ nanoparticles. Energy Environ Sci 5:7539–7542
8. Krebs FC, Tromholt T, Jorgensen M (2010) Upscaling of polymer solar cell fabrication using full roll-to-roll processing. Nanoscale 2:873–886
9. O'Regan B, Gratzel M (1991) A low-cost, high-efficiency solar cell based on dye-sensitized colloidal TiO_2 films. Nature 353:737–740
10. Murakoshi K, Kogure R, Wada Y, Yanagida S (1997) Solid state dye-sensitized TiO_2 solar cell with polypyrrole as hole transport layer. Chem Lett 5:471–472
11. Bach U, Lupo D, Comte P, Moser JE, Weissortel F, Salbeck J, Spreitzer H, Gratzel M (1998) Solid-state dye-sensitized mesoporous TiO_2 solar cells with high photon-to-electron conversion efficiencies. Nature 395:583–585
12. Clifford JN, Palomares E, Nazeeruddin MK, Grätzel M, Durrant JR (2007) Dye dependent regeneration dynamics in dye sensitized nanocrystalline solar cells: evidence for the formation of a ruthenium

bipyridyl cation/iodide intermediate. J Phys Chem C 111:6561–6567

13. Snaith HJ (2010) Estimating the maximum attainable efficiency in dye-sensitized solar cells. Adv Funct Mater 20:13–19

14. Burschka J, Dualeh A, Kessler F, Baranoff E, Cevey-Ha N-L, Yi C, Nazeeruddin MK, Grätzel M (2011) Tris(2-(1H-pyrazol-1-yl)pyridine)cobalt(III) as p-type dopant for organic semiconductors and its application in highly efficient solid-state dye-sensitized solar cells. J Am Chem Soc 133:18042–18045

15. Snaith HJ, Schmidt-Mende L (2007) Advances in liquid-electrolyte and solid-state dye-sensitized solar cells. Adv Mater 19:3187–3200

16. Kruger J, Plass R, Gratzel M, Cameron P, Peter L (2003) Charge transport and back reaction in solid-state dye-sensitized solar cells: a study using intensity-modulated photovoltage and photocurrent spectroscopy. J Phys Chem B 107:7536–7539

17. Schmidt-Mende L, Grätzel M (2006) TiO_2 pore-filling and its effect on the efficiency of solid-state dye-sensitized solar cells. Thin Solid Films 500:296–301

18. Kroeze JE, Hirata N, Schmidt-Mende L, Orizu C, Ogier SD, Carr K, Gratzel M, Durrant JR (2006) Parameters influencing charge separation in solid-state dye-sensitized solar cells using novel hole conductors. Adv Funct Mater 16:1832–1838

19. Ding IK, Tetreault N, Brillet J, Hardin BE, Smith EH, Rosenthal SJ, Sauvage F, Gratzel M, McGehee MD (2009) Pore-filling of spiro-OMeTAD in solid-state dye sensitized solar cells: quantification, mechanism, and consequences for device performance. Adv Funct Mater 19:2431–2436

20. Docampo P, Hey A, Guldin S, Gunning R, Steiner U, Snaith HJ (2012) Pore filling of spiro-OMeTAD in solid-state dye-sensitized solar cells determined via optical reflectometry. Adv Funct Mater 22:5010–5019

21. Snaith HJ, Gratzel M (2007) Electron and hole transport through mesoporous TiO_2 infiltrated with spiro-OMeTAD. Adv Mater 19:3643–3647

22. Hauch A, Georg A (2001) Diffusion in the electrolyte and charge-transfer reaction at the platinum electrode in dye-sensitized solar cells. Electrochim Acta 46:3457–3466

23. Kopidakis N, Benkstein KD, van de Lagemaat J, Frank AJ (2003) Transport-limited recombination of photocarriers in dye-sensitized nanocrystalline TiO_2 solar cells. J Phys Chem B 107:11307–11315

24. Orenstein J, Kastner M (1981) Photocurrent transient spectroscopy: measurement of the density of localized states in a -As_2Se_3. Phys Rev Lett 46:1421–1424

25. Tiedje T, Rose A (1981) A physical interpretation of dispersive transport in disordered semiconductors. Solid State Commun 37:49–52

26. Peter LM (2007) Dye-sensitized nanocrystalline solar cells. Phys Chem Chem Phys 9:2630–2642

27. Nelson J, Chandler RE (2004) Random walk models of charge transfer and transport in dye sensitized systems. Coord Chem Rev 248:1181–1194

28. Nelson J (1999) Continuous-time random-walk model of electron transport in nanocrystalline TiO_2 electrodes. Phys Rev B 59:15374–15380

29. Bisquert J, Vikhrenko VS (2004) Interpretation of the time constants measured by kinetic techniques in nanostructured semiconductor electrodes and Dye-sensitized solar cells. J Phys Chem B 108:2313–2322

30. Bedja I, Kamat PV, Hua X, Lappin AG, Hotchandani S (1997) Photosensitization of nanocrystalline ZnO films by Bis(2,2′-bipyridine) (2,2;-bipyridine-4,4′-dicarboxylic acid)ruthenium(II). Langmuir 13:2398–2403

31. Keis K, Magnusson E, Lindstrom H, Lindquist S-E, Hagfeldt A (2002) A 5% efficient photoelectrochemical solar cell based on nanostructured ZnO electrodes. Solar Energ Mater Solar Cell 73:51–58

32. Memarian N, Concina I, Braga A, Rozati SM, Vomiero A, Sberveglieri G (2011) Hierarchically assembled ZnO nanocrystallites for high-efficiency dye-sensitized solar cells. Angew Chem Int Ed 50:12321–12325

33. Kay A, Gratzel M (2002) Dye-sensitized core-shell nanocrystals: improved efficiency of mesoporous tin oxide electrodes coated with a thin layer of an insulating oxide. Chem Mater 14:2930–2935

34. Tennakone K, Bandara J, Bandaranayake PKM, Kumara GRA, Konno A (2001) Enhanced efficiency of a dye-sensitized solar cell made from MgO-coated nanocrystalline SnO_2. Jpn J Appl Phys Part 2-Lett 40: L732–L734

35. Look DC, Reynolds DC, Sizelove JR, Jones RL, Litton CW, Cantwell G, Harsch WC (1998) Electrical properties of bulk ZnO. Solid State Commun 105:399–401

36. Nagasawa M, Shionoya S, Makishima S (1965) Vapor reaction growth of SnO_2 single crystals and their properties. Jpn J Appl Phys 4:195–202

37. Forro L, Chauvet O, Emin D, Zuppiroli L, Berger H, Levy F (1994) High mobility n-type charge carriers in large single crystals of anatase (TiO_2). J Appl Phys 75:633–635

38. Bouclé J, Ackermann J (2012) Solid-state dye-sensitized and bulk heterojunction solar cells using TiO_2 and ZnO nanostructures: recent progress and new concepts at the borderline. Polym Int 61:355–373

39. Boucharef M, Bin CD, Boumaza MS, Colas M, Snaith HJ, Ratier B, Bouclé J (2010) Solid-state dye-sensitized solar cells based on ZnO nanocrystals. Nanotechnology 21:205203

40. O'Regan B, Schwartz DT, Zakeeruddin SM, Grätzel M (2000) Electrodeposited nanocomposite n–p heterojunctions for solid-state dye-sensitized photovoltaics. Adv Mater 12:1263–1267

41. Docampo P, Tiwana P, Sakai N, Miura H, Herz L, Murakami T, Snaith HJ (2012) Unraveling the function of an MgO inter layer in both electrolyte and

solid-state SnO_2 based dye-sensitized solar cells. J Phys Chem C 116:22840–22846

42. Docampo P, Snaith HJ (2011) Obviating the requirement for oxygen in SnO_2-based solid-state dye-sensitized solar cells. Nanotechnology 22(22):225403

43. Snaith HJ, Ducati C (2010) SnO_2-based dye-sensitized hybrid solar cells exhibiting near unity absorbed photon-to-electron conversion efficiency. Nano Lett 10:1259–1265

44. Tiwana P, Docampo P, Johnston MB, Herz LM, Snaith HJ (2012) The origin of an efficiency improving "light soaking" effect in SnO_2 based solid-state dye-sensitized solar cells. Energy Environ Sci 5:9566–9573

45. Sadoughi G, Sivaram V, Gunning R, Docampo P, Bruder I, Pschirer N, Irajizad A, Snaith HJ (2013) Enhanced electronic contacts in SnO_2-dye-P3HT based solid state dye sensitized solar cells. Phys Chem Chem Phys 15:2075–2080

46. Docampo P, Guldin S, Stefik M, Tiwana P, Orilall MC, Hüttner S, Sai H, Wiesner U, Steiner U, Snaith HJ (2010) Control of solid-state dye-sensitized solar cell performance by block-copolymer-directed TiO_2 synthesis. Adv Funct Mater 20:1787–1796

47. Willis RL, Olson C, O'Regan B, Lutz T, Nelson J, Durrant JR (2002) Electron dynamics in nanocrystalline ZnO and TiO_2 films probed by potential step chronoamperometry and transient absorption spectroscopy. J Phys Chem B 106:7605–7613

48. Fukai Y, Kondo Y, Mori S, Suzuki E (2007) Highly efficient dye-sensitized SnO_2 solar cells having sufficient electron diffusion length. Electrochem Commun 9:1439–1443

49. Quintana M, Edvinsson T, Hagfeldt A, Boschloo G (2006) Comparison of dye-sensitized ZnO and TiO_2 solar cells: studies of charge transport and carrier lifetime. J Phys Chem C 111:1035–1041

50. Tiwana P, Docampo P, Johnston MB, Snaith HJ, Herz LM (2011) Electron mobility and injection dynamics in mesoporous ZnO, SnO_2, and TiO_2 films used in dye-sensitized solar cells. ACS Nano 5:5158–5166

51. Li K-L, Xie Z-B, Adams S (2010) Fast charge transport of titania nanotube arrays in dye-sensitized solar cells. Z Kristallogr 225:173–179

52. Kopidakis N, Neale NR, Zhu K, van de Lagemaat J, Frank AJ (2005) Spatial location of transport-limiting traps in TiO_2 nanoparticle films in dye-sensitized solar cells. Appl Phys Lett 87(20):202106

53. Nakade S, Matsuda M, Kambe S, Saito Y, Kitamura T, Sakata T, Wada Y, Mori H, Yanagida S (2002) Dependence of TiO_2 nanoparticle preparation methods and annealing temperature on the efficiency of dye-sensitized solar cells. J Phys Chem B 106:10004–10010

54. Nakade S, Saito Y, Kubo W, Kitamura T, Wada Y, Yanagida S (2003) Influence of TiO_2 nanoparticle size on electron diffusion and recombination in dye-sensitized TiO_2 solar cells. J Phys Chem B 107:8607–8611

55. Docampo P, Guldin S, Steiner U, Snaith HJ (2012) Triblock terpolymer directed self-assembly of mesoporous TiO_2 – high performance photoanodes for solid state dye-sensitized solar cells. Adv Energy Mater 2(6):676–682

56. Zhu K, Neale NR, Miedaner A, Frank AJ (2007) Enhanced charge-collection efficiencies and light scattering in dye-sensitized solar cells using oriented TiO_2 nanotubes arrays. Nano Lett 7:69–74

57. Ohsaki Y, Masaki N, Kitamura T, Wada Y, Okamoto T, Sekino T, Niihara K, Yanagida S (2005) Dye-sensitized TiO_2 nanotube solar cells: fabrication and electronic characterization. Phys Chem Chem Phys 7(24):4157–63

58. Docampo P, Guldin S, Steiner U, Snaith HJ (2013) Charge transport limitations in self-assembled TiO_2 photoanodes for dye-sensitized solar cells. J Phys Chem Lett 4:698–703

59. Gubbala S, Chakrapani V, Kumar V, Sunkara MK (2008) Band-edge engineered hybrid structures for dye-sensitized solar cells based on SnO_2 nanowires. Adv Funct Mater 18:2411–2418

60. Martinson ABF, Góes M r S, Fabregat-Santiago F, Bisquert J, Pellin MJ, Hupp JT (2009) Electron transport in dye-sensitized solar cells based on ZnO nanotubes: evidence for highly efficient charge collection and exceptionally rapid dynamics†. J Phys Chem A 113:4015–4021

61. Law M, Greene LE, Johnson JC, Saykally R, Yang P (2005) Nanowire dye-sensitized solar cells. Nat Mater 4:455–459

62. Xu C, Wu J, Desai UV, Gao D (2012) High-efficiency solid-state dye-sensitized solar cells based on TiO_2-coated ZnO nanowire arrays. Nano Lett 12(5):2420–2424

63. Plank NOV, Howard I, Rao A, Wilson MWB, Ducati C, Mane RS, Bendall JS, Louca RRM, Greenham NC, Miura H et al (2009) Efficient ZnO nanowire solid-state dye-sensitized solar cells using organic dyes and core-shell nanostructures. J Phys Chem C 113:18515–18522

64. Schlur L, Carton A, Lévêque P, Guillon D, Pourroy G (2013) Optimization of a New ZnO nanorods hydrothermal synthesis method for solid state dye sensitized solar cells applications. J Phys Chem C 117:2993–3001

65. Plank NOV, Snaith HJ, Ducati C, Bendall JS, Schmidt-Mende L, Welland ME (2008) A simple low temperature synthesis route for ZnO–MgO core-shell nanowires. Nanotechnology 19:465603

66. Tétreault N, Horváth E, Moehl T, Brillet J, Smajda R, Bungener S, Cai N, Wang P, Zakeeruddin SM, Forró L et al (2010) High-efficiency solid-state dye-sensitized solar cells: fast charge extraction through self-assembled 3D fibrous network of crystalline TiO_2 nanowires. ACS Nano 4:7644–7650

67. Lee TH, Sun DZ, Zhang X, Sue HJ, Cheng X (2009) Solid-state dye-sensitized solar cell based on semiconducting nanomaterials. J Vac Sci Technol B 27:3073–3077

68. Desai UV, Xu C, Wu J, Gao D (2012) Solid-state dye-sensitized solar cells based on ordered ZnO nanowire arrays. Nanotechnology 23:205401
69. Selk Y, Minnermann M, Oekermann T, Wark M, Caro J (2011) Solid-state dye-sensitized ZnO solar cells prepared by low-temperature methods. J Appl Electrochem 41:445–452
70. Abate A, Leijtens T, Pathak S, Teuscher J, Avolio R, Errico ME, Kirkpatrik J, Ball JM, Docampo P, McPherson I et al (2013) Lithium salts as "redox active" p-type dopants for organic semiconductors and their impact in solid-state dye-sensitized solar cells. Phys Chem Chem Phys 15:2572–2579
71. Margulis GY, Hardin BE, Ding IK, Hoke ET, McGehee MD (2013) Parasitic absorption and internal quantum efficiency measurements of solid-state dye sensitized solar cells. Adv Energ Mater 3(7):959–966
72. Snaith HJ, Moule AJ, Klein C, Meerholz K, Friend RH, Grätzel M (2007) Efficiency enhancements in solid-state hybrid solar cells via reduced charge recombination and increased light capture. Nano Lett 7:3372–3376
73. Crossland EJW, Noel N, Sivaram V, Leijtens T, Alexander-Webber JA, Snaith HJ (2013) Mesoporous TiO$_2$ single crystals delivering enhanced mobility and optoelectronic device performance. Nature 495:215–219
74. Crossland EJW, Kamperman M, Nedelcu M, Ducati C, Wiesner U, Smilgies DM, Toombes GES, Hillmyer MA, Ludwigs S, Steiner U et al (2008) A bicontinuous double gyroid hybrid solar cell. Nano Lett 9:2807–2812
75. Crossland EJW, Nedelcu M, Ducati C, Ludwigs S, Hillmyer MA, Steiner U, Snaith HJ (2008) Block copolymer morphologies in dye-sensitized solar cells: probing the photovoltaic structure–function relation. Nano Lett 9:2813–2819
76. Nedelcu M, Lee J, Crossland EJW, Warren SC, Orilall MC, Guldin S, Huttner S, Ducati C, Eder D, Wiesner U et al (2009) Block copolymer directed synthesis of mesoporous TiO$_2$ for dye-sensitized solar cells. Soft Matter 5:134–139
77. Guldin S, Docampo P, Stefik M, Kamita G, Wiesner U, Snaith HJ, Steiner U (2012) Layer-by-layer formation of block-copolymer-derived TiO$_2$ for solid-state dye-sensitized solar cells. Small 8:432–440
78. Brendel JC, Lu Y, Thelakkat M (2010) Polymer templated nanocrystalline titania network for solid state dye sensitized solar cells. J Mater Chem 20:7255–7265
79. Miyasaka T (2011) Toward printable sensitized mesoscopic solar cells: light-harvesting management with thin TiO$_2$ films. J Phys Chem Lett 2:262–269
80. Khazraji AC, Hotchandani S, Das S, Kamat PV (1999) Controlling dye (Merocyanine-540) aggregation on nanostructured TiO$_2$ films. An organized assembly approach for enhancing the efficiency of photosensitization. J Phys Chem B 103:4693–4700

81. Yum J-H, Chen P, Grätzel M, Nazeeruddin MK (2008) Recent developments in solid-state dye-sensitized solar cells. ChemSusChem 1:699–707
82. Gui EL, Kang AM, Pramana SS, Yantara N, Mathews N, Mhaisalkar S (2012) Effect of TiO$_2$ mesoporous layer and surface treatments in determining efficiencies in antimony sulfide-(Sb$_2$S$_3$) sensitized solar cells. J Electrochem Soc 159:B247–B250
83. Itzhaik Y, Niitsoo O, Page M, Hodes G (2009) Sb$_2$S$_3$-sensitized nanoporous TiO$_2$ solar cells. J Phys Chem C 113:4254–4256
84. Nezu S, Larramona G, Choné C, Jacob A, Delatouche B, Péré D, Moisan C (2010) Light soaking and gas effect on nanocrystalline TiO$_2$/Sb$_2$S$_3$/CuSCN photovoltaic cells following extremely thin absorber concept. J Phys Chem C 114:6854–6859
85. Chang JA, Rhee JH, Im SH, Lee YH, Kim H-J, Seok SI, Nazeeruddin MK, Gratzel M (2010) High-performance nanostructured inorganic–organic heterojunction solar cells. Nano Lett 10:2609–2612
86. Im SH, Lim C-S, Chang JA, Lee YH, Maiti N, Kim H-J, Nazeeruddin MK, Grätzel M, Seok SI (2011) Toward interaction of sensitizer and functional moieties in hole-transporting materials for efficient semiconductor-sensitized solar cells. Nano Lett 11:4789–4793
87. Yong Hui L, Sang Hyuk I, Jeong Ah C, Jong-Heun L, Sang Il S (2012) CdSe-sensitized inorganic–organic heterojunction solar cells: The effect of molecular dipole interface modification and surface passivation. Org Electron 13:975–979
88. Larramona G, Choné C, Jacob A, Sakakura D, Delatouche B, Péré D, Cieren X, Nagino M, Bayón R (2006) Nanostructured photovoltaic cell of the type titanium dioxide, cadmium sulfide thin coating, and copper thiocyanate showing high quantum efficiency. Chem Mater 18:1688–1696
89. Lévy-Clément C, Tena-Zaera R, Ryan MA, Katty A, Hodes G (2005) CdSe-sensitized p-CuSCN/nanowire n-ZnO heterojunctions. Adv Mater 17:1512–1515
90. Snaith HJ, Stavrinadis A, Docampo P, Watt AAR (2011) Lead-sulphide quantum-dot sensitization of tin oxide based hybrid solar cells. Solar Energy 85:1283–1290
91. Lee H, Leventis HC, Moon S-J, Chen P, Ito S, Haque SA, Torres T, Nüesch F, Geiger T, Zakeeruddin SM et al (2009) PbS and CdS quantum dot-sensitized solid-state solar cells: "old concepts, new results". Adv Funct Mater 19:2735–2742
92. Ball JM, Lee MM, Hey A, Snaith H (2013) Low-temperature processed mesosuperstructured to thin-film perovskite solar cells. Energy Environ Sci 6:1739–1743
93. Kojima A, Teshima K, Shirai Y, Miyasaka T (2009) Organometal halide perovskites as visible-light sensitizers for photovoltaic cells. J Am Chem Soc 131:6050–6051
94. Lee MM, Teuscher J, Miyasaka T, Murakami TN, Snaith HJ (2012) Efficient hybrid solar cells based

on meso-superstructured organometal halide perovskites. Science 338:643–647

95. Kim H-S, Lee C-R, Im J-H, Lee K-B, Moehl T, Marchioro A, Moon S-J, Humphry-Baker R, Yum J-H, Moser JE et al (2012) Lead iodide perovskite sensitized all-solid-state submicron thin film mesoscopic solar cell with efficiency exceeding 9%. Sci Rep 2:591

Solid State Electrochemistry, Electrochemistry Using Solid Electrolytes

Ulrich Guth
Kurt-Schwabe-Institut für Mess- und Sensortechnik e.V. Meinsberg, Waldheim, Germany
FB Chemie und Lebensmittelchemie, Technische Universität Dresden, Dresden, Germany

The electrochemistry of cells with ▶ *solid electrolytes* is in many aspects similar to the electrochemistry with aqueous solutions or ▶ *ionic liquids*, but we have to take into account special conditions. Solid electrolyte cells can be built up with all ionic conductors provided that the transference number for ionic species is nearly one. Very common and widely used in industry and scientific labs are cells with electrolytes having an ionic conductivity based on mobile oxygen ions (more precisely oxide ions).

The specific characteristic of solid electrolyte cells can be summarized as follows:

- The solid electrolyte is only under certain conditions regarding temperature and partial pressure a pure ionic conductor. Generally, such solids are so-called *mixed conductors*. (see entry ▶ Mixed Conductors, Determination of Electronic and Ionic Conductivity (Transport Numbers)). For example, the oxide ionic conductor is not fixed in its chemical composition because under low partial pressure oxygen is released and under high partial pressure oxygen is incorporated. As the result electronic defects (excess electrons and holes) are formed. The higher the temperature and the more extreme the oxygen partial pressure

(high and low), the more electronic carriers are generated. Therefore the transference number of ionic charge carriers is a function temperature and partial pressure. Generally accepted is that solid conductors which transference number is >0.99 are called as ▶ *solid electrolytes*.

- Unlike classical electrolytes mostly only one ion is mobile, whereas the counterion is fixed in the sublattice. Sometimes this phenomenon is called as unipolare ionic conductivity.

- The conductivity for ions as well as for electronic carriers is an activated process. Hence, such cells operate at elevated temperature. The activation energies for the ionic transport depend on the defect structure of the electrolyte. It can be distinguished between so-called super ionic conductors like α-AgI in which the number of free places for mobile ions in the lattice is much higher than the number of mobile ions and other, for practical purpose more interesting electrolytes based on chemical disorder. That means, the defects are formed by the chemical composition, e.g., by doping with aliovalent ions. The activation energy of ionic conductivity is small for super ion conductors ($\sim$30 kJ/mol and less) and much higher for solids with chemical disorder ($\sim$100 kJ/mol and more).

- The most used solid electrolyte is *yttria*-stabilized zirconia, where the tetravalent zirconium ion (Zr^{4+}) in ZrO_2 is partially substituted by the three-valent yttrium (Y^{3+}). The yttrium ion needs only 1.5 oxide ions. Therefore a part of the oxide lattice is not occupied by oxygen ions. From the point of defect chemistry (see entry "▶ Kröger-Vinks Notation of Point Defects"), such defects are called as oxygen vacancies (more precisely oxide ion vacancies). Due to the incorporation of lower valent yttrium ions instead of zirconium ions, oxide ion vacancies are generated in the cubic fluorite-type lattice. For 8 mol-% Y_2O_3 the composition of such substituted zirconia is $Zr_{0.84}Y_{0.16}O_{1.92}$.

- With increasing temperature the transport of oxide ions (O^{2-}) becomes more and more possible. As a result the electrical (ionic) conductivity increases exponentially with increasing temperature. On the other hand, the crystal structure of the substituted zirconia becomes stable over a wide range of temperature. Therefore this modified zirconia is denoted as yttria-stabilized zirconia (YSZ) that means ceramic stabilized. Yttria-stabilized zirconia as the stabilized ceramic was prepared and used as an electrical conductor firstly by W. Nernst in 1899 [1]. The mechanism of ionic conductivity in such mixed oxides was described by C. Wagner in 1934 [2]. Besides yttria other oxides like CaO, MgO, Yb_2O_3, and Sc_2O_3 are used to stabilize the structure and to increase the electrical conductivity.

- Due to the high operating temperature, typically between 400 °C and 1,500 °C, and the presence of oxygen electrode, materials have to be chemically stable under these conditions; they must not evaporate. Only precious metals like platinum, gold, and silver as well as oxide systems (perovskites) having a high electric conductivity fulfil these requirements.
- Electrochemical cells using solid electrolytes are widely applied as high-temperature fuel cells (SOFC) [3], electrochemical sensors [4, 5] for *combustion control* (see entry ▶ Combustion Control Sensors, Electrochemical), lambda probes, and as ceramic membranes for oxygen separation, for hydrogengeneration by water *vapour electrolyses* and selective oxidation of hydrocarbons [6]. In the recent 50 years, sensors with solid electrolytes have conquered a big market especially in automotive application and in a lot of high-temperature processes [7–9]. The application of such sensors are extended to other biological processes in order to achive in situ information.

Cross-References

▶ Solid Electrolytes Cells, Electrochemical Cells with Solid Electrolytes in Equilibrium
▶ Solid Oxide Fuel Cells, Direct Hydrocarbon Type
▶ Solid Oxide Fuel Cells, Introduction

References

1. Nernst W (1899) On the electrolytic conductivity of solids at high temperatures (in German) Über die elektrolytische Leitung fester Körper bei sehr hohen Temperaturen. Z Elektrochem 6:41–43
2. Wagner C (1943) On the mechanism of the electrical current conductivity in the Nernst glower (in German: Über den Mechanismus der elektrischen Stromleitung im Nernststift). Naturwissenschaften 31:265–268
3. Möbius HH (1997) On the history of solid electrolyte fuel cells. J Solid State Electrochem 1:2
4. Peters H, Möbius HH (1958) Electrochemical investigation of the equilibria $CO + 0,5\, O_2 = CO2$ und $C + CO_2 = 2\, CO$ (in German: Elektrochemische Untersuchung der Gleichgewichte $CO + 0,5\, O_2 = CO2$ und $C + CO_2 = 2\, CO$). Z phys Chem (Leipzig) 209:298–309
5. Weisbart J, Ruka R (1961) Oxygen Gauge. Rev sci Instrum 32:563–595
6. Gellings PJ, Bouwmeester HJM (1997) The CRC handbook of solid state electrochemistry. CRC Press, Boca Raton/New York/London/Tokyo
7. Möbius HH (1991) Solid state electrochemical potentiometric sensors for gas analysis. In: Göpel W, Hesse J, Zemel JN (eds) Sensors a comprehensive survey. VCH, New York/Weinheim
8. Guth U (2008) Gas sensors. In: Bard AJ, Inzelt G, Scholz F (eds) Electrochemical dictionary, 2 extendedth edn. Springer, Heidelberg/Dordrecht/London/New York, pp 400–406, 2012
9. Fischer WA, Janke D (1970) Solid electrolyte cells with gaseous reference electrodes for measurement of oxygen activity in iron melts (in German: Festelektrolytzellen mit gasförmigen Vergleichselektroden zur Messung der Sauerstoffaktivität in Eisenschmelzen). Arch Eisenhüttenw 41:1027–1033

Solvents and Solutions

Rudolf Holze
AG Elektrochemie, Institut für Chemie,
Technische Universität Chemnitz, Chemnitz,
Germany

Electrochemistry as the science of the chemistry of charged species in bulk volumes and at interfaces couples the flow of electricity carried by electrons (in electronic or hole conductors) with the flow of ions (in ionic conductors) by interfacial electrode reactions. Although there are numerous materials known as ionic conductors, they can be classified into a few types:

- Solid electrolytes (inorganic crystalline and amorphous materials, solid polymers)
- Molten salts and ionic liquids, the latter being molten salts liquid already at temperatures $T < 100 \,°C$
- Electrolyte solutions.

The last one shall be characterized with emphasis on technological aspects below.

An electrolyte solution (frequently the term "solution" is forgotten; this may possibly result in confusion) is always composed of a solvent and an electrolyte. Electrolytes are either potential electrolytes, materials which are present as molecular compounds when not yet brought into a solvent (e.g., HCl or acetic acid), or true electrolytes, where the ions are already present before dissolution (e.g., salts like NaCl). In both classes strong and weak electrolytes can be found. The former ones are completely dissociated in solution (at least at low and moderate concentrations), whereas the latter ones are only partially dissociated with the degree of dissociation depending on the concentration of the electrolyte (Ostwald's dilution law). Typical examples of the former type are HCl (a potential electrolyte) or KCl, of the latter class butyric acid or phenol (weak electrolytes) or AgCl. Reasons for being a weak electrolyte may either be low solubility (as in case of AgCl) or rather strong covalent bonds (possibly accompanied with a rather low gain in energy upon solvation (hydration)) as with butyric acid. This distinction vanishes almost when the reason for dissociation and dissolution is understood: The driving force is the release of energy by solvation of the formed ions. This gain is large enough to overcompensate the lattice energy keeping the ions in a salt crystal or to split covalent bonds in, e.g., HCl. Strong covalent bonds or large lattice energies will not be compensated by the energy gain during solvation; in case of weakly solvated species, the electrolyte becomes even weaker.

Solvents are numerous and different in properties. A common criterion is "inorganic" or "organic." In the former class, water is certainly the most popular solvent; less common and popular are thionyl chloride ($SOCl_2$), sulfuryl chloride (SO_2Cl_2), or liquid ammonia (NH_3). Into the latter class belong all organic compounds being liquid around room temperature. Common examples are tetrahydrofuran, acetonitrile, or propylene carbonate. Beyond this traditional approach, a classification based on solvent and in particular solvent molecule properties is more suitable. The most basic property is the capability of a solvent to release a proton: Protic solvents can dissociate themselves already without an electrolyte being added into protons and an anion like methanol or water:

$$H_2O \rightarrow H^+ + OH^-$$

Many acids, both organic and inorganic ones, are also protic solvents. Solvents, which cannot release protons, are called aprotic. Numerous organic compounds and some inorganic ones belong into this class.

Further characterization (and possibly classification) is possible based on other molecular properties. The most fundamental problem very closely related to the capability of any liquid to form an electrolyte solution is the polarity of a molecule. Dissolution is based on the energy gain during the formation of solvated ions composed of ions from the electrolyte and molecules forming a solvation shell. Higher polarity (larger dipole moment) will support higher degrees of solvation, formation of solvation shells with more solvent molecules. In addition the charge density of the ion will influence solvation. Ions with low charge density (large molecular ions with only a single charge) will be less strongly solvated, whereas small ions with high charge density (e.g., Mg^{2+} or Al^{3+}) will be solvated more strongly. Ion-dipole interactions as the driving force of solvation and thus of dissolution are of smaller importance with ions having small charge densities. In this case further properties are influential. Strong intermolecular interactions like hydrogen bonds in water are not desirable; they will keep the solvent from forming an even weakly bound solvation shell and will thus limit solubility. When using solvents with low polarity and small dipole moment, consequently electrolytes composed of ions with low charge density are preferably used: singly charged large monoatomic ions or large polyatomic ions again with only a single charge (e.g., tetraalkylammonium ions).

Properties not considered so far are those which are related to a particular application. The stability of a solvent molecule versus electrooxidation and electroreduction is of utmost importance in the use of solvents (and implicitly electrolyte solutions) in electrochemical synthesis procedures and electrochemical energy technology. In a battery, an accumulator, or a supercapacitor, the solution must not be decomposed either by oxidation at the positive mass or by reduction at the negative mass. In case of decomposition because of, e.g., overcharge reaction, products will decrease the energy efficiency; even more dangerous is the formation of gaseous decomposition products resulting in overpressure and finally explosion and leakage of the device. Stability can be based on thermodynamics, i.e., the oxidation or reduction potential value is simply beyond the electrode potential encountered at the active masses. Because many popular solvents hardly meet this requirement, kinetic limitations are of importance, too. In case of the lead-acid accumulator, corrosion of the thermodynamically unstable lead electrode is avoided because the cathodic reaction associated with the anodic corrosion of lead is the hydrogen evolution in the strongly battery acid. This reaction is extremely slow at the lead surface resulting in very small corrosion. In case of basically all lithium- and lithium-ion-based batteries and accumulators, formation of a solid electrolyte interface (SEI) keeps the solvent (and sometimes the electrolyte) from reacting with the active mass, in particular the negative one and thus suppresses self-discharge by direct chemical reaction between the active mass and the electrolyte solution.

In synthetic processes this requirement may be eased somewhat when divided cells are used wherein electrolyte solutions in both half cells may employ different solvents. Molecules particularly stable versus oxidation (to be used in the anode compartment) are acetonitrile, dimethyl sulfoxide, nitromethane, or dichloromethane,

Solvents and Solutions, Fig. 1 Examples of hydrophilic cations

whereas dimethylformamide, diethyl ether, tetrahydrofuran, or N-methylpyrrolidone is stable at rather negative electrode potentials.

The same stability criterion has to be applied to the electrolyte. It must be stable towards the solvent; in addition it should participate only in the intended reactions (in energy storage and conversion systems) or in a particular way (e.g., large cations directing the adsorption of acrylonitrile at the cathode in the Baizer process of electrochemical production of adipodinitrile from acrylonitrile). In electrosynthetic procedures further specific interactions with compounds involved in the electrode reaction (both educts and products) need attention. A concentration of educts high enough to permit operation of the synthesis at an economically attractive rate is required; depending on the properties of the educt, rather different features of the solvent and the electrolyte are required. Electrolyte ions can be classified in various ways as outlined below (Figs. 1–3). Hydrophilic cations in an aqueous solution will be highly solvated, and accordingly they will drag water molecules towards the cathode and thus into the electrochemical double layer. Lipophilic cations will hardly be solvated; thus, they will be adsorbed at the cathode without attached solvent molecules. The substantial difference in structure and composition of the double layer may subsequently influence the electrode reactions similar to the effect of ion size observed with the Baizer process (see above).

A major property beyond stability and the capability to dissolve an electrolyte is the ionic conductance (the specific conductance given with respect to a standard sample volume of 1 cm^3 size

Solvents and Solutions, Fig. 2 Examples of lipophilic cations

Solvents and Solutions, Fig. 3 Examples of neutral ions (the term neutrality does not pertain to the electric charge)

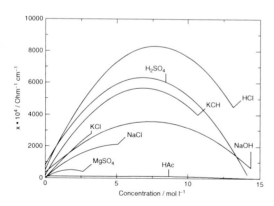

Solvents and Solutions, Fig. 4 Specific conductances (conductivities) of selected aqueous electrolyte solutions

is also called conductivity; both terms are frequently confused. The same confusion exists with the respective reciprocal values resistivity and resistance). In all known applications, a high conductance is desirable. In those few cases of, e.g., electrochemical reactors with multiple electrodes connected electrically in series and supplied from a single electrolyte solution reservoir, parasitic currents may result. They would indeed be smaller with a less conducting solution, but the gain thus achieved would be no match of the overall loss; thus in this case, other engineering solutions must be found.

Conductance is most directly related to concentration, but not in a straightforward way. As depicted below an increasing concentration initially results in the expected rise of conductance, but beyond a maximum value further increase in concentration results in no gain in conductance, actually conductance may decrease.

In some cases (e.g., NaCl, KCl) the limit of solubility has been reached, and further addition of electrolyte does not yield more dissociated charge carriers. In many other cases, solubility is not the cause. Instead incomplete dissolution resulting in the formation of uncharged (and thus nonconducting) ion pairs and enhanced viscosity, which decreases ionic mobility and thus conductance, are the causes [1, 2]. Conductivity data on numerous electrolyte solutions have been collected elsewhere [3].

Future Directions

In most practical uses of electrolyte solutions, stability and conductivity are of equally large importance; actually the conductivity of the ionically conducting phase as well as of the other materials in an electrochemical energy storage and conversion system is of potentially limiting importance [4]. Consequently studies will aim at compounds and mixtures showing higher conductance because of lower viscosity and other improvements of system properties with particular attention to selectivity of the transport processes regarding the ions involved in the electrode reaction. Solvation properties will need attention with respect to transport as well as solubility of electrolytes.

Cross-References

▶ Electrolytes, Classification
▶ Electrolytes, Thermodynamics
▶ Ionic Liquids
▶ Solid Electrolytes

References

1. Wright MR (2007) An introduction to aqueous electrolyte solutions. Wiley, Chichester
2. Bockris JO'M, Reddy AKN (1998) Modern electrochemistry 1, 2nd edn. Kluwer, Dordrecht
3. Holze R Landolt-Börnstein: numerical data and functional relationships in science and technology, new series, group IV: physical chemistry, vol 9: Electrochemistry, subvolume B: ionic conductivities of liquid systems. In: Martienssen W, Lechner MD (eds) Springer-Verlag, Berlin, in preparation
4. Park M, Zhang X, Chung M, Less GB, Sastry AM (2010) A review of conduction phenomena in Li-ion batteries. J Power Sources 195:7904

Specific Ion Effects, Evidences

Werner Kunz
Institut für Biophysik, Fachbereich Physik,
Johann Wolfgang Goethe-Universität Frankfurt
am Main, Frankfurt am Main, Germany

History and Importance

Probably the first scientific study on specific salt effects was performed by Jean Luis Poiseuille in 1847 [1]. He discovered that some salts increase the viscosity of water, whereas others decrease it. In the first half of the twentieth century, the investigations on specific ion effects on viscosity were further refined by Jones and Dole in 1929 [2] and Cox and Wolfenden in 1934 [3]. Based on these studies, Frank and Evans [4] proposed the expressions "water structure-maker" and "water-structure breaker," a concept that recently turned out to be slightly misleading, at least for simple 1–1 electrolytes in water.

Most popular is Franz Hofmeister's work carried out essentially in the 1880s and 1890s in Prague together with some colleagues and students [5]. He was essentially interested in physiological effects of salts. For example, he studied the influence of salts on protein precipitation and on swelling processes. From this work the so-called Hofmeister series was deduced. It should be mentioned that Hofmeister only made classifications of entire salts and not of separate ions. Such ion series were introduced much later, for example, by Voet and Pearson [6, 7], based on the lyotropic series and classifications according to the "softness" and "hardness."

The series of ions is usually as follows:

The ions on the left-hand side of Fig. 1 increase protein stability whereas the ions on the right-hand side decrease it. Many other properties of ionic solutions roughly follow also this scheme. However, several points should be stressed:

(a) Whereas the anionic series are widely respected, the ions in the cation series often have different places depending on the particular system or surface next to the ions. Further, anion specificity is usually more pronounced than cation specificity.

(b) As far as protein properties are concerned, usually, a reversal of the series is concerned above a certain concentration. More strictly: At very dilute solutions only the electrostatics matter, then at higher concentration the direct Hofmeister series is followed, then the inverse one (small anions bind stronger than big ones), and at very high concentrations all salts are salting-out (stabilizing). And all this depends also on the isoelectric point of the protein.

(c) Also bell-shaped curves with a maximum of the effect for the intermediate ("Hofmeister neutral" ions) are observed [8].

(d) Some ions are difficult to classify because of their Janus-type nature. This is the case, for example, of guanidinium. Its structure classifies it close to ammonium, but it is a strong denaturant.

(e) Organic ions are more difficult to classify. In the case of quaternary ammonium ions, the series goes as indicated by the white arrow in Fig. 1, which is counterintuitive at first sight. However, here the underlying

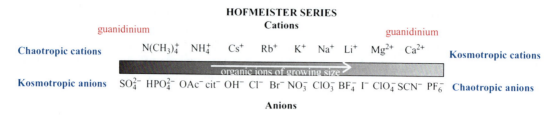

Specific Ion Effects, Evidences, Fig. 1 Typical Hofmeister ion series for salt effects on proteins

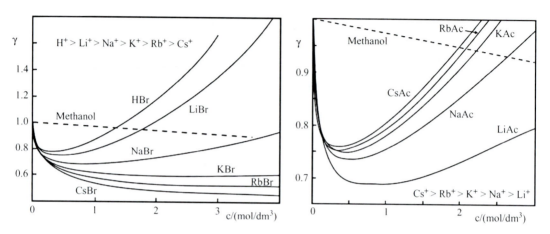

Specific Ion Effects, Evidences, Fig. 2 Activity coefficients of some aqueous electrolyte solutions as compared to the activity coefficient of methanol [10] (Reproduced with permission)

interactions are different and hydrophobicity plays a major role.

(f) The series is not universal. There are some important phenomena that do not follow at all these series. An example is the gas bubble-bubble coalescence in salt water. Here, a completely different classification has been found, but still not fully explained [9].

(g) It should be noted that the cation and anion series in Fig. 1 go in opposite directions. This will be discussed later on in this entry.

There is an enormous amount of examples of specific ion effects, from thermodynamic and transport properties of aqueous electrolyte solutions to effects at the water-air interface, in biological systems (e.g., the nervous system and ion channels) and in colloidal interactions. A compilation, although not complete, is given in a recent monograph [10].

Here only two examples are given and discussed. The first concerns the activity coefficients of salts in aqueous solutions, as shown in Fig. 2.

Clearly, on the left side, a Hofmeister series is concerned with activity coefficients increasing from CsBr to HBr at comparable concentrations above about 0.2 M. When bromide is exchanged by acetate as shown on the right side, the series is reversed. An intuitive explanation is given by Collins. He argued that small ions with high charge density form ion pairs more easily with counterions that have also a high charge density. The reason is the strong electrostatic attraction. Similarly, pairs of ions with low charge density also form readily ion pairs, because the ions lose easily their hydration shells. By contrast, an ion with a high charge density will not pair with a counterion of low charge density. The borderline between high and low charge density is given by water, which in one case binds stronger to ions than to other water molecules and in the opposite case (ions of low charge density) binds less strongly to the ions than to another water molecule. This "law of matching water affinities" [11] turned out to be very general, as long as ion pairing is the dominant driving force. In the above-mentioned example, bromide is of low charge density. It forms more easily ion pairs with Cs⁺ than, for example, with Na⁺ and consequently, the activity coefficient of CsBr is lower than that of NaCl. In the opposite case, the carboxylate group is of high charge density and therefore pairs more easily with Li⁺ than with Na⁺. Roughly speaking, the similarity of ion size is important. It is also interesting to see that at very small concentrations, ion specificity does not show up. Here, all is determined by the electrostatic attraction as described by Debye and Hückel. It depends on the charge of every ion, but not on their seizes.

The second example shown here is the surface tension of water as a function of added electrolytes, see Fig. 3.

Following features are noteworthy:

(a) Compared to the absolute value of the surface tension of pure water (72.8 mN/m) at 25 °C, the effects are small.
(b) Some salts increase the surface tension of water, others decrease it. Acids decrease it: H_3O^+ ions seemingly have a certain propensity toward the air-water interface. In general, it is expected that all nonorganic ions increase the surface tension, which means that the interface is depleted of ions. However, this is not the case. In a pioneering work, Jungwirth and Tobias could show (first by molecular dynamics simulation, the result of which was confirmed later on by experiments) that ions of low charge density can be significantly concentrated in the very first layer at the air-water interface, followed by a depletion zone more inside the bulk [13]. The integral of the ion distributions perpendicular to the interface would still give an overall depletion, but *locally* a significant adsorption can be found. The reason is again in agreement with Collins' ideas. The ions of low charge density easily lose their hydration shells, and it is energetically slightly more profitable to be situated close to the air interface.

If this is true, why do hydronium ions go to the surface? They are certainly of high charge density. The reason is that the geometry of their

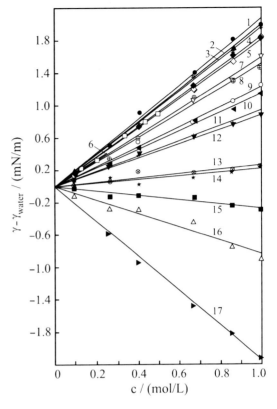

Specific Ion Effects, Evidences, Fig. 3 Effect of electrolyte concentration on the change in surface tension relative to water in 1:1 electrolytes. Experimental error in data points is ±0.1 mN m⁻¹. HCl (■), LiCl (▲), NaCl (●), KCl (♦), CaCl (⊕), NaF (□), NaI (○), NH₄Cl (▽), NaBr (□), HNO₃ (△), (CH₂)₄NCl (+), NH₄NO₃ (◄), HClO₄ (▶), NaClO₂ (□), LiClO₄ (⊗), NaClO₄ (□), KOH (#) [12]

hydration shell is not very favorable in the three-dimensional bulk phase. The hydronium ion fits better into the network structure of water near the interface [13]. Another option could be that the water surface spontaneously polarizes to become slightly negative.

In general and as a rule of thumb, it can be expected that ions of low charge density such as Cs⁺ are attracted by neutral hydrophobic surfaces. The release of hydration water of the ion and of the hydrophobic surface is the driving force. The entropic gain through the detachment of water molecules from both can be higher than the hydration enthalpy. This can be even more important than the electrostatic interaction between a negatively charged headgroup and Cs^+, or in other words, the release of unfavorable hydration water is energetically more relevant than electrostatic interactions. Water is the discriminating factor and consequently ion-water interactions are at least as important as ion-ion interactions. As a consequence, a correct modeling of water (and its polarizability) is of major importance whenever specific ion effects are to be considered.

The behavior of ions close to interfaces follows general features, c.f. Schwierz et al [14]. It holds also for inorganic surfaces, such as metals and oxides [15].

Collins' concept of matching water affinities can also be applied to the interactions of ions with charges headgroups (surfactants, polyelectrolytes) and is very useful for the interpretation of critical micellar concentrations, phase diagrams of colloidal systems, ion effects near membranes, etc. [16]. For example, it explains why the cation and anion series in Fig. 1 are reversed. In proteins, cations mostly interact with strongly hydrated headgroups such as carboxylates. Their interaction with corresponding cations (sodium to calcium) is much stronger than their interactions with weakly charged ions such as ammonium or potassium. By contrast, the relevant positively charged headgroups of proteins are all of low charge density (mainly ammonium groups). They preferably interact with the anions on the right-hand side, whereas their interactions with ions of high charge density, such as sulfate, is less favorable.

Within proteins, the detailed geometry in the confined medium plays a crucial role. This can even lead to significant attractions of negatively charged phosphate ions by negatively charged pockets.

Whereas most of the older literature presents specific ion effects on a macroscopic level, recent investigations highlight the specific molecular distributions of ions and their specific hydration in aqueous phases or near interfaces [13, 17]. Most useful are X-ray and neutron scattering experiments [18], terahertz and IR spectroscopy [19], dielectric relaxation spectroscopy [20], and NMR [21] combined with computer simulations [22]. For example, the studies by Bakker and his group showed that single-charged ions influence the water dynamics mostly in their first hydration shell. As a consequence, they cannot be considered as making or breaking structures. However, their studies were made in very concentrated solutions so that nearly all water molecules are in the first hydration shell. From dielectric relaxation studies, it can be concluded that the water dynamics is influenced by the presence of ions well beyond the first hydration shell.

Nevertheless, the term "cosmotropic" only means that the ion is energetically strongly hydrated and "chaotropic" that it is energetically weakly hydrated. Note that this classification is not in phase with Pearson's concept of "hard and "soft" ions." Ag^+ and Hg^{2+} are soft ions, but their enthalpies of hydration -483 and $-1,853\,kJ\,mol^{-1}$ respectively are more negative than those of the hard ions of similar size, Na^+ and Ca^{2+} with values of -416 and $-1,602$ kJ mol^{-1}, respectively. As a first approximation, cosmotropic ions have high charge densities, and chaotropic ions have low charge densities. Compared to this property, the polarizability of the ions plays only a minor role. Especially, as far as bulk properties are concerned, dispersion forces are negligible. If ever, the ion polarizability has (as well as their charge density) an influence on the water polarization, but there is no direct ion-ion interaction force between ions because of the ion polarizabilities [23].

A combination of the molecular picture with the macroscopic properties allows now for a general description of specific ion effects. As pointed out before, the mentioned concepts are very valuable to describe ion pairing effects. Other effects of course require other concepts and models. This is, for example, the case, when ion partitioning prevails or when hydrophobic interactions dominate, as pointed out by Leontidis [24].

As far as specific ion effects in nonaqueous solutions (including Ionic Liquids) are concerned, the literature is scarce and no general conclusions have been drawn so far. Mostly, solubility data are given and sometimes also conductivity or osmotic coefficients. A collection of data can be found in [25], but there is no review on specific ion effects in nonaqueous systems.

Specific Ion Effects, Evidences

In most cases, the solubility of ions in such systems (including in Ionic Liquids) is much lower than in water so that much less data are available. Exceptions are, for example, AgCl that dissolves well in acetonitrile and uranyl nitrate that dissolves in diethylether, both being examples for specific ion effects (on solubility) in nonaqueous solvents.

Future Directions

Today we are not far from a general picture of specific ion effects, especially the Hofmeister series is no longer a mystery. However, the temperature dependence of thermodynamic properties is still a challenge [26]. For example, it is still impossible to precisely predict the influence of salts on the water vapor pressure, probably, because the interactions involved are finely balanced and strongly depend also on the change in water behavior with temperature. The problem is further complicated in multicomponent systems like sea water, where numerous charged species interact and are also in equilibrium with their precipitated solid state. The relative diffusions and the time evolution of such complex systems are still very difficult to model or to predict. Similarly, ions in confined media behave highly specifically, for instance, in biological ion channels. In such systems, of course, the chemical environment plays a crucial role so that every system is specific on its own.

Cross-References

▶ Ion Properties
▶ Ions at Solid-Liquid Interfaces
▶ Polyelectrolytes, Films-Specific Ion Effects in Thin Films
▶ Polyelectrolytes, Properties
▶ Salting-In and Salting-Out
▶ Specific Ion Effects, Theory

References

1. Poiseuille JL (1847) Sur le mouvement des liquides de nature différente dans les tubes de très petits diamètres. Ann Chim Phys 21:76–110
2. Jones G, Dole M (1929) Viscosity of aqueous solutions of strong electrolytes with special reference to barium chloride. J Am Chem Soc 51:2950–2964
3. Cox WM, Wolfenden JH (1934) Viscosity of strong electrolytes measured by a differential method. Proc R Soc Lond A145:475–488
4. Frank HS, Evans MW (1945) Free volume and entropy in condensed systems. III. Entropy in binary liquid mixtures; partial molal entropy in dilute solutions; structure and thermodynamics in aqueous electrolytes. J Chem Phys 13:507–532
5. Hofmeister F (1888) Ueber die wasserentziehende Wirkung der Salze. Naunyn Schmiedebergs Arch Pharmakol Exp Pathol 25:1–30; for English translation see: Kunz W, Henle J, Ninham BW (2004) Zur Lehre von der Wirkung der Salze (About the science of the effect of salts): Franz Hofmeister's historical papers. Curr Opin Colloid Interface Sci 9:3
6. Voet A (1937) Quantitative lyotropy. Chem Rev 20:169–179
7. Pearson RG (1963) Hard and soft acids and bases. J Am Chem Soc 85:3533–3539
8. Bauduin P, Renoncourt A, Touraud D, Kunz W, Ninham BW (2004) Hofmeister effect on enzymatic catalysis and colloidal structures. Curr Opin Colloid Interface Sci 9:43–47
9. Henry CL, Craig VSJ (2010) The link between ion specific bubble Coalescence and Hofmeister effects is the partitioning of ions within the interface. Langmuir 26:6478–6483
10. Kunz W (ed) (2010) Specific ion effects. World Scientific, Singapore
11. Collins KD (2004) Ions from the Hofmeister series and osmolytes: effects on proteins in solution and in the crystallization process. Methods 34:300–311
12. Weissenborn PK, Pugh RJ (1995) Surface tension and bubble coalescence phenomena of aqueous solutions of electrolytes. Langmuir 11:1422–1426
13. Pavel J, Tobias DJ (2006) Specific ion effects at the air/water interface. Chem Rev 106:1259–1281
14. Schwierz N, Horinek D, Netz RR (2010) Reversed anionic Hofmeister series: the interplay of surface charge and surface polarity. Langmuir 26:7370–7379
15. Lyklema J (2003) Lyotropic sequences in colloid stability revisited. Adv Colloid Interface Sci 100–102:1–12
16. Vlachy N, Jagoda-Cwiklik B, Vácha R, Touraud D, Jungwirth P, Kunz W (2009) Hofmeister series and specific interaction of charged headgroups with aqueous ions. Adv Colloid Surf Sci 146:42–47
17. Zhang Y, Cremer PS (2006) Interactions between macromolecules and ions: the Hofmeister series. Curr Opin Chem Biol 10:658–663; see also Ref. [32] of the entry "Electrolytes: History"
18. Neilson GW, Enderby JE (1979) Neutron and x-ray diffraction studies of concentrated aqueous electrolyte solutions. Ann Rep Prog Chem Sect C: Phys Chem 76:185–220

19. Tielrooij KJ, Garcia-Araez N, Bonn M, Bakker HJ (2010) Cooperativity in ion hydration. Science 328:1006–1009
20. Buchner R, Barthel J (2001) Dielectric relaxation in solutions. Ann Rep Prog Chem Sect C: Phys Chem 97:349–382
21. Fabre H, Kamenka N, Khan A, Lindblom G, Lindman B, Tiddy GJT (1980) Self-diffusion and NMR studies of chloride and bromide ion binding in aqueous hexadecyltrimethylammonium salt solutions. J Phys Chem 84:3428–3433
22. Koelsch P, Viswanath P, Motschmann H, Shapovalov VL, Brezesinski G, Möhwald H, Horinek D, Netz RR, Giewekemeyer K, Salditt T, Schollmeyer H, von Klitzing R, Daillant J, Guenoun P (2007) Specific ion effects in physicochemical and biological systems: simulations, theory and experiments. Colloid Surf A 303:110–136
23. Serr A, Netz RR (2006) Polarizabilities of hydrated and free ions derived from DFT calculations. Int J Quantum Chem 106:2960–2974
24. Leontidis E (2010) Phospholipid aggregates as model systems to understand ion-specific effects: experiments and models. In: Kunz W (ed) Specific ion effects. World Scientific, Singapore, pp 55–84
25. Barthel J, Krienke H, Kunz W (1998) Physical chemistry of electrolyte solutions. Modern aspects. Springer, New York
26. Drzymala J, Lyklema J (2012) Surface tension of aqueous electrolyte solutions. Thermodynamics. J Phys Chem A 116:6465–6472

Specific Ion Effects, Theory

Dominik Horinek
Institute of Physical and Theoretical Chemistry, University of Regensburg, Regensburg, Germany

Description

Ion-specific phenomena are observed when ions of the same valency behave distinctly in many bulk liquids and in interfacial systems [1]. The primary ion-specific interactions in an electrolyte solution even at high dilution are the ion-solvent interactions. One of the earliest continuum models for ion specificity is the Born solvation model, which assigns the free energy of solvation of an ion of charge q and radius R in a liquid with dielectric constant ε_r as

$$\Delta G_{Born} = q^2/8p\varepsilon_0 R[1 - 1/\varepsilon_r]$$

The Born radius R is an ion-specific parameter that depends on the solvation properties of the ion and is not equal to the ionic radius derived from crystal structures of salts. For ions of the same charge, the Born radius increases with increasing molecular weight: Among the halide ions, fluoride has a smaller Born radius than iodide. An astonishing general observation is that anions appear to be better solvated than cations: The Born radii of cations are significantly larger than their Pauling radii [2]. A more sophisticated description of ionic solutions includes molecular details of the solvent. This is achieved in atomistic simulations with explicit inclusion of the solvent. Typically, a force field is assigned that describes the forces acting in the system as a function of the system's conformation. Such force fields describe the inter- and intramolecular interactions. For the understanding of ion specificity, the intermolecular, nonbonding terms are important, which are commonly described as a pairwise sum of van der Waals interactions

$$V(r_{ij}) = C_{12}/r_{ij}^{12} - C_6/r_{ij}^6$$

with numerical parameters C_{12} and C_6 that depend on the two atom types, and electrostatic Coulomb interactions

$$V(r_{ij}) = 1/4\pi\varepsilon_0 q_i q_j/r_{ij}$$

with the static partial atomic charges q_i and q_j. When ion specificity of simple mononuclear ions is studied, the partial charges are given by the valency of the ion, and the ion specificity is accounted for by the van der Waals parameters [3]. In some force fields, polarizable terms are added, that account for electronic polarization effects of the ion in an electric field. The difficulty of the simulation approach lies in the classical force fields: Their mathematical form and

Specific Ion Effects, Theory, Fig. 1 Anion-cation potentials of mean force showing primary minima (contact pairs), secondary minima (solvent shared pairs), and weak minima at larger separations corresponding to solvent separated pairs. The differently strong appearance of minima and barriers in the potentials of mean force is a clear sign of ion specificity in electrolyte solutions. Simulation results taken from the chapter of J. Dzubiella et al. in Ref. [4]

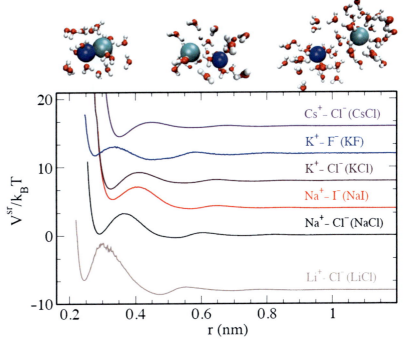

numerical parameters are only an approximate description of the real forces. Their major strength lies in the qualitative and semiquantitative elucidation of mechanisms.

Molecular dynamics simulations of solvated ions give predictions of specific thermodynamic solvation properties like free energies and entropies of solvation as well as dynamic properties like ion conductivities. More complex behavior is observed for properties like the osmotic pressure or the activity coefficient of a salt solution. In addition to the ion-water interactions, ion-ion interactions are important. Much of these specific effects can be understood by looking at the ion-ion potentials of mean force, which are linked to thermodynamic properties along the lines of the liquid state theories of Kirkwood and Buff or McMillan and Mayer. Figure 1 shows the short-ranged part of ion-ion potentials of mean force that were obtained by molecular dynamics simulations [4].

In Fig. 1, the "like-seeks-like" rule can be recognized: Small anions have a strong tendency to form contact pairs with small cations, large anions tend to pair with large cations, but large ions do not like to pair with small ions.

A qualitative theoretical explanation of this model based on ionic surface charge densities or continuum electrostatics is possible, but quantitative interpretations are difficult, even more so when molecular ions like SCN^- are studied, which cannot be approximated by spheres with an isotropic surface charge density.

The last field of ion specificity is the most complex one: ion effects in inhomogeneous systems where interfaces are present: macroscopic interfaces, colloidal systems, or polymer solutions. In such systems, the presence of salt affects a whole range of properties: interfacial tensions, colloidal coagulation, protein stability, and protein activity, just to name a few. It is remarkable that the ion specificity for all these properties follows the same order, which is expressed in the Hofmeister series, and that sometimes this order is completely reversed.

The key to the understanding of ion specificity in such systems lies in the affinity of the ions to the interface, which can be described by a potential of mean force. There are several attempts to explain these potentials of mean force by simple analytical expressions, which

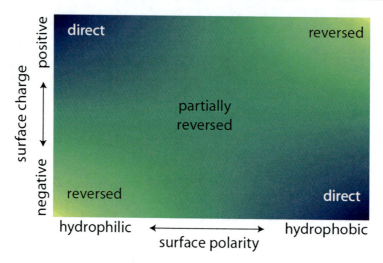

Specific Ion Effects, Theory, Fig. 2 Hofmeister series for salt-induced coagulation of colloidal particles as a function of surface polarity and surface charge: A direct series is observed when the potency for inducing coagulation is $I^- < Cl^- < F^-$, and a reversed series is observed when $F^- < Cl^- < I^-$, and other ordering correspond to partial reversals in the middle region of the diagram

was first attempted by Boström and Ninham with the inclusion of dispersion coefficients [5], but so far, there is no generally accepted theory that is free of undetermined parameters. Such potentials of mean force are again accessible by molecular dynamics simulations.

The ion-specific potentials of mean force can be included in an extension of the Poisson-Boltzmann equation

$$\varepsilon \frac{d}{dz} \varepsilon(z) \frac{d}{dz} \Phi(z) = -\sum_i q_i c_0 e^{-\left(V_i^{PMF}(z) + q_i \Phi(z)\right)/k_B T}$$

For $V_i^{PMF}(z) = 0$, the standard Poisson-Boltzmann equation is recovered. For a given set of $V_i^{PMF}(z)$, this equation yields the electrostatic potential and ionic concentration profiles. The solution is an ion-specific analogy of the Gouy-Chapman model. An important aspect of the inclusion of single-ion potentials of mean force is the possibility of non-monotonic concentration profiles, while plain electrostatic screening always leads to exponentially decaying concentration profiles.

The simplest and most studied surface is the air/water interface [6], where concentration profiles of the ions yield the surface excesses Γ of the involved ions, which are thermodynamically linked to the surface tension increment $d\gamma/da$ by the Gibbs adsorption isotherm [7]

$$d\gamma/da = -kT\Sigma_i \Gamma_i.$$

It is generally observed that large ions like iodide yield a weaker increase in surface tension than small ions, but the strength of this effect largely depends on the employed force field or model potential.

A similar approach gives a qualitative account of stability of colloids with different surface charge and polarity [8]. First, the specific ion enrichment depends on the ion type and on the surface polarity, which determine the ion affinities to the surface: At nonpolar surfaces, large ions have a higher surface affinity than small ions, but they have a lower affinity to polar surfaces. Next, the specific adsorption, double layer formation, and the bare surface charge are combined in the extended Poisson-Boltzmann equation and yield the electrostatic stabilization. Along the lines of DLVO theory, this electrostatic stabilization is added to the van der Waals attraction of the colloids, and a prediction for the critical coagulation concentrations for different salts is obtained. This approach gives a double reversal of the Hofmeister series when the polarity and charge of the surface are varied, as seen in Fig. 2.

Future Directions

Atomistic molecular dynamics simulations reveal a detailed molecular view of electrolyte solutions and interfaces that goes far beyond simple continuum theories. This view has been started with studies of the air/water interface and is currently extended to more complex solid interfaces, colloidal systems, and back to biopolymer solutions, where the whole endeavor of ion specificity began with the classical studies of Franz Hofmeister. Quantum chemical simulations, which have the potential to give more reliable predictions of ion density profiles than classical force field simulations, become increasingly feasible with increasing computational resources [9].

Cross-References

▶ DLVO Theory
▶ Ion Properties
▶ Ions at Biological Interfaces
▶ Ions at Solid-Liquid Interfaces
▶ Specific Ion Effects, Evidences

References

1. Kunz W (ed) (2010) Specific ion effects. World Scientific, Singapore
2. Reif MM, Hünenberger PH (2011) Computation of methodology-independent single-ion solvation properties from molecular simulations. IV. Optimized Lennard-Jones interaction parameter sets for the alkali and halide ions in water. J Phys Chem 134:144104
3. Horinek D, Mamatkulov SI, Netz RR (2009) Rational design of ion force fields based on thermodynamic solvation properties. J Chem Phys 130:124507
4. Kalcher I, Dzubiella J (2009) Structure- thermodynamics relation of electrolyte solutions. J Chem Phys 130:134507
5. Boström M, Williams DRM, Ninham BW (2001) Specific ion effects: why DLVO theory fails for biology and colloid science. Phys Rev Lett 87:168103
6. Jungwirth P, Tobias DJ (2006) Specific ion effects at the air/water interface. Chem Rev 109:1259
7. Netz RR, Horinek D (2012) Progress in modeling of ion effects at the vapor/water interface. Ann Rev Phys Chem 63:18

8. Schwierz N, Horinek D, Netz RR (2010) Reversed anionic Hofmeister series: the interplay of surface charge and surface polarity. Langmuir 26:7370
9. Baer MD, Mundy CJ (2011) Towards an understanding of the specific ion effect using density functional theory. J Phys Chem Lett 2:1088

Spectroelectrochemistry, Potential of Combining Electrochemistry and Spectroscopy

Werner Mäntele
Institut für Biophysik, Fachbereich Physik, Johann Wolfgang Goethe-Universität Frankfurt am Main, Frankfurt am Main, Germany

Introduction

Redox processes play an essential role in biological energy conversions. In photosynthesis and in the respiratory chain, for example, electron transfer chains within and between protein complexes couple to proton or ion gradients across the lipid membrane which in turn are used to form chemical energy in the form of adenosinetriphosphate, ATP. Some of these protein complexes simple act as "electron shuttles," i.e., they commute between membrane-spanning complexes transferring an electron. Others span the membrane and contain multiple redox centers with complex electron transfer mechanisms and catalytic functions.

Our knowledge on these biological electron transfer chains nowadays relies on detailed X-ray structures that have emerged in the past 10–20 years. Studies on the reactions and mechanisms rely mostly on spectroscopic analysis, since most of the cofactors exhibit strong electronic transitions in the visible and UV spectral range. Vibrational spectroscopy, which can give answers on the role of the protein scaffold and of amino acid side chains, has rarely been used until the late 1980s because of the lack of suitable technologies.

Electrochemistry has been used in the study of biological electron transfer processes mainly as

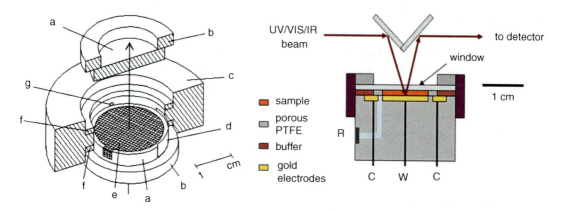

Spectroelectrochemistry, Potential of Combining Electrochemistry and Spectroscopy, Fig. 1 Design of transmission cells according to [2, 3] or reflection cells according to [4]. Windows can be CaF$_2$ for the UV to Mid-IR range or ZnS for the Visible to the extended IR range. *Left*: (**a**) Windows, (**b**) Window holder, (**c**) Cell body, (**d**) Pt counter electrode, (**e**) Gold grid working electrode; (**g**) Capillary to reference electrode *Right*: W, working electrode; C, counter electrode, R, reference electrode

redox chemistry, where the redox midpoint potentials of proteins and other small redox-active molecules have been determined by oxidative or reductive chemical titrations. The pioneering work of Armstrong et al. [1] has indicated that direct electrochemistry, i.e., electron transfer reactions of redox proteins at the surface of suitable electrodes without the use of mediators, is much more elegant and can yield precise electrochemical parameters such as the midpoint potential E_m, the number of transferred electrons n, or the number of protons coupled with electron transfer. If cyclic voltammetry is used, additional thermodynamic parameters are obtained.

Electrochemistry Meets Spectroscopy

Although many attempts have been made to establish spectroelectrochemical cells for spectral analysis in combination with electrode redox control, this method suffered from inadequate equilibration of the redox proteins with the electrode potential. This is mostly due to the Nernst diffusion layer with a thickness of some tens of micrometers which requires stirring for cells with longer optical pathlength.

We have been able to develop OTTLE (**O**ptically **T**ransparent **T**hin-**L**ayer **E**lectrochemical) cells with pathlengths as low as a few micrometers for the spectroelectrochemical study of biological redox compounds [2, 3]. The very low thickness combines two advantages: (1) The entire cell volume is within the Nernst layer and thus in equilibrium with the electrode potential, without the need of stirring. (2) A pathlength of some µm (5–20 µm) yields sufficient transmission even in the water absorbance range in the infrared spectral region and thus allows the entire spectral range from the UV (190 nm) to the mid-IR (20 µm) to be used for spectral analysis. All cell versions are three-electrode configurations with the working electrode in the measuring beam, the counter electrode outside the beam and the reference electrode outside the current path between working and counter electrode. While the OTTLE cells described in [2] and in [3] were transmission cells either with grid or gold mesh electrodes, an OTTLE cell using a solid gold electrode in reflection was developed later [4]. Figure 1 shows both cell principles.

Both in transmission and reflection cells, the protein in buffer solution forms a layer of some µm thickness between two windows or between a window and the electrode, resp.

Gold grid electrodes usually employ an electrodeposited gold mesh (transmission 30–60 %), while reflection cells use a polished electrode as a mirror, not necessarily from gold, but also from polished glassy carbon or others.

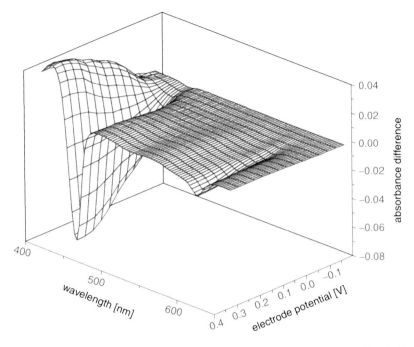

Spectroelectrochemistry, Potential of Combining Electrochemistry and Spectroscopy, Fig. 2 Electrochemical redox titration of the cytochrome c oxidase from *Paracoccus denitrificans*

If metal electrodes are used, the use of surface modifiers or "promotors" is essential. Surface modifiers are bifunctional molecules which attach to the metal surface typically with a sulphur-containing group and present a charged or polar side to the solution which allows the protein or enzyme to interact at a distance close enough for rapid electron transfer but distant enough to prevent denaturation. Examples for promoters are given in [1].

Equilibration of the redox protein is faster in the reflection cell, because the diffusional path length is only half the optical pathlength, which enables a rapid switching of the molecule between different redox states. Although direct electrochemistry is preferable, it turns out that redox cofactors are sometimes deeply buried in proteins and enzymes so that mediators are required to speed up the equilibration [4]. However, they can be kept at very low concentrations and thus do not interfere with the optical measurement. Under these conditions, amperometric, coulometric, or cyclovoltammetric experiments in combination with UV/VIS/IR spectroscopy, but also with fluorescence or circular dichroism spectroscopy are possible.

Insight into Redox Mechanisms from Electrochemical Redox Titrations

Electrochemical titrations of redox proteins with spectroscopic monitoring bear the advantage that the sample is *not* successively diluted with redox reagents, that titrations can be performed in cycles, and that the electrode potential (and thus the ox/red equilibrium of the sample) can be set at mV precision. This precision offers the possibility to analyze redox proteins with several cofactors that may be coupled and that may exhibit cooperativity. As an example, Fig. 2 shows the complete redox titration of cytochrome c oxidase, the terminal enzyme of the respiratory chain. Provided that fast equilibration (typically within seconds) at the electrode occurs, a full spectral data set can be obtained within some minutes.

There is multiple advantages of precise midpoint potentials for the analysis of redox proteins [5]. First, the coupling of different redox cofactors in one protein can be analyzed from the titration curves. In the case of the titration of the cytochrome c oxidase, heme a and heme a_3 can be sufficiently separated to determine individual redox potentials or cooperativity. This is also the case for multiheme cytochrome c proteins. Second, the precision of the redox midpoint potentials allows the influence of site-directed mutations by structural alterations or by electrostatic interactions to be determined in detail. Third, the possibility for a redox clamping of a protein at an electrode at mV precision offers to study in detail electron transfer by time-resolved optical techniques.

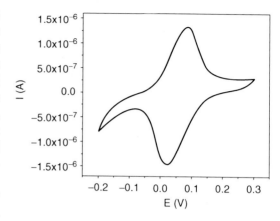

Spectroelectrochemistry, Potential of Combining Electrochemistry and Spectroscopy, Fig. 3 Cyclovoltammogram of horse heart cytochrome c in a thin-layer OTTLE cell [4]

Cyclic Voltammograms of Proteins in Thin-layer Spectroelectrochemical Cells

Cyclovoltammetry is typically performed by dipping a working electrode into a solution or suspension of the redox-active sample. Under these conditions, the electrode current is diffusion controlled and only a small fraction of the sample material that is in diffusional exchange with the electrode is involved in the reaction. The separation of the anodic peak ($E_{p,a}$) and the cathodic peak ($E_{p,c}$) depends on the scan speed, and the midpoint potential of a reversible electron transfer reaction is calculated as the average of $E_{p,a}$ and $E_{p,c}$. Cyclovoltammograms for thin-layer OTTLE cells differ significantly. If the layer thickness is in the order of the Nernst layer (<100 μm), the entire cell volume is involved in the reaction because of fast diffusional transport to the electrode. Consequently, the anodic and the cathodic peak are hardly separated, and are at identical potentials under ideal conditions.

Figure 3 shows a thin-layer cyclovoltammogram of horse heart cytochrome c taken in the reflection cell (Fig. 1) at low scan rates (1 mV/s). The reaction is completely reversible and the absence of additional chemical reactions is evident. The redox midpoint potential can be calculated from ($E_{p,a}$ + $E_{p,c}$)/2 to +52 mV, in perfect agreement with that extracted from the spectroscopic data analogous to those presented in Fig. 2.

An interesting option is the combination of cyclovolt-ammetry with rapid scanning UV/VIS or infrared spectroscopy. This technique generates a series of spectra in the course of a cyclovoltammogram and allows a very rapid screening for potential reversible or irreversible chemical reactions following electron transfer.

Why Combine IR Spectroscopy with Electrochemistry?

Spectroelectrochemistry as a combination of electrochemistry and UV/VIS spectroscopy is self-evident because of the direct relation of electron transfer with changes in electronic orbitals and thus with spectral changes. In redox proteins, with a few exceptions, the redox activity is caused by cofactors such as hemes, quinines, iron sulfur centers or metal centers, depending on the function and on the potential range. The role of the protein backbone and the amino acid side chains is the binding and precise orientation of these cofactors and the fine-tuning of their optical and redox properties with polarity or charges. In this context, the protein must able to react to redox transitions with conformational

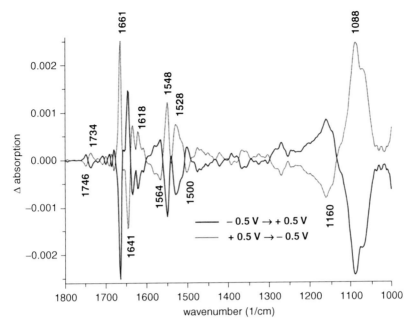

Spectroelectrochemistry, Potential of Combining Electrochemistry and Spectroscopy, Fig. 4 Redox-induced infrared difference spectra of the cytochrome c oxidase from *Paracoccus denitrificans*

changes, dipole formation/reorientation or charge displacements in order to contribute reorganization energy for the electron transfer. Indeed, electron transfer between cofactors in proteins may lead to charge displacements at amino acids which can stabilize a charged state of a cofactor.

It is this aspect that makes the combination of infrared (IR) spectroscopy with electrochemical reactions so attractive. The infrared spectrum in the wavelength range from 4 to 20 μm reflects the structural properties of a polypeptide molecule, both for backbone and the side chain conformations. However, the potential of IR spectroscopy is better exploited with reaction-induced IR difference spectroscopy.

Figure 4 shows the electrochemically-induced IR difference spectra of the cytochrome c oxidase from *Paracoccus denitrificans*. The heme difference spectra of this redox enzyme were already discussed in Fig. 2. For the spectra shown in Fig. 4, the protein sample was first equilibrated at −0.5 V, then at +0.5 V, and finally again at −0.5 V versus Ag/AgCl/KCl. The difference spectra calculated from these states represent the transitions from the reduced state of all cofactors (copper, heme a, copper heme a3) to the oxidized state and reverse. These transitions are fully reversible as indicated by the perfect symmetry and by numerous isosbestic points.

The strong signals in the amide I and II range (around 1,660 and 11,550 cm^{-1}) indicate the response of the protein scaffold to the redox transitions. The small bands at 1,734 cm^{-1}/ 1,746 cm^{-1} have been assigned to a glutamic acid side chain (Glu 278) and represents the proton coupling to electron transfer.

Along this line, the combination of electrochemistry and IR spectroscopy has been extremely successful in the identification and unraveling of molecular mechanisms of biological electron transfer and of catalysis coupled to redox reactions.

Future Directions

The combination of electrochemistry and spectroscopy provides a vast number of applications in chemistry and biochemistry. The precise selection and clamping of redox states by direct or mediated electrochemistry together with the high analytical selectivity of spectroscopic techniques, in particular infrared spectroscopy, demonstrates this potential. The spectroelectrochemical cells developed up to now can be miniaturized to accomodate sample volumes of just a few 100 nL in very thin layers. These can be microstructured cells in combination with microreactors and/or microseparation devices.

In combination with infrared lasers, in particular tunable quantum cascade lasers (QCLs), very fast redox switching combined with IR monitoring appears possible. This opens the possibility to use direct electrochemistry for the triggering of fast chemical and biochemical reactions, and the fingerprint selectivity of infrared spectroscopy for the analysis of the reaction details.

Cross-References

▶ Infrared Spectroelectrochemistry

References

1. Armstrong FA, Hill HAO, Walton NJ (1988) Direct electrochemistry of redox proteins. Acc Chem Res 21:407–413
2. Mäntele W, Wollenweber A, Rashwan F, Heinze J, Nabedryk E, Berger G, Breton J (1988) Fourier-transform infrared spectroelectrochemistry of the bacteriochlorophyll-a anion radical. Photochem Photobiol 47:451–455
3. Moss D, Nabedryk E, Breton J, Mäntele W (1990) Redox-linked conformational changes in proteins detected by a combination of infrared-spectroscopy and protein electrochemistry – evaluation of the technique with cytochrome-c. Eur J Biochem 187:565–572
4. Bernad S, Mäntele W (2006) An innovative spectroelectrochemical reflection cell for rapid protein electrochemistry and ultraviolet/visible/infrared spectroscopy. Anal Biochem 351:214–218
5. Baymann F, Moss DA, Mäntele W (1991) An electrochemical assay for the characterization of redox proteins from biological electron-transfer chains. Anal Biochem 199:269–274

Sulfur Removal from Crude Oil and its Fraction

Ahmad Hammad, Zaki Yusuf and Nayif A. Rasheedi
Research and Development Center, Saudi Aramco, Dhahran, Saudi Arabia

Introduction

Growing environmental awareness around the world for the reduction of particulate matter, SO_x, and NO_x emissions from transportation vehicles is being reciprocated by the introduction of ever-increasingly stringent legislation. As the maximum sulfur (S) concentration has already been limited to 10 ppm levels in many industrialized countries for highway transportation fuel, imposed legislation has put the petroleum refining industry under severe constraints to supply sweetened fuel in a cost-effective manner. On the same note, with ever-declining feed quality with consequential increase in refractory S compounds, challenges to produce fuel by deep hydrodesulfurization (HDS) are becoming steeper while ensuring higher fuel standards. Additionally, traditional catalysts (Co-Mo/Al_2O_3) are not as effective during deep HDS for removing residual refractory S compounds. Moreover, H_2S formed during HDS, along with the nitrogenous and metal-organic compounds present in the feedstock, has inhibitory effects on these conventional HDS catalysts. Furthermore, the need for the removal of refractory S compounds present in crude fractions demands increased severity, namely, high temperatures (350–400 °C), to attain lower S levels in the final product, which in turn leads to coke formation on the catalyst surface, impeding S removal kinetics. On the same note, it is also a requirement to reduce the aromatic constituents of fuel to curtail particulate emissions. Deep hydrogenation of aromatics is favored only at lower temperatures (<300 °C), due to the exothermic nature of the reaction. Therefore, the reduction of

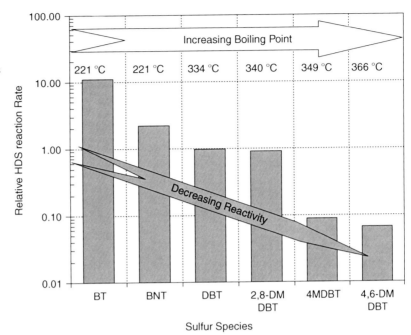

Sulfur Removal from Crude Oil and its Fraction, Fig. 1 Reactivities of different sulfur compounds and their boiling points as a function of diesel sulfur species [8] (catalyst: CoMo/Al$_2$O$_3$; T = 350 °C; P ≈ 100 atm)

S content in fuel should be met with other specifications to meet both environmental and fuel standards [1–8].

Sulfur Chemistry

Various types of S species are reported to be present in different crude fractions. Among them, the relative amounts of thiophenic S species vary significantly with respect to different refinery hydrocarbon (HC) streams. The relative rates of reactivities of various S species are shown in Fig. 1 as a function of their boiling points [8]. Dibenzothiophene (DBT) and its derivatives β-DBTs (4,6 DMDBT, 4,6 DEMDBT, etc.) are significantly less reactive than benzothiophenes (BT), due to their stereochemical hindrance that prevents them from interacting with the active sites of conventional catalysts.

Novel HDS Catalysts and Processes

Many research entities have been searching for different alternative solutions, including diverse catalyst preparation protocols, superior performing promoters, and unsupported novel catalyst series, to achieve deep HDS. Although some of the new-generation unsupported catalysts (e.g., MoWNi NEBULA®) have been commercialized for deep HDS processes, challenges still remain due to high material cost and a high hydrogen pressure requirement. Also, in other cases, nano-sized MoS$_2$ catalysts were capable of utilizing H$_2$S as an autocatalysis agent, thus reversing the inhibitory effect of H$_2$S on HDS. Additionally, novel reactor design and configurations with the right type of catalyst deployment have been proposed to capitalize on the special features of each group of catalysts to cater to the needs of various feed types and their constituents [8–23]. In general, for both conventional and ultra-deep HDS processes, hydrogen is produced ex situ by steam methane reforming (SMR) and water-gas shift reaction (WGSR). The higher total reactor pressure and high hydrogen partial pressure requirements in HDS reactors demand higher capital investments, due to the need for heavier reactors integrated with compressors and high-pressure piping networks in association with the SMR hydrogen plant.

In Situ Hydrogen Generation Process

The electrochemical in situ hydrogen generation route is an alternative route that eliminates the need for an external hydrogen source for HDS. The process involves parallel reactors for batch operation. For continuous operation, parallel reactors can also be employed. In both cases, the residence time of the reactor is governed by the deployment of optimal surface area of the electrocatalysts in relation to reactor configuration. A simplified PFD of such a process for batch operation is shown in Fig. 2.

In this process, preheated HC stream is sent to an externally heated batch reactor integrated with an electrochemical cell. In addition to HC and water, 0.25 M sulfuric acid (H_2SO_4) is introduced as a charge carrier in the reaction media. As a typical example, the temperature of the reaction is kept around 250 °C, while the pressure remains at around 40 atm. The air is removed from the reactor before any potential or current is applied to the media. An adequate amount of atomic hydrogen is electrolytically generated over the working electrode (cathode) by applying constant current (e.g., 0.03 amps) for its optimal, almost instantaneous, and in situ utilization during the desulfurization of S compounds on the electrode surface, which also serves as an electrocatalyst. Water is the only source of hydrogen for these desulfurization reactions. The correct selection of the electrocatalyst helps enhance the desulfurization reaction, while requiring a maximum of ~10 atm hydrogen partial pressure and thus further improving the overall process economics. The total pressure in the reactor never exceeds 50 atm during the reaction. The reaction is followed by an online process monitoring system. At the end of the batch reaction, the HC-water mixture is sent for separation and stabilization. Low-MW HC is fractionated from the vapor stream, while the separated sour gases are sent for further treatment and elemental S recovery.

Representative Results

In the current study, a batch reaction was carried out by maintaining reaction conditions mentioned in the preceding section, with Au and Pt acting as the working and counter electrode, respectively. Figure 3 shows the gas phase composition of various reaction products from an in situ hydrogen generation process (iHGP). The x-axis represents time (in hours) in Fig. 3, representing the duration of the experimental run for each batch. Each bar graph on the left side of Fig. 3 represents a stack of product gases corresponding to the batch experimental run. Four separate batch reactions are shown on Fig. 3 at the left side of the slide. The duration of the four (4) runs was 4, 10, 16, and 24 hours (hrs), respectively. The left y-axis corresponding to the bar graph provides quantitative information about 16 gaseous product concentrations in the gas phase. The height or thickness of each color represents the relative concentration of various product gases at the end of the batch reactions. The right y-axis provides information about the total % S removal in the liquid phase, determined after the completion of each run. It is evident from Fig. 3 that H_2S is the primary reaction product in the gas phase. Other S compounds, namely, SO_2 and COS, are also observed from the gas analysis. It is also evident from Fig. 3 that the total maximum % S removal is reached after ~4 h of batch operation. As the reaction progressed with time, the % S removal decreased as the gas phase S products moved into the liquid phase by other reverse reactions. The role of molecular hydrogen (H_2) is shown in Fig. 4. The x-axis represents three different types of electrolytes. Both y-axis parameters are identical to that of Fig. 3. During this study, as shown in Fig. 4, the electrochemical desulfurization reaction was carried out with the addition of various electrolytes: 0.25 M H_2SO_4, a combination of 0.1 wt% dodecyl trimethyl ammonium bromide (DTAB, $C_{15}H_{34}N.Br$) and 0.25 M H_2SO_4, and standalone 0.1 wt% DTAB. The results demonstrate that despite the presence of a high concentration of

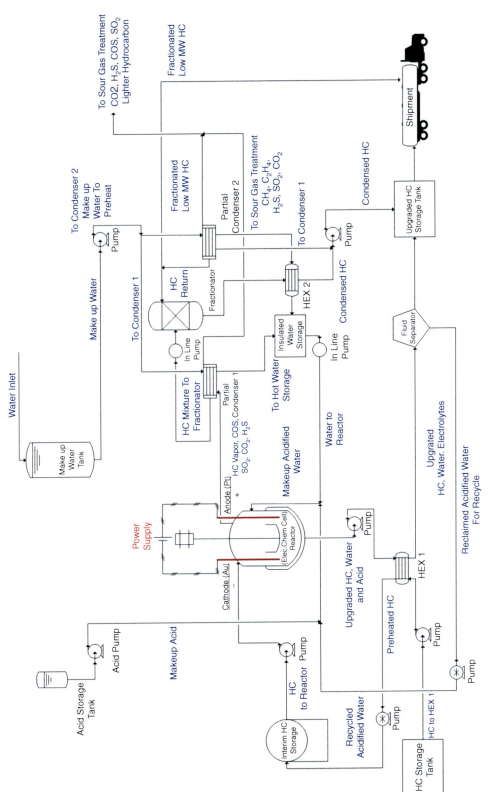

Sulfur Removal from Crude Oil and its Fraction, Fig. 2 Simplified PFD of in situ electrochemical hydrodesulfurization of crude fractions

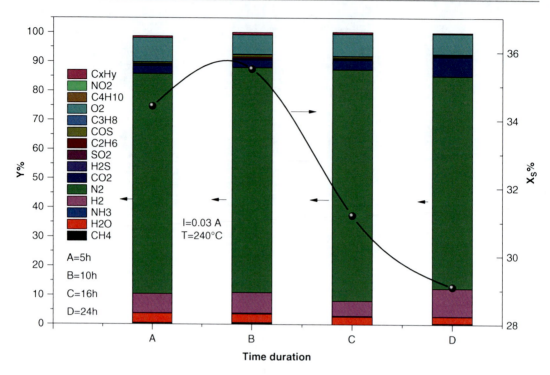

Sulfur Removal from Crude Oil and its Fraction, Fig. 3 Typical gas phase composition

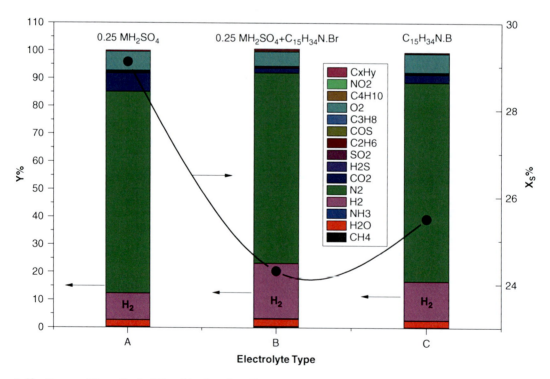

Sulfur Removal from Crude Oil and its Fraction, Fig. 4 The role of hydrogen

Sulfur Removal from Crude Oil and its Fraction

molecular hydrogen – as evident from "case b" and "case c" – the % S removal was not as high as observed in "case a," which involved 0.25 M H_2SO_4, the sole electrolyte in the reaction media.

Reaction Mechanism

As noted earlier from Fig. 4, many side reactions may be taking place during the iHGP. The major reactions involving H_2S formation are proposed below to describe the reaction mechanism for S removal. Atomic hydrogen is formed and remains adsorbed ($H^{\bullet}_{(ad)}$) on the cathode surface (Au) from the electrolysis of water given in Eq. 1 after the application of the right potential/current. The adsorbed $H^{\bullet}_{(ad)}$, in general, converts into molecular hydrogen (H_2), as shown by Eqs. 2 and/or 3. The depleted $H^+_{(aq)}$ gets replenished by the electrolysis of water during the desulfurization reactions.

$$H^+_{(aq)} + e^- \rightarrow H^{\bullet}_{(ad)} \qquad (1)$$

Volmer

$$H^{\bullet}_{(ad)} + H^{\bullet}_{(ad)} \rightarrow H_2 \qquad (2)$$

Tafel

$$H^{\bullet}_{(ad)} + H^+_{(aq)} + e^- \rightarrow H_2 \qquad (3)$$

Heyrovsky

During the desulfurization reactions, the S compounds also get competitively adsorbed on the cathode surface (Au) active sites adjacent to adsorbed $H^{\bullet}_{(ad)}$ atoms given by Eqs. 4 and 5 to react to form S-free adsorbed HC molecules on the active sites, with concurrent release of H_2S followed by the release of S-free adsorbed HC molecules from the active sites of the electrodes. This mechanism is also being supported by Fig. 4, which shows that, despite the presence of large quantities of molecular H_2, higher desulfurization could not be achieved. Therefore, it is proposed

that it is the atomic aqueous hydrogen ion $H^+_{(aq)}$ that is contributing to the formation of atomic adsorbed $H^{\bullet}_{(ad)}$ during the desulfurization reactions and not the molecular H_2.

$$R\text{-}S\text{-}S\text{-}R' \rightarrow (R\text{-}S\text{-}S\text{-}R')_{(ad)} \qquad (4)$$

Adsorption

$$(R\text{-}S\text{-}S\text{-}R')_{(ad)} + 6H^{\bullet}_{(ad)} \rightarrow (R\text{-}H + R'\text{-}H)_{(ad)} + 2H_2S \qquad (5)$$

Hydrogenation

$$(R\text{-}H + R'\text{-}H)_{(ad)} \rightarrow R\text{-}H + R'\text{-}H \qquad (6)$$

Desorption

Advantages

The advantage of in situ HDS is that it is carried out in milder conditions, at lower temperature ($\sim$250 °C) and pressure ($\sim$40 atm) with lower H partial pressure. The process also has the ability to accommodate various types of low-cost electrocatalysts for targeting polyaromatic molecules during the electrocatalytic reactions, thereby reducing the total cost.

Future Directions

Major work is needed for designing new-generation batch and continuous reactors to suit the needs of different feedstocks during in situ HDS processes (iHGP). Novel electrocatalyst developments should also be undertaken with a wide array of materials to target the tenacious sulfur, nitrogenous, and aromatic molecules. An example would be sputter deposited customized electrocatalysts that have the potential to target specific classes of molecules with improved selectivity at reduced cost [24, 25].

Cross-References

▶ Electrochemical Removal of H_2S

References

1. Xiang C, Chai Y, Liu Y, Liu C (2008) Mutual influences of hydrodesulfurization of dibenzothiophene and hydrodenitrogenation of indole over NiMoS/γ-Al_2O_3 catalyst. J Fuel Chem Technol 36(6):684–690
2. Orozco EO, Vrinat M (1998) Kinetics of dibenzothiophene hydrodesulfurization over MoS_2 supported catalysts: modelization of the H_2S partial pressure effect. Appl Catal Gen 170:195–206
3. Egorova M, Prins R (2004) Hydrodesulfurization of dibenzothiophene and 4,6-dimethyldibenzothiophene over sulfided NiMo/γ-Al_2O_3, CoMo/γ-Al_2O_3, and Mo/γ-Al_2O_3 catalysts. J Catal 225:417–427
4. Granados-Aguilar AS, Viveros-García T, Pérez-Cisneros ES (2008) Thermodynamic analysis of a reactive distillation process for deep hydrodesulfurization of diesel: effect of the solvent and operating conditions. Chem Eng J 143:210–219
5. Kabe T, Aoyama Y, Wang D, Ishihara A, Qian W, Hosoya M, Zhang Q (2001) Effects of H_2S on hydrodesulfurization of dibenzothiophene and 4,6-dimethyldibenzothiophene on alumina-supported NiMo and NiW catalysts. Appl Catal Gen 209:237–247
6. Manoli JM, Costa PD, Brun M, Vrinat M, Maugé F, Potvin C (2004) Hydrodesulfurization of 4,6-dimethyldibenzothiophene over promoted (Ni, P) alumina-supported molybdenum carbide catalysts: activity and characterization of active sites. J Catal 221:365–377
7. Song C (2000) Introduction to chemistry of diesel fuels. In: Song C, Hsu CS, Mochida I (eds) Chemistry of diesel fuels. Taylor & Francis, New York
8. Stanislaus A, Marafi A, Rana MS (2010) Recent advances in the science and technology of ultralow sulfur diesel (ULSD) production. Catal Today 153:1–68
9. Pawelec B, Castaño P, Zepeda TA (2008) Morphological investigation of nanostructured CoMo catalysts. Appl Surf Sci 254:4092–4102
10. Eijsbouts S, Plantenga F, Leliveld B, Inoue Y, and Fujita K (2003) STARS and NEBULA – new generations of hydroprocessing catalysts for the production of ultra low sulfur diesel. http://www.anl.gov/PCS/acsfuel/preprint%20archive/Files/48_2_New%20York_10-03_0583.pdf. Accessed 28 Nov 2011. Prepr Pap.-Am Chem Soc, Div Fuel Chem 48(2):494
11. Farag H, El-Hendawy ANA, Sakanishi K, Kishida M, Mochida I (2009) Catalytic activity of synthesized nanosized molybdenum disulfide for the hydrodesulfurization of dibenzothiophene: effect of H_2S partial pressure. Appl Catal Environ 91:189–197
12. Roukoss C, Laurenti D, Devers E, Marchand K et al (2009) Hydrodesulfurization catalysts: promoters, promoting methods and support effect on catalytic activities. C R Chim 12:683–691
13. Vakros J, Papadopouloua C, Lycourghiotisa A, Kordulisa C (2011) Hydrodesulfurization catalyst bodies with various Co and Mo profiles. Appl Catal Gen 399:211–220
14. Wang H, Prins R (2009) Hydrodesulfurization of dibenzothiophene, 4,6-dimethyldibenzothiophene, and their hydrogenated intermediates over Ni–MoS_2/γ-Al_2O_3. J Catal 264:31–43
15. Deng Z, Wang T, Wang Z (2010) Hydrodesulfurization of diesel in a slurry reactor. ChemEng Sci 65:480–486
16. Lee JJ, Kim H, Koh JH, Sang AJ, Moon H (2005) Performance of CoMoS/Al_2O_3 prepared by sonochemical and chemical vapor deposition methods in the hydrodesulfurization of dibenzothiophene and 4,6-dimethyldibenzothiophene. Appl Catal Environ 58:89–95
17. Farag H et al (2006) Autocatalysis-like behavior of hydrogen sulfide on hydrodesulfurization of polyaromatic thiophenes over a synthesized molybdenum sulfide catalyst. Appl Catal Gen 314:114–122
18. Farag H, Sakanishi K, Kouzu M, Matsumura A et al (2003) Dual character of H_2S as promoter and inhibitor for hydrodesulfurization of dibenzothiophene. Catal Commun 4:321–326
19. Lee SI, Cho A, Koh JH, Oh SH, Moon SH (2011) Co promotion of sonochemically prepared MoS_2/Al_2O_3 by impregnation with $Co(acac)_2 \cdot 2H_2O$. Appl Catal Environ 101:220–225
20. Pawelec B, Fierro JLG et al (2008) Influence of the acidity of nanostructured CoMo/P/Ti-HMS catalysts on the HDS of 4,6-DMDBT reaction pathways. Appl Catal Environ 80:1–14
21. Babich IV, Moulijn JA (2003) Science and technology of novel processes for deep desulfurization of oil refinery streams: a review. Fuel 82:607–631
22. Cui X, Yao D, Li H, Yang J, Hu D (2010) Nano-magnetic particles as multifunctional microreactor for deep desulfurization. J Hazard Mater. doi:10.1016/j.jhazmat.2011.11.063
23. Huirache-Acuña R, Albiter MA, Ornelas C, Paraguay-Delgado F, Martínez-Sánchez R, Alonso-Nuñez G (2006) Appl Catal Gen 308:134–142
24. Slavcheva E, Radev I, Bliznakov S (2007) Sputtered iridium oxide films as electrocatalysis for water splitting via PEM electrolysis: Electrochimica. ACTA 52:3889–3894
25. Carmo M, Fritz DL, Mergel J (2013) A Comprehensive review on water electrolysis. Int J Hydrogen Energ 38:4901–4934

Super Capacitors

Masanobu Chiku
Department of Applied Chemistry, Graduate
School of Engineering, Osaka Prefecture
University, Osaka, Japan

Introduction

Electrical energy can be basically stored in two different ways. First, electrical energy can be stored as chemical energy which becomes available through the faradaic reaction of the active material. This type of energy storage device is called a battery. Second, electrical energy can be stored as electrostatically adsorbed positive and negative charges, that is, to say non-faradaic reaction occurs on the electrode. Conventional capacitors use non-faradaic reaction to store the electrical energy.

There is a major difference between faradaic and non-faradaic reactions. Non-faradaic reactions involve no chemical reaction on the electrodes, but faradaic reactions involve chemical reactions including phase transition of active materials. Therefore, the cycle life of battery reactions is limited to several thousand cycles or less, due to irreversible chemical reactions and irreversible phase changes of active materials. On the other hand, the cycle life of capacitors is over $10^5 \sim 10^6$ cycles.

Super capacitors, which are also called electrochemical capacitors or ultra capacitors, have far more capacity than conventional capacitors. For example, electrical double-layer capacitors have $10 \sim 100$ F g^{-1} capacity.

Classification of Capacitors

Table 1 shows the type of capacitors and the method used to store electrical energy, i.e., faradaic or non-faradaic electrostatic reactions.

The electrostatic capacitor uses organic polymer, mica, glass, ceramic, and anodic oxide film as dielectric material. On the other hand, electrolytic capacitors use aluminum or tantalum metal oxide film formed by electrolytic techniques as dielectric materials. These types of capacitors are used as elements on electrical circuits and have small capacity. So these types of capacitors are not classified as super capacitors.

The following types of capacitors are classified as super capacitors. Electrical double-layer capacitors are a type of capacitor which uses charge separation at the interface between electrode and electrolyte. This type of capacitors realizes large capacities – up to several hundred F g^{-1} – by using electrode materials with large surfaces such as activated carbon. In electrical double-layer capacitors, solutions of sulfuric acid or potassium hydroxide are used as aqueous electrolytes, and aprotic organic solvents with alkylammonium salt are used as nonaqueous electrolytes. The capacities of electrical double-layer capacitors are generically up to 200 F g^{-1} with organic electrolytes and 300 F g^{-1} with aqueous electrolytes [1].

There are also super capacitors using faradaic reactions. Pseudo capacitors, which are also called redox capacitors, use redox reactions on electrode materials like ruthenium oxide or polyaniline. Batteries also use redox reactions of active mass during charging and discharging. In a battery, the electrode maintains a nearly constant potential during charging and discharging. In the case of the pseudo capacitor, electrode potential changes during charging and discharging, as it does in an electrical double-layer capacitor. Because faradaic reactions have larger capacitance than non-faradaic reactions, pseudo capacitors have very large capacitances. For example, pseudo capacitors with ruthenium oxide electrodes have more than 1,000 F g^{-1} capacitance with aqueous electrolyte [2]. Charging and discharging of pseudo capacitors is accompanied by the transfer of materials on the electrode/electrolyte interface. Thus, high rate charging and discharging of pseudo capacitors will require innovative electrode and electrolyte materials or configuration of the capacitor cells.

Super Capacitors, Table 1 Classification of capacitors

Name	Mechanism	Type	Classification
Electrostatic capacitor	Electrostatic	Non-faradaic	Conventional capacitor
Electrolytic capacitor	Electrostatic	Non-faradaic	Conventional capacitor
Electrical double-layer capacitor	Electrical double layer	Non-faradaic	Super capacitor
Pseudo capacitor	Redox reaction	Faradaic	Super capacitor
Hybrid capacitor	Redox reaction	Faradaic	Super capacitor

Hybrid capacitors use a combination of battery-type and capacitor-type electrodes. For example, the negative electrode is a battery-type electrode and the positive electrode is an electrical double-layer capacitor. Comparing to pseudo capacitors, the potential of battery-type electrodes is constant during charging and discharging.

Comparison with Secondary Battery

From the view point of charge/discharge capacity, super capacitors are positioned between electrostatic capacitors and secondary batteries. That is, the charge/discharge capacity (energy density) of super capacitors is greater than electrostatic capacitors but smaller than secondary batteries. The operation voltage of super capacitors is almost same as secondary batteries, but charging/discharging speed is much faster than secondary batteries. Thus, the power density of super capacitors is greater than secondary batteries. Turning to the cycle characteristics of super capacitors, electrical double-layer capacitors are possible which cycle an infinite time – at least – in principle. On the other hand, pseudocapacitors have better cycle characteristics than secondary batteries but poorer cycle characteristics than electrical double-layer capacitors.

Commercially Available Super Capacitor

The electrochemical double-layer capacitor is already commercially available from several companies, and their energy densities are about 10 Wh L^{-1} [3].

Lithium-ion capacitor is one of the most promising commercially available super capacitor due to its high energy density and high specific power. As the negative electrode, a conventional graphite electrode for lithium-ion battery is used and activated carbon electrode is used as positive electrode. Thus, the lithium-ion capacitor belongs to the group of hybrid capacitor. The cell voltage of lithium-ion capacitors is 3.8 V with fully charged state and 2.2 V with fully discharged state. The lithium-ion capacitor is already commercially available, and Japanese companies, including JM Energy Corp., Asahi Kasei FDK Energy, and TAIYO YUDEN Co., Ltd, took initiative to develop the practical lithium-ion capacitor [3]. Commercial lithium-ion capacitor shows the energy density about 20 Wh L^{-1}.

Application of Super Capacitors

Super capacitors have different characteristics with rechargeable batteries as mentioned above, so super capacitors are used in different applications with rechargeable batteries. The major applications for super capacitors are high-pulse power and short-term power hold [4]. There are several applications for super capacitors like memory backup, portable power supplies, electric vehicles, and hybrid electric vehicles. Recent attracting applications are load leveling of renewable energy applications like photovoltaic or wind energy. Rechargeable batteries for such applications need to be replaced every 1–3 years. On the other hand, life duration of super capacitors in such application is around

Super Capacitors

20 years, which is similar to the life span of the photovoltaic panels.

Future Perspectives

Electrochemical double-layer capacitors with activated carbon electrodes are commercially available from several companies. In the case of hybrid capacitors, lithium-ion capacitors which consist of the anode of a lithium-ion battery and a cathode of activated carbon are commercially available, as mentioned above. These super capacitors show great promise as next generation power sources for electric devices and electric vehicles. More innovation is needed before pseudo capacitors are suitable for commercialization.

Advanced carbon materials such as graphene are promising active materials for electrical double-layer capacitors [5]. Graphene have high specific surface area, but the graphene-based materials usually show smaller surface area than the theoretical value, i.e., 2,630 m^2 g^{-1}. However, recent progress in activation process for graphene-based materials realized the electrode which is mostly made of single graphene layers with high surface area around 3,000 m^2 g^{-1} [6]. Graphene has not only large surface area but high electrical conductivity, so compositions of these carbon materials and pseudo capacitive materials such as ruthenium oxide, nickel oxide, or manganese oxide are attracting attentions for several researchers [5].

Templated carbon materials are also considered as future electrode materials for super capacitors [7]. Templated carbon materials are made by using nanostructured and/or self-assembled materials as template for carbon materials. Several materials are used as template, for example, mesoporous silica, silica colloidal crystals, zeolite, and anodic aluminum oxide film are used as template for ordered mesoporous carbon, carbon inverse opal, zeolite template carbon, and anodic aluminum oxide template carbon, respectively. These carbon materials show lower surface area than activated carbons, but their uniformly ordered highly porous structures are suitable for combining with pseudo capacitive materials such as metal oxides.

Cross-References

▶ Electrolytes for Electrochemical Double Layer Capacitors
▶ Ionic Liquids for Supercapacitors

References

1. Zhang LL, Zhao XS (2009) Carbon-based materials as supercapacitor electrodes. Chem Soc Rev 38:2520–2531
2. Simon P, Gogotsi Y (2008) Materials for electrochemical capacitors. Nat Mater 7:845–854
3. Naoi K, Ishimoto S, Miyamoto J, Naoi W (2012) Second generation 'nanohybrid supercapacitor': evolution of capacitive energy storage devices. Energy Environ Sci 5:9363–9373
4. Shukla AK, Banerjee A, Rabikumar MK, Jalajakshi A (2012) Electrochemical capacitors: technical challenges and prognosis for future markets. Electrochim Acta 84:165–173
5. Huang X, Zeng Z, Fan Z, Liu J, Zhang H (2012) Graphene-based electrodes. Adv Mater 24:5979–6004
6. Zhu YW, Murali S, Stoller MD, Ganesh KJ, Cai WW, Ferreira PJ, Pirkle A, Wallace RM, Cychosz KA, Thommes M, Su D, Stach EA, Ruoff RS (2011) Carbon-based supercapacitors produced by activation of graphene. Science 332:1537–1541
7. Nishihara H, Kyotani T (2012) Templated nanocarbons for energy storage. Adv Mater 24:4473–4498

T

Thermal Effects in Electrochemical Systems

Adam Z. Weber
Lawrence Berkeley National Laboratory,
Berkeley, CA, USA

Introduction

Most electrochemical systems do not operate thermodynamically reversibly as there is also some heat generation or input required. However, unlike thermal engines, efficiencies for electrochemical systems can be quite high. Most of the heat generation in these systems is due to irreversible losses or overpotentials (kinetic or mass transfer) and the passage of current (ohmic or IR losses). In some systems, the heating can become significant enough to cause thermal runaway events such as shorting in lithium batteries or hydrogen crossover through membrane pinholes in polymer-electrolyte fuel cells. In this chapter, we discuss the treatment of thermal effects from a macroscopic view through understanding and exploring the governing equations. The focus is on electrochemical systems that are involved with storage and conversion. Thus, solid-state systems such as photovoltaics, where incident photons are absorbed to promote hole/electron separation within a semiconductor band structure and which release heat, or thermoelectrics, where the Peltier and Seebeck effects give rise to

temperature changes due to flow of electricity and vice versa, are not discussed, respectively, although many of the descriptions below can be applied to such devices [1]. Below, first the nature of the thermal effects and then the transport of energy are discussed.

Thermodynamics

Heat generation and thermal effects are due to both irreversible and reversible heat generation. From the first law of thermodynamics,

$$\Delta H = Q - W_s \tag{1}$$

one sees that the net release of energy is in the form of heat (Q) or work (W_s). The work performed by the system is simply the power as expressed by

$$P = IV \tag{2}$$

where I is the current and V is the cell voltage. The net energy due to the electrochemical reaction is the difference between the heat of formation of the products and reactants, ΔH, which can be converted to an electrochemical potential, resulting in the enthalpy potential,

$$U_H = \frac{\Delta H}{z_i F} \tag{3}$$

G. Kreysa et al. (eds.), *Encyclopedia of Applied Electrochemistry*, DOI 10.1007/978-1-4419-6996-5,
© Springer Science+Business Media New York 2014

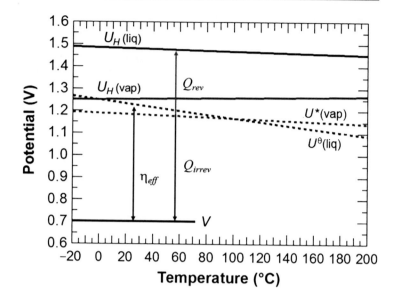

Thermal Effects in Electrochemical Systems, Fig. 1 Thermodynamic potentials and definitions of a hydrogen fuel cell, where the heat lost (Q) is composed of reversible and irreversible losses and the cell is operating at a potential V; the efficiency is also shown

where z_i is the charge number of species i and F is Faraday's constant. Thus, the expression for the heat released becomes

$$Q = i(U_H - V) \quad (4)$$

If the cell potential equals the enthalpy potential, there is no net heat loss, which is why the enthalpy potential is often termed the thermoneutral potential. However, the enthalpy energy is not fully accessible as it is composed of both a reversible or entropic ($Q_{rev} = T\Delta S$) component, as well as irreversible components.

The maximum energy that is convertible to electrical energy is related to the free energy of the reactive species:

$$\Delta G = \Delta H - T\Delta S \quad (5)$$

The potential corresponding to the Gibbs free energy is defined as the equilibrium potential:

$$U^0 = \frac{\Delta G}{z_i F} \quad (6)$$

Usually, most devices are operated not at the reference conditions. To account for this change, Eqs. 5 and 6 are combined and differentiated with respect to temperature, yielding the change in equilibrium potential with temperature of

$$U = U^0 + \Delta U = U^0 + \frac{\Delta S}{z_i F}(T - T^0) \quad (7)$$

As an example, Fig. 1 shows the equilibrium and enthalpy potentials as calculated from handbooks [2, 3] for the electrochemical reaction of hydrogen with oxygen. Depending on if the product water is vapor or liquid, one arrives at different potentials due to the latent heat and free energy difference between liquid (U^0) and vapor (U^*) water (the two potentials are related logarithmically by the vapor pressure of water, which is why they cross at 100 °C). Thus, as the cell temperature increases, the amount of usable work from the fuel decreases (assuming the generated heat is expelled to the atmosphere), while the amount of heat generated increases for a given operating voltage or potential. Also, Fig. 1 clearly shows how significant the latent heat of water can be (~200 mV).

For a cell that operates at a voltage, V, below the equilibrium and enthalpy potentials, the net heat lost from the cell is given by

$$Q_{loss} = Q_{rev} + Q_{irrev}$$
$$= i(U_H - U^0) + i(U^0 - V) \quad (8)$$

which is summarized in Fig. 1. Thus, if the cell operates at a potential between the equilibrium

Thermal Effects in Electrochemical Systems

and enthalpy potentials, the process is endothermic, whereas if the operating potential is above or below both potentials, the process is exothermic.

Also as shown in Fig. 1, for fuel cells, the efficiency of the cell, η_{eff}, is typically defined relative to the maximum free energy available for electrical work:

$$\eta_{\text{eff}} = 1 - \frac{V}{U} \tag{9}$$

One must also be cognizant of whether the efficiency is defined in terms of the equilibrium or enthalpy values, and what the reference state is for the calculation (i.e., vapor or liquid water). This is especially important when comparing different fuel cells as well as with fuel cells to other systems. For example, solid oxide fuel cells operate at temperatures (600–900 °C) where the heat generated can be recovered to electrical energy, thereby making efficiency greater than 100 % possible using the definition above. Thus, it is more advisable to use the heating value or enthalpy or enthalpy potential of the fuel as the metric for efficiency since this also allows for a better comparison among technologies (e.g., combustion engines to fuel cells).

Thermal Balance and Transport

Equation 8 can be used to determine the net heat generated by the system. If one assumes that the cell or system is internally isothermal for that time, then an overall heat balance can be used to determine the temperature rise. This assumption is often used in modeling of battery systems, where one calculates temperature as a function of time but with each time being at an isothermal temperature. Such a lumped parameter estimate has also been used in modeling other electrochemical cells (e.g., fuel cells) since it provides an estimate of the temperature increase in a given segment or time without worrying about the complicated and nonlinear effects of actual thermal gradients. Mathematically, such an energy balance takes the form of

$$\frac{dT}{dt} = \frac{1}{m_{cell}\hat{C}_P^{cell}}(Q - Q_{sens} - Q_{loss}) \tag{10}$$

where m_{cell} and $\hat{C}_P^{cell}$ are the mass and heat capacity of the cell, respectively. The energies on the right side of the equation, from left to right, correspond to the theoretical heat production (see Eq. 8), the heat transferred to the coolant and reactants (including phase change and latent heat processes), and the heat lost to the environment from the surface of the cell or stack:

$$Q_{loss} = UA(T_{cell} - T_{ambient}) \tag{11}$$

where U is the overall heat transfer coefficient and A is the effective interfacial surface area.

If one desires to know the temperature distribution throughout the cell, then an overall energy balance and transport equation must be utilized. This is often critical for identification of failure points that are geometry specific or perhaps in understanding phenomena that are due to the temperature gradient (e.g., phase-change-induced flow in fuel cells [4]).

The conservation of thermal energy can be written as

$$\begin{aligned}
\rho_k \hat{C}_{p_k} & \left(\frac{\partial T_k}{\partial t} + \mathbf{v}_k \cdot \nabla T_k \right) \\
& + \left(\frac{\partial \ln \rho_k}{\partial \ln T_k} \right)_{p_k, x_{i,k}} \left(\frac{\partial p_k}{\partial t} + \mathbf{v}_k \cdot \nabla p_k \right) \\
& = Q_{k,p} - \nabla \cdot \mathbf{q}_k - \boldsymbol{\tau} : \nabla \mathbf{v}_k \\
& + \sum_i \bar{H}_{i,k} \nabla \cdot \mathbf{J}_{i,k} - \sum_i \bar{H}_{i,k} \Re_{i,k}
\end{aligned} \tag{12}$$

In the above expression, the first term represents the accumulation and convective transport of enthalpy, where $\hat{C}_{p_k}$ is the heat capacity of phase k which is a combination of the various components of that phase. The second term is energy due to reversible work. For condensed phases, this term is negligible. The first two terms on the right side of Eq. 12 represent the net heat input by conduction and interphase transfer.

The first is due to heat transfer between two phases:

$$Q_{k,p} = h_{k,p} a_{k,p} (T_p - T_k) \tag{13}$$

where $h_{k,p}$ is the heat transfer coefficient between phases k and p per interfacial area. Most often but not always this term is used as a boundary condition since it occurs only at the edges (see Eq. 11). The second term is due to the heat flux in phase k:

$$\mathbf{q}_k = \sum_i \bar{H}_{i,k} \mathbf{J}_{i,k} - k_{T_k}^{\text{eff}} \nabla T_k \tag{14}$$

where $\bar{H}_{i,k}$ is the partial molar enthalpy of species i in phase k, $\mathbf{J}_{i,k}$ is the flux density of species i relative to the mass-average velocity of phase k

$$\mathbf{J}_{i,k} = \mathbf{N}_{i,k} - c_{i,k} \mathbf{v}_k \tag{15}$$

and $k_{T_k}^{\text{eff}}$ is the effective thermal conductivity of phase k. The third term on the right side of Eq. 12 represents viscous dissipation, the heat generated by viscous forces, where τ is the stress tensor. This term is often small for electrochemical systems and for most cases can be neglected unless highly viscous fluids are used such as some nonaqueous or high salt-content solvents in batteries. The fourth term on the right side comes from enthalpy changes due to diffusion. Finally, the last term represents the change in enthalpy due to reaction

$$\sum_i \bar{H}_{i,k} \mathfrak{R}_{i,k} = - \sum_h a_{1,k} i_{h,1-k} \left(\eta_{s_{h,1-k}} + \Pi_h \right)$$
$$- \sum_{p \neq k} \Delta H_l a_{k,p} r_{l,k-p}$$
$$- \sum_g \Delta H_g R_{g,k}$$
$$\tag{16}$$

The above reaction terms include homogeneous reactions, interfacial reactions (e.g., phase change), and interfacial electron-transfer reactions. As discussed above, the irreversible heat generation is represented by the activation overpotential, $\eta_{s_{h,1-k}}$, and the reversible heat generation is represented by the Peltier coefficient, Π_h (see Fig. 1). Upon comparison to Eqs. 7 and 8, the Peltier coefficient for charge-transfer reaction h can be expressed as

$$\Pi_h \approx \frac{T}{n_h F} \sum_i s_{i,k,h} \bar{S}_{i,k} = T \frac{\Delta S_h}{n_h F} \tag{17}$$

where ΔS_h is the entropy of reaction h. The above equation neglects transported entropy (hence the approximate sign), and the summation includes all species that participate in the reaction (e.g., electrons, ions, neutral products and reactants). While the entropy of half-reactions that occur at electrodes are not truly obtainable since they involve knowledge of the activity of an uncoupled ion, Peltier coefficients can be measured through calorimetry.

It is often the case that in electrochemical systems there is enough intimate contact between the various phases that local equilibrium can be assumed such that all of the phases have the same temperature at a given point. Assuming this eliminates the phase dependences in the above equations and allows for a single thermal energy equation to be written. Neglecting those phenomena that are minor as mentioned above and summing over the phases result in

$$\sum_k \rho_k \hat{C}_{p_k} \frac{\partial T}{\partial t} = - \sum_k \rho_k \hat{C}_{p_k} \mathbf{v}_k \cdot \nabla T + \nabla$$
$$\cdot \left(k_T^{\text{eff}} \nabla T \right) + \sum_k \frac{\mathbf{i}_k \cdot \mathbf{i}_k}{\kappa_k^{\text{eff}}}$$
$$+ \sum_h i_h (\eta_h + \Pi_h)$$
$$- \sum_h \Delta H_h r_h \tag{18}$$

where the expression for Joule or ohmic heating has been substituted in from the combination of Eq. 14 into Eq. 12 and subsequent cancellation with the fourth term in the right side of Eq. 12, i.e.,

$$- \sum_i \mathbf{J}_{i,k} \cdot \nabla \bar{H}_{i,k} = - \mathbf{i}_k \cdot \nabla \Phi_k = \frac{\mathbf{i}_k \cdot \mathbf{i}_k}{\kappa_k^{\text{eff}}} \tag{19}$$

In Eq. 18, the first term on the right side is energy transport due to convection, the second is energy transport due to conduction, the third is ohmic heating, the fourth is reaction heats, and the last represents reactions in the bulk which include such things as phase change. As noted above, heat lost to the surroundings (i.e., Eq. 13) is only accounted for at the boundaries of the cell. This energy balance is often what is used for understanding and predicting temperature distributions in electrochemical systems, although the general form (Eq. 12) or significant terms can always be added and also one does not need to assume local equilibrium among the phases.

It should also be noted that the total heat generation is still given by Eq. 8, since upon summation over all of the charge-transfer reactions the total heat generation is

$$Q = \sum_{k-p} \sum_{h} a_{k,p} i_{h,k-p} (U_{H_h} - V) \qquad (21)$$

which can be used in lumped energy balances (e.g., Eq. 10) if one only cares about the total heat generation and not its distribution.

Future Directions

The above discussion outlines and discusses heat generation and transport within electrochemical systems. The presentation and derivation of the equations are meant as a guide to initial understanding of such thermal effects. While the lessons are generally valid, there are nuances in terms of properties, aggregation of cells, and including such effects as thermal runaway. In particular, it is noted that large battery systems typically require complex thermal management systems in order to prevent catastrophic failures if something goes awry. In addition, thermal issues often impact durability by increasing degradation rates or may cause problems at very low temperatures in terms of performance. These discussions are beyond the scope of this chapter and can readily be analyzed and learned about in the literature and throughout this encyclopedia.

References

1. Newman J, Thomas-Alyea KE (2004) Electrochemical systems, 3rd edn. Wiley, New York
2. Perry RH, Green DW (1997) Perry's chemical engineers' handbook, 7th edn. McGraw-Hill, New York
3. Weast RC (1979) CRC handbook of chemistry and physics, 59th edn. CRC Press, Boca Raton
4. Weber AZ, Newman J (2006) Coupled thermal and water management in polymer electrolyte fuel cells. J Electrochem Soc 153:A2205–A2214

Thermodynamic Properties of Ionic Solutions - MSA and NRTL Models

Jean-Pierre Simonin
Laboratoire PECSA, UMR CNRS 7195,
Université Paris, Paris, France

Introduction

Numerous models are available in the literature for the description of the thermodynamic properties of ionic solutions [1–8]. In this entry, we focus on two particular models: the mean spherical approximation (MSA) and the electrolyte nonrandom two liquid model (NRTL).

The Mean Spherical Approximation (MSA)

The MSA was first introduced by Percus and Yevick [9] as an approximate way of introducing hard-core effects on the distribution of charged particles. Then, Waisman and Lebowitz applied this approximation to electrolytes [10]. They obtained the solution to the Ornstein-Zernike equation [8] in the case of ions having the same diameter, in the primitive model of solutions in which ions are regarded as charged spheres immersed in a continuum (the solvent) of relative permittivity ε. The treatment was later extended to the case of ions of different diameters by Blum [11] and Blum and Høye [12].

The MSA solution for a mixture of ions and dipoles (a non-primitive model simulating an aqueous electrolyte), with all particles having the same diameter, has been found independently by Blum [13] and Adelman and Deutsch [14]. The case of equisized ions and dipoles of different size was later studied by Blum et al. [15] and by Høye and Lomba [16]. Explicit expressions for the thermodynamic quantities were obtained.

The MSA is fundamentally connected to the Debye-Hückel (DH) theory [7, 8], in which the linearized Poisson-Boltzmann equation is solved for a central ion surrounded by a neutralizing ionic cloud. In the DH framework, the main simplifying assumption is that the ions in the cloud are point ions. These ions are supposed to be able to approach the central ion to some minimum distance, the distance of closest approach. The MSA is the solution of the same linearized Poisson-Boltzmann equation but with finite size for all ions. The mathematical solution of the proper boundary conditions of this problem is more complex than for the DH theory. However, it is tractable and the MSA leads to analytical expressions. The latter shares with the DH theory the remarkable simplicity of being a function of a single screening parameter, generally denoted by Γ. For an arbitrary (neutral) mixture of ions, this parameter satisfies a simple equation which can be easily solved numerically by iterations. Its expression is explicit in the case of equisized ions (restricted case) [12]. One has

$$\Gamma = \frac{\sqrt{1 + 2\kappa\sigma} - 1}{2\sigma}$$

in which κ is the classical DH screening parameter and σ is the common ion diameter. Besides, the expressions of the thermodynamic excess functions in the MSA are formally similar to those of the DH theory [12]. In the restricted case, the contributions to the osmotic and mean salt activity coefficient are respectively given by

$$\Delta\Phi^{MSA} = -\frac{\Gamma^3}{3\pi\rho_t}$$

$$\Delta\ln\gamma_{\pm}^{MSA} = -\frac{1}{4\pi}\frac{\Gamma}{1 + \Gamma\sigma}\frac{\kappa^2}{\rho_t}$$

in which ρ_t is the total number density of ions (number of ions per volume unit).

As compared to the MSA, better statistical mechanical approximations exist, such as the HNC equation and its improvements [17]. However, these equations need be solved numerically for every individual system, and it is sometimes very difficult to achieve convergence of the numerical algorithm.

The MSA has been applied to the representation of deviations from ideality in ionic solutions. Effective ionic radii have been determined for the calculation of osmotic coefficients for solutions up to 1 mol L^{-1} [18], for concentrated aqueous solutions [19] and for the computation of activity coefficients in ionic mixtures [20]. In these studies, for a given salt, a unique hard sphere diameter was determined for the whole concentration range. Also, thermodynamic data were fitted with the use of one linearly density-dependent parameter (a hard-core size or dielectric parameter), up to moderate concentrations, by least-squares refinement [21–24] or with a nonlinearly varying cation size [25] in very concentrated electrolytes. The deviations from ideality for a great number of strong electrolytes have been described by using a concentration-dependent cation size and solution permittivity to very high concentrations [26, 27]. In this approach, the osmotic and activity coefficients of salt are thermodynamically consistent in the sense that they satisfy the Gibbs-Duhem relationship. This was achieved by applying a suitable conversion to these quantities [27]. Next, associating electrolytes have been treated within the framework of the binding MSA [28] or associative MSA [29], by introducing a mass action law for the formation of the pair (ionic or chemical) [30–32] or of a trimer [33].

The Electrolyte Nonrandom Two Liquid Model (NRTL)

The NRTL was proposed originally by Renon and Prausnitz [1, 34] for the thermodynamics of liquid mixtures. The NRTL equation has relationship to Wilson's equation. It was established

Thermodynamic Properties of Ionic Solutions - MSA and NRTL Models

based on the concept of local composition and nonrandomness of mixing. In the case of a binary mixture of species A and B, the relation assumed in NRTL between the local fractions of particles is [34]

$$\frac{x_{J(I)}}{x_{K(I)}} = \frac{x_J}{x_K} \frac{\exp\left(-\alpha_{JI}\tau_{JI}\right)}{\exp\left(-\alpha_{KI}\tau_{KI}\right)}$$

in which x_M is the mole fraction of species M ($=$ A or B) in the mixture, $x_{M(I)}$ is the mole fraction of species M in the immediate vicinity of species I (local mole fraction, with I $=$ A or B), $\tau_{MI} = (g_{MI}-g_{II})/RT$ (with R the gas constant and T the temperature), g_{MI} is an energy parameter for the M-I interaction, and α_{MI} is the nonrandomness parameter (typically of the order of 0.3). The resulting equation satisfied by these local fractions of species has similarity [34] with that of the quasichemical theory of Guggenheim [35].

On the other hand, the local composition assumed in NRTL does not satisfy the symmetry condition [36, 37], $x_A x_{B(A)} = x_B x_{A(B)}$, for two species A and B. This condition simply expresses the fact that the number of A-B pairs must be the same according to whether it is calculated by considering central A or central B species. Besides, it may be noticed that the NRTL equation for a binary mixture involves *two* energy parameters, τ_{AB} and τ_{BA}. This feature is in contrast with lattice models [35] with only nearest-neighbor interaction (as is the case in NRTL) in which the interchange energy, $\Delta g = 2g_{AB}-g_{AA}-g_{BB} = \tau_{AB} + \tau_{BA}$, is the only energy parameter which may appear in the final expressions [38].

The molar excess Gibbs energy, $\overline{G}^{NRTL}$, for a binary mixture A + B is given by the simple relation [1, 34]

$$\frac{\overline{G}^{NRTL}}{RT} = x_A x_B \left[\tau_{BA} \frac{G_{BA}}{x_A + x_B G_{BA}} + \tau_{AB} \frac{G_{BA}}{x_B + x_A G_{AB}} \right]$$

with $G_{IJ} = \exp(-\alpha_{IJ}\tau_{IJ})$, from which the activity coefficients of the species can be computed [1].

NRTL has been widely used in the chemical engineering community [1]. Unlike Wilson's equation, it is applicable to partially miscible as well as completely miscible systems. It has been used to parameterize the thermodynamic properties of liquid mixtures. It has also been employed to represent deviations from ideality in electrolyte solutions in combination with a term for long-range electrostatic interactions. The NRTL term was introduced to account for short-range interactions. NRTL has been associated with the Debye-Hückel (DH) contribution [39], with the Pitzer-DH term [40, 41], and with the MSA [42]. These models have been applied to the prediction of the thermodynamics of multicomponent aqueous electrolytes [43] and of mixed-solvent electrolytes [42, 44] by using the values of parameters adjusted for the binary systems. No higher-order parameters were introduced. The effect of ion hydration and association has been accounted for in various ways [45–47].

Lately, this type of model has experienced renewed interest for the correlation of liquid-liquid equilibria for systems containing (room temperature) ionic liquids [48, 49].

Future Directions

In the future, the MSA may be used to describe the properties of ions interacting through Coulomb plus Yukawa interaction potentials. The Yukawa potential may account for dispersion forces or for the interaction between colloids in solution [50]. Another field for which the MSA could be suitable is that of polyelectrolytes [51] or chains carrying local charges.

The NRTL is an engineering model that has been widely used because of its versatility, of the simplicity of its equations, and of its capability to accurately represent deviations from ideality in many systems. Other models of the same class have been proposed in the literature, such as UNIQUAC [1]. These models are based on a combination of phenomenological and empirical arguments. Engineering models improving on these models would likely involve more complicated developments, without however

Cross-References

▸ Activity Coefficients
▸ Electrolytes, History
▸ Electrolytes, Thermodynamics

necessarily producing a significant gain in accuracy. Decisive improvements will likely emerge, and actually are beginning to emerge, from molecular models, such as those based on SAFT equation [52].

References

1. Prausnitz JM, Lichtenthaler RN, Gomes de Azevedo E (1999) Molecular thermodynamics of fluid phase equilibria. Prentice Hall, Upper Saddle River
2. Maurer G (1983) Electrolyte solutions. Fluid Phase Equil 13:269–293
3. Renon H (1986) Electrolyte solutions. Fluid Phase Equil 30:181–195
4. Cabezas H Jr, O'Connell JP (1993) Some uses and misuses of thermodynamic models for dilute liquid solutions. Ind Eng Chem Res 32:2892–2904
5. Loehe JR, Donohue MD (1997) Recent advances in modeling thermodynamic properties of aqueous strong electrolyte systems. AIChE J 43:180–195
6. Anderko A, Wang P, Rafal M (2002) Electrolyte solutions: from thermodynamic and transport property models to the simulation of industrial processes. Fluid Phase Equil 194–197:123–142
7. Lee LL (2008) Molecular thermodynamics of electrolyte solutions. World Scientific, Hackensack
8. Barthel J, Krienke H, Kunz W (1998) Physical chemistry of electrolyte solutions. Modern aspects. Springer, New York
9. Percus JK, Yevick GJ (1964) Hard-core insertion in the many-body problem. Phys Rev 136:B290–B296
10. Waisman E, Lebowitz JL (1970) Exact solution of an integral equation for the structure of a primitive model of electrolytes. J Chem Phys 52:4307–4309
11. Blum L (1975) Mean spherical model for asymmetric electrolytes. Mol Phys 30:1529–1535
12. Blum L, Høye JS (1977) Mean spherical model for asymmetric electrolytes. 2. Thermodynamic properties and the pair correlation function. Mol Phys 30:1529–1535
13. Blum L (1974) Solution of a model for the solvent-electrolyte interactions in the mean spherical approximation. J Chem Phys 61:2129–2133
14. Adelman S, Deutch JM (1974) Exact solution of the mean spherical model for strong electrolytes in polar solvents. J Chem Phys 60:3935–3949
15. Høye JS, Lomba E (1988) Mean spherical approximation (MSA) for a simple model of electrolytes. I. Theoretical foundations and thermodynamics. J Chem Phys 88:5790–5797
16. Blum L, Vericat F, Fawcett WR (1992) On the mean spherical approximation for hard ions and dipoles. J Chem Phys 96:3039–3044, and Erratum (1994) J Chem Phys 101:10197
17. Hansen JP, McDonald IR (2006) Theory of simple liquids. Academic, Amsterdam
18. Ebeling W, Scherwinski K (1983) On the estimation of theoretical individual activity coefficients of electrolytes. 1. Hard-Sphere Model. Z Phys Chem 264:1–14
19. Watanasiri S, Brulé MR, Lee LL (1982) Prediction of thermodynamic properties of electrolytic solutions using the mean spherical approximation. J Phys Chem 86:292–294
20. Corti HR (1987) Prediction of activity coefficients in aqueous electrolyte mixtures using the mean spherical approximation. J Phys Chem 91:686–689
21. Triolo R, Grigera JR, Blum L (1976) Simple electrolytes in the mean spherical approximation. J Phys Chem 80:1858–1861
22. Triolo R, Blum L, Floriano MA (1976) Simple electrolytes in the mean spherical approximation. 2. Study of a refined model. J Phys Chem 82:1368–1370
23. Triolo R, Blum L, Floriano MA (1977) Simple electrolytes in the mean spherical approximation. III A workable model for aqueous solutions. J Chem Phys 67:5956–5959
24. Fawcett WR, Tikanen AC (1996) Role of solvent permittivity in estimation of electrolyte activity coefficients on the basis of the mean spherical approximation. J Phys Chem 100:4251–4255
25. Sun T, Lénard JL, Teja AS (1994) A simplified mean spherical approximation for the prediction of the osmotic coefficient of aqueous electrolyte solutions. J Phys Chem 98:6870–6875
26. Simonin JP, Blum L, Turq P (1996) Real ionic solutions in the mean spherical approximation. 1. Simple salts in the primitive model. J Phys Chem 100:7704–7709
27. Simonin JP (1997) Real ionic solutions in the mean spherical approximation. 2. Pure strong electrolytes up to very high concentrations, and mixtures, in the primitive model. J Phys Chem B 101:4313–4320
28. Bernard O, Blum L (1996) Binding mean spherical approximation for pairing ions: an exponential approximation and thermodynamics J Chem Phys 104:4746–4754
29. Kalyuzhnyi YV, Holovko M (1998) Thermodynamics of the associative mean spherical approximation for the fluid of dimerizing particles. J Chem Phys 108:3709–3715
30. Simonin JP, Bernard O, Blum L (1998) Real ionic solutions in the mean spherical approximation. 3. Osmotic and activity coefficients for associating electrolytes in the primitive model. J Phys Chem B 102:4411–4417

31. Simonin JP, Bernard O, Blum L (1999) Ionic solutions in the binding mean spherical approximation: thermodynamic properties of mixtures of associating electrolytes. J Phys Chem B 103:699–704
32. Ruas A, Moisy P, Simonin JP, Bernard O, Dufrêche JF, Turq P (2005) Lanthanide salts solutions: representation of osmotic coefficients within the binding mean spherical approximation. J Phys Chem B109:5243–5248
33. Ruas A, Bernard O, Caniffi B, Simonin JP, Turq P, Blum L, Moisy P (2006) Uranyl(VI) nitrate salts: modelling thermodynamic properties using the BIMSA theory and determination of "fictive" binary data. J Phys Chem B110:3435–3443
34. Renon H, Prausnitz JM (1968) Local compositions in thermodynamic excess functions for liquid mixtures. AIChE J 14:135–144
35. Guggenheim EA (1952) Mixtures. Clarendon, London
36. Flemr V (1976) A note on excess Gibbs energy equations based on local composition concept. Coll Czech Chem Commun 41:3347–3349
37. McDermott C, Ashton N (1977) Note on the definition of local composition. Fluid Phase Equil 1:33–35
38. Madden WG (1990) On the internal energy at lattice polymer interfaces. J Chem Phys 92:2055–2060
39. Cruz JL, Renon H (1978) A new thermodynamic representation of binary electrolyte solutions nonideality in the whole range of concentrations. AIChE J 24:817–830
40. Chen CC, Britt HI, Boston JF, Evans LB (1982) Local composition model for excess Gibbs energy of electrolyte systems. AIChE J 28:588–596
41. Kolker A, de Pablo JJ (1995) Thermodynamic modeling of concentrated aqueous electrolyte and nonelectrolyte solutions. AIChE J 41:1563–1571
42. Papaiconomou N, Simonin JP, Bernard O, Kunz W (2002) MSA-NRTL model for the description of the thermodynamic properties of electrolyte solutions. Phys Chem Chem Phys 4:4435–4443
43. Chen CC, Evans LB (1986) A local composition model for the excess Gibbs energy of aqueous electrolyte systems. AIChE J 32:444–454
44. Mock B, Evans LB, Chen CC (1986) Thermodynamic representation of phase equilibria of mixed-solvent electrolyte systems. AIChE J 32:1655–1664
45. Chen CC, Mathias PM, Orbey H (1999) Use of hydration and dissociation chemistries with the electrolyte-NRTL model. AIChE J 45:1576–1585
46. Simonin JP, Krebs S, Kunz W (2006) Inclusion of ionic hydration and association in the MSA-NRTL model for a description of the thermodynamic properties of aqueous ionic solutions: application to solutions of associating acids. Ind Eng Chem Res 45:4345–4354
47. Simonin JP, Bernard O, Krebs S, Kunz W (2006) Modelling of the thermodynamic properties of ionic solutions using a stepwise solvation-equilibrium model. Fluid Phase Equil 242:176–188
48. Simoni LD, Lin Y, Brennecke JF, Stadtherr MA (2008) Modeling liquid-liquid equilibrium of ionic liquid systems with NRTL, electrolyte-NRTL, and UNIQUAC. Ind Eng Chem Res 47:256–272
49. Maia FM, Rodríguez O, Macedo EA (2010) LLE for (water + ionic liquid) binary systems using [Cxmim][BF4] (x = 6, 8) ionic liquids. Fluid Phase Equil 296:184–191
50. Yu YX, Jin L (2008) Thermodynamic and structural properties of mixed colloids represented by a hard-core two-Yukawa mixture model fluid: Monte Carlo simulations and an analytical theory. J Chem Phys 128:014901, 1–13
51. Bernard O, Blum L (2000) Thermodynamics of a model for flexible polyelectrolytes in the binding mean spherical approximation. J Chem Phys 112:7227–7237
52. Paricaud P, Galindo A, Jackson G (2002) Recent advances in the use of the SAFT approach in describing electrolytes, liquid crystals and polymers. Fluid Phase Equil 194–197:87–96

Three-Dimensional Electrode

Luisa M. Abrantes
Departamento de Química e Bioquímica, FCUL, Lisbon, Portugal

Introduction

The designation covers those electrodes displaying a very high surface area per unit volume, A_e, caused by no planarity. Usually, they also cause conditions of good turbulence at their interface with the electrolyte, enhancing the mass transfer process of the electroactive species towards the electrode surface. Both characteristics strongly improve the electrochemical reaction rate as expressed by the equation

$$I_1 = z\,F\,A_e V_e\,K_m\,c$$

where I_1 is the limiting current, z the number of electrons involved in the electrode reaction, F the Faraday constant, V_e the three-dimensional electrode volume, K_m the mass transfer coefficientand c the concentration of the reactant in solution.

This kind of electrodes allows the design of compact electrochemical reactors, being particularly suited to operative conditions demanding low current densities, namely for processing very dilute solutions or for accomplishing specific requirements of reaction selectivity. Notwithstanding, the modelling of electrochemical reactors bearing three-dimensional electrodes is a complex task and significant simplifications are needed to reach an easy computational processing. Some examples of research work related to this subject can be found in literature [1–3].

Classification

A comprehensive classification of the large variety of three-dimensional electrodes has been given by Pletcher and Walsh [4], taking into account the electrodic motion and the different available forms, from intrinsic three-dimensional structures to beds of electronically conducting particles.

As static electrodes, packed bed electrodes [5, 6], consisting of a restrained bed of regular or irregular conducting particles in continuous intimate contact, and porous electrodes built from a broad range of materials, have been extensively used. These materials include reticulated vitreous carbon RVC [7], carbon/graphite and metal felts, fibres, meshes and foams [8–13], stacked metal screens and grids [14, 15]; nowadays these traditional options are extended by emerging structures based on conducting polymer composites [16–18], modified substrates with conducting nanoparticles [19–21] or nanostructured arrays [22–24].

The moving electrodes include fluidized bed electrodes [25, 26], formed by individual carbon or metal particles fluidized by an appropriate flow of electrolyte, and also Circulating bed electrodes [26, 27] where the geometrical configuration promotes two differentiated zones concerning the relative motion of particles and solution: one with both circulating in the same direction, the other with countercurrent stream of each phase. Several less relevant dynamic electrodes may also be considered, namely tumbled, pulsed flow and vibrated/beds [28–30].

Design Options

Electrode geometry is only one of the multiple factors involved in the design of the electrochemical reactor [31]. According to the specific requirements of a given process, the eventual selection of three-dimensional electrodes has to cope with decisions regarding the operation mode (batch or continuous), shape of the cell (e.g. rectangular, cylindrical, profiled), conditions of solution stream (trickle or flooded), directions of current and electrolyte flows (flow-through or flow-by), electrical connections (mono- or bipolar). As a consequence, possible design alternatives shall be contrasted before retaining the apparently best configuration, paying particular attention to the potential and current distribution over the entire electrode matrix.

Before the economic-financial feasibility study which is on the basis of the final decision to implement a process, it is necessary to evaluate the technical/technological viability of the possible routes. In electrochemical engineering practice, it is common to consider a set of figures of merit [32] and to establish comparisons with the same parameters of competitive alternatives. Unfortunately, the literature concerning the above mentioned matters on the employment of three dimensional electrodes is scarce and out-dated.

Applications

Profiting from the ability for processing very dilute solutions, traditional major applications have been associated to pollution abatement, namely metal removal and organics destruction from contaminated effluents [6, 27, 33–35]; inorganic and organic syntheses have also been reported [12, 36–39]. With the advent of nanomaterials and nanotechnology, the research in batteries [22, 40, 41], fuel cells [42–44], supercapacitors [45–47], solar cells [48–50] biosensors [51–53], medical implantable devices [54–56] and analytical systems [57–59] has reached an astonishing rapid progress.

Future Directions

The knowledge of conventional three-dimensional electrode behaviour is well established and the most usual technological applications are at a significant level of maturity. So, in the very near future, discounting a few works aiming at incremental enhancements of existing processes or testing some novel materials, it a research slow down in the developed countries is predictable whereas applied studies envisaging specific needs, e.g. pollution abatement, will continue active in the emerging economies of the world.

The following years will certainly reveal an exponential growth of new three-dimensional electrodes exploring nanoscale structures. One of the priority areas of research will be energy storage, where the great challenge is to improve the capability of batteries and capacitors for storing more energy and to increase the life of charge/discharge cycles without disturbing the stability of the interface electrode/electrolyte. Since batteries are used in a large spectrum of electromechanical devices with their miniaturization being compromised by the dimensions of available power sources, such kind of electrodes will bring in an obvious progress.

Other major domains of future development will be multiple nanoelectrode arrays applied in dye-sensitized solar cells, fuel cells and sensing technologies. Aiming to restore lost body functions of disabled people, the extensive research (both in vitro and in vivo) already registered in medical sciences will progress further. As three-dimensional electrode architectures are used to connect neurons to electronic circuits, they will play a relevant role in the new generations of advanced (hi-tech) bionic prosthetics and implants.

Cross-References

▶ Electrode
▶ Macroscopic Modeling of Porous Electrodes

References

1. Pérez OG, Bisang JM (2010) Theoretical and experimental study of electrochemical reactors with three-dimensional bipolar electrodes. J Appl Electrochem 40:709–718
2. Scott K, Sun Y-P (2007) Approximate analytical solutions for models of three-dimensional electrodes by Adomian's decomposition method. In: Vayenas C, White R, Gamboa-Aldeco ME (eds) Modern aspects of electrochemistry 41. Springer, New York, pp 221–304
3. Scott K, Argyropoulos P (2004) A current distribution model for a porous fuel cell electrode. J Electroanal Chem 567:103–109
4. Pletcher D, Walsh FC (1992) Three dimensional electrodes. In: Genders D, Weinberg N (eds) Electrochemistry for a cleaner environment. The Electrosynthesis Company Inc, New York
5. Wendt H, Kreysa G (1999) Electrochemical engineering: science and technology in chemical and other industries. Springer, Berlin
6. Rajeshwar K, Ibanez J (1997) Environmental electrochemistry, fundamentals and applications in pollution abatement. Academic, San Diego
7. Friedrich JH, Ponce-de-Léon C, Reade GW, Walsh FC (2004) Reticulated vitreous carbon as an electrode material. J Electroanal Chem 561:203–217
8. Vatistas N, Marconi PF, Bartolozzi M (1991) Mass-transfer study of the carbon felt electrode. Electrochim Acta 36:339–343
9. Tricoli V, Vatistas N, Marconi PF (1993) Removal of silver using graphite-felt electrodes. J Appl Electrochem 23:390–392
10. Abda M, Oren Y (1993) Removal of cadmium and associated contaminants from aqueous wastes by fibrous carbon electrodes. Water Res 27:1535–1544
11. Marracino JM, Couret F, Langlois S (1987) A first investigation of flow-through porous electrodes made of metallic felts or foams. Electrochim Acta 32:1303–1309
12. Yang KS, Mull G, Moulijn JA (2007) Electrochemical generation of hydrogen peroxide using surface area-enhanced Ti-mesh electrodes. Electrochim Acta 52:6304–6309
13. Ponce-de-Léon C, Kulak A, Williams S, Jiménez IM, Walsh FC (2011) Improvements in direct borohydride fuel cells using three-dimensional electrodes. Catal Today 170:148–154
14. Brown CJ, Pletcher D, Walsh FC, Hammond JK, Robinson D (1994) Studies of three-dimensional electrodes in FM 01-LC laboratory electrolyser. J Appl Electrochem 24:95–106
15. Hu J-L, Chou WL, Wang CT, Kuo YM (2008) Colour and COD removal using three-dimensional stacked Pt/Ti screen anode. Environ Eng Sci 25:1009–1016
16. Ferreira VC, Melato AI, Silva AF, Abrantes LM (2011) Attachment of noble metal nanoparticles to conducting polymers containing sulphur-preparation

conditions for enhanced electrocatalytic activity. Electrochim Acta 56:3567–3574

17. Singh N (2010) Studies on conducting polymer and conducting polymerinorganic composite electrodes prepared via a new cathodic polymerization method. Dissertation, Purdue University

18. Frackowiak E, Khomenko V, Jurewicz K, Lota K, Béguin F (2006) Supercapacitors based on conducting polymers/nanotubes composites. J Power Sources 153:413–418

19. Celebanska A, Lesniewski A, Paszewsky M, Niedziolka MJ, Jonsson JN, Opallo M (2011) Gold three-dimensional film electrode prepared from oppositely charged nanoparticles. Electrochem Commun 13:1170–1173

20. Murata K, Kajiya K, Nukaga M, Suga Y, Watanabe T, Nakamura N, Ohno H (2010) A simple fabrication method for three-dimensional gold nanoparticle electrodes and their application for the study of the direct electrochemistry of Citochrome C. Electroanalysis 22:185–190

21. Ferreira VC, Silva F, Abrantes LM (2010) Electrochemical and morphological characterization of new architectures containing self-assembled monolayers and Au-NPs. J Phys Chem C 114:7710–7716

22. Wang J, Du N, Zhang H, Yu J, Yang D (2011) Cu-Sn core-shell nanowire arrays as three-dimensional electrodes for lithium-ion batteries. J Phys Chem C 115:23620–23624

23. Ahn H-J, Moon WJ, Seong TY, Wang D (2009) Three-dimensional nanostructured carbon - nanotube array/PtRu nanoparticle electrodes for micro-fuel cells. Electrochem Commun 11:635–638

24. Teixidor GT, Zaouk RB, Park BY, Madou MJ (2008) Fabrication and characterization of three-dimensional carbon electrodes for lithium-ion batteries. J Power Sources 18:730–740

25. Coueuret F (1980) The fluidized bed electrode for the continuous recovery of metals. J Appl Electrochem 10:687–696

26. Thilakavathi R, Balasubramanian N, Ahmed-Basha C (2009) Modeling electrowinning process in an expanded bed electrode. J Hazard Mater 162:154–160

27. Ferreira BK (2008) Three-dimensional electrodes for the removal of metals from dilute solutions: a review. Miner Process Extr Metall Rev 29:330–371

28. Al-Shammari AA, Rahman SU, Chin D-T (2004) An oblique rotating barrel electrochemical reactor for removal of copper ions from waste water. J Appl Electrochem 34:447–453

29. Belmant C, Cognet P, Berlan J, Lacoste G, Fabre P-L, Jud J-M (1998) Application of an electrochemical pulsed flow reactor to electroorganic synthesis. Part I: Reduction of acetophenone. J Appl Electrochem 28:185–191

30. Sedahmed GH, El-Abd MZ, Zatout AA, El-Taweel YA, Zaki MM (1994) Mass-transfer behaviour of

electrochemical reactor employing vibrating screen electrodes. J Electrochem Soc 141:437–440

31. Walsh FC (1993) A first course in electrochemical engineering. The Electrochemical Consultancy, Romsey, Hants

32. Pletcher D, Walsh FC (1990) Industrial electrochemistry, 2nd edn. Chapman and Hall, London

33. Reade GW, Nahle AH, Bond P, Friedrich JM, Walsh FC (2004) Removal of cupric ions from acidic sulphate solution using reticulated vitreous carbon rotating cylinder electrodes. J Chem Technol Biotechnol 79:935–945

34. Recio FJ, Herrasti P, Sirés I, Kulak AN, Bavykin DV, Ponce-de-Léon C, Walsh FC (2011) The preparation of PbO_2 coatings on reticulated vitreous carbon for the electro-oxidation of organic pollutants. Electrochim Acta 56:5158–5165

35. Liu Z, Shi W, Li Y, Zhu S (2011) Continuous electrochemical oxidation of phenol using a three-dimensional electrode reactor. Appl Mech Mater 71–78:2169

36. Gieng EL, Oloman CW (2003) Electrosynthesis of hydrogen peroxide in acidic solutions by mediated oxygen reduction in a three-phase (aqueous/organic/gaseous) system. Part I: emulsion, stricture, electrode kinetics and batch electrolysis. J Appl Electrochem 33:655–663

37. Lütz S, Vuorilehto K, Liese A (2007) Process development for the electroenzymatic synthesis of (R-) methylphenylsulfoxide by use of a 3-dimensional electrode. Biotechnol Bioeng 98:523–534

38. Szanto D, Trinidad P, Walsh F (1998) Evaluation of carbon electrodes and electrosynthesis of coumestan and catecholamine derivatives in the FM 01-LC electrolyser. J Appl Electrochem 28:251–258

39. Roquero P, Ghenem-Lakhal A, Cognet P, Lacoste G, Berlan J, Fabre P-L, Duverneuil P (1996) A new reactor for industrial organic electrosynthesis. Chem Eng Sci 51:1847–1855

40. Arthur TS, Bates DJ, Cirigliano N, Johnson DC, Malati P, Mosby JM, Perre E, Rawlo MT, Prieto AL, Dunn B (2011) Three-dimensional electrodes and battery architectures. MRS Bull 36:523–531

41. Armand M, Tarascon JM (2008) Building better batteries. Nature 451:652–657

42. Wang H, Ye F, Wang C, Yang J (2011) Ultrafine Pt nanoclusters for the direct methanol fuel cell reactions. J Clust Sci 22:173–181

43. Yi L, Hu B, Song Y, Wang X, Zou G, Yi W (2011) Studies of electrochemical performance of carbon supported Pt-Cu nanoparticles as anode catalysts for direct borohydride-hydrogen peroxide fuel cell. J Power Sources 196:9924–9930

44. Uhm S, Jeong B, Lee J (2011) A facile route for preparation of non-noble CNF cathode catalysts in alkaline ethanol fuel cells. Electrochim Acta 56:9186–9190

45. Dhawale DS, Vinu A, Lokhande CD (2011) Stable nanostructured polyaniline electrode for

supercapacitor application. Electrochim Acta 56:9482–9487

46. Yu G, Hu L, Liu N, Wang H, Vosgueritchian M, Yang Y, Cui Y, Bao Z (2011) Enhancing the supercapacitor performance of graphene/MnO_2 nanostructered electrodes by conductive wrapping. Nano Lett 11:4438–4442

47. Lang XY, Yuan HT, Iwasa Y, Chen MW (2011) Three-dimensional nanoporous gold for electrochemical supercapacitors. Scr Mater 64:923–926

48. Trevisan R, Döbbelin M, Boix PP, Barea EM, Tena-Zaera R, Mora-Seró I, Bisquert J (2011) Pedot nanotube arrays as high-performing counter electrodes for dye sensitized solar cells. Study of the interactions among electrolytes and counter electrodes. Adv Energ Mater 1:781–784

49. Gu D, Baumgart H, Namkoong G (2011) New template based nanoelectrode arrays for organic/inorganic photovoltaic applications. Phys Stat Solidi Rapid Res Lett 5:104–106

50. Rühle S, Shalom M, Zaban A (2010) Quantum-dot sensitized solar cells. Chem Phys Chem 11:2290–2304

51. Iost RM, da Silva WC, Madurro JM, Madurro AG, Ferreira LF, Crespilho FN (2011) Recent advances in nano-based electrochemical biosensors: applications in diagnosis and monitoring of diseases. Front Biosci (Elite Ed) 3:663–689

52. Siangproh W, Dungchai W, Rattanarat P, Chailapakul Q (2011) Nanoparticle-based electrochemical detection in conventional and miniaturized systems and their bioanalytical applications: a review. Anal Chim Acta 690:10–25

53. Li X-R, Xu J-J, Chen H-Y (2011) Potassium-doped carbon nanotubes toward the direct electrochemistry of cholesterol oxidase and its application in highly sensitive cholesterol biosensor. Electrochim Acta 56:9378–9385

54. de Assis Jr ED, Nguyen-Vu TDB, Arumugam PU, Chen H, Cassel AM, Andrews RJ, Yang CY, Li J (2009) High efficient electrical stimulation of hippocampal slices with vertically aligned carbon nanofibres microbrush array. Biomed Microdevices 11:801–808

55. Linsmeier CE, Prinz CN, Petterson LME, Caroff P, Samuelson L, Shouenborg J, Montelius L, Danielsen N (2009) Nanowire biocompatibility in the brain – looking for a needle in a 3D stack. Nano Lett 9:4184–4190

56. Li J, Andrews RJ (2007) Trimodal nanoelectrode array for precise deep brain stimulation: prospects of a new technology based on carbon nanofibers arrays. Acta Neurochir Suppl 97:537–545

57. Panchompoo J, Aldous L, Dowing C, Crossley A, Compton RG (2011) Facile synthesis of Pd nanoparticle modified carbon black for electroanalysis:application to the detection of hydrazine. Electroanalysis 23:1568–1578

58. Peña RC, Bertotti M, Brett CM (2011) Methylene blue/multiwall carbon nanotube modified electrode for the amperometric determination of hydrogen peroxide. Electroanalysis 23:2290–2296

59. Ugo P, Moretto LM, Silvestrini M, Pereira FC (2010) Nanoelectrode emsembles for the direct voltammetric determination of trace iodide in water. Int J Environ Anal Chem 90:747–759

TiO$_2$ Photocatalyst

Hiroshi Irie
Yamanashi University, Yamanashi Prefecture, Japan

Introduction

Photocatalysis has become a common word, and various products using photocatalytic functions have been commercialized. Among many candidates for photocatalysts, titanium dioxide (TiO_2) is almost the only material for the industrial use at present and also probably in the future. This is because TiO_2 has the most efficient photoactivity, the highest stability, the lowest cost, and the safety to humans and environment. There are two types of photochemical reactions proceeding on its surface irradiated with ultraviolet (UV) light. One is the photo-induced redox reactions of adsorbed substances, and the other is photo-induced hydrophilic conversion of TiO_2 itself. The combination of these two functions opened various novel applications of TiO_2. In the last decade, numerous studies have performed to modify TiO_2 so that it is sensitive to visible light. Here the progress of the research of TiO_2 photocatalysis is reviewed as well as future prospects of this field.

Water Splitting with a TiO$_2$ Electrode

The solar water splitting evolving hydrogen (H_2) and oxygen (O_2) was demonstrated for the first time in *Nature* in 1972 with a rutile TiO_2 single crystal which was exposed to ultraviolet (UV) light and was connected to a platinum (Pt) counter electrode [1]. In those days, crude oil prices ballooned suddenly, and the future lack of crude oil was a serious concern. Therefore, this report attracted the attention not only of electrochemists but also of many scientists in a broad area, and numerous related studies have been reported in the 1970s.

When the surface of the rutile TiO_2 electrode was irradiated with light consisting of wavelengths shorter than its bandgap, about 415 nm (3.0 eV), photocurrent flowed from the Pt electrode to the TiO_2 electrode through the external circuit. The direction of the current revealed that the oxidation reaction (O_2 evolution) occurs at the TiO_2 electrode and the reduction reaction (H_2 evolution) at the Pt electrode. This observation shows that water can be decomposed, using UV light, into O_2 and H_2, without the application of an external voltage, according to the following scheme.

$$TiO_2 + hv \rightarrow e^- + H^+ \\ \text{(at the } TiO_2 \text{ electrode)} \tag{1}$$

$$2H_2O + 4h^+ \rightarrow O_2 + 4H^+ \\ \text{(at the Pt electrode)} \tag{2}$$

$$2H^+ + 2e^- \rightarrow H_2 \tag{3}$$

The overall reaction is
$$2H_2O + 4hv \rightarrow O_2 + 2H_2 \tag{4}$$

When a semiconductor electrode is in contact with an electrolyte solution, thermodynamic equilibration occurs at the interface. This may result in the formation of a space-charge layer within a thin surface region of the semiconductor, in which the electronic energy bands are generally bent upwards and downwards in the cases of the n- and p-type semiconductors, respectively. The thickness of the space-charge layer is usually of the order of $1–10^3$ nm, depending on the carrier density and dielectric constant of the semiconductor. If this electrode receives photons with energies greater than that of the material bandgap, E_G, electron–hole pairs are generated and separated in the space-charge layer. In the case of an n-type semiconductor, the electric field existing across the space-charge layer drives photo-generated holes toward the interfacial region (i.e., solid–liquid) and electrons toward the interior of the electrode and from there to the electrical connection to the external circuit. The reverse process occurs at a p-type semiconductor electrode.

Powdered TiO_2 Photocatalysis Toward Environmental Purification

Photocatalytic water splitting has been studied intensively with powdered anatase TiO_2 suspensions, with Pt deposited on TiO_2 as a cocatalyst. Although there were several experiments for the simultaneous evolution of H_2 and O_2 in the powder systems, the reaction efficiency was very low, due to the recombination to regenerate water molecules through the back reaction. To solve this problem, Kawai and Sakata added organic compounds to the aqueous suspension of platinized TiO_2. In this case, water is reduced, producing H_2 at the Pt sites, and the organic compounds are oxidized instead of water by photo-generated holes at the TiO_2 sites. The H_2 production proceeds surprisingly efficiently, with a quantum yield of more than 50 % in the presence of ethanol [2]. Most organic compounds can enhance the H_2 production efficiency. This is because the redox potential for photo-generated holes is ~ 3 V versus SHE, and most organic compounds can be oxidized finally into carbon dioxide (CO_2). After the discovery, TiO_2 photocatalysis drew the attention of many people again as one of the promising methods for H_2 production. However, even though the reaction efficiency is very high, TiO_2 can absorb only the UV light contained in a solar spectrum, which is only about 3 %.

Instead, the research shifted to the utilization of the strong photo-generated oxidation power of TiO_2 for the decomposition of pollutants. The first such reports were those of Frank and Bard in 1977, in which they described the decomposition of cyanide in the presence of aqueous TiO_2 suspensions [3]. In the 1980s, detoxications of various harmful compounds in both water and air were demonstrated using powdered TiO_2 actively as potential purification methods of wastewater and polluted air. Then, the purpose shifted to the oxidation reaction of harmful compounds, the reduction reaction was not necessarily H_2 production. The platinization of TiO_2 was not necessary in this case, and TiO_2 powder itself was used under ambient condition. Now, both the reduction and oxidation sites are located on the TiO_2 surface, and the reduction of adsorbed oxygen molecules proceeds on the TiO_2 surface. As already mentioned, the holes (h^+) generated in TiO_2 were highly oxidizing, and most compounds were essentially oxidized completely, i.e., each constitution element of the compounds was oxidized to its final oxidation state. In addition, various forms of active oxygen, such as O_2^-, $\cdot OH$, $HO_2\cdot$, and $O\cdot$, produced by the following processes, may be responsible for the decomposition reactions.

$$e^- + O_2 \rightarrow O_2^-(ad) \qquad (5)$$

$$O_2^-(ad) + H^+ \rightarrow HO_2 \cdot (ad) \qquad (6)$$

$$h^+ + H_2O \rightarrow \cdot OH(ad) + H^+ \qquad (7)$$

$$h^+ + O_2^-(ad) \rightarrow 2 \cdot O(ad) \qquad (8)$$

Many research studies have been performed on the purification of wastewater and polluted air utilizing TiO_2 photocatalysis up to now.

Photo-Induced Super-Hydrophilicity on Filmed TiO_2 Surface

The surface wettability is generally evaluated by the water contact angle (CA). The CA (θ) is defined as the angle between the solid surface and the tangent line of the liquid phase at the interface of the solid–liquid–gas phases. In 1995, Wananabe and coworkers discovered by chance the marked change in the water wettability of the TiO_2 surface before and after UV light irradiation [4]. After the discovery of this phenomenon, the application range of TiO_2 coating has been largely widened.

A TiO_2 thin film exhibits an initial CA of several tens of degrees depending on the surface conditions mainly surface roughness. When this surface is exposed to UV light, water starts to exhibit a decreasing CA, that is, it tends to spread out flat instead of beading up. Finally, the CA reaches almost $0°$. At this stage, the surface becomes completely non-water repellant and is termed "super-hydrophilic." The surface retains a CA of a few degrees for a day or two under the ambient condition without being exposed to UV light. Then, the CA slowly increases, and the surface becomes the initial less-hydrophilic state again. At this point, the super-hydrophilicity can be recovered simply by exposing the surface again to UV light.

How can we explain this phenomenon? Before explaining its mechanism, let me briefly review a classical theoretical treatment for the wettability of the flat solid surface. It is commonly evaluated in terms of the CA, which is given by either Young equation [5] or Girifalco-Good (G-G) equation [6]. By combining Young and G-G equations, the CA can be simply expressed as follows:

$$\cos \theta = c\gamma_S^{1/2} - 1 \qquad (9)$$

where γ_S is a surface free energy per unit area of the solid and c is a constant. Equation 9 shows that the CA decreases simply with increasing γ_S. Therefore, it can be considered that the highly hydrophilic state with a CA of $0°$ is achieved by the generation of some states with large surface energies when irradiated with UV light.

Opinions on the mechanism underlying super-hydrophilicity differ; some researchers propose that super-hydrophilicity originates from photocatalytic oxidative decomposition, which removes contaminants from the TiO_2 surface [7–9]. This is because metal oxides have large surface energies in general. However, it is important to note that a completely clean surface cannot be obtained by the photocatalytic reaction because the surface is easily contaminated by airborne stains under the ambient condition. In contrast, others hypothesize that the mechanism involves a photo-induced surface structural change of TiO_2, which increases the surface energy of TiO_2 [10–12]. The present author supports the latter mechanism. That is, the photo-generated holes produced in the bulk of TiO_2 diffuse to the surface and are trapped at lattice oxygen sites. Most trapped holes are consumed to react with the adsorbed organics directly, or adsorbed water, producing OH radicals. However, a small portion of the trapped hole may react with TiO_2 itself, breaking the bond between the lattice titanium and oxygen ions by the coordination of water molecules at the titanium site. The coordinated water molecules release a proton for charge compensation, and then a new OH group forms, leading to the increase in the number of OH groups at the surface. It is considered that the singly coordinated new OH groups produced by UV light irradiation are thermodynamically less stable compared to the initial doubly coordinated OH groups. Therefore, the surface energy of the TiO_2 surface covered with the thermodynamically less stable OH groups is higher than that of the TiO_2 surface covered with the initial OH groups. Because a water droplet is substantially larger than the hydrophilic (or hydrophobic) domains, it instantaneously spreads completely on such a surface, thereby resembling a two-dimensional capillary phenomenon.

The finding of the photo-induced super-hydrophilicity has markedly widened the application range of TiO_2-coated materials. That is, the stains adsorbed on the TiO_2 surface can easily be washed by water, because water soaks between stain and the highly hydrophilic TiO_2 surface. In other words, this has removed the limitation of the cleaning function of the TiO_2 photocatalysis, that is, the function is limited by the number of photons. Even though the number of photons is not sufficient to decompose the adsorbed stains, the surface is maintained clean when water is supplied there. Another function blessed with the photo-induced super-hydrophilicity is the antifogging function. The fogging of the surfaces of mirrors and glasses occurs when steam cools down on these surfaces to form many water droplets. On a super-hydrophilic surface, no water drops are formed. Instead, a uniform thin film of water is formed on the surface. This uniform water film prevents the fogging. Once the surface turns into the super-hydrophilic state, it remains unchanged for several days or 1 week. Thus, we expect that various glass products, mirrors, and eyeglasses, for example, can be imparted with antifogging functions using this technology, with simple processing and at a low cost [13].

Visible-Light-Sensitive TiO_2

The current area of interest in this field has been the modification of TiO_2 sensitive to visible light. One approach was so-called dye-sensitization, $H_2[PtCl_6]$ (or $PtCl_4$)-modified TiO_2 system, reported by Kisch and coworkers [14]. Another approach was to substitute Cr, Fe, or Ni for a Ti site or to form Ti^{3+} sites by introducing an oxygen vacancy in TiO_2 [15–17]. However, these approaches were not widely accepted due to the lack of reproducibility and chemical stability. In 2001, Asahi and coworkers reported visible-light-sensitive nitrogen-doped TiO_2 has attracted considerable attention [18]. After that, numerous studies have investigated such photocatalysts, not only nitrogen but also sulfur, carbon, etc., as anionic dopants, because anion-doping into TiO_2 can be used to control the density of states (DOSs) of its valence band (VB) (the potential of VB top is ca. 3.0 V vs. SHE, pH = 0) [19–21]. In fact, N-doping into TiO_2 forms the isolated state originated from N 2p above the VB top (the potential of the N 2p state is ca. 2.3 V vs. SHE, pH = 0) [19]. Upon irradiation with visible light, electrons in the N 2p states are excited to the conduction band (CB) (the potential of CB bottom is ca. –0.2 V vs. SHE, pH = 0), and holes are produced simultaneously in the N 2p states. The produced holes decompose organic substances, and the electrons in the CB are consumed by oxygen molecule through a one-electron reduction reaction ($O_2 + H^+ + e^- \rightarrow HO_2$, –0.046 V vs. SHE) [22]. It should be noted, however, that anion-doped TiO_2 usually shows a lower photocatalytic activity under visible light than UV light and the quantum efficiency (QE) of N-doped TiO_2 under visible light is one or two orders of magnitude smaller than that under UV light. This is due to the oxidation power and

TiO₂ Photocatalyst

mobility of the photo-generated holes in the isolated state, which are lower than those in the VB of TiO_2 [19]. However, if the DOSs of the CB can be controlled to either form the isolated state below the CB or narrow the bandgap by shifting the CB bottom positively so that TiO_2 can absorb visible light, it is self-evident that the photo-excited electrons will accumulate in either the isolated state or the newly constructed CB as they rarely reduce oxygen through a one-electron reduction on TiO_2 surface. Subsequently, the photo-excited holes will recombine with the accumulated electrons.

Very recently, we have investigated TiO_2 powders with grafted metal ions (Cu(II) and Fe(III)); these powders were found to be sensitive to visible light [23–25]. In this system, we have proposed that visible light initiates interfacial charge transfer (IFCT), after the suggestions made in the literatures as to the photo-induced IFCT between the continuous energy levels of solids and discrete one of molecular species on the surface as theoretically formulated by Creutz et al. [26, 27] That is, electrons in the VB of TiO_2 are directly transferred to Cu(II) (Fe(III)), forming Cu(I) (Fe(II)). The holes produced in the VB are then capable of decomposing organic substances. Although the reduction reaction mechanism is not clearly understood, it may possibly proceed by multi-electron reduction (two-electron reduction: $O_2 + 2H^+ + 2e^- \rightarrow H_2O_2$, 0.68 V; or four-electron reduction: $O_2 + 2H_2O + 4H^+ + 4e^- \rightarrow 4H_2O$, 1.23 V vs. SHE) [22]. Consequently, this system functions catalytically and exhibits oxidative decomposition activity. It is noteworthy that the holes in the VB with the strong oxidative power contribute to the decomposition of organic substances.

The discovery that O_2 reduction can be induced by the photo-produced Cu(I) (Fe(II)) on TiO_2 releases the TiO_2 photocatalyst from the restriction that VB control is required to gain visible-light sensitivity and enables one to control the CB of TiO_2 for this purpose. Based on this concept, we also developed a novel and efficient visible-light-sensitive CB-controlled TiO_2 photocatalyst, $Cu(II)/W_xGa_{2x}Ti_{1-3x}O_2$ [28].

Future Perspectives

TiO_2 photocatalysis has become a real practical technology after the middle of the 1990s, particularly in the field of building materials outdoors. In the late 2000s, visible-light-sensitive TiO_2 has developed, so TiO_2 photocatalysis would become real practical applications indoors. We believe that TiO_2 photocatalysis is one of the good examples how, on the time scale of tens of years, basic scientific knowledge can be developed into a technological field and can produce a new industry.

Cross-References

▶ Photocatalyst
▶ Photoelectrochemistry, Fundamentals and Applications
▶ Photoelectrochemical Disinfection of Air (TiO_2)

References

1. Fujishima A, Honda K (1972) Electrochemical photolysis of water at a semiconductor electrode. Nature 238:37–38
2. Kawai T, Sakata T (1980) Conversion of carbohydrate into hydrogen fuel by a photocatalytic process. Nature 286:474–476
3. Kraeutler B, Bard AJ (1977) Photoelectrosynthesis of ethane from acetate ion at an n-type titanium dioxide electrode. The photo-Kolbe reaction. J Am Chem Soc 99:7729–7731
4. Wang R, Hashimoto K, Fujishima A, Chikuni M, Kojima E, Kitamura A, Shimohigoshi M, Watanabe T (1997) Light-induced amphiphilic surfaces. Nature 388:431–432
5. Hiemenz PC (ed) (1986) Principles of colloid and surface chemistry. Marcel Dekker, New York
6. Girifalco LA, Goog RJ (1957) A theory for the estimation of surface and interfacial energies. I. Derivation and application to interfacial tension. J Phys Chem 61:904–909
7. White JM, Szanyi J, Henderson MA (2003) The photon-driven hydrophilicity of titania: a model study using $TiO_2(110)$ and adsorbed trimethyl acetate. J Phys Chem B 107:9029–9033
8. Wang C, Groenzin H, Schltz MJ (2003) Molecular species on nanoparticulate anatase TiO_2 film detected by sum frequency generation: trace hydrocarbons and hydroxyl groups. Langmuir 19:7330–7334
9. Zubkov T, Stahl D, Thompson TL, Panayotov D, Diwald O, Yates JT Jr (2005) Ultraviolet light-induced hydrophilicity effect on $TiO_2(110)(1'1)$.

Dominant role of the photooxidation of adsorbed hydrocarbons causing wettability by water droplets. J Phys Chem B 109:15454–15462

10. Sakai N, Fujishima A, Watanabe T, Hashimoto K (2003) Quantitative evaluation of the photoinduced hydrophilic conversion properties of TiO_2 thin film surfaces by the reciprocal of contact angles. J Phys Chem B 107:1028–1035

11. Sakai N, Fujishima A, Watanabe T, Hashimoto K (2001) Enhanced of the photoinduced hydrophilic conversion rate of TiO_2 film electrode surfaces by anodic polarization. J Phys Chem B 105:3023–3026

12. Hashimoto K, Irie H, Fujishima A (2005) TiO_2 photocatalysis: a historical overview and future prospects. Jpn J Appl Phys 44:8269–8285

13. Fujishima A, Hashimoto K, Watanabe T (1999) TiO_2 photocatalysis: fundamentals and applications. BKC, Tokyo

14. Kisch H, Zang L, Lange C, Maier WF, Antonius C, Meissner D (1998) Modified, amorphous titania – a hybrid semiconductor for detoxification and current generation by visible light. Angew Chem Int Ed 37:3034–3036

15. Borgarello E, Kiwi J, Gratzel M, Pelizzetti E, Visca M (1982) Visible light induced water cleavage in colloidal solutions of chromium-doped titanium dioxide particles. J Am Chem Soc 104:2996–3002

16. Linsebigler AL, Lu GQ, Yates JT Jr (1995) Photocatalysis on TiO_2 surfaces: principles, mechanisms, and selected results. Chem Rev 95:735–758

17. Hoffmann MR, Martin ST, Choi W, Bahnemann DW (1995) Environmental applications of semiconductor photocatalysis. Chem Rev 95:69–96

18. Asahi R, Morikawa T, Ohwaki T, Aoki K, Taga Y (2001) Visible-light photocatalysis in nitrogen-doped titanium oxides. Science 293:269–271

19. Irie H, Watanabe Y, Hashimoto K (2003) Nitrogen-concentration dependence on photocatalytic activity of $TiO_{2-x}N_x$ powders. J Phys Chem B 107:5483–5486

20. Umebayashi T, Yamaki T, Itoh H, Asai K (2002) Band-gap narrowing of titanium dioxide by sulfur doping. Appl Phys Lett 81:454–456

21. Ohno T, Mitsui T, Matsumura M (2003) Photocatalytic activity of S-doped TiO_2 photocatalyst under visible light. Chem Lett 32:364–365

22. Bard AJ, Parsons R, Jordan J (1985) Standard potentials in aqueous solution. Marcel Dekker, New York

23. Irie H, Miura S, Kamiya K, Hashimoto K (2008) Efficient visible light sensitive photocatalysts; grafting Cu(II) ions onto TiO_2 and WO_3 photocatalysts. Chem Phys Lett 457:202–205

24. Irie H, Kamiya K, Shibanuma T, Miura S, Tryk DA, Yokoyama T, Hashimoto K (2009) Visible light-sensitive Cu(II)-grafted TiO_2 photocatalysts: activities and x-ray absorption fine structure analyses. J Phys Chem C 113:10761–10766

25. Yu H, Irie H, Shimodaira Y, Hosogi Y, Kuroda Y, Miyauchi M, Hashimoto K (2010) An efficient visible-light-sensitive Fe(III)-grafted TiO_2 photocatalyst. J Phys Chem C 114:16481–16487

26. Creutz C, Brunschwig BS, Sutin N (2005) Interfacial charge-transfer absorption: semiclassical treatment. J Phys Chem B 109:10251–10260

27. Creutz C, Brunschwig BS, Sutin N (2006) Interfacial charge-transfer absorption: 3. Application to semiconductor – molecule assemblies. J Phys Chem B 110:25181–25190

28. Yu H, Irie H, Hashimoto K (2010) Conduction band energy level control of titanium dioxide towards an efficient visible light-sensitive photocatalyst. J Am Chem Soc 132:6898–6899

Transference Numbers of Ions in Electrolytes

Sandra Zugmann[1,2,3] and Heiner Jakob Gores[4]
[1]EVA Fahrzeugtechnik, Munich, Germany
[2]Institució Catalana de Recerca i Estudis Avançats (ICREA), Institute for Bioengineering of Catalonia (IBEC), Barcelona, Spain
[3]Electrical and Computer Engineering, University of California Davis, Davis, CA, USA
[4]Institute of Physical Chemistry, Münster Electrochemical Energy Technology (MEET), Westfälische Wilhelms-Universität Münster (WWU), Münster, Germany

Basics

Complete characterization of an electrolyte requires various parameters, especially when charge transport is described or modeled. An electrolyte with n species needs $n(n-1)/2$ concentration-dependent transport properties, where n is the number of independent species [1]. In literature, conductivity of electrolytes receives the majority of attention because it is easy to measure. But for a full characterization of the electrolyte, other properties such as ionic transference numbers and ion mobilities, structure formation by ion association, and solvation, as well as bulk properties such as viscosity, are required as well [2–7].

Transference numbers describe directly the charge transport and accordingly the current transport of a specific ion. The cation transference number t_+ is given by Eq. 1

Transference Numbers of Ions in Electrolytes

$$t_+ = \frac{I_+}{I_0} = \frac{Q_+}{Q_0} = \frac{u_+}{u_0} = \frac{\lambda_+}{\Lambda_0} = \frac{D_+}{D_+ + D_-} \quad (1)$$

with $I_+, Q_+, u_+, \lambda_+,$ and D_+ as the current, charge, mobility, ionic conductivity, and diffusion coefficient of the cation, the subscript 0 indicates the value for the whole electrolyte. For the anionic transference number t_-, an analogous expression is valid and $t_+ + t_- = 1$. Generally, liquid electrolytes show both cationic and anionic transference numbers deviating from 1 in contrast to solid electrolytes where many technically important electrolytes are known where only the cations $t_+ = 1$ or the anions $t_- = 1$ are mobile. If the electrolyte is not ideal and contains, e.g., ion associates, the simple definition according to Eq. 1 is invalid. The transference number has to be defined in a much more general way, as "the transference number of a cation- or anion-constituent R as the net number of Faradays carried by that constituent in the direction of the cathode or anode, respectively, across a reference plane fixed with respect to the solvent, when one Faraday of electricity passes across the plane" [8]. According to this definition, the transference number can even be negative, e.g., if a cation A^+ is carried as an ion-constituent $R = AX_2^-$ (anionic triple ion) to the anode.

Measurement Methods

Several experimental methods for measuring transference numbers are known. The most often used methods are presented here in a short overview.

Moving Boundary Method

Two different solutions with different density (test solution and indicator solution but with same anion in the same solvent) build a stable boundary in the measurement cell [9]. By applying a defined current I for a known time t, this boundary is moving and the covered distance and the associated volume V are measured, and the transference number of the cation t_+ is calculated from the transported amount of electrolyte

in that volume [10], F Faraday constant, and c molar concentration of the electrolyte.

$$t_+ = \frac{c \cdot V \cdot F}{I \cdot t} \quad (2)$$

Hittorf's Method

In Hittorf's method [10, 11], a known quantity of charge $Q = (I\,t)$ passes through a symmetrical cell M|MX(solvent)|M, where the metal M is both anodically dissolved and cathodically plated. The part of the cell containing the salt MX and the solvent (MX(solvent)) is divided into three different sections. After electrolysis the cell is separated into the different electrode compartments and the change of moles of the metal ion Δn_M in the anode compartment is analyzed [12, 13]. Gain of cations M^{z+} (charge z^+) by electrochemical oxidation of the metal is $Q/(z^+ \cdot F)$, loss by migration out of the compartment is $t_+ Q/(z^+ \cdot F)$. The net gain is $\Delta n_M = (1-t_+) Q/(z^+F) = t_-\,Q/(z^+F)$; it equilibrates the mol number of anions migrating to the anode. Equation 3 yields the transference numbers of the cation and the anion

$$t_- = 1 - t_+ = \frac{\Delta n_M \cdot z^+ F}{I \cdot t} \quad (3)$$

In useful variant of Hittorf's method, the solvent that is containing a small amount of (ideally) noninteracting molecules is analyzed as well. This experiment suggested by Nernst yields specific solvation numbers for anions and cations and can also contribute to a better precision of transference numbers determined by Hittorf's method as cations and anions generally transport different number of solvent molecules.

Emf Measurements (Electromotive Force Measurements)

Two test series have to be done, including measurements in two nearly identical cells, an emf cell with transference and a cell without transference; the cell without transference includes a salt bridge. Both cells have different molal concentrations in their half cells that is the reason for the cell potential (electromotive force). The emf of the cell with transference E_{trans} includes the anionic transference number, whereas the cell without transference is just the well-known Nernst equation for concentration cells [11] with potential E. From these two series, the transference number t_i can be easily determined from the slope of a plot E_{trans} vs. E or for a single set of measurements at an equal concentration in the two half cells difference just by the ratio [10].

$$t_i = \frac{dE_{trans}}{dE} \approx \frac{E_{trans}}{E} \qquad (4)$$

The type of the transference number t_i depends on the setup of the cells and the related cell reaction [10]. In favorable cases, it is possible that t_i is related to a single ion; however, also complex ions may contribute to t_i. For a discussion, see [10], p. 111 ff.

Determination by Conductivity

The correlation of transference number and conductivity is given in Eq. 1. To measure the cationic conductivity, a reference substance has to be defined with two nearly identical ionic conductances for the cation and the anion, such as tetrabutylammonium tetrabutylborate [14]. The measured molar conductivity can be simply divided by two, and thus, the ionic conductivities are determined. The next step is to measure a salt with the already known anion and the required cation. Thus, the ion conductivity and the transference number of the cation can be calculated as well.

$$\lambda_+ + \lambda_- = \Lambda; \frac{\Lambda}{2} = \lambda_+ = \lambda_- \qquad (5)$$

Potentiostatic Polarization Method

By applying a constant potential ΔV, the current decreases until a steady state is reached that is only due to the cations. The transference number though is determined by dividing the cationic current I_{ss} through the initial current I_0. Because

electrode surfaces vary with time, the resistances before R_0 and after the polarization R_{ss} are measured and included into the equation [15]:

$$t_+ = \frac{I_{ss}(\Delta V - I_0 R_0)}{I_0 (\Delta V - I_{ss} R_{ss})} \qquad (6)$$

Galvanostatic Polarization Method

Applying a constant current for a known time leads to a concentration gradient that is measured indirectly by observing the relaxation potential after current interruption [16]. Combining the galvanostatic polarization experiment, determination of salt diffusion coefficient and concentration dependence of the potential by emf measurements enables to calculate the cationic transference number [17].

$$t_+ = 1 - \frac{z^+ v^+ mFc\sqrt{\pi \cdot D}}{4\left(\frac{d\Phi}{d\ln c}\right)} \qquad (7)$$

where v^+ is the stoichiometric number of cations per mole salt added to a solution, c is the bulk concentration of the salt, D is the salt diffusion coefficient, $d\Phi/d\ln c$ is the concentration dependence of the potential Φ, and m is the slope of a plot of the cell potential at the time of current interruption obtained from $\Delta\Phi$ vs. $jt_i^{1/2}$ fits, where j is the current density and t_i the polarization time.

Determination by NMR (Nuclear Magnetic Resonance Spectroscopy)

Transference numbers can be calculated using self-diffusion coefficients of cation and anion, see Eq. 1, that are measured, e.g., by pfg NMR [18]. This calculation is only valid for completely dissociated salts, an important limitation of the procedure.

Transference numbers in aqueous electrolytes are already well known for many systems. In the last few years, the focus was shifting to organic electrolytes for lithium-ion batteries where charge transport plays a very important role. The next section shows some recent results and encountered problems.

Lithium Transference Numbers

The capability for fast charge and mass flow in the system is an important criterion for electrolytes usable in high-power lithium-ion batteries. Concentration gradients that develop across a cell during current flow are defined by the value of the lithium transference number. Sufficiently high currents and resulting high concentration gradients may lead to precipitation of the salt at the anode or depletion of electrolyte at the cathode with resulting cell failure. Low or negative transference numbers result in poor high-rate performance and limit the cell's power output. In contrary, high lithium-ion transference numbers can significantly reduce the effects of concentration polarization and thus decrease the potential loss in the cell [19]. Therefore, knowledge about this transport property can give access to rational optimization of electrolytes for lithium-ion batteries. The rough rule, "the higher the transference number the better," is valid. It is known from theoretical considerations that transference numbers higher than 0.5 at infinite dilution increase with increasing concentration, whereas those lower than 0.5 decrease with increasing concentration [20]. For the most part, lithium ions in organic solvents have transference numbers lower than 0.5 due to their strong ion solvent interaction resulting in a big solvent shell entailing a continuous decrease up to higher concentrations as mentioned above.

The number of precise methods to measure transference numbers in liquid aqueous electrolytes is quite acceptable [9–14]. Various methods are already used for more than a hundred years, such as the moving boundary, Hittorf's method, or the indirect determination of transference numbers by conductivity measurements. In contrast, accurate data for nonaqueous liquid electrolytes, especially with respect to lithium salts, are very rare. In literature, the most often used methods are the potentiostatic polarization method and determination by NMR [21, 22]. Interestingly, the first was developed for solid electrolytes; the latter is only valid for ideal solutions. To measure concentrated electrolyte solutions, other methods such as the galvanostatic polarization method and the emf method are recommended. To sum up our experience, every method has several assumptions and therefore resulting restrictions that have to be kept in mind. The potentiostatic polarization method can be used for a fast electrochemical estimation in screening experiments, but it is restricted to binary, ideal electrolytes. The galvanostatic polarization method is the best method and can even be used for concentrated solutions, but it combines three different measurements that are very time-consuming [23]. The emf method requires an adequate nonaqueous salt bridge, and it assumes ion transference numbers that are not varying with concentration, which is not the case, see above. For multicomponent systems over wide concentration ranges, the determination by NMR is a useful method. However, the diffusive transport of both ion pairs and free ions is detected. Often, electrochemical methods and NMR measurements show contradicting results for concentration dependence of cationic transference numbers as expected [24]. Whereas electrochemical methods yield values decreasing with concentration in accordance with electrostatic theories, NMR measurements show increasing transference numbers with increasing concentration [25]. This observation can be explained by ion–ion interaction leading to ion-pair formation (ion association) [25, 26] that leads to decreasing transference numbers of lithium ions for electrochemical methods.

Future Directions

Further studies on ion-pair formation have to be done for a reliable understanding of ion transport in multicomponent solutions and concentration dependence of transference numbers for nonaqueous electrolyte solutions, solid polymer electrolytes, and associated gels. In transport models for lithium-ion batteries, transference numbers are often introduced as constants entailing inexact results. Although the transference number is already known for a long time as a very important parameter in transport phenomena, the underlying processes are not yet fully understood. A broad study based on various electrochemical and spectroscopic methods at multicomponent systems is necessary for filling the obvious gaps of knowledge.

Cross-References

▶ Conductivity of Electrolytes
▶ Electrolytes for Rechargeable Batteries
▶ Ion Mobilities
▶ Ion Properties
▶ Ionic Liquids
▶ Solid Electrolytes

References

1. Onsager L (1945) Theories and problems of liquid diffusion. Ann NY Acad Sci 46:241–256
2. Gores HJ, Schweiger HG, Multerer M (2007) Advanced materials and methods for lithium ion batteries. In: Zhang SS (ed) Ch 11 Optimizing the conductivity of electrolytes for lithium ion-cells, Research Signpost Special Review Books, Kerala, pp 257–277
3. Nyman A, Behm M, Lindbergh G (2008) Electrochemical characterisation and modelling of the mass transport phenomena in LiPF6-EC-EMC electrolyte. Electrochim Acta 53:6356–6365
4. Petrowsky M, Frech R (2008) Concentration dependence of ionic transport in dilute organic electrolyte solutions. J Phys Chem B 112:8285–8290
5. Newman J, Tomas-Alyea KE (2004) Electrochemical systems. Wiley, Hoboken
6. Thomas KE, Darling RM, Newman J (2002) Advances in lithium-ion batteries. In: van Schalkwijk WA, Scrosati B (eds) Ch 12 Mathematical Modeling of Lithium Batteries Kluwer Academic/Plenum, New York
7. Gores HJ, Barthel J, Zugmann S, Moosbauer D, Amereller M, Hartl R, Maurer A (2011) Handbook of battery materials. In: Daniel C (ed) 2nd edn, Ch 17, Liquid Nonaqueous Electrolytes, Wiley VCH, Weinheim, pp 525–626
8. Spiro M (1973) Physical chemistry of organic solvent systems. In: Convington AK, Dickinson T, Fernandez-Prini R (eds) Ch 5, Part 2, Transference Numbers, Plenum Press, London
9. MacInnes DA, Longsworth L (1932) Transference numbers by the method of moving boundaries. Chem Rev 11:171–230
10. Robinson RA, Stokes RH (2003) Electrolyte solutions, 2nd rev edn. Dover, New York
11. Spiro M (1971) Physical methods of chemistry: electrochemical methods. In: Weissberger A, Rossiter B (eds) vol 1, Ch IV Determination of transference numbers, Wiley-Interscience, New York
12. Hittorf W (1901) Bemerkungen über die Bestimmungen der Überführungszahlen. Z Phys Chem 39:612
13. Hittorf W (1903) Über die Wanderung der Ionen während der Elektrolyse. Ann Phys 165:239
14. Barthel J, Wachter R, Gores HJ (1979) Modern aspects of electrochemistry. In: Conway B, Bockris J (eds) vol 13, Ch 1. Springer/Plenum, New York, pp 1–79
15. Evans JW, Vincent CA, Bruce PG (1987) Electrochemical measurement of transference numbers in polymer electrolytes. Polymer 28:2324–2328
16. Ma Y, Doyle M, Fuller TF, Doeff MM, de Jonghe LC, Newman J (1995) The measurement of a complete set of transport properties for a concentrated solid polymer electrolyte solution. J Electrochem Soc 142:1859–1868
17. Hafezi H, Newman J (2000) Verification and analysis of transference number measurements by the galvanostatic polarization method. J Electrochem Soc 147:3036–3042
18. Saito Y, Yamamoto H, Nakamura O, Kageyama H, Ishikawa H, Miyoshi T, Matsuoka M (1999) Determination of ionic self-diffusion coefficients of lithium electrolytes using the pulsed field gradient NMR. J Power Sources 81–82:772–776
19. Scrosati B (2005) Power sources for portable electronics and hybrid cars: lithium batteries and fuel cells. Chem Rec 5:286–297
20. Barthel J, Stroeder U, Iberl L, Hammer H (1982) The temperature dependence of the properties of electrolyte solutions. IV. Determination of cationic transference numbers in methanol, ethanol, propanol, and acetonitrile at various temperatures. Ber Bunsen-Ges 86:636–645
21. Li LF, Lee HS, Li H, Yang XQ, Nam KW, Yoon WS, McBreen J, Huang XJ (2008) New electrolytes for lithium ion batteries using LiF salt and boron based anion receptors. J Power Sources 184:517–521
22. Froemling T, Kunze M, Schoenhoff M, Sundermeyer J, Roling B (2008) Enhanced lithium transference numbers in ionic liquid electrolytes. J Phys Chem B 112:12985–12990
23. Zugmann S, Moosbauer D, Amereller M, Schreiner C, Wudy F, Schmitz R, Schmitz RW, Isken P, Dippel C, Mueller RA, Kunze M, Lex-Balducci A, Winter M, Gores HJ (2011) Electrochemical characterization of electrolytes for lithium-ion batteries based on lithium difluoromono(oxalato)borate. J Power Sources 196:1417–1424
24. Zhao J, Wang L, He X, Wan C, Jiang C (2008) Determination of lithium-ion transference numbers in LiPF6-PC solutions based on electrochemical polarization and NMR measurements. J Electrochem Soc 155:A292–A296

25. Zugmann S, Fleischmann M, Amereller M, Gschwind RM, Wiemhoefer H, Gores HJ (2011) Measurement of transference numbers for lithium ion electrolytes via four different methods, a comparative study. Electrochim Acta 56:3926–3933
26. Zugmann S, Fleischmann M, Amereller M, Gschwind RM, Winter M, Gores HJ (2011) Salt diffusion coefficients, concentration dependence of cell potentials, and transference numbers of lithium difluoromono (oxalato)borate-based solutions. J Chem Eng Data 56:4786–4789

Transport in Concentrated Solutions

Wesley A. Henderson[1], Daniel Seo[1] and Oleg Borodin[2]
[1]Department of Chemical and Biomolecular Engineering, North Carolina State University, Raleigh, NC, USA
[2]Electrochemistry Branch, Sensor and Electron Devices Directorate, U.S. Army Research Laboratory, Adelphi, MD, USA

Predictive capabilities for the transport properties (i.e., viscosity and conductivity) of concentrated electrolytes remain limited [1, 2]. This is generally attributed to the ionic association of the ions – i.e., as ion pair and aggregate solvates – which occurs in electrolyte solutions. Unfortunately, the general lack of understanding of what factors govern the formation of these solvates and how these are correlated to transport properties such as viscosity and conductivity has severely hampered the tailored formulation of electrolytes for specific applications.

Mass transfer in an electrolyte for a particular dissolved ion is determined by three terms which correspond to migration (conduction), diffusion, and convection [3]:

$$\mathbf{N}_i = -z_i u_i F C_i \nabla \Phi - D_i \nabla C_i + C_i \mathbf{v}$$

where $\mathbf{N}_i$ is the flux of the ion i. Conduction therefore occurs for charged species (ions) in an electric field $(-\nabla \Phi)$ (z_i is the charge on the ion, u_i is the mobility, F is Faraday's constant, C_i is the concentration, and Φ is the electrostatic potential).

The driving force for diffusion is a concentration gradient (∇C_i) (D_i is the diffusion coefficient of the ion), while a net velocity of the fluid $(\mathbf{v})$ results in convection. Some or all of these modes may govern mass transport in electrolytes, but the focus is generally on ionic conductivity.

The current density of an electrolyte solution is

$$\mathbf{I} = F \Sigma z_i \mathbf{N}_i$$
$$i$$

If diffusion and convection are absent, then the current is proportional to the ionic conductivity (κ)[3]:

$$\mathbf{I} = -\kappa \nabla \Phi$$

where

$$\kappa = F^2 \Sigma z_i^2 u_i C_i$$
$$i$$

A knowledge of the species present in an electrolyte (i.e., uncoordinated solvent, uncoordinated ions, solvated ions, ion pair, and aggregate clusters), as well as the mobility and dynamics associated with the formation of these species, is therefore necessary to identify the origins of variations in transport properties such as viscosity and ionic conductivity. Unfortunately, the empirical equations widely utilized to evaluate electrolyte mixtures do not provide this information about solution structure. This is the principal reason why theories about concentrated electrolytes remain inadequate for describing and predicting concentrated electrolyte interactions and properties.

Solution Structure: Solvation Versus Ionic Association

Solvation interactions can occur for both cations and anions. Solvation may involve the formation of a coordination bond between a solvent molecule and an ion. This occurs, for example, when a lithium salt is dissolved and the electron lone

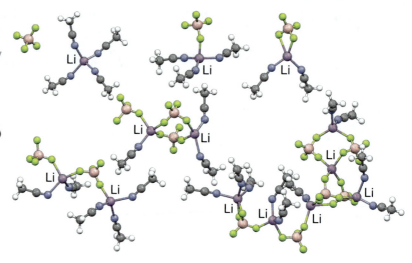

Transport in Concentrated Solutions, Fig. 1 Examples of uncoordinated anions, fully solvated Li⁺ cations, and various CIP and AGG solvate species found in MD simulations of (AN)$_n$-LiBF$_4$ mixtures (Li purple, N blue, B tan, F light green)

pairs on the donor atoms of the solvent molecules coordinate the electron-deficient alkali metal (i.e., Li⁺) cation. But solvation may also take the form of longer-range interactions between the lone pairs and the positively charged center of cations for which steric or other factors prevent the close approach of the solvent molecules (i.e., the solvation of tetraalkylammonium cations). In contrast, anion solvation typically occurs via hydrogen-bond formation between the solvent molecules and anion. Thus, anion solvation is generally only present for salts dissolved in protic solvents such as water, alcohols, and amines. In aprotic solvents – such as acetonitrile (AN) and ethylene carbonate (EC) – therefore, anion solvation does not occur, and the ionic association interactions are therefore predicated upon the competition which occurs between the solvent molecules and anions for coordination to the cations (in the case of alkali metal cations).

Based upon such interactions, electrolytes can be categorized as either "strong" or "weak." Strong electrolytes are those in which the ions (cations and anions) are fully solvated and thus interact only weakly through long-range electrostatic forces. To a great extent, the solvent molecules shield the ions from one another. Thus, the ions may be regarded as independent of one another. This is often the case for dilute aqueous electrolytes. As the salt concentration increases, however, the solvates come in closer proximity to one another (thus, increasing the electrostatic interactions), and ion pair or aggregate species may form.

For weak electrolytes, ionic association occurs – even for dilute mixtures [4, 5]. The various species are often denoted as solvent-separated ion pair (SSIP), contact ion pair (CIP), and aggregate (AGG) solvates. The SSIP solvates consist of uncoordinated anions and fully solvated cations. In contrast, CIP solvates consist of cation-anion pairs with solvated cations, while AGG solvates have multiple ions (anions and solvated cations) clustered together – depending upon the ion composition, these solvates may have a net positive, neutral, or negative charge. Examples of these various ionic species are noted in Fig. 1. These have been extracted from molecular dynamics (MD) simulations of (AN)$_n$-LiBF$_4$ mixtures [4, 5].

The types of solvates present and their distribution are key determinants in the resulting electrolyte properties. It is noteworthy that the solvates continuously transform from one type of coordination to another with continuous exchanges of solvent molecules and anions in the cation's coordination shell [6]. The structure of the solvent is one important factor in the solvate formation, in terms of both its ability to donate electron density through donor atoms and steric effects which impact the packing of the solvent molecules and anions around the cations. The structure of the anions is another critical factor. For lithium salts in aprotic

Transport in Concentrated Solutions 2093

Transport in Concentrated Solutions, Fig. 2 Snapshots of MD simulations of (AN)$_n$-LiX mixtures with (*left*) LiBF$_4$ and (*right*) LiClO$_4$ with concentrations of (*top*) $n = 30$, (*middle*) $n = 20$, and (*bottom*) $n = 10$ (*Li purple, N blue, B tan, F light green, Cl dark green,* and *O red*) (uncoordinated AN molecules have been removed)

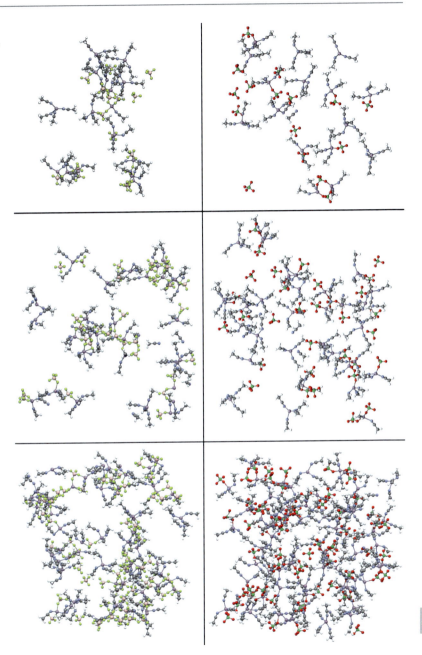

solvents, increasing ionic association is found in the order [4, 5]

$$PF_6^- > TFSI^- \geq ClO_4^- > BF_4^- > CF_3SO_3^- > CF_3CO_2^-$$

where TFSI$^-$ is the N(SO$_2$CF$_3$)$_2^-$ anion. This ordering is evident in Fig. 2 which shows snapshots of the solvates found in MD simulations of (AN)$_n$-LiX mixtures with LiBF$_4$ and LiClO$_4$. To facilitate the viewing of the solvates, all of the uncoordinated AN solvent molecules have been removed. Thus, although it appears that the simulations with LiClO$_4$ have more solvent than those with LiBF$_4$, this is not the case. Instead, the LiClO$_4$ mixtures have a higher

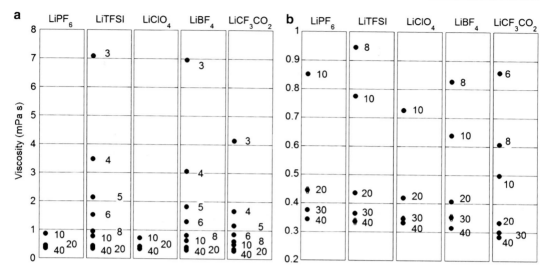

Transport in Concentrated Solutions, Fig. 3 (a) Viscosity of isothermal (AN)$_n$-LiX mixtures at 60 °C (AN/LiX (n) noted in plots) and (b) the same data expanded for the dilute mixtures alone [6]

average solvation number than the LiBF$_4$ mixtures, resulting in a greater fraction of SSIP and CIP solvates. Significant differences are noted for these salts, even though the anions are both tetrahedral and approximately of the same size. Other factors which influence the solution structure and electrolyte properties include the temperature and salt concentration [4, 5].

Viscosity of Concentrated Electrolytes

Electrolyte viscosity data is frequently modeled using the Jones-Dole (JD) equation [7–9]. This equation can be extended to include a third term for concentrated electrolytes [10–13]:

$$\eta_r = \eta/\eta_0 = 1 + A\left(C^{1/2}\right) + B(C) + D(C^2)$$

where η_r, η, and η_0 are the relative viscosity of a solution, viscosity of the solution, and viscosity of the pure solvent, respectively; C is the concentration; and A, B, and D are fitted coefficients. The A-term is typically associated with a reference ion and its ionic atmosphere, but for concentrated electrolytes (>0.05 M) with organic solvents, this term is often negligible [13].

The B-term is associated with ion-solvent interactions and volume effects. The D-term is employed for electrolytes with high salt concentrations (0.5–2.0 M or higher) and is associated with ion-ion and/or ion-solvent interactions [13]. This form of analysis, however, simply results in values for the fitted parameters without any practical indication of the significance of these values and their origin.

If the solution structure is known, then a mechanistic rather than quantitative explanation for the property variations is possible. For example, it might be expected that the most dissociated salt (i.e., LiPF$_6$) would result in the lowest viscosity for (AN)$_n$-LiX mixtures, but this is not what is found (Fig. 3). Instead, for the relatively dilute mixtures, the viscosity decreases with increasing salt association. Scrutiny of the simulations (Fig. 2) suggests that this occurs because the amount of uncoordinated solvent increases with increasing ionic association (i.e., the anions displace the solvent molecules in the Li$^+$ cation's coordination shell as most of the cations have fourfold coordination). The uncoordinated solvent may serve as a lubricant between the solvates. Given that the solution viscosity is the response to shear of the entire mixture, rather than the charged species alone (as is the case for the ionic conductivity), the lower number of larger solvates and increased amount of uncoordinated solvent with increasing ionic association does aptly explain the trend of decreasing viscosity with increasing ionic association [6].

Ionic Conductivity of Concentrated Electrolytes

The theory of electrolyte ionic conduction is well established for strong electrolytes in which the dissociated solvated ions act largely independently of one another. No general theory for conduction in weak electrolytes (concentrated electrolytes or aprotic electrolytes which lack anion solvation) exists, however, which is widely accepted. This is due to a lack of understanding of how the ions interact with each other and how this, in turn, influences the conduction behavior.

Conduction equations are often defined in terms of molar conductivity Λ (S cm^2 mol^{-1}) as the conductivity of electrolyte solutions is dependent upon the salt concentration. Thus, Λ is related to the conductivity κ or σ (S cm^{-1}) by

$$\Lambda = \kappa/C$$

where C (mol dm^{-3}) is the salt concentration.

For strong electrolytes (i.e., low salt concentration in an aqueous solution), Λ is frequently described by Kohlrausch's law [14–16]:

$$\Lambda = \Lambda_0 - KC^{1/2}$$

where Λ_0 is the limiting molar conductivity (at infinite dilution) and K is an empirical constant. For fully dissociated ions, Λ_0 is equal to

$$\Lambda_0 = v_+\lambda_{0+} + v_-\lambda_{0-}$$

where v_+ and v_- are the moles of cations and anions, respectively, from the dissociation of 1 mol of the dissolved electrolyte salt and λ_{0+} and λ_{0-} are the limiting molar conductivity values for the individual ions. This has been expanded upon to give the Debye-Hückel-Onsager equation [14–16]

$$\Lambda = \Lambda_0 - (A + B\Lambda_0)C^{1/2}$$

where A and B are constants that depend upon temperature, the ion charges, and the dielectric constant and viscosity of the solvent. For weak electrolytes where ion-ion interactions are nonnegligible, however, both Kohlrausch's law and the Debye-Hückel-Onsager equation become ineffective.

To evaluate concentrated electrolytes, association constants are often calculated. CIP solvates can be defined as

$$\mathrm{C^+}_S + \mathrm{A^-} \leftrightarrow [\mathrm{C^+A^-}]_S$$

where $\mathrm{C^+}_S$ and $\mathrm{A^-}$ are solvated cations and anions, respectively, and $[\mathrm{C^+A^-}]_S$ is a solvated ion pair. The corresponding association constant for ion pairs (K_A) is frequently determined using conductivity measurements of symmetrical electrolytes at low salt concentration [17–19]. Different equations may be used. For example, the Fernández-Prini expansion [20, 21] of the Fuoss-Hsia equation [22] (an extension of the Fuoss-Onsager equation [23, 24]) gives

$$\Lambda = \Lambda_0 - S(C\gamma)^{1/2} + E(C\gamma)\ln(C\gamma) + J_1(C\gamma)$$
$$- J_2(C\gamma)^{3/2} - K_A(C\gamma)f_\pm^2\Lambda$$
$$S = \alpha\Lambda_0 + \beta \quad E = E_1\Lambda_0 - E_2$$
$$J_1 = \sigma_1\Lambda_0 + \sigma_2 \quad J_2 = \sigma_3\Lambda_0 + \sigma_4$$

where γ is the degree of dissociation and $f_\pm$ is the mean molar activity coefficient for an ion pair. The σ, β, E_1, E_2, σ_1, σ_2, σ_3, and σ_4 coefficients are all dependent on the solvent properties (ϵ_r, η) and the ion charges. The σ_1, σ_2, σ_3, and σ_4 coefficients also depend upon the distance R of closest approach of the ions. The mean activity coefficient may be calculated with the Debye-Hückel equation [25]

$$\ln f_\pm = -A(C\gamma)^{1/2}/\left[1 + BR(C\gamma)^{1/2}\right]$$

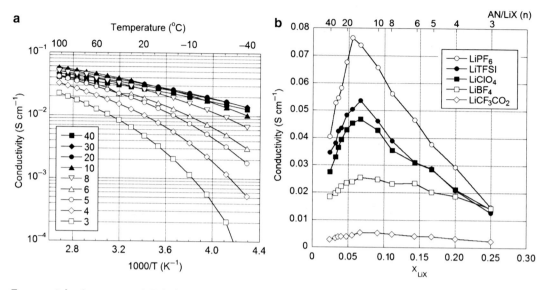

Transport in Concentrated Solutions, Fig. 4 Ionic conductivity of (a) $(AN)_n$-LiTFSI mixtures at variable temperatures (the AN/LiTFSI (n) ratio is indicated) and (b) isothermal $(AN)_n$-LiX mixtures at 60 °C (AN/LiX (n) noted at the top of the plot) [6]

K_A can then be evaluated using the mass action equation [25]

$$K_A = (1-\gamma)/C\gamma^2 f_\pm^2$$

where the activity coefficient for the ion pair is assumed to be unity. Note that this assumes that only ion pairs are present in solution.

This form of analysis has been extended to include an association constant for triple-ion aggregates (K_T) [26] – i.e., $[C^+A^-C^+]^+_S$ and $[A^-C^+A^-]^-_S$ – and some authors have further extended this to include quadruple-ion formation [27–30]. Even assuming that all of the assumptions made in such an evaluation are valid, ultimately the association constant parameters obtained provide little insight into the solution interactions and fail to describe property variations with concentration, especially for high salt concentrations.

A knowledge of the solution structure, however, enables the various factors which govern ionic conduction to be probed in depth. These include the types of solvates present (Fig. 1) and their population, their distribution (i.e., proximity to one another), and the dynamics associated with solvate formation (i.e., solvent and anion exchange in the Li$^+$ cation coordination shell) [6]. For the $(AN)_n$-LiX electrolytes, a 1 M solution corresponds to 15–17 AN molecules per Li$^+$ cation [6]. As for many aprotic solvent-based electrolytes with lithium salts, this is the concentration with the highest ionic conductivity (Fig. 4). The type and population of solvate species typically do not change notably between the $n = 30$ and $n = 20$ mixtures, but there is a sizable increase in the conductivity (Fig. 4). This can be attributed principally to an increase in the number of available charge carriers, which remain well dispersed (Fig. 2). For concentrations greater than $n = 20$, the number of ions continues to increase, but an increase in the association of the ions is also observed (the extent of which depends upon the anion present) which decreases the number of charge carriers and their mobility to some extent. In addition, the separation between the solvates decreases resulting in greater electrostatic interactions between the charged species. Thus, a maximum in the conductivity is observed. Increasing the salt concentration further leads to a decline in the conductivity (Fig. 4) as the solvate species become more associated (with a significant decline in uncoordinated anions and fully solvated Li^+ cations), the exchange rate of solvent molecules and anions tends to decrease, and electrostatic interactions between neighboring solvates strongly influence their mobility.

Empirical equations with fitted parameters have thus far been the principal means for electrolyte evaluation. Utilizing these fitted parameters, however, provides limited insight as to how differences in solvent and/or ion structure, composition, and temperature translate into wide variations in measureable transport properties. Further, the complexity of the solution structure for weak electrolytes, even in dilute solutions, suggests that the assumptions implicit in the current understanding of the significance of the fitted parameters may be overly simplistic. The identification of the molecular-level solution structure within electrolytes, however, enables the many determinate factors for electrolyte transport properties to be parsed out and their relevance evaluated. A general lack of this information has greatly hampered the development of a broader understanding of electrolyte structure–property correlations to date, especially for concentrated electrolytes for which ion-ion interactions are prevalent and strongly influence the property characteristics. Improved experimental characterization methods and the complementary use of MD simulations, however, offer a promising route forward to surmount the barriers which have impeded progress in the understanding of weak electrolytes.

Cross-References

▶ Electrolytes, Classification
▶ Electrolytes for Rechargeable Batteries

References

1. Dufrêche J-F, Bernard O, Durand-Vidal S, Turq P (2005) Analytical theories of transport in concentrated electrolyte solutions from the MSA. J Phys Chem B 109:9873–9884. doi:10.1021/jp050387y
2. Anderko A, Wang P, Rafal M (2002) Electrolyte solutions: from thermodynamic and transport property models to the simulation of industrial processes. Fluid Phase Equil 194/197:123–142. doi:10.1016/S0378-3812(01)00645-8
3. Newman JS (1991) Electrochemical systems, 2nd edn. Prentice-Hall/Wiley, Englewood Cliffs
4. Seo DM, Borodin O, Han S-D, Ly Q, Boyle PD, Henderson WA (2012) Electrolyte solvation and ionic association. I. Acetonitrile-lithium salt mixtures: intermediate and highly associated salts. J Electrochem Soc 159:A553–A565. doi:10.1149/2.jes112264
5. Seo DM, Borodin O, Han S-D, Boyle PD, Henderson WA (2012) Electrolyte solvation and ionic association. II. Acetonitrile-lithium salt mixtures: highly dissociated salts. J Electrochem Soc 159:A1489–A1500. doi:10.1149/2.035209jes
6. Seo DM, Borodin O, Balogh D, O'Connell M, Ly Q, Han SD, Passerini S, Henderson WA (2012) Electrolyte solvation and ionic association. III. Acetonitrile-lithium salt mixtures–transport properties. J Electrochem Soc under-review
7. Jones G, Dole M (1929) The viscosity of aqueous solutions of strong electrolytes with special reference to barium chloride. J Am Chem Soc 51:2950–2964. doi:10.1021/ja01385a012
8. Dole M (1984) Debye's contribution to the theory of the viscosity of strong electrolytes. J Phys Chem 88:6468–6469. doi:10.1021/j150670a003
9. Kaminsky M (1957) Experimentelle untersuchungen über die konzentrations- und temperaturabhängigkeit der zähigkeit wäßriger lösungen starker elektrolyte. Z Phys Chem 12:206–231. doi:10.1524/zpch.1957.12.3_4.206
10. Martinus N, Sinclair CD, Vincent CA (1977) The extended Jones-Dole equation. Electrochim Acta 22:1183–1187. doi:10.1016/0013-4686(77)80059-5
11. Lencka MM, Anderko A, Sanders SJ, Young RD (1998) Modeling viscosity of multicomponent electrolyte solutions. Int J Thermophys 19:367–378. doi:10.1023/A:1022501108317
12. Wang P, Anderko A, Young RD (2004) Modeling viscosity of concentrated and mixed-solvent electrolyte systems. Fluid Phase Equil 226:71–82. doi:10.1016/j.fluid.2004.09.008
13. Lemordant D, Montigny B, Chagnes A, Caillon-Caravanier M, Blanchard F, Bosser G, Carré B, Willmann P (2002) Viscosity-conductivity relationships in concentrated lithium salt-organic solvent electrolytes. In: Kumagai N, Komaba S, Wakihara M (eds) Materials chemistry in lithium batteries. Research Signpot, Trivandrum
14. Fuoss RM (1978) Review of the theory of electrolytic conductance. J Soln Chem 7:771–782. doi:10.1007/BF00643581
15. Petrowsky M, Frech R (2008) Concentration dependence of ionic transport in dilute organic electrolyte solutions. J Phys Chem B 112:8285–8290. doi:10.1021/jp801146k
16. Aihara Y, Sugimoto K, Price WS, Hayamizu K (2000) Ionic conduction and self-diffusion near

infinitesimal concentration in lithium salt-organic solvent electrolytes. J Chem Phys 113:1981–1991. doi:10.1063/1.482004

17. Fuoss RM (1978) Paired ions: dipolar pairs as subset of diffusion pairs. Proc Natl Acad Sci 75:16–20

18. Barthel J, Gores HJ (1994) In: Mamontov G, Popov AI (eds) Chemistry of nonaqueous electrolyte solutions, current progress. New York, VCH

19. Barthel J, Gores HJ (1999) In: Besenhard JO (ed) Handbook of battery materials, 1st edn. Wiley, Darmstadt

20. Fernández-Prini R (1969) Conductance of electrolyte solutions. A modified expression for its concentration dependence. Trans Faraday Soc 65:3311–3313. doi:10.1039/TF9696503311

21. Fernández-Prini R (1973) In: physical chemistry of organic solvent systems, convington AK, Dickinson T (eds). Plenum, London

22. Fuoss RM, Hsia KL (1967) Association of 1–1 salts in water. Proc Natl Acad Sci 57:1550–1557

23. Fuoss RM, Onsager L (1957) Conductance of unassociated electrolytes. J Phys Chem 61:668–682. doi:10.1021/j150551a038

24. Accascina F, Kay RL, Kraus CA (1959) The Fuoss-Onsager conductance equation at high concentration. Proc Natl Acad Sci 45:804–807

25. Popovych O, Tomkins RPT (1981) Nonaqueous solution chemistry. Wiley, New York

26. Salomon M, Uchiyama MC (1987) Treatment of triple ion formation. J Solut Chem 16:21–30. doi:10.1007/BF00647011

27. Hojo M, Miyauchi Y, Nakatani I, Mizobuchi T, Imai Y (1990) Conductometric studies on the triple ion and quadrupole formations from lithium and tributylammonium trifluoroacetates in protophobic aprotic solvents. Chem Lett 6:1035–1038. doi:10.1246/cl.1990.1035

28. Hojo M, Miyauchi Y, Imai Y (1990) Triple ion and quadrupole formations from trialkylammonium sulfonates and nitrate in protophobic aprotic solvents with higher dielectric constants. Bull Chem Soc Jpn 63:3288–3295

29. Salomon M (1990) Complexation of silver salts in pyridine. Electrochim Acta 35:509–513. doi:10.1016/0013-4686(90)87037-3

30. Miyauchi Y, Hojo M, Moriyama H, Imai Y (1992) Conductometric identification of triple-ion and quadrupole formation by the coordination forces from lithium trifluoroacetate and lithium pentafluoropropionate in protophobic aprotic solvents. J Chem Soc Faraday trans 88:3175–3182. doi:10.1039/FT9928803175

U

UV–Vis Spectroelectrochemistry

Anwar-ul-Haq Ali Shah
Institute of Chemical Sciences, University of
Peshawar, Peshawar, Pakistan

Introduction

Spectroscopy is a powerful technique that finds
applications in all branches of science and
technology. It is the investigation of interaction
of electromagnetic waves and matter. The electric field strength of the electric component,
which is perpendicular to the magnetic component, of an electromagnetic wave at certain time t
is given by:

$$E = E_o \sin 2\pi\nu t$$

where E_o is the amplitude and ν is the frequency.
When satisfying Bohr's frequency condition,
transfer of energy occurs between the interacting
electromagnetic field and a molecule. The
quantity of energy transfer (ΔE) is transition
dependent. Since molecules in a sample of matter
exist in a variety of states, transition between the
states occurs due to absorption or emission of
a fixed quantity of energy. When a sample is
irradiated with electromagnetic waves of
different frequencies, different types of transitions are investigated. UV–vis spectroscopy
probes the electronic transitions of molecules
that absorb light in the ultraviolet and visible

region of the electromagnetic spectrum and is
considered a reliable and accurate analytical
technique for qualitative as well as quantitative
analysis of samples. An optical spectrometer
records the wavelength at which absorption
occurs, together with the degree of absorption at
each wavelength. The resulting spectrum is
presented as graph of absorbance versus wavelength [1]. The coupling of UV–vis spectroscopic
technique to the electrochemical systems creates
the field of UV–vis spectroelectrochemistry.

Principle

In a typical UV–vis spectroelectrochemical
experiment, a UV–vis light beam is directed
through an electrode in an electrochemical cell,
and the changes in absorbance, resulting from the
species generated or consumed in the electrode
process or being present at electrode–solution
interface, are measured (Fig. 1).

With optically transparent electrodes (OTE),
molecular adsorbates, polymer films, or other
modifying layers attached to the electrode
surface or being present in the phase adjacent to
the electrode can be studied. With opaque
electrode materials, internal or external reflection
may be applied. Glass, quartz, or plastic
substrates coated with a thin layer of semiconductors (indium-doped tin oxide) or conducting
metals (gold, platinum) are often used as OTE.
The optically transparent electrode is immersed
as working electrode in a standard cuvette,

G. Kreysa et al. (eds.), *Encyclopedia of Applied Electrochemistry*, DOI 10.1007/978-1-4419-6996-5,
© Springer Science+Business Media New York 2014

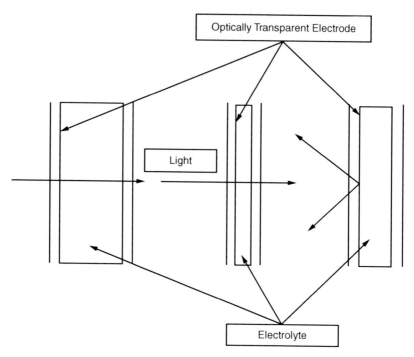

UV–Vis Spectroelectrochemistry, Fig. 1 Schematic of the interaction of electromagnetic radiation with optically transparent electrode immersed in an electrolytic solution

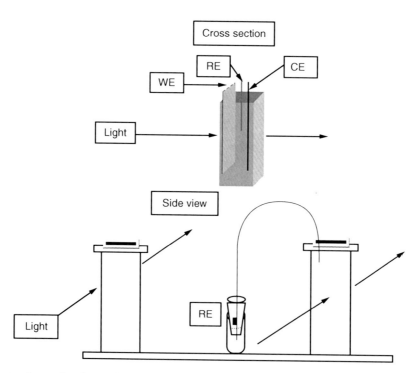

UV–Vis Spectroelectrochemistry, Fig. 2 Cross section and side view of the simple UV-Vis spectroelectrochemical cell

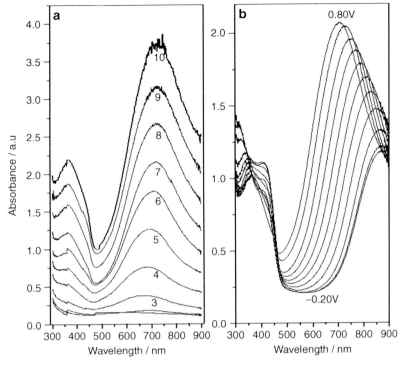

UV–Vis Spectroelectrochemistry, Fig. 3 (a) UV–vis spectra obtained at different time intervals (as indicated, in minutes) after applying an electrode potential of $E_{SCE} = 0.90$ V in 0.5 M H_2SO_4 solution containing 10 mM aniline. (b) UV–vis spectra of a polyaniline-coated ITO glass electrode, obtained at different electrode potential values, ranging from $E_{SCE} = -0.20$ to 0.80 V at every 0.10 V

containing electrolyte and a wire as an auxiliary electrode, inserted in a conventional double beam spectrophotometer with an unmodified OTE in the reference beam. Reference electrode is connected to the cuvette with a salt bridge (see below). Fast scanning instruments are utilized when acquisition of many spectra is needed over short time scales (Fig. 2).

Spectroelectrochemical measurements can be carried out in various modes such as absorbance vs. wavelength or absorbance versus time as the potential is stepped. The resulting spectrum shows only the absorptions, caused by electrochemically generated species, which depend on the magnitude of the applied electrode potential (see example below) and are considered very interesting for studying optoelectronic properties of the species under investigations (Fig. 3).

It is also possible to study opaque or bulk electrode materials in the transmission mode by designing them in the form of minigrids or fine mesh. While passing through the cell, the light will interact with electrolyte solution inside the openings as well as the species adsorbed or deposited on the grid. In the reflectance mode, light is either reflected from the working electrode at an angle greater than critical angle (internal reflectance) [2] or guided with the necessary optical accessory towards the interface through a window in the cell, and the reflected beam is then guided towards the detector (external reflection). Beyond the single reflection (specular reflection or specular reflectance spectroscopy), multiple reflections are possible in the attenuated total reflectance (ATR) mode where an evanescent wave penetrates the film to a thickness of ca. one wavelength or so at each reflection point, so film thickness is often matched to the wavelength of light employed [3]. A typical setup is shown below (Fig. 4):

The ATR element is coated with the electrode material under investigation; besides ITO other

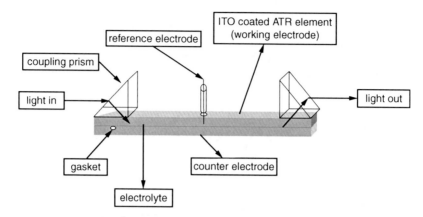

UV-Vis Spectroelectrochemistry, Fig. 4 Diagram of the ATR spectroelectrochemical cell demonstrating the placement of the optically transparent electrode and the coupling prisms thereon

materials which can be sputtered or evaporated on the ATR crystal are applicable [4]. Evaluation of optical constants of metals and other materials whose surface properties may differ markedly from the bulk is among the important applications of specular reflectance spectroscopy [5]. Electronic band structure and surface crystallographic features of thin film of metals sputtered on the ATR crystal can be studied in a method called electroreflectance spectroscopy. In this method the desired information is obtained from differential spectra obtained by rapid electrode potential modulation employing light at alternatively parallel and perpendicular polarization orientation to the plane of incidence.

Future Directions

It is obvious that spectroelectrochemical techniques are becoming common and versatile for electrochemists in a variety of scientific and applied fields such as electrocatalysis, electrochromism, sensors, fuel cell, and solar energy conversion. In situ UV–vis and other spectroelectrochemical (transmission and reflectance) techniques play very important role in understanding of electrochemical processes in variety of objects ranging from single molecule to polymers and nanomaterials. The most important challenges which electrochemists face are the design of cell geometries, proper selection of electrode materials, and chemically selective films for deposition on the electrode material for particular application.

Cross-References

▶ Ellipsometry
▶ Infrared Spectroelectrochemistry
▶ Raman Spectroelectrochemistry
▶ Spectroelectrochemistry, Potential of Combining Electrochemistry and Spectroscopy

References

1. Harvey D (2001) Modern analytical chemistry, 1st edn. McGraw-Hill, New York, pp 397–398
2. Zavarine IS, Kubiak CP (2001) A versatile variable temperature thin layer reflectancespectroelectrochemical cell. J Electroanal Chem 495:106–109
3. Richardson JN, Aguilar Z, Kaval N, Andria SE, Shtoyko T, Seliskar CJ, Heineman WR (2003) Optical and electrochemical evaluation of colloidal Au nanoparticle-ITO hybrid optically transparent electrodes and their application to attenuated total reflectance spectroelectrochemistry. Electrochim Acta 48:4291–4299
4. Holze R (2007) Surface and interface analysis: An electrochemists toolbox. Springer, Heidelberg
5. Purushothaman BK, Pelsozy M, Morrison PW Jr, Lvovich VF, Martin HB (2012) In situ infrared attenuated total reflectance spectroelectrochemical study of lubricant degradation. J Appl Electrochem 42:111–120

V

Voltammetry of Adsorbed Proteins

Julea Butt
School of Chemistry, University of East Anglia,
Norwich, UK

Introduction

Redox active proteins are vital constituents of living cells where they play key roles in both biosynthetic and respiratory metabolisms. For example, these proteins underpin the synthesis of amino acids and nucleotides, support respiration and photosynthesis, and may contribute to the perception of, and response to, oxidative stress. These properties of redox proteins lend themselves to application in biotechnology. Redox proteins define the selectivity and specificity of amperometric biosensors and can provide efficient catalysts for the synthesis of high-value chemicals and renewable fuels. As a consequence redox proteins are much studied, both for insight into their biological activity and with a view to advancing novel biotechnologies. Voltammetry allows for rapid definition of a protein's redox chemistry. It quantifies the number of redox couples accessible to a protein in addition to their reduction potentials and electron stoichiometry. Voltammetry can also detect and quantify more complex redox-driven events, such as ligand binding and redox catalysis. The resolving power of the method is optimal when the entire sample is adsorbed as a (sub-) monolayer film on an electrode with which the protein undergoes direct and facile electron exchange, Fig. 1. Under these conditions the relatively sluggish process of protein diffusion, which dominates the voltammetric response of proteins in solution, makes minimal contribution to the data that is collected. As a consequence the kinetic and thermodynamic parameters describing processes intrinsic to the protein's redox chemistry are resolved with high fidelity [1, 2]. Voltammetry of adsorbed proteins is sometimes termed protein-film voltammetry (PFV) or protein-film electrochemistry (PFE).

Practical Considerations

The voltammetry of adsorbed proteins is typically performed using a three-electrode cell configuration that incorporates working, reference, and counter electrodes. The protein of interest is adsorbed onto the working electrode. When the natural properties of the protein are to be studied or exploited, it is important that adsorption occurs with minimal perturbation of the protein's natural functionality and/or via interactions that mimic those employed by the protein in its cellular context. Working electrode materials that have proved particularly versatile in this regard include metal oxides such as SnO_2, graphite, and gold electrodes modified with self-assembled monolayers (SAMs) of pure, or mixtures of, $-OH$, $-NH_3^+$, $-COO^-$, and $-CH_3$ terminated alkanethiol compounds. Both water-soluble and

G. Kreysa et al. (eds.), *Encyclopedia of Applied Electrochemistry*, DOI 10.1007/978-1-4419-6996-5,
© Springer Science+Business Media New York 2014

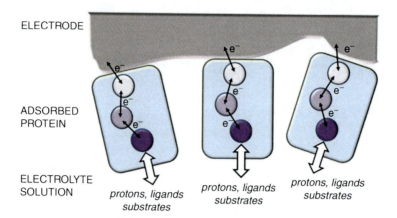

Voltammetry of Adsorbed Proteins, Fig. 1 Cartoon illustrating the voltammetry of adsorbed redox proteins. The proteins are represented by blue oblongs and their redox active cofactors by circles. The voltammetry quantifies electron transfer (*small arrows*) in addition to interactions between the adsorbed proteins and ions or molecules in the electrolyte (*large arrows*). Some electrodes, for example, alumina polished graphite, present a rough surface for protein adsorption as shown here. Other electrodes are atomically flat, such as template-stripped gold

membrane-associated proteins can be adsorbed onto electrodes for voltammetric analysis. Furthermore, membrane proteins may be adsorbed from a detergent solubilized suspension or proteoliposomes, and, when adsorbed, the proteins may be incorporated within electrode-supported bilayers.

Many proteins adsorb spontaneously onto electrodes in electroactive orientations through favorable electrostatic interactions. When using graphite or metal-oxide electrodes to study acidic proteins, it may be necessary to facilitate protein adsorption by including positively charged co-adsorbates such as neomycin, polylysine, or polymyxin to promote adsorption onto these negatively charged electrode surfaces. Alternatively the protein may be covalently attached to the electrode, directly through modification of its surface residues or indirectly through a cofactor such as flavin with which the protein then associates.

In cyclic voltammetry the potential of the working electrode is swept back and forth between two preset values that define the potential "window" for investigation. The flow of current at the working electrode is plotted against working electrode potential in a cyclic voltammogram that immediately maps out the redox activity of the sample. The rate of potential change is termed the scan rate. Typical values for this parameter fall between 1×10^{-3} and 100×10^3 V s^{-1}. This dynamic range provides access to both thermodynamic *and* kinetic descriptions of redox chemistry. Square-wave or differential-pulse voltammetry may provide better separation of faradaic and capacitive currents than cyclic voltammetry. However, the latter is generally preferred since the resulting voltammogram is more easily interpreted to afford an overview of the protein's redox behavior as described below.

Non-Turnover Voltammetry

Many proteins support redox events of the type described by the half-reactions below:

$$Ox + ne^- \leftrightarrow Red \qquad (1)$$

$$Ox + ne^- \leftrightarrow H + me^- \leftrightarrow Red \qquad (2)$$

$$Ox + ne^- + L \leftrightarrow Red:L \qquad (3)$$

$$Ox + ne^- \leftrightarrow Red \leftrightarrow Red^* \qquad (4)$$

where *Ox* is the oxidized species, $n(m)$ is the electron stoichiometry of the couple, *L* is a ligand from the electrolyte solution, *H* is the

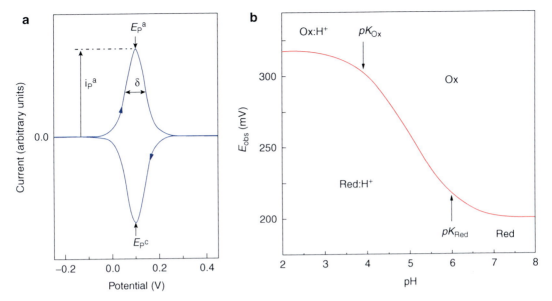

Voltammetry of Adsorbed Proteins, Fig. 2 Non-turnover behavior during cyclic voltammetry at scan rates where the protein oxidation state is in equilibrium with the electrode potential at all times in the experiment. (a) Cyclic voltammogram of a surface confined redox center undergoing redox transformation with a midpoint potential of 0.1 V and $n = 1$. The peak potentials of the reductive, E_P^a, and oxidative, E_P^a, peaks are indicated. For the oxidative peak the peak current, i_p^a, and half-height width, δ, are shown. Arrowheads indicate the scan direction. (b) The pH dependence of the observed reduction potential for an $n = 2$ redox couple with $pK_{Ox} = 4$, $pK_{Red} = 6$ and a reduction potential at alkaline pH of 200 mV; see text for details. Within the graph, the predominant species at each pH and potential is indicated

semi-reduced species, *Red* is the reduced form, and *Red** differs from *Red* in its chemical configuration. Processes (1)–(4) are termed non-turnover events since they do not describe the catalytic redox transformation of molecules in solution. The redox properties of monoheme cytochromes *c*, ferredoxins, and cupredoxins are generally described by non-turnover events. The properties of redox enzymes are also described by non-turnover processes when they occur in the absence of substrates or when the cyclic voltammetric scan rate is sufficiently fast to outrun the rate-limiting event(s) of catalysis.

The interactions between redox centers within a protein and in adjacent protein molecules are generally minimal. As a consequence, a single redox couple undergoing a reversible non-turnover event gives rise to a well-defined pair of peaks during cyclic voltammetry, Figs. 2a and 3. The peaks correspond to reduction (negative current) and oxidation (positive current) of the redox center. When the scan rate is sufficiently low, the oxidation state of the protein at all times in the experiment will be defined by the working electrode potential. This is to say that the protein is being studied under equilibrium conditions and the cyclic voltammogram provides a thermodynamic description of the redox process(es), Fig. 2a. The peak separation, $E_p^a - E_p^c$, is zero and the peak potentials are equal to the reduction potential, E_m, of the couple. The peak current magnitudes for the oxidative (i_p^a) and reductive waves (i_p^c) are equal, and the electron stoichiometry (n) of the couple is derived from the half-height width, δ, of either peak from the relationship $\delta = 3.53RT/nF$ (V). The peak area quantifies the number of moles of electrons exchanged between the protein and electrode. In combination with knowledge of n, this allows the amount of adsorbed, electroactive protein to be calculated.

When a ligand binds with greater affinity to one oxidation state of the protein than another, as implied by (3), then the observed reduction potential, E_{obs}, varies in a systematic manner with the concentration of L, $[L]$, according to:

$$E_{obs} = E_m + \frac{2.303RT}{nF} \log \frac{\left(1 + \frac{[L]}{K_{red}}\right)}{\left(1 + \frac{[L]}{K_{ox}}\right)}$$

where E_m is the reduction potential of the couple in the absence of L binding to either oxidation state. The dissociation constants describing L binding to the reduced and oxidized state are K_{red} and K_{ox}, respectively. When $K_{red} < K_{ox}$, the reduced form of the cofactor is stabilized by L such that the presence of L results in $E_{obs} > E_m$. This is illustrated in Fig. 2b for $L = H^+$ and the redox couple $Ox + 2e^- \leftrightarrow Red$ where both oxidation states can bind a single proton.

When the scan rate is sufficiently low that the system is at equilibrium throughout the experiment, the peak current magnitudes vary in direct proportion to the scan rate according to $i_p = n^2F^2\upsilon A\Gamma_{TOT}/4RT$ where υ is is the scan rate, A is the surface area of the electrode, and Γ_{TOT} is the total electroactive surface coverage of the protein. If the protein contains more than one redox center, or a single center that has more than one accessible oxidation state, the voltammetric response is simply the sum of the response from each couple. From the equation for peak current, it is apparent that the magnitude of the peaks from two $n = 1$ couples in the same protein will be equal. The magnitude of the peak from an $n = 2$ couple will be four times that from an $n = 1$ couple.

When the scan rate is sufficiently high to preclude sample equilibration with the electrode potential, then one, or both, of the peak currents will fail to achieve the magnitude predicted from the behavior displayed at lower scan rates; for examples, see Fig. 3. In such conditions, the peaks become smeared across the potential axis in a manner that allows for kinetic resolution of the underlying events. If the oxidative and reductive peaks are displaced equally, but in opposite directions, about the reduction potential, then interfacial electron transfer is the rate-limiting step, Fig. 3a. The variation of peak separation with scan rate, sometimes termed a trumpet plot, allows quantification of the standard heterogeneous rate constant for electron transfer.

When the oxidative and reductive peak currents differ in their magnitudes, the cyclic voltammetry reports on the rate of chemical and electron transfer processes, Fig. 3b. Such a situation will arise for ligand binding/release as described by (Eq. 3) or the chemical rearrangement of Red to Red^* in (Eq. 4). In these cases simulation, or modeling, of the voltammetric response through an appropriate choice of mechanism will resolve rate constants for the chemical steps, in addition to that for heterogeneous electron transfer.

Catalytic Voltammetry

When the redox activity of an adsorbed protein is coupled to catalytic redox transformation of a molecule in solution, termed the substrate, then catalytic voltammetry is observed. Electrons may be relayed between the active site and electrode by the action of one or more ancillary redox centers, as implied by the cartoon in Fig. 1. Alternatively, direct electron exchange between the electrode and the catalytic center may occur. In either case, when the electrode potential provides sufficient driving force for catalysis, there will be a sustained flow of current due to the net transfer of electrons between the electrode and molecules in solution, Fig. 4. This situation is in contrast to the discrete peaks that arise from cyclic voltammetry of non-turnover events, Figs. 2 and 3. A negative (positive) catalytic current describes catalytic reduction (oxidation). The catalytic current magnitude quantifies the rate of electron transfer through the adsorbed enzyme and so the catalytic rate. Voltammetry of a solution of the substrate should always be performed in the absence of enzyme to confirm that the catalytic currents arise from the activity of the enzyme rather than direct catalytic transformation of the substrate by the electrode surface.

Catalytic voltammetry reports on the intrinsic properties of the enzyme when the response is free from limitation by (a) relatively slow rate of interfacial electron transfer and (b) mass transport of substrate to the adsorbed enzyme from the bulk of the electrolyte solution. The latter is

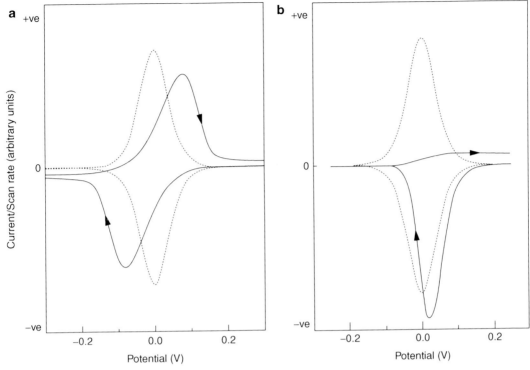

Voltammetry of Adsorbed Proteins, Fig. 3 Illustrative cyclic voltammograms from non-turnover processes measured at scan rates that are sufficiently high to preclude equilibration with the electrode potential. (**a**) Interfacial electron transfer is rate limiting when the peaks (*black lines*) are offset symmetrically from the midpoint potential of the redox couple. (**b**) Information on the rates of chemical events coupled to the electron transfer is available when the oxidative and reductive peaks (*black lines*) differ in their magnitude and width. In both panels the arrowheads indicate the scan direction and the response under equilibrium conditions is shown (*broken red lines*)

usually ensured by rapid rotation of the electrode during measurements. During steady-state catalysis, the variation of the catalytic current at high overpotential (i_{cat}) will vary systematically with substrate concentration. Typically i_{cat} will increase to approach a maximum value (i_{max}) as the substrate concentration is raised and as described by the Michaelis-Menten equation, $i_{cat} = i_{max}$ [substrate]/(K_M + [substrate]) where K_M is the Michaelis constant. This behavior provides the basis for the operation of amperometric biosensors. For fundamental studies of the enzyme, this allows the Michaelis constant, K_M, to be defined for a range of applied potential (driving force) and for comparison to the behavior of the enzyme in solution. The variation of K_M with the concentration of a second molecule allows mechanisms of enzyme activation and inhibition to be defined through standard biochemical approaches. If the amount of electroactive enzyme is known, see above, the current magnitude can be converted into finite rate constants describing catalysis.

In cyclic voltammetry a sigmoidal variation of catalytic activity with electrochemical potential, Fig. 4, is expected from consideration of the Nernst equation as the active site is swept between its oxidized and reduced states. More complex variations of activity with electrochemical potential are frequently observed [1, 2]. These may be independent of the direction of potential change and scan rate. Alternatively they may display a hysteresis that is repeated over successive cycles or seen only on the first scan. Such behaviors arise from the chemical events, reversible or irreversible, that are coupled

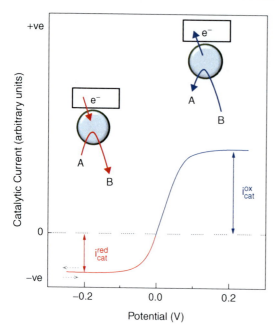

Voltammetry of Adsorbed Proteins, Fig. 4 Schematic of a cyclic voltammogram that could arise from an adsorbed redox enzyme engaged in steady-state catalysis. Positive catalytic currents (*blue*) correspond to oxidation of B and negative catalytic currents (*red*) correspond to reduction of A. The ratio of the maximum catalytic currents for each process, $i^{red}_{cat} : i^{ox}_{cat}$, reveals the catalytic bias of the enzyme under the experimental conditions. Broken arrows indicate the scan direction. Inset: cartoons illustrating the direction of electron exchange between the electrode (*rectangle*), enzyme (*circle*) and substrate (A or B) during each of the catalytic processes

to redox transformations of centers within the enzymes. It is important to recognize that these behaviors can report on the diffusion of molecules (inhibitors and substrates) into the active sites of redox enzymes [1]. Thus, the information afforded by the voltammetry of adsorbed proteins extends beyond redox events to descriptions of physicochemical processes that limit the rate of electron transfer [1].

Advantages and Disadvantages

A number of approaches, in addition to the voltammetry of adsorbed protein, can provide thermodynamic descriptions of a protein's redox chemistry. These include the voltammetry of proteins in solution and potentiometric titration of proteins monitored by a spectroscopy that reports on the oxidation state of the protein. Voltammetry of adsorbed proteins offers a number of advantages. Foremost are the very low sample requirements, typically less than 10 picomoles of protein per electrode, and the ability to resolve redox chemistry without the need for readily detectable spectroscopic change in the sample. There is access to a broad and continuous potential window, approximately +1 to −1 V vs the standard hydrogen electrode, that is not limited by the availability of suitable chemical titrants that may interact specifically with the protein and alter its redox properties. In addition the method facilitates the rapid collection of data across a broad range of experimental conditions. Simply by moving the electrode, with its adsorbed protein, into a solution of the desired composition achieves the rapid, and complete, transfer of sample to a second set of conditions.

While the redox chemistry of metal- and flavin-based cofactors may be readily detected by voltammetry, that associated with thiol-disulfides and amino acids is not. It is also important to be aware that voltammetry by itself provides no direct insight into the chemical identity of the redox couple. This can be overcome by using electrodes that allow for simultaneous spectroscopic and voltammetric analysis of adsorbed proteins. Examples include Ag electrodes that allow for surface-enhanced resonance Raman spectroscopy (SERRS) and mesoporous nanocrystalline SnO_2 electrodes that allow for electronic absorption or magnetic circular dichroism spectroscopies [3]. Another consideration is the need for a redox center to be positioned within ca. 14 Å of the electrode, and so the surface of the protein, for facile interfacial electron exchange. As a consequence, proteins with only buried redox centers are not routinely addressed by direct electrochemical methods.

Voltammetry of adsorbed redox enzymes provides much information that is not readily accessible to other methods. Catalytic rate is resolved as a function of electrochemical potential, that is to say the driving force for the reaction being catalyzed. If the substrates for both oxidative and reductive transformation are present in a single experiment, the bias of the enzyme is immediately revealed. In cyclic voltammetry, this is the ratio of i^{red}_{cat} to i^{ox}_{cat} in Fig. 4. The catalytic parameters i_{max} (equivalent to V_{max} and proportional to k_{cat}) in addition to K_M can be resolved as a function of electrochemical potential. The analysis can be performed at the reduction potential of a redox dye that is used in a spectroscopic solution-phase assay of enzyme activity. Alternatively the analysis may be performed at a potential close to that which the enzyme is thought to experience in its cellular context. The kinetic parameters typically vary over orders of magnitude as a function of potential. This is an important consideration if the parameters are to be used in numerical models of metabolic flux [2].

Future Directions

Ongoing advances aim to expand the range of systems that can be addressed and enhance the mechanistic insights that are available. These aims are frequently addressed in parallel. Direct insight into the chemical and structural consequences of redox events within proteins is provided by voltammetry with simultaneous spectroscopy [3]. Here the opportunities are expanding rapidly. Examples include surface-enhanced infrared absorption at gold electrodes and Fourier transform infrared spectroscopy of high-surface area porous, or network, electrodes. The assembly of multicomponent interfaces, such as bilayers incorporating membrane proteins, can be monitored with electrodes integrated with quartz crystal microbalance technologies [4]. Ideally, the adsorption of proteins in an optimal orientation for facile interfacial electron exchange will become routine. Tools to deliver such bespoke adsorption will arise as information on protein structure is combined with advances in protein engineering and

techniques for electrode surface modification. Analysis of the data arising from these studies will be facilitated by access to user-friendly software, akin to that available for studies of molecules in solution, which tests the ability of various reaction mechanisms to simulate and/or model the voltammetric response of the adsorbed protein.

Cross-References

▶ AFM Studies of Biomolecules
▶ Biofilms, Electroactive
▶ Biosensors, Electrochemical
▶ Cofactor Substitution, Mediated Electron Transfer to Enzymes
▶ Cyclic Voltammetry
▶ Direct Electron Transfer to Enzymes
▶ Enzymatic Electrochemical Biosensors
▶ Infrared Spectroelectrochemistry
▶ Scanning Tunneling Microscopy Studies of Immobilized Biomolecules
▶ Spectroelectrochemistry, Potential of Combining Electrochemistry and Spectroscopy
▶ UV–Vis Spectroelectrochemistry

References

1. Leger C, Bertrand P (2008) Direct electrochemistry of redox enzymes as a tool for mechanistic studies. Chem Rev 108:2379–2438. doi:10.1021/cr0680742
2. Gates AJ, Kemp GL, To CY, Mann J, Marritt SJ, Mayes AG, Richardson DJ, Butt JN (2011) The relationship between redox enzyme activity and electrochemical potential—cellular and mechanistic implications from protein film electrochemistry. Phys Chem Chem Phys 13:7720–7731. doi:10.1039/C0CP02887H
3. Ash PA, Vincent KA (2012) Spectroscopic analysis of immobilised redox enzymes under direct electrochemical control. Chem Commun 48:1400–1409. doi:10.1039/C1CC15871F
4. Jeuken LJC (2009) Electrodes for integral membrane enzymes. Nat Prod Rep 26:1234–1240. doi:10.1039/B903252E

W

Wastewater Treatment (Microbial Bioelectrochemical) and Production of Value-Added By-Products

Ilje Pikaar[1], Bernardino Virdis[1,2],
Stefano Freguia[1] and Jurg Keller[1]
[1]Advanced Water Management Centre
(AWMC), The University of Queensland,
Brisbane, QLD, Australia
[2]Centre for Microbial Electrosynthesis
(CEMES), The University of Queensland,
Brisbane, QLD, Australia

Fundamentals of Bioelectrochemical Systems

A bioelectrochemical system (BES) is an electrochemical device used to convert electrical energy into chemical energy and vice versa. A BES consists of an anode and a cathode compartment, often separated by an ion-selective membrane. The anode is the site of the oxidation reaction which liberates electrons to the electrode and protons to the electrolyte; the cathode is the site of the reduction reaction, which consumes the electrons to reduce a final electron acceptor. To maintain electroneutrality of the system, protons (or other cations) need to migrate to the cathode through the ion-selective membrane. Depending on the half-cell potentials of the electrodes, a BES can be operated either as a microbial fuel cell (MFC), in which electric energy is generated, or as a microbial

electrolysis cell (MEC), in which an input of electric energy is necessary to drive a cathodic reaction otherwise thermodynamically unfavorable [1]. In Fig. 1, a schematic overview of these two configurations is presented.

The catalytic reactions occurring in BES rely on electrochemically active microorganisms, capable of performing extracellular electron transfer (EET), a process by which electrons are transported out of the microorganism cell envelope towards a solid-state electron acceptor (i.e., anode) or from a solid-state electron donor (i.e., cathode) into a microorganism cell [2]. Mechanisms of EET are still hotly debated among researchers. The strategies used by microorganisms that enable EET are still poorly understood. However, it is widely acknowledged that membrane-bound proteins, such as c-type cytochromes (cyt-c), play a key role in electron transfer reactions [3]. EET mechanisms that have been proposed include (i) the use of soluble components that "shuttle" electrons from the cell towards the electrode [4], (ii) direct interaction between the bacterial cell and the electrode by means of outer membrane cytochromes, and (iii) the use of electrically conductive appendages, often referred to as *nanowires* [5, 6]. Electron shuttles can be secondary metabolites produced by the microorganism itself or compounds of abiotic origin, such as humic acids. The relative importance and factors controlling which of the abovementioned EET mechanisms would be dominant in an electroactive community are yet to be fully understood; however, it is

G. Kreysa et al. (eds.), *Encyclopedia of Applied Electrochemistry*, DOI 10.1007/978-1-4419-6996-5,
© Springer Science+Business Media New York 2014

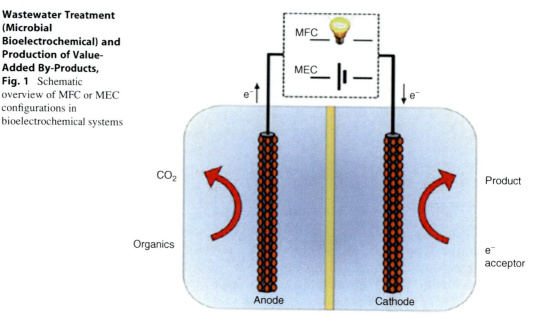

Wastewater Treatment (Microbial Bioelectrochemical) and Production of Value-Added By-Products, Fig. 1 Schematic overview of MFC or MEC configurations in bioelectrochemical systems

likely that different mechanisms would be needed simultaneously to allow electron transfer over distances larger than a bacterial cell [3].

Microbial Bioelectrochemical Wastewater Treatment

Traditionally, wastewater has always been regarded as a source of pollution that requires treatment prior to disposal into the environment. Recently, wastewater has been increasingly viewed as a renewable resource of water, with nutrients and energy contained in organic matter as chemical bonds. This energy can potentially be converted directly into electricity in a MFC, be used to drive the production of a wide range of commodity chemicals, or be used for bioremediation purposes, including the removal of (recalcitrant) pollutants or for denitrification.

BES for Energy Production from Wastewater

A lot of research has focused on the removal of organics with concomitant electricity production in MFCs. Numerous studies showed the feasibility of this concept with and without an ion-selective membrane, mainly using synthetic media containing acetate, glucose, and butyrate as electron donors, but also with domestic and industrial wastewater (e.g., digester effluent, brewery, and paper-recycling water) [1, 7]. At the cathode, the reduction of oxygen, which is provided by air sparging or diffusion, is normally carried out by either chemical or biological catalysis. It is also possible to reduce nitrate to nitrogen gas at the cathode, which results in a lower energy recovery but enables the simultaneous removal of organics and nitrogen from wastewater [8]. Over the last few years, the performance of MFCs has improved significantly, approaching current densities that would be suitable for full-scale applications, even though most of these studies make use of highly buffered, synthetic media with acetate as a carbon source [9, 10]. So far, however, only low current densities and Coulombic efficiencies have been achieved using actual wastewater as feed [7]. This is most likely related to (i) the presence of a range of complex organic compounds that are more difficult to break down by bacteria, (ii) the possible concomitantly occurring side reactions, such as methanogenesis, and (iii) the low buffer capacity of the wastewater, resulting in acidification of the anode biofilms. Although the feasibility of simultaneous wastewater treatment and electricity generation has been shown, large-scale implementation is most likely not economically viable, even with prospective low-cost materials, mainly due to our current cheap energy economy [11] which keeps the value of electricity quite low. Therefore, in recent years, research has broadened significantly beyond electricity generation and is now focusing on the application of BES for both the removal of (recalcitrant) pollutants and the production of value-added products.

BES for Environmental Remediation

Conventional methods for the removal of recalcitrant compounds such as dyes, pesticides, heavy metals, and phenolic compounds normally involve physical and chemical techniques such as ozonation, UV/H2O2, carbon adsorption, electrochemical treatment, and membrane filtration [12]. Although all of these methods have been demonstrated, they also carry a number of disadvantages, namely, the formation of toxic by-products, high costs, and high energy consumption. All of these disadvantages could theoretically be avoided in a BES. The working principle is based on the biocatalyzed oxidation of organics at the anode, with the concurrent reduction of pollutants at the cathode, using either an abiotic or a biocatalyzed cathodic reaction. Several studies have showed the suitability of BES for the removal of recalcitrant compounds, such as the biocatalyzed anodic removal of heterocyclic organic compounds, the cathodic removal of dyes [13] and chlorinated compounds [14], and the removal/recovery of heavy metals/metalloids [15] such as chromium, copper, uranium, and selenium using either biotic or abiotic cathodes, as outlined in Huang et al. 2010 [12]. Pollutant removal can be achieved in both MFC and MEC configurations, although the removal rate can be significantly increased with the addition of a small amount of energy [12]. Currently, the removal of recalcitrant compounds in BES is still in its initial phase, with most studies being performed at a laboratory scale using defined synthetic media. Before it can become a mature and cost-effective technology that can be implemented in practice at full scale, significant challenges lay ahead (see Future Directions).

BES for the Production of Added-Value Products

The disadvantage of MFCs for electricity generation is their high cost-to-revenue ratio. Reported current densities for bench scale MFC are still below the target required to support application development. However, when the electrons liberated during the oxidation reaction are used to generate products with higher revenue than electricity, the process may be economically valuable [11]. Liu et al. [16] provided the first account of the use of BES for the production of products other than electricity by coupling acetate oxidation with the cathodic production of hydrogen. This process is very similar to water electrolysis, with the very important difference that acetate oxidation at a bioanode occurs at lower potentials than that required for water oxidation to O_2 (i.e., -0.2 to -0.3 V instead of $+0.82$ V versus standard hydrogen electrode, SHE). Hence, considering that H_2 evolution requires a theoretical redox potential of -0.414 V vs SHE, the process can be driven with a much lower energy input than through water electrolysis (i.e., 1.2 V). After this preliminary study, numerous studies have been conducted to further optimize this process. Most research focused on reducing the hydrogen evolution overpotential by using either metal catalysts (e.g., platinum) [17] or microorganisms to catalyze the cathodic reaction [18]. The use of biocathodes is preferable, as they avoid the need for expensive catalysts and allow the application of cheap cathode materials such as carbon or graphite.

These studies have broadened the possible application of BES outside of the electricity production area. For example, the feasibility of the production of other biofuels, including ethanol via the biocatalyzed reduction of acetate [19], methane through the biocathodic reduction of carbon dioxide [20], and organic acids by the biocatalyzed reduction of carbon dioxide [21], has recently been shown. In addition to the production of biofuels, other demonstrated applications include the production of chemicals such as hydrogen peroxide from partial oxygen reduction [22] and caustic soda [23]. For the generation of hydrogen peroxide, carbon-based materials are used, as they are cheap and have a high overpotential for the reduction of oxygen to water. The air/oxygen is supplied either by sparging it into the liquid or by diffusion using a gas diffusion electrode (GDE). The advantage of a GDE is that high oxygen partial pressure can be obtained at the electrode surface where the reactions take place. Caustic soda is produced at the cathode, as most commercially available ion-exchange membranes are not highly selective towards protons, which allows for other cations such as sodium ions, which are normally present at much higher concentrations than protons, to be also transferred. As protons are consumed at the cathode by the reductive reaction, alkaline caustic solutions can be produced [23].

Future Directions

Considerations for Practical Applications: Scaling Up

So far, mainly lab-scale BES using mostly synthetic media have been used, with only a limited number of studies at (small) pilot scale using real wastewater. In order to become a suitable system for practical implementation, BESs have to be scaled up by at least two orders of magnitude. However, the scaling up to practical implementation will not be straightforward, as it will require significant efforts to address all the technological and microbial challenges associated with it. Technological challenges include the limited alkalinity of typical wastewaters, electrical losses through electrodes and electrolytes, and the development of suitable process configurations and reactor designs that enable scale-up and control strategies to ensure stable operation.

For most applications, a two-compartment cell configuration using a cation-exchange membrane (CEM) or anion-exchange membrane (AEM) to separate the anode and cathode is needed to avoid the losses of efficiency resulting from mixing of the two electrolytes. However, such configuration does incur some drawbacks. First, during the oxidation of organics, 4 moles of protons are generated per mol of COD oxidized. To avoid a pH drop below physiological conditions, the alkalinity of the wastewater should be at least 30 meq/L for a low-strength wastewater of 250 mg/L COD and proportionally much higher for high-strength wastewaters (625 meq/L required for a COD of 5,000 mg/L). However, the typical alkalinity of most wastewater streams ranges between 5 and 30 meq/L; hence, a significant shortfall of alkalinity can be expected in most systems. This is often avoided in experimental studies by the addition of high concentrations of pH buffer (typically phosphate), which is not realistic, however, in any practical situation. In some applications, this may not be a major challenge, since only part of the COD is oxidized in the anode, while in the cathode, hydroxide ions are deliberately accumulated to generate an alkali solution (sodium/potassium hydroxide) as the product stream [23]. Recirculation of the anodic flow to the cathode and back was proposed to help balance the alkalinity requirements, while still allowing the incoming COD to be oxidized in the anode [24]. Secondly, under normal wastewater conditions, that is, a neutral pH range, CEMs predominantly transport ions other than protons, due to the low concentration of protons. Due to this imperfect proton transport and the fact that the oxidation reactions in the anode produce protons, and the reduction reactions in the cathode consumes protons, the pH in the cathode chamber can increase by several units. As shown by the Nernst equation, a voltage loss of 60 mV results per pH unit difference between anode and cathode, thus compromising the overall cell performance.

Electrical losses are caused by overpotential losses at the electrode, ohmic losses (e.g., wastewater conductivity and electrode resistivity) and membrane pH gradients. The ohmic losses especially are of major concern due to the low conductivity of the wastewater (1–10 mS/cm) typically observed. For example, at a current density of 10 A/m^2 and a conductivity of 1 mS/cm, the ohmic losses would be 1 V/cm anode-to-cathode distance, as noted by Rozendal et al. (2008) [11]. Hence, the design of a BES should aim to minimize the anode-to-cathode spacing. Normally, cheap carbon-based materials such as graphite felt, granules, or fibers are used as the electrode material. Although graphite has a high electrical

Microbial Bioelectrochemical Wastewater Treatment

resistivity (1,375 $\mu\Omega$/cm), the voltage loss caused by electrode resistivity is negligible in laboratory-scale systems due to the small absolute current flowing through the electrodes. However, the electrode ohmic losses can be several hundreds of millivolts at full scale as much higher absolute currents flow through the electrode. These losses can be reduced by using highly conductive current collectors such as stainless steel or titanium [25]; however, it should be noted that this will result in additional investment costs. Alternatively, the need for current collectors can be avoided when using a bipolar stack design. Bipolar plates are normally made of graphite and connect the anode of one cell with the cathode of the adjacent cell and, by doing so, minimize the travel distance for the electrons, thus reducing ohmic losses. Considering the above, it is likely that the future design of a full-scale BES will involve the use of multiple BES modules using bipolar plates operating in series. Although bipolar plates are common for PEM fuel cells, to the best knowledge of the authors, their feasibility for BES has only been shown in two studies [26, 27]. This is most likely related to the fact that bipolar plates require a series connection, and at this stage no suitable control strategy exists to avoid cell reversal in stacked cells when operating in series [28]. Cell reversal (i.e., one or more of the cells operate as an electrolysis cell) occurs when one cell of the stack cannot deliver the current that is demanded from the system [29]. As a consequence, the anode potentials can reach high positive potentials that are detrimental to the microorganisms (or even cause oxidation of carbon-based electrode materials) and will ultimately result in failure of the whole system. So far, this cell-reversal phenomenon has been poorly studied, without any suitable process control strategy to avoid its occurrence. Recently, it has been proposed that cell reversal can be avoided by using an electronic circuit in which multiple BESs charge multiple capacitors in parallel while discharging them in series [30]. However, this method can only be used in an MFC configuration and not in MECs, which severely limits its practical application.

Microbial challenges include metabolic diversity, substrate competition with other bacteria or Archaea (e.g., methanogens), exposure to high concentrations of recalcitrant compounds, and biofilm pH gradients. Real wastewater typically contains a large variety of both soluble and particulate organic materials that are often difficult to degrade, e.g., cellulose. This implies that in real wastewater applications, lower cell performances may be expected. Indeed, several studies using real wastewater showed a reduction in cell performance compared to synthetic media that is well buffered and using mostly readily biodegradable substrates, typically acetate. Fermenters and methanogens compete with electrochemically active bacteria at the anode for the organics present in the wastewater [31]; this parasite process lowers the electron recovery and hence the Coulombic efficiency. Many of the abovementioned issues can be overcome, but some significant efforts and innovation may be required to achieve suitable solutions that work in practice. Currently there is only limited research focusing on these aspects, although they are likely the main constraints for the applications of BES in the future.

Potential Practical Applications

Since its first discovery, BES has been rapidly evolving from the initial concept of MFC, which focuses on the production of electricity from the treatment of waste streams to the broader concept of BES, which also encompasses bioremediation, microbial electrolysis, and microbial electrosynthesis. Production of biofuels such as hydrogen, methane, ethanol, or even bulk chemicals such as caustic and hydrogen peroxide has already been proven. In addition, a wide range of other products can theoretically be produced while using wastewater as a source of redox equivalents. For example, the biocatalyzed reduction of butyrate to butanol and the reduction of glycerol to 1,3-propanediol, mainly used as a building block in the production of polymers, have recently gained attention [1, 32]. Glycerol is a by-product of biodiesel production, and it is generated in increasingly large quantities and often with impurities. Therefore, the utilization and "upgrading" of such a low-value by-product using a BES is very attractive. Recently, the use of BES for desalination in so-called microbial desalination cells and the generation of hydrogen from fresh and salt water using microbial reverse-electrodialysis cells have also been proposed [33, 34]. Both applications have the draw-back that they require enormous amounts of wastewater with easily biodegradable organics, saline water and fresh water all at one location, which limits the practical applicability to few suitable locations.

To summarize, BES for wastewater treatment may become attractive as a method for the recovery of energy from the treatment of waste streams, especially when value-added by-products can be generated simultaneously. However, due to the intrinsic limitations detailed above and the limited economy of scale of their modular design, it will remain highly unlikely that BES will completely replace existing wastewater treatment systems any time soon.

Cross-References

▶ Wastewater Treatment by Electrocoagulation
▶ Wastewater Treatment, Electrochemical Design Concepts

References

1. Rabaey K, Rozendal RA (2010) Microbial electrosynthesis - revisiting the electrical route for microbial production. Nat Rev Microbiol 8(10):706–716
2. Logan BE, Hamelers B, Rozendal R, Schröder U, Keller J, Freguia S, Aelterman P, Verstraete W, Rabaey K (2006) Microbial fuel cells: methodology and technology. Environ Sci Technol 40(17):5181–5192
3. Borole AP, Reguera G, Ringeisen B, Wang ZW, Feng Y, Kim BH (2011) Electroactive biofilms: current status and future research needs. Energ Environ Sci 4(12):4813–4834
4. Gralnick JA, Newman DK (2007) Extracellular respiration. Mol Microbiol 65(1):1–11
5. Gorby YA, Yanina S, McLean JS, Rosso KM, Moyles D, Dohnalkova A, Beveridge TJ, Chang IS, Kim BH, Kim KS, Culley DE, Reed SB, Romine MF, Saffarini DA, Hill EA, Shi L, Elias DA, Kennedy DW, Pinchuk G, Watanabe K, Ishii S, Logan B, Nealson KH, Fredrickson JK (2006) Electrically conductive bacterial nanowires produced by Shewanella oneidensis strain MR-1 and other microorganisms. Proc Natl Acad Sci USA 103(30):11358–11363
6. Reguera G, McCarthy KD, Mehta T, Nicoll JS, Tuominen MT, Lovley DR (2005) Extracellular electron transfer via microbial nanowires. Nature 435(7045):1098–1101
7. Ahn Y, Logan BE (2010) Effectiveness of domestic wastewater treatment using microbial fuel cells at ambient and mesophilic temperatures. Bioresour Technol 101(2):469–475
8. Virdis B, Rabaey K, Yuan Z, Keller J (2008) Microbial fuel cells for simultaneous carbon and nitrogen removal. Water Res 42(12):3013–3024
9. Fan Y, Hu H, Liu H (2007) Enhanced coulombic efficiency and power density of air-cathode microbial fuel cells with an improved cell configuration. J Power Sources 171(2):348–354
10. Chen S, Hou H, Harnisch F, Patil SA, Carmona-Martinez AA, Agarwal S, Zhang Y, Sinha-Ray S, Yarin AL, Greiner A, Schröder U (2011) Electrospun and solution blown three dimensional carbon fiber nonwovens for application as electrodes in microbial fuel cells. Energ Environ Sci 4(4):1417–1421
11. Rozendal RA, Hamelers HVM, Rabaey K, Keller J, Buisman CJN (2008) Towards practical implementation of bioelectrochemical wastewater treatment. Trends Biotechnol 26(8):450–459
12. Huang L, Cheng S, Chen G (2011) Bioelectrochemical systems for efficient recalcitrant wastes treatment. J Chem Technol Biotechnol 86(4):481–491
13. Mu Y, Rabaey K, Rozendal RA, Yuan Z, Keller J (2009) Decolorization of azo dyes in bioelectrochemical systems. Environ Sci Technol 43(13):5137–5143
14. Pham H, Boon N, Marzorati M, Verstraete W (2009) Enhanced removal of 1,2-dichloroethane by anodophilic microbial consortia. Water Res 43(11):2936–2946
15. Heijne AT, Liu F, Weijden RVD, Weijma J, Buisman CJN, Hamelers HVM (2010) Copper recovery combined with electricity production in a microbial fuel cell. Environ Sci Technol 44(11):4376–4381
16. Liu H, Grot S, Logan BE (2005) Electrochemically assisted microbial production of hydrogen from acetate. Environ Sci Technol 39(11):4317–4320
17. Logan BE, Call D, Cheng S, Hamelers HVM, Sleutels THJA, Jeremiasse AW, Rozendal RA (2008) - Microbial electrolysis cells for high yield hydrogen gas production from organic matter. Environ Sci Technol 42(23):8630–8640
18. Rozendal RA, Jeremiasse AW, Hamelers HVM, Buisman CJN (2008) Hydrogen production with a microbial biocathode. Environ Sci Technol 42(2):629–634
19. Steinbusch KJJ, Hamelers HVM, Schaap JD, Kampman C, Buisman CJN (2010) Bioelectrochemical ethanol production through mediated acetate reduction by mixed cultures. Environ Sci Technol 44(1):513–517
20. Cheng S, Xing D, Call DF, Logan BE (2009) Direct biological conversion of electrical current into

methane by electromethanogenesis. Environ Sci Technol 43(10):3953–3958

21. Nevin KP, Woodard TL, Franks AE, Summers ZM, Lovley DR (2010) Microbial electrosynthesis: feeding microbes electricity to convert carbon dioxide and water to multicarbon extracellular organic compounds. mBio 1(2)

22. Rozendal RA, Leone E, Keller J, Rabaey K (2009) Efficient hydrogen peroxide generation from organic matter in a bioelectrochemical system. Electrochem Commun 11(9):1752–1755

23. Rabaey K, Bützer S, Brown S, Keller J, Rozendal RA (2010) High current generation coupled to caustic production using a lamellar bioelectrochemical system. Environ Sci Technol 44(11):4315–4321

24. Freguia S, Rabaey K, Yuan Z, Keller J (2008) Sequential anode–cathode configuration improves cathodic oxygen reduction and effluent quality of microbial fuel cells. Water Res 42(6–7):1387–1396

25. Logan B, Cheng S, Watson V, Estadt G (2007) Graphite fiber brush anodes for increased power production in air-cathode microbial fuel cells. Environ Sci Technol 41(9):3341–3346

26. Dekker A, Ter Heijne A, Saakes M, Hamelers HVM, Buisman CJN (2009) Analysis and improvement of a scaled-up and stacked microbial fuel cell. Environ Sci Technol 43(23):9038–9042

27. Shin SH, Choi Y, Na SH, Jung S, Kim S (2006) Development of bipolar plate stack type microbial fuel cells. B Kor Chem Soc 27(2):281–285

28. Aelterman P, Rabaey K, Pham HT, Boon N, Verstraete W (2006) Continuous electricity generation at high voltages and currents using stacked microbial fuel cells. Environ Sci Technol 40(10):3388–3394

29. Oh SE, Logan BE (2007) Voltage reversal during microbial fuel cell stack operation. J Power Sources 167(1):11–17

30. Kim Y, Hatzell MC, Hutchinson AJ, Logan BE (2011) Capturing power at higher voltages from arrays of microbial fuel cells without voltage reversal. Energ Environ Sci 4(11):4662–4667

31. Freguia S, Rabaey K, Yuan Z, Keller J (2008) Syntrophic processes drive the conversion of glucose in microbial fuel cell anodes. Environ Sci Technol 42(21):7937–7943

32. Rabaey K, Wise A, Johnstone AJ, Virdis B, Freguia S, Rozendal RA (2011) Bioelectrochemical conversion of glycerol to 1,3-propanediol. ACS National Meeting Book of Abstracts

33. Kim Y, Logan BE (2011) Series assembly of microbial desalination cells containing stacked electrodialysis cells for partial or complete seawater desalination. Environ Sci Technol 45(13):5840–5845

34. Kim Y, Logan BE (2011) Hydrogen production from inexhaustible supplies of fresh and salt water using microbial reverse-electrodialysis electrolysis cells. P Natl Acad Sci USA 108(39):16176–16181

Wastewater Treatment by Electrocoagulation

Yunny Meas Vong[1] and Darlene G. Garey[2]
[1]CIDETEQ, Parque Tecnológico Querétaro, México, Estado de Querétaro, México
[2]CIDETEQ, Centro de Investigación y Desarrollo Tecnológico en Electroquímica Parque Tecnológico Querétaro, Pedro Escobedo, Edo. Querétaro, México

Introduction

Various electrochemical wastewater treatment methods have been reported to reduce chemical use and to better remove persistent pollutants, as compared to traditional physical, chemical or biological treatment processes. Associated costs of excessive energy consumption inherent in such methods, however, have created a disincentive for industrial use. Electrocoagulation (EC) has been proven to be effective in removing inorganic contaminants, hydrocarbons, pathogens, colloids, dissolved organics, suspensions, colorants and pigments. A new reactor has been designed to use electrocoagulation in a process to achieve both technical and economic feasibility for application in industry.

The process involves *in situ* generation of coagulants by the electrodissolution of a sacrificial anode rather than with the direct addition of chemical coagulants [1]. The electrochemical reactor promotes destabilization of the pollutants (electrocoagulation), formation of larger solid agglomerates by aggregation (electroflocculation) and separation of the solids by settling or flotation [2]. The same mechanisms involved with chemical coagulants are thus exploited with electrocoagulation [3].

The consumable anode, normally iron or aluminum, introduces metal ions that destabilize particles by neutralizing their charges, thereby reducing electrostatic particle repulsion. Generation of hydroxide species at a cathode leads to

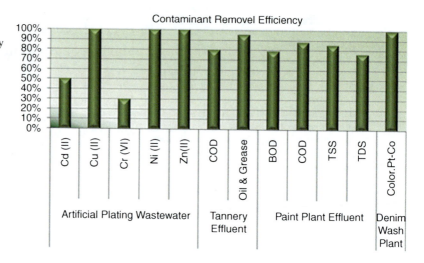

Wastewater Treatment by Electrocoagulation, Fig. 1 Removal efficiency for different effluent types

polymerization of the cations into Fe(OH)$_3$ or Al(OH)$_3$ that can entrap or adsorb the pollutants:

$$Fe \rightarrow Fe^{2+} \rightarrow Fe^{3+} \rightarrow Fe(OH)_3 \quad (1)$$

$$Al \rightarrow Al^{3+} \rightarrow Al(OH)_3 \quad (2)$$

This new reactor has proven effective in the removal of a variety of pollutants from various industrial sources. As can be seen in Fig. 1 below, treatment reduced or removed entirely certain heavy metals, COD, BOD, suspended and dissolved solids, oil by I Want This and grease and color from wastewaters of various industries:

Most notable were the complete removal of Cu, Ni and Zn and the nearly complete removal of color and oils and greases. All others were significantly reduced, with the exception of Cd (II) and Cr(IV), both metals particularly difficult to treat by conventional means.

Concepts

In an electrocoagulation reactor, electric current is applied to a sacrificial anode and a cathode in a processing tank. Reactions at the (aluminum or iron) anode and the cathode would be:

a.) On an aluminum anode :

$$Al \rightarrow Al^{3+} + 3\ e^- \quad (3)$$

On the cathode:

$$2\ H_2O + 2\ e^- \rightarrow H_2I + 2\ OH^- \quad (4)$$

In the solution:

$$Al^{3+} + 3\ OH^- \rightarrow Al(OH)_3 \downarrow \quad (5)$$

b.) On an iron anode:

$$Fe \rightarrow Fe^{2+} + 2\ e^- \quad (6)$$

$$Fe^{2+} \rightarrow Fe^{3+} + e^- \quad (7)$$

$$Fe \rightarrow Fe^{3+} + 3\ e^- \quad (8)$$

On the cathode:

$$2\ H_2O + 2\ e^- \rightarrow H_2I + 2\ OH^- \quad (9)$$

In the solution:

$$Fe^{2+} + 2\ OH^- \rightarrow Fe(OH)_2 \downarrow \quad (10)$$

$$Fe^{3+} + 3\ OH^- \rightarrow Fe(OH)_3 \downarrow \quad (11)$$

Coagulant is generated by the dissolution of metal from the anode, with the simultaneous formation of hydroxyl ions and hydrogen gas at the cathode. This process produces the corresponding aluminum or iron hydroxides.

Charged particles in colloidal suspension are then neutralized and destabilized by mutual collision with the metallic hydroxide ions (i.e., electrocoagulation).

Destabilized or "coagulated" particles are then brought together to form a larger agglomeration of "floc" solids by physical mixing (i.e., electroflocculation). Whereas coagulation is enhanced by strong stirring that produces turbulence to bring hydroxide and metal ions into contact with pollutants, flocculation is facilitated by mild stirring that allows the action of Van der Waals forces to attract pollutants to the growing flocs.

Separation of these solids then occurs by sedimentation or flotation. The H_2 gas generated on the cathode helps to float flocculated particles to the water surface (i.e., electroflotation).

Reactor and Process Design

Laboratory Scale Studies

Initial feasibility studies were performed in batch processes on a laboratory scale. Reactor design was evaluated to identify the influence of parameters of interest, interrelationships, useful ranges for key inputs (conductivity, current density, electrolysis time, etc) and efficiencies of pollutant removal.

The reactor employed aluminum electrodes supplied by DC power. Chemical oxygen demand (COD), color, conductivity, pH and turbidity were monitored during the process. Aluminum dissolved at the anode, with water electrolyzed at the cathode to produce aluminum hydroxide and hydrogen gas. These reactions and hydroxide precipitation resulted in changes in solution pH during the process. Aluminum solubility was affected by solution pH, which was found to affect COD and color removal.

Analyses indicated over 95 % removal of COD, color and turbidity from a sample of industrial wastewater containing hydrocarbons and penetrant dye. Results were used to optimize reactor conditions for final pH, conductivity, current densities and electrolysis time. Laboratory investigations also identified optimal electrode surface area and inter-electrode distance. These parameters lead to decreased current density and cell potential and result in reduction of energy consumption.

Pilot Plant Prototype

A transparent scale model prototype was created to encompass a continuous treatment process. Such processes not only generally require smaller reactors but are also often preferred in industrial applications. Optimization results from the laboratory testing were incorporated into the design of an electrochemical cell reactor of 3.75l L volume. A clarification chamber, sludge conditioning tank and horizontal plate pressure filter were added in series for electroflocculation and separation (electroflotation/settling) following the electrocoagulation process. The pilot system had the capacity to treat 1.2 m^3 of wastewater per day.

It was demonstrated that removal efficiencies increased with strong turbulence in the electrolysis cell followed by flow relaxation in the subsequent clarification chamber. As in the laboratory studies, COD, color and turbidity were significantly reduced. It was also found that partial recirculation of sludge supernatant enhanced coagulation and increased removal efficiencies without increasing energy consumption. This also aided in reduction of the amount of sludge generated. These improvements were incorporated into the design of an industrial scale process.

Industrial Scale Unit

The electrochemical cell was scaled up to a volume of 23 L, exhibiting the capacity to treat 6 m^3 of wastewater per day. Electrodes of parallel aluminum plates were used, incorporating potential for polarity inversion in order to avoid incrustation. Based on the pilot unit results, the clarifier was modified with internal conduits serving as transition zones to turn turbulent to laminar flow and enhance electroflocculation. Sludge from sedimentation and electroflotation was sent to a conditioning tank and then to a horizontal plate pressure filter to produce a dry filter cake. Clear effluent was filtered with sand and activated carbon for reuse in industrial processes.

It was seen that the pH range must be controlled to ensure minimal solubility of the hydroxide formed. The minimum solubility of aluminum hydroxide was found to be at pH 5.5–7.5, optimally around 6.5 (note – a final pH between 6 and 8 is preferred for many industrial reuse and discharge applications).

Results showed a 95–98 % reduction in COD and color for hydrocarbon and dye contaminated wastewater. Sludge generation was relatively low, compared to traditional chemical treatments. Energy consumption was minimized, producing treated water of desired quality for reuse at relatively low cost – and therefore attractive to industry.

Advantages

The electrocoagulation process can be used to treat a wide range of wastewaters and is effective in particular in removing colloidal and suspended solids, colorants and pigments. Inorganic contaminants, dissolved organics and pathogens can be eliminated with EC, as well.

The EC process is characterized by compact size, ease of operation, low sludge generation, high water recovery, and avoidance of chemical addition.

Even including the equipment cost, potential return on investment (ROI) could be less than one year, due to reduced costs for chemicals and sludge disposal.

Future Directions

Although success has been repeatedly demonstrated for use of electrocoagulation in the treatment of industrial wastewaters, there is room for process improvement. Areas suggested for further examination include the following:

a. Pursue the reduction of technical problems, specifically that of electrode scaling or deposition. This can be a particular issue with hard water of high mineral content (principally calcium and/or magnesium). A solution may

Wastewater Treatment by Electrocoagulation, Table 1 Comparison of different treatment strategies for a synthetic contaminated aqueous solution of 1 mM phenol and 0.1 mM direct yellow dye 52

Process	% color removal	% phenol degradation
Electrocoagulation*	90	5
Fenton**	20	>95
Combined Electrocoagulation-Fenton***	95	>95

Experimental conditions for Electrocoagulation*: 200 A. cm^{-2}, pH = 6.1; Fenton**: 0.05 mM Fe(II) + 10.7 mM $[H_2O_2]$, pH = 6.1; Combined electrocoagulation-Fenton***: under O_2 saturation.

be to intermittently reverse electrode polarity (e.g., 1 h in one direction, followed by 1 h in the other). Performance may thus be optimized for particular EC systems and wastewaters.

b. Investigate the use of other anode materials. Magnesium and zinc, for example, are less toxic than aluminum in waters slated for discharge and may be therefore preferable, should acceptable performance be obtained.

c. Study the use of carbon as the cathode in order to produce peroxide and therefore better oxidize organics.

d. Investigate the performance of iron anode/carbon cathode combinations in relation to Fenton reactions [4, 5]. These could exploit a synergy of oxidizing organics as well as dissolving them during the EC process (Table 1).

$$Fe \rightarrow Fe^{2+} \tag{12}$$

$$Fe^{2+} + H_2O_2 \rightarrow Fe^{3+} + {}^*OH + OH^- \tag{13}$$

Further investigate relationships between the formation of Fe(II) and Fe(III) and their derivatives during the electrocoagulation process. Iron of a higher state of valence should be more optimal for precipitation, as well as produce a more neutral treated water (Fig. 2).

Such investigations may serve to further enhance the successes seen with electrocoagulation

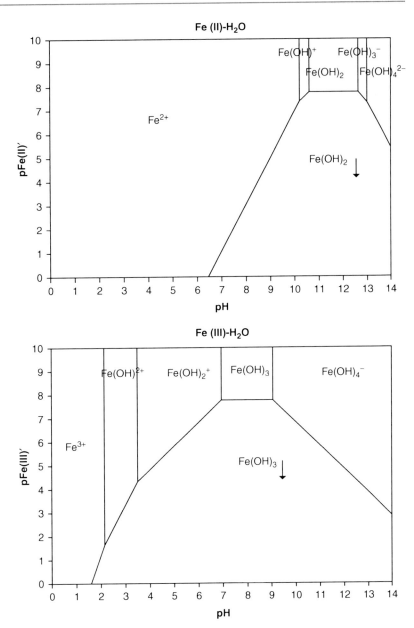

Wastewater Treatment by Electrocoagulation, Fig. 2 Predominance zone diagrams of Fe(II) and Fe(III)

for wastewater treatment, while working toward solving some of the technical issues experienced in the past.

Cross-References

▶ Wastewater Treatment, Electrochemical Design Concepts

References

1. Zhao H, Peng J, Xue A, Ni J (2009) Distribution and transformation of Al species in organic silicate aluminum hybrid coagulants. Compos Sci Technol 69(10):1629
2. Essadki AH, Gourich G, Vial C, Delmas H, Bennajah M (2009) Defluoridation of drinking water by electrocoagulation/electroflotation in a stirred tank reactor with a comparative performance to an external-loop airlift reactor. J Hazard Mater 168:1325

3. Gregor JE, Nokes CJ, Fenton E (1997) Optimising natural organic matter removal from low turbidity waters by controlled pH adjustment of aluminium coagulation. Water Res 31:2949
4. Peralta JM, Mejia S, Godinez LA, Meas Y (2005) Fenton approaches for water purification technologies. In: Palomar M (ed) Applications of analytical chemistry in environmental research. Research Signpost, Kerala
5. Brillas E, Sires I, Oturan MA (2009) Electro-fenton process and related electrochemical technologies Based on fenton's reaction chemistry. Chem Rev 109(12):6570–6631

Wastewater Treatment by Electroflocculation

Vivian Robinson
ETP Semra Pty Ltd, Canterbury, NSW, Australia

Introduction

Electroflocculation is a process wherein flocculating metal ions are electrolytically added to polluted water at an anode, and gas micro bubbles are released at a cathode. The flocculating metal ions adhere to pollutants in the water, increasing their size, and the gas micro bubbles capture the flocculated pollutants and float them to the surface, from where they can be easily removed. By the appropriate choice of electrode materials, this process can remove a wide variety of pollutants without the need for chemicals or filters, with pH adjustment to near neutral at discharge excluded.

The Process

The Electrochemical Reactions
The main electrochemical reactions involved are:

$$\text{Anode 1} \qquad Al - 3e^- \rightarrow Al^{3+} \qquad (1)$$

$$\text{Anode 2} \qquad Fe - 3e^- \rightarrow Fe^{3+} \qquad (2)$$

$$\text{Cathode} \quad 2H_2O + 2e^- \rightarrow 2(OH)^- + H_2 \quad (3)$$

Both the Al and Fe trivalent ions form as proton donors, $Al(H_2O)_6^{3+}$ and $Fe(H_2O)_6^{3+}$, which are only stable at low pH [1–3]. At higher pH, they lose protons to form the monomers $Al(H_2O)_5(OH)^{2+}$, $Al(H_2O)_4(OH)_2^+$, $Al(OH)_3$, and $Al(OH)_4^-$ progressively as the pH increases. Some of these monomers are unstable and try to form the OH-bridged dimer $Al(H_2O)_4(OH)_2(H_2O)_4Al^{4+}$. These unstable monomers react best at a pH in the range of $5 < pH < 7$. This process goes much faster when there are pollutants in the water onto which the dimer can easily form, binding to the pollutant at the same time. The Fe^{3+} reactions are similar. These are the standard chemical trivalent ion flocculation processes used in water treated by both chemical and electrocoagulation processes.

By supplying the ions electrolytically, the cathodic reaction (3) generates hydrogen gas micro bubbles, which capture the flocculated pollutants and float them to the surface. Under the right conditions, this combination of processes captures the pollutants as a stable surface floc layer that is easily separated from the treated water. Reaction (3) also means that the water's pH will increase with greater treatment dosing, and its pH must be pre-adjusted such that it is always near neutral at discharge.

The flocculation reactions described above attach Fe or Al oxides to the pollutants. At near neutral pH, these oxides have very low solubility product constants [4, 5]. This binding to insoluble oxides effectively makes most captured pollutants non-leachable and suitable for landfill. It has been noted that many pollutants captured in this manner also formed acid resistant compounds that passed TCLP toxicity leach tests [6].

Engineering Requirements
For the process to occur, electrodes of the appropriate material must be placed in water with an electric potential applied across the electrodes to cause an electric current to flow between them. The electrochemical reactions occur instantly upon the passage of electric current. Typically, the flocculating process, whereby the monomers form the bridged dimers,

requires many hours to go to completion. This is often referred to as the "incubation" time. In operation, a convenient time must be established between the ultimate pollution removal and the practical completion required for economic operation to the required standard.

For the process to work properly, the water needs to be treated using an anode current density less than 100 A/sq m, with equal cathode area. Above 100 A/sq m, the micro bubbles become too large to produce a stable floc. With "typical" treatment being 100 amp hours per kL for moderately polluted water, that requires at least 1 sq m each of anode and cathode per kL and a current of 100 Amps for a treatment time of 1 h or 10 sq m of anode/cathode for a treatment time of 6 min. These large electrode surface areas mean that low voltages can pass the required current and that large areas of metal are used in electrode fabrication. The two beneficial effects of this are that the voltage required to power the electrochemical reaction is low, typically 2–4 V, hence power consumption is also low, and electrode replacement is an infrequent occurrence, typically several months to yearly intervals. Electrolytic dosing times of less than a few minutes usually imply too great a current density for stable surface floc formation and hence easy pollutant removal.

The water's electrical conductivity has a major impact upon power requirements and process time. The lower the conductivity, the greater the voltage and hence power required, or the longer time needed to process it. Otherwise, conductivity has no detected effect upon the treated water quality, even with salinities over ten times greater than that of seawater. The pollutant type also affects the required treatment dose. Larger pollutants, such as clay, require less treatment per unit weight of pollutant than do smaller pollutants, such as starch. For a given pollutant type, the treatment required is linearly proportional to the amount of pollutant to be removed.

When applied correctly, the result is that the pollutants are captured by the trivalent metal ions, most of which float to the surface, with the heaviest sinking to the bottom of the reaction

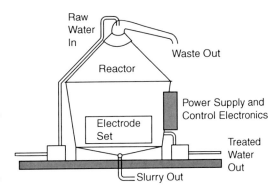

Wastewater Treatment by Electroflocculation, Fig. 1 Schematic illustration of a batch treatment electroflocculation facility

vessel. Those that float to the surface form a stable floc, which can be easily removed. This enables the cleaned water to flow out of the system without the need for other separation techniques, such as filters. In this manner, electroflocculation is like a chemical-free dissolved air flotation (DAF) system or a membrane-free filter. As a filter, it can clean to nanofiltration level, handling some pollutant loadings well over 20,000 mg/L without clogging.

Practical Implementation

There are two separate mechanisms for treating water by electroflocculation: batch and continuous flow. Both systems involve a reaction chamber into which is placed a set of electrodes. In a batch system, a single chamber holds all the electrodes and water, as illustrated in Fig. 1. The water is pumped into the reactor, the current for the electrochemical reactions is passed, and the pollutants float to the surface. In this situation, they are best removed by raising the water level, which forces the floc out the top chute. The water rests in the chamber for a predetermined time before it is pumped out. With batch processes requiring time for the water to be pumped in and out, there is a practical volume limit of about 10 kL per batch, beyond which pump in/out times may be too long to be practical.

Continuous flow uses multiple chambers, as illustrated in Fig. 2. The first two chambers each contain the electrode sets, and the remainder are

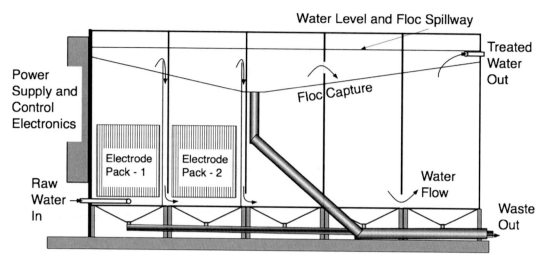

Wastewater Treatment by Electroflocculation, Fig. 2 Schematic illustration of a continuous flow electroflocculation facility

rest chambers. The water is pumped into the first chamber where one electrochemical reaction occurs, flowing out the top into the bottom of the second chamber. There, the second electrochemical reaction occurs, again generating surface floc, before the water exits again at the top. From there it moves under laminar flow through the rest chambers. The low current density and laminar flow ensure that the captured pollutants form a stable floc layer on the surface, from where it is easily removed by simple mechanical means. Different numbers of electrode sets and/or chambers can be used for different applications, depending upon the nature of the pollutant and the desired result.

Continuous flow electroflocculation can be scaled to any desired size. For practical limitations, a 40,000 L system that can typically clean 1 mL of water per day could be transported in a 40 ft shipping container. Larger volumes are handled by multiple systems or large plants built on site.

Electroflocculation water treatment systems need to be automated with microprocessor control. This is required to make sure that changing conditions do not prevent successful operation. It is also required to adequately remove the pollutants, including those that may settle to the bottom of the chambers.

With appropriate microprocessor control, the one set of hardware can handle a wide variety of pollutants, with different pollutants, pollution levels, and desired outcomes being handled by different electronic settings. In practice, if the systems are treating water of uniform quality, they operate for months at a time with only minor maintenance between the electrode changes that are required as a result of reactions (1) and (2) eroding the anodes until there is no longer sufficient metal left to pass current. When raw water quality varies significantly, operator input to adjust the treatment settings is recommended for efficient operation. This could be automated by the use of adequate input water monitoring techniques.

Note that typically about 5 % of the water being treated is removed when the wastes are removed from the electroflocculation facility. With the appropriate posttreatment collection, it is possible to separate the pollutants from the water and return the removed water for further processing if required. For example, emulsified fats, oils, and greases are liberated and float to the surface. They can be skimmed off and sent for recycling. The dried weight of other removed pollutants is typically the dried weight of the pollutant plus an additional 1–10 % due to the added metal oxide. This sinks to the bottom when removed from the electroflocculation facility and

Wastewater Treatment by Electroflocculation

the micro bubbles broken. The pollutants can be separated by appropriate techniques and, in most cases, the waste can be dried and sent to a landfill as a non-leachable product. In this manner, no liquid waste needs to be removed from the site. In many situations, the treated water is either recycled or reused. When recycling, losses due to evaporation are replaced by fresh water, keeping salinity buildup to a minimum.

Results and Environmental Benefits

The following is a list of some contaminants that have been removed by electroflocculation [7]:
- Suspended solids down to the size of large dissolved molecules, including clay, asbestos, and starch
- Organic molecules with a molecular weight greater than ≈ 300 g/mol
- Free and emulsified fats, oils, greases, and hydrocarbons, including benzene ring structures and many halogenated hydrocarbons
- "Sticky" substances, such as glues, polymers, and monomers
- Cations with $Z > 23$ and some lower Z, such as Mg and Ca
- Anions with $Z > 33$ and some with lower Z, such as F
- Chemicals such as cyanide, arsenate, and phosphate
- Pathogens such as algae, bacteria, and viruses
- BOD/COD
- Many dye molecules, tannins, and humic substances
- Soaps and detergents (MBAS)

It should be noted that removal rates are typically >99 % for most pollutants, a figure that requires sufficient treatment dosage being applied, with the treated water resting for a sufficient time. Many BOD and COD pollutants that consist of small organic molecules of molecular weight less than 150 g/Mol, such as C_2H_5OH, CH_3COOH, and similar, are difficult to remove by this process. This broad array of pollutants covers many of those of interest in the treatment of wastewater. In removing these pollutants, electroflocculation liberates emulsified FOGs, separates most suspended solids, and captures the dissolved molecules by forming insoluble metal oxide bonds with them. Low voltages for the electrochemical reactions and low pressure pumping mean that electricity consumption is minimal.

These features give electroflocculation a number of environmental advantages. It has found many applications in the food processing industry, removing FOGs, suspended solids, and BOD. It has also shown itself to be an ideal pretreatment for reverse osmosis (RO), almost irrespective of the original water source. In many mining applications, water salinity prevents wastewater disposal to the environment. Desalination by RO is expensive because pollutants often clog the RO membranes. Electroflocculation removes those pollutants that cause problems for RO membranes, making RO desalination a more viable proposition economically.

Future Directions

A big advantage of electroflocculation is its ability to float pollutants to the surface as an insoluble compound between the soluble pollutant and the flocculating Fe and Al metal ions. This offers a unique opportunity for reducing the pollution in many existing large water bodies, such as dams and lakes. Appropriately designed electroflocculation systems could be floated over parts of the water body, powered by either solar or DC wind generators if required. The low voltages required means minimum power consumption for maximum pollutant removal. Whenever power is activated, the electroflocculation reaction captures the pollutants and floats them to the surface. Note that the rising micro bubbles draw in and circulate water from great distances away from the electrode sets. At the surface, they could be either constrained and removed from the water, to be dried and sent to a landfill, or simply be allowed to spread over the surface, where they will sink to the bottom when the bubbles eventually burst. There they will form an insoluble precipitate that will not leach into the water and will always settle at the bottom, even when stirred up (it is recommended that radioactive wastes, oily contaminants, and highly toxic pollutants be removed rather than retained in the water body).

In this manner, electroflocculation offers an excellent opportunity for environmental restoration by in situ cleaning of large bodies of water without the need to add chemicals or the need to treat and discharge the water. The only by-products will be inert common aluminum and iron oxides that are chemically bound in an insoluble bond to the pollutants that have been either removed or settled to the bottom. With the appropriately designed equipment placed in the large water body, all that would be noticed is that the water slowly became clean and restored to a usable condition.

Note that the minor anode reaction $2H_2O - 4e^- \rightarrow 4H^+ + O_2$ oxygenates the water, making it a more suitable environment for aquatic life to flourish. The residual Al level is the saturated Al level at the water's pH, which if near neutral is very low [4, 5]. Provided the water is not overtreated too much, residual Al levels remain low, avoiding Al toxicity issues.

Cross-References

▶ Wastewater Treatment, Electrochemical Design Concepts

References

1. Phillips CSG, Williams RJP (1966) Inorganic chemistry, vol 1. Oxford University Press, Oxford, Ch 14
2. Phillips CSG, Williams RJP (1966) Inorganic chemistry, vol 2. Oxford University Press, Oxford, p 524
3. Morrison TI, Reis AH, Knapp GS, Fradin FY, Chen H, Klippert E (1978) Extended x-ray absorption fine structure studies of the hydrolytic polymerisation of iron(III). 1. Structural characterisation of the μ-dihydroxo-actaaquodriiron(III) dimer. J Am Chem Soc 100:3262
4. Baes CF Jr, Mesmer RE (1976) The hydrolysis of cations. Wiley Interscience, New York, Ch 6
5. Lide R (ed) (1999) CRC handbook of chemistry and physics. The Chemical Rubber Co. http://www.ktf-split.hr/periodni/en/abc/kpt.html
6. US EPA Report No. EPA/540/R-96/502 (1998)
7. Robinson VNE (1999) AWWA Regional Conference. AWWA, Albury, p 181

Wastewater Treatment by Electrogeneration of Strong Oxidants Using Borondoped Diamond (BDD)

Karine Groenen Serrano
Laboratory of Chemical Engineering, University of Paul Sabatier, Toulouse, France

Introduction

Chemical industries are making significant efforts to limit pollution by the improvement of processes and the development of extensive recycling. In spite of these efforts, numerous wastewaters still contain residual persistent organic pollutants (POPs). The extreme variety of the chemicals involved, their concentration, their flow and their chemical, biological and ecological properties, explains the efforts being made in the research for treatment technologies suited to every situation. In this context, electrochemical technologies represent an interesting alternative. There are two possible approaches: direct or indirect electrochemical oxidation. In the case of direct electrochemical oxidation (DEO), the refractory compound is oxidized in the neighborhood of the electrode. The process is then limited by mass transfer and is suitable for concentrated wastewater. For dilute solutions, another way is to use the electrode surface to generate a strong oxidant stable enough to diffuse through the solution and react chemically with the refractory compound. Moreover, the electrosynthesis of powerful oxidants allows selective reactions to be performed for organic synthesis. Table 1 summarizes the standard potentials of different strong oxidants.

To electrogenerate a strong oxidant, the anode material is a key parameter: it requires high water discharge overpotential to avoid the competitive reaction of oxygen formation.

$$2H_2O \rightarrow O_2 + 4H^+ + 4e \quad (E^\circ = 1.23V/ENH)$$

$$(1)$$

Wastewater Treatment by Electrogeneration of Strong Oxidants Using Borondoped Diamond (BDD), Table 1 List of strong oxidants in aqueous solution

Oxidant	Reaction	E°/V vs NHE	
ClO^-	$Cl^- + 2OH^- \rightarrow ClO^- + H_2O + 2e$	0.84	(1)
$HClO$	$Cl^- + H_2O \rightarrow HClO + H^+ + 2e$	1.48	(2)
MnO_4^-	$Mn^{2+} + 4H_2O \rightarrow MnO_4^- + 8H^+ + 5e$	1.51	(3)
O_3	$3H_2O \rightarrow O_3 + 6H^- + 6e$	1.51	(4)
$C_2O_6^{2-}$	$2HCO_3^- \rightarrow C_2O_6^{2-} + 2H^+ + 2e$	1.80	(5)
$Ag(II)$	$Ag^+ + NO_3^- \rightarrow AgNO_3^+ + e$	1.93	(6)
$S_2O_5^{2-}$	$2SO_4^{2-} \rightarrow S_2O_5^{2-} + 2e$	2.01	(7)
	$2HSO_4^- \rightarrow S_2O_5^{2-} + 2H^+ + 2e$	2.12	(8)
$P_2O_5^{2-}$	$2PO_4^{3-} \rightarrow P_2O_5^{2-} + 2e$	2.07	(9)
FeO_4^{2-}	$Fe^{3+} + 4H_2O \rightarrow FeO_4^{2-} + 8H^+ + 3e$	2.20	(10)

Boron doped diamond (BDD) exhibits properties that distinguish it from conventional electrodes, such as: (i) an extremely wide electrochemical potential window allowing the generation of hydroxyl radicals at the electrode (Eq. 2)

$$H_2O \rightarrow OH^\bullet + H^+ + e \ (E^\circ = 2.74V/ENH) \tag{2}$$

(ii) a very low double layer capacitance and background current, and (iii) corrosion stability in highly aggressive media (acid and basic solutions). Thanks to these properties, BDD has been widely studied these last decades for strong oxidants (peroxo, oxo compounds, metal ions, ozone, and chlorine species) for wastewater treatment [1–3].

Peroxo Compounds

Peroxodisulfate

Persulfuric acid is widely used as an oxidant in many applications. In wastewater treatment, it is used for cyanide removal from effluents. Other applications concern dye oxidation, fiber whitening, radical polymerization, and measurement of total organic compounds. Persulfuric acid is an important intermediate in the electrochemical production of hydrogen peroxide.

Industrial production of peroxodisulfate is achieved by electrolysis of sulfuric acid on Pt anodes. The $S_2O_8^{2-}$ ions are formed from the oxidation of SO_4^{2-} and HSO_4^- species at very high potentials (Table 1, Eqs.7 and 8). In 2000, Michaud et al demonstrated that BDD anodes are suitable to produce peroxodisulfate [4]. Serrano et al discussed the mechanism of formation of peroxodisulfate [5]. Only HSO_4^- and molecular H_2SO_4 react with OH^- radicals to form sulfate radicals, per the following reactions (Eqs. 3–5):

$$HSO_4^- + OH^\bullet \rightarrow SO_4^{-\bullet} + H_2O \tag{3}$$

$$H_2SO_4 + OH^\bullet \rightarrow SO_4^{-\bullet} + H_3O^+ \tag{4}$$

$$SO_4^{-\bullet} + SO_4^{-\bullet} \rightarrow S_2O_8^{2-} \tag{5}$$

As shown in Fig. 1, the maximum current efficiency, close to 100 %, is obtained in a 2M H_2SO_4 solution (Fig. 1):

It is interesting to note that the maximum value of the current efficiency using a BDD anode is higher than those obtained in industrial electrosynthesis on platinum electrodes (about 70–75 %).

Peroxodiphosphate

Peroxophosphates have a wide variety of applications, among them wastewater treatment and treatment of water polluted by petroleum. Due to its similar properties, peroxodiphosphate

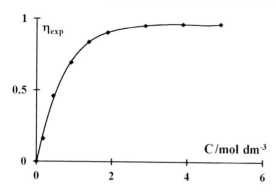

Wastewater Treatment by Electrogeneration of Strong Oxidants Using Borondoped Diamond (BDD), Fig. 1 Plot of current efficiency versus sulfuric acid concentration, C, for electrolysis using a boron-doped diamond electrode. i = 23 mA cm^{-2}, T = 9 °C. [5]

can also be used in other processes as a substitute for peroxodisulfate: it is more environmentally friendly because phosphate can be removed from aqueous wastes more easily than sulfate.

Nowadays, peroxodiphosphoric acid is not produced commercially, and only tetrapotassium peroxophosphate has been produced on a commercial scale. The synthesis of $K_4P_2O_8$ can be carried out on a platinum anode in alkaline conditions, but the current efficiency is low. The addition of certain reagents (fluoride, thiocyanate or nitrate) blocks the oxygen evolution sites of the anode in the process. Consequently, the direct oxidation of phosphate to peroxodiphosphate is favored over the process of water oxidation, and higher current efficiencies (70–75 %) are obtained. However, some of the reagents can be toxic or highly corrosive to platinum. Thus, in order to obtain impurity-free peroxodiphosphate, extensive purification is required. This greatly increases manufacturing costs.

In this context, Canizares et al [6] demonstrated that the production of peroxodiphosphate by electrolysis with conductive diamond in K_3PO_4 under alkaline conditions (pH = 12.5, i = 1,250 A/m^2, T = 25 °C) leads to the synthesis of a high quality product with good efficiency and conversion yield (over 70 %) with high stability. However, they observed that electrolysis at high K_3PO_4 concentration damaged the surface of the diamond electrode, and small corrosion circles appeared. They recommended the use of 1M K_3PO_4.

Moreover, the electrolysis of phosphate solutions at acidic pH leads to the generation of peroxomonophosphate [7]. Nevertheless, the process is less efficient than the generation of peroxodiphosphate in alkaline solutions, due to the lower chemical stability of the peroxomono compound.

Peroxycarbonate

Peroxycarbonate is a biodegradable bleaching agent. Saha et al [8] reported a method for the production of peroxycarbonate on BDD in 1M Na_2CO_3 solution. The current efficiency of the formation of sodium peroxycarbonate was found to depend significantly on the current density, anode material, and electrolyte concentration. The maximum current efficiency for producing sodium peroxycarbonate was 82 % at a current density of 0.05 A cm^{-2} after 30 min of electrolysis in 1 M Na_2CO_3 solution.

Another way to generate peroxycarbonate is to remove CO_2 from the atmosphere by dissolving it in NaOH [9].

$$CO_2 + OH^{-\bullet} \rightarrow HCO_3 - \qquad (6)$$

$$2HCO_3^- + 2OH^{\bullet} \rightarrow C_2O_6^{2-} + 2H_2O \qquad (7)$$

The highest current efficiency of 70 % for producing peroxycarbonate in the anode chamber was found at 5 °C with a current density of 0.05 A cm^{-2} after 30 min of electrolysis.

Oxo Compounds

Ferrate

A range of studies cited by Jiang and Lloyd [10] have shown that various types of organics can be oxidized by ferrate (VI), FeO_4^{2-}, effectively. The electrochemical method is one of the most promising to synthesize ferrate: The iron (or iron based alloy) anode is dissolved and then oxidized to form K_2FeO_4 in a highly concentrated KOH electrolyte (Eq. 8).

$$Fe + 8OH^- \rightarrow FeO_4^{2-} + 4H_2O + 6e \quad (8)$$

Moreover, during the oxidation process, ferrate (VI) ion is reduced to Fe(III) ions or ferric hydroxide, and this simultaneously generates a coagulant in a single dosing and mixing unit process.

The efficiency of the electrochemical route is highly dependent on the nature of the anode material and the composition of the electrolyte. A production yield of 40 % was obtained at 30 °C at 3 mA cm^{-2}. The maximal concentration of ferrate obtainable is limited by the moderate stability of ferrate species and deactivation of the anode surface.

In this context, Rodrigo and coworkers studied the use of diamond-based electrodes to electrolyze Fe(OH)$_3$ solutions in 10M KOH. The current efficiency obtained was very low (<0.8 %). The use of iron-powder bed as iron raw material increases the availability of oxidizable iron species and slightly improves the efficiency of the electrosynthesis with BDD [11]. Lee et al [12] observed that electrochemical generation of ferrate in acidic solution is possible using a BDD electrode. However, the electrochemically generated ferrate undergoes rapid decomposition to produce oxygen and Fe^{3+}.

Permanganate

Mn(III) and Mn(VII) are strong oxidants which have been used for the destruction of organic pollutants, as well as for synthetic or analytical purposes.

The oxidation of Mn^{2+} to MnO$_4^-$ has been investigated with BDD electrodes [13]. Current efficiencies of 37 % were obtained in potentiostatic conditions. In these experimental conditions, a MnO$_2$ film is formed on the electrode surface, which reduces the current efficiency of permanganate production. The presence of Bi(III) in the Mn(II) solution increases the current efficiency (about 10 % more than that obtained without Bi(III)). Bi(III) is oxidized to Bi(V), which acts as an electrocatalyst in MnO$_4^-$ production.

Disinfection

Chlorine Species

Chlorination is the most common method of disinfection, thanks to its low cost. Cl$^-$ can be oxidized to yield active chlorine species (dissolved chlorine, hypochlorous acid HclO, and hypochlorite ions ClO$^-$) with beneficial disinfecting effects (Eqs. 1 and 2) [14]. Chlorine-mediated electrolysis is particularly suitable for the treatment of wastewater with high sodium chloride concentrations, such as olive oil wastewater, and textile and tannery effluents. The main drawback of this indirect oxidation is the likely formation of chlorinated organic byproducts.

The most used electrode materials for *in situ* generation of active chlorine are dimensionally stable anodes based on RuO$_2$ or IrO$_2$ that have good electrocatalytic properties for chlorine evolution, as well as long-term chemical and electrochemical stability.

In 2000, Ferro et al [15] reported that in dilute chloride media and at neutral to weakly alkaline pH, a Faraday yield of about 65 % was found for chlorine production using a BDD anode. The voltammogram suggested that other chlorinated species could be present in solution. Bergmann et al [16, 17], Polcaro et al [18], and Rodrigo et al [19] studied the oxidation products during electrolysis of water containing chloride ions using different experimental conditions and reactor configurations. They highlighted the formation of chlorate and perchlorate (ClO$_3^-$, ClO$_4^-$), which is highly relevant because of the effect of these compoundson human health. It is concluded that the electrolysis of chloride-containing solutions using a BDD anode must be carefully controlled.

Ozone

Ozone is a strong oxidant used for water treatment and disinfection purposes. In many applications, it replaces chlorine; it has the main advantage of avoiding the formation of unwanted chlorinated byproducts. An increasingly important O$_3$ application is its use as a disinfectant in purified water loops for the pharmaceutical and electronic industries, where extreme standards of purification are needed. Today for most applications, ozone is produced by corona discharge, which is based on passing an electric discharge through dry oxygen or air. The low concentrations of O_3 obtained have limited its viability as an oxidant for many potential applications. Thus, the development of electrochemical ozone production (EOP) technology is of interest. The main advantages of EOP are that it is a low dc voltage technology, and ozone can be produced directly in water. During water electrolysis, oxygen evolution is the main rival reaction to ozone production. Thermodynamically, oxygen evolution is strongly favored versus ozone production. Therefore, ozone production is only possible if high oxygen overvoltage anode materials are used (Eq. 4).

A significant contribution to EOP technology was made by Stucki et al [20]. These authors developed an electrochemical ozone generator that evolves ozone directly into a stream of water at a porous PbO_2 anode in contact with a solid polymer electrolyte (SPE). The current efficiency is close to 14 %. In conventional electrochemical reactors, O_3 is produced on BDD with a low current efficiency of a few percent. Kraft et al [21] combined diamond anodes formed on niobium expanded metal substrate and SPE technology; a current efficiency reaching 24 % was achieved with an applied current of 2A and a 40L/h flow rate at 20 °C. In parallel, Arihara et al [22] developed a cell with a perforated conductive diamond plate and a Pt mesh cathode onto a polymer electrolyte membrane. They attained high current efficiency reaching 29 % with an applied current of 1 A and a flow rate of over 40 L/h at 20 °C. This EOP technology is well suited to small productions of ozone (until 50 m^3h^{-1}).

Silver (II) for Nuclear Wastewater

Ag(II) is a strong oxidant which can be used for the treatment of organics in waste and the gasses (NOx, SO_2), as well as the dissolution of plutonium dioxide in the nuclear industry. Ag(II) can be generated electrochemically or chemically. The efficiency of the chemical process is low,

so the electrochemical route is therefore more convenient. The electrochemical process was developed in the 1980s in the nuclear industry. Because silver (II) is unstable in aqueous solution, it is regenerated electrochemically *in situ* in concentrated nitric acid, in the form of $AgNO_3^+$ (Eq. 6). Because of its high redox potential, Ag (II) can chemically attack the refractory compounds, and the reduced form Ag (I) is oxidized in a closed cycle. This electrochemical process is called Mediated Electrochemical Oxidation (MEO) (see Fig. 2). A porous separator or an ion exchange membrane have to be used to prevent parasitic reactions such as Ag(II) reduction and the chemical reaction between Ag(II) and HNO_2 produced at the cathode to reform $AgNO_3$.

Platinum or Ti/Pt are the anode materials generally used because of their high oxygen overpotential and good corrosion resistance in nitric acid. During galvanostatic electrolysis, the oxidant concentration increases with the electrical charge until the concentration becomes stationary. This stationary state results from a dynamic equilibrium between the rate of electroregeneration and the rate of competitive reactions such as hydrolysis of Ag(II). Arnaud et al [23] reported that electrode deactivation occurs on various anode materials, resulting from the slow formation of an unstable layer, probably related to AgO formation.

In this context, Panizza et al [24] studied the potential of the BDD anode for Ag(II) electrogeneration. They obtained an efficiency of 80 % in potentiostatic conditions, avoiding the hydroxyl radical formation. Racaud et al [25] compared the performance of BDD and Ti/Pt in the process in galvanostatic conditions. They highlighted that BDD is a very promising anode; the generation rate and the Faraday yield are on the same order of magnitude as Ti/Pt. Further studies are necessary totest the durability of BDD in nitric acid media.

Future Directions

The BDD anode easily produces active oxidants and thus opens new possibilities in the field of

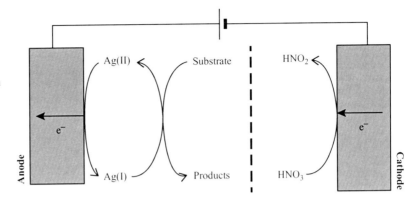

Wastewater Treatment by Electrogeneration of Strong Oxidants Using Borondoped Diamond (BDD), Fig. 2 Mediated Electrochemical Oxidation (*MEO*) in a divided cell. Example of the *Ag(II)* process

wastewater treatment, in particular for wastewater containing biorefractory organic species at low concentrations. Indeed, the ability to form highly reactive hydroxyl radicals allows the transfer of oxygen atoms to sulfate, phosphate or carbonate ions to form peroxocompounds. The process is particularly efficient for the electrosynthesis of persulfate.

The BDD anode has a very important role to play in the disinfection of water. Indeed, percarbonate persulfate and ozone can be generated by electrolysis directly from the water without adding chemicals. Note that the presence of chloride ions can provoke the formation of chloride species, which are dangerous for health.

The BDD anode prepared on a silicon substrate shows high electrochemical and chemical stability in drastic conditions. However, the brittleness of silicon and the size of electrodes hinder the industrial development of the process and the fields of application. BDD had been deposited on metal substrates such as niobium, tantalum and tungsten. The use of refractory metals provides a good electrical conductivity, mechanical strength and electrochemical inertness for easy formation of a protective film on the electrode surface by passivation. Nevertheless, the high cost of Nb, Ta, and W substrates prevents large-scale use. Additional efforts must be made to produce strong, long-lasting electrodes, with a size on the order of a square meter and at a more affordable price.

Cross-References

▶ Boron-Doped Diamond for Green Electro-organic Synthesis
▶ Wastewater Treatment, Electrochemical Design Concepts

References

1. Panizza M, Cerisola G (2005) Application of diamond electrodes to electrochemical processes. Electrochim Acta 51:191–199
2. Kraft A (2007) Doped diamond: A compact review on a new, versatile electrode material. Int J Electrochem Sci 2:355–385
3. Panizza M, Cerisola G (2009) Direct and mediated anodic oxidation of organic pollutants. Chem Rev 109:6541–6569
4. Michaud PA, Mahé E, Haenni W, Perret A, Comninellis C (2000) Preparation of peroxodisulfuric acid using boron–doped diamond thin film electrodes. Electrochem Solid State Lett 3(2):77–79
5. Serrano K, Michaud PA, Comninellis C, Savall A (2002) Electrochemical preparation of peroxodisulfuric acid using boron doped diamond thin film electrodes. Electrochim Acta 48(4):431–436
6. Canizares P, Larrondo F, Lobato J, Rodrigo MA, Saez C (2005) Electrochemical synthesis of peroxodiphosphate using boron-doped diamond anodes. J Electrochem Soc 152(11):D191–D196
7. Weiss E, Saez C, Groenen Serrano K, Canizares P, Savall A, Rodrigo M (2008) Electrochemical synthesis of peroxomonophosphate using boron doped diamond anodes. J Appl Electrochem 38(1):93–100
8. Saha MS, Furuta T, Nishiki Y (2003) Electrochemical synthesis of sodium peroxycarbonate at boron-doped diamond electrodes. Electrochem Solid-State Lett 6(7):D5–D7

9. Saha MS, Furuta T, Nishiki Y (2004) Conversion of carbon dioxide to peroxycarbonate at borondoped diamond electrode. Electrochem Commun 6:201–204
10. Jiang JQ, Lloyd B (2002) Progress in the development and use of ferrate (vi) salt as an oxidant and coagulant for water and wastewater treatment Water Res 36:1397–1408
11. Saez C, Rodrigo MA, Canizares P (2008) Electrosynthesis of ferrates with diamond anodes. AIChE 54(6):1601–1607
12. Lee J, Tryk DA, Fujishima A, Park SM (2002) Electrochemical generation of ferrate in acidic media at boron-doped diamond electrodes. Chem Commun 5:486–487
13. Lee J, Einaga Y, Fujishima A, Park SM (2004) Electrochemical oxidation of Mn2+ on boron-doped diamond electrodes with Bi3+ used as an electron transfer mediator. J Electrochem Soc 151(8): E265–E270
14. Kraft A (2008) Electrochemical water disinfection: a short review. Platin Met Rev 52(3):177–185
15. Ferro S, De Battisti A, Duo I, Comninellis C, Haenni W, Perret A (2000) Chlorine evolution at highly boron-doped diamond electrodes. J Electrochem Soc 147(7):2614–2619
16. Bergmann MEH, Rollin J (2007) Product and by-product formation in disinfection electrolysis of drinking water using boron-doped diamond anodes. Catal Today 124:198–203
17. Bergmann MEH, Rollin J, Iourtchouk T (2009) The occurrence of perchlorate during drinking water electrolysis using BDD electrodes. Electrochim Acta 54:2102–2107
18. Vacca A, Mascia M, Palmas S, Da Pozzo A (2011) Electrochemical treatment of water containing chlorides under non-ideal flow conditions with BDD anodes. J Appl Electrochem 41(9):1087–1097
19. Sanchez-Carretero A, Saez C, Canizares P, Rodrigo MA (2011) Electrochemical production of perchlorates using conductive diamond electrolyses. Chem Eng J 166:710–714
20. Stucki S, Theis G, Kötz R, Devantay H, Christen HJ (1985) In situ production of ozone in water using a membrel electrolyser. J Electrochem Soc 132:367–371
21. Kraft A, Stadelmann M, Wünsche M, Blaschke M (2006) Electrochemical ozone production using diamond anodes and a solid polymer electrolyte. Electrochem Commun 8.883–886
22. Arihara K, Terashima C, Fujishima A (2007) Electrochemical Production of high-concentration ozone-water using freestanding perforated diamond electrodes. J Electrochem Soc 154(4):E71–E75
23. Arnaud O, Eysseric C, Savall A (1999) Optimising the anode material for electrogeneration of Ag(II). Inst Chem Eng Symp Ser 45:229–238
24. Panizza M, Duo I, Michaud PA, Cerisola G, Comninellis C (2000) Electrochemical generation of silver (II) at boron-doped diamond electrodes. Electrochem Solid State Lett 3(12):550–551
25. Racaud Ch, Savall A, Rondet Ph, Bertrand N, Groenen Serrano K (2012) New electrodes for silver (II) electrogeneration: comparison between Ti/Pt, Nb/Pt, and Nb/BDD. Chemical Engineering Journal, (211-212): 53–59

Wastewater Treatment, Electrochemical Design Concepts

José M. Bisang
Programa de Electroquímica Aplicada e Ingeniería Electroquímica (PRELINE), Facultad de Ingeniería Química, Universidad Nacional del Litoral, Santa Fe, Santa Fe, Argentina

Introduction

The increasing requirements of legal limitations for environmental protection demand the development of reliable and cost-effective processes for the treatment of effluents with small concentration of dangerous species. Electrochemistry can offer an interesting proposal for the removal of a contaminant and also its transformation in a product of commercial value. Several flow configurations are possible. One of them is the use of a single-pass electrochemical reactor to process the final effluent. However, the more frequently arrangement is to couple the reactor with a reservoir, where the effluent may be stored. Both configurations are sketched in Fig. 1. Thus, for the practical use of the electrochemical technique it is very important the calculation of the reactor outlet concentration, the specific energy consumption of the process and figures of merit to compare the performance of different electrochemical reactors.

Flow Arrangements

The concentration in the reactor will be reported according to the dispersion model, DM, because

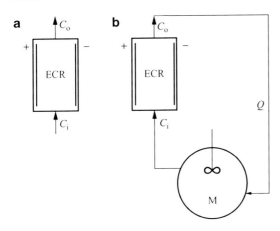

Wastewater Treatment, Electrochemical Design Concepts, Fig. 1 Flow arrangements. (**a**) Single pass. (**b**) Electrochemical reactor, ECR, coupled with a reservoir, M

several electrochemical equipments can be represented by this mathematical treatment and also the simplified models, stirred tank and plug flow, can be obtained as limiting cases.

Single-Pass Through an Electrochemical Reactor

The reactor outlet concentration, C_o, is related to the reactor inlet concentration, C_i, by the following relationship [1]:

$$\frac{C_o}{C_i} = \frac{4\alpha e^{Pe/2}}{(\alpha+1)^2 e^{\alpha Pe/2} - (\alpha-1)^2 e^{-\alpha Pe/2}} \quad (1)$$

where

$$\alpha = \sqrt{1 + \frac{4\beta}{Pe}} \quad (2)$$

and

$$\beta = \frac{k a_e \tau_R}{\varepsilon} \quad (3)$$

The reactor space time, τ_R, is defined as

$$\tau_R = \frac{\varepsilon L}{u} \quad (4)$$

and the Peclet number, Pe, is

$$Pe = \frac{uL}{\varepsilon D_L} \quad (5)$$

a_e being the specific surface area, D_L the dispersion coefficient, L the electrode length, u the mean superficial fluid velocity, and ε the porosity.

The kinetic constant, k, is given by

$$k = \frac{k_f}{1 + Da} \quad (6)$$

k_f is the electrochemical rate constant which is a function of potential and temperature and Da the Damköhler number defined as k_f/k_m, where k_m is the mass-transfer coefficient. When Da = 0 the kinetic expression is truncated to a Tafel equation for a first-order reaction, and high values of the Damköhler number indicate an approach to mass-transfer-controlled rates.

For high values of the Peclet number, Eq. 1 is

$$\left.\frac{C_o}{C_i}\right|_{PF} = e^{-\beta} \quad (7)$$

valid for an electrochemical reactor according to the plug flow model, PF.

Likewise, for Pe → 0 Eq. 1 is simplified to

$$\left.\frac{C_o}{C_i}\right|_{ST} = \frac{1}{1+\beta} \quad (8)$$

Equation 8 corresponds to the stirred tank model, ST, for a single-pass electrochemical reactor.

Electrochemical Reactor Coupled with a Reservoir

The concentration in the reservoir, related to its initial value, in the Laplace plane is given as [2]

$$\bar{C}_i(s) = \frac{1 + (s + \beta R)^{-1}\left[1 - \frac{4\sigma e^{Pe/2}}{(\sigma+1)^2 e^{\sigma Pe/2} - (\sigma-1)^2 e^{-\sigma Pe/2}}\right]}{s + 1 - \frac{4\sigma e^{Pe/2}}{(\sigma+1)^2 e^{\sigma Pe/2} - (\sigma-1)^2 e^{-\sigma Pe/2}}} \quad (9)$$

where s is the Laplace variable, and

$$R = \frac{\tau_M}{\tau_R} \tag{10}$$

where τ_M is the reservoir space time, and

$$\sigma = \sqrt{1 + 4(\beta R + s)(RPe)^{-1}} \tag{11}$$

The inversion of Eq. 9 from the complex s plane to the time plane can be made by a numerical Laplace transform inversion method. However, for high values of R, the dimensionless concentration in the reservoir is given by [3]

$$C_i(T)|_{DM,R\to\infty}$$
$$= \exp\left\{-\left[1 - \frac{4\alpha e^{Pe/2}}{(\alpha+1)^2 e^{\alpha Pe/2} - (\alpha-1)^2 e^{-\alpha Pe/2}}\right]T\right\} \tag{12}$$

being

$$T = \frac{t}{\tau_M} \tag{13}$$

where t is the time. When the Pe number approaches zero, Eq. 12 yields

$$C_i(T)|_{ST,R\to\infty} = e^{-[\beta/(1+\beta)]T} \tag{14}$$

and at high Pe number, Eq. 12 is simplified to

$$C_i(T)|_{PF,R\to\infty} = e^{-(1-e^{-\beta})T} \tag{15}$$

Equations 14 and 15 were reported by Walker and Wragg [4].

For small β values, the exponential function can be expressed as

$$e^\beta \approx 1 + \beta \tag{16}$$

Introducing Eq. 16 into Eq. 15 gives Eq. 14 as a limiting case of Eq. 15.

For $\beta \ll 1$ Eq. 14 is simplified to

$$C_i(T) = e^{-\beta T} \tag{17}$$

Taking into account Eq. 3 and assuming that the reservoir volume is approximately the total volume of electrolyte, Eq. 17 yields

$$C_i(t)|_{BR} = e^{-k a_e t} \tag{18}$$

valid for a batch reactor, BR.

The above equations show that the conversion increases when the dimensionless number β is enlarged. This can be achieved following two strategies:

1. Increase the kinetic constant
2. Increase the specific surface area with the use of three-dimensional electrodes.

The proposed reactors for the effluents treatment are focused on these strategies or a combination of both. In case of mass-transfer-controlled reactions, the increase in the kinetic constant can be achieved by two different actions: the first is the movement of the electrode and the second one consists in the modification of specific hydrodynamic aspects of the reactor.

Concerning to the movement of the electrode can be cited the rotating cylinder electrode [5–8], the electrochemical pump cell in monopolar [9] and bipolar [10] electrical connection, and the multiplate bipolar stack [11].

The modification of hydrodynamic aspects is exploited in the "falling-film cell" [12], where the electrolyte flows as a thin film in the channel between an inclined plane plate and a sheet of expanded metal which work as electrodes. Other proposal is to include turbulence promoters in the interelectrode gap in conventional parallel plate electrochemical reactors [13–16], or the use of expanded metal electrodes immersed in a fluidized bed of small glass beads, called Chemelec cell [17]. Likewise, the Metelec cell [18] incorporates a cylindrical foil cathode concentric arranged around an inner anode, with a helical turbulent electrolyte flow between the electrodes. The electrochemical hydrocyclone cell [19] makes use of the good mass-transfer conditions due to the helical downward accelerated flow in a modified conventional hydrocyclone.

Other strategy to enlarge β is to increase the specific surface area of the electrode with the use

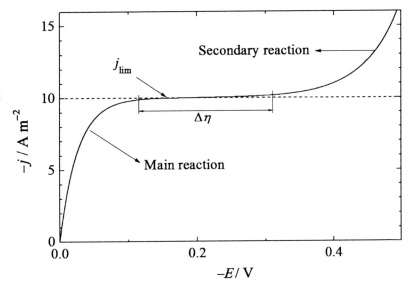

Wastewater Treatment, Electrochemical Design Concepts, Fig. 2 Schematic representation of a microkinetic polarization curve showing the admitted range of overpotential for the main reaction and the secondary reaction

of three-dimensional structures. Several materials were proposed such as steel wool used in the Zadra [20] and IMT cells [21], carbon particles [22] as in the Enviro-Cell or metallic particles [23], woven wire meshes called extended surface electrolysis [24–26], expanded metal meshes [27, 28], metallic foams [29–32], carbon clothes or felts [33–36], and reticulated vitreous carbon [37, 38]. The three-dimensional electrodes are frequently arranged with a packed bed configuration [22]. In case of meshes it was proposed to roll a flexible sandwich of electrodes and separators around a central axis and placing the roll in a container, known as Swiss-roll electrolysis [39]. Likewise, particles in a three-dimensional electrode can be arranged as a fluidized bed [40, 41], a circulating fluidized bed [42], or as circulating bed electrodes [43].

To increase the efficiency for the electrochemical treatment of effluents rotating three-dimensional electrodes were also tested. Thus, a rotating packed bed electrode [44] and a vertically moving particle bed electrode were proposed [45, 46]. Plater barrels were also adapted for the removal of metals [47, 48]. Likewise, rotating cylinder electrodes made of reticulated vitreous carbon [49], expanded metal sheets [50], woven wire meshes [51] and wedge wire screens [52] were also examined.

Optimal Bed Depth in Three-Dimensional Electrodes

The use of three-dimensional electrodes requires that the microkinetic polarization curve of the main reaction, sketched in Fig. 2, shows a potential range of width $\Delta \eta$; within this the current density approaches the limiting current density. Thus, the optimal bed depth in the direction of current flow is introduced as a new important design parameter, for which the whole bed is working under limiting current conditions. This means that at each point the local overpotential lies within the $\Delta \eta$ range, and the full limiting current density is realized.

The following geometries can be identified.

Rectangular Arrangement with a Counter Electrode Placed Opposite the Feeder Side of the Bed Electrode

The optimum bed depth, L_{op}, can be calculated using the equations [53, 54]:

$$L_{op} = \frac{1}{\rho_p} \left[\frac{2(\rho_p + \rho_s)\Delta \eta}{j_{lim} A_s} \right]^{0.5} \text{ for } \rho_p > \rho_s \quad (19)$$

and

$$L_{op} = \frac{1}{\rho_s} \left[\frac{2(\rho_p + \rho_s)\Delta\eta}{j_{lim}A_s} \right]^{0.5} \text{ for } \rho_p < \rho_s \quad (20)$$

where A_s is the surface area per unit volume of electrode, ρ_p is the resistivity of the solid phase, ρ_s the resistivity of the solution phase, and j_{lim} the limiting current density. When $\rho_p = \rho_s$ the largest value of the optimum bed depth is achieved, and Eqs. 19 and 20 give the same result:

$$L_{op} = \left(\frac{4\Delta\eta}{j_{lim}A_s\rho_s} \right)^{0.5} \text{ for } \rho_p = \rho_s \quad (21)$$

The case of a three-dimensional electrode with an isopotential metal phase can be obtained from Eq. 20 as [55]

$$L_{op} = \left(\frac{2\Delta\eta}{j_{lim}A_s\rho_s} \right)^{0.5} \text{ for } \rho_p = 0 \quad (22)$$

Cylindrical Arrangement with Outer Counter Electrode and Inner Current Feeder
In this case is valid [56]

$$\Delta\eta - \frac{j_{lim}A_s\rho_s r_i^2}{2} \left\{ \frac{\lambda^2 - 1}{2} - (\delta\lambda^2 + 1) \right.$$

$$\left. \ln\left[\frac{\lambda}{\left(\frac{\delta\lambda^2+1}{\delta+1} \right)^{0.5}} \right] \right\} = 0 \text{ for } \delta < \frac{2\ln(\lambda) - \lambda^2 + 1}{\lambda^2 - 1 - 2\lambda^2 \ln(\lambda)}$$

$$(23)$$

or

$$\Delta\eta + \frac{j_{lim}A_s\rho_s r_i^2}{2} \left[\frac{\delta(\lambda^2 - 1)}{2} - \frac{\delta\lambda^2 + 1}{2} \right.$$

$$\left. \ln\left(\frac{\delta\lambda^2 + 1}{\delta + 1} \right) \right] = 0 \text{ for } \delta > \frac{2\ln(\lambda) - \lambda^2 + 1}{\lambda^2 - 1 - 2\lambda^2 \ln(\lambda)}$$

$$(24)$$

where

$$\lambda = r_e/r_i \quad (25)$$

being r_e and r_i the external and internal radius of the three-dimensional electrode and

$$\delta = \rho_p/\rho_s \quad (26)$$

For the case of packed bed electrodes, $\rho_p = 0$, isopotential metal phase, and Eq. 23 is simplified to

$$\Delta\eta - \frac{j_{lim}A_s\rho_s r_i^2}{2} \left[\frac{\lambda^2 - 1}{2} - \ln(\lambda) \right] = 0 \quad (27)$$

Cylindrical Arrangement with Inner Counter Electrode and Outer Current Feeder

$$\Delta\eta - \frac{j_{lim}A_s\rho_s r_i^2}{2} \left\{ \delta \frac{\lambda^2 - 1}{2} - (\delta + \lambda^2) \right.$$

$$\left. \ln\left[\frac{\lambda}{\left(\frac{\delta+\lambda^2}{\delta+1} \right)^{0.5}} \right] \right\} = 0 \text{ for } \delta > \frac{2\lambda^2 \ln(\lambda) - \lambda^2 + 1}{\lambda^2 - 1 - 2\ln(\lambda)}$$

$$(28)$$

or [56]

$$\Delta\eta + \frac{j_{lim}A_s\rho_s r_i^2}{2} \left[\frac{\lambda^2 - 1}{2} - \frac{\delta + \lambda^2}{2} \ln\left(\frac{\delta + \lambda^2}{\delta + 1} \right) \right]$$

$$= 0 \text{ for } \delta < \frac{2\lambda^2 \ln(\lambda) - \lambda^2 + 1}{\lambda^2 - 1 - 2\ln(\lambda)}$$

$$(29)$$

For the case of packed bed electrodes, $\rho_p = 0$, isopotential metal phase, and Eq. 29 is simplified to

$$\Delta\eta + \frac{j_{lim}A_s\rho_s r_i^2}{2} \left[\frac{\lambda^2 - 1}{2} - \lambda^2 \ln(\lambda) \right] = 0 \quad (30)$$

Bipolar Rectangular Arrangement with an Isopotential Metal Phase
The optimal bed depth of the cathodic region, L_c, can be calculated as [57]

$$L_c = -\frac{I^*}{SA_s j_{lim}} + \sqrt{\left(\frac{I^*}{SA_s j_{lim}} \right)^2 + \frac{2\Delta\eta}{j_{lim}A_s\rho_s}}$$

$$(31)$$

where S is the separator area and I^* is the leakage current given by

$$I^* = \frac{S}{\rho_s} \frac{dE}{dx}\bigg|_{\text{inactive region}} \quad (32)$$

The evaluation of the leakage current requires the calculation of the potential distribution in the electrode, E. When $I^* = 0$, Eq. 31 is simplified to Eq. 22 valid for a monopolar three-dimensional electrode.

Figures of Merit for an Electrochemical Reactor

The specific energy consumption, E_s, is defined as:

$$E_s = \frac{\text{Required electrical power}}{\text{Produced mass}} = \frac{UnF}{M\phi} \quad (33)$$

where U is the applied cell voltage, n is the number of electrons interchanged, F is the Faraday constant, M is the atomic weight, and ϕ is the current efficiency. The specific energy consumption must be as small as possible.

An appropriate figure of merit to compare the performance of electrochemical reactors for effluents treatment is the normalized space velocity, s_n, which depends on the model assumed for the equipment. Thus, for a continuous stirred tank is given by

$$s_n|_{ST} = \frac{ka_e}{9} = \frac{Qx}{9V_R(1-x)} \quad (34)$$

where Q is the volumetric flow rate, V_R is the reactor volume and x is the conversion, and for the plug flow model is valid:

$$s_n|_{PF} = \frac{ka_e}{2.3026} = -\frac{Q}{V_R} \log(1-x) \quad (35)$$

Future Directions

During the last years, many works were related to the treatment of effluents with heavy metals as contaminants. However, the research can be extended to the processing of gases with the use of two-phase electrochemical reactors, and also the investigation linked to the removal of organic compounds must be intensified. In this last case, the combination of the electrochemical procedure with other disciplines, such as microbiology, catalysis, and photochemistry, can be a good strategy to solve the problem of complex effluents. In these cases new models will be necessary to represent the equipments.

Cross-References

▶ Electrochemical Cell Design for Water Treatment
▶ Electrochemical Cells, Current and Potential Distributions

References

1. Fahidy TZ (1985) Principles of electrochemical reactor analysis, Ch 6. Elsevier, Amsterdam, p 137
2. Colli AN, Bisang JM (2011) Generalized study of the temporal behaviour in recirculating electrochemical reactor systems. Electrochim Acta 58:406–416
3. Mustoe LH, Wragg AA (1978) Concentration-time behaviour in a recirculating electrochemical reactor system using a dispersed plug-flow model. J Appl Electrochem 8:467–472
4. Walker ATS, Wragg AA (1977) The modelling of concentration-time relationships in recirculating electrochemical reactor systems. Electrochim Acta 22:1129–1134
5. Gabe DR (1974) The rotating cylinder electrode. J Appl Electrochem 4:91–108
6. Gabe DR, Walsh FC (1983) The rotating cylinder electrode: a review of development. J Appl Electrochem 13:3–22
7. Gabe DR, Wilcox GD, González-García J, Walsh FC (1998) The rotating cylinder electrode: its continued development and application. J Appl Electrochem 28:759–780
8. John Low CT, Ponce de Leon C, Walsh FC (2005) The rotating cylinder electrode (RCE) and its application to the electrodeposition of metals. Aust J Chem 58:246–262
9. Jansson REW, Ashworth GA (1977) The continuous deposition of metal powders in a pump cell. J Appl Electrochem 7:309–314
10. Jansson REW, Marshall RJ (1978) The rotating electrolyser. II Transport properties and design equations. J Appl Electrochem 8:287–291

11. Marshall RJ, Walsh FC (1985) A review of some recent electrolytic cell designs. Surface Technology 24:45–77
12. Coeuret F, Legrand J (1985) Mass transfer at the electrodes of the "falling-film cell". J Appl Electrochem 15:181–190
13. Hammond JK, Robinson D, Walsh FC (1991) Mass transport studies in filterpress monopolar (FM-type) electrolysers. I. Pilot scale studies in the FM21-SP reactor. Dechema Monogr 123:279–297
14. Brown CJ, Pletcher D, Walsh FC, Hammond JK, Robinson D (1992) Local mass transport effects in the FM01-LC laboratory electrolyser. J Appl Electrochem 22:613–619
15. Brown CJ, Pletcher D, Walsh FC, Hammond JK, Robinson D (1993) Studies of space-averaged mass transport in the FM01-LC laboratory electrolyser. J Appl Electrochem 23:38–43
16. Carlsson L, Sandegren B, Simonsson D, Rihovsky M (1983) Design and performance of a modular, multi-purpose electrochemical reactor. J Electrochem Soc 130:342–346
17. Tyson AG (1984) An electrochemical cell for cadmium recovery and recycling. Plating and Surface Finishing (December):44–47
18. Walsh FC, Wilson G (1986) The electrolytic removal of gold from spent electroplating liquors. Trans Inst Met Finish 64:55–61
19. Dhamo N (1994) An electrochemical hydrocyclone cell for the treatment of dilute solutions: approximate plug-flow model for electrodeposition kinetics. J Appl Electrochem 24:745–750
20. Zadra JB, Engel AL, Heinen HJ (1952) Process for recovering gold and silver from activated carbon by leaching and electrolysis. Report of the Bureau of Mines, RI 4843, 39 pp
21. Elges CH III, Wroblewski MD, Eisela JA (1984) Direct electrowinning of gold. Proc Electrochem Soc 84:501–512
22. Kreysa G (1978) Festbettelektrolyse-ein Verpharen zur Reinigung metallhaltiger Abwässer. Chem Ing Tech 50:332–337
23. Kreysa G (1978) Kinetic behaviour of packed and fluidised bed electrodes. Electrochim Acta 23:1354–1359
24. Williams JM, Olson MC (1976) Extended surface electrolysis for trace metal removal: testing a commercial scale system. AIChE Symposium Series73, pp. 119–131
25. Keating KB, Williams JM (1983) Extended-surface electrolysis. Chem Eng 90:61–62
26. Bisang JM, Bogado F, Rivera MO, Dorbessan OL (2004) Electrochemical removal of arsenic from technical grade phosphoric acid. J Appl Electrochem 34:375–381
27. Leroux F, Coeuret F (1985) Flow-by electrodes of ordered sheets of expanded metal. I. Mass transfer and current distribution. Electrochim Acta 30:159–165
28. Leroux F, Coeuret F (1985) Flow-by electrodes of ordered sheets of expanded metal. I. Potential

distribution for the diffusional regime. Electrochim Acta 30:167–172
29. Langlois S, Coeuret F (1989) Flow-through and flow-by porous electrodes of nickel foam. I. Material characterization. J Appl Electrochim 19:43–50
30. Langlois S, Coeuret F (1989) Flow-through and flow-by porous electrodes of nickel foam. II. Diffusion-convective mass transfer between the electrolyte and the foam. J Appl Electrochim 19:61–60
31. Montillet A, Comiti J, Legrand J (1994) Application of metallic foams in electrochemical reactors of the filter-press type: part II Mass transfer performance. J Appl Electrochim 24:384–389
32. Panizza M, Solisio C, Cerisola G (1999) Electrochemical remediation of copper (II) from an industrial effluent. Part II. Three-dimensional foam electrode. Resour Conserv Recy 27:299–307
33. Abda M, Gavra Z, Oren Y (1991) Removal of chromium from aqueous solutions by treatment with fibrous carbon electrodes: column effects. J Appl Electrochem 21:734–739
34. Vatistas N, Marconi PF, Bartolozzi M (1991) Mass-transfer study of the carbon felt electrode. Electrochim Acta 36:339–343
35. Carta R, Palmas S, Polcaro AM, Tola G (1991) Behaviour of a carbon felt flow by electrodes Part. I: mass transfer characteristics. J Appl Electrochem 21:793–798
36. Polcaro AM, Palmas S (1992) Flow-by porous electrode for removal and recovery of heavy metal pollutants. Inst Chem Eng Symp Ser 127:85–96
37. Konicek MG, Platek G (1983) Reticulate electrode cell removes heavy metals from rinse waters. New Mater New Processes 2:232–235
38. Walsh FC, Pletcher D, Whyte I, Millington JP (1992) Electrolytic removal of cupric ions from dilute liquors using reticulated vitreous carbon cathodes. J Chem Technol Biotechnol 55:147–155
39. Robertson PM, Scholder B, Theis G, Ibl N (1978) Construction and properties of the Swiss-roll electrolysis cell and its application to waste water treatment. Chem Ind (London) 13:459–465
40. van der Heiden G, Raats CMS, Boon HF (1978) Fluidized bed electrolysis for removal or recovery of metals from dilute solutions. Chem Ind (London) 13:465–468
41. Coeuret F (1980) The fluidized bed electrode for the continuous recovery of metals. J Appl Electrochem 10:687–696
42. Goodridge F, Vance CJ (1977) The electrowinning of zinc using a circulating bed electrode. Electrochim Acta 22:1073–1076
43. Scott K (1988) A consideration of circulating bed electrodes for the recovery of metal from dilute solutions. J Appl Electrochem 18:504–510
44. Kreysa G (1983) Elektrochemie mit dreidimensionalen electroden. Chem Ing Tech 55:23–30
45. Bouzek K, Chmelíková R, Paidar M, Bergmann H (2003) Study of mass transfer in a vertically moving

particle bed electrode. J Appl Electrochem 33:205–215
46. Bouzek K, Bergmann H (2003) Mathematical simulation of a vertically moving particle bed electrochemical cell. J Appl Electrochem 33:839–851
47. Zhou CD, Chin DT (1995) Mass transfer and particle motion in a barrel plater. J Electrochem Soc 142:1933–1942
48. Al-Shammari AA, Rahman SU, Chin DT (2004) An oblique rotating barrel electrochemical reactor for removal of copper ions from wastewater. J Appl Electrochem 34:447–453
49. Nahlé AH, Reade GW, Walsh FC (1995) Mass transport to reticulated vitreous carbon cylinder electrodes. J Appl Electrochem 25:450–455
50. Grau JM, Bisang JM (2005) Mass transfer studies at rotating cylinder electrodes of expanded metal. J Appl Electrochem 35:285–291
51. Grau JM, Bisang JM (2006) Mass transfer studies at packed bed rotating cylinder electrodes of woven-wire meshes. J Appl Electrochem 36:759–763
52. Grau JM, Bisang JM (2007) Electrochemical removal of cadmium from dilute aqueous solutions using a rotating cylinder electrode of wedge wire screens. J Appl Electrochem 37:275–282
53. Kreysa G (1983) Modellierung von Fest- und Wirbelbettelektroden. Dechema Monogr 94:123–137
54. G. Kreysa (1987) Cells with three-dimensional electrodes. In: 'Ullmann's encyclopedia of industrial chemistry' A9, VCH Verlagsgesellschaft, Weinheim, pp. 204–209
55. Armstrong RD, Brown OR, Giles RD, Harrison JA (1968) Factors in the design of electrochemical reactors. Nature 219:94
56. Kreysa G, Jüttner K, Bisang JM (1993) Cylindrical three-dimensional electrodes under limiting current conditions. J Appl Electrochem 23:707–714
57. González Pérez O, Bisang JM (2011) Theoretical and experimental study of electrochemical reactors with three-dimensional bipolar electrodes. J Appl Electrochem 40:709–718

Water Treatment by Adsorption on Carbon and Electrochemcial Regeneration

Nigel W. Brown
Daresbury Innovation Centre, Arvia Technology Ltd., Daresbury, UK

Introduction

Society is increasingly demanding that water quality must achieve ever higher standards through the introduction of more stringent legislation. This stems from heightened public awareness, concern over long-term environmental and health effects, and improved analytical techniques that are detecting lower concentrations of chemicals. Some of the most difficult effluents to treat are those containing low and trace quantities of toxic, non-biodegradable, or colored organics.

Water Treatment Using Adsorption

Adsorption is an attractive route for removal of these pollutants, as, not only can very low discharge consents be achieved [1], but unit designs are simple and low cost, with high removal efficiency and availability [2]. However, it is only a concentration process, and, once the adsorbent is fully loaded, it must be disposed of (by landfill or incineration) or regenerated. Activated carbon is the most widely used adsorbent, as high-pollutant loadings are possible. However, the increasing costs of waste disposal have meant that, in order to be economically viable and to satisfy increasingly stringent environmental regulation, complex and costly regeneration is required. Analysis of the whole life costs of water treatment by adsorption indicates that most of the treatment costs are associated with regeneration [3]. Thermal regeneration is the most widely used industrial regeneration process. However, this is a high-energy/high-cost process, often requiring transportation to off-site specialist regenerators, and results in a 5–10 % material loss [1].

Electrochemical Treatment of Water

Direct electrochemical treatment of water to remove dissolved organic contaminants is an area where significant research has been undertaken over many years, with many positive benefits such as the use of a "clean" reagent, the electron [4]; the ease of automation [5]; operation at atmospheric pressures and relatively low temperatures [6]; increased efficiencies provided by the use of compact bipolar electrochemical reactors; and the large surface areas of three-dimensional electrodes [7] and almost instantaneous start-up and shutdown [8].

However, this research has also identified a number of disadvantages in its application, in particular:

- The low conductivity of many wastewaters results in high cell voltages and/or narrow electrode compartments [9].
- It's difficult to treat low concentrations of organic contaminants due to mass transport limitations and side reactions [8, 10].
- A large charge is often required to achieve complete mineralization of organics when the concentration of organics in the effluent is high [11].
- Electrochemical treatment of chloride-containing wastes can result in the generation of more toxic chlorinated organic compounds [9].

These disadvantages have restricted the application of electrochemical water treatment to a relatively small number of niche applications.

Electrochemical Regeneration of Adsorbents

While the treatment of wastewater in electrochemical cells has been the subject of many reports, electrochemical regeneration of adsorbents has not been widely studied. Electrochemical regeneration refers to the regeneration of loaded adsorbent inside an electrolytic cell. The regeneration involves desorption and/or destruction of the adsorbed compounds, restoring the adsorptive capacity. In comparison with electrochemical water treatment, adsorption followed by electrochemical regeneration has, in principle, the potential to address most of the disadvantages discussed above. Mass transport limitations are avoided since contaminants are concentrated on the adsorbent, which is then used as the electrode. During regeneration, conditions can be controlled to ensure high conductivity, and any breakdown products are not produced directly in the treated effluent.

Typically, the performance of the regeneration process is characterized in terms of regeneration efficiency (g), defined as [12]

$$\gamma = 100 \; q_r/q_i \qquad (1)$$

where q_r and q_i are the adsorptive capacities of regenerated and fresh adsorbent under identical adsorption conditions.

The first reference to electrochemical regeneration of activated carbon was reported by *Owen and Barry* in 1972 [13]. By placing a bed of loaded GAC in an electrochemical cell with a sodium chloride electrolyte, they achieved regeneration efficiencies between 37 % and 61 % (compared with virgin activated carbon) when treating a municipal secondary effluent. They found that the electrochemical regeneration of activated carbon resulted in carbon losses of 2–3 % per cycle, significantly less than those occurring in thermal regeneration. They also noted that the production of oxidizing agents within the electrochemical cell would result in sterilization of the adsorbent, controlling the potential problem of bacterial growth. Their cost analysis concluded that electrochemical regeneration of activated carbon is potentially a much less costly method of regenerating carbon compared with thermal regeneration. However, they were unable to identify any relationship between the regeneration efficiency and the operating parameters.

Despite their positive cost analysis, the next reference to the electrochemical regeneration of activated carbon was not until 1980, when *Doniat et al.* [14] proposed this process in their patent. They achieved regeneration efficiencies of 75–85 % (compared with virgin GAC) using granular activated carbon to treat a primary domestic effluent. No further reduction in regeneration efficiency (compared with virgin GAC) was observed during six successive adsorption/regeneration cycles. The regeneration time required was 4–16 h, and they noted that the activated carbon loss per cycle was less than 1 %. Between 1984 and 1991, a number of papers were published in the Soviet Journal of Water Chemistry and Technology [15–18].

Unfortunately, the data presented in these papers is very limited.

In 1994, *Narbaitz and Cen* [12] were the first researchers to report a systematic assessment of the electrochemical regeneration of granular activated carbon. They demonstrated the technical feasibility of regenerating activated carbon after loading with phenol, both cathodically and anodically. Their work was undertaken in an undivided cell using either sodium chloride or sulfate as the electrolyte. Their work suggested that cathodic regeneration resulted in higher regeneration efficiencies, although in this case some phenol was detected in the electrolyte. The regeneration efficiencies were affected by the particle size and the electrolyte type and concentration, but not the phenol loading.

Cathodic regeneration resulted in the initial desorption of the phenol and its subsequent destruction. Desorption was due to the increase in pH around the cathode resulting from the electrochemical treatment. The authors believed that the electrochemical effects would be largely confined to the external surface of the activated carbon. Since most adsorption in activated carbon occurs on its large internal surface area, desorption due to changes in the local environment is the most likely mechanism for regeneration. Data on anodic regeneration presented by *Boudenne and Cerclier* [19] supports this hypothesis, suggesting that electrochemical oxidation occurs only at the external surface. They observed that electrochemical oxidation of adsorbed chlorophenol occurred on a nonporous carbon black, but not on activated carbon. Using their slurry electrode, low contact times occur between the carbon particles and the electrode. This short period would be insufficient for desorption of the organics from the internal pores of the GAC but would allow for electrochemical oxidation from the external surface of the carbon black.

Narbaitz and Cen [12] achieved regeneration efficiencies of up to 95 % during the first regeneration, and subsequent regenerations only reduced the regeneration efficiency by 2 % per cycle with no apparent carbon loss. This result represents an improvement over the regeneration results quoted by *Doniat et al.* [14], who found a more significant fall in regeneration efficiency. However, *Doniat et al.* used anodic oxidation rather than cathodic desorption of the phenol. This result is consistent with data reported in the literature on the effect of chemical oxidation or reduction of activated carbon prior to phenol adsorption [20, 21]. Chemical reduction was found to cause an increase in adsorptive capacity, and chemical oxidation resulted in a reduction.

The findings of *Narbaitz and Cen* [12] have been confirmed by *Zhang et al.* at the Xiamen University [22, 23], who reported that cathodic regeneration is achieved by desorption followed by reaction. They confirmed that activated carbon regeneration efficiency increased with increasing electrolyte (sodium chloride) concentration, regeneration current intensity, and regeneration time. In general, sodium chloride has been used as the electrolyte. However, this could lead to the production of chlorinated hydrocarbons with greater toxicities than the original effluent [24], and sodium sulfate has sometimes been used to avoid this possibility [12].

The available literature on the electrochemical regeneration of activated carbon suggests that recovery of the activated carbon capacity can be achieved by cathodic desorption of the adsorbate followed by electrochemical oxidation at the anode. Oxidation is likely to be achieved by indirect oxidation within the bulk phase.

Electrochemical Regeneration of Nonporous Carbon Adsorbents

An alternative approach is to adsorb onto a nonporous material which would eliminate intra-particle diffusion. If this material had a high electrical conductivity, it would facilitate its electrochemical regeneration. The use of such a material should significantly reduce the time required to achieve both equilibrium and regeneration, at the expense of greatly reduced adsorbent capacity due to the lack of internal surface area.

This is an approach that has been undertaken at the University of Manchester [25–27] using anodic regeneration of a nonporous carbon

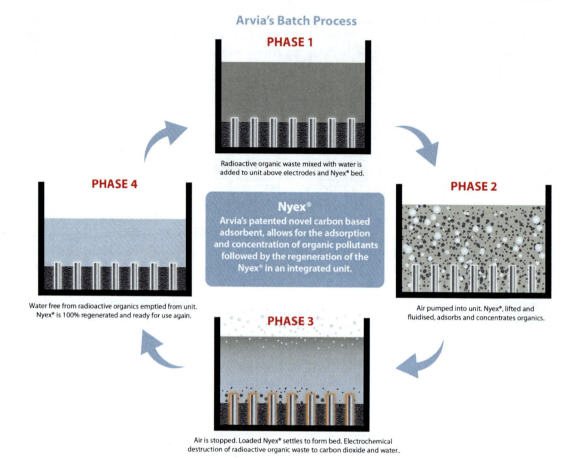

Water Treatment by Adsorption on Carbon and Electrochemcial Regeneration, Fig. 1 Use of the Arvia™ Process in batch operation for the treatment of radioactive organic waste using adsorption coupled with electrochemical regeneration (Source: Arvia Technology Ltd)

material based on an intercalated compound of graphite, developed under the trade name Nyex™. This adsorbent material is highly electrically conducting, dense, and nonporous. This process is now commercialized as the Arvia™ Process (Fig. 1) by Arvia Technology Ltd, a spin-out from the University of Manchester. Electrochemical regeneration of the adsorbent is achieved through the anodic oxidation of adsorbed organic contaminants. Regeneration efficiencies of 100 % over a number of cycles have been obtained by passing a charge of 25 C g^{-1} through a bed of adsorbent particles, at a current density of 20 mA cm^{-2} for 10 min [26]. The energy required for the removal of organics from an industrial wastewater was 27 kWh per kg COD (chemical oxygen demand) removed [27]. The process can be operated in either a batch [28] or continuous [29] mode within a single unit, where adsorption, separation, and regeneration occur either continuously within different zones of the unit or sequentially in batch operation.

Compared to activated carbon, the graphite adsorbent used by the group in Manchester has a very low specific surface area of around 3 m^2 g^{-1} [25]. Consequently, the adsorption capacity of this adsorbent is relatively low. For example, a Nyex™ adsorbent was reported to have a loading capacity for atrazine of only around 0.7 mg g^{-1} [25] and for a crystal violet dye only 2 mg g^{-1} [26]. However, 100 % regeneration efficiencies were achieved after only

10–20 min of anodic treatment [26]. The high conductivity of the adsorbent leads to a relatively low cell voltage, and the energy consumption is low. This rapid and simple electrochemical regeneration opens up the possibility of on-site regeneration, with circulation of the adsorbent between water treatment and regeneration processes, as demonstrated by Mohammed et al. in 2011 [29].

In their earlier studies, the group in Manchester added a sodium chloride electrolyte to the adsorbent during the regeneration process [25–27]. However, recent work has shown that the high conductivity of the adsorbent enables effective regeneration to be achieved without the need to add electrolyte to the anode compartment [29–31].

In addition, Nyex^{TM} has been shown to preferentially remove chlorinated compounds [27]. Hence even if chloride is present in the water and toxic chlorinated compounds are formed, then they are unlikely to be released, as they will be preferentially adsorbed. Analysis of breakdown products in the treated water has suggested that oxidation products are observed in the treated water only from indirect oxidation of unadsorbed organics [30]. Thus, with appropriate design of the treatment system, breakdown products can be eliminated from the treated water.

The electrochemical regeneration of the Nyex^{TM} results in the destruction of the organics through anodic oxidation. While there may be some indirect oxidation within the bulk phase, it is believed that the majority of the oxidation occurs on the surface of the Nyex^{TM} particles which act as the anode. Three separate mechanisms are likely to be involved [30]:

- Direct electron transfer – oxidation of the adsorbed organic on the surface through the transfer of an electron from the organic to the anode.
- Hydroxyl radical oxidation – a number of electrode materials have been shown to oxidize organic contaminants via the generation of hydroxyl radicals. These extremely reactive radicals react at or very close to the surface of the electrode.

- Surface oxidation via oxidized surface species – the generation of oxidized species on the electrode surface (possibly generated through the oxidation of the surface by hydroxyl radicals [32]. While these will oxidize some of the organics in the effluent, their lower oxidation potentials mean that some of the breakdown products (e.g., oxalic and maleic acid) may remain in solution.

While the exact method of oxidation is not fully understood, the observation of the removal of high concentrations of COD (70,000 mg l^{-1}) to below the detection limit [33] suggests that hydroxyl radicals are likely to be involved.

Industrial Applications

Although adsorption onto activated carbon is widely used, to date no industrial applications of electrochemical regeneration of GAC have been installed. However, adsorption coupled with electrochemical regeneration has been demonstrated at an industrial scale by Arvia Technology Ltd at Trawsfynydd nuclear decommissioning site (Fig. 2) [32]. This project investigated the destruction of "orphan" radioactive oils which have no existing disposal option. By creating an emulsion of the oils, the organic component was adsorbed using a Nyex^{TM} adsorbent and oxidized by anodic electrochemical regeneration. The radioactive elements were preferentially released into the water phase, where they were removed in the site's existing active effluent treatment plant. While not strictly for water treatment, this was the first commercial application of adsorption and electrochemical regeneration technology in industry.

Conclusions

Electrochemical regeneration of activated carbon has been shown to be effective, with regeneration efficiencies on the order of 80–95 % being achieved. However, long regeneration periods with relatively low current densities on the order of 1–2 mA cm^{-2} are required. For organic

Water Treatment by Adsorption on Carbon and Electrochemcial Regeneration, Fig. 2 Commercial scale application of the Arvia™ Process at Trawsfynydd nuclear decommissioning site for the removal and destruction of radioactive oils

contaminants on activated carbon, the performance of anodic regeneration of activated carbon has been found to be limited, with better performance obtained using cathodic regeneration. In spite of this, there have been no industrial applications of electrochemical regeneration of activated carbon, presumably because of the high capital and operating costs associated with the low current density and long treatment times required.

In recent years, an alternative, graphite-based adsorbent has been developed, which has a low adsorption capacity but which can be rapidly regenerated by anodic treatment. The principle of on-site regeneration, with adsorbent circulating between water treatment and regeneration processes, has been demonstrated for water treatment with a small prototype.

Future Directions

Industrial application for the treatment of radioactive oils has recently been demonstrated. The potential for large-scale water treatment applications is clear, and the success of the technology will depend on the overall process economics. However, the benefits of on-site, chemical-free treatment, and the relative simplicity of the process, indicate that the prospects are good. Further work is needed to develop reliable, large-scale process equipment and to demonstrate the long-term performance.

Cross-References

▶ Activated Carbons
▶ Electrochemical Cell Design for Water Treatment
▶ Electrochemical Reactor Design for the Oxidation of Organic Pollutants
▶ Organic Pollutants, Direct and Mediated Anodic Oxidation
▶ Organic Pollutants, Direct Electrochemical Oxidation
▶ Organic Pollutants in Water, Direct Electrochemical Oxidation Using PbO_2
▶ Organic Pollutants in Water Using BDD, Direct and Indirect Electrochemical Oxidation
▶ Organic Pollutants, Oxidation on Active and Non-Active Anodes

References

1. McKay G (1996) Use of adsorbents for the removal of pollutants from wastewaters. CRC Press, London/Boca Raton
2. Sanghi R, Bhattacharya B (2002) Review on decolorisation of aqueous dye solutions by low cost adsorbents. Coloration Technol 118:256–269
3. EPA (1989) Technologies for upgrading existing or designing new drinking water treatment facilities. United State, EPA technology transfer report EPA/625/4-89/023
4. Juttner K, Galla U, Schmieder H (2000) Electrochemical approaches to environmental problems in the process industry. Electrochim Acta 45(15–16):2575–2594
5. Torres RA, Sarria V, Torres W, Peringer P, Pulgarin C (2003) Electrochemical treatment of industrial wastewater containing 5-amino-6-methyl-2-benzimidazolone: toward electrochemical-biological coupling. Water Res 37:3118–3124
6. Serikawa RM, Isaka M, Su Q, Usui T, Nishimura T, Sato H, Hamada S (2000) Wet electrolytic oxidation of organic pollutants in wastewater treatment. J Appl Electrochem 30(7):875–883
7. Comninellis C, Nerini A (1995) Anodic oxidation of phenol in the presence of NaCl for wastewater treatment. J Appl Electrochem 25(1):23–28
8. Steele DF, Richardson D, Campbell JD, Graig DR, Quinn JD (1990) The low temperature destruction of organic waste by electrochemical oxidation. Process Safety Environ Prot 68:115–121
9. Murphy OJ, Hitchens GD, Kaba L, Verostko CE (1992) Direct electrochemical oxidation of organics for wastewater treatment. Water Res 26(4):443–451
10. Wang C-T (2003) Decolorization of congo red with three dimensional flow-by packed-bed electrodes. J Environ Sci Health Part A Toxic/Hazard Subst Envirom Eng A38(2):399–413
11. Canizares P, Garcia-Gomez J, Saez C, Rodrigo MA (2003) Electrochemical oxidation of several chlorophenols on diamond electrodes part I. Reaction mechanism. J Appl Electrochem 33:917–927
12. Narbaitz RM, Cen JQ (1994) Electrochemical regeneration of granular activated carbon. Water Res 28(8):1771–1778
13. Owen PH, Barry JP (1972) Electrochemical carbon regeneration. PB 239156, Environics, Huntington Beach, California
14. Doniat D, Corajoud J-M, Mosetti J, Porta A (1980) Process for regenerating contaminated activated carbon. US Patent 4,217,191
15. Slavinskii AS, Velikaya LP, Karimova AM, Baturin AP (1984) Electrochemical regeneration of activated carbon saturated with p-nitrotoluene. Sov J Water Chem Technol 6(6):40–43
16. Sveshnikova DA, Ramazanov AK, Aliev ZM, Shakhnazarov TA, Kamalutdinova IA, Babaev ME (1987) Electrochemical regeneration of activated carbons. Sov J Water Chem Technol 9(5):115–117
17. Sheveleva IV, Khabalov VV (1990) Electrochemical regeneration of unwoven carbon fibre after saturation with phenol. Sov J Water Chem Technol 12(3):23–26
18. Lazareva LP, Lisitskaya IG, Gorchakova NK, Khabalov VV (1991) Investigation of patterns of electrochemical regeneration of carbon sorbents after adsorption of dyes. Sov J Water Chem Technol 13(11):26–29
19. Boudenne JL, Cerclier O (1999) Performance of carbon black-slurry electrodes for 4- chlorophenol oxidation. Water Res 33(2):494–504
20. Coughlin RW, Ezra FS, Tan RN (1968) Influence of chemisorbed oxygen in adsorption onto carbon from aqueous solution. J Colloid Interface Sci 28(3/4):386–396
21. Nevskaia DM, Guerrero-Ruiz A (2001) Comparative study of the adsorption from aqueous solutions and the desorption of phenol and nonylphenol substrates on activated carbons. J Colloid Interface Sci 234(2):316–321
22. Zhang H, Ye L, Zhong H (2002) Regeneration of phenol saturated activated carbon in an electrochemical reactor. J Chem Technol Biotechnol 77:1246–1250
23. Zhang HP (2002) Regeneration of exhausted activated carbon by electrochemical method. Chem Eng J 85(1):81–85
24. Clifford AL, Dong DF, Mumby TA, Rogers DJ (1997) Chemical and electrochemical regeneration of active carbon. US Patent 5,702,587
25. Brown NW, Roberts EPL, Chasiotis A, Cherdron T, Sanjhrajka N (2004) Atrazine removal using adsorption and electrochemical regeneration. Water Res 38:3067–3074
26. Brown NW, Roberts EPL, Garforth AA, Dryfe RAW (2004) Electrochemical regeneration of a carbon based adsorbent loaded with crystal violet dye. Electrochim Acta 49:3269–3281
27. Brown NW, Roberts EPL (2007) Electrochemical pre-treatment of effluents containing chlorinated compounds using an adsorbent. J Appl Electrochem 37:1329–1335
28. Mohammed FM (2011) Modelling and design of water treatment processes using adsorption and electrochemical regeneration. Ph.D. thesis, University of Manchester, Manchester
29. Mohammed FM, Roberts EPL, Hill A, Campen AC, Brown NW (2011) Continuous water treatment by adsorption and electrochemical regeneration. Water Res 45:3065–3074
30. Hussain SN (2011) Water treatment using graphite adsorbents with electrochemical regeneration. Ph.D. thesis, University of Manchester, Manchester
31. Asghar HMA (2011) Development of graphitic adsorbents for water treatment using adsorption and electrochemical regeneration. Ph.D. thesis, University of Manchester, Manchester

32. Brown NW, Campen AK, Robinson P, Eaton D (2011) On-site active oil trial at Magnox Ltd (Trawsfynydd site) of the Arvia^{TM} process. Report no 2010AT11-Final October 2011, Arvia Technology Ltd, Daresbury Innovation Centre, Daresbury
33. Comninellis C (1994) Electrocatalysis in the electrochemical conversion/combustion of organic pollutants for waste-water treatment. Electrochim Acta 39(11–12):1857–1862

Water Treatment with Electrogenerated Fe(VI)

Juan Manuel Peralta-Hernández
Centro de Innovación Aplicada en Tecnologías Competitivas, Guanajuato, Mexico

Description

Water quality regulations around the world are becoming increasingly prevalent and stringent due to increasing social and governmental concern for the environment. One of the greatest issues in water quality involves the handling of effluent that contains soluble organic pollutants that are either toxic or nonbiodegradable [1]. In recent times, a novel process for the treatment of organic pollutants in wastewater has been developed using iron in the +6 oxidation state. The ferrate ion Fe(VI) has been suggested to strongly oxidize and degrade several organic compounds [2–12]. Commonly, ferrate is used in wastewater treatment with various Fe(VI) salts, but the potassium salt (K_2FeO_4) is the most widely studied compound among the family of ferrates. Fe(VI) reduces rapidly and exothermally to Fe^{3+} and oxygen in strong acids, while the oxygen ligands of Fe(VI) exchange very slowly with water at pH 10. The reduction potential of Fe(VI) is 2.2 and 0.7 V in acidic and alkaline media, respectively [13]. Nevertheless, the use of solid ferrate salts has several notable disadvantages. The reaction of the ferrate salt strongly depends on the reagent's initial concentration, solution pH, and temperature. Moreover, the production of ferrate salts requires large quantities of chemicals and

Water Treatment with Electrogenerated Fe(VI), Table 1 Some research that used electrochemically generated Fe(VI) for wastewater treatment

Pollutant	Electrolytic solution	% remove	Ref
2 Industrial liquid wastes	15 M NaOH	COD/gL^{-1} Liquid1: 19.1 Liquid2: 2 9	[17]
Chromium	NaOH (3 and 5 M)	Chromium removed 90.0	[18]
Acidic yellow 36	0.1 M HClO4	C/CO 94.0	[19]

also involves several reaction steps [14]. An interesting alternative in the synthesis of ferrate ion was proposed by Bouzek et al. [6, 15]. This electrochemical method is based on alkaline hydroxide solutions using anodic iron dissolution.

Anode reaction:

$$Fe + 8OH^- \rightarrow FeO_4^{2-} + 4H_2O + 6e^-$$

Cathode reaction:

$$6H_2O \rightarrow 3H_2 + 6OH^- - 6e^-$$

Híveš et al. [16] published results using a similar system where different working electrodes were tested. Fewer reports are available on the electrochemical generation of ferrate ions in acidic conditions.

In light of this being known, several studies have been carried out to degrade organic compounds during wastewater treatment, as can be seen in Table 1.

Lee et al. [20] reported the use of a boron-doped diamond (BDD) working electrode, perchloric acid ($HClO_4$) as a support electrolyte, and ferrous sulfate iron salt to promote Fe(VI) generation according to the following reaction:

Study of the properties of BDD electrodes has increased recently for industrial applications, especially in the area of wastewater treatment. The attractiveness of BDD electrodes stems from their very high overpotential and ability to allow oxygen and hydrogen evolution to occur in a wide potential window [21].

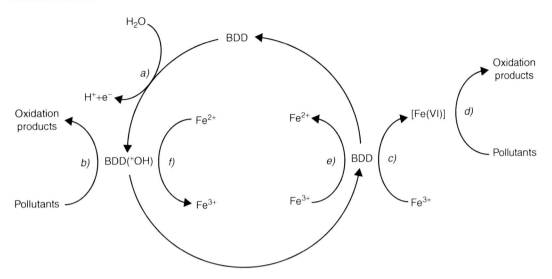

Water Treatment with Electrogenerated Fe(VI), Fig. 1 Mechanism proposed by M. Villanueva-Rodríguez et al. for the electrochemical oxidation of organic compounds with simultaneous ferrate ion formation in the system [19]

The reactivity of Fe(VI), electrochemically generated in acidic conditions over BDD anodes, was recently tested by M. Villanueva-Rodríguez [22], who reported experimental results to carry out the comparison of the effectiveness of ferrate ion Fe(VI) and the typical electrooxidation process in the removal of color from solutions containing the common azo dye, acid yellow 36 (AY-36). Experiments were performed using a batch electrochemical cell operated under two conditions of interest: electrochemical synthesis of ferrate ions Fe(VI) and a common electrooxidation process. In this research, the authors proposed a general mechanism for the electrochemical oxidation of organic compounds with simultaneous ferrate ion generation in the system, as shown in Fig. 1: (a) water discharge to hydroxyl radicals, (b) oxidation of organic compounds by means of ˙OH, (c) formation of ferrate ions, (d) oxidation of organic compounds by ferrate ions, (e) electrochemical regeneration of Fe^{3+} to Fe^{2+}, and (f) scavenger action of Fe^{2+} excess over hydroxyl radicals.

Future Directions

Important development has been made on understanding the chemistry of ferrate ion, many authors now has ben publish papers related to this specie, however the next step is the understand the real mechanism the attack of ferrate ion to organic compounds for degradation.

Cross-References

▶ Wastewater Treatment, Electrochemical Design Concepts

References

1. Alsheyab M, Jiang J-Q, Stanford C (2010) Electrochemical generation of ferrate (VI): determination of optimum conditions. Desalination 254:175–178
2. Sánchez-Carretero A, Rodrigo MA, Cañizares P, Sáez C (2010) Electrochemical synthesis of ferrate in presence ultrasound using boron doped diamond anodes. Electrochem Commun 12:644–646
3. Jeannot C, Malaman B, Gérardin R, Oulladiaf B (2002) Synthesis, crystal and magnetic structures of the sodium ferrate (IV) Na_4FeO_4 studied by neutron diffraction and Mössbauer techniques. J Solid State Chem 165:266–277
4. Thompson GW, Ockerman LT, Schreyer JM (1951) Preparation and purification of potassium ferrate VI. J Am Chem Soc 73:1379–1381
5. Williams DH, Rile JT (1974) Preparation and alcohol oxidation studies of the ferrate(VI) ion, FeO_4^{2-}. Inorg Chim Acta 8:177–183

6. Bouzek K, Rousar I, Taylor MA (1996) Influence of anode material on current yield during ferrate(VI) production by anodic iron dissolution part II: current efficiency during anodic dissolution of white cast iron to ferrate (v0 in concentrated alkali hydroxide solutions. J Appl Electrochem 26:925–931
7. Bouzek K, Schmidt MJ, Wragg AA (1999) Influence of anode material composition on the stability of electrochemically prepared ferrate (VI) solutions. J Chem Technol Biotechnol 74:1188–1194
8. Alsheyab M, Jiang J-Q, Stanford C (2009) Electrochemical production of ferrate and its potential on-line application for wastewater treatment. A review. J Environ Manage 90:1350–1356
9. Sharma VK, Burnett CR, O'Connor DB, Cabelli D (2002) Iron(VI) and iron(V) oxidation of thiocyanate. Environ Sci Tech 36:4182–4186
10. Read JF, Bewick SA, Graves CR, MacPherson JM, Salah JC, Theriault A, Wyand AEH (2000) The kinetics and mechanism of the oxidation of s-methyl-l-cysteine, l-cysteine and l-cysteine by potassium ferrate. Inorg Chim Acta 303:244–255
11. Ma J, Liu W (2002) Effectiveness and mechanism of potassium ferrate(VI) preoxidation for algae removal by coagulation. Water Res 36:871–878
12. Sharma VK, Mishra SK, Nesnas N (2006) Oxidation of sulfonamide antimicrobials by ferrate(VI) $[Fe^{VI}O_4^{2-}]$. Environ Sci Technol 40:7222–7227
13. Noorhasan NN, Sharma VK (2008) Kinetics of the reaction of aqueous iron(VI) $(Fe^{VI}O_4^{2-})$ with ethylenediaminetetraacetic acid. Dalton Trans 14:1883–1887
14. Lee Y, Cho M, Kim JY, Yoon J (2004) Chemistry of ferrate (Fe(VI)) in aqueous solution and its applications as a green chemical. J Ind Eng Chem 10:161–171
15. Bouzek K, Rousar I (1996) Influence of anode material on current yields during ferrate(vi) production by anodic iron dissolution part I: current efficiency during anodic dissolution of grey cast iron to ferrate (vi) in concentrated alkali hydroxide solutions. J Appl Electrochem 26:919–923
16. Híveš J, Benová M, Bouzek K, Sharma VK (2006) Electrochemical formation of ferrate(VI) in a molten NaOH–KOH system. Electrochem Commun 8:1737–1740
17. Lescuras-Darrou V, Lapicque F, Valentin G (2002) Electrochemical ferrate generation for waste water treatment using cast irons with high silicon contents. J Appl Electrochem 32:57–63
18. Sylvester P et al. (2001) Ferrate treatment for removing chromium from high-level radioactive tank waste. Environ Sci Technol 35:216–221
19. Villanueva-Rodríguez M, Hernández-Ramírez A et al. (2009) Enhancing the electrochemical oxidation of acid-yellow 36 azo dye using boron-doped diamond electrodes by addition of ferrous ion. J Hazard Mater 167:1226–1230
20. Lee J, Tryk DA, Fujishima A, Park S-M (2002) Electrochemical generation of ferrate in acidic media at boron-doped diamond electrodes. Chem Commun 5:486–487
21. Cruz-González K, Torres-López O, García-León A, Guzmán-Mar JL, Reyes LH, Hernández-Ramírez A, Peralta-Hernández JM (2010) Determination of optimum operating parameters for Acid Yellow 36 decolorization by electro-Fenton process using BDD cathode. Chemical Engineering Journal 160:199–206
22. Villanueva-Rodríguez M, Bandala ER, Quiroz MA, Guzmán-Mar JL, Peralta-Hernández JM, Hernández-Ramírez A (2011) Azo dyes degradation by electrogenerated ferrate ion using BDD electrodes. Sustain Environ Res 21(5):337–340.

Water Vapor Electrolysis

Ulrich Guth
Kurt-Schwabe-Institut für Mess- und Sensortechnik e.V. Meinsberg, Waldheim, Germany
FB Chemie und Lebensmittelchemie, Technische Universität Dresden, Dresden, Germany

Hydrogen as an Energy Carrier

Hydrogen plays an important role in the so-called hydrogen economy (technology). The term expresses an energy concept in which hydrogen serves as energy storage and fuel for combustion in engines or fuel cells. Hydrogen can be generated by a steam reforming process or by electrolysis based on renewable electrical energy. It can be stored as a gas in high pressure gas cylinders, in submontane caverns, liquefied in cryogen tanks, and can be distributed by pipelines like natural gas. This concept, that is an alternative to the current generation and use of electric energy, can be applied only if the generation of hydrogen is cheap and its handling is safe. The idea of hydrogen technology was born in the beginning of the 1970s of the last century due to the oil price shock and was widely pursued in science and technology [1, 2]. As early as 1969, based on the experiences of SOFC developments, it was proposed to produce hydrogen by water vapor electrolysis at high temperature

Theoretical Background

Well known and widely used is the electrolysis of water in aqueous solution at normal temperatures. The water vapor electrolysis cell can be understood as a reverse fuel cell. From thermodynamic considerations this electrolysis has the advantage that the Gibbs energy of reaction $\Delta_r G^0$ and hence the equilibrium voltage U_{eq} (emf), which is nearly equal to the decomposition voltage, decreases with increasing temperature (Table 1). The decomposition voltage depends only slightly on pressure.

The high temperature electrolysis cells produce pure gases without electrolyte aerosols (KOH) at high temperatures and can be combined in one unit with a high temperature solid oxide fuel cell (SOFC) in order to utilize completely fuel and heat.

The volume of hydrogen which can be generated by electrolysis is given by Faraday's law in consideration of the ideal gas equation:

$$It = n z F = \frac{pv}{RT} zF = \frac{zF}{V} v \tag{1}$$

where I is the current, t the time, n the amount of substances, z the number of electrons, F the Faraday constant, v the volume and V the molar volume. For a flowing system all values are time dependent. The current I which flows according to Ohm's law is inversely proportional to the cell resistance R_{cell}:

$$I = \frac{U - U_{eq}}{R_{cell}} \tag{2}$$

The voltage to be applied is the difference between electrolysis voltage U and equilibrium voltage $U_{eq} = -E$. R_{cell} includes all ohmic and non-ohmic resistances. The composition of the water-hydrogen gas changes during the electrolysis and is dependent on electrode length. The cell cannot be extended to any length because of the IR drop across the electronic conducting electrode material. Therefore, it is necessary to connect single cells in series. The choice of the suitable length of a single cell is an optimization problem. According to an optimization approach by Archer [7], who first described this problem, the resistance of a tube cell can be calculated

$$
\begin{aligned}
R_{cell} &= \frac{\sigma_{se} d_{se} + r_{pk} + r_{pa}}{A} \varphi + R_0 \\
&= \frac{r}{A} \varphi + R_0
\end{aligned} \tag{3}
$$

where A is the electrode area and R_0 the resistance of the connecting part between two single cells. φ is a factor which expresses how the resistance increases due to IR drop in the electrode material lengthwise of the cell.

$$\varphi = lB \left(\frac{1 + \cosh\ lB}{2 \sinh\ lB} + lB/4 \right) \tag{4}$$

$$B = \frac{\rho_a / d_a + \rho_c / d_c}{\sigma_{se} d_{se} + r_{pa} + r_{pc}} \tag{5}$$

B has the dimension of $[l/\text{length}]$ and is the ratio of resistance-layer thickness parameter ρ/d of anode and cathode, respectively, and the sum of solid electrolyte resistance and polarization resistances. R_{cell} decreases with increasing cell length up to a minimum and then increases slightly. By minimization of the cell resistance of a single cell a high hydrogen flux per cell is obtained and hence a low number of single cells for a plant at a given hydrogen production. The power density, the hydrogen flux per area as a measure for the utilization of electrolyte, and electrode materials have also to be taken into account. The power density is inversely proportional to $R_{cell}A$. This value is zero at cell length $l = 0$. If both the number of single cells and the cell connections have the same importance in fabrication, then the expression $R_{cell}^2 A$ should have a minimum [7, 8]. Up to now there are

(800–1,000 °C) by means of solid electrolyte cells [3, 4]. It was also known that oxygen can be pumped into a gas or out of a gas using solid electrolyte cells [5, 6].

W 2150

Water Vapor Electrolysis, Table 1 Thermodynamic standard values for the hydrogen, oxygen, and water equilibrium

$H_2 + \frac{1}{2} O_2 \rightleftharpoons H_2O\ (g)$			
Temperature T/°C	$\Delta_r H^0 / kJmol^{-1}$	$\Delta_r G^0 / kJmol^{-1}$	$|U_{eq}|/V = |E|/V$
25	−241.8	−228.6	1.18
100	−242.6	−225.2	1.17
500	−246.2	−203.52	1.05
1,000		−185.33	0.96

different models in SOFC branch to optimize the performance of cells, cell stacks, and cell bundles respecting materials, geometry, dimensions, temperature, fuel, etc. [9].

Provided the values are independent on the cell length, then the volume flux of hydrogen results in:

$$\frac{v_{H_2}}{t} = \frac{VI}{zF} = \frac{V(U - U_{eq})}{zFR_{cell}}$$
$$= \frac{RT}{zF} \frac{1}{p} \frac{(U - U_{eq})}{R_{cell}} \qquad (6)$$

The equilibrium voltage U_{eq} can be calculated according to the composition of the gas in terms of partial pressure p_{H_2O} and p_{H_2} using the potential function for the H_2O,H_2-electrode ($\rightarrow$ *Solid Electrolyte Cells*):

$$-E/mV = U_{eq}/mV = -1290.6$$
$$+ \left[0.2924 - 0.0992 \log \left(\frac{p_{H_2O}}{p_{H_2}} \right) \right] T/K \qquad (5)$$

Modes of Operation

The high temperature electrolysis can be performed in two different modes: cells with oxygen anode and cells with depolarized anodes. Both are possible with current SOFC cell arrangements.

Oxygen Anode

In this mode pure oxygen and pure hydrogen are generated according to the electrode reaction (*Kröger-Vinks Notation of Point Defects* is used) [8]:

$$H_2O\ (g) + V_O^{\circ\circ}(se) + 2e^- \text{(electrode material)} \rightleftharpoons H_2\ (g)$$
$$+ O_O(se) \text{ cathode reaction}$$

$$O_O(se) \rightleftharpoons 1/2\ O_2(g) + 2e^- \text{(electrode material)}$$
$$+ V_O^{\circ\circ}\ (se) \text{ anode reaction}$$

$$H_2O\ (g) \rightleftharpoons 1/2\ O_2(g) + H_2(g) \text{ cell reaction}$$

For the H_2O,H_2-electrode cheap nickel (nickel cermets, i.e., Ni/YSZ) can be used as electrode material, whereas for the oxygen anode a redox stable material like $La_{0.8}Sr_{0.2}MnO_{3-\delta}$ (LSGM) has to be applied. The advantage to produce pure gases is to balance to the drawback of high electrolysis voltage and hence a high energy consumption. The demands on the electrode material concerning the chemical stability against solid electrolyte (se), thermal expansion coefficient similar to solid electrolyte, and high electronic conductivity are the same as for solid oxide fuel cells (SOFC). All technologies which were developed for planar and tube-shaped SOFC like screen printing, tape casting, sputtering, dip coating, and spray pyrolysis can be utilized for electrolysis cells as well [10]. As a solid electrolyte yttria stabilized zirconia (YSZ), which is commonly used in SOFC and well investigated concerning aging effects, can be used. On the other hand, other zirconia-based electrolytes such as scandia- (0.3 S/cm at 1,000 °C) and ytterbia- doped zirconia can be taken into account due to the higher ionic conductivity as compared

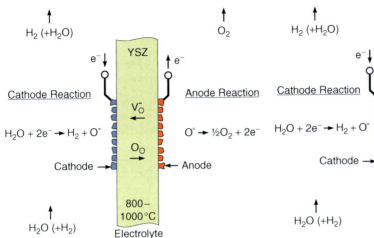
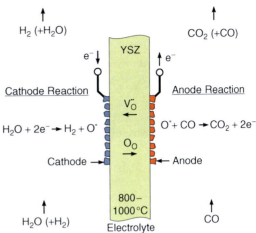

Water Vapor Electrolysis, Fig. 1 Water vapor electrolysis with an oxygen anode

Water Vapor Electrolysis, Fig. 2 Water vapor electrolysis with depolarized anode

with that of YSZ (0.13 S/cm). ZrO_2 doped with 8 mol% scandia is attractive for intermediate temperature (600–800 °C) because of long-term stability which whereas it shows aging effects at 1,000 °C. Figure 1 presents a schema of an electrolysis cell with oxygen anode.

Depolarized Anode

In this mode the anode gas is a reducing fuel gas (cheap flue gas which contains CO and other combustibles). The electrode reactions can be written as follows:

$$H_2O\,(g) + V_O^{\circ\circ}(se) + 2e^-(\text{electrode material}) \rightleftarrows H_2\,(g)$$
$$+\, O_O(se)\;\text{cathode reaction}$$

$$CO\,(g) + O_O(se) \rightleftarrows CO_2(g) + 2e^-(\text{electrode material})$$
$$+\, V_O^{\circ\circ}(se)\;\text{anode reaction}$$

$$H_2O\,(g) + CO\,(g) \rightleftarrows CO_2(g)$$
$$+\, H_2(g)\;\text{cell }reaction$$

That is the water gas shift reaction. The decomposition voltage is remarkably smaller as for cells with oxygen anode and tends to about zero. The applied voltage serves to overcome the IR drops in electrolyte and electrodes and the polarization. Nickel (or nickel cermets) may also be used as electrode materials for the anode. In Fig. 2 schematic view of a cell with fuel gas (depolarized) anode is shown.

Whether the high temperature electrolysis is applied in praxis and in which mode depends on terms of investments and amortization and above all on the marketing condition.

Cross-References

▶ Solid Electrolytes Cells, Electrochemical Cells with Solid Electrolytes in Equilibrium

References

1. Anonymous (1972) Hydrogen: likely fuel of the future. Chem Eng News 50(26):14–17, 50(27):16–18, 50(28):17–19
2. Bockris JOM (1974) Hydrogen economy. Science 176:1323
3. Spacil HS, Tedmon CS Jr (1969) Electrochemical dissociation of water vapor in solid oxide electrolyte cells. I. Thermodynamics and cell characteristics. J Electrochem Soc 116:1618–1626
4. Spacil HS, Tedmon CS Jr (1969) Electrochemical dissociation of water vapor in solid oxide electrolyte cells. II. Materials, fabrication, and properties. J Electrochem Soc 116:1627–1633

5. Yuan D, Kröger FA (1969) Stabilized Zirconia as an oxygen pump. J Electrochem Soc 116:594–600
6. Alcock CB, Zador S (1972) Electrolytic removal of oxygen from gases by means of solid electrolyte. J Appl Electrochem 189(2):289–299
7. Archer DA (1963) Technical Document Report AD 277224 and AD 412789 (ASD-TDR −63-448)
8. Guth U, Möbius HH (1985) Hydrogen production by means of high temperature electrolysis of water vapour in solid electrolyte cells (in German). Wiss Z Univ Greifswald 34:4–8
9. Khaleel MA, Selman JR (2003) Cell, stack and system modelling. In: Singhal CS, Kendall K (eds) High temperature solid oxide fuel cells: fundamentals, design and applications. Elsevier, Oxford, pp 291–331
10. Singhal CS, Kendall K (eds) (2003) High temperature solid oxide fuel cells: fundamentals, desingn and applications. Elsevier, Oxford

X

X-Ray Absorption and Scattering Methods

Dirk Lützenkirchen-Hecht
Fachbereich C- Abteilung Physik,
Wuppertal, Germany

Introduction

Studies of the electronic and the atomic structure of electrodes at the surface and in the bulk material are of fundamental importance e.g., in corrosion research, battery technology, electrocatalysis or electrodeposition. Therefore, a large variety of different in situ and ex situ techniques have been successfully introduced to the investigation of electrochemical interfaces in the past. Due to possible changes of the electrode structure related to the loss of potential control after a removal of the electrode from the electrolyte, in-situ techniques are in general favored compared to ex situ methods such as photoelectron or Auger-electron spectroscopies and electron diffraction despite many successful studies using these vacuum based approaches (see, e.g. [1]). In contrast, especially photon-in photon-out techniques using X-rays appear very promising due to the relatively weak interactions of hard X-rays and matter – this way the electrochemical reactions are not influenced by the X-ray probe while useful information about the state of the sample can be achieved. In this contribution, we will focus on X-ray scattering techniques and X-ray spectroscopic measurements.

X-Ray Diffraction (XRD)

The physical basis for X-ray diffraction is Braggs law, i.e.

$$n\lambda = 2d \cos \Theta \qquad (1)$$

where λ is the X-ray wavelength, d a lattice spacing in the crystalline sample of interest and Θ the scattering angle between the impinging radiation and the considered lattice planes. A constructive interference can thus only occur, if the optical path difference between the X-rays which are diffracted from neighboring lattice planes is equal to an integer multiple of the X-ray wavelength λ. Due to the distribution of the lattice atoms in the unit cell, each crystal structure is represented by a set of characteristic lattice planes, and thus the resulting diffraction pattern is representative for the considered crystal structure. In more detail, structure factors F_{hkl} which include the positions of the atoms in the crystal structure under investigation as well as the occupation of these positions by different types of atoms can be calculated according to

$$F_{hkl} = \sum f_n \exp(2\pi i \mathbf{Q} * \mathbf{R_n}) \qquad (2)$$

with $\mathbf{R_n}$ the lattice vector of the nth atom and $\mathbf{Q}$ the reciprocal lattice vector which represents scattering processes at the hkl-lattice planes [2]. The summation includes all the atoms in the unit cell, and f_n is the atom form factor, which represents the scattering power of an individual atom.

G. Kreysa et al. (eds.), *Encyclopedia of Applied Electrochemistry*, DOI 10.1007/978-1-4419-6996-5,
© Springer Science+Business Media New York 2014

Basically, f_n is determined by the number and the distribution of the electrons belonging to a single atom [2]. Values for f_n are tabulated e.g. in Ref. [3]. The diffracted intensity then is given by $I \sim |F_{hkl}|^2$, thus not only the positions of the diffraction peaks are representative for the crystal structure, but also their intensities. This is in principle the basis for extremely detailed in-situ investigations of electrode materials in electrochemical environments.

As an example, the structure of the passive layers on iron single crystals formed in slightly alkaline solution has been studied using synchrotron radiation X-ray diffraction [4]. The structure of those passive layers has been controversially discussed for decades. While electrochemical experiments as well as ex-situ surface analytical experiments using electron diffraction suggest that the film consists of a duplex structure with an inner layer of Fe_3O_4 and an outer layer of Fe_2O_3 (maghemite) [5–7], other studies using e.g. Raman, Mößbauer and X-ray absorption spectroscopies lead to the conclusion that the passive film is amorphous or has an extremely small grain size [8–12]. According to these experiments, the presence of structures similar to those of disordered iron hydroxides or oxyhydroxides seems to be likely, while a spinel-type structure was deduced from scanning tunnelling microscopy studies [13]. First of all, the existence of well defined diffraction peaks in the XRD experiments clearly evidenced the presence of a well defined crystalline structure, and a detailed analysis of a large number of indexed, symmetrically inequivalent diffraction peaks in terms of their structure factors enabled the structure identification. In contrast to the phases described above, a spinel structure with randomly distributed vacancies on octahedral and tetrahedral Fe lattice sites and octahedral interstitial sites matches best the experimental data (for more details see [4]).

Diffraction techniques are of key importance for the understanding of the functionality and the improvement of battery materials. In recent years, numerous studies have appeared making use of the feasibility to determine the electrode structure during battery operation using X-ray or neutron diffraction. Here we will consider an application of Braggs law using energy dispersive X-ray diffraction: While monochromatic X-rays are used in the conventional setup and the Bragg angle is varied according to Eq. 1, a polychromatic X-ray beam combined with a fixed geometry is used in the case of energy dispersive diffraction, i.e. the X-ray wavelength $\lambda = hc/E$ (h, Planck´s constant; c, speed of light; E, photon energy) is varied instead of the Bragg angle [14]. Here we will consider the lithium-titanium-spinel compound $Li[Li_{1/3}Ti_{5/3}]O_4$, the main feature of which are the minimal variations of the cubic unit cell that accompany lithium insertion and extraction, coining the expression of a zero strain material [15]. This property is of fundamental impact in any battery application, because lattice strain is one of the main causes of electrode capacity fading [16]. In-situ energy dispersive X-ray diffraction clearly demonstrates that the positions of the Bragg peaks do not shift substantially during charge/discharge of the spinel compound, suggesting that the lithium insertion processes do not induce any phase transitions [17]. According to the high-quality diffraction data, only small variations (<1 %) of the lattice parameters are resulting [17]. Energy dispersive diffraction experiments directly profit from the high intensity and spectral characteristics of synchrotron radiation that are ideal for in situ studies, making even time-resolved experiments during the charging/discharging of batteries feasible [17, 18].

Coming back to the detailed analysis of diffraction patterns, we note that such efforts can be in practice more complicated for real samples for different reasons. First of all, the crystallites (grains) inside a polycrystalline sample might have a preferred orientation (texture), and accordingly, the Bragg reflexes of all other orientations are extremely suppressed in their intensity compared to those expected from calculated structure factors. Such a behavior can be expected e.g. in the case of epitaxially grown thin films that adopt the structure or at least the orientation of the substrate. This is observed e.g. for the passive films on iron discussed above [4], and in part also for those on Ni [19] but also for electrodeposited metal films.

Furthermore, the presence of defects in a crystal lattice may also alter the diffraction pattern: Depending on the type and the concentration of the defects, systematic peak broadening, peak shifts as well as peak splitting may be observed, and stress and strain may also influence the diffractograms [20, 21]. Thus the detailed analysis of measured peak positions, their widths and intensities can be used for the identification of the defects existing in a particular sample.

When finite-size effects are included in the diffraction peak evaluation, the peak widths are found to be inversely related to the dimension of the diffracting regions of the crystal; thus, the line broadening provide information about the particle size D of small crystallites, for example by application of the Scherrer formula:

$$D = K\lambda/(\Delta\Theta \cos\Theta) \qquad (3)$$

where K is the Scherrer constant, which depends on the shape of the crystallites under investigation and in most cases is close to 0.9, and $\Delta\Theta$ is the line width at half maximum (in radians) after correction for instrumental broadening (see e.g. [21]). A more sophisticated Fourier analyses of the line shape may allow determining the particle size distribution [21, 22]. As a conclusion for the practical data analysis of real samples, it is by far not trivial to separate the different contributions.

All the diffraction techniques described so far are in principle bulk-sensitive techniques only, and thus the scattering intensities are in general extremely small if thin surface layers are investigated. Therefore the use of high-intensity synchrotron radiation as well as state-of-the art detector equipment suited for single photon counting is recommended in practice. Scattering contributions from the substrate can substantially be reduced by using the asymmetric grazing incidence geometry: By using incidence angles smaller than the critical angle of total reflection, the penetration depth of the X-rays can be reduced to some few nm [23], and thus the scattered intensities are originating from a near-surface region only [24]. However, the Bragg peaks related to surface layers still have small intensities. Furthermore, due to the break of the

symmetry at the surface of any single crystal, there are special Bragg peaks that are sensitive only to the near-surface structure of the electrodes (surface X-ray scattering, see e.g. [25]). Moreover, the morphology of the surface substantially influence the intensity distribution between the diffraction peaks so that a detailed picture of electrode surfaces can be gained. For example, the intensity between the Bragg peaks is generally smaller for a rough surface compared to a perfectly sharp interface [25]. A detailed review on the application of surface X-ray scattering in electrochemistry has been given recently [26]. Just to mention a few examples, the adsorption behavior of organic molecules [27], the reconstruction of metal surfaces under potential variation in an electrolyte [28] and the distribution of surface water have been measured [29]. Further studies deal with electrocatalysis [30] and with the underpotential deposition of metals – where the deposition of monolayers may occur for potentials positive to the thermodynamic equilibrium [31, 32].

X-Ray Absorption Spectroscopy (EXAFS/XANES)

X-ray diffraction experiments are in general restricted to investigations of crystalline or even single-crystalline electrode materials, but they can hardly be used for the investigation of highly disordered or even amorphous samples. In the case of corrosion research, species dissolved in the electrolytes are of particular interest; especially their valence and coordination (e.g. complexation) are important to understand corrosion or inhibition processes in detail. For those situations, the X-ray absorption fine-structure spectroscopy (XAFS) technique may be an alternative and complimentary method. In a XAFS experiment, the absorption of X-rays within the sample is measured as a function of the X-ray energy $E = h\nu$ in the vicinity of an absorption edge of an clement in the sample, i.e. an energy, where core electrons can be excited to unoccupied states above the Fermi or vacuum level. The energy of the absorption edge is characteristic for the absorbing material and accordingly the measured edges are signatures of the atomic species present in a material, with the intensity of the absorption being directly linked to the concentration of the considered element.

Depending on the ligands of the considered X-ray absorbing atom, the binding energies of the core electrons are slightly changed, and thus a precise measurement of the absorption edge position may yield a sensitive measurement of the oxidation state of the X-ray absorbing element [33, 34]. This is an important characteristic especially in battery research [35, 36], corrosion science [37, 38], electrocatalysis and the chemistry of fuel cells [39, 40], electrodeposition or underpotential deposition [41]. As can already be anticipated from the fact that the transition of the excited photoelectron can only occur into unoccupied states, it is clear that the energy region close to the edge is intimately connected to the density of unoccupied states of the absorbing element [33, 42]. For higher photon energies, the photoelectron has enough kinetic energy to propagate freely through the material, undergoing multiple scattering events at neighboring atoms. Thereby, the shape of the edge and its absorption fine structure close to the edge (XANES) are highly sensitive also to the coordination of the absorbing atoms [43]. While the complexity of the involved quantum mechanical processes makes a full theoretical treatment on an ab initio basis very difficult [43], it is quite convenient to use the XANES as a fingerprint technique: Near-edge spectra are measured for a number of standard compounds of the same element with known valence and crystal structure and these are compared to the spectrum of the actual sample [8, 36, 37, 44]. Certain compounds such as Cr^{6+} or Mo^{6+} give rise for distinct pre-edge peaks so that these species can easily be identified qualitatively and quantitatively [37, 38].

For samples in an electrochemical environment, however, the absorber element is typically present in more than one chemical form. For example, a battery only consists of a single phase material if it is fully charged or discharged, but all intermediate situations are in best case

superpositions of both states [35, 36]. Similarly only fractions of a catalyst material are modified during operation, and a metal is in general not fully present as an oxide or hydroxide during a corrosion or passivation process. In all those situations, a quantitative XANES data analysis may thus be difficult and not straightforward. For these problems, the application of sophisticated mathematical techniques such as the principal component analysis (PCA, [45]) is recommended. A PCA is able to specify whether or not a chosen reference compound contributes to the spectrum of the actual sample, and thereby the application of the PCA yields the number and type of reference compounds within the sample [46, 47]. In combination with a least square fit, the PCA may also provide the concentrations of all the identified phases [45–48]. This capability is useful for the investigation of many electrochemical reactions, where e.g. reaction intermediates have to be identified or the course of a reaction may be followed quantitatively as a function of polarization potential or reaction time [35, 36, 48]. Making use of the extended X-ray absorption fine structure (EXAFS), even the detailed atomic configuration in the vicinity of the X-ray absorbing element can be quantified irrespective of the state of matter of the sample [43]. Especially this property makes XAFS an ideal tool for in situ investigations of all kinds of electrochemical processes.

Conclusions and Future Directions

Both X-ray scattering and X-ray spectroscopic methods may provide a detailed, microscopic picture of the electrode in contact with the electrolyte, and all the information may be obtained in situ. However, it should be kept in mind that the described X-ray methods always average over an extended volume of the investigated samples. Future efforts should therefore especially include the use of focussed X-ray beams in order to combine X-ray scattering and X-ray absorption techniques with X-ray microscopy, making laterally resolved structural investigations feasible. Using focussing mirrors or X-ray lenses, the nm domain seems to be accessible, and thus all localized electrochemical phenomena such as corrosion, heterogeneous electrocatalysis or the growth of micro- and nanostructures may be investigated. Another important aspect is the feasibility of time-resolved experiments: Due to the large photon flux at modern synchrotron sources, the time for an individual experiment is not any more limited by the photon flux of the source. Thus, extremely fast investigations of dynamic reactions are possible [48], reaching a resolution in the ms domain, which is important for many electrochemical processes.

References

1. Marcus P, Mansfeld F (eds) (2005) Analytical methods in corrosion science and engineering. CRC Taylor and Francis, Boca Raton
2. Kittel C (1996) Introduction to solid state physics, 7th edn. Wiley, New York
3. Macgillavry CH, Rieck GD (1962) International tables for X-ray crystallography, Vol III: mathematical tables. Kynoch Press, Birmingham
4. Davenport AJ, Oblonsky LJ, Ryan MP, Toney MF (2000) The structure of the passive film that forms on iron in aqueous environments. J Electrochem Soc 147:2162–2173. doi:10.1149/1.1393502
5. Haupt S, Strehblow H-H (1987) Corrosion, layer formation and oxide reduction of passive iron in alkaline solution, a combined electrochemical and surface analytical study. Langmuir 3:873–885. doi:10.1021/la00078a003
6. Nagayama M, Cohen M (1962) The anodic oxidation of iron in a neutral solution. J Electrochem Soc 109:781–790. doi:10.1149/1.2425555
7. Foley CL, Kruger J, Bechtoldt CJ (1967) Electron diffraction studies of active, passive and transpassive oxide films formed on iron. J Electrochem Soc 114:994–1001. doi:10.1149/1.2424199
8. Davenport AJ, Sansone M (1995) High resolution in situ XANES investigation of the nature of the passive film on iron in a pH 8.4 borate buffer. J Electrochem Soc 142:725–730. doi:10.1149/1.2048525
9. Long GG, Kruger J, Black DR, Kuriyama M (1983) Structure of passive films on iron using a new surface-exafs technique. J Electroanal Chem 150:603–610. doi:10.1016/S0022-0728(83)80239-3
10. Rubim JC, Dünnwald J (1989) Enhanced Raman scattering from passive films on silver-coated iron electrodes. J Electroanal Chem 258:327–344. doi:10.1016/0022-0728(89)85118-6
11. Gui J, Devine TM (1991) In situ vibrational spectra of the passive film on iron in buffered borate solution. Corros Sci 32:1105–1124. doi:10.1016/0010-938X(91)90096-8
12. O'Grady WE (1980) Moessbauer study of the passive oxide film on iron. J Electrochem Soc 127:555–563. doi:10.1149/1.2129711
13. Ryan MP, Newman RC, Thompson GE (1995) An STM study of the passive film formed on iron in borate buffer solution. J Electrochem Soc 142: L177–L179. doi:10.1149/1.2050035
14. Giessen BC, Gordon GE (1968) X-ray diffraction: new high-speed technique based on X-ray spectrography. Science 159:973–975. doi:10.1126/science.159.3818.973-a
15. Ohzuku T, Ueda A, Yamamota N (1995) Zero-strain insertion material of $Li[Li_{1/3}Ti_{5/3}]O_4$ for rechargeable lithium cells. J Electrochem Soc 142:1431–1435. doi:10.1149/1.2048592
16. Scrosati B (2000) Recent advances in lithium ion battery materials. Electrochim Acta 45:2461–2466. doi:10.1016/S0013-4686(00)00333-9
17. Panero S, Reale P, Ronci F, Scrosati B, Perfetti P, Rossi Albertini V (2001) Refined, in-situ EDXD structural analysis of the $Li[Li_{1/3}Ti_{5/3}]O_4$ electrode under lithium insertion-extraction. Phys Chem Chem Phys 3:845–847. doi:10.1039/B008703N
18. Rowles MR, Styles MJ, Madsen IC, Scarlett NVY, McGregor K, Riley DP, Snook GA, Urban AJ, Connolley T, Reinhard C (2012) Quantification of passivation layer growth in inert anodes for molten salt electrochemistry by in situ energy-dispersive diffraction. J Appl Crystallogr 45:28–37. doi:10.1107/S0021889811044104
19. Magnussen OM, Scherer J, Ocko BM, Behm RJ (2000) In situ X-ray scattering study of the passive film on Ni(111) in sulfuric acid solution. J Phys Chem B 104:1222–1226. doi:10.1021/jp993615v
20. Paterson MS (1952) X-ray diffraction by face centered cubic crystals with deformation faults. J Appl Phys 23:805–811. doi:10.1063/1.1702312
21. Klug HP, Alexander LE (1974) X-ray diffraction procedures. Wiley, New York
22. Warren BE, Averbach BL (1952) The separation of cold-work distortion and particle size broadening in X-ray patterns. J Appl Phys 23:497. doi:10.1063/1.1702234
23. Parratt LG (1954) Surface studies of solids by total reflection of X-rays. Phys Rev 95:359–369. doi:10.1103/PhysRev.95.359
24. Toney MF, Brennan S (1989) Structural depth profiling of iron oxide thin films using grazing incidence asymmetric Bragg X-ray diffraction. J Appl Phys 65:4763–4768. doi:10.1063/1.343230
25. Robison IK, Tweet DJ (1992) Surface X-ray diffraction. Rep Prog Phys 55:599–651. doi:10.1088/0034-4885/55/5/002
26. Nagy Z, You H (2002) Applications of surface X-ray scattering to electrochemistry problems. Electrochim Acta 47:3037–3055. doi:10.1016/S0013-4686(02)00223-2

27. Wandlowski T, Ocko BM, Magnussen OM, Wu S, Lipkowski J (1996) The surface structure of Au(111) in the presence of organic adlayers: a combined electrochemical and surface X-ray scattering study. J Electroanal Chem 409:155–164. doi:10.1016/0022-0728(95)04479-5

28. Ocko BM, Wang J, Davenport A, Isaacs H (1990) In situ x-ray reflectivity and diffraction studies of the Au(001) reconstruction in an electrochemical cell. Phys Rev Lett 65:1466–1469. doi:10.1103/PhysRevLett.65.1466

29. Toney MF, Howard JN, Richer J, Borges GL, Gordon JG, Melroy OR, Wiesler DG, Yee D, Sorensen LB (1995) Distribution of water molecules at Ag(111)/electrolyte interface as studied with surface X-ray scattering. Surf Sci 335:326–332. doi:10.1016/0039-6028(95)00455-6

30. Stamenkovic VR, Fowler B, Mun BS, Wang G, Ross PN, Lucas CA, Marković NM (2007) Improved oxygen reduction activity on Pt₃Ni(111) via increased surface site availability. Science 315:493–497. doi:10.1126/science.1135941

31. Kondo T, Morita J, Okamura M, Saito T, Uosaki K (2002) In situ structural study on underpotential deposition of Ag on Au(111) electrode using surface X-ray scattering technique. J Electroanal Chem 532:201–205. doi:10.1016/S0022-0728(02)00705-2

32. Tamura K, Wang JX, Adžic RR, Ocko BM (2004) Kinetics of monolayer Bi electrodeposition on Au(111): surface X-ray scattering and current transients. J Phys Chem B 108:1992–1998. doi:10.1021/jp0368435

33. Lengeler B (1989) X-ray absorption and reflection in materials science. Adv Solid State Physics 29:53–73. doi:10.1007/BFb0108007

34. Ressler T, Wong J, Roos J (1999) Manganese speciation in exhaust particulates of automobiles using MMT-containing gasoline. J Synchrotron Radiat 6:656–658. doi:10.1107/S0909049598015623

35. Deb A, Bergmann U, Cairns EJ, Cramer SP (2004) X-ray absorption spectroscopy study of the Li$_x$FePO$_4$ cathode during cycling using a novel electrochemical in situ reaction cell. J Synchrotron Radiat 11:497–504. doi:10.1107/S0909049504024641

36. Wagemaker M, Lützenkirchen-Hecht D, van Well AA, Frahm R (2004) Atomic and electronic bulk versus surface structure: lithium intercalation in Anatase TiO$_2$. J Phys Chem B 108:12456–12464. doi:10.1021/jp048567f

37. Schmuki P, Virtanen S, Davenport AJ, Vitus CM (1996) Transpassive dissolution of Cr and sputter deposited Cr oxides studied by in situ X-ray near edge spectroscopy. J Electrochem Soc 143:3997–4005. doi:10.1149/1.1837327

38. Lützenkirchen-Hecht D, Frahm R (2001) Corrosion of Mo in KOH: time resolved XAFS investigations. J Phys Chem B 105:9988–9993. doi:10.1021/jp003414n

39. Russell AE, Rose A (2004) X-ray absorption spectroscopy of low temperature fuel cell catalysts. Chem Rev 104:4613–4635. doi:10.1021/cr020708r

40. Friebel D, Miller DJ, O'Grady CP, Anniyev T, Bargar J, Bergmann U, Ogasawara H, Wikfeldt KT, Pettersson LGM, Nilsson A (2011) In situ X-ray probing reveals fingerprints of surface platinum oxide. Phys Chem Chem Phys 13:262–266. doi:10.1039/c0cp01434f

41. Herrero E, Buller LJ, Abruna HD (2001) Underpotential deposition at single crystal surfaces of Au, Pt, Ag and other materials. Chem Rev 101:1897–1930. doi:10.1021/cr9600363

42. Fuggle JC, Inglesfield JE (1992) Unoccupied electronic states: fundamentals for XANES, EELS, IPS and BIS, vol 69, Topics in applied physics. Springer, Berlin

43. Rehr JJ, Albers RC (2000) Theoretical approaches to x-ray absorption fine structure. Rev Mod Phys 72:621–654. doi:10.1103/RevModPhys.72.621

44. Davenport AJ, Sansone M, Bardwell JA, Aldykiewicz JA Jr, Taube M, Vitus CM (1994) In situ multielement XANES study of formation and reduction of the oxide film on stainless steel. J Electrochem Soc 141:L6–L8. doi:10.1149/1.2054720

45. Malinowski ER, Howery DG (1980) Factor analysis in chemistry. Wiley, New York

46. Wasserman SR (1997) The analysis of mixtures: application of principal component analysis to XAS spectra. J Phys IV (France) 7:C2-203–C2-205. doi:10.1051/jp4/1997163

47. Ressler T, Wong J, Roos J, Smith IL (2000) Speciation of manganese particles in exhaust fumes of cars utilizing MMT. Environ Sci Technol 34:950–958. doi:10.1021/es990787x

48. Grunwaldt J-D, Lützenkirchen-Hecht D, Richwin M, Grundmann S, Clausen BS, Frahm R (2001) Piezo X-ray absorption spectroscopy for the investigation of solid-state transformations in the millisecond range. J Phys Chem B 105:5161–5168. doi:10.1021/jp010092u

X-Ray Diffraction Methods

Deepak Dubal and Rudolf Holze
AG Elektrochemie, Institut für Chemie, Technische Universität Chemnitz, Chemnitz, Germany

X-Ray Diffraction

Besides the use of X-ray absorption for the elucidation of adsorbates and adsorbate-surface interaction X-ray diffraction is a tool useful in particular for the determination of structural data

X-Ray Diffraction Methods

on a very high level of precision provided the presence of a minimum level of structuring. The advent of synchrotron radiation providing an intense source of electromagnetic radiation in the range of X-rays has greatly stimulated the application of X-ray diffraction methods. A broad overview of experimental approaches and recent results has been provided elsewhere [1–5].

A wave front of electromagnetic radiation arriving at an array of atoms acting as scattering centers causes these centers to emit (scatter) spherical waves of electromagnetic radiation of the same energy like the incoming radiation. Depending on the wavelength of the employed radiation, the a ngle of incidence, and the spatial arrangement of the scatterers, the emitted radiation of the different scatterers will interfere constructively or destructively. With a crystalline sample containing a considerable number of scatterers, the conditions for constructive interference will be fulfilled only when the angle of incidence relative to the crystallographic plane, and the scatterers are located in is matched very precisely. The relationship is given by the Braggs equation $n\lambda = 2\,d_{hkl}\sin\vartheta$ with n being the order of the reflection, λ the wavelength of the used radiation, d_{hkl} being the crystallographic spacing, and ϑ the angle of diffraction (i.e., the angle of incidence). From the number and intensities found at different values of values of ϑ, the crystallographic data of the sample under investigation can be derived. In order to obtain a diffractogram a crystal is illuminated with X-rays of a fixed wavelength λ. The angle of incidence is varied by turning the crystal, and the diffracted intensity is measured. (for further details, see [6, 7]).

X-ray diffraction as described above is done in an angular dispersive way (ADXD). The diffracted X-ray intensity is displayed as a function of the scattering angle 2ϑ. This diffractogram is correct only for the single wavelength used during its acquisition. Any change of the incident wavelength causes a nonlinear stretching of the diffractogram. A more general display is obtained by displaying the scattered intensity as a function of the scattering parameter q. Further details and advantages of this approach

have been discussed elsewhere [8, 9]. Consequently two different ways to obtain a diffractogram are possible: use of light with a fixed wavelength and measurement of the diffracted intensity as a function of the scattering angle or use of polychromatic light (Bremsstrahlung) and detection of the diffracted radiation with a solid-state detector connected to a multichannel analyzer.

A setup includes a source for X-rays (usually a fixed or a rotating anode source, more recently synchrotron radiation has become an attractive choice as a source of radiation), a monochromator or at least a filter to select the desired wavelength, the spectroelectrochemical cell, and the detector (for an overview on detectors see [3]).

The cell has to fulfill three general requirements:

- The path for the incident and the reflected beam should not be obstructed.
- The number of scattering or absorbing particles beyond the sample under investigation in the beam path should be kept at a minimum.
- The electrochemical characteristics of the cell (like e.g., even current distribution, reliable potential control, low iR drop) should be as perfect as possible.

For X-ray diffraction experiments two basic designs are possible: the Bragg (or reflection) and the Laue (or transmission) mode. In electrochemical investigations the former is better suited for studies of adsorbates or of other features parallel to the electrode surface, whereas the second mode is suitable for thick films or layers. In both cases a cell window as transparent as possible for X-rays with sufficient stability towards this radiation is needed. Most commonly thin polymer foils (Mylar® or Melinex®) are used. A typical design of a cell of the Laue type as depicted below shows the X-ray passing through two polymer film windows and the electrolyte solution (Fig. 1).

The working electrode is coated onto one of the windows. In order to keep scattering from the electrolyte solution low, one window is mounted on the end of a hollow syringe barrel. For electrochemical measurements it is retracted to provide acceptable current distribution, and for X-ray measurements the barrel is moved as close as possible towards the fixed window. An obvious drawback of the cell design is the poor electrochemical properties of the cell in the latter position. A spectroelectrochemical cell of the Bragg (reflection) type as used for the investigation of materials which are chemically or electrochemically deposited onto a gold film sputtered before onto a porous membrane foil has been described [10]. The cross-section as displayed below shows the Prussian blue-coated membrane assembly with its coating towards a polyethylene film transparent and amorphous for X-rays (Fig. 2).

The electrolyte penetrates the porous membrane, via the porous glass body, and the glass fiber paper connection is provided with the silver chloride-coated silver plate acting both as a reference and a counter electrode. The electrochemical behavior of the cell as demonstrated with cyclic voltammetry is fairly close to standard electrochemical cells. Results show changes of lattice constants of Prussian blue as a function of the electrode potential and the transferred charge. Because of the penetration depth of X-rays and the need for a minimum of crystallinity, this approach is not exactly surface sensitive; it probes instead the interphase between the current collector (i.e., the gold layer) and the electrolyte solution. Results obtained during charging/discharging of TiS_2 (as proposed for use in secondary lithium batteries [11]) involving formation of $LiTiS_2$ are displayed below; the employed cell is depicted thereafter (Figs. 3 and 4).

Upon discharge lithium is intercalated. This results in shifts of the (101), (002), and (100) Bragg peaks. Detailed studies of the various shifts as a function of the state of charge/discharge reveal further information on the mechanism of the different phase transitions [12]. Reactions of lithium with S_8 in a secondary Li/S cell have been tracked with in situ X-ray diffraction [14]. The electrochemical reaction of lithium with crystalline silicon was monitored [15]. Various Li_xSi_y phases were identified. The transition from the crystalline into the amorphous state in the first lithium intercalation was observed [16]. For alternate cell constructions, see [17], and for a cell used for studies of ion insertion into electrode materials, see [18]. A complete cell not modified in any way has been used with high-energy X-rays from a synchrotron source for simultaneous measurement of diffraction patterns of cell components of a lithium ion battery; recorded diffraction peaks were assigned to cathode and anode material [19]. A cell particularly suitable for

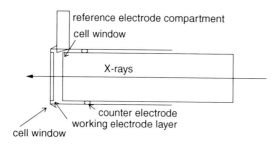

X-Ray Diffraction Methods, Fig. 1 Spectroelectrochemical cell for X-ray diffraction studies in transmission mode

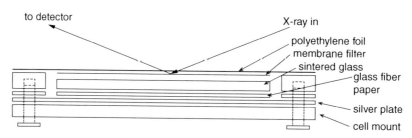

X-Ray Diffraction Methods, Fig. 2 Spectroelectrochemical cell (for better identification of components in exploded arrangement) for in situ X-ray diffraction studies according to [10]

measurements under ultrapure conditions has been described [20]. Basically the same design has been employed in a study of the solid-state electrochemistry of PbO and Pb(OH)Cl (laurionite) [21]. The particles of red PbO (litharge) were attached to a paraffin-impregnated graphite (PIG) rod used as a working electrode. X-ray diffraction patterns obtained at various electrode potentials (see Fig. 5 below) show peaks indicative of PbO, a mix of PbO and Pb, and finally of Pb as a function of reduction potential and time (Fig. 5).

Reflexes of graphite were also observed, they could be distinguished easily from those of the electrochemically active material. The reduction proceeds entirely as a solid state reaction, no evidence of solution phase intermediates was observed.

Improved surface specificity can be obtained by making measurements at grazing angle of incidence. The structure of small two-dimensional clusters and linear nanostructures of copper and cadmium deposits on Pt(533) at characteristic electrode potentials in the upd range has been studied with GIXD combined with GIXAFS [22]. Using X-rays provided by a synchrotron (the method is now named SR-GIXRD), the corrosion of mild steel in the presence of carbon dioxide containing brine electrolyte solutions was studied. The mechanism of corrosion, in particular corrosion products present as thin films like $Fe_2O_2CO_3$, $Fe_2O_2CO_3$, and $Fe_2(OH)_2CO_3$, was identified with SR-GIXRD [23]. Investigations of iron chalcogenide glasses suitable for ion-selective electrodes ISE with a combination of electrochemical impedance measurements and GIXRD have been reported together with a description of a suitable electrochemical cell [24, 25]. Selective dissolution of various crystallographic surfaces of iron chalcogenides associated with electrode potential shifts of these materials employed in ion-selective electrodes has been monitored with SR-GIXRD [26]; results were found to be in agreement with those obtained with AFM.

Cell designs and examples discussed above pertain to X-ray diffraction performed in the angular dispersive mode, i.e., the angle of incidence is varied by, e.g., turning the electrochemical cell with respect to the radiation source.

As already mentioned energy dispersive measurements are possible with the spectroelectrochemical cell staying in place, whereas the wavelength of the incident radiation is varied. An additional advantage of EDXD is the high intensity of X-ray radiation in the employed energy range with an upper limit only given by the power supply of the X-ray tube. The strong X-ray absorption of standard K- and L-lines in the

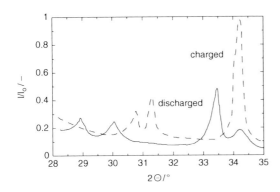

X-Ray Diffraction Methods, Fig. 3 In situ X-ray diffraction pattern of TiS_2 recorded during charging/discharging experiments (Based on data in [12], see also [13])

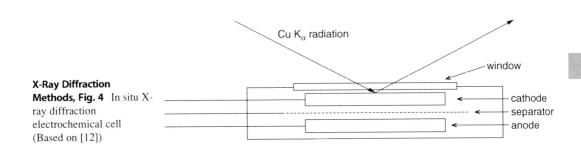

X-Ray Diffraction Methods, Fig. 4 In situ X-ray diffraction electrochemical cell (Based on [12])

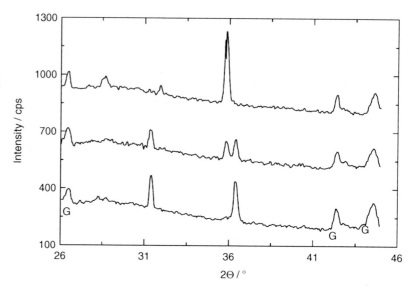

X-Ray Diffraction Methods, Fig. 5 In situ X-ray diffraction pattern of PbO and its reduction products: before reduction, at $E_{Ag/AgCl} = -0.5$ V (*top*); during reduction at $E_{Ag/AgCl} = -1$ V; after reduction, at $E_{Ag/AgCl} = -1.3$ V (*bottom*) (Based on data by [21])

range of lower energies caused by electrochemical cell components is no problem, because usually employed construction materials show only weak absorption at higher wavelengths [9]. The poor resolution of the solid-state detector causes a broadening of the peaks in the measured diffractogram [9]. Results reported so far deal with intercalation materials as used in batteries [9].

Application of time-resolved high-energy X-ray diffraction on platinum nanoparticles in fuel electrodes have been described [27]. Results indicate severe surface reconstruction of the nanoparticle surface showing at least three types of Pt-O bonds (adsorbed OH, adsorbed atomic O and amorphous PtO_x) under oxidative conditions.

X-ray diffraction has also been used in studies of solid electrolytes as reviewed elsewhere [28]. Diffraction techniques (including X-ray as well as neutron diffraction) as applied in electrolyte solution studies are described in [29].

Surface X-Ray Diffraction (SXD)

X-ray diffraction discussed so far as applied to the elucidation of the crystallographic structure of thick layers, films or other three-dimensional samples can also be applied to basically two-dimensional structures on surfaces or at interfaces based on various instrumental advances, in particular the use of synchrotron radiation. The slightly gradual distinction implies possible overlap between both groups of methods. This way information on the two-dimensional periodicity of surface layers can be obtained. The required surface sensitivity can be obtained primarily by employing an angle of incidence less than or in the order of the critical angle. This results in a significantly reduced depth of penetration of the X-rays, scattering from the bulk of the sample is also diminished. Under these conditions the ratio of the signal from the surface vs. the signal from the bulk (or substrate) can be enhanced by a factor of $2/\theta_c$ (with θ_c being the critical angle of total external reflection in radians.). θ_c is about 0.5° (or 9 mrad) at X-ray frequencies typical for metals in contact with transparent incident phases. These conditions are also known as total reflection Bragg diffraction [30]. If grazing angle conditions (i.e. angle of incidence θ about 3.5° or lower) are employed the method is also named grazing incidence X-ray diffraction GIXRD. Further enhancement of general and surface sensitivity can be achieved by applying an electrode potential modulation procedure combined with an appropriate data treatment. This way only potential dependent features in the diffractogram will become visible whereas features (Bragg reflexes) of the bulk not affected by the electrode potential will cancel out. Because of fluctuations in the intensity of the X-ray source and other experimental components it is insufficient to record complete diffractograms sequentially; instead a slow potential modulation (with the frequency set to a value resulting in a time constant sufficiently longer than that of the occurring electrochemical process) is applied. Using computer-based data acquisition the respective data are stored and treated; for further details and an overview see [5].

The electrochemical cells described in the preceding section can be used. A cell design with a significantly reduced radiation absorption of the electrolyte solution film as used for specular X-ray reflectivity measurements can also be used. Electrode potentials are selected previously based on standard electrochemical experiments (e.g. cyclic voltammetry) with respect to well-defined changes of the electrode/solution interface (e.g. potential steps between potentials of complete desorption and maximum adsorption). Control of the potentiostat and the X-ray diffractometer as well as data acquisition, storage and manipulation are done with a suitably programmed computer.

Typical examples include studies of underpotential deposition of various metals on metallic substrates. The structure of the upd layer [31, 32], the position of adsorbed anions and water molecules on top of the upd layer and the respective bond angles and lengths could be elucidated [33, 34]. Surface reconstruction caused by e.g. weakly adsorbed hydrogen [35], surface expansion effects of low-index platinum and gold surfaces correlated with adsorption/desorption of solution species [36] and crystallographic transformations in layers of $NiOH_2$ have been studied [37]. In a corrosion study of a $Cu_3Au(111)$ alloy single crystal used as a model system initial stages of corrosion, in particular structure formation, have been elucidated with surface XRD [38]. Evidence of diffusion of gold atoms into an upd-Te layer on top of

a Au(111) has been reported [31]. In a study of the water layer structure on a Cu(111) electrode surface during hydrogen evolution combining SXD and IRRAS showed a closest packing-like stacked structure with significantly different inter- and intra-layer specific oxygen-oxygen nearest neighbor distances [39]; the infrared spectra indicated the presence of both free and hydrogen-bonded hydroxyl ions.

Surface Differential X-Ray Diffraction (SDD)

The scattered X-ray from crystal planes or a thin metal film can interfere coherently with scattered X-rays from an adsorbate layer. This phenomenon is called surface differential diffraction SDD or Bragg peak interference. The theory has been described elsewhere in detail [40, 41]. From SDD measurements information on the relative amount of material in the adsorbate layer (coverage) and the distance between atoms in the top layer of the substrate and in the adsorbate layer as a function of electrode potential can be derived.

The experimental setup uses a working electrode prepared from mica coated with the metal to be investigated (e.g. silver) by vapor deposition. Because of the mode of growth a certain monocrystalline orientation will result (e.g. (111) in case of silver). The working electrode is attached to an electrochemical cell equipped with counter and reference electrode and the standard ancillary equipment for electrochemical measurements. The X-ray of well-defined wavelength from a standard laboratory source or from a synchrotron (selected with e.g. a Si(220) monochromator) impinges on the working electrode through the mica substrate. Diffracted intensity is collected with a multichannel analyzer; it can be done in real time during electrode potential scans. Investigated systems include predominantly metal adlayers (upd layers) on metal surfaces [41, 42]. Data on mode of layer growth, adsorbate-atom and substrate-atom distance and relaxation of involved atom layers were obtained.

Future Directions

In studies of solid materials like active masses in electrochemical energy storage and conversion devices knowledge of structural changes is of growing importance. Application of X-ray diffraction methods in particular under in situ conditions will thus increase.

Cross-References

▶ Insertion Electrodes for Li Batteries
▶ Electrode
▶ Solid State Electrochemistry, Electrochemistry Using Solid Electrolytes
▶ X-Ray Absorption and Scattering Methods

References

1. (2002) Special issue of Electrochim Acta 47:3035
2. Robinson J (1990) In: Gutiérrez C, Melendres C (eds) Spectroscopic and diffraction techniques in interfacial electrochemistry, vol 320, NATO ASI series C. Kluwer, Dordrecht, p 313
3. Robinson J (1988) In: Gale RJ (ed) Spectroelectrochemistry. Plenum Press, New York, p 9
4. Holze R (2007) Surface and interface analysis: an electrochemists toolbox. Springer, Heidelberg
5. Zegenhagen J, Renner F (2006) Nachr Chem 54:847
6. Massa W (1996) Kristallstrukturbestimmung. Teubner Verlag, Stuttgart
7. Clegg W (2001) Crystal structure analysis. Oxford University Press, Oxford
8. Caminiti R, Rossi Albertini V (1999) Int Rev Phys Chem 18:263
9. Rossi Albertini V, Perfetti P, Ronci F, Scrosati B (2001) Chem Mater 13:450
10. Ikeshoji T, Iwasaki T (1988) Inorg Chem 27:1123
11. Whittingham MS (1976) Science 192:1126
12. Chianelli RR, Scanlon JC, Rao BML (1979) J Solid State Chem 29:323
13. Chianelli RR, Scanlon JC, Rao BML (1978) J Electrochem Soc 125:1563
14. Puglisi V, Simoneau M, Geronov Y 210. Electrochemical society meeting, Cancun, 29.10.–03.11.2006, Extended abstract #164
15. Li J, Dahn JR 210. Electrochemical society meeting, Cancun, 29.10.–03.11.2006, Extended abstract #261
16. Li J, Dahn JR (2007) J Electrochem Soc 154:A156
17. Nishizawa M, Uchida I (1998) Denki Kagaku 66:991
18. Bergström O, Gustafsson T, Thomas JO (1998) J Appl Crystallogr 31:103
19. Li ZG, Harlow RL, Gao F, Lin P, Miao R, Liang L (2003) J Electrochem Soc 150:A1171
20. Koop T, Schindler W, Kazimirov A, Scherb G, Zegenhagen J, Schulz T, Feidenhans'l R, Kirschner J (1998) Rev Sci Instrum 69:1840
21. Meyer B, Ziemer B, Scholz F (1995) J Electroanal Chem 392:79
22. Prinz H, Strehblow HH (2002) Electrochim Acta 47:3093
23. De Marco R, Jiang Z-T, Pejcic B, Poinen E (2005) J Electrochem Soc 152:B389
24. de Marco R, Pejcic B, Prince K, van Riessen A (2003) Analyst 128:742
25. De Marco R, Jiang Z-T, Martizano J, Lowe A, Pejcic B, van Riessen A (2006) Electrochim Acta 51:5920
26. de Marco R, Jiang ZT, Pejcic B, van Riessen A (2006) Electrochim Acta 51:4886
27. Imai H, Izumi K, Kubo Y, Kato K, Imai Y Extended abstracts of the 209th meeting of the electrochemical society Spring Denver, Colorado, USA, 07.05.–11.05.2006, Ext. Abstr. #302
28. Wagner JB Jr (1991) In: Varma R, Selman JR (eds) Techniques for characterization of electrodes and electrochemical processes. Wiley, New York, p 3
29. Enderby JE (1991) In: Varma R, Selman JR (eds) Techniques for characterization of electrodes and electrochemical processes. Wiley, New York, p 327
30. Marra WC, Eisenberger P, Cho AY (1979) J Appl Phys 50:6927
31. Kawamura H, Takahasi M, Hojo N, Miyake M, Murase K, Tamura K, Uosaki K, Awakura Y, Mizuki J (2002) J Electrochem Soc 149:C83
32. Fleischmann M, Mao BW (1988) J Electroanal Chem 247:297
33. Nakamura M, Endo O, Ohta T, Ito M, Yoda Y (2002) Surf Sci 514:227
34. Herrero E, Glazier S, Abruna HD (1998) J Phys Chem B 102:9825
35. Fleischmann M, Mao BW (1988) J Electroanal Chem 247:311
36. Lucas CA (2002) Electrochim Acta 47:3065
37. Fleischmann M, Oliver A, Robinson J (1986) Electrochim Acta 31:899
38. Renner FU, Stierle A, Dosch H, Kolb DM, Lee T-L, Zegenhagen J (2006) Nature 439:707
39. Ito M, Yamazaki M (2006) Phys Chem Chem Phys 8:3623
40. Rayment T, Thomas RK, Bomchil G, White JW (1981) Mol Phys 43:601
41. Chabala ED, Rayment T (1994) Langmuir 10:4324
42. Ramadan AR, Chabala ED, Rayment T (1999) Phys Chem Chem Phys 1:1591

Z

ZEBRA Batteries

Hikari Sakaebe
Research Institute for Ubiquitous Energy
Devices, National Institute of Advanced
Industrial Science and Technology (AIST),
Ikeda, Osaka, Japan

Introduction

ZEBRA battery is actually a Z E B R A (*Zeolite Battery Research Africa*) battery, that is, sodium-nickel chloride cell. This battery consists of a liquid Na negative electrode and $NiCl_2$ separated by β-alumina solid electrolyte with Na^+ conduction. Total cell reaction is as follows:

$$2Na + NiCl_2 \Longleftrightarrow 2NaCl + Ni \ V$$
$$= 2.58 \ V \text{ at } 300 \ °C$$

This battery system was invented by Johan Coetzer at CSIR (Council for Scientific and Industrial Research) in South Africa. Energy density of ZEBRA battery is relatively high (theoretical 788 Wh/kg, practical 90 Wh/kg) among the rechargeable battery. The first patent was applied in 1978. Then this battery system was developed by BETA Research and Development Ltd. in England, and the development was continued. Afterward this was integrated into the joint venture of AEG (later Daimler) and Anglo American Corporation 10 years later. The company AEG Anglo Batteries GmbH funded by joint venture

started the pilot line production of ZEBRA batteries in 1994. After the merger of Daimler and Chrysler, AEG Anglo Batteries GmbH was terminated. MES-DEA succeeded the ZEBRA technology and industrialized the battery. At the time, production capacity was 2,000 battery packs per year in a building designed for a capacity of 30,000 battery packs per year. We can find detailed review in references [1, 2].

Composition of the Battery

Cell Chemistry

At a charged state, positive and negative electrode could be $NiCl_2$ and Na metal. It is quite difficult to handle these materials in a production scale, and the cell is usually built with NaCl and Ni metal as discharged products. In order to maintain the sufficient utilization of Na electrode, Al powder is incorporated in the positive electrode side to form $NaAlCl_4$ that can be an electrolyte when the electrode materials are melted. Operation temperature is at around 270–350 °C. Sodium ion conductivity has a practical value ($\geq 0.2 - 1 \ cm^{-1}$) at 260 °C and is temperature dependent with a positive gradient [3]. Thus, the operational temperature of ZEBRA batteries has been chosen for the range above. Current collector for positive electrode is Cu wire with Ni plating.

Cell structure is schematically shown in Fig. 1. Outside of β-alumina electrolyte case, Na electrode is incorporated in the cell can tube

G. Kreysa et al. (eds.), *Encyclopedia of Applied Electrochemistry*, DOI 10.1007/978-1-4419-6996-5,
© Springer Science+Business Media New York 2014

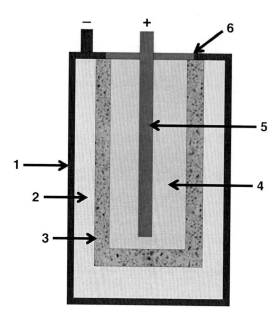

ZEBRA Batteries, Fig. 1 Basic cell structure of ZEBRA battery. *1* Cell can for the Na negative electrode, *2* Na negative electrode, *3* ceramic electrode tube made of β-alumina, *4* mixture for the positive electrode (NiCl₂ + NaAlCl₄), *5* current collector for positive electrode, *6* thermal compression bond

(stainless steel with Ni plating), and $NiCl_2$ and $NaAlCl_4$ are inside the β-alumina tube. Outer cell case is a current collector for Na negative electrode. Molten Na can be partly isolated in sodium compartment if discharge depth becomes so deep and Na loses electronic contact with current collector. And thus 100 % utilization for Na is difficult.

Because of these phenomena, potential drops suddenly at the end of the discharge. This behavior is not desirable for practical use, and Al powder takes part in the additional electrode reaction to alleviate the potential drop.

$$Al + 4NaCl \Longleftrightarrow 3Na + NaAlCl_4 \quad Voc = 1.58\ V$$

At the same time, Al helps $NaAlCl_4$ melt to penetrate into the pores in electrode. Iron sulfide is added in order to stabilize the Ni particle size. The plateau can be found in the potential region in 2–1.6 V.

This salt liquefies at 154 °C, and in the liquid state it is conductive for sodium ions. It has several functions that play very important roles for the ZEBRA battery technology. Concretely, the salt conducts Na^+, prevents the cracks of β-alumina, keeps the safety, and protects the cell from overcharge and overdischarge.

Positive electrode of the cell is theoretically a mixture of Ni and $NiCl_2$. Na^+ conduction in the electrode is not enough, and the liquid salt $NaAlCl_4$ conducts the Na^+ between the solid electrolyte ceramic surface and the reaction zone inside the positive electrode bulk during the cell operation. With this function, all chemical species in positive electrode can be utilized.

Ceramic is a quite brittle material. By several triggers like a vibration, stress, and so on, cracks may be developed. In this case, through the crack, the liquid salt $NaAlCl_4$ becomes in contact with the liquid sodium (the melting point of sodium is 90 °C) to form a salt and Al metal:

$$NaAlCl_4 + 3Na \rightarrow 4NaCl + Al$$

Small cracks in the ceramic electrolyte can be closed by formed salt and Al. When crack is larger, formed Al short-circuits between positive and negative electrode. This cell loses voltages, but still the whole system can be operated as long as failed cell was within 5–10 % of the total cells. The battery controller detects this and adjusts all operative parameters. In this meaning, ZEBRA battery is failure tolerant to some extent.

The same reaction helps to passivate the positive electrode material ($NiCl_2$). In this sequence, some energy is released, but the total energy can be reduced almost 30 % of the normal and full reaction in discharging.

The charge capacity of the ZEBRA cell is governed by the quantity of NaCl available in the positive electrode. During an operation, the liquid salt $NaAlCl_4$ is a Na reserve as shown in the following reaction:

$$2NaAlCl_4 + Ni \leftrightarrow 2\ Na + 2\ AlCl_3 + NiCl_2$$

This corresponds to an overcharge reaction and requires a higher voltage than 3.05 V.

When the cell is overdischarged, the same reaction with the case of cell failure occurs at 1.58 V.

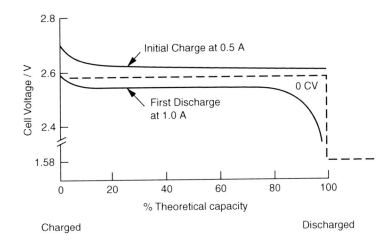

ZEBRA Batteries, Fig. 2 First charge and discharge curves of Na/NiCl₂ cell without additives assembled in discharged state (Reproduced figure from Ref. [4])

Battery Design

Cells are connected in parallel and in series combination. Dustmann at MES-DEA S. A introduced some example [1]. Types of battery vary with one to five parallel strings, up to 220 cells in series and 100–500 cells in one battery pack. The standard battery type Z5C (32 or 64 Ah as nominal capacity and 557 or 278.6 V as OCV, respectively) has 216 cells arranged in one (OCV = 557 V) or two (OCV = 278 V) strings. Cooling plate separates each second cell, through which ambient air is circulated. Cooling power is 1.6–2 kW. Cells are surrounded by a double-walled vacuum insulation for thermal insulation. The thickness of walls is typically 25 mm. This configuration has a heat conductivity of only 0.006 W/mK and is stable for up to 1,000 °C.

Battery System Design

The complete system ready for assembly is equipped with the ohmic heater, the fan for cooling, and the battery management interface (BMI) [1]. BMI controls the thermal issues and electrical connection through a main circuit breaker that can disconnect the battery from outside. The BMI measures and supervises a lot of parameters: voltage, current, status of charge, and insulation resistance of negative and positive pole to ground. It controls the charger by a dedicated pulse width modulated (PWM) signal. A CAN bus is used for the communication between the BMI and devices. All battery data are monitored for diagnostic with a notebook.

Performance

There is no side reaction in the cell chemistry of ZEBRA battery, and therefore, the charge and discharge cycle has almost 100 % efficiency. This is due to the ceramic electrolyte. Figure 2 shows a first charge and discharge curves of Na/NiCl₂ cell assembled in a discharged state [4]. This cell contains no additives like FeS.

Discharge curves of cell in practical battery system are slightly different from Fig. 2. Figure 3 is an example for the discharge curves of ML/3I 32 Ah cell reported by J.L. Sudworth at Beta Research and Development Ltd. [5]. At around 2.35 V, Fe species is involved in the cell reaction:

$$FeS + 2NaCl + Ni \rightarrow FeCl_2 + NiS + 2Na \quad V = 2.37\ V$$

$$FeCl_2 + 2Na \rightarrow Fe + 2Na \quad V = 2.35\ V$$

Charging of ZEBRA cells and batteries is at a 6 h rate for normal charge and 1 h rate for

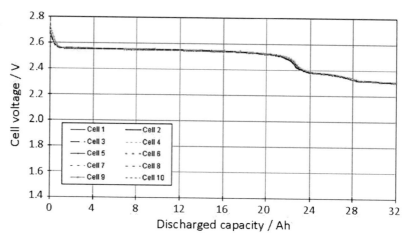

ZEBRA Batteries, Fig. 3 Discharge curves of the ML/3I 32 Ah cells at 1.5 A (Reproduced from the figure in Ref. [5])

fast charge. The voltage limitation differs depending on the mode. Limitation is set per cell and 2.67 V and 2.85 V for normal and fast charge, respectively. Fast charging mode does not allow the charge over 80 % SOC. Regenerative breaking is limited to 3.1 V and 60 A per cell so that high regenerative breaking rates are possible. The peak power during discharge is independent of SOC when it is defined as the power at 67 % OCV. It is generally reported that the power density of this battery is 150 W/kg. Specific power depends on the operation mode of positive electrode. Detailed mechanism is described in references [2, 3, 5].

Battery life is specified as calendar and cycle life. Factors that limit the life are corrosion, resistance rise, and capacity loss [5]. Without electrochemical operation, cell slowly degrades, but the calendar life of 11 years is demonstrated [1]. The cycle life is measured by the accumulation of all discharged charge measured in Ah divided by the nominal capacity in Ah. Each cycle for nominal capacity is equivalent to a 100 % discharge cycle. The expected cycle life is up to 3,500 cycles [1] from module tests and 1,450 cycles from battery testing [1] that simulated all real-life operation conditions. The thermal insulation was stable for more than 15 years in the report of C. H. Dustmann [1].

Battery Safety

Battery safety is so important for mobile and vehicle applications. Especially for vehicles, on the road, accident likely becomes heavy, and the crash accident should not bring more danger by release of the energy stored in the cells. And various tests are usually conducted. In ZEBRA battery case, test results were reported. Crash of an operative battery against a pole with 50 km/h, overcharge test, overdischarge test, short circuit test, vibration test, external fire test, and submersion of the battery in water have been specified and performed [6]. The ZEBRA battery did pass all these tests owing to its four-barrier safety concept [7, 8]: chemical aspects, cell case, thermal structure, and battery controller.

As commented in "Cell Chemistry," mechanical damage in a ceramic electrolyte is recovered even in the mechanical damage. And liquid electrolyte reacted with the liquid sodium forms layer of salt and aluminum equal to the overcharge reaction. This reaction reduces the thermal load by about 1/3 compared to the total electrochemically stored energy. Cell case was closed by TCB to stand up to 900 °C. Thermal insulation material is made out of foamed SiO_2. This is thermally stable over 1,000 °C, and the heat conductivity is very low (only 0.006 W/mK) in the vacuum insulation box. Even when the vacuum environment

ZEBRA Batteries

is broken, heat conductivity increased only by a factor of three, and this is still possible to insulate the heat. Also as stated in "Battery System," BMI senses the thermal failure and controls the cell operation.

Applications

The ZEBRA battery system has been designed for mainly electric vehicles. This application requires both of power and energy. ZEBRA battery has a sufficient balance of power and energy and a candidate system of the practical EV. Other applications are electric vans, buses, and hybrid buses with ZEV range. For these applications, ZEBRA battery requires some devices to maintain the temperature for quick start. Once the battery is cooled down to room temperature, it usually takes 1–2 days to be operated. In this meaning, continuous operation like a stationary use may be favorable. Telecommunication or marine use is also a possible application. Hybrid system concept of solid oxide fuel cell and ZEBRA battery for automotive application is designed, and possibility of the system is discussed by D. J. L. Brett et al. [9]. This kind of hybrid system may improve energy efficiency making use of the heat from fuel cells.

References

1. Dustmann CH (2004) Advances in ZEBRA batteries. J Power Sources 127:85–92
2. Sudworth JL, Galloway RC (2009) Secondary batteries – high temperature systems – sodium-nickel chloride. In: Jürgen Garche, et al. Encyclopedia of electrochemical power sources. Amsterdam, The Nederland Elsevier BV, p 312
3. Sudworth JL, Tilley AR (1985) The sodium sulphur battery. Chapman & Hall, London
4. Bones RJ, Teagle DA, Brooker SD, Cullen FL (1989) Development of a Ni, $NiCl_2$ positive electrode for a liquid sodium (ZEBRA) battery cell. J Electrochem Soc 136:1274–1277
5. Sudworth JL (2001) The sodium/nickel chloride (ZEBRA) battery. J Power Sources 100:149–163
6. Böhm H, Bull RN, Prassek A (1989) ZEBRA's response to the new EUCAR/USABC abuse test procedures, EVS-15, Brussels, 29 Sept–3 Oct
7. Zyl AV, Dustmann CH (1995) Safety aspects of ZEBRA high energy batteries. evt95, Paris, 13–15 Nov 57
8. Trickett D, Current Status of Health and Safety Issues of Sodium/Metal Chloride (ZEBRA) Batteries (1998) National renewable energy laboratory report, TP-460-25553
9. Brett DJL, Aguiar P, Brandon NP, Bull RN, Galloway RC, Hayes GW, Lillie K, Mellors C, Smith C, Tilley AR (2006) Concept and system design for a ZEBRA battery–intermediate temperature solid oxide fuel cell hybrid vehicle. Concept and system design for a ZEBRA battery–mediate. J Power Sources 157:782–798